Student Solutions Manual

Randy Gallaher Kevin Bodden

Lewis and Clark Community College

Intermediate Algebra

for College Students

■ Seventh Edition

Algebra

for College Students

■ Third Edition

Allen R.
Angel

PEARSON
Prentice
Hall

Upper Saddle River, NJ 07458

Vice President and Editorial Director, Mathematics: Christine Hoag
Executive Editor: Paul Murphy
Senior Managing Editor: Linda Behrens
Project Manager: Dawn Nuttall
Project Manager, Production: Debbie Ryan
Editorial Assistant: Georgina Brown
Supplement Cover Manager: Paul Gourhan
Supplement Cover Designer: Victoria Colotta
Operations Specialist: Ilene Kahn
Senior Operations Supervisor: Diane Peirano

Pearson Prentice Hall
Pearson Education, Inc.
Upper Saddle River, NJ 07458

Printed in the United States of America

10 9 8 7 6 5 4 3 2 1

ISBN 13: 978-0-13-238405-6 Standalone

ISBN 10: 0-13-238405-1 Standalone
ISBN 13: 978-0-13-238359-2 Component

ISBN 10: 0-13-238359-4 Component

Pearson Education Ltd., *London*
Pearson Education Australia Pty. Ltd., *Sydney*
Pearson Education Singapore, Pte. Ltd.
Pearson Education North Asia Ltd., *Hong Kong*
Pearson Education Canada, Inc., *Toronto*
Pearson Educación de Mexico, S.A. de C.V.
Pearson Education—Japan, *Tokyo*
Pearson Education Malaysia, Pte. Ltd.

Table of Contents
Intermediate Algebra for College Students, 7e

Chapter 1 Basic Concepts

Chapter 2 Equations and Inequalities

Chapter 3 Graphs and Functions

Chapter 4 Systems of Equations and Inequalities

Chapter 9 Exponential and Logarithmic Functions

Chapter 10 Conic Sections

Chapter 11 Sequences, Series, and the Binomial Theorem

ADDITIONAL CHAPTERS FOR *Algebra for College Students, 3e*

NOTE: See the above *Intermediate Algebra for College Students, 7e* Table of Contents for Chapters 1-9 of *Algebra for College Students*, 3e. Chapters 1-9 are exactly the same in both texts.

Chapter 10 Characteristics of Functions and Their Graphs

Chapter 11 Conic Sections

Chapter 12 Sequences, Series, and Probability

Chapter 1

Exercise Set 1.1

1 – 9. Answers will vary.

11. Answers will vary.

13. Do all the homework and preview the new material to be covered in class.

15. **(1)** Carefully write down any formulas or ideas that you need to remember.

(2) Look over the entire exam quickly to get an idea of its length and to make sure that no pages are missing. You will need to pace yourself to make sure that you complete the entire exam. Be prepared to spend more time on problems worth more points.

(3) Read the test directions carefully.

(4) Read each problem carefully. Answer each question completely and make sure you have answered the specific question asked.

(5) Starting with number 1, work each question in order. If you come across a problem that you are not sure of, do not spend too much time on it. Continue working the problems that you understand. After completing all other questions, come back to finish those questions you are not sure of. Do not spend too much time on any one question.

(6) Attempt each problem. You may be able to earn at least partial credit.

(7) Work carefully and write clearly so that your instructor can read your work. Also, it is easy to make mistakes when your writing is unclear.

(8) Check your work and your answers if you have time.

(9) Do not be concerned if others finish the test before you. Do not be disturbed if you are the last to finish. Use all your extra time to check your work.

17. The more you put into the course, the more you will get out of it.

19. Answers will vary.

Exercise Set 1.2

1. A variable is a letter used to represent various numbers.

3. A set is a collection of objects.

5. The null or empty set is a set that contains no elements.

7. The five inequality symbols are:
$<$, is less than
$\le$, is less than or equal to
$>$, is greater than
$\ge$, is greater than or equal to
$\ne$, is not equal to

9. $\{4, 5, 6\}$

11. Every integer is also a rational number because it can be written with a denominator of 1.

13. True; the set of whole numbers contains the set of natural numbers.

15. True; $2 = \frac{2}{1}$ is both a rational number and an integer.

17. False; $\frac{1}{2}$ is a rational number but is not an integer.

19. True; this is how the rational and irrational numbers are defined.

21. True; there are no integers between π and 4.

23. $5 > 3$

25. $0 > -2$

27. $-1 > -1.01$

29. $-5 < -3$

31. $-14.98 > -14.99$

33. $1.7 < 1.9$

35. $-\pi > -4$

37. $-\frac{7}{8} > -\frac{10}{11}$

39. $A = \{0\}$

41. $C = \{18, 20\}$

43. $E = \{0, 1, 2\}$

45. $H = \{0, 7, 14, 21, 28, \ldots\}$

47. $J = \{1, 2, 3, 4, \ldots\}$ or $J = N$

49. **a.** 4 is a natural number.

b. 4 and 0 are whole numbers.

c. –2, 4 and 0 are integers.

d. $-2, 4, \frac{1}{2}, \frac{5}{9}, 0, -1.23,$ and $\frac{78}{79}$ are rational numbers.

e. $\sqrt{2}$ and $\sqrt{8}$ are irrational numbers.

f. $-2, 4, \frac{1}{2}, \frac{5}{9}, 0, \sqrt{2}, \sqrt{8}, -1.23,$ and $\frac{78}{79}$ are real numbers.

51. $A \cup B = \{1, 2, 3, 4, 5, 6\}$
$A \cap B = \{\ \}$ or $\varnothing$

53. $A \cup B = \{-4, -3, -2, -1, 0, 1, 3\}$
$A \cap B = \{-3, -1\}$

55. $A \cup B = \{2, 4, 6, 8, 10\}$
$A \cap B = \{\ \}$ or $\varnothing$

57. $A \cup B = \{0, 5, 10, 15, 20, 25, 30\}$
$A \cap B = \{\ \}$ or $\varnothing$

59. $A \cup B = \{-1, 0, 1, e, i, \pi\}$
$A \cap B = \{-1, 0, 1\}$

61. The set of natural numbers.

63. The set of whole number multiples of three.

65. The set of odd integers.

67. **a.** Set A is the set of all x such that x is a natural number less than 7.

b. $A = \{1, 2, 3, 4, 5, 6\}$

69. $\{x | x \geq 0\}$

71. $\{z | z \leq 2\}$

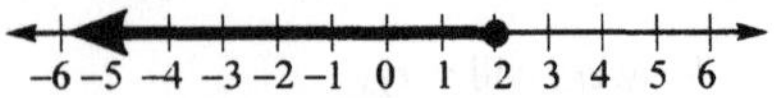

73. $\{p | -6 \leq p < 3\}$

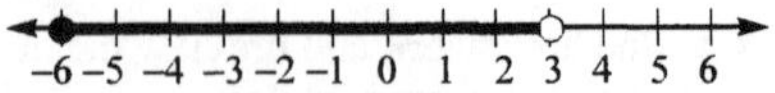

75. $\{q | q > -3 \text{ and } q \in N\}$

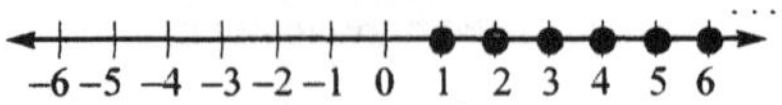

77. $\{r | r \leq \pi \text{ and } r \in W\}$

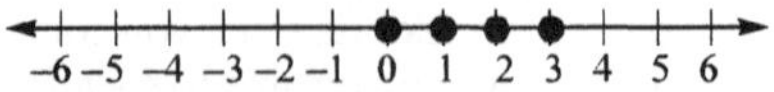

79. $\{x | x \geq 1\}$

81. $\{x | x \leq 4 \text{ and } x \in I\}$ or $\{x | x < 5 \text{ and } x \in I\}$

83. $\{x | -3 < x \leq 5\}$

85. $\{x | -2.5 \leq x < 4.2\}$

87. $\{x | -3 \leq x \leq 1 \text{ and } x \in I\}$

89. Yes; the set of natural numbers is a subset of the set of whole numbers

91. No; the set of whole numbers is not a subset of the set of natural numbers.

93. Yes; the set of rational numbers is a subset of the set of real numbers.

95. No; the set of rational numbers is not a subset of the set of integers.

97. Answers may vary.
Possible answer: $\left\{\frac{3}{2}, \frac{4}{3}, \frac{5}{4}, \frac{6}{5}, \frac{7}{6}\right\}$

99. Answers may vary.
Possible answer: $A = \{2, 4, 5, 6, 9\}$,
$B = \{4, 5, 8, 9\}$
Therefore, $A \cup B = \{2, 4, 5, 6, 8, 9\}$ and
$A \cap B = \{4, 5, 9\}$

101. a. The set of top 6 finishers in the Pocono 500 *or* the Ford 500 is: {Johnson, Mayfield, Labonte, Gordon, Busch, Earnhardt Jr., Biffle, Stewart, Gaughan}

b. Part (a) represents the union because it asks for the drivers in either category.

c. The set of top 6 finishers in the Pocono 500 *and* the Ford 500 is: {Johnson, Gordon, Busch}

d. Part (c) represents the intersection because it asks for the drivers in both categories.

103. a. The set of students who had a grade of A on the first *or* second tests is: {Albert, Carmen, Frank, Linda, Barbara, Jason, David, Earl, Kate, Ingrid}

b. Part (a) represents the union because it asks for the students in either category.

c. The set of students who had a grade of A on the first *and* second tests is: {Frank, Linda}

d. Part (c) represents the intersection because it asks for the students in both categories.

105. a. The set of the five most populous countries in 2005 *or* 2050 is: {China, India, United States, Indonesia, Brazil, Nigeria}

b. The set of the five most populous countries in 1950 *or* 2050 is: {China, India, United States, Russia, Japan, Indonesia, Nigeria}

c. The set of the five most populous countries in 1950 *and* 2050 is: {China, India, United States}

d. The set of the five most populous countries in 2005 *and* 2050 is: {China, India, United States, Indonesia}

e. The set of the five most populous countries in 1950 *and* 2005 *and* 2050is: {China, India, United States}

107. a. A = {Alex, James}
B = {Alex, James, George, Connor}
C = {Alex, Stephen}
D = {Alex, George, Connor}

b. $A \cap B \cap C \cap D = \{\text{Alex}\}$

c. Only Alex met all the requirements to receive the Wolf Badge.

109. a. $A = \{1,3,4,5,6,7\}$

b. $B = \{2,3,4,6,8,9\}$

c. $A \cup B = \{1,2,3,4,5,6,7,8,9\}$

d. $A \cap B = \{3,4,6\}$

111. a. The set $\{x | x > 1 \text{ and } x \in N\}$ does not contain fractions and decimal numbers while the set $\{x | x > 1\}$ does contain fractions and decimal number.

b. {2, 3, 4, 5, ...}

c. No, it is not possible to list all real numbers greater than 1 in roster form.

113.

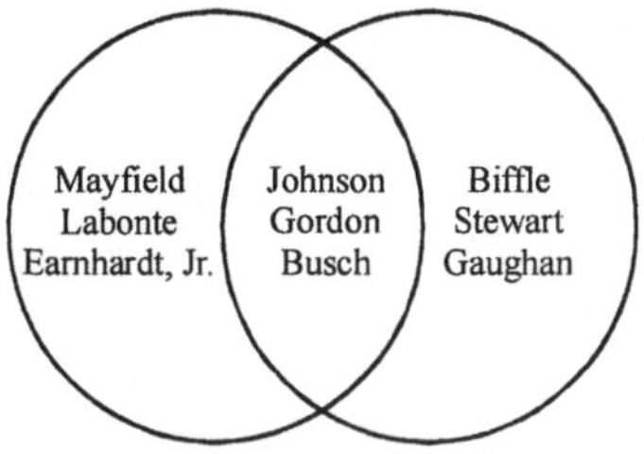

Exercise Set 1.3

1. Additive inverses or opposites are two numbers that sum to 0.

3. No, the absolute value of every real number is not always a positive number because $|0|$ is not positive.

5. Since a and $-a$ are the same distance from 0 on a number line, $|a| = |-a|$, is true for all real numbers, $\mathbb{R}$.

7. Since $|6| = 6$ and $|-6| = 6$, the desired values for a are 6 and −6.

9. Since $|a| \geq 0$ for all real numbers, there are no values of a for which $|a| = -9$.

11 – 13. Answers will vary.

15. $-\frac{a}{b}$ or $\frac{-a}{b}$

17. **a.** $a+b=b+a$

b. Answers will vary.

19. In general, $a+(b\cdot c)\neq(a+b)\cdot(a+c)$. To see this, consider $2+(3\cdot 4)$ and $(2+3)\cdot(2+4)$. The left side is $2+(3\cdot 4)=2+12=14$ and the right side is $(2+3)\cdot(2+4)=5\cdot 6=30$.

21. $|5|=5$

23. $|-7|=7$

25. $\left|-\frac{7}{8}\right|=\frac{7}{8}$

27. $|0|=0$

29. $-|-7|=-7$

31. $-\left|\frac{5}{9}\right|=-\frac{5}{9}$

33. $|-9|=9$ and $|9|=9$, so $|-9|=|9|$.

35. $|-8|=8$ and $-8=-8$, so $|-8|>-8$.

37. $|-\pi|=\pi\approx 3.14$ and $-3=-3$, so $|-\pi|>-3$.

39. $|-7|=7$ and $-|2|=-2$, so $|-7|>-|2|$.

41. $-(-3)=3$ and $-|-3|=-3$, so $-(-3)>-|-3|$.

43. $|19|=19$ and $|-25|=25$, so $|19|<|-25|$.

45. $-|5|, -2, -1, |-3|, 4$

47. $-32, -|4|, 4, |-7|, 15$

49. $-|-6.5|, -6.1, |-6.3|, |6.4|, 6.8$

51. $-2, \frac{1}{3}, \left|-\frac{1}{2}\right|, \left|\frac{3}{5}\right|, \left|-\frac{3}{4}\right|$

53. $7+(-4)=3$

55. $-12+(-10)=-22$

57. $-9-(-5)=-9+5=-4$

59. $\frac{4}{5}-\frac{6}{7}=\frac{28}{35}-\frac{30}{35}=\frac{28-30}{35}=-\frac{2}{35}$

61. $-14.21-(-13.22)=-14.21+13.22=-0.99$

63.
$$\begin{aligned}10-(-2.31)+(-4.39)&=10+2.31-4.39\\&=12.31-4.39\\&=7.92\end{aligned}$$

65.
$$\begin{aligned}9.9-|8.5|-|17.6|&=9.9-8.5-17.6\\&=1.4-17.6\\&=-16.2\end{aligned}$$

67. $|17-12|-|3|=|5|-|3|=5-3=2$

69.
$$\begin{aligned}-|-3|-|7|+(6+|-2|)&=-|-3|-|7|+(6+2)\\&=-|-3|-|7|+8\\&=-3-7+8\\&=-10+8\\&=-2\end{aligned}$$

71.
$$\begin{aligned}\left(\frac{3}{5}+\frac{3}{4}\right)-\frac{1}{2}&=\left(\frac{12}{20}+\frac{15}{20}\right)-\frac{1}{2}\\&=\left(\frac{12+15}{20}\right)-\frac{1}{2}\\&=\frac{27}{20}-\frac{1}{2}\\&=\frac{27}{20}-\frac{10}{20}\\&=\frac{27-10}{20}\\&=\frac{17}{20}\end{aligned}$$

73. $-5\cdot 8=-40$

75. $-4\left(-\frac{5}{16}\right)=\left(-\frac{4}{1}\right)\left(-\frac{5}{16}\right)=\frac{(-4)(-5)}{(1)(16)}=\frac{20}{16}=\frac{5}{4}$

77.
$$\begin{aligned}(-1)(-2)(-1)(2)(-3)&=2(-1)(2)(-3)\\&=-2(2)(-3)\\&=-4(-3)\\&=12\end{aligned}$$

79. $(-1.1)(3.4)(8.3)(-7.6) = -3.74(8.3)(-7.6)$
$= -31.042(-7.6)$
$= 235.9192$

81. $-55 \div (-5) = \dfrac{-55}{-5} = 11$

83. $-\dfrac{5}{9} \div \dfrac{-5}{9} = \dfrac{-5}{9} \cdot \dfrac{9}{-5} = \dfrac{-45}{-45} = 1$

85. $\left(-\dfrac{3}{4}\right) \div |-16| = -\dfrac{3}{4} \div 16 = -\dfrac{3}{4} \cdot \dfrac{1}{16} = -\dfrac{3 \cdot 1}{4 \cdot 16} = -\dfrac{3}{64}$

87. $\left|-\dfrac{7}{6}\right| \div \left|\dfrac{-1}{2}\right| = \dfrac{7}{6} \div \dfrac{1}{2} = \dfrac{7}{6} \cdot \dfrac{2}{1} = \dfrac{7 \cdot 2}{6 \cdot 1} = \dfrac{14}{6} = \dfrac{7}{3}$

89. $10 - 14 = 10 + (-14) = -4$

91. $7 - (-13) = 7 + 13 = 20$

93. $3\left(-\dfrac{2}{3}\right)\left(-\dfrac{5}{2}\right) = -\dfrac{6}{3}\left(-\dfrac{5}{2}\right) = \dfrac{30}{6} = 5$

95. $-14.4 - (-9.6) - 15.8 = -14.4 + 9.6 - 15.8$
$= -4.8 - 15.8$
$= -20.6$

97. $9 - (6 - 5) - (-2 - 1) = 9 - (1) - (-3)$
$= 9 - 1 + 3$
$= 8 + 3$
$= 11$

99. $-|12| \cdot \left|\dfrac{-1}{2}\right| = -(12)\left(\dfrac{1}{2}\right) = -6$

101. $\left|\dfrac{-9}{4}\right| \div \left|\dfrac{-4}{9}\right| = \dfrac{9}{4} \div \dfrac{4}{9} = \dfrac{9}{4} \cdot \dfrac{9}{4} = \dfrac{9 \cdot 9}{4 \cdot 4} = \dfrac{81}{16}$

103. $5 - |-7| + 3 - |-2| = 5 - 7 + 3 - 2$
$= -2 + 3 - 2$
$= 1 - 2$
$= -1$

105. $\left(-\dfrac{3}{5} - \dfrac{4}{9}\right) - \left(-\dfrac{2}{3}\right) = \left(-\dfrac{3}{5} - \dfrac{4}{9}\right) + \dfrac{2}{3}$
$= \left(-\dfrac{27}{45} - \dfrac{20}{45}\right) + \dfrac{2}{3}$
$= \left(\dfrac{-27 - 20}{45}\right) + \dfrac{2}{3}$
$= -\dfrac{47}{45} + \dfrac{2}{3}$
$= -\dfrac{47}{45} + \dfrac{30}{45}$
$= \dfrac{-47 + 30}{45}$
$= \dfrac{-17}{45}$ or $-\dfrac{17}{45}$

107. $(25 - |32|)(-7 - 4) = (25 - 32)(-7 - 4)$
$= (-7)(-11)$
$= 77$

109. $-10 - 29 = -10 + (-29) = -39$

111. $7(3)(0)(-15.2) = 0$

113. $r + s = s + r$; Commutative property of addition

115. $b \cdot 0 = 0$; Multiplicative property of zero

117. $(x + 3) + 6 = x + (3 + 6)$; Associative property of addition

119. $x = 1 \cdot x$; Identity property of multiplication

121. $2(xy) = (2x)y$; Associative property of multiplication

123. $4(x + y + 2) = 4x + 4y + 8$; Distributive property

125. $5 + 0 = 5$; Identity property of addition

127. $3 + (-3) = 0$; Inverse property of addition

129. $-(-x) = x$; Double negative property

131. -6 is the additive inverse
$\dfrac{1}{6}$ is the multiplicative inverse

133. $\frac{22}{7}$ is the additive inverse

$-\frac{7}{22}$ is the multiplicative inverse

135. The change in temperature is
$45° - (-4°) = 45° + 4° = 49°\text{ F}$

137. Final depth is $-358.9 + 210.7 = -148.2$ ft or 148.2 ft below the starting point.

139. $-69.7 - (-79.8) = -69.7 + 79.8 = 10.1$
The difference in the temperatures is 10.1°F.

141. $100 \cdot \$30.30 = \$3030$
$100 \cdot \$42.37 = \$4237$
$\$4237 - \$3030 = \$1207$
Ron had a gain of $2080.00.

143. Answers will vary.

145. First-year income $\approx \$52{,}000$
First-year expenditures $\approx 28{,}000$
$52{,}000 - 28{,}000 = 24{,}000$
The average first year profit will be approximately $24,000.

147.
$$\begin{aligned}&1+2-3+4+5-6+7+8-9+10+11-12+13+14-15+16+17-18+19+20-21+22+23-24\\&=(1+2-3)+(4+5-6)+(7+8-9)+(10+11-12)+(13+14-15)+(16+17-18)+(19+20-21)+(22+23-24)\\&=0+3+6+9+12+15+18+21\\&=84\end{aligned}$$

149.
$$\begin{aligned}&\frac{(1)(-2)(3)(-4)(5)\ldots(97)(-98)}{(-1)(2)(-3)(4)(-5)\ldots(-97)(98)}\\&=\frac{(1)(-2)}{(-1)(2)}\cdot\frac{(3)(-4)}{(-3)(4)}\cdot\ldots\cdot\frac{(97)(-98)}{(-97)(98)}\\&=\left(\frac{-2}{-2}\right)\left(\frac{-12}{-12}\right)\cdot\ldots\cdot\left(\frac{-9506}{-9506}\right)\\&=1\cdot1\cdot\ldots\cdot1\\&=1\end{aligned}$$

150. True; the set of real numbers contains the irrational numbers.

151. The set of natural numbers is {1, 2, 3, 4, ...}

152. **a.** 3, 4, –2, and 0 are integers.

b. $3, 4, -2, \frac{5}{6}$, and 0 are rational numbers.

c. $\sqrt{11}$ is an irrational number.

d. $3, 4, -2, \frac{5}{6}, \sqrt{11}$, and 0 are real numbers.

153. **a.** $A \cup B = \{1, 4, 7, 9, 12, 15\}$

b. $A \cap B = \{4, 7\}$

154. $\{x \mid -4 < x \le 6\}$

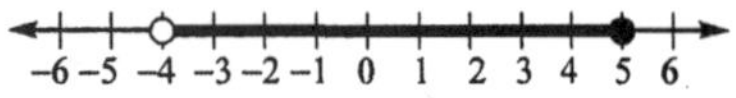

Exercise Set 1.4

1. **a.** In the expression a^n, a is called the base.

b. In the expression a^n, n is called the exponent.

3. **a.** In the expression $\sqrt[n]{a}$, n is called the index.

b. In the expression $\sqrt[n]{a}$, a is called the radicand.

5. The principal square root of a positive number radicand is the positive number whose square equals the radicand.

7. An odd root of a negative number will be negative because a negative number raised to an odd power is a negative number.

9. Parentheses, exponents and roots, multiplication and division from left to right, addition and subtraction from left to right.

11. **a.** Answers will vary.

b.
$$\begin{aligned}16 \div 2^2 + 6\cdot 4 - 24 \div 6 &= 16 \div 4 + 6\cdot 4 - 24 \div 6\\&= 4 + 24 - 4\\&= 28 - 4\\&= 24\end{aligned}$$

13. $3^2 = 3 \cdot 3 = 9$

15. $-3^2 = -(3)(3) = -9$

17. $(-3)^2 = (-3)(-3) = 9$

19. $-\left(\frac{3}{5}\right)^4 = -\left(\frac{3}{5}\right)\left(\frac{3}{5}\right)\left(\frac{3}{5}\right)\left(\frac{3}{5}\right) = -\frac{81}{625}$

21. $\sqrt{49} = 7$ since $7 \cdot 7 = 49$

23. $-\sqrt{36} = -6$ since $6 \cdot 6 = 36.$

25. $\sqrt[3]{-27} = -3$ since $(-3)(-3)(-3) = -27$

27. $\sqrt[3]{0.001} = 0.1$ since $(0.1)(0.1)(0.1) = 0.001$

29. $(0.35)^4 \approx 0.015$

31. $\left(-\frac{13}{12}\right)^8 \approx 1.897$

33. $(6.721)^{5.9} \approx 76{,}183.335$

35. $\sqrt[3]{26} \approx 2.962$

37. $\sqrt[5]{362.65} \approx 3.250$

39. $-\sqrt[3]{\frac{20}{53}} \approx -0.723$

41. **a.** x^2 becomes $3^2 = 3 \cdot 3 = 9$

b. $-x^2$ becomes $-3^2 = -3 \cdot 3 = -9$

43. **a.** x^2 becomes $10^2 = 10 \cdot 10 = 100$

b. $-x^2$ becomes $-10^2 = -10 \cdot 10 = -100$

45. **a.** x^2 becomes $(-1)^2 = (-1)(-1) = 1$

b. $-x^2$ becomes $-(-1)^2 = -(-1)(-1) = -1$

47. **a.** x^2 becomes $\left(\frac{1}{3}\right)^2 = \left(\frac{1}{3}\right)\left(\frac{1}{3}\right) = \frac{1}{9}$

b. $-x^2$ becomes $-\left(\frac{1}{3}\right)^2 = -\left(\frac{1}{3}\right)\left(\frac{1}{3}\right) = -\frac{1}{9}$

49. **a.** x^3 becomes $3^3 = 3 \cdot 3 \cdot 3 = 27$

b. $-x^3$ becomes $-3^3 = -(3 \cdot 3 \cdot 3) = -27$

51. **a.** x^3 becomes $(-5)^3 = (-5)(-5)(-5) = -125$

b. $-x^3$ becomes $-(-5)^3 = -(-5)(-5)(-5) = 125$

53. **a.** x^3 becomes $(-2)^3 = (-2)(-2)(-2) = -8$

b. $-x^3$ becomes $-(-2)^3 = -(-2)(-2)(-2) = 8$

55. **a.** x^3 becomes $\left(\frac{2}{5}\right)^3 = \left(\frac{2}{5}\right)\left(\frac{2}{5}\right)\left(\frac{2}{5}\right) = \frac{8}{125}$

b. $-x^3$ becomes $-\left(\frac{2}{5}\right)^3 = -\left(\frac{2}{5}\right)\left(\frac{2}{5}\right)\left(\frac{2}{5}\right) = -\frac{8}{125}$

57. $4^2 + 2^3 - 2^2 - 3^3 = 16 + 8 - 4 - 27 = 24 - 31 = -7$

59.
$$\begin{aligned} -2^2 - 2^3 + 1^{10} + (-2)^3 &= -4 - 8 + 1 + (-8) \\ &= -4 - 8 + 1 - 8 \\ &= -12 + 1 - 8 \\ &= -11 - 8 \\ &= -19 \end{aligned}$$

61.
$$\begin{aligned} (1.5)^2 - (3.9)^2 + (-2.1)^3 &= 2.25 - 15.21 - 9.261 \\ &= -12.96 - 9.261 \\ &= -22.221 \end{aligned}$$

63.
$$\begin{aligned} \left(-\frac{1}{2}\right)^4 - \left(\frac{1}{2}\right)^2 + \left(-\frac{1}{2}\right)^3 &= \frac{1}{16} - \frac{1}{4} - \frac{1}{8} \\ &= \frac{2}{32} - \frac{8}{32} - \frac{4}{32} \\ &= -\frac{10}{32} \\ &= -\frac{5}{16} \end{aligned}$$

65. $3 + 5 \cdot 8 = 3 + 40 = 43$

67. $18 - 6 \div 6 + 8 = 18 - 1 + 8 = 17 + 8 = 25$

69.
$$\begin{aligned} \frac{3}{4} \div \frac{1}{2} - 2 + 5 \div 10 &= \frac{3}{4} \cdot \frac{2}{1} - 2 + 5 \div 10 \\ &= \frac{3}{2} - 2 + 5 \div 10 \\ &= \frac{3}{2} - 2 + \frac{1}{2} \\ &= \frac{3}{2} - \frac{4}{2} + \frac{1}{2} \\ &= 0 \end{aligned}$$

71.
$$\begin{aligned}\frac{1}{2}\cdot\frac{2}{3}\div\frac{3}{4}-\frac{1}{6}\cdot\left(-\frac{1}{3}\right)&=\frac{1}{3}\div\frac{3}{4}-\frac{1}{6}\cdot\left(-\frac{1}{3}\right)\\&=\frac{1}{3}\cdot\frac{4}{3}-\frac{1}{6}\cdot\left(-\frac{1}{3}\right)\\&=\frac{8}{18}+\frac{1}{18}\\&=\frac{9}{18}\\&=\frac{1}{2}\end{aligned}$$

73.
$$\begin{aligned}10\div\left[(3+2^2)-(2^4-8)\right]&=10\div\left[(3+4)-(16-8)\right]\\&=10\div\left[7-8\right]\\&=10\div(-1)\\&=-10\end{aligned}$$

75.
$$\begin{aligned}5\left(\sqrt[3]{27}+\sqrt[5]{32}\right)\div\frac{\sqrt{100}}{2}&=5(3+2)\div\frac{\sqrt{100}}{2}\\&=5(5)\div\frac{10}{2}\\&=25\div 5\\&=5\end{aligned}$$

77.
$$\begin{aligned}\{[(12-15)-3]-2\}^2&=\{[(-3)-3]-2\}^2\\&=[(-6)-2]^2\\&=(-8)^2\\&=64\end{aligned}$$

79.
$$\begin{aligned}4\left[5(16-6)\div(25\div 5)^2\right]^2&=4\left[5(10)\div(5)^2\right]^2\\&=4[50\div 25]^2\\&=4[2]^2\\&=4\cdot 4\\&=16\end{aligned}$$

81.
$$\begin{aligned}\frac{4-(2+3)^2-6}{4(3-2)-3^2}&=\frac{4-(5)^2-6}{4(1)-3^2}\\&=\frac{4-25-6}{4(1)-9}\\&=\frac{4-25-6}{4-9}\\&=\frac{-27}{-5}\\&=\frac{27}{5}\end{aligned}$$

83.
$$\begin{aligned}\frac{8+4\div 2\cdot 3+4}{5^2-3^2\cdot 2-7}&=\frac{8+4\div 2\cdot 3+4}{25-9\cdot 2-7}\\&=\frac{8+2\cdot 3+4}{25-18-7}\\&=\frac{8+6+4}{25-18-7}\\&=\frac{14+4}{7-7}\\&=\frac{18}{0}\text{ which is undefined}\end{aligned}$$

85.
$$\begin{aligned}\frac{8-[4-(3-1)^2]}{5-(-3)^2+4\div 2}&=\frac{8-[4-(2)^2]}{5-(-3)^2+4\div 2}\\&=\frac{8-(4-4)}{5-9+4\div 2}\\&=\frac{8-0}{5-9+4\div 2}\\&=\frac{8}{5-9+2}\\&=\frac{8}{-4+2}\\&=\frac{8}{-2}\\&=-4\end{aligned}$$

87.
$$\begin{aligned}-2|-3|-\sqrt{36}\div|2|+3^2&=-2(3)-6\div 2+3^2\\&=-2(3)-6\div 2+9\\&=-6-3+9\\&=-9+9\\&=0\end{aligned}$$

89.
$$\begin{aligned}\frac{6-|-4|-4|8-5|}{5-6\cdot 2\div|-6|}&=\frac{6-|-4|-4|3|}{5-6\cdot 2\div|-6|}\\&=\frac{6-4-4\cdot 3}{5-6\cdot 2\div 6}\\&=\frac{6-4-12}{5-12\div 6}\\&=\frac{2-12}{5-2}\\&=\frac{-10}{3}\text{ or }-\frac{10}{3}\end{aligned}$$

91. $\frac{2}{5}\left[\sqrt[3]{27}-|-9|+4-3^2\right]^2 = \frac{2}{5}[3-(9)+4-9]^2$
$= \frac{2}{5}(3-9+4-9)^2$
$= \frac{2}{5}(-6+4-9)^2$
$= \frac{2}{5}(-2-9)^2$
$= \frac{2}{5}(-11)^2$
$= \frac{2}{5}(121)$
$= \frac{242}{5}$

93. $\frac{24-5-4^2}{|-8|+4-2(3)}+\frac{4-(-3)^2+|4|}{3^2-4\cdot 3+|-7|}$
$= \frac{24-5-16}{8+4-2(3)}+\frac{4-(9)+4}{9-4\cdot 3+7}$
$= \frac{24-5-16}{8+4-6}+\frac{4-9+4}{9-12+7}$
$= \frac{19-16}{12-6}+\frac{-5+4}{-3+7}$
$= \frac{3}{6}+\frac{-1}{4}$
$= \frac{2}{4}+\frac{-1}{4}$
$= \frac{2-1}{4}$
$= \frac{1}{4}$

95. Substitute 2 for x:
$5x^2+4x = 5(2)^2+4(2)$
$= 5\cdot 4+4(2)$
$= 20+8$
$= 28$

97. Substitute −1 for x:
$-9x^2+3x-29 = -9(-1)^2+3(-1)-29$
$= -9\cdot 1+3(-1)-29$
$= -9-3-29$
$= -12-29$
$= -41$

99. Substitute −4 for x:
$16(x+5)^3-25(x+5)$
$= 16[(-4)+5]^3-25[(-4)+5]$
$= 16(1)^3-25(1)$
$= 16\cdot 1-25\cdot 1$
$= 16-25$
$= -9$

101. Substitute 1 for x and −3 for y:
$6x^2+3y^2-15 = 6(1)^2+3(-3)^3-15$
$= 6(1)+3(-27)-15$
$= 6+(-81)-15$
$= -75-15$
$= -90$

103. Substitute 4 for a and −1 for b:
$3(a+b)^2+4(a+b)-6$
$= 3[4+(-1)]^2+4[4+(-1)]-6$
$= 3(3)^2+4(3)-6$
$= 3(9)+4(3)-6$
$= 27+12-6$
$= 39-6$
$= 33$

105. Substitute 4 for x:
$-8-\{x-[2x-(x-3)]\}$
$= -8-\{4-[2\cdot 4-(4-3)]\}$
$= -8-\{4-[2\cdot 4-1]\}$
$= -8-\{4-[8-1]\}$
$= -8-[4-(7)]$
$= -8-(4-7)$
$= -8-(-3)$
$= -8+3$
$= -5$

107. Substitute 6 for a, −11 for b, and 3 for c:
$\frac{-b+\sqrt{b^2-4ac}}{2a} = \frac{-(-11)+\sqrt{(-11)^2-4(6)(3)}}{2(6)}$
$= \frac{11+\sqrt{121-72}}{12}$
$= \frac{11+\sqrt{49}}{12}$
$= \frac{11+7}{12} = \frac{18}{12} = \frac{3}{2}$

109. The expression is $\frac{7y-14}{2}$.

Now substitute 6 for y:

$$\frac{7y-14}{2}=\frac{7(6)-14}{2}=\frac{42-14}{2}=\frac{28}{2}=14$$

111. The expression is $6(3x+6)-9$.

Now substitute 3 for x:

$$\begin{aligned}6(3x+6)-9&=6(3\cdot 3+6)-9\\&=6(9+6)-9\\&=6(15)-9\\&=90-9\\&=81\end{aligned}$$

113. The expression is $\left(\frac{x+3}{2y}\right)^2-3$

Now substitute 5 for x and 2 for y:

$$\begin{aligned}\left(\frac{x+3}{2y}\right)^2-3&=\left(\frac{5+3}{2\cdot 2}\right)^2-3\\&=\left(\frac{5+3}{4}\right)^2-3\\&=\left(\frac{8}{4}\right)^2-3\\&=2^2-3\\&=4-3\\&=1\end{aligned}$$

115. a. Substitute 3 for x:

$\text{distance}=8.2x=8.2(3)=24.6$

Frank can travel 24.6 miles in 3 hours.

b. Substitute 7 for x:

$\text{distance}=8.2x=8.2(7)=57.4$

Frank can travel 57.4 miles in 7 hours.

117. a. Substitute 2 for x:

$$\begin{aligned}\text{height}&=-16x^2+72x+22\\&=-16(2)^2+72(2)+22\\&=-16(4)+72(2)+22\\&=-64+144+22\\&=80+22\\&=102\end{aligned}$$

After 2 seconds, the baseball will be 102 feet above the ground.

b. Substitute 4 for x:

$$\begin{aligned}\text{height}&=-16x^2+72x+22\\&=-16(4)^2+72(4)+22\\&=-16(16)+72(4)+22\\&=-256+288+22\\&=32+22\\&=54\end{aligned}$$

After 4 seconds, the baseball will be 54 feet above the ground.

119. a. 2007 is represented by $x=5$; substitute 5 for x:

$$\begin{aligned}\text{spending}&=26.865x+488.725\\&=26.865(5)+488.725\\&=134.325+488.725\\&=623.050\end{aligned}$$

The amount each consumer will spend on holiday gifts in 2007 will be \$623.05.

b. 2015 is represented by $x=13$; substitute 13 for x:

$$\begin{aligned}\text{spending}&=26.865x+488.725\\&=26.865(13)+488.725\\&=349.245+488.725\\&=837.970\end{aligned}$$

The amount each consumer will spend on holiday gifts in 2015 will be \$837.97.

121. a. 2000 is represented by $x=8$; substitute 8 for x:

$$\begin{aligned}\text{number of trips}&=0.065x^2-0.39x+8.47\\&=0.065(8)^2-0.39(8)+8.47\\&=0.065(64)-0.39(8)+8.47\\&=4.16-3.12+8.47\\&=9.51\end{aligned}$$

The number of trips made by public transportation in 2000 was approximately 9.51 billion.

b. 2010 is represented by $x=18$; substitute 18 for x:

$$\begin{aligned}\text{number of trips}&=0.065x^2-0.39x+8.47\\&=0.065(18)^2-0.39(18)+8.47\\&=0.065(324)-0.39(18)+8.47\\&=21.06-7.02+8.47\\&=22.51\end{aligned}$$

The number of trips made by public transportation in 2010 will be approximately 22.51 billion.

123. a. 2010 is represented by $x = 8$; substitute 8 for x:

$$\begin{aligned}\text{sales} &= 13.5x + 189.83\\ &= 13.5(8) + 189.83\\ &= 108 + 189.83\\ &= 297.83\end{aligned}$$

The sales from auctions in 2010 will be approximately \$297.83.

b. 2018 is represented by $x = 16$; substitute 16 for x:

$$\begin{aligned}\text{sales} &= 13.5x + 189.83\\ &= 13.5(16) + 189.83\\ &= 216 + 189.83\\ &= 405.83\end{aligned}$$

The sales from auctions in 2018 will be approximately \$405.83.

125. a. Substitute 10 for x:
percent of children

$$\begin{aligned}&= 0.23x^2 - 1.98x + 4.42\\ &= 0.23(10)^2 - 1.98(10) + 4.42\\ &= 0.23(100) - 1.98(10) + 4.42\\ &= 23 - 19.8 + 4.42\\ &= 7.62\end{aligned}$$

The percent of all 10-year-olds who are latchkey kids is 7.62%.

b. Substitute 14 for x.
percent of children

$$\begin{aligned}&= 0.23x^2 - 1.98x + 4.42\\ &= 0.23(14)^2 - 1.98(14) + 4.42\\ &= 0.23(196) - 1.98(14) + 4.42\\ &= 45.08 - 27.72 + 4.42\\ &= 21.78\end{aligned}$$

The percent of all 14-year-olds who are latchkey kids is 21.78%.

127. a. 1991 is represented by $x = 1$; substitute 1 for x:

$$\begin{aligned}\text{sales} &= 0.062x^2 + 0.020x + 1.18\\ &= 0.062(1)^2 + 0.020(1) + 1.18\\ &= 0.062 + 0.020 + 1.18\\ &= 1.262\end{aligned}$$

The amount of sales of organically grown food in 1991 was about \$1.262 billion.

b. 2007 is represented by $x = 17$; substitute 17 for x:

$$\begin{aligned}\text{sales} &= 0.062x^2 + 0.020x + 1.18\\ &= 0.062(17)^2 + 0.020(17) + 1.18\\ &= 17.918 + 0.34 + 1.18\\ &= 19.438\end{aligned}$$

The amount of sales of organically grown food in 2001 was about \$19.438 billion.

129. a. $A \cap B = \{b, c, f\}$

b. $A \cup B = \{a, b, c, d, f, g, h\}$

130. $|a| = |-a|$ for all real numbers or $\mathbb{R}$.

131. Since $|a| = \begin{cases} a & \geq 0 \\ -a & a < 0 \end{cases}$ then $|a| = a$ for $a \geq 0$.

132. $|a| = 6$ for $a = 6$ or $a = -6$ since $|6| = 6$ and $|-6| = 6$.

133. $-|6|, -4, -|-2|, 0, |-5|$

134. Associative property of addition

Mid-Chapter Test: 1.1 – 1.4

1. Answers will vary.

2. $A \cup B = \{-3, -2, -1, 0, 1, 2, 3, 5\}$
$A \cap B = \{-1, 1\}$

3. $D = \{0, 5, 10, 15, \ldots\}$ is the set of whole number multiples of 5.

4. $\{x \mid x \geq 3\}$

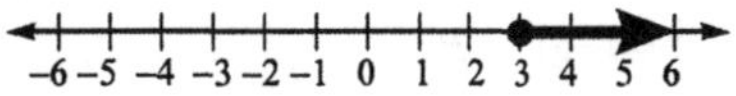

5. $\frac{3}{5} > \frac{4}{9}$

6. $\{x \mid -5 \leq x < 2\}$

7. No, W is not a subset of N because $0 \in W$, but $0 \notin N$.

8. $-15, |-6|, 7, |-17|$

9. $7-2.3-(-4.5)=7-2.3+4.5$
$=4.7+4.5$
$=9.2$

10. $\left(\frac{2}{5}+\frac{1}{3}\right)-\frac{1}{2}=\left(\frac{6}{15}+\frac{5}{15}\right)-\frac{1}{2}$
$=\frac{11}{15}-\frac{1}{2}$
$=\frac{22}{30}-\frac{15}{30}$
$=\frac{7}{30}$

11. $(5)(-2)(3.2)(-8)=-10(3.2)(-8)$
$=-32(-8)$
$=256$

12. $\left|-\frac{8}{13}\right|\div(-2)=\frac{8}{13}\div(-2)$
$=\frac{8}{13}\cdot\left(\frac{1}{-2}\right)$
$=\frac{8(1)}{13(-2)}$
$=\frac{8}{-26}$
$=-\frac{4}{13}$

13. $(7-|-2|)-(-8+|16|)=(7-2)-(-8+16)$
$=5-8$
$=5+(-8)$
$=-3$

14. $5(x+y)=5x+5y$ illustrates the distributive property

15. $\sqrt{0.81}=0.9$ because $(0.9)(0.9)=0.81$.

16. a. $x^2=(-6)^2=36$

b. $-x^2=-(-6)^2=-36$

17. a. Grouping symbols, exponents and roots, multiplication and division from left to right, addition and subtraction from left to right.

b. $4-2\cdot 3^2=4-2\cdot 9=4-18=-14$

18. $5\cdot 4\div 10+2^5-8=5\cdot 4\div 10+32-8$
$=20\div 10+32-8$
$=2+32-8$
$=34-8$
$=26$

19. $\frac{1}{4}\left\{\left[(12\div 4)^2-7\right]^3\div 2\right\}^2=\frac{1}{4}\left\{\left[(3)^2-7\right]^3\div 2\right\}^2$
$=\frac{1}{4}\left\{\left[9-7\right]^3\div 2\right\}^2$
$=\frac{1}{4}\left\{\left[2\right]^3\div 2\right\}^2$
$=\frac{1}{4}\{8\div 2\}^2$
$=\frac{1}{4}\{4\}^2$
$=\frac{1}{4}\cdot 16$
$=4$

20. $\frac{\sqrt{16}+\left(\sqrt{49}-6\right)^4}{\sqrt[3]{-27}-\left(4-3^2\right)}=\frac{\sqrt{16}+(7-6)^4}{\sqrt[3]{-27}-(4-9)}$
$=\frac{\sqrt{16}+(1)^4}{\sqrt[3]{-27}-(-5)}$
$=\frac{4+1}{-3-(-5)}$
$=\frac{4+1}{-3+5}$
$=\frac{5}{2}$

Exercise Set 1.5

1. a. $a^m\cdot a^n=a^{m+n}$

b. Answers will vary.

3. a. $a^0=1,\ a\neq 0$

b. Answers will vary.

5. a. $(ab)^m=a^m b^m$

b. Answers will vary.

7. a. $\left(\frac{a}{b}\right)^m=\frac{a^m}{b^m},\ b\neq 0$

b. Answers will vary.

9. $x^{-1} = 5, \frac{1}{x} = 5, \frac{1}{x} = \frac{5}{1}$

   Cross multiply: $5x = 1$

   Solve for x: $x = \frac{1}{5}$

11. a. The opposite of x is $-x$.

   The reciprocal of x is $\frac{1}{x}$.

   b. x^{-1} or $\frac{1}{x}$

   c. $-x$

13. $2^3 \cdot 2^2 = 2^{3+2} = 3^5 = 32$

15. $\frac{3^7}{3^5} = 3^{7-5} = 3^2 = 9$

17. $9^{-2} = \frac{1}{9^2} = \frac{1}{81}$

19. $\frac{1}{5^{-3}} = 2^3 = 8$

21. $15^0 = 1$

23. $(2^3)^2 = 2^{3 \cdot 2} = 2^6 = 64$

25. $(2 \cdot 4)^2 = 2^2 \cdot 4^2 = 4 \cdot 16 = 64$

27. $\left(\frac{4}{7}\right)^2 = \frac{4^2}{7^2} = \frac{16}{49}$

29. a. $3^{-2} = \frac{1}{3^2} = \frac{1}{9}$

   b. $(-3)^{-2} = \frac{1}{(-3)^2} = \frac{1}{9}$

   c. $-3^{-2} = -\frac{1}{3^2} = -\frac{1}{9}$

   d. $-(-3)^{-2} = -\frac{1}{(-3)^2} = -\frac{1}{9}$

31. a. $\left(\frac{1}{2}\right)^{-1} = \left(\frac{2}{1}\right)^1 = 2$

   b. $\left(-\frac{1}{2}\right)^{-1} = \left(-\frac{2}{1}\right)^1 = -2$

   c. $-\left(\frac{1}{2}\right)^{-1} = -\left(\frac{2}{1}\right)^1 = -2$

   d. $-\left(-\frac{1}{2}\right)^{-1} = -\left(-\frac{2}{1}\right)^1 = -(-2) = 2$

33. a. $5x^0 = 5 \cdot x^0 = 5 \cdot 1 = 5$

   b. $-5x^0 = -5 \cdot x^0 = -5 \cdot 1 = -5$

   c. $(-5x)^0 = 1$

   d. $-(-5x)^0 = -1 \cdot (-5x)^0 = -1 \cdot 1 = -1$

35. a. $3xyz^0 = 3 \cdot x \cdot y \cdot z^0 = 3 \cdot x \cdot y \cdot 1 = 3xy$

   b. $(3xyz)^0 = 1$

   c. $3x(yz)^0 = 3x \cdot (yz)^0 = 3x \cdot 1 = 3x$

   d. $3(xyz)^0 = 3 \cdot (xyz)^0 = 3 \cdot 1 = 3$

37. $7y^{-3} = 7 \cdot \frac{1}{y^3} = \frac{7}{y^3}$

39. $\frac{9}{x^{-4}} = 9x^4$

41. $\frac{2a}{b^{-3}} = 2ab^3$

43. $\frac{13m^{-2}n^{-3}}{2} = \frac{13}{2m^2n^3}$

45. $\frac{5x^{-2}y^{-3}}{z^{-4}} = \frac{5z^4}{x^2y^3}$

47. $\frac{9^{-1}x^{-1}}{y} = \frac{1}{9^1x^1y} = \frac{1}{9xy}$

49. $2^5 \cdot 2^{-7} = 2^{5+(-7)} = 2^{-2} = \frac{1}{2^2} = \frac{1}{4}$

51. $x^6 \cdot x^{-4} = x^{6+(-4)} = x^2$

53. $\frac{8^5}{8^3} = 8^{5-3} = 8^2 = 64$

55. $\dfrac{7^{-5}}{7^{-3}} = 7^{-5-(-3)} = 7^{-5+3} = 7^{-2} = \dfrac{1}{7^2} = \dfrac{1}{49}$

57. $\dfrac{m^{-6}}{m^5} = m^{-6-5} = m^{-11} = \dfrac{1}{m^{11}}$

59. $\dfrac{5w^{-2}}{w^{-7}} = 5w^{-2-(-7)} = 5w^{-2+7} = 5w^5$

61.
$$\begin{aligned} 3a^{-2}\cdot 4a^{-6} &= 3\cdot 4\cdot a^{-2}\cdot a^{-6} \\ &= 12a^{-2+(-6)} \\ &= 12a^{-8} \\ &= \frac{12}{a^8} \end{aligned}$$

63.
$$\begin{aligned} \left(-3p^{-2}\right)\left(-p^3\right) &= (-3)(-1)p^{-2}\cdot p^3 \\ &= 3p^{-2+3} \\ &= 3p^1 \\ &= 3p \end{aligned}$$

65.
$$\begin{aligned} \left(5r^2s^{-2}\right)\left(-2r^5s^2\right) &= 5(-2)r^2\cdot r^5\cdot s^{-2}\cdot s^2 \\ &= -10r^{2+5}\cdot s^{-2+2} \\ &= -10r^7s^0 \\ &= -10r^7 \end{aligned}$$

67.
$$\begin{aligned} \left(2x^4y^7\right)\left(4x^3y^{-5}\right) &= 2\cdot 4\cdot x^4\cdot x^3\cdot y^7\cdot y^{-5} \\ &= 8x^{4+3}y^{7+(-5)} \\ &= 8x^7y^2 \end{aligned}$$

69. $\dfrac{33x^5y^{-4}}{11x^3y^2} = \left(\dfrac{33}{11}\right)\dfrac{x^{5-3}}{y^{2-(-4)}} = \dfrac{3x^2}{y^6}$

71. $\dfrac{9xy^{-4}z^3}{-3x^{-2}yz} = \left(\dfrac{9}{-3}\right)\dfrac{x^{1-(-2)}z^{3-1}}{y^{1-(-4)}} = -\dfrac{3x^3z^2}{y^5}$

73. a. $4(a+b)^0 = 4\cdot 1 = 4$

b.
$$\begin{aligned} 4a^0+4b^0 &= 4\cdot a^0+4\cdot b^0 \\ &= 4\cdot 1+4\cdot 1 \\ &= 4+4 \\ &= 8 \end{aligned}$$

c. $(4a+4b)^0 = 1$

d.
$$\begin{aligned} -4a^0+4b^0 &= -4\cdot a^0+4\cdot b^0 \\ &= -4\cdot 1+4\cdot 1 \\ &= -4+4 \\ &= 0 \end{aligned}$$

75. a. $4^{-1}-3^{-1} = \dfrac{1}{4}-\dfrac{1}{3} = \dfrac{3}{12}-\dfrac{4}{12} = -\dfrac{1}{12}$

b. $4^{-1}+3^{-1} = \dfrac{1}{4}+\dfrac{1}{3} = \dfrac{3}{12}+\dfrac{4}{12} = \dfrac{7}{12}$

c.
$$\begin{aligned} 2\cdot 4^{-1}+3\cdot 5^{-1} &= 2\cdot\frac{1}{4}+3\cdot\frac{1}{5} \\ &= \frac{2}{4}+\frac{3}{5} \\ &= \frac{10}{20}+\frac{12}{20} \\ &= \frac{22}{20} \\ &= \frac{11}{10} \text{ or } 1\frac{1}{10} \end{aligned}$$

d.
$$\begin{aligned} (2\cdot 4)^{-1}+(3\cdot 5)^{-1} &= 8^{-1}+15^{-1} \\ &= \frac{1}{8}+\frac{1}{15} \\ &= \frac{15}{120}+\frac{8}{120} \\ &= \frac{23}{120} \end{aligned}$$

77. $\left(3^2\right)^2 = 3^{2\cdot 2} = 3^4 = 81$

79. $\left(3^2\right)^{-2} = 3^{2\cdot(-2)} = 3^{-4} = \dfrac{1}{81}$

81. $\left(b^{-3}\right)^{-2} = b^{(-3)(-2)} = b^6$

83. $(-c)^3 = (-1\cdot c)^3 = (-1)^3\cdot c^3 = -1\cdot c^3 = -c^3$

85. $\left(-4x^{-3}\right)^2 = (-4)^2\cdot\left(x^{-3}\right)^2 = 16\cdot x^{-6} = \dfrac{16}{x^6}$

87. $5^{-1}+2^{-1} = \dfrac{1}{5}+\dfrac{1}{2} = \dfrac{2}{10}+\dfrac{5}{10} = \dfrac{7}{10}$

89. $$\begin{aligned}3\cdot 4^{-2}+9\cdot 8^{-1} &= 3\cdot\frac{1}{4^2}+9\cdot\frac{1}{8^1}\\ &= 3\cdot\frac{1}{16}+9\cdot\frac{1}{8}\\ &= \frac{3}{16}+\frac{9}{8}\\ &= \frac{3}{16}+\frac{18}{16}\\ &= \frac{21}{16}\text{ or }1\frac{5}{16}\end{aligned}$$

91. $$\left(\frac{4b}{3}\right)^{-2}=\left(\frac{3}{4b}\right)^2=\frac{3^2}{(4b)^2}=\frac{9}{16b^2}$$

93. $$\begin{aligned}\left(4x^2y^{-2}\right)^2 &= (4)^2\left(x^2\right)^2\left(y^{-2}\right)^2\\ &= (4)^2x^{2\cdot 2}y^{(-2)\cdot 2}\\ &= 16x^4y^{-4}\\ &= \frac{16x^4}{y^4}\end{aligned}$$

95. $$\begin{aligned}\left(5p^2q^{-4}\right)^{-3} &= 5^{-3}p^{2(-3)}q^{(-4)(-3)}\\ &= 5^{-3}p^{-6}q^{12}\\ &= \frac{q^{12}}{5^3p^6}\\ &= \frac{q^{12}}{125p^6}\end{aligned}$$

97. $$\begin{aligned}\left(-3g^{-4}h^3\right)^{-3} &= (-3)^{-3}g^{(-4)(-3)}h^{3\cdot(-3)}\\ &= (-3)^{-3}g^{12}h^{-9}\\ &= \frac{g^{12}}{(-3)^3h^9}\\ &= \frac{g^{12}}{-27h^9}\\ &= -\frac{g^{12}}{27h^9}\end{aligned}$$

99. $$\left(\frac{3j}{4k^2}\right)^2=\frac{(3j)^2}{\left(4k^2\right)^2}=\frac{3^2j^2}{4^2k^{2\cdot 2}}=\frac{9j^2}{16k^4}$$

101. $$\begin{aligned}\left(\frac{2r^4s^5}{r^2}\right)^3 &= \left(2r^4r^{-2}s^5\right)^3\\ &= \left(2r^2s^5\right)^3\\ &= 2^3r^{2\cdot 3}s^{5\cdot 3}\\ &= 8r^6s^{15}\end{aligned}$$

103. $$\begin{aligned}\left(\frac{4xy}{y^3}\right)^{-3} &= \left(\frac{y^3}{4xy}\right)^3\\ &= \left(\frac{y^3y^{-1}}{4x}\right)^3\\ &= \frac{\left(y^2\right)^3}{(4x)^3}\\ &= \frac{y^{2\cdot 3}}{4^3x^3}\\ &= \frac{y^6}{64x^3}\end{aligned}$$

105. $$\begin{aligned}\left(\frac{5x^{-2}y}{x^{-5}}\right)^3 &= \left(5x^{-2}x^5y\right)^3\\ &= \left(5x^3y\right)^3\\ &= 5^3x^{3\cdot 3}y^3\\ &= 125x^9y^3\end{aligned}$$

107. $$\begin{aligned}\left(\frac{10x^2y}{5xz}\right)^{-3} &= \left(\frac{5xz}{10x^2y}\right)^3\\ &= \left(\frac{z}{2x^2x^{-1}y}\right)^3\\ &= \left(\frac{z}{2xy}\right)^3\\ &= \frac{z^3}{2^3x^3y^3}\\ &= \frac{z^3}{8x^3y^3}\end{aligned}$$

109. $$\left(\frac{x^8y^{-2}}{x^{-2}y^3}\right)^2=\left(\frac{x^{8-(-2)}}{y^{3-(-2)}}\right)^2=\left(\frac{x^{10}}{y^5}\right)^2=\frac{\left(x^{10}\right)^2}{\left(y^5\right)^2}=\frac{x^{20}}{y^{10}}$$

111. $\left(\frac{4x^{-1}y^{-2}z^{3}}{2xy^{2}z^{-3}}\right)^{-2}=\left(\frac{2z^{3-(-3)}}{x^{1-(-1)}y^{2-(-2)}}\right)^{-2}$

$=\left(\frac{2z^{6}}{x^{2}y^{4}}\right)^{-2}$

$=\left(\frac{x^{2}y^{4}}{2z^{6}}\right)^{2}$

$=\frac{(x^{2})^{2}(y^{4})^{2}}{2^{2}(z^{6})^{2}}$

$=\frac{x^{4}y^{8}}{4z^{12}}$

113. $\left(\frac{-a^{3}b^{-1}c^{-3}}{4ab^{3}c^{-4}}\right)^{-3}=\left(\frac{-a^{3-1}c^{-3-(-4)}}{4b^{3-(-1)}}\right)^{-3}$

$=\left(\frac{-a^{2}c^{1}}{4b^{4}}\right)^{-3}$

$=\left(\frac{4b^{4}}{-a^{2}c}\right)^{3}$

$=\frac{4^{3}b^{4\cdot 3}}{-a^{2\cdot 3}c^{3}}$

$=-\frac{64b^{12}}{a^{6}c^{3}}$

115. $\frac{(3x^{-4}y^{2})^{3}}{(2x^{3}y^{5})^{3}}=\frac{3^{3}x^{-4\cdot 3}y^{2\cdot 3}}{2^{3}x^{3\cdot 3}y^{5\cdot 3}}$

$=\frac{3^{3}x^{-12}y^{6}}{2^{3}x^{9}y^{15}}$

$=\frac{27}{8x^{9-(-12)}y^{15-6}}$

$=\frac{27}{8x^{21}y^{9}}$

117. $x^{2a}\cdot x^{5a+3}=x^{2a+5a+3}=x^{7a+3}$

119. $w^{2a-5}\cdot w^{3a-2}=w^{2a-5+3a-2}=w^{5a-7}$

121. $\frac{x^{2w+3}}{x^{w-4}}=x^{2w+3-(w-4)}=x^{2w+3-w+4}=x^{w+7}$

123. $(x^{3p+5})(x^{2p-3})=x^{3p+5+2p-3}=x^{5p+2}$

125. $x^{-m}(x^{3m+2})=x^{-m}x^{3m+2}=x^{-m+3m+2}=x^{2m+2}$

127. $\frac{30m^{a+b}n^{b-a}}{6m^{a-b}n^{a+b}}=\left(\frac{30}{6}\right)\frac{m^{a+b-(a-b)}}{n^{a+b-(b-a)}}$

$=\frac{5m^{a+b-a+b}}{n^{a+b-b+a}}$

$=\frac{5m^{2b}}{n^{2a}}$

129. **a.** $x^{4}>x^{3}$ when $x<0$ or $x>1$

b. $x^{4}<x^{3}$ when $0<x<1$

c. $x^{4}=x^{3}$ when $x=0$ or $x=1$

d. x^{4} is not greater than x^{3} when $0\le x\le 1$

131. **a.** $(-1)^{n}=1$ for any even number n because an even number of negative factors is positive.

b. $(-1)^{n}=-1$ for any odd number n because an odd number of negative factors is negative.

133. **a.** Yes, $\left(-\frac{2}{3}\right)^{-2}=\left(\frac{2}{3}\right)^{-2}$. By the negative exponent rule, $\left(-\frac{2}{3}\right)^{-2}=\left(-\frac{3}{2}\right)^{2}=\frac{9}{4}$ and $\left(\frac{2}{3}\right)^{-2}=\left(\frac{3}{2}\right)^{2}=\frac{9}{4}$.

b. Yes. $x^{-2}=\frac{1}{x^{2}}$, $x\ne 0$, and $(-x)^{-2}=\frac{1}{(-x)^{2}}=\frac{1}{x^{2}}$, $x\ne 0$.

135. Let a represent the unknown exponent,

$\left(\frac{x^{2}y^{-2}}{x^{-3}y^{a}}\right)^{2}=x^{10}y^{2}$

$\left(x^{2-(-3)}y^{-2-a}\right)^{2}=x^{10}y^{2}$

$\left(x^{5}y^{-2-a}\right)^{2}=x^{10}y^{2}$

$x^{10}y^{2(-2-a)}=x^{10}y^{2}$

$x^{10}y^{-4-2a}=x^{10}y^{2}$

Thus, $-4-2a=2$

$2a=-4-2$

$2a=-6$

$a=-3$

137. Let a and b represent the unknown exponents.

$$\left(\frac{x^a y^5 z^{-2}}{x^4 y^b z}\right)^{-1} = \frac{x^5 z^3}{y^2}$$

$$\left(\frac{y^{5-b}}{x^{4-a} z^{1-(-2)}}\right)^{-1} = \frac{x^5 z^3}{y^2}$$

$$\frac{x^{4-a} z^3}{y^{5-b}} = \frac{x^5 z^3}{y^2}$$

Thus, $4-a=5$

$a=-1$

and $5-b=2$

$b=3$

139. $\left(\frac{x^{5/8}}{x^{1/4}}\right)^3 = \left(x^{5/8-1/4}\right)^3$

$= \left(x^{3/8}\right)^3$

$= x^{(3/8)\cdot 3}$

$= x^{9/8}$

141. $\frac{x^{1/2} y^{-3/2}}{x^5 y^{5/3}} = \frac{1}{x^{5-1/2} y^{5/3-(-3/2)}}$

$= \frac{1}{x^{9/2} y^{19/6}}$

144. **a.** $A \cup B = \{1, 2, 3, 4, 5, 6, 8\}$

b. $A \cap B = \{\ \}$ or $\varnothing$

145. $\{x \mid -3 \le x < 2\}$

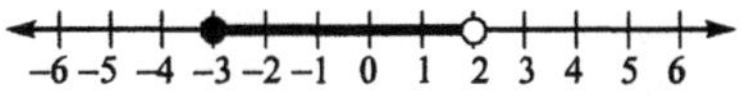

146. $8+|12|\div|-3|-4\cdot 2^2 = 8+12\div 3-4\cdot 2^2$

$= 8+12\div 3-4\cdot 4$

$= 8+4-16$

$= 12-16$

$= -4$

147. $\sqrt[3]{-125} = -5$ because $(-5)(-5)(-5) = -125$

Exercise Set 1.6

1. The form of a number in scientific notation is a number greater than or equal to 1 and less than 10 multiplied by a power of 10.

3. $1\times 10^{-2} = 0.01$ and $1\times 10^{-3} = 0.001$, so 1×10^{-2} is greater than 1×10^{-3}.

5. $3700 = 3.7\times 10^3$

7. $0.041 = 4.1\times 10^{-2}$

9. $760{,}000 = 7.6\times 10^5$

11. $0.00000186 = 1.86\times 10^{-6}$

13. $5{,}780{,}000 = 5.78\times 10^6$

15. $0.000106 = 1.06\times 10^{-4}$

17. $3.1\times 10^4 = 31{,}000$

19. $2.13\times 10^{-5} = 0.0000213$

21. $9.17\times 10^{-1} = 0.917$

23. $8\times 10^6 = 8{,}000{,}000$

25. $2.03\times 10^5 = 203{,}000$

27. $1\times 10^6 = 1{,}000{,}000$

29. $\left(4\times 10^5\right)\left(6\times 10^2\right) = (4\times 6)\left(10^5\times 10^2\right)$

$= 24\times 10^7$

$= 240{,}000{,}000$

31. $\frac{8.4\times 10^{-6}}{4\times 10^{-4}} = \left(\frac{8.4}{4}\right)\times 10^{-6-(-4)} = 2.1\times 10^{-2} = 0.021$

33. $\frac{6.75\times 10^{-3}}{2.5\times 10^2} = \left(\frac{6.75}{2.5}\right)\times 10^{-3-2}$

$= 2.7\times 10^{-5}$

$= 0.000027$

35. $\left(8.2\times 10^5\right)\left(1.4\times 10^{-2}\right) = (8.2\times 1.4)\left(10^5\times 10^{-2}\right)$

$= 11.48\times 10^3$

$= 11{,}480$

37. $\dfrac{1.68\times10^{4}}{5.6\times10^{7}}=\left(\dfrac{1.68}{5.6}\right)\times10^{4-7}=0.3\times10^{-3}=0.0003$

39. $(9.1\times10^{-4})(7.4\times10^{-4})=(9.1\times7.4)(10^{-4}\times10^{-4})$
$=67.34\times10^{-8}$
$=0.0000006734$

41. $(0.03)(0.0005)=(3\times10^{-2})(5\times10^{-4})$
$=(3\times5)(10^{-2}\times10^{-4})$
$=15\times10^{-6}$
$=1.5\times10^{-5}$

43. $\dfrac{35{,}000{,}000}{7000}=\dfrac{3.5\times10^{7}}{7.0\times10^{3}}$
$=\left(\dfrac{3.5}{7}\right)\times10^{7-3}$
$=0.5\times10^{4}$
$=5.0\times10^{3}$

45. $\dfrac{0.00069}{23{,}000}=\dfrac{6.9\times10^{-4}}{2.3\times10^{4}}=\left(\dfrac{6.9}{2.3}\right)\times10^{-4-4}=3.0\times10^{-8}$

47. $(47{,}000)(35{,}000{,}000)=(4.7\times10^{4})(3.5\times10^{7})$
$=(4.7\times3.5)(10^{4}\times10^{7})$
$=16.45\times10^{11}$
$=1.645\times10^{12}$

49. $\dfrac{1008}{0.0021}=\dfrac{1.008\times10^{3}}{2.1\times10^{-3}}$
$=\left(\dfrac{1.0082}{2.1}\right)\times10^{3-(-3)}$
$=0.48\times10^{6}$
$=4.8\times10^{5}$

51. $\dfrac{0.00153}{0.00051}=\dfrac{1.53\times10^{-3}}{5.1\times10^{-4}}$
$=\left(\dfrac{1.53}{5.1}\right)\times10^{-3+4}$
$=0.3\times10^{1}$
$=3.0\times10^{0}$

53. $(4.78\times10^{9})(1.96\times10^{5})=(4.78\times1.96)(10^{9}\times10^{5})$
$=9.3688\times10^{14}$
$\approx9.369\times10^{14}$

55. $(7.23\times10^{-3})(1.46\times10^{5})$
$=(7.23\times1.46)(10^{-3}\times10^{5})$
$=10.5558\times10^{2}$
$=1.05558\times10^{3}$
$\approx1.056\times10^{3}$

57. $\dfrac{4.36\times10^{-4}}{8.17\times10^{-7}}=\left(\dfrac{4.36}{8.17}\right)\times10^{-4-(-7)}$
$=0.5337\times10^{3}$
$=5.337\times10^{2}$

59. $(4.89\times10^{15})(6.37\times10^{-41})$
$=(4.89\times6.37)(10^{15}\times10^{-41})$
$=31.1493\times10^{-26}$
$=3.11493\times10^{-25}$
$\approx3.115\times10^{-25}$

61. $(8.32\times10^{3})(9.14\times10^{-31})$
$=(8.32\times9.14)(10^{3}\times10^{-31})$
$=76.0448\times10^{-28}$
$=7.60448\times10^{-27}$
$\approx7.604\times10^{-27}$

63. $\dfrac{1.50\times10^{35}}{4.5\times10^{-26}}=\left(\dfrac{1.50}{4.5}\right)\times10^{35-(-26)}$
$\approx0.3333\times10^{61}$
$\approx3.333\times10^{60}$

65. 850 million $=850{,}000{,}000=8.5\times10^{8}$

67. 2.4 million $=2{,}400{,}000=2.4\times10^{6}$

69. 52.8 billion $=52{,}800{,}000{,}000=5.28\times10^{10}$

71. 9.1 trillion $=9{,}100{,}000{,}000{,}000=9.1\times10^{12}$

73. $0.00001=1\times10^{-5}$

75. $0.0000158=1.58\times10^{-5}$

77. $0.000000001 = 1\times10^{-9}$

79. a. $\frac{1}{10} = 10^{-1}$; subtract 1 from the exponent.

b. $\frac{1}{100} = 10^{-2}$; subtract 2 from the exponent.

c. $\frac{1}{1\text{ million}} = \frac{1}{1{,}000{,}000} = 10^{-6}$; subtract 6 from the exponent.

d. $\frac{6.58\times10^{-4}}{1\text{ million}} = 6.58\times10^{-4}\times10^{-6} = 6.58\times10^{-10}$

81. a. $5.25\times10^4 - 4.25\times10^4 = 1\times10^4$

It is off by 1×10^4 or 10,000.

b.
$$\begin{aligned}5.25\times10^5 - 5.25\times10^4 &= 52.5\times10^4 - 5.25\times10^4\\ &= 47.25\times10^4\\ &= 4.725\times10^5\end{aligned}$$

It is off by 4.725×10^5 or 472,500.

c. The error in part b is more serious because 472,500 is greater than 10,000.

83.
$$\begin{aligned}\frac{93{,}000{,}000}{3100} &= \frac{9.3\times10^7}{3.1\times10^3}\\ &= \left(\frac{9.3}{3.1}\right)\times10^{7-3}\\ &= 3.0\times10^4\\ &= 30{,}000\end{aligned}$$

It will take 30,000 hours to reach the sun.

85. a.
$$\begin{aligned}&6.536\times10^9 - 2.995\times10^8\\ &= 6.536\times10^9 - 0.2995\times10^1\times10^8\\ &= 6.536\times10^9 - 0.2995\times10^9\\ &= 6.2365\times10^9\\ &= 6{,}236{,}500{,}000\end{aligned}$$

About 6.2365×10^9 or 6,236,500,000 people lived outside of the United States on September 1, 2006.

b.
$$\begin{aligned}\frac{2.995\times10^8}{6.536\times10^9} &= \left(\frac{2.995}{6.536}\right)\times10^{8-9}\\ &\approx 0.46\times10^{-1}\\ &\approx 0.046\\ &\approx 4.6\%\end{aligned}$$

About 4.6% of the world's population lived in the United States on September 1, 2006.

87. a.
$$\begin{aligned}11.728\text{ trillion} &= 11.728\times10^{12}\\ &= 1.1728\times10^1\times10^{12}\\ &= 1.1728\times10^{13}\end{aligned}$$

$$\begin{aligned}296.5\text{ million} &= 296.5\times10^6\\ &= 2.965\times10^2\times10^6\\ &= 2.965\times10^8\end{aligned}$$

b.
$$\begin{aligned}\frac{1.1728\times10^{13}}{2.965\times10^8} &= \left(\frac{1.1728}{2.965}\right)\times10^{13-8}\\ &\approx 0.3955481\times10^5\\ &\approx 39{,}554.81\end{aligned}$$

The GDP *per capita* was about \$39,554.81.

89. $\frac{1.29\times10^9}{9.8\times10^6} = \left(\frac{1.29}{9.8}\right)\times10^{9-6} \approx 0.132\times10^3 \approx 132$

The population density of China in 2005 was approximately 132 people per square kilometer.

91. a.
$$\begin{aligned}4.2\times10^9\times5\% &= 4.2\times10^9\times0.05\\ &= 4.2\times10^9\times5\times10^{-2}\\ &= (4.2\times5)\left(10^9\times10^{-2}\right)\\ &= 21\times10^7\\ &= 2.1\times10^8\end{aligned}$$

About 2.1×10^8 pounds are recycled.

b.
$$\begin{aligned}4.2\times10^9\times95\% &= 4.2\times10^9\times0.95\\ &= 4.2\times10^9\times9.5\times10^{-1}\\ &= (4.2\times9.5)\left(10^9\times10^{-1}\right)\\ &= 39.9\times10^8\\ &= 3.99\times10^9\end{aligned}$$

About 3.99×10^9 pounds are not recycled.

93. a. 1290 million − 296 million = 994 million. About 994 million more people lived in China than in the United States.

b.
$$\begin{aligned}\frac{1290\text{ million}}{6{,}446{,}000{,}000} &= \frac{1.290\times10^9}{6.446\times10^9}\\ &= \left(\frac{1.290}{6.446}\right)\times10^{9-9}\\ &\approx 0.2001\times10^0\\ &= 20.01\%\end{aligned}$$

Approximately 20.01% of the worlds population lived in China.

c. $\frac{1290 \text{ million}}{3.70\times10^6}=\frac{1.290\times10^9}{3.70\times10^6}$

$=\left(\frac{1.290}{3.70}\right)\times10^{9-6}$

$\approx 0.3486\times10^3$

≈ 348.6

The population density in China was about 348.6 people per square mile.

d. $\frac{296 \text{ million}}{3.62\times10^6}=\frac{2.96\times10^8}{3.62\times10^6}$

$=\left(\frac{2.96}{3.62}\right)\times10^{8-6}$

$\approx 0.818\times10^2$

≈ 81.8

The population density in the United States was about 81.8 people per square mile.

95. a. $0.03\cdot(2.3\times10^{12})=(3\times10^{-2})(2.3\times10^{12})$

$=(3\times2.3)\times10^{-2+12}$

$=6.9\times10^{10}$

The outlay in FY 2004 for law enforcement and general government was $\$6.9\times10^{10}$.

b. $0.36\cdot(2.3\times10^{12})=(3.6\times10^{-1})(2.3\times10^{12})$

$=(3.6\times2.3)\times10^{-1+12}$

$=8.28\times10^{11}$

The outlay in FY 2004 for Social Security, Medicare, and other retirement was $\$8.28\times10^{11}$.

c. $100\%-7\%=93\%$

$0.93\cdot(2.3\times10^{12})=(9.3\times10^{-1})(2.3\times10^{12})$

$=(9.3\times2.3)\times10^{-1+12}$

$=21.39\times10^{11}$

$=2.139\times10^{12}$

The total outlay in FY 2004 for all programs other than the net interest on the nation debt was $\$2.139\times10^{12}$.

97. a. $14.0+16.9+9.6+9.8+10.0=60.3$ million

$=60.3\times10^6$

$=6.03\times10^7$

The total land area of the five largest countries is 6.03×10^7 square kilometers.

b. 14.0 million $-$ 9.6 million $=$ 4.4 million

$=4.4\times10^6$

Antarctica has 4.4×10^6 more square kilometers of land than the United States.

Chapter 1 Review Exercises

1. {4, 5, 6, 7, 8}

2. {0, 3, 6, 9,...}

3. Yes, the set of natural numbers is a subset of the set of whole numbers.

4. Yes, the set of rational numbers is a subset of the set of real numbers.

5. No, the set of rational numbers is not a subset of the set of irrational numbers.

6. Yes, the set of irrational numbers is a subset of the set of real numbers.

7. 4 and 6 are natural numbers.

8. 4, 6 and 0 are whole numbers.

9. –2, 4, 6 and 0 are integers.

10. $-2, 4, 6, \frac{1}{2}, 0, \frac{15}{27}, -\frac{1}{5},$ and 1.47 are rational numbers.

11. $\sqrt{7}$ and $\sqrt{3}$ are irrational numbers.

12. $-2, 4, 6, \frac{1}{2}, \sqrt{7}, \sqrt{3}, 0, \frac{15}{27}, -\frac{1}{5},$ and 1.47 are real numbers.

13. False. $\frac{0}{1}=0$ *is* a real number.

14. True. 0, –2, and 4 can be written with a denominator of 1.

15. True. Division by 0 is undefined.

16. True. The set of real numbers contains both the rational and irrational numbers.

17. $A\cup B=\{1, 2, 3, 4, 5, 6, 8, 10\}$

$A\cap B=\{2, 4, 6\}$

18. $A\cup B=\{2, 3, 4, 5, 6, 7, 8, 9\}$

$A\cap B=\{\ \}$ or $\varnothing$

19. $A\cup B=\{1, 2, 3, 4,...\}$

$A\cap B=\{\ \}$ or $\varnothing$

20. $A \cup B = \{3, 4, 5, 6, 9, 10, 11, 12\}$
$A \cap B = \{9, 10\}$

21. $\{x \mid x > 5\}$

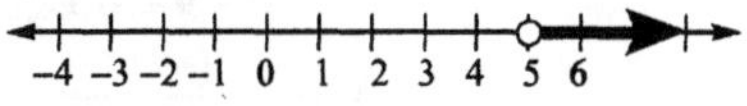

22. $\{x \mid x \le -2\}$

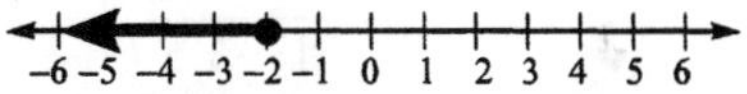

23. $\{x \mid -1.3 < x \le 2.4\}$

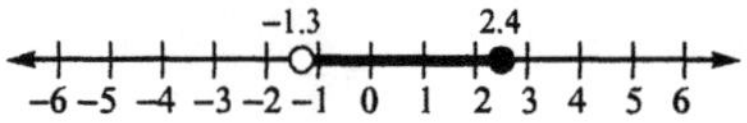

24. $\left\{x \middle| \frac{2}{3} \le x < 4 \text{ and } x \in N\right\}$

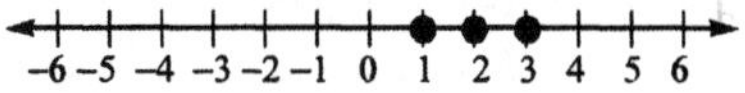

25. $-3 < 0$

26. $-4 < -3.9$

27. $1.06 < 1.6$

28. $|-8| = 8$

29. $|-4| = 4$ and $|-10| = 10$, so $|-4| < |-10|$.

30. $|-9| = 9$, so $13 > |-9|$.

31. $\left|-\frac{2}{3}\right| = \frac{2}{3} = \frac{10}{15}$ and $\frac{3}{5} = \frac{9}{15}$, so $\left|-\frac{2}{3}\right| > \frac{3}{5}$.

32. $-|-2| = -2$, so $-|-2| > -6$.

33. $-\pi, -3, 3, \pi$

34. $0, \frac{3}{5}, 2.7, |-3|$

35. $-2, 3, |-5|, |-10|$

36. $-7, -3, |-3|, |-7|$

37. $-4, -|-3|, 5, 6$

38. $-2, 0, |1.6|, |-2.3|$

39. Distributive property

40. Commutative property of multiplication

41. Associative property of addition

42. Identity property of addition

43. Associative property of multiplication

44. Double negative property

45. Multiplicative property of zero

46. Inverse property of addition

47. Inverse property of multiplication

48. Identity property of multiplication

49.
$$\begin{aligned} 8 + 3^2 - \sqrt{36} \div 2 &= 8 + 9 - 6 \div 2 \\ &= 8 + 9 - 3 \\ &= 17 - 3 \\ &= 14 \end{aligned}$$

50.
$$\begin{aligned} -4 \div (-2) + 16 - \sqrt{81} &= -4 \div (-2) + 16 - 9 \\ &= 2 + 16 - 9 \\ &= 18 - 9 \\ &= 9 \end{aligned}$$

51.
$$\begin{aligned} (7-9) - (-3+5) + 15 &= (-2) - 2 + 15 \\ &= -2 + (-2) + 15 \\ &= -4 + 15 \\ &= 11 \end{aligned}$$

52.
$$\begin{aligned} 2|-7| - 4|-6| + 5 &= 2(7) - 4(6) + 5 \\ &= 14 - 24 + 5 \\ &= -10 + 5 \\ &= -5 \end{aligned}$$

53.
$$\begin{aligned} (6-9) \div (9-6) + 2 &= -3 \div 3 + 2 \\ &= -1 + 2 \\ &= 1 \end{aligned}$$

54.
$$\begin{aligned} |6-3| \div 3 + 4 \cdot 8 - 12 &= |3| \div 3 + 4 \cdot 8 - 12 \\ &= 3 \div 3 + 4 \cdot 8 - 12 \\ &= 1 + 32 - 12 \\ &= 33 - 12 \\ &= 21 \end{aligned}$$

55. $\sqrt{9} + \sqrt[3]{64} + \sqrt[5]{32} = 3 + 4 + 2 = 9$

56. $3^2-6\cdot 9+4\div 2^2-5=9-6\cdot 9+4\div 4-5$
$=9-54+1-5$
$=-45+1-5$
$=-44-5$
$=-49$

57. $4-(2-9)^0+3^2\div 1+3=4-(-7)^0+3^2\div 1+3$
$=4-1+9\div 1+3$
$=4-1+9+3$
$=3+9+3$
$=12+3$
$=15$

58. $5^2+(-2+2^2)^3+1=5^2+(-2+4)^3+1$
$=5^2+(2)^3+1$
$=25+8+1$
$=33+1$
$=34$

59. $-3^2+14\div 2\cdot 3-6=-9+14\div 2\cdot 3-6$
$=-9+7\cdot 3-6$
$=-9+21-6$
$=12-6$
$=6$

60. $\left\{\left[(12\div 4)^2-1\right]^2\div 16\right\}^3=\left\{\left[(3)^2-1\right]^2\div 16\right\}^3$
$=\left[(9-1)^2\div 16\right]^3$
$=\left[8^2\div 16\right]^3$
$=(64\div 16)^3$
$=4^3$
$=64$

61. $\dfrac{9+7\div(3^2-2)+6\cdot 8}{\sqrt{81}+\sqrt{1}-10}=\dfrac{9+7\div(9-2)+6\cdot 8}{\sqrt{81}+\sqrt{1}-10}$
$=\dfrac{9+7\div 7+6\cdot 8}{9+1-10}$
$=\dfrac{9+1+48}{10-10}$
$=\dfrac{58}{0}$ which is undefined

62. $\dfrac{-(5-7)^2-3(-2)+|-6|}{18-9\div 3\cdot 5}=\dfrac{-(-2)^2-3(-2)+|-6|}{18-9\div 3\cdot 5}$
$=\dfrac{-4-3(-2)+|-6|}{18-9\div 3\cdot 5}$
$=\dfrac{-4-3(-2)+6}{18-9\div 3\cdot 5}$
$=\dfrac{-4+6+6}{18-3\cdot 5}$
$=\dfrac{-4+6+6}{18-15}$
$=\dfrac{8}{3}$

63. Substitute 2 for x:
$2x^2+3x+8=2(2)^2+3(2)+8$
$=2(4)+3(2)+8$
$=8+6+8$
$=14+8$
$=22$

64. Substitute −3 for a and −4 for b:
$5a^2-7b^2=5(-3)^2-7(-4)^2$
$=5(9)-7(16)$
$=45-112$
$=-67$

65. **a.** 1976 is represented by $x=7$, so substitute 7 for x:
dollars spent $=50.86x^2-316.75x+541.48$
$=50.86(7)^2-316.75(7)+541.48$
$=50.86(49)-316.75(7)+541.48$
$=2492.14-2217.25+541.48$
$=816.37$
In 1976, the amount spent was approximately \$816.37 million.

b. 2008 is represented by $x=15$, so substitute 15 for x:
dollars spent $=50.86x^2-316.75x+541.48$
$=50.86(15)^2-316.75(15)+541.48$
$=50.86(225)-316.75(15)+541.48$
$=11,443.5-4751.25+541.48$
$=7233.73$
In 2008, the amount spent will be approximately \$7,233.73 million.

66. **a.** 1980 is represented by $x = 4$, so substitute 4 for x:

$$\text{freight hauled} = 14.04x^2 + 1.96x + 712.05$$
$$= 14.04(4)^2 + 1.96(4) + 712.05$$
$$= 14.04(16) + 1.96(4) + 712.05$$
$$= 224.64 + 7.84 + 712.05$$
$$= 944.53$$

In 1980, the amount of freight hauled by trains was approximately 944.53 ton-miles.

b. 2010 is represented by $x = 10$, so substitute 10 for x:

$$\text{freight hauled} = 14.04x^2 + 1.96x + 712.05$$
$$= 14.04(10)^2 + 1.96(10) + 712.05$$
$$= 14.04(100) + 1.96(10) + 712.05$$
$$= 1404 + 19.6 + 712.05$$
$$= 2135.65$$

In 2010, the amount of freight hauled by trains will be approximately 2135.65 ton-miles.

67. $2^3 \cdot 2^2 = 2^{3+2} = 2^5 = 32$

68. $x^2 \cdot x^3 = x^{2+3} = x^5$

69. $\dfrac{a^{12}}{a^4} = a^{12-4} = a^8$

70. $\dfrac{y^{12}}{y^5} = y^{12-5} = y^7$

71. $\dfrac{b^7}{b^{-2}} = b^{7-(-2)} = b^{7+2} = b^9$

72. $c^3 \cdot c^{-6} = c^{3+(-6)} = c^{-3} = \dfrac{1}{c^3}$

73. $5^{-2} \cdot 5^{-1} = 5^{-2+(-1)} = 5^{-3} = \dfrac{1}{5^3} = \dfrac{1}{125}$

74. $8x^0 = 8(1) = 8$

75. $\left(-9m^3\right)^2 = (-9)^2\left(m^3\right)^2 = 81m^{3\cdot 2} = 81m^6$

76. $\left(\dfrac{4}{7}\right)^{-1} = \left(\dfrac{7}{4}\right)^1 = \dfrac{7}{4}$

77. $\left(\dfrac{2}{3}\right)^{-3} = \left(\dfrac{3}{2}\right)^3 = \dfrac{3^3}{2^3} = \dfrac{27}{8}$

78. $\left(\dfrac{x}{y^2}\right)^{-1} = \left(\dfrac{y^2}{x}\right)^1 = \dfrac{y^2}{x}$

79. $\left(5xy^3\right)\left(-3x^2y\right) = (5\cdot(-3))\cdot x^{1+2}\cdot y^{3+1}$
$$= -15x^3y^4$$

80. $\left(2v^3w^{-4}\right)\left(7v^{-6}w\right) = (2\cdot 7)\cdot v^{3+(-6)}\cdot w^{-4+1}$
$$= 14v^{-3}w^{-3}$$
$$= \frac{14}{v^3w^3}$$

81. $\dfrac{6x^{-3}y^5}{2x^2y^{-2}} = \left(\dfrac{6}{2}\right)\dfrac{y^{5-(-2)}}{x^{2-(-3)}} = \dfrac{3y^7}{x^5} = \dfrac{3y^7}{x^5}$

82. $\dfrac{12x^{-3}y^{-4}}{4x^{-2}y^5} = \left(\dfrac{12}{4}\right)\dfrac{1}{x^{-2-(-3)}y^{5-(-4)}} = \dfrac{3}{xy^9}$

83. $\dfrac{g^3h^{-6}j^{-9}}{g^{-2}h^{-1}j^5} = \dfrac{g^{3-(-2)}}{h^{-1-(-6)}j^{5-(-9)}} = \dfrac{g^5}{h^5j^{14}}$

84. $\dfrac{21m^{-3}n^{-2}}{7m^{-4}n^2} = \left(\dfrac{21}{7}\right)\dfrac{m^{-3-(-4)}}{n^{2-(-2)}} = \dfrac{3m}{n^4}$

85. $\left(\dfrac{4a^2b}{a}\right)^3 = (4a^{2-1}b)^3 = (4ab)^3 = 4^3a^3b^3 = 64a^3b^3$

86. $\left(\dfrac{x^5y}{-3y^2}\right)^2 = \left(\dfrac{x^5}{-3y^{2-1}}\right)^2$
$$= \left(\frac{x^5}{-3y}\right)^2$$
$$= \frac{(x^5)^2}{(-3)^2\,y^2}$$
$$= \frac{x^{5\cdot 2}}{9y^2}$$
$$= \frac{x^{10}}{9y^2}$$

87. $\left(\dfrac{p^3q^{-1}}{p^{-4}q^5}\right)^2 = \left(\dfrac{p^{3-(-4)}}{q^{5-(-1)}}\right)^2 = \left(\dfrac{p^7}{q^6}\right)^2 = \dfrac{p^{7\cdot 2}}{q^{6\cdot 2}} = \dfrac{p^{14}}{q^{12}}$

88. $\left(\dfrac{-2ab^{-3}}{c^2}\right)^3=\left(\dfrac{-2a}{b^3c^2}\right)^3$
$=\dfrac{(-2a)^3}{(b^3c^2)^3}$
$=\dfrac{(-2)^3\cdot a^3}{b^{3\cdot 3}\cdot c^{2\cdot 3}}$
$=-\dfrac{8a^3}{b^9c^6}$

89. $\left(\dfrac{5xy^3}{z^2}\right)^{-2}=\left(\dfrac{z^2}{5xy^3}\right)^2$
$=\dfrac{(z^2)^2}{5^2x^2(y^3)^2}$
$=\dfrac{z^{2\cdot 2}}{25x^2y^{3\cdot 2}}$
$=\dfrac{z^4}{25x^2y^6}$

90. $\left(\dfrac{9m^{-2}n}{3mn}\right)^{-3}=\left(\dfrac{3mn}{9m^{-2}n}\right)^3$
$=\left[\left(\dfrac{3}{9}\right)m^{1-(-2)}n^{1-1}\right]^3$
$=\left(\dfrac{m^3n^0}{3}\right)^3$
$=\left(\dfrac{m^3\cdot 1}{3}\right)^3$
$=\dfrac{(m^3)^3}{3^3}$
$=\dfrac{m^9}{27}$

91. $\left(-2m^2n^{-3}\right)^{-2}=(-2)^{-2}\left(m^2\right)^{-2}\left(n^{-3}\right)^{-2}$
$=(-2)^{-2}m^{2\cdot(-2)}n^{-3\cdot(-2)}$
$=(-2)^{-2}m^{-4}n^6$
$=\dfrac{n^6}{(-2)^2m^4}$
$=\dfrac{n^6}{4m^4}$

92. $\left(\dfrac{15x^5y^{-3}z^{-2}}{-3x^4y^{-4}z^3}\right)^4=\left[\left(\dfrac{15}{-3}\right)\dfrac{x^{5-4}y^{-3-(-4)}}{z^{3-(-2)}}\right]^4$
$=\left[\dfrac{-5xy}{z^5}\right]^4$
$=\dfrac{(-5)^4x^4y^4}{z^{5\cdot 4}}$
$=\dfrac{625x^4y^4}{z^{20}}$

93. $\left(\dfrac{2x^{-1}y^5z^4}{3x^4y^{-2}z^{-2}}\right)^{-2}=\left(\dfrac{2y^{5-(-2)}z^{4-(-2)}}{3x^{4-(-1)}}\right)^{-2}$
$=\left(\dfrac{2y^7z^6}{3x^5}\right)^{-2}$
$=\left(\dfrac{3x^5}{2y^7z^6}\right)^2$
$=\dfrac{3^2x^{5\cdot 2}}{2^2y^{7\cdot 2}z^{6\cdot 2}}$
$=\dfrac{9x^{10}}{4y^{14}z^{12}}$

94. $\left(\dfrac{8x^{-2}y^{-2}z}{-x^4y^{-4}z^3}\right)^{-1}=\left(\dfrac{8y^{-2-(-4)}}{-x^{4-(-2)}z^{3-1}}\right)^{-1}$
$=\left(\dfrac{8y^2}{-x^6z^2}\right)^{-1}=\left(\dfrac{-x^6z^2}{8y^2}\right)^1=-\dfrac{x^6z^2}{8y^2}$

95. $0.0000742=7.42\times10^{-5}$

96. $460{,}000=4.6\times10^5$

97. $183{,}000=1.83\times10^5$

98. $0.000001=1.0\times10^{-6}$

99. $\left(25\times10^{-3}\right)\left(1.2\times10^6\right)=(25\times1.2)\times10^{-3+6}$
$=30\times10^3$
$=3\times10^4$
$=30{,}000$

101. $\dfrac{4{,}000{,}000}{0.02}=\dfrac{4\times10^6}{2\times10^{-2}}$
$=\left(\dfrac{4}{2}\right)\times10^{6-(-2)}$
$=2\times10^8$
$=200{,}000{,}000$

100. $\dfrac{27\times10^3}{9\times10^5}=\left(\dfrac{27}{9}\right)\times10^{3-5}=3\times10^{-2}=0.03$

102. $(0.004)(500{,}000)=\left(4\times10^{-3}\right)\left(5\times10^{5}\right)$
$=(4\times5)\times10^{-3+5}$
$=20\times10^{2}$
$=2\times10^{3}$
$=2000$

103. a. $\$2.86\times10^7-\$2.69\times10^7$
$=(\$2.86-\$2.69)\times10^7$
$=\$0.17\times10^7$
$=\$1.7\times10^6$
SBC Communications spent $\$1.7\times10^6$ or $1,700,000 more than Netflix.

b. $\$2.69\times10^7-\$2.23\times10^7$
$=(\$2.69-\$2.23)\times10^7$
$=\$0.46\times10^7$
$=\$4.6\times10^6$
Netflix spent $\$4.6\times10^6$ or $4,600,000 more than Dell Computers.

c. $\dfrac{\$2.86\times10^7}{\$2.23\times10^7}=\left(\dfrac{\$2.86}{\$2.23}\right)\times10^{7-7}$
$\approx1.28\times10^0$
≈1.28
The amount spent by SBC Communications was approximately 1.28 times larger than the amount spent by Dell Computers.

104. a. $1.4\times10^{10}=14{,}000{,}000{,}000$

b. $1.4\times10^{10}=14\times10^9=14$ billion
Voyager 1 has traveled 14 billion kilometers.

c. $\dfrac{1.4\times10^{10}}{28}=\dfrac{1.4\times10^{10}}{2.8\times10^1}$
$=\left(\dfrac{1.4}{2.8}\right)\times10^{10-1}$
$=0.5\times10^9$
$=5.0\times10^8$
Voyager 1 has averaged 5.0×10^8 or 500,000,000 kilometers per year.

d. $(1.4\times10^{10})\cdot(0.6)=(1.4\times10^{10})\cdot(6\times10^{-1})$
$=(1.4\times6)\times10^{10-1}$
$=8.4\times10^9$
Voyager 1 has traveled about 8.4×10^9 miles or 8,400,000,000 miles.

Chapter 1 Practice Test

1. $A=\{6, 7, 8, 9, \ldots\}$

2. False. For example, π is a real number but it is not a rational number.

3. True. This is how the set of real numbers is defined.

4. $-\dfrac{3}{5}, 2, -4, 0, \dfrac{19}{12}, 2.57$, and -1.92 are rational numbers.

5. $-\dfrac{3}{5}, 2, -4, 0, \dfrac{19}{12}, 2.57, \sqrt{8}, \sqrt{2}$, and -1.92 are real numbers.

6. $A\cup B=\{5, 7, 8, 9, 10, 11, 14\}$
$A\cap B=\{8, 10\}$

7. $A\cup B=\{1, 3, 5, 7, \ldots\}$
$A\cap B=\{3, 5, 7, 9, 11\}$

8. $\{x\,|\,-2.3\le x<5.2\}$

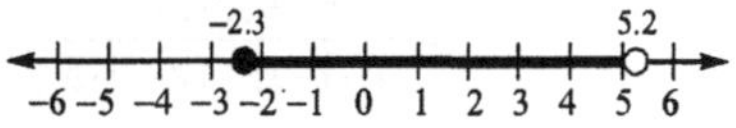

9. $\left\{x\,\middle|\,-\dfrac{5}{2}<x<\dfrac{6}{5} \text{ and } x\in I\right\}$

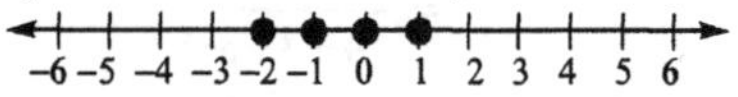

10. $-|4|, -2, |3|, 9$

11. Associative property of addition

12. Commutative property of addition

13. $\{6-[7-3^2\div(3^2-2\cdot3)]\}$
$=\{6-[7-3^2\div(9-2\cdot3)]\}$
$=\{6-[7-3^2\div(9-6)]\}$
$=\{6-[7-3^2\div3]\}$
$=\{6-[7-9\div3]\}$
$=\{6-[7-3]\}$
$=\{6-4\}$
$=2$

14. $$\begin{aligned}2^4+4^2\div 2^3\cdot\sqrt{25}+7&=16+16\div 8\cdot\sqrt{25}+7\\&=16+16\div 8\cdot 5+7\\&=16+2\cdot 5+7\\&=16+10+7\\&=33\end{aligned}$$

15. $$\begin{aligned}\frac{-3|4-8|\div 2+6}{-\sqrt{36}+18\div 3^2+4}&=\frac{-3|-4|\div 2+6}{-\sqrt{36}+18\div 3^2+6}\\&=\frac{-3(4)\div 2+6}{-\sqrt{36}+18\div 3^2+4}\\&=\frac{-3(4)\div 2+6}{-6+18\div 9+4}\\&=\frac{-12\div 2+6}{-6+2+4}\\&=\frac{-6+6}{-6+2+4}\\&=\frac{0}{0}\text{ which is indeterminate}\end{aligned}$$

16. $$\begin{aligned}\frac{-6^2+3\left(4-|6|\right)\div 6}{4-(-3)+12\div 4\cdot 5}&=\frac{-6^2+3(4-6)\div 6}{4-(-3)+12\div 4\cdot 5}\\&=\frac{-6^2+3(-2)\div 6}{4-(-3)+12\div 4\cdot 5}\\&=\frac{-36+3(-2)\div 6}{4-(-3)+12\div 4\cdot 5}\\&=\frac{-36+(-6)\div 6}{4+3+3\cdot 5}\\&=\frac{-36+(-1)}{4+3+15}\\&=\frac{-37}{7+15}\\&=\frac{-37}{22}\\&=-\frac{37}{22}\end{aligned}$$

17. Substitute 2 for x and 3 for y:

$$\begin{aligned}-x^2+2xy+y^2&=-(2)^2+2(2)(3)+(3)^2\\&=-4+2(2)(3)+9\\&=-4+12+9\\&=8+9\\&=17\end{aligned}$$

18. a. Substitute 1 for t:

$$\begin{aligned}h&=-16t^2+120t+200\\&=-16(1)^2+120(1)+200\\&=-16(1)+120(1)+200\\&=-16+120+200\\&=304\end{aligned}$$

After 1 second, the cannonball is 304 feet above sea level.

b. Substitute 5 for t:

$$\begin{aligned}h&=-16t^2+120t+200\\&=-16(5)^2+120(5)+200\\&=-16(25)+120(5)+200\\&=-400+600+200\\&=400\end{aligned}$$

After 5 seconds, the cannonball is 400 feet above sea level.

19. $3^{-2}=\frac{1}{3^2}=\frac{1}{9}$

20. $\left(\frac{4m^{-3}}{n^2}\right)^2=\left(\frac{4}{m^3n^2}\right)^2=\frac{4^2}{m^{3\cdot 2}n^{2\cdot 2}}=\frac{16}{m^6n^4}$

21. $\frac{24a^2b^{-3}c^0}{30a^3b^2c^{-2}}=\left(\frac{24}{30}\right)\frac{c^{0-(-2)}}{a^{3-2}b^{2-(-3)}}=\frac{4c^2}{5ab^5}$

22. $$\begin{aligned}\left(\frac{-3x^3y^{-2}}{x^{-1}y^5}\right)^{-3}&=\left(\frac{-3x^{3-(-1)}}{y^{5-(-2)}}\right)^{-3}\\&=\left(\frac{-3x^4}{y^7}\right)^{-3}\\&=\left(\frac{y^7}{-3x^4}\right)^3\\&=\frac{y^{7\cdot 3}}{(-3)^3x^{4\cdot 3}}\\&=-\frac{y^{21}}{27x^{12}}\end{aligned}$$

23. $389{,}000{,}000=3.89\times 10^8$

24. $$\begin{aligned}\frac{3.12\times 10^6}{1.2\times 10^{-2}}&=\left(\frac{3.12}{1.2}\right)\times 10^{6-(-2)}\\&=2.6\times 10^8\\&=260{,}000{,}000\end{aligned}$$

25. a. $9.2 \text{ billion} = 9{,}000{,}000{,}000 = 9.2\times10^9$

b. Ages 0 – 14:

$$\begin{aligned}0.195\cdot(9.2\times10^9) &= (1.95\times10^{-1})(9.2\times10^9)\\ &= (1.95\times9.2)\times10^{-1+9}\\ &= 17.97\times10^8\\ &= 1.797\times10^9\end{aligned}$$

Ages 15 – 64:

$$\begin{aligned}0.631\cdot(9.2\times10^9) &= (6.31\times10^{-1})(9.2\times10^9)\\ &= (6.31\times9.2)\times10^{-1+9}\\ &= 58.052\times10^8\\ &= 5.8052\times10^9\end{aligned}$$

Ages 65 and older:

$$\begin{aligned}0.174\cdot(9.2\times10^9) &= (1.74\times10^{-1})(9.2\times10^9)\\ &= (1.74\times9.2)\times10^{-1+9}\\ &= 16.008\times10^8\\ &= 1.6008\times10^9\end{aligned}$$

Chapter 2

Exercise Set 2.1

1. The terms of an expression are the parts added.

3. a. The coefficient of $\frac{x+y}{4}$ or $\frac{1}{4}(x+y)$ is $\frac{1}{4}$.

b. The coefficient of $-(p+3)$ or $-1(p+3)$ is -1.

c. The coefficient of $-\frac{3(x+2)}{5}$ or $-\frac{3}{5}(x+2)$ is $-\frac{3}{5}$.

5. a. Like terms have the same variables and exponents.

b. No; $3x$ and $3x^2$ are not like terms because the exponent on x is different for each term.

7.
$$\begin{aligned} 2x+3&=x+5\\ 2(4)+3&=4+5\\ 8+3&=4+5\\ 11&\neq 9 \end{aligned}$$
No, 4 is not a solution to the equation because substituting 4 for x results in a false equation.

9. The addition property of equality states that if $a = b$, then $a + c = b + c$.

11. a. An identity has infinitely many solutions

b. Its solution set is $\mathbb{R}$.

13. a. Answers will vary.

b.
$$\begin{aligned} 5x-2(x-4)&=2(x-2)\\ 5x-2x+8&=2x-4\\ 3x+8&=2x-4\\ 3x-2x+8&=2x-2x-4\\ x+8&=-4\\ x+8-8&=-4-8\\ x&=-12 \end{aligned}$$

15. Symmetric property

17. Transitive property

19. Reflexive property

21. Addition property

23. Multiplication property

25. Multiplication property

27. $5c^3$ is degree three since the exponent is 3.

29. $3ab$ is degree two since $3ab$ can be written as $3a^1b^1$ and the sum of the exponents is $1 + 1 = 2$.

31. The degree of 6 is zero since 6 can be written as $6x^0$.

33. $-5r$ is degree one since $-5r$ can be written as $-5r^1$.

35. $5a^2b^4c$ is degree seven since $5a^2b^4c$ can be written as $5a^2b^4c^1$ and the sum of the exponents is $2 + 4 + 1 = 7$.

37. $3x^5y^6z$ is degree 12 since $3x^5y^6z$ can be written as $3x^5y^6z^1$ and the sum of the exponents is $5 + 6 + 1 = 12$.

39. $7r+3b-11x+12y$ cannot be simplified since all the terms are "unlike".

41. $5x^2-11x+10x-5$
Combine like terms
$=5x^2-x-5$

43. $10.6c^2-2.3c+5.9c-1.9c^2$
$=10.6c^2-1.9c^2-2.3c+5.9c$
Combine like terms
$=8.7c^2+3.6c$

45. w^3+w^2-w+1 cannot be further simplified since all of the terms are "unlike".

47. $8pq-9pq+p+q$
Combine like terms
$=-pq+p+q$

49. $12\left(\frac{1}{6}+\frac{d}{4}\right)+5d$
Distributive property
$=12\cdot\frac{1}{6}+12\cdot\frac{d}{4}+5d$
$=\frac{12}{6}+\frac{12d}{4}+5d$
$=2+3d+5d$
Combine like terms
$=2+8d=8d+2$

51. $3\left(x+\frac{1}{2}\right)-\frac{1}{3}x+5$
Distributive property
$=3x+\frac{3}{2}-\frac{1}{3}x+5$
$=3x-\frac{1}{3}x+\frac{3}{2}+5$
$=\frac{9}{3}x-\frac{1}{3}x+\frac{3}{2}+\frac{10}{2}$
Combine like terms
$=\frac{8}{3}x+\frac{13}{2}$

53. $4-[6(3x+2)-x]+4$
Distributive property
$=4-[18x+12-x]+4$
Combine like terms
$=4-[17x-12]+4$
Distributive property
$=4-17x-12+4$
$=4-12+4-17x$
Combine like terms
$=-4-17x$
$=-17x-4$

55. $9x-[3x-(5x-4y)]-2y$
Distributive property
$=9x-[3x-5x+4y]-2y$
Combine like terms
$=9x-[-2x+4y]-2y$
Distributive property
$=9x+2x-4y-2y$
Combine like terms
$=11x-6y$

57. $5b-\{7[2(3b-2)-(4b+9)]-2\}$
Distributive property inside [ ]
$=5b-\{7[6b-4-4b-9]-2\}$
$=5b-\{7[6b-4b-4-9]-2\}$
Combine like terms
$=5b-[7(2b-13)-2]$
Distributive property
$=5b-[14b-91-2]$
Combine like terms
$=5b-(14b-93)$
Distributive property
$=5b-14b+93$
Combine like terms
$=-9b+93$

59. $-\{[2rs-3(r+2s)]-2(2r^2-s)\}$
Distributive property
$=-\{[2rs-3r-6s]-4r^2+2s\}$
$=-\{2rs-3r-6s-4r^2+2s\}$
$=-(2rs-3r-6s+2s-4r^2)$
Combine like terms
$=-(2rs-3r-4s-4r^2)$
Distributive property
$=-2rs+3r+4s+4r^2$
$=4r^2-2rs+3r+4s$

61.
$$5a-1=14$$
$$5a-1+1=14+1$$
$$5a=15$$
$$\frac{5a}{5}=\frac{15}{5}$$
$$a=3$$

63. $5x-9=3(x-2)$
$$5x-9=3x-6$$
$$5x-9+9=3x-6+9$$
$$5x=3x+3$$
$$5x-3x=3x+3-3x$$
$$2x=3$$
$$\frac{2x}{2}=\frac{3}{2}$$
$$x=\frac{3}{2}$$

65. $4x-8=-4(2x-3)+4$
$$4x-8=-8x+12+4$$
$$4x-8=-8x+16$$
$$4x-8+8x=-8x+16+8x$$
$$12x-8=16$$
$$12x-8+8=16+8$$
$$12x=24$$
$$\frac{12x}{12}=\frac{24}{12}$$
$$x=2$$

67. $-6(z-1)=-5(z+2)$
$$-6z+6=-5z-10$$
$$-6z+6+6z=-5z-10+6z$$
$$6=1z-10$$
$$6+10=1z-10+10$$
$$16=z$$

69. $-3(t-5)=2(t-5)$
$$-3t+15=2t-10$$
$$-3t+15-2t=2t-10-2t$$
$$-5t+15=-10$$
$$-5t+15-15=-10-15$$
$$-5t=-25$$
$$\frac{-5t}{-5}=\frac{-25}{-5}$$
$$t=5$$

71. $3x+4(2-x)=4x+5$
$$3x+8-4x=4x+5$$
$$-x+8=4x+5$$
$$-x+8-4x=4x+5-4x$$
$$-5x+8=5$$
$$-5x+8-8=5-8$$
$$-5x=-3$$
$$\frac{-5x}{-5}=\frac{-3}{-5}$$
$$x=\frac{3}{5}$$

73. $2-(x+5)=4x-8$
$$2-x-5=4x-8$$
$$-x-3=4x-8$$
$$-x-3-4x=4x-8-4x$$
$$-5x-3=-8$$
$$-5x-3+3=-8+3$$
$$-5x=-5$$
$$\frac{-5x}{-5}=\frac{-5}{-5}$$
$$x=1$$

75. $p-(p+4)=4(p-1)+2p$
$$p-p-4=4p-4+2p$$
$$-4=6p-4$$
$$-4+4=6p-4+4$$
$$0=6p$$
$$\frac{0}{6}=\frac{6p}{6}$$
$$0=p$$

77. $-3(y-1)+2y=4(y-3)$
$$-3y+3+2y=4y-12$$
$$-y+3=4y-12$$
$$-y+3+y=4y-12+y$$
$$3=5y-12$$
$$3+12=5y-12+12$$
$$15=5y$$
$$\frac{15}{5}=\frac{5y}{5}$$
$$3=y$$

79. $6-(n+3)=3n+5-2n$
$$6-n-3=3n+5-2n$$
$$3-n=n+5$$
$$3-n+n=n+5+n$$
$$3=2n+5$$
$$3-5=2n+5-5$$
$$-2=2n$$
$$\frac{-2}{2}=\frac{2n}{2}$$
$$-1=n$$
$$n=-1$$

81. $4(2x-2)-3(x+7)=-4$
$$8x-8-3x-21=-4$$
$$5x-29=-4$$
$$5x-29+29=-4+29$$
$$5x=25$$
$$\frac{5x}{5}=\frac{25}{5}$$
$$x=5$$

83. $-4(3-4x)-2(x-1)=12x$
$$-12+16x-2x+2=12x$$
$$14x-10=12x$$
$$14x-10-14x=12x-14x$$
$$-10=-2x$$
$$\frac{-10}{-2}=\frac{-2x}{-2}$$
$$5=x$$
$$x=5$$

85. $5(a+3)-a=-(4a-6)+1$
$$5a+15-a=-4a+6+1$$
$$4a+15=-4a+7$$
$$4a+15+4a=-4a+7+4a$$
$$8a+15=7$$
$$8a+15-15=7-15$$
$$8a=-8$$
$$\frac{8a}{8}=\frac{-8}{8}$$
$$a=-1$$

87. $5(x-2)-14x=x-5$
$$5x-10-14x=x-5$$
$$-9x-10=x-5$$
$$-9x-10+9x=x-5+9x$$
$$-10=10x-5$$
$$-10+5=10x-5+5$$
$$-5=10x$$
$$\frac{-5}{10}=\frac{10x}{10}$$
$$-\frac{1}{2}=x$$
$$x=-\frac{1}{2}$$

89. $2[3x-(4x-6)]=5(x-6)$
$$2(3x-4x+6)=5x-30$$
$$6x-8x+12=5x-30$$
$$-2x+12=5x-30$$
$$-2x+12+2x=5x-30+2x$$
$$12=7x-30$$
$$12+30=7x-30+30$$
$$42=7x$$
$$\frac{42}{7}=\frac{7x}{7}$$
$$6=x$$
$$x=6$$

91. $4\{2-[3(c+1)-2(c+1)]\}=-2c$
$$4\{2-[3c+3-2c-2]\}=-2c$$
$$4\{2-[c+1]\}=-2c$$
$$4\{2-c-1\}=-2c$$
$$4\{1-c\}=-2c$$
$$4-4c=-2c$$
$$4-4c+2c=-2c+2c$$
$$-2c+4=0$$
$$-2c+4-4=0-4$$
$$-2c=-4$$
$$\frac{-2c}{-2}=\frac{-4}{-2}$$
$$c=2$$

93. $-\{4(d+3)-5[3d-2(2d+7)]-8\}=-10d-6$
$$-\{4(d+3)-5[3d-4d-14]-8\}=-10d-6$$
$$-\{4(d+3)-5[-d-14]-8\}=-10d-6$$
$$-\{4d+12+5d+70-8\}=-10d-6$$
$$-\{9d+74\}=-10d-6$$
$$-9d-74=-10d-6$$
$$-9d-74+10d=-10d-6+10d$$
$$d-74=-6$$
$$d-74+74=-6+74$$
$$d=68$$

95. $\frac{s}{4} = -16$

$4\left(\frac{s}{4}\right) = 4(-16)$

$s = -64$

97. $\frac{4x-2}{3} = -6$

$3\left(\frac{4x-2}{3}\right) = 3(-6)$

$4x - 2 = -18$

$4x - 2 + 2 = -18 + 2$

$4x = -16$

$\frac{4x}{4} = \frac{-16}{4}$

$x = -4$

99. $\frac{3}{4}t + \frac{7}{8}t = 39$

$8\left(\frac{3}{4}t + \frac{7}{8}t\right) = 8(39)$

$6t + 7t = 312$

$13t = 312$

$\frac{13t}{13} = \frac{312}{13}$

$t = 24$

101. $\frac{1}{2}(x-2) = \frac{1}{3}(x+2)$

$6\left[\frac{1}{2}(x-2)\right] = 6\left[\frac{1}{3}(x+2)\right]$

$3(x-2) = 2(x+2)$

$3x - 6 = 2x + 4$

$3x - 6 - 2x = 2x + 4 - 2x$

$x - 6 = 4$

$x - 6 + 6 = 4 + 6$

$x = 10$

103. $4 - \frac{3}{4}a = 7$

$4 - \frac{3}{4}a - 4 = 7 - 4$

$-\frac{3}{4}a = 3$

$-4\left(-\frac{3}{4}a\right) = -4(3)$

$3a = -12$

$\frac{3a}{3} = \frac{-12}{3}$

$a = -4$

105. $\frac{1}{2} = \frac{4}{5}x - \frac{1}{4}$

$20\left(\frac{1}{2}\right) = 20\left(\frac{4}{5}x - \frac{1}{4}\right)$

$10 = 16x - 5$

$10 + 5 = 16x - 5 + 5$

$15 = 16x$

$\frac{15}{16} = \frac{16x}{16}$

$\frac{15}{16} = x$

107. $\frac{1}{4}(x+3) = \frac{1}{3}(x-2) + 1$

$12\left[\frac{1}{4}(x+3)\right] = 12\left[\frac{1}{3}(x-2) + 1\right]$

$3(x+3) = 4(x-2) + 12$

$3x + 9 = 4x - 8 + 12$

$3x + 9 = 4x + 4$

$3x + 9 - 4x = 4x + 4 - 4x$

$-x + 9 = 4$

$-x + 9 - 9 = 4 - 9$

$-x = -5$

$x = 5$

109. $0.4n + 4.7 = 5.1n$

$0.4n + 4.7 - 0.4n = 5.1n - 0.4n$

$4.7 = 4.7n$

$\frac{4.7}{4.7} = \frac{4.7n}{4.7}$

$1.00 = n$ or $n = 1.00$

111.
$$\begin{aligned}4.7x-3.6(x-1)&=4.9\\4.7x-3.6x+3.6&=4.9\\1.1x+3.6&=4.9\\1.1x+3.6-3.6&=4.9-3.6\\1.1x&=1.3\\\frac{1.1x}{1.1}&=\frac{1.3}{1.1}\\x&\approx1.18\end{aligned}$$

113.
$$\begin{aligned}&5(z+3.41)=-7.89(2z-4)-5.67\\&5z+17.05=-15.78z+31.56-5.67\\&5z+17.05=-15.78z+25.89\\&15.78z+5z+17.05=-15.78z+25.89+15.78z\\&20.78z+17.05=25.89\\&20.78z+17.05-17.04=25.89-17.04\\&20.78z=8.85\\&\frac{20.78z}{20.78}=\frac{8.85}{20.78}\\&z\approx0.43\end{aligned}$$

115.
$$\begin{aligned}0.6(500-2.4x)&=3.6(2x-4000)\\300-1.44x&=7.2x-14,400\\300-1.44x+1.44x&=7.2x-14,400+1.44x\\300&=8.64x-14,400\\300+14,400&=8.64x-14,400+14,400\\14,700&=8.64x\\\frac{14,700}{8.64}&=\frac{8.64x}{8.64}\\1701.39&\approx x\\x&\approx1701.39\end{aligned}$$

117.
$$\begin{aligned}1000(7.34q+14.78)&=100(3.91-4.21q)\\7340q+14780&=391-421q\\7340q+14780+421q&=391-421q+421q\\7761q+14780&=391\\7761q+14780-14780&=391-14780\\7761q&=-14389\\\frac{7761q}{7761}&=\frac{-14389}{7761}\\q&=-1.85\end{aligned}$$

119.
$$\begin{aligned}3(y+3)-4(2y-7)&=-5y+2\\3y+9-8y+28&=-5y+2\\-5y+37&=-5y+2\\-5y+37+5y&=-5y+2+5y\\37&=2\end{aligned}$$
The solution set is $\varnothing$.
The equation is a contradiction.

121.
$$\begin{aligned}4(2x-3)+15&=-6(x-4)+12x-21\\8x-12+15&=-6x+24+12x-21\\8x+3&=6x+3\\8x+3-3&=6x+3-3\\8x&=6x\\8x-6x&=6x-6x\\2x&=0\\x&=0\end{aligned}$$
The solution set is $\{0\}$.
The equation is conditional.

123.
$$\begin{aligned}4-\left(\frac{2}{3}x+2\right)&=2\left(-\frac{1}{3}x+1\right)\\4-\frac{2}{3}x-2&=-\frac{2}{3}x+2\\-\frac{2}{3}x+2&=-\frac{2}{3}x+2\\-\frac{2}{3}x+2+\frac{2}{3}x&=-\frac{2}{3}x+2+\frac{2}{3}x\\2&=2\end{aligned}$$
The solution set is $\mathbb{R}$.
The equation is an identity.

125.
$$\begin{aligned}6(x-1)&=-3(2-x)+3x\\6x-6&=-6+3x+3x\\6x-6&=-6+6x\\6x-6-6x&=-6+6x-6x\\-6&=-6\end{aligned}$$
The solution set is $\mathbb{R}$.
The equation is an identity.

127.
$$\begin{aligned}0.8z-0.3(z+10)&=0.5(z+1)\\0.8z-0.3z-3.0&=0.5z+0.5\\0.5z-3.0&=0.5z+0.5\\0.5z-3.0-0.5z&=0.5z+0.5-0.5z\\-3.0&=0.5\end{aligned}$$
The solution set is $\varnothing$.
The equation is a contradiction.

129. a. For 2008, substitute 8 for t.

$$P = 0.82t + 78.5$$
$$P = 0.82(8) + 78.5$$
$$P = 6.56 + 78.5$$
$$P = 85.06$$
$$P \approx 85$$

In 2008, the population density will be about 85 people per square mile.

b. Substitute 100 for P and solve for t.

$$P = 0.82t + 78.5$$
$$100 = 0.82t + 78.5$$
$$100 - 78.5 = 0.82t + 78.5 - 78.5$$
$$21.5 = 0.82t$$
$$\frac{21.5}{0.82} = \frac{0.82t}{0.82}$$
$$26.22 \approx t$$

The population density should reach 100 people per square mile during the year 2026.

131. a. For 2006, substitute 2 for x.

$$M = -1.26x + 61.48$$
$$= -1.26(2) + 61.48$$
$$= -2.52 + 61.48$$
$$= 58.96$$

The percent of total cars sold in the United States made by American automakers was 58.96%.

b.

$$M = -1.26x + 61.48$$
$$53.92 = -1.26x + 61.48$$
$$-7.56 = -1.26x$$
$$x = \frac{-7.56}{-1.26}$$
$$x = 6$$
$$6 + 2004 = 2010$$

The percent of total cars sold in the United States made by American automakers will be about 53.92% in 2010.

133. a. For 1941, substitute 1 for x.

$$t = 2.405 - 0.005x$$
$$t = 2.405 - 0.005(1)$$
$$t = 2.405 - 0.005$$
$$t = 2.4$$

The winning time for male runners in the Boston Marathon in 1941 is estimated to have been about 2.4 hours.

b. For 2005, substitute 65 for x.

$$t = 2.405 - 0.005x$$
$$t = 2.405 - 0.005(65)$$
$$t = 2.405 - 0.325$$
$$t = 2.08$$

The winning time for male runners in the Boston Marathon in 2005 is estimated to have been about 2.08 hours.

135. Answers may vary. Possible answer:

$$2x + 3 = 8$$
$$14x = 35$$
$$x = \frac{5}{2}$$

All three equations can be written in the form $2x = 5$.

137. Answers may vary. One possible answer is $x + 5 = x + 3$. Make sure that the variable terms "cancel" and leave a false statement.

139. Answers may vary. One possible answer is $\frac{5}{2}p + 7 = 6 + 2p + 4$.

141. $2(a + 5) + n = 4a - 8$

Substitute -2 for a and solve for n.

$$2(-2+5) + n = 4(-2) - 8$$
$$2(3) + n = -8 - 8$$
$$6 + n = -16$$
$$6 + n - 6 = -16 - 6$$
$$n = -22$$

143. $*\Delta - \square = \odot$ for Δ

$$*\Delta - \square + \square = \odot + \square$$
$$*\Delta = \odot + \square$$
$$\frac{*\Delta}{*} = \frac{\odot + \square}{*}$$
$$\Delta = \frac{\odot + \square}{*}$$

145. $\odot\square + \Delta = \otimes$ for $\odot$

$$\odot\square + \Delta - \Delta = \otimes - \Delta$$
$$\odot\square = \otimes - \Delta$$
$$\frac{\odot\square}{\square} = \frac{\otimes - \Delta}{\square}$$
$$\odot = \frac{\otimes - \Delta}{\square}$$

147. a. Answers will vary.

b. The definition of absolute value is

$$|a| = \begin{cases} a \text{ if } a \ge 0 \\ -a \text{ if } a < 0 \end{cases}$$

148. a. $-3^2 = -(3\cdot 3) = -9$

b. $(-3)^2 = (-3)(-3) = 9$

149. $\sqrt[3]{-125} = -5$ since $(-5)^3 = -125$

150. $\left(-\frac{2}{7}\right)^2 = \left(-\frac{2}{7}\right)\left(-\frac{2}{7}\right) = \frac{4}{49}$

Exercise Set 2.2

1. A formula is an equation that is a mathematical model of a real-life situation.

3. 1. Understand
2. Translate
3. Carry out
4. Check
5. Answer

5. a.
$$16 = 2l + 2(3)$$
$$16 = 2l + 6$$
$$16 - 6 = 2l + 6 - 6$$
$$10 = 2l$$
$$\frac{10}{2} = \frac{2l}{2}$$
$$5 = l$$

b.
$$P = 2l + 2w$$
$$P - 2w = 2l + 2w - 2w$$
$$P - 2w = 2l$$
$$\frac{P - 2w}{2} = \frac{2l}{2}$$
$$\frac{P - 2w}{2} = l$$
$$l = \frac{P - 2w}{2}$$

c. No; the same procedure was used for each solution.

d.
$$l = \frac{P - 2w}{2}$$
$$l = \frac{16 - 2(3)}{2}$$
$$l = \frac{16 - 6}{2}$$
$$l = \frac{10}{2}$$
$$l = 5$$
They are the same.
The formula and the equation are equivalent when $P = 16$ and $w = 3$.

7.
$$E = IR$$
$$= (63)(100)$$
$$= 6300$$

9.
$$R = R_1 + R_2$$
$$= 100 + 200$$
$$= 300$$

11.
$$A = \pi r^2$$
$$= \pi(8)^2$$
$$= \pi(64)$$
$$\approx 201.06$$

13.
$$\bar{x} = \frac{x_1 + x_2 + x_3}{3}$$
$$= \frac{40 + 90 + 80}{3}$$
$$= \frac{210}{3}$$
$$= 70$$

15.
$$A = P + Prt$$
$$= 160 + 160(0.05)(2)$$
$$= 160 + 16$$
$$= 176$$

17.
$$m = \frac{y_2 - y_1}{x_2 - x_1}$$
$$= \frac{4 - (-3)}{-2 - (-6)}$$
$$= \frac{4 + 3}{-2 + 6}$$
$$= \frac{7}{4}$$

19. $$\begin{aligned} R_T &= \frac{R_1R_2}{R_1+R_2} \\ &= \frac{100\cdot 200}{100+200} \\ &= \frac{20{,}000}{300} \\ &\approx 66.67 \end{aligned}$$

21. $$\begin{aligned} x &= \frac{-b+\sqrt{b^2-4ac}}{2a} \\ &= \frac{-(-5)+\sqrt{(-5)^2-4(2)(-12)}}{2(2)} \\ &= \frac{5+\sqrt{25+96}}{4} \\ &= \frac{5+\sqrt{121}}{4} \\ &= \frac{5+11}{4} \\ &= \frac{16}{4} \\ &= 4 \end{aligned}$$

23. $$\begin{aligned} A &= p\left(1+\frac{r}{n}\right)^{nt} \\ &= 100\left(1+\frac{0.06}{1}\right)^{1\cdot 3} \\ &= 100(1.06)^3 \\ &= 100(1.191016) \\ &\approx 119.10 \end{aligned}$$

25. $$\begin{aligned} 3x+y &= 5 \\ 3x+y-3x &= 5-3x \\ y &= 5-3x \end{aligned}$$
or
$$y = -3x+5$$

27. $$\begin{aligned} x-7y &= 13 \\ x-7y-x &= -x+13 \\ -7y &= -x+13 \\ \frac{-7y}{-7} &= \frac{-x+13}{-7} \\ y &= \frac{1}{7}x-\frac{13}{7} \end{aligned}$$

29. $$\begin{aligned} 6x-2y &= 16 \\ 6x-2y-6x &= -6x+16 \\ -2y &= -6x+16 \\ \frac{-2y}{-2} &= \frac{-6x+16}{-2} \\ y &= 3x-8 \end{aligned}$$

31. $$\begin{aligned} \frac{3}{4}x-y &= 5 \\ 4\left(\frac{3}{4}x-y\right) &= 4\cdot 5 \\ 3x-4y &= 20 \\ 3x-4y-3x &= -3x+20 \\ -4y &= -3x+20 \\ \frac{-4y}{-4} &= \frac{-3x+20}{-4} \\ y &= \frac{3}{4}x-5 \end{aligned}$$

33. $$\begin{aligned} 3(x-2)+3y &= 6x \\ 3x-6+3y &= 6x \\ 3x-6+3y-3x &= 6x-3x \\ -6+3y &= 3x \\ -6+3y+6 &= 3x+6 \\ 3y &= 3x+6 \\ \frac{3y}{3} &= \frac{3x+6}{3} \\ y &= x+2 \end{aligned}$$

35. $$\begin{aligned} y+1 &= -\frac{4}{3}(x-9) \\ 3[y+1] &= 3\left[-\frac{4}{3}(x-9)\right] \\ 3y+3 &= -4(x-9) \\ 3y+3 &= -4x+36 \\ 3y+3-3 &= -4x+36-3 \\ 3y &= \\ \frac{3y}{3} &= \frac{-4x+33}{3} \\ y &= -\frac{4}{3}+11 \end{aligned}$$

37. $d = rt$

$\frac{d}{r} = \frac{rt}{r}$

$\frac{d}{r} = t$ or $t = \frac{d}{r}$

39. $C = \pi d$

$\frac{C}{\pi} = \frac{\pi d}{\pi}$

$\frac{C}{\pi} = d$ or $d = \frac{C}{\pi}$

41. $P = 2l + 2w$

$P - 2w = 2l + 2w - 2w$

$P - 2w = 2l$

$\frac{P-2w}{2} = \frac{2l}{2}$

$\frac{P-2w}{2} = l$ or $l = \frac{P-2w}{2}$

43. $V = lwh$

$\frac{V}{lw} = \frac{lwh}{lw}$

$\frac{V}{lw} = h$ or $h = \frac{V}{lw}$

45. $A = P + Prt$

$A - P = P + Prt - P$

$A - P = Prt$

$\frac{A-P}{Pt} = \frac{Prt}{Pt}$

$\frac{A-P}{Pt} = r$ or $r = \frac{A-P}{Pt}$

47. $V = \frac{1}{3}lwh$

$3V = 3\left(\frac{1}{3}lwh\right)$

$3V = lwh$

$\frac{3V}{wh} = \frac{lwh}{wh}$

$\frac{3V}{wh} = l$ or $l = \frac{3V}{wh}$

49. $y = mx + b$

$y - b = mx + b - b$

$y - b = mx$

$\frac{y-b}{x} = \frac{mx}{x}$

$\frac{y-b}{x} = m$ or $m = \frac{y-b}{x}$

51. $y - y_1 = m(x - x_1)$

$\frac{y-y_1}{x-x_1} = \frac{m(x-x_1)}{x-x_1}$

$\frac{y-y_1}{x-x_1} = m$ or $m = \frac{y-y_1}{x-x_1}$

53. $z = \frac{x-\mu}{\sigma}$

$\sigma z = \sigma\left(\frac{x-\mu}{\sigma}\right)$

$\sigma z = x - \mu$

$\sigma z - x = x - \mu - x$

$\sigma z - x = -\mu$

$x - \sigma z = \mu$

55. $P_1 = \frac{T_1 P_2}{T_2}$

$T_2 P_1 = T_2\left(\frac{T_1 P_2}{T_2}\right)$

$T_2 P_1 = T_1 P_2$

$\frac{T_2 P_1}{P_1} = \frac{T_1 P_2}{P_1}$

$T_2 = \frac{T_1 P_2}{P_1}$

57. $A = \frac{1}{2}h(b_1 + b_2)$

$2A = 2\left[\frac{1}{2}h(b_1 + b_2)\right]$

$2A = h(b_1 + b_2)$

$\frac{2A}{b_1 + b_2} = \frac{h(b_1 + b_2)}{b_1 + b_2}$

$\frac{2A}{b_1 + b_2} = h$ or $h = \frac{2A}{b_1 + b_2}$

59. $S = \frac{n}{2}(f + l)$

$2S = 2\left[\frac{n}{2}(f + l)\right]$

$2S = n(f + l)$

$\frac{2S}{f + l} = \frac{n(f + l)}{f + l}$

$\frac{2S}{f + l} = n$ or $n = \frac{2S}{f + l}$

61. $C = \frac{5}{9}(F - 32)$

$\frac{9}{5}C = \frac{9}{5} \cdot \frac{5}{9}(F - 32)$

$\frac{9}{5}C = F - 32$

$\frac{9}{5}C + 32 = F - 32 + 32$

$\frac{9}{5}C + 32 = F$ or $F = \frac{9}{5}C + 32$

63. $F = \frac{km_1m_2}{d^2}$

$Fd^2 = d^2\left(\frac{km_1m_2}{d^2}\right)$

$Fd^2 = km_1m_2$

$\frac{Fd^2}{km_2} = \frac{km_1m_2}{km_2}$

$\frac{Fd^2}{km_2} = m_1$ or $m_1 = \frac{Fd^2}{km_2}$

65. a. Let d = U.S. dollars, and p = Mexican pesos. Then $p = 9.11d$.

b. Solve $p = 9.11d$ for d.

$p = 9.11d$

$\frac{p}{9.11} = \frac{9.11d}{9.11}$

$\frac{p}{9.11} = d$ or $d = \frac{p}{9.11}$

c. Answers will vary.

67. $i = prt$

$= 1100(0.07)(4)$

$= 308$

Ken must pay \$308 in simple interest.

69. $i = prt$

$4875 = (20000)(.0375)t$

$4875 = 750t$

$\frac{4875}{750} = \frac{750t}{750}$

$6.5 = t$

The length of the loan was 6.5 years.

71. a. $A = \pi r^2$

$= \pi(1)^2$

≈ 3.14 square inches

b. $A = \pi r^2$

$= \pi(5)^2$

$= 25\pi$

≈ 78.54 square inches.

73. a. 6 inches is 0.5 feet.

$V = lwh$

$= 15(10)(0.5)$

$= 75$ cubic feet

b. $\frac{75}{27} \approx 2.78$ cubic yards

c. To get 2.78 cubic yards of concrete, 3 cubic yards must be purchased.

$3(\$35) = \$105$

75. The volume of a cylinder is given by $V = \pi r^2 h$. Note that the radius is half the diameter so the radius is 2.5 inches.

$V = \pi r^2 h$

$= \pi(2.5)^2(6.25)$

≈ 122.72 cubic inches

For the volume of the box:

$V = lwh$

$= (7)(5)(3.5)$

$= 122.5$ cubic inches

The cylinder has greater volume by about $122.72 - 122.5 = 0.22$ cubic inches.

77. $A = p\left(1+\frac{r}{n}\right)^{nt}$

$= 10{,}000\left(1+\frac{0.06}{4}\right)^{4\cdot 2}$

$= 10{,}000(1.015)^8$

$= 11{,}264.93$

Beth will have \$11,264.93 in her account.

79. Note that 36 months is 3 years so $t = 3$.

$A = p\left(1+\frac{r}{n}\right)^{nt}$

$= 4390\left(1+\frac{0.041}{2}\right)^{(2)(3)}$

$= 4390(1+0.0205)^6$

$= 4390(1.0205)^6$

≈ 4958.41

The certificate will be worth \$4958.41 after 36 months.

81. $T_f = T_a(1-F)$

$0.035 = T_a(1-0.15)$

$0.035 = T_a(0.85)$

$\frac{0.035}{0.85} = T_a$

$T_a \approx 0.0412$

The equivalent taxable rate is 4.12%.

83. a. The 4.6% tax-free rate for Anthony:

$T_f = T_a(1-F)$

$0.046 = T_a(1-0.35)$

$0.046 = T_a(0.65)$

$\frac{0.046}{0.65} = T_a$

$T_a \approx 0.0708 = 7.08\%$

b. The 4.6% tax-free rate for Angelo:

$T_f = T_a(1-F)$

$0.046 = T_a(1-0.28)$

$0.046 = T_a(0.72)$

$\frac{0.046}{0.72} = T_a$

$T_a \approx 0.0639 = 6.39\%$

85. $w = 0.02c$

a. $c = 2600 - 2400 = 200$

$w = 0.02(200)$

$= 4$

Her weekly weight loss is 4 pounds per week.

b. $2 = 0.02c$

$\frac{2}{0.02} = c$

$c = 100$

$2400 + 100 = 2500$

She would have to burn 2500 calories per day to lose 2 pounds in a week.

87. a. $S = 100 - a$

b. $S = 100 - 60 = 40$

A 60-year-old should keep 40% in stocks.

89. $r = \frac{\frac{s}{t}}{\frac{t}{u}} = \frac{s}{t} \div \frac{t}{u} = \frac{s}{t}\cdot\frac{u}{t} = \frac{su}{t^2}$

In simplified form, it is $r = \frac{su}{t^2}$

a. $r = \frac{su}{t^2}$

$rt^2 = t^2\left(\frac{su}{t^2}\right)$

$rt^2 = su$

$\frac{rt^2}{u} = \frac{su}{u}$

$\frac{rt^2}{u} = s$ or $s = \frac{rt^2}{u}$

b. $r = \frac{su}{t^2}$

$rt^2 = t^2\left(\frac{su}{t^2}\right)$

$rt^2 = su$

$\frac{rt^2}{s} = \frac{su}{s}$

$\frac{rt^2}{s} = u$ or $u = \frac{rt^2}{s}$

90. $-\sqrt{3^2+4^2}+|3-4|-6^2 = -\sqrt{9+16}+|-1|-36$
$= -\sqrt{25}+1-36$
$= -5+1-36$
$= -40$

91. $\dfrac{7+9\div(2^3+4\div4)}{|3-7|+\sqrt{5^2-3^2}} = \dfrac{7+9\div(8+4\div4)}{|3-7|+\sqrt{25-9}}$
$= \dfrac{7+9\div(8+1)}{|-4|+\sqrt{16}} = \dfrac{7+9\div9}{4+4}$
$= \dfrac{7+1}{4+4} = \dfrac{8}{8}$
$= 1$

92. $a^3-3a^2b+3ab^2-b^3$ with $a=-2$ and $b=3$
$(-2)^3-3(-2)^2(3)+3(-2)(3)^2-(3)^3$
$-8-3(4)(3)+3(-2)(9)-27$
$-8-(12)(3)+(-6)(9)-27$
$-8-36+-54-27$
-125

93. $\frac{1}{4}t+\frac{1}{2}=1-\frac{1}{8}t$
$8\left(\frac{1}{4}t+\frac{1}{2}\right)=8\left(1-\frac{1}{8}t\right)$
$2t+4=8-t$
$2t+4+t=8-t+t$
$3t+4=8$
$3t+4-4=8-4$
$3t=4$
$\frac{3t}{3}=\frac{4}{3}$
$t=\frac{4}{3}$

Exercise Set 2.3

1. $x-3$

3. $v+6$

5. $d+2$

7. $19.95y$

9. $0.096x$

11. Let x = the length of the first piece in feet. Then the second piece has length $12-x$ feet.
x; $12-x$

13. Let w = the width of the rectangle in meters. Then the length is $w+29$ meters.
w; $w+29$

15. Let p = the dollar amount of Max's portion. Then Lora's portion is $165-p$.
p; $165-p$

17. Let z = the speed at which Betty can jog. Then Nora's speed is $z+1.3$.
z; $z+1.3$

19. Let e = the original cost of the electricity. The increase is $0.22e$, so the new cost is the original cost plus the increase, $e+0.22e$ or $1.22e$.
e; $e+0.22e$

21. Let x = measure of angle B
$4x$ = measure of angle A
measure of angle A + measure of angle B = 90°
$4x+x=90$
$5x=90$
$x=18$
Angle B is 18° and angle A is $4\times18°=72°$.

23. $B=4A$
$A+B=180$
$A+4A=180$
$5A=180$
$A=36$
$B=4A=4(36)=144$
The measure of angle A is 36° and B is 144°.

25. Let x=smallest angle, then
$x+20$=second angle
$2x$=third angle
$x+x+20+2x=180$
$4x+20=180$
$4x=160$
$x=40$
$x+20=40+20=60$
$2x=2(40)=80$
The measures of the angles are 40°, 60°, and 80°.

27. Let x = the cost of the regular subscription.

$$x - 0.25x = 24$$
$$0.75x = 24$$
$$\frac{0.75x}{0.75} = \frac{24}{0.75}$$
$$x = 32$$

The original cost of the subscription was \$32.

29. Let x = number of rides.

$$1.80x = 45$$
$$\frac{1.80x}{1.80} = \frac{45}{1.80}$$
$$x = 25 \text{ rides}$$

Kate would need to ride the bus 25 times per month.

31. Let n = number of miles.

$$0.20n + 35 = 80$$
$$0.20n = 45$$
$$\frac{0.20n}{0.20} = \frac{45}{0.20}$$
$$n = 225 \text{ miles}$$

Tanya can drive 225 miles in one day.

33. Let x = the number of golfing trips.

The cost of a social membership:

$$50x + 25x + 1775 = 75x + 1775$$

The cost of a golf membership:

$$25x + 2425$$

Set these two expressions equal to each other and solve for x.

$$75x + 1775 = 25x + 2425$$
$$75x + 1775 - 25x = 25x + 2425 - 25x$$
$$50x + 1775 = 2425$$
$$50x + 1775 - 1775 = 2425 - 1775$$
$$50x = 650$$
$$x = 13$$

She must go golfing 13 times per year for the two options to cost the same amount.

35. Let t = number of trips, then

$2.50t$ = cost for one trip without pass

$$0.50t + 20 = 2.50t$$
$$20 = 2.00t$$
$$10 = t$$

The Morgans would have to go more than 10 times for the cost of the monthly pass to be worthwhile.

37. Let r = the monthly rent in 2006. Then $r + 0.075r = 1.075r$ is the monthly rent in 2007.

$$r + 0.075r = 1720$$
$$1.075r = 1720$$
$$\frac{1.075r}{1.075} = \frac{1720}{1.075}$$
$$r = 1600$$

The monthly rent in 2006 was \$1600.

39. Let s = sales from the northwest district (in millions of dollars), then $s + 0.31$ is the sales from the southeast district (in millions of dollars).

$$s + s + 0.31 = 4.6$$
$$2s + 0.31 = 4.6$$
$$2s = 4.29$$
$$s = 2.145$$
$$2.145 + 0.31 = 2.455$$

The sales from the northwest district were \$2.145 million and the sales from the southeast district were \$2.455 million.

41. Let p = the average personal income in 1980. Then $2.32p$ is the average personal income in 2004.

$$p + 2.32p = 29{,}367$$
$$3.32p = 29{,}367$$
$$\frac{3.32p}{3.32} = \frac{29{,}367}{3.32}$$
$$p = 8845.48$$

The average personal income in 1980 was \$8845.48.

43. Let w = the minimum wage in 1980. Then $w + 0.6613w = 1.6613w$ is the minimum wage in 2005.

$$1.6631w = 5.15$$
$$\frac{1.6631w}{1.6631} = \frac{5.15}{1.6631}$$
$$w \approx 3.10$$

The minimum wage in 1980 was \$3.10 per hour.

45. Let x = number of grasses. Then $2x - 5$ = number of weeds and $2x + 2$ is the number of trees.

$$x + (2x - 5) + (2x + 2) = 57$$
$$5x - 3 = 57$$
$$5x = 60$$
$$\frac{5x}{5} = \frac{60}{5}$$
$$x = 12$$

There are 12 grasses, $2(12) - 5 = 24 - 5 = 19$ weeds, and $2(12) + 2 = 24 + 2 = 26$ trees.

47. Let x = price of lunch. Note that the tip is 15% of cost of the *meal plus the tax*. So the tip is $0.15(x + 0.07x)$.

$$x + 0.07x + 0.15(x + 0.07x) = 20.00$$
$$x + 0.07x + 0.15x + 0.0105x = 20.00$$
$$1.2305x = 20.00$$
$$x \approx 16.25$$

The maximum price of the lunch she can order is \$16.25.

49. a. Let x = number of months for the total payments to be the same.

$$563.50x = 538.30x + 0.02(70,000) + 200$$
$$563.50x = 538.30x + 1400 + 200$$
$$563.50x = 538.30x + 1600$$
$$25.2x = 1600$$
$$\frac{25.2x}{25.2} = \frac{1600}{25.2}$$
$$x \approx 63.49$$

It takes about 63.49 months (5.29 years) for the total payments to be the same.

b. If they plan to keep the house for 30 years, the lower total cost would be with the mortgage from First National.

51. a. Let x = number of months (or monthly payments) necessary for the accumulated payments under the original mortgage plan to equal the accumulated payments and closing cost under the other plan.

$$510x = 420.50x + 2500$$
$$89.50x = 2500$$
$$\frac{89.50x}{89.50} = \frac{2500}{89.50}$$
$$x \approx 28$$

In about 28 months or 2.33 years, he would have paid the same amount under either plan.

b. Yes, any time after 2.33 years makes the refinancing worth it.

53. Let n = the number of medals won by Germany. Then $n+1$ = the number won by Australia, $2n-4$ = the number won by Russia, $n+15$ = the number won by China, and $2n+7$ = the number won by the United States.

$$n + (n+1) + (2n-4) + (n+15) + (2n+7) = 355$$
$$7n + 19 = 355$$
$$7n = 336$$
$$n = 48$$

Germany won 48 medals, Australia won $48 + 1 = 49$ medals, Russia won $2(48) - 4 = 92$ medals, China won $48 + 15 = 63$ medals, and the United States won $2(48) + 7 = 103$ medals.

55. Let x = number of animals. Then $x + 100,000$ = number of plants, $x + 290,000$ = number of non-beetle insects, and $2x - 140,000$ = number of beetles.

$$x + (x + 100,000) + (x + 290,000) + (2x - 140,000)$$
$$= 5x + 250,000$$
$$5x + 250,000 = 1,500,000$$
$$5x = 1,250,000$$
$$\frac{5x}{5} = \frac{1,250,000}{5}$$
$$x = 250,000$$

There are 250,000 animal species. 250,000 + 100,000 = 350,000 plant species, 250,000 + 290,000 = 540,000 non-beetle insect species, and 2(250,000) – 140,000 = 360,000 beetle species.

57. Let s = the length of the smaller side. The lengths of the other two sides are $(s + 3)$ and $(2s - 3)$. The perimeter is the sum of the sides.

$$s + (s+3) + (2s-3) = 36$$
$$4s = 36$$
$$s = 9$$

The length of the smaller side is 9 in. The lengths of the other two sides are $(9+3) = 12$ in. and $(2(9)-3) = 18 - 3 = 15$ in.

59. Let s = the length of the smaller side. The lengths of the other two sides are $(2s + 4)$ and $(3s - 4)$. The perimeter is the sum of the sides.

$$s+(2s+4)+(3s-4)=60$$
$$6s=60$$
$$s=10$$

The length of the smaller side is 10 ft. The lengths of the other two sides are $2(10)+4=24$ ft. and $3(10)-4=26$ ft.

61. Let x = length of one side of the square. Since there are 7 sides, the total perimeter is $7x$.

$$7x=91$$
$$\frac{7x}{7}=\frac{91}{7}$$
$$x=13$$

The dimensions of each square will be 13 meters by 13 meters.

63. Let h = height of each bookshelf. Then $h + 3$ is the width.

$$2h+4(h+3)=30$$
$$2h+4h+12=30$$
$$6h=18$$
$$h=3$$
$$h+3=6$$

The width is 6 feet and the height is 3 feet.

65. Let p = the original price of the calculator.

$$p-0.10p-5=49$$
$$0.90p-5=49$$
$$0.90p=54$$
$$\frac{0.90p}{0.90}=\frac{54}{0.90}$$
$$p=60$$

The original price of the calculator was $60.

67. Let x = number of paintings to be sold. The break-even point occurs when sales equals expenses.

$$500x=1350+0.10(500x)$$
$$500x=1350+500x$$
$$450x=1350$$
$$\frac{4500x}{450}=\frac{1350}{450}$$
$$x=3$$

The break-even point occurs when 3 paintings are sold.

69. Let x = energy cost.

$$(9.75+73)-(20+x)=46.75$$
$$9.75+73-20-x=46.75$$
$$62.75-x=46.75$$
$$-x=-16$$
$$x=16$$

The energy cost is $16.

71. **a.** Let x = the fifth score.

$$\frac{88+92+97+96+x}{5}=90$$

b. Answers may vary

c.
$$\frac{373+x}{5}=90$$
$$5\left(\frac{373+x}{5}\right)=5(90)$$
$$373+x=450$$
$$x=77$$

Paula needs a score of 77 on the fifth test.

73. Answers will vary.

75. Let x = number of miles driven.

$$3(28)+0.15x+0.04[3(28)+0.15x]=121.68$$

Original Charge 4% Sales Tax

$$84+0.15x+0.04(84+0.15x)=121.68$$
$$84+0.15x+3.36+0.006x=121.68$$
$$87.36+0.156x=121.68$$
$$0.156x=34.32$$
$$\frac{0.156x}{0.156}=\frac{34.32}{0.156}$$
$$x=220$$

Martin drove a total of 220 miles during the three days.

78. $2+\left|-\frac{3}{5}\right|=2+\frac{3}{5}=\frac{10}{5}+\frac{3}{5}=\frac{13}{5}$

79. $-6.4-(-3.7)=-6.4+3.7=-2.7$

80. $\left|-\frac{5}{8}\right|\div\left|-4\right|=\frac{5}{8}\div 4=\frac{5}{8}\cdot\frac{1}{4}=\frac{5}{32}$

81. $5-\left|-3\right|-\left|12\right|=5-3-12=2-12=-10$

82. $\left(2x^4y^{-6}\right)^{-3}=2^{-3}\left(x^4\right)^{-3}\left(y^{-6}\right)^{-3}$
$=\frac{1}{8}x^{-12}y^{18}$
$=\frac{y^{18}}{8x^{12}}$

Mid-Chapter Test: 2.1 – 2.3

1. The degree of $6x^5y^7$ is 12 because the sum of the exponents is $5+7=12$.

2. $3x^2+7x-9x+2x^2-11$
$=3x^2+2x^2+7x-9x-11$
$=5x^2-2x-11$

3. $2(a-1.3)+4(1.1a-6)+17$
$=2a-2.6+4.4a-24+17$
$=2a+4.4a-2.6-24+17$
$=6.4a-9.6$

4. $7x-9=5x-21$
$7x-9-5x=5x-21-5x$
$2x-9=-21$
$2x-9+9=-21+9$
$2x=-12$
$\frac{2x}{2}=\frac{-12}{2}$
$x=-6$

5. $\frac{3}{4}y+\frac{1}{2}=\frac{7}{8}y-\frac{5}{4}$
$\frac{3}{4}y+\frac{1}{2}-\frac{7}{8}y=\frac{7}{8}y-\frac{5}{4}-\frac{7}{8}y$
$\frac{6}{8}y+\frac{1}{2}-\frac{7}{8}y=-\frac{5}{4}$
$-\frac{1}{8}y+\frac{1}{2}=-\frac{5}{4}$
$-\frac{1}{8}y+\frac{1}{2}-\frac{1}{2}=-\frac{5}{4}-\frac{1}{2}$
$-\frac{1}{8}y=-\frac{5}{4}-\frac{2}{4}$
$-\frac{1}{8}y=-\frac{7}{4}$
$-8\left(-\frac{1}{8}y\right)=-8\left(-\frac{7}{4}\right)$
$y=14$

6. $3p-2(p+6)=4(p+1)-5$
$3p-2p-12=4p+4-5$
$p-12=4p-1$
$p-12-p=4p-1-p$
$-12=3p-1$
$-12+1=3p-1+1$
$-11=3p$
$\frac{-11}{3}=\frac{3p}{3}$
$-\frac{11}{3}=p$

7. $0.6(a-3)-3(0.4a+2)=-0.2(5a+9)-4$
$0.6a-1.8-1.2a-6=-a-1.8-4$
$-0.6a-7.8=-a-5.8$
$-0.6a-7.8+a=-a-5.8+a$
$0.4a-7.8=-5.8$
$0.4a-7.8+7.8=-5.8+7.8$
$0.4a=2$
$\frac{0.4a}{0.4}=\frac{2}{0.4}$
$a=5$

8. $4x+15-9x=-7(x-2)+2x+1$
$4x+15-9x=-7x+14+2x+1$
$-5x+15=-5x+15$
$-5x+15+5x=-5x+15+5x$
$15=15$
The equation is an identity. The solution set is $\mathbb{R}$, the set of all real numbers.

9. $-3(3x+1)=-\left[4x+(6x-5)\right]+x+7$
$-9x-3=-\left[4x+6x-5\right]+x+7$
$-9x-3=-\left[10x-5\right]+x+7$
$-9x-3=-10x+5+x+7$
$-9x-3=-9x+12$
$-9x-3+9x=-9x+12+9x$
$-3=12$
The equation is a contradiction. The solution set is $\varnothing$, the empty set.

10. $A=\frac{1}{2}hb$

$A=\frac{1}{2}(10)(16)$

$=5(16)$

$=80$

11. $R_T=\frac{R_1R_2}{R_1+R_2}$

$R_T=\frac{(100)(50)}{100+50}$

$=\frac{5000}{150}$

$=\frac{100}{3}$

12. $y=7x+13$

$y-13=7x+13-13$

$y-13=7x$

$\frac{y-13}{7}=\frac{7x}{7}$

$\frac{y-13}{7}=x$ or $x=\frac{y-13}{7}$

13. $A=\frac{2x_1+x_2+x_3}{n}$

$nA=n\left(\frac{2x_1+x_2+x_3}{n}\right)$

$nA=2x_1+x_2+x_3$

$nA-2x_1-x_2=2x_1+x_2+x_3-2x_1-x$

$nA-2x_1-x_2=x_3$

or

$x_3=nA-2x_1-x_2$

14. $A=P\left(1+\frac{r}{n}\right)^{nt}$

$=700\left(1+\frac{0.06}{4}\right)^{4\cdot 5}$

$=700(1.015)^{20}$

$=942.80$

The certificate of deposit will be worth $942.80 after 5 years.

15. $A=2B+6$

$A+B=90$

$2B+6+B=90$

$3B+6=90$

$3B=84$

$B=28$

Angle A measures $2(28)+6=62°$ and angle B measures $28°$.

16. Let d = the number of days.

$15+1.75d=32.50$

$15+1.75d-15=32.50-15$

$1.75d=17.5$

$\frac{1.75d}{1.75}=\frac{17.5}{1.75}$

$d=10$

Tom rented the ladder for 10 days.

17. Let x = the length of the shortest side. Then the length of the longest side is $4x$ and the length of the last side is $x+10$.

$x+4x+x+10=100$

$6x+10=100$

$6x=90$

$x=15$

The sides of the triangle have lengths of 15 feet, $15+10=25$ feet, and $4(15)=60$ feet.

18. Let r = the tax rate (as a decimal). Then the total tax is given by $36r$.

$36+36r=37.62$

$36r=1.62$

$r=\frac{1.62}{36}=0.45$

The sales tax rate was 4.5%.

19. Let n = the number of months, then $52n$ is the total increase in population.

$3613+52n=5693$

$52n=2080$

$n=\frac{2080}{52}=40$

40 months ago the population was 3613.

20. Mary is incorrect. To obtain an equivalent equation, both sides of the equation must be multiplied by the *same* non-zero constant. She should multiply both sides of the equation by the least common multiple of the denominators, 12.

$$\frac{1}{2}x+\frac{1}{3}=\frac{1}{4}x-\frac{1}{2}$$
$$12\left(\frac{1}{2}x+\frac{1}{3}\right)=12\left(\frac{1}{4}x-\frac{1}{2}\right)$$
$$6x+4=3x-6$$
$$3x+4=-6$$
$$3x=-10$$
$$x=-\frac{10}{3}$$

Exercise Set 2.4

1.

	Rate	Time	Distance
Don	5	1.2	5(1.2)
Judy	4.5	1.2	4.5(1.2)

distance=5(1.2)+4.5(1.2)
=6+5.4
=11.4

The distance around the lake is 11.4 miles

3. Let t = time in hours

Balloon	Rate	Time	Distance
1	14	t	$14t$
2	11	t	$11t$

distance apart = balloon 1 dist.–balloon 2 dist.

$$12=14t-11t$$
$$12=3t$$
$$4=t$$

It will take 4 hours for the balloons to be 12 miles apart.

5. Let t = the time each are gleaning

	Rate	Time	Distance
Rodney	0.15	t	$0.15t$
Dennis	0.10	t	$0.10t$

$$0.15t+0.10t=1.5$$
$$0.25t=1.5$$
$$t=6$$

Rodney and Dennis will meet after 6 hours.

7. a. Let r = Wayne's speed

	Rate	Time	Distance
Mary	$2r$	3	$(2r)(3)$
Wayne	r	3	$3r$

After 3 hours, Laura is 18 miles ahead of Wayne:

$$(2r)(3)=3r+18$$
$$6r=3r+18$$
$$3r=18$$
$$r=6$$

Wayne's speed is 6 miles per hour.

b. Mary's speed is 2(6) = 12 miles per hour.

9. a. Let t = time needed for Kristen to catch up with Luis

	Rate	Time	Distance
Luis	4	$t+0.75$	$4(t+0.75)$
Kristen	24	t	$24t$

$$24t=4(t+0.75)$$
$$24t=4t+3$$
$$20t=3$$
$$t=0.15$$
$$0.15 \text{ hours }=(0.15)(60)=9 \text{ minutes}$$

It will take Kristen 9 minutes to catch up with Luis.

b. When Kristen catches up with Luis (after 0.15 hours), Kristen will have traveled a distance of (24)(0.15) = 3.6 miles. She will be 3.6 miles from their house.

11. Let t = time of operation for smaller machine

	Rate	Time	Amount
Smaller machine	400	t	$400t$
Larger machine	600	$t+2$	$600(t+2)$

$$400t+600(t+2)=15{,}000$$
$$400t+600t+1200=15{,}000$$
$$1000t=13{,}800$$
$$t=13.8$$

The smaller machine operated for 13.8 hours.

13. Let t = time needed for Linda's husband to catch up with her. Note that 15 minutes = 0.25 hours.

	Rate	Time	Distance
Husband	50	t	$50t$
Linda	35	$t+0.25$	$35(t+0.25)$

Since the distances traveled are the same,

$$50t=35(t+0.25)$$
$$50t=35t+8.75$$
$$15t=8.75$$
$$t=0.58\bar{3} \text{ hours or 35 minutes}$$

It took Linda's husband 35 minutes to catch her.

15. Let x = amount invested at 3%. Assume that the interest is earned over a one-year period.

Account	Principal	Rate	Time	Interest
3%	x	0.03	1	$0.03x$
4.1%	$30{,}000-x$	0.041	1	$0.041(30{,}000-x)$

The total interest is \$1091.73.

$$0.03x+0.041(30000-x)=1091.73$$
$$0.03x+1230-0.041x=1091.73$$
$$-0.011x+1230=1091.73$$
$$-.011x=-138.27$$
$$x=12570$$

Thus, \$12,570 was invested at 3% and the remaining amount of \$30,000 – \$12,570 = \$17,430 was invested at 4.1%.

17. Let x = pounds of Kona coffee

Item	Cost	Pounds	Total
Kona	6.20	x	$6.20x$
Amaretto	5.80	18	5.80(18)
Mixture	6.10	$18+x$	$6.10(18+x)$

$$6.20x+5.80(18)=6.10(18+x)$$
$$6.2x+104.4=109.8+6.1x$$
$$0.1x=5.4$$
$$x=54$$

She should mix 54 pounds of Kona coffee with the amaretto coffee.

19. a. Let x = the number of shares of Johnson & Johnson stock.

	No. of Shares	Price/Share	Total Value
J & J	x	56.88	$56.88x$
AOL	$2x$	27.36	$27.36(2x)$

The total amount invested is $250,000.

$$56.88x + 27.36(2x) = 250000$$
$$56.88x + 54.72x = 250000$$
$$111.6x = 250000$$
$$x \approx 2200 \text{ (to the nearest hundred)}$$

Don should buy 2200 shares of Johnson & Johnson and 2(2200) = 4400 shares of AOL.

b. Don's total purchase is

$$(2200)(\$56.88) + (4400)(\$27.36)$$
$$\$125,136 + \$120,384$$
$$\$245,520$$

Don will have $250,000 - $245,520 = $4,480 left over after the purchase.

21. Let x = ounces of 12% solution

Solution	Strength	Ounces	Acid
Mail	12%	x	$0.12x$
Store	5%	40	0.05(40)
Mixture	8%	$x+40$	$0.08(x+40)$

$$0.12x + 0.05(40) = 0.08(x+40)$$
$$0.12x + 2 = 0.08x + 3.2$$
$$0.04x = 1.2$$
$$x = 30$$

She should mix 30 ounces of the 12% vinegar.

23. Let x = number of teaspoons of 30% sauce.

Sauce	Strength	Teaspoons	Acid
#1	30%	x	$0.30x$
#2	80%	$4-x$	$0.80(4-x)$
Mixture	45%	4	$0.45(4)$

$$0.30x + 0.80(4-x) = 0.45(4)$$
$$0.30x + 3.2 - 0.80x = 1.8$$
$$-0.50x = -1.4$$
$$x = 2.8$$

She should use 2.8 teaspoons of the 30% sauce and $4 - 2.8 = 1.2$ teaspoons of the 80% sauce.

25. Let x = strength of the second solution.

Solution	Strength of Solution	No. of Milliliters	Amount
20%	0.20	200	0.20(200)
Unknown	x	100	$x(100)$
25%	0.25	300	0.25(300)

$$0.20(200) + x(100) = 0.25(300)$$
$$40 + 100x = 75$$
$$100x = 35$$
$$x = \frac{35}{100} = 0.35 \text{ or } 35\%$$

The strength of the unknown solution is 35%.

27. Let x = number of pounds of the striped orange slices.

Type	Cost	No. of Pounds	Amount
Orange Slices	\$1.29	x	$1.29x$
Strawberry Leaves	\$1.79	$12 - x$	$1.79(12 - x)$
Mixture	$\frac{17.48}{12} = \$1.46$	12	17.48

$$1.29x + 1.79(12 - x) = 17.48$$
$$1.29x + 21.48 - 1.79x = 17.48$$
$$-0.50x + 21.48 = 17.48$$
$$-0.50x = -4$$
$$x = \frac{-4}{-0.50} = 8$$

8 pounds of the orange slices should be mixed with 12 – 8 = 4 pounds of the strawberry leaves to produce the desired mixture.

29. Let t = time (in hours) before they meet.

	Rate	Time	Distance
Julie	45	t	$45t$
Kamilia	50	t	$50t$

Their combined distances are 2448 miles.

$$45t + 50t = 2448$$
$$95t = 2448$$
$$t \approx 25.77$$

They will meet after about 25.77 hours.

31. Let x = time needed for both pumps to empty the pool.

Pump	Rate	Time	Amount Pumped
1	10	t	$10t$
2	20	t	$20t$

The total amount of water pumped is 15,000 gallons.

$$10t + 20t = 15{,}000$$
$$30t = 15{,}000$$
$$t = \frac{15{,}000}{30} = 500 \text{ minutes}$$

It will take the pumps 500 minutes or $\frac{500}{60} = 8\frac{1}{3}$ hours to empty the pool.

33. Let x = amount of pure antifreeze to be added.

Type	Strength of Solution	No. of Quarts	Amount
Pure	1.00	x	$1.00x$
20%	0.20	10	0.20(10)
Mixture	0.50	$x+10$	$0.50(x+10)$

$$1.00x + 0.20(10) = 0.50(x+10)$$
$$1.00x + 2 = 0.50x + 5$$
$$0.50x = 3$$
$$x = \frac{3}{0.50} = 6$$

Dureen Kelly should add 6 quarts of pure antifreeze to 10 quarts of 20% antifreeze to produce a mixture (solution) of 16 quarts of 50% antifreeze.

35. a. Let t = time before the jets meet.

	Rate	Time	Distance
Jet	800	t	$800t$
Refueling Plane	520	$t+2$	$520(t+2)$

The distances traveled are equal.

$$800t = 520(t+2)$$
$$800t = 520t + 1040$$
$$280t = 1040$$
$$t = \frac{26}{7} \approx 3.7143$$

The two planes will meet in approximately 3.71 hours.

b. $800t = 800(26/7) = 2971.43$

The refueling will take place approximately 2971.43 miles from the base.

37. Let x = number of small paintings sold.

Item	Cost	Number	Amount
Small	60	x	$60x$
Large	180	$12-x$	$180(12-x)$

$$60x + 180(12-x) = 1200$$
$$60x + 2160 - 180x = 1200$$
$$-120x = -960$$
$$x = 8$$

Joseph DeGuizman sold 8 small paintings and $12-8=4$ large paintings.

39. Let x = amount of 80% solution needed.

Solution	Strength of Solution	No. of Ounces	Amount of Alcohol
80%	0.80	x	$0.80x$
Water	0	$128-x$	$0(128-x)$
6%	0.06	128	0.06(128)

$$0.80x + 0(128-x) = 0.06(128)$$
$$0.80x = 7.68$$
$$x = \frac{7.68}{0.80} = 9.6$$

Herb should combine 9.6 ounces of the 80% solution with $128 - 9.6 = 118.4$ ounces of water to produce the desired solution.

41. Let x = amount of sirloin needed.

Type	Fat	No. of Ounces	Total Fat
Sirloin	1.2	x	$1.2x$
Veal	0.3	$64-x$	$0.3(64-x)$
Mixture	0.8	64	0.8(64)

$$1.2x + 0.3(64-x) = 0.8(64)$$
$$1.2x + 19.2 - 0.3x = 51.2$$
$$0.9x + 19.2 = 51.2$$
$$0.9x = 32$$
$$x = \frac{32}{0.9} \approx 35.6$$

Lori must combine about 35.6 ounces of sirloin with about $64 - 35.6 = 28.4$ ounces of veal to produce the desired meatloaf.

43. Let r = the rate George rides his bike.

Type	Rate	Time	Distance
Bike	r	$\frac{3}{4}$	$\frac{3}{4}r$
Car	$r+14$	$\frac{1}{6}$	$\frac{1}{6}(r+14)$

Both distances are the same.

$$\frac{1}{6}(r+14)=\frac{3}{4}r$$
$$12\left[\frac{1}{6}(r+14)\right]=12\left(\frac{3}{4}r\right)$$
$$2(r+14)=3(3r)$$
$$2r+28=9r$$
$$28=7r$$
$$\frac{28}{7}=r \text{ or } r=4$$

The distance is $\frac{3}{4}r=\frac{3}{4}(4)=3$ miles.

45. Let x = amount of water that needs to evaporate.

$$64(37)=45(64-x)$$
$$2368=2880-45x$$
$$-512=-45x$$
$$\frac{-512}{-45}=x \text{ or } x\approx 11.4$$

About 11.4 ounces of water must evaporate to raise the salinity of the solution.

47. Answers will vary.

49. It is possible to determine the times for the 2nd and 3rd parts of the trip.

2nd Part: $t=\frac{d}{r}=\frac{31}{90}\approx 0.344$ hour

3rd Part: $t=\frac{d}{r}=\frac{68}{45}\approx 1.511$ hours

The time for the first part (Paris to Calais) is $3.000-0.344-1.511=1.145$ hours. The distance is
(130 mph)(1.145 hours) $\approx$ 149 miles

51. Let x be the amount of 20% solution which must be drained. Then, $16-x$ is the amount remaining.

$$0.20(16-x)+1.00x=0.50(16)$$
$$3.2-0.20x+1.00x=8$$
$$3.2+0.80x=8$$
$$0.80x=4.8$$
$$x=\frac{4.8}{0.80}=6$$

Thus, 6 quarts must be drained before adding the same amount of antifreeze.

52.
$$\frac{2.52\times10^{17}}{3.6\times10^{4}}=\frac{2.52}{3.6}\times\frac{10^{17}}{10^{4}}$$
$$=0.7\times10^{17-4}$$
$$=0.7\times10^{13}$$
$$=7.0\times10^{-1}\times10^{13}$$
$$=7.0\times10^{-1+13}$$
$$=7.0\times10^{12}$$

53.
$$0.6x+0.22=0.4(x-2.3)$$
$$0.6x+0.22=0.4x-0.92$$
$$0.6x-0.4x+0.22=0.4x-0.4x-0.92$$
$$0.2x+0.22=-0.92$$
$$0.2x+0.22-0.22=-0.92-0.22$$
$$0.2x=-1.14$$
$$\frac{0.2x}{0.2}=\frac{-1.14}{0.2}$$
$$x=-5.7$$

54.
$$\frac{2}{3}x+8=x+\frac{25}{4}$$
$$12\left(\frac{2}{3}x\right)+12(8)=12(x)+12\left(\frac{25}{4}\right)$$
$$8x+96=12x+75$$
$$8x-8x+96=12x-8x+75$$
$$96=4x+75$$
$$96-75=4x+75-75$$
$$21=4x$$
$$\frac{21}{4}=\frac{4x}{4}$$
$$\frac{21}{4}=x$$

55. $\frac{3}{5}(x-2)=\frac{2}{7}(2x+3y)$

$$35\left[\frac{3}{5}(x-2)\right]=35\left[\frac{2}{7}(2x+3y)\right]$$
$$21(x-2)=10(2x+3y)$$
$$21x-42=20x+30y$$
$$21x-20x-42=20x-20x+30y$$
$$x-42=30y$$
$$\frac{x-42}{30}=\frac{30y}{30}$$
$$\frac{x-42}{30}=y \text{ or } y=\frac{x-42}{30}$$

56. Let x be the distance driven in one day.

$$30+0.14x=16+0.24x$$
$$30=16+0.10x$$
$$14=0.10x$$
$$\frac{14}{0.10}=x$$
$$x=140 \text{ miles}$$

The costs are the same when 140 miles are driven per day.

Exercise Set 2.5

1. It is necessary to change the direction of the inequality symbol when multiplying or dividing by a negative number.

3. **a.** Use open circles when the endpoints are not included.

b. Use closed circles when the endpoints are included.

c. Answers may vary. One possible answer is $x>4$.

d. Answers may vary. One possible answer is $x\geq 4$.

5. $a<x<b$ means $a<x$ and $x<b$.

7. **a.** 

b. $(-2, \infty)$

c. $\{x|x>-2\}$

9. **a.** 

b. $(-\infty, \pi]$

c. $\{w|w\leq\pi\}$

11. **a.** 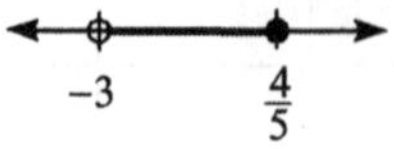

b. $\left(-3, \frac{4}{5}\right]$

c. $\left\{q\middle|-3<q\leq\frac{4}{5}\right\}$

13. **a.** 

b. $(-7, -4]$

c. $\{x|-7<x\leq -4\}$

15. $x-9>-6$

$$x>3$$

17. $3-x<-4$

$$-x<-7$$

Reverse the inequality

$$\frac{-x}{-1}>\frac{-7}{-1}$$
$$x>7$$

19.
$$4.7x-5.48\geq 11.44$$
$$4.7x-5.48+5.48\geq 11.44+5.48$$
$$4.7x\geq 16.92$$
$$\frac{4.7x}{4.7}\geq\frac{16.92}{4.7}$$
$$x\geq 3.6$$

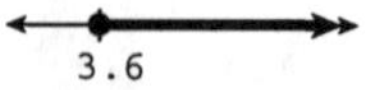

21. $4(x+2) \le 4x+8$

$4x+8 \le 4x+8$

$8 \le 8$

Since this is a true statement, the solution is the entire real number line.

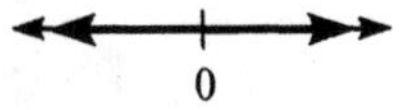

23. $5b-6 \ge 3(b+3)+2b$

$5b-6 \ge 3b+9+2b$

$5b-6 \ge 5b+9$

$-6 \ge 9$

Since this is a false statement, there is no solution.

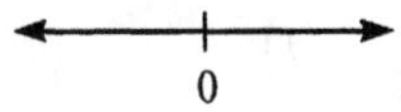

25. $2y-6y+8 \le 2(-2y+9)$

$-4y+8 \le -4y+18$

$8 \le 18$

Since this is a true statement, the solution is all real numbers.

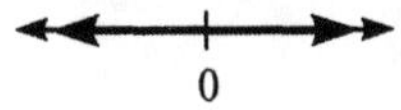

27. $4+\dfrac{4x}{3} < 6$

$\dfrac{4x}{3} < 2$

$3\left(\dfrac{4x}{3}\right) < 3(2)$

$4x < 6$

$\dfrac{4x}{4} < \dfrac{6}{4}$

$x < \dfrac{3}{2}$

$\left(-\infty, \dfrac{3}{2}\right)$

29. $\dfrac{v-5}{3} - v \ge -3(v-1)$

$3\left[\dfrac{v-5}{3} - v\right] \ge 3\left[-3(v-1)\right]$

$v-5-3v \ge -9(v-1)$

$-5-2v \ge -9v+9$

$7v-5 \ge 9$

$7v \ge 14$

$v \ge 2 \Rightarrow [2,\infty)$

31. $\dfrac{t}{3} - t + 7 \le -\dfrac{4t}{3} + 8$

$3\left(\dfrac{t}{3} - t + 7\right) \le 3\left(-\dfrac{4t}{3} + 8\right)$

$t-3t+21 \le -4t+24$

$-2t+21 \le -4t+24$

$2t \le 3$

$\dfrac{2t}{2} \le \dfrac{3}{2}$

$t \le \dfrac{3}{2} \Rightarrow \left(-\infty, \dfrac{3}{2}\right]$

33. $-3x+1 < 3\left[(x+2)-2x\right]-1$

$-3x+1 < 3\left[x+2-2x\right]-1$

$-3x+1 < 3\left[2-x\right]-1$

$-3x+1 < 6-3x-1$

$-3x+1 < 5-3x$

$1 < 5 \Rightarrow$ a true statement

The solution set is $(-\infty, \infty)$.

35. $-2 \le t+3 < 4$

$-2-3 \le t+3-3 < 4-3$

$-5 \le t < 1$

$[-5,1)$

37. $-15 \le -3z \le 12$

Divide by -3 and reverse inequalities.

$$\frac{-15}{-3} \ge \frac{-3z}{-3} \ge \frac{12}{-3}$$

$$5 \ge z \ge -4$$

$$-4 \le z \le 5$$

$$[-4, 5]$$

39. $4 \le 2x - 4 < 7$

$$4 + 4 \le 2x - 4 + 4 < 7 + 4$$

$$8 \le 2x < 11$$

$$\frac{8}{2} \le \frac{2x}{2} < \frac{11}{2}$$

$$4 \le x < \frac{11}{2}$$

$$\left[4, \frac{11}{2}\right)$$

41. $14 \le 2 - 3g < 15$

$$14 - 2 \le 2 - 3g - 2 < 15 - 2$$

$$12 \le -3g < 13$$

Divide by -3 and reverse inequalities.

$$\frac{12}{-3} \ge \frac{-3g}{-3} > \frac{13}{-3}$$

$$-4 \ge g > -\frac{13}{3}$$

$$-\frac{13}{3} < g \le -4$$

$$\left(-\frac{13}{3}, -4\right]$$

43. $5 \le \frac{3x+1}{2} < 11$

$$2(5) \le 2\left(\frac{3x+1}{2}\right) < 2(11)$$

$$10 \le 3x + 1 < 22$$

$$10 - 1 \le 3x + 1 - 1 < 22 - 1$$

$$9 \le 3x < 21$$

$$\frac{9}{3} \le \frac{3x}{3} < \frac{21}{3}$$

$$3 \le x < 7$$

$$\{x \mid 3 \le x < 7\}$$

45. $-6 \le -3(2x - 4) < 12$

$$-6 \le -6x + 12 < 12$$

$$-6 - 12 \le -6x + 12 - 12 < 12 - 12$$

$$-18 \le -6x < 0$$

Divide by -6 and reverse inequalities

$$\frac{-18}{-6} \ge \frac{-6x}{-6} > \frac{0}{-6}$$

$$3 \ge x > 0$$

$$0 < x \le 3$$

$$\{x \mid 0 < x \le 3\}$$

47. $0 \le \frac{3(u-4)}{7} \le 1$

$$7(0) \le 7\left(\frac{3(u-4)}{7}\right) \le 7(1)$$

$$0 \le 3(u - 4) \le 7$$

$$0 \le 3u - 12 \le 7$$

$$0 + 12 \le 3u - 12 + 12 \le 7 + 12$$

$$12 \le 3u \le 19$$

$$\frac{12}{3} \le \frac{3u}{3} \le \frac{19}{3}$$

$$4 \le u \le \frac{19}{3}$$

$$\left\{u \,\middle|\, 4 \le u \le \frac{19}{3}\right\}$$

49. $c \le 1$

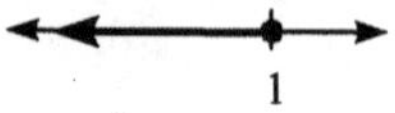

$c > -3$

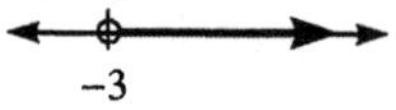

$c \le 1$ and $c > -3 \;\Rightarrow\; -3 < c \le 1$

$$\{c \mid -3 < c \le 1\}$$

51. $x < 2$

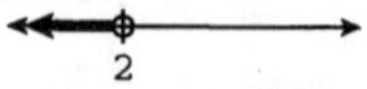

$x > 4$

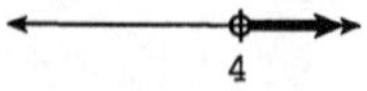

$x < 2$ and $x > 4$

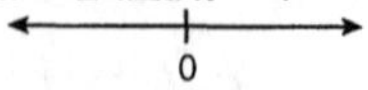

There is no overlap so the solution is the empty set, $\varnothing$.

53. $x+1<3$ and $x+1>-4$

$x<2$ and $x>-5$

$x>-5$

$x<2$

$x<-2$ and $x>-5$ which is $-5<x<2$ or

$\{x|-5<x<2\}$

55. $2s+3<7$ or $-3s+4\le -17$

$2s<4$ or $-3s\le -21$

$\frac{2s}{2}<\frac{4}{2}$ or $\frac{-3s}{-3}\ge\frac{-21}{-3}$

$s<2$ $\quad s\ge 7$

$s<2$ or $s\ge 7$ which is $(-\infty,2)\cup[7,\infty)$.

57. $4x+5\ge 5$ and $3x-7\le -1$

$4x\ge 0$ and $3x\le 6$

$x\ge 0$ and $x\le 2$

$x\ge 0$

$x\le 2$

$x\ge 0$ and $x\le 2$ which is $0\le x\le 2$

In interval notation: [0, 2]

59. $4-r<-2$ or $3r-1<-1$

$-r<-6$ $\quad 3r<0$

$r>6$ $\quad r<0$

$r>6$

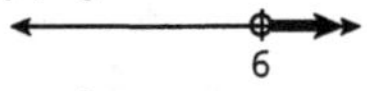

$r<0$

$r>6$ or $r<0$

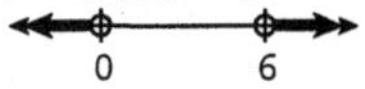

In interval notation: $(-\infty,0)\cup(6,\infty)$

61. $2k+5>-1$ and $7-3k\le 7$

$2k>-6$ $\quad -3k\le 0$

$\frac{2k}{2}>\frac{-6}{2}$ $\quad \frac{-3k}{-3}\ge\frac{0}{-3}$

$k>-3$ $\quad k\ge 0$

$k>-3$ and $k\ge 0\ \Rightarrow\ k\ge 0$

In interval notation: $[0,\infty)$

63. **a.** $l+g\le 130$

b. $g=2w+2d$

$l+g\le 130$

$l+2w+2d\le 130$

c. $l=40,\ w=20.5$

$l+2w+2d\le 130$

$40+2(20.5)+2d\le 130$

$40+41+2d\le 130$

$81+2d\le 130$

$2d\le 49$

$d\le 24.5$

The maximum depth is 24.5 inches.

65. Let x be the maximum number of boxes.

$70x\le 800$

$x\le\frac{800}{70}$

$x\le 11.43$

The maximum number of boxes is 11.

67. Let x = the number of minutes she talks beyond the first 20 minutes.

$0.99+0.07x\le 5.00$

$0.07x\le 4.01$

$\frac{0.07x}{0.07}\le\frac{4.01}{0.07}$

$x\le 57$ (to nearest whole number)

She can talk for 57 minutes beyond the first 20 minutes for a total of 77 minutes.

69. To make a profit, the cost must be less than the revenue: cost < revenue.

$10,025 + 1.09x < 6.42x$

$10,025 < 5.33x$

$\frac{10,025}{5.33} < x$

$1880.86 < x$

She needs to sell a minimum of 1881 books to make a profit.

71. Let x = the number of additional ounces beyond the first ounce.

$0.39 + 0.24x \le 10.00$

$0.24x \le 9.61$

$\frac{0.24x}{0.24} \le \frac{9.61}{0.24}$

$x \le 40$ (rounded down)

The maximum weight is 41 ounces (the first ounce plus up to 40 additional ounces).

73. Let x be the amount of sales in dollars.

$300 + 0.10x > 400 + 0.08x$

$0.10x > 100 + 0.08x$

$0.02x > 100$

$x > \frac{100}{0.02}$

$x > 5000$

She will earn more by plan 1 if her weekly sales total more than \$5000.

75. Let x be the minimum score for the sixth exam.

$\frac{66 + 72 + 90 + 49 + 59 + x}{6} \ge 60$

$\frac{336 + x}{6} \ge 60$

$6\left(\frac{336 + x}{6}\right) \ge 6(60)$

$336 + x \ge 360$

$x \ge 24$

She must make a 24 or higher on the sixth exam to pass the course.

77. Let x be the score on the fifth exam.

$80 \le \frac{85 + 92 + 72 + 75 + x}{5} < 90$

$80 \le \frac{324 + x}{5} < 90$

$5(80) \le 5\left(\frac{324 + x}{5}\right) < 5(90)$

$400 \le 324 + x < 450$

$76 \le x < 126$

To receive a final grade of B, Ms. Mahoney must score 76 or higher on the fifth exam. That is, the score must be

$76 \le x \le 100$ (maximum grade is 100).

79. a. The taxable income of \$78,221 places a married couple filing jointly in the 25% tax bracket. The tax is \$8,180.00 plus 25% of the taxable income over \$59,400.

The tax is

$\$8,180.00 + 0.25(\$78,221 - \$59,400)$

$\$8,180.00 + 0.25(\$18,821)$

$\$8,180.00 + \$4,705.25$

$\$12,885.25$

They will owe \$12,885.25 in taxes.

b. The taxable income of \$301,233 places a married couple filing jointly in the 33% tax bracket. The tax is \$40,915.50 plus 33% of the taxable income over \$182,800.

The tax is

$\$40,915.5 + 0.33(\$301,233 - \$182,800)$

$\$40,915.5 + 0.33(\$118,433.00)$

$\$40,915.5 + \$39,082.89$

$\$79,998.39$

They will owe \$79,998.39 in taxes.

81. a. $v(t) \ge 0$

$-32t + 96 \ge 0$

$-32t \ge -96$

$t \le 3$

The object is traveling upward on the interval $[0, 3]$.

b.
$$v(t) \le 0$$
$$-32t + 96 \le 0$$
$$-32t \le -96$$
$$t \ge 3$$
The object is traveling downward on the interval $[3,10]$.

83. a.
$$v(t) \ge 0$$
$$-9.8t + 49 \ge 0$$
$$-9.8t \ge -49$$
$$t \le 5$$
The object is traveling upward on the interval $[0,5]$.

b.
$$v(t) \le 0$$
$$-9.8t + 49 \le 0$$
$$-9.8t \le -49$$
$$t \ge 5$$
The object is traveling downward on the interval $[5,13]$.

85. a.
$$v(t) \ge 0$$
$$-32t + 320 \ge 0$$
$$-32t \ge -320$$
$$t \le 10$$
The object is traveling upward on the interval $[0,8]$ (note: we restricted t to the interval $[0,8]$).

b. From part (a) we saw that the object was moving upward for all values of t on the interval $[0,8]$. Therefore, there are no values for t in that interval where the object is traveling downward.

87. Let x be the value of the third reading.
$$7.2 < \frac{7.48 + 7.15 + x}{3} < 7.8$$
$$7.2 < \frac{14.63 + x}{3} < 7.8$$
$$3(7.2) < 3\left(\frac{14.63 + x}{3}\right) < 3(7.8)$$
$$21.6 < 14.63 + x < 23.4$$
$$6.97 < x < 8.77$$
Any value between 6.97 and 8.77 would result in a normal pH reading.

89. a. The goal is greater than 6000 for all five months shown and the enlisted number is greater than 4000 for all five months shown except for April. Thus, the goal is greater than 6000 and the number enlisted is greater than 4000 for the months January, February, March, and May.

b. March, April, May

c. April

91. Answers may vary. The three components cover all possible total medical costs.

First piece: 0 if $c \le \$100$
There is a deductible of \$100 so for total costs less than or equal to \$100, the patient covers all costs.

Second piece: $0.80(c - 100)$ if $\$100 < c \le \$2100$
If total costs are more than \$100, but no more than \$2100, the patient pays the \$100 deductible and Blue Cross/Blue Shield pays 80% of the remaining amount.

Third piece: $c - 500$ if $c > \$2100$
The maximum out-of-pocket cost for the patient is \$500. For the second piece, the maximum cost to the patient occurs when $c = \$2100$. Since $0.80(2100 - 100) = 500$, this corresponds to the maximum amount a patient will have to pay. Thus, for any total medical cost over \$2100, the patient must pay the \$500 limit and Blue Cross/Blue Shield will pay the remaining amount.

93. a. From the chart, the 10th percentile is approximately 17.5 pounds and the 90th percentile is approximately 23.5 pounds. Therefore, 80% of the weights for 9 month old boys are in the interval $[17.5, 23.5]$ (in pounds).

b. From the chart, the 10th percentile is approximately 23.5 pounds and the 90th percentile is approximately 31 pounds. Therefore, 80% of the weights for 21 month old boys are in the interval $[23.5, 31]$ (in pounds).

c. From the chart, the 10th percentile is approximately 27.2 pounds and the 90th percentile is approximately 36.5 pounds. Therefore, 80% of the weights for 36 month old boys are in the interval $[27.2, 36.5]$ (in pounds).

95. First find the average of 82, 90, 74, 76, and 68.

$$\frac{82+90+74+76+68}{5} = \frac{390}{5} = 78$$

This represents $\frac{2}{3}$ of the final grade.

Let x be the score from the final exam. Since this represents $\frac{1}{3}$ of the final grade, the inequality is

$$80 \le \frac{2}{3}(78) + \frac{1}{3}x < 90$$

$$3(80) < 3\left[\frac{2}{3}(78) + \frac{1}{3}x\right] < 3(90)$$

$$240 \le 2(78) + x < 270$$
$$240 \le 156 + x < 270$$
$$84 \le x < 114$$

Stephen must score at least 84 points on the final exam to have a final grade of B. The range is $84 \le x \le 100$.

97. a. Answers may vary. One possible answer is: Write $x < 2x + 3 < 2x + 5$ as $x < 2x + 3$ and $2x + 3 < 2x + 5$

b. Solve each of the inequalities.

$$x < 2x + 3 \quad \text{and} \quad 2x + 3 < 2x + 5$$
$$-x < 3 \qquad\qquad 3 < 5$$
$$x > -3 \qquad\qquad \text{All real numbers}$$

The final answer is $x > -3$ or $(-3, \infty)$.

99. a. $A \cup B = \{1, 2, 3, 4, 5, 6, 8, 9\}$

b. $A \cap B = \{1, 8\}$

100. a. 4 is a counting number.

b. 0 and 4 are whole numbers.

c. $-3, 4, \frac{5}{2}, 0$ and $-\frac{13}{29}$ are rational numbers.

d. $-3, 4, \frac{5}{2}, \sqrt{7}, 0$ and $-\frac{13}{29}$ are real numbers.

101. Associative property of addition.

102. Commutative property of addition

103.

$$R = L + (V - D)r$$
$$R = L + Vr - Dr$$
$$R - L + Dr = Vr$$
$$\frac{R - L + Dr}{r} = V \text{ or } V = \frac{R - L + Dr}{r}$$

Exercise Set 2.6

1. $|x| = a, a > 0$

Set $x = a$ or $x = -a$.

3. $|x| < a, a > 0$

Write $-a < x < a$.

5. $|x| > a, a > 0$

Write $x < -a$ or $x > a$.

7. The solution to $|x| > 0$ is all real numbers except 0. The absolute value of any real number, except 0, is greater than 0, i.e., positive.

9. Set $x = y$ or $x = -y$.

11. If $a \ne 0$, and $k > 0$,

a. $|ax + b| = k$ has 2 solutions.

b. $|ax + b| < k$ has an infinite number of solutions.

c. $|ax + b| > k$ has an infinite number of solutions.

13. **a.** $|x| = 5, \{-5, 5\}$, D

b. $|x| < 5, \{x | -5 < x < 5\}$, B

c. $|x| > 5, \{x | x < -5 \text{ or } x > 5\}$, E

d. $|x| \le 5, \{x | -5 \le x \le 5\}$, C

e. $|x| \ge 5, \{x | x \le -5 \text{ or } x \ge 5\}$, A

15. $|a| = 2$

$a = 2$ or $a = -2$

The solution set is $\{-2, 2\}$.

17. $|c| = \frac{1}{2}$

$c = \frac{1}{2}$ or $c = -\frac{1}{2}$

The solution set is $\left\{-\frac{1}{2}, \frac{1}{2}\right\}$.

19. $|d| = -\frac{5}{6}$

There is no solution since the right side is a negative number and the absolute value can never be equal to a negative number. The solution set $\varnothing$.

21. $|x+5| = 8$

$x+5 = 8$ $\qquad$ $x+5 = -8$

$x = 3$ or $\qquad$ $x = -13$

The solution set is $\{-13, 3\}$.

23. $|4.5q + 31.5| = 0$

$4.5q + 31.5 = 0$

$4.5q = -31.5$

$q = -7$

The solution set is $\{-7\}$.

25. $|5-3x| = \frac{1}{2}$

$5-3x = \frac{1}{2}$ or $5-3x = -\frac{1}{2}$

$-3x = \frac{1}{2} - 5$ $\qquad$ $-3x = -\frac{1}{2} - 5$

$-3x = -\frac{9}{2}$ $\qquad$ $-3x = -\frac{11}{2}$

$-\frac{1}{3}(3x) = -\frac{1}{3}\left(-\frac{9}{2}\right)$ $\qquad$ $-\frac{1}{3}(-3x) = -\frac{1}{3}\left(-\frac{11}{2}\right)$

$x = \frac{3}{2}$ $\qquad$ $x = \frac{11}{6}$

The solution set is $\left\{\frac{3}{2}, \frac{11}{6}\right\}$.

27. $\left|\frac{x-3}{4}\right| = 5$

$\frac{x-3}{4} = 5$ or $\frac{x-3}{4} = -5$

$4\left(\frac{x-3}{4}\right) = 4(5)$ $\qquad$ $4\left(\frac{x-3}{4}\right) = 4(-5)$

$x-3 = 20$ $\qquad$ $x-3 = -20$

$x = 23$ $\qquad$ $x = -17$

The solution set is $\{-17, 23\}$.

29. $\left|\frac{x-3}{4}\right| + 8 = 8$

$\left|\frac{x-3}{4}\right| = 0$

$\frac{x-3}{4} = 0$

$4\left(\frac{x-3}{4}\right) = 4(0)$

$x-3 = 0$

$x = 3$

The solution set is $\{3\}$.

31. $|w| < 11$

$-11 < w < 11$

The solution set is $\{w | -11 < w < 11\}$.

33. $|q+5| \le 8$

$-8 \le q+5 \le 8$

$-8-5 \le q+5-5 \le 8-5$

$-13 \le q \le 3$

The solution set is $\{q \mid -13 \le q \le 3\}$.

35. $|5b-15| < 10$

$-10 < 5b-15 < 10$

$-10+15 < 5b-15+15 < 10+15$

$5 < 5b < 25$

$\frac{5}{5} < \frac{5b}{5} < \frac{25}{5}$

$1 < b < 5$

The solution set is $\{b \mid 1 < b < 5\}$.

37. $|2x+3|-5 \le 10$

$|2x+3| \le 15$

$-15 \le 2x+3 \le 15$

$-15-3 \le 2x+3-3 \le 15-3$

$-18 \le 2x \le 12$

$\frac{-18}{2} \le \frac{2x}{2} \le \frac{12}{2}$

$-9 \le x \le 6$

The solution set is $\{x \mid -9 \le x \le 6\}$

39. $|3x-7|+8 < 14$

$|3x-7| < 6$

$-6 < 3x-7 < 6$

$-6+7 < 3x-7+7 < 6+7$

$1 < 3x < 13$

$\frac{1}{3} < \frac{3x}{3} < \frac{13}{3}$

$\frac{1}{3} < x < \frac{13}{3}$

The solution set is $\left\{x \,\middle|\, \frac{1}{3} < x < \frac{13}{3}\right\}$.

41. $|2x-6|+5 \le 1$

$|2x-6| \le -4$

There is no solution since the right side is negative whereas the left side is non-negative; zero or a positive number is never less than a negative number. The solution set is $\varnothing$.

43. $\left|\frac{1}{2}j+4\right| < 7$

$-7 < \frac{1}{2}j+4 < 7$

$-7-4 < \frac{1}{2}j+4-4 < 7-4$

$-11 < \frac{1}{2}j < 3$

$2(-11) < 2\left(\frac{1}{2}j\right) < 2(3)$

$-22 < j < 6$

The solution set is $\{j \mid -22 < j < 6\}$.

45. $\left|\frac{x-3}{2}\right| - 4 \le -2$

$\left|\frac{x-3}{2}\right| \le 2$

$-2 \le \frac{x-3}{2} \le 2$

$2(-2) \le 2\left(\frac{x-3}{2}\right) \le 2(2)$

$-4 \le x-3 \le 4$

$-4+3 \le x-3+3 \le 4+3$

$-1 \le x \le 7$

The solution set is $\{x \mid -1 \le x \le 7\}$.

47. $|y| > 2$

$y < -2$ or $y > 2$

The solution set is $\{y \mid y < -2 \text{ or } y > 2\}$.

49. $|x+4| > 5$

$x+4 < -5$ or $x+4 > 5$

$x < -9$ $\quad$ $x > 1$

The solution set is $\{x \mid x < -9 \text{ or } x > 1\}$.

51. $|7-3b|>5$

$7-3b<-5$ or $7-3b>5$

$-3b<-12$ $\qquad$ $-3b>-2$

$\frac{-3b}{-3}>\frac{-12}{-3}$ $\qquad$ $\frac{-3b}{-3}<\frac{-2}{-3}$

$b>4$ $\qquad$ $b<\frac{2}{3}$

The solution set is $\left\{b \middle| b<\frac{2}{3} \text{ or } b>4\right\}$.

53. $|2h-5|>3$

$2h-5<-3$ or $2h-5>3$

$2h<2$ $\qquad$ $2h>8$

$h<\frac{2}{2}$ $\qquad$ $h>\frac{8}{2}$

$h<1$ $\qquad$ $h>4$

The solution set is $\{h \mid h<1 \text{ or } h>4\}$.

55. $|0.1x-0.4|+0.4>0.6$

$|0.1x-0.4|>0.2$

$0.1x-0.4<-0.2$ or $0.1x-0.4>0.2$

$0.1x<0.2$ $\qquad$ $0.1x>0.6$

$x<\frac{0.2}{0.1}$ $\qquad$ $x>\frac{0.6}{0.1}$

$x<2$ $\qquad$ $x>6$

The solution set is $\{x \mid x<2 \text{ or } x>6\}$.

57. $\left|\frac{x}{2}+4\right|\ge 5$

$\frac{x}{2}+4\le -5$ or $\frac{x}{2}+4\ge 5$

$2\left(\frac{x}{2}+4\right)\le 2(-5)$ $\qquad$ $2\left(\frac{x}{2}+4\right)\ge 2(5)$

$x+8\le -10$ $\qquad$ $x+8\ge 10$

$x\le -18$ $\qquad$ $x\ge 2$

The solution set is $\{x \mid x\le -18 \text{ or } x\ge 2\}$.

59. $|7w+3|-12\ge -12$

$|7w+3|\ge 0$

Observe that the absolute value of a number is always greater than or equal to 0. Thus, the solution is the set of real numbers, or $\mathbb{R}$.

61. $|4-2x|>0$

$4-2x<0$ or $4-2x>0$

$-2x<-4$ $\qquad$ $-2x>-4$

$x>\frac{-4}{-2}$ $\qquad$ $x<\frac{-4}{-2}$

$x>2$ $\qquad$ $x<2$

The solution set is $\{x \mid x<2 \text{ or } x>2\}$.

63. $|3p-5|=|2p+10|$

$3p-5=-(2p+10)$ or $3p-5=2p+10$

$3p-5=-2p-10$ $\qquad$ $p-5=10$

$5p-5=-10$ $\qquad$ $p=15$

$5p=-5$

$p=-1$

The solution set is $\{-1, 15\}$.

65. $|6x|=|3x-9|$

$6x=-(3x-9)$ or $6x=3x-9$

$6x=-3x+9$ $\qquad$ $3x=-9$

$9x=9$ $\qquad$ $x=\frac{-9}{3}$

$x=\frac{9}{9}$ $\qquad$ $x=-3$

$x=1$

The solution set is $\{-3, 1\}$.

67. $\left|\frac{2r}{3}+\frac{5}{6}\right|=\left|\frac{r}{2}-3\right|$

$\frac{2r}{3}+\frac{5}{6}=-\left(\frac{r}{2}-3\right)$ or $\frac{2r}{3}+\frac{5}{6}=\frac{r}{2}-3$

$\frac{2r}{3}+\frac{5}{6}=-\frac{r}{2}+3$ $\qquad$ $6\left(\frac{2r}{3}+\frac{5}{6}\right)=6\left(\frac{r}{2}-3\right)$

$6\left(\frac{2r}{3}+\frac{5}{6}\right)=6\left(-\frac{r}{2}+3\right)$ $\qquad$ $4r+5=3r-18$

$4r+5=-3r+18$ $\qquad$ $r+5=-18$

$7r+5=18$ $\qquad$ $r=-23$

$7r=13$

$r=\frac{13}{7}$

The solution set is $\left\{-23, \frac{13}{7}\right\}$.

69. $\left|-\frac{3}{4}m+8\right|=\left|7-\frac{3}{4}m\right|$

$-\frac{3}{4}m+8=-\left(7-\frac{3}{4}m\right)$ or $-\frac{3}{4}m+8=7-\frac{3}{4}m$

$-\frac{3}{4}m+8=-7+\frac{3}{4}m$ $\quad$ $-\frac{3}{4}m+8=7-\frac{3}{4}m$

$-\frac{6}{4}m=-15$ $\quad$ $8=7$ False!

$m=10$

The solution set is $\{10\}$.

71. $|h|=1$

$h=1$ or $h=-1$

The solution set is $\{-1, 1\}$.

73. $|q+6|>2$

$q+6<-2$ or $q+6>2$

$q<-8$ $\quad$ $q>-4$

The solution set is $\{q \mid q<-8 \text{ or } q>-4\}$.

75. $|2w-7|\le 9$

$-9\le 2w-7\le 9$

$-9+7\le 2w-7+7\le 9+7$

$-2\le 2w\le 16$

$\frac{-2}{2}\le\frac{2w}{2}\le\frac{16}{2}$

$-1\le w\le 8$

The solution set is $\{w \mid -1\le w\le 8\}$.

77. $|5a-1|=9$

$5a-1=-9$ or $5a-1=9$

$5a=-8$ $\quad$ $5a=10$

$a=-\frac{8}{5}$ $\quad$ $a=2$

The solution set is $\left\{-\frac{8}{5}, 2\right\}$.

79. $|5x+2|>0$

$5+2x<0$ or $5+2x>0$

$2x<-5$ $\quad$ $2x>-5$

$x<-\frac{5}{2}$ $\quad$ $x>-\frac{5}{2}$

The solution set is $\left\{x \mid x<-\frac{5}{2} \text{ or } x>-\frac{5}{2}\right\}$.

81. $|4+3x|\le 9$

$-9\le 4+3x\le 9$

$-13\le 3x\le 5$

$-\frac{13}{3}\le x\le\frac{5}{3}$

The solution set is $\left\{x \mid -\frac{13}{3}\le x\le\frac{5}{3}\right\}$.

83. $|3n+8|-4=-10$

$|3n+8|=-6$

Since the right side is negative and the left side is non-negative, there is no solution since the absolute value can never equal a negative number. The solution set is $\varnothing$.

85. $\left|\frac{w+4}{3}\right|+5<9$

$\left|\frac{w+4}{3}\right|<4$

$-4<\frac{w+4}{3}<4$

$3(-4)<3\left(\frac{w+4}{3}\right)<3(4)$

$-12<w+4<12$

$-16<w<8$

The solution set is $\{w \mid -16<w<8\}$.

87. $\left|\frac{3x-2}{4}\right|-\frac{1}{3}\ge-\frac{1}{3}$

$\left|\frac{3x-2}{4}\right|\ge 0$

Since the absolute value of a number is always greater than or equal to zero, the solution is the set of all real numbers or $\mathbb{R}$.

89. $|2x-8|=\left|\frac{1}{2}x+3\right|$

$2x-8=-\left(\frac{1}{2}x+3\right)$ or $2x-8=\frac{1}{2}x+3$

$2x-8=-\frac{1}{2}x-3$ $\qquad \frac{3}{2}x-8=3$

$\frac{5}{2}x-8=-3$ $\qquad \frac{3}{2}x=11$

$\frac{5}{2}x=5$ $\qquad \frac{2}{3}\left(\frac{3}{2}x\right)=\frac{2}{3}(11)$

$\frac{2}{5}\left(\frac{5}{2}x\right)=\frac{2}{5}(5)$ $\qquad x=\frac{22}{3}$

$x=2$

The solution set is $\left\{2,\frac{22}{3}\right\}$.

91. $|2-3x|=\left|4-\frac{5}{3}x\right|$

$2-3x=-\left(4-\frac{5}{3}x\right)$ or $2-3x=4-\frac{5}{3}x$

$2-3x=-4+\frac{5}{3}x$ $\qquad -3x=2-\frac{5}{3}x$

$-3x=-6+\frac{5}{3}x$ $\qquad -\frac{4}{3}x=2$

$-\frac{14}{3}x=-6$ $\qquad -\frac{3}{4}\left(-\frac{4}{3}x\right)=-\frac{3}{2}(2)$

$\left(-\frac{3}{14}\right)\left(-\frac{14}{3}\right)x=\left(-\frac{3}{14}\right)(-6)$ $\qquad x=-\frac{3}{2}$

$x=\frac{9}{7}$

The solution set is $\left\{-\frac{3}{2},\frac{9}{7}\right\}$.

93. a. $|t-0.089|\le 0.004$

$-0.004\le t-0.089\le 0.004$

$-0.004+0.089\le t-0.089+0.089\le 0.004+0.089$

$0.085\le t\le 0.093$

The solution is $[0.085, 0.093]$.

b. 0.085 inches

c. 0.093 inches

95. a. $|d-160|\le 28$

$-28\le d-160\le 28$

$-28+160\le d-160+160\le 28+160$

$132\le d\le 188$

The solution is [132, 188]

b. The submarine can move between 132 feet and 188 feet below sea level, inclusive.

97. $\{-5,5\}$ is the solution set to the equation $|x|=5$.

99. $\{x|x\le -5 \text{ or } x\ge 5\}$ is the solution set of $|x|\ge 5$.

101. $|ax+b|\le 0$

$0\le ax+b\le 0$

which is the same as

$ax+b=0$

$ax=-b$

$x=-\frac{b}{a}$

103. a. Set $ax+b=-c$ or $ax+b=c$ and solve each equation for x.

b. $ax+b=-c$ or $ax+b=c$

$ax=-c-b$ $\qquad ax=c-b$

$x=\frac{-c-b}{a}$ $\qquad x=\frac{c-b}{a}$

The solution is $x=\frac{-c-b}{a}$ or $x=\frac{c-b}{a}$.

105. a. Write $ax+b<-c$ or $ax+b>c$ and solve each inequality for x.

b. $ax+b<-c$ or $ax+b>c$

$ax<-c-b$ $\qquad ax>c-b$

$x<\frac{-c-b}{a}$ $\qquad x>\frac{c-b}{a}$

The solution is $x<\frac{-c-b}{a}$ or $x>\frac{c-b}{a}$.

107. $|x-4|=|4-x|$

$x-4=-(4-x)$ or $x-4=4-x$

$x-4=-4+x$ $\qquad 2x-4=4$

$0=0$ $\qquad 2x=8$

True $\qquad x=4$

Since the first statement is always true all real values work. The solution set is $\mathbb{R}$.

109. $|x| = x$

By definition $|x| = \begin{cases} x, & x \geq 0 \\ -x, & x < 0 \end{cases}$

Thus, $|x| = x$ when $x \geq 0$

The solution set is $\{x | x \geq 0\}$.

111. $|x+1| = 2x-1$

$$\begin{aligned} x+1 &= -(2x-1) \quad \text{or} \quad & x+1 &= 2x-1 \\ x+1 &= -2x+1 & 1 &= x-1 \\ 3x+1 &= 1 & 2 &= x \\ 3x &= 0 \\ x &= 0 \end{aligned}$$

Checking both possible solutions, only $x = 2$ checks. The solution set is $\{2\}$.

113. $|x-4| = -(x-4)$

By the definition, $|x-4| = \begin{cases} x-4, & x-4 \geq 0 \\ -(x-4), & x-4 \leq 0 \end{cases}$

$= \begin{cases} x-4, & x \geq 4 \\ -(x-4), & x \leq 4 \end{cases}$

Thus, $|x-4| = -(x-4)$ for $x \leq 4$.

The solution set is $\{x | x \leq 4\}$.

115. $x + |-x| = 8$

For $x \geq 0$, $x + |-x| = 8$

$$\begin{aligned} x + x &= 8 \\ 2x &= 8 \\ x &= 4 \end{aligned}$$

For $x < 0$, $x + |-x| = 8$

$$\begin{aligned} x - x &= 8 \\ 0 &= 8 \text{ False} \end{aligned}$$

The solution set is $\{4\}$.

117. $x - |x| = 8$

For $x \geq 0$, $x - |x| = 8$

$$\begin{aligned} x - x &= 8 \\ 0 &= 8 \text{ False} \end{aligned}$$

For $x < 0$, $x - |x| = 8$

$$\begin{aligned} x - (-x) &= 8 \\ x + x &= 8 \\ 2x &= 8 \\ x &= 4 \quad \text{Contradicts } x < 0 \end{aligned}$$

There are no values of x, so the solution set is $\varnothing$.

119.

$$\begin{aligned} \frac{1}{3} + \frac{1}{4} \div \frac{2}{5}\left(\frac{1}{3}\right)^2 &= \frac{1}{3} + \frac{1}{4} \div \frac{2}{5} \cdot \frac{1}{9} \\ &= \frac{1}{3} + \frac{1}{4} \cdot \frac{5}{2} \cdot \frac{1}{9} \\ &= \frac{1}{3} + \frac{5}{72} \\ &= \frac{1}{3} \cdot \frac{24}{24} + \frac{5}{72} \\ &= \frac{24}{72} + \frac{5}{72} \\ &= \frac{29}{72} \end{aligned}$$

120. Substitute 1 for x and 3 for y.

$$\begin{aligned} 4(x+3y) - 5xy &= 4(1+3\cdot 3) - 5(1)(3) \\ &= 4(1+9) - 5(1)(3) \\ &= 4(10) - 5(1)(3) \\ &= 40 - 15 \\ &= 25 \end{aligned}$$

121. Let x be the time needed to swim across the lake. Then $1.5 - x$ is the time needed to make the return trip.

	Rate	Time	Distance
First Trip	2	x	$2x$
Return Trip	1.6	$1.5 - x$	$1.6(1.5 - x)$

The distances are the same.

$$\begin{aligned} 2x &= 1.6(1.5 - x) \\ 2x &= 2.4 - 1.6x \\ 3.6x &= 2.4 \\ x &= \frac{2.4}{3.6} = \frac{2}{3} \end{aligned}$$

The total distance across the lake is

$2x = 2\left(\frac{2}{3}\right) = \frac{4}{3}$ or 1.33 miles.

122. $3(x-2) - 4(x-3) > 2$

$$\begin{aligned} 3x - 6 - 4x + 12 &> 2 \\ -x + 6 &> 2 \\ -x &> -4 \\ \frac{-x}{-1} &< \frac{-4}{-1} \\ x &< 4 \end{aligned}$$

The solution set is $\{x | x < 4\}$.

Chapter 2 Review Exercises

1. $15a^3b^5$ has degree eight since the sum of the exponents is 3 + 5 = 8.

2. $-5x$ has degree one since $-5x$ can be written as $-5x^1$ and the only exponent is 1.

3. $-21xyz^5$ has degree seven since $-21xyz^5$ can be written as $-21x^1y^1z^5$ and the sum of the exponents is 1 + 1 + 5 = 7.

4. $a(a+3)-4(a-1)$
$$a^2+3a-4a+4$$
$$a^2-a+4$$

5. $x^2+2xy+6x^2-13=x^2+6x^2+2xy-13$
$$=7x^2+2xy-13$$

6. b^2+b-9 cannot be simplified since there are no like terms.

7. $2[-(x-y)+3x]-5y+10$
$=2[-x+y+3x]-5y+10$
$=2[2x+y]-5y+10$
$=4x+2y-5y+10$
$=4x-3y+10$

8. $5(c+4)-2c=-(c-4)$
$$5c+20-2c=-c+4$$
$$3c+20=-c+4$$
$$4c+20=4$$
$$4c=-16$$
$$c=-4$$

9. $3(x+1)-3=4(x-5)$
$$3x+3-3=4x-20$$
$$3x+0=4x-20$$
$$3x=4x-20$$
$$3x-4x=4x-4x-20$$
$$-x=-20$$
$$\frac{-1x}{-1}=\frac{-20}{-1}$$
$$x=20$$

10. $3+\frac{x}{2}=\frac{5}{6}$
$$6(3)+6\left(\frac{x}{2}\right)=6\left(\frac{5}{6}\right)$$
$$18+3x=5$$
$$18-18+3x=5-18$$
$$3x=-13$$
$$\frac{3x}{3}=\frac{-13}{3}$$
$$x=-\frac{13}{3}$$

11. $\frac{1}{2}(3t+4)=\frac{1}{3}(4t+1)$
$$6\left(\frac{1}{2}(3t+4)\right)=6\left(\frac{1}{3}(4t+1)\right)$$
$$3(3t+4)=2(4t+1)$$
$$9t+12-8t=8t+2-8t$$
$$t+12=2$$
$$t+12-12=2-12$$
$$t=-10$$

12. $2\left(\frac{x}{2}-4\right)=3\left(x+\frac{1}{3}\right)$
$$x-8=3x+1$$
$$x-8+8=3x+1+8$$
$$x=3x+9$$
$$x-3x=3x-3x+9$$
$$-2x=9$$
$$\frac{-2x}{-2}=\frac{9}{-2}$$
$$x=-\frac{9}{2}$$

13. $3x-7=9x+8-6x$
$$3x-7=3x+8$$
$$3x-3x-7=3x-3x+8$$
$$-7=8$$
This is a false statement which means there is no solution, or $\varnothing$.

14.
$$2(x-6)=5-\{2x-[4(x-2)-9]\}$$
$$2x-12=5-\{2x-[4x-8-9]\}$$
$$2x-12=5-\{2x-[4x-17]\}$$
$$2x-12=5-\{2x-4x+17\}$$
$$2x-12=5-\{-2x+17\}$$
$$2x-12=5+2x-17$$
$$2x-12=2x-12$$
$$2x-12-2x=2x-12-2x$$
$$12=12$$
Since this is a true statement, the solution set is all real numbers, or $\mathbb{R}$.

15.
$$m=\frac{y_2-y_1}{x_2-x_1}=\frac{(4)-(-3)}{(-8)-(6)}=\frac{4+3}{-8-6}=\frac{7}{-14}=-\frac{1}{2}$$

16.
$$x=\frac{-b+\sqrt{b^2-4ac}}{2a}=\frac{-10+\sqrt{(10)^2-4(8)(-3)}}{2(8)}=\frac{-10+\sqrt{100+96}}{16}=\frac{-10+\sqrt{196}}{16}=\frac{-10+14}{16}=\frac{4}{16}=\frac{1}{4}$$

17.
$$h=\frac{1}{2}at^2+v_0t+h_0$$
$$=\frac{1}{2}(-32)(1)^2+0(2)+85$$
$$=\frac{1}{2}(-32)(1)+0+85$$
$$=-16(1)+0+85$$
$$=-16+0+85$$
$$=69$$

18. $z=\dfrac{\overline{x}-\mu}{\frac{\sigma}{\sqrt{n}}}=\dfrac{50-54}{\frac{5}{\sqrt{25}}}=\dfrac{50-54}{\frac{5}{5}}=\dfrac{50-54}{1}=-4$

19. $E=IR$
$$\frac{E}{I}=\frac{IR}{I}$$
$$\frac{E}{I}=R \text{ or } R=\frac{E}{I}$$

20.
$$P=2l+2w$$
$$P-2l=2l-2l+2w$$
$$P=2l=2w$$
$$\frac{P-2l}{2}=\frac{2w}{2}$$
$$\frac{P-2l}{2}=w \text{ or } w=\frac{P-2l}{2}$$

21.
$$A=\pi r^2h$$
$$\frac{A}{\pi r^2}=\frac{\pi r^2}{\pi r^2}$$
$$\frac{A}{\pi r^2}=h \text{ or } h=\frac{A}{\pi r^2}$$

22.
$$A=\frac{1}{2}bh$$
$$2(A)=2\left(\frac{1}{2}bh\right)$$
$$2A=bh$$
$$\frac{2A}{b}=\frac{bh}{b}$$
$$\frac{2A}{b}=h \text{ or } h=\frac{2A}{b}$$

23.
$$y = mx + b$$
$$y - b = mx + b - b$$
$$y - b = mx$$
$$\frac{y-b}{x} = \frac{mx}{x}$$
$$\frac{y-b}{x} = m \text{ or } m = \frac{y-b}{x}$$

24.
$$2x - 3y = 5$$
$$2x - 2x - 3y = -2x + 5$$
$$-3y = -2x + 5$$
$$\frac{-3y}{-3} = \frac{-2x+5}{-3}$$
$$y = \frac{-2x+5}{-3} \text{ or } y = \frac{2x-5}{3}$$

25.
$$R_T = R_1 + R_2 + R_3$$
$$R_T - R_1 - R_3 = R_1 + R_2 + R_3 - R_1 - R_3$$
$$R_T - R_1 - R_3 = R_2$$
$$\text{or } R_2 = R_T - R_1 - R_3$$

26.
$$S = \frac{3a+b}{2}$$
$$2(S) = 2\left(\frac{3a+b}{2}\right)$$
$$2S = 3a + b$$
$$2S - b = 3a + b - b$$
$$2S - b = 3a$$
$$\frac{2S-b}{3} = \frac{3a}{3}$$
$$\frac{2S-b}{3} = a \text{ or } a = \frac{2S-b}{3}$$

27.
$$K = 2(d + l)$$
$$K = 2d + 2l$$
$$K - 2d = 2d - 2d + 2l$$
$$K - 2d = 2l$$
$$\frac{K-2d}{2} = \frac{2l}{2}$$
$$\frac{D-2d}{2} = l \text{ or } l = \frac{K-2d}{2}$$

28. Let x be the original price.
$$x - 0.75x = 7.50$$
$$0.25x = 7.50$$
$$\frac{0.25x}{0.25} = \frac{7.50}{0.25}$$
$$x = 30$$
The original price was \$30.

29. Let x be the number of years for the population to reach 5800.
$$4750 + 350x = 7200$$
$$350x = 2450$$
$$x = \frac{2450}{350}$$
$$x = 7$$
It will take 7 years for the population to grow from 4750 people to 7200 people.

30. Let x be the amount of sales.
$$300 + 0.06x = 708$$
$$0.06x = 408$$
$$\frac{0.06x}{0.06} = \frac{408}{0.06}$$
$$x = 6800$$
Celeste's sales must be \$6800 to earn \$708 in a week.

31. Let x be the number of miles she drives.
$$3(24.99) = 3(19.99) + 0.10x$$
$$74.97 = 59.97 + 0.10x$$
$$15.00 = 0.10x$$
$$\frac{15.00}{0.10} = \frac{0.10x}{0.10}$$
$$150 = x$$
The costs would be the same if she drives 150 miles.

32. Let x be the regular price.
$$x - 0.40x - 20 = 136$$
$$0.60x - 20 = 136$$
$$0.60x = 156$$
$$\frac{0.60x}{0.60} = \frac{156}{0.60}$$
$$x = 260$$
The regular price was \$260.

33. Let x = the amount invested at 3.5%. Then $5000 - x$ is the amount invested at 4.0%.

Account	Principal	Rate	Time	Interest
3.5%	x	0.035	1	$0.035x$
4.0%	$5000 - x$	0.04	1	$0.04(5000 - x)$

$$\begin{aligned} 0.035x + 0.04(5000 - x) &= 187.15 \\ 0.035x + 200 - 0.04x &= 187.15 \\ -0.005x + 200 &= 187.15 \\ -0.005x &= -12.85 \\ x &= \frac{-12.85}{-0.005} \\ x &= 2570 \end{aligned}$$

Thus, Mr. Olden invested \$2,570 at 3.5% and \$5,000 – \$2,570 = \$2,430 at 4.0%.

34. Let x = the amount of 20% solution.

Solution	Strength of Solution	No. of Gallons	Amount
20%	0.20	x	$0.20x$
60%	0.06	$250 - x$	$0.60(250 - x)$
Mixture	0.30	250	0.30(250)

$$\begin{aligned} 0.20x + 0.60(250 - x) &= 0.30(250) \\ 0.20x + 150 - 0.60x &= 75 \\ -0.40x + 150 &= 75 \\ -0.40x &= -75 \\ x &= \frac{-75}{-0.40} \\ x &= 187.5 \end{aligned}$$

Dale must combine 187.5 gallons of the 20% solution with 250 – 187.5 = 62.5 gallons of the 60% solution to obtain the 30% solution.

35. Let t be the amount of time needed.

Type	Rate	Time	Distance
One Train	60	t	$60t$
Other Train	80	t	$80t$

The total distance is 910 miles.

$$\begin{aligned} 60t + 80t &= 910 \\ 140t &= 910 \\ t &= \frac{910}{140} = \frac{13}{2} = 6\frac{1}{2} \end{aligned}$$

In $6\frac{1}{2}$ hours, the trains are 910 miles apart.

36. a. Let x be the speed of Shuttle 1. Then $x + 300$ is the speed of Shuttle 2.

Type	Rate	Time	Distance
Shuttle 1	x	5.5	$5.5x$
Shuttle 2	$x + 300$	5.0	$5.0(x + 300)$

The distances are the same.

$$5.5x = 5.0(x+300)$$
$$5.5x = 5.0x + 1500$$
$$0.5x = 1500$$
$$x = \frac{1500}{0.5} = 3000$$

The speed of Shuttle 1 is 3000 mph.

b. The distance is 5.5(3000) = 16,500 miles.

37. Let x be the amount of \$6.00 coffee needed. Then $40 - x$ is the amount of \$6.80 coffee needed.

Item	Cost per Pound	No. of Pounds	Total Value
\$6.00 Coffee	\$6.00	x	$6.00x$
\$6.80 Coffee	\$6.80	$40 - x$	$6.80(40 - x)$
Mixture	\$6.50	40	6.50(40)

$$6.00x + 6.80(40 - x) = 6.50(40)$$
$$6.00x + 272 - 6.80x = 260$$
$$-0.80x + 272 = 260$$
$$-0.80x = -12$$
$$x = \frac{-12}{-0.80} = 15$$

Mr. Tomlins needs to combine 15 pounds of \$6.00 coffee with 40 – 15 = 25 pounds of \$6.80 coffee to produce the mixture.

38. Let x = the original price of the telephone.

$$x - 0.20x = 28.80$$
$$0.80x = 28.80$$
$$x = \frac{28.80}{0.80} = 36$$

The original price of the telephone was \$36.

39. Let x be the time spent jogging. Then $4 - x$ is the time spent walking.

Trip	Rate	Time	Distance
Jogging	7.2	x	$7.2x$
Walking	2.4	$4 - x$	$2.4(4 - x)$

a. The distances are the same.

$$7.2x = 2.4(4 - x)$$
$$7.2x = 9.6 - 2.4x$$
$$9.6x = 9.6$$
$$x = \frac{9.6}{9.6} = 1$$

Thus, Nicolle jogged for 1 hour and walked for 4 – 1 = 3 hours.

b. The distance one-way is 7.2(1) = 7.2 miles. The total distance is twice this value or 2(7.2) = 14.4 miles.

40. Let x be the measure of the smallest angle. The measure of the other two angles are $x + 25$ and $2x - 5$.

$$x+(x+25)+(2x-5)=180$$
$$4x+20=180$$
$$4x=160$$
$$x=\frac{160}{4}=40$$

The measures of the angles are 40°, 40 + 25 = 65°, and 2(40) – 5= 80 – 5 = 75°.

41. Let x be the flow rate of the smaller hose.

Type	Rate	Time	Amount (No. of Gallons)
Smaller	r	3	$3r$
Larger	$1.5r$	5	$5(1.5r)$

The total number of gallons of water is 3150 gallons.

$$3r+5(1.5r)=3150$$
$$3r+7.5r=3150$$
$$10.5r=3150$$
$$r=\frac{3150}{10.5}=300$$

The flow rate for the smaller hose is 300 gallons per hour and the flow rate for the larger hose is 1.5(300) = 450 gallons per hour.

42. Let x = measure of one of the angles. Then the other angle measure is $2x - 30$. The sum of the measures of complementary angles is 90°.

$$x+(2x-30)=90$$
$$3x-30=90$$
$$3x=120$$
$$x=\frac{120}{3}$$
$$x=40$$

The measures of the angles are 40° and 2(40°) – 30° = 50°.

43. Let x be the amount of 20% solution.

Solution	Strength of Solution	No. of Ounces	Amount
20%	0.20	x	$0.20x$
6%	0.06	10	0.06(10)
Mixture	0.12	$x + 10$	$0.12(x + 10)$

$$0.20x+0.06(10)=0.12(x+10)$$
$$0.20x+0.6=0.12x+1.2$$
$$0.08x+0.6=1.2$$
$$0.08x=0.6$$
$$x=\frac{0.6}{0.08}=7.5$$

The clothier must combine 7.5 ounces of the 20% solution with 10 ounces of the 6% solution to obtain the 12% solution.

44. Let x be the amount invested at 10%. Then $12{,}000 - x$ is the amount invested at 6%.

Acct	Principal	Rate	Time	Interest
10%	x	0.10	1	$0.10x$
6%	$12{,}000 - x$	0.06	1	$0.06(12{,}000 - x)$

$$0.10x=0.06(12{,}000-x)$$
$$0.10x=720-0.06x$$
$$0.16x=720$$
$$x=\frac{720}{0.16}=4500$$

Thus, David invested $4500 at 10% and 12000 – 4500 = $7500 at 6%.

45. Let x be the number of visits. The cost of the first plan = cost of second plan gives the equation

$$40+1(x)=25+4(x)$$
$$40+x=25+4x$$
$$15+x=4x$$
$$15=3x$$
$$\frac{15}{3}=x \quad \text{or } x=5$$

Jeff needs to make more than 5 visits for the first plan to be advantageous.

46. Let x be the speed of the faster train. Then $x - 10$ is the speed of the slower train.

Train	Rate	Time	Distance
Faster	x	3	$3x$
Slower	$x - 10$	3	$3(x - 10)$

$$3x+3(x-10)=270$$
$$3x+3x-30=270$$
$$6x-30=270$$
$$6x=300$$
$$x=\frac{300}{6}=50$$

The speed of the faster train is 50 mph and the speed of the slower train is 40 mph.

47. $3z+9\le 15$

$$3z\le 6$$
$$z\le 2$$

48. $8-2w>-4$

$$-2w>-12$$
$$\frac{-2w}{-2}<\frac{-12}{-2}$$
$$w<6$$

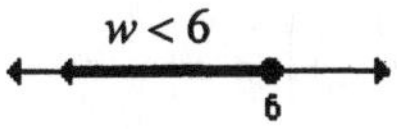

49. $2x+1>6$

$$2x>5$$
$$x>\frac{5}{2}$$

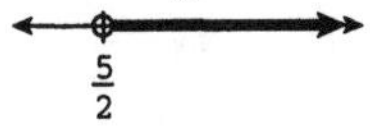

50. $26\le 4x+5$

$$21\le 4x$$
$$\frac{21}{4}\le x$$

51. $\frac{4x+3}{3}>-5$

$$3\left(\frac{4x+3}{3}\right)>3(-5)$$
$$4x+3>-15$$
$$4x>-18$$
$$x>\frac{-18}{4}$$
$$x>-\frac{9}{2}$$

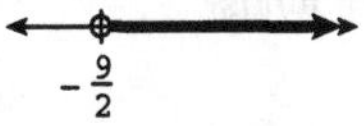

52. $2(x-1)>3x+8$

$$2x-2>3x+8$$
$$2x-10>3x$$
$$-10>x$$

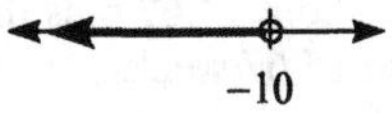

53. $-4(x-2)\ge 6x+8-10x$

$$-4x+8\ge -4x+8$$
$$8\ge 8 \quad \text{a true statement}$$

The solution is all real numbers.

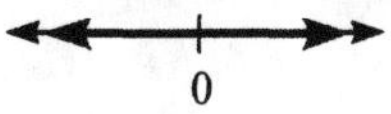

54. $\frac{x}{2}+\frac{3}{4}>x-\frac{x}{2}+1$

$$4\left(\frac{x}{2}+\frac{3}{4}\right)>4\left(x-\frac{x}{2}+1\right)$$
$$2x+3>4x-2x+4$$
$$2x+3>2x+4$$
$$3>4$$

This is a contradiction, so the solution is { }.

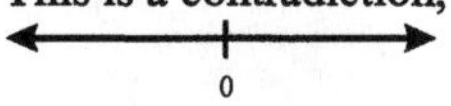

55. Let x be the maximum number of 40-pound boxes. Since the maximum load is 560 pounds, the total weight of Bob, Kathy, and the boxes must be less than or equal to 560 pounds.

$$300+40x\le 560$$
$$40x\le 260$$
$$x\le\frac{260}{40}$$
$$x\le 6.5$$

The maximum number of boxes that Joseph can carry in the canoe is 6.

56. Let x be the number of additional minutes (beyond 3 minutes) of the phone call.

$$4.50+0.95x \le 8.65$$
$$0.95x \le 4.15$$
$$x \le \frac{4.15}{0.95}$$
$$x \le 4.4$$

The customer can talk for 3 minutes plus an additional 4 minutes for a total of 7 minutes.

57. Let x be the number of weeks (after the first week) needed to lose 27 pounds.

$$5+1.5x \ge 27$$
$$1.5x \ge 22$$
$$x \ge \frac{22}{1.5}$$
$$x \ge 14\frac{2}{3} \approx 14.67$$

The number of weeks is about 14.67 plus the initial week for a total of 15.67 weeks.

58. Let x be the grade from the 5th exam. The inequality is

$$80 \le \frac{94+73+72+80+x}{5} < 90$$
$$80 \le \frac{319+x}{5} < 90$$
$$5(80) \le 5\left(\frac{319+x}{5}\right) < 5(90)$$
$$400 \le 319+x < 450$$
$$400-319 \le 319+x < 450-319$$
$$81 \le x < 131$$

(must use 100 here since it is not possible to score 131)

Thus, Patrice needs to score 81 or higher on the 5th exam to receive a B.

$$\{x \mid 81 \le x \le 100\}$$

59. $1 < x-4 < 7$

$$1+4 < x-4+4 < 7+4$$
$$5 < x < 11$$
$$(5, 11)$$

60. $8 < p+11 \le 16$

$$8-11 < p+11-11 \le 16-11$$
$$-3 < p \le 5$$
$$(-3, 5]$$

61. $3 < 2x-4 < 12$

$$3+4 < 2x-4+4 < 12+4$$
$$7 < 2x < 16$$
$$\frac{7}{2} < \frac{2x}{2} < \frac{16}{2}$$
$$\frac{7}{2} < x < 8$$
$$\left(\frac{7}{2}, 8\right)$$

62. $-12 < 6-3x < -2$

$$-12-6 < 6-6-3x < -2-6$$
$$18 < -3x < -8$$
$$\frac{-18}{-3} > \frac{-3x}{-3} > \frac{-8}{-3}$$
$$6 > x > \frac{8}{3}$$
$$\frac{8}{3} < x < 6$$
$$\left(\frac{8}{3}, 6\right)$$

63. $-1 < \frac{5}{9}x+\frac{2}{3} \le \frac{11}{9}$

$$9(-1) < 9\left(\frac{5}{9}x+\frac{2}{3}\right) \le 9\left(\frac{11}{9}\right)$$
$$-9 < 5x+6 \le 11$$
$$-9-6 < 5x+6-6 \le 11-6$$
$$-15 < 5x \le 5$$
$$\frac{-15}{5} < \frac{5x}{5} \le \frac{5}{5}$$
$$-3 < x \le 1$$
$$(-3, 1]$$

64. $-8 < \frac{4-2x}{3} < 0$

$3(-8) < 3\left(\frac{4-2x}{3}\right) < 3(0)$

$-24 < 4-2x < 0$

$-24-4 < 4-4-2x < 0-4$

$-28 < -2x < -4$

$\frac{-28}{-2} > \frac{-2x}{-2} > \frac{-4}{-2}$

$14 > x > 2$

$2 < x < 14$

(2, 14)

65. $h \le 1$ and $7h - 4 > -25$

$h \le 1$ and $7h > -21$

$h \le 1$ and $h > -3$

$h \le 1$

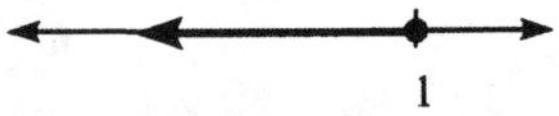

$h > -3$

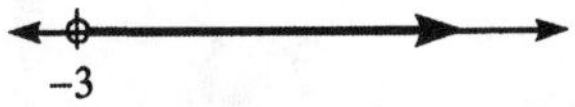

$x \le 1$ and $x > -3$ which is $-3 < h \le 1$.

The solution set is $\{h \mid -3 < h \le 1\}$.

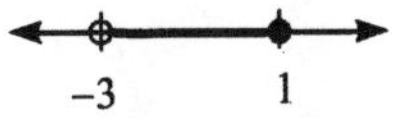

66. $2x - 1 > 5$ or $3x - 2 \le 10$

$2x > 6$ or $3x \le 12$

$x > 3$ or $x \le 4$

$x > 3$

$x \le 4$

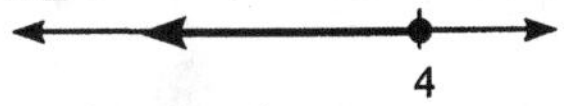

$x > 3$ or $x \le 4$

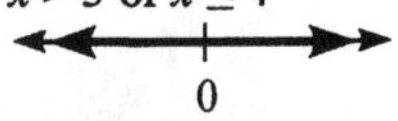

which is the entire real number line or $\mathbb{R}$.

67. $4x - 5 < 11$ and $-3x - 4 \ge 8$

$4x < 16$ and $-3x \ge 12$

$x < 4$ and $x \le -4$

$x \le 4$

$x \le -4$

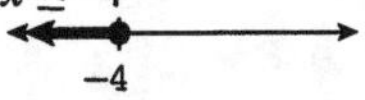

$x \le -4$ and $x < 4$ which is $x \le -4$

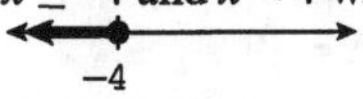

$\{x \mid x \le -4\}$

68. $\frac{7-2g}{3} \le -5$ or $\frac{3-g}{9} > 1$

$7 - 2g \le -15$ or $3 - g > 9$

$-2g \le -22$ or $-g > 6$

$g \ge 11$ or $g < -6$

$g < -6$

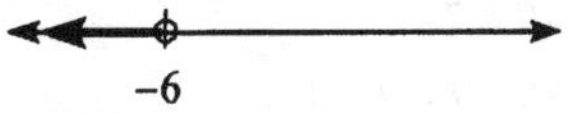

$g \ge 11$

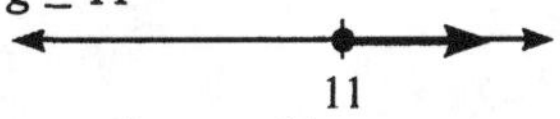

$g < -6$ or $g \ge 11$

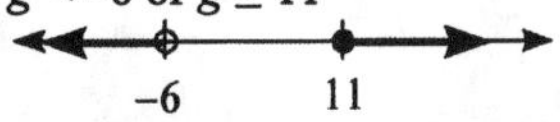

$\{g \mid g < -6 \text{ or } g \ge 11\}$

69. $|a| = 2$

$a = 2$ or $a = -2$

The solution set is $\{-2, 2\}$.

70. $|x| < 8$

$-8 < x < 8$

The solution set is $\{x \mid -8 < x < 8\}$.

71. $|x| \ge 9$

$x \le -9$ or $x \ge 9$

The solution set is $\{x \mid x \le -9 \text{ or } x \ge 9\}$.

72. $|l + 5| = 13$

$l + 5 = -13$ or $l + 5 = 13$

$l = -18$ $\quad l = 8$

The solution set is $\{-18, 8\}$.

73. $|x-2| \geq 5$

$x-2 \leq -5$ or $x-2 \geq 5$

$x \leq -3$ $\quad$ $x \geq 7$

The solution set is $\{x | x \leq -3 \text{ or } x \geq 7\}$.

74. $|4-2x| = 5$

$4-2x=5$ or $4-2x=-5$

$-2x=1$ $\quad$ $-2x=-9$

$x=\frac{1}{-2}$ $\quad$ $x=\frac{-9}{-2}$

$x=-\frac{1}{2}$ $\quad$ $x=\frac{9}{2}$

The solution set is $\left\{-\frac{1}{2}, \frac{9}{2}\right\}$.

75. $|-2q+9| < 7$

$-7 < -2q+9 < 7$

$-7-9 < -2q+9-9 < 7-9$

$-16 < -2q < -2$

$\frac{-16}{-2} > \frac{-2q}{-2} > \frac{-2}{-2}$

$8 > q > 1$

$1 < q < 8$

The solution set is $\{q | 1 < q < 8\}$.

76. $\left|\frac{2x-3}{5}\right| = 1$

$\frac{2x-3}{5}=1$ or $\frac{2x-3}{5}=-1$

$2x-3=5$ $\quad$ $2x-3=-5$

$2x=8$ $\quad$ $2x=-2$

$x=4$ $\quad$ $x=-1$

The solution set is {–1, 4}.

77. $\left|\frac{x-4}{3}\right| < 6$

$-6 < \frac{x-4}{3} < 6$

$3(-6) < 3\left(\frac{x-4}{3}\right) < 3(6)$

$-18 < x-4 < 18$

$-14 < x < 22$

The solution set is $\{x | -14 < x < 22\}$.

78. $|4d-1| = |6d+9|$

$4d-1=-(6d+9)$ or $4d-1=6d+9$

$4d-1=-6d-9$ $\quad$ $4d-10=6d$

$10d-1=-9$ $\quad$ $-10=2d$

$10d=-8$ $\quad$ $-5=d$

$d=-\frac{4}{5}$

The solution set is $\left\{-5, -\frac{4}{5}\right\}$.

79. $|2x-3|+4 \geq -17$

$|2x-3| \geq -21$

Since the right side is negative and the left side is non-negative, the solution is the entire real number line since the absolute value of a number is always greater than a negative number. The solution set is all real numbers, or $\mathbb{R}$.

80. $|3c+8|-6 \leq 1$

$|3c+8| \leq 7$

$-7 \leq 3c+8 \leq 7$

$-7-8 \leq 3c+8-8 \leq 7-8$

$-15 \leq 3c \leq -1$

$\frac{-15}{3} \leq \frac{3c}{3} \leq \frac{-1}{3}$

$-5 \leq c \leq -\frac{1}{3}$

$\left[-5, -\frac{1}{3}\right]$

81. $3 < 2x-5 \leq 11$

$3+5 < 2x-5+5 \leq 11+5$

$8 < 2x \leq 16$

$\frac{8}{2} < \frac{2x}{2} \leq \frac{16}{2}$

$4 < x \leq 8$

The solution is (4, 8].

82. $-6 \le \dfrac{3-2x}{4} < 5$

$$4(-6) \le 4\left(\frac{3-2x}{4}\right) < 4(5)$$
$$-24 \le 3 - 2x < 20$$
$$-27 \le -2x < 17$$
$$\frac{-27}{-2} \ge \frac{-2x}{-2} > \frac{17}{-2}$$
$$\frac{27}{2} \ge x > -\frac{17}{2}$$
$$-\frac{17}{2} < x \le \frac{27}{2}$$

The solution is $\left(-\dfrac{17}{2}, \dfrac{27}{2}\right]$.

83. $2p-5<7$ or $9-3p \le 15$

$$2p<12 \quad \text{or} \quad -3p \le 6$$
$$p<6 \quad \text{or} \quad p \ge -2$$
$$-2 \le p < 6$$

The solution is $[-2, 6)$.

84. $x-3 \le 4$ or $2x-5>7$

$$x-3+3 \le 4+3 \qquad 2x-5+5>7+5$$
$$x \le 7 \qquad 2x>12$$
$$x>6$$

The solution is $(-\infty, \infty)$.

85. $-10 < 3(x-4) \le 18$

$$-10 < 3x - 12 \le 18$$
$$-10 + 12 < 3x - 12 + 12 \le 18 + 12$$
$$2 < 3x \le 30$$
$$\frac{2}{3} < x \le 10$$

The solution is $\left(\dfrac{2}{3}, 10\right]$.

Chapter 2 Practice Test

1. $-3a^2bc^4$ is degree seven since $-3a^2bc^4$ can be written as $-3a^2b^1c^4$ and the sum of the exponents is $2 + 1 + 4 = 7$.

2. $2p - 3q + 2pq - 6p(q - 3) - 4p$

$$= 2p - 3q + 2pq - 6pq + 18p - 4p$$
$$= (2p + 18p - 4p) - 3q + (2pq - 6pq)$$
$$= 16p - 3q - 4pq$$

3. $7q-\{2[3-4(q+7)]+5q\}-8$

$$=7q-\{2[3-4q-28]+5q\}-8$$
$$=7q-\{2(-25-4q)+5q\}-8$$
$$=7q-(-50-8q+5q)-8$$
$$=7q-(-3q-50)-8$$
$$=7q+3q+50-8$$
$$=10q+42$$

4. $7(d+2)=3(2d-4)$

$$7d+14=6d-12$$
$$7d+14-6d=6d-12-6d$$
$$d+14=-12$$
$$d+14-14=-12-14$$
$$d=-26$$

5. $\dfrac{r}{12}+\dfrac{1}{3}=\dfrac{4}{9}$

$$36\left(\frac{r}{12}+\frac{1}{3}\right)=36\left(\frac{4}{9}\right)$$
$$3r+12=16$$
$$3r+12-12=16-12$$
$$3r=4$$
$$\frac{3r}{3}=\frac{4}{3}$$
$$r=\frac{4}{3}$$

6. $-2(x+3)=4\{3[x-(3x+7)]+2\}$
$-2x-6=4\{3[x-3x-7]+2\}$
$-2x-6=4\{3[-2x-7]+2\}$
$-2x-6=4\{-6x-21+2\}$
$-2x-6=4\{-6x-19\}$
$-2x-6=-24x-76$
$-2x-6+24x=-24x-76+24x$
$22x-6=-76$
$22x-6+6=-76+6$
$22x=-70$
$\frac{22x}{22}=\frac{-70}{22}$
$x=-\frac{35}{11}$

7. $7x-6(2x-4)=3-(5x-6)$
$7x-12x+24=3-5x+6$
$-5x+24=-5x+9$
$-5x+24+5x=-5x+9+5x$
$24=9$
This is a false statement which means there is no solution. $\varnothing$

8. $-\frac{1}{2}(4x-6)=\frac{1}{3}(3-6x)+2$
$-2x+3=1-2x+2$
$-2x+3=-2x+3$
$-2x+3+2x=-2x+3+2x$
$3=3$
This is always true which means the solution is any real number or $\mathbb{R}$.

9. $S_n=\frac{a_1(1-r^n)}{1-r}$
$$S_3=\frac{3\left[1-\left(\frac{1}{3}\right)^3\right]}{1-\frac{1}{3}}=\frac{3\left[1-\frac{1}{27}\right]}{1-\frac{1}{3}}=\frac{3\left(\frac{26}{27}\right)}{\frac{2}{3}}$$
$$=\frac{\frac{26}{9}}{\frac{2}{3}}=\frac{26}{9}\cdot\frac{3}{2}=\frac{13}{3}$$

10. $c=\frac{a-5b}{2}$
$2(c)=2\left(\frac{a-5b}{2}\right)$
$2c=a-5b$
$2c-a=a-a-5b$
$2c-a=-5b$
$\frac{2c-a}{-5}=\frac{-5b}{-5}$
$\frac{2c-a}{-5}=b$ or $b=\frac{a-2c}{5}$

11. $A=\frac{1}{2}h(b_1+b_2)$
$2(A)=2\left[\frac{1}{2}h(b_1+b_2)\right]$
$2A=h(b_1+b_2)$
$2A=hb_1+hb_2$
$2A-hb_1=hb_1-hb_1+hb_2$
$2A-hb_1=hb_2$
$\frac{2A-hb_1}{h}=\frac{hb_2}{h}$
$\frac{2A-hb_1}{h}=b_2$ or $b_2=\frac{2A-hb_1}{h}$

12. Let x be the cost of the clubs before tax, then $0.07x$ is the tax.
$x+0.07x=668.75$
$1.07x=668.75$
$x=\frac{668.75}{1.07}$
$x=625$
The cost of the clubs before tax is \$625.

13. Let x = the number of visits Jay can make.
$240+2x=400$
$2x=160$
$x=80$
Bill can visit the health club 80 times.

14. Let x = the number of hours in which the will be 147 miles apart.

Person	Rate	Time	Distance
Jeffrey	15	x	$15x$
Roberto	20	x	$20x$

The total distance is the sum of the distances they traveled.

$$15x + 20x = 147$$
$$35x = 147$$
$$x = \frac{147}{35}$$
$$x = 4.2$$

In 4.2 hours, the cyclists will be 147 miles apart.

15. Let x be the amount of 12% solution.

Solution	Strength of Solution	No. of Liters	Amount of Salt
12%	0.12	x	$0.12x$
25%	0.25	10	0.25(10)
20%	0.20	$x + 10$	$0.20(x + 10)$

$$0.12x + 0.25(10) = 0.20(x + 10)$$
$$0.12x + 2.50 = 0.20x + 2.00$$
$$0.12x + 0.50 = 0.20x$$
$$0.50 = 0.08x$$
$$\frac{0.50}{0.08} = x$$
$$6.25 = x$$

Combine 6.25 liters of the 12% solution with 10 liters of the 25% solution to obtain the mixture.

16. Let x be the amount invested at 8%. Then $12{,}000 - x$ is the amount invested at 7%.

Account	Principal	Rate	Interest
8%	x	0.08	$0.08x$
7%	$12{,}000 - x$	0.07	$0.07(12{,}000 - x)$

The total interest is $910.

$$0.08x + 0.07(12{,}000 - x) = 910$$
$$0.08x + 840 - 0.07x = 910$$
$$0.01x + 840 = 910$$
$$0.01x = 70$$
$$x = \frac{70}{0.01}$$
$$x = 7000$$

Thus, $7000 was invested at 8% and the remaining amount of 12000 – 7000 = $5000 was invested at 7%.

17. $3(2q + 4) < 5(q - 1) + 7$

$$6q + 12 < 5q - 5 + 7$$
$$6q + 12 < 5q + 2$$
$$q + 12 < 2$$
$$q < -10$$

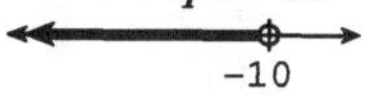

18. $\frac{6 - 2x}{5} \geq -12$

$$5\left(\frac{6 - 2x}{5}\right) \geq 5(-12)$$
$$6 - 2x \geq -60$$
$$-2x \geq -66$$
$$\frac{-2x}{-2} \leq \frac{-66}{-2}$$
$$x \leq 33$$

19. $x - 3 \leq 4$ and $2x + 1 > 10$

$$x - 3 + 3 \leq 4 + 3 \qquad 2x + 1 - 1 > 10 - 1$$
$$x \leq 7 \qquad 2x > 9$$
$$x > \frac{9}{2}$$

The solution is $\left(\frac{9}{2}, 7\right]$.

20. $7 \le \frac{2u-5}{3} < 9$

$3(7) \le 3\left(\frac{2u-5}{3}\right) < 3(9)$

$21 \le 2u-5 < 27$

$21+5 \le 2u-5+5 < 27+5$

$26 \le 2u < 32$

$13 \le u < 16$

The solution is [13, 16).

21. $|2b+5| = 9$

$2b+5 = -9$ or $2b+5 = 9$

$2b = -14$ $\quad$ $2b = 4$

$b = -7$ $\quad$ $b = 2$

The solution set is $\{-7, 2\}$.

22. $|2x-3| = \left|\frac{1}{2}x-10\right|$

$2x-3 = -\left(\frac{1}{2}x-10\right)$ or $2x-3 = \frac{1}{2}x-10$

$2x-3 = -\frac{1}{2}x+10$ $\quad$ $\frac{3}{2}x-3 = -10$

$\frac{5}{2}x-3 = 10$ $\quad$ $\frac{3}{2}x = -7$

$\frac{5}{2}x = 13$ $\quad$ $\frac{2}{3}\left(\frac{3}{2}x\right) = \frac{2}{3}(-7)$

$\frac{2}{5}\left(\frac{5}{2}x\right) = \frac{2}{5}(13)$ $\quad$ $x = -\frac{14}{3}$

$x = \frac{26}{5}$

The solution set is $\left\{-\frac{14}{3}, \frac{26}{5}\right\}$.

23. $|4z+12| = 0$

$4z+12 = 0$

$4z = -12$

$z = -3$

The solution set is $\{-3\}$.

24. $|2x-3|+6 > 11$

$|2x-3| > 5$

$2x-3 < -5$ or $2x-3 > 5$

$2x < -2$ $\quad$ $2x > 8$

$x < -1$ $\quad$ $x > 4$

The solution set is $\{x | x < -1 \text{ or } x > 4\}$.

25. $\left|\frac{2x-3}{8}\right| \le \frac{1}{4}$

$-\frac{1}{4} \le \frac{2x-3}{8} \le \frac{1}{4}$

$8\left(-\frac{1}{4}\right) \le 8\left(\frac{2x-3}{8}\right) \le 8\left(\frac{1}{4}\right)$

$-2 \le 2x-3 \le 2$

$1 \le 2x \le 5$

$\frac{1}{2} \le x \le \frac{5}{2}$

The solution set is $\left\{x \middle| \frac{1}{2} \le x \le \frac{5}{2}\right\}$.

Chapter 2 Cumulative Review Test

1. a. $A \cup B = \{1, 2, 3, 5, 7, 9, 11, 13, 15\}$

b. $A \cap B = \{3, 5, 7, 11, 13\}$

2. a. Commutative property of addition

b. Associative property of multiplication

c. Distributive property

3. $-4^3 + (-6)^2 \div (2^3 - 2)^2$

$= -4^3 + (-6)^2 \div (8-2)^2$

$= -4^3 + (-6)^2 \div (6)^2$

$= -64 + 36 \div 36$

$= -64 + 1$

$= -63$

4. Substitute −1 for *a* and − 2 for *b*.

$$a^2b^3+ab^2-3b$$
$$=(-1)^2(-2)^3+(-1)(-2)^2-3(-2)$$
$$=(1)(-8)+(-1)(4)-3(-2)$$
$$=-8+(-4)-(-6)$$
$$=-8+(-4)+6$$
$$=-12+6$$
$$=-6$$

5. $$\frac{8-\sqrt[3]{27}\cdot 3\div 9}{|-5|-[5-(12\div 4)]^2}=\frac{8-\sqrt[3]{27}\cdot 3\div 9}{|-5|-[5-3]^2}$$
$$=\frac{8-\sqrt[3]{27}\cdot 3\div 9}{|-5|-2^2}$$
$$=\frac{8-3\cdot 3\div 9}{5-4}$$
$$=\frac{8-9\div 9}{5-4}$$
$$=\frac{8-1}{5-4}$$
$$=\frac{7}{1}$$
$$=7$$

6. $$(5x^4y^3)^{-2}=\left(\frac{1}{5x^4y^3}\right)^2$$
$$=\frac{1^2}{5^2x^{4\cdot 2}y^{3\cdot 2}}$$
$$=\frac{1}{25x^8y^6}$$

7. $$\left(\frac{4m^2n^{-4}}{m^{-3}n^2}\right)^2=\left(\frac{4m^{2-(-3)}}{n^{2-(-4)}}\right)^2$$
$$=\left(\frac{4m^5}{n^6}\right)^2$$
$$=\frac{4^2m^{5\cdot 2}}{n^{6\cdot 2}}$$
$$=\frac{16m^{10}}{n^{12}}$$

8. $$\frac{5.704\times 10^5}{1.045\times 10^3}=\frac{5.704}{1.045}\times 10^{5-3}$$
$$\approx 5.458\times 10^2$$
$$\approx 545.8$$

The land area of Alaska is about 545.8 times larger than that of Rhode Island.

9. $$-3(y+7)=2(-2y-8)$$
$$-3y-21=-4y-16$$
$$y-21=-16$$
$$y=5$$

10. $$1.2(x-3)=2.4x-4.98$$
$$1.2x-3.6=2.4x-4.98$$
$$1.2x=2.4x-1.38$$
$$-1.2x=-1.38$$
$$x=\frac{-1.38}{-1.2}$$
$$x=1.15$$

11. $$\frac{2m}{3}-\frac{1}{6}=\frac{4}{9}m$$
$$18\left(\frac{2m}{3}-\frac{1}{6}\right)=18\left(\frac{4}{9}m\right)$$
$$12m-3=8m$$
$$4m-3=0$$
$$4m=3$$
$$m=\frac{3}{4}$$

12. A conditional equation is true only under specific conditions. An identity is true for all values of the variable. A contradiction is never true.
Answers may vary. One possible answer is:
$3x + 4 = 13$ is a conditional linear equation.
$3(x + 7) = 2(x + 10) + x + 1$ is an identity.
$3x + 4 = 3x + 8$ is a contradiction.

13. $$x=\frac{-b+\sqrt{b^2-4ac}}{2a}$$
$$=\frac{-(-8)+\sqrt{(-8)^2-4(3)(-3)}}{2(3)}$$
$$=\frac{-(-8)+\sqrt{64+36}}{6}=\frac{-(-8)+\sqrt{100}}{6}$$
$$=\frac{8+10}{6}=\frac{18}{6}=3$$

14. $y - y_1 = m(x - x_1)$

$\frac{y - y_1}{m} = \frac{m(x - x_1)}{m}$

$\frac{y - y_1}{m} = x - x_1$

$\frac{y - y_1}{m} + x_1 = x$

$x = \frac{y - y_1}{m} + x_1$ or $x = \frac{y - y_1 + mx_1}{m}$

15. a. $-4 < \frac{5x - 2}{3} < 2$

$3(-4) < 3\left(\frac{5x - 2}{3}\right) < 3(2)$

$-12 < 5x - 2 < 6$

$-12 + 2 < 5x - 2 + 2 < 6 + 2$

$-10 < 5x < 8$

$\frac{-10}{5} < \frac{5x}{5} < \frac{8}{5}$

$-2 < x < \frac{8}{5}$

b. $\left\{x \middle| -2 < x < \frac{8}{5}\right\}$

c. $\left(-2, \frac{8}{5}\right)$

16. $|3h - 1| = 8$

$3h - 1 = -8$ or $3h - 1 = 8$

$3h = -7$ $\quad$ $3h = 9$

$h = -\frac{7}{3}$ $\quad$ $h = 3$

Solution is $\left\{-\frac{7}{3}, 3\right\}$.

17. $|2x - 4| - 6 \geq 18$

$|2x - 4| \geq 24$

$2x - 4 \leq -24$ or $2x - 4 \geq 24$

$2x \leq -20$ $\quad$ $2x \geq 28$

$x \leq -10$ $\quad$ $x \geq 14$

The solution set is $\{x | x \leq -10 \text{ or } x \geq 14\}$.

18. Let x be the original price.

$x - 0.40x = 21$

$0.60x = 21$

$x = \frac{21}{0.60}$

$x = 35$

The original price was $35.

19. Let x be the speed of the car traveling south. Then $x + 20$ is the speed of the car traveling north.

Car	Rate	Time	Distance
South	x	3	$3x$
North	$x + 20$	3	$3(x + 20)$

The total distance is 300 miles.

$3x + 3(x + 20) = 300$

$3x + 3x + 60 = 300$

$6x + 60 = 300$

$6x = 240$

$x = \frac{240}{6}$

$x = 40$

The speed of the car traveling south is 40 mph and the speed of the car traveling north is 40 + 20 = 60 mph.

20. Let x = the number of pounds of cashews. Then $40 - x$ is the number of pounds of peanuts.

	Cost	Pounds	Cost
cashews	6.50	x	$6.50x$
peanuts	2.50	$40 - x$	$2.50(40 - x)$
mixture	4.00	40	4.00(40)

$6.50x + 2.50(40 - x) = 4.00(40)$

$6.50x + 100 - 2.50x = 160$

$4.00x + 100 = 160$

$4.00x = 60$

$x = 15$

Molly should combine 15 pounds of cashews with 40 lbs – 15 lbs = 25 lbs of peanuts.

Chapter 3

Exercise Set 3.1

1. a. The graph of any linear equation looks like a straight line.

b. Two points are needed to graph a linear equation. Two points uniquely determine a straight line.

3. If a set of points is collinear, they are in a straight line.

5. $A(3, 1)$, $B(-6, 0)$, $C(2, -4)$, $D(-2, -4)$, $E(0, 3)$, $F(-8, 1)$, $G\left(\frac{3}{2}, -1\right)$

7.

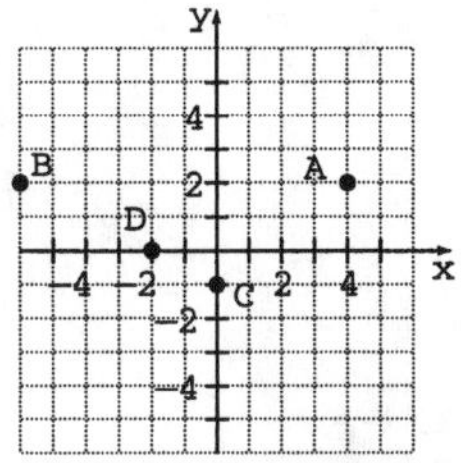

9. I

11. IV

13. II

15. III

17.
$$\begin{aligned} y &= 2x-5 \\ 21 &\stackrel{?}{=} 2(2)-5 \\ 21 &\stackrel{?}{=} 4-5 \\ 21 &= -1 \quad \text{False} \end{aligned}$$
No, (2, 21) is not a solution to $y = 2x - 5$.

19.
$$\begin{aligned} y &= |x|+3 \\ (-2) &\stackrel{?}{=} |-4|+3 \\ -2 &\stackrel{?}{=} 4+3 \\ -2 &= 7 \quad \text{False} \end{aligned}$$
No, (–4, –2) is not a solution to $y = |x|+3$.

21.
$$\begin{aligned} s &= 2r^2-r-5 \\ 5 &\stackrel{?}{=} 2(-2)^2-(-2)-5 \\ 5 &\stackrel{?}{=} 8+2-5 \\ 5 &= 5 \quad \text{True} \end{aligned}$$
Yes, (–2, 5) is a solution to $s = 2r^2-r-5$.

23.
$$\begin{aligned} -a^2+2b^2 &= -2 \\ -(2)^2+2(1)^2 &\stackrel{?}{=} -2 \\ -(4)+2(1) &\stackrel{?}{=} -2 \\ -4+2 &\stackrel{?}{=} -2 \\ -2 &= -2 \quad \text{True} \end{aligned}$$
Yes, (2, 1) is a solution to $-a^2+2b^2=-2$.

25.
$$\begin{aligned} 2x^2+6x-y &= 0 \\ 2\left(\frac{1}{2}\right)^2+6\left(\frac{1}{2}\right)-\left(\frac{5}{2}\right) &\stackrel{?}{=} 0 \\ 2\left(\frac{1}{4}\right)+6\left(\frac{1}{2}\right)-\left(\frac{5}{2}\right) &\stackrel{?}{=} 0 \\ \frac{1}{2}+3-\frac{5}{2} &\stackrel{?}{=} 0 \\ 1 &= 0 \quad \text{False} \end{aligned}$$
No, $\left(\frac{1}{2}, \frac{5}{2}\right)$ is not a solution to $2x^2+4x-y=0$.

27.

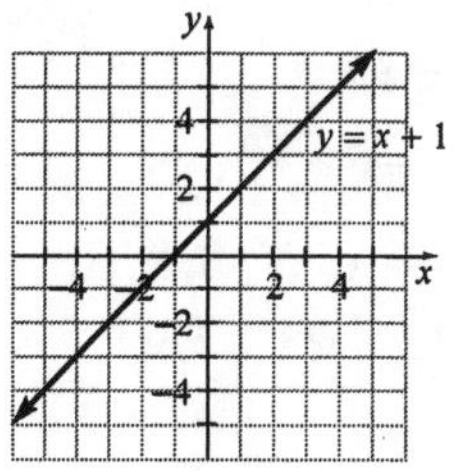

29.

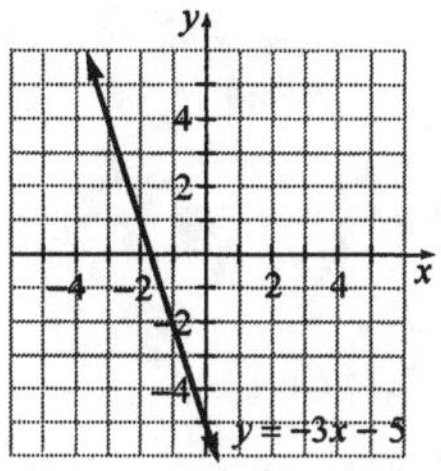

31.

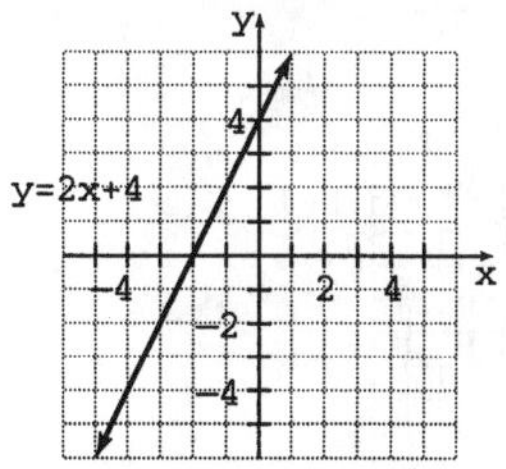

33.

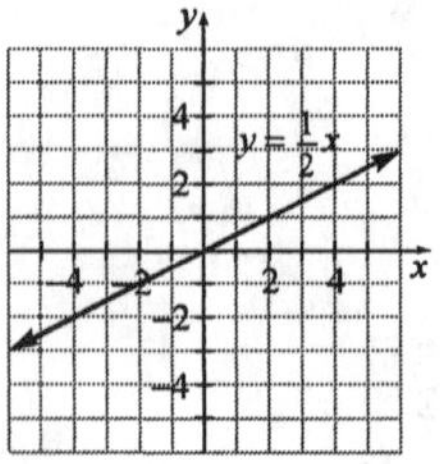

35.

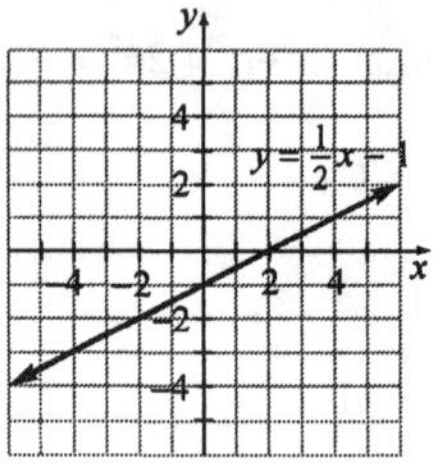

37.

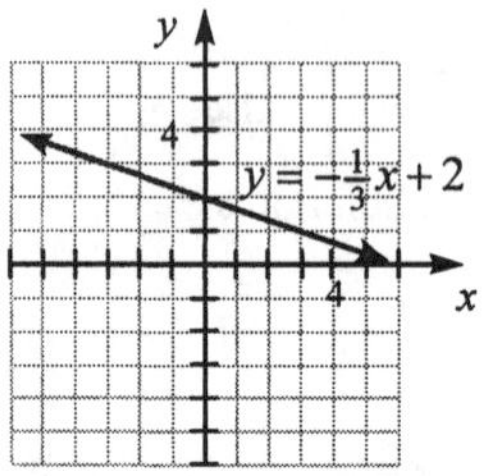

39.

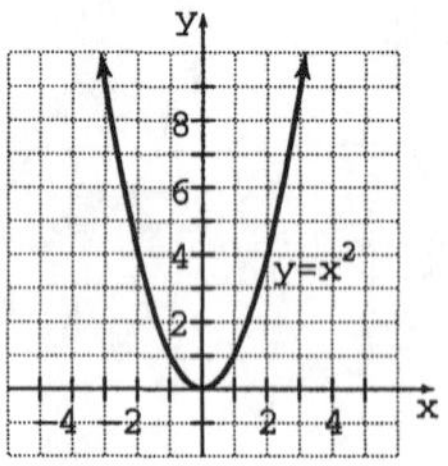

41.

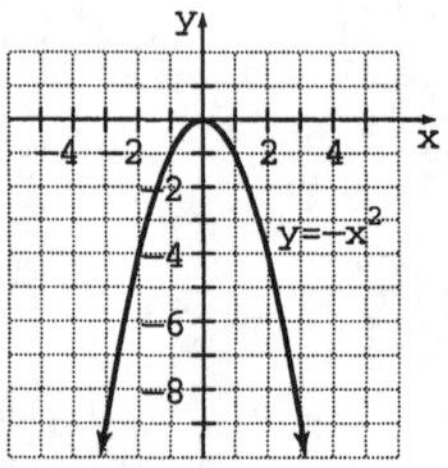

43.

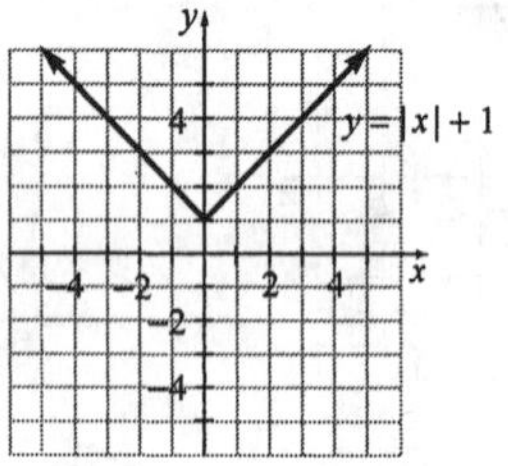

45.

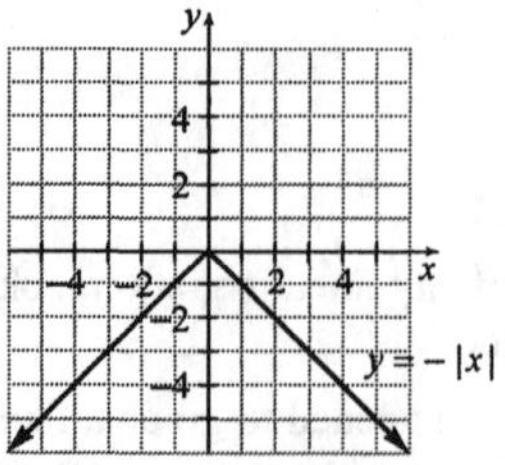

47.

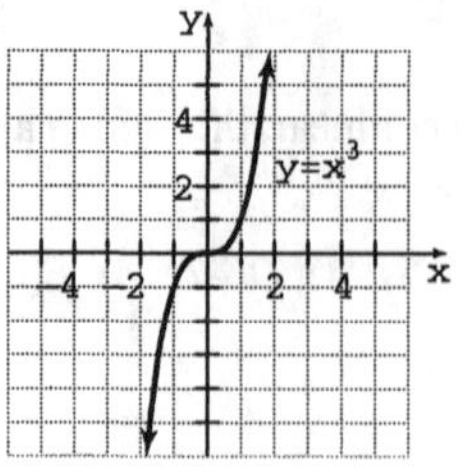

49.

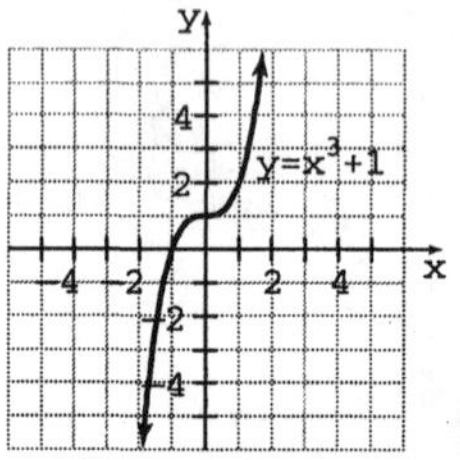

51.

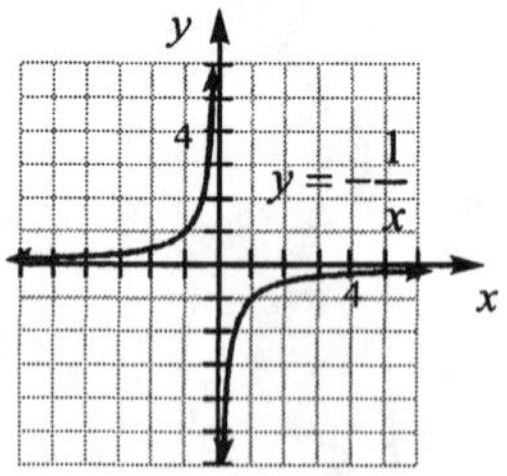

53.

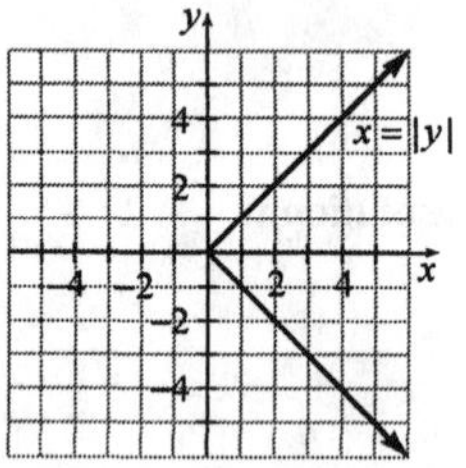

55.

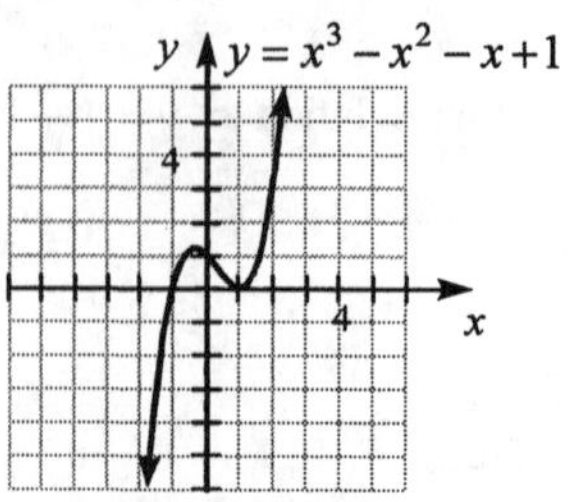

57.

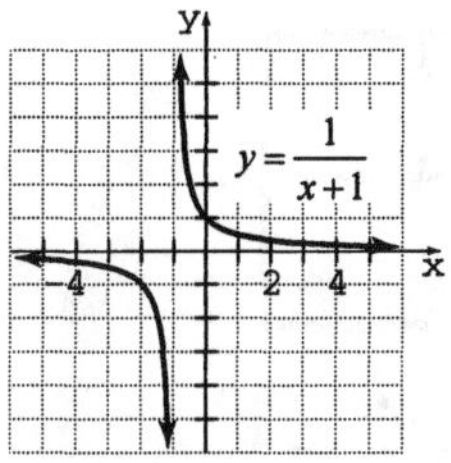

59.

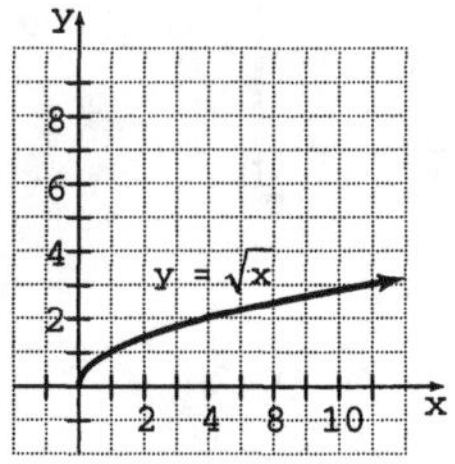

61.

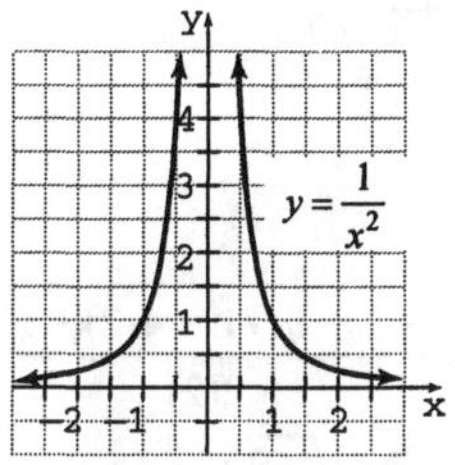

63. $y=\dfrac{x^2}{x+1}$

$$\left(\frac{1}{12}\right)\stackrel{?}{=}\frac{\left(\frac{1}{3}\right)^2}{\left(\frac{1}{3}\right)+1}$$

$$\frac{1}{12}\stackrel{?}{=}\frac{\frac{1}{9}}{\frac{4}{3}}$$

$$\frac{1}{12}\stackrel{?}{=}\frac{1}{9}\cdot\frac{3}{4}$$

$$\frac{1}{12}=\frac{1}{12}\quad\text{True}$$

Yes, $\left(\frac{1}{3},\frac{1}{12}\right)$ is on the graph of the equation $y=\dfrac{x^2}{x+1}$.

65. a.

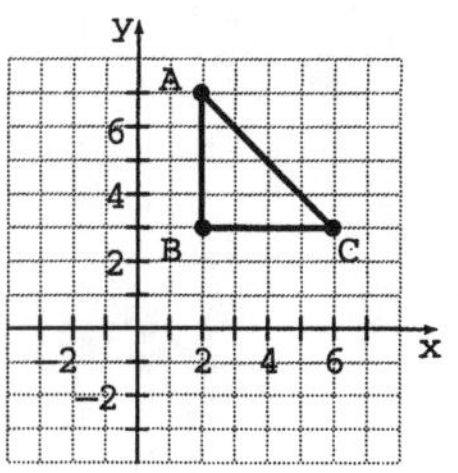

b. Area $=\frac{1}{2}bh=\frac{1}{2}(4)(4)=8$

The area is 8 square units.

67. a. The average length of the golf course at the majors in 1980 was about 6875 yards.

b. The average length of the golf course at the majors in 2005 was about 7300 yards.

c. The average length of the golf course at the majors was greater than 7000 yards for the years 1990, 2000, and 2005.

d. No, the average length of the golf courses in the majors from 1995 to 2005 does not appear to be linear. The increase in length is not the same from 1995 to 2000 as it is from 2000 to 2005. That is, the increase in the length is not constant.

69.

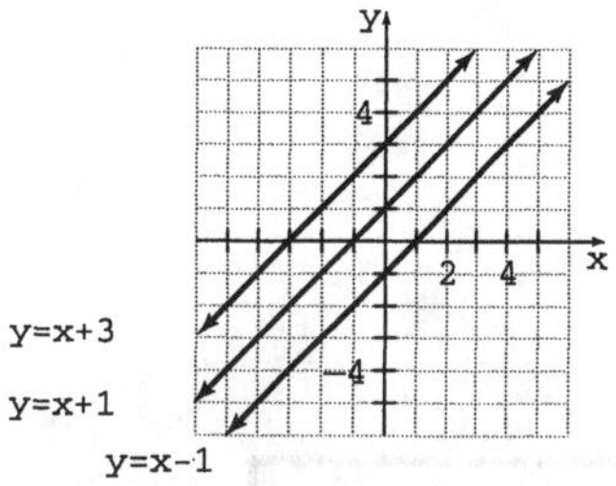

a. Each graph crosses the y-axis at the point corresponding to the constant term in the graph's equation.

b. Yes, all the equations seem to have the same slant or slope.

71.

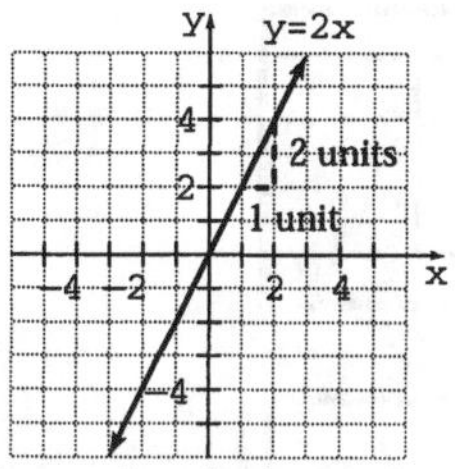

For each unit change in x, y changes 2 units. Thus, the rate of change of y with respect to x is 2.

73.

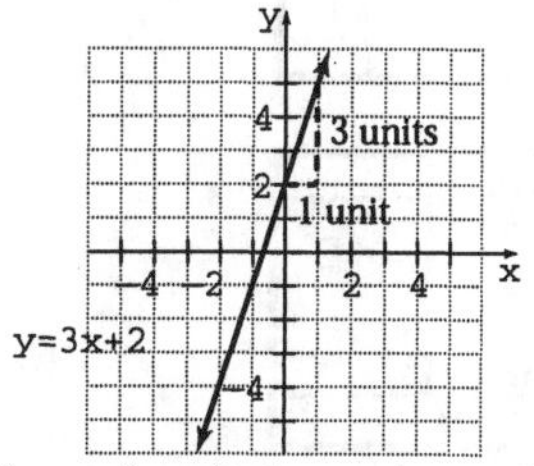

For each unit change in x, y changes 3 units. Thus, the rate of change of y with respect to x is 3.

75. Answers may vary. One possible answer is the points (4, –3) and (5, 1).

Starting at (3, –7): For a unit change, *x* changes from 3 to 3 + 1 = 4. At the same time, *y* changes from –7 to –7 + 4 = –3. So, (4, –3) is a solution to the equation.

Starting at (4, –3): For a unit change, *x* changes from 4 to 4 + 1 = 5. At the same time, *y* changes from –3 to –3 + 4 = 1. So, (5, 1) is a solution to the equation.

77. c

79. a

81. d

83. b

85. b

87. d

89. b

91. d

93.

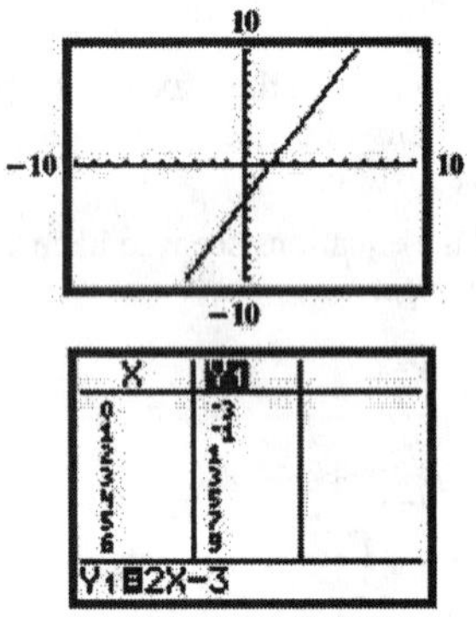

95.

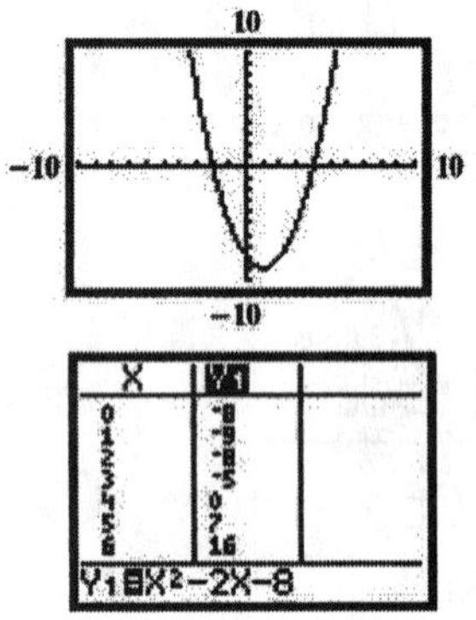

97.

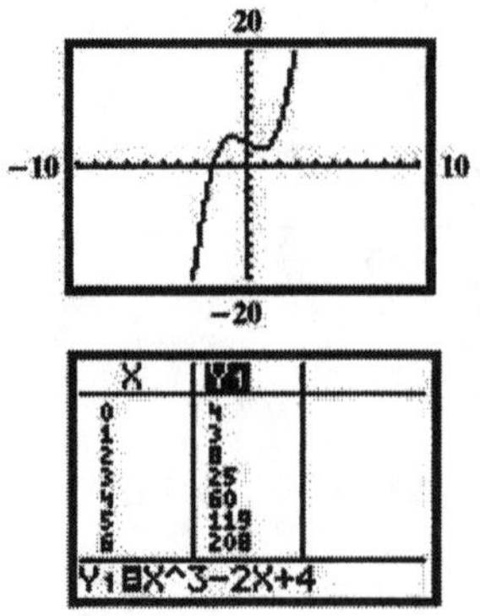

99.

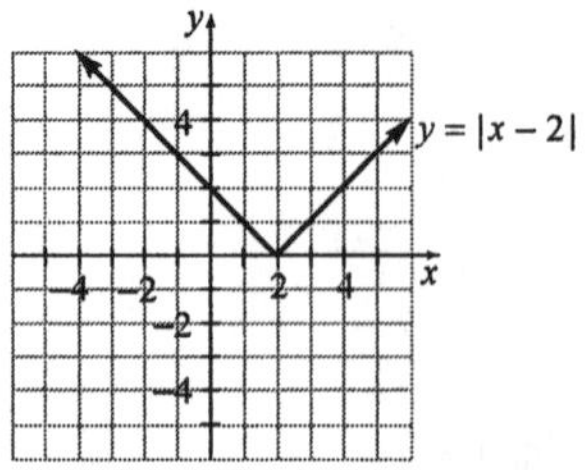

103.
$$\frac{-b+\sqrt{b^2-4ac}}{2a}=\frac{-(7)+\sqrt{(7)^2-4(2)(-15)}}{2(2)}$$
$$=\frac{-7+\sqrt{169}}{4}$$
$$=\frac{-7+13}{4}=\frac{3}{2}$$

104. Let *x* be the number of miles driven.
The cost for renting from Hertz is
$y_1 = 30 + 0.14x$.
The cost for renting from National Automobile Rental Agency is $y_2 = 16 + 0.24x$. The costs are equal when
$$y_1 = y_2$$
$$30 + 0.14x = 16 + 0.24x$$
$$14 = 0.10x$$
$$x = \frac{14}{0.10}$$
$$x = 140$$
You would have to drive 140 miles to make the costs equal.

105.
$$-1 \le \frac{4-3x}{2} < 5$$
$$2(-1) \le 4-3x < 2(5)$$
$$-2 \le 4-3x < 10$$
$$-2-4 \le -3x < 10-4$$
$$-6 \le -3x < 6$$
$$\frac{-6}{-3} \ge x > \frac{6}{-3}$$
$$2 \ge x > -2$$
$$-2 < x \le 2$$
The solution set is $\{x|-2 < x \le 2\}$.

106. $|3x+2| > 7$
$$3x+2 < -7 \text{ or } 3x+2 > 7$$
$$3x < -9 \qquad 3x > 5$$
$$x < -3 \qquad x > \frac{5}{3}$$
The solution set is $\left\{x \middle| x < -3 \text{ or } x > \frac{5}{3}\right\}$.

Exercise Set 3.2

1. A function is a correspondence between a first set of elements, the domain, and a second set of elements, the range, such that each element of the domain corresponds to exactly one element in the range.

3. Yes, all functions are also relations. A function is a set of ordered pairs so it is a relation.

5. If each vertical line drawn through any part of the graph intersects the graph in at most one point, the graph represents a function.

7. The range is the set of values for the dependent variable.

9. Domain: $\{x|\, x \ne 0\}$
   The denominator cannot be zero.
   Range: $\{y|\, y \ne 0\}$
   All values of y except $y = 0$ are represented in the function.

11. Domain: $\mathbb{R}$ or $(-\infty, \infty)$
    There are no restrictions on values of x that can be used.
    Range: $\{y|y \ge 0\}$
    The absolute value of any number is never negative.

13. If y depends on x, then x is the independent variable.

15. **a.** Yes, the relation is a function.
    b. Domain: {3, 5, 11}, Range: {6, 10, 22}

17. **a.** Yes, the relation is a function.
    b. Domain: {Cameron, Tyrone, Vishnu}, Range:{3, 6}

19. **a.** No, the relation is not a function.
    b. Domain: {1990, 2001, 2002}, Range: {20, 34, 37}

21. **a.** The relation is a function.
    b. Domain: {1, 2, 3, 4, 5}, Range: {1, 2, 3, 4, 5}

23. **a.** The relation is a function.
    b. Domain: {1, 2, 3, 4, 5, 7}, Range: {–1, 0, 2, 4, 9}

25. **a.** The relation is not a function.
    b. Domain: {1, 2, 3}, Range: {1, 2, 4, 5, 6}

27. **a.** The relation is not a function.
    b. Domain: {0, 1, 2}, Range: {–7, –1, 2, 3}

29. **a.** The graph represents a function.
    b. Domain: $\mathbb{R}$, Range: $\mathbb{R}$
    c. $x = 2$

31. **a.** The graph does not represent a function.
    b. Domain: $\{x|\, 0 \le x \le 2\}$, Range: $\{y|-3 \le y \le 3\}$
    c. $x \approx 1.5$

33. **a.** The graph represents a function.
    b. Domain: $\mathbb{R}$, Range: $\{y|y \ge 0\}$
    c. $x = -3$ or $x = -1$

35. **a.** The graph represents a function.
    b. Domain: {–1, 0, 1, 2, 3}, Range: {–1, 0, 1, 2, 3}
    c. $x = 2$

37. a. The graph does not represent a function.

b. Domain: $\{x|x \geq 2\}$, Range: $\mathbb{R}$

c. $x = 3$

39. a. The graph represents a function.

b. Domain: $\{x|-2 \leq x \leq 2\}$,

Range: $\{y|-1 \leq y \leq 2\}$

c. $x = -2$ or $x = 2$

41. a. $f(2) = -2(2) + 7 = -4 + 7 = 3$

b. $f(-3) = -2(-3) + 7 = 6 + 7 = 13$

43. a. $h(0) = (0)^2 - (0) - 6 = -6$

b. $h(-1) = (-1)^2 - (-1) - 6 = 1 + 1 - 6 = -4$

45. a.
$$\begin{aligned} r(1) &= -(1)^3 - 2(1)^2 + (1) + 4 \\ &= -(1) - 2(1) + (1) + 4 \\ &= -1 - 2 + 1 + 4 \\ &= 2 \end{aligned}$$

b.
$$\begin{aligned} r(-2) &= -(-2)^3 - 2(-2)^2 + (-2) + 4 \\ &= -(-8) - 2(4) + (-2) + 4 \\ &= 8 - 8 - 2 + 4 \\ &= 2 \end{aligned}$$

47. a. $h(6) = |5 - 2(6)| = |5 - 12| = |-7| = 7$

b. $h\left(\frac{5}{2}\right) = \left|5 - 2\left(\frac{5}{2}\right)\right| = |5 - 5| = 0$

49. a. $s(-3) = \sqrt{(-3) + 3} = \sqrt{0} = 0$

b. $s(6) = \sqrt{(6) + 3} = \sqrt{9} = 3$

51. a. $g(0) = \dfrac{(0)^3 - 2}{(0) - 2} = \dfrac{-2}{-2} = 1$

b. $g(0) = \dfrac{(2)^3 - 2}{2 - 2} = \dfrac{8 - 2}{0} = \dfrac{6}{0}$ undefined

53. a. $A(4) = 6(4) = 24$
The area is 24 square feet.

b. $A(6.5) = 6(6.5) = 39$
The area is 39 square feet.

55. a. $A(r) = \pi r^2$

b. $A(12) = \pi(12)^2 = 144\pi \approx 452.4$
The area is about 452.4 square yards.

57. a. $C(F) = \frac{5}{9}(F - 32)$

b. $C(-31) = \frac{5}{9}(-31 - 32) = \frac{5}{9}(-63) = -35$
The Celsius temperature that corresponds to $-31°F$ is $-35°C$.

59. a.
$$\begin{aligned} T(3) &= -0.03(3)^2 + 1.5(3) + 14 \\ &= -0.27 + 4.5 + 14 \\ &= 18.23 \end{aligned}$$
The temperature is 18.23°C.

b.
$$\begin{aligned} T(12) &= -0.03(12)^2 + 1.5(12) + 14 \\ &= -4.32 + 18 + 14 \\ &= 27.68 \end{aligned}$$
The temperature is 27.68°C.

61. a.
$$\begin{aligned} T(4) &= -0.02(4)^2 - 0.34(4) + 80 \\ &= -0.32 - 1.36 + 80 \\ &= 78.32 \end{aligned}$$
The temperature is 78.32°.

b.
$$\begin{aligned} T(12) &= -0.02(12)^2 - 0.34(12) + 80 \\ &= -2.88 - 4.08 + 80 \\ &= 73.04 \end{aligned}$$
The temperature is 73.04°.

63. a.
$$\begin{aligned} T(6) &= \frac{1}{3}(6)^3 + \frac{1}{2}(6)^2 + \frac{1}{6}(6) \\ &= 72 + 18 + 1 \\ &= 91 \end{aligned}$$
If the base is 6 by 6 oranges, then there will be 91 oranges in the pyramid.

b.
$$\begin{aligned} T(8) &= \frac{1}{3}(8)^3 + \frac{1}{2}(8)^2 + \frac{1}{6}(8) \\ &= \frac{512}{3} + 32 + \frac{4}{3} \\ &= 204 \end{aligned}$$
If the base is 8 by 8 oranges, then there will be 204 oranges in the pyramid.

65. Answers will vary. One possible interpretation: The person warms up slowly, possibly by walking for 5 minutes, then begins jogging slowly over a period of 5 minutes. For the next 15 minutes, the person jogs at a steady pace. For the next 5 minutes, he walks slowly and his heart rate decreases to his normal resting heart rate. The rate stays the same for the next 5 minutes.

67. Answers will vary. One possible interpretation: The man walks on level ground, about 30 feet above sea level, for 5 minutes. For the next 5 minutes he walks uphill to 45 feet above sea level. For 5 minutes he walks on level ground then walks quickly downhill for 3 minutes to an elevation of 20 feet above sea level. For 7 minutes he walks on level ground. Then he walks quickly uphill.

69. Answers may vary. One possible interpretation: A woman drives in stop-and-go traffic for five minutes. Then she drives on the highway for fifteen minutes, gets off onto a country road for a few minutes, stops for a couple of minutes, and returns to stop-and-go traffic.

71. **a.** Yes, both graphs pass the vertical line test.

b. The independent variable is the year.

c. $f(2005) = \$218,600$

d. $g(2005) = \$865,000$

e.
$$\text{percent increase } = \frac{218,600 - 89,300}{89,300} \approx 1.448 \approx 144.8\%$$

The percent increase from 1988 through 2005 was about 144.8%.

73. **a.** Yes, both graphs pass the vertical line test.

b. $f(1998) \approx 6.0$ million viewers.

c. $g(1998) \approx 4.4$ million viewers.

d. Yes, both graphs are approximately linear from 1998 to 2005. The number of Today Show viewers is remained relatively constant, while the number of Good Morning America viewers increased at a relatively constant rate.

e. The two shows will have the same number of viewers around the year 2006 or 2007.

75. **a.**

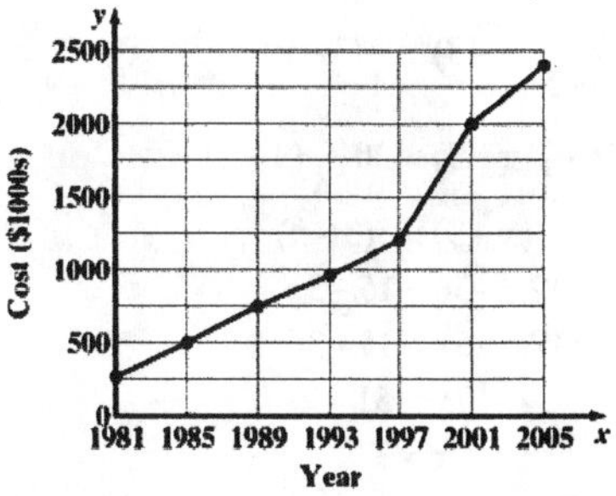

b. No, the points don't lie on a straight line.

c. The cost of a 30-second commercial in 2004 was about $2,300,000.

77. **a.**

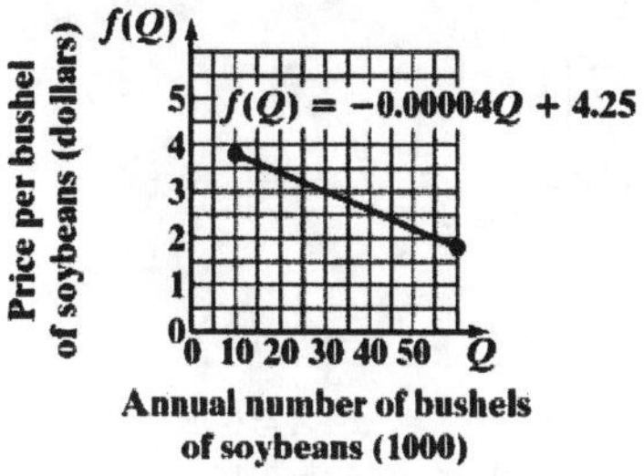

b.
$$\begin{aligned} f(40,000) &= -0.00004(40,000) + 4.25 \\ &= -1.6 + 4.25 \\ &= 2.65 \end{aligned}$$

The cost of a bushel of soybeans if 40,000 bushels are produced is approximately $2.65 per bushel.

80.
$$\begin{aligned} 3x - 2 &= \frac{1}{3}(3x - 3) \\ 3x - 2 &= \frac{1}{3}(3x) - \frac{1}{3}(3) \\ 3x - 2 &= x - 1 \\ 3x - x &= -1 + 2 \\ 2x &= 1 \\ x &= \frac{1}{2} \end{aligned}$$

The solution set is $\left\{\frac{1}{2}\right\}$.

81.
$$\begin{aligned} E &= a_1 p_1 + a_2 p_2 + a_3 p_3 \\ E - a_1 p_1 - a_3 p_3 &= a_2 p_2 \\ p_2 &= \frac{E - a_1 p_1 - a_3 p_3}{a_2} \end{aligned}$$

82. $\frac{3}{5}(x-3) > \frac{1}{4}(3-x)$

$$20 \cdot \frac{3}{5}(x-3) > 20 \cdot \frac{1}{4}(3-x)$$
$$12(x-3) > 5(3-x)$$
$$12x - 36 > 15 - 5x$$
$$12x + 5x > 15 + 36$$
$$17x > 51$$
$$x > 3$$

a. 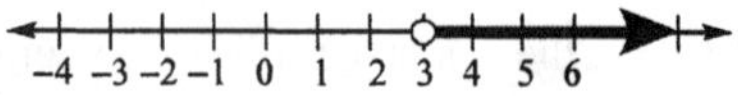

b. $(3, \infty)$

c. $\{x | x > 3\}$

83. $\left|\frac{x-4}{3}\right| + 9 = 11$

$$\left|\frac{x-4}{3}\right| = 2$$

$\frac{x-4}{3} = -2$ or $\frac{x-4}{3} = 2$

$x - 4 = -6$ $\quad$ $x - 4 = 6$

$x = -2$ $\quad$ $x = 10$

The solution set is $\{-2, 10\}$.

Exercise Set 3.3

1. The standard form of a linear equation is $ax + by = c$, where a, b, and c are real numbers, and a and b are not both 0.

3. To find the x-intercept, set $y = 0$ and solve for x. To find the y-intercept, set $x = 0$ and solve for y.

5. The graph of $x = a$, for any real number a, will be a vertical line.

7. The graph of $f(x) = b$, for any real number b, will be a horizontal line.

9. To solve an equation in one variable, graph both sides of the equation. The solution is the x-coordinate of the point of intersection.

11. $y = -2x + 5$

$$2x + y = 5$$

13. $3(x-2) = 4(y-5)$

$$3x - 6 = 4y - 20$$
$$3x - 4y = -14$$

15. $y = -2x + 1$

For the y-intercept, set $x = 0$ and solve for y:
$$y = -2x + 1$$
$$y = -2(0) + 1$$
$$y = 0 + 1$$
$$y = 1$$
The y-intercept is at (0, 4).
For the x-intercept, set $y = 0$ and solve for x:
$$y = -2x + 1$$
$$0 = -2x + 1$$
$$2x = 1$$
$$x = \frac{1}{2}$$
The x-intercept is at $\left(\frac{1}{2}, 0\right)$.

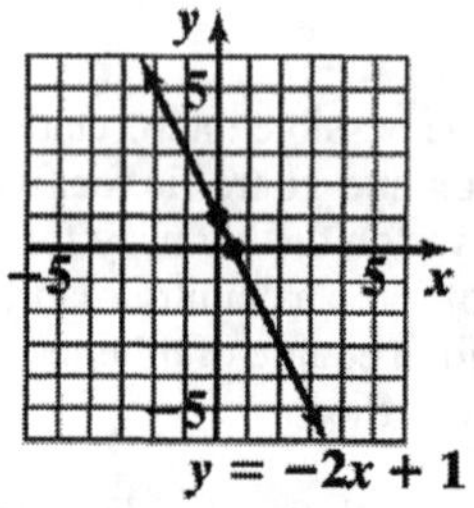

17. $f(x) = 2x + 3$

For the y-intercept, set $x = 0$ and solve for y:
$$f(x) = 2x + 3$$
$$f(0) = 2(0) + 3$$
$$f(0) = 0 + 3$$
$$f(0) = 3$$
The y-intercept is at (0, 3).
For the x-intercept, set $f(x) = 0$ and solve for x:
$$f(x) = 2x + 3$$
$$0 = 2x + 3$$
$$-3 = 2x$$
$$-\frac{3}{2} = x$$
The x-intercept is at $\left(-\frac{3}{2}, 0\right)$.

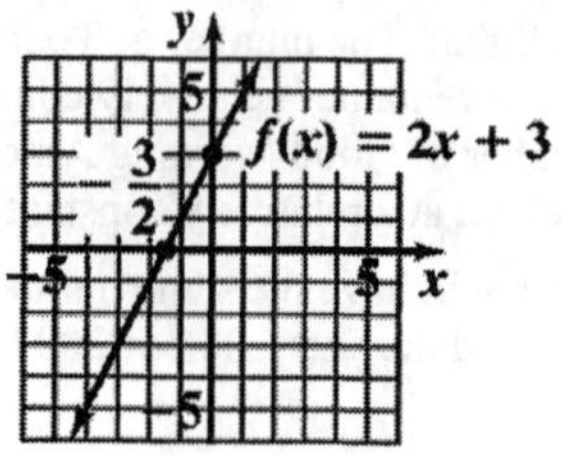

19. $2y = 4x + 6$

For the y-intercept, set $x = 0$ and solve for y:

$2y = 4x + 6$
$2y = 4(0) + 6$
$2y = 0 + 6$
$y = 3$

The y-intercept is at (0, 3).

For the x-intercept, set $y = 0$ and solve for y:

$$2y = 4x+6$$
$$2(0) = 4x+6$$
$$0 = 4x+6$$
$$-6 = 4x$$
$$\frac{-6}{4} = x$$
$$-\frac{3}{2} = x$$

The x-intercept is at $\left(-\frac{3}{2}, 0\right)$.

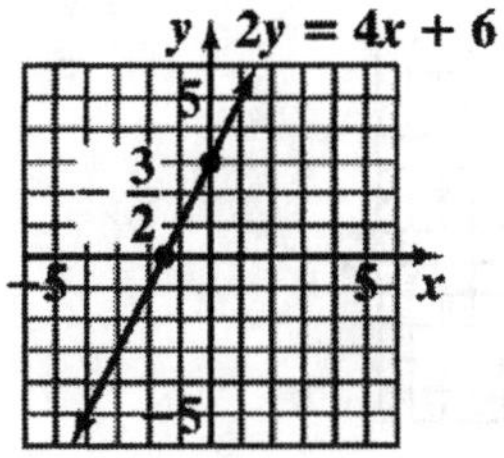

21. $\frac{4}{3}x = y-3$

For the y-intercept, set $x = 0$ and solve for y:

$$\frac{4}{3}(0) = y-3$$
$$0 = y-3$$
$$3 = y$$

The y-intercept is at (0, 3).

For the x-intercept, set $y = 0$ and solve for x:

$$\frac{4}{3}x = (0)-3$$
$$\frac{4}{3}x = -3$$
$$3\left(\frac{4}{3}x\right) = 3(-3)$$
$$4x = -9$$
$$x = -\frac{9}{4}$$

The x-intercept is at $\left(-\frac{9}{4}, 0\right)$.

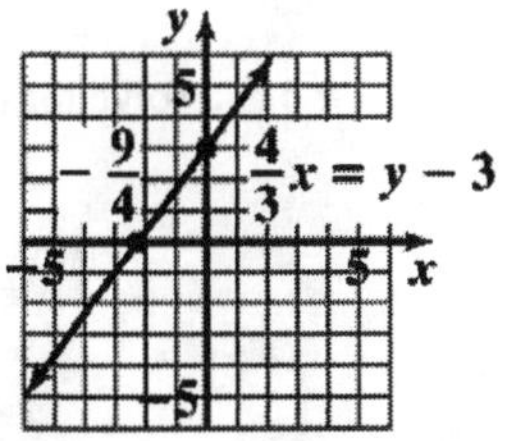

23. $15x+30y = 60$

For the y-intercept, set $x = 0$ and solve for y:

$$15(0)+30y = 60$$
$$0+30y = 60$$
$$30y = 60$$
$$y = 2$$

The y-intercept is at (0, 2).

For the x-intercept, set $y = 0$ and solve for x:

$$15x+30(0) = 60$$
$$15x = 60$$
$$x = 4$$

The x-intercept is at (4, 0).

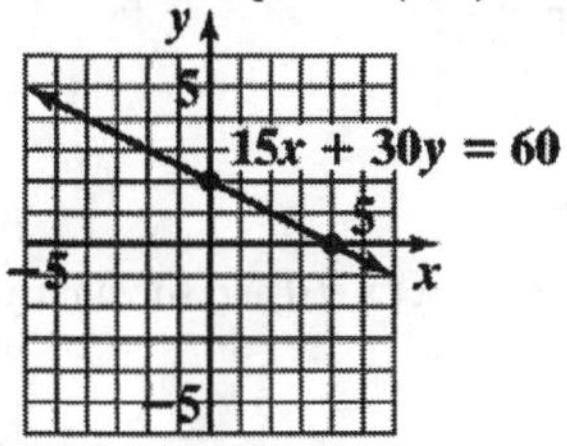

25. $0.25x+0.50y = 1.00$

For the y-intercept, set $x = 0$ and solve for y:

$$0.25(0)+0.50y = 1.00$$
$$0+0.50y = 1.00$$
$$0.50y = 1.00$$
$$y = 2$$

The y-intercept is at (0, 2).

For the x-intercept, set $y = 0$ and solve for x:

$$0.25x+0.50(0) = 1.00$$
$$0.25x+0 = 1.00$$
$$0.25x = 1.00$$
$$x = 4$$

The x-intercept is (4, 0).

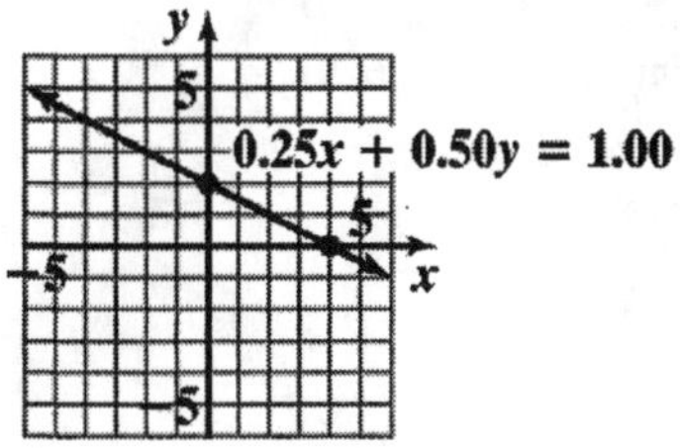

27. $120x - 360y = 720$

For the y-intercept, set $x = 0$ and solve for y:

$$120(0) - 360y = 720$$
$$0 - 360y = 720$$
$$-360y = 720$$
$$y = -2$$

The y-intercept is at $(0, -2)$.
For the x-intercept, set $y = 0$ and solve for x:

$$120x - 360(0) = 720$$
$$120x - 0 = 720$$
$$120x = 720$$
$$x = 6$$

The x-intercept is 6 and the point is $(6, 0)$.

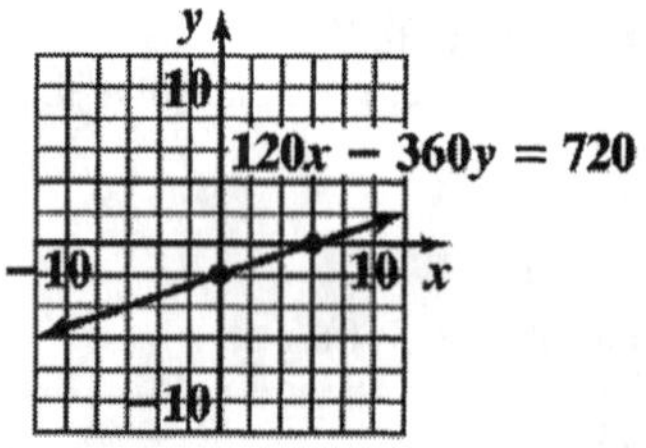

29. $\frac{1}{3}x + \frac{1}{4}y = 12$

For the y-intercept, set $x = 0$ and solve for y:

$$\frac{1}{3}(0) + \frac{1}{4}y = 12$$
$$0 + \frac{1}{4}y = 12$$
$$4\left(\frac{1}{4}x\right) = 4(12)$$
$$x = 48$$

The y-intercept is at $(0, 48)$.
For the x-intercept, set $y = 0$ and solve for x:

$$\frac{1}{3}x + \frac{1}{4}(0) = 12$$
$$\frac{1}{3}x + 0 = 12$$
$$3\left(\frac{1}{3}x\right) = 3(12)$$
$$x = 36$$

The x-intercept is at $(36, 0)$.

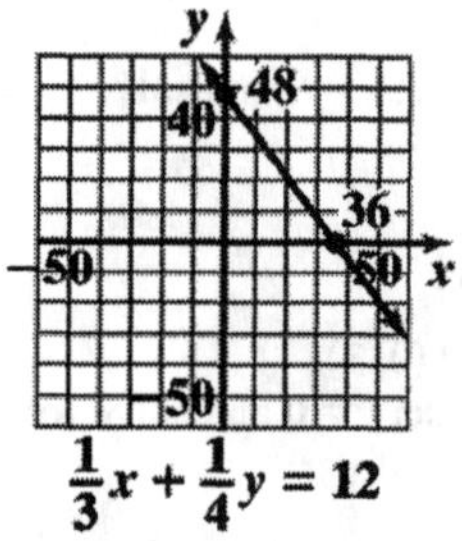

31. The equation is $y = -2x$. To graph, plot a few points.

x	Calculation	y
0	$-2(0)$	0
1	$-2\ (1)$	-2
-1	$-2\ (-1)$	2

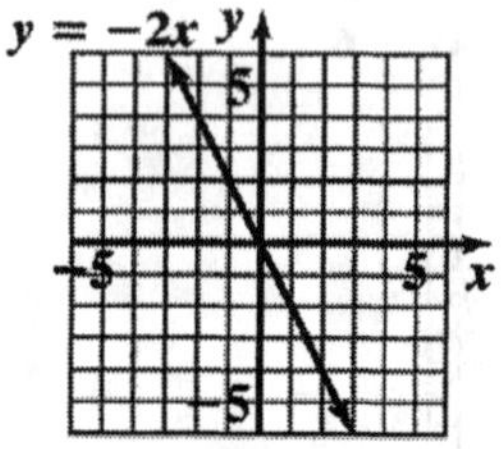

33. The equation is $f(x) = \frac{1}{3}x$. To graph, plot a few points. Pick multiples of 3 for easier calculation.

x	Calculation	$f(x)$
0	$\frac{1}{3}\ (0)$	0
3	$\frac{1}{3}\ (3)$	1
-3	$\frac{1}{3}\ (-3)$	-1

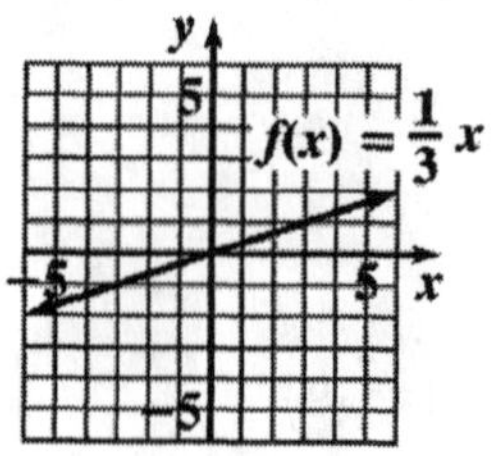

35. Solve for y and make a table of values.

$2x+4y=0$

$4y=-2x$

$y=\frac{-2}{4}x$ or $y=-\frac{1}{2}x$

x	Calculation	y
0	$-\frac{1}{2}(0)$	0
2	$-\frac{1}{2}(2)$	-1
-2	$-\frac{1}{2}(-2)$	1

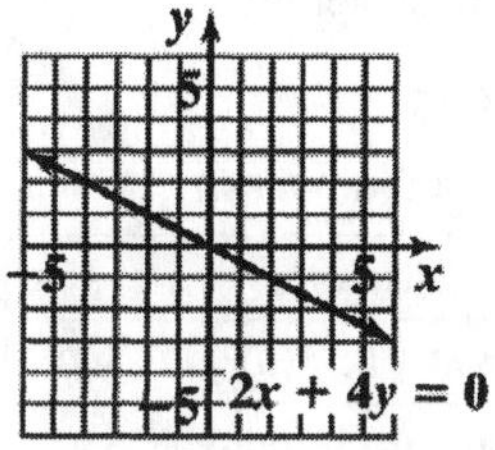

37. Solve for y and make a table of values.

$6x-9y=0$

$-9y=-6x$

$y=\frac{-6}{-9}x$ or $y=\frac{2}{3}x$

x	Calculation	y
0	$\frac{2}{3}(0)$	0
3	$\frac{2}{3}(3)$	2
-3	$\frac{2}{3}(-3)$	-2

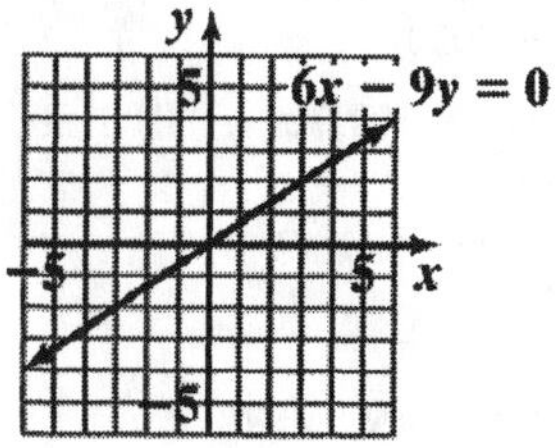

39. $y=4$

This is a horizontal line 4 units above the x-axis.

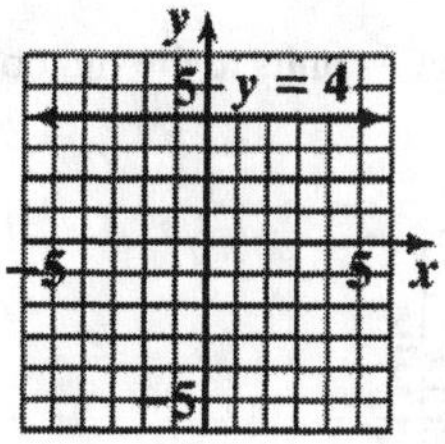

41. $x=-4$

This is a vertical line 4 units to the left of the y-axis.

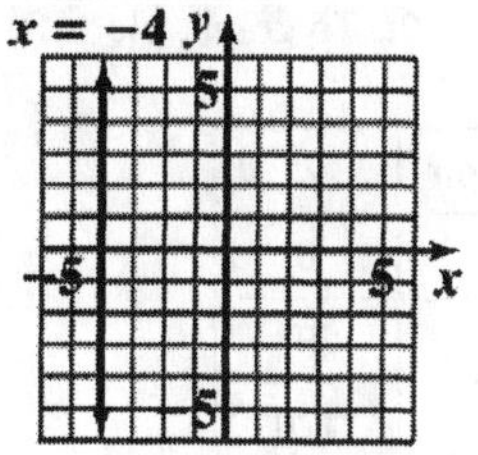

43. $y=-1.5$

This is a horizontal line 1.5 units below the x-axis.

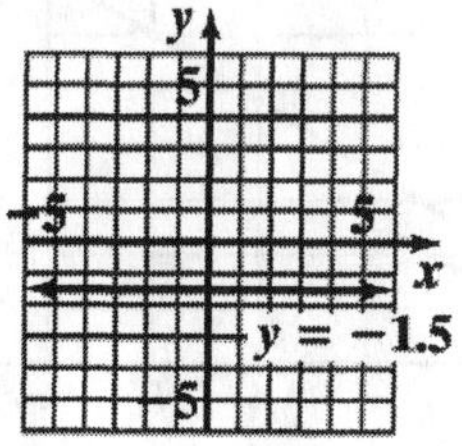

45. $x=0$

This is a vertical line corresponding to the y-axis.

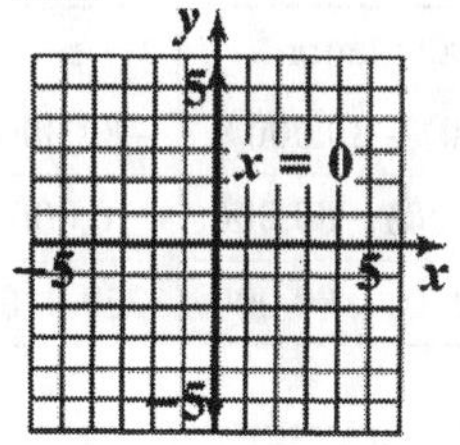

47. $x = \frac{5}{2}$

This is a vertical line $\frac{5}{2}$ units to the right of the *y*-axis.

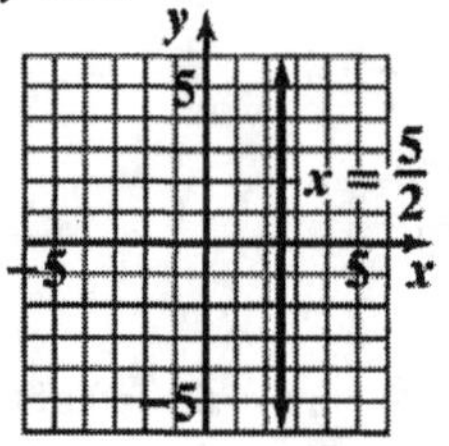

49. The equation is $d = 30t$. To graph, plot a few points.

t	Calculation	d
0	30(0)	0
1	30(1)	30
4	30(4)	120

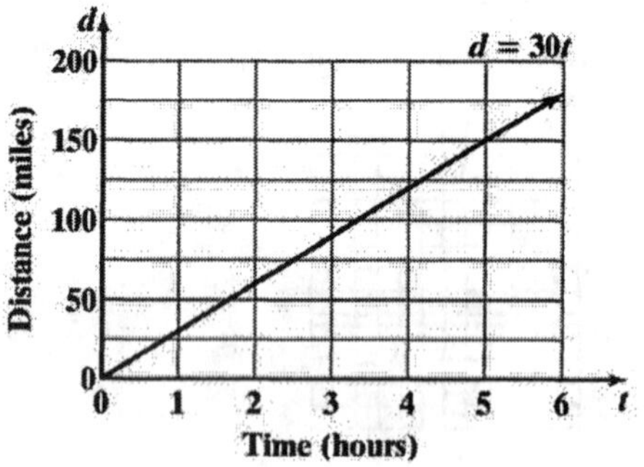

51. a. $p = 60x - 80{,}000$. To graph, plot a few points.

x	Calculation	p
0	60(0) – 80,000	–80,000
2500	60(2500) – 80,000	70,000
5000	60(5000) – 80,000	220,000

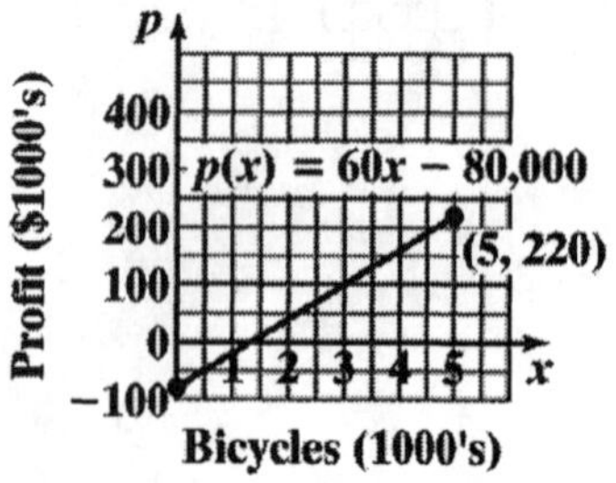

b. To break even, the profit would be zero. That is, set $p = 0$ and solve for x:

$$0 = 60x - 80{,}000$$
$$80{,}000 = 60x$$
$$x = \frac{-80{,}000}{-60} \approx 1333$$

The company must sell about 1,300 bicycles to break even.

c. To earn a profit of \$150,000, set $p = 150{,}000$ and solve for x:

$$150{,}000 = 60x - 80{,}000$$
$$230{,}000 = 60x$$
$$x = \frac{230{,}000}{60} \approx 3833$$

The company must sell about 3,800 bicycles to make a \$150,000 profit.

53. a. $s(x) = 500 + 0.15x$

b. To graph, plot a few points.

x	Calculation	s
0	500 + 0.15(0)	500
1000	500 + 0.15(1000)	650
5000	500 + 0.15(5000)	1250

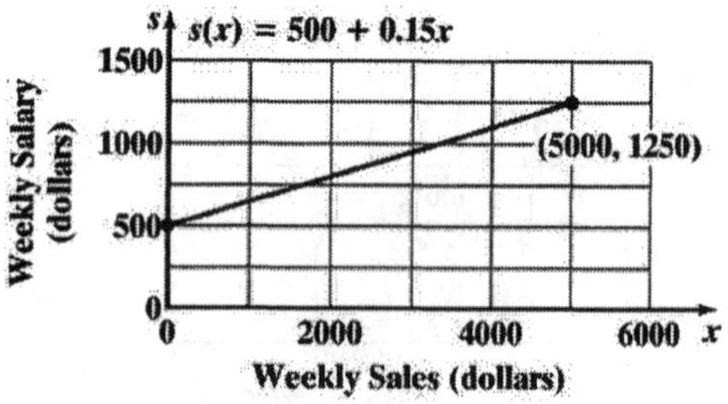

c. For weekly sales of \$3000,

$$s(3000) = 500 + 0.15(3000)$$
$$= 500 + 450$$
$$= 950$$

Her salary is \$950.

d. For a salary of \$1100, set $s = 1100$ and solve for x.

$$1100 = 500 + 0.15x$$
$$600 = 0.15x$$
$$4000 = x$$

Her weekly sales are \$4000.

55. a. There is only one y-value for each x-value. (It passes the vertical line test.)

b. The independent variable is length. The dependent variable is weight.

c. Yes, the graph of weight versus length is approximately linear.

d. The weight of the average girl who is 85 centimeters long is 11.5 kilograms.

e. The average length of a girl with a weight of 7 kilograms is 65 centimeters.

f. For a girl 95 centimeters long, the weights 12.0 – 15.5 kilograms are considered normal.

g. As the lengths increase, the normal range of weights increases. Yes, this is expected. As the girl grows, it is reasonable that her weight would increase with her length.

57. The x- and y-intercepts of a graph will be the same when the graph goes through the origin.

59. Answers may vary. One possible answer is, $f(x) = 4$ is a function whose graph has no x-intercept but has a y-intercept of (0, 4).

61. The x- and y-intercepts will both be (0, 0).

63. a.

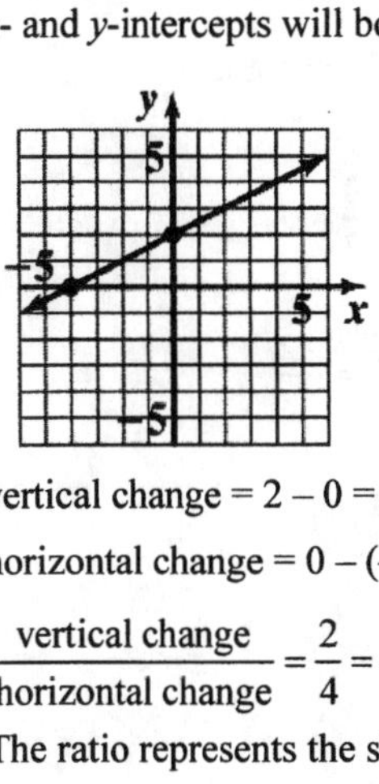

b. vertical change = 2 – 0 = 2

c. horizontal change = 0 – (–4) = 4

d. $\dfrac{\text{vertical change}}{\text{horizontal change}} = \dfrac{2}{4} = \dfrac{1}{2}$

The ratio represents the slope of the line.

65. Graph $Y_1 = 2x + 5$ and $Y_2 = 8x - 1$, and find the intersection.

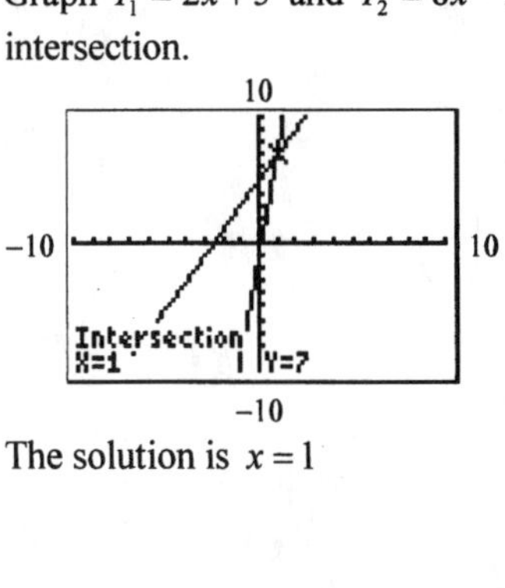

The solution is $x = 1$

67. Graph $Y_1 = 0.3(x + 5)$ and $Y_2 = -0.6(x + 2)$, and find the intersection.

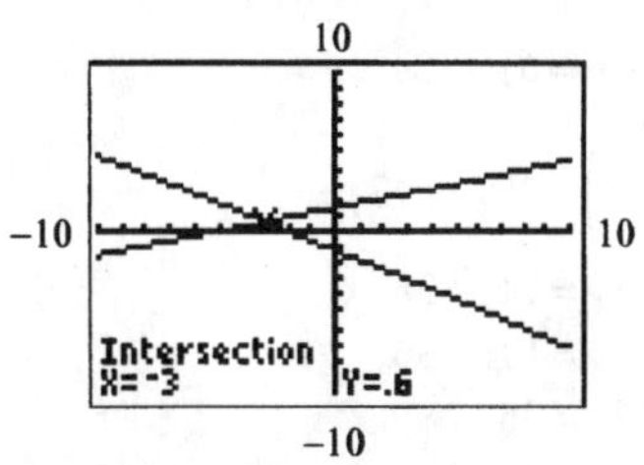

The solution is $x = -3$.

69. Let $Y_1 = 2(x + 3.2)$.

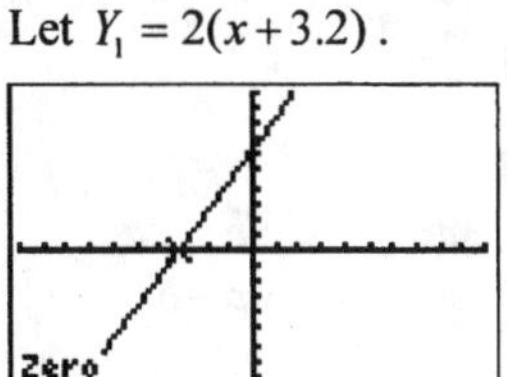

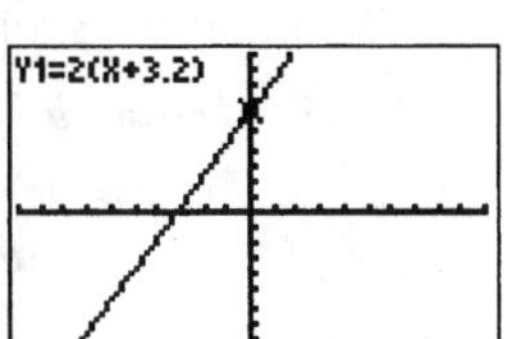

Window: –10,10,1,–10,10,1

The x-intercept is (–3.2, 0), and the y-intercept is (0, 6.4).

71. To use the graphing calculator, we must rewrite the equation in the form $y = f(x)$.

$$-4x - 3.2y = 8$$
$$-3.2y = 4x + 8$$
$$y = \frac{4x + 8}{-3.2}$$
$$y = -1.5625x - 2.5$$

Let $Y_1 = -1.5625x - 2.5$.

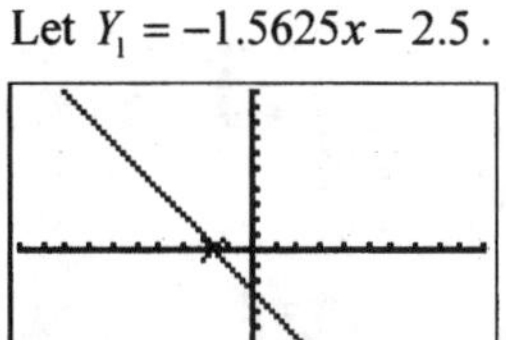

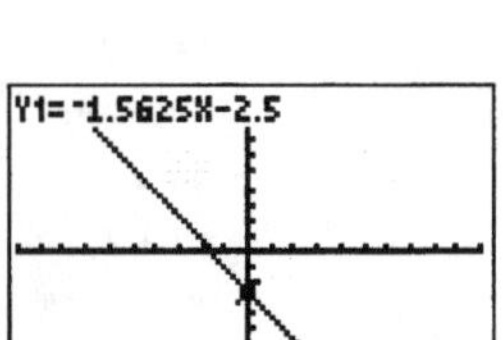

Window: –10,10,1,–10,10,1

The x-intercept is (–1.6, 0), and the y-intercept is (0, –2.5).

73.
$$\begin{aligned} 4\{2-3[(1-4)-5]\}-8 &= 4\{2-3[(-3)-5]\}-8 \\ &= 4\{2-3[-8]\}-8 \\ &= 4\{2+24\}-8 \\ &= 4\{26\}-8 \\ &= 104-8 \\ &= 96 \end{aligned}$$

74.
$$\frac{1}{3}y-3y=6(y+2)$$
$$\frac{1}{3}y-3y=6y+12$$
$$3\left(\frac{1}{3}y-3y\right)=3(6y+12)$$
$$y-9y=18y+36$$
$$-8y=18y+36$$
$$-26y=36$$
$$y=\frac{36}{-26}=-\frac{18}{13}$$

75. a. Answers will vary.

b. $|x-a|=b$

$x-a=-b$ or $x-a=b$

$x=a-b$ or $x=a+b$

76. a. Answers will vary.

b. $|x-a|<b$

$-b<x-a<b$

$a-b<x<a+b$

77. a. Answers will vary.

b. $|x-a|>b$

$x-a<-b$ or $x-a>b$

$x<a-b$ or $x>a+b$

78. $|x-4|=|2x-2|$

$x-4=2x-2$ or $x-4=-(2x-2)$

$x-2x=-2+4$ $x-4=-2x+2$

$-x=2$ $x+2x=2+4$

$x=-2$ $3x=6$

$x=-2$ $x=2$

The solution set is {–2, 2}.

Exercise Set 3.4

1. Select two points on the line. Then find $\frac{\Delta y}{\Delta x}$, the ratio of the vertical change (or rise) to the horizontal change (or run) between the two points.

3. The line rises going from left to right.

5. The horizontal change on a vertical line is zero, and we cannot divide by zero. So the slope is undefined.

7. To get the slope-intercept form from the standard form, solve for y in terms of x.

9. a. If a graph is translated down 6 units, it is lowered or moved down 6 units.

b. If the y-intercept is $(0,-3)$ and the graph is translated down 5 units, the new y-intercept will be at $y=-3-5=-8$. The new y-intercept is (0, –8).

11. When the slope is given as a rate of change it means the change in y for a unit change in x.

13. $m=\frac{11-5}{0-3}=\frac{6}{-3}=-2$

15. $m=\frac{4-2}{1-5}=\frac{2}{-4}=-\frac{1}{2}$

17. $m=\frac{1-5}{1-(-3)}=\frac{-4}{4}=-1$

19. $m=\frac{-6-2}{4-4}=\frac{-8}{0}$, undefined

21. $m=\frac{4-4}{-1-(-3)}=\frac{0}{2}=0$

23. $m=\frac{-3-3}{9-0}=\frac{-6}{9}=-\frac{2}{3}$

25.
$$\frac{b-2}{4-3}=1$$
$$\frac{b-2}{1}=1$$
$$b-2=1$$
$$b=3$$

27.
$$\frac{k-0}{1-5}=\frac{1}{2}$$
$$\frac{k}{-4}=\frac{1}{2}$$
$$-4\left(\frac{k}{-4}\right)=-4\left(\frac{1}{2}\right)$$
$$k=-2$$

29.
$$\frac{-4-2}{3-x}=2$$
$$\frac{-6}{3-x}=2$$
$$(3-x)\left(\frac{-6}{3-x}\right)=(3-x)(2)$$
$$-6=6-2x$$
$$-12=-2x$$
$$6=x$$

31.
$$\frac{2-(-4)}{r-12}=-\frac{1}{2}$$
$$\frac{6}{r-12}=-\frac{1}{2}$$
$$2(r-12)\left(\frac{6}{r-12}\right)=2(r-12)\left(-\frac{1}{2}\right)$$
$$2\cdot 6=(r-12)(-1)$$
$$12=-r+12$$
$$r+12=12$$
$$r=0$$

33. The slope is negative and y decreases 6 units when x increases 2 units. Thus, $m = -\frac{6}{2} = -3$. The line crosses the y-axis at 0 so $b = 0$. Hence, $m = -3$ and $b = 0$, and the equation of the line is $y = -3x + 0$ or $y = -3x$.

35. The slope is negative and y decreases 1 unit when x increases 3 units. Thus, $m = -\frac{1}{3}$. The line crosses the y-axis at 2 so $b = 2$. Hence, $m=-\frac{1}{3}$ and $b = 2$, and the equation of the line is $y=-\frac{1}{3}x+2$.

37. The slope is undefined since the change in x is 0. The equation of this vertical line is $x = -2$.

39. The line is horizontal so $m = 0$. The line crosses the y-axis at 3 so $b = 3$. Hence, $m = 0$ and $b = 3$, and the equation of the line is $y = 3$.

41. The slope is negative and y decreases 15 units when x increases 10 units. Thus, $m=-\frac{15}{10}=-\frac{3}{2}$. The line crosses the y-axis at 15 so $b = 15$. Hence, $m=-\frac{3}{2}$ and $b = 15$, and the equation of the line is $y=-\frac{3}{2}x+15$.

43. The equation $y = -x + 2$ is given in slope-intercept form . The slope is -1 and the y-intercept is (0, 2).

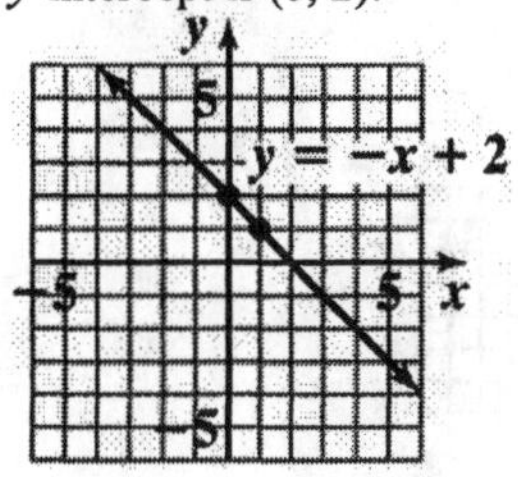

45.
$$5x+15y=30$$
$$15y=-5x+30$$
$$y=\frac{-5x+30}{15}$$
$$y=-\frac{1}{3}x+2$$

The slope is $-\frac{1}{3}$ and the y-intercept is (0, 2).

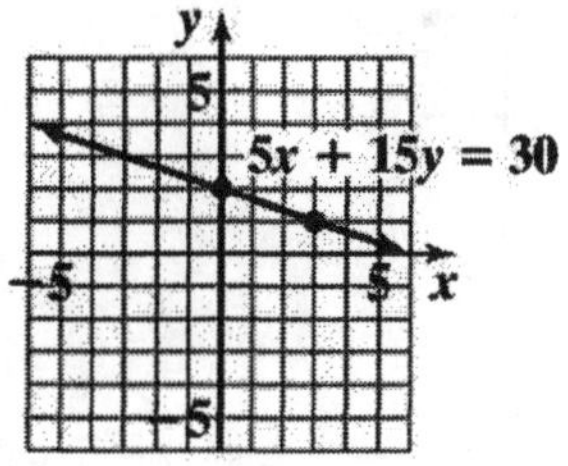

47.
$$-50x+20y=40$$
$$20y=50x+40$$
$$y=\frac{50x+40}{20}$$
$$y=\frac{5}{2}x+2$$

The slope is $\frac{5}{2}$ and the y-intercept is (0, 2).

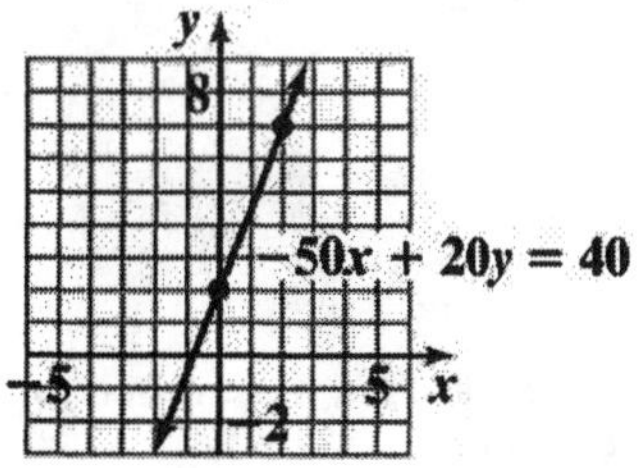

49. $f(x) = -2x + 1$
The slope is -2 and the y-intercept is (0, 1).

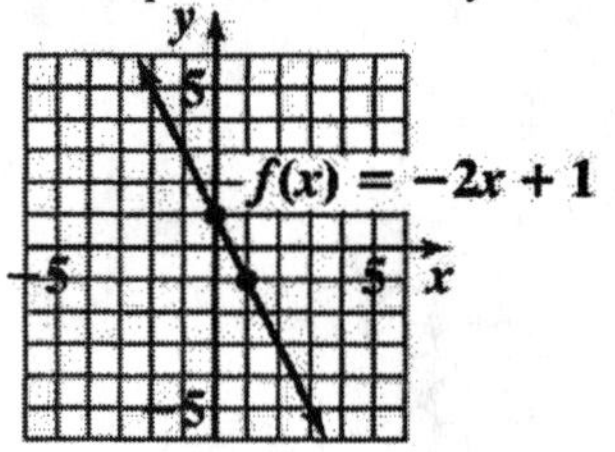

51. $h(x) = -\frac{3}{4}x + 2$

The slope is $-\frac{3}{4}$ and the y-intercept is (0, 2).

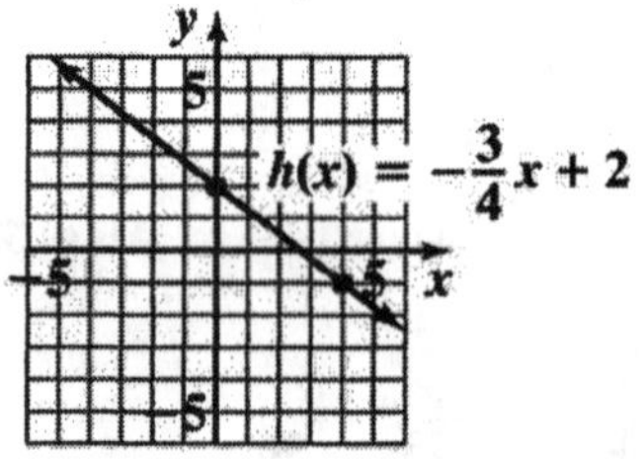

53. **a.** 2

b. 4

c. 1

d. 3

55. If the slopes are the same and the y-intercepts are different, the lines are parallel.

57. Begin with $y = mx + b$. If $m = \frac{4}{3}$ and (6, 3) is a point on the graph, then

$$y = \frac{4}{3}x + b$$
$$3 = \frac{4}{3}(6) + b$$
$$3 = 8 + b$$
$$-5 = b$$

The y-intercept is (0, –5).

59. **a.** The slope is 3 and the y-intercept is (0, 1), so the equation is $y = 3x + 1$.

b. The slope is 3 and the y-intercept is (0, –5), so the equation is $y = 3x - 5$.

61. **a.** The slope of the translated graph is 1.

b. If the y-intercept $b = -1$ is translated up 5 units, the y-intercept of the translated graph is at $y = -1 + 5 = 4$. The new y-intercept is (0, 4).

c. Using $m = 1$ and $b = 4$, the equation of the translated graph is $y = x + 4$.

63. First, rewrite the equation in the slope-intercept form by solving for y in terms of x.

$$3x - 2y = 6$$
$$-2y = -3x + 6$$
$$y = \frac{-3x + 6}{-2}$$
$$y = \frac{3}{2}x - 3$$

Thus, $m = \frac{3}{2}$ and $b = -3$. If the graph is translated down 4 units, then the y-intercept of the translated graph is at $y = -3 - 4 = -7$. Therefore, the equation of the translated graph is $y = \frac{3}{2}x - 7$.

65. $m = \frac{2-4}{-4-6} = \frac{-2}{-10} = \frac{1}{5}$

Thus, for a unit change in x, y changes $\frac{1}{5}$ or 0.2 unit.

67. **a.** Let $(x_1, y_1) = (2005, 19.7)$ and $(x_2, y_2) = (2006, 31.0)$. Then $m = \frac{31.0 - 19.7}{2006 - 2005} = \frac{11.3}{1} = 11.3$.

b. positive

c. Let $(x_1, y_1) = (2004, 7.3)$ and $(x_2, y_2) = (2008, 35.6)$. Then $m = \frac{35.6 - 7.3}{2008 - 2004} = \frac{28.3}{4} = 7.075$.
The average rate of change in digital TV sales from 2004 to 2008 is 7.075 million sales per year.

69. a, b.

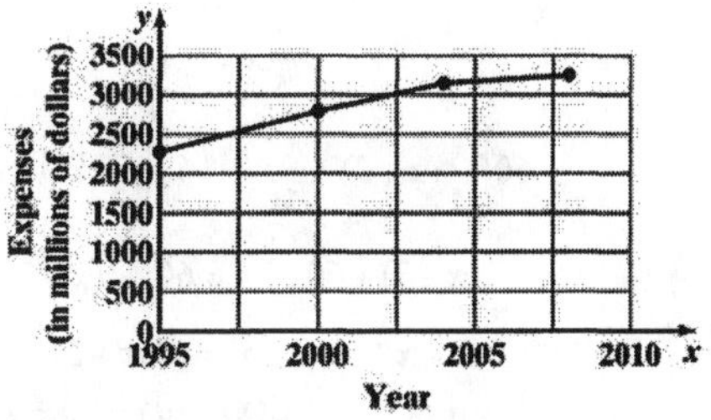

c. From 1995 to 2000,

$$m = \frac{2876 - 2257}{2000 - 1995} = \frac{619}{5} = 123.8$$

From 2000 to 2004,

$$m = \frac{3133 - 2876}{2004 - 2000} = \frac{257}{4} = 64.25$$

From 2004 to 2008,

$$m = \frac{3260 - 3133}{2008 - 2004} = \frac{127}{4} = 31.75$$

d. The greatest average rate of change occurred during the period 1995 to 2000, because the largest slope corresponds to these years.

71. a. If x is the number of years after age 20, two points on the graph are (0, 200) and (50, 150).

The slope is $m = \frac{150 - 200}{50 - 0} = \frac{-50}{50} = -1$ and the y-intercept is (0, 200), so $b = 200$. Thus, the equation for the line is $h(x) = -1x + 200$, or $h(x) = -x + 200$.

b. Note that a 34-year-old man is represented by $x = 14$. Now, $h(14) = -(14) + 200 = 186$. Therefore, the maximum recommended heart rate for a 34-year-old man is 186 beats per minute.

73. a. Note that $t = 0$ represents 1997 and $t = 7$ represents 2004. Thus, two ordered pairs of the function are (0, 159.5) and (7, 294.9). The slope of the line is

$$m = \frac{294.9 - 159.50}{2004 - 1997} = \frac{135.4}{7} \approx 19.34$$

The y-intercept is (0, 159.5), so $b = 159.5$. The linear function is $M(t) \approx 19.34t + 159.5$.

b. Note that 2003 is represented by $t = 6$:

$$\begin{aligned} M(6) &= 19.34(6) + 159.5 \\ &= 116.04 + 159.5 \\ &= 275.54 \end{aligned}$$

The function indicates that Medicaid spending for 2003 was \$275.54 billion. This estimate is similar to the spending shown in the graph.

c. Note that 2010 is represented by $t = 13$:

$$\begin{aligned} M(13) &= 19.34(13) + 159.5 \\ &= 251.42 + 159.5 \\ &= 410.92 \end{aligned}$$

The function indicates that Medicaid spending for 2010 will be \$410.92 billion.

d.

$$\begin{aligned} M(t) &= 340 \\ 19.34t + 159.5 &= 340 \\ 19.34t &= 180.5 \\ t &= \frac{180.5}{19.34} \approx 9.3 \end{aligned}$$

Note that $t = 9$ represents the year 2006. Thus, according to the function, Medicaid spending reached \$340 billion during the year 2006.

75. a. Note that $t = 0$ represents 2001 and $t = 3$ represents 2004. Thus, two ordered pairs of the function are (0, 19.4) and (3, 16.1). The slope of the line is

$$m = \frac{16.1 - 19.4}{3 - 0} = \frac{-3.3}{3} = -1.1$$

The y-intercept is (0, 19.4), so , $b = 19.4$. The linear function is $P(t) = -1.1t + 19.4$.

b. Negative. The negative slope occurs because the percent of teenagers who used illicit drugs decreased from 2001 through 2004.

c. Note that 2003 is represented by $t = 2$:

$$P(2) = -1.1(2) + 19.4 = -2.2 + 19.4 = 17.2$$

According to the function, 17.2% of teenagers used illicit drugs in 2003. This estimate is close to the percent shown in the graph.

d. Note that 2010 is represented by $t = 9$:

$$P(9) = -1.1(9) + 19.4 = -9.9 + 19.4 = 9.5$$

According to the function, 9.5% of teenagers will be using illicit drugs in 2010.

77. a. Note that $t = 0$ represents 1995 and that $t = 9$ represents 2004. Thus, two ordered pairs of the function are (0, 110500) and (9, 185200). The slope of the line is

$$m = \frac{185{,}200 - 110{,}500}{9 - 0} = \frac{74{,}700}{9} = 8300$$

The y-intercept is (0, 110500), so $b = 110{,}500$. The linear function is $P(t) = 8300t + 110{,}500$.

b. Note that $t = 5$ represents 2000:

$$\begin{aligned} P(5) &= 8300(5) + 110{,}500 \\ &= 41{,}500 + 110{,}500 \\ &= 152{,}000 \end{aligned}$$

The median home sale price in the year 2000 was about \$152,000.

c. Note that $t = 15$ represents 2010:

$$\begin{aligned} P(15) &= 8300(15) + 110{,}500 \\ &= 124{,}500 + 110{,}500 \\ &= 235{,}000 \end{aligned}$$

The median home sale price in the year 2010 will be about \$235,000.

d.
$$\begin{aligned} 8300t + 110{,}500 &= 200{,}000 \\ 8300t &= 89{,}500 \\ t &= \frac{89{,}500}{8300} \approx 10.8 \end{aligned}$$

Now $t \approx 10.8$ corresponds to the year 2005. Thus, assuming the trend continues, the median home sale price will reach \$200,000 near the end of the year 2005.

79. The y-intercept of $y = 3x + 6$ is 6. On the screen, the y-intercept is not 6. The y-intercept is wrong.

81. The slope of $y = \frac{1}{2}x + 4$ is $\frac{1}{2}$. On the screen, the slope is not $\frac{1}{2}$. The slope is wrong.

83. There are 91 steps and the total vertical distance is 1292.2 in. Thus, the average height of a step is $\frac{1292.2}{91} = 14.2$ inches. If the slope is 2.21875 and the average height, or "rise", is 14.2 inches, the average width, or "run", is found as follows:

$$\begin{aligned} \text{slope} &= \frac{\text{rise}}{\text{run}} \\ m &= \frac{\text{height}}{\text{width}} \\ 2.21875 &= \frac{14.2}{\text{width}} \\ \text{width} &= \frac{14.2}{2.21875} = 6.4 \end{aligned}$$

The average width is 6.4 inches.

86.
$$\begin{aligned} \frac{-6^2 - 32 \div 2 \div |-8|}{5 - 3 \cdot 2 - 4 \div 2^2} &= \frac{-36 - 32 \div 2 \div 8}{5 - 6 - 4 \div 4} \\ &= \frac{-36 - 16 \div 8}{5 - 6 - 1} \\ &= \frac{-36 - 2}{-1 - 1} \\ &= \frac{-38}{-2} \\ &= 19 \end{aligned}$$

87.
$$\begin{aligned} \frac{1}{4}(x+3) + \frac{1}{5}x &= \frac{2}{3}(x-2) + 1 \\ 60\left[\frac{1}{4}(x+3) + \frac{1}{5}x\right] &= 60\left[\frac{2}{3}(x-2) + 1\right] \\ 60 \cdot \frac{1}{4}(x+3) + 60 \cdot \frac{1}{5}x &= 60 \cdot \frac{2}{3}(x-2) + 60 \cdot 1 \\ 15(x+3) + 12x &= 40(x-2) + 60 \\ 15x + 45 + 12x &= 40x - 80 + 60 \\ 27x + 45 &= 40x - 20 \\ -13x + 45 &= -20 \\ -13x &= -65 \\ x &= 5 \end{aligned}$$

88.
$$\begin{aligned} 2.6x - (-1.4x + 3.4) &= 6.2 \\ 2.6x + 1.4x - 3.4 &= 6.2 \\ 4.0x - 3.4 &= 6.2 \\ 4.0x &= 9.6 \\ x &= 2.4 \end{aligned}$$

89. Let r = the rate of the second, slower train, in miles per hour. Then $r + 15$ is the rate of the first train in miles per hour. The first train travels for a total of 6 hours at which time it is a distance of $6(r + 15)$ miles from Chicago. The second train travels for 3 hours at which time it is a distance of $3r$ miles from Chicago. They are 270 miles apart, so

$$\begin{aligned} 6(r+15) - 3r &= 270 \\ 6r + 90 - 3r &= 270 \\ 3r + 90 &= 270 \\ 3r &= 180 \\ r &= 60 \end{aligned}$$

The second train travels at $r = 60$ miles per hour. The first train travels at $r + 15 = 60 + 15 = 75$ miles per hour.

90. a. $|2x+1| > 5$

$$\begin{aligned} 2x + 1 < -5 \quad &\text{or} \quad 2x + 1 > 5 \\ 2x < -6 \quad &\text{or} \quad 2x > 4 \\ x < \frac{-6}{2} \quad &\text{or} \quad x > \frac{4}{2} \\ x < -3 \quad &\text{or} \quad x > 2 \end{aligned}$$

Solution: $x < -3$ or $x > 2$.

b. $|2x+1| < 5$

$$\begin{aligned} -5 < 2x &+ 1 < 5 \\ -6 < 2&x < 4 \\ \frac{-6}{2} < &\,x < \frac{4}{2} \\ -3 < &\,x < 2 \end{aligned}$$

Mid-Chapter Test: 3.1-3.4

1. The point $(-3.5, -4.2)$ lies in quadrant III.

2. 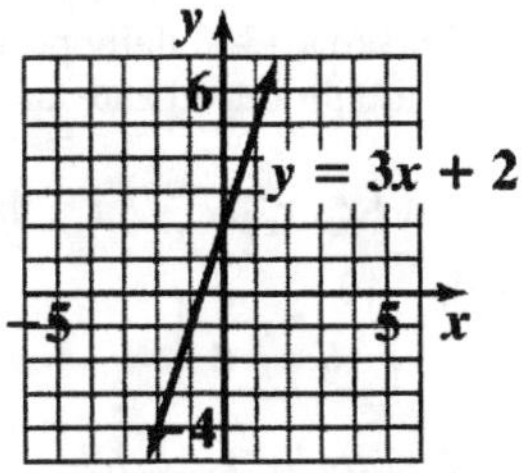

3. 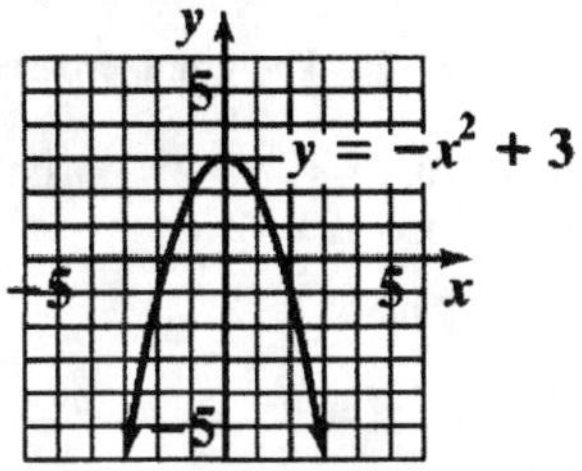

4. 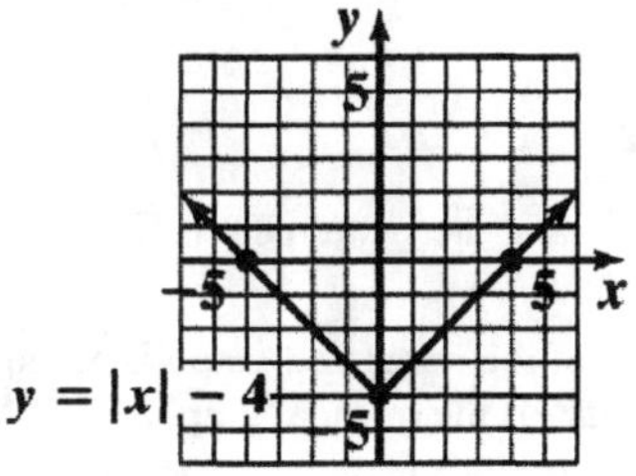

5. 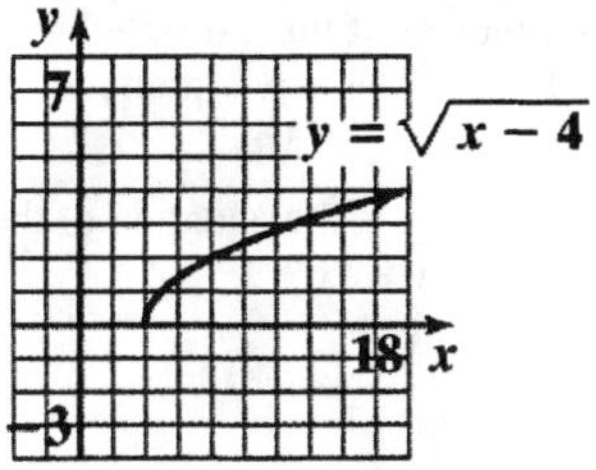

6. **a.** A relation is any set of ordered pairs.

   b. A function is a correspondence between a first set of elements, the domain, and a second set of elements, the range, such that each element of the domain corresponds to exactly one element in the range.

   c. No, every relation is not a function. A relation can have two ordered pairs with the same first element but a function cannot.

   d. Yes, every function is also a relation. A function is a set of ordered pairs so it is a relation.

7. The relation is a function.
   Domain: $\{1, 2, 7, -5\}$
   Range: $\{5, -3, -1, 6\}$

8. The relation is not a function.
   Domain: $\{x \mid -2 \le x \le 2\}$
   Range: $\{y \mid -4 \le y \le 4\}$

9. The relation is a function.
   Domain: $\{x \mid -5 \le x \le 3\}$
   Range: $\{y \mid -1 \le y \le 3\}$

10. $$\begin{aligned} g(-2) &= 2(-2)^2 + 8(-2) - 13 \\ &= 2(4) + 8(-2) - 13 \\ &= 8 - 16 - 13 \\ &= -21 \end{aligned}$$

11. $$\begin{aligned} h(3) &= -6(3)^2 + 3(3) + 150 \\ &= -6(9) + 3(3) + 150 \\ &= -54 + 9 + 150 \\ &= 105 \end{aligned}$$
    At $t = 3$ seconds, the apple is 105 feet above the ground.

12. $$\begin{aligned} 7(x+3) + 2y &= 3(y-1) + 18 \\ 7x + 21 + 2y &= 3y - 3 + 18 \\ 7x + 21 + 2y &= 3y + 15 \\ 7x + 21 + 2y - 21 &= 3y + 15 - 21 \\ 7x + 2y &= 3y - 6 \\ 7x + 2y - 3y &= 3y - 6 - 3y \\ 7x - y &= -6 \end{aligned}$$

13. $x + 3y = -3$
    For the y-intercept, set $x = 0$ and solve for y:
    $$\begin{aligned} (0) + 3y &= -3 \\ 3y &= -3 \\ y &= -1 \end{aligned}$$
    The y-intercept is at $(0, -1)$.
    For the x-intercept, set $y = 0$ and solve for x:
    $$\begin{aligned} x + 3(0) &= -3 \\ x &= -3 \end{aligned}$$
    The y-intercept is at $(-3, 0)$.

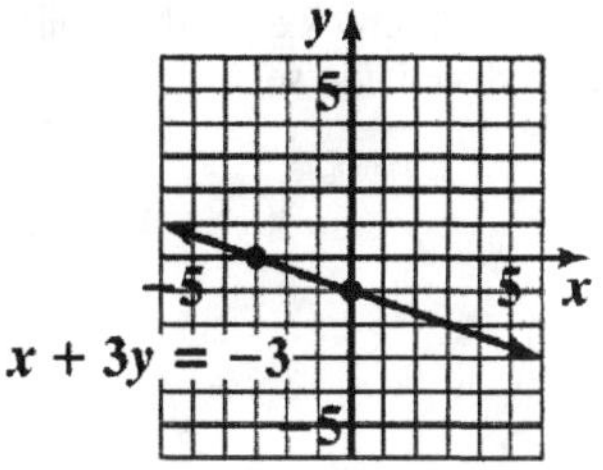

14. $x = -4$
This is a vertical line 4 units to the left of the y-axis.

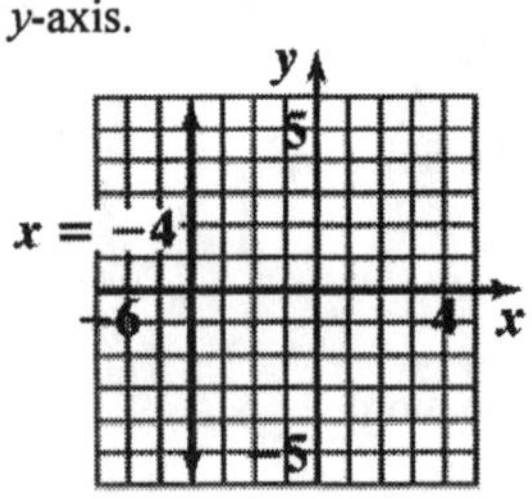

15. $y = 5$
This is a horizontal line 5 units above the x-axis.

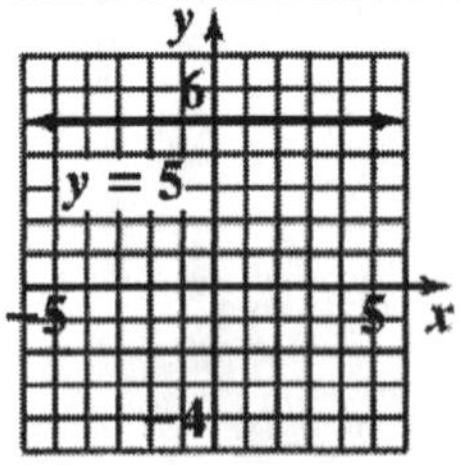

16. a. Let $x = 0$: $p(0) = 30(0) - 660$
$= 0 - 660$
$= -660$

Let $x = 40$: $p(40) = 30(40) - 660$
$= 1200 - 660$
$= 540$

Plots the points $(0, -660)$ and $(40, 540)$ and connect to obtain the graph.

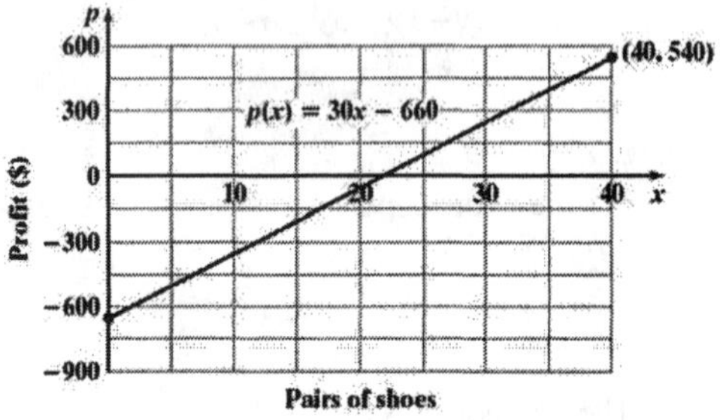

b.
$$p(x) = 0$$
$$30x - 660 = 0$$
$$30x = 660$$
$$x = 22$$
The company will break even if 22 pairs of shoes are produced and sold.

c.
$$p(x) = 360$$
$$30x - 660 = 360$$
$$30x = 1020$$
$$x = 34$$
The company will make a \$360 daily profit if 34 pairs of shoes are produced and sold.

17. Let $(x_1, y_1) = (9, -3)$ and $(x_2, y_2) = (-7, 8)$, then
$$m = \frac{y_2 - y_1}{x_2 - x_1} = \frac{8 - (-2)}{-7 - 9} = \frac{10}{-16} = -\frac{5}{8}.$$

18. The line contains the points $(-1, 4)$ and $(2, -2)$, so $m = \frac{-2 - 4}{2 - (-1)} = \frac{-6}{3} = -2$. The graph crosses the y-axis at the point (0, 2), so $b = 2$. Thus, the equation of the line is $y = -2x + 2$.

19.
$$-3x + 2y = 18$$
$$2y = 3x + 18$$
$$y = \frac{3x + 18}{2}$$
$$y = \frac{3}{2}x + 9$$
The slope is $\frac{3}{2}$ and the y-intercept is (0, 9).

20. a. The slope of the translated graph is 5, the same as that of the original graph.

b. If the y-intercept $b = -3$ is translated up 4 units, the y-intercept of the translated graph is at $y = -3 + 4 = 1$. The new y-intercept is (0, 1).

c. Using $m = 5$ and $b = 1$, the equation of the translated graph is $y = 5x + 1$.

Exercise Set 3.5

1. The point-slope form of linear equation is $y - y_1 = m(x - x_1)$ where m is the slope of the line and (x_1, y_1) is a point on the line.

3. Two lines are perpendicular if their slopes are negative reciprocals, or if one line is horizontal and the other is vertical.

5. $y - y_1 = m(x - x_1)$
$y - 1 = 2(x - 3)$
$y - 1 = 2x - 6$
$y = 2x - 5$

7. $y - y_1 = m(x - x_1)$
$y - (-1) = -\frac{1}{2}(x - 4)$
$y + 1 = -\frac{1}{2}x + 2$
$y = -\frac{1}{2}x + 1$

9. $y - y_1 = m(x - x_1)$
$y - (-5) = \frac{1}{2}(x - (-1))$
$y + 5 = \frac{1}{2}(x + 1)$
$y + 5 = \frac{1}{2}x + \frac{1}{2}$
$y = \frac{1}{2}x - \frac{9}{2}$

11. $m = \frac{9 - (-3)}{-6 - 2} = \frac{12}{-8} = -\frac{3}{2}$

Use $m = -\frac{3}{2}$ and $(x_1, y_1) = (2, -3)$.
$y - y_1 = m(x - x_1)$
$y - (-3) = -\frac{3}{2}(x - 2)$
$y + 3 = -\frac{3}{2}x + 3$
$y = -\frac{3}{2}x$

13. $m = \frac{-2 - (-3)}{6 - 4} = \frac{-2 + 3}{2} = \frac{1}{2}$

Use $m = \frac{1}{2}$ and $(x_1, y_1) = (4, -3)$.
$y - y_1 = m(x - x_1)$
$y - (-3) = \frac{1}{2}(x - 4)$
$y + 3 = \frac{1}{2}x - 2$
$y = \frac{1}{2}x - 5$

15. $m_1 = \frac{2 - 0}{0 - 2} = \frac{2}{-2} = -1$; $m_2 = \frac{3 - 0}{0 - 3} = \frac{3}{-3} = -1$

Since their slopes are equal, l_1 and l_2 are parallel.

17. $m_1 = \frac{7 - 6}{5 - 4} = \frac{1}{1} = 1$; $m_2 = \frac{4 - (-1)}{1 - (-1)} = \frac{5}{2}$

Since their slopes are different and since the product of their slopes is not −1, l_1 and l_2 are neither parallel nor perpendicular.

19. $m_1 = \frac{-2 - 2}{-1 - 3} = \frac{-4}{-4} = 1$; $m_2 = \frac{-1 - 0}{3 - 2} = \frac{-1}{1} = -1$

Since the product of their slopes is −1, the lines are perpendicular.

21. $y = \frac{1}{5}x + 9$, so $m_1 = \frac{1}{5}$

$y = -5x + 2$, so $m_2 = -5$

Since the product of their slopes is −1, the lines are perpendicular.

23.

$4x + 2y = 8$
$2y = -4x + 8$
$y = -2x + 4$
$m_1 = -2$

$8x = 4 - 4y$
$4y = -8x + 4$
$y = -2x + 1$
$m_2 = -2$

Since their slopes are equal and their y-intercepts are different, the lines are parallel.

25.

$2x - y = 4$
$-y = -2x + 4$
$y = 2x - 4$
$m_1 = 2$

$-x + 4y = 4$
$4y = x + 4$
$y = \frac{1}{4}x + 1$
$m_2 = \frac{1}{4}$

Since their slopes are different and since the product of their slopes is not −1, the lines are neither parallel nor perpendicular.

27.

$y = \frac{1}{2}x - 6$
$m_1 = \frac{1}{2}$

$-4y = 8x + 15$
$y = \frac{8x + 15}{-4}$
$y = -2x - \frac{15}{4}$
$m_2 = -2$

Since the product of their slopes is −1, the lines are perpendicular.

29. $y=\frac{1}{2}x+6$ $\qquad -2x+4y=8$

$$m_1=\frac{1}{2} \qquad 4y=2x+8$$

$$y=\frac{2x+8}{4}$$

$$y=\frac{1}{2}x+2$$

$$m_2=\frac{1}{2}$$

Since their slopes are equal and their y-intercepts are different, the lines are parallel.

31. $x-2y=-9$ $\qquad y=x+6$

$$-2y=-x-9 \qquad m_2=1$$

$$y=\frac{-x-9}{-2}$$

$$y=\frac{1}{2}x+\frac{9}{2}$$

$$m_1=\frac{1}{2}$$

Since their slopes are different and since the product of their slopes is not -1, the lines are neither parallel nor perpendicular.

33. The slope of the given line, $y = 2x + 4$, is $m_1=2$. So $m_2=2$. Now use the point-slope form with $m = 2$ and $(x_1, y_1)=(2, 5)$ to obtain the slope-intercept form.

$$y-y_1=m(x-x_1)$$
$$y-5=2(x-2)$$
$$y-5=2x-4$$
$$y=2x+1$$

35. Find the slope of the given line.

$$2x-5y=7$$
$$-5y=-2x+7$$
$$y=\frac{-2x+7}{-5}$$
$$y=\frac{2}{5}x-\frac{7}{5}$$

$m_1=\frac{2}{5}$, so $m_2=\frac{2}{5}$. Now use the point-slope form with $m=\frac{2}{5}$ and $(x_1, y_1)=(-3,-5)$ to obtain the standard form.

$$y-y_1=m(x-x_1)$$
$$y-(-5)=\frac{2}{5}(x-(-3))$$
$$y+5=\frac{2}{5}(x+3)$$
$$y+5=\frac{2}{5}x+\frac{6}{5}$$
$$5(y+5)=5\left(\frac{2}{5}x+\frac{6}{5}\right)$$
$$5y+25=2x+6$$
$$-2x+5y=-19$$
$$2x-5y=19$$

37. Find the slope of the line with the given intercepts:

$m=\frac{5-0}{0-3}=-\frac{5}{3}$. The y-intercept (0, 5), so $b = 5$.

Thus, the slope-intercept form of the equation is

$y=-\frac{5}{3}x+5$.

39. The slope of the given line $y=\frac{1}{3}x+1$ is $m_1=\frac{1}{3}$. So m_2 is the negative reciprocal, or

$m_2=-\frac{1}{m_1}=-\frac{1}{\frac{1}{3}}=-1\cdot 3=-3$. Use the point-slope form with $m=-3$ and $(x_1, y_1)=(5,-2)$ to obtain the function notation.

$$y-y_1=m(x-x_1)$$
$$y-(-2)=-3(x-5)$$
$$y+2=-3x+15$$
$$y=-3x+13$$
$$f(x)=-3x+13$$

41. Find the slope of the line with the given intercepts:

$m_1=\frac{-3-0}{0-2}=\frac{-3}{-2}=\frac{3}{2}$. So m_2 is the negative reciprocal, or $m_2=-\frac{1}{m_1}=-\frac{1}{\frac{3}{2}}=-\frac{2}{3}$. Now use the point-slope form with $m=-\frac{2}{3}$ and $(x_1, y_1)=(6, 2)$ and obtain the slope-intercept form.

$$y-y_1=m(x-x_1)$$
$$y-2=-\frac{2}{3}(x-6)$$
$$y-2=-\frac{2}{3}x+4$$
$$y=-\frac{2}{3}x+6$$

43. a. To find the function, use the points (2.5, 210) and (6, 370) to determine the slope.

$$m=\frac{370-210}{6-2.5}=\frac{160}{3.5}\approx 45.7$$

Now use the point-slope form with $m = 45.7$ and $(s_1, C_1) = (2.5, 210)$

$$C-C_1=m(s-s_1)$$
$$C-210=45.7(s-2.5)$$
$$C-210=45.7s-114.25$$
$$C=45.7s+95.75$$
$$C(s)\approx 45.7s+95.8$$

b. $C(5)\approx 45.7(5)+95.8=228.5+95.8=324.3$

The average person will burn about 324.3 calories.

45. a. To find the function, use the points (200, 50) and (300, 30) to determine the slope: $m=\frac{30-50}{300-200}=\frac{-20}{100}=-0.20$

Now use the point-slope form with $m=-0.20$ and $(p_1, d_1)=(200,50)$

$$d-d_1=m(p-p_1)$$
$$d-50=-0.20(p-200)$$
$$d-50=-0.20p+40$$
$$d=-0.20p+90$$
$$d(p)=-0.20p+90$$

b. $d(260)=-0.20(260)+90=-52+90=38$

When the price is \$260, the demand will be 38 DVD players.

c.
$$d(p)=45$$
$$-0.20p+90=45$$
$$-0.20p=-45$$
$$p=225$$

In order to have a demand of 45 DVD players, the price should be \$225.

47. a. To find the function, use the points (2, 130) and (4, 320) to determine the slope:

$$m=\frac{320-130}{4-2}=\frac{190}{2}=95.$$

Now use the point-slope form with $m = 95$ and $(p_1, s_1)=(2,130)$.

$$s-s_1=m(p-p_1)$$
$$s-130=95(p-2.00)$$
$$s-130=95p-190$$
$$s=95p-60$$
$$s(p)=95p-60$$

b. $s(2.80)=95(2.80)-60=266-60=206$

If the price is \$2.80, the supply will be 206 kites.

c.
$$s(p)=255$$
$$95p-60=225$$
$$95p=285$$
$$p=3$$

For the supply to be 225 kites, the price should be \$3.00.

49. a. To find the function, use the points (80, 1000) and (200, 2500) to determine the slope:

$$m=\frac{2500-1000}{200-80}=\frac{1500}{120}=12.5.$$

Now use the point-slope form with $m=12.5$ and $(t_1, i_1)=(80, 1000)$.

$$i-i_1=m(t-t_1)$$
$$i-1000=12.5(t-80)$$
$$i-1000=12.5t-1000$$
$$i=12.5t$$
$$i(t)=12.5t$$

b. $i(120)=12.5(120)=1500$

If 120 tickets are sold, the income will be \$1500.

c.
$$i(t)=2200$$
$$12.5t=2200$$
$$t=176$$

If the income is \$2200, then 176 tickets were sold.

51. a. To find the function, use the points (2000, 30) and (4000, 50) to determine the slope:

$$m=\frac{50-30}{4000-2000}=\frac{20}{2000}=0.01.$$

Now use the point-slope form with $m = 0.01$ and $(w_1, r_1)=(2000, 30)$.

$$r-r_1=m(w-w_1)$$
$$r-30=0.01(w-2000)$$
$$r-30=0.01w-20$$
$$r=0.01w+10$$
$$r(w)=0.01w+10$$

b.
$$r(3613)=0.01(3613)+10$$
$$=36.13+10$$
$$=46.13$$

If the weight of the care is 3613 pounds, then the registration fee will be \$46.13.

c.
$$\begin{aligned} r(w) &= 60 \\ 0.01w+10 &= 60 \\ 0.01w &= 50 \\ w &= 5000 \end{aligned}$$
If the registration fee is \$60, then the weight of the car is 5000 pounds.

53. a. To find the function, use the points (50, 36.0) and (70, 18.7) to determine the slope:
$m = \dfrac{18.7-36.0}{70-50} = \dfrac{-17.3}{20} = -0.865$.
Now use the point-slope form with $m = -0.865$ and $(a_1, y_1) = (50, 36.0)$:
$$\begin{aligned} y - y_1 &= m(a-a_1) \\ y-36.0 &= -0.865(a-50) \\ y-36.0 &= -0.865a+43.25 \\ y &= -0.865a+79.25 \\ y(a) &= -0.865a+79.25 \end{aligned}$$

b.
$$\begin{aligned} y(37) &= -0.865(37)+79.25 \\ &= -32.005+79.25 \\ &= 47.245 \end{aligned}$$
The additional life expectancy will be about 47.2 years.

c.
$$\begin{aligned} y(a) &= 25 \\ -0.865a+79.25 &= 25 \\ -0.865a &= -54.25 \\ a &\approx 62.7 \end{aligned}$$
In order to have an additional life expectancy of 25 years, one would need to be currently about 62.7 years old.

55. a. To find the function, use the points (18, 14) and (36, 17.4) to determine the slope:
$m = \dfrac{17.4-14}{36-18} = \dfrac{3.4}{18} \approx 0.189$.
Now use the point-slope form with $m = 0.189$ and $(a_1, w_1) = (18, 14)$:
$$\begin{aligned} w - w_1 &= m(a-a_1) \\ w-14 &= 0.189(a-18) \\ w-14 &= 0.189a-3.402 \\ w &= 0.189a+10.598 \\ w(a) &\approx 0.189a+10.6 \end{aligned}$$

b.
$$\begin{aligned} w(22) &= 0.189(22)+10.6 \\ &= 4.158+10.6 \\ &= 14.758 \end{aligned}$$
A 22-month-old boy who is in the 95th percentile will weigh about 14.758 kg. This is close to the weight shown in the graph.

58.
$$\begin{aligned} 6-\frac{1}{2}x &> 2x+5 \\ 2\left(6-\frac{1}{2}x\right) &> 2(2x+5) \\ 12-x &> 4x+10 \\ -5x &> -2 \\ \frac{-5x}{-5} &< \frac{-2}{-5} \\ x &< \frac{2}{5} \end{aligned}$$
The solution is $\left(-\infty, \frac{2}{5}\right)$.

59. When multiplying or dividing both sides of an inequality by a negative number, you must reverse the direction of the inequality symbol.

60. a. A relation is any set of ordered pairs.

b. A function is a correspondence between a first set of elements, the domain, and a second set of elements, the range, such that each element of the domain corresponds to exactly one element in the range.

c. Answers will vary.

61. Domain: $\{3, 4, 5, 6\}$; Range: $\{-4, -1, 2, 7\}$

Exercise Set 3.6

1. Yes, $f(x)+g(x) = (f+g)(x)$ for all values of x. This is how addition of functions is defined.

3. $f(x)/g(x) = (f/g)(x)$ provided $g(x) \neq 0$. This is because division by zero is undefined.

5. No, $(f-g)(x) \neq (g-f)(x)$ for all values of x since subtraction is not commutative. For example, if $f(x) = x^2+1$ and $g(x) = x$, then
$$\begin{aligned} (f-g)(x) &= f(x)-g(x) \\ &= (x^2+1)-(x) \\ &= x^2-x+1 \\ (g-f)(x) &= g(x)-f(x) \\ &= (x)-(x^2+1) \\ &= -x^2+x-1 \end{aligned}$$
So, $(f-g)(x) \neq (g-f)(x)$.

7. a. $(f+g)(-2) = f(-2)+g(-2) = (-3)+5 = 2$

b. $(f-g)(-2) = f(-2)-g(-2) = (-3)-5 = -8$

c. $(f\cdot g)(-2) = f(-2)\cdot g(-2) = (-3)\cdot 5 = -15$

d. $(f/g)(-2) = f(-2)/g(-2) = (-3)/5 = -\frac{3}{5}$

9. a. $(f+g)(x) = f(x)+g(x)$
$= (x+5)+(x^2+x)$
$= x^2+2x+5$

b. $(f+g)(a) = a^2+2a+5$

c. $(f+g)(2) = (2)^2+2(2)+5$
$= 4+2(2)+5$
$= 4+4+5$
$= 13$

11. a. $(f+g)(x) = f(x)+g(x)$
$= (-3x^2+x-4)+(x^3+3x^2)$
$= x^3+x-4$

b. $(f+g)(a) = a^3+a-4$

c. $(f+g)(2) = (2)^3+(2)-4$
$= 8+2-4$
$= 6$

13. a. $(f+g)(x) = f(x)+g(x)$
$= (4x^3-3x^2-x)+(3x^2+4)$
$= 4x^3-x+4$

b. $(f+g)(a) = 4a^3-a+4$

c. $(f+g)(2) = 4(2)^3-(2)+4$
$= 4(8)-2+4$
$= 32-2+4$
$= 34$

15. $f(2) = (2)^2-4 = 4-4 = 0$
$g(2) = -5(2)+3 = -10+3 = -7$
$f(2)+g(2) = 0+(-7) = -7$

17. $f(4) = (4)^2-4 = 16-4 = 12$
$g(4) = -5(4)+3 = -20+3 = -17$
$f(4)-g(4) = 12-(-17) = 29$

19. $f(3) = (3)^2-4 = 9-4 = 5$
$g(3) = -5(3)+3 = -15+3 = -12$
$f(3)\cdot g(3) = 5(-12) = -60$

21. $f\left(\frac{3}{5}\right) = \left(\frac{3}{5}\right)^2 - 4 = \frac{9}{25} - \frac{100}{25} = -\frac{91}{25}$

$g\left(\frac{3}{5}\right) = -5\left(\frac{3}{5}\right)+3 = -3+3 = 0$

$\frac{f\left(\frac{3}{5}\right)}{g\left(\frac{3}{5}\right)} = \frac{-\frac{91}{25}}{0}$ which is undefined.

23. $f(-3) = (-3)^2-4 = 9-4 = 5$
$g(-3) = -5(-3)+3 = 15+3 = 18$
$g(-3)-f(-3) = 18-5 = 13$

25. $f(0) = (0)^2-4 = 0-4 = -4$
$g(0) = -5(0)+3 = 0+3 = 3$
$g(0)/f(0) = 3/-4 = -\frac{3}{4}$

27. $(f+g)(x) = f(x)+g(x)$
$= (2x^2-x)+(x-6)$
$= 2x^2-6$

29. $(f+g)(x) = 2x^2-6$
$(f+g)(2) = 2(2)^2-6$
$= 2(4)-6$
$= 8-6$
$= 2$

31. $(f-g)(-2) = f(-2)-g(-2)$
$= \left(2\cdot(-2)^2-(-2)\right)-\left((-2)-6\right)$
$= (8+2)-(-8)$
$= 10+8$
$= 18$

33. $(f\cdot g)(0) = f(0)\cdot g(0)$
$= \left(2\cdot 0^2-0\right)\cdot(0-6)$
$= (0)(-6)$
$= 0$

35. $(f/g)(-1) = f(-1)/g(-1)$
$= \left(2\cdot(-1)^2-(-1)\right)/\left((-1)-6\right)$
$= (2+1)/(-7)$
$= 3/(-7)$
$= -\frac{3}{7}$

37. $(g/f)(5) = g(5)/f(5)$
$= (5-6)/\left(2\cdot 5^2 - 5\right)$
$= (-1)/(50-5)$
$= (-1)/45$
$= -\dfrac{1}{45}$

39. $(g-f)(x) = g(x) - f(x)$
$= (x-6) - (2x^2 - x)$
$= x - 6 - 2x^2 + x$
$= -2x^2 + 2x - 6$

41. $(f+g)(0) = f(0) + g(0) = 2 + 1 = 3$

43. $(f\cdot g)(2) = f(2)\cdot g(2) = 4\cdot(-1) = -4$

45. $(g-f)(-1) = g(-1) - f(-1) = 2 - 1 = 1$

47. $(g/f)(4) = g(4)/f(4) = 1/0$, which is undefined

49. $(f+g)(-2) = f(-2) + g(-2) = 1 + (-1) = 0$

51. $(f\cdot g)(1) = f(1)\cdot g(1) = 0\cdot 1 = 0$

53. $(f/g)(4) = f(4)/g(4) = -3/1 = -3$

55. $(g/f)(2) = g(2)/f(2) = 2/(-1) = -2$

57. a. Frank contributed \$1000 in 2004.

b. $1400 - 600 = 800$
In 2006, Sharon contributed about \$800 more than Frank.

c. $1600 + 2000 + 1400 + 900 + 2000 = 7900$
Frank and Sharon contributed about \$7900 over the five-year period.

d. $(F+S)(2005) = \$900$

59. a. The import of crude oil to China was the greatest in 2003. That year approximately 1.8 million barrels per day were imported.

b. In 1998 and 2001, the import of crude oil decrease from the year before.

c. $I(2002) \approx 1.4$ million barrels

d. $5.8 - 1.8 = 4.0$
In 2003, about 4.0 million barrels of crude oil were produced in China per day.

61. a. About 20 houses were sold in Fuller in the summer of 2006.

b. $28 - 20 = 8$
About 8 houses were sold in Fuller at other times in 2006.

c. $Y(2005) \approx 30 - 18 = 12$ houses

d. $(S+Y)(2003) \approx 23$ houses

63. a.

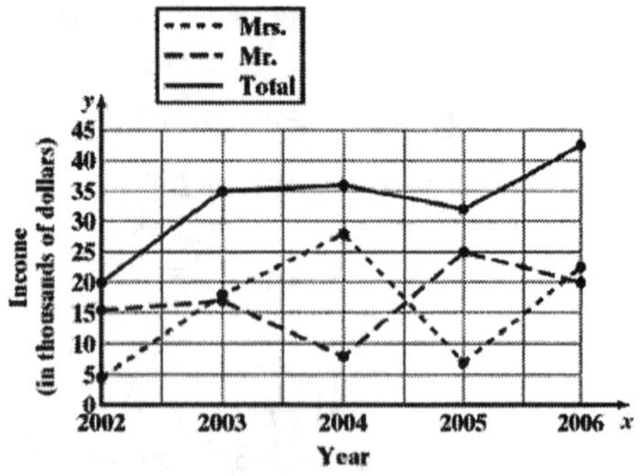

b.

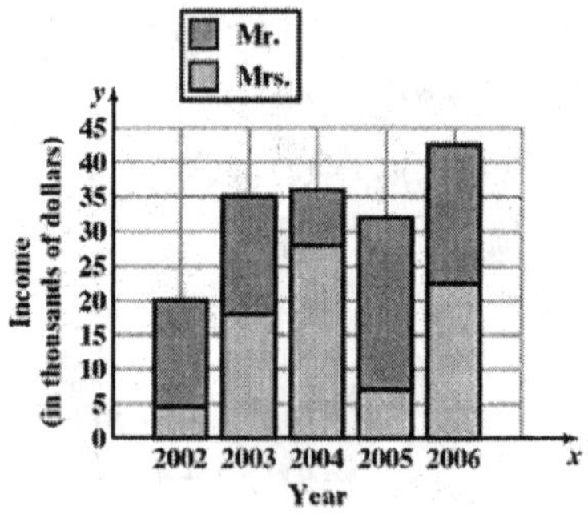

c.

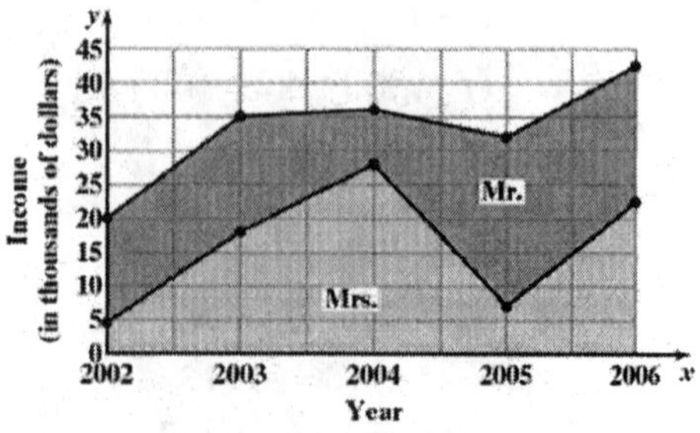

65. a.

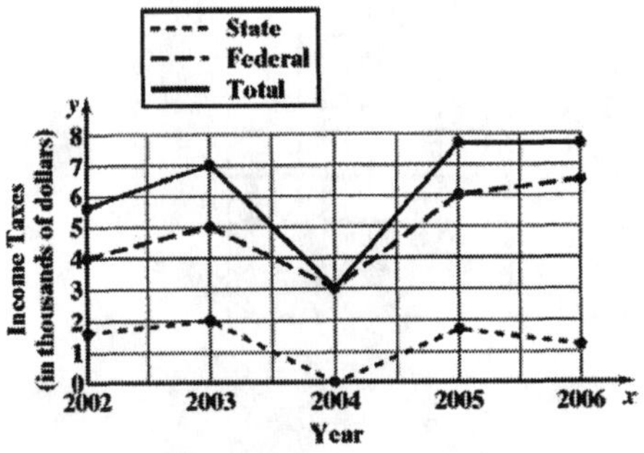

b.

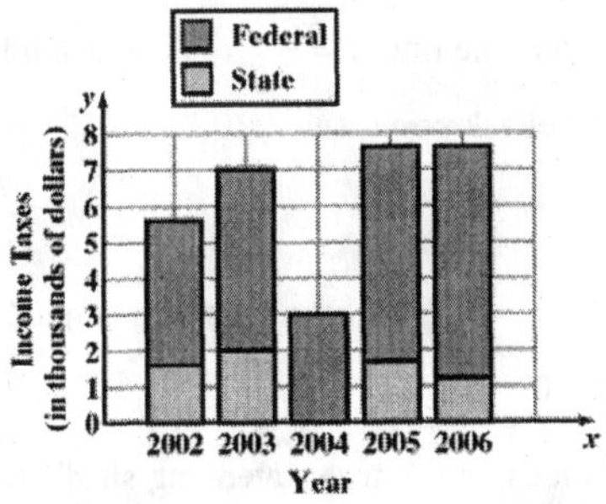

c.

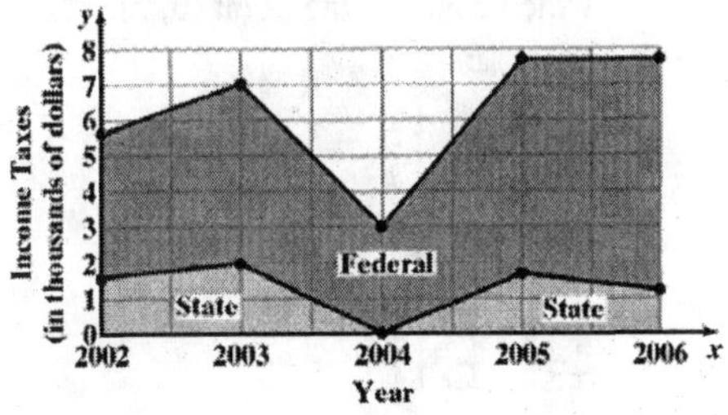

67. If $(f+g)(a)=0$, then, $f(a)$ and $g(a)$ must either be opposites or both be equal to 0.

69. If $(f-g)(a)=0$, then $f(a)=g(a)$.

71. If $(f/g)(a)<0$, then $f(a)$ and $g(a)$ must have opposite signs.

73.

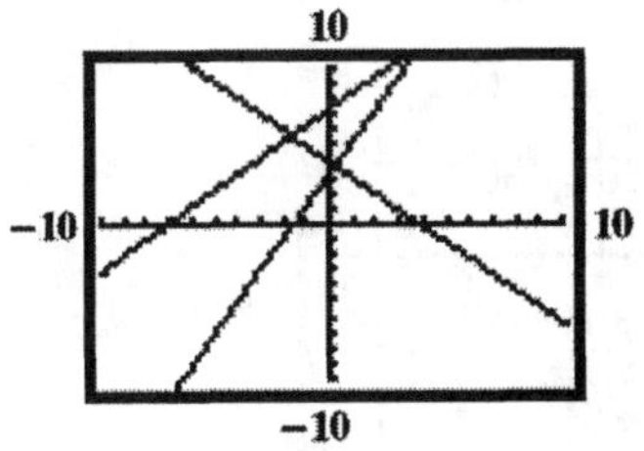

75.

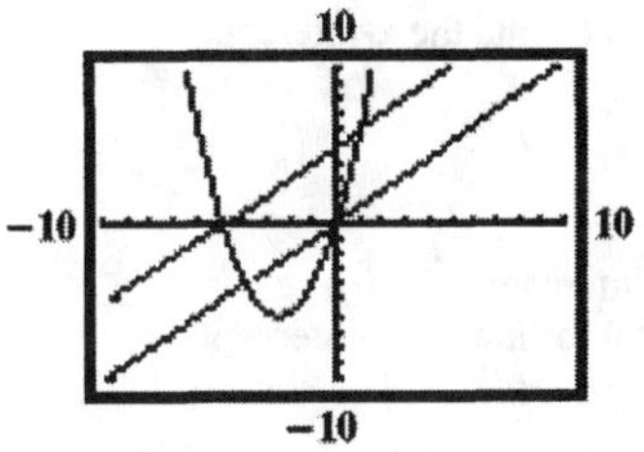

78. $(-4)^{-3}=\dfrac{1}{(-4)^3}=\dfrac{1}{-64}=-\dfrac{1}{64}$

79. $2{,}960{,}000=2.96\times10^6$

80.

$$A=\frac{1}{2}bh$$
$$2\cdot A=2\cdot\frac{1}{2}bh$$
$$2A=bh$$
$$\frac{2A}{b}=\frac{bh}{b}$$
$$\frac{2A}{b}=h \text{ or } h=\frac{2A}{b}$$

81. Let the pre-tax cost of the washing machine be x.

$$x+0.06x=477$$
$$1.06x=477$$
$$x=450$$

The pre-tax cost of the washing machine was \$450.

82.

x	y
−3	1
−2	0
−1	−1
0	−2
1	−1
2	0
3	1

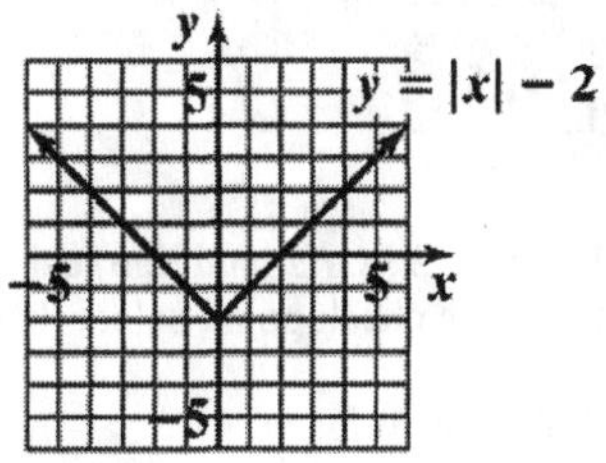

83. Set $y = 0$ to find the x-intercept.

$$3x - 4(0) = 12$$
$$3x = 12$$
$$x = 4$$

The x-intercept is (4, 0).
Set $x = 0$ to find the y-intercept.

$$3(0) - 4y = 12$$
$$-4y = 12$$
$$y = -3$$

The y-intercept is (0, –3).

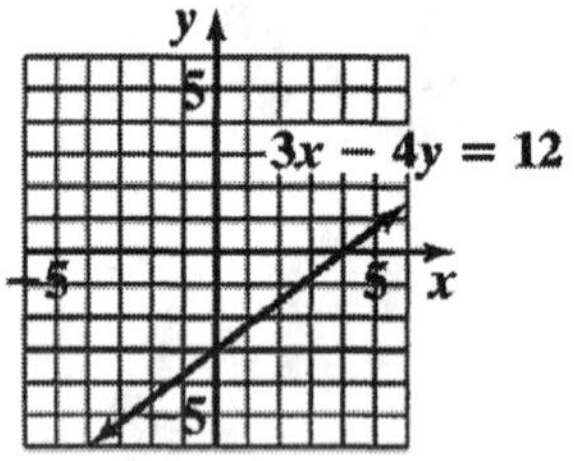

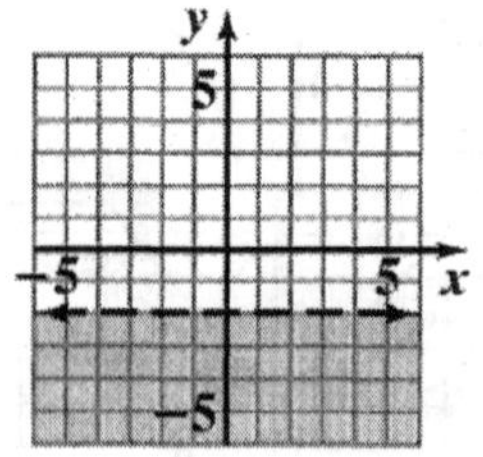

Exercise Set 3.7

1. The inequalities > and < do not include the corresponding equation. The points on the line satisfy only the equation.

3. (0, 0) cannot be used as a test point if the line passes through the origin.

5. $x > 1$
Graph the line $x = 1$ (vertical line) using a dashed line. For the check point, select (0, 0):
$x > 1$
$0 > 1$ ← Substitute 0 for x
Since this is a false statement, shade the region which does not contain (0, 0).

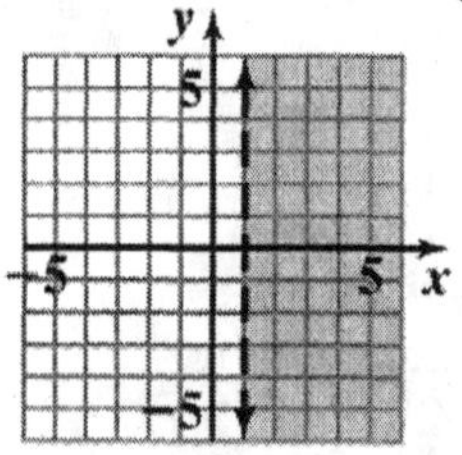

7. $y < -2$
Graph the line $y = -2$ (horizontal line) using a dashed line. For the check point, select (0, 0).
$y < -2$
$0 < -2$ ← Substitute 0 for y.
Since this is a false statement, shade the region which does not contain (0, 0).

9. $y \geq -\frac{1}{2}x$

Graph the line $y = -\frac{1}{2}x$ using a solid line. For the check point, select (0, 2).

$y \geq -\frac{1}{2}x$

$2 \geq -\frac{1}{2}(0)$ ← Substitute 0 for x and 2 for y

$2 \geq 0$

Since this is a true statement, shade the region which contains the point (0, 2).

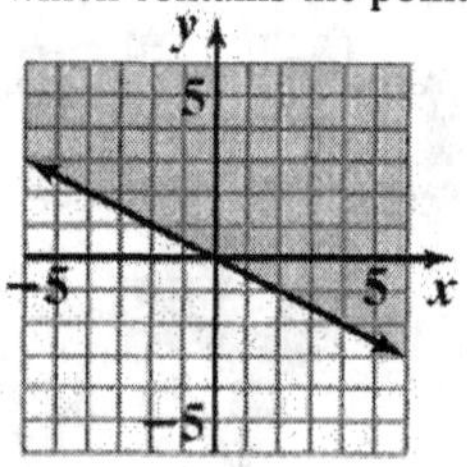

11. $y < 2x + 1$
Graph the line $y = 2x + 1$ using a dashed line. For the check point, select (0, 0).
$y < 2x + 1$
$0 < 2(0) + 1$ ← Substitute 0 for x and y
$0 < 1$
Since this is a true statement, shade the region which contains the point (0, 0).

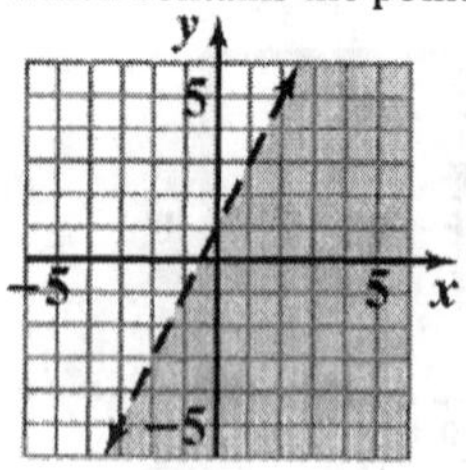

13. $y > 2x - 1$
Graph the line $y = 2x - 1$ using a dashed line.
For the check point, select (0, 0).
$y > 2x - 1$
$0 > 2(0) - 1 \leftarrow$ Substitute 0 for x and y
$0 > -1$
Since this is a true statement, shade the region which contains the point (0, 0).

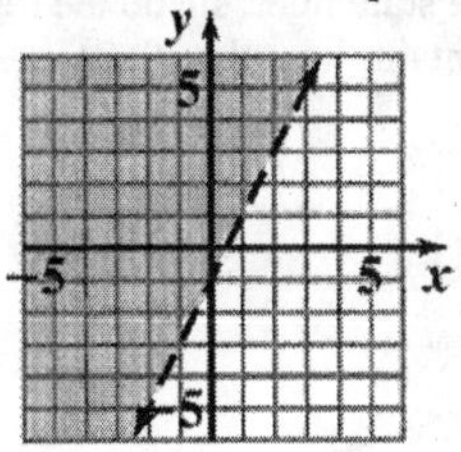

15. $y \geq \frac{1}{2}x - 3$

Graph the line $y = \frac{1}{2}x - 3$ using a solid line. For the check point, select (0, 0).
$y \geq \frac{1}{2}x - 3$
$0 \geq \frac{1}{2}(0) - 3 \leftarrow$ Substitute 0 for x and y
$0 \geq -3$
Since this is a true statement, shade the region which contains the point (0, 0).

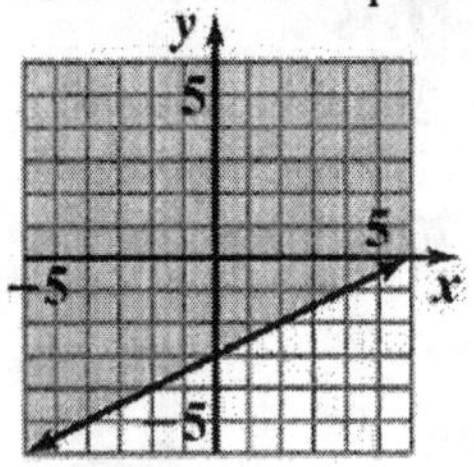

17. $2x + 3y > 6$
Graph the line $2x + 3y = 6$ using a dashed line.
For the check point, select (0, 0).
$2x + 3y > 6$
$2(0) + 3(0) > 6 \leftarrow$ Substitute 0 for x and y
$0 > 6$
Since this is a false statement, shade the region which does not contain the point (0, 0).

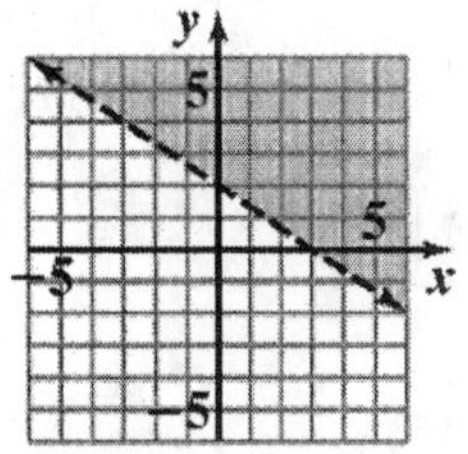

19. $y \leq -3x + 5$
Graph the line $y = -3x + 5$ using a solid line.
For the check point, select (0, 0).
$y \leq -3x + 5$
$0 \leq -3(0) + 5 \leftarrow$ Substitute 0 for x and y
$0 \leq 5$
Since this is a true statement, shade the region which contains the point (0, 0).

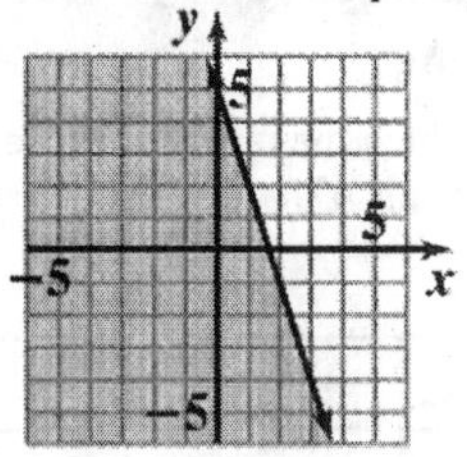

21. $2x + y < 4$
Graph the line $2x + y = 4$ using a dashed line.
For the check point, select (0, 0).
$2x + y < 4$
$2(0) + 0 < 4 \leftarrow$ Substitute 0 for x and y
$0 < 4$
Since this is a true statement, shade the region which contains the point (0, 0).

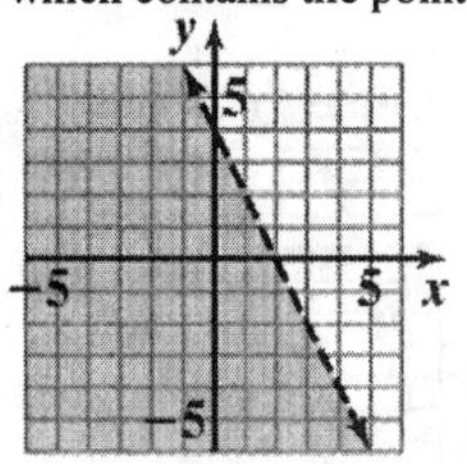

23. $10 \geq 5x - 2y$
Graph the line $10 = 5x - 2y$ using a solid line.
For the check point, select (0, 0).
$10 \geq 5x - 2y$
$10 \geq 5(0) - 2(0) \leftarrow$ Substitute 0 for x and y
$10 \geq 0$
Since this is a true statement, shade the region which contains the point (0, 0).

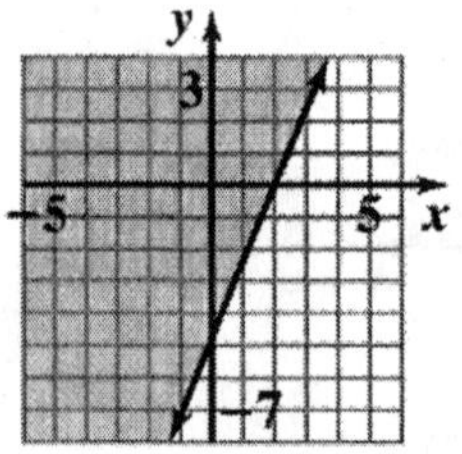

25. a,b.

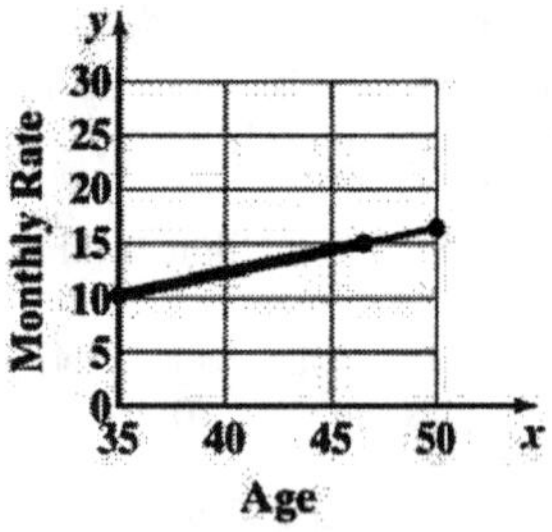

c. The age at which the rate first exceeds \$15 per month is 47.

27. a,b.

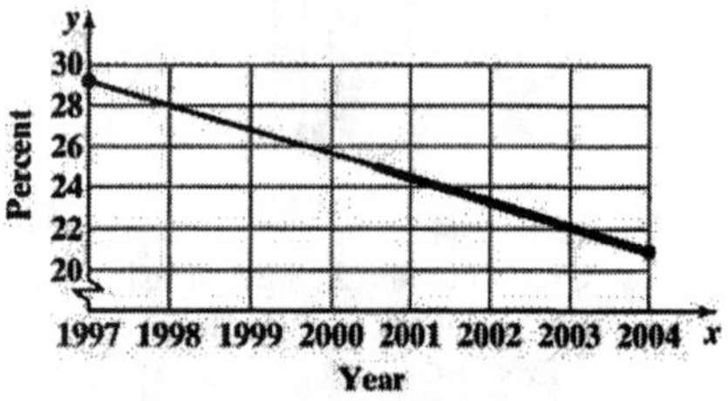

c. The year 2003 is first year in which the percentage of Americans 18 and older who smoke was less than or equal to 23%.

29. a.

$f(x) = 2x - 4$

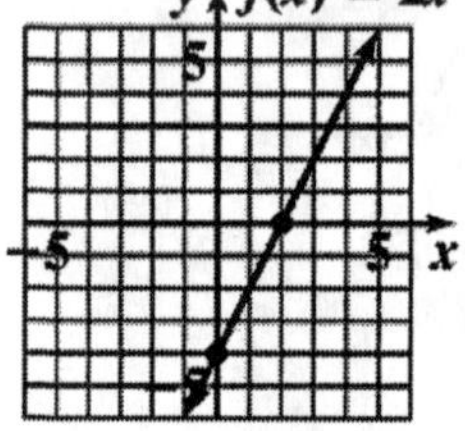

b.

$f(x) = 2x - 4$

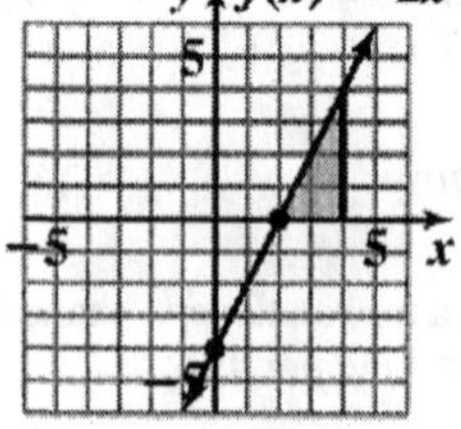

31. $y < |x|$

Graph the equation $y = |x|$ using a dashed line. For the check point, select (0, 2).

$y < |x|$

$2 < (0) \leftarrow$ Substitute 0 for x and 2 for y

$2 < 0$

Since this is a false statement, shade the region which does not contain the point (0, 2).

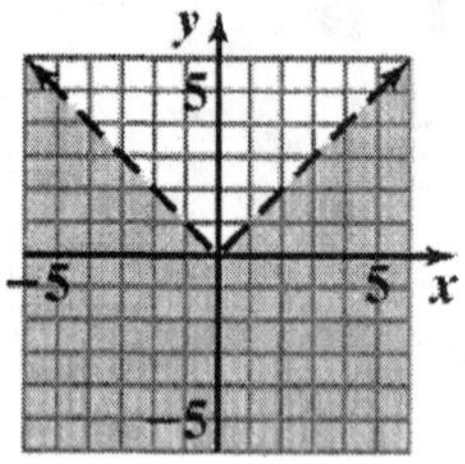

33. $y < x^2 - 4$

Graph the equation $y = x^2 - 4$ using a dashed line. For the check point, select (0, 0).

$y < x^2 - 4$

$0 < 0^2 - 4 \leftarrow$ Substitute 0 for x and y

$0 < -4$

Since this is a false statement, shade the region which does not contain the point (0, 0).

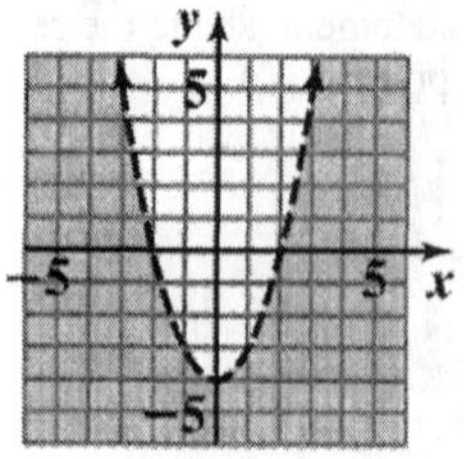

34.

$$9 - \frac{5x}{3} = -6$$

$$-\frac{5x}{3} = -15$$

$$3\left(-\frac{5x}{3}\right) = 3(-15)$$

$$-5x = -45$$

$$x = \frac{-45}{-5}$$

$$x = 9$$

35. $C = \bar{x} + Z\frac{\sigma}{\sqrt{n}}$

$C = 80 + 1.96\frac{3}{\sqrt{25}}$

$C = 80 + 1.96\left(\frac{3}{5}\right)$

$C = 80 + 1.176$

$C = 81.176$

36. Let x be the original cost of the CD. The first week, the price was reduced by 10%, and the second week, the price was reduced an additional $2.

$$\begin{aligned} x - 0.10x - 2.00 &= 12.15 \\ 0.90x - 2.00 &= 12.15 \\ 0.90x &= 14.15 \\ x &= \frac{14.15}{0.90} \\ x &\approx 15.72 \end{aligned}$$

The original cost of the CD was $15.72.

37. $f(-1) = -(-1)^2 + 5 = -1 + 5 = 4$

38. $2x - y = 4$

$-y = -2x + 4$

$y = 2x - 4$

So $m_1 = 2$. The slope of a line perpendicular to this one is $m_2 = -\frac{1}{m_1} = -\frac{1}{2}$. Use the point-slope form with $m = -\frac{1}{2}$ and $(x_1, y_1) = (8, -2)$.

$y - y_1 = m(x - x_1)$

$y - (-2) = -\frac{1}{2}(x - 8)$

$y + 2 = -\frac{1}{2}x + 4$

$y = -\frac{1}{2}x + 2$

39. $(x_1, y_1) = (-2, 7)$ and $(x_2, y_2) = (2, -1)$

$m = \frac{-1 - 7}{2 - (-2)} = \frac{-8}{4} = -2$

Chapter 3 Review Exercises

1.

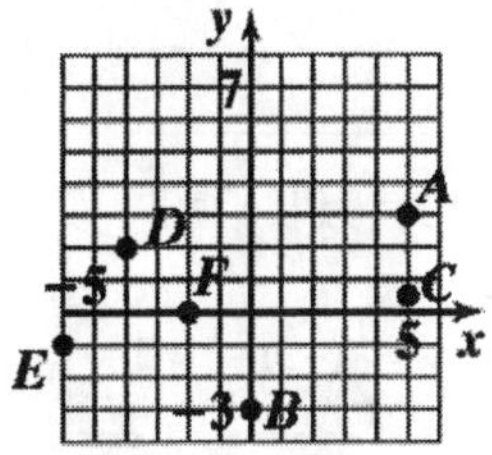

2. $y = \frac{1}{2}x$

x	y
–2	$y = \frac{1}{2}(-2) = -1$
0	$y = \frac{1}{2}(0) = 0$
2	$y = \frac{1}{2}(2) = 1$

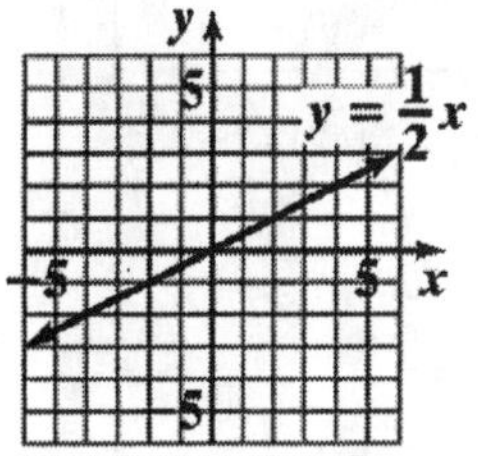

3. $y = -2x - 1$

x	y
0	$y = -2(0) - 1 = -1$
1	$y = -2(1) - 1 = -3$
2	$y = -2(2) - 1 = -4$

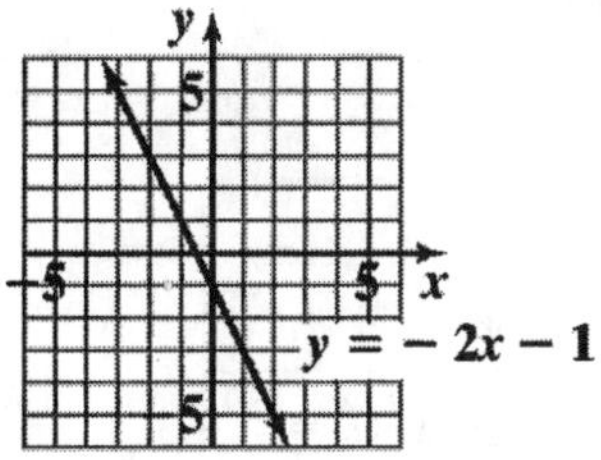

4. $y=\frac{1}{2}x+3$

x	y
0	$y=\frac{1}{2}(0)+3=3$
–2	$y=\frac{1}{2}(-2)+3=2$
–4	$y=\frac{1}{2}(-4)+3=1$

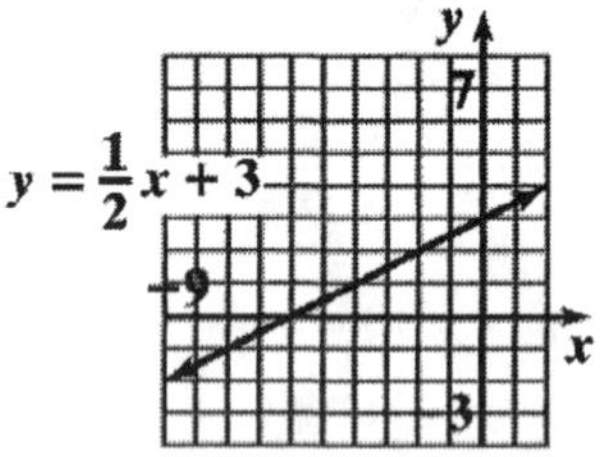

5. $y=-\frac{3}{2}x+1$

x	y
–2	$y=-\frac{3}{2}(-2)+1=4$
0	$y=-\frac{3}{2}(0)+1=1$
2	$y=-\frac{3}{2}(2)+1=-2$

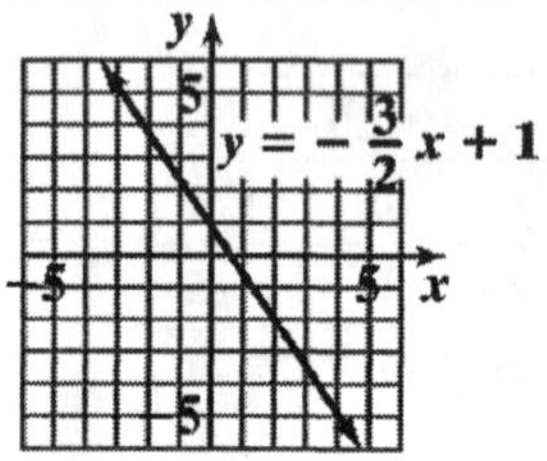

6. $y=x^2$

x	y
–3	$y=(-3)^2=9$
–1	$y=(-1)^2=1$
0	$y=0^2=0$
2	$y=2^2=4$

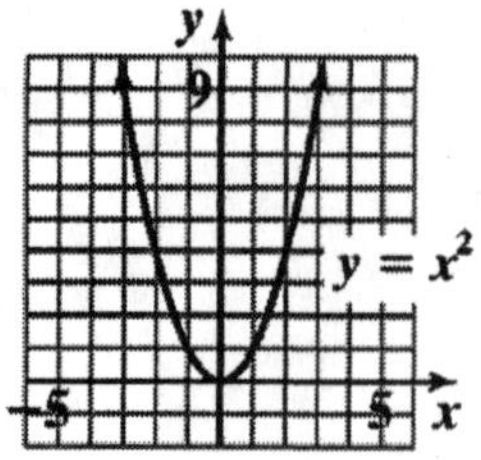

7. $y=x^2-1$

x	y
–3	$y=(-3)^2-1=8$
–1	$y=(-1)^2-1=0$
0	$y=0^2-1=0$
1	$y=1^2-1=0$
2	$y=2^2-1=3$

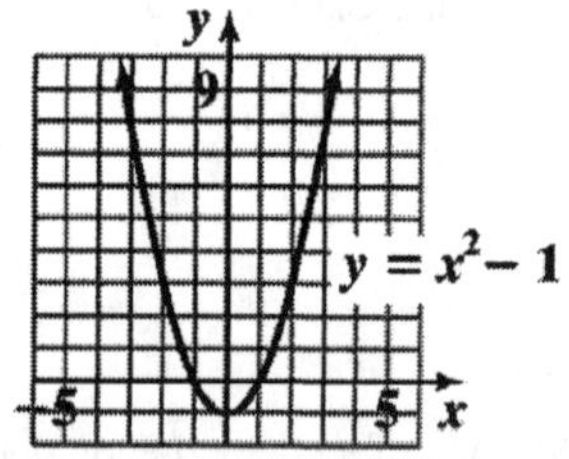

8. $y=|x|$

x	y
–4	$y=\lvert -4\rvert=4$
–1	$y=\lvert -1\rvert=1$
0	$y=\lvert 0\rvert=0$
2	$y=\lvert 2\rvert=2$

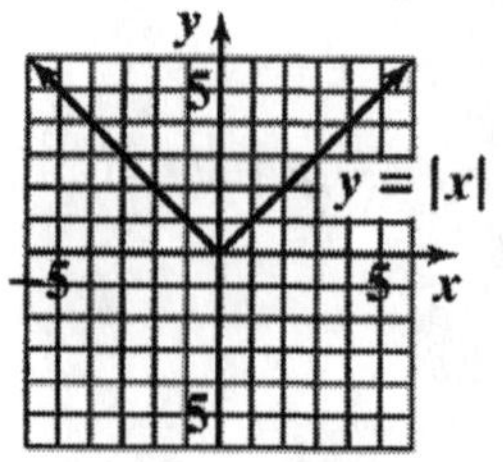

9. $y = |x| - 1$

x	y
–4	$y = \|-4\| - 1 = 3$
–1	$y = \|-1\| - 1 = 0$
0	$y = \|0\| - 1 = -1$
2	$y = \|2\| - 1 = 1$

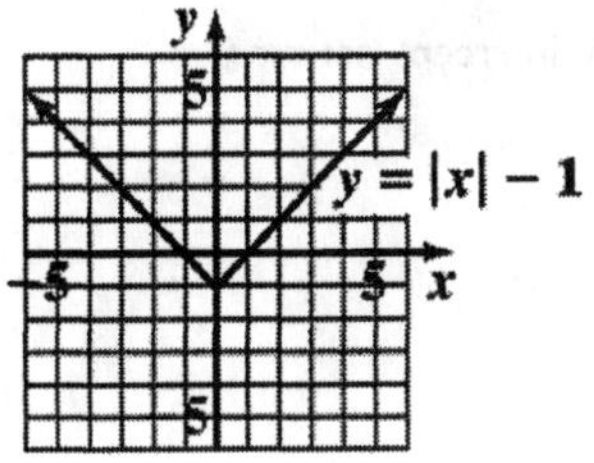

10. $y = x^3$

x	y
–2	$y = (-2)^3 = -8$
–1	$y = (-1)^3 = -1$
0	$y = 0^3 = 0$
1	$y = 1^3 = 1$
2	$y = 2^3 = 8$

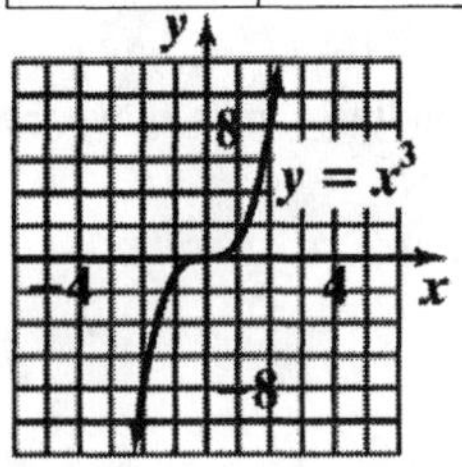

11. $y = x^3 + 4$

x	y
–2	$y = (-2)^3 + 4 = -4$
–1	$y = (-1)^3 + 4 = 3$
0	$y = 0^3 + 4 = 4$
1	$y = 1^3 + 4 = 5$

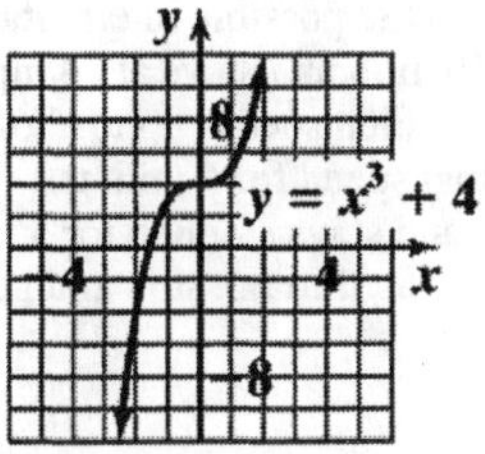

12. A function is a correspondence between a first set of elements, the domain, and a second set of elements, the range, such that each element of the domain corresponds to exactly one element in the range.

13. No, every relation is not a function. For example, {(4, 2), (4, –2)} is a relation but not a function.

Yes, every function is a relation because it is a set of ordered pairs.

14. Yes, each member of the domain corresponds to exactly one member of the range.

15. No, the domain element 2 corresponds to more than one member of the range (5 and –2).

16. a. Yes, the relation is a function.

b. Domain: $\mathbb{R}$; Range: $\mathbb{R}$

17. a. Yes, the relation is a function.

b. Domain: $\mathbb{R}$; Range: $\{y \mid y \le 0\}$

18. a. No, the relation is not a function.

b. Domain: $\{x \mid -3 \le x \le 3\}$;

Range: $\{y \mid -3 \le y \le 3\}$

19. a. No, the relation is not a function.

b. Domain: $\{x \mid -2 \le x \le 2\}$;

Range: $\{y \mid -1 \le y \le 1\}$

20. $f(x) = -x^2 + 3x - 4$

a. $f(2) = -(2)^2 + 3(2) - 4 = -4 + 6 - 4 = -2$

b. $f(h) = -h^2 + 3h - 4$

21. $g(t) = 2t^3 - 3t^2 + 6$

a. $g(-1) = 2(-1)^3 - 3(-1)^2 + 6 = -2 - 3 + 6 = 1$

b. $g(a) = 2a^3 - 3a^2 + 6$

22. Answers will vary. One possible interpretation: Car speeds up to 50 mph and stays at 50 mph for about 11 minutes. It then speeds up to about 68 mph and stays at that speed for 5 minutes. It then stops quickly and stays stopped for 5 minutes. It then travels through stop and go traffic for 5 minutes.

23. $N(x) = 40x - 0.2x^2$

a. $N(30) = 40(30) - 0.2(30)^2$
$= 1200 - 180$
$= 1020$
1020 baskets of apples are produced by 20 trees.

b. $N(50) = 40(50) - 0.2(50)^2$
$= 2000 - 500$
$= 1500$
1500 baskets of apples are produced by 50 trees.

24. $h(t) = -16t^2 + 196$

a. $h(1) = -16(1)^2 + 196 = -16 + 196 = 180$
After 1 second, the height of the ball is 180 feet.

b. $h(3) = -16(3)^2 + 196 = -144 + 196 = 52$
After 3 seconds, the height of the ball is 52 feet.

25. $3x - 4y = 6$
To find the x-intercept, set $y = 0$:
$3x - 4(0) = 6$
$3x = 6$
$x = 2$
To find the y-intercept , set $x = 0$.
$3(0) - 4y = 6$
$-4y = 6$
$y = \frac{6}{-4} = -\frac{3}{2}$

The intercepts are (2, 0) and $\left(0, -\frac{3}{2}\right)$.

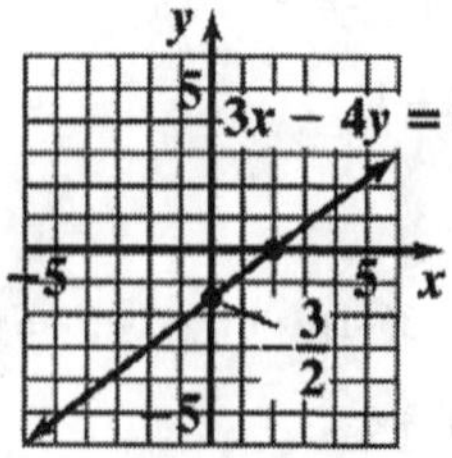

26. $\frac{1}{3}x = \frac{1}{8}y + 10$
To find the x-intercept, set $y = 0$.
$\frac{1}{3}x = \frac{1}{8}(0) + 10$
$\frac{1}{3}x = 10$
$3\left(\frac{1}{3}x\right) = 3(10)$
$x = 30$
To find the y-intercept, set $x = 0$.
$\frac{1}{3}(0) = \frac{1}{8}y + 10$
$0 = \frac{1}{8}y + 10$
$-10 = \frac{1}{8}y$
$8(-10) = 8\left(\frac{1}{8}y\right)$
$-80 = y$
The intercepts are (30, 0) and (0, –80).

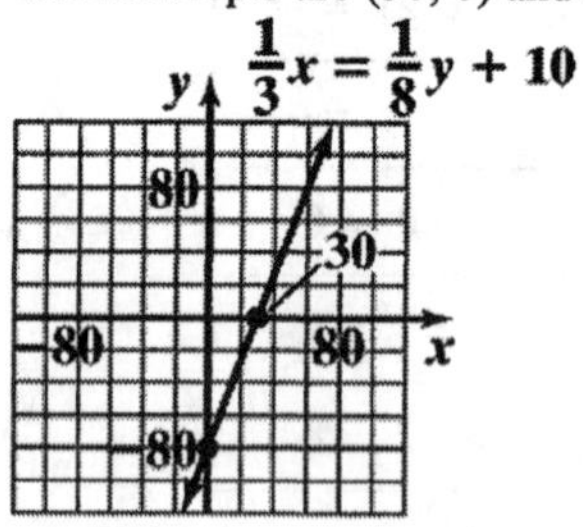

27. $f(x) = 4$ is a horizontal line 4 units above the x-axis.

x	y
–2	4
0	4
2	4

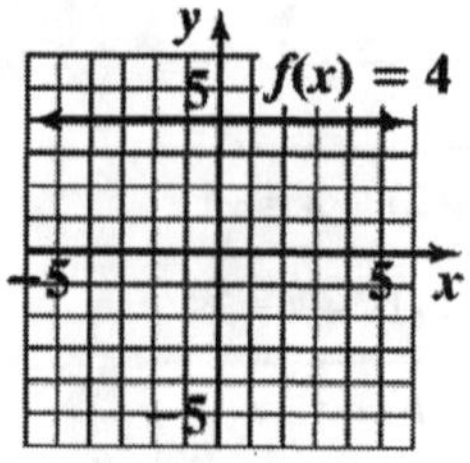

28. $x = -2$ is a vertical line 2 units to the left of the y-axis.

x	y
−2	2
−2	0
−2	−2

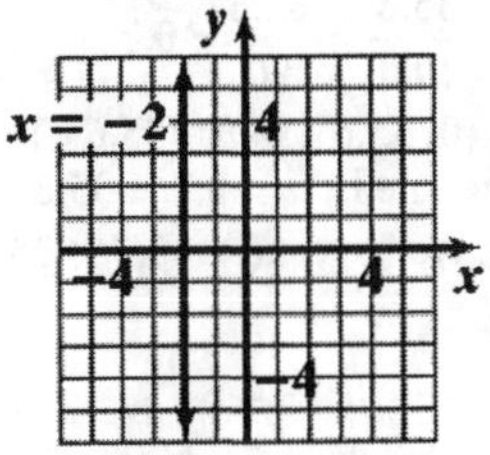

29. a. $p(x) = 0.1x - 5000$

x	p
0	$p = 0.1(0) - 5000 = -5000$
50,000	$p = 0.1(50{,}000) - 5000 = 0$
100,000	$p = 0.1(100{,}000) - 5000 = 5000$

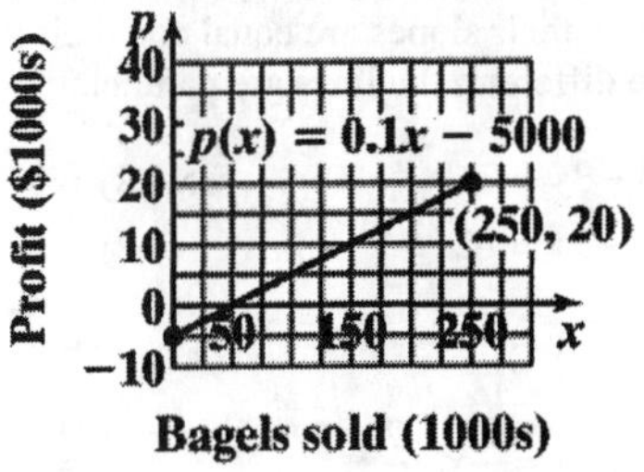

b. Approximately 50,000 bagels are sold when the company breaks even.

c. If the company has $22,000 profit, then approximately 270,000 bagels were sold.

30. The principle is $12,000 and the time is one year. Use the decimal form of the interest rate.
$I = 12{,}000r$

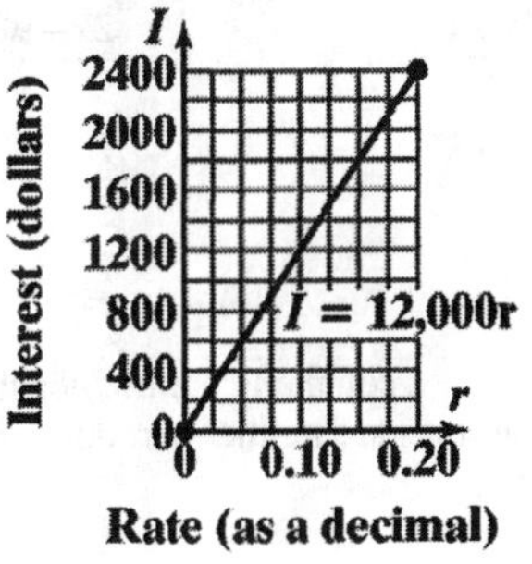

31. $y = \frac{1}{2}x - 5$

$m = \frac{1}{2}, \; b = -5$

The slope is $\frac{1}{2}$, and the y-intercept is $(0, -5)$.

32. $f(x) = -2x + 3$
$m = -2, \; b = 3$
The slope is -2, and the y-intercept is $(0, 3)$.

33. $3x + 5y = 13$
Solve for y.

$$5y = -3x + 13$$

$$y = \frac{-3x + 13}{5}$$

$$y = -\frac{3}{5}x + \frac{13}{5}$$

$$m = -\frac{3}{5}, \; b = \frac{13}{5}$$

The slope is $-\frac{3}{5}$, and the y-intercept is $\left(0, -\frac{13}{5}\right)$.

34. $3x + 4y = 10$
Solve for y.

$$4y = -3x + 10$$

$$y = \frac{-3x + 10}{4}$$

$$y = -\frac{3}{4}x + \frac{5}{2}$$

$$m = -\frac{3}{4}, b = \frac{5}{2}$$

The slope is $-\frac{3}{4}$, and the y-intercept is $\left(0, \frac{5}{2}\right)$.

35. $x = -7$ is a vertical line, so the slope undefined and there is no y-intercept.

36. $f(x) = 8$ is a horizontal line, so the slope is 0 and the y-intercept is (0, 8).

37. Let $(x_1, y_1) = (2, -5)$ and $(x_2, y_2) = (6, 7)$. Then

$$m = \frac{y_2 - y_1}{x_2 - x_1} = \frac{7 - (-5)}{6 - 2} = \frac{12}{4} = 3.$$

38. Let $(x_1, y_1) = (-2, 3)$ and $(x_2, y_2) = (4, 1)$. Then

$$m = \frac{y_2 - y_1}{x_2 - x_1} = \frac{1 - 3}{4 - (-2)} = \frac{-2}{6} = -\frac{1}{3}.$$

39. This is a horizontal line, so the slope is 0. The y-intercept is (0, 3), so the equation is $y = 3$.

40. This is a vertical line, so the slope is undefined. The x-intercept is (2, 0), so the equation is $x = 2$.

41. The x-intercept is (4, 0) and the y-intercept is (0, 2), so $m = \frac{2-0}{0-4} = \frac{2}{-4} = -\frac{1}{2}$ and $b = 2$. Thus, the equation is $y = -\frac{1}{2}x + 2$.

42. **a.** The slope of the translated graph is the same as the slope of the original graph: $m = -2$.

b. The y-intercept of the translated graph is the 4 less than the y-intercept of the original graph: $5 - 4 = 1$. The y-intercept of the translated graph is (0, 1).

c. Since $m = -2$ and $b = 1$, the equation of the translated graph is $y = -2x + 1$.

43. Use the point-slope form.

$$y - y_1 = m(x - x_1)$$
$$y - (-4) = \frac{2}{3}(x - (-6))$$
$$y + 4 = \frac{2}{3}(x + 6)$$
$$y + 4 = \frac{2}{3}x + 4$$
$$y = \frac{2}{3}x$$

Since $b = 0$, the y-intercept is (0, 0).

44. **a.**

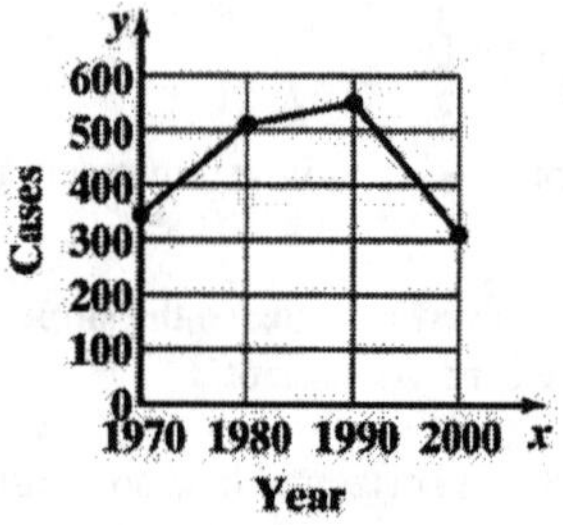

b. 1970 to 1980:

$$m_1 = \frac{510-346}{1980-1970} = \frac{164}{10} = 16.4$$

1980 to 1990:

$$m_2 = \frac{552-510}{1990-1980} = \frac{42}{10} = 4.2$$

1990 to 2000:

$$m_3 = \frac{317-552}{2000-1990} = \frac{-235}{10} = -23.5$$

c. The number of reported cases of typhoid fever increased the most during the 10-year period of 1970 – 1980.

45. Let t represent the number of years after 1980. Then 1980 is represented by $t = 0$ and 2070 is represented by $t = 90$. Therefore, the points (0, 35.6) and (90, 98.2) are on the line. Find the slope: $m = \frac{98.2-35.6}{2070-1980} = \frac{62.6}{90} \approx 0.7$.

The t-intercept is (0, 35,6), so $b = 35.6$. Thus, the equation of the line is: $n = 0.7t + 35.6$. With function notation, we have $n(t) = 0.7t + 35.6$.

46.

$$2x - 3y = 10$$
$$-3y = -2x + 10$$
$$\frac{-3y}{-3} = \frac{-2x+10}{-3}$$
$$y = \frac{2}{3}x - \frac{10}{3}$$
$$m_1 = \frac{2}{3}$$

$$y = \frac{2}{3}x - 5$$
$$m_2 = \frac{2}{3}$$

Since their slopes are equal and their y-intercepts are different, the lines are parallel.

47.

$$2x - 3y = 7$$
$$-3y = -2x + 7$$
$$y = \frac{-2x+7}{-3}$$
$$y = \frac{2}{3}x - \frac{7}{3}$$
$$m_1 = \frac{2}{3}$$

$$-3x - 2y = 8$$
$$-2y = 3x + 8$$
$$y = \frac{3x+8}{-2}$$
$$y = -\frac{3}{2}x - 4$$
$$m_2 = -\frac{3}{2}$$

Since the slopes are negative reciprocals, the lines are perpendicular.

48.

$$4x - 2y = 13$$
$$-2y = -4x + 13$$
$$y = \frac{-4x+13}{-2}$$
$$y = 2x - \frac{13}{2}$$
$$m_1 = 2$$

$$-2x + 4y = -9$$
$$4y = 2x - 9$$
$$y = \frac{2x-9}{4}$$
$$y = \frac{1}{2}x - \frac{9}{4}$$
$$m_2 = \frac{1}{2}$$

Since the slopes not equal and are not negative reciprocals, the lines are neither parallel nor perpendicular.

49. Use the point-slope form with $m=\frac{1}{2}$ and $(x_1, y_1)=(4, 9)$.

$$y-9=\frac{1}{2}(x-4)$$
$$y-9=\frac{1}{2}x-2$$
$$y=\frac{1}{2}x+7$$

50. First, find the slope: $m=\frac{-6-1}{4-(-3)}=\frac{-7}{7}=-1$.

Now, use the point-slope form with $m=-1$ and $(x_1, y_1)=(-3,1)$.

$$y-1=-1(x-(-3))$$
$$y-1=-1(x+3)$$
$$y-1=-x-3$$
$$y=-x-2$$

51. The slope of the line $y=-\frac{2}{3}x+1$ is $-\frac{2}{3}$. Since the new line is parallel to this line, its slope is also $-\frac{2}{3}$. Use the slope-intercept form with $m=-\frac{2}{3}$ and y-intercept of (0, 6).

$$y=-\frac{2}{3}x+6$$

52. $-2y=-5x+7$

$$y=\frac{-5x+7}{-2}$$
$$y=\frac{5}{2}x-\frac{7}{2}$$

The slope of this line is $\frac{5}{2}$. The new line is parallel to this line, so its slope is also $\frac{5}{2}$. Use the point-slope form with $m=\frac{5}{2}$ and $(x_1, y_1)=(2, 8)$.

$$y-8=\frac{5}{2}(x-2)$$
$$y-8=\frac{5}{2}x-5$$
$$y=\frac{5}{2}x+3$$

53. The slope of the line $y=\frac{3}{5}x+5$ is $\frac{3}{5}$. Since the new line is perpendicular to this line, its slope is $-\frac{5}{3}$. Use the point-slope form with $m=-\frac{5}{3}$ and $(x_1, y_1)=(-3, 1)$.

$$y-1=-\frac{5}{3}(x-(-3))$$
$$y-1=-\frac{5}{3}(x+3)$$
$$y-1=-\frac{5}{3}x-5$$
$$y=-\frac{5}{3}x-4$$

54. $4x-2y=8$

$$-2y=-4x+8$$
$$y=\frac{-4x+8}{-2}$$
$$y=2x-4$$

The slope of this line is 2. Since the new line is perpendicular to this line, its slope is $-\frac{1}{2}$. Use the point-slope form with $m=-\frac{1}{2}$ and $(x_1, y_1)=(4, 5)$.

$$y-5=-\frac{1}{2}(x-4)$$
$$y-5=-\frac{1}{2}x+2$$
$$y=-\frac{1}{2}x+7$$

55. $m_1=\frac{-3-3}{0-5}=\frac{-6}{-5}=\frac{6}{5}$

$m_2=\frac{-2-(-1)}{2-1}=\frac{-2+1}{2-1}=\frac{-1}{1}=-1$

Since the slopes are not equal and are not negative reciprocals, the lines are neither parallel nor perpendicular.

56. $m_1=\frac{3-2}{2-3}=\frac{1}{-1}=-1$

$m_2=\frac{4-1}{1-4}=\frac{3}{-3}=-1$

Since the slopes are equal, the lines are parallel.

57. $m_1 = \frac{6-3}{4-7} = \frac{3}{-3} = -1$

$m_2 = \frac{3-2}{6-5} = \frac{1}{1} = 1$

Since the slopes are negative reciprocals, the lines are perpendicular.

58. $m_1 = \frac{3-5}{2-(-3)} = \frac{3-5}{2+3} = \frac{-2}{5} = -\frac{2}{5}$

$m_2 = \frac{2-(-2)}{-1-(-4)} = \frac{2+2}{-1+4} = \frac{4}{3}$

Since the slopes are not equal and are not negative reciprocals, the lines are neither parallel nor perpendicular.

59. a. First, find the slope of the linear function using the points (35, 10.76) and (50, 19.91):

$m = \frac{19.91-10.76}{50-35} = \frac{9.15}{15} = 0.61$.

Use the point-slope form with $m = 0.61$ and $(a_1, r_1) = (35, 10.76)$:

$$\begin{aligned} r - r_1 &= m(a - a_1) \\ r - 10.76 &= 0.61(a - 35) \\ r - 10.76 &= 0.61a - 21.35 \\ r &= 0.61a - 10.59 \\ r(a) &= 0.61a - 10.59 \end{aligned}$$

b. $$\begin{aligned} r(40) &= 0.61(40) - 10.59 \\ &= 24.40 - 10.59 \\ &= 13.81 \end{aligned}$$

The monthly rate for a 40-year-old man is about \$13.81.

60. a. First, find the slope of the linear function using the points (30, 489) and (50, 525):

$m = \frac{525-489}{50-30} = \frac{36}{20} = 1.8$.

Use the point-slope form with $m = 1.8$ and $(r_1, C_1) = (30, 489)$:

$$\begin{aligned} C - C_1 &= m(r - r_1) \\ C - 489 &= 1.8(r - 30) \\ C - 489 &= 1.8r - 54 \\ C &= 1.8r + 435 \\ C(r) &= 1.8r + 435 \end{aligned}$$

b. $C(40) = 1.8(40) + 435 = 72 + 435 = 507$

When a person swims at 40 yards per minute for one hour, he or she will burn 507 calories.

c. $$\begin{aligned} C(r) &= 600 \\ 1.8r + 435 &= 600 \\ 1.8r &= 165 \\ r &\approx 91.7 \end{aligned}$$

To burn 600 calories in 1 hour, the person needs to swim at a speed of about 91.7 yards per minute.

61. $$\begin{aligned} (f+g)(x) &= f(x) + g(x) \\ &= (x^2 - 3x + 4) + (2x - 5) \\ &= x^2 - x - 1 \end{aligned}$$

62. $(f+g)(x) = x^2 - x - 1$

$(f+g)(4) = (4)^2 - 4 - 1 = 16 - 4 - 1 = 11$

63. $$\begin{aligned} (g-f)(x) &= (2x-5) - (x^2 - 3x + 4) \\ &= 2x - 5 - x^2 + 3x - 4 \\ &= -x^2 + 5x - 9 \end{aligned}$$

64. $(g-f)(x) = -x^2 + 5x - 9$

$$\begin{aligned} (g-f)(-1) &= -(-1)^2 + 5(-1) - 9 \\ &= -1 - 5 - 9 \\ &= -15 \end{aligned}$$

65. $$\begin{aligned} (f \cdot g)(-1) &= f(-1) \cdot g(-1) \\ &= \left((-1)^2 - 3(-1) + 4\right) \cdot (2(-1) - 5) \\ &= (1 + 3 + 4) \cdot (-2 - 5) \\ &= 8(-7) \\ &= -56 \end{aligned}$$

66. $$\begin{aligned} (f \cdot g)(3) &= f(3) \cdot g(3) \\ &= \left(3^2 - 3(3) + 4\right) \cdot (2(3) - 5) \\ &= (9 - 9 + 4) \cdot (6 - 5) \\ &= 4(1) \\ &= 4 \end{aligned}$$

67. $$\begin{aligned} (f/g)(1) &= \frac{f(1)}{g(1)} \\ &= \frac{1^2 - 3(1) + 4}{2(1) - 5} \\ &= \frac{1 - 3 + 4}{2 - 5} \\ &= \frac{2}{-3} \\ &= -\frac{2}{3} \end{aligned}$$

68. $(f/g)(2) = \dfrac{f(2)}{g(2)}$
$= \dfrac{2^2 - 3(2) + 4}{2(2) - 5}$
$= \dfrac{4 - 6 + 4}{4 - 5}$
$= \dfrac{2}{-1}$
$= -2$

69. **a.** The projected female population worldwide in 2050 is about 4.6 billion.

b. 2.9 billion − 0.8 billion = 2.1 billion
The projected number of women 15–49 years of age in 2050 about 2.1 billion.

c. 3.5 billion − 2.7 billion = 0.8 billion
The number of women who are projected to be in the 50 years and older age group in 2010 is about 0.8 billion.

d. 2002: 3.1 billion − 2.5 billion = 0.6 billion
2010: 3.5 billion − 2.7 billion = 0.8 billion
$\dfrac{0.8 \text{ billion} - 0.6 \text{ billion}}{0.6 \text{ billion}} = \dfrac{0.2 \text{ billion}}{0.6 \text{ billion}} \approx 0.33$
There is approximately a 33% increase in the projected number of women 50 years and older from 2002 to 2010.

70. **a.** Ginny's total retirement income in 2006 was about $47,000.

b. Ginny's pension income in 2005 was about $28,000.

c. 25,000 − 22,000 = 3000
Ginny's interest and dividend income in 2003 was about $3000.

71. $y \geq -5$
Graph the line $y = -5$ using a solid line. For the check point, select (0, 0).
$y \geq -5$
$0 \geq -5 \leftarrow$ Substitute 0 for y
Since this is a true statement, shade the region which contains (0, 0).

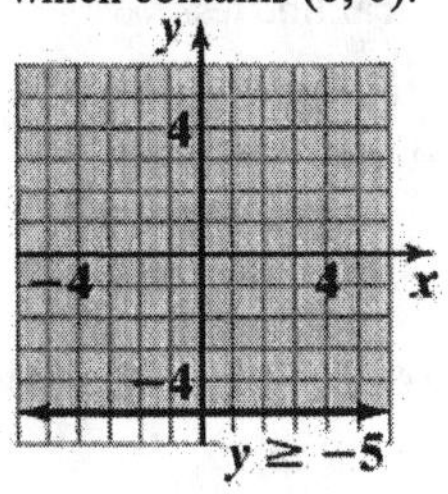

72. $x < 4$
Graph the line $x = 4$ using a dashed line. For the check point, select (0, 0).
$x < 4$
$0 < 4 \leftarrow$ Substitute 0 for x
Since this is a true statement, shade the region which contains (0, 0).

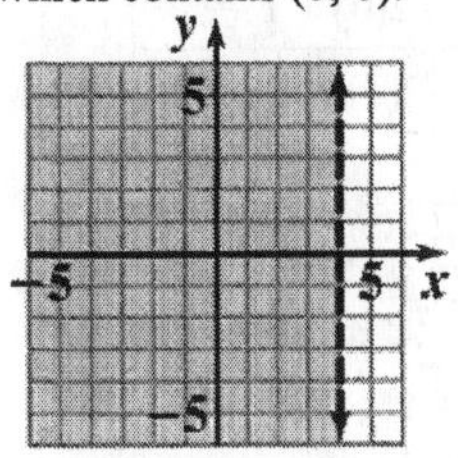

73. $y \leq 4x - 3$
Graph the line $y = 4x - 3$ using a solid line. For the check point, select (0, 0).
$y \leq 4x - 3$
$0 \leq 4(0) - 3 \leftarrow$ Substitute 0 for x and y
$0 \leq -3$
Since this is a false statement, shade the region which does not contain (0, 0).

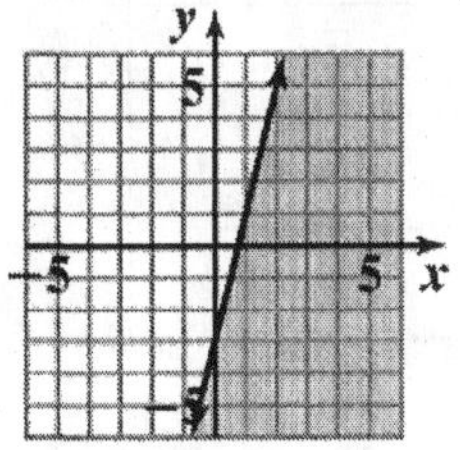

74. $y < \dfrac{1}{3}x - 2$

Graph the line $y = \dfrac{1}{3}x - 2$ using a dashed line.
For the check point, select (0, 0).
$y < \dfrac{1}{3}x - 2$
$0 < \dfrac{1}{3}(0) - 2 \leftarrow$ Substitute 0 for x and y
$0 < -2$
Since this is a false statement, shade the region which does not contain (0, 0).

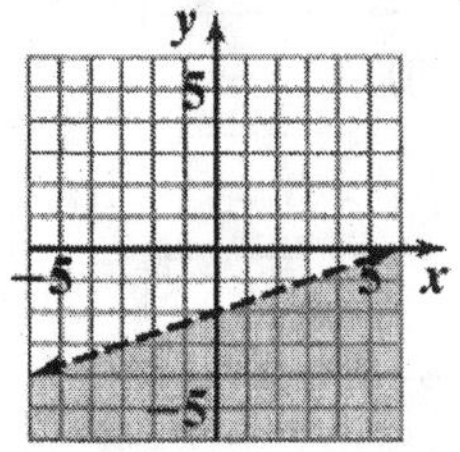

Chapter 3 Practice Test

1. $y = -2x + 1$

x	y
-1	$y = -2(-1) + 1 = 3$
0	$y = -2(0) + 1 = 1$
1	$y = -2(1) + 1 = -1$

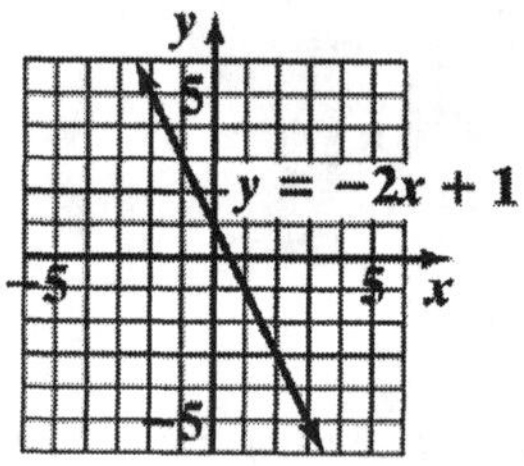

2. $y = \sqrt{x}$

x	y
-1	$y = \sqrt{-1}$ undefined
0	$y = \sqrt{0} = 0$
1	$y = \sqrt{1} = 1$
4	$y = \sqrt{4} = 2$

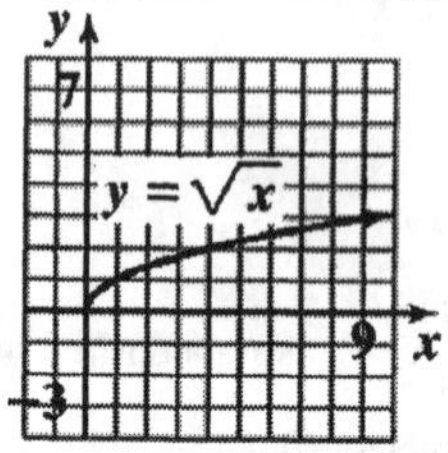

3. $y = x^2 - 4$

x	y
-2	$y = (-2)^2 - 4 = 0$
-1	$y = (-1)^2 - 4 = -3$
0	$y = (0)^2 - 4 = -4$
1	$y = (1)^2 - 4 = -3$
2	$y = (2)^2 - 4 = 0$

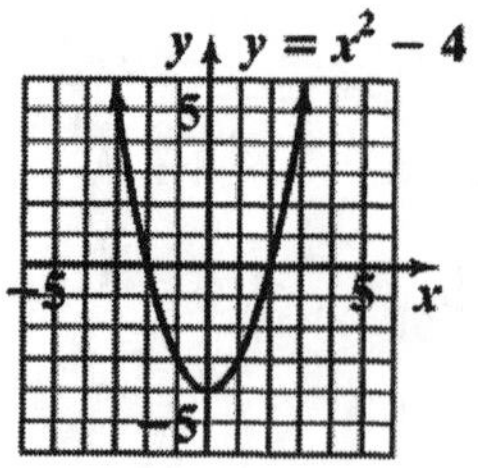

4. $y = |x|$

x	y
-3	$y = |-3| = 3$
0	$y = |0| = 0$
4	$y = |4| = 4$

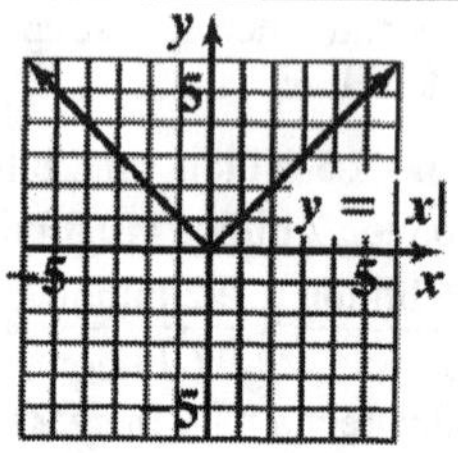

5. A function is a correspondence between a first set of elements, the domain, and a second set of elements, the range, such that each element of the domain corresponds to exactly one element in the range.

6. Yes, the set of ordered pairs represents a function because each member in the domain corresponds with exactly one member in the range.

7. Yes, the graph represents a function because it passes the vertical line test.
   Domain: $\mathbb{R}$
   Range: $\{y | y \le 4\}$

8. No, the graph does not represent a function because it fails the vertical line test.
   Domain: $\{x | -3 \le x \le 3\}$
   Range: $\{y | -2 \le y \le 2\}$

9. $f(x) = 3x^2 - 6x + 5$
   $f(-2) = 3(-2)^2 - 6(-2) + 5 = 12 + 12 + 5 = 29$

10. $-20x + 10y = 40$

To find the x-intercept, set $y = 0$:

$$-20x + 10(0) = 40$$
$$-20x = 40$$
$$x = -2$$

The x-intercept is $(-2, 0)$.

To find the y-intercept, set $x = 0$:

$$-20(0) + 10y = 40$$
$$10y = 40$$
$$y = 4$$

The y-intercept is $(0, 4)$.

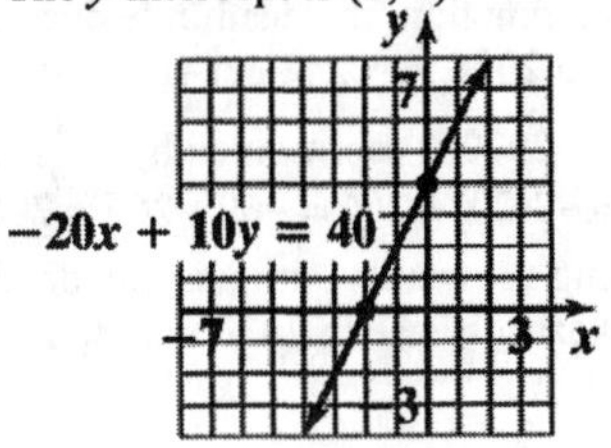

11. $\frac{x}{5} - \frac{y}{4} = 1$

To find the x-intercept, set $y = 0$:

$$\frac{x}{5} - \frac{0}{4} = 1$$
$$\frac{x}{5} = 1$$
$$x = 5$$

The x-intercept is $(5, 0)$.

To find the y-intercept, set $x = 0$:

$$\frac{0}{5} - \frac{y}{4} = 1$$
$$-\frac{y}{4} = 1$$
$$-4\left(-\frac{y}{4}\right) = -4(1)$$
$$y = -4$$

The y-intercept is $(0, -4)$.

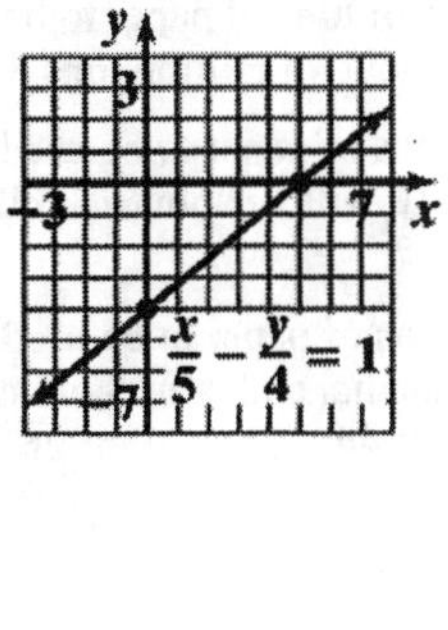

12. The graph of $f(x) = -3$ is a horizontal line 3 units below the x-axis.

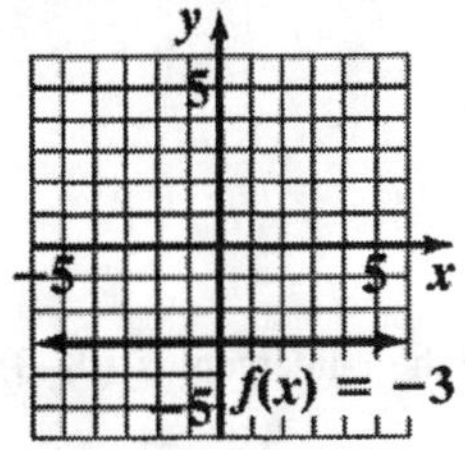

13. The graph of $x = 4$ is a vertical line 4 units to the right of the y-axis.

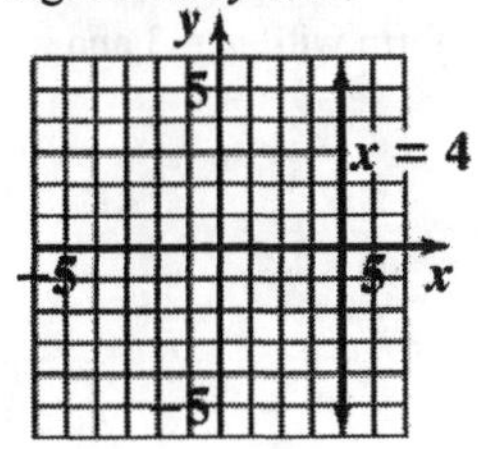

14. a. $p(x) = 10.2x - 50{,}000$

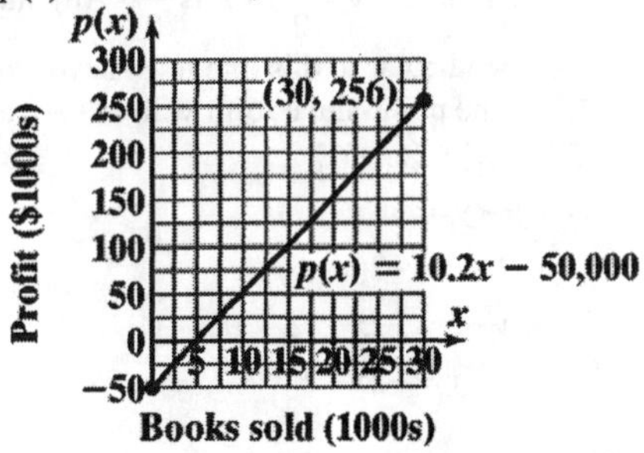

b. The company breaks even when $p(x) = 0$.

$$10.2x - 50{,}000 = 0$$
$$10.2x = 50{,}000$$
$$x = \frac{50{,}000}{10.2} = 4900$$

The company breaks even when it sells 4900 books.

c. $10.2x - 50{,}000 = 100{,}000$

$$10.2x = 150{,}000$$
$$x = \frac{150{,}000}{10.2} \approx 14{,}700$$

The company needs to sell about 14,700 books to break even.

15. $4x-3y=15$

$$-3y=-4x+15$$
$$y=\frac{-4x+15}{-3}$$
$$y=\frac{4}{3}x-5$$

The slope is $\frac{4}{3}$ and the y-intercept is $(0,-3)$.

16. $m=\frac{5-2}{4-3}=\frac{3}{1}=3$

Use the point-slope form with $m=3$ and $(x_1, y_1)=(3, 2)$.

$$y-y_1=m(x-x_1)$$
$$y-2=3(x-3)$$
$$y-2=3x-9$$
$$y=3x-7$$

17. The slope of $y=\frac{1}{2}x+1$ is $\frac{1}{2}$. Any line perpendicular to this line must have slope -2. Use the point-slope form with $m=-2$ and $(x_1, y_1)=(6,-5)$.

$$y-y_1=m(x-x_1)$$
$$y-(-5)=-2(x-6)$$
$$y+5=-2x+12$$
$$y=-2x+7$$

18. The year 2000 correspond to $t=0$, and the year 2050 corresponds to $t=50$. Thus, the points (0, 274.634) and (50, 419.854) lie on the graph.

$$m=\frac{419.854-274.634}{50-0}=\frac{145.220}{50}=2.9044$$

The y-intercept is (0, 274.634), so $b=274.634$. Thus, the equation is $p(t)=2.9044t+274.634$.

19. Write each equation in slope-intercept form by solving for y.

$$2x-3y=12 \qquad\qquad 4x+10=6y$$
$$-3y=-2x+12 \qquad\qquad \frac{4x+10}{6}=y$$
$$y=\frac{-2x+12}{-3} \qquad\qquad \frac{4}{6}x+\frac{10}{6}=y$$
$$y=\frac{2}{3}x-4 \qquad\qquad y=\frac{2}{3}x+\frac{5}{3}$$

Since the slopes are the same and the y-intercepts are different, the lines are parallel.

20. a. Let t be the number of years since 2000. Then 2000 is represented by $t=0$ and 2010 is represented by $t=10$. Find the slope of the linear function using the points (0, 266) and (10, 236).

$$m=\frac{236-266}{10-0}=\frac{-30}{10}=-3$$

The y-intercept is (0, 266), so $b=266$. Thus, the equation is $r(t)=-3t+266$.

b. Note that 2006 is represented by $t=6$:

$r(6)=-3(6)+266=-18+266=248$

From the function, the death rate due to heart disease in 2006 was 248 per 100,000.

c. Note that 2020 is represented by $t=20$:

$r(20)=-3(20)+266=-60+266=206$

Assuming the trend continues, the death rate due to heart disease in 2020 will be 206 per 100,000.

21.
$$\begin{aligned}(f+g)(3)&=f(3)+g(3)\\&=\left(2(3)^2-(3)\right)+\left((3)-6\right)\\&=\left(2(9)-3\right)+(3-6)\\&=(18-3)+(3-6)\\&=15+(-3)\\&=12\end{aligned}$$

22.
$$\begin{aligned}(f/g)(-1)&=\frac{f(-1)}{g(-1)}\\&=\frac{2(-1)^2-(-1)}{(-1)-6}\\&=\frac{2+1}{-7}\\&=-\frac{3}{7}\end{aligned}$$

23. $f(a)=2a^2-a$

24. a. The total number of tons of paper to be used in 2010 will be about 44 million tons.

b. The number of tons of paper to be used by businesses in 2010 will be about 18 million tons.

c. The number of tons of paper to be used for reference, print media, and household use in 2010 will be about 44 – 18 = 26, or 26 million tons.

25. Graph the line $y = 3x - 2$ using a dashed line. For the check point, select (0, 0).
$y < 3x - 2$
$0 < 3 \cdot 0 - 2 \leftarrow$ Substitute 0 for x and y
$0 < -2$
Since the statement is false, shade the region that does not contain (0, 0).

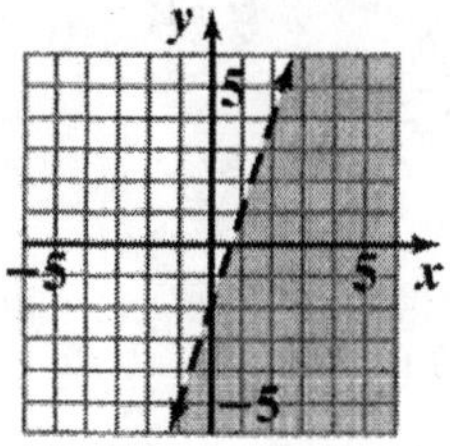

Chapter 3 Cumulative Review Test

1. a. $A \cap B = \{3, 5, 7\}$

b. $A \cup B = \{1, 2, 3, 5, 7, 9, 11, 14\}$

2. a. None of the numbers are natural numbers.

b. $-6, -4, \frac{1}{3}, 0, \sqrt{3}, 4.67, \frac{37}{2},$ and $-\sqrt{5}$ are real numbers.

3.
$$\begin{aligned} 10 - \{3[6 - 4(6^2 \div 4)]\} &= 10 - \{3[6 - 4(36 \div 4)]\} \\ &= 10 - \{3[6 - 4(9)]\} \\ &= 10 - \{3[6 - 36]\} \\ &= 10 - \{3[-30]\} \\ &= 10 - \{-90\} \\ &= 10 + 90 \\ &= 100 \end{aligned}$$

4. $$\left(\frac{5x^2}{y^{-3}}\right)^2 = (5x^2y^3)^2 = 5^2(x^2)^2(y^3)^2 = 25x^4y^6$$

5. $$\left(\frac{3x^4y^{-2}}{6xy^3}\right)^3 = \left(\frac{x^3}{2y^5}\right)^3 = \frac{(x^3)^3}{2^3(y^5)^3} = \frac{x^9}{8y^{15}}$$

6. a.
$$\begin{aligned} \text{Commercial Sector} &= 14\% \text{ of } 2.18 \times 10^{13} \\ &= 0.14 \times 2.18 \times 10^{13} \\ &= 0.3052 \times 10^{13} \\ &= 3.052 \times 10^{12} \end{aligned}$$
Commercial sector consumption of natural gas in 2003 was 3.052×10^{12} cubic feet.

b.
$$\begin{aligned} \text{Industrial Sector} &= 37\% \text{ of } 2.18 \times 10^{13} \\ &= 0.37 \times 2.18 \times 10^{13} \\ &= 0.8066 \times 10^{13} \\ &= 8.066 \times 10^{12} \end{aligned}$$
$$\begin{aligned} \text{Transportation Sector} &= 3\% \text{ of } 2.18 \times 10^{13} \\ &= 0.03 \times 2.18 \times 10^{13} \\ &= 0.0654 \times 10^{13} \\ &= 6.54 \times 10^{11} \end{aligned}$$
$$\begin{aligned} &8.066 \times 10^{12} - 6.54 \times 10^{11} \\ &= 8.066 \times 10^{12} - 0.654 \times 10^{12} \\ &= (8.066 - 0.654) \times 10^{12} \\ &= 7.412 \times 10^{12} \end{aligned}$$
In 2003, the industrial sector consumed 7.412×10^{12} cubic feet more of natural gas than the transportation sector.

c.
$$\begin{aligned} &2.18 \times 10^{13} + 10\% \text{ of } 2.18 \times 10^{13} \\ &= 110\% \text{ of } 2.18 \times 10^{13} \\ &= 1.1 \times 2.18 \times 10^{13} \\ &= 2.398 \times 10^{13} \end{aligned}$$
Consumption of natural gas in 2006 will be 2.398×10^{13} cubic feet.

7.
$$\begin{aligned} 2(x+4) - 5 &= -3[x - (2x+1)] \\ 2x + 8 - 5 &= -3[x - 2x - 1] \\ 2x + 3 &= -3[-x - 1] \\ 2x + 3 &= 3x + 3 \\ 3 &= x + 3 \\ 0 &= x \end{aligned}$$

8.
$$\begin{aligned} \frac{4}{5} - \frac{x}{3} &= 10 \\ 15\left(\frac{4}{5} - \frac{x}{3}\right) &= 15(10) \\ 12 - 5x &= 150 \\ -5x &= 138 \\ x &= -\frac{138}{5} \end{aligned}$$

9.
$$\begin{aligned} 7x - \{4 - [2(x-4)] - 5\} &= 7x - \{4 - [2x - 8] - 5\} \\ &= 7x - \{4 - 2x + 8 - 5\} \\ &= 7x - \{-2x + 7\} \\ &= 7x + 2x - 7 \\ &= 9x - 7 \end{aligned}$$

10. $A = \frac{1}{2}h(b_1 + b_2)$

$2A = h(b_1 + b_2)$

$\frac{2A}{h} = b_1 + b_2$

$\frac{2A}{h} - b_2 = b_1$

$b_1 = \frac{2A}{h} - b_2$

11. Let x be amount of 15% hydrogen peroxide solution.

Solution	Strength	Amount	Salt
15%	0.15	x	$0.15x$
4%	0.04	10	0.04(10)
Mixture	0.10	$x + 10$	$0.10(x + 10)$

$0.15x + 0.04(10) = 0.10(x + 10)$

$0.15x + 0.4 = 0.10x + 1$

$0.15x - 0.10x = 1 - 0.4$

$0.05x = 0.6$

$x = \frac{0.6}{0.05}$

$x = 12$

12 gallons of the 15% hydrogen peroxide solution must be added.

12. $4(x-4) < 8(2x+3)$

$4x - 16 < 16x + 24$

$4x - 16x < 24 + 16$

$-12x < 40$

$\frac{-12x}{-12} > \frac{40}{-12}$

$x > -\frac{10}{3}$

The solution set is $\left\{x \middle| x > -\frac{10}{3}\right\}$.

13. $-1 < 3x - 7 < 11$

$-1 + 7 < 3x - 7 + 7 < 11 + 7$

$6 < 3x < 18$

$\frac{6}{3} < \frac{3x}{3} < \frac{18}{3}$

$2 < x < 6$

The solution set is $\{x |\ 2 < x < 6\}$.

14. $|3x + 5| = |2x - 10|$

$3x + 5 = 2x - 10$ or $3x + 5 = -(2x - 10)$

$3x - 2x = -10 - 5$ $\quad$ $3x + 5 = -2x + 10$

$x = -15$ $\quad$ $3x + 2x = 10 - 5$

$5x = 5$

$x = 1$

The solution set is $\{-15, 1\}$.

15. $|2x - 1| \le 3$

$-3 \le 2x - 1 \le 3$

$-3 + 1 \le 2x - 1 + 1 \le 3 + 1$

$-2 \le 2x \le 4$

$\frac{-2}{2} \le \frac{2x}{2} \le \frac{4}{2}$

$-1 \le x \le 2$

The solution set is $\{x | -1 \le x \le 2\}$.

16. Set up a table of values to find some points on the line $y = -\frac{3}{2}x - 4$.

x	y
–4	$-\frac{3}{2}(-4) - 4 = 2$
–2	$-\frac{3}{2}(-2) - 4 = -1$
0	$-\frac{3}{2}(0) - 4 = -4$

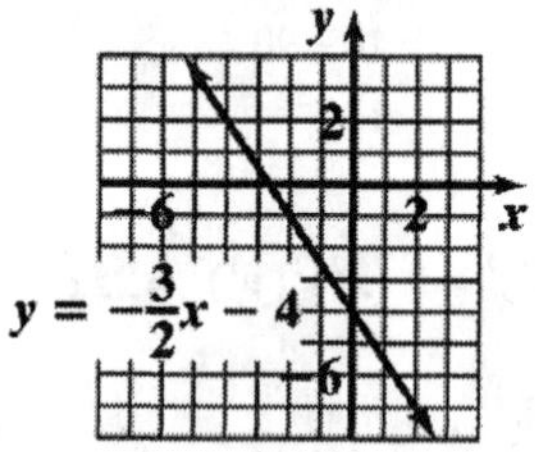

17. a. The graph is not a function because it fails the vertical line test.

b. Domain: $\{x | x \le 2\}$

Range: $\mathbb{R}$

18. Let $(x_1, y_1) = (-5, 3)$ and $(x_2, y_2) = (4, -1)$:

$m = \frac{y_2 - y_1}{x_2 - x_1} = \frac{-1 - 3}{4 - (-5)} = \frac{-4}{9} = -\frac{4}{9}$

19. Write each equation in slope-intercept form by solving for y.

$$2x-5y=8 \qquad 5x-2y=12$$
$$-5y=-2x+8 \qquad -2y=-5x+12$$
$$y=\frac{-2x+8}{-5} \qquad y=\frac{-5x+12}{-2}$$
$$y=\frac{2}{5}x-\frac{8}{5} \qquad y=\frac{5}{2}x-6$$

Since the slopes are $\frac{2}{5}$ and $\frac{5}{2}$ which are neither equal nor negative reciprocals, the lines are neither parallel nor perpendicular.

20. $(f+g)(x)=f(x)+g(x)$

$$=(x^2+3x-2)+(4x-9)$$
$$=x^2+7x-11$$

Chapter 4

Exercise Set 4.1

1. The solution to a system of linear equations is the point(s) that satisfy all equations in the system.

3. A dependent system of equations is a system of equations that has an infinite number of solutions.

5. A consistent system of equations has a solution.

7. Compare the slopes and y-intercepts of the equations. If the slopes are different, the system is consistent. If the slopes and y-intercepts are the same, the system is dependent. If the slopes are the same and the y-intercepts are different, the system is inconsistent.

9. You will get a true statement, like $0 = 0$.

11. $y = 2x + 4$ and $y = 2x - 1$

a. (0, 4) does not satisfy the second equation since the left side is 4, whereas the right side is $2(0) - 1 = 0 - 1 = -1$.

b. (3, 10) does not satisfy the second equation since the left side is 10, whereas the right side is
$2(3) - 1 = 6 - 1 = 5$

13. $x + y = 25$ and $0.25x + 0.45y = 7.50$

a. (5, 20) does not satisfy the second equation and is therefore not a solution to the system.

1st equation:	2nd equation:
$(5)+(20)=25$	$0.25(5)+0.45(20)=7.50$
$25=25$	$10.25 \neq 7.50$

b. (18.75, 6.25) satisfies both equations and is therefore a solution to the system.

1st equation:	2nd equation:
$(18.75)+(6.25)=25$	$0.25(18.75)+0.45(6.25)=7.50$
$25=25$	$7.50=7.50$

15.
$$x+2y-z=-5$$
$$2x-y+2z=8$$
$$3x+3y+4z=5$$

a. (3, 1, –2) does not satisfy the first equation since the left side is $3+2(1)-(-2)=3+2+2=7$, whereas the right side is –5. Thus, $(3,1,-2)$ is not a solution to the system.

b. (1, –2, 2) satisfies all three equations. For the first equation, the left side is $1+2(-2)-(2)=1-4-2=-5$ and the right side is –5. For the second equation, the left side is $2(1)-(-2)+2(2)=2+2+4=8$ and the right side is 8. Finally, for the third equation, the left side is $3(1)+3(-2)+4(2)=3-6+8=5$ and the right side is 5. Therefore, $(1,-2,2)$ is a solution to the system.

17. Write each equation in slope-intercept form.

$-7x+3y=1$	$3y+12=-6x$
$3y=7x+1$	$3y=-6x-12$
$y=\frac{7}{3}x+\frac{1}{3}$	$y=-2x-4$

The slope of the first line is $\frac{7}{3}$ and the slope of the second line is -2. The slopes are different so the lines intersect to produce one solution. This is a consistent system.

19. Multiply the first equation through by 12.

$$\frac{x}{3}+\frac{y}{4}=1 \quad \text{and} \quad 4x+3y=12$$
$$12\left(\frac{x}{3}+\frac{y}{4}\right)=12(1)$$
$$4x+3y=12$$

Since both equations are identical, the line is the same for both of them to produce an infinite number of solutions. This is a dependent system.

21. Write each equation in slope-intercept form.

$$3x-3y=9 \qquad\qquad 2x-2y=-4$$
$$-3y=-3x+9 \qquad\qquad -2y=-2x-4$$
$$y=x-3 \qquad\qquad y=x+2$$

Since the slope of each line is 1, but the y-intercepts are different ($b=-3$ for the first equation, $b=2$ for the second equation), the two lines are parallel and produce no solution. This is an inconsistent system.

23. Write each equation in slope-intercept form.

$$y=\frac{3}{2}x+\frac{1}{2} \qquad\qquad 3x-2y=-\frac{5}{2}$$
$$-2y=-3x-\frac{5}{2}$$
$$y=\frac{3}{2}x+\frac{5}{4}$$

Since the slope of each line is $\frac{3}{2}$, but the y-intercepts are different ($b=\frac{1}{2}$ for the first equation, $b=\frac{5}{4}$ for the second equation) the two lines are parallel and produce no solution. This is an inconsistent system.

25. Graph the equations $y = x + 5$ and $y = -x + 3$.

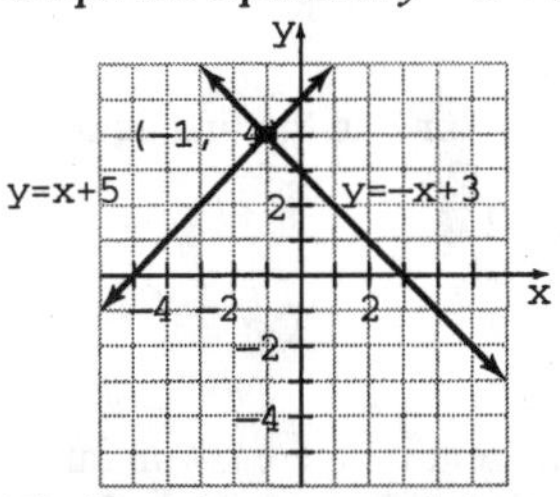

The lines intersect and the point of intersection is (−1, 4). This is a consistent system.

27. Graph the equations $y=4x-1$ and $3y=12x+9$.

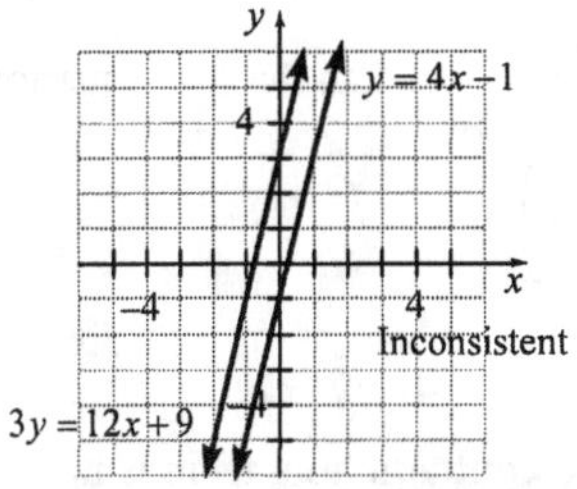

The lines are parallel. The system is inconsistent and there is no solution.

29. Graph the equations $2x+3y=6$ and $4x=-6y+12$.

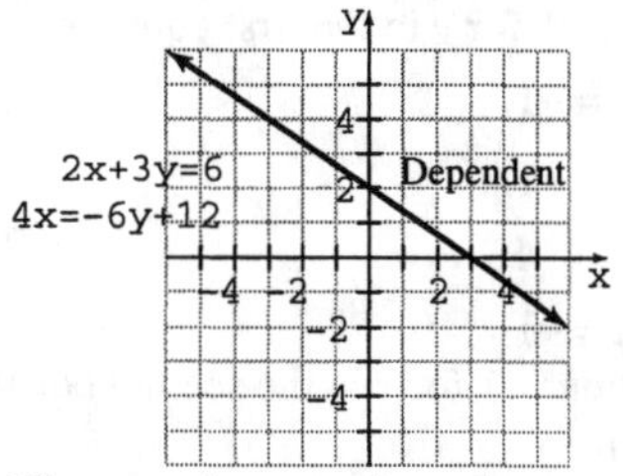

The equations produce the same line. The system is dependent and there are an infinite number of solutions.

31. Graph the equations $5x+3y=13$ and $x=2$.

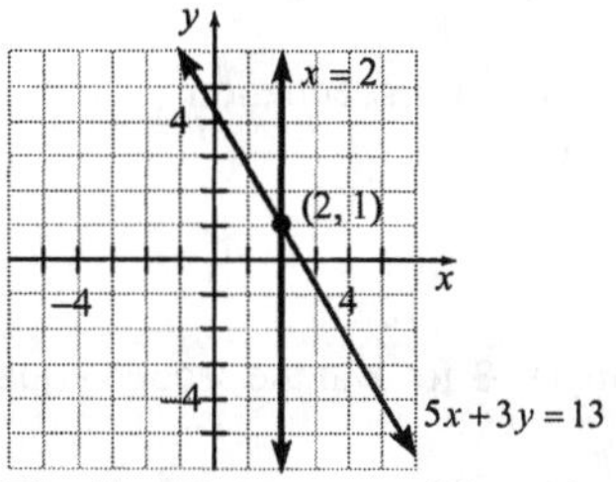

The lines intersect and the point of intersection is (2, 1). This is a consistent system.

33. Graph the equations $y=-5x+5$ and $y=2x-2$.

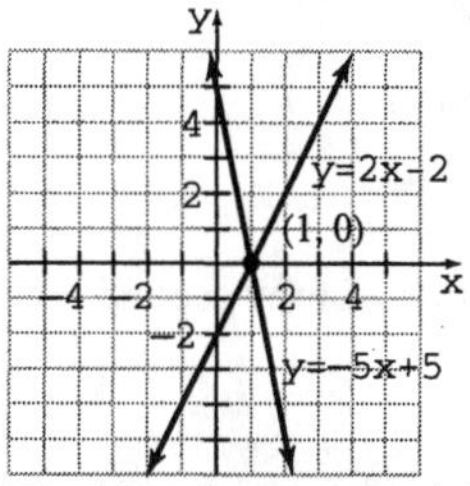

The lines intersect and the point of intersection is (1, 0). This is a consistent system.

35. Graph the equations $x-\frac{1}{2}y=-2$ and $2y=4x-6$.

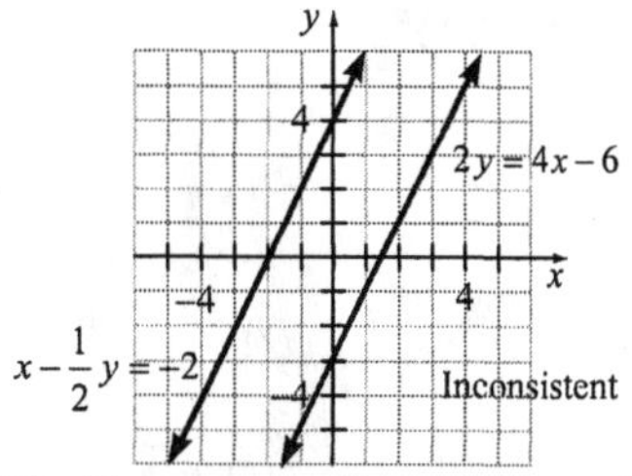

The lines are parallel and do not intersect. The system is inconsistent and there is no solution.

37. $x+3y=-1$

$y=x+1$

Substitute $x+1$ for y in the first equation.

$$x+3(x+1)=-1$$
$$x+3x+3=-1$$
$$4x=-4$$
$$x=-1$$

Now, substitute -1 for x in the second equation.

$$y=(-1)+1$$
$$=0$$

The solution is (–1, 0).

39. $x=2y+3$

$y=x$

Substitute x for y in first equation.

$$x=2(x)+3$$
$$-x=3$$
$$x=-3$$

Now, substitute -3 for x in the second equation.

$$x=y$$
$$-3=y$$

The solution is (–3, –3).

41. $a+3b=5$

$2a-b=3$

Solve the first equation for a.

$$a+3b=5$$
$$a=-3b+5$$

Substitute $-3b+5$ for a in the second equation.

$$2(-3b+5)-b=3$$
$$-6b+10-b=3$$
$$-7b=-7$$
$$b=1$$

Now, substitute 1 for b in the first equation.

$$a+3(1)=5$$
$$a=2$$

The solution is (2, 1).

43. $5x+6y=6.7$

$3x-2y=0.1$

Solve the second equation for y:

$$3x-2y=0.1$$
$$-2y=-3x+0.1$$
$$y=\frac{3}{2}x-0.05$$

Now substitute $\frac{3}{2}x-0.05$ for y in the first equation and solve for x:

$$5x+6\left(\frac{3}{2}x-0.05\right)=6.7$$
$$5x+9x-0.3=6.7$$
$$14x=7.0$$
$$x=0.5$$

Substitue $x=0.5$ into the equation for y:

$$y=\frac{3}{2}x-0.05$$
$$=\frac{3}{2}(0.5)-0.05$$
$$=0.7$$

The solution is $(0.5, 0.7)$.

45. $a-\frac{1}{2}b=2$

$b=2a-4$

Substitute $2a-4$ for b in the first equation.

$$a-\frac{1}{2}(2a-4)=2$$
$$a-a+2=2$$
$$2=2$$

Since this is an identity, there are an infinite number of solutions. This is a dependent system.

47. $5x-2y=-7$

$y=\frac{5}{2}x+1$

Substitute $\frac{5}{2}x+1$ for y in the first equation.

$$5x-2\left(\frac{5}{2}x+1\right)=-7$$
$$5x-5x-2=-7$$
$$-2\neq-7$$

Since this is a contradiction, there is no solution. This is an inconsistent system.

49. $5x-4y=-7$

$x-\frac{3}{5}y=-2$

First, solve the second equation for x.

$x-\frac{3}{5}y=-2$

$x=\frac{3}{5}y-2$

Now substitute $\frac{3}{5}y-2$ for x in the first equation.

$5\left(\frac{3}{5}y-2\right)-4y=-7$

$3y-10-4y=-7$

$-10-y=-7$

$-y=3$

$y=-3$

Finally, substitute –3 for y in the equation $x=\frac{3}{5}y-2$.

$x=\frac{3}{5}(-3)-2$

$x=-\frac{9}{5}-2$

$x=-\frac{9}{5}-\frac{10}{5}$

$x=-\frac{19}{5}$

The solution is $\left(-\frac{19}{5},-3\right)$.

51. $\frac{1}{2}x-\frac{1}{3}y=2$

$\frac{1}{4}x+\frac{2}{3}y=6$

Solve the first equation for y.

$\frac{1}{2}x-\frac{1}{3}y=2$

$-\frac{1}{3}y=-\frac{1}{2}x+2$

$y=\frac{3}{2}x-6$

Substitute $\frac{3}{2}x-6$ for y in the second equation.

$\frac{1}{4}x+\frac{2}{3}\left(\frac{3}{2}x-6\right)=6$

$\frac{1}{4}x+x-4=6$

$\frac{5}{4}x-4=6$

$\frac{5}{4}x=10$

$x=\frac{4}{5}(10)=8$

Substitute 8 for x in the equation $y=\frac{3}{2}x-6$.

$y=\frac{3}{2}(8)-6=12-6=6$

The solution is (8, 6).

53.

$$\begin{aligned} x+y&=9\\ \underline{x-y}&\underline{=-3}\\ \text{Add: }\ 2x\quad&=6\\ x\quad&=3\end{aligned}$$

Substitute 3 for x in the first equation.

$(3)+y=9$

$y=6$

The solution is (3, 6).

55.

$$\begin{aligned} 4x-3y&=1\\ \underline{5x+3y}&\underline{=-10}\\ \text{Add: }\ 9x\quad&=-9\\ x&=-1\end{aligned}$$

Substitute –1 for x in the second equation.

$5(-1)+3y=-10$

$-5+3y=-10$

$3y=-5$

$y=-\frac{5}{3}$

The solution is $\left(-1,-\frac{5}{3}\right)$.

57. $10m - 2n = 6$

$-5m + n = -8$

To eliminate m, multiply the second equation by 2 and then add.

$10m - 2n = 6$

$2[-5m + n = -8]$

gives

$$\begin{aligned} 10m - 2n &= 6 \\ -10m + 2n &= -16 \\ \hline \text{Add:} \quad 0 &= -10 \end{aligned}$$

Since $0 = -10$ is a false statement, the system has no solution. It is an inconsistent system.

59. $2c - 5d = 1$

$-4c + 10d = 6$

To eliminate c, multiply the first equation by 2 and then add.

$2[2c - 5d = 1]$

$-4c + 10d = 6$

gives

$$\begin{aligned} 4c - 10d &= 2 \\ -4c + 10d &= 6 \\ \hline \text{Add:} \quad 0 &= 8 \end{aligned}$$

Since $0 = 8$ is a false statement, the system has no solution. It is an inconsistent system.

61. $7p - 3q = 4$

$2p + 5q = 7$

To eliminate q, multiply the first equation by 5, multiply the second equation by 3, and then add:

$5[7p - 3q = 4]$

$3[2p + 5q = 7]$

gives

$$\begin{aligned} 35p - 15q &= 20 \\ 6p + 15q &= 21 \\ \hline \text{Add:} \quad 41p \quad &= 41 \\ p &= 1 \end{aligned}$$

Substitute 1 for p in the second equation.

$$\begin{aligned} 2(1) + 5q &= 7 \\ 2 + 5q &= 7 \\ 5q &= 5 \\ q &= 1 \end{aligned}$$

The solution is (1, 1).

63. $5a - 10b = 15$

$a = 2b + 3$

Write the system in standard form.

$5a - 10b = 15$

$a - 2b = 3$

To eliminate a, multiply the second equation by –5 and then add.

$5a - 10b = 15$

$-5[a - 2b = 3]$

gives

$$\begin{aligned} 5a - 10b &= 15 \\ -5a + 10b &= -15 \\ \hline \text{Add:} \quad 0 &= 0 \end{aligned}$$

Since this is an identity, there are an infinite number of solutions. This is a dependent system.

65.

$$\begin{aligned} 2x - y &= 8 \\ 3x + y &= 6 \\ \hline \text{Add:} \quad 5x \quad &= 14 \\ x &= \frac{14}{5} \end{aligned}$$

Substitute $\frac{14}{5}$ for x in the second equation.

$$\begin{aligned} 3x + y &= 6 \\ 3\left(\frac{14}{5}\right) + y &= 6 \\ \frac{42}{5} + y &= 6 \\ y &= 6 - \frac{42}{5} \\ y &= \frac{30}{5} - \frac{42}{5} \\ y &= -\frac{12}{5} \end{aligned}$$

The solution is $\left(\frac{14}{5}, -\frac{12}{5}\right)$.

67. $3x-4y=5$

$2x=5y-3$

Write the system in standard form.

$3x-4y=5$

$2x-5y=-3$

To eliminate x, multiply the first equation by –2 and the second equation by 3 and then add.

$-2[3x-4y=5]$

$3[2x-5y=-3]$

gives

$$\begin{aligned} -6x+8y&=-10\\ \underline{6x-15y}&\underline{=-9} \end{aligned}$$

Add: $-7y=-19$

$$y=\frac{-19}{-7}=\frac{19}{7}$$

Substitute $\frac{19}{7}$ for y in the second equation.

$2x=5y-3$

$2x=5\left(\frac{19}{7}\right)-3$

$2x=\frac{95}{7}-3$

$2x=\frac{74}{4}$

$x=\frac{1}{2}\left(\frac{74}{7}\right)=\frac{37}{7}$

The solution is $\left(\frac{37}{7},\frac{19}{7}\right)$.

69. $0.2x-0.5y=-0.4$

$-0.3x+0.4y=-0.1$

To eliminate x, multiply the first equation by 3 and the second equation by 2 and then add.

$3[0.2x-0.5y=-0.4]$

$2[-0.3x+0.4y=-0.1]$

gives

$$\begin{aligned} 0.6x-1.5y&=-1.2\\ \underline{-0.6x+0.8y}&\underline{=-0.2} \end{aligned}$$

Add: $-0.7y=-1.4$

$$y=\frac{-1.4}{-0.7}=2$$

Now, substitute 2 for y in the first equation.

$0.2x-0.5y=-0.4$

$0.2x-0.5(2)=-0.4$

$0.2x-1=-0.4$

$0.2x=0.6$

$$x=\frac{0.6}{0.2}=3$$

The solution is (3, 2).

71. $2.1m-0.6n=8.4$

$-1.5m-0.3n=-6.0$

To eliminate n, multiply the second equation by –2 and then add.

$2.1m-0.6n=8.4$

$-2[-1.5m-0.3n=-6.0]$

gives

$$\begin{aligned} 2.1m-0.6n&=8.4\\ \underline{3.0m+0.6n}&\underline{=12.0} \end{aligned}$$

Add: $5.1m=20.40$

$$m=\frac{20.40}{5.1}=4$$

Substitute 4 for m into the second equation.

$-1.5m-0.3n=-6.0$

$-1.5(4)-0.3n=-6.0$

$-6.0-0.3n=-6.0$

$-0.3n=0$

$n=0$

The solution is (4, 0).

73. $\frac{1}{2}x-\frac{1}{3}y=1$

$\frac{1}{4}x-\frac{1}{9}y=\frac{2}{3}$

To clear fractions, multiply the first equation by 6 and the second equation by 36:

$6\left[\frac{1}{2}x-\frac{1}{3}y=1\right]$

$36\left[\frac{1}{4}x-\frac{1}{9}y=\frac{2}{3}\right]$

gives

$3x-2y=6$

$9x-4y=24$

To eliminate y, multiply the first equation by –2 and then add.

$-2[3x-2y=6]$

$9x-4y=24$

gives

$$\begin{aligned} -6x+4y &= -12 \\ 9x-4y &= 24 \\ \text{Add: } 3x \quad &= 12 \\ x &= 4 \end{aligned}$$

Substitute 4 for x in the equation $3x - 2y = 6$.

$$\begin{aligned} 3(4)-2y &= 6 \\ 12-2y &= 6 \\ 6-2y &= 0 \\ 6 &= 2y \\ 3 &= y \end{aligned}$$

The solution is (4, 3).

75. $\frac{1}{3}x = 4 - \frac{1}{4}y$

$3x = 4y$

Write the system in standard form.

$\frac{1}{3}x + \frac{1}{4}y = 4$

$3x - 4y = 0$

To clear fractions and to eliminate y, multiply the first equation by 48 and the second equation by 3 and then add:

$48\left[\frac{1}{3}x + \frac{1}{4}y = 4\right]$

$3[3x - 4y = 0]$

gives

$$\begin{aligned} 16x+12y &= 192 \\ 9x-12y &= 0 \\ \text{Add: } 25x \quad &= 192 \\ x &= \frac{192}{25} \end{aligned}$$

Substitute $\frac{192}{25}$ for x in the second equation.

$$\begin{aligned} 3x &= 4y \\ 3\left(\frac{192}{25}\right) &= 4y \\ \frac{576}{25} &= 4y \\ \frac{1}{4}\left(\frac{576}{25}\right) &= \frac{1}{4}(4y) \\ \frac{144}{25} &= y \end{aligned}$$

The solution is $\left(\frac{192}{25}, \frac{144}{25}\right)$.

77. a.-c. Answers will vary. The system should involve a variable that has a coefficient of 1. One example:

$2x + 3y = 20$

$y = 5x + 1$

This system will be easier to solve by substitution because one of the equations is already solved for a variable.

$$\begin{aligned} 2x+3y &= 20 \\ 2x+3(5x+1) &= 20 \\ 2x+15x+3 &= 20 \\ 17x &= 17 \\ x &= 1 \end{aligned}$$

Then, $y = 5(1) + 1 = 6$. The solution is $(1, 6)$.

79. The system of equations is

$y = 1000t + 38{,}000$

$y = 45{,}500 + 500t$

Substitute $1000t + 38{,}000$ for y in the second equation and solve for t.

$$\begin{aligned} 1000t+38{,}000 &= 45{,}500+500t \\ 500t+38{,}000 &= 45{,}500 \\ 500t &= 7{,}500 \\ t &= 15 \end{aligned}$$

The salaries will be the same in the year 2006+15=2021.

Substitute $t = 15$ in the first equation and solve for y.

$y = 1000(15) + 38{,}000 = 53{,}000$

In 2021, both salaries will be $53,000.

81. Multiply the first equation by 2. The resulting equations are identical.

83. a. If a system has more than one solution, then it must have an infinite number of solutions. All of these solutions must lie on the same line.

b. Slope is $m = \frac{11-3}{-6-(-4)} = \frac{8}{-2} = -4$. Use the point-slope form with $m = -4$ and $(x_1, y_1) = (-4, 3)$.

$$\begin{aligned} y - y_1 &= m(x - x_1) \\ y-3 &= -4(x-(-4)) \\ y-3 &= -4(x+4) \\ y-3 &= -4x-16 \\ y &= -4x-13 \end{aligned}$$

The y-intercept is (0, –13).

c. Yes, the graph of a non-vertical line is a function.

85. Answers may vary. One example is

$$\begin{aligned} x+y&=1 \\ 2x+2y&=2 \end{aligned}.$$

(multiply the first equation by 2)

87. a. Answers may vary. One example is

$$\begin{aligned} x+y&=7 \\ x-y&=-3 \end{aligned}.$$

b. Choose coefficients for *x* and *y*, then use the given coordinates to find the constants.

89. $Ax+4y=-8$

$3x-By=21$

Since the solution is (2, –3), substitute 2 for *x* and –3 for *y*.

$A(2)+4(-3)=-8$

$3(2)-B(-3)=21$

or

$2A-12=-8$

$6+3B=21$

This is a system of two equations in the two unknowns *A* and *B*. To solve, solve each equation for the unknown variable. In the first equation, solve for *A*.

$$\begin{aligned} 2A-12&=-8 \\ 2A&=4 \\ A&=\frac{4}{2}=2 \end{aligned}$$

In the second equation, solve for *B*.

$$\begin{aligned} 6+3B&=21 \\ 3B&=15 \\ B&=\frac{15}{3}=5 \end{aligned}$$

Thus, $A=2$ and $B=5$.

91. $f(x)=mx+b$

Substitute (2, 6) and (–1, –6) into the equation to get a system.

$6=2m+b$

$-6=-m+b$

Multiply the second equation by –1 and add.

$$\begin{aligned} 6&=2m+b \\ 6&=m-b \\ \hline 12&=3m \\ 4&=m \end{aligned}$$

Substitute 4 for *m* in the first equation and solve for *b*.

$$\begin{aligned} 6&=2(4)+b \\ -2&=b \end{aligned}$$

Thus, $m=4$ and $b=-2$.

93. The system is dependent or one graph is not in the viewing window.

95. $\dfrac{x+2}{2}-\dfrac{y+4}{3}=4$

$\dfrac{x+y}{2}=\dfrac{1}{2}+\dfrac{x-y}{3}$

Start by writing each equation in standard form after clearing fractions.

$$6\left(\frac{x+2}{2}\right)-6\left(\frac{y+4}{3}\right)=6(4)$$

$$6\left(\frac{x+y}{2}\right)=6\left(\frac{1}{2}\right)+6\left(\frac{x-y}{3}\right)$$

$$3(x+2)-2(y+4)=24$$

$$3(x+y)=3+2(x-y)$$

$$3x+6-2y-8=24$$

$$3x+3y=3+2x-2y$$

$$\begin{aligned} 3x-2y&=26 \quad (1) \\ x+5y&=3 \quad (2) \end{aligned}$$

To eliminate *x*, multiply equation (2) by –3 and then add to equation (1).

$3x-2y=26$

$-3[x+5y=3]$

gives

$$\begin{aligned} 3x-2y&=26 \\ -3x-15y&=-9 \\ \hline \text{Add:} \quad -17y&=17 \\ y&=\frac{17}{-17}=-1 \end{aligned}$$

Now, substitute –1 for *y* in equation (2).

$$\begin{aligned} x+5(-1)&=3 \\ x-5&=3 \\ x&=8 \end{aligned}$$

The solution is (8, –1).

97. Rewrite the system using the hint.

$$3\cdot\frac{1}{a}+4\cdot\frac{1}{b}=-1$$
$$\frac{1}{a}+6\cdot\frac{1}{b}=2$$

Now let $x=\frac{1}{a}$ and $y=\frac{1}{b}$.

$$3x+4y=-1\ (1)$$
$$x+6y=2\ (2)$$

Multiply equation (2) by –3 and add.

$$3x+4y=-1$$
$$-3[x+6y=2]$$

gives

$$3x+4y=-1$$
$$-3x-18y=-6$$

Add: $-14y=-7$

$$y=\frac{-7}{-14}=\frac{1}{2}$$

Substitute $\frac{1}{2}$ for y in equation (2).

$$x+6\left(\frac{1}{2}\right)=2$$
$$x+3=2$$
$$x=-1$$

Now find the values of a and b.

$$x=\frac{1}{a}$$
$$-1=\frac{1}{a}$$
$$a=-1$$
$$y=\frac{1}{b}$$
$$\frac{1}{2}=\frac{1}{b}$$
$$b=2$$

The solution is (–1, 2).

99. $4ax+3y=19$

$-ax+y=4$

Solve the second equation for y.

$$-ax+y=4$$
$$y=ax+4$$

Substitute $ax+4$ for y in the first equation.

$$4ax+3y=19$$
$$4ax+3(ax+4)=19$$
$$4ax+3ax+12=19$$
$$7ax+12=19$$
$$7ax=7$$
$$x=\frac{7}{7a}=\frac{1}{a}$$

Now substitute $\frac{1}{a}$ for x in the equation $y=ax+4$.

$$y=a\left(\frac{1}{a}\right)+4$$
$$y=1+4$$
$$y=5$$

The solution is $\left(\frac{1}{a},5\right)$.

103. Rational numbers can be expressed as quotients of two integers. Irrational numbers cannot.

104. a. Yes, the set of real numbers includes the set of rational numbers.

b. Yes, the set of real numbers includes the set of irrational numbers.

105.

$$\frac{1}{2}(x-7)=\frac{3}{4}(2x+1)$$
$$4\left(\frac{1}{2}(x-7)\right)=4\left(\frac{3}{4}(2x+1)\right)$$
$$2(x-7)=3(2x+1)$$
$$2x-14=6x+3$$
$$-4x=17$$
$$x=-\frac{17}{4}$$

106. $|x-6|=|6-x|$

$$x-6=6-x \quad \text{or} \quad x-6=-(6-x)$$
$$2x-6=6 \qquad x-6=-6+x$$
$$2x=12 \qquad -6=-6$$
$$x=6$$

The second piece is always true, so the solution is all real numbers, or $\mathbb{R}$.

107. $A = p\left(1+\frac{r}{n}\right)^{n\cdot t} = 500\left(1+\frac{0.04}{2}\right)^{2\cdot 1}$
$= 500(1.02)^2 = 500(1.0404)$
$= 520.20$

108. No, the points $(-3,4)$ and $(-3,-1)$ have the same first coordinate but different second coordinates.

109. $f(x) = x+3$ and $g(x) = x^2-9$
$(f/g)(x) = \frac{x+3}{x^2-9}$
$(f/g)(3) = \frac{(3)+3}{(3)^2-9} = \frac{6}{0}$ undefined

Exercise Set 4.2

1. The graph will be a plane.

3.
$x = 1$
$2x - y = 4$
$-3x + 2y - 2z = 1$
Substitute 1 for *x* in the second equation.
$2(1) - y = 4$
$2 - y = 4$
$-y = 2$
$y = -2$
Substitute 1 for *x* and –2 for *y* in the third equation.
$-3(1) + 2(-2) - 2z = 1$
$-3 - 4 - 2z = 1$
$-7 - 2z = 1$
$-2z = 8$
$z = -4$
The solution is (1, –2, –4).

5.
$5x - 6z = -17$
$3x - 4y + 5z = -1$
$2z = -6$
Solve the third equation for *z*.
$2z = -6$
$z = -3$
Substitute –3 for *z* in the first equation.
$5x - 6z = -17$
$5x - 6(-3) = -17$
$5x + 18 = -17$
$5x = -35$
$x = -7$
Substitute –7 for *x* and –3 for *z* in the second equation.
$3x - 4y + 5z = -1$
$3(-7) - 4y + 5(-3) = -1$
$-21 - 4y - 15 = -1$
$-4y - 36 = -1$
$-4y = 35$
$y = \frac{35}{-4} = -\frac{35}{4}$
The solution is $\left(-7, -\frac{35}{4}, -3\right)$.

7.
$x + 2y = 6$
$3y = 9$
$x + 2z = 12$
Solve the second equation for *y*.
$3y = 9$
$y = 3$
Substitute 3 for *y* in the first equation.
$x + 2y = 6$
$x + 2(3) = 6$
$x + 6 = 6$
$x = 0$
Substitute 0 for *x* in the third equation.
$x + 2z = 12$
$0 + 2z = 12$
$2z = 12$
$z = 6$
The solution is (0, 3, 6).

9. $x-2y=-3$ (1)

$3x+2y=7$ (2)

$2x-4y+z=-6$ (3)

To eliminate y between equations (1) and (2), add equations (1) and (2).

$$\begin{array}{r} x-2y=-3 \\ 3x+2y=7 \\ \hline \text{Add:}\quad 4x=4 \end{array}$$

$$x=1$$

Substitute 1 for x in equation (1).

$$(1)-2y=-3$$
$$-2y=-4$$
$$y=2$$

Substitute 1 for x, and 2 for y in equation (3).

$$2(1)-4(2)+z=-6$$
$$2-8+z=-6$$
$$-6+z=-6$$
$$z=0$$

The solution is (1, 2, 0).

11. $2y+4z=2$ (1)

$x+y+2z=-2$ (2)

$2x+y+z=2$ (3)

To eliminate x between equations (2) and (3), multiply equation (2) by –2 and then add.

$$-2[x+y+2z=-2]$$
$$2x+y+z=2$$

gives

$$\begin{array}{r} -2x-2y-4z=4 \\ 2x+y+z=2 \\ \hline \text{Add:}\quad -y-3z=6\quad (4) \end{array}$$

To eliminate y between equations (1) and (4), multiply equation (4) by 2 and then add.

$$2y+4z=2$$
$$2[-y-3z=6]$$

gives

$$\begin{array}{r} 2y+4z=2 \\ -2y-6z=12 \\ \hline \text{Add:}\quad -2z=14 \end{array}$$

$$z=-7$$

Substitute –7 for z in equation (1).

$$2y+4(-7)=2$$
$$2y-28=2$$
$$2y=30$$
$$y=15$$

Substitute 15 for y, and –7 for z in equation (3).

$$2x+(15)+(-7)=2$$
$$2x+8=2$$
$$2x=-6$$
$$x=-3$$

The solution is (–3, 15, –7).

13. $3p+2q=11$ (1)

$4q-r=6$ (2)

$6p+7r=4$ (3)

To eliminate r between equations (2) and (3), multiply equation (2) by 7 and add to equation (3).

$$7[4q-r=6]$$
$$6p+7r=4$$

gives

$$\begin{array}{r} 28q-7r=42 \\ 6p+7r=4 \\ \hline \text{Add:}\quad 6p+28q=46\quad (4) \end{array}$$

Equations (1) and (4) are two equations in two unknowns. To eliminate p, multiply equation (1) by –2 and add to equation (4).

$$-2[3p+2q=11]$$
$$6p+28q=46$$

gives

$$\begin{array}{r} -6p-4q=-22 \\ 6p+28q=46 \\ \hline \text{Add:}\quad 24q=24 \end{array}$$

$$q=1$$

Substitute 1 for q in equation (1).

$$3p+2q=11$$
$$3p+2(1)=11$$
$$3p+2=11$$
$$3p=9$$
$$p=3$$

Substitute 1 for q in equation (2).

$$4q-r=6$$
$$4(1)-r=6$$
$$4-r=6$$
$$r=-2$$

The solution is (3, 1, –2).

15. $p+q+r=4$ (1)

$p-2q-r=1$ (2)

$2p-q-2r=-1$ (3)

To eliminate q between equations (1) and (3), simply add.

$$\begin{array}{r} p+q+\ r=\ 4 \\ \underline{2p-q-2r=-1} \\ \text{Add: } 3p \quad -r=\ 3 \end{array} \quad (4)$$

To eliminate q between equations (1) and (2), multiply equation (1) by 2 and then add.

$2[p+q+r=4]$

$p-2q-r=1$

gives

$$\begin{array}{r} 2p+2q+2r=8 \\ \underline{p-2q-\ r=1} \\ \text{Add: } 3p \quad +r=9 \end{array} \quad (5)$$

Equations (4) and (5) are two equations in two unknowns.

$3p-r=3$

$3p+r=9$

To eliminate r, simply add these two equations.

$$\begin{array}{r} 3p-r=\ 3 \\ \underline{3p+r=\ 9} \\ \text{Add: } 6p \quad =12 \\ p=2 \end{array}$$

Substitute 2 for p in equation (5).

$3p+r=9$

$3(2)+r=9$

$6+r=9$

$r=3$

Substitute 2 for p and 3 for r in equation (1).

$p+q+r=4$

$2+q+3=4$

$q+5=4$

$q=-1$

The solution is (2, –1, 3).

17. $2x-2y+3z=5$ (1)

$2x+y-2z=-1$ (2)

$4x-y-3z=0$ (3)

To eliminate y between equations (2) and (3), simply add.

$$\begin{array}{r} 2x+y-2z=-1 \\ \underline{4x-y-3z=\ 0} \\ \text{Add: } 6x \quad -5z=-1 \end{array} \quad (4)$$

To eliminate y between equations (1) and (2), multiply equation (2) by 2 and then add.

$2x-2y+3z=5$

$2[2x+y-2z=-1]$

gives

$$\begin{array}{r} 2x-2y+3z=\ 5 \\ \underline{4x+2y-4z=-2} \\ \text{Add: } 6x \quad -z=\ 3 \end{array} \quad (5)$$

Equations (4) and (5) are two equations in two unknowns.

$6x-5z=-1$

$6x-z=3$

To eliminate x, multiply equation (5) by –1 and then add.

$6x-5z=-1$

$-1[6x-z=3]$

gives

$$\begin{array}{r} 6x-5z=-1 \\ \underline{-6x+z=-3} \\ \text{Add: } -4z=\ -4 \\ z=\ 1 \end{array}$$

Substitute 1 for z in equation (5).

$6x-z=3$

$6x-1=3$

$6x=4$

$x=\frac{4}{6}=\frac{2}{3}$

Substitute $\frac{2}{3}$ for x and 1 for z in equation (2).

$2x+y-2z=-1$

$2\left(\frac{2}{3}\right)+y-2(1)=-1$

$\frac{4}{3}+y-2=-1$

$y-\frac{2}{3}=-1$

$y=-1+\frac{2}{3}=-\frac{1}{3}$

The solution is $\left(\frac{2}{3}, -\frac{1}{3}, 1\right)$.

19. $r-2s+t=2$ (1)

$2r+3s-t=-3$ (2)

$2r-s-2t=1$ (3)

To eliminate t between equations (1) and (2), add the two equations.

$$\begin{array}{r} r-2s+t=2 \\ \underline{2r+3s-t=-3} \\ \text{Add:}\quad 3r+s=-1 \quad (4) \end{array}$$

To eliminate t between equations (2) and (3), multiply equation (2) by -2 and then add.

$-2[2r+3s-t=-3]$

$2r-s-2t=-1$

gives

$$\begin{array}{r} -4r-6s+2t=6 \\ \underline{2r-s-2t=1} \\ \text{Add:}\quad -2r-7s=7 \quad (5) \end{array}$$

Equations (4) and (5) make a system of two equations in two variables. To eliminate r, multiply equation (4) by 2 and equation (5) by 3, then add the equations.

$2[3r+s=-1]$

$3[-2r-7s=7]$

gives:

$$\begin{array}{r} 6r+\ 2s=-2 \\ \underline{-6r-21s=21} \\ \text{Add:}\quad -19s=19 \\ s=-1 \end{array}$$

Substitute -1 for s into equation (4) and solve for r.

$3r+s=-1$

$3r-1=-1$

$3r=0$

$r=0$

Finally, substitute 0 for r and -1 for s into equation (1)

$r-2s+t=2$

$(0)-2(-1)+t=2$

$2+t=2$

$t=0$

The solution is (0, –1, 0).

21. $2a+2b-c=2$ (1)

$3a+4b+c=-4$ (2)

$5a-2b-3c=5$ (3)

To eliminate c between equations (1) and (2), simply add.

$$\begin{array}{r} 2a+2b-c\ =2 \\ \underline{3a+4b+c=-4} \\ \text{Add:}\ 5a+6b\quad =-2 \quad (4) \end{array}$$

To eliminate c between equations (2) and (3), multiply equation (2) by 3 and then add.

$3[3a+4b+c=-4]$

$5a-2b-3c=5$

gives

$$\begin{array}{r} 9a+12b+3c=-12 \\ \underline{5a-\ 2b-3c=\quad 5} \\ \text{Add:}\ 14a+10b\qquad =-7 \quad (5) \end{array}$$

Equations (4) and (5) are two equations in two unknowns.

$5a+6b=-2$

$14a+10b=-7$

To eliminate b, multiply equation (4) by –5 and multiply equation (5) by 3 and then add.

$-5[5a+6b=-2]$

$3[14a+10b=-7]$

gives

$$\begin{array}{r} -25a-30b=\ 10 \\ \underline{42a+30b=-21} \\ \text{Add:}\ 17a\qquad =-11 \\ a=-\frac{11}{17} \end{array}$$

Substitute $-\frac{11}{17}$ for a in equation (4).

$5a+6b=-2$

$5\left(-\frac{11}{17}\right)+6b=-2$

$-\frac{55}{17}+6b=-2$

$6b=-2+\frac{55}{17}$

$b=\frac{1}{6}\cdot\frac{21}{17}=\frac{7}{34}$

Substitute $-\frac{11}{17}$ for a and $\frac{7}{34}$ for b in equation (2).

$$3a+4b+c=-4$$
$$3\left(-\frac{11}{17}\right)+4\left(\frac{7}{34}\right)+c=-4$$
$$-\frac{33}{17}+\frac{14}{17}+c=-4$$
$$-\frac{19}{17}+c=-4$$
$$c=-4+\frac{19}{17}$$
$$c=-\frac{49}{17}$$

The solution is $\left(-\frac{11}{17}, \frac{7}{34}, -\frac{49}{17}\right)$.

23. $-x+3y+z=0\ (1)$
$-2x+4y-z=0\ (2)$
$3x-y+2z=0\ (3)$

To eliminate z between equations (1) and (2), simply add.

$$\begin{aligned}&-x+3y+z=0\\&\underline{-2x+4y-z=0}\\\text{Add: }&-3x+7y\quad=0\quad(4)\end{aligned}$$

To eliminate z between equations (2) and (3), multiply equation (2) by 2 and then add.

$2[-2x+4y-z]=0$
$3x-y+2z=0$

gives

$$\begin{aligned}&-4x+8y-2z=0\\&\underline{3x-\ \ y+2z=0}\\\text{Add: }&-x+7y\quad=0\quad(5)\end{aligned}$$

Equations (4) and (5) are two equations in two unknowns.

$-3x+7y=0$
$-x+7y=0$

To eliminate y, multiply equation (4) by –1 and then add.

$-1[-3x+7y=0]$
$-x+7y=0$

gives

$$\begin{aligned}&3x-7y=0\\&\underline{-x+7y=0}\\\text{Add: }&2x\quad=0\\&x=0\end{aligned}$$

Substitute 0 for x in equation (5).

$$\begin{aligned}-x+7y&=0\\-0+7y&=0\\7y&=0\\y&=0\end{aligned}$$

Finally, substitute 0 for x and 0 for y into equation (1).

$$\begin{aligned}-x+3y+z&=0\\-0+3(0)+z&=0\\0+z&=0\\z&=0\end{aligned}$$

The solution is (0, 0, 0).

25. $-\frac{1}{4}x+\frac{1}{2}y-\frac{1}{2}z=-2\quad(1)$
$\frac{1}{2}x+\frac{1}{3}y-\frac{1}{4}z=2\quad(2)$
$\frac{1}{2}x-\frac{1}{2}y+\frac{1}{4}z=1\quad(3)$

To clear fractions, multiply equation (1) by 4, equation (2) by 12, and equation (3) by 4.

$$4\left(-\frac{1}{4}x+\frac{1}{2}y-\frac{1}{2}z=-2\right)$$
$$12\left(\frac{1}{2}x+\frac{1}{3}y-\frac{1}{4}z=2\right)$$
$$4\left(\frac{1}{2}x-\frac{1}{2}y+\frac{1}{4}z=1\right)$$

gives

$-x+2y-2z=-8\ (4)$
$6x+4y-3z=24\ (5)$
$2x-2y+z=4\quad(6)$

To eliminate y between equations (4) and (6), simply add.

$$\begin{aligned}&-x+2y-2z=-8\\&\underline{2x-2y+\ \ z=\ \ 4}\\\text{Add: }&x\qquad -z=-4\quad(7)\end{aligned}$$

To eliminate y between equations (5) and (6), multiply equation (6) by 2 and then add to equation (5).

$6x+4y-3z=24$
$2[2x-2y+z=4]$

gives

$$\begin{aligned}&6x+4y-3z=24\\&\underline{4x-4y+2z=\ \ 8}\\\text{Add: }&10x\qquad -z=32\quad(8)\end{aligned}$$

Equations (7) and (8) are two equations in two unknowns.

$x - z = -4$

$10x - z = 32$

To eliminate z, multiply equation (7) by -1 and then add.

$-1[x - z = -4]$

$10x - z = 32$

gives

$$\begin{array}{r} -x + z = 4 \\ \underline{10x - z = 32} \\ \text{Add: } 9x \quad = 36 \end{array}$$

$$x = \frac{36}{9} = 4$$

Substitute 4 for x in equation (7).

$x - z = -4$

$4 - z = -4$

$-z = -8$

$z = 8$

Finally, substitute 4 for x and 8 for z in equation (4).

$-x + 2y - 2z = -8$

$-4 + 2y - 2(8) = -8$

$-4 + 2y - 16 = -8$

$2y - 20 = -8$

$2y = 12$

$y = \frac{12}{2} = 6$

The solution is (4, 6, 8).

27. $x - \frac{2}{3}y - \frac{2}{3}z = -2$ (1)

$\frac{2}{3}x + y - \frac{2}{3}z = \frac{1}{3}$ (2)

$-\frac{1}{4}x + y - \frac{1}{4}z = \frac{3}{4}$ (3)

To clear fractions, multiply equation (1) by 3, equation (2) by 3, and equation (3) by 4. The resulting system is

$3\left(x - \frac{2}{3}y - \frac{2}{3}z = -2\right)$

$3\left(\frac{2}{3}x + y - \frac{2}{3}z = \frac{1}{3}\right)$

$4\left(-\frac{1}{4}x + y - \frac{1}{4}z = \frac{3}{4}\right)$

gives

$3x - 2y - 2z = -6$ (4)

$2x + 3y - 2z = 1$ (5)

$-x + 4y - z = 3$ (6)

To eliminate x between equations (4) and (6), multiply equation (6) by 3 and then add.

$3x - 2y - 2z = -6$

$3[-x + 4y - z = 3]$

gives

$$\begin{array}{r} 3x - 2y - 2z = -6 \\ \underline{-3x + 12y - 3z = 9} \\ \text{Add: } \quad 10y - 5z = 3 \quad (7) \end{array}$$

To eliminate x between equations (5) and (6), multiply equation (6) by 2 and then add.

$2x + 3y - 2z = 1$

$2[-x + 4y - z = 3]$

gives

$$\begin{array}{r} 2x + 3y - 2z = 1 \\ \underline{-2x + 8y - 2z = 6} \\ \text{Add: } \quad 11y - 4z = 7 \quad (8) \end{array}$$

Equations (7) and (8) are two equations in two unknowns.

$10y - 5z = 3$

$11y - 4z = 7$

To eliminate z, multiply equation (7) by -4 and equation (8) by 5 and then add.

$-4[10y - 5z = 3]$

$5[11y - 4z = 7]$

gives

$$\begin{array}{r} -40y + 20z = -12 \\ \underline{55y - 20z = 35} \\ \text{Add: } 15y \quad = 23 \end{array}$$

$$y = \frac{23}{15}$$

Substitute $\frac{23}{15}$ for y into equation (7).

$10y - 5z = 3$

$10\left(\frac{23}{15}\right) - 5z = 3$

$\frac{46}{3} - 5z = 3$

$-5z = -\frac{37}{3}$

$z = \left(-\frac{1}{5}\right)\left(-\frac{37}{3}\right) = \frac{37}{15}$

Substitute $\frac{23}{15}$ for y and $\frac{37}{15}$ for z in equation (6).

$$-x+4y-z=3$$
$$-x+4\left(\frac{23}{15}\right)-\frac{37}{15}=3$$
$$-x+\frac{92}{15}-\frac{37}{15}=3$$
$$-x+\frac{11}{3}=3$$
$$-x=-\frac{2}{3}$$
$$x=\frac{2}{3}$$

The solution is $\left(\frac{2}{3},\frac{23}{15},\frac{37}{15}\right)$.

29. Multiply each equation by 10.

$$10(0.2x+0.3y+0.3z=1.1)$$
$$10(0.4x-0.2y+0.1z=0.4)$$
$$10(-0.1x-0.1y+0.3z=0.4)$$

gives

$$2x+3y+3z=11\ (1)$$
$$4x-2y+z=4\ (2)$$
$$-x-y+3z=4\ (3)$$

To eliminate z between equations (1) and (2), multiply equation (2) by –3 and then add.

$$2x+3y+3z=11$$
$$-3[4x-2y+z=4]$$

gives

$$\begin{array}{rl} 2x+3y+3z= & 11 \\ -12x+6y-3z= & -12 \\ \hline \text{Add: } -10x+9y \quad = & -1 \quad (4) \end{array}$$

To eliminate z between equations (1) and (3) multiply equation (1) by –1 and then add.

$$-1[2x+3y+3z=11]$$
$$-x-y+3z=4$$

gives

$$\begin{array}{rl} -2x-3y-3z= & -11 \\ -x-\ y+3z= & 4 \\ \hline \text{Add: } -3x-4y \quad = & -7 \quad (5) \end{array}$$

Equations (4) and (5) are two equations in two unknowns.

$$-10x+9y=-1$$
$$-3x-4y=-7$$

To eliminate x, multiply equation (4) by –3 and equation (5) by 10.

$$-3[-10x+9y=-1]$$
$$10[-3x-4y=-7]$$

gives

$$\begin{array}{rl} 30x-27y= & 3 \\ -30x-40y= & -70 \\ \hline \text{Add: } -67y= & -67 \end{array}$$
$$y=1$$

Substitute 1 for y in equation (4).

$$-10x+9y=-1$$
$$-10x+9(1)=-1$$
$$-10x=-10$$
$$x=1$$

Substitute 1 for x and 1 for y in equation (1).

$$2x+3y+3z=11$$
$$2(1)+3(1)+3z=11$$
$$5+3z=11$$
$$3z=6$$
$$z=2$$

The solution is (1, 1, 2).

31. $2x+y+2z=1\ (1)$

$$x-2y-z=0\ (2)$$
$$3x-y+z=2\ (3)$$

To eliminate z between equations (2) and (3), add.

$$\begin{array}{rl} x-2y-z= & 0 \\ 3x-\ y+z= & 2 \\ \hline \text{Add: } 4x-3y \quad = & 2 \quad (4) \end{array}$$

To eliminate z between equations (1) and (2), multiply equation (2) by 2 and then add.

$$2x+y+2z=1$$
$$2[x-2y-z=0]$$

gives

$$\begin{array}{rl} 2x+\ y+2z= & 1 \\ 2x-4y-2z= & 0 \\ \hline \text{Add: } 4x-3y \quad = & 1 \quad (5) \end{array}$$

Equations (4) and (5) are two equations in two unknowns.

$$4x-3y=2$$
$$4x-3y=1$$

To eliminate x, multiply equation (4) by –1 and then add.

$$-1[4x-3y=2]$$
$$4x-3y=1$$

gives

$$\begin{array}{rl} -4x+3y= & -2 \\ 4x-3y= & 1 \\ \hline \text{Add: } 0= & -1 \quad \text{False} \end{array}$$

Since this is a false statement, there is no solution and the system is inconsistent.

33. $x - 4y - 3z = -1$ (1)
$-3x + 12y + 9z = 3$ (2)
$2x - 10y - 7z = 5$ (3)

To eliminate x between equations (1) and (2), multiply equation (1) by 3 and then add.

$3[x - 4y - 3z = -1]$
$-3x + 12y + 9z = 3$

gives

$3x - 12y - 9z = -3$
$\underline{-3x + 12y + 9z = 3}$

Add: $0=0$

Since $0 = 0$ is a true statement, we suspect that the system is dependent and therefore has infinitely many solutions. However, we could still encounter a contradiction if we used a different pair of equations.

To eliminate x from equations (1) and (3), multiply equation (1) by -2 and add.

$-2[x - 4y - 3z = -1]$
$2x - 10y - 7z = 5$

gives

$-2x + 8y + 6z = 2$
$\underline{2x - 10y - 7z = 5}$

Add: $-2y - z = 7$

To eliminate x from equations (2) and (3), multiply equation (2) by 2 and equation (3) by 3, then add.

$2[-3x + 12y + 9z = 3]$
$3[2x - 10y - 7z = 5]$

gives

$-6x + 24y + 18z = 6$
$\underline{6x - 30y - 21z = 15}$

Add: $-6y - 3z = 21$

Since neither of the remaining pairs yields a contradiction, the system is indeed dependent and has an infinite number of solutions.

35. $x + 3y + 2z = 6$ (1)
$x - 2y - z = 8$ (2)
$-3x - 9y - 6z = -7$ (3)

To eliminate x between equations (1) and (3), multiply equation (1) by 3 and then add.

$3[x + 3y + 2z = 6]$
$-3x - 9y - 6z = -7$

gives

$3x + 9y + 6z = 18$
$\underline{-3x - 9y - 6z = -7}$

Add: $0 = 11$

Since $0 = 11$ is a false statement, the system is inconsistent.

37. No point is common to all three planes. Therefore, the system is inconsistent.

39. One point is common to all three planes. There is one solution and the system is consistent.

41. **a.** Yes, if two or more of the planes are parallel, there will be no solution.

b. Yes, three planes may intersect at a single point.

c. No, the possibilities are no solution, one solution, or infinitely many solutions.

43. $Ax + By + Cz = 1$

Substitute $(-1, 2, -1)$, $(-1, 1, 2)$, and $(1, -2, 2)$ into the equation forming three equations in the three unknowns A, B, and C.

$A(-1) + B(2) + C(-1) = 1$
$A(-1) + B(1) + C(2) = 1$
$A(1) + B(-2) + C(2) = 1$

gives

$-A + 2B - C = 1$ (1)
$-A + B + 2C = 1$ (2)
$A - 2B + 2C = 1$ (3)

To eliminate A between equations (1) and (2), multiply equation (2) by -1 and then add.

$-A + 2B - C = 1$
$-1[-A + B + 2C = 1]$

gives

$-A + 2B - C = 1$
$\underline{A - B - 2C = -1}$

Add: $B - 3C = 0$ (4)

To eliminate A between equations (1) and (3), simply add.

$-A+2B\ \ -C=1$

$\underline{A-2B+2C=1}$

Add: $C=2$

Substitute 2 for C in equation (4).

$B-3(2)=0$

$B-6=0$

$B=6$

Substitute 6 for B and 2 for C in equation (1).

$-A+2B-C=1$

$-A+2(6)-(2)=1$

$-A+12-2=1$

$-A+10=1$

$-A=-9$

$A=9$

The equation is $9x+6y+2z=1$.

45. One example is

$x+y+z=10$

$x+2y+z=11$

$x+y+2z=16$

Choose coefficients for x, y, and z, then use the given coordinates to find the constants.

47. a. $y=ax^2+bx+c$

For the point $(1, -1)$,
let $y=-1$ and $x=1$.

$-1=a(1)^2+b(1)+c$

$-1=a+b+c$ (1)

For the point $(-1, -5)$,
let $y=-5$ and $x=-1$.

$-5=a(-1)^2+b(-1)+c$

$-5=a-b+c$ (2)

For the point $(3, 11)$,
let $y=11$ and $x=3$.

$11=a(3)^2+b(3)+c$

$11=9a+3b+c$ (3)

Equations (1), (2), and (3) give us a system of three equations.

$a+b+c=-1$ (1)

$a-b+c=-5$ (2)

$9a+3b+c=11$ (3)

To eliminate a and c between equations (1) and (2) multiply equation (2) by -1 and then add.

$a+b+c=-1$

$-1[a-b+c=-5]$

gives

$a+b+c=-1$

$\underline{-a+b-c=\ 5}$

Add: $2b\ \ =\ 4$

$b=2$

Substitute 2 for b in equations (1) and (3).

Equation (1) becomes

$a+b+c=-1$

$a+2+c=-1$

$a+c=-3$ (4)

Equation (3) becomes

$9a+3b+c=11$

$9a+3(2)+c=11$

$9a+c=5$ (5)

Equations (4) and (5) are two equations in two unknowns. To eliminate c, multiply equation (4) by -1 and then add.

$-1[a+c=-3]$

$9a+c=5$

gives

$-a-c=\ 3$

$\underline{9a+c=5}$

Add: $8a\ \ =8$

$a=1$

Finally, substitute 1 for a in equation (4).

$a+c=-3$

$1+c=-3$

$c=-4$

Thus, $a=1$, $b=2$, and $c=-4$.

b. The quadratic equation is $y=x^2+2x-4$. This is the equation determined by the values found in part **a.**

49. $3p+4q=11$ (1)

$2p+r+s=9$ (2)

$q-s=-2$ (3)

$p+2q-r=2$ (4)

To eliminate r between equations (2) and (4), simply add.

$$\begin{array}{l} 2p \quad +r+s=9 \\ \underline{p+2q-r \quad =2} \end{array}$$

Add: $3p+2q \quad +s=11$ (5)

To eliminate s between equations (3) and (5), simply add.

$$\begin{array}{l} q-s=-2 \\ \underline{3p+2q+s=11} \end{array}$$

Add: $3p+3q \quad = 9$ (6)

Equations (1) and (6) give us a system of two equations in two unknowns.
To eliminate p, multiply equation (6) by -1 and then add.

$3p+4q=11$

$-1[3p+3q=9]$

gives

$$\begin{array}{l} 3p+4q=11 \\ \underline{-3p-3q=-9} \end{array}$$

Add: $\quad q=2$

Substitute 2 for q in equation (3).

$q-s=-2$

$2-s=-2$

$-s=-4$

$s=4$

Substitute 2 for q in equation (1).

$3p+4q=11$

$3p+4(2)=11$

$3p+8=11$

$3p=3$

$p=1$

Finally, substitute 1 for p and 4 for s in equation (2).

$2p+r+s=9$

$2(1)+r+4=9$

$r+6=9$

$r=3$

The solution is (1, 2, 3, 4).

51. Let t be the time for Margie.

Then, $t-\frac{1}{6}$ is the time for David.

	rate	time	distance
David	5	t	$5t$
Margie	3	$t+\frac{1}{6}$	$3\left(t+\frac{1}{6}\right)$

a. The distances traveled are the same.

$3\left(t+\frac{1}{6}\right)=5t$

$3t+\frac{1}{2}=5t$

$\frac{1}{2}=2t$

$\frac{1}{4}=t$ or $t=15$ minutes

b. The distance is

$5t=5\left(\frac{1}{4}\right)=\frac{5}{4}=1\frac{1}{4}$

or 1.25 miles.

52. $\left|4-\frac{2}{3}x\right|>5$

$4-\frac{2}{3}x>5$ or $4-\frac{2}{3}x<-5$

$-\frac{2}{3}x>1$ $\qquad -\frac{2}{3}x<-9$

$\left(-\frac{3}{2}\right)\left(-\frac{2}{3}\right)x<\left(-\frac{3}{2}\right)(1)$ $\qquad \left(-\frac{3}{2}\right)\left(-\frac{2}{3}\right)x>\left(-\frac{3}{2}\right)(-9)$

$x<-\frac{3}{2}$ $\qquad x>\frac{27}{2}$

The solution is $\left\{x \middle| x<-\frac{3}{2} \text{ or } x>\frac{27}{2}\right\}$.

53. $\left|\frac{3x-4}{2}\right|+1<7$

$\left|\frac{3x-4}{2}\right|<6$

$$-6<\frac{3x-4}{2}<6$$
$$2(-6)<2\left(\frac{3x-4}{2}\right)<2(6)$$
$$-12<3x-4<12$$
$$-12+4<3x-4+4<12+4$$
$$-8<3x<16$$
$$-\frac{8}{3}<\frac{3x}{3}<\frac{16}{3}$$
$$-\frac{8}{3}<x<\frac{16}{3}$$

The solution is $\left\{x\middle|-\frac{8}{3}<x<\frac{16}{3}\right\}$.

54. $\left|3x+\frac{1}{5}\right|=-5$

There is no solution since the right side is a negative number and the left side is non-negative and it is not possible for a non-negative quantity to be equal to a negative number. The solution is $\varnothing$.

Exercise Set 4.3

1. Let x = the land area of Georgia and y = the land area of Ireland.

$$x+y=139{,}973$$
$$y=x+573$$

Substitute $x+573$ for y in the first equation.

$$x+(x+573)=139{,}973$$
$$2x+573=139{,}973$$
$$2x=139{,}400$$
$$x=69{,}700$$

Substitute 69,700 for x in the second equation.

$$y=69{,}700+573$$
$$y=70{,}273$$

The land area of Georgia is 69,700 km^2 and the land area of Ireland is 70,273 km^2.

3. Let F = grams of fat in fries and H = grams of fat in hamburger

$$F=3H+4$$
$$F-H=46$$

Substitute $3H+4$ for F in the second equation.

$$F-H=46$$
$$3H+4-H=46$$
$$2H=42$$
$$H=21$$

Substitute 21 for H in the first equation.

$$F=3H+4$$
$$F=3(21)+4$$
$$F=63+4$$
$$F=67$$

The hamburger has 21 grams of fat and the fries have 67 grams of fat.

5. Let h = the cost of a hot dog and s = the cost of a soda.

$$2h+3s=7$$
$$4h+2s=10$$

Multiply the first equation by –2 and then add.

$$-2[2h+3s=7]$$
$$4h+2s=10$$

gives

$$-4h-6s=-14$$
$$\underline{4h+2s=10}$$
$$\text{Add:}\quad -4s=-4$$
$$s=1$$

Substitute 1 for s in the first equation.

$$2h+3(1)=7$$
$$2h+3=7$$
$$2h=4$$
$$h=2$$

Each hot dog costs \$2 and each soda costs \$1.

7. Let x = the number of photos on the 128MB card and y = the number on the 516MB card.

$y = 4x$

$x + y = 360$

Substitute $4x$ for y in the second equation.

$x + 4x = 360$

$5x = 360$

$x = 72$

Substitute 72 for x in the first equation.

$y = 4(72)$

$y = 288$

The 128-megabyte memory card can store 72 photos and the 516-megabyte card can store 288 photos.

9. Let x = the measure of the larger angle and y = the measure of the smaller angle.

$x + y = 90$

$x = 2y + 15$

Substitute $2y + 15$ for x in the first equation.

$x + y = 90$

$2y + 15 + y = 90$

$3y + 15 = 90$

$3y = 75$

$y = 25$

Now, substitute 25 for y in the second equation.

$x = 2y + 15$

$x = 2(25) + 15$

$x = 50 + 15$

$x = 65$

The two angles measure 25° and 65°.

11. Let A and B be the measures of the two angles.

$A + B = 180$

$A = 3B - 28$

Substitute $3B - 28$ for A in the first equation.

$A + B = 180$

$3B - 28 + B = 180$

$4B - 28 = 180$

$4B = 208$

$B = 52$

Now substitute 52 for B in the second equation.

$A = 3B - 28$

$A = 3(52) - 28$

$A = 128$

The two angles measure 52° and 128°.

13. Let t = team's rowing speed in still water and c = speed of current

$t + c = 15.6$

$t - c = 8.8$

Add the equations to eliminate variable c.

$$\begin{array}{r} t + c = 15.6 \\ t - c = \ \ 8.8 \\ \hline 2t \quad = 24.4 \end{array}$$

$t = 12.2$

Substitute 12.2 for t in the first equation.

$(12.2) + c = 15.6$

$c = 3.4$

The team's speed in still water is 12.2 mph and the speed of the current is 3.4 mph.

15. Let x = the weekly salary and y = the commission rate.

$x + 4000y = 660$

$x + 6000y = 740$

Multiply the first equation by –1 and then add.

$-1[x + 4000y = 660]$

$x + 6000y = 740$

gives

$$\begin{array}{rr} -x - 4000y = & -660 \\ x + 6000y = & 740 \\ \hline \text{Add:} \quad 2000y = & 80 \end{array}$$

$$y = \frac{80}{2000} = 0.04$$

Substitute 0.04 for y in the first equation.

$x + 4000y = 660$

$x + 4000(0.04) = 660$

$x + 160 = 660$

$x = 500$ dollars

His weekly salary is \$500 and the commission rate is 4%.

17. Let x = the amount of 5% solution and y = the amount of 30% solution.

$x + y = 3$

$0.05x + 0.30y = 0.20(3)$

Solve the first equation for x.

$x = 3 - y$

Substitute $3 - y$ for x in the second equation.

$0.05x + 0.30y = 0.20(3)$

$0.05(3 - y) + 0.30y = 0.6$

$0.15 - 0.05y + 0.30y = 0.6$

$0.25y = 0.45$

$y = 1.8$

Substitute 1.8 for y in the first equation.

$x + y = 3$

$x + 1.8 = 3$

$x = 1.2$

Pola should mix 1.2 ounces of the 5% solution with 1.8 ounces of the 30% solution.

19. Let x = gallons of concentrate (18% solution) and y = gallons of water (0% solution).

$x + y = 200$

$0.18x + 0y = 0.009(200)$

Solve the second equation for x.

$0.18x + 0y = 0.009(200)$

$0.18x = 1.8$

$x = 10$

Substitute 10 for x in the first equation.

$x + y = 200$

$10 + y = 200$

$y = 190$

The mixture should contain 10 gallons of concentrate and 190 gallons of water.

21. Let x = pounds of birdseed and y = pounds of sunflower seeds

$0.59x + 0.89y = 0.76(40)$

$x + y = 40$

Solve the second equation for y.

$y = 40 - x$

Substitute $40 - x$ for y in the first equation.

$0.59x + 0.89y = 0.76(40)$

$0.59x + 0.89(40 - x) = 30.4$

$0.59x + 35.6 - 0.89x = 30.4$

$-0.3x = -5.2$

$x = 17\frac{1}{3}$

Substitute $17\frac{1}{3}$ for x in the second equation.

$x + y = 40$

$17\frac{1}{3} + y = 40$

$y = 22\frac{2}{3}$

Angela Leinenbach should mix $17\frac{1}{3}$ pounds of birdseed at \$0.59 per pound with $22\frac{2}{3}$ pounds of sunflower seeds at \$0.89 per pound.

23. Let x = the price of an adult ticket and y = the price of a child ticket.

$3x + 4y = 159$

$2x + 3y = 112$

Multiply the first equation by 2 and the second equation by -3, then add.

$2[3x + 4y = 159]$

$-3[2x + 3y = 112]$

gives

$6x + 8y = 318$

$-6x - 9y = -336$

Add: $-y = -18$

$y = 18$

Substitute 18 for y in the second equation.

$2x + 3(18) = 112$

$2x + 54 = 112$

$2x = 58$

$x = 29$

Adult tickets cost \$29 and child tickets cost \$18.

25. Let x = amount invested at 5% and y = amount invested at 6%

$x + y = 10{,}000$

$0.05x + 0.06y = 540$

Solve the first equation for y.

$y = 10{,}000 - x$

Substitute $10{,}000 - x$ for y in the second equation.

$0.05x + 0.06y = 540$

$0.05x + 0.06(10{,}000 - x) = 540$

$0.05x + 600 - 0.06x = 540$

$-0.01x = -60$

$x = 6000$

Substitute 6000 for x in the first equation.

$x + y = 10{,}000$

$6000 + y = 10{,}000$

$y = 4000$

Mr. and Mrs. Gamton invested \$6000 at 5% and \$4000 at 6%.

27. Let x = the amount of the whole milk (3.25% fat) and y = the amount of the skim milk (0% fat).
$x + y = 260$
$0.0325x + 0y = 0.02(260)$
Solve the second equation for x.
$0.0325x + 0y = 0.02(260)$
$0.0325x = 5.2$
$x = 160$
Now substitute 160 for x in the first equation.
$(160) + y = 260$
$y = 100$
Becky needs to mix 160 gallons of the whole milk with 100 gallons of skim milk to produce 260 gallons of 2% fat milk.

29. Let x = pounds of *Season's Choice* birdseed at \$1.79/lb and y = pounds of *Garden Mix* birdseed at \$1.19/lb
$1.79x + 1.19y = 28.00$
$x + y = 20$
Solve the second equation for y: $y = 20 - x$
Substitute $20 - x$ for y in the first equation.
$1.79x + 1.19(20 - x) = 28.00$
$1.79x + 23.80 - 1.19x = 28.00$
$0.60x = 4.20$
$x = 7$
Substitute 7 for x in the second equation.
$(7) + y = 20$
$y = 13$
The Carters should buy 7 pounds *of Season's Choice* and 13 pounds of *Garden Mix.*

31. Let x = the rate of the slower car and y = the rate of the faster car.
$4x + 4y = 420$
$y = x + 5$
Substitute $x + 5$ for y in the first equation.
$4x + 4y = 420$
$4x + 4(x + 5) = 420$
$4x + 4x + 20 = 420$
$8x + 20 = 420$
$8x = 400$
$x = 50$
Now substitute 50 for x in the second equation.
$y = x + 5$
$y = 50 + 5 = 55$
The rate of the slower car is 50 mph and the rate of the faster car is 55 mph.

33. Let x = the amount of time traveled at 65 mph and y = the amount of time traveled at 50 mph.
$x + y = 11.4$
$65x + 50y = 690$
Solve the first equation for x
$x = 11.4 - y$
Now substitute $11.4 - y$ into x in the second equation.
$65(11.4 - y) + 50y = 690$
$741 - 65y + 50y = 690$
$-15y = -51$
$y = 3.4$
Substitute 3.4 for y in the first equation.
$x + (3.4) = 11.4$
$x = 8$
Cabrina traveled for 8 hours at 65 mph and Dabney traveled for 3.4 hours at 50 mph.

35. Let x = the number of grams of Mix A and y = the number of grams of Mix B.
$0.1x + 0.2y = 20$
$0.06x + 0.02y = 6$
To solve, multiply the second equation by -10 and then add.
$0.1x + 0.2y = 20$
$-10[0.06x + 0.02y = 6]$
gives
$0.1x + 0.2y = 20$
$-0.6x - 0.2y = -60$
Add: $-0.5x = -40$
$x = \frac{-40}{-0.5} = 80$
Now substitute 80 for x in the first equation.
$0.1x + 0.2y = 20$
$0.1(80) + 0.2y = 20$
$8 + 0.2y = 20$
$0.2y = 12$
$y = \frac{12}{0.2} = 60$
The scientist should feed each animal 80 grams of Mix A and 60 grams of Mix B.

37. Let x = the amount of the first alloy and y = the amount of the second alloy.

$0.7x + 0.4y = 0.6(300)$

$0.3x + 0.6y = 0.4(300)$

To solve, multiply the first equation by 3 and the second equation by –2 and then add.

$3[0.7x + 0.4y = 0.6(300)]$

$-2[0.3x + 0.6y = 0.4(300)]$

gives

$$\begin{aligned} 2.1x + 1.2y &= 540 \\ -0.6x - 1.2y &= -240 \end{aligned}$$

Add: $1.5x = 300$

$$x = \frac{300}{1.5} = 200$$

Now substitute 200 for x in the first equation.

$$\begin{aligned} 0.7x + 0.4y &= 0.6(300) \\ 0.7(200) + 0.4y &= 0.6(300) \\ 140 + 0.4y &= 180 \\ 0.4y &= 40 \\ y &= \frac{40}{0.4} = 100 \end{aligned}$$

200 grams of the first alloy should be combined with 100 grams of the second alloy to produce the desired mixture.

39. Let t = the number of years since 2002 and y = the number of returns filed. We can use the given functions to form a system of equations.

Paper $y = -2.73t + 58.37$

Online $y = 1.95t + 10.58$

Subsitute $-2.73t + 58.37$ for y in the second equation.

$$\begin{aligned} -2.73t + 58.37 &= 1.95t + 10.58 \\ 58.37 &= 4.68t + 10.58 \\ 47.79 &= 4.68t \\ \frac{47.79}{4.68} &= t \quad \text{or} \quad t \approx 10.2 \end{aligned}$$

Thus, if the trends continue, the number of paper Form 1040 returns will be about the same as the number of online Form 1040, 1040A, 1040EZ returns in 2012 (2002 + 10 = 2012).

41. Let x = speed of Melissa's car and y = speed of Tom's car

$x = y + 15$

$\frac{150}{x} = \frac{120}{y}$ (travel times are equal)

Substitute $y + 15$ for x in the second equation.

$$\begin{aligned} \frac{150}{y+15} &= \frac{120}{y} \\ 150y &= 120y + 1800 \\ 30y &= 1800 \\ y &= 60 \end{aligned}$$

Substitute 60 for y in the first equation.

$$\begin{aligned} x &= y + 15 \\ x &= 60 + 15 \\ x &= 75 \end{aligned}$$

Tom traveled at 60 mph and Melissa traveled at 75 mph.

43. Let x = pieces of personal mail

y = number of bills and statements

z = number of advertisements

$x + y + z = 24$

$y = 2x - 2$

$z = 5x + 2$

Substitute $2x - 2$ for y and $5x + 2$ for z in the first equation.

$$\begin{aligned} x + y + z &= 24 \\ x + 2x - 2 + 5x + 2 &= 24 \\ 8x &= 24 \\ x &= 3 \end{aligned}$$

Substitute 3 for x in the second and third equations.

$y = 2x - 2$ $\quad z = 5x + 2$

$y = 2(3) - 2$ $\quad z = 5(3) + 2$

$y = 4$ $\quad z = 17$

An average American household receives 3 pieces of personal mail, 4 bills and statements, and 17 advertisements per week.

45. Let x = the number of bowl game appearances by Alabama, y = the number of bowl game appearnces by Tennessee, and z = the number of bowl game appearances by Texas.

$$x+y+z=141$$
$$x=z+8$$
$$y+z=x+37$$

Solve the third equation for y.

$$y+z=x+37$$
$$y=x+37-z$$

Substitute $x+37-z$ for y in the first equation.

$$x+(x+37-z)+z=141$$
$$2x+37=141$$
$$2x=104$$
$$x=52$$

Substitute 52 for x in the second equation.

$$52=z+8$$
$$44=z$$

Substitute 44 for z and 52 for x in the third equation.

$$y+44=52+37$$
$$y+44=89$$
$$y=45$$

Through 2004, Alabama has appeared in 52 bowl games, Tennessee has appeared in 45 bowl games, and Texas has appeared in 44 bowl games.

47. Let x = the number of top-ten finishes for Vijay Singh, y = the number of top-ten finishes for Tiger Woods, and z = the number of top-ten finishes for Phil Mickelson.

$$x+y+z=191$$
$$y=z+8$$
$$x=z+12$$

Substitute $z+8$ for y and $z+12$ for x in the first equation.

$$(z+12)+(z+8)+z=191$$
$$3z+20=191$$
$$3z=171$$
$$z=57$$

Substitute 57 for z in the second equation.

$$y=57+8=65$$

Substitute 57 for z in the third equation.

$$x=57+12=69$$

Vijay Singh had 69 top-ten finishes, Tiger Woods had 65 top-ten finishes, and Phil Mickelson had 57 top-ten finishes.

49. Let x = the amount of snowfall (inches) in Haverhill, y = the amount of snowfall in Plymouth, and z = the amount of snowfall in Salem.

$$x+y+z=112.5$$
$$y=z$$
$$y=x+1.5$$

Substitute $x+1.5$ for both y and z in the first equation.

$$x+(x+1.5)+(x+1.5)=112.5$$
$$3x+3=112.5$$
$$3x=109.5$$
$$x=36.5$$

Substitute 36.5 for x in the third equation.

$$y=36.5+1.5$$
$$y=38$$

Haverhill received 36.5 inches of snow while Plymouth and Salem each received 38 inches.

51. Let F = the number of Super Bowls held in Florida, L = the number held in Louisiana, and C = the number held in California.

$$F+L+C=32 \quad (1)$$
$$F=L+3 \quad (2)$$
$$F+L=2C-1 \quad (3)$$

Substitute $L+3$ for F in the first and third equations.

$$(L+3)+L+C=32 \qquad (L+3)+L=2C-1$$
$$2L+C+3=32 \qquad 2L+3=2C-1$$
$$2L+C=29 \qquad 2L-2C=-4$$

We now have two equations in two unknowns.

$$2L+C=29 \quad (4)$$
$$2L-2C=-4 \quad (5)$$

Eliminate L by multiplying equation (5) by -1 and adding to equation (4).

$$2L+C=29$$
$$-1[2L-2C=-4]$$

gives

$$2L+C=29$$
$$\underline{-2L+2C=4}$$

Add: $3C=33$

$$C=11$$

Substitute 11 for C in equation (4).

$$2L+11=29$$
$$2L=18$$
$$L=9$$

Substitute 9 for L in equation (2).

$F = 9 + 3$

$F = 12$

Florida hosted 12 Super Bowls, Louisiana hosted 9 Super Bowls, and California hosted 11 Super Bowls.

53. Let x be the smallest angle measure, y the second smallest, and z the largest.

$x + y + z = 180$

$x = \frac{2}{3}y$

$z = 3y - 30$

Substitute $\frac{2}{3}y$ for x and $3y - 30$ for z in the first equation.

$$x + y + z = 180$$
$$\frac{2}{3}y + y + 3y - 30 = 180$$
$$\frac{14}{3}y - 30 = 180$$
$$\frac{14}{3}y = 210$$
$$y = 45$$

Substitute 45 for y in the second equation.

$x = \frac{2}{3}y$

$x = \frac{2}{3}(45)$

$x = 30$

Substitute 45 for y in the third equation.

$z = 3y - 30$

$z = 3(45) - 30$

$z = 135 - 30$

$z = 105$

The three angles are 30°, 45°, and 105°.

55. Let x = the amount invested at 3%, y = the amount invested at 5%, and z = the amount invested at 6%.

$$y = 2x \quad (1)$$
$$x + y + z = 10{,}000 \; (2)$$
$$0.03x + 0.05y + 0.06z = 525 \quad (3)$$

Substitute $2x$ for y in equation (2).

$x + y + z = 10{,}000$

$x + 2x + z = 10{,}000$

$3x + z = 10{,}000 \quad (4)$

Substitute $2x$ for y in equation (3).

$$0.03x + 0.05y + 0.06z = 525$$
$$0.03x + 0.05(2x) + 0.06z = 525$$
$$0.03x + 0.10x + 0.06z = 525$$
$$0.13x + 0.06z = 525 \; (5)$$

Equations (4) and (5) are a system of two equations in two unknowns.

$3x + z = 10{,}000$

$0.13x + 0.06z = 525$

To eliminate z, multiply equation (4) by -3 and equation (5) by 50 and add.

$-3[3x + z = 10{,}000]$

$50[0.13x + 0.06x = 525]$

gives

$$-9x - 3z = -30{,}000$$
$$6.5x + 3z = 26{,}250$$

Add: $-2.5x = -3750$

$$x = \frac{-3750}{-2.5} = 1500$$

Substitute 1500 for x in equation (4).

$3x + z = 10{,}000$

$3(1500) + z = 10{,}000$

$4500 + z = 10{,}000$

$z = 5500$

Substitute 1500 for x and 5500 for z in equation (2).

$x + y + z = 10{,}000$

$1500 + y + 5500 = 10{,}000$

$y + 7000 = 10{,}000$

$y = 3000$

Tam invested \$1500 at 3%, \$3000 at 5%, and \$5500 at 6%.

57. Let x = the amount of the 10% solution, y = the amount of the 12% solution, and z = the amount of the 20% solution.

$$x + y + z = 8 \quad (1)$$
$$0.10x + 0.12y + 0.20z = (0.13)8 \; (2)$$
$$z = x - 2 \quad (3)$$

Substitute $x - 2$ for z in equation (1).

$x + y + z = 8$

$x + y + (x - 2) = 8$

$2x + y - 2 = 8$

$2x + y = 10 \quad (4)$

Substitute $x - 2$ for z in equation (2).

$$0.10x + 0.12y + 0.20z = (0.13)8$$
$$0.10x + 0.12y + 0.20(x - 2) = (0.13)8$$
$$0.10x + 0.12y + 0.20x - 0.40 = 1.04$$
$$0.30x + 0.12y = 1.44 \quad (5)$$

Equations (4) and (5) are a system of two equations in two unknowns.

$$2x + y = 10$$
$$0.30x + 0.12y = 1.44$$

To solve, multiply equation (5) by 100 and equation (4) by –12 and then add.

$$-12[2x + y = 10]$$
$$100[0.30x + 0.12y = 1.44]$$

gives

$$-24x - 12y = -120$$
$$30x + 12y = 144$$
$$\text{Add: } 6x = 24$$
$$x = 4$$

Substitute 4 for x in equation (4).

$$2x + y = 10$$
$$2(4) + y = 10$$
$$8 + y = 10$$
$$y = 2$$

Finally, substitute 4 for x in equation (3).

$$z = x - 2$$
$$z = 4 - 2$$
$$z = 2$$

The mixture consists of 4 liters of the 10% solution, 2 liters of the 12% solution, and 2 liters of the 20% solution.

59. Let x = the number of children's chairs, y = the number of standard chairs, and z = the number of executive chairs.

$$5x + 4y + 7z = 154 \ (1)$$
$$3x + 2y + 5z = 94 \ (2)$$
$$2x + 2y + 4z = 76 \ (3)$$

To eliminate y between equations (1) and (2), multiply equation (2) by –2 and add.

$$5x + 4y + 7z = 154$$
$$-2[3x + 2y + 5z = 94]$$

gives

$$5x + 4y + 7z = 154$$
$$-6x - 4y - 10z = -188$$
$$\text{Add: } -x - 3z = -34 \quad (4)$$

To eliminate y between equations (2) and (3), multiply equation (3) by –1 and add.

$$3x + 2y + 5z = 94$$
$$-1[2x + 2y + 4z = 76]$$

gives

$$3x + 2y + 5z = 94$$
$$-2x - 2y - 4z = -76$$
$$\text{Add: } x + z = 18 \quad (5)$$

Equations (4) and (5) are a system of two equations in two unknowns. To eliminate x, simply add.

$$-x - 3z = -34$$
$$x + z = 18$$
$$\text{Add: } -2z = -16$$
$$z = \frac{-16}{-2} = 8$$

Substitute 8 for z in equation (5).

$$x + z = 18$$
$$x + 8 = 18$$
$$x = 10$$

Substitute 10 for x and 8 for z in equation (3).

$$2x + 2y + 4z = 76$$
$$2(10) + 2y + 4(8) = 76$$
$$20 + 2y + 32 = 76$$
$$2y + 52 = 76$$
$$2y = 24$$
$$y = 12$$

The Donaldson Furniture Company should produce 10 children's chairs, 12 standard chairs, and 8 executive chairs.

61. $I_A + I_B + I_C = 0 \ (1)$

$$-8I_B + 10I_C = 0 \ (2)$$
$$4I_A - 8I_B = 6 \ (3)$$

To eliminate I_A between equations (1) and (3), multiply equation (1) by –4 and add.

$$-4[I_A + I_B + C = 0]$$
$$4I_A - 8I_B = 6$$

gives

$$-4I_A - 4I_B - 4I_C = 0$$
$$4I_A - 8I_B = 6$$
$$\text{Add: } -12I_B - 4I_C = 6$$
$$\text{or } -6I_B - 2I_C = 3 \quad (4)$$

Equations (4) and (2) are a system of two equations in two unknowns.

$$-8I_B + 10I_C = 0$$
$$-6I_B - 2I_C = 3$$

Multiply equation (4) by 5 and add this result to equation (2).

$-8I_B + 10I_C = 0$

$5[-6I_B - 2I_C = 3]$

gives

$$\begin{aligned} -8I_B + 10I_C &= 0 \\ -30I_B - 10I_C &= 15 \end{aligned}$$

Add: $-38I_B \qquad = 15$

$$I_B = \frac{15}{-38} = -\frac{15}{38}$$

Substitute $-\frac{15}{38}$ for I_B in equation (2).

$$-8I_B + 10I_C = 0$$

$$-8\left(-\frac{15}{38}\right) + 10I_C = 0$$

$$\frac{120}{38} + 10I_C = 0$$

$$10I_C = -\frac{120}{38}$$

$$\frac{1}{10}(10I_C) = \frac{1}{10}\left(-\frac{120}{38}\right)$$

$$I_C = -\frac{12}{38} = -\frac{6}{19}$$

Finally, substitute $-\frac{15}{38}$ for I_B in equation (3).

$$4I_A - 8I_B = 6$$

$$4I_A - 8\left(-\frac{15}{38}\right) = 6$$

$$4I_A + \frac{120}{38} = 6$$

$$4I_A = 6 - \frac{120}{38}$$

$$4I_A = 6 - \frac{60}{19}$$

$$4I_A = \frac{114}{19} - \frac{60}{19}$$

$$4I_A = \frac{54}{19}$$

$$\frac{1}{4}(4I_A) = \frac{1}{4}\left(\frac{54}{19}\right)$$

$$I_A = \frac{27}{38}$$

The current in branch A is $\frac{27}{38}$, the current in branch B is $-\frac{15}{38}$ and the current in branch C is $-\frac{6}{19}$.

64. Substitute –2 for x and 5 for y.

$$\begin{aligned} \frac{1}{2}x + \frac{2}{5}xy + \frac{1}{8}y &= \frac{1}{2}(-2) + \frac{2}{5}(-2)(5) + \frac{1}{8}(5) \\ &= -1 - 4 + \frac{5}{8} \\ &= -5 + \frac{5}{8} \\ &= -\frac{35}{8} \end{aligned}$$

65. $4 - 2\left[(x-5) + 2x\right] = -(x+6)$

$$\begin{aligned} 4 - 2(x - 5 + 2x) &= -x - 6 \\ 4 - 2(3x - 5) &= -x - 6 \\ 4 - 6x + 10 &= -x - 6 \\ -6x + 14 &= -x - 6 \\ -6x + x &= -6 - 14 \\ -5x &= -20 \\ x &= 4 \end{aligned}$$

66. Use the vertical line test. If a vertical line cannot be drawn to intersect the graph in more than one point, the graph represents a fimction.

67. The slope is

$$m = \frac{-4-(-8)}{6-2} = \frac{-4+8}{6-2} = \frac{4}{4} = 1$$

Use the point-slope form with $m = 1$ and $(x_1, y_1) = (6, -4)$.

$$\begin{aligned} y - y_1 &= m(x - x_1) \\ y - (-4) &= 1(x - 6) \\ y + 4 &= x - 6 \\ y &= x - 10 \end{aligned}$$

Mid-Chapter Test: 4.1 – 4.3

1. a. $7x - y = 13$ $\quad$ $2x + 3y = 9$

$-y = -7x + 13$ $\quad$ $3y = -2x + 9$

$y = 7x - 13$ $\quad$ $y = \dfrac{-2x+9}{3}$

$y = -\dfrac{2}{3}x + 3$

b. The slopes are different so the lines will cross. Therefore, the system is consistent.

c. Because the slopes are different, the graphs will cross exactly once. Therefore, the system will have exactly one solution.

2. The equations are already in slope-intercept form. Graph $y = 2x$ and $y = -x + 3$.

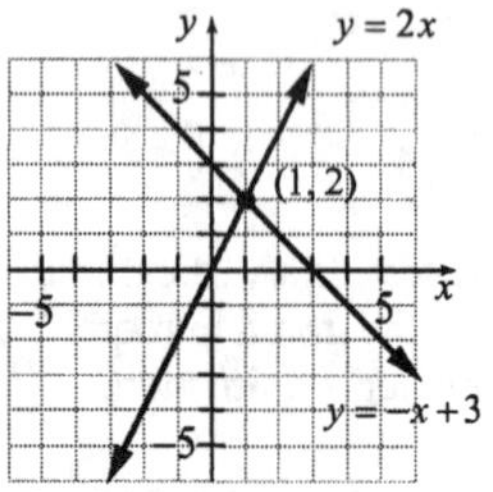

The lines intersect and the intersection point is $(1, 2)$. The system is consistent.

3. Start by writing each equation in slope-intercept form.

$x + y = -4$ $\quad$ $3x - 2y = 3$

$y = -x - 4$ $\quad$ $-2y = -3x + 3$

$y = \dfrac{-3x+3}{-2}$

$y = \dfrac{3}{2}x - \dfrac{3}{2}$

Graph the equations.

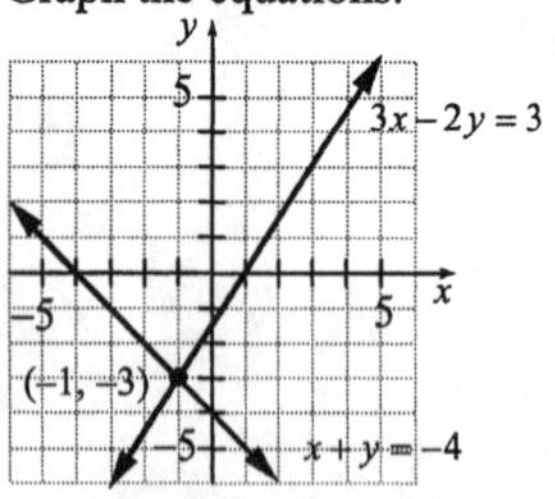

The lines intersect and the intersection point is $(-1, -3)$. The system is consistent.

4. $2x + 5y = -3$

$x - 2y = -6$

Solve the second equation for *x*.

$x - 2y = -6$

$x = 2y - 6$

Substitute $2y - 6$ for *x* in the first equation.

$2(2y - 6) + 5y = -3$

$4y - 12 + 5y = -3$

$9y = 9$

$y = 1$

Substitute 1 for *y* in the equation for *x*.

$x = 2y - 6$

$x = 2(1) - 6$

$x = -4$

The solution is $(-4, 1)$.

5. $4x - 3y = 8$

$2x + y = -1$

Solve the second equation for *y*.

$2x + y = -1$

$y = -2x - 1$

Substitute $-2x - 1$ for *y* in the first equation.

$4x - 3(-2x - 1) = 8$

$4x + 6x + 3 = 8$

$10x = 5$

$x = \dfrac{1}{2}$

Substitute $\dfrac{1}{2}$ for *x* in the equation for *y*.

$y = -2x - 1$

$y = -2\left(\dfrac{1}{2}\right) - 1$

$y = -1 - 1 = -2$

The solution is $\left(\dfrac{1}{2}, -2\right)$.

6. $x = 4y - 19$

$7x + 5y = -1$

Write the equations in standard form.

$x - 4y = -19$

$7x + 5y = -1$

Multiply the first equation by -7 and add to the second equation.

$-7[x-4y=-19]$

$7x+5y=-1$

gives

$$\begin{array}{r} -7x+28y=133 \\ \underline{7x+\ \ 5y=\ \ -1} \end{array}$$

Add: $33y=132$

$y=4$

Substitute 4 for y in the first equation.

$x=4y-19$

$x=4(4)-19$

$x=-3$

The solution is $(-3,4)$.

7. $3x+4y=3$

$9x+5y=\dfrac{11}{2}$

Multiply the first equation by -3 and add to the second.

$-3[3x+4y=3]$

$9x+5y=\dfrac{11}{2}$

gives

$$\begin{array}{r} -9x-12y=-9 \\ \underline{9x+\ \ 5y=\dfrac{11}{2}} \end{array}$$

Add: $-7y=-\dfrac{7}{2}$

$y=\dfrac{1}{2}$

Substitute $\dfrac{1}{2}$ for y in the first equation.

$3x+4y=3$

$3x+4\left(\dfrac{1}{2}\right)=3$

$3x+2=3$

$3x=1$

$x=\dfrac{1}{3}$

The solution is $\left(\dfrac{1}{3},\dfrac{1}{2}\right)$.

8. $\dfrac{1}{3}a-\dfrac{1}{4}b=-1$

$\dfrac{1}{2}a+\dfrac{1}{6}b=5$

Clear the fractions by multiplying the first equation by 12 and the second equation by 6.

$12\left[\dfrac{1}{3}a-\dfrac{1}{4}b=-1\right]$

$6\left[\dfrac{1}{2}a+\dfrac{1}{6}b=5\right]$

gives

$4a-3b=-12 \quad (1)$

$3a+b=30 \quad (2)$

Multiply equation (2) by 3 and add.

$4a-3b=-12$

$3[3a+b=30]$

gives

$$\begin{array}{r} 4a-3b=-12 \\ \underline{9a+3b=90} \end{array}$$

Add: $13a=78$

$a=6$

Substitute 6 for a in equation (2).

$3a+b=30$

$3(6)+b=30$

$18+b=30$

$b=12$

The solution is $(6,12)$.

9. $3m-2n=1$

$n=\dfrac{3}{2}m-7$

Substitute $\dfrac{3}{2}m-7$ for n in the first equation.

$3m-2n=1$

$3m-2\left(\dfrac{3}{2}m-7\right)=1$

$3m-3m+14=1$

$14=1$

This is a contradiction so the system has no solution. The system is inconsistent.

10. $8x-16y=24$

$x=2y+3$

Substitute $2y+3$ for x in the first equation.

$$8x-16y=24$$
$$8(2y+3)-16y=24$$
$$16y+24-16y=24$$
$$24=24$$

The result is an identity. The system is dependent and has an infinite number of solutions.

11. $x+y+z=2 \quad (1)$

$2x-y+2z=-2 \quad (2)$

$3x+2y+6z=1 \quad (3)$

To eliminate y between equations (1) and (2), simply add.

$$x+y+z=2$$
Add: $$2x-y+2z=-2$$
$$3x+3z=0 \quad (4)$$

To eliminate y between equations (2) and (3), multiply equation (2) by 2 and then add.

$$4x-2y+4z=-4$$
$$3x+2y+6z=1$$
Add: $$7x+10z=-3 \quad (5)$$

Equations (4) and (5) are two equations in two unknowns.

$$3x+3z=0 \quad (4)$$
$$7x+10z=-3 \quad (5)$$

To eliminate x, multiply equation (4) by -7, multiply equation (5) by 3 and then add.

$$-21x-21z=0$$
$$21x+30z=-9$$
Add: $$9z=-9$$
$$z=-1$$

Substitute -1 for z in equation (4).

$$3x+3(-1)=0$$
$$3x-3=0$$
$$3x=3$$
$$x=1$$

Substitute 1 for x and -1 for z in equation (1).

$$x+y+z=2$$
$$1+y+(-1)=2$$
$$y=2$$

The solution is $(1,2,-1)$.

12. $2x-y-z=1 \quad (1)$

$3x+5y+2z=12 \quad (2)$

$-6x-4y+5z=3 \quad (3)$

To eliminate x between equations (1) and (2), multiply equation (1) by -3 and equation (2) by 2, then add.

$$-6x+3y+3z=-3$$
$$6x+10y+4z=24$$
Add: $$13y+7z=21 \quad (4)$$

To eliminate x between equations (2) and (3), multiply equation (2) by 2 and then add.

$$6x+10y+4z=24$$
$$-6x-4y+5z=3$$
Add: $$6y+9z=27 \quad (5)$$

Equations (4) and (5) are two equations in two unknowns.

$$13y+7z=21 \quad (4)$$
$$6y+9z=27 \quad (5)$$

To eliminate z, multiply equation (4) by -9, multiply equation (5) by 7 and then add.

$$-117y-63z=-189$$
$$42y+63z=189$$
Add: $$-75y=0$$
$$y=0$$

Substitute 0 for y in equation (4).

$$13(0)+7z=21$$
$$7z=21$$
$$z=3$$

Substitute 0 for y and 3 for z in equation (1).

$$2x-y-z=1$$
$$2x-0-3=1$$
$$2x-3=1$$
$$2x=4$$
$$x=2$$

The solution is $(2,0,3)$.

13. The system contains three variables. Any solution to the system must be an ordered triple. That is, there needs to be a value for each of the three variables.

$$x+y+z=4 \quad (1)$$
$$-x+2y+2z=5 \quad (2)$$
$$7x+5y-z=-2 \quad (3)$$

To eliminate x between equations (1) and (2), simply add.

$$x+\ y+\ z=4$$
$$\underline{-x+2y+2z=5}$$

Add: $3y+3z=9$ (4)

To eliminate x between equations (2) and (3), multiply equation (2) by 7 and then add.

$$-7x+14y+14z=35$$
$$\underline{7x+\ 5y-\ z=-2}$$

Add: $19y+13z=33$ (5)

Equations (4) and (5) are two equations in two unknowns.

$$3y+3z=9 \quad (4)$$
$$19y+13z=33 \quad (5)$$

To eliminate z, multiply equation (4) by –13, multiply equation (5) by 3 and then add.

$$-39y-39z=-117$$
$$\underline{57y+39z=99}$$

Add: $18y=-18$

$$y=-1$$

Substitute -1 for y in equation (4).

$$3(-1)+3z=9$$
$$-3+3z=9$$
$$3z=12$$
$$z=4$$

Substitute -1 for y and 4 for z in equation (1).

$$x+y+z=4$$
$$x-1+4=4$$
$$x+3=4$$
$$x=1$$

The solution is $(1,-1,4)$.

14. Let x = the number of pounds of cashews and y = the number of pounds of pecans.

$$x+y=15 \quad \rightarrow \quad x+y=15$$
$$12x+6y=15(10) \qquad 12x+6y=150$$

Solve the first equation for y.

$$x+y=15$$
$$y=-x+15$$

Substitute $-x+15$ for y in the second equation.

$$12x+6(-x+15)=150$$
$$12x-6x+90=150$$
$$6x+90=150$$
$$6x=60$$
$$x=10$$

Substitute 10 for x in the first equation.

$$x+y=15$$
$$10+y=15$$
$$y=5$$

William should mix 10 pounds of cashews with 5 pounds of pecans.

15. Let x, y, and z be the three numbers in increasing order.

$$x+y+z=32$$
$$z=4x$$
$$x+y=z-8$$

Solve the third equation for y.

$$x+y=z-8$$
$$y=-x+z-8$$

Substitute $-x+z-8$ for y in the first equation.

$$x+y+z=32$$
$$x-x+z-8+z=32$$
$$2z-8=32$$
$$2z=40$$
$$z=20$$

Substitute 20 for z in the second equation.

$$20=4x$$
$$5=x$$

Substitute 5 for x and 20 for z in the equation for y.

$$y=-x+z-8$$
$$y=-5+20-8$$
$$y=7$$

The three numbers are 5, 7, and 20.

Exercise Set 4.4

1. A square matrix has the same number of rows and columns.

3. $-\frac{1}{2}R_2$; the next step is to change the –2 in the second row to 1 by multiplying the second row of numbers by $-\frac{1}{2}$.

5. Switch row (2) and row (3) in order to continuing placing ones along the diagonal.

7. Dependent (assuming no contradictions exist)

9. $$\left[\begin{array}{cc|c} 5 & -10 & -25 \\ 3 & -7 & -4 \end{array}\right] \Rightarrow \left[\begin{array}{cc|c} 1 & -2 & -5 \\ 3 & -7 & -4 \end{array}\right]\frac{1}{5}R_1$$

11. $$\left[\begin{array}{ccc|c} 4 & 7 & 2 & -1 \\ 3 & 2 & 1 & -5 \\ 1 & 1 & 3 & -8 \end{array}\right] \Rightarrow \left[\begin{array}{ccc|c} 1 & 1 & 3 & -8 \\ 3 & 2 & 1 & -5 \\ 4 & 7 & 2 & -1 \end{array}\right] \begin{array}{l} \text{switch} \\ \\ R_1 \text{ and } R_3 \end{array}$$

13. $$\left[\begin{array}{cc|c} 1 & 3 & 12 \\ -4 & 11 & -6 \end{array}\right] \Rightarrow \left[\begin{array}{cc|c} 1 & 3 & 12 \\ 0 & 23 & 42 \end{array}\right] 4R_1 + R_2$$

15. $$\left[\begin{array}{ccc|c} 1 & 0 & 8 & \frac{1}{4} \\ 5 & 2 & 2 & -2 \\ 6 & -3 & 1 & 0 \end{array}\right] \Rightarrow$$

$$\left[\begin{array}{ccc|c} 1 & 0 & 8 & \frac{1}{4} \\ 0 & 2 & -38 & -\frac{13}{4} \\ 6 & -3 & 1 & 0 \end{array}\right] -5R_1 + R_2$$

17. $x + 3y = 3$

$-x + y = -3$

$$\left[\begin{array}{cc|c} 1 & 3 & 3 \\ -1 & 1 & -3 \end{array}\right]$$

$$\left[\begin{array}{cc|c} 1 & 3 & 3 \\ 0 & 4 & 0 \end{array}\right] R_1 + R_2$$

$$\left[\begin{array}{cc|c} 1 & 3 & 3 \\ 0 & 1 & 0 \end{array}\right] \frac{1}{4}R_2$$

The system is

$x + 3y = 3$

$y = 0$

Substitute 0 for y in the first equation.

$x + 3y = 3$

$x + 3(0) = 3$

$x + 0 = 3$

$x = 3$

The solution is (3, 0).

19. $x + 3y = -2$

$-2x - 7y = 3$

$$\left[\begin{array}{cc|c} 1 & 3 & -2 \\ -2 & -7 & 3 \end{array}\right]$$

$$\left[\begin{array}{cc|c} 1 & 3 & -2 \\ 0 & -1 & -1 \end{array}\right] 2R_1 + R_2$$

$$\left[\begin{array}{cc|c} 1 & 3 & -2 \\ 0 & 1 & 1 \end{array}\right] -R_2$$

The system is

$x + 3y = -2$

$y = 1$

Substitute 1 for y in the first equation.

$x + 3(1) = -2$

$x + 3 = -2$

$x = -5$

The solution is $(-5, 1)$.

21. $5a - 10b = -10$

$2a + b = 1$

$$\left[\begin{array}{cc|c} 5 & -10 & -10 \\ 2 & 1 & 1 \end{array}\right]$$

$$\left[\begin{array}{cc|c} 1 & -2 & -2 \\ 2 & 1 & 1 \end{array}\right] \frac{1}{5}R_1$$

$$\left[\begin{array}{cc|c} 1 & -2 & -2 \\ 0 & 5 & 5 \end{array}\right] -2R_1 + R_2$$

$$\left[\begin{array}{cc|c} 1 & -2 & -2 \\ 0 & 1 & 1 \end{array}\right] \frac{1}{5}R_2$$

The system is

$a - 2b = -2$

$b = 1$

Substitute 1 for b in the first equation.

$a - 2(1) = -2$

$a - 2 = -2$

$a = 0$

The solution is (0, 1).

23. $2x-5y=-6$

$-4x+10y=12$

$\left[\begin{array}{cc|c} 2 & -5 & -6 \\ -4 & 10 & 12 \end{array}\right]$

$\left[\begin{array}{cc|c} 1 & -\frac{5}{2} & -3 \\ -4 & 10 & 12 \end{array}\right] \frac{1}{2}R_1$

$\left[\begin{array}{cc|c} 1 & -\frac{5}{2} & -3 \\ 0 & 0 & 0 \end{array}\right] 4R_1+R_2$

Since the last row contains all 0's, this is a dependent system of equations.

25. $12x+2y=2$

$6x-3y=-11$

$\left[\begin{array}{cc|c} 12 & 2 & 2 \\ 6 & -3 & -11 \end{array}\right]$

$\left[\begin{array}{cc|c} 1 & \frac{1}{6} & \frac{1}{6} \\ 6 & -3 & -11 \end{array}\right] \frac{1}{12}R_1$

$\left[\begin{array}{cc|c} 1 & \frac{1}{6} & \frac{1}{6} \\ 0 & -4 & -12 \end{array}\right] -6R_1+R_2$

$\left[\begin{array}{cc|c} 1 & \frac{1}{6} & \frac{1}{6} \\ 0 & 1 & 3 \end{array}\right] -\frac{1}{4}R_2$

The system is

$x+\frac{1}{6}y=\frac{1}{6}$

$y=3$

Substitute 3 for y in the first equation.

$x+\frac{1}{6}y=\frac{1}{6}$

$x+\frac{1}{6}(3)=\frac{1}{6}$

$x+\frac{3}{6}=\frac{1}{6}$

$x=-\frac{2}{6}$

$x=-\frac{1}{3}$

The solution is $\left(-\frac{1}{3},3\right)$.

27. $-3x+6y=5$

$2x-4y=7$

$\left[\begin{array}{cc|c} -3 & 6 & 5 \\ 2 & -4 & 7 \end{array}\right]$

$\left[\begin{array}{cc|c} 1 & -2 & -\frac{5}{3} \\ 2 & -4 & 7 \end{array}\right] -\frac{1}{3}R_1$

$\left[\begin{array}{cc|c} 1 & -2 & -\frac{5}{3} \\ 0 & 0 & \frac{31}{3} \end{array}\right] -2R_1+R_2$

Since the last row contains zeros on the left and a nonzero number on the right, this is an inconsistent system and there is no solution.

29. $12x-8y=6$

$-3x+4y=-1$

$\left[\begin{array}{cc|c} 12 & -8 & 6 \\ -3 & 4 & -1 \end{array}\right]$

$\left[\begin{array}{cc|c} 1 & -\frac{2}{3} & \frac{1}{2} \\ -3 & 4 & -1 \end{array}\right] \frac{1}{12}R_1$

$\left[\begin{array}{cc|c} 1 & -\frac{2}{3} & \frac{1}{2} \\ 0 & 2 & \frac{1}{2} \end{array}\right] 3R_1+R_2$

$\left[\begin{array}{cc|c} 1 & -\frac{2}{3} & \frac{1}{2} \\ 0 & 1 & \frac{1}{4} \end{array}\right] \frac{1}{2}R_2$

The system is

$x-\frac{2}{3}y=\frac{1}{2}$

$y=\frac{1}{4}$

Substitute $\frac{1}{4}$ for y in the first equation.

$x-\frac{2}{3}y=\frac{1}{2}$

$x-\frac{2}{3}\left(\frac{1}{4}\right)=\frac{1}{2}$

$x-\frac{1}{6}=\frac{1}{2}$

$x=\frac{1}{2}+\frac{1}{6}$

$x=\frac{4}{6}$

$x=\frac{2}{3}$

The solution is $\left(\frac{2}{3},\frac{1}{4}\right)$.

31. $10m = 8n + 15$

$16n = -15m - 2$

Write the system in standard form.

$10m - 8n = 15$

$15m + 16n = -2$

$$\left[\begin{array}{cc|c} 10 & -8 & 15 \\ 15 & 16 & -2 \end{array}\right]$$

$$\left[\begin{array}{cc|c} 1 & -\frac{4}{5} & \frac{3}{2} \\ 15 & 16 & -2 \end{array}\right] \frac{1}{10}R_1$$

$$\left[\begin{array}{cc|c} 1 & -\frac{4}{5} & \frac{3}{2} \\ 0 & 28 & -\frac{49}{2} \end{array}\right] -15R_1 + R_2$$

$$\left[\begin{array}{cc|c} 1 & -\frac{4}{5} & \frac{3}{2} \\ 0 & 1 & -\frac{7}{8} \end{array}\right] \frac{1}{28}R_2$$

The system is

$m - \frac{4}{5}n = \frac{3}{2}$

$n = -\frac{7}{8}$

Substitute $-\frac{7}{8}$ for n in the first equation.

$$m - \frac{4}{5}n = \frac{3}{2}$$

$$m - \frac{4}{5}\left(-\frac{7}{8}\right) = \frac{3}{2}$$

$$m + \frac{7}{10} = \frac{3}{2}$$

$$m = \frac{3}{2} - \frac{7}{10}$$

$$m = \frac{15}{10} - \frac{7}{10}$$

$$m = \frac{8}{10}$$

$$m = \frac{4}{5}$$

The solution is $\left(\frac{4}{5}, -\frac{7}{8}\right)$.

33. $x - 3y + 2z = 5$

$2x + 5y - 4z = -3$

$-3x + y - 2z = -11$

$$\left[\begin{array}{ccc|c} 1 & -3 & 2 & 5 \\ 2 & 5 & -4 & -3 \\ -3 & 1 & -2 & -11 \end{array}\right]$$

$$\left[\begin{array}{ccc|c} 1 & -3 & 2 & 5 \\ 0 & 11 & -8 & -13 \\ -3 & 1 & -2 & -11 \end{array}\right] -2R_1 + R_2$$

$$\left[\begin{array}{ccc|c} 1 & -3 & 2 & 5 \\ 0 & 11 & -8 & -13 \\ 0 & -8 & 4 & 4 \end{array}\right] 3R_1 + R_3$$

$$\left[\begin{array}{ccc|c} 1 & -3 & 2 & 5 \\ 0 & 1 & -\frac{8}{11} & -\frac{13}{11} \\ 0 & -8 & 4 & 4 \end{array}\right] \frac{1}{11}R_2$$

$$\left[\begin{array}{ccc|c} 1 & -3 & 2 & 5 \\ 0 & 1 & -\frac{8}{11} & -\frac{13}{11} \\ 0 & 0 & -\frac{20}{11} & -\frac{60}{11} \end{array}\right] 8R_2 + R_3$$

$$\left[\begin{array}{ccc|c} 1 & -3 & 2 & 5 \\ 0 & 1 & -\frac{8}{11} & -\frac{13}{11} \\ 0 & 0 & 1 & 3 \end{array}\right] -\frac{11}{20}R_3$$

The system is

$x - 3y + 2z = 5$

$y - \frac{8}{11}z = -\frac{13}{11}$

$z = 3$

Substitute 3 for z in the second equation.

$$y - \frac{8}{11}(3) = -\frac{13}{11}$$

$$y - \frac{24}{11} = -\frac{13}{11}$$

$$y = \frac{11}{11}$$

$$y = 1$$

Substitute 1 for y and 3 for z in the first equation.

$$x-3(1)+2(3)=5$$
$$x-3+6=5$$
$$x+3=5$$
$$x=2$$

The solution is (2, 1, 3).

35. $x+2y=5$

$y-z=-1$

$2x-3z=0$

Write the system in standard form.

$$x+2y+0z=5$$
$$0x+y-z=-1$$
$$2x+0y-3z=0$$

$$\left[\begin{array}{ccc|c} 1 & 2 & 0 & 5 \\ 0 & 1 & -1 & -1 \\ 2 & 0 & -3 & 0 \end{array}\right]$$

$$\left[\begin{array}{ccc|c} 1 & 2 & 0 & 5 \\ 0 & 1 & -1 & -1 \\ 0 & -4 & -3 & -10 \end{array}\right] -2R_1+R_3$$

$$\left[\begin{array}{ccc|c} 1 & 2 & 0 & 5 \\ 0 & 1 & -1 & -1 \\ 0 & 0 & -7 & -14 \end{array}\right] 4R_2+R_3$$

$$\left[\begin{array}{ccc|c} 1 & 2 & 0 & 5 \\ 0 & 1 & -1 & -1 \\ 0 & 0 & 1 & 2 \end{array}\right] -\tfrac{1}{7}R_3$$

The system is

$$x+2y=5$$
$$y-z=-1$$
$$z=2$$

Substitute 2 for z in the second equation.

$$y-z=-1$$
$$y-2=-1$$
$$y=1$$

Substitute 1 for y in the first equation.

$$x+2y=5$$
$$x+2(1)=5$$
$$x=3$$

The solution is (3, 1, 2).

37. $x-2y+4z=5$

$-3x+4y-2z=-8$

$4x+5y-4z=-3$

$$\left[\begin{array}{ccc|c} 1 & -2 & 4 & 5 \\ -3 & 4 & -2 & -8 \\ 4 & 5 & -4 & -3 \end{array}\right]$$

$$\left[\begin{array}{ccc|c} 1 & -2 & 4 & 5 \\ 0 & -2 & 10 & 7 \\ 4 & 5 & -4 & -3 \end{array}\right] 3R_1+R_2$$

$$\left[\begin{array}{ccc|c} 1 & -2 & 4 & 5 \\ 0 & -2 & 10 & 7 \\ 0 & 13 & -20 & -23 \end{array}\right] -4R_1+R_3$$

$$\left[\begin{array}{ccc|c} 1 & -2 & 4 & 5 \\ 0 & 1 & -5 & -\frac{7}{2} \\ 0 & 13 & -20 & -23 \end{array}\right] -\tfrac{1}{2}R_2$$

$$\left[\begin{array}{ccc|c} 1 & -2 & 4 & 5 \\ 0 & 1 & -5 & -\frac{7}{2} \\ 0 & 0 & 45 & \frac{45}{2} \end{array}\right] -13R_2+R_3$$

$$\left[\begin{array}{ccc|c} 1 & -2 & 4 & 5 \\ 0 & 1 & -5 & -\frac{7}{2} \\ 0 & 0 & 1 & \frac{1}{2} \end{array}\right] \tfrac{1}{45}R_3$$

The system is

$$x-2y+4z=5$$
$$y-5z=-\frac{7}{2}$$
$$z=\frac{1}{2}$$

Substitute $\frac{1}{2}$ for z in the second equation.

$$y-5\left(\frac{1}{2}\right)=-\frac{7}{2}$$
$$y-\frac{5}{2}=-\frac{7}{2}$$
$$y=-1$$

Substitute -1 for y and $\frac{1}{2}$ for z in the first equation.

$$x-2(-1)+4\left(\frac{1}{2}\right)=5$$
$$x+2+2=5$$
$$x+4=5$$
$$x=1$$

The solution is $\left(1, -1, \frac{1}{2}\right)$.

39. $2x-5y+z=1$
$3x-5y+z=3$
$-4x+10y-2z=-2$

$$\left[\begin{array}{ccc|c} 2 & -5 & 1 & 1 \\ 3 & -5 & 1 & 3 \\ -4 & 10 & -2 & -2 \end{array}\right]$$

$$\left[\begin{array}{ccc|c} 1 & -\frac{5}{2} & \frac{1}{2} & \frac{1}{2} \\ 3 & -5 & 1 & 3 \\ -4 & 10 & -2 & -2 \end{array}\right]\frac{1}{2}R_1$$

$$\left[\begin{array}{ccc|c} 1 & -\frac{5}{2} & \frac{1}{2} & \frac{1}{2} \\ 0 & \frac{5}{2} & -\frac{1}{2} & \frac{3}{2} \\ -4 & 10 & -2 & -2 \end{array}\right]-3R_1+R_2$$

$$\left[\begin{array}{ccc|c} 1 & -\frac{5}{2} & \frac{1}{2} & \frac{1}{2} \\ 0 & \frac{5}{2} & -\frac{1}{2} & \frac{3}{2} \\ 0 & 0 & 0 & 0 \end{array}\right]4R_1+R_3$$

Since there is a row of all zeros, the system is dependent.

41. $4p-q+r=4$
$-6p+3q-2r=-5$
$2p+5q-r=7$

$$\left[\begin{array}{ccc|c} 4 & -1 & 1 & 4 \\ -6 & 3 & -2 & -5 \\ 2 & 5 & -1 & 7 \end{array}\right]$$

$$\left[\begin{array}{ccc|c} 1 & -\frac{1}{4} & \frac{1}{4} & 1 \\ -6 & 3 & -2 & -5 \\ 2 & 5 & -1 & 7 \end{array}\right]\frac{1}{4}R_1$$

$$\left[\begin{array}{ccc|c} 1 & -\frac{1}{4} & \frac{1}{4} & 1 \\ 0 & \frac{3}{2} & -\frac{1}{2} & 1 \\ 2 & 5 & -1 & 7 \end{array}\right]6R_1+R_2$$

$$\left[\begin{array}{ccc|c} 1 & -\frac{1}{4} & \frac{1}{4} & 1 \\ 0 & \frac{3}{2} & -\frac{1}{2} & 1 \\ 0 & \frac{11}{2} & -\frac{3}{2} & 5 \end{array}\right]-2R_1+R_3$$

$$\left[\begin{array}{ccc|c} 1 & -\frac{1}{4} & \frac{1}{4} & 1 \\ 0 & 1 & -\frac{1}{3} & \frac{2}{3} \\ 0 & \frac{11}{2} & -\frac{3}{2} & 5 \end{array}\right]\frac{2}{3}R_2$$

$$\left[\begin{array}{ccc|c} 1 & -\frac{1}{4} & \frac{1}{4} & 1 \\ 0 & 1 & -\frac{1}{3} & \frac{2}{3} \\ 0 & 0 & \frac{1}{3} & \frac{4}{3} \end{array}\right]-\frac{11}{2}R_2+R_3$$

$$\left[\begin{array}{ccc|c} 1 & -\frac{1}{4} & \frac{1}{4} & 1 \\ 0 & 1 & -\frac{1}{3} & \frac{2}{3} \\ 0 & 0 & 1 & 4 \end{array}\right]3R_3$$

The system is

$$x-\frac{1}{4}y+\frac{1}{4}z=1$$
$$y-\frac{1}{3}z=\frac{2}{3}$$
$$z=4$$

Substitute 4 for z in the second equation.

$$y-\frac{1}{3}z=\frac{2}{3}$$
$$y-\frac{1}{3}(4)=\frac{2}{3}$$
$$y-\frac{4}{3}=\frac{2}{3}$$
$$y=\frac{6}{3}$$
$$y=2$$

Substitute 2 for y and 4 for z in the first equation.

$$x-\frac{1}{4}y+\frac{1}{4}z=1$$
$$x-\frac{1}{4}(2)+\frac{1}{4}(4)=1$$
$$x-\frac{1}{2}+1=1$$
$$x+\frac{1}{2}=1$$
$$x=\frac{1}{2}$$

The solution is $\left(\frac{1}{2}, 2, 4\right)$.

43. $2x-4y+3z=-12$
$3x-y+2z=-3$
$-4x+8y-6z=10$

$$\left[\begin{array}{ccc|c} 2 & -4 & 3 & -12 \\ 3 & -1 & 2 & -3 \\ -4 & 8 & -6 & 10 \end{array}\right]$$

$$\left[\begin{array}{ccc|c} 1 & -2 & \frac{3}{2} & -6 \\ 3 & -1 & 2 & -3 \\ -4 & 8 & -6 & 10 \end{array}\right]\frac{1}{2}R_1$$

$$\left[\begin{array}{ccc|c} 1 & -2 & \frac{3}{2} & -6 \\ 0 & 5 & -\frac{5}{2} & 15 \\ -4 & 8 & -6 & 10 \end{array}\right] -3R_1 + R_2$$

$$\left[\begin{array}{ccc|c} 1 & -2 & \frac{3}{2} & -6 \\ 0 & 5 & -\frac{5}{2} & 15 \\ 0 & 0 & 0 & -14 \end{array}\right] 4R_1 + R_3$$

Since the last row contains zeros on the left and a nonzero number on the right, the system is inconsistent and there is no solution.

45. $5x-3y+4z=22$

$-x-15y+10z=-15$

$-3x+9y-12z=-6$

$$\left[\begin{array}{ccc|c} 5 & -3 & 4 & 22 \\ -1 & -15 & 10 & -15 \\ -3 & 9 & -12 & -6 \end{array}\right]$$

$$\left[\begin{array}{ccc|c} 1 & -\frac{3}{5} & \frac{4}{5} & \frac{22}{5} \\ -1 & -15 & 10 & -15 \\ -3 & 9 & -12 & -6 \end{array}\right] \frac{1}{5}R_1$$

$$\left[\begin{array}{ccc|c} 1 & -\frac{3}{5} & \frac{4}{5} & \frac{22}{5} \\ 0 & -\frac{78}{5} & \frac{54}{5} & -\frac{53}{5} \\ -3 & 9 & -12 & -6 \end{array}\right] R_1 + R_2$$

$$\left[\begin{array}{ccc|c} 1 & -\frac{3}{5} & \frac{4}{5} & \frac{22}{5} \\ 0 & -\frac{78}{5} & \frac{54}{5} & -\frac{53}{5} \\ 0 & \frac{36}{5} & -\frac{48}{5} & \frac{36}{5} \end{array}\right] 3R_1 + R_3$$

$$\left[\begin{array}{ccc|c} 1 & -\frac{3}{5} & \frac{4}{5} & \frac{22}{5} \\ 0 & 1 & -\frac{9}{13} & \frac{53}{78} \\ 0 & \frac{36}{5} & -\frac{48}{5} & \frac{36}{5} \end{array}\right] -\frac{5}{78}R_2$$

$$\left[\begin{array}{ccc|c} 1 & -\frac{3}{5} & \frac{4}{5} & \frac{22}{5} \\ 0 & 1 & -\frac{9}{13} & \frac{53}{78} \\ 0 & 0 & -\frac{60}{13} & \frac{30}{13} \end{array}\right] -\frac{36}{5}R_2 + R_3$$

$$\left[\begin{array}{ccc|c} 1 & -\frac{3}{5} & \frac{4}{5} & \frac{22}{5} \\ 0 & 1 & -\frac{9}{13} & \frac{53}{78} \\ 0 & 0 & 1 & -\frac{1}{2} \end{array}\right]$$

The system is

$$x-\frac{3}{5}y+\frac{4}{5}z=\frac{22}{5}$$
$$y-\frac{9}{13}z=\frac{53}{78}$$
$$z=-\frac{1}{2}$$

Substitute $-\frac{1}{2}$ for z in the second equation.

$$y-\frac{9}{13}z=\frac{53}{78}$$
$$y-\frac{9}{13}\left(-\frac{1}{2}\right)=\frac{53}{78}$$
$$y+\frac{9}{26}=\frac{53}{78}$$
$$y=\frac{53}{78}-\frac{27}{78}$$
$$y=\frac{26}{78}$$
$$y=\frac{1}{3}$$

Substitute $\frac{1}{3}$ for y and $-\frac{1}{2}$ for z in the first equation.

$$x-\frac{3}{5}y+\frac{4}{5}z=\frac{22}{5}$$
$$x-\frac{3}{5}\left(\frac{1}{3}\right)+\frac{4}{5}\left(-\frac{1}{2}\right)=\frac{22}{5}$$
$$x-\frac{1}{5}-\frac{2}{5}=\frac{22}{5}$$
$$x-\frac{3}{5}=\frac{22}{5}$$
$$x=\frac{25}{5}$$
$$x=5$$

The solution is $\left(5, \frac{1}{3}, -\frac{1}{2}\right)$.

47. No, this is the same as switching the order of the equations.

49. Let x = smallest angle
y = remaining angle
z = largest angle

$$z = x + 55$$
$$z = y + 20$$
$$x + y + z = 180$$

Write the system in standard form:

$$x - z = -55$$
$$y - z = -20$$
$$x + y + z = 180$$

$$\left[\begin{array}{ccc|c} 1 & 0 & -1 & -55 \\ 0 & 1 & -1 & -20 \\ 1 & 1 & 1 & 180 \end{array}\right]$$

$$\left[\begin{array}{ccc|c} 1 & 0 & -1 & -55 \\ 0 & 1 & -1 & -20 \\ 0 & 1 & 2 & 235 \end{array}\right] -1R_1 + R_3$$

$$\left[\begin{array}{ccc|c} 1 & 0 & -1 & -55 \\ 0 & 1 & -1 & -20 \\ 0 & 0 & 3 & 255 \end{array}\right] -1R_2 + R_3$$

$$\left[\begin{array}{ccc|c} 1 & 0 & -1 & -55 \\ 0 & 1 & -1 & -20 \\ 0 & 0 & 1 & 85 \end{array}\right] \tfrac{1}{3}R_3$$

The system is

$$x - z = -55$$
$$y - z = -20$$
$$z = 85$$

Substitute 85 for z in the second equation.

$$y - z = -20$$
$$y - 85 = -20$$
$$y = 65$$

Substitute 85 for z in the first equation.

$$x - z = -55$$
$$x - 85 = -55$$
$$x = 30$$

The angles are 30°, 65°, and 85°.

51. Let x = amount Chiquita controls,
y = amount Dole controls,
z = amount Del Monte controls

$$x = z + 12$$
$$y = 2z - 3$$
$$x + y + z = 65$$

Write the system in standard form.

$$x - z = 12$$
$$y - 2z = -3$$
$$x + y + z = 65$$

$$\left[\begin{array}{ccc|c} 1 & 0 & -1 & 12 \\ 0 & 1 & -2 & -3 \\ 1 & 1 & 1 & 65 \end{array}\right]$$

$$\left[\begin{array}{ccc|c} 1 & 0 & -1 & 12 \\ 0 & 1 & -2 & -3 \\ 0 & 1 & 2 & 53 \end{array}\right] -1R_1 + R_3$$

$$\left[\begin{array}{ccc|c} 1 & 0 & -1 & 12 \\ 0 & 1 & -2 & -3 \\ 0 & 0 & 4 & 56 \end{array}\right] -1R_2 + R_3$$

$$\left[\begin{array}{ccc|c} 1 & 0 & -1 & 12 \\ 0 & 1 & -2 & -3 \\ 0 & 0 & 1 & 14 \end{array}\right] \tfrac{1}{4}R_3$$

The system is

$$x - z = 12$$
$$y - 2z = -3$$
$$z = 14$$

Substitute 14 for z in the second equation.

$$y - 2z = -3$$
$$y - 2(14) = -3$$
$$y - 28 = -3$$
$$y = 25$$

Substitute 14 for z in the first equation.

$$x - z = 12$$
$$x - 14 = 12$$
$$x = 26$$

Thus, Del Monte controls 14% of the bananas, Dole controls 25% and Chiquita controls 26%, with the remaining 100% − 65% = 35% being controlled by "Other".

53. a. $A \cup B = \{1, 2, 3, 4, 5, 6, 9, 10\}$

b. $A \cap B = \{4, 6\}$

54. a.

b. $\{x \mid -1 < x \le 4\}$

c. $(-1, 4]$

55. A graph is the set of points whose coordinates satisfy an equation.

56. $f(x) = -2x^2 + 3x - 6$

$$\begin{aligned} f(-5) &= -2(-5)^2 + 3(-5) - 6 \\ &= -50 - 15 - 6 \\ &= -71 \end{aligned}$$

Exercise Set 4.5

1. Answers will vary.

3. If $D = 0$ and either D_x, D_y, or D_z is not equal to 0, the system is inconsistent.

5. $x = \dfrac{D_x}{D} = \dfrac{12}{4} = 3$

$y = \dfrac{D_y}{D} = \dfrac{-2}{4} = -\dfrac{1}{2}$

The solution is $\left(3, -\dfrac{1}{2}\right)$.

7. $\begin{vmatrix} 2 & 4 \\ 1 & 5 \end{vmatrix} = (2)(5) - (1)(4) = 10 - 4 = 6$

9. $\begin{vmatrix} \frac{1}{2} & 3 \\ 2 & -4 \end{vmatrix} = \dfrac{1}{2}(-4) - (2)(3) = -2 - 6 = -8$

11. $\begin{vmatrix} 3 & 2 & 0 \\ 0 & 5 & 3 \\ -1 & 4 & 2 \end{vmatrix} = 3\begin{vmatrix} 5 & 3 \\ 4 & 2 \end{vmatrix} - 0\begin{vmatrix} 2 & 0 \\ 4 & 2 \end{vmatrix} + (-1)\begin{vmatrix} 2 & 0 \\ 5 & 3 \end{vmatrix}$

$$\begin{aligned} &= 3(10 - 12) - 0(4 - 0) - 1(6 - 0) \\ &= 3(-2) - 0(4) - 1(6) \\ &= -6 - 0 - 6 \\ &= -12 \end{aligned}$$

13. $\begin{vmatrix} 2 & 3 & 1 \\ 1 & -3 & -6 \\ -4 & 5 & 9 \end{vmatrix}$

$$\begin{aligned} &= 2\begin{vmatrix} -3 & -6 \\ 5 & 9 \end{vmatrix} - 1\begin{vmatrix} 3 & 1 \\ 5 & 9 \end{vmatrix} + (-4)\begin{vmatrix} 3 & 1 \\ -3 & -6 \end{vmatrix} \\ &= 2[-27 - (-30)] - 1(27 - 5) - 4[-18 - (-3)] \\ &= 2(3) - 1(22) - 4(-15) \\ &= 6 - 22 + 60 \\ &= 44 \end{aligned}$$

15. $x + 3y = 1$

$-2x - 3y = 4$

To solve, first calculate D, D_x, and D_y.

$D = \begin{vmatrix} 1 & 3 \\ -2 & -3 \end{vmatrix} = (1)(-3) - (-2)(3) = -3 - (-6) = 3$

$D_x = \begin{vmatrix} 1 & 3 \\ 4 & -3 \end{vmatrix} = (1)(-3) - (4)(3) = -15$

$D_y = \begin{vmatrix} 1 & 1 \\ -2 & 4 \end{vmatrix} = (1)(4) - (-2)(1) = 4 - (-2) = 6$

$x = \dfrac{D_x}{D} = \dfrac{-15}{3} = -5$ and $y = \dfrac{D_y}{D} = \dfrac{6}{3} = 2$

The solution is $(-5, 2)$.

17. $-x - 2y = 2$

$x + 3y = -6$

To solve, first calculate D, D_x, and D_y.

$D = \begin{vmatrix} -1 & -2 \\ 1 & 3 \end{vmatrix} = (-1)(3) - (1)(-2) = -3 + 2 = -1$

$$\begin{aligned} D_x &= \begin{vmatrix} 2 & -2 \\ -6 & 3 \end{vmatrix} \\ &= (2)(3) - (-6)(-2) \\ &= 6 - 12 \\ &= -6 \end{aligned}$$

$D_y = \begin{vmatrix} -1 & 2 \\ 1 & -6 \end{vmatrix} = (-1)(-6) - (1)(2) = 6 - 2 = 4$

$x = \dfrac{D_x}{D} = \dfrac{-6}{-1} = 6$ and $y = \dfrac{D_y}{D} = \dfrac{4}{-1} = -4$

The solution is $(6, -4)$.

19. $6x = 4y + 7$

$8x - 1 = -3y$

Rewrite the system in standard form:

$6x - 4y = 7$

$8x + 3y = 1$

Now calculate D, D_x, and D_y.

$$D = \begin{vmatrix} 6 & -4 \\ 8 & 3 \end{vmatrix} = (6)(3) - (8)(-4) = 18 + 32 = 50$$

$$D_x = \begin{vmatrix} 7 & -4 \\ 1 & 3 \end{vmatrix} = (7)(3) - (1)(-4) = 21 + 4 = 25$$

$$D_y = \begin{vmatrix} 6 & 7 \\ 8 & 1 \end{vmatrix} = (6)(1) - (8)(7) = 6 - 56 = -50$$

$$x = \frac{D_x}{D} = \frac{25}{50} = \frac{1}{2} \text{ and } y = \frac{D_y}{D} = \frac{-50}{50} = -1$$

The solution is $\left(\frac{1}{2}, -1\right)$.

21. $5p - 7q = -21$

$-4p + 3q = 22$

To solve, first calculate D, D_p, and D_q.

$$D = \begin{vmatrix} 5 & -7 \\ -4 & 3 \end{vmatrix} = (5)(3) - (-4)(-7)$$
$$= 15 - 28$$
$$= -13$$

$$D_p = \begin{vmatrix} -21 & -7 \\ 22 & 3 \end{vmatrix} = (-21)(3) - (22)(-7)$$
$$= -63 - (-154)$$
$$= 91$$

$$D_q = \begin{vmatrix} 5 & -21 \\ -4 & 22 \end{vmatrix} = (5)(22) - (-4)(-21)$$
$$= 110 - 84$$
$$= 26$$

$$p = \frac{D_p}{D} = \frac{91}{-13} = -7 \text{ and } q = \frac{D_q}{D} = \frac{26}{-13} = -2$$

The solution is (–7, –2).

23. $x + 5y = 3$

$2x - 6 = -10y$

Rewrite the system in standard form:

$x + 5y = 3$

$2x + 10y = 6$

To solve, first calculate D, D_x, and D_y.

$$D = \begin{vmatrix} 1 & 5 \\ 2 & 10 \end{vmatrix} = (1)(10) - (2)(5) = 10 - 10 = 0$$

$$D_x = \begin{vmatrix} 3 & 5 \\ 6 & 10 \end{vmatrix} = (3)(10) - (6)(5) = 30 - 30 = 0$$

$$D_y = \begin{vmatrix} 1 & 3 \\ 2 & 6 \end{vmatrix} = (1)(6) - (2)(3) = 6 - 6 = 0$$

Since $D = 0$, $D_x = 0$, and $D_y = 0$, the system is dependent so there are an infinite number of solutions.

25. $3r = -4s - 6$

$3s = -5r + 1$

Rewrite the system in standard form.

$3r + 4s = -6$

$5r + 3s = 1$

Now calculate D, D_r, and D_s.

$$D = \begin{vmatrix} 3 & 4 \\ 5 & 3 \end{vmatrix} = (3)(3) - (5)(4) = 9 - 20 = -11$$

$$D_r = \begin{vmatrix} -6 & 4 \\ 1 & 3 \end{vmatrix} = (-6)(3) - (1)(4) = -18 - 4 = -22$$

$$D_s = \begin{vmatrix} 3 & -6 \\ 5 & 1 \end{vmatrix} = (3)(1) - (5)(-6) = 3 + 30 = 33$$

$$r = \frac{D_r}{D} = \frac{-22}{-11} = 2 \text{ and } s = \frac{D_s}{D} = \frac{33}{-11} = -3$$

The solution is (2, –3).

27. $5x - 5y = 3$

$-x + y = -4$

To solve, first calculate D, D_x, and D_y.

$$D = \begin{vmatrix} 5 & -5 \\ -1 & 1 \end{vmatrix} = (5)(1) - (-1)(-5) = 5 - 5 = 0$$

$$D_x = \begin{vmatrix} 3 & -5 \\ -4 & 1 \end{vmatrix} = (3)(1) - (-4)(-5) = 3 - 20 = -17$$

Since $D = 0$ and $D_x \neq 0$, the system is inconsistent, so there is no solution.

29. $6.3x - 4.5y = -9.9$

$-9.1x + 3.2y = -2.2$

Here, you can work with decimals in the determinants. If you do not want to use decimals, then you need to multiply each equation by 10 to clear the decimals.

First, calculate D, D_x, and D_y.

$$D = \begin{vmatrix} 6.3 & -4.5 \\ -9.1 & 3.2 \end{vmatrix} = (6.3)(3.2) - (-9.1)(-4.5)$$

$$= 20.16 - 40.95 = -20.79$$

$$D_x = \begin{vmatrix} -9.9 & -4.5 \\ -2.2 & 3.2 \end{vmatrix} = (-9.9)(3.2) - (-2.2)(-4.5)$$

$$= -31.68 - 9.90 = -41.58$$

$$D_y = \begin{vmatrix} 6.3 & -9.9 \\ -9.1 & -2.2 \end{vmatrix} = (6.3)(-2.2) - (-9.1)(-9.9)$$

$$= -13.86 - 90.09 = -103.95$$

$$x = \frac{D_x}{D} = \frac{-41.58}{-20.79} = 2 \text{ and}$$

$$y = \frac{D_y}{D} = \frac{-103.95}{-20.79} = 5$$

The solution is (2, 5).

31. $x + y + z = 3$

$0x - 3y + 4z = 15$

$-3x + 4y - 2z = -13$

To solve, first calculate D, D_x, D_y, and D_z.

$$D = \begin{vmatrix} 1 & 1 & 1 \\ 0 & -3 & 4 \\ -3 & 4 & -2 \end{vmatrix} \text{ (using first column)}$$

$$= 1\begin{vmatrix} -3 & 4 \\ 4 & -2 \end{vmatrix} - 0\begin{vmatrix} 1 & 1 \\ 4 & -2 \end{vmatrix} + (-3)\begin{vmatrix} 1 & 1 \\ -3 & 4 \end{vmatrix}$$

$$= 1(6-16) - 0(-2-4) - 3(4+3)$$

$$= 1(-10) - 0(-6) - 3(7)$$

$$= -10 - 0 - 21$$

$$= -31$$

$$D_x = \begin{vmatrix} 3 & 1 & 1 \\ 15 & -3 & 4 \\ -13 & 4 & -2 \end{vmatrix} \text{ (using first row)}$$

$$= 3\begin{vmatrix} -3 & 4 \\ 4 & -2 \end{vmatrix} - 1\begin{vmatrix} 15 & 4 \\ -13 & -2 \end{vmatrix} + 1\begin{vmatrix} 15 & -3 \\ -13 & 4 \end{vmatrix}$$

$$= 3(6-16) - 1(-30+52) + 1(60-39)$$

$$= 3(-10) - 1(22) + 1(21)$$

$$= -30 - 22 + 21$$

$$= -31$$

$$D_y = \begin{vmatrix} 1 & 3 & 1 \\ 0 & 15 & 4 \\ -3 & -13 & -2 \end{vmatrix} \text{ (using first column)}$$

$$= 1\begin{vmatrix} 15 & 4 \\ -13 & -2 \end{vmatrix} - 0\begin{vmatrix} 3 & 1 \\ -13 & -2 \end{vmatrix} + (-3)\begin{vmatrix} 3 & 1 \\ 15 & 4 \end{vmatrix}$$

$$= 1(-30+52) - 0(-6+13) - 3(12-15)$$

$$= 1(22) - 0(7) - 3(-3)$$

$$= 22 - 0 + 9$$

$$= 31$$

$$D_z = \begin{vmatrix} 1 & 1 & 3 \\ 0 & -3 & 15 \\ -3 & 4 & -13 \end{vmatrix} \text{ (using first column)}$$

$$= 1\begin{vmatrix} -3 & 15 \\ 4 & -13 \end{vmatrix} - 0\begin{vmatrix} 1 & 3 \\ 4 & -13 \end{vmatrix} + (-3)\begin{vmatrix} 1 & 3 \\ -3 & 15 \end{vmatrix}$$

$$= 1(39-60) - 0(-13-12) - 3(15+9)$$

$$= 1(-21) - 0(-25) - 3(24)$$

$$= -21 - 0 - 72$$

$$= -93$$

$$x = \frac{D_x}{D} = \frac{-31}{-31} = 1,\ y = \frac{D_y}{D} = \frac{31}{-31} = -1, \text{ and}$$

$$z = \frac{D_z}{D} = \frac{-93}{-31} = 3$$

The solution is $(1, -1, 3)$.

33. $3x-5y-4z=-4$

$4x+2y+0z=1$

$0x+6y-4z=-11$

To solve, first calculate D, D_x, D_y, and D_z.

$$D=\begin{vmatrix}3&-5&-4\\4&2&0\\0&6&-4\end{vmatrix} \text{ (using the first row)}$$

$$=3\begin{vmatrix}2&0\\6&-4\end{vmatrix}-(-5)\begin{vmatrix}4&0\\0&-4\end{vmatrix}+(-4)\begin{vmatrix}4&2\\0&6\end{vmatrix}$$

$=3(-8-0)+5(-16-0)-4(24-0)$

$=3(-8)+5(-16)-4(24)$

$=-24-80-96$

$=-200$

$$D_x=\begin{vmatrix}-4&-5&-4\\1&2&0\\-11&6&-4\end{vmatrix} \text{ (using the first row)}$$

$$=-4\begin{vmatrix}2&0\\6&-4\end{vmatrix}-(-5)\begin{vmatrix}1&0\\-11&-4\end{vmatrix}+(-4)\begin{vmatrix}1&2\\-11&6\end{vmatrix}$$

$=-4(-8-0)+5(-4+0)-4(6+22)$

$=-4(-8)+5(-4)-4(28)$

$=32-20-112$

$=-100$

$$D_y=\begin{vmatrix}3&-4&-4\\4&1&0\\0&-11&-4\end{vmatrix} \text{ (using the first row)}$$

$$=3\begin{vmatrix}1&0\\-11&-4\end{vmatrix}-(-4)\begin{vmatrix}4&0\\0&-4\end{vmatrix}+(-4)\begin{vmatrix}4&1\\0&-11\end{vmatrix}$$

$=3(-4+0)+4(-16-0)-4(-44-0)$

$=3(-4)+4(-16)-4(-44)$

$=-12-64+176$

$=100$

$$D_z=\begin{vmatrix}3&-5&-4\\4&2&1\\0&6&-11\end{vmatrix} \text{ (using the first row)}$$

$$=3\begin{vmatrix}2&1\\6&-11\end{vmatrix}-(-5)\begin{vmatrix}4&1\\0&-11\end{vmatrix}+(-4)\begin{vmatrix}4&2\\0&6\end{vmatrix}$$

$=3(-22-6)+5(-44-0)-4(24-0)$

$=3(-28)+5(-44)-4(24)$

$=-84-220-96$

$=-400$

$x=\frac{D_x}{D}=\frac{-100}{-200}=\frac{1}{2}$, $y=\frac{D_y}{D}=\frac{100}{-200}=-\frac{1}{2}$, and

$z=\frac{D_z}{D}=\frac{-400}{-200}=2$

The solution is $\left(\frac{1}{2},-\frac{1}{2},2\right)$.

35. $x+4y-3z=-6$

$2x-8y+5z=12$

$3x+4y-2z=-3$

To solve, first calculate D, D_x, D_y, and D_z

$$D=\begin{vmatrix}1&4&-3\\2&-8&5\\3&4&-2\end{vmatrix} \text{ (using the first row)}$$

$$=1\begin{vmatrix}-8&5\\4&-2\end{vmatrix}-4\begin{vmatrix}2&5\\3&-2\end{vmatrix}+(-3)\begin{vmatrix}2&-8\\3&4\end{vmatrix}$$

$=1(16-20)-4(-4-15)-3(8+24)$

$=1(-4)-4(-19)-3(32)$

$=-4+76-96$

$=-24$

$$D_x=\begin{vmatrix}-6&4&-3\\12&-8&5\\-3&4&-2\end{vmatrix} \text{ (using the first row)}$$

$$=-6\begin{vmatrix}-8&5\\4&-2\end{vmatrix}-4\begin{vmatrix}12&5\\-3&-2\end{vmatrix}+(-3)\begin{vmatrix}12&-8\\-3&4\end{vmatrix}$$

$=-6(16-20)-4(-24+15)-3(48-24)$

$=-6(-4)-4(-9)-3(24)$

$=24+36-72$

$=-12$

$$D_y=\begin{vmatrix}1&-6&-3\\2&12&5\\3&-3&-2\end{vmatrix} \text{ (using the first row)}$$

$$=1\begin{vmatrix}12&5\\-3&-2\end{vmatrix}-(-6)\begin{vmatrix}2&5\\3&-2\end{vmatrix}+(-3)\begin{vmatrix}2&12\\3&-3\end{vmatrix}$$

$=1(-24+15)+6(-4-15)-3(-6-36)$

$=1(-9)+6(-19)-3(-42)$

$=-9-114+126$

$=3$

$$D_z = \begin{vmatrix} 1 & 4 & -6 \\ 2 & -8 & 12 \\ 3 & 4 & -3 \end{vmatrix} \text{ (using the first row)}$$

$$= 1\begin{vmatrix} -8 & 12 \\ 4 & -3 \end{vmatrix} - 4\begin{vmatrix} 2 & 12 \\ 3 & -3 \end{vmatrix} + (-6)\begin{vmatrix} 2 & -8 \\ 3 & 4 \end{vmatrix}$$

$$= 1(24-48) - 4(-6-36) - 6(8+24)$$

$$= 1(-24) - 4(-42) - 6(32)$$

$$= -24 + 168 - 192$$

$$= -48$$

$x = \frac{D_x}{D} = \frac{-12}{-24} = \frac{1}{2}$, $y = \frac{D_y}{D} = \frac{3}{-24} = -\frac{1}{8}$, and

$z = \frac{D_z}{D} = \frac{-48}{-24} = 2$.

The solution is $\left(\frac{1}{2}, -\frac{1}{8}, 2\right)$.

37. $a - b + 2c = 3$

$a - b + c = 1$

$2a + b + 2c = 2$

To solve, first calculate D, D_a, D_b, and D_c.

$$D = \begin{vmatrix} 1 & -1 & 2 \\ 1 & -1 & 1 \\ 2 & 1 & 2 \end{vmatrix} \text{ (using first column)}$$

$$= 1\begin{vmatrix} -1 & 1 \\ 1 & 2 \end{vmatrix} - 1\begin{vmatrix} -1 & 2 \\ 1 & 2 \end{vmatrix} + 2\begin{vmatrix} -1 & 2 \\ -1 & 1 \end{vmatrix}$$

$$= 1(-2-1) - 1(-2-2) + 2(-1+2)$$

$$= 1(-3) - 1(-4) + 2(1)$$

$$= -3 + 4 + 2$$

$$= 3$$

$$D_a = \begin{vmatrix} 3 & -1 & 2 \\ 1 & -1 & 1 \\ 2 & 1 & 2 \end{vmatrix} \text{ (using first column)}$$

$$= 3\begin{vmatrix} -1 & 1 \\ 1 & 2 \end{vmatrix} - 1\begin{vmatrix} -1 & 2 \\ 1 & 2 \end{vmatrix} + 2\begin{vmatrix} -1 & 2 \\ -1 & 1 \end{vmatrix}$$

$$= 3(-2-1) - 1(-2-2) + 2(-1+2)$$

$$= 3(-3) - 1(-4) + 2(1)$$

$$= -9 + 4 + 2$$

$$= -3$$

$$D_b = \begin{vmatrix} 1 & 3 & 2 \\ 1 & 1 & 1 \\ 2 & 2 & 2 \end{vmatrix} \text{ (using first column)}$$

$$= 1\begin{vmatrix} 1 & 1 \\ 2 & 2 \end{vmatrix} - 1\begin{vmatrix} 3 & 2 \\ 2 & 2 \end{vmatrix} + 2\begin{vmatrix} 3 & 2 \\ 1 & 1 \end{vmatrix}$$

$$= 1(2-2) - 1(6-4) + 2(3-2)$$

$$= 1(0) - 1(2) + 2(1)$$

$$= 0 - 2 + 2$$

$$= 0$$

$$D_c = \begin{vmatrix} 1 & -1 & 3 \\ 1 & -1 & 1 \\ 2 & 1 & 2 \end{vmatrix} \text{ (using first column)}$$

$$= 1\begin{vmatrix} -1 & 1 \\ 1 & 2 \end{vmatrix} - 1\begin{vmatrix} -1 & 3 \\ 1 & 2 \end{vmatrix} + 2\begin{vmatrix} -1 & 3 \\ -1 & 1 \end{vmatrix}$$

$$= 1(-2-1) - 1(-2-3) + 2(-1+3)$$

$$= 1(-3) - 1(-5) + 2(2)$$

$$= -3 + 5 + 4$$

$$= 6$$

$a = \frac{D_a}{D} = \frac{-3}{3} = -1$, $b = \frac{D_b}{D} = \frac{0}{3} = 0$, and

$c = \frac{D_c}{D} = \frac{6}{3} = 2$

The solution is (–1, 0, 2).

39. $a + 2b + c = 1$

$a - b + 3c = 2$

$2a + b + 4c = 3$

To solve, first calculate D, D_a, D_b, and D_c.

$$D = \begin{vmatrix} 1 & 2 & 1 \\ 1 & -1 & 3 \\ 2 & 1 & 4 \end{vmatrix} \text{ (using first column)}$$

$$= 1\begin{vmatrix} -1 & 3 \\ 1 & 4 \end{vmatrix} - 1\begin{vmatrix} 2 & 1 \\ 1 & 4 \end{vmatrix} + 2\begin{vmatrix} 2 & 1 \\ -1 & 3 \end{vmatrix}$$

$$= 1(-4-3) - 1(8-1) + 2(6+1)$$

$$= 1(-7) - 1(7) + 2(7)$$

$$= -7 - 7 + 14 = 0$$

$$D_a = \begin{vmatrix} 1 & 2 & 1 \\ 2 & -1 & 3 \\ 3 & 1 & 4 \end{vmatrix} \text{ (using first column)}$$

$$= 1\begin{vmatrix} -1 & 3 \\ 1 & 4 \end{vmatrix} - 2\begin{vmatrix} 2 & 1 \\ 1 & 4 \end{vmatrix} + 3\begin{vmatrix} 2 & 1 \\ -1 & 3 \end{vmatrix}$$

$$= 1(-4-3) - 2(8-1) + 3(6+1)$$

$$= 1(-7) - 2(7) + 3(7)$$

$$= -7 - 14 + 21 = 0$$

$$D_b = \begin{vmatrix} 1 & 1 & 1 \\ 1 & 2 & 3 \\ 2 & 3 & 4 \end{vmatrix} \text{ (using first row)}$$

$$= 1\begin{vmatrix} 2 & 3 \\ 3 & 4 \end{vmatrix} - 1\begin{vmatrix} 1 & 3 \\ 2 & 4 \end{vmatrix} + 1\begin{vmatrix} 1 & 2 \\ 2 & 3 \end{vmatrix}$$

$$= 1(8-9) - 1(4-6) + 1(3-4)$$

$$= 1(-1) - 1(-2) + 1(-1)$$

$$= -1 + 2 - 1 = 0$$

$$D_c = \begin{vmatrix} 1 & 2 & 1 \\ 1 & -1 & 2 \\ 2 & 1 & 3 \end{vmatrix} \text{ (using first column)}$$

$$= 1\begin{vmatrix} -1 & 2 \\ 1 & 3 \end{vmatrix} - 1\begin{vmatrix} 2 & 1 \\ 1 & 3 \end{vmatrix} + 2\begin{vmatrix} 2 & 1 \\ -1 & 2 \end{vmatrix}$$

$$= 1(-3-2) - 1(6-1) + 2(4+1)$$

$$= 1(-5) - 1(5) + 2(5)$$

$$= -5 - 5 + 10 = 0$$

Since $D = 0$, $D_a = 0$, $D_b = 0$, and $D_c = 0$, there are an infinite number of solutions to the system and it is a dependent system.

41. $1.1x + 2.3y - 4.0z = -9.2$

$-2.3x + 0y + 4.6z = 6.9$

$0x - 8.2y - 7.5z = -6.8$

Here, you can work with decimals in the determinants. If you do not want to use decimals, then you need to multiply each equation by 10 to clear the decimals. To solve, first calculate D, D_x, D_y, and D_z.

$$D = \begin{vmatrix} 1.1 & 2.3 & -4.0 \\ -2.3 & 0 & 4.6 \\ 0 & -8.2 & -7.5 \end{vmatrix} = 1.1\begin{vmatrix} 0 & 4.6 \\ -8.2 & -7.5 \end{vmatrix} - (-2.3)\begin{vmatrix} 2.3 & -4.0 \\ -8.2 & -7.5 \end{vmatrix} + 0\begin{vmatrix} 2.3 & -4.0 \\ 0 & 4.6 \end{vmatrix} \text{ (using first column)}$$

$$= 1.1(0 + 37.72) + 2.3(-17.25 - 32.8) + 0(10.58 - 0)$$

$$= 1.1(37.72) + 2.3(-50.05) + 0(10.58)$$

$$= 41.492 - 115.115 + 0$$

$$= -73.623$$

$$D_x = \begin{vmatrix} -9.2 & 2.3 & -4.0 \\ 6.9 & 0 & 4.6 \\ -6.8 & -8.2 & -7.5 \end{vmatrix} = -9.2\begin{vmatrix} 0 & 4.6 \\ -8.2 & -7.5 \end{vmatrix} - 6.9\begin{vmatrix} 2.3 & -4.0 \\ -8.2 & -7.5 \end{vmatrix} + (-6.8)\begin{vmatrix} 2.3 & -4.0 \\ 0 & 4.6 \end{vmatrix} \text{ (using first column)}$$

$$= -9.2(0 + 37.72) - 6.9(-17.25 - 32.8) - 6.8(10.58 - 0)$$

$$= -9.2(37.72) - 6.9(-50.05) - 6.8(10.58)$$

$$= -347.024 + 345.345 - 71.944$$

$$= -73.623$$

$$D_y = \begin{vmatrix} 1.1 & -9.2 & -4.0 \\ -2.3 & 6.9 & 4.6 \\ 0 & -6.8 & -7.5 \end{vmatrix} = 1.1\begin{vmatrix} 6.9 & 4.6 \\ -6.8 & -7.5 \end{vmatrix} - (-2.3)\begin{vmatrix} -9.2 & -4.0 \\ -6.8 & -7.5 \end{vmatrix} + 0\begin{vmatrix} -9.2 & -4.0 \\ 6.9 & 4.6 \end{vmatrix} \quad \text{(using first column)}$$

$= 1.1(-51.75 + 31.28) + 2.3(69 - 27.2) + 0(-42.32 + 27.6)$

$= 1.1(-20.47) + 2.3(41.8) + 0(-14.72)$

$= -22.517 + 96.14 + 0$

$= 73.623$

$$D_z = \begin{vmatrix} 1.1 & 2.3 & -9.2 \\ -2.3 & 0 & 6.9 \\ 0 & -8.2 & -6.8 \end{vmatrix} = 1.1\begin{vmatrix} 0 & 6.9 \\ -8.2 & -6.8 \end{vmatrix} - (-2.3)\begin{vmatrix} 2.3 & -9.2 \\ -8.2 & -6.8 \end{vmatrix} + 0\begin{vmatrix} 2.3 & -9.2 \\ 0 & 6.9 \end{vmatrix} \quad \text{(using first column)}$$

$= 1.1(0 + 56.58) + 2.3(-15.64 - 75.44) + 0(15.87 - 0)$

$= 1.1(56.58) + 2.3(-91.08) + 0(15.87)$

$= 62.238 - 209.484 + 0$

$= -147.246$

$x = \frac{D_x}{D} = \frac{-73.623}{-73.623} = 1$, $y = \frac{D_y}{D} = \frac{73.623}{-73.623} = -1$, and $z = \frac{D_z}{D} = \frac{-147.246}{-73.623} = 2$

The solution is $(1, -1, 2)$.

43. $-6x + 3y - 12z = -13$

$5x + 2y - 3z = 1$

$2x - y + 4z = -5$

To solve, first calculate D, D_x, D_y, and D_z.

$$D = \begin{vmatrix} -6 & 3 & -12 \\ 5 & 2 & -3 \\ 2 & -1 & 4 \end{vmatrix} \quad \text{(using the first row)}$$

$$= -6\begin{vmatrix} 2 & -3 \\ -1 & 4 \end{vmatrix} - 3\begin{vmatrix} 5 & -3 \\ 2 & 4 \end{vmatrix} + (-12)\begin{vmatrix} 5 & 2 \\ 2 & -1 \end{vmatrix}$$

$= -6(8 - 3) - 3(20 + 6) - 12(-5 - 4)$

$= -6(5) - 3(26) - 12(-9)$

$= -30 - 78 + 108$

$= 0$

$$D_x = \begin{vmatrix} -13 & 3 & -12 \\ 1 & 2 & -3 \\ -5 & -1 & 4 \end{vmatrix} \quad \text{(using the first row)}$$

$$= -13\begin{vmatrix} 2 & -3 \\ -1 & 4 \end{vmatrix} - 3\begin{vmatrix} 1 & -3 \\ -5 & 4 \end{vmatrix} + (-12)\begin{vmatrix} 1 & 2 \\ -5 & -1 \end{vmatrix}$$

$= -13(8 - 3) - 3(4 - 15) - 12(-1 + 10)$

$= -13(5) - 3(-11) - 12(9)$

$= -65 + 33 - 108$

$= -140$

Since $D = 0$ and $D_x = -140 \neq 0$, there is no solution to the system and it is inconsistent.

45. $2x+\frac{1}{2}y-3z=5$

$-3x+2y+2z=1$

$4x-\frac{1}{4}y-7z=4$

To clear the system of fractions, multiply the first equation by 2 and the third equation by 4.

$4x+y-6z=10$

$-3x+2y+2z=1$

$16x-y-28z=16$

To solve, first calculate D, D_x, D_y, and D_z.

$$D=\begin{vmatrix}4 & 1 & -6\\ -3 & 2 & 2\\ 16 & -1 & -28\end{vmatrix} \text{ (use the first row)}$$

$$=4\begin{vmatrix}2 & 2\\ -1 & -28\end{vmatrix}-1\begin{vmatrix}-3 & 2\\ 16 & -28\end{vmatrix}+(-6)\begin{vmatrix}-3 & 2\\ 16 & -1\end{vmatrix}$$

$=4(-56+2)-1(84-32)-6(3-32)$

$=4(-54)-1(52)-6(-29)$

$=-216-52+174$

$=-94$

$$D_x=\begin{vmatrix}10 & 1 & -6\\ 1 & 2 & 2\\ 16 & -1 & -28\end{vmatrix} \text{ (use the first row)}$$

$$=10\begin{vmatrix}2 & 2\\ -1 & -28\end{vmatrix}-1\begin{vmatrix}1 & 2\\ 16 & -28\end{vmatrix}+(-6)\begin{vmatrix}1 & 2\\ 16 & -1\end{vmatrix}$$

$=10(-56+2)-1(-28-32)-6(-1-32)$

$=10(-54)-1(-60)-6(-33)$

$=-540+60+198$

$=-282$

$$D_y=\begin{vmatrix}4 & 10 & -6\\ -3 & 1 & 2\\ 16 & 16 & -28\end{vmatrix} \text{ (use the first row)}$$

$$=4\begin{vmatrix}1 & 2\\ 16 & -28\end{vmatrix}-10\begin{vmatrix}-3 & 2\\ 16 & -28\end{vmatrix}+(-6)\begin{vmatrix}-3 & 1\\ 16 & 16\end{vmatrix}$$

$=4(-28-32)-10(84-32)-6(-48-16)$

$=4(-60)-10(52)-6(-64)$

$=-240-520+384$

$=-376$

$$D_z=\begin{vmatrix}4 & 1 & 10\\ -3 & 2 & 1\\ 16 & -1 & 16\end{vmatrix} \text{ (use the first row)}$$

$$=4\begin{vmatrix}2 & 1\\ -1 & 16\end{vmatrix}-1\begin{vmatrix}-3 & 1\\ 16 & 16\end{vmatrix}+10\begin{vmatrix}-3 & 2\\ 16 & -1\end{vmatrix}$$

$=4(32+1)-1(-48-16)+10(3-32)$

$=4(33)-1(-64)+10(-29)$

$=132+64-290$

$=-94$

$x=\frac{D_x}{D}=\frac{-282}{-94}=3$, $y=\frac{D_y}{D}=\frac{-376}{-94}=4$,

$z=\frac{D_z}{D}=\frac{-94}{-94}=1$

The solution is (3, 4, 1).

47. $0.3x-0.1y-0.3z=-0.2$

$0.2x-0.1y+0.1z=-0.9$

$0.1x+0.2y-0.4z=1.7$

To clear decimals multiply each equation by 10.

$3x-y-3x=-2$

$2x-y+z=-9$

$x+2y-4z=17$

To solve, first calculate D, D_x, D_y, and D_z.

$$D=\begin{vmatrix}3 & -1 & -3\\ 2 & -1 & 1\\ 1 & 2 & -4\end{vmatrix} \text{ (use first column)}$$

$$=3\begin{vmatrix}-1 & 1\\ 2 & -4\end{vmatrix}-2\begin{vmatrix}-1 & -3\\ 2 & -4\end{vmatrix}+1\begin{vmatrix}-1 & -3\\ -1 & 1\end{vmatrix}$$

$=3(4-2)-2(4+6)+1(-1-3)$

$=3(2)-2(10)+1(-4)$

$=6-20-4$

$=-18$

$$D_x=\begin{vmatrix}-2 & -1 & -3\\ -9 & -1 & 1\\ 17 & 2 & -4\end{vmatrix} \text{ (use first column)}$$

$$=-2\begin{vmatrix}-1 & 1\\ 2 & -4\end{vmatrix}-(-9)\begin{vmatrix}-1 & -3\\ 2 & -4\end{vmatrix}+17\begin{vmatrix}-1 & -3\\ -1 & 1\end{vmatrix}$$

$=-2(4-2)+9(4+6)+17(-1-3)$

$=-2(2)+9(10)+17(-4)$

$=-4+90-68$

$=18$

$$D_y = \begin{vmatrix} 3 & -2 & -3 \\ 2 & -9 & 1 \\ 1 & 17 & -4 \end{vmatrix} \text{ (use first column)}$$

$$= 3\begin{vmatrix} -9 & 1 \\ 17 & -4 \end{vmatrix} - 2\begin{vmatrix} -2 & -3 \\ 17 & -4 \end{vmatrix} + 1\begin{vmatrix} -2 & -3 \\ -9 & 1 \end{vmatrix}$$

$$= 3(36-17) - 2(8+51) + 1(-2-27)$$

$$= 3(19) - 2(59) + 1(-29)$$

$$= 57 - 118 - 29$$

$$= -90$$

$$D_z = \begin{vmatrix} 3 & -1 & -2 \\ 2 & -1 & -9 \\ 1 & 2 & 17 \end{vmatrix} \text{ (use first column)}$$

$$= 3\begin{vmatrix} -1 & -9 \\ 2 & 17 \end{vmatrix} - 2\begin{vmatrix} -1 & -2 \\ 2 & 17 \end{vmatrix} + 1\begin{vmatrix} -1 & -2 \\ -1 & -9 \end{vmatrix}$$

$$= 3(-17+18) - 2(-17+4) + 1(9-2)$$

$$= 3(1) - 2(-13) + 1(7)$$

$$= 3 + 26 + 7$$

$$= 36$$

$$x = \frac{D_x}{D} = \frac{18}{-18} = -1,\ y = \frac{D_y}{D} = \frac{-90}{-18} = 5, \text{ and}$$

$$z = \frac{D_z}{D} = \frac{36}{-18} = -2$$

The solution is (–1, 5, –2).

49. $\begin{vmatrix} a_1 & b_1 \\ a_2 & b_2 \end{vmatrix} = a_1b_2 - a_2b_1$

$\begin{vmatrix} a_2 & b_2 \\ a_1 & b_1 \end{vmatrix} = a_2b_1 - a_1b_2$

The second result is the negative of the first result. Thus, the second determinant has the opposite sign.

51. 0; $\begin{vmatrix} a & b \\ a & b \end{vmatrix} = (a)(b) - (a)(b) = ab - ab = 0$

53. 0; for example

$$\begin{vmatrix} 0 & 0 & 0 \\ a & b & c \\ d & e & f \end{vmatrix} = 0\begin{vmatrix} b & c \\ e & f \end{vmatrix} - 0\begin{vmatrix} a & c \\ d & f \end{vmatrix} + 0\begin{vmatrix} a & b \\ d & e \end{vmatrix}$$

$$= 0 - 0 + 0 = 0$$

55. Yes, the determinant will become the opposite of the original value.

57. No; the value of the new determinant will be the same as the original value. Each time a row (or column) is multiplied by -1, the value of the determinant changes sign. Since two rows were multiplied by -1, the value of the determinant changes sign twice, which yields the original value.

59. Yes; the value of the new determinant will be double the original value. That is, we would multiply the original determinant by 2 as well.

61. $\begin{vmatrix} 4 & 6 \\ -2 & y \end{vmatrix} = 32$

$$(4)(y) - (-2)(6) = 32$$

$$4y + 12 = 32$$

$$4y = 20$$

$$y = \frac{20}{4} = 5$$

63. $\begin{vmatrix} 4 & 7 & y \\ 3 & -1 & 2 \\ 4 & 1 & 5 \end{vmatrix} = -35$

$$4\begin{vmatrix} -1 & 2 \\ 1 & 5 \end{vmatrix} - 3\begin{vmatrix} 7 & y \\ 1 & 5 \end{vmatrix} + 4\begin{vmatrix} 7 & y \\ -1 & 2 \end{vmatrix} = -35$$

$$4(-5-2) - 3(35-y) + 4(14+y) = -35$$

$$4(-7) - 3(35-y) + 4(14+y) = -35$$

$$-28 - 105 + 3y + 56 + 4y = -35$$

$$-77 + 7y = -35$$

$$7y = 42$$

$$y = 6$$

65. a. To eliminate *y*, multiply the first equation by b_2 and the second equation by $-b_1$ and then add.

$$b_2[a_1x + b_1y = c_1]$$

$$-b_1[a_2x + b_2y = c_2]$$

gives

$$a_1b_2x + b_1b_2y = c_1b_2$$

$$\underline{-a_2b_1x - b_1b_2y = -c_2b_1}$$

Add: $(a_1b_2 - a_2b_1)x = c_1b_2 - c_2b_1$

$$x = \frac{c_1b_2 - c_2b_1}{a_1b_2 - a_2b_1}$$

b. To eliminate x, multiply the first equation by $-a_2$ and the second equation by a_1 and then add.

$-a_2[a_1x+b_1y=c_1]$

$a_1[a_2x+b_2y=c_2]$

gives

$$-a_1a_2x-a_2b_1y=-a_2c_1$$
$$a_1a_2x+a_1b_2y=a_1c_2$$

Add: $(a_1b_2-a_2b_1)y=a_1c_2-a_2c_1$

$$y=\frac{a_1c_2-a_2c_1}{a_1b_2-a_2b_1}$$

66. $3(x-2)<\frac{4}{5}(x-4)$

$$5[3(x-2)]<5\left[\frac{4}{5}(x-4)\right]$$
$$15(x-2)<4(x-4)$$
$$15x-30<4x-16$$
$$11x-30<-16$$
$$11x<14$$
$$x<\frac{14}{11}$$

$\left(-\infty, \frac{14}{11}\right)$

67. $3x+4y=8$

Solve for y.

$4y=-3x+8$

$y=-\frac{3}{4}x+2$

x	y
−4	$y=-\frac{3}{4}(-4)+2=3+2=5$
0	$y=-\frac{3}{4}(0)+2=0+2=2$
4	$y=-\frac{3}{4}(4)+2=-3+2=-1$

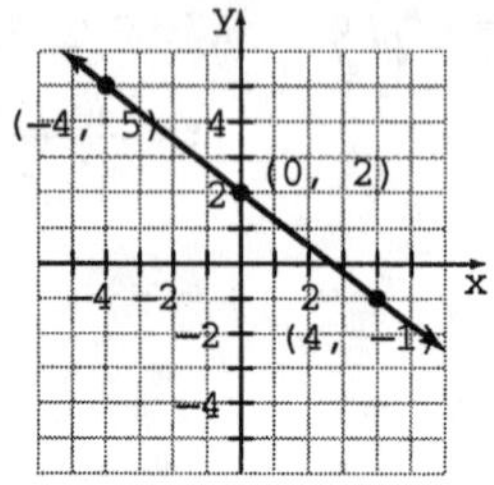

68. $3x+4y=8$

For the x-intercept, let $y=0$.

$$3x+4y=8$$
$$3x+4(0)=8$$
$$3x+0=8$$
$$3x=8$$
$$x=\frac{8}{3}=2\frac{2}{3}$$

For the y-intercept, let $x=0$.

$$3x+4y=8$$
$$3(0)+4y=8$$
$$0+4y=8$$
$$4y=8$$
$$y=2$$

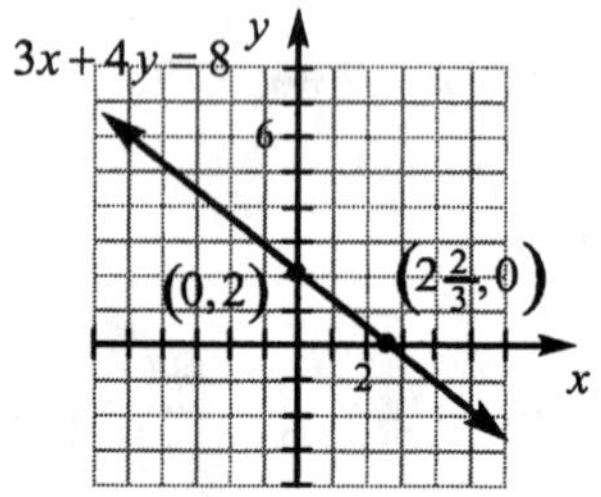

69. $3x+4y=8$

Solve for y.

$4y=-3x+8$

$y=-\frac{3}{4}x+2$

The slope is $-\frac{3}{4}$ and the y-intercept is 2.

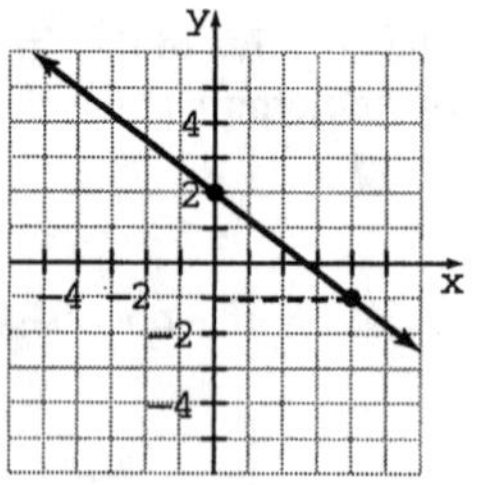

Exercise Set 4.6

1. Answers will vary.

3. Yes; the point of intersection of the boundary lines is in the solution set if the inequalities are both non-strict.

5. $2x-y<4$

   $y\geq -x+2$

   For $2x-y<4$, graph the line $2x-y=4$ using a dashed line. For the check point, select (0, 0):

   $2x-y<4$

   $2(0)-(0)<4$

   $0<4$ True

   Since this is a true statement, shade the region which contains the point (0, 0). This is the region "above" the line.

   For $y\geq -x+2$, graph the line $y=-x+2$ using a solid line. For the check point, select (0, 0):

   $y\geq -x+2$

   $(0)\geq -(0)+2$

   $0\geq 2$ False

   Since this is a false statement, shade the region which does not contain the point (0, 0). This is the region "above" the line. To obtain the final region, take the intersection of the above two regions.

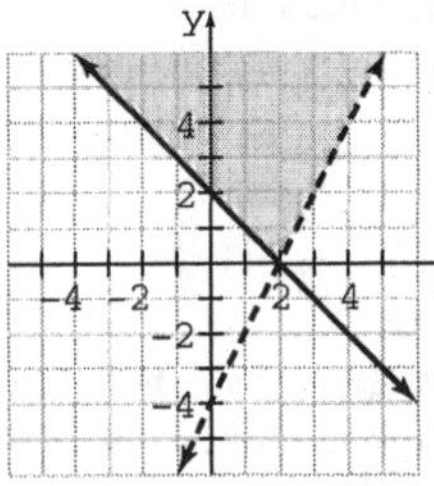

7. $y<3x-2$

   $y\leq -2x+3$

   For $y<3x-2$, graph the line $y=3x-2$ using a dashed line. For the check point, select (0, 0):

   $y<3x-2$

   $(0)<3(0)-2$

   $0<-2$ False

   Since this is a false statement, shade the region which does not contain the point (0, 0). This is the region "below" the line.

   For $y\leq -2x+3$, graph the line $y=-2x+3$ using a solid line. For the check point, use (0, 0):

   $y\leq -2x+3$

   $(0)\leq -2(0)+3$

   $0\leq 3$ True

   Since this is a true statement, shade the region which contains the point (0, 0). This is the region "below" the line. To obtain the final region, take the intersection of the above two regions.

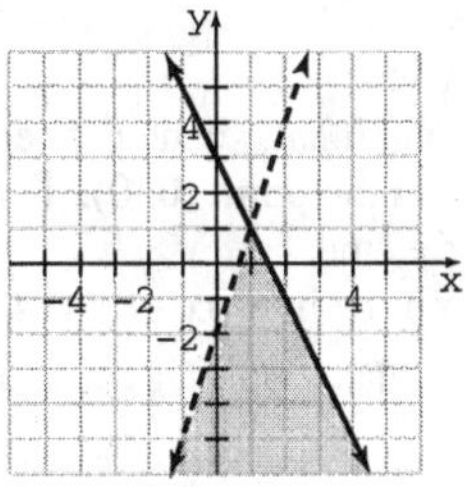

9. $y<x$

   $y\geq 3x+2$

   For $y<x$, graph the line $y=x$ using a dashed line. For the check point, select (1, 0):

   $y<x$

   0<1 True

   Since this is a true statement, shade the region which contains the point (1, 0). This is the region "below" the line.

   For $y\geq 3x+2$, graph the line $y=3x+2$ using a solid line. For the check point, select (0, 0):

   $y\geq 3x+2$

   $(0)\geq 3(0)+2$

   $0\geq 2$ False

   Since this is a false statement, shade the region which does not contain the point (0, 0). This is the region "above" the line. To obtain the final region, take the intersection of the above two regions.

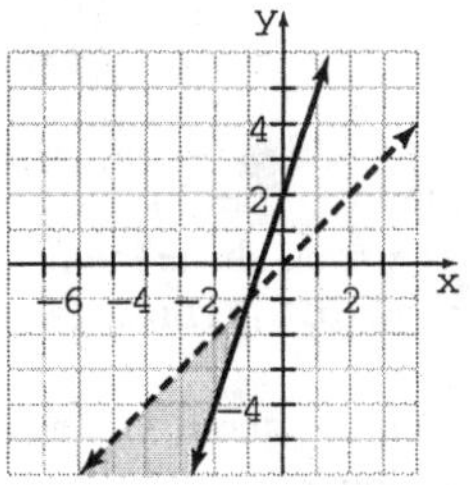

11. $-2x+3y<-5$

$3x-8y>4$

For $-2x+3y<-5$, graph the line $-2x+3y=-5$ using a dashed line. For the check point, select (0, 0):

$-2x+3y<-5$

$-2(0)+3(0)<-5$

$0<-5$ False

Since this is a false statement, shade the region which does not contain the point (0, 0). This is the region "below" the line.

For $3x-8y>4$, graph the line $3x-8y=4$ using a dashed line. For the check point, select (0, 0):

$3x-8y>4$

$3(0)-8(0)>4$

$0>4$ False

Since this is a false statement, shade the region which does not contain the point (0, 0). This is the region "below" the line. To obtain the final region, take the intersection of the above two regions.

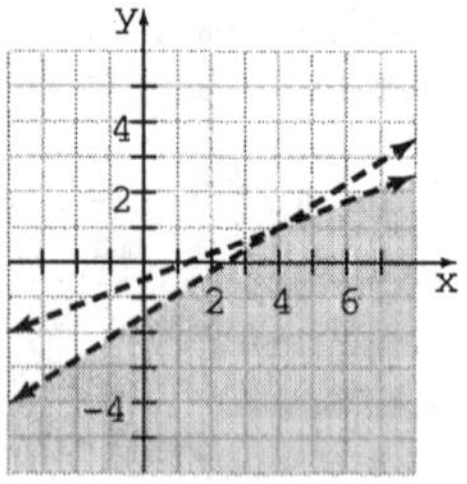

13. $-4x+5y<20$

$x\geq-3$

For $-4x+5y<20$, graph the line $-4x+5y=20$ using a dashed line. For the check point, select (0, 0):

$-4x+5y<20$

$-4(0)+5(0)<20$

$0<20$ True

Since this is a true statement, shade the region which contains the point (0, 0). This is the region "below" the line.

For $x\geq-3$, the graph is the line $x=-3$ along with the region to the right of $x=-3$. To obtain the final region, take the intersection of the above two regions.

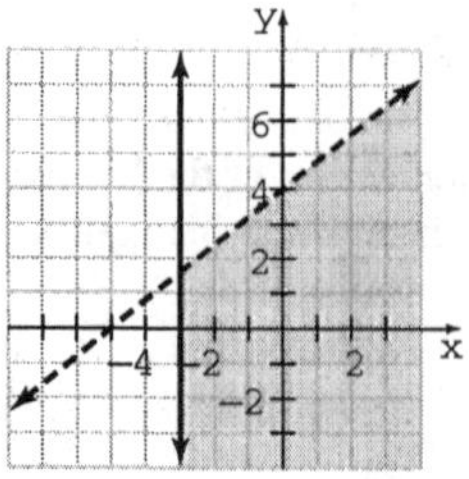

15. $x\leq4$

$y\geq-2$

For $x\leq4$, the graph is the line $x=4$ along with the region to the left of $x=4$.

For $y\geq-2$, the graph is the line $y=-2$ along with the region above the line $y=-2$. To obtain the final region, take the intersection of the above two regions.

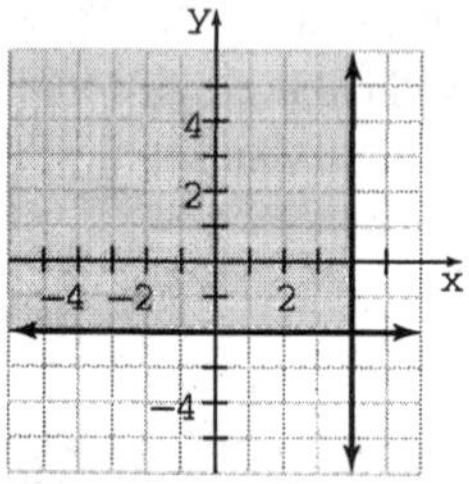

17. $5x+2y>10$

$3x-y>3$

For $5x+2y>10$, graph the line $5x+2y=10$ using a dashed line. For the check point, select (0, 0):

$5x+2y>10$

$5(0)+2(0)>10$

$0>10$ False

Since this is a false statement, shade the region which does not contain the point (0, 0). This is the region "above" the line.

For $3x-y>3$, graph the line $3x-y=3$ using a dashed line. For the check point, select (0, 0):

$3x-y>3$

$3(0)-0>3$

$0>3$ False

Since this is a false statement, shade the region which does not contain the point (0, 0). This is the region "below" the line. To obtain the final region, take the intersection of the above two regions.

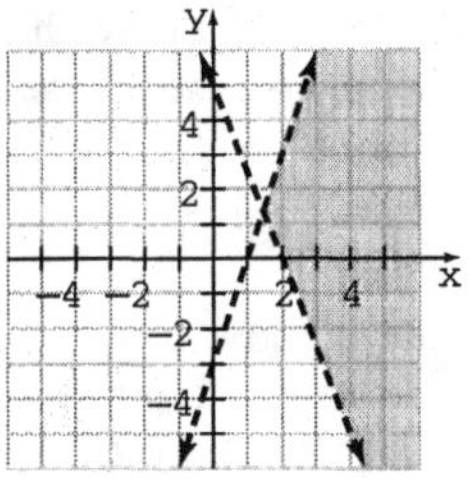

19. $-2x > y + 4$

$-x < \frac{1}{2}y - 1$

For $-2x > y + 4$, graph the line $-2x = y + 4$ using a dashed line. For the check point, select (0, 0):

$-2x > y + 4$

$-2(0) > 0 + 4$

$0 > 4$ False

Since this is a false statement, shade the region which does not contain the point (0, 0). This is the region "below" the line.

For $-x < \frac{1}{2}y - 1$, graph the line $-x = \frac{1}{2}y - 1$ using a dashed line. For the check point, select (0, 0):

$-x < \frac{1}{2}y - 1$

$-0 < \frac{1}{2}(0) - 1$

$0 < -1$ False

Since this is a false statement, shade the region which does not contain the point (0, 0). This is the region "above" the line. To obtain the final region take the intersection of the above two regions. Since the regions do not overlap, the final result is the empty set which means there is no solution.

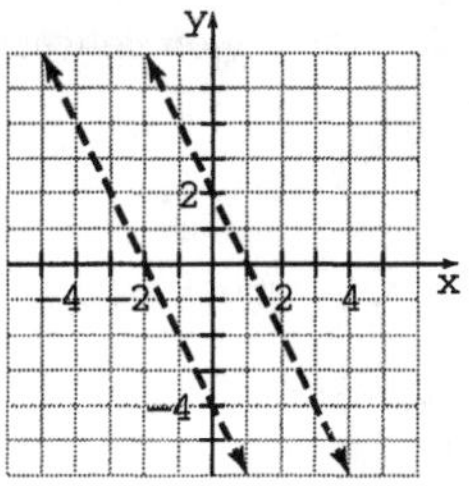

21. $y < 3x - 4$

$6x \geq 2y + 8$

Solve the second inequality for y.

$6x \geq 2y + 8$

$6x - 2y > 8$

$-2y \geq -6x + 8$

$y \leq 3x - 4$

The second inequality is now identical to the first except that the second inequality includes the line $y = 3x - 4$.

For $y < 3x - 4$, graph the line $y = 3x - 4$ using a dashed line. For the check point, select (0, 0):

$y < 3x - 4$

$0 < 3(0) - 4$

$0 < -4$ False

Since this is a false statement, shade the region which does not contain the point (0, 0). This is the region "below" the line.

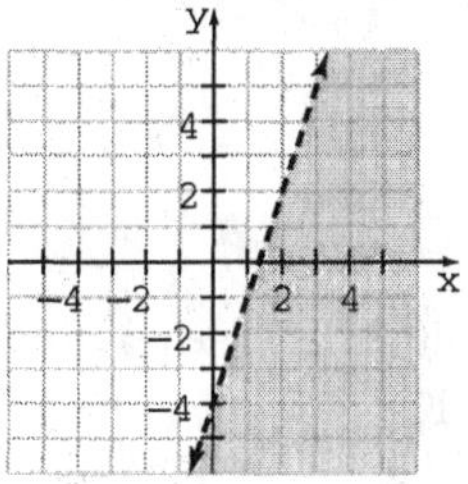

23. $x \geq 0$

$y \geq 0$

$2x + 3y \leq 6$

$4x + y \leq 4$

The first two inequalities indicate that the region must be in the first quadrant. For $2x + 3y \leq 6$, the graph is the line $2x + 3y = 6$ along with the region below this line. For $4x + y \leq 4$, the graph is the line $4x + y = 4$ along with the region below this line. To obtain the final region, take the intersection of these regions.

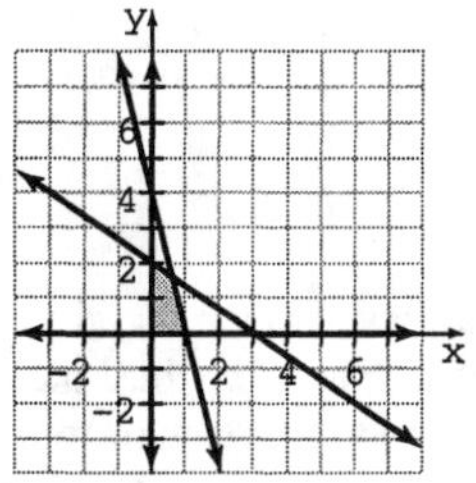

25. $x \geq 0$

$y \geq 0$

$2x + 3y \leq 8$

$4x + 2y \leq 8$

The first two inequalities indicate that the region must be in the first quadrant. For $2x + 3y \leq 8$, the graph is the line $2x + 3y = 8$ along with the region below this line. For $4x + 2y \leq 8$, the graph is the line $4x + 2y = 8$ along with the region below this line. To obtain the final region, take the intersection of these regions.

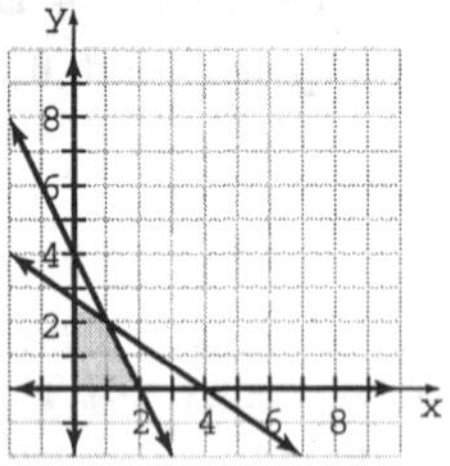

27. $x \geq 0$

$y \geq 0$

$3x + y \leq 9$

$2x + 5y \leq 10$

The first two inequalities indicate that the region must be in the first quadrant. For $3x + y \leq 9$, the graph is the line $3x + y = 9$ along with the region below this line. For $2x + 5y \leq 10$, the graph is the line $2x + 5y = 10$ along with the region below this line. To obtain the final region, take the intersection of these regions.

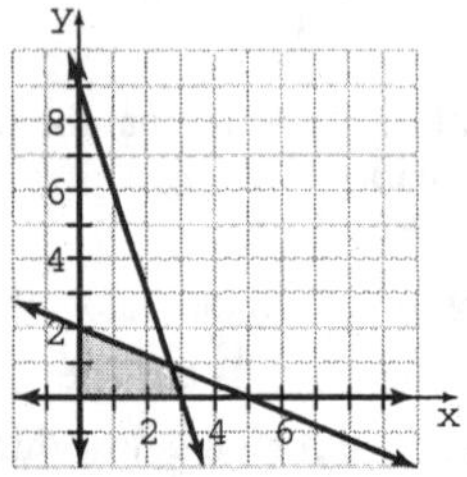

29. $x \geq 0$

$y \geq 0$

$x \leq 4$

$x + y \leq 6$

$x + 2y \leq 8$

The first two inequalities indicate that the region must be in the first quadrant. The third inequality indicates that the region must be on or to the left of the line $x = 4$. For $x + y \leq 6$, the graph is the line $x + y = 6$ along with the region below this line. For $x + 2y \leq 8$, the graph is the line $x + 2y = 8$ along with the region below the line. To obtain the final region, take the intersection of these regions.

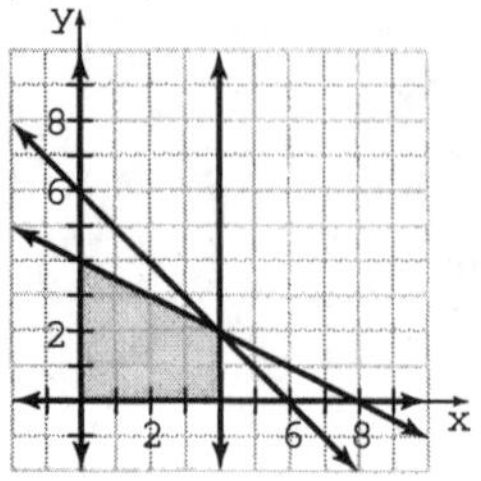

31. $x \geq 0$

$y \geq 0$

$x \leq 15$

$30x + 25y \leq 750$

$10x + 40y \leq 800$

The first two inequalities indicate that the region must be in the first quadrant. The third inequality indicates that the region must be on or to the left of the line $x = 15$. For $30x + 25y \leq 750$ the graph is the line $30x + 25y = 750$ along with the region below this line. For $10x + 40y \leq 800$, the graph is the line $10x + 40y = 800$ along with the region below the line. To obtain the final region, take the intersection of these regions.

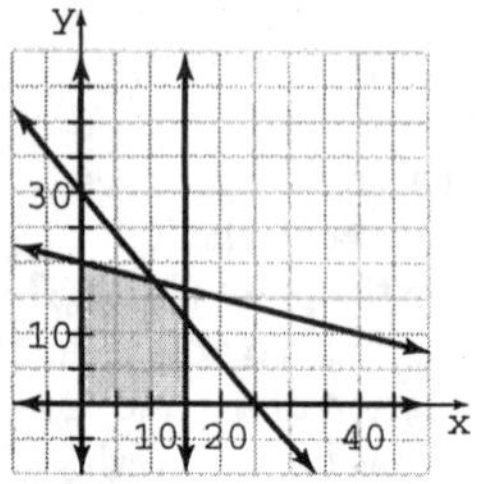

33. $|x| < 2$

For $|x| < 2$, the graph is the region between the dashed lines $x = -2$ and $x = 2$.

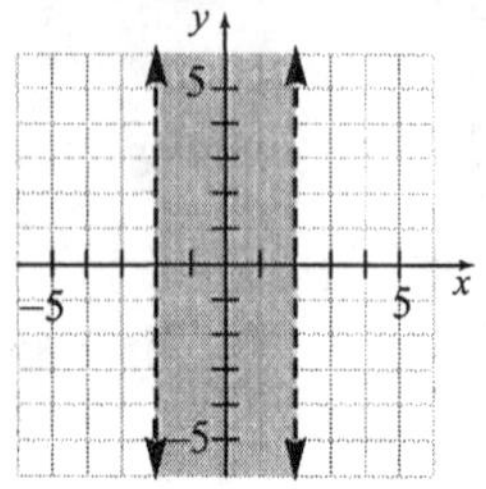

35. $|y-2| \le 4$

$-4 \le y-2 \le 4$

$-2 \le y \le 6$

For $|y-2| \le 4$, the graph is the solid lines $y=-2$ and $y=6$ along with the region between these lines.

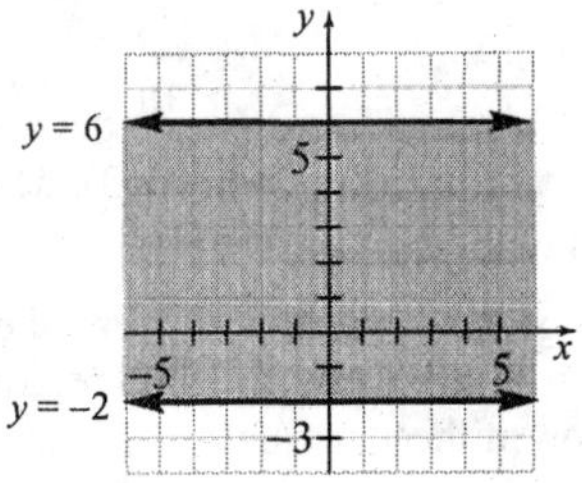

37. $|y| > 2$

$y \le x+3$

For $|y| > 2$, the graph is the region above the dashed line $y = 2$ along with the region below the dashed line $y = -2$. For $y \le x + 3$, the graph is the region below the solid line $y = x + 3$. To obtain the final region, take the intersection of these regions.

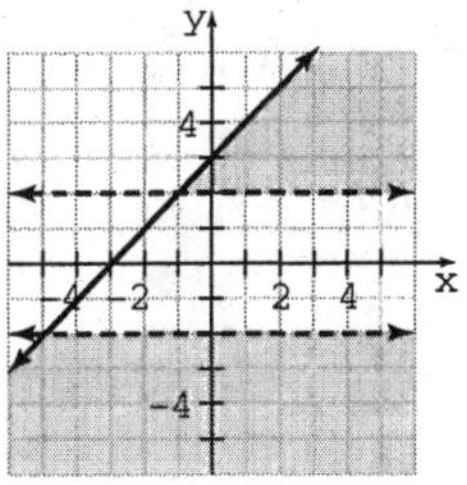

39. $|y| < 4$

$y \ge -2x+2$

For $|y| < 4$, the graph is the region between the dashed lines $y = -4$ and $y = 4$. For $y \ge -2x + 2$, the graph is the region above the solid line $y = -2x + 2$. To obtain the final region, take the intersection of these regions.

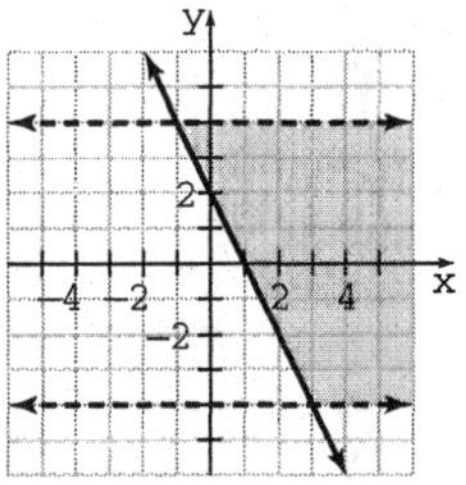

41. $|x+2| < 3$

$|y| > 4$

$|x+2| < 3$ can be written as

$-3 < x+2 < 3$

$-5 < x < 1$

For $|x+2| < 3$, the graph is the region between the dashed lines $x = -5$ and $x = 1$. For $|y| > 4$, the graph is the region above the dashed line $y = 4$ along with the region below the dashed line $y = -4$. To obtain the final region, take the intersection of these regions.

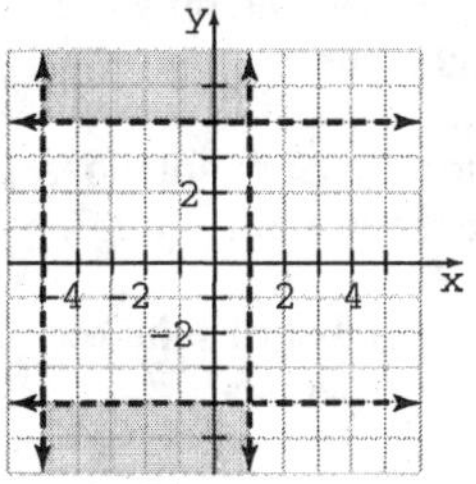

43. $|x-3| \le 4$

$|y+2| \le 1$

$|x-3| \le 4$ can be written as

$-4 \le x-3 \le 4$

$-1 \le x \le 7$

For $|x-3| \le 4$, the graph is the region between the solid lines $x = -1$ and $x = 7$.

$|y+2| \le 1$ can be written as

$-1 \le y+2 \le 1$

$-3 \le y \le -1$

For $|y+2| \le 1$, the graph is the region between the solid lines $y = -3$ and $y = -1$. To obtain the final region, take the intersection of these regions.

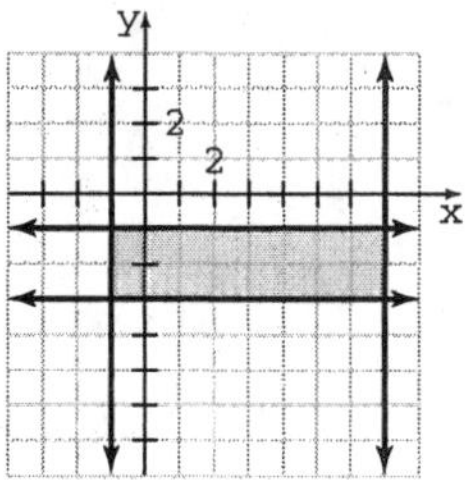

45. **a.** Region A; we are looking for the region that lies below the graph for the paper returns and above the graph for the electronic returns.

b. Region B; we are looking for the region that lies above the graph for the paper returns and below the graph for the electronic returns.

47. Yes; if the boundary lines are parallel, there may be no solution. For example, the system $y < x$ and $y > x + 1$ has no solution.

49. There are no solutions. Opposite sides of the same line are being shaded and only one inequality includes the line.

51. There are an infinite number of solutions. Both inequalities include the line $5x - 2y = 3$.

53. There are an infinite number of solutions. The lines are not parallel or identical.

55. $y \ge x^2$

$y \le 4$

For $y \ge x^2$, graph the equation $y = x^2$ using a solid line. For the check point, select (0, 1).

$y \ge x^2$

$(1) \ge (0)^2$

$1 \ge 0$ True

Since this is a true statement, shade the region which contains the point (0, 1). This is the region "above" the graph of $y \ge x^2$.

For $y \le 4$, the graph is the region below the solid line $y = 4$. To obtain the final region, take the intersection of these regions.

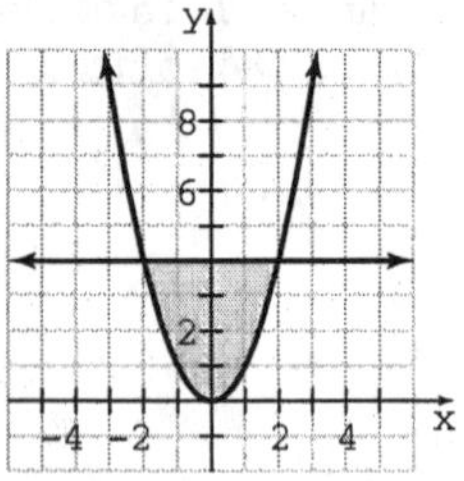

57. $y < |x|$

$y < 4$

For $y < |x|$, graph the equation $y = |x|$ using a dashed line. For the check point, select (0, 3).

$y < |x|$

$3 < |0|$

$3 < 0$ False

Since this is a false statement, shade the region which does not contain the point (0, 3). This is the region below the graph of $y = |x|$.

For $y < 4$, the graph is the region below the dashed line $y = 4$. To obtain the final region, take the intersection of these regions.

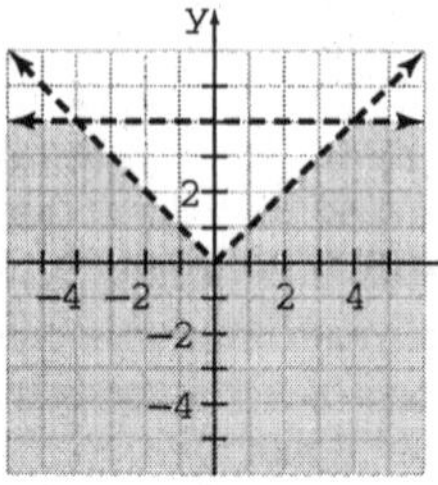

59.
$$\begin{aligned} f_1d_1 + f_2d_2 &= f_3d_3 \\ f_1d_1 - f_1d_1 + f_2d_2 &= f_3d_3 - f_1d_1 \\ f_2d_2 &= f_3d_3 - f_1d_1 \\ \frac{f_2d_2}{d_2} &= \frac{f_3d_3 - f_1d_1}{d_2} \\ f_2 &= \frac{f_3d_3 - f_1d_1}{d_2} \end{aligned}$$

60. Domain: {–1, 0, 4, 5}
Range: (–5, –2, 2, 3}

61. Domain: $\mathbb{R}$
Range: $\mathbb{R}$

62. Domain: $\mathbb{R}$
Range: $\{y \mid y \ge -1\}$

Chapter 4 Review Exercises

1. Write each equation in slope-intercept form.

$2x-3y=-1$ $\quad -3y=-2x-1 \quad y=\frac{2}{3}x+\frac{1}{3}$

$-4x+6y=1$ $\quad 6y=4x+1 \quad y=\frac{2}{3}x+\frac{1}{6}$

Since the slope of each line is $\frac{2}{3}$ but the y-intercepts are different ($b=\frac{1}{3}$ for first equation, $b=\frac{1}{6}$ for second equation), the two lines are parallel and produce no solution. This is an inconsistent system.

2. Write each equation in slope-intercept form.

$4x-5y=8$ $\quad -5y=-4x+8 \quad y=\frac{4}{5}x-\frac{8}{5}$

$3x+4y=9$ $\quad 4y=-3x+9 \quad y=-\frac{3}{4}x+\frac{9}{4}$

Since the slope of the first line is $\frac{4}{5}$ and the slope of the second line is $-\frac{3}{4}$, the slopes are different so that the lines intersect to produce one solution. This is a consistent system.

3. Write each equation in slope-intercept form.

$y=\frac{1}{3}x+4$ is already in this form.

$x+2y=8$

$2y=-x+8$

$y=-\frac{1}{2}x+4$

Since the slope of the first line is $\frac{1}{3}$ and the slope of the second line is $-\frac{1}{2}$, the slopes are different so that the lines intersect to produce one solution. This is a consistent system.

4. Write each equation in slope-intercept form.

$6x=5y-8$ $\quad 6x+8=5y \quad \frac{6x+8}{5}=y \quad \frac{6}{5}x+\frac{8}{5}=y$

$4x=6y+10$ $\quad 4x-10=6y \quad \frac{4x-10}{6}=y \quad \frac{2}{3}x-\frac{5}{3}=y$

Since the slope of the first line is $\frac{6}{5}$ and the slope of the second line is $\frac{2}{3}$, the slopes are different so that the lines intersect to produce one solution. This is a consistent system.

5. Graph the equations $y=x+3$ and $y=2x+5$.

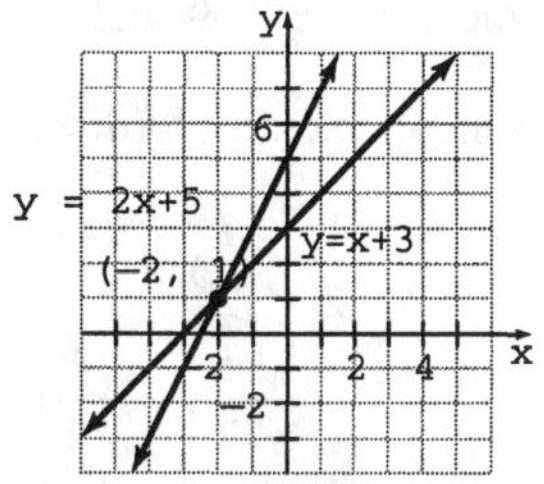

They intersect. The point of intersection is (–2, 1).

6. Graph the equations $x=-5$ and $y=3$.

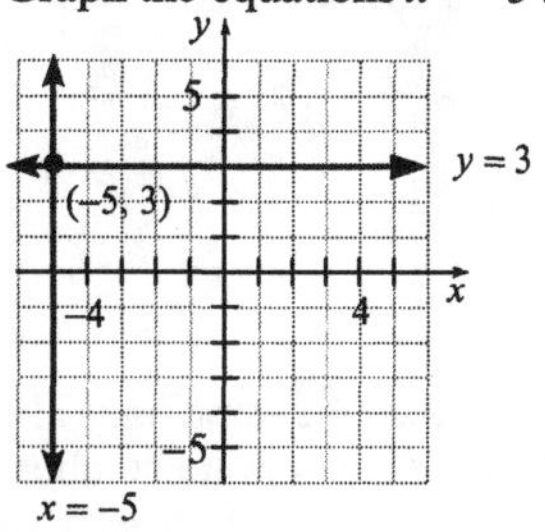

The lines intersect and the point of intersection is (–5, 3).

7. Graph the equations $3x+3y=12$ and $2x-y=-4$. In slope-intercept form we have:

$$\begin{aligned} 3x+3y&=12 \\ 3y&=-3x+12 \\ y&=-x+4 \end{aligned} \qquad \begin{aligned} 2x-y&=-4 \\ -y&=-2x-4 \\ y&=2x+4 \end{aligned}$$

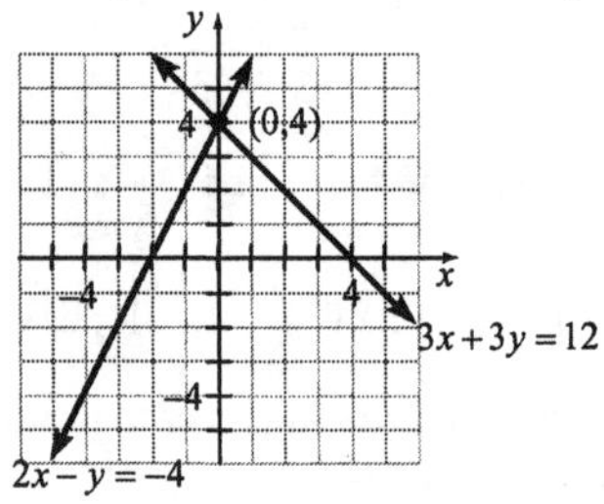

They intersect. The point of intersection is (0, 4).

8. Graph the equations $3y-3x=-9$ and $\frac{1}{2}x-\frac{1}{2}y=\frac{3}{2}$. In slope-intercept form we have:

$$\begin{aligned} 3y-3x&=-9 \\ 3y&=3x-9 \\ y&=x-3 \end{aligned} \qquad \begin{aligned} \frac{1}{2}x-\frac{1}{2}y&=\frac{3}{2} \\ -\frac{1}{2}y&=-\frac{1}{2}x+\frac{3}{2} \\ y&=x-3 \end{aligned}$$

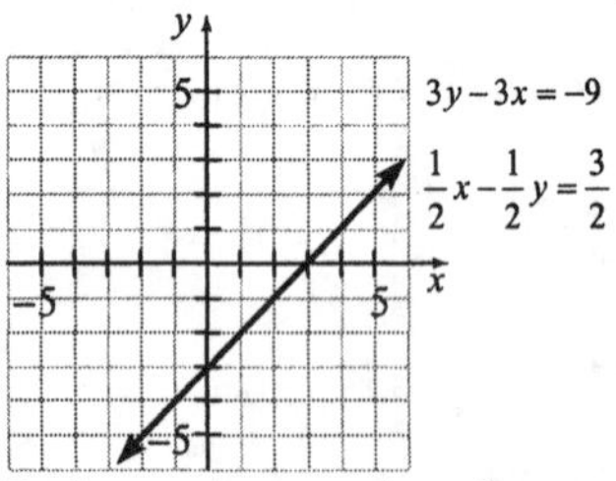

Both equations produce the same line. This is a dependent system with an infinite number of solutions.

9. $y=-4x+2$

$y=3x-12$

Substitute $3x-12$ for y in the first equation.

$$\begin{aligned} y&=-4x+2 \\ 3x-12&=-4x+2 \\ 7x-12&=2 \\ 7x&=14 \\ x&=2 \end{aligned}$$

Now, substitute 2 for x in the first equation.

$$\begin{aligned} y&=-4x+2 \\ y&=-4(2)+2 \\ y&=-8+2 \\ y&=-6 \end{aligned}$$

The solution is (2, –6).

10. $4x-3y=-1$

$y=2x+1$

Substitute $2x+1$ for y in the first equation.

$$\begin{aligned} 4x-3(2x+1)&=-1 \\ 4x-6x-3&=-1 \\ -2x-3&=-1 \\ -2x&=2 \\ x&=-1 \end{aligned}$$

Now, substitute –1 for x in the second equation.

$$\begin{aligned} y&=2x+1 \\ y&=2(-1)+1 \\ y&=-2+1 \\ y&=-1 \end{aligned}$$

The solution is (–1, –1).

11. $a=2b-8$

$2b-5a=0$

Substitute $2b-8$ for a in the second equation.

$$\begin{aligned} 2b-5a&=0 \\ 2b-5(2b-8)&=0 \\ 2b-10b+40&=0 \\ -8b+40&=0 \\ -8b&=-40 \\ b&=\frac{-40}{-8}=5 \end{aligned}$$

Now, substitute 5 for b in the first equation.

$$\begin{aligned} a&=2b-8 \\ a&=2(5)-8 \\ a&=10-8 \\ a&=2 \end{aligned}$$

The solution is (2, 5).

12. $2x+y=12$

$\frac{1}{2}x-\frac{3}{4}y=1$

First multiply the second equation by 4 to eliminate fractions.

$2x+y=12$

$4\left(\frac{1}{2}x-\frac{3}{4}y=1\right)$

gives

$2x+y=12 \quad (1)$

$2x-3y=4 \quad (2)$

Now, solve the first equation for y.

$2x+y=12$

$y=-2x+12$

Substitute $-2x+12$ for y in the second equation.

$$\begin{aligned} 2x-3y&=4 \\ 2x-3(-2x+12)&=4 \\ 2x+6x-36&=4 \\ 8x-36&=4 \\ 8x&=40 \\ x&=\frac{40}{8}=5 \end{aligned}$$

Finally, substitute 5 for x in the equation for y.

$y=-2x+12.$

$y=-2(5)+12$

$y=-10+12$

$y=2$

The solution is (5, 2).

13. Start by adding the two equations.

$$\begin{array}{lrl} & x-2y &= 5 \\ & \underline{2x+2y} &\underline{= 4} \\ \text{Add:} & 3x \quad &= 9 \\ & x &= 3 \end{array}$$

Substitute 3 for x in the first equation.

$(3)-2y=5$

$-2y=2$

$y=-1$

The solution is (3, –1).

14. Start by adding the two equations.

$$\begin{array}{lrl} & -2x- \;\; y &= 5 \\ & \underline{2x+\;2y} &\underline{=\;6} \\ \text{Add:} & y &= 11 \end{array}$$

Substitute 11 for y in the first equation.

$-2x-(11)=5$

$-2x=16$

$x=-8$

The solution is (–8, 11).

15. $2a+3b=7$

$a-2b=-7$

To eliminate a, multiply the second equation by –2 and then add.

$2a+3b=7$

$-2[a-2b=-7]$

gives

$$\begin{array}{lrl} & 2a+3b &= 7 \\ & \underline{-2a+4b} &\underline{=14} \\ \text{Add:} & 7b &= 21 \\ & b &= 3 \end{array}$$

Substitute 3 for b in the second equation.

$a-2(3)=-7$

$a-6=-7$

$a=-1$

The solution is (–1, 3).

16. $0.4x-0.3y=1.8$

$-0.7x+0.5y=-3.1$

To eliminate y, multiply the first equation by 5 and the second equation by 3 and then add.

$5[0.4x-0.3y=1.8]$

$3[-0.7x+0.5y=-3.1]$

gives

$$\begin{array}{lrl} & 2.0x-1.5y &= \;9 \\ & \underline{-2.1x+1.5y} &\underline{= \;-9.3} \\ \text{Add:} & -0.1x \qquad &= -0.3 \\ & x &= \frac{-0.3}{-0.1}=3 \end{array}$$

Substitute 3 for x in the first equation.

$0.4(3)-0.3y=1.8$

$1.2-0.3y=1.8$

$-0.3y=0.6$

$y=\frac{0.6}{-0.3}$

$y=-2$

The solution is (3, –2).

17. $4r-3s=8$

$2r+5s=8$

To eliminate r, multiply the second equation by -2 and then add.

$4r-3s=8$

$-2[2r+5s=8]$

gives

$$\begin{array}{r} 4r-\ \ 3s=\ \ \ 8 \\ \underline{-4r-10s=-16} \end{array}$$

Add: $-13s=-8$

$$s=\frac{-8}{-13}=\frac{8}{13}$$

Substitute $\frac{8}{13}$ for s in the first equation.

$$4r-3s=8$$
$$4r-3\left(\frac{8}{13}\right)=8$$
$$4r-\frac{24}{13}=8$$
$$4r=8+\frac{24}{13}$$
$$4r=\frac{104}{13}+\frac{24}{13}$$
$$4r=\frac{128}{13}$$
$$r=\frac{128}{13}\cdot\frac{1}{4}$$
$$r=\frac{32}{13}$$

The solution is $\left(\frac{32}{13}, \frac{8}{13}\right)$.

18. $-2m+3n=15$

$3m+3n=10$

To eliminate n, multiply the second equation by -1 and then add.

$-2m+3n=15$

$-1[3m+3n=10]$

gives

$$\begin{array}{r} -2m+3n=\ \ 15 \\ \underline{-3m-3n=-10} \end{array}$$

Add: $-5m \quad = \quad 5$

$$m=\frac{5}{-5}=-1$$

Substitute -1 for m in the second equation.

$$3m+3n=10$$
$$3(-1)+3n=10$$
$$-3+3n=10$$
$$3n=13$$
$$n=\frac{13}{3}$$

The solution is $\left(-1, \frac{13}{3}\right)$.

19. $x+\frac{3}{5}y=\frac{11}{5}$

$x-\frac{3}{2}y=-2$

To clear fractions and to eliminate x, multiply the first equation by 10 and the second equation by -10 and then add.

$$10\left(x+\frac{3}{5}y=\frac{11}{5}\right)$$
$$-10\left(x-\frac{3}{2}y=-2\right)$$

gives

$$\begin{array}{rl} 10x+\ \ 6y=22 & (1) \\ \underline{-10x+15y=20} & \underline{(2)} \end{array}$$

Add: $21y=42$

$$y=2$$

Now substitute 2 for y in the equation (1).

$$10x+6y=22$$
$$10x+6(2)=22$$
$$10x+12=22$$
$$10x=10$$
$$x=1$$

The solution is (1, 2).

20. $4x+4y=16$

$y=4x-3$

Write the system in standard form.

$4x+4y=16$

$-4x+y=-3$

To eliminate x, add the two equations:

$$\begin{array}{r} 4x+4y=16 \\ \underline{-4x+\ \ y=-3} \end{array}$$

Add: $5y=13$

$$y=\frac{13}{5}$$

Substitute $\frac{13}{5}$ for y in the first equation.

$$x+y=4$$
$$x+\frac{13}{5}=4$$
$$x=4-\frac{13}{5}$$
$$x=\frac{20}{5}-\frac{13}{5}$$
$$x=\frac{7}{5}$$

The solution is $\left(\frac{7}{5}, \frac{13}{5}\right)$.

21. $$y=-\frac{3}{4}x+\frac{5}{2}$$
$$x+\frac{5}{4}y=\frac{7}{2}$$

Write the system in standard form.

$$\frac{3}{4}x+y=\frac{5}{2}$$
$$x+\frac{5}{4}y=\frac{7}{2}$$

To clear fractions and to eliminate x, multiply the first equation by 16 and the second equation by -12 and then add.

$$16\left[\frac{3}{4}x+y=\frac{5}{2}\right]$$
$$-12\left[x+\frac{5}{4}y=\frac{7}{2}\right]$$

gives

$$\begin{aligned} 12x+16y&=40\\ -12x-15y&=-42\\ \hline \text{Add:}\quad y&=-2 \end{aligned}$$

Now, substitute -2 for y in the equation $x+\frac{5}{4}y=\frac{7}{2}$ and then solve for x.

$$x+\frac{5}{4}y=\frac{7}{2}$$
$$x+\frac{5}{4}(-2)=\frac{7}{2}$$
$$x-\frac{5}{2}=\frac{7}{2}$$
$$x=\frac{5}{2}+\frac{7}{2}$$
$$x=\frac{12}{2}$$
$$x=6$$

The solution is $(6, -2)$.

22. $$2x-5y=12$$
$$x-\frac{4}{3}y=-2$$

To clear fractions and to eliminate x, multiply the first equation by -3 and the second equation by 6 and then add.

$$-3[2x-5y=12]$$
$$6\left[x-\frac{4}{3}y=-2\right]$$

gives

$$\begin{aligned} -6x+15y&=-36\\ 6x-8y&=-12\\ \hline \text{Add:}\quad 7y&=-48 \end{aligned}$$
$$y=-\frac{48}{7}$$

Now substitute $-\frac{48}{7}$ for y in the first equation.

$$2x-5y=12$$
$$2x-5\left(-\frac{48}{7}\right)=12$$
$$2x+\frac{240}{7}=12$$
$$2x=12-\frac{240}{7}$$
$$2x=\frac{84}{7}-\frac{240}{7}$$
$$2x=-\frac{156}{7}$$
$$x=\frac{1}{2}\left(-\frac{156}{7}\right)=-\frac{78}{7}$$

The solution is $\left(-\frac{78}{7}, -\frac{48}{7}\right)$.

23. $2x+y=4$

$3x+\frac{3}{2}y=6$

To eliminate y, multiply the first equation by 3 and the second equation by −2, then add.

$3[2x+y=4]$

$-2\left[3x+\frac{3}{2}y=6\right]$

gives

$6x+3y=12$

$\underline{-6x-3y=-12}$

Add: $0=0$ True

The system has an infinite number of solutions.

24. $2x=4y+5$

$2y=x-7$

Write the system in standard form.

$2x-4y=5$

$-x+2y=-7$

To eliminate x, multiply the second equation by 2 and then add.

$2x-4y=5$

$2[-x+2y=-7]$

gives

$2x-4y=\ \ 5$

$\underline{-2x+4y=-14}$

Add: $0=-9$ False

Since this a false statement, there is no solution to the system. The system is inconsistent.

25. $x-2y-4z=13$ (1)

$3y+2z=-2$ (2)

$5z=-20$ (3)

Solve equation (3) for z.

$5z=-20$

$z=-4$

Substitute −4 for z in equation (2).

$3y+2(-4)=-2$

$3y-8=-2$

$3y=6$

$y=2$

Substitute −4 for z and 2 for y in equation (1).

$x-2(2)-4(-4)=13$

$x-4+16=13$

$x+12=13$

$x=1$

The solution is (1, 2, −4).

26. $2a+b-2c=5$

$3b+4c=1$

$3c=-6$

Solve the third equation for c.

$3c=-6$

$c=-2$

Substitute −2 for c in the second equation.

$3b+4(-2)=1$

$3b-8=1$

$3b=9$

$b=3$

Substitute 3 for b and −2 for c in the first equation.

$2a+(3)-2(-2)=5$

$2a+3+4=5$

$2a+7=5$

$2a=-2$

$a=-1$

The solution is (−1, 3, −2).

27. $x+2y+3z=3$ (1)

$-2x-3y-z=5$ (2)

$3x+3y+7z=2$ (3)

To eliminate x between equations (1) and (2), multiply equation (1) by 2 and then add.

$2[x+2y+3z=3]$

$-2x-3y-z=5$

gives

$2x+4y+6z=6$

$\underline{-2x-3y\ -z=5}$

Add: $y+5z=11$ (4)

To eliminate x between equations (2) and (3), multiply equation (2) by 3 and equation (3) by 2, then add.

$3[-2x-3y-z=5]$

$2[3x+3y+7z=2]$

gives

$$-6x-9y-\ 3z=15$$
$$\underline{6x+6y+14z=\ 4}$$
Add: $-3y+11z=19$ (5)

Equations (4) and (5) are two equations in two unknowns.

$y+5z=11$ (4)

$-3y+11z=19$ (5)

To eliminate y, multiply equation (4) by 3 and then add.

$3[y+5z=11]$

$-3y+11z=19$

gives

$$3y+15z=33$$
$$\underline{-3y+11z=19}$$
Add: $26z=52$

$$z=2$$

Substitute 2 for z in equation (4).

$$y+5(2)=11$$
$$y+10=11$$
$$y=1$$

Finally, substitute 1 for y and 2 for z in equation (1).

$$x+2(1)+3(2)=3$$
$$x+2+6=3$$
$$x+8=3$$
$$x=-5$$

The solution is (–5, 1, 2).

28. $-x-4y+2z=1$ (1)

$2x+2y+z=0$ (2)

$-3x-2y-5z=5$ (3)

To eliminate y between equations (1) and (2), multiply equation (2) by 2 and add.

$-x-4y+2z=1$

$2[2x+2y+z=0]$

gives

$$-x-4y+2z=1$$
$$\underline{4x+4y+2z=0}$$
Add: $3x\quad +4z=1$ (4)

To eliminate y between equations (2) and (3), simply add.

$$2x+2y+z=0$$
$$\underline{-3x-2y-5z=5}$$
Add: $-x\quad -4z=5$ (5)

Equations (4) and (5) are two equations in two unknowns.

$3x+4z=1$ (4)

$-x-4z=5$ (5)

To eliminate z, simply add equations (4) and (5).

$$3x+4z=1$$
$$\underline{-x-4z=5}$$
Add: $2x\quad =6$

$$x=3$$

Substitute 3 for x in equation (4).

$$3(3)+4z=1$$
$$9+4z=1$$
$$4z=-8$$
$$z=-2$$

Finally, substitute 3 for x and –2 for z in equation (1).

$$-(3)-4y+2(-2)=1$$
$$-3-4y-4=1$$
$$-7-4y=1$$
$$-4y=8$$
$$y=-2$$

The solution is (3, –2, –2).

29. $3y-2z=-4$ (1)

$3x-5z=-7$ (2)

$2x+y=6$ (3)

To eliminate y between equations (1) and (3), multiply equation (3) by –3 and then add.

$3y-2z=-4$

$-3[2x+y=6]$

gives

$$3y-2z=-4$$
$$\underline{-6x-3y\quad =-18}$$
Add: $-6x\quad -2z=-22$

or

$-3x-z=-11$ (4)

Equations (4) and (2) are two equations in two unknowns. To eliminate x, simply add.

$$3x-5z=-7$$
$$\underline{-3x-\ z=-11}$$
Add: $-6z=-18$

$$z=3$$

Substitute 3 for z in equation (2).

$3x-5z=-7$

$3x-5(3)=-7$

$3x-15=-7$

$3x=8$

$x=\frac{8}{3}$

Substitute $\frac{8}{3}$ for x in equation (3).

$2x+y=6$

$2\left(\frac{8}{3}\right)+y=6$

$\frac{16}{3}+y=6$

$y=6-\frac{16}{3}$

$y=\frac{18}{3}-\frac{16}{3}$

$y=\frac{2}{3}$

The solution is $\left(\frac{8}{3},\frac{2}{3},3\right)$.

30. $a+2b-5c=19$ (1)

$2a-3b+3c=-15$ (2)

$5a-4b-2c=-2$ (3)

To eliminate a between equations (1) and (2), multiply equation (1) by -2 and then add.

$-2[a+2b-5c=19]$

$2a-3b+3c=-15$

gives

$-2a-4b+10c=-38$

$\underline{2a-3b+\ 3c=-15}$

Add: $-7b+13c=-53$ (4)

To eliminate a between equations (1) and (3), multiply equation (1) by -5 and then add.

$-5[a+2b-5c=19]$

$5a-4b-2c=-2$

gives

$-5a-10b+25c=-95$

$\underline{5a-\ 4b-\ 2c=-2}$

Add: $-14b+23c=-97$ (5)

Equations (4) and (5) are two equations in two unknowns.

$-7b+13c=-53$

$-14b+23c=-97$

To eliminate b, multiply equation (4) by -2 and then add.

$-2[-7b+13c=-53]$

$-14b+23c=-97$

gives

$14b-26c=\ 106$

$\underline{-14b+23c=-97}$

Add: $-3c=9$

$c=-3$

Substitute -3 for c in equation (4).

$-7b+13c=-53$

$-7b+13(-3)=-53$

$-7b-39=-53$

$-7b=-14$

$b=2$

Finally, substitute 2 for b and -3 for c in equation (1).

$a+2b-5c=19$

$a+2(2)-5(-3)=19$

$a+4+15=19$

$a+19=19$

$a=0$

The solution is (0, 2, –3).

31. $x-y+3z=1$ (1)

$-x+2y-2z=1$ (2)

$x-3y+z=2$ (3)

To eliminate x between equations (1) and (2), simply add.

$x-y+3z=1$

$\underline{-x+2y-2z=1}$

Add: $y\ +z=2$ (4)

To eliminate x between equations (2) and (3), simply add.

$-x+2y-2z=1$

$\underline{x-3y+\ z=2}$

Add: $-y-\ z=3$ (5)

Equations (4) and (5) are two equations in two unknowns. To eliminate y and z simply add.

$y+z=2$

$\underline{-y-z=3}$

Add: $0=5$ False

Since this is a false statement, there is no solution to the system. This is an inconsistent system.

32. $-2x+2y-3z=6 \quad (1)$

$4x-y+2z=-2 \, (2)$

$2x+y-z=4 \quad (3)$

To eliminate x between equations (1) and (2), multiply equation (1) by 2 and then add.

$2[-2x+2y-3z=6]$

$4x-y+2z=-2$

gives

$-4x+4y-6z=12$

$\underline{4x-\ \ y+2z=-2}$

Add: $\quad 3y-4z=10 \quad (4)$

To eliminate x between equations (1) and (3), simply add.

$-2x+2y-3z=\ 6$

$\underline{2x+\ \ y-\ \ z=\ 4}$

Add: $\quad 3y-4z=10 \quad (5)$

Equations (4) and (5) are two equations in two unknowns.

$3y-4z=10$

$3y-4z=10$

Since they are identical, there are an infinite number of solutions. This is a dependent system.

33. Let x = Jennifer's age and y = Luan's age.

$y=x+10$

$x+y=66$

Substitute $x + 10$ for y in the second equation.

$x+(x+10)=66$

$2x+10=66$

$2x=56$

$x=28$

Now substitute 28 for x in the first equation.

$y=(28)+10$

$y=38$

Jennifer is 28 years old and Luan is 38 years old.

34. Let x = the speed of the plane in still air and y = the speed of the wind.

$x+y=560$

$x-y=480$

To eliminate y, simply add.

$x+y=560$

$\underline{x-y=480}$

Add: $2x \quad =1040$

$x \quad =520$

Substitute 520 for x in the first equation.

$x+y=560$

$(520)+y=560$

$y=40$

The speed of the plane in still air is 520 mph and the speed of the wind is 40 mph.

35. Let x = the amount of 20% acid solution and y = the amount of 50% acid solution.

$x+y=6$

$0.2x+0.5y=0.4(6)$

To clear decimals, multiply the second equation by 10.

$x+y=6$

$2x+5y=24$

Solve the first equation for y.

$x+y=6$

$y=-x+6$

Substitute $-x + 6$ for y in the second equation.

$2x+5y=24$

$2x+5(-x+6)=24$

$2x-5x+30=-24$

$-3x+30=24$

$-3x=-6$

$x=\frac{-6}{-3}=2$

Finally, substitute 2 for x in the equation $y=-x+6$.

$y=-x+6$

$y=-2+6$

$y=4$

Sally should combine 2 liters of solution A with 4 liters of solution B.

36. Let x = the number of adult tickets and y = the number of children's tickets.

$x+y=650$

$15x+11y=8790$

To solve, multiply the first equation by –11 and then add.

$-11[x+y=650]$

$15x+11y=8790$

gives

$-11x-11y=-7150$

$15x+11y=8790$

Add: $4x = 1640$

$x=\frac{1640}{4}=410$

Substitute 410 for x in the first equation.

$x+y=650$

$410+y=650$

$y=240$

Thus, 410 adult tickets and 240 children's tickets were sold.

37. Let x = age at first time and y = age at second time.

$y=2x-5$

$x+y=118$

Substitute $2x-5$ for y in the second equation.

$x+y=118$

$x+2x-5=118$

$3x-5=118$

$3x=123$

$x=41$

Substitute 41 for x in the first equation.

$y=2x-5$

$y=2(41)-5$

$y=82-5$

$y=77$

His ages were 41 years and 77 years.

38. Let x = the amount invested at 7%, y = the amount invested at 5%, and z = the amount invested at 3%.

$x+y+z=40{,}000$ (1)

$y=x-5000$ (2)

$0.07x+0.05y+0.03z=2300$ (3)

Substitute $x-5000$ for y in equations (1) and (3).
Equation (1) becomes

$x+y+z=40{,}000$

$x+x-5000+z=40{,}000$

$2x+z=45{,}000$ (4)

Equation (3) becomes

$0.07x+0.05y+0.03z=2300$

$0.07x+0.05(x-5000)+0.03z=2300$

$0.07x+0.05x-250+0.03z=2300$

$0.12x+0.03z=2550$ (5)

Equation (4) and (5) are a system of two equations in two unknowns. Solve equation (4) for z.

$2x+z=45{,}000$

$z=-2x+45{,}000$

Substitute $-2x+45{,}000$ for z in equation (5).

$0.12x+0.03z=2550$

$0.12x+0.03(-2x+45{,}000)=2550$

$0.12x-0.06x+1350=2550$

$0.06x=1200$

$x=20{,}000$

Now substitute 20,000 for x in equation (2).

$y=x-5000$

$y=20{,}000-5000=15{,}000$

Finally, substitute 20,000 for x and 15,000 for y in equation (1).

$x+y+z=40{,}000$

$20{,}000+15{,}000+z=40{,}000$

$35{,}000+z=40{,}000$

$z=5000$

Thus, \$20,000 was invested at 7%, \$15,000 at 5%, and \$5000 at 3%.

39. $x+5y=1$

$-2x-8y=-6$

$\left[\begin{array}{cc|c} 1 & 5 & 1 \\ -2 & -8 & -6 \end{array}\right]$

$\left[\begin{array}{cc|c} 1 & 5 & 1 \\ 0 & 2 & -4 \end{array}\right] 2R_1 + R_2$

$\left[\begin{array}{cc|c} 1 & 5 & 1 \\ 0 & 1 & -2 \end{array}\right] \frac{1}{2}R_2$

The system is

$x+5y=1$

$y=-2$

Substitute –2 for y in the first equation.

$x+5(-2)=1$

$x-10=1$

$x=11$

The solution is (11, –2).

40. $2x-5y=1$

$2x+4y=10$

$\left[\begin{array}{cc|c} 2 & -5 & 1 \\ 2 & 4 & 10 \end{array}\right]$

$\left[\begin{array}{cc|c} 1 & -\frac{5}{2} & \frac{1}{2} \\ 2 & 4 & 10 \end{array}\right] \frac{1}{2}R_1$

$\left[\begin{array}{cc|c} 1 & -\frac{5}{2} & \frac{1}{2} \\ 0 & 9 & 9 \end{array}\right] -2R_1 + R_2$

$\left[\begin{array}{cc|c} 1 & -\frac{5}{2} & \frac{1}{2} \\ 0 & 1 & 1 \end{array}\right] \frac{1}{9}R_2$

The system is

$x-\frac{5}{2}y=\frac{1}{2}$

$y=1$

Substitute 1 for y in the first equation.

$x-\frac{5}{2}(1)=\frac{1}{2}$

$x-\frac{5}{2}=\frac{1}{2}$

$x=\frac{6}{2}$

$x=3$

The solution is (3, 1).

41. $3y=6x-12$

$4x=2y+8$

Write the system in standard form.

$-6x+3y=-12$

$4x-2y=8$

$\left[\begin{array}{cc|c} -6 & 3 & -12 \\ 4 & -2 & 8 \end{array}\right]$

$\left[\begin{array}{cc|c} 1 & -\frac{1}{2} & 2 \\ 4 & -2 & 8 \end{array}\right] -\frac{1}{6}R_1$

$\left[\begin{array}{cc|c} 1 & -\frac{1}{2} & 2 \\ 0 & 0 & 0 \end{array}\right] -4R_1 + R_2$

Since the last row is all zeros, the system is dependent. There are an infinite number of solutions.

42. $2x-y-z=5$

$x+2y+3z=-2$

$3x-2y+z=2$

$\left[\begin{array}{ccc|c} 2 & -1 & -1 & 5 \\ 1 & 2 & 3 & -2 \\ 3 & -2 & 1 & 2 \end{array}\right]$

$\left[\begin{array}{ccc|c} 1 & -\frac{1}{2} & -\frac{1}{2} & \frac{5}{2} \\ 1 & 2 & 3 & -2 \\ 3 & -2 & 1 & 2 \end{array}\right] \frac{1}{2}R_1$

$\left[\begin{array}{ccc|c} 1 & -\frac{1}{2} & -\frac{1}{2} & \frac{5}{2} \\ 0 & \frac{5}{2} & \frac{7}{2} & -\frac{9}{2} \\ 3 & -2 & 1 & 2 \end{array}\right] -1R_1 + R_2$

$\left[\begin{array}{ccc|c} 1 & -\frac{1}{2} & -\frac{1}{2} & \frac{5}{2} \\ 0 & \frac{5}{2} & \frac{7}{2} & -\frac{9}{2} \\ 0 & -\frac{1}{2} & \frac{5}{2} & -\frac{11}{2} \end{array}\right] -3R_1 + R_3$

$\left[\begin{array}{ccc|c} 1 & -\frac{1}{2} & -\frac{1}{2} & \frac{5}{2} \\ 0 & 1 & \frac{7}{5} & -\frac{9}{5} \\ 0 & -\frac{1}{2} & \frac{5}{2} & -\frac{11}{2} \end{array}\right] \frac{2}{5}R_2$

$\left[\begin{array}{ccc|c} 1 & -\frac{1}{2} & -\frac{1}{2} & \frac{5}{2} \\ 0 & 1 & \frac{7}{5} & -\frac{9}{5} \\ 0 & 0 & \frac{16}{5} & -\frac{32}{5} \end{array}\right] \frac{1}{2}R_2 + R_3$

$\left[\begin{array}{ccc|c} 1 & -\frac{1}{2} & -\frac{1}{2} & \frac{5}{2} \\ 0 & 1 & \frac{7}{5} & -\frac{9}{5} \\ 0 & 0 & 1 & -2 \end{array}\right] \frac{5}{16}R_3$

The system is

$$x-\frac{1}{2}y-\frac{1}{2}z=\frac{5}{2}$$
$$y+\frac{7}{5}z=-\frac{9}{5}$$
$$z=-2$$

Substitute –2 for z in the second equation.

$$y+\frac{7}{5}z=-\frac{9}{5}$$
$$y+\frac{7}{5}(-2)=-\frac{9}{5}$$
$$y-\frac{14}{5}=-\frac{9}{5}$$
$$y=\frac{5}{5}=1$$

Substitute 1 for y and –2 for z in the first equation.

$$x-\frac{1}{2}y-\frac{1}{2}z=\frac{5}{2}$$
$$x-\frac{1}{2}(1)-\frac{1}{2}(-2)=\frac{5}{2}$$
$$x-\frac{1}{2}+1=\frac{5}{2}$$
$$x+\frac{1}{2}=\frac{5}{2}$$
$$x=\frac{4}{2}=2$$

The solution is (2, 1, –2).

43. $3a-b+c=2$

$2a-3b+4c=4$

$a+2b-3c=-6$

$$\left[\begin{array}{ccc|c} 3 & -1 & 1 & 2 \\ 2 & -3 & 4 & 4 \\ 1 & 2 & -3 & -6 \end{array}\right]$$

$$\left[\begin{array}{ccc|c} 1 & -\frac{1}{3} & \frac{1}{3} & \frac{2}{3} \\ 2 & -3 & 4 & 4 \\ 1 & 2 & -3 & -6 \end{array}\right]\frac{1}{3}R_1$$

$$\left[\begin{array}{ccc|c} 1 & -\frac{1}{3} & \frac{1}{3} & \frac{2}{3} \\ 0 & -\frac{7}{3} & \frac{10}{3} & \frac{8}{3} \\ 1 & 2 & -3 & -6 \end{array}\right]-2R_1+R_2$$

$$\left[\begin{array}{ccc|c} 1 & -\frac{1}{3} & \frac{1}{3} & \frac{2}{3} \\ 0 & -\frac{7}{3} & \frac{10}{3} & \frac{8}{3} \\ 0 & \frac{7}{3} & -\frac{10}{3} & -\frac{20}{3} \end{array}\right]-1R_1+R_3$$

$$\left[\begin{array}{ccc|c} 1 & -\frac{1}{3} & \frac{1}{3} & \frac{2}{3} \\ 0 & 1 & -\frac{10}{7} & -\frac{8}{7} \\ 0 & \frac{7}{3} & -\frac{10}{3} & -\frac{20}{3} \end{array}\right]-\frac{3}{7}R_2$$

$$\left[\begin{array}{ccc|c} 1 & -\frac{1}{3} & \frac{1}{3} & \frac{2}{3} \\ 0 & 1 & -\frac{10}{7} & -\frac{8}{7} \\ 0 & 0 & 0 & -4 \end{array}\right]$$

Since the last row has all zeros on the left side and a nonzero number on the right side, the system is inconsistent and has no solution.

44. $x+y+z=3$

$3x+4y=-1$

$y-3z=-10$

$$\left[\begin{array}{ccc|c} 1 & 1 & 1 & 3 \\ 3 & 4 & 0 & -1 \\ 0 & 1 & -3 & -10 \end{array}\right]$$

$$\left[\begin{array}{ccc|c} 1 & 1 & 1 & 3 \\ 0 & 1 & -3 & -10 \\ 0 & 1 & -3 & -10 \end{array}\right]-3R_1+R_2$$

$$\left[\begin{array}{ccc|c} 1 & 1 & 1 & 3 \\ 0 & 1 & -3 & -10 \\ 0 & 0 & 0 & 0 \end{array}\right]-1R_2+R_3$$

Since the last row is all zeros and there are no contradictions, the system is dependent. There are an infinite number of solutions.

45. $7x-8y=-10$

$-5x+4y=2$

To solve, first calculate D, D_x, and D_y.

$$D=\begin{vmatrix} 7 & -8 \\ -5 & 4 \end{vmatrix}=(7)(4)-(-5)(-8)$$
$$=28-40=-12$$
$$D_x=\begin{vmatrix} -10 & -8 \\ 2 & 4 \end{vmatrix}=(-10)(4)-(2)(-8)$$
$$=-40+16=-24$$
$$D_y=\begin{vmatrix} 7 & -10 \\ -5 & 2 \end{vmatrix}=(7)(2)-(-5)(-10)$$
$$=14-50=-36$$
$$x=\frac{D_x}{D}=\frac{-24}{-12}=2 \text{ and } y=\frac{D_y}{D}=\frac{-36}{-12}=3$$

The solution is (2, 3).

46. $x+4y=5$

$5x+3y=-9$

To solve, first calculate D, D_x, and D_y.

$$D=\begin{vmatrix}1 & 4\\5 & 3\end{vmatrix}$$
$$=(1)(3)-(5)(4)$$
$$=3-20$$
$$=-17$$

$$D_x=\begin{vmatrix}5 & 4\\-9 & 3\end{vmatrix}$$
$$=(5)(3)-(-9)(4)$$
$$=15+36$$
$$=51$$

$$D_y=\begin{vmatrix}1 & 5\\5 & -9\end{vmatrix}$$
$$=(1)(-9)-(5)(5)$$
$$=-9-25$$
$$=-34$$

$$x=\frac{D_x}{D}=\frac{51}{-17}=-3 \text{ and } y=\frac{D_y}{D}=\frac{-34}{-17}=2$$

The solution is (–3, 2).

47. $9m+4n=-1$

$7m-2n=-11$

To solve, first calculate D, D_m, and D_n.

$$D=\begin{vmatrix}9 & 4\\7 & -2\end{vmatrix}$$
$$=(9)(-2)-(7)(4)$$
$$=-18-28$$
$$=-46$$

$$D_m=\begin{vmatrix}-1 & 4\\-11 & -2\end{vmatrix}$$
$$=(-1)(-2)-(-11)(4)$$
$$=2+44$$
$$=46$$

$$D_n=\begin{vmatrix}9 & -1\\7 & -11\end{vmatrix}$$
$$=(9)(-11)-(7)(-1)$$
$$=-99+7$$
$$=-92$$

$$m=\frac{D_m}{D}=\frac{46}{-46}=-1 \text{ and } n=\frac{D_n}{D}=\frac{-92}{-46}=2.$$

The solution is (–1, 2).

48. $p+q+r=5$

$2p+q-r=-5$

$3p+2q-3r=-12$

To solve, calculate D, D_p, D_q, and D_r.

$$D=\begin{vmatrix}1 & 1 & 1\\2 & 1 & -1\\3 & 2 & -3\end{vmatrix}$$
$$=1\begin{vmatrix}1 & -1\\2 & -3\end{vmatrix}-1\begin{vmatrix}2 & -1\\3 & -3\end{vmatrix}+1\begin{vmatrix}2 & 1\\3 & 2\end{vmatrix}$$
$$=1(-3+2)-1(-6+3)+1(4-3)$$
$$=1(-1)-1(-3)+1(1)$$
$$=-1+3+1=3$$

$$D_p=\begin{vmatrix}5 & 1 & 1\\-5 & 1 & -1\\-12 & 2 & -3\end{vmatrix}$$
$$=5\begin{vmatrix}1 & -1\\2 & -3\end{vmatrix}-1\begin{vmatrix}-5 & -1\\-12 & -3\end{vmatrix}+1\begin{vmatrix}-5 & 1\\-12 & 2\end{vmatrix}$$
$$=5(-3+2)-1(15-12)+1(-10+12)$$
$$=5(-1)-1(3)+1(2)$$
$$=-5-3+2=-6$$

$$D_q=\begin{vmatrix}1 & 5 & 1\\2 & -5 & -1\\3 & -12 & -3\end{vmatrix}$$
$$=1\begin{vmatrix}-5 & -1\\-12 & -3\end{vmatrix}-5\begin{vmatrix}2 & -1\\3 & -3\end{vmatrix}+1\begin{vmatrix}2 & -5\\3 & -12\end{vmatrix}$$
$$=1(15-12)-5(-6+3)+1(-24+15)$$
$$=1(3)-5(-3)+1(-9)$$
$$=3+15-9=9$$

$$D_r=\begin{vmatrix}1 & 1 & 5\\2 & 1 & -5\\3 & 2 & -12\end{vmatrix}$$
$$=1\begin{vmatrix}1 & -5\\2 & -12\end{vmatrix}-1\begin{vmatrix}2 & -5\\3 & -12\end{vmatrix}+5\begin{vmatrix}2 & 1\\3 & 2\end{vmatrix}$$
$$=1(-12+10)-1(-24+15)+5(4-3)$$
$$=1(-2)-1(-9)+5(1)$$
$$=-2+9+5=12$$

$$p=\frac{D_p}{D}=\frac{-6}{3}=-2,\ q=\frac{D_q}{D}=\frac{9}{3}=3, \text{ and}$$
$$r=\frac{D_r}{D}=\frac{12}{3}=4$$

The solution is (–2, 3, 4).

49. $-2a+3b-4c=-7$

$2a+b+c=5$

$-2a-3b+4c=3$

To solve, calculate D, D_a, D_b, and D_c.

$$D=\begin{vmatrix}-2 & 3 & -4\\ 2 & 1 & 1\\ -2 & -3 & 4\end{vmatrix}$$

$$=-2\begin{vmatrix}1 & 1\\ -3 & 4\end{vmatrix}-3\begin{vmatrix}2 & 1\\ -2 & 4\end{vmatrix}+(-4)\begin{vmatrix}2 & 1\\ -2 & -3\end{vmatrix}$$

$$=-2(4+3)-3(8+2)-4(-6+2)$$
$$=-2(7)-3(10)-4(-4)$$
$$=-14-30+16$$
$$=-28$$

$$D_a=\begin{vmatrix}-7 & 3 & -4\\ 5 & 1 & 1\\ 3 & -3 & 4\end{vmatrix}$$

$$=-7\begin{vmatrix}1 & 1\\ -3 & 4\end{vmatrix}-3\begin{vmatrix}5 & 1\\ 3 & 4\end{vmatrix}+(-4)\begin{vmatrix}5 & 1\\ 3 & -3\end{vmatrix}$$

$$=-7(4+3)-3(20-3)-4(-15-3)$$
$$=-7(7)-3(17)-4(-18)$$
$$=-49-51+72$$
$$=-28$$

$$D_b=\begin{vmatrix}-2 & -7 & -4\\ 2 & 5 & 1\\ -2 & 3 & 4\end{vmatrix}$$

$$=-2\begin{vmatrix}5 & 1\\ 3 & 4\end{vmatrix}-(-7)\begin{vmatrix}2 & 1\\ -2 & 4\end{vmatrix}+(-4)\begin{vmatrix}2 & 5\\ -2 & 3\end{vmatrix}$$

$$=-2(20-3)+7(8+2)-4(6+10)$$
$$=-2(17)+7(10)-4(16)$$
$$=-34+70-64$$
$$=-28$$

$$D_c=\begin{vmatrix}-2 & 3 & -7\\ 2 & 1 & 5\\ -2 & -3 & 3\end{vmatrix}$$

$$=-2\begin{vmatrix}1 & 5\\ -3 & 3\end{vmatrix}-3\begin{vmatrix}2 & 5\\ -2 & 3\end{vmatrix}+(-7)\begin{vmatrix}2 & 1\\ -2 & -3\end{vmatrix}$$

$$=-2(3+15)-3(6+10)-7(-6+2)$$
$$=-2(18)-3(16)-7(-4)$$
$$=-36-48+28$$
$$=-56$$

$a=\frac{D_a}{D}=\frac{-28}{-28}=1$, $b=\frac{D_b}{D}=\frac{-28}{-28}=1$, and

$c=\frac{D_c}{D}=\frac{-56}{-28}=2$

The solution is (1, 1, 2).

50. $y+3z=4$

$-x-y+2z=0$

$x+2y+z=1$

To solve, first calculate D, D_x, D_y, and D_z.

$$D=\begin{vmatrix}0 & 1 & 3\\ -1 & -1 & 2\\ 1 & 2 & 1\end{vmatrix}$$

$$=0\begin{vmatrix}-1 & 2\\ 2 & 1\end{vmatrix}-(-1)\begin{vmatrix}1 & 3\\ 2 & 1\end{vmatrix}+1\begin{vmatrix}1 & 3\\ -1 & 2\end{vmatrix}$$

$$=0(-1-4)+1(1-6)+1(2+3)$$
$$=0(-5)+1(-5)+1(5)$$
$$=0-5+5$$
$$=0$$

$$D_x=\begin{vmatrix}4 & 1 & 3\\ 0 & -1 & 2\\ 1 & 2 & 1\end{vmatrix}$$

$$=4\begin{vmatrix}-1 & 2\\ 2 & 1\end{vmatrix}-0\begin{vmatrix}1 & 3\\ 2 & 1\end{vmatrix}+1\begin{vmatrix}1 & 3\\ -1 & 2\end{vmatrix}$$

$$=4(-1-4)-0(1-6)+1(2+3)$$
$$=4(-5)-0(-5)+1(5)$$
$$=-20+0+5$$
$$=-15$$

Since $D=0$ and $D_x=-15$, the system is inconsistent and has no solution.

51. $-x+3y>6$

$2x-y\le 2$

For $-x+3y>6$, graph the line $-x+3y=6$ using a dashed line. For the check point, select (0, 0):

$-x+3y>6$

$-0+3(0)>6$

$0>6$ False

Since this is a false statement, shade the region which does not contain the point (0, 0).

For $2x-y\le 2$, graph the line $2x-y=2$ using a solid line. For the check point, select (0, 0):

$2x-y\le 2$

$2(0)-0\le 2$

$0\le 2$ True

Since this is a true statement, shade the region which contains the point (0, 0).
To obtain the final region, take the intersection of the above two regions.

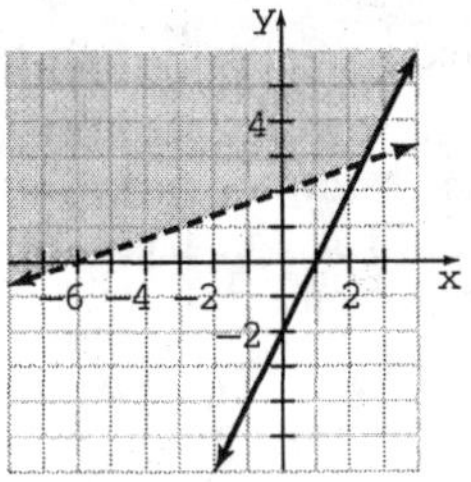

52. $5x - 2y \leq 10$

$3x + 2y > 6$

For $5x - 2y \leq 10$, graph the line $5x - 2y = 10$ using a solid line.
For the check point, select (0, 0):

$5x - 2y \leq 10$

$5(0) - 2(0) \leq 10$

$0 \leq 10$ True

Since this is a true statement, shade the region which contains the point (0, 0).
For $3x + 2y > 6$, graph the line $3x + 2y = 6$ using a dashed line. For the check point, select (0, 0):

$3x + 2y > 6$

$3(0) + 2(0) > 6$

$0 > 6$ False

Since this is a false statement, shade the region which does not contain the point (0, 0). To obtain the final region, take the intersection of the above two regions.

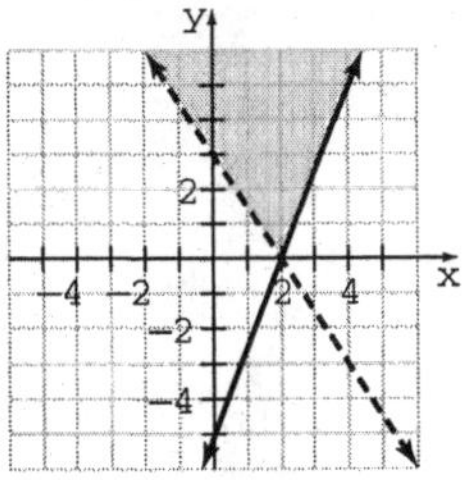

53. $y > 2x + 3$

$y < -x + 4$

For $y > 2x + 3$, graph the line $y = 2x + 3$ using a dashed line. For the check point, select (0, 0):

$y > 2x + 3$

$0 > 2(0) + 3$

$0 > 3$ False

Since this is a false statement, shade the region which does not contain the point (0, 0).
For $y < -x + 4$, graph the line $y = -x + 4$ using a dashed line. For the check point, select (0, 0):

$y < -x + 4$

$0 < -0 + 4$

$0 < 4$ True

Since this is a true statement, shade the region which contains the point (0, 0). To obtain the final region, take the intersection of the above two regions.

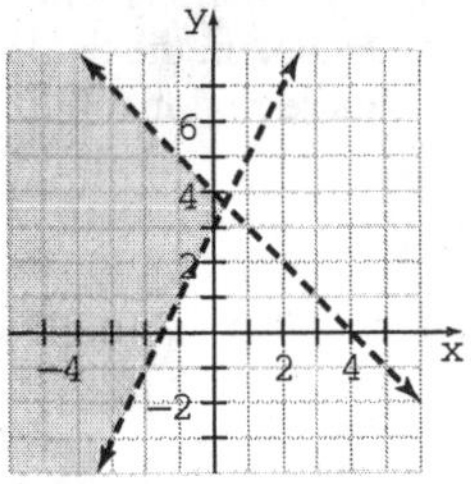

54. $x > -2y + 4$

$y < -\frac{1}{2}x - \frac{3}{2}$

For $x > -2y + 4$, graph the line $x = -2y + 4$ using a dashed line. For the check point, select (0, 0):

$x > -2y + 4$

$0 > -2(0) + 4$

$0 > 4$ False

Since this is a false statement, shade the region which does not contain the point (0, 0).

For $y < -\frac{1}{2}x - \frac{3}{2}$, graph the line $y = -\frac{1}{2}x - \frac{3}{2}$ using a dashed line. For the check point, select (0, 0):

$y < -\frac{1}{2}x - \frac{3}{2}$

$0 < -\frac{1}{2}(0) - \frac{3}{2}$

$0 < -\frac{3}{2}$ False

Since this is a false statement, shade the region which does not contain the point
(0, 0). To obtain the final region, take the intersection of the above two regions. The regions do not overlap, so there are no solutions.

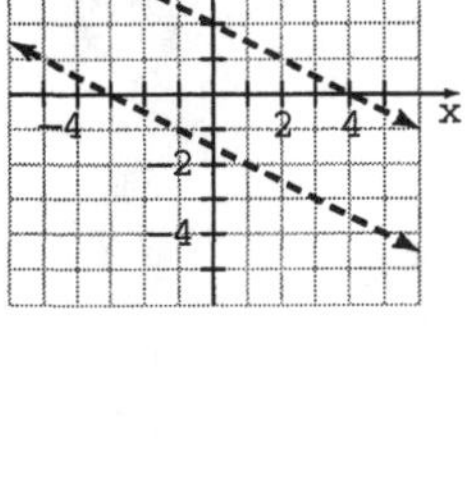

55. $x \geq 0$

$y \geq 0$

$x + y \leq 6$

$4x + y \leq 8$

The first two inequalities indicate that the solution must be in the first quadrant. For $x + y \leq 6$, the graph is the line $x + y = 6$ along with the region below this line. For $4x + y \leq 8$, the graph is the line $4x + y = 8$ along with the region below this line. To obtain the final region, take the intersection of these regions.

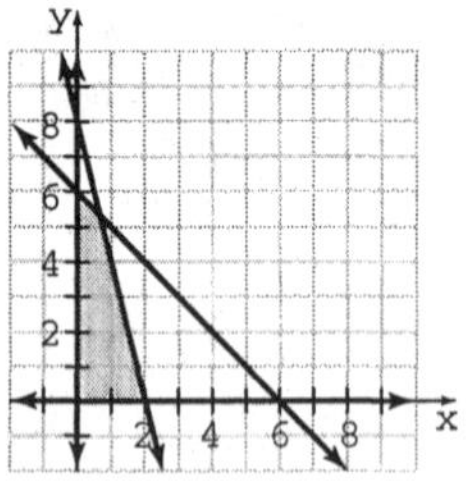

56. $x \geq 0$

$y \geq 0$

$2x + y \leq 6$

$4x + 5y \leq 20$

The first two inequalities indicate that the solution must be in the first quadrant. For $2x + y \leq 6$, graph the line $2x + y = 6$ along with the region below this line. For $4x + 5y \leq 20$, graph the line $4x + 5y = 20$ along with the region below this line. To obtain the final region, take the intersection of these regions.

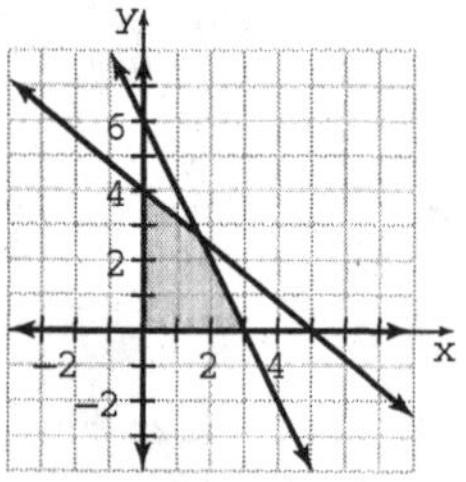

57. $|x| \leq 3$

$|y| > 2$

For $|x| \leq 3$, the graph is the region between the solid lines $x = -3$ and $x = 3$. For $|y| > 2$, the graph is the region above the dashed line $y = 2$ along with the region below the dashed line $y = -2$. To obtain the final region, take the intersection of these regions.

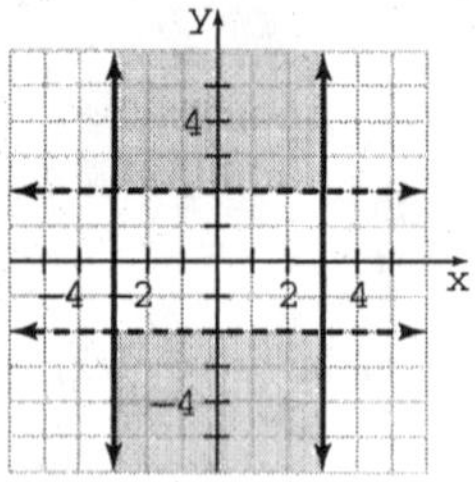

58. $|x| > 4$

$|y - 2| \leq 3$

For $|x| > 4$, the graph is the region to the left of dashed line $x = -4$ along with the region to the right of the dashed line $x = 4$.

$|y - 2| \leq 3$ can be written as

$-3 \leq y - 2 \leq 3$

$-1 \leq y \leq 5$

For $|y - 2| \leq 3$, the graph is the region between the solid lines $y = -1$ and $y = 5$. To obtain the final region, take the intersection of these regions.

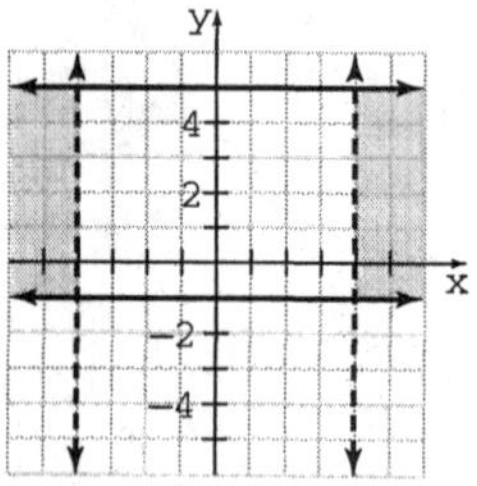

Chapter 4 Practice Test

1. Answers will vary.

2. Write both equations in slope-intercept form.

$$5x+2y=4 \qquad 6x=3y-7$$
$$2y=-5x+4 \qquad -3y=-6x-7$$
$$y=\frac{-5x+4}{2} \qquad y=\frac{-6x-7}{-3}$$
$$y=-\frac{5}{2}x+2 \qquad y=2x+\frac{7}{3}$$

Since the slope of the first line is $-\frac{5}{2}$ and the slope of the second line is 2, the slopes are different so that the lines intersect to produce one solution. This is a consistent system.

3. Write both equations in slope-intercept form.

$$5x+3y=9 \qquad 2y=-\frac{10}{3}x+6$$
$$3y=-5x+9$$
$$y=\frac{-5x+9}{3} \qquad \frac{1}{2}(2y)=\frac{1}{2}\left(-\frac{10}{3}x+6\right)$$
$$y=-\frac{5}{3}x+3 \qquad y=-\frac{5}{3}x+3$$

Since the equations are identical, there is an infinite number of solutions and this is a dependent system.

4. Write both equations in slope-intercept form.

$$5x-4y=6 \qquad -10x+8y=-10$$
$$-4y=-5x+6 \qquad 8y=10x-10$$
$$y=\frac{-5x+6}{-4} \qquad y=\frac{10x-10}{8}$$
$$y=\frac{5}{4}x-\frac{3}{2} \qquad y=\frac{5}{4}x-\frac{5}{4}$$

Since the slope of each line is $\frac{5}{4}$, but the y-intercepts are different $\left(b=-\frac{3}{2}\text{ for the first equation, } b=-\frac{5}{4}\text{ for the second equation}\right)$, the two lines are parallel and produce no solution. This is an inconsistent system.

5. Graph the equations $y = 3x - 2$ and $y = -2x + 8$.

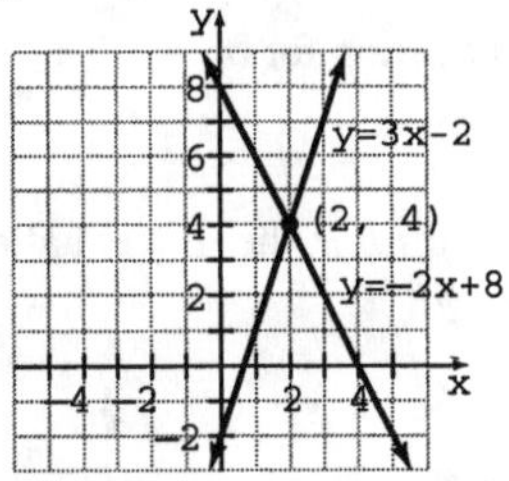

The lines intersect and the point of intersection is (2, 4).

6. Graph the equations $y = -x + 6$ and $y = 2x + 3$.

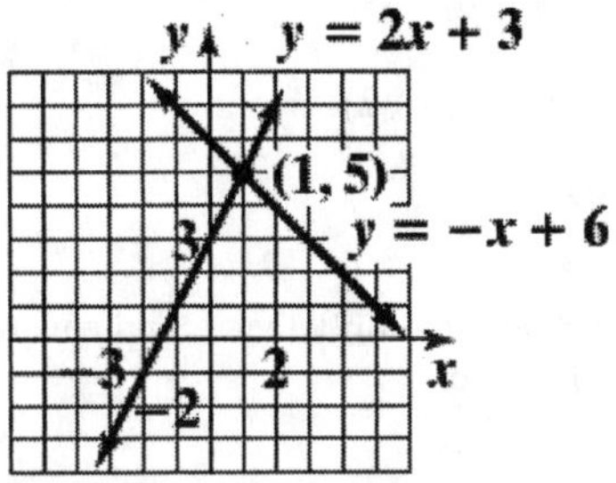

The lines intersect and the point of intersection is (1, 5).

7. $y=4x-3$

$y=5x-4$

Substitute $4x-3$ for y in the second equation.

$$4x-3=5x-4$$
$$-x-3=-4$$
$$-x=-1$$
$$x=1$$

Substitute 1 for x in the first equation.

$$y=4x-3$$
$$y=4(1)-3$$
$$y=4-3$$
$$y=1$$

The solution is (1, 1).

8. $4a+7b=2$

$5a+b=-13$

Solve the second equation for b.

$5a+b=-13$

$b=-5a-13$

Substitute $-5a-13$ for b in the first equation.

$4a+7(-5a-13)=2$

$4a-35a-91=2$

$-31a-91=2$

$-31a=93$

$a=-3$

Substitute -3 for a in the equation $b=-5a-13$.

$b=-5(-3)-13$

$b=2$

The solution is (–3, 2).

9. $8x+3y=8$

$6x+y=1$

To eliminate y, multiply the second equation by –3 and then add.

$8x+3y=8$

$-3[6x+y=1]$

gives

$$\begin{array}{r} 8x+3y=8 \\ \underline{-18x-3y=-3} \\ -10x \quad =5 \end{array}$$

$$x=\frac{5}{-10}=-\frac{1}{2}$$

Substitute $-\frac{1}{2}$ for x in the second equation.

$6\left(-\frac{1}{2}\right)+y=1$

$-3+y=1$

$y=4$

The solution is $\left(-\frac{1}{2}, 4\right)$.

10. $0.3x=0.2y+0.4$

$-1.2x+0.8y=-1.6$

Write the system in standard form.

$0.3x-0.2y=0.4$

$-1.2x+0.8y=-1.6$

To eliminate x, multiply the first equation by 4 and then add.

$4[0.3x-0.2y=0.4]$

$-1.2x+0.8y=-1.6$

gives

$$\begin{array}{r} 1.2x-0.8y=\ \ 1.6 \\ \underline{-1.2x+0.8y=-1.6} \end{array}$$

Add: $0=0$ True

Since this is a true statement, there are an infinite number of solutions and this is a dependent system.

11. $\frac{3}{2}a+b=6$

$a-\frac{5}{2}b=-4$

To clear fractions, multiply both equations by 2.

$2\left[\frac{3}{2}a+b=6\right]$

$2\left[a-\frac{5}{2}b=-4\right]$

gives

$3a+2b=12$

$2a-5b=-8$

Now, to eliminate b, multiply the first equation by 5 and the second equation by 2 and then add.

$5[3a+2b=12]$

$2[2a-5b=-8]$

gives

$$\begin{array}{r} 15a+10b=\ \ 60 \\ \underline{4a-10b=-16} \end{array}$$

Add: $19a \quad =44$

$$a=\frac{44}{19}$$

Substitute $\frac{44}{19}$ for a in the first equation.

$\frac{3}{2}a+b=6$

$\frac{3}{2}\left(\frac{44}{19}\right)+b=6$

$\frac{66}{19}+b=6$

$b=6-\frac{66}{19}$

$b=\frac{114}{19}-\frac{66}{19}$

$b=\frac{48}{19}$

The solution is $\left(\frac{44}{19}, \frac{48}{19}\right)$.

12. $x+y+z=2\,(1)$

$-2x-y+z=1\,(2)$

$x-2y-z=1\,(3)$

To eliminate z between equations (1) and (3) simply add.

$$\begin{array}{r} x+\ y+z=2 \\ \underline{x-2y-z=1} \end{array}$$

Add: $2x-\ y\quad =3\quad (4)$

To eliminate z between equations (2) and (3) simply add.

$$\begin{array}{r} -2x-\ y+z=1 \\ \underline{x-2y-z=1} \end{array}$$

Add: $-x-3y\quad =2\quad (5)$

Equations (4) and (5) are two equations in two unknowns.

$2x-y=3$

$-x-3y=2$

To eliminate x, multiply equation (5) by 2 and then add.

$2x-y=3$

$2[-x-3y=2]$

gives

$$\begin{array}{r} 2x-\ y=3 \\ \underline{-2x-6y=4} \end{array}$$

Add: $\quad -7y=7$

$$y=\frac{7}{-7}=-1$$

Substitute -1 for y in equation (4).

$2x-y=3$

$2x-(-1)=3$

$2x+1=3$

$2x=2$

$$x=\frac{2}{2}=1$$

Finally, substitute 1 for x and -1 for y in equation (1).

$x+y+z=2$

$1-1+z=2$

$0+z=2$

$z=2$

The solution is $(1, -1, 2)$.

13. $-2x+3y+7z=5$

$3x-2y+z=-2$

$x-6y+9z=-13$

The augmented matrix is $\left[\begin{array}{rrr|r} -2 & 3 & 7 & 5 \\ 3 & -2 & 1 & -2 \\ 1 & -6 & 9 & -13 \end{array}\right]$

14. $\left[\begin{array}{rrr|r} 6 & -2 & 4 & 4 \\ 4 & 3 & 5 & 6 \\ 2 & -1 & 4 & -3 \end{array}\right]$

$\left[\begin{array}{rrr|r} 6 & -2 & 4 & 4 \\ 0 & 5 & -3 & 12 \\ 2 & -1 & 4 & -3 \end{array}\right] -2R_3+R_2$

15. $2x+7y=1$

$3x+5y=7$

$\left[\begin{array}{rr|r} 2 & 7 & 1 \\ 3 & 5 & 7 \end{array}\right]$

$\left[\begin{array}{rr|r} 1 & \frac{7}{2} & \frac{1}{2} \\ 3 & 5 & 7 \end{array}\right] \frac{1}{2}R_1$

$\left[\begin{array}{rr|r} 1 & \frac{7}{2} & \frac{1}{2} \\ 0 & -\frac{11}{2} & \frac{11}{2} \end{array}\right] -3R_1+R_2$

$\left[\begin{array}{rr|r} 1 & \frac{7}{2} & \frac{1}{2} \\ 0 & 1 & -1 \end{array}\right] -\frac{2}{11}R_2$

The system is

$$x+\frac{7}{2}y=\frac{1}{2}$$

$$y=-1$$

Substitute -1 for y in the first equation.

$$x+\frac{7}{2}(-1)=\frac{1}{2}$$

$$x-\frac{7}{2}=\frac{1}{2}$$

$$x=\frac{8}{2}=4$$

The solution is $(4, -1)$.

16. $x-2y+z=7$
$-2x-y-z=-7$
$4x+5y-2z=3$

$$\left[\begin{array}{ccc|c}1&-2&1&7\\-2&-1&-1&-7\\4&5&-2&3\end{array}\right]$$

$$\left[\begin{array}{ccc|c}1&-2&1&7\\0&-5&1&7\\4&5&-2&3\end{array}\right]2R_1+R_2$$

$$\left[\begin{array}{ccc|c}1&-2&1&7\\0&-5&1&7\\0&13&-6&-25\end{array}\right]-4R_1+R_3$$

$$\left[\begin{array}{ccc|c}1&-2&1&7\\0&1&-\frac{1}{5}&-\frac{7}{5}\\0&13&-6&-25\end{array}\right]-\tfrac{1}{5}R_2$$

$$\left[\begin{array}{ccc|c}1&-2&1&7\\0&1&-\frac{1}{5}&-\frac{7}{5}\\0&0&-\frac{17}{5}&-\frac{34}{5}\end{array}\right]-13R_2+R_3$$

$$\left[\begin{array}{ccc|c}1&-2&1&7\\0&1&-\frac{1}{5}&-\frac{7}{5}\\0&0&1&2\end{array}\right]-\tfrac{5}{17}R_3$$

The system is
$x-2y+z=7$
$y-\frac{1}{5}z=-\frac{7}{5}$
$z=2$

Substitute 2 for *z* in the second equation.
$y-\frac{1}{5}z=-\frac{7}{5}$
$y-\frac{1}{5}(2)=-\frac{7}{5}$
$y-\frac{2}{5}=-\frac{7}{5}$
$y=-1$

Substitute –1 for *y* and 2 for *z* in the first equation.
$x-2y+z=7$
$x-2(-1)+2=7$
$x+2+2=7$
$x+4=7$
$x=3$

The solution is (3, –1, 2).

17. $\begin{vmatrix}3&-1\\5&-2\end{vmatrix}=(3)(-2)-(5)(-1)$
$=-6-(-5)$
$=-6+5$
$=-1$

18. $\begin{vmatrix}8&2&-1\\3&0&5\\6&-3&4\end{vmatrix}=8\begin{vmatrix}0&5\\-3&4\end{vmatrix}-3\begin{vmatrix}2&-1\\-3&4\end{vmatrix}+6\begin{vmatrix}2&-1\\0&5\end{vmatrix}$
$=8(0+15)-3(8-3)+6(10-0)$
$=8(15)-3(5)+6(10)$
$=120-15+60$
$=165$

19. $4x+3y=-6$
$-2x+5y=16$

To solve, first calculate D, D_x, and D_y.

$D=\begin{vmatrix}4&3\\-2&5\end{vmatrix}$
$=(4)(5)-(-2)(3)$
$=20+6$
$=26$

$D_x=\begin{vmatrix}-6&3\\16&5\end{vmatrix}$
$=(-6)(5)-(16)(3)$
$=-30-48$
$=-78$

$D_y=\begin{vmatrix}4&-6\\-2&16\end{vmatrix}$
$=(4)(16)-(-2)(-6)$
$=64-12$
$=52$

$x=\frac{D_x}{D}=\frac{-78}{26}=-3$ and $y=\frac{D_y}{D}=\frac{52}{26}=2$.

The solution is (–3, 2).

20. $2r-4s+3t=-1$
$-3r+5s-4t=0$
$-2r+s-3t=-2$

To solve, first calculate D, D_r, D_s, and D_t.

$$D = \begin{vmatrix} 2 & -4 & 3 \\ -3 & 5 & -4 \\ -2 & 1 & -3 \end{vmatrix}$$
$$= 2\begin{vmatrix} 5 & -4 \\ 1 & -3 \end{vmatrix} - (-4)\begin{vmatrix} -3 & -4 \\ -2 & -3 \end{vmatrix} + 3\begin{vmatrix} -3 & 5 \\ -2 & 1 \end{vmatrix}$$
$$= 2(-15+4)+4(9-8)+3(-3+10)$$
$$= 2(-11)+4(1)+3(7)$$
$$= -22+4+21$$
$$= 3$$

$$D_r = \begin{vmatrix} -1 & -4 & 3 \\ 0 & 5 & -4 \\ -2 & 1 & -3 \end{vmatrix}$$
$$= -1\begin{vmatrix} 5 & -4 \\ 1 & -3 \end{vmatrix} - (-4)\begin{vmatrix} 0 & -4 \\ -2 & -3 \end{vmatrix} + 3\begin{vmatrix} 0 & 5 \\ -2 & 1 \end{vmatrix}$$
$$= -1(-15+4)+4(0-8)+3(0+10)$$
$$= -1(-11)+4(-8)+3(10)$$
$$= 11-32+30$$
$$= 9$$

$$D_s = \begin{vmatrix} 2 & -1 & 3 \\ -3 & 0 & -4 \\ -2 & -2 & -3 \end{vmatrix}$$
$$= 2\begin{vmatrix} 0 & -4 \\ -2 & -3 \end{vmatrix} - (-1)\begin{vmatrix} -3 & -4 \\ -2 & -3 \end{vmatrix} + 3\begin{vmatrix} -3 & 0 \\ -2 & -2 \end{vmatrix}$$
$$= 2(0-8)+1(9-8)+3(6-0)$$
$$= 2(-8)+1(1)+3(6)$$
$$= -16+1+18$$
$$= 3$$

$$D_t = \begin{vmatrix} 2 & -4 & -1 \\ -3 & 5 & 0 \\ -2 & 1 & -2 \end{vmatrix}$$
$$= 2\begin{vmatrix} 5 & 0 \\ 1 & -2 \end{vmatrix} - (-4)\begin{vmatrix} -3 & 0 \\ -2 & -2 \end{vmatrix} + (-1)\begin{vmatrix} -3 & 5 \\ -2 & 1 \end{vmatrix}$$
$$= 2(-10-0)+4(6-0)-1(-3+10)$$
$$= 2(-10)+4(6)-1(7)$$
$$= -20+24-7$$
$$= -3$$

$$r = \frac{D_r}{D} = \frac{9}{3} = 3,\ s = \frac{D_s}{D} = \frac{3}{3} = 1,$$
$$t = \frac{D_t}{D} = \frac{-3}{3} = -1$$

The solution is (3, 1, –1).

21. Let x = the number of pounds of sunflower seeds and y = the number of pounds of bird seed.

$$x + y = 20$$
$$0.49x + 0.89y = 0.73(20)$$

Solve the first equation for y.

$$x + y = 20$$
$$y = -x + 20$$

Substitute $-x + 20$ for y in the second equation.

$$0.49x + 0.89(-x+20) = 0.73(20)$$
$$0.49x - 0.89x + 17.8 = 14.6$$
$$-0.4x = -3.2$$
$$x = 8$$

Substitute 8 for x in the equation for y.

$$y = -x + 20$$
$$y = -(8) + 20$$
$$y = 12$$

Thus, 8 pounds of sunflower seeds should be mixed with 12 pounds of the gourmet bird seed to obtain the desired mixture.

22. Let x = amount of 6% solution
y = amount of 15% solution

$$x + y = 10$$
$$0.06x + 0.15y = 0.09(10)$$

The system can be written as

$$x + y = 10$$
$$6x + 15y = 90$$

Solve the first equation for y.

$y = 10 - x$

Substitute $10 - x$ for y in the second equation.

$$6x + 15y = 90$$
$$6x + 15(10 - x) = 90$$
$$6x + 150 - 15x = 90$$
$$-9x = -60$$
$$x = \frac{-60}{-9} = \frac{20}{3} = 6\frac{2}{3}$$

Substitute $6\frac{2}{3}$ for x into $y = 10 - x$

$$y = 10 - 6\frac{2}{3}$$
$$y = 3\frac{1}{3}$$

She should mix $6\frac{2}{3}$ liters of 6% solution and $3\frac{1}{3}$ liters of 15% solution.

23. Let x = smallest number
y = remaining number
z = largest number

$$x+y+z=29$$
$$z=4x$$
$$y=2x+1$$

Substitute $2x + 1$ for y and $4x$ for z in the first equation.

$$x+y+z=29$$
$$x+2x+1+4x=29$$
$$7x=28$$
$$x=4$$

Substitute 4 for x in the third equation.

$$y=2(4)+1$$
$$y=9$$

Substitute 4 for x in the second equation.

$$z=4(4)$$
$$z=16$$

The three numbers are 4, 9, and 16.

24. $3x+2y<9$

$-2x+5y\le 10$

For $3x + 2y < 9$, graph the line $3x + 2y = 9$ using a dashed line. For the check point, select (0, 0).

$$3x+2y<9$$
$$3(0)+2(0)<9$$
$$0<9 \quad \text{True}$$

Since this is a true statement, shade the region which contains the point (0, 0). this is the region "below" the line.

For $-2x + 5y \le 10$, graph the line $-2x + 5y = 10$ using a solid line. For the check point, select (0, 0).

$$-2x+5y\le 10$$
$$-2(0)+5(0)\le 10$$
$$0\le 10 \quad \text{True}$$

Since this is a true statement, shade the region which contains the point (0, 0). this is the region "below" the line. To obtain the final region, take the intersection of the above two regions.

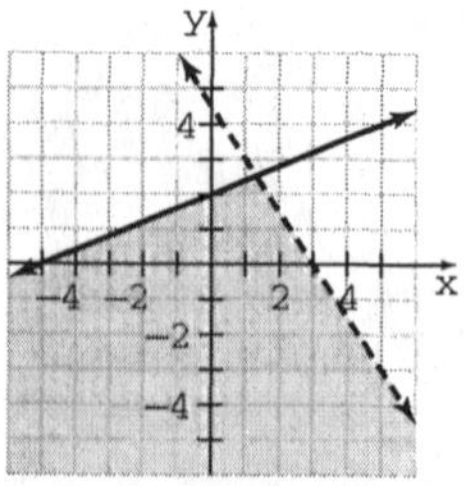

25. $|x|>3$

$|y|\le 1$

For $|x|>3$, the graph is the region to the left of the dashed line $x = -3$ along with the region to the right of the dashed line $x = 3$.

For $|y|\le 1$, the graph is the region between the solid lines $y = -1$ and $y = 1$. To obtain the final region, take the intersection of these regions.

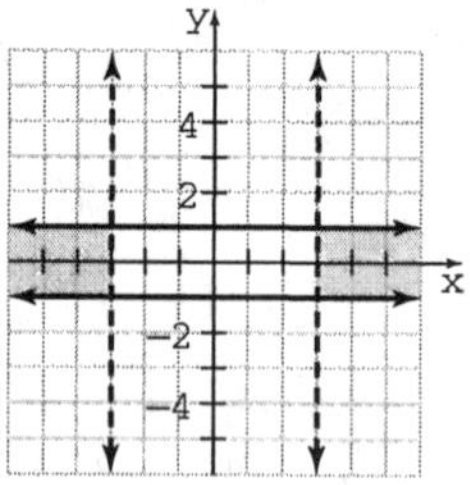

Chapter 4 Cumulative Review Test

1. $48\div\left\{4\left[3+\left(\frac{5+10}{5}\right)^2\right]-32\right\}$

$$48\div\left\{4\left[3+\left(\frac{15}{5}\right)^2\right]-32\right\}$$
$$48\div\left\{4\left[3+(3)^2\right]-32\right\}$$
$$48\div\{4[3+9]-32\}$$
$$48\div\{4[12]-32\}$$
$$48\div\{48-32\}$$
$$48\div\{16\}$$
$$3$$

2. a. 9 and 1 are natural numbers.

b. $\frac{1}{2}$, –4, 9, 0, –4.63, and 1 are rational numbers.

c. $\frac{1}{2}$, –4, 9, 0, $\sqrt{3}$, –4.63, and 1 are real numbers.

3. $-|-8|,\ -1,\ \frac{5}{8},\ \frac{3}{4},\ |-4|,\ |-12|$

4. $-[3-2(x-4)]=3(x-6)$
$-[3-2x+8]=3(x-6)$
$-(-2x+11)=3(x-6)$
$2x-11=3x-18$
$-11=x-18$
$7=x$
$x=7$

5. $\frac{2}{3}x-\frac{5}{6}=2$
$6\left(\frac{2}{3}x-\frac{5}{6}\right)=6(2)$
$4x-5=12$
$4x=17$
$x=\frac{17}{4}$

6. $|2x-3|-5=4$
$|2x-3|=9$
$2x-3=-9$ or $2x-3=9$
$2x=-6$ $\quad$ $2x=12$
$x=-3$ $\quad$ $x=6$
The solution set is $\{-3,6\}$.

7. $M=\frac{1}{2}(a+x)$
$2[M]=2\left[\frac{1}{2}(a+x)\right]$
$2M=a+x$
$2M-a=x$ or $x=2M-a$

8. $0<\frac{3x-2}{4}\le 8$
$4(0)<4\left(\frac{3x-2}{4}\right)\le 4(8)$
$0<3x-2\le 32$
$0+2<3x-2+2\le 32+2$
$2<3x\le 34$
$\frac{2}{3}<\frac{3x}{3}\le\frac{34}{3}$
$\frac{2}{3}<x\le\frac{34}{3}$
The solution set is $\left\{x\,\middle|\,\frac{2}{3}<x\le\frac{34}{3}\right\}$.

9. $\left(\frac{3x^2y^{-2}}{y^3}\right)^{-2}=\left(\frac{y^3}{3x^2y^{-2}}\right)^2$
$=\left(\frac{y^5}{3x^2}\right)^2$
$=\frac{y^{5\cdot 2}}{3^2x^{2\cdot 2}}$
$=\frac{y^{10}}{9x^4}$

10. $2y=3x-8$
$y=\frac{3x-8}{2}$
$y=\frac{3}{2}x-4$

x	y
0	$y=\frac{3}{2}(0)-4=0-4=-4$
2	$y=\frac{3}{2}(2)-4=3-4=-1$
4	$y=\frac{3}{2}(4)-4=6-4=2$

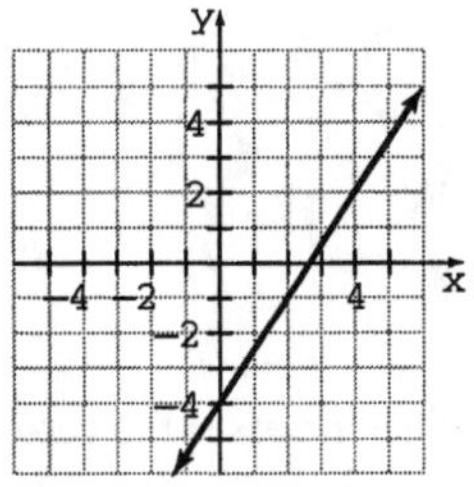

11. $2x-3y=8$
To find the slope, solve for y.
$2x-3y=8$
$-3y=-2x+8$
$y=\frac{-2x+8}{-3}$
$y=\frac{2}{3}x-\frac{8}{3}$

The slope of this line is $\frac{2}{3}$. Since the new line is parallel to this line, its slope is also $\frac{2}{3}$. Use the point-slope form with $m=\frac{2}{3}$ and

$(x_1, y_1) = (2, 3)$.

$$y - y_1 = m(x - x_1)$$
$$y - 3 = \frac{2}{3}(x - 2)$$
$$y - 3 = \frac{2}{3}x - \frac{4}{3}$$
$$y = \frac{2}{3}x - \frac{4}{3} + 3$$
$$y = \frac{2}{3}x - \frac{4}{3} + \frac{9}{3}$$
$$y = \frac{2}{3}x + \frac{5}{3}$$

12. $6x - 3y < 12$
Graph the line $6x - 3y = 12$ using a dashed line. For the check point, select (0, 0).

$$6x - 3y < 12$$
$$6(0) - 3(0) < 12$$
$$0 - 0 < 12$$
$$0 < 12 \quad \text{True}$$

Since this is a true statement, shade the region containing the point (0, 0).

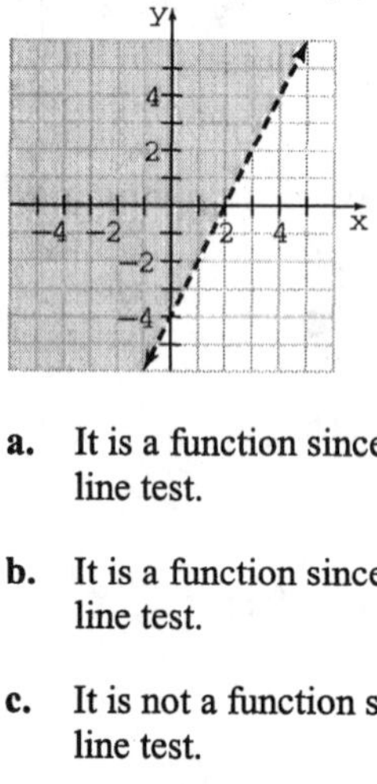

13. a. It is a function since it passes the vertical line test.

b. It is a function since it passes the vertical line test.

c. It is not a function since it fails the vertical line test.

14. a. $f(x) = \dfrac{x+3}{x^2-9}$

$$f(-4) = \frac{(-4)+3}{(-4)^2 - 9}$$
$$f(-4) = \frac{-1}{16-9}$$
$$f(-4) = -\frac{1}{7}$$

b. $f(h) = \dfrac{h+3}{h^2-9}$

c. $f(3) = \dfrac{(3)+3}{(3)^2-9}$

$$f(3) = \frac{6}{9-9}$$
$$f(3) = \frac{6}{0} \quad \text{undefined}$$

15. $3x + y = 6$
$y = 4x - 1$

Substitute $4x - 1$ for y in the first equation.

$$3x + y = 6$$
$$3x + 4x - 1 = 6$$
$$7x - 1 = 6$$
$$7x = 7$$
$$x = \frac{7}{7} = 1$$

Substitute 1 for x in the second equation.

$$y = 4x - 1$$
$$y = 4(1) - 1$$
$$y = 4 - 1$$
$$y = 3$$

The solution is (1, 3).

16. $2p + 3q = 11$
$-3p - 5q = -16$

To eliminate p, multiply the first equation by 3 and the second equation by 2 and then add.

$$3[2p + 3q = 11]$$
$$2[-3p - 5q = -16]$$

gives

$$6p + 9q = 33$$
$$-6p - 10q = -32$$

Add: $-q = 1$

$$q = -1$$

Substitute -1 for q in the first equation.

$$2p + 3(-1) = 11$$
$$2p - 3 = 11$$
$$2p = 14$$
$$p = 7$$

The solution is (7, –1).

17. $x-2y=0$ (1)
$2x+z=7$ (2)
$y-2z=-5$ (3)
To eliminate z between equations (2) and (3), multiply equation (2) by 2 and then add.
$2[2x+z=7]$
$y-2z=-5$
gives
$$\begin{array}{l} 4x \quad +2z=14 \\ \underline{\quad y-2z=-5} \end{array}$$
Add: $4x+y \quad =9$ (4)
Equations (4) and (1) are two equations in two unknowns:
$x-2y=0$
$4x+y=9$
To eliminate y, multiply equation (4) by 2 and then add.
$x-2y=0$
$2[4x+y=9]$
gives
$$\begin{array}{l} x-2y=0 \\ \underline{8x+2y=18} \end{array}$$
Add: $9x \quad =18$
$$x=\frac{18}{9}=2$$
Substitute 2 for x in equation (4).
$4x+y=9$
$4(2)+y=9$
$8+y=9$
$y=1$
Finally, substitute 2 for x in equation (2).
$2x+z=7$
$2(2)+z=7$
$4+z=7$
$z=3$
The solution is (2, 1, 3).

18. Let x be the measure of the smallest angle. Then $9x$ is the measure of the largest angle and $x + 70$ is the measure of the remaining angle. The sum of the measures of the three angles is 180°.
$x+(x+70)+9x=180$
$11x+70=180$
$11x=110$
$x=10$
$x + 70 = 10 + 70 = 80$; $9x = 9(10) = 90$
The three angles are 10°, 80°, and 90°.

19. Let t be the time for Judy to catch up to Dawn.

	rate	time	distance
Judy	6	t	$6t$
Mark	4	$t+\frac{1}{2}$	$4\left(t+\frac{1}{2}\right)$

$$6t=4\left(t+\frac{1}{2}\right)$$
$6t=4t+2$
$2t=2$
$$t=\frac{2}{2}=1$$
It takes 1 hour for Judy to catch up to Mark.

20. Let x = the number of \$20 tickets sold and y = the number of \$16 tickets sold. The system is
$x+y=1000$
$20x+16y=18,400$
Solve the first equation for y.
$x+y=1000$
$y=-x+1000$
Substitute $-x+1000$ for y in the second equation.
$20x+16(-x+1,000)=18,400$
$20x-16x+16,000=18,400$
$4x+16,000=18,400$
$4x=2400$
$$x=\frac{2400}{4}=600$$
Substitute 600 for x in the equation for y.
$y=-x+1000$
$y=-(600)+1000$
$y=400$
Thus, 600 \$20 tickets and 400 \$16 tickets were sold for the concert.

Chapter 5

Exercise Set 5.1

1. The terms are the parts that are added or subtracted.

3. A polynomial is a finite sum of terms in which all variables have whole number exponents and no variable appears in a denominator.

5. The leading coefficient is the coefficient of the leading term.

7. **a.** The degree of a polynomial is the same as that of the highest-degree term.

   b. The degree of $-4x^4+6x^3y^4+z^5$ is the same as the degree of the highest-degree term $(6x^3y^4)$. The degree is 3 + 4 = 7.

9. **a.** A polynomial is linear if its degree is 0 or 1.

   b. Answers will vary. One example is $x + 4$

11. **a.** A polynomial is cubic if it has degree 3 and is in one variable.

    b. Answers will vary. One example is x^3+x-4.

13. Answers will vary. One example is x^5+x+1.

15. Since the polynomial –6 has only one term, it is a monomial.

17. Since the polynomial $7z$ has only term, it is a monomial.

19. Since $5z^{-3}$ has a negative exponent, it is not a polynomial.

21. Since $3x^{1/2}+2xy$ has a fractional exponent, it is not a polynomial.

23. $-5+2x-x^2=-x^2+2x-5$; degree 2

25. $9y^2+3xy+10x^2=10x^2+3xy+9y^2$; degree 2

27. $-2x^4+5x^2-4$ is already in descending order; degree 4

29. $x^4+3x^6-2x-13=3x^6+x^4-2x-13$

    a. $3x^6$: The degree of the polynomial is 6.

    b. $3x^6$: The leading coefficient is 3.

31. $4x^2y^3+6xy^4+9xy^5=9xy^5+6xy^4+4x^2y^3$

    a. $9xy^5$ or $9x^1y^5$: The degree of the polynomial is 1 + 5 = 6.

    b. $9xy^5$: The leading coefficient is 9.

33. $-\frac{1}{3}m^4n^5p^8+\frac{3}{5}m^3p^6-\frac{5}{9}n^4p^6q$

    a. $-\frac{1}{3}m^4n^5p^8$: The degree of the polynomial is 4 + 5 + 8 = 17.

    b. $-\frac{1}{3}m^4n^5p^8$: The leading coefficient is $-\frac{1}{3}$.

35. $P(x)=x^2-6x+5$

    $P(2)=2^2-6(2)+5=4-12+5=-3$

37. $P(x)=2x^2-3x-6$

    $P\left(\frac{1}{2}\right)=2\left(\frac{1}{2}\right)^2-3\left(\frac{1}{2}\right)-6=\frac{1}{2}-\frac{3}{2}-6=-7$

39. $P(x)=0.2x^3+1.6x^2-2.3$

    $$\begin{aligned}P(0.4)&=0.2(0.4)^3+1.6(0.4)^2-2.3\\&=0.0128+0.256-2.3\\&=-2.0312\end{aligned}$$

41. $$\begin{aligned}(x^2+3x-1)+(6x-5)&=x^2+3x-1+6x-5\\&=x^2+3x+6x-1-5\\&=x^2+9x-6\end{aligned}$$

43. $$\begin{aligned}(x^2-8x+11)-(5x+9)&=x^2-8x+11-5x-9\\&=x^2-8x-5x+11-9\\&=x^2-13x+2\end{aligned}$$

45. $$\begin{aligned}&(4y^2+9y-1)-(2y^2+10)\\&=4y^2+9y-1-2y^2-10\\&=4y^2-2y^2+9y-1-10\\&=2y^2+9y-11\end{aligned}$$

47. $\left(-\frac{5}{9}a+6\right)+\left(-\frac{2}{3}a^2-\frac{1}{4}a-1\right)$
$=-\frac{5}{9}a+6-\frac{2}{3}a^2-\frac{1}{4}a-1$
$=-\frac{2}{3}a^2-\frac{5}{9}a-\frac{1}{4}a+6-1$
$=-\frac{2}{3}a^2-\frac{20}{36}a-\frac{9}{36}a+6-1$
$=-\frac{2}{3}a^2-\frac{29}{36}a+5$

49. $(1.4x^2+1.6x-8.3)-(4.9x^2+3.7x+11.3)$
$=1.4x^2+1.6x-8.3-4.9x^2-3.7x-11.3$
$=1.4x^2-4.9x^2+1.6x-3.7x-8.3-11.3$
$=-3.5x^2-2.1x-19.6$

51. $\left(-\frac{1}{3}x^3+\frac{1}{4}x^2y+8xy^2\right)+\left(-x^3-\frac{1}{2}x^2y+xy^2\right)$
$=-\frac{1}{3}x^3+\frac{1}{4}x^2y+8xy^2-x^3-\frac{1}{2}x^2y+xy^2$
$=-\frac{1}{3}x^3-x^3+\frac{1}{4}x^2y-\frac{1}{2}x^2y+8xy^2+xy^2$
$=-\frac{4}{3}x^3-\frac{1}{4}x^2y+9xy^2$

53. $(3a-6b+5c)-(-2a+4b-8c)$
$=3a-6b+5c+2a-4b+8c$
$=3a+2a-6b-4b+5c+8c$
$=5a-10b+13c$

55. $(3a^2b-6ab+5b^2)-(4ab-6b^2-5a^2b)$
$=3a^2b-6ab+5b^2-4ab+6b^2+5a^2b$
$=3a^2b+5a^2b-6ab-4ab+5b^2+6b^2$
$=8a^2b-10ab+11b^2$

57. $(8r^2-5t^2+2rt)+(-6rt+2t^2-r^2)$
$=8r^2-5t^2+2rt-6rt+2t^2-r^2$
$=8r^2-r^2+2rt-6rt-5t^2+2t^2$
$=7r^2-4rt-3t^2$

59. $6x^2-5x-\left[3x-(4x^2-9)\right]$
$=6x^2-5x-\left[3x-4x^2+9\right]$
$=6x^2-5x-3x+4x^2-9$
$=6x^2+4x^2-5x-3x-9$
$=10x^2-8x-9$

61. $5w-6w^2-\left[(3w-2w^2)-(4w+w^2)\right]$
$=5w-6w^2-\left[3w-2w^2-4w-w^2\right]$
$=5w-6w^2-\left[3w-4w-2w^2-w^2\right]$
$=5w-6w^2-(-w-3w^2)$
$=5w-6w^2+w+3w^2$
$=-6w^2+3w^2+5w+w$
$=-3w^2+6w$

63. $(7x+8)-(4x-11)=7x+8-4x+11$
$=7x-4x+8+11$
$=3x+19$

65. $(-2x^2+4x-12)+(-x^2-2x)$
$=-2x^2+4x-12-x^2-2x$
$=-2x^2-x^2+4x-2x-12$
$=-3x^2+2x-12$

67. $(-5.2a^2-9.6a)-(0.2a^2-3.9a+26.4)$
$=-5.2a^2-9.6a-0.2a^2+3.9a-26.4$
$=-5.2a^2-0.2a^2-9.6a+3.9a-26.4$
$=-5.4a^2-5.7a-26.4$

69. $\left(-\frac{1}{2}x^2y+xy^2+\frac{3}{5}\right)-\left(5x^2y+\frac{5}{9}\right)$
$=-\frac{1}{2}x^2y+xy^2+\frac{3}{5}-5x^2y-\frac{5}{9}$
$=-\frac{1}{2}x^2y-5x^2y+xy^2+\frac{3}{5}-\frac{5}{9}$
$=-\frac{11}{2}x^2y+xy^2+\frac{2}{45}$

71. $(3x^{2r}-7x^r+1)+(2x^{2r}-3x^r+2)$
$=3x^{2r}-7x^r+1+2x^{2r}-3x^r+2$
$=3x^{2r}+2x^{2r}-7x^r-3x^r+1+2$
$=5x^{2r}-10x^r+3$

73. $\left(x^{2s}-8x^s+6\right)-\left(2x^{2s}-4x^s-13\right)$
$=x^{2s}-8x^s+6-2x^{2s}+4x^s+13$
$=x^{2s}-2x^{2s}-8x^s+4x^s+6+13$
$=-x^{2s}-4x^s+19$

75. $\left(7b^{4n}-5b^{2n}+1\right)-\left(3b^{3n}-b^{2n}\right)$
$=7b^{4n}-5b^{2n}+1-3b^{3n}+b^{2n}$
$=7b^{4n}-3b^{3n}-5b^{2n}+b^{2n}+1$
$=7b^{4n}-3b^{3n}-4b^{2n}+1$

77. The perimeter of a square is $P=4s$.
$P=4\left(x^2+2x+6\right)=4x^2+8x+24$

79. The perimeter of a triangle is $P=s_1+s_2+s_3$.
$P=\left(x^2+3x+1\right)+\left(x^2+2x+5\right)+\left(x^2-x+13\right)$
$=3x^2+4x+19$

81. The perimeter is the sum of the sides.
$P=(5x-1)+\left(x^2+8\right)+(4x+1)+\left(x^2+3x+1\right)$
$=2x^2+12x+9$

83. No, the sum of two trinomials is not always a trinomial. Explanations will vary. One possible example: $(x^2+3x+4)+(2x^2-3x-5)=3x^2-1$.

85. No, the sum of two quadratic polynomials is not always a quadratic polynomial. Explanations will vary. One possible example:
$(2x^2+3x+4)+(-2x^2+x+1)=4x+5$.

87. $A(s)=s^2$
$A(12)=(12)^2=144$
The area is 144 square meters.

89. $A(r)=\pi r^2$
$A(6)=\pi(6)^2\ A=36\pi\approx 113.10$
The area is about 113.10 square inches.

91. $h=P(t)=-16t^2+1250$
$h=P(6)=-16(6)^2+1250=674$
The object is 674 feet from the ground.

93. $c(n)=\frac{1}{2}\left(n^2-n\right)$
$c(15)=\frac{1}{2}(15^2-15)=\frac{1}{2}(225-15)=\frac{1}{2}(210)=105$
There are 105 committees possible.

95. $A(t)=650+24t$

a. Note 2007 is represented by $t=1$:
$A(1)=650+24(1)=650+24=674$
In 2007, Jorge will have \$674 in the account.

b. Note 2021 is represented by $t=15$:
$A(15)=650+24(15)=650+360=1010$
In 2021, Jorge will have \$1010 in the account.

97. a. $R(x)=2x^2-60x$, $C(x)=8050-420x$
$P(x)=R(x)-C(x)$
$P(x)=\left(2x^2-60x\right)-\left(8050-420x\right)$
$P(x)=2x^2-60x0-8050+420x$
$P(x)=2x^2+360x-8050$

b. $P(100)=2(100)^2+360(100)-8050$
$=20,000+36,000-8050$
$=47,950$
The profit is \$47,950.

99. $y=x^2+3x-4$ is graph (c).
The coefficient of the leading term is positive, so the graph opens up. The y-intercept is (0,–4).

101. $y=-x^3+2x-6$ is graph (c).
The coefficient of the leading term is negative, so that eliminates graphs (a) and (b).

103. $E(t)=7t^2-7.8t+81.2$

a. Note 2004 is represented by $t=3$:
$E(3)=7(3)^2-7.8(3)+81.2$
$=63-23.4+81.2$
$=120.8$
According to the function, the expenditure by oil companies in 2004 was \$120.8 billion.

b. Yes, the graph indicates that the expenditure was about \$120 billion, which supports the answer from part (a).

c. Note 2007 is represented by $t=6$:
$E(6)=7(6)^2-7.8(6)+81.2$
$=252-46.8+81.2$
$=286.4$
According to the function, the expenditure by oil companies in 2007 would be \$286.4 billion.

105. $C(t) = 0.31t^2 + 0.59t + 9.61$
$t = 2012 - 1997 = 15$
$C(15) = 0.31(15)^2 + 0.59(15) + 9.61$
$= 69.75 + 8.85 + 9.61$
$= 88.21$ thousand or $88,210$
The cost in 2012 is $88,210.

107. a. $y_1 = x^3;\ y_2 = x^3 - 3x^2 - 3$

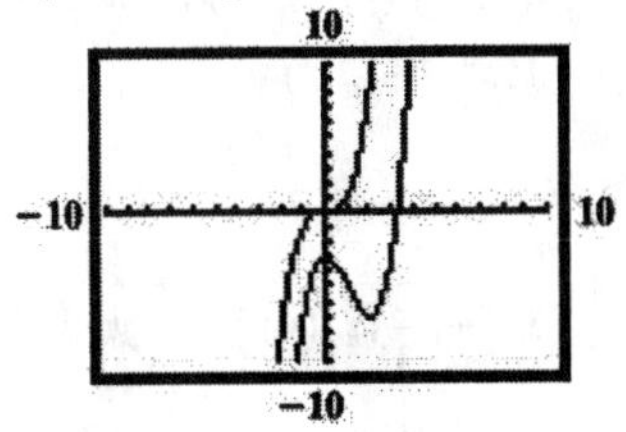

b. increase

c. Answers will vary.

d. decrease

e. Answers will vary.

109. $y = -x^4 + 3x^3 - 5$ is graph (b).
The leading coefficient is negative, therefore eliminating graph (c). The y-intercept is (0,–5).

113. $\sqrt[4]{81} = 3$ since $3^4 = 81$

114.
$$1 = \frac{8}{5}x - \frac{1}{2}$$
$$10(1) = 10\left(\frac{8}{5}x - \frac{1}{2}\right)$$
$$10 = 16x - 5$$
$$15 = 16x$$
$$\frac{15}{16} = x$$

115. Let t be the number of hours for the two machines to produce 540 buckets.
$40t + 50t = 540$
$90t = 540$
$t = 6$
It will take the two machines 6 hours to produce 540 buckets.

116. Let $(x_1, y_1) = (10, -4)$ and $(x_2, y_2) = (-1, -2)$.
$$m = \frac{y_2 - y_1}{x_2 - x_1} = \frac{-2-(-4)}{-1-10} = \frac{-2+4}{-11} = \frac{2}{-11} = -\frac{2}{11}$$

117.
$-4s + 3t = 16$ (1)
$4t - 2u = 2$ (2)
$-s + 6u = -2$ (3)

To eliminate the u variable in equation (3), multiply equation (2) by 3 and add the result to equation (3):
$3[4t - 2u = 2]$
$-s + 6u = -2$
gives:
$12t - 6u = 6$
$-s + 6u = -2$
$-s + 12t = 4$ (4)

Equations (1) and (4) form a system of equations in s and t. Solve this system.
$-4s + 3t = 16$ (1)
$-s + 12t = 4$ (4)

Multiply equation (4) by –4 and add the result to equation (1).
$-4[-s + 12t = 4]$
$-4s + 3t = 16$
gives:
$4s - 48t = -16$
$-4s + 3t = 16$
$-45t = 0$
$t = 0$

Substitute 0 for t in equation (4) and solve for s.
$-s + 12(0) = 4$
$-s = 4$
$s = -4$

Substitute 0 for t in equation (2) and solve for u.
$4(0) - 2u = 2$
$-2u = 2$
$u = -1$

The solution is $(-4, 0, -1)$.

Exercise Set 5.2

1. Answers will vary.

3. a. Answers will vary.

b. $(4 + x)(x^2 - 6x + 3)$
$= 4(x^2 - 6x + 3) + x(x^2 - 6x + 3)$
$= 4x^2 - 24x + 12 + x^3 - 6x^2 + 3x$
$= x^3 - 2x^2 - 21x + 12$

5. a. Answers will vary.
   b. Answers will vary. One example is $(x+4)(x-4)$
   c. Answers will vary.
   d. Answers will vary. One example is x^2-16

7. Yes, the product of two first-degree polynomials will always be a second-degree polynomial. We obtain the degree of the product by adding the degrees of the factors.

9. $$\begin{aligned}(4xy)(6xy^4) &= 4\cdot 6\cdot x\cdot x\cdot y\cdot y^4\\ &= 24x^{1+1}y^{1+4}\\ &= 24x^2y^5\end{aligned}$$

11. $$\begin{aligned}\left(\frac{5}{9}x^2y^5\right)\left(\frac{1}{5}x^5y^3z^2\right) &= \frac{5}{9}\cdot\frac{1}{5}\cdot x^2\cdot x^5\cdot y^5\cdot y^3\cdot z^2\\ &= \frac{1}{9}x^{2+5}y^{5+3}z^2\\ &= \frac{1}{9}x^7y^8z^2\end{aligned}$$

13. $$\begin{aligned}&-3x^2y(-2x^4y^2+5xy^3+4)\\ &=(-3x^2y)(-2x^4y^2)+(-3x^2y)(5xy^3)+(-3x^2y)(4)\\ &=6x^6y^3-15x^3y^4-12x^2y\end{aligned}$$

15. $$\begin{aligned}&\frac{2}{3}yz(3x+4y-12y^2)\\ &=\left(\frac{2}{3}yz\right)(3x)+\left(\frac{2}{3}yz\right)(4y)+\left(\frac{2}{3}yz\right)(-12y^2)\\ &=2xyz+\frac{8}{3}y^2z-8y^3z\end{aligned}$$

---

17. $$\begin{aligned}0.3(2x^2-5x+11y) &= 0.3(2x^2)+0.3(-5x)+0.3(11y)\\ &=0.6x^2-1.5x+3.3y\end{aligned}$$

19. $$\begin{aligned}0.3a^5b^4(9.5a^6b-4.6a^4b^3+1.2ab^5) &= 0.3a^5b^4(9.5a^6b)+0.3a^5b^4(-4.6a^4b^3)+0.3a^5b^4(1.2ab^5)\\ &=(0.3)(9.5)a^{5+6}b^{4+1}+0.3(-4.6)a^{5+4}b^{4+3}+0.3(1.2)a^{5+1}b^{4+5}\\ &=2.85a^{11}b^5-1.38a^9b^7+0.36a^6b^9\end{aligned}$$

21. $$\begin{aligned}(4x-6)(3x-5) &= (4x)(3x)+(4x)(-5)+(-6)(3x)+(-6)(-5)\\ &=12x^2-20x-18x+30\\ &=12x^2-38x+30\end{aligned}$$

23. $$\begin{aligned}(4-x)(3+2x^2) &= (4)(3)+(4)(2x^2)+(-x)(3)+(-x)(2x^2)\\ &=12+8x^2-3x-2x^3\\ &=-2x^3+8x^2-3x+12\end{aligned}$$

25. $$\begin{aligned}\left(\frac{1}{2}x+2y\right)\left(2x-\frac{1}{3}y\right) &= \left(\frac{1}{2}x\right)(2x)+\left(\frac{1}{2}x\right)\left(-\frac{1}{3}y\right)+(2y)(2x)+(2y)\left(-\frac{1}{3}y\right)\\ &=x^2-\frac{1}{6}xy+4xy-\frac{2}{3}y^2\\ &=x^2+\frac{23}{6}xy-\frac{2}{3}y^2\end{aligned}$$

27. $$\begin{aligned}(0.3a+0.5b)(0.3a-0.5b) &= (0.3a)(0.3a)+(0.3a)(-0.5b)+(0.5b)(0.3a)+(0.5b)(-0.5b)\\ &=0.09a^2-0.15ab+0.15ab-0.25b^2\\ &=0.09a^2-0.25b^2\end{aligned}$$

29.
$$\begin{array}{rl}
x^2+3x+1 & \\
x-4 & \\ \hline
-4x^2-12x-4 & \leftarrow \text{Multiply top expression by } -4 \\
x^3+3x^2+x & \leftarrow \text{Multiply top expression by } x \\ \hline
x^3-x^2-11x-4 & \leftarrow \text{Add like terms}
\end{array}$$

31.
$$\begin{array}{rl}
2a^2-ab+2b^2 & \\
a-3b & \\ \hline
-6a^2b+3ab^2-6b^3 & \leftarrow \text{Multiply top expression by } a \\
2a^3-a^2b+2ab^2 & \leftarrow \text{Multiply top expression by } x \\ \hline
2a^3-7a^2b+5ab^2-6b^3 & \leftarrow \text{Add like terms}
\end{array}$$

33.
$$\begin{array}{rl}
x^3-x^2+3x+7 & \\
x+1 & \\ \hline
x^3-x^2+3x+7 & \leftarrow \text{Multiply top expression by } 1 \\
x^4-x^3+3x^2+7x & \leftarrow \text{Multiply top expression by } x \\ \hline
x^4 \quad +2x^2+10x+7 & \leftarrow \text{Add like terms}
\end{array}$$

35.
$$\begin{array}{rl}
5x^3+4x^2-6x+2 & \\
x+5 & \\ \hline
25x^3+20x^2-30x+10 & \leftarrow \text{Multiply top expression by } 5 \\
5x^4+4x^3-6x^2+2x & \leftarrow \text{Multiply top expression by } x \\ \hline
5x^4+29x^3+14x^2-28x+10 & \leftarrow \text{Add like terms}
\end{array}$$

37.
$$\begin{array}{rl}
3m^2-2m+4 & \\
m^2-3m-5 & \\ \hline
-15m^2+10m-20 & \leftarrow \text{Multiply top expression by } -5 \\
-9m^3+6m^2-12m & \leftarrow \text{Multiply top expression by } -3m \\
3m^4-2m^3+4m^2 & \leftarrow \text{Multiply top expression by } m^2 \\ \hline
3m^4-11m^3-5m^2-2m-20 & \leftarrow \text{Add like terms}
\end{array}$$

39.
$$\begin{aligned}
(2x-1)^3 &= (2x-1)(2x-1)(2x-1) \\
&= \left[(2x)^2-2(2x)(1)+1^2\right](2x-1) \\
&= (4x^2-4x+1)(2x-1)
\end{aligned}$$

To complete the solution, multiply vertically

$$\begin{array}{rl}
4x^2-4x+1 & \\
2x-1 & \\ \hline
-4x^2+4x-1 & \leftarrow \text{Multiply top expression by } -1 \\
8x^3-8x^2+2x & \leftarrow \text{Multiply top expression by } 2x \\ \hline
8x^3-12x^2+6x-1 & \leftarrow \text{Add like terms}
\end{array}$$

41.

$$\begin{array}{rl} 5r^2 - rs + 2s^2 & \\ 2r^2 - s^2 & \\ \hline -5r^2s^2 + rs^3 - 2s^4 & \leftarrow \text{Multiply top expression by } -s^2 \\ 10r^4 - 2r^3s + 4r^2s^2 \qquad\qquad & \leftarrow \text{Multiply top expression by } 2r^2 \\ \hline 10r^4 - 2r^3s - r^2s^2 + rs^3 - 2s^4 & \leftarrow \text{Add like terms} \end{array}$$

---

43. $(x+2)(x+2) = (x+2)^2$
$= (x)^2 + 2(x)(2) + (2)^2$
$= x^2 + 4x + 4$

45. $(2x-7)(2x-7) = (2x-7)^2$
$= (2x)^2 - 2(2x)(7) + (7)^2$
$= 4x^2 - 28x + 49$

47. $(4x-3y)^2 = (4x)^2 - 2(4x)(3y) + (3y)^2$
$= 16x^2 - 24xy + 9y^2$

49. $(5m^2 + 2n)(5m^2 - 2n) = (5m^2)^2 - (2n)^2$
$= 25m^4 - 4n^2$

51. $[y + (4-2x)]^2$
$= (y)^2 + 2(y)(4-2x) + (4-2x)^2$
$= y^2 + 2y(4-2x) + (4)^2 - 2(4)(2x) + (2x)^2$
$= y^2 + 8y - 4xy + 16 - 16x + 4x^2$

53. $[5x + (2y+1)]^2$
$= (5x)^2 + 2(5x)(2y+1) + (2y+1)^2$
$= 25x^2 + 10x(2y+1) + (2y)^2 + 2(2y)(1) + (1)^2$
$= 25x^2 + 20xy + 10x + 4y^2 + 4y + 1$

55. $[a + (b+4)][a - (b+4)]$
$= (a)^2 - (b+4)^2$
$= (a)^2 - [(b)^2 + 2(b)(4) + (4)^2]$
$= a^2 - (b^2 + 8b + 16)$
$= a^2 - b^2 - 8b - 16$

57. $2xy(x^2 + xy + 12y^2)$
$= 2xy(x^2) + 2xy(xy) + 2xy(12y^2)$
$= 2x^3y + 2x^2y^2 + 24xy^3$

59. $\frac{1}{2}xy^2(4x^2 + 3xy - 7y^4)$
$= \frac{1}{2}xy^2(4x^2) + \frac{1}{2}xy^2(3xy) + \frac{1}{2}xy^2(-7y^4)$
$= 2x^3y^2 + \frac{3}{2}x^2y^3 - \frac{7}{2}xy^6$

61. $-\frac{3}{5}xy^3z^2\left(-xy^2z^5 - 5xy + \frac{1}{6}xz^7\right)$
$= -\frac{3}{5}xy^3z^2(-xy^2z^5) - \frac{3}{5}xy^3z^2(-5xy) - \frac{3}{5}xy^3z^2\left(\frac{1}{9}xz^7\right)$
$= \frac{3}{5}x^2y^5z^7 + 3x^2y^4z^2 - \frac{1}{15}x^2y^3z^9$

63. $(3a+4)(7a-6)$
$= 3a(7a) + 3a(-6) + 4(7a) + 4(-6)$
$= 21a^2 - 18a + 28a - 24$
$= 21a^2 + 10a - 24$

65. $\left(8x + \frac{1}{5}\right)\left(8x - \frac{1}{5}\right) = (8x)^2 - \left(\frac{1}{5}\right)^2 = 64x^2 - \frac{1}{25}$

67. $\left(x - \frac{1}{2}y\right)^3 = \left(x - \frac{1}{2}y\right)\left[\left(x - \frac{1}{2}y\right)\left(x - \frac{1}{2}y\right)\right]$
$= \left(x - \frac{1}{2}y\right)\left[(x)^2 - 2(x)\left(\frac{1}{2}y\right) + \left(\frac{1}{2}y\right)^2\right]$
$= \left(x - \frac{1}{2}y\right)\left(x^2 - xy + \frac{1}{4}y^2\right)$
$= x(x^2) + x(-xy) + x\left(\frac{1}{4}y^2\right) - \frac{1}{2}y(x^2)$
$\quad - \frac{1}{2}y(-xy) - \frac{1}{2}y\left(\frac{1}{4}y^2\right)$
$= x^3 - x^2y + \frac{1}{4}xy^2 - \frac{1}{2}x^2y + \frac{1}{2}xy^2 - \frac{1}{8}y^3$
$= x^3 - \frac{3}{2}x^2y + \frac{3}{4}xy^2 - \frac{1}{8}y^3$

69. $(x+3)(2x^2+4x-3)$
$= x(2x^2)+x(4x)+x(-3)+3(2x^2)+3(4x)$
$+3(-3)$
$= 2x^3+4x^2-3x+6x^2+12x-9$
$= 2x^3+10x^2+9x-9$

71. $(2p-3q)(3p^2+4pq-2q^2)$
$= 2p(3p^2)+2p(4pq)+2p(-2q^2)-3q(3p^2)$
$-3q(4pq)-3q(-2q^2)$
$= 6p^3+8p^2q-4pq^2-9p^2q-12pq^2+6q^3$
$= 6p^3-p^2q-16pq^2+6q^3$

73. $[(3x+2)+y][(3x+2)-y]$
$= (3x+2)^2-y^2$
$= (3x)^2+2(3x)(2)+2^2-y^2$
$= 9x^2+12x+4-y^2$

75. $(a+b)(a-b)(a^2-b^2)$
$= (a^2-b^2)(a^2-b^2)$
$= (a^2)^2-2(a^2)(b^2)+(b^2)^2$
$= a^4-2a^2b^2+b^4$

77. $(x-4)(6+x)(2x-8)$
$= [x(6+x)-4(6+x)](2x-8)$
$= (6x+x^2-24-4x)(2x-8)$
$= (x^2+2x-24)(2x-8)$
$= x^2(2x-8)+2x(2x-8)-24(2x-8)$
$= 2x^3-8x^2+4x^2-16x-48x+192$
$= 2x^3-4x^2-64x+192$

79. a. $(f\cdot g)(x) = f(x)\cdot g(x)$
$= (x-5)(x+6)$
$= x^2+6x-5x-30$
$= x^2+x-30$

b. $(f\cdot g)(4) = (4)^2+(4)-30$
$= 16+4-30$
$= -10$

81. a. $(f\cdot g)(x) = f(x)\cdot g(x)$
$= (2x^2+6x-4)(5x+3)$
$= 2x^2(5x+3)+6x(5x+3)-4(5x+3)$
$= 10x^3+6x^2+30x^2+18x-20x-12$
$= 10x^3+36x^2-2x-12$

b. $(f\cdot g)(4) = 10(4)^3+36(4)^2-2(4)-12$
$= 640+576-8-12$
$= 1196$

83. a. $(f\cdot g)(x) = f(x)\cdot g(x)$
$= (-x^2+3x)(x^2+2)$
$= -x^4-2x^2+3x^3+6x$
$= -x^4+3x^3-2x^2+6x$

b. $(f\cdot g)(4) = -(4)^4+3(4)^3-2(4)^2+6(4)$
$= -256+192-32+24$
$= -72$

85. The area is the sum of the three component areas:
$(x)(x)+2(x)+3(x) = x^2+2x+3x = x^2+5x$

87. The total area is the sum of the areas of the large square and the small square: x^2+y^2

89. a. The sum of the areas of the four pieces is
$x(x)+4(x)+x(3)+4(3) = x^2+4x+3x+12$
$= x^2+7x+12$

b. length × width $= (x+5)(x+3)$
$= x^2+5x+3x+15$
$= x^2+8x+15$

91. length × width $= (6+x)(6-x)$
$= 6^2-x^2$
$= 36-x^2$

93. a. The area of the large rectangle is
$(2x+3)(x+4) = 2x^2+8x+3x+12$
$= 2x^2+11x+12$
The area of the smaller rectangle is
$(2x)(x) = 2x^2$
The area of the shaded portion is the area of the larger rectangle – area of the smaller rectangle.
shaded portion $= (2x^2+11x+12)-2x^2$
$= 11x+12$

b. Since the area of the shaded portions is 67 square inches, set $11x + 12$ equal to 67 and then solve:

$$\begin{aligned}11x+12&=67\\11x&=55\\x&=5\end{aligned}$$

Now, the dimensions of the larger rectangle are $2(5)+3=10+3=13$ inches by $5 + 4 = 9$ inches. The area is $(13)(9) = 117$ square inches. Also, the dimensions of the smaller rectangle are $2(5) = 10$ inches by 5 inches. The area is $(10)(5) = 50$ square inches.

95. $x^2-49=x^2-7^2=(x+7)(x-7)$

97. $x^2+12x+36=x^2+2(x)(6)+(6)^2$
$=(x+6)^2$ or $(x+6)(x+6)$

99. $a(x-n)^3=a(x-n)(x-n)(x-n)$

101. a. Answers will vary. One example is: Observe that the length is $a + b$ and the width is $a + b$. Since the area is the product of the length and the width, this gives area = $(a+b)(a+b)=(a+b)^2$

b. The sum of the area of the four pieces is

$$\begin{aligned}&(a)(a)+(a)(b)+(b)(a)+(b)(b)\\&=a^2+ab+ab+b^2\\&=a^2+2ab+b^2\end{aligned}$$

c. $(a+b)^2=a^2+2ab+b^2$

d. The are the same

103. $A=P\left(1+\dfrac{r}{n}\right)^{nt}$

a. $A=P\left(1+\dfrac{r}{1}\right)^{1\cdot t}=P(1+r)^t$

b. $A=1000(1+0.06)^2=1000(1.06)^2=1123.6$
The amount is \$1123.60.

105. $P(n)=n(n-1)$

a. $P(11)=11(11-1)=11(10)=110$ ways

b. $P(n)=n(n-1)$
$P(n)=n^2-n$

c. $P(11)=(11)^2-(11)=121-11=110$ ways

d. Yes, the results to parts (a) and (c) are the same because the formulas are equivalent.

107. $f(x)=x^2+3x+4$

$$\begin{aligned}f(a+b)&=(a+b)^2-3(a+b)+5\\&=a^2+2ab+b^2-3a-3b+5\end{aligned}$$

109.

$$\begin{aligned}3x^t\left(5x^{2t-1}+6x^{3t}\right)&=\left(3x^t\right)\left(5x^{2t-1}\right)+\left(3x^t\right)\left(6x^{3t}\right)\\&=15x^{t+(2t-1)}+18x^{t+3t}\\&=15x^{3t-1}+18x^{4t}\end{aligned}$$

111.

$$\begin{aligned}&\left(6x^m-5\right)\left(2x^{2m}-3\right)\\&=\left(6x^m\right)\left(2x^{2m}\right)+\left(6x^m\right)(-3)+(-5)\left(2x^{2m}\right)\\&\quad+(-5)(-3)\\&=12x^{m+2m}-18x^m-10x^{2m}+15\\&=12x^{3m}-18x^m-10x^{2m}+15\end{aligned}$$

113. $\left(y^{a-b}\right)^{a+b}=y^{(a-b)(a+b)}=y^{a^2-b^2}$

115. First, find

$$\begin{aligned}(x-3y)^2&=(x)^2-2(x)(3y)+(3y)^2\\&=x^2-6xy+9y^2\end{aligned}$$

Then,

$$(x-3y)^4=\left[(x-3y)^2\right]^2=\left(x^2-6xy+9y^2\right)^2$$

$$\begin{array}{r}x^2-6xy+9y^2\\x^2-6xy+9y^2\\\hline 9x^2y^2-54xy^3+81y^4\\-6x^3y+36x^2y^2-54xy^3\quad\quad\\x^4-6x^3y+9x^2y^2\quad\quad\quad\quad\quad\quad\\\hline x^4-12x^3y+54x^2y^2-108xy^3+81y^4\end{array}$$

117. a. Answers will vary. Possible answer. Graph $y_1=\left(x^2+2x+3\right)(x+2)$ and $y_2=x^3+4x^2+7x+6$.
If they coincide, then the multiplication is correct.

b.

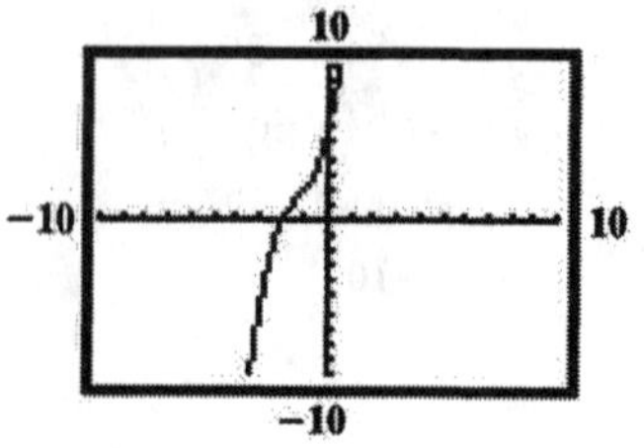

It is correct.

119. $[(y+1)-(x+2)]^2$

$$= (y+1)^2 - 2(y+1)(x+2) + (x+2)^2$$
$$= y^2 + 2y + 1 - 2(xy + 2y + x + 2) + x^2 + 4x + 4$$
$$= y^2 + 2y + 1 - 2xy - 4y - 2x - 4 + x^2 + 4x + 4$$
$$= y^2 - 2y - 2xy + 2x + x^2 + 1$$

121. $\frac{4}{5} - \left(\frac{3}{4} - \frac{2}{3}\right) = \frac{4}{5} - \frac{3}{4} + \frac{2}{3}$

$$= \frac{4\cdot 12}{5\cdot 12} - \frac{3\cdot 15}{4\cdot 15} + \frac{2\cdot 20}{3\cdot 20}$$
$$= \frac{48}{60} - \frac{45}{60} + \frac{40}{60}$$
$$= \frac{43}{60}$$

122. $\left(\frac{2r^4s^5}{r^2}\right)^3 = \left(2r^{4-2}s^5\right)^3 = \left(2r^2s^5\right)^3 = 8r^6s^{15}$

123.

$$-12 < 3x - 5 \le -1$$
$$-12 + 5 < 3x - 5 + 5 \le -1 + 5$$
$$-7 < 3x \le 4$$
$$\frac{-7}{3} < \frac{3x}{3} \le \frac{4}{3}$$
$$-\frac{7}{3} < x \le \frac{4}{3}$$

The solution set is $\left(-\frac{7}{3}, \frac{4}{3}\right]$.

124. $g(x) = -x^2 - 2x + 3$

$$g\left(\frac{1}{2}\right) = -\left(\frac{1}{2}\right)^2 + 2\left(\frac{1}{2}\right) + 3$$
$$= -\frac{1}{4} + 1 + 3$$
$$= -\frac{1}{4} + 4$$
$$= \frac{15}{4}$$

Exercise Set 5.3

1. a. Answers will vary.

b. $\dfrac{5x^4 - 6x^3 - 4x^2 - 12x + 1}{3x}$

$$= \frac{5x^4}{3x} - \frac{6x^3}{3x} - \frac{4x^2}{3x} - \frac{12x}{3x} + \frac{1}{3x}$$
$$= \frac{5}{3}x^3 - 2x^2 - \frac{4}{3}x - 4 + \frac{1}{3x}$$

3. Yes. Explanations will vary.

5. Place the polynomials in descending order of the variable.

7. a. Answers will vary.

b.

$$\begin{array}{r|rrr} 5 & 1 & 3 & -4 \\ & & 5 & 40 \\ \hline & 1 & 8 & 36 \end{array} \quad \leftarrow \text{Remainder}$$

Thus, $\dfrac{x^2 + 3x - 4}{x - 5} = x + 8 + \dfrac{36}{x - 5}$

9. To find the remainder, use synthetic division:

$$\begin{array}{r|rrr} -2 & 1 & 11 & 21 \\ & & -2 & -18 \\ \hline & 1 & 9 & 3 \end{array}$$

The remainder is 3. Since the remainder is not zero, $x + 2$ is not a factor.

11. $\frac{x^9}{x^7} = x^{9-7} = x^2$

13. $\frac{a^{11}}{a^7} = a^{11-7} = a^4$

15. $\frac{z^{16}}{z^8} = z^{18-8} = z^8$

17. $\frac{12r^7s^{10}}{3rs^8} = \frac{12}{3}r^{7-1}s^{10-8} = 4r^6s^2$

19. $\frac{15x^{18}y^{19}}{3x^{10}y^8} = \frac{15}{3}x^{18-10}y^{19-8} = 5x^8y^{11}$

21. $\frac{4x+18}{2} = \frac{4x}{2} + \frac{18}{2} = 2x + 9$

23. $$\frac{4x^2+2x}{2x}=\frac{4x^2}{2x}+\frac{2x}{2x}=2x+1$$

25. $$\frac{5y^3+6y^2-12y}{3y}=\frac{5y^3}{3y}+\frac{6y^2}{3y}-\frac{12y}{3y}=\frac{5}{3}y^2+2y-4$$

27. $$\frac{4x^5-6x^4+12x^3-8x^2}{4x^2}=\frac{4x^5}{4x^2}-\frac{6x^4}{4x^2}+\frac{12x^3}{4x^2}-\frac{8x^2}{4x^2}=x^3-\frac{3}{2}x^2+3x-2$$

29. $$\frac{8x^2y^2-10xy^3-5y}{2y^2}=\frac{8x^2y^2}{2y^2}-\frac{10xy^3}{2y^2}-\frac{5y}{2y^2}=4x^2-5xy-\frac{5}{2y}$$

31. $$\frac{9x^2y-12x^3y^2+15y^3}{2xy^2}=\frac{9x^2y}{2xy^2}-\frac{12x^3y^2}{2xy^2}+\frac{15y^3}{2xy^2}=\frac{9x}{2y}-6x^2+\frac{15y}{2x}$$

33. $$\frac{3xyz+6xyz^2-9x^3y^5z^7}{6xy}=\frac{3xyz}{6xy}+\frac{6xyz^2}{6xy}-\frac{9x^3y^5z^7}{6xy}=\frac{z}{2}+z^2-\frac{3}{2}x^2y^4z^7$$

35.
$$\begin{array}{r}x+2\\ x+1\overline{)x^2+3x+2}\\ \underline{x^2+x}\quad\leftarrow x(x+1)\\ 2x+2\\ \underline{2x+2}\quad\leftarrow 2(x+1)\\ 0\quad\leftarrow \text{Remainder}\end{array}$$

Thus, $\dfrac{x^2+3x+2}{x+1}=x+2$

37.
$$\begin{array}{r}2x+4\\ 3x+2\overline{)6x^2+16x+8}\\ \underline{6x^2+4x}\quad\leftarrow 2x(3x+2)\\ 12x+8\\ \underline{12x+8}\quad\leftarrow 4(3x+2)\\ 0\quad\leftarrow \text{Remainder}\end{array}$$

Thus, $\dfrac{6x^2+16x+8}{3x+2}=2x+4$

39.
$$\begin{array}{r}3x+2\\ 2x-1\overline{)6x^2+x-2}\\ \underline{6x^2-3x}\quad\leftarrow 3x(2x-1)\\ 4x-2\\ \underline{4x-2}\quad\leftarrow 2(2x-1)\\ 0\quad\leftarrow \text{Remainder}\end{array}$$

Thus, $\dfrac{6x^2+x-2}{2x-1}=3x+2$

41.
$$\begin{array}{r}x+5\\ x+1\overline{)x^2+6x+3}\\ \underline{x^2+x}\quad\leftarrow x(x+1)\\ 5x+3\\ \underline{5x+5}\quad\leftarrow 5(x+1)\\ -2\quad\leftarrow \text{Remainder}\end{array}$$

Thus, $\dfrac{x^2+6x+3}{x+1}=x+5-\dfrac{2}{x+1}$

43.
$$\begin{array}{r}2b+5\\ b-2\overline{)2b^2+b-8}\\ \underline{2b^2-4b}\quad\leftarrow 2b(b-2)\\ 5b-8\\ \underline{5b-10}\quad\leftarrow 5(b-2)\\ 2\quad\leftarrow \text{Remainder}\end{array}$$

Thus, $\dfrac{2b^2+b-8}{b-2}=2b+5+\dfrac{2}{b-2}$

45.
$$\begin{array}{r}4x+9\\ 2x-3\overline{)8x^2+6x-25}\\ \underline{8x^2-12x}\quad\leftarrow 4x(2x-3)\\ 18x-25\\ \underline{18x-27}\quad\leftarrow 9(2x-3)\\ 2\quad\leftarrow \text{Remainder}\end{array}$$

Thus, $\dfrac{8x^2+6x-25}{2x-3}=4x+9+\dfrac{2}{2x-3}$

47.
$$\begin{array}{r}2x+6\\ 2x-6\overline{)4x^2+0x-36}\\ \underline{4x^2-12x}\quad\leftarrow 2x(2x-6)\\ 12x-36\\ \underline{12x-36}\quad\leftarrow 5(2x-6)\\ 0\quad\leftarrow \text{Remainder}\end{array}$$

Thus, $\dfrac{4x^2-36}{2x-6}=2x+6$

49.

$$x+1\overline{\big)\,x^3+3x^2+5x+4}\quad\text{quotient: } x^2+2x+3$$

$$\underline{x^3+x^2} \leftarrow x^2(x+1)$$

$$2x^2+5x$$

$$\underline{2x^2+2x} \leftarrow 2x(x+1)$$

$$3x+4$$

$$\underline{3x+3} \leftarrow 3(x+1)$$

$$1 \leftarrow \text{Remainder}$$

Thus, $\dfrac{x^3+3x^2+5x+4}{x+1}=x^2+2x+3+\dfrac{1}{x+1}$

51.

$$2y+3\overline{\big)\,4y^3+12y^2+7y-9}\quad\text{quotient: } 2y^2+3y-1$$

$$\underline{4y^3+6y^2} \leftarrow 2y^2(2y+3)$$

$$6y^2+7y$$

$$\underline{6y^2+9y} \leftarrow 3y(2y+3)$$

$$-2y-9$$

$$\underline{-2y-3} \leftarrow -1(2y+3)$$

$$-6 \leftarrow \text{Remainder}$$

Thus, $\dfrac{4y^3+12y^2+7y-9}{2y+3}=2y^2+3y-1-\dfrac{6}{2y+3}$

53.

$$2a-1\overline{\big)\,4a^3+0a^2-5a+0}\quad\text{quotient: } 2a^2+a-2$$

$$\underline{4a^3-2a^2} \leftarrow 2a^2(2a-1)$$

$$2a^2-5a$$

$$\underline{2a^2-a} \leftarrow a(2a-1)$$

$$-4a+0$$

$$\underline{-4a+2} \leftarrow -2(2a-1)$$

$$-2 \leftarrow \text{Remainder}$$

Thus, $\dfrac{4a^2-5a}{2a-1}=2a^2+a-2-\dfrac{2}{2a-1}$

55.

$$x^2+0x-2\overline{\big)\,3x^5+0x^4+0x^3+2x^2-12x-4}\quad\text{quotient: } 3x^3\ \ +6x+2$$

$$\underline{3x^5+0x^4-6x^3} \leftarrow 3x^3(x^2+0x-2)$$

$$6x^3+2x^2-12x$$

$$\underline{6x^3+0x^2-12x} \leftarrow 6x(x^2+0x-2)$$

$$2x^2+0x-4$$

$$\underline{2x^2+0x-4} \leftarrow 2(x^2+0x-2)$$

$$0 \leftarrow \text{Remainder}$$

Thus, $\dfrac{3x^5+2x^2-12x-4}{x^2-2}=3x^3+6x+2$

57.

$$3x^3-8x^2+0x-5\overline{\big)\,3x^4+4x^3-32x^2-5x-20}\quad\text{quotient: } x+4$$

$$\underline{3x^4-8x^3+\ 0x^2-5x} \leftarrow x(3x^3-8x^2+0x-5)$$

$$12x^3-32x^2+0x-20$$

$$\underline{12x^3-32x^2+0x-20} \leftarrow 4(3x^3-8x^2+0x-5)$$

$$0 \leftarrow \text{Remainder}$$

Thus, $\dfrac{3x^4+4x^3-32x^2-5x-20}{3x^3-8x^2-5}=x+4$

59.

$$\begin{array}{r l} & 2c^2-6c+3 \\ c^2-c+5\,\big) & 2c^4-8c^3+19c^2-33c+15 \\ & \underline{2c^4-2c^3+10c^2} \quad \leftarrow 2c^2(c^2-c+5) \\ & -6c^3+9c^2-33c \\ & \underline{-6c^3+6c^2-30c} \quad \leftarrow -6c(c^2-c+5) \\ & 3c^2-3c+15 \\ & \underline{3c^2-3c+15} \leftarrow 3(c^2-c+5) \\ & 0 \leftarrow \text{Remainder} \end{array}$$

Thus $\dfrac{2c^4-8c^3+19c^2-33c+15}{c^2-c+5}=2c^2-6c+3$

61.

$$\begin{array}{r|rrr} -1 & 1 & 7 & 6 \\ & & -1 & -6 \\ \hline & 1 & 6 & 0 \end{array}$$

Thus, $\dfrac{x^2+7x+6}{x+1}=x+6$

63.

$$\begin{array}{r|rrr} -2 & 1 & 5 & 6 \\ & & -2 & -6 \\ \hline & 1 & 3 & 0 \end{array}$$

Thus, $\dfrac{x^2+5x+6}{x+2}=x+3$

65.

$$\begin{array}{r|rrr} 4 & 1 & -11 & 28 \\ & & 4 & -28 \\ \hline & 1 & -7 & 0 \end{array}$$

Thus, $\dfrac{x^2-11x+28}{x-4}=x-7$

67.

$$\begin{array}{r|rrr} 3 & 1 & 5 & -14 \\ & & 3 & 24 \\ \hline & 1 & 8 & 10 \end{array}$$

Thus, $\dfrac{x^2+5x-12}{x-3}=x+8+\dfrac{10}{x-3}$

69.

$$\begin{array}{r|rrr} 4 & 3 & -7 & -10 \\ & & 12 & 20 \\ \hline & 3 & 5 & 10 \end{array}$$

Thus, $\dfrac{3x^2-7x-10}{x-4}=3x+5+\dfrac{10}{x-4}$

71.

$$\begin{array}{r|rrrr} 1 & 4 & -3 & 2 & 0 \\ & & 4 & 1 & 3 \\ \hline & 4 & 1 & 3 & 3 \end{array}$$

Thus, $\dfrac{4x^3-3x^2+2x}{x-1}=4x^2+x+3+\dfrac{3}{x-1}$

73.

$$\begin{array}{r|rrrr} -3 & 3 & 7 & -4 & 16 \\ & & -9 & 6 & -6 \\ \hline & 3 & -2 & 2 & 10 \end{array}$$

Thus, $\dfrac{3c^3+7c^2-4c+16}{c+3}=3c^2-2c+2+\dfrac{10}{c+3}$

75.

$$\begin{array}{r|rrrrr} 1 & 1 & 0 & 0 & 0 & -1 \\ & & 1 & 1 & 1 & 1 \\ \hline & 1 & 1 & 1 & 1 & 0 \end{array}$$

Thus, $\dfrac{y^4-1}{y-1}=y^3+y^2+y+1$

77.

$$\begin{array}{r|rrrrr} -4 & 1 & 0 & 0 & 0 & 16 \\ & & -4 & 16 & -64 & 256 \\ \hline & 1 & -4 & 16 & -64 & 272 \end{array}$$

Thus, $\dfrac{x^4+16}{x+4}=x^3-4x^2+16x-64+\dfrac{272}{x+4}$

79.

$$\begin{array}{r|rrrrrr} -1 & 1 & 1 & 0 & 0 & 0 & -9 \\ & & -1 & 0 & 0 & 0 & 0 \\ \hline & 1 & 0 & 0 & 0 & 0 & -9 \end{array}$$

Thus, $\dfrac{x^5+x^4-9}{x+1}=x^4-\dfrac{9}{x+1}$

81.

$$\begin{array}{r|rrrrrr} -1 & 1 & 4 & 0 & 0 & 0 & -14 \\ & & -1 & -3 & 3 & -3 & 3 \\ \hline & 1 & 3 & -3 & 3 & -3 & -11 \end{array}$$

Thus, $\dfrac{b^5+4b^4-14}{b+1}=b^4+3b^3-3b^2+3b-3-\dfrac{11}{b+1}$

83.

$$\begin{array}{r|rrrr} \frac{1}{3} & 3 & 2 & -4 & 1 \\ & & 1 & 1 & -1 \\ \hline & 3 & 3 & -3 & 0 \end{array}$$

Thus, $\dfrac{3x^3+2x^2-4x+1}{x-\frac{1}{3}}=3x^2+3x-3$

85.

$$\begin{array}{r|rrrrr} \frac{1}{2} & 2 & -1 & 2 & -3 & 7 \\ & & 1 & 0 & 1 & -1 \\ \hline & 2 & 0 & 2 & -2 & 6 \end{array}$$

Thus, $\dfrac{2x^4-x^3+2x^2-3x+1}{x-\frac{1}{2}}=2x^3+2x-2+\dfrac{6}{x-\frac{1}{2}}$

87. To find the remainder, use synthetic division:

$$\begin{array}{r|rrr} 2 & 4 & -5 & 6 \\ & & 8 & 6 \\ \hline & 4 & 3 & 12 \end{array}$$

The remainder is 12.

89. To find the remainder, use synthetic division:

$$\begin{array}{r|rrrr} 2 & 1 & -2 & 4 & -8 \\ & & 2 & 0 & 8 \\ \hline & 1 & 0 & 4 & 0 \end{array}$$

The remainder is 0 which means that $x-2$ is a factor of x^3-2x^2+4x-8.

91. To find the remainder, use synthetic division.

$$\begin{array}{r|rrrr} \frac{1}{2} & -2 & -6 & 2 & -4 \\ & & -1 & -\frac{7}{2} & -\frac{3}{4} \\ \hline & -2 & -7 & -\frac{3}{4} & -\frac{19}{4} \end{array} \quad \text{or } -4.75$$

The remainder is $-\frac{19}{4}$ or -4.75.

93. $\text{Width} = \dfrac{\text{area}}{\text{length}} = \dfrac{6x^2-8x-8}{2x-4}$

$$\begin{array}{rll} & 3x+2 & \\ 2x-4 \overline{)} & 6x^2-8x-8 & \\ & \underline{6x^2-12x} & \leftarrow 3x(2x-4) \\ & 4x-8 & \\ & \underline{4x-8} & \leftarrow 2(2x-4) \\ & 0 & \leftarrow \text{Remainder} \end{array}$$

Thus the width is $3x + 2$.

95. To compare the areas, compare the lengths and widths of the two figures (rectangles).
Length: observe that $12x + 24$ is six times $2x + 4$
Width: observe that $\frac{1}{2}x+4$ is $\frac{1}{2}$ of $x + 8$

The area of the larger rectangle is $\frac{1}{2}(6)=3$ times the area of the smaller rectangle.

97. No, it is not possible to divide a binomial by a monomial and obtain a monomial as a quotient. Explanations will vary.

99. If the remainder is 0, then $x-a$ is a factor.

101. $\dfrac{P(x)}{x-4}=x+2$

$P(x)=(x-4)(x+2)=x^2-2x-8$

103. $\dfrac{P(x)}{x+4}=x+5+\dfrac{6}{x+4}$

$$\begin{aligned} P(x) &= (x+4)\left(x+5+\frac{6}{x+4}\right) \\ &= x(x+4)+5(x+4)+\frac{6}{x+4}(x+4) \\ &= x^2+4x+5x+20+6 \\ &= x^2+9x+26 \end{aligned}$$

105.

$$\begin{array}{rll} & 2x^2+3xy-y^2 & \\ x-2y \overline{)} & 2x^3-x^2y-7xy^2+2y^3 & \\ & \underline{2x^3-4x^2y} & \leftarrow 2x^2(x-2y) \\ & 3x^2y-7xy^2 & \\ & \underline{3x^2y-6xy^2} & \leftarrow 3xy(x-2y) \\ & -xy^2+2y^3 & \\ & \underline{-xy^2+2y^3} & \leftarrow -y^2(x-2y) \\ & 0 & \leftarrow \text{Remainder} \end{array}$$

Thus, $\dfrac{2x^3-x^2y-7xy^2+2y^3}{x-2y}=2x^2+3xy-y^2$

107.

$$\begin{array}{rll} & x+\frac{5}{2} & \\ 2x-3 \overline{)} & 2x^2+2x-2 & \\ & \underline{2x^2-3x} & \leftarrow x(2x-3) \\ & 5x-2 & \\ & \underline{5x-\frac{15}{2}} & \leftarrow \frac{5}{2}(2x-3) \\ & \frac{11}{2} & \leftarrow \text{Remainder} \end{array}$$

Thus, $\dfrac{2x^2+2x-2}{2x-3}=x+\dfrac{5}{2}+\dfrac{11}{2(2x-3)}$

109. Two of the dimensions are r and $2r + 2$. This product is $r(2r+2)=2r^2+2r$. To find the third side, divide the volume by the above product. That is, the third side is

$$w=\frac{\text{volume}}{\text{product of two sides}}=\frac{2r^3+4r^2+2r}{2r^2+2r}$$

$$\begin{array}{rll} & r+1 & \\ 2r^2+2r \overline{)} & 2r^3+4r^2+2r & \\ & \underline{2r^3+2r^2} & \leftarrow r(2r^2+2r) \\ & 2r^2+2r & \\ & \underline{2r^2+2r} & \leftarrow 1(2r^2+2r) \\ & 0 & \leftarrow \text{remainder} \end{array}$$

Thus, the third side is $w = r + 1$.

111. The polynomial is the product of $x-3$ with x^2-3x+4 plus the remainder of 5:

$$\begin{array}{rl} x^2-3x+4 & \\ x-3 & \\ \hline -3x^2+9x-12 & \leftarrow -3(x^2-3x+4) \\ x^3-3x^2+4x \quad & \leftarrow x(x^2-3x+4) \\ \hline x^3-6x^2+13x-12 & \end{array}$$

Now, add on the remainder 5 to get $x^3-6x^2+13x-7$ for the polynomial.

113. $\dfrac{4x^{n+1}+2x^n-3x^{n-1}-x^{n-2}}{2x^n}$

$$= \frac{4x^{n+1}}{2x^n}+\frac{2x^n}{2x^n}-\frac{3x^{n-1}}{2x^n}-\frac{x^{n-2}}{2x^n}$$
$$= 2x^{n+1-n}+x^{n-n}-\frac{3}{2}x^{n-1-n}-\frac{1}{2}x^{n-2-n}$$
$$= 2x+x^0-\frac{3}{2}x^{-1}-\frac{1}{2}x^{-2}$$
$$= 2x+1-\frac{3}{2x}-\frac{1}{2x^2}$$

115. Let $P(x)=x^{100}+x^{99}+\ldots+x+1$. To determine if $x-1$ is a factor, compute $P(1)$. If it is 0, then $x-1$ is a factor. Now,

$$P(1)=1^{100}+1^{99}+1^{98}+1^{97}+\ldots+1^2+1^1+1$$
$$=1+1+1+1+\ldots+1+1+1=101$$

Note: Here you are adding up the number 1 a total of 101 times. Since $P(1)=101$, which is not zero, then $x-1$ is not a factor of $x^{100}+x^{99}+\ldots+x+1$.

117. Let $P(x)=x^{99}+x^{98}+\ldots+x^1+1$. To determine if $x+1$ is a factor, compute $P(-1)$. If it is 0, then $x+1$ is a factor. Now,

$$P(-1)=(-1)^{99}+(-1)^{98}+(-1)^{97}+(-1)^{96}+\ldots+(-1)^3+(-1)^2+(-1)^1+1$$
$$=(-1)+1+(-1)+1+\ldots+(-1)+1+(-1)+1$$
$$=0+0+\ldots+0+0$$
$$=0$$

Since $P(-1)=0$, then $x+1$ is a factor of $x^{99}+x^{98}+\ldots+x+1$.

119. $\dfrac{8.45\times10^{23}}{4.225\times10^{13}}=\dfrac{8.45}{4.225}\times\dfrac{10^{23}}{10^{13}}$

$$=2.0\times10^{25-15}$$
$$=2.0\times10^{10}$$

120. Let x be the measure of the smallest angle. Then $2x$ is the measure of the second angle, and $x+60$ is the measure of the third angle.

$$x+2x+x+60=180$$
$$4x+60=180$$
$$4x=120$$
$$x=30$$

The smallest angle measures $30°$. The second angle measures $2(30°)=60°$. The third angle measures $30°+60°=90°$.

121. $\left|\dfrac{5x-3}{2}\right|+4=8$

$$\left|\frac{5x-3}{2}\right|=4$$
$$\frac{5x-3}{2}=-4 \quad\text{or}\quad \frac{5x-3}{2}=4$$
$$5x-3=-8 \qquad 5x-3=8$$
$$x=-1 \qquad x=\frac{11}{5}$$

The solution set is $\left\{-1,\dfrac{11}{5}\right\}$.

122. $f(x)=x^2-4,\quad g(x)=-5x+3$

$$f(6)=(6)^2-4=36-4=32$$
$$g(6)=-5(6)+3=-30+3=-27$$
$$f(6)\cdot g(6)=(32)(-27)=-864$$

123. $(6r+5s-t)+(-3r-2s-7t)$

$$=6r-3r+5s-2s-t-7t$$
$$=3r+3s-8t$$

Exercise Set 5.4

1. The first step in *any* factoring problem is to determine whether all the terms contain a greatest common factor and, if so, factor it out.

3. a. Answers will vary.

b. The greatest common factor is $2x^2y$.

c. $6x^2y^5-2x^3y+12x^9y^3$

$$=2x^2y\left(3y^4-x+6x^7y^2\right)$$

5. $12(x-4)^3 = 3\cdot 4(x-4)^3$

$6(x-4)^6 = 3\cdot 2(x-4)^6$

$3(x-4)^9 = 3\cdot (x-4)^9$

The GCF of 12, 6, and 3 is 3.

The lowest power of $(x-4)$ is 3.

Therefore, the GCF of the terms is $3(x-4)^3$.

7. a. Answers will vary.

b. $6x^3 - 2xy^3 + 3x^2y^2 - y^5$
$= 2x(3x^2 - y^3) + y^2(3x^2 - y^3)$
$= (3x^2 - y^3)(2x + y^2)$

9. $7n + 14 = 7\cdot n + 7\cdot 2 = 7(n+2)$

11. $2x^2 - 4x + 10 = 2\cdot x^2 + 2\cdot(-2x) + 2(5)$
$= 2(x^2 - 2x + 5)$

13. $12y^2 - 16y + 28 = 4\cdot 3y^2 - 4\cdot 4y + 4\cdot 7$
$= 4(3y^2 - 4y + 7)$

15. $9x^4 - 3x^3 + 11x^2 = x^2\cdot 9x^2 + x^2(-3x) + x^2\cdot 11$
$= x^2(9x^2 - 3x + 11)$

17. $-24a^7 + 9a^6 - 3a^2$
$= -3a^2\cdot 8a^5 - 3a^2(-3a^4) - 3a^2\cdot 1$
$= -3a^2(8a^5 - 3a^4 + 1)$

19. $3x^2y + 6x^2y^2 + 3xy = 3xy\cdot x + 3xy\cdot 2xy + 3xy\cdot 1$
$= 3xy(x + 2xy + 1)$

21. $80a^5b^4c - 16a^4b^2c^2 + 24a^2c$
$= 8a^2c\cdot 10a^3b^4 + 8a^2c(-2a^2b^2c) + 8a^2c(3)$
$= 8a^2c(10a^3b^4 - 2a^2b^2c + 3)$

23. $9p^4q^5r - 3p^2q^2r^2 + 12pq^5r^3$
$= 3pq^2r\cdot 3p^3q^3 + 3pq^2r(-pr) + 3pq^2r(4q^3r^2)$
$= 3pq^2r(3p^3q^3 - pr + 4q^3r^2)$

25. $-22p^2q^2 - 16pq^3 + 26r$
$= -2\cdot 11p^2q^2 - 2\cdot 8pq^3 - 2(-13r)$
$= -2(11p^2q^2 + 8pq^3 - 13r)$

27. $-8x + 4 = -4(2x) - 2(-1) = -4(2x-1)$

29. $-x^2 - 4x + 22 = -(x^2 + 4x - 22)$

31. $-3r^2 - 6r + 9 = -3(r^2 + 2r - 3)$

33. $-6r^4s^3 + 4r^2s^4 + 2rs^5 = -2(3r^4s^3 - 2r^2s^4 - rs^5)$
$= -2rs^3(3r^3 - 2rs - s^2)$

35. $-a^4b^2c + 5a^3bc^2 + a^2b$
$= -a^2b(a^2bc) - a^2b(-5ac^2) - a^2b(-1)$
$= -a^2b(a^2bc - 5ac^2 - 1)$

37. $x(a+3) + 1(a+3) = (a+3)(x+1)$

39. $7x(x-4) + 2(x-4)^2 = (x-4)[7x + 2(x-4)]$
$= (x-4)(7x + 2x - 8)$
$= (x-4)(9x-8)$

41. $(x-2)(3x+5) - (x-2)(5x-4)$
$= (x-2)[(3x+5) - (5x-4)]$
$= (x-2)(3x + 5 - 5x + 4)$
$= (x-2)(-2x+9)$
$= -(x-2)(2x-9)$

43. $(2a+4)(a-3) - (2a+4)(2a-1)$
$(2a+4)[(a-3) - (2a-1)]$
$= (2a+4)(a - 3 - 2a + 1)$
$= (2a+4)(-a-2)$
$= -2(a+2)(a+2)$ or $-2(a+2)^2$

45. $x^2 + 4x - 5x - 20 = x(x+4) - 5(x+4)$
$= (x+4)(x-5)$

47. $8y^2 - 4y - 20y + 10 = 2(4y^2 - 2y - 10y + 5)$
$= 2[2y(2y-1) - 5(2y-1)]$
$= 2(2y-1)(2y-5)$

49. $am + an + bm + bn = a(m+n) + b(m+n)$
$= (m+n)(a+b)$

51. $x^3-3x^2+4x-12=x^2(x-3)+4(x-3)$
$=(x-3)(x^2+4)$

53. $10m^2-12mn-25mn+30n^2$
$=2m(5m-6n)-5n(5m-6n)$
$=(5m-6n)(2m-5n)$

55. $5a^3+15a^2-10a-30=5[a^3+3a^2-2a-6]$
$=5[a^2(a+3)-2(a+3)]$
$=5(a+3)(a^2-2)$

57. $c^5-c^4+c^3-c^2=c^2(c^3-c^2+c-1)$
$=c^2[c^2(c-1)+1(c-1)]$
$=c^2(c-1)(c^2+1)$

59. $A_1-A_2=6x(2x+1)-5(2x+1)$
$=(2x+1)(6x-5)$

61. $A_1-A_2=(3x^2+12x)-(2x-8)$
$=3x(x+4)-2(x+4)$
$=(x+4)(3x-2)$

63. $V_1-V_2=9x(3x+2)-5(3x+2)$
$=(3x+2)(9x-5)$

65. a. $h(t)=-16t^2+80t$
$h(3)=-16(3)^2+80(3)$
$=-16(9)+80(3)$
$=-144+240$
$=96$
The height after 3 seconds is 96 feet.

b. $h(t)=-16t^2+80t$
$=-16t(t)-16t(-5)$
$=-16t(t-5)$

c. $h(3)=-16(3)(3-5)$
$=-16(3)(-2)$
$=-48(-2)$
$=96$ feet

67. a. $A=\pi r^2+2rl$
$=\pi(20)^2+2(20)(40)$
$=400\pi+1600$
≈ 2856.64
The area is about 2856.64 ft^2.

b. $A=\pi r^2+2rl=r(\pi r)+r(2l)=r(\pi r+2l)$

c. $A=20(\pi\cdot 20+2\cdot 40)$
$=20(20\pi+80)$
≈ 2856.64 ft^2

69. a. $A(t)=975-75t$
$A(6)=975-75(6)=975-450=525$
After 6 months, Fred still owed $525.

b. $A(t)=975-75t=75\cdot 13-75\cdot t=75(13-t)$

c. $A(6)=75(13-6)=75(7)=\$525$

71. a. $(x+0.06x)-0.06(x+0.06x)$
$=(1-0.06)(x+0.06x)$
$=(0.94)(1.06x)$

b. Upon multiplication, $(0.94)(1.06x)=0.9964x$. This represents 99.64% of the 2005 price which means this sale price is slightly lower than the price of the 2005 model.

73. a. Final price is
$(x+0.15x)-0.20(x+0.15x)$
$=0.80(x+0.15x)$

b. $(x+0.15x)-0.20(x+0.15x)$
$=(1-0.20)(x+0.15x)$
$=0.80(1.15x)$
$=0.92x$
The sale price is 92% of the regular price.

75. $5a(3x+2)^5+4(3x+2)^4$
$=5a(3x+2)\cdot(3x+2)^4+4\cdot(3x+2)^4$
$=(3x+2)^4[5a(3x+2)+4]$
$=(3x-2)^4(15ax+10a+4)$

77. $4x^2(x-3)^3-6x(x-3)^2+4(x-3)$
$=2(x-3)\left[2x^2(x-3)^2-3x(x-3)+2\right]$
$=2(x-3)\left[2x^2\left(x^2-6x+9\right)-3x\left(x-3\right)+2\right]$
$=2(x-3)\left(2x^4-12x^3+18x^2-3x^2+9x+2\right)$
$=2(x-3)\left(2x^4-12x^3+15x^2+9x+2\right)$

79. $ax^2+2ax-3a+bx^2+2bx-3b$
$=ax^2+bx^2+2ax+2bx-3a-3b$
$=x^2\left(a+b\right)+2x\left(a+b\right)-3\left(a+b\right)$
$=\left(a+b\right)\left(x^2+2x-3\right)$

81. $x^{6m}-2x^{4m}=x^{4m}\cdot x^{2m}-x^{4m}\cdot 2=x^{4m}\left(x^{2m}-2\right)$

83. $3x^{4m}-2x^{3m}+x^{2m}=x^{2m}\cdot 3x^{2m}-x^{2m}\cdot 2x^{m}+x^{2m}\cdot 1$
$=x^{2m}\left(3x^{2m}-2x^{m}+1\right)$

85. $a^rb^r+c^rb^r-a^rd^r-c^rd^r$
$=b^r\left(a^r+c^r\right)-d^r\left(a^r+c^r\right)$
$=\left(a^r+c^r\right)\left(b^r-d^r\right)$

87. **a.** Yes
$6x^3-3x^2+9x=3x\left(2x^2\right)+3x\left(-x\right)+3x\left(3\right)$
$=3x\left(2x^2-x+3\right)$

b. 0

c. Answers will vary.

89. **a.** They should be the same graph.

b. $y_1=8x^3-16x^2-4x$
$y_2=4x(2x^2-4x-1)$

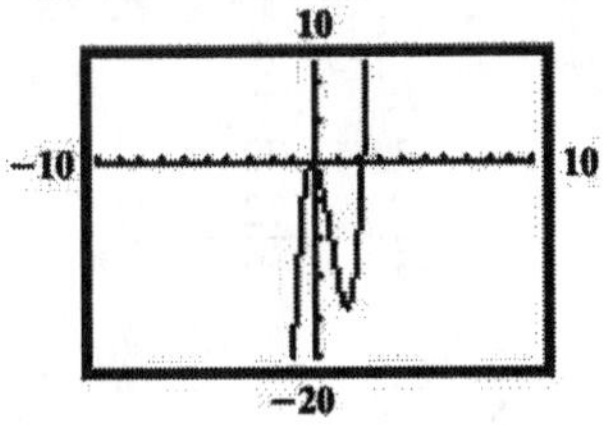

c. Answers will vary.

d. The factoring is not correct.

91. $$\frac{\left(\left|\frac{1}{2}\right|-\left|-\frac{1}{3}\right|\right)^2}{-\left|\frac{1}{3}\right|\cdot\left|-\frac{2}{5}\right|}=\frac{\left(\frac{1}{2}-\frac{1}{3}\right)^2}{-\left(\frac{1}{3}\right)\left(\frac{2}{5}\right)}=\frac{\left(\frac{3}{6}-\frac{2}{6}\right)^2}{-\left(\frac{1}{3}\right)\left(\frac{2}{5}\right)}$$
$$=\frac{\left(\frac{1}{6}\right)^2}{-\left(\frac{1}{3}\right)\left(\frac{2}{5}\right)}=\frac{\frac{1}{36}}{-\frac{2}{15}}$$
$$=\frac{1}{36}\cdot\left(-\frac{15}{2}\right)=-\frac{15}{72}=-\frac{5}{24}$$

92. $3(2x-4)+3(x+1)=9$
$6x-12+3x+3=9$
$9x-9=9$
$9x=18$
$x=2$

93. This graph is a translation of the parent graph $y=x^2$. The parent graph is shifted down one unit.

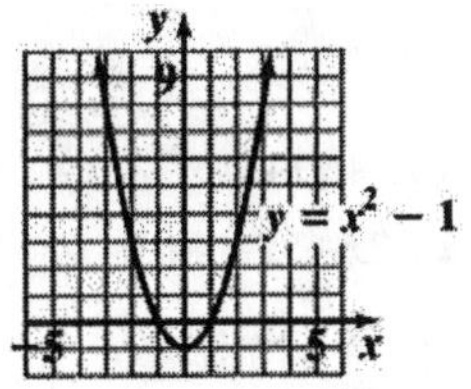

94. Let x be the walking time and y be the jogging time.
$x+y=0.9 \Rightarrow y=0.9-x$
$3x+5y=3.5$
Substitute $0.9-x$ into y in the second equation.
$3x+5\left(0.9-x\right)=3.5$
$3x+4.5-5x=3.5$
$-2x=-1$
$x=0.5$
Jason jogs for 0.9 – 0.5 = 0.4 hours every day.

95. $\left(7a-3\right)\left(-2a^2-4a+1\right)$
$=7a\left(-2a^2\right)+7a\left(-4a\right)+7a\left(1\right)-3\left(-2a^2\right)$
$-3\left(-4a\right)-3\left(1\right)$
$=-14a^3-28a^2+7a+6a^2+12a-3$
$=-14a^3-22a^2+19a-3$

Mid-Chapter Test: 5.1-5.4

1. $5x^4 - 1.5x^3 + 2z - 7$; degree 4

2. $P(x) = 8x^2 - 7x + 3$

$$P\left(\frac{1}{2}\right) = 8\left(\frac{1}{2}\right)^2 - 7\left(\frac{1}{2}\right) + 3$$
$$= 8\left(\frac{1}{4}\right) - 7\left(\frac{1}{2}\right) + 3$$
$$= 2 - \frac{7}{2} + 3$$
$$= 5 - \frac{7}{2} = \frac{10}{2} - \frac{7}{2} = \frac{3}{2} \text{ or } 1\frac{1}{2}$$

3. $(2n^2 - n - 12) + (-3n^2 - 6n + 8)$
$= 2n^2 - 3n^2 - n - 6n - 12 + 8$
$= -n^2 - 7n - 4$

4. $(-9x^2y + 4xy) - (7x^2y - 10xy)$
$= -9x^2y + 4xy - 7x^2y + 10xy$
$= -16x^2y + 14xy$

5. $(2x^2 - x + 4) + (3x^2 + 2x + 3) + (4x^2 - 5x + 6)$
$= 2x^2 + 3x^2 + 4x^2 - x + 2x - 5x + 4 + 3 + 6$
$= 9x^2 - 4x + 13$

6. $2x^5(3xy^4 + 5x^2 - 7x^3y)$
$= 2x^5(3xy^4) + 2x^5(5x^2) + 2x^5(-7x^3y)$
$= 6x^6y^4 + 10x^7 - 14x^8y$

7. $(7x - 6y)(3x + 2y)$
$= 7x(3x) + 7x(2y) - 6y(3x) - 6y(2y)$
$= 21x^2 + 14xy - 18xy - 12y^2$
$= 21x^2 - 4xy - 12y^2$

8. $(3x + 1)(2x^3 - x^2 + 5x + 9)$
$= 3x(2x^3) + 3x(-x^2) + 3x(5x) + 3x(9)$
$\quad + 1(2x^3) + 1(-x^2) + 1(5x) + 1(9)$
$= 6x^4 - 3x^3 + 15x^2 + 27x + 2x^3 - x^2 + 5x + 9$
$= 6x^4 - x^3 + 14x^2 + 32x + 9$

9. $\left(8p - \frac{1}{5}\right)\left(8p + \frac{1}{5}\right) = (8p)^2 - \left(\frac{1}{5}\right)^2 = 64p^2 - \frac{1}{25}$

10. $(4m - 3n)(3m^2 + 2mn - 6n^2)$
$= 4m(3m^2) + 4m(2mn) + 4m(-6n^2) - 3n(3m^2)$
$\quad - 3n(2mn) - 3n(-6n^2)$
$= 12m^3 + 8m^2n - 24mn^2 - 9m^2n - 6mn^2 + 18n^3$
$= 12m^3 - m^2n - 30mn^2 + 18n^3$

11. $x^2 - 14x + 49 = (x)^2 - 2(x)(7) + (7)^2$
$= (x - 7)^2$

12. $$\frac{4x^4y^3 + 6x^2y^2 - 11x}{2x^2y^2} = \frac{4x^4y^3}{2x^2y^2} + \frac{6x^2y^2}{2x^2y^2} - \frac{11x}{2x^2y^2}$$
$$= 2x^2y + 3 - \frac{11}{2xy^2}$$

13.
$$\begin{array}{r l}
 & 3x + 5 \\
4x + 1 \overline{)\,12x^2 + 23x + 7} & \\
\underline{12x^2 + 3x} & \leftarrow x^2(x+1) \\
20x + 7 & \\
\underline{20x + 5} & \leftarrow 2x(x+1) \\
2 & \leftarrow \text{Remainder}
\end{array}$$

Thus, $\dfrac{12x^2 + 23x + 7}{4x + 1} = 3x + 5 + \dfrac{2}{4x + 1}$

14.
$$\begin{array}{r l}
 & y^2 + y + 5 \\
2y - 3 \overline{)\,2y^3 - y^2 + 7y - 10} & \\
\underline{2y^3 - 3y^2} & \leftarrow y^2(2y-3) \\
2y^2 + 7y & \\
\underline{2y^2 - 3y} & \leftarrow y(2y-3) \\
10y - 10 & \\
\underline{10y - 15} & \leftarrow 5(2y-3) \\
5 & \leftarrow \text{Remainder}
\end{array}$$

Thus, $\dfrac{2y^3 - y^2 + 7y - 10}{2y - 3} = y^2 + y + 5 + \dfrac{5}{2y - 3}$

15.
$$\begin{array}{r|r r r}
-8 & 1 & -1 & -72 \\
 & & -8 & 72 \\
\hline
 & 1 & -9 & 0
\end{array}$$

Thus, $\dfrac{x^2 - x - 72}{x + 8} = x - 9$

16.

2	3	−2	−14	11	2
		6	8	−12	−2
	3	4	−6	−1	0

Thus,

$$\frac{3a^4-2a^3-14a^2+11a+2}{a-2}=3a^3+4a^2-6a-1$$

17. $32b^3c^3+16b^2c+24b^5c^4$
$=8b^2c\cdot4bc^2+8b^2c\cdot2+8b^2c\cdot3b^3c^3$
$=8b^2c\left(4bc^2+2+3b^3c^3\right)$

18. $7b(2x+9)-3c(2x+9)=(2x+9)(7b-3c)$

19. $2b^4-b^3c+4b^3c-2b^2c^2$
$=b^2\left(2b^2-bc+4bc-2c^2\right)$
$=b^2\left[b(2b-c)+2c(2b-c)\right]$
$=b^2(2b-c)(b+2c)$

20. $5a(3x-2)^5-4(3x-2)^6$
$=(3x-2)^5\left[5a-4(3x-2)\right]$
$=(3x-2)^5\left(5a-12x+8\right)$

Exercise Set 5.5

1. Factor out the greatest common factor if there is one.

3. a. Answers will vary.

b. $6x^2-x-12$
Observe that $6(-12)=-72$. The two numbers whose product is –72 and whose sum is 1 are –8 and 9, since $(-8)(9)=-72$ and $-8+9=1$.. Then the middle term, *x*, can be written as $x=-8x+9x$ and the factorization is
$6x^2+x-12=6x^2-8x+9x-12$
$=2x(3x-4)+3(3x-4)$
$=(3x-4)(2x+3)$

5. No. You must first factor out the GCF.
$2x^2+8x+6=2(x^2+4x+3)$
$=2(x+3)(x+1)$

7. No. You must first factor out the GCF.
$3x^3+6x^2-24x=3x(x^2+2x-8)$
$=3x(x+4)(x-2)$

9. Both are +.

11. One is +, one is –.

13. $x^2+7x+12$
The two numbers whose product is 12 and whose sum is 7 are 3 and 4, since (3)(4) = 12 and 3 + 4 = 7. Thus, $x^2+7x+12=(x+3)(x+4)$.

15. b^2+8b-9
The two numbers whose product is -9 and whose sum is 8 are –1 and 9, since (–1)(9) = –9 and
(–1) + 9 = 8. Thus, $b^2+8b-9=(b-1)(b+9)$.

17. z^2+4z+4
The two numbers whose product is 4 and whose sum is 4 are 2 and 2, since (–6)(–6) = 4 and –6 + (–6) = 4. Thus,
$z^2+4z+4=(z+2)(z+2)=(z+2)^2$.

19. $r^2+24r+144$
The two numbers whose product is 144 and whose sum is 24 are 12 and 12, since (12)(12) = 144 and 12 + (12) = 24. Thus,
$r^2+24r+144=(r+12)(r+12)=(r+12)^2$.

21. $x^2+30x-64$
The two numbers whose product is -64 and whose sum is 30 are –2 and 32, since (–2)(32) = –64 and –2 + 32 = 30. Thus, $x^2+30x-64=(x-2)(x+32)$.

23. $x^2-13x-30$
The two numbers whose product is –30 and whose sum is –13 are 2 and –15, since (2)(–15) = –30 and 2 – 15 = –13. Thus, $x^2-13x-30=(x+2)(x-15)$.

25. $-a^2+18a-45$ can be written as $-(a^2-18a+45)$.
Now, factor $a^2-18a+45$. The two numbers whose product is 45 and whose sum is –18 are –3 and –15, since (–3)(–15) = 45 and –3 + (–15) = –18. Thus, $-(a^2-18a+45)=-(a-3)(a-15)$.

27. $x^2+xy+7y^2$
Notice that there are no two rational numbers whose product is 7 and whose sum is 1. Thus, the polynomial is prime and cannot be factored.

29. Factor out the common factor -2:
$-2m^2-14m-20=-2(m^2+7m+10)$. Now, factor $m^2+7m+10$. The two numbers whose product is 10 and whose sum is 7 are 5 and 2, since (5)(2) = 10 and 5 + 2 = 7. Thus,
$m^2+7m+10=(m+2)(m+5)$, and
$-2m^2-14m-20=-2(m+2)(m+5)$.

31. Factor out the common factor 4:
$4r^2+12r-16=4(r^2+3r-4)$. Now, factor r^2+3r-4. The two numbers whose product is –4 and whose sum is 3 are 4 and –1, since (4)(–1) = –4 and 4 + (–1) = 3. Thus,
$r^2+3r-4=(r+4)(r-1)$, and
$4r^2+12r-16=4(r+4)(r-1)$.

33. Factor out the common factor x:
$x^3+3x^2-18x=x(x^2+3x-18)$. Now, factor $x^2+3x-18$. The two numbers whose product is –18 and whose sum is 3 are –3 and 6 since (–3)(6) = –18 and –3 + 6 = 3. Thus,
$x^2+3x-18=(x-3)(x+6)$, and
$x^3+3x^2-18x=x(x-3)(x+6)$.

35. $5a^2-8a+3$
Observe that (5)(3) = 15. The two numbers whose product is 15 and whose sum is –8 are –5 and –3, since (–5)(–3) = 15 and (–5) + (–3) = –8. Now the middle term, $-8a$, can be written as $-5a-3a$ and the factorization is

$$\begin{aligned}5a^2-8a+3&=5a^2-5a-3a+3\\&=5a(a-1)-3(a-1)\\&=(a-1)(5a-3)\end{aligned}$$

37. $3x^2-3x-6=3\left(x^2-x-2\right)$
Now factor $\left(x^2-x-2\right)$. The two numbers whose product is –2 and sum is –1 are –2 and 1, since (–2)(–1) = –2 and –2 + 1 = –1. Thus,
$3x^2-3x-6=3\left(x^2-x-2\right)=3\left(x-2\right)\left(x+1\right)$

39. $6c^2-13c-63$
Observe that 6(–63) = –378. The two numbers whose product is –378 and whose sum is –13 are –27 and 14, since (–27)(14) = –378 and –27 + 14 = –13. Now the middle term $-13c$ can be written as $-27c+14c$ and the factorization is

$$\begin{aligned}6c^2-13c-63&=6c^2-27c+14c-63\\&=3c(2c-9)+7(2c-9)\\&=(3c+7)(2c-9)\end{aligned}$$

41. $8b^2-2b-3$
Observe that (8)(–3) = –24. The two numbers whose product is –24 and whose sum is –2 are –6 and 4, since (–6)(4) = –24 and –6 + 4 = –2. Now the middle term, $-2b$, can be written as $-6b+4b$ and the factorization is

$$\begin{aligned}8b^2-2b-3&=8b^2-6b+4b-3\\&=2b(4b-3)+1(4b-3)\\&=(4b-3)(2b+1)\end{aligned}$$

43. $6c^2+11c-10$
Observe that (6)(–10) = –60. The two numbers whose product is –60 and whose sum is 11 are 15 and –4, since (15)(–4) = –60 and 15 + (–4) = 11. Now the middle term, $11c$, can be written as $15c+(-4c)$ and the factorization is

$$\begin{aligned}6c^2+11c-10&=6c^2+15c-4c-10\\&=3c(2c+5)-2(2c+5)\\&=(2c+5)(3c-2)\end{aligned}$$

45. $16p^2-16pq-12q^2=4\left(4p^2-4pq-3q^2\right)$
[4 is a common factor]
Now, factor $4p^2-4pq-3q^2$. Observe that $(4)(-3)=-12$. The two numbers whose product is –12 and whose sum is –4 are –6 and 2, since $(-6)(2)=-12$ and $-6+2=-4$. Now, the middle term, $-4pq$, can be written as $-6pq+2pq$ and the factorization is

$$\begin{aligned}4p^2-4pq-3q^2&=4p^2-6pq+2pq-3q^2\\&=2p(2p-3q)+q(2p-3q)\\&=(2p+q)(2p-3q)\end{aligned}$$

Thus, $16p^2-16pq-12q^2=4(2p+q)(2p-3q)$.

47. $4x^2 + 4xy + 9y^2$

Observe that (4)(9) = 36. Notice that there are no two rational numbers whose product is 36 and whose sum is 4. Therefore, the polynomial is prime and cannot be factored.

49. $18a^2 + 18ab - 8b^2 = 2\left(9a^2 + 9ab - 4b^2\right)$

Now, factor $9a^2 + 9ab - 4b^2$. Observe that (9)(–4) = –36. The two numbers whose product is –36 and whose sum is 9 are 12 and –3, since (12)(–3) = –36 and 12 + (–3) = 9. Then the middle term, 9*ab*, can be written as 9*ab* = 12*ab* – 3*ab* and the factorization is

$$\begin{aligned} 9a^2 + 9ab - 4b^2 &= 9a^2 + 12ab - 3ab - 4b^2 \\ &= 3a(3a+4b) - b(3a+4b) \\ &= (3a+4b)(3a-b) \end{aligned}$$

Thus, $18a^2 + 18ab - 8b^2 = 2(3a+4b)(3a-b)$.

51. $8x^2 + 30xy - 27y^2$

Observe that 8(–27) = –216. The two numbers whose product is –216 and whose sum is 30 are 36 and –6, since (36)(–6) = –216 and 36 – 6 = 30. The middle term can be written as $36xy - 6xy = 30xy$ and the factorization is

$$\begin{aligned} 8x^2 + 30xy - 27y^2 &= 8x^2 + 36xy - 6xy - 27y^2 \\ &= 4x(2x+9y) - 3(2x+9y) \\ &= (4x-3y)(2x+9y) \end{aligned}$$

53. $100b^2 - 90b + 20 = 10(10b^2 - 9b + 2)$

Now, factor $10b^2 - 9b + 2$. Observe that (10)(2) = 20. The two numbers whose product is 20 and whose sum is –9 are –4 and –5, since (–4)(–5) = 20 and –4 + (–5) = –9. Then the middle term, –9*b*, can be written as –4*b* – 5*b* and the factorization is

$$\begin{aligned} 10b^2 - 9b + 2 &= 10b^2 - 4b - 5b + 2 \\ &= 2b(5b-2) - 1(5b-2) \\ &= (5b-2)(2b-1) \end{aligned}$$

Thus, $100b^2 - 90b + 20 = 10(5b-2)(2b-1)$.

55. $a^3b^5 - a^2b^5 - 12ab^5 = ab^5(a^2 - a - 12)$

Now, factor $a^2 - a - 12$. The two numbers whose product is –12 and whose sum is –1 are –4 and 3, since (–4)(3) = –12 and –4 + 3 = –1. Thus, $a^2 - ab - 12 = (a-4)(a+3)$, and

$a^3b^5 - a^2b^5 - 12ab^5 = ab^5(a-4)(a+3)$.

57. $3b^4c - 18b^3c^2 + 27b^2c^3 = 3b^2c(b^2 - 6bc + 9c^2)$

Now factor $b^2 - 6bc + 9c^2$. The two numbers whose product is 9 and whose sum is –6 are –3 and –3, since (–3)(–3) = 9 and –3 + (–3) = –6. Thus, $b^2 - 6bc + 9c^2 = (b-3c)(b-3c)$, and

$$\begin{aligned} 3b^4c - 18b^3c^2 + 27b^2c^3 &= 3b^2c(b-3c)(b-3c) \\ &= 3b^2c(b-3c)^2 \end{aligned}$$

59. $8m^8n^3 + 4m^7n^4 - 24m^6n^5 = 4m^6n^3(2m^2 + mn - 6n^2)$

Now factor $2m^2 + mn - 6n^2$. Observe that (2)(–6) = –12. The two numbers whose product is –12 and whose sum is 1 are 4 and –3, since (4)(–3) = –12 and 4 + (–3) = 1. Then the middle term, *mn*, can be written as 4*mn* – 3*mn* and the factorization is

$$\begin{aligned} 2m^2 + mn - 6n^2 &= 2m^2 + 4mn - 3mn - 6n^2 \\ &= 2m(m+2n) - 3n(m+2n) \\ &= (2m-3n)(m+2n) \end{aligned}$$

Thus, $8m^8n^3 + 4m^7n^4 - 24m^6n^5$

$= 4m^6n^3(m+2n)(2m-3n)$

61. $30x^2 - x - 20$

Observe that (30)(–20) = –600. The two numbers whose product is –600 and whose sum is –1 are –25 and 24, since (–25)(24) = –600 and –25 + 24 = –1. Then the middle term, –*x*, can be written as –25*x* + 24*x* and the factorization is

$$\begin{aligned} 30x^2 - x - 20 &= 30x^2 - 25x + 24x - 20 \\ &= 5x(6x-5) + 4(6x-5). \\ &= (6x-5)(5x+4) \end{aligned}$$

63. $8x^4y^5 + 24x^3y^5 - 32x^2y^5 = 8x^2y^5\left(x^2 + 3x - 4\right)$

Now, factor $x^2 + 3x - 4$. The two numbers whose product is –4 and whose sum is 3 are 4 and –1, since (4)(–1) = –4 and 4 + (–1) = 3. Thus, $8x^4y^5 + 24x^3y^5 - 32x^2y^5 = 8x^2y^5(x+4)(x-1)$.

65. $x^4+x^2-6=(x^2)^2+x^2-6$

$=y^2+y-6$ ← Replace x^2 by y

$=(y+3)(y-2)$ ← Use 3 and –2 since $(3)(-2)=-6$, $3+(-2)=1$

$=(x^2+3)(x^2-2)$ ← Replace y by x^2

67. $b^4+9b^2+20=(b^2)^2+9b^2+20$

$=u^2+9u+20$ ← Replace b^2 by u

$=(u+4)(u+5)$ ← Use 4 and 5 since $(4)(5)=20, 4+5=9$

$=(b^2+4)(b^2+5)$ ← Replace u by b^2

69. $6a^4+5a^2-25=6(a^2)^2+5a^2-25$

$=6w^2+5w-25$ ← Replace a^2 by w

$=6w^2-10w+15w-25$ ← Use $-10w+15w$ for $5w$ since $(-10)(15)=-150$, $-10+15=5$

$=2w(3w-5)+5(3w-5)$ ← Factor by grouping

$=(2w+5)(3w-5)$

$=(2a^2+5)(3a^2-5)$ ← Replace w by a^2

71. $4(x+1)^2+8(x+1)+3=4y^2+8y+3$ ← Replace $x+1$ by y

$=4y^2+6y+2y+3$ ← Use $6y+2y$ for $8y$ since $(6)(2)=12$, and $6+2=8$

$=2y(2y+3)+1(2y+3)$

$=(2y+1)(2y+3)$

$=[2(x+1)+1][2(x+1)+3]$ ← Replace y by $x+1$

$=(2x+2+1)(2x+2+3)$

$=(2x+3)(2x+5)$

73. $6(a+2)^2-7(a+2)-5=6y^2-7y-5$ ← Replace $a+2$ by y

$=6y^2+3y-10y-5$ ← Use $3y-10y$ for $-7y$ since $(3)(-10)=-30$ and $3+(-10)=-7$

$=3y(2y+1)-5(2y+1)$

$=(3y-5)(2y+1)$

$=[3(a+2)-5][2(a+2)+1]$ ← Replace y by $a+2$

$=(3a+6-5)(2a+4+1)$

$=(3a+1)(2a+5)$

75. $x^2y^2+9xy+14=(xy)^2+9xy+14$

$=w^2+9w+14$ ← Replace xy by w

$=(w+2)(w+7)$ ← Use 2 and 7 since $(2)(7)=14$, $2+7=9$

$=(xy+2)(xy+7)$ ← Replace w by xy

77. $2x^2y^2-9xy-11=2(xy)^2-9xy-11$

$=2w^2-9w-11 \quad \leftarrow$ Replace xy by w

$=2w^2+2w-11w-11 \leftarrow$ Use $2w-11w$ for $-9w$ since $(2)(-11)=-22$ and $2+(-11)=-9$

$=2w(w+1)-11(w+1)$

$=(2w-11)(w+1)$

$=(2xy-11)(xy+1) \quad \leftarrow$ Replace w by xy

79. $2y^2(2-y)-7y(2-y)+5(2-y)$

$=(2-y)(2y^2-7y+5) \quad \leftarrow 2-y$ is a common factor

$=(2-y)(2y^2-2y-5y+5) \leftarrow$ Use $-2y-5y$ for $-7y$ since $(-2)(-5)=10$ and $-2+(-5)=-7$

$=(2-y)[2y(y-1)-5(y-1)] \leftarrow$ Factor by grouping

$=(2-y)(y-1)(2y-5)$

81. $2p^2(p-4)+7p(p-4)+6(p-4)$

$=(p-4)(2p^2+7p+6) \quad \leftarrow p-4$ is a common factor

$=(p-4)(2p^2+4p+3p+6) \leftarrow$ Use $4p+3p$ for $7p$ since $(4)(3)=12$ and $4+3=7$

$=(p-4)[2p(p+2)+3(p+2)] \leftarrow$ Factor by grouping

$=(p-4)(2p+3)(p+2)$

83. $a^6-7a^3-30=(a^3)^2-7a^3-30$

$=b^2-7b-30 \quad \leftarrow$ Replace a^3 by b

$=(b-10)(b+3) \quad \leftarrow$ Use -10 and 3 since $(-10)(3)=-30$ and $-10+3=-7$

$=(a^3-10)(a^3+3) \leftarrow$ Replace b by a^3

85. $x^2(x+5)+3x(x+5)+2(x+5)=(x+5)(x^2+3x+2) \quad \leftarrow x+5$ is a common factor

$=(x+5)(x+2)(x+1) \quad \leftarrow$ Use 2 and 1 since $(2)(1)=2$ and $2+1=3$

87. $5a^5b^2-8a^4b^3+3a^3b^4$

$=a^3b^2(5a^2-8ab+3b^2) \quad \leftarrow a^3b^2$ is a common factor

$=a^3b^2[5a^2-5ab-3ab+3b^2] \leftarrow$ Use $-5ab-3ab$ for $8ab$ since $(-5)(-3)=-15, -5+(-3)=-8$

$=a^3b^2[5a(a-b)-3b(a-b)] \leftarrow$ Factor by grouping

$=a^3b^2(5a-3b)(a-b)$

---

89. $A_1-A_2=(x+5)(x+2)-(2)(2)$

$=x^2+2x+5x+10-4$

$=x^2+7x+6$

$=(x+6)(x+1)$

91. $A_1-A_2=(x+4)(x+5)-(2)(1)$

$=x^2+5x+4x+20-3$

$=x^2+9x+18$

$=(x+6)(x+3)$

93. To find the polynomial multiply the factors.

$$\begin{aligned}(2x+3y)(x-4y) &= 2x(x-4y)+3y(x-4y)\\ &= 2x^2-8xy+3xy-12y^2\\ &= 2x^2-5xy-12y^2\end{aligned}$$

95. To find the other factor, simply divide $x^2+4x-21$ by $x-3$.

$$x-3\overline{)x^2+4x-21}\quad \text{quotient } x+7$$

$\underline{x^2-3x}$ ← $x(x-3)$

$7x-21$

$\underline{7x-21}$ ← $7(x-3)$

0 ← Remainder

The other factor is $x+7$.

97. a. Answers will vary.

b. $$\begin{aligned}30x^2+23x-40 &= 30x^2-25x+48x-40\\ &= 5x(6x-5)+8(6x-5)\\ &= (5x+8)(6x-5)\end{aligned}$$

$$\begin{aligned}49x^2-98x+13 &= 49x^2-7x-91x+13\\ &= 7x(7x-1)-13(7x-1)\\ &= (7x-13)(7x-1)\end{aligned}$$

99. To factor $2x^2+bx-5$, the factors must be of the form $(2x-5)(x+1)$ which gives $2x^2-3x-5$ or $(2x-1)(x+5)$ which gives $2x^2+9x-5$ or $(2x+5)(x-1)$ which gives $2x^2+3x-5$ or $(2x+1)(x-5)$ which gives $2x^2-9x-5$. Therefore $b=\pm3,\ \pm9$.

101. To factor x^2+bx+5, the factors must be of the form $(x+1)(x+5)$ which gives x^2+6x+5 or $(x-1)(x-5)$ which gives x^2-6x+5. Therefore, $b=6$ or -6.

103. a. $b^2-4ac=(-8)^2-4(1)(15)=64-60=4$ a perfect square

b. $$\begin{aligned}x^2-8x+15 &= x^2-5x-3x+15\\ &= x(x-5)-3(x-5)\\ &= (x-3)(x-5)\end{aligned}$$

105. a. $b^2-4ac=(-4)^2-4(1)(6)=16-24=-8$ not a perfect square

b. Not factorable

107. Answers will vary. One example is x^2+2x+1

109. $$\begin{aligned}&4a^{2n}-4a^n-15\\ &=4(a^n)^2-4a^n-15\\ &=4y^2-4y-15 \qquad \leftarrow \text{Replace } a^n \text{ by } y\\ &=4y^2-10y+6y-15\\ &=2y(2y-5)+3(2y-5)\\ &=(2y+3)(2y-5)\\ &=(2a^n+3)(2a^n-5) \qquad \leftarrow \text{Replace } y \text{ by } a^n\end{aligned}$$

111. $$\begin{aligned}&x^2(x+y)^2-7xy(x+y)^2+12y^2(x+y)^2\\ &=(x+y)^2(x^2-7xy+12y^2)\\ &=(x+y)^2(x^2-3xy-4xy+12y^2)\\ &=(x+y)^2[x(x-3y)-4y(x-3y)]\\ &=(x+y)^2(x-4y)(x-3y)\end{aligned}$$

113. $$\begin{aligned}&x^{2n}+3x^n-10\\ &=(x^n)^2+3(x^n)-10\\ &=y^2+3y-10 \qquad \leftarrow \text{Replace } x^n \text{ with } y\\ &=y^2-2y+5y-10\\ &=y(y-2)+5(y-2)\\ &=(y+5)(y-2)\\ &=(x^n+5)(x^n-2) \qquad \leftarrow \text{Replace } y \text{ with } x^n\end{aligned}$$

115. a. Answers will vary.

b.

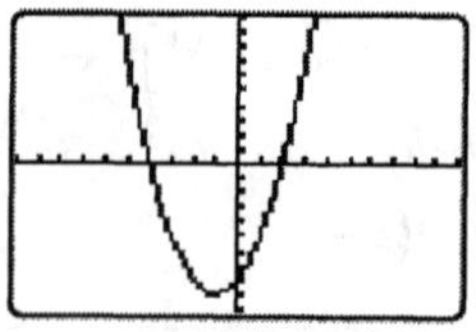

−10,10,1,−10,10,1

The graphs of $y_1=x^2+2x-8$ and $y_2=(x+4)(x-2)$ are the same. The factoring is correct.

117.
$$F=\frac{9}{5}C+32$$
$$F-32=\frac{9}{5}C$$
$$\frac{5}{9}(F-32)=\frac{5}{9}\left(\frac{9}{5}C\right)$$
$$\frac{5}{9}(F-32)=C \text{ or } C=\frac{5}{9}(F-32)$$

118. Since the equation $y=-3x+4$ is in slope-intercept form, we know that the slope is –3 and the y-intercept is 4. Use these to graph the line.

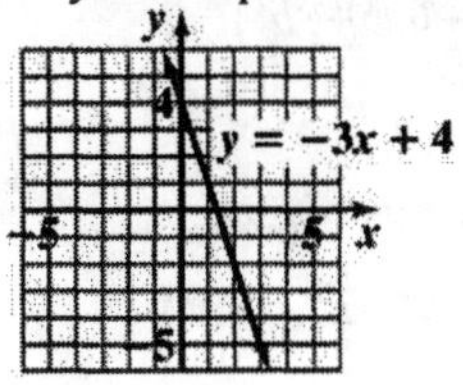

119.
$$\begin{vmatrix}3 & -2 & -1\\ 2 & 3 & -2\\ 1 & -4 & 1\end{vmatrix}$$
$$=3\begin{vmatrix}3 & -2\\ -4 & 1\end{vmatrix}-(-2)\begin{vmatrix}2 & -2\\ 1 & 1\end{vmatrix}+(-1)\begin{vmatrix}2 & 3\\ 1 & -4\end{vmatrix}$$
$$=3(3-8)+2(2+2)-1(-8-3)$$
$$=3(-5)+2(4)-1(-11)$$
$$=-15+8+11$$
$$=4$$

120.
$$\left[(x+y)+6\right]^2=(x+y)^2+2\cdot 6\cdot(x+y)+(6)^2$$
$$=(x+y)^2+12(x+y)+36$$
$$=x^2+2xy+y^2+12x+12y+36$$

121.
$$2x^3+4x^2-5x-10=2x^2(x+2)-5(x+2)$$
$$=(x+2)(2x^2-5)$$

Exercise Set 5.6

1. a. Answers will vary.

b. $x^2-16=x^2-4^2=(x+4)(x-4)$

3. Answers will vary.

5. $a^3+b^3=(a+b)(a^2-ab+b^2)$

7. No. $(x+7)(x-7)=x^2-7x+7x-49$
$$=x^2-49$$
$$\neq x^2+14x-49$$

9. No. $(x-9)^2=(x-9)(x-9)$
$$=x^2-9x-9x+81$$
$$=x^2-18x+81$$
$$\neq x^2-81$$

11. $x^2-81=(x)^2-(9)^2=(x+9)(x-9)$

13. $a^2-100=(a)^2-(10)^2=(a+10)(a-10)$

15. $1-49b^2=(1)^2-(7b)^2=(1+7b)(1-7b)$

17. $25-16y^4=(5)^2-(4y^2)^2=(5+4y^2)(5-4y^2)$

19. $\frac{1}{100}-y^2=\left(\frac{1}{10}\right)^2-(y)^2=\left(\frac{1}{10}+y\right)\left(\frac{1}{10}-y\right)$

21. $x^2y^2-121c^2=(xy)^2-(11c)^2$
$$=(xy+11c)(xy-11c)$$

23. $0.4x^2-0.9=(0.2x)^2-(0.3)^2$
$$=(0.2x+0.3)(0.2x-0.3)$$

25. $36-(x-6)^2=(6)^2-(x-6)^2$
$$=\left[6+(x-6)\right]\left[6-(x-6)\right]$$
$$=(6+x-6)(6-x+6)$$
$$=x(12-x)$$

27. $a^2-(3b+2)^2=\left[a+(3b+2)\right]\left[a-(3b+2)\right]$
$$=(a+3b+2)(a-3b-2)$$

29. $x^2+10x+25=x^2+2(x)(5)+5^2=(x+5)^2$

31. $49-14t+t^2=7^2-2(7)(t)+t^2=(7-t)^2$

33. $36p^2q^2+12pq+1=(6pq)^2+2(6pq)(1)+1^2$
$$=(6pq+1)^2$$

35. $0.81x^2-0.36x+0.04$
$$=(0.9x)^2-2(0.9)(0.2)x+(0.2)^2$$
$$=(0.9x-0.2)^2$$

37. $y^4+4y^2+4=(y^2)^2+2(y^2)(2)+2^2=(y^2+2)^2$

39. $(a+b)^2+6(a+b)+9=(a+b)^2+2(a+b)(3)+3^2$
$=[(a+b)+3]^2$
$=(a+b+3)^2$

41. $(y-3)^2+8(y-3)+16=(y-3)^2+2(y-3)(4)+4^2$
$=[(y-3)+4]^2$
$=(y-3+4)^2$
$=(y+1)^2$

43. $x^2+6x+9-y^2=(x+3)^2-(y)^2$
$=[(x+3)+y][(x+3)-y]$
$=(x+3+y)(x+3-y)$

45. $25-(x^2+4x+4)=(5)^2-(x+2)^2$
$=[5+(x+2)][5-(x+2)]$
$=(5+x+2)(5-x-2)$
$=(x+7)(3-x)$

47. $9a^2-12ab+4b^2-9$
$=(3a-2b)^2-(3)^2$
$=[(3a-2b)+3][(3a-2b)-3]$
$=(3a-2b+3)(3a-2b-3)$

49. $y^4-6y^2+9=(y^2)^2-2(y^2)(3)+3^2=(y^2-3)^2$

51. $a^3+125=(a)^3+(5)^3$
$=(a+5)[a^2-a(5)+5^2]$
$=(a+5)(a^2-5a+25)$

53. $64-a^3=(4)^3-(a)^3$
$=(4-a)[4^2+4(a)+a^2]$
$=(4-a)(16+4a+a^2)$

55. $p^3-27a^3=(p)^3-(3a)^3$
$=(p-3a)((p)^2+(p)(3a)+(3a)^2)$
$=(p-3a)(p^2+3ap+9a^2)$

57. $27y^3-8x^3=(3y)^3-(2x)^3$
$=(3y-2x)[(3y)^2+(3y)(2x)+(2x)^2]$
$=(3y-2x)(9y^2+6xy+4x^2)$

59. $16a^3-54b^3$
$=2(8a^3-27b^3)$
$=2[(2a)^3-(3b)^3]$
$=2(2a-3b)[(2a)^2+(2a)(3b)+(3b)^2]$
$=2(2a-3b)(4a^2+6ab+9b^2)$

61. $y^6+x^9=(y^2)^3+(x^3)^3$
$=(y^2+x^3)[(y^2)^2-(y^2)(x^3)+(x^3)^2]$
$=(y^2+x^3)(y^4-x^3y^2+x^6)$

63. $(x+1)^3+1$
$=(x+1)^3+(1)^3$
$=[(x+1)+1][(x+1)^2-(x+1)(1)+1^2]$
$=(x+2)(x^2+2x+1-x-1+1)$
$=(x+2)(x^2+x+1)$

65. $(a-b)^3-27$
$=(a-b)^3-(3)^3$
$=[(a-b)-3][(a-b)^2+(a-b)(3)+3^2]$
$=(a-b-3)(a^2-2ab+b^2+3a-3b+9)$

67. $b^3-(b+3)^3$
$=[b-(b+3)][b^2+b(b+3)+(b+3)^2]$
$=(b-b-3)(b^2+b^2+3b+b^2+6b+9)$
$=(-3)(3b^2+9b+9)$
$=(-3)\cdot 3(b^2+3b+3)$
$=-9(b^2+3b+3)$

69. $a^4-4b^4=(a^2)^2-(2b^2)^2=(a^2+2b^2)(a^2-2b^2)$

71. $49-64x^2y^2=(7)^2-(8xy)^2=(7+8xy)(7-8xy)$

73. $(x+y)^2-16=(x+y)^2-4^2$
$=[(x+y)+4][(x+y)-4]$
$=(x+y+4)(x+y-4)$

75. $x^3-64=(x)^3-(4)^3$
$=(x-4)\left[x^2+x(4)+4^2\right]$
$=(x-4)(x^2+4x+16)$

77. $9x^2y^2+24xy+16=(3xy)^2+2(3xy)(4)+(4)^2$
$=(3xy+4)^2$

79. $a^4+2a^2b^2+b^4=(a^2)^2+2(a^2)(b^2)+(b^2)^2$
$=(a^2+b^2)^2$

81. $x^2-2x+1-y^2=(x-1)^2-(y)^2$
$=[(x-1)+y][(x-1)-y]$
$=(x-1+y)(x-1-y)$

83. $(x+y)^3+1$
$=(x+y)^3+(1)^3$
$=[(x+y)+1]\left[(x+y)^2-(x+y)(1)+1^2\right]$
$=(x+y+1)(x^2+2xy+y^2-x-y+1)$

85. $(m+n)^2-(2m-n)^2$
$=[(m+n)+(2m-n)][(m+n)-(2m-n)]$
$=(3m)(-m+2n)$
$=3m(-m+2n)$

87. $V_1-V_2=(3x)^3-(2)^3$
$=(3x-2)\left((3x)^2+(3x)(2)+(2)^2\right)$
$=(3x-2)(9x^2+6x+4)$

89. $V_1-V_2=(6a)^3-(b)^3$
$=(6a-b)\left((6a)^2+(6a)(b)+(b)^2\right)$
$=(6a-b)(36a^2+6ab+b^2)$

91. $V_1+V_2=(4x)^3+(3a)^3$
$=(4x+3a)\left((4x)^2-(4x)(3a)+(3a)^2\right)$
$=(4x+3a)(16x^2-12ax+9a^2)$

93. Area of larger square is $(a)(a)=a^2$.
Area of smaller square is $(b)(b)=b^2$.

a. Area of shaded region is a^2-b^2.

b. In factored form, the area is
$a^2-b^2=(a+b)(a-b)$.

95. Volume of larger solid is $(6a)(a)(a)=6a^3$.
Volume of smaller solid is $(6a)(b)(b)=6ab^2$.

a. Volume of shaded region is $6a^3-6ab^2$.

b. In factored form, the volume is
$6a^3-6ab^2=6a(a^2-b^2)=6a(a+b)(a-b)$.

97. Volume of larger sphere is $\frac{4}{3}\pi R^3$.

Volume of smaller sphere is $\frac{4}{3}\pi r^3$.

a. Volume of shaded region is $\frac{4}{3}\pi R^3-\frac{4}{3}\pi r^3$.

b. In factored form the volume is
$\frac{4}{3}\pi R^3-\frac{4}{3}\pi r^3=\frac{4}{3}\pi(R^3-r^3)$
$=\frac{4}{3}\pi(R-r)(R^2+Rr+r^2)$.

99. Express $4x^2+bx+9$ as $(2x)^2+bx+(3)^2$. Now,
$bx=2(2x)(3)$ or $bx=-2(2x)(3)$
$bx=12$ $\quad$ $bx=-12x$
$b=12$ $\quad$ $b=-12$

101. Express c as a^2. Then $25x^2+20x+c$ becomes
$25x^2+20x+a^2$ or $(5x)^2+20x+(a)^2$. But,
$20x=2(5x)(a)$
$20x=10xa$
$\frac{20x}{10x}=a$
$2=a$
Now, $a=2$ and $c=a^2=(2)^2=4$.

103. a. Find an expression whose square is $25x^2-30x+9$.

b. $A(x)=[s(x)]^2$

$A(x)=25x^2-30x+9=(5x-3)^2$

Therefore $s(x)=5x-3$

c. $s(2)=5(2)-3=10-3=7$

105. $x^4+64=(x^4+16x^2+64)-16x^2$

$=(x^2+8)^2-(4x)^2$

$(x^2+8+4x)(x^2+8-4x)$

$=(x^2+4x+8)(x^2-4x+8)$

107. $P(x)=x^2$

$P(a+h)-P(a)=(a+h)^2-a^2$

$=a^2+2ah+h^2-a^2$

$=2ah+h^2$

$=h(2a+h)$

109. a. The area is $4\cdot 4=16$.

b. The sum is $x^2+8x+16$

c. $x^2+8x+16=(x+4)^2$

111. $64x^{4a}-9y^{6a}=(8x^{2a})^2-(3y^{3a})^2$

$=(8x^{2a}+3y^{3a})(8x^{2a}-3y^{3a})$

113. $a^{2n}-16a^n+64=(a^n)^2-2(a^n)(8)+(8)^2$

$=(a^n-8)^2$

115. $x^{3n}-8=(x^n)^3-2^3$

$=(x^n-2)\left[(x^n)^2+(x^n)(2)+(2)^2\right]$

$=(x^n-2)(x^{2n}+2x^n+4)$

117. $y_1=2x^2-18$

$y_2=2(x+3)(x-3)$

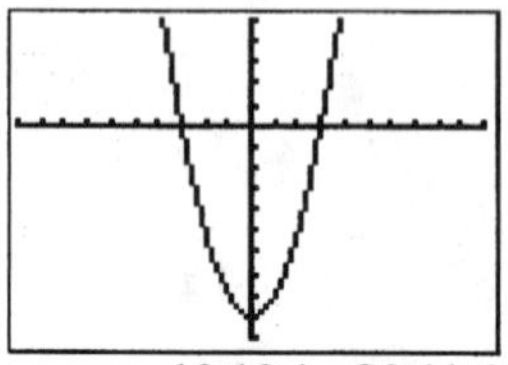

$-10,10,1,-20,10,2$

The factoring is correct because the graphs are the same.

119. a. $x^6-1=(x^3)^2-1^2=(x^3+1)(x^3-1)$

b. $x^6-1=(x^2)^3-1^3$

$=(x^2-1)\left[(x^2)^2+(x^2)(1)+(1)^2\right]$

$=(x^2-1)(x^4+x^2+1)$

c. x^6-1

$=(x^3-1)(x^3+1)$

$=(x-1)(x^2+x+1)(x+1)(x^2-x+1)$

$=(x-1)(x+1)(x^2+x+1)(x^2-x+1)$

$=(x^2-1)(x^4+x^2+1)$ upon multiplication

121. $-2[3x-(2y-1)-5x]+3y$

$=-2[3x-2y+1-5x]+3y$

$=-2[-2x-2y+1]+3y$

$=4x+4y-2+3y$

$=4x+7y-2$

122. $f(x)=x^2-3x+6$ and $g(x)=5x-2$

$(g-f)(x)=(5x-2)-(x^2-3x+6)$

$=5x-2-x^2+3x-6$

$=-x^2+8x-8$

$(g-f)(-1)=-(-1)^2+8(-1)-8$

$=-1-8-8$

$=-17$

123. Let x be the measure of the smallest angle. Then $2x$ will be the measure of the largest angle, and $x+10$ will be the measure of the remaining angle.

$x+2x+(x+10)=90$

$4x+10=90$

$4x=80$

$x=20$

So, $2x=2(20)=40$ and $x+10=20+10=30$.

The measures of the angles are 20°, 30°, and 40°.

124. $45y^{12}+60y^{10}=15y^{10}(3y^2+4)$

125. $12x^2-9xy+4xy-3y^2$

$=3x(4x-3y)+y(4x-3y)$

$=(4x-3y)(3x+y)$

Exercise Set 5.7

1. Answers will vary. Possible answers follow.

   a. Factor out the GCF other than 1 if one is present. If the remaining polynomial is the difference of two squares, difference of two cubes, or the sum of two cubes, use the following factoring formulas.
   $a^2-b^2=(a+b)(a-b)$
   $a^3-b^3=(a-b)(a^2+ab+b^2)$
   $a^3+b^3=(a+b)(a^2-ab+b^2)$

   b. Factor out the GCF other than 1 if one is present. If the remaining polynomial is a perfect square, use the special factoring formulas:
   $a^2+2ab+b^2=(a+b)^2$
   $a^2-2ab+b^2=(a-b)^2$
   If the remaining polynomial is not a perfect square, factor using trial and error, grouping, or substitution.

   c. Factor out the GCF other than 1 if one is present. Try to factor the remaining polynomial using by grouping. If that does not work, check to see if three of the terms is are the square of a binomial.

---

3. $3x^2-75=3(x^2-25)$ ← 3 is a common factor
   $=3(x+5)(x-5)$ ← Factor x^2-25 as difference of squares

5. $10s^2+19s-15=10s^2+25s-6s-15$ ← Use $19s=25s-6s$
   $=5s(2s+5)-3(2s+5)$
   $=(5s-3)(2s+5)$

7. $6x^3y^2+10x^2y^3+14x^2y^2=2x^2y^2(3x+5y+7)$ ← $2x^2y^2$ is a common factor

9. $0.8x^2-0.072=0.8(x^2-0.09)$ ← 0.8 is a common factor
   $=0.8(x^2-0.3^2)$ ← Difference of two squares
   $=0.8(x+0.3)(x-0.3)$

11. $6x^5-54x=6x(x^4-9)$ ← $6x$ is a common factor
    $=6x\left[(x^2)^2-(3)^2\right]$ ← Difference of two squares
    $=6x(x^2+3)(x^2-3)$

13. $3x^6-3x^5+12x^5-12x^4=3x^4(x^2-x+4x-4)$ ← $3x^4$ is a common factor
    $=3x^4[x(x-1)+4(x-1)]$ ← Factor by grouping
    $=3x^4(x-1)(x+4)$

15. $5x^4y^2+20x^3y^2+15x^3y^2+60x^2y^2=5x^2y^2(x^2+4x+3x+12)$ ← $5x^2y^2$ is a common factor
    $=5x^2y^2[x(x+4)+3(x+4)]$ ← Factor by grouping
    $=5x^2y^2(x+4)(x+3)$

17. $x^4-x^2y^2=x^2(x^2-y^2)$ ← x^2 is a common factor
    $=x^2(x+y)(x-y)$ ← Difference of two squares

19. $x^7y^2 - x^4y^2 = x^4y^2\left(x^3 - 1\right)$ ← x^4y^2 is a common factor

$= x^4y^2(x-1)\left(x^2+x+1\right)$ ← Difference of two cubes

21. $x^5 - 16x = x\left(x^4 - 16\right)$ ← x is a common factor

$= x\left[\left(x^2\right)^2 - (4)^2\right]$ ← Difference of two squares

$= x\left(x^2+4\right)\left(x^2-4\right)$ ← Difference of two squares on $x^2 - 4$

$= x\left(x^2+4\right)(x+2)(x-2)$

23. $4x^6 + 32y^3 = 4\left(x^6 + 8y^3\right)$ ← 2 is a common factor

$= 4\left[\left(x^2\right)^3 + (2y)^3\right]$ ← Sum of two cubes

$= 4\left(x^2+2y\right)\left[\left(x^2\right)^2 - \left(x^2\right)(2y) + (2y)^2\right]$ ← Simplify

$= 4\left(x^2+2y\right)\left(x^4 - 2x^2y + 4y^2\right)$

25. $5(a+b)^2 - 20 = 5\left[(a+b)^2 - 4\right]$ ← 2 is a common factor

$= 5[(a+b)+2][(a+b)-2]$ ← Difference of two squares

$= 5(a+b+2)(a+b-2)$ ← Simplify

27. $6x^2 + 36xy + 54y^2 = 6\left(x^2 + 6xy + 9y^2\right)$ ← 6 is a common factor

$= 6\left[x^2 + 2(x)(3y) + (3y)^2\right]$ ← Perfect square trinomial

$= 6(x+3y)^2$

29. $(x+2)^2 - 4 = [(x+2)-2][(x+2)+2]$ ← Difference of two squares

$= (x+2-2)(x+2+2)$ ← Simplify

$= x(x+4)$

31. $6x^2 + 24xy - 3xy - 12y^2 = 3(2x^2 + 12xy - xy - 6y^2$ ← 3 is a common factor

$= 3[2x(x+4y) - y(x+4y)]$ ← Factor by grouping

$= 3(x+4y)(2x-y)$

33. $(y+5)^2 + 4(y+5) + 4 = (y+5)^2 + 2(y+5)(2) + (2)^2$ ← Perfect square trinomial

$= [(y+5)+2]^2$

$= (y+7)^2$

35. $b^4 + 2b^2 + 1 = \left(b^2\right)^2 + 2\left(b^2\right)(1) + 1^2$ ← Perfect square trinomial

$= \left(b^2+1\right)^2$

37. $x^3+\frac{1}{64}=x^3+\left(\frac{1}{4}\right)^3$ ← Sum of two cubes

$=\left(x+\frac{1}{4}\right)\left[x^2-x\left(\frac{1}{4}\right)+\left(\frac{1}{4}\right)^2\right]$

$=\left(x+\frac{1}{4}\right)\left[x^2-\frac{1}{4}x+\frac{1}{16}\right]$

39. $6y^3+14y^2+4y=2y(3y^2+7y+2)$ ← $2y$ is a common factor

$=2y(3y^2+6y+y+2)$ ← Use $7y=6y+y$

$=2y[3y(y+2)+1(y+2)]$

$=2y(3y+1)(y+2)$

41. $a^3b-81ab^3=ab(a^2-81b^2)$ ← ab is a common factor

$=ab\left[a^2-(9b)^2\right]$ ← Difference of two squares

$=ab(a+9b)(a-9b)$

43. $49-(x^2+2xy+y^2)=7^2-(x+y)^2$ ← Perfect square trinomial

$=[7+(x+y)][7-(x+y)]$ ← Difference of two squares

$=(7+x+y)(7-x-y)$ ← Simplify

45. $24x^2-34x+12=2(12x^2-17x+6)$ ← 2 is a common factor

$=2(12x^2-8x-9x+6)$ ← Use $-17x=-8x-9x$

$=2[4x(3x-2)-3(3x-2)]$ ← Factor by grouping

$=2(3x-2)(4x-3)$

47. $18x^2+39x-15=3(6x^2+13x-5)$ ← 3 is a common factor

$=3(6x^2-2x+15x-5)$ ← Use $13x=-2x+15x$

$=3[2x(3x-1)+5(3x-1)]$ ← Factor by grouping

$=3(3x-1)(2x+5)$

49. $x^4-16=(x^2)^2-(4)^2$ ← Difference of two squares

$=(x^2+4)(x^2-4)$ ← Difference of two squares (x^2-4)

$=(x^2+4)(x+2)(x-2)$

51. $5bc-10cx-7by+14xy=5c(b-2x)-7y(b-2x)$ ← Factor by grouping

$=(b-2x)(5c-7y)$ ← $(b-2x)$ is a common factor

53. $3x^4 - x^2 - 4 = 3(x^2)^2 - x^2 - 4$

$= 3(x^2)^2 - 4x^2 + 3x^2 - 4$ ← Use $-x^2 = -4x^2 + 3x^2$

$= x^2(3x^2 - 4) + 1(3x^2 - 4)$ ← Factor by grouping

$= (3x^2 - 4)(x^2 + 1)$ ← $(3x^2 - 4)$ is a common factor

55. $z^2 - (x^2 - 12x + 36) = z^2 - (x-6)^2$ ← Perfect square trinomial

$= [z + (x-6)][z - (x-6)]$ ← Difference of two squares

$= (z + x - 6)(z - x + 6)$ ← Simplify

57. $2(y+4)^2 + 5(y+4) - 12 = 2x^2 + 5x - 12$ ← Replace $y+4$ with x

$= 2x^2 + 8x - 3x - 12$ ← Use $5x = 8x - 3x$

$= 2x(x+4) - 3(x+4)$ ← Factor by grouping

$= (2x-3)(x+4)$ ← $x+4$ is a common factor

$= [2(y+4) - 3][(y+4) + 4]$ ← Replace x with $y+4$

$= (2y + 8 - 3)(y+8)$ ← Simplify

$= (2y+5)(y+8)$ ← Simplify

59. $a^2 + 12ab + 36b^2 - 16c^2 = (a+6b)^2 - (4c)^2$ ← Perfect square trinomial

$= [(a+6b) + 4c][(a+6b) - 4c]$ ← Difference of two squares

$= (a + 6b + 4c)(a + 6b - 4c)$ ← Simplify

61. $10x^4y + 25x^3y - 15x^2y = 5x^2y(2x^2 + 5x - 3)$ ← $3x^2y$ is a common factor

$= 5x^2y(2x^2 + 6x - x - 3)$ ← Use $5x = 6x - x$

$= 5x^2y[2x(x+3) - 1(x+3)]$ ← Factor by grouping

$= 5x^2y(x+3)(2x-1)$ ← $x+3$ is a common factor

63. $x^4 - 2x^2y^2 + y^4 = (x^2)^2 - 2(x^2)(y^2) + (y^2)^2$ ← Perfect square trinomial

$= (x^2 - y^2)^2$ ← Difference of two squares

$= [(x+y)(x-y)]^2$

$= (x+y)^2(x-y)^2$

---

65. $a^2 + b^2$ is not factorable, (e)

67. $a^2 + 2ab + b^2 = (a+b)^2$, (d)

69. $a^3 - b^3 = (a-b)(a^2 + ab + b^2)$, (f)

71. $a^3 + b^3 = (a+b)(a^2 - ab + b^2)$, (c)

73. $P = 2(x^2 + 2) + 2(5x + 4)$

$= 2x^2 + 4 + 10x + 8$

$= 2x^2 + 10x + 12$

$= 2(x^2 + 5x + 6)$

$= 2(x+3)(x+2)$

75. $A = x^2 + 4x + 4x + (3)(4)$
$= x^2 + 8x + 12$
$= (x+6)(x+2)$

77. $A = y^2 - (3)(3)$
$= y^2 - 9$
$= (y+3)(y-3)$

79. $V = (5x)^3 - (3)^3$
$= (5x-3)\left((5x)^2 + (5x)(3) + (3)^2\right)$
$= (5x-3)(25x^2 + 15x + 9)$

81. The area of larger rectangle is $a(a + b)$.
The area of center rectangle is $b(a + b)$.

a. The area of shaded region is
$a(a+b) - b(a+b) = a^2 - b^2$.

b. In factored form, the area is
$a(a+b) - b(a+b) = (a-b)(a+b)$

83. a. The area of the shaded region is the sum of the areas of the three regions. This is
$(a)(a) + (2b)(a) + (b)(b) = a^2 + 2ab + b^2$.

b. In factored form, the area is
$a^2 + 2ab + b^2 = (a+b)^2$.

85. The area of the left side is $b(a-b)$.
The area of the right side is $b(a-b)$.
The area of the front side is $a(a-b)$.
The area of the back side is $a(a-b)$.

a. The surface area is $2a(a-b) + 2b(a-b)$.

b. In factored form, the area is
$2a(a-b) + 2b(a-b) = 2(a+b)(a-b)$.

87. a. Answers will vary.

b. Answers will vary.

89. a. $x^{-3} - 2x^{-4} - 3x^{-5} = x^{-5}(x^2 - 2x - 3)$

b. $x^{-5}(x^2 - 2x - 3) = x^{-5}(x-3)(x+1)$

91. a. $5x^{1/2} + 2x^{-1/2} - 3x^{-3/2}$
$= 5x^{-3/2}x^2 + 2x^{-3/2}x^1 - 3x^{-3/2}$
$= x^{-3/2}(5x^2 + 2x - 3)$

b. $x^{-3/2}(5x^2 + 2x - 3) = x^{-3/2}(5x-3)(x+1)$

92. $6(x+4) - 4(3x+3) = 6$
$6x + 24 - 12x - 12 = 6$
$-6x + 12 = 6$
$-6x = -6$
$x = 1$

93. $\left|\dfrac{6+2z}{3}\right| > 2$

$\dfrac{6+2z}{3} < -2$ or $\dfrac{6+2z}{3} > 2$
$6 + 2z < -6$ $\quad$ $6 + 2z > 6$
$2z < -12$ $\quad$ $2z > 0$
$z < -6$ $\quad$ $z > 0$
$\{z \mid z < -6 \text{ or } z > 0\}$

94. Let x be the number of pounds of coffee at \$5.20/lb and y be the number of pounds of coffee at \$6.30/lb. The problem situation is described by the following system of equations.
$x + y = 30$ (1)
$5.20x + 6.30y = 170$ (2)
From equation (1) we have $y = 30 - x$.
Substitute $30 - x$ for y in equation (2).
$5.20x + 6.30(30 - x) = 170$
$5.20x + 189 - 6.30x = 170$
$-1.10x = -19$
$x \approx 17.3$
John should mix 17.3 pounds of coffee at \$5.20/lb with 30 – 17.3 = 12.7 pounds of coffee at \$6.30/lb.

95.
$$\begin{array}{r} x^2 - x + 4 \\ \underline{5x + 4} \\ 4x^2 - 4x + 16 \\ \underline{5x^3 - 5x^2 + 20x } \\ 5x^3 - x^2 + 16x + 16 \end{array}$$

96. $2x^3 + 6x^2 - 5x - 15$
$2x^2(x+3) - 5(x+3)$
$(x+3)(2x^2 - 5)$

Exercise Set 5.8

1. The degree of a polynomial function is the same as the degree of the leading term.

3. $ax^2+bx+c=0$ is the standard form of a quadratic equation.

5. a. The zero factor property only holds when one side of the equation is 0.

b.
$$(x+3)(x+4)=2$$
$$x^2+7x+12=2$$
$$x^2+7x+10=0$$
$$(x+5)(x+2)=0$$
$$x+5=0 \quad \text{or} \quad x+2=0$$
$$x=-5 \qquad x=-2$$
The solutions are –2 and –5.

7. a. Answers will vary. Possible answer: Factor the polynomial, set the factors equal to 0, and then solve.

b.
$$-x-20=-12x^2$$
$$12x^2-x-20=0$$
$$(3x-4)(4x+5)=0$$
$$3x-4=0 \quad \text{or} \quad 4x+5=0$$
$$3x=4 \quad \text{or} \quad 4x=-5$$
$$x=\frac{4}{3} \quad \text{or} \quad x=-\frac{5}{4}$$
The solutions are $\frac{4}{3}, -\frac{5}{4}$.

9. a. The two shorter sides of a right triangle are called the legs.

b. The longest side of a right triangle is called the hypotenuse.

11. –8 and –2; answers will vary.

13. Yes, the graph of the function may be such that the graph never crosses the x-axis.

15. Yes, the graph of the function may be such that the graph touches the x-axis at two points.

17.
$$x(x+3)=0$$
$$x=0 \quad \text{or} \quad x+3=0$$
$$x=-3$$
The solutions are 0, –3.

19.
$$4x(x-1)=0$$
$$4x=0 \quad \text{or} \quad x-1=0$$
$$x=0 \qquad x=1$$
The solutions are 0, 1.

21.
$$2(x+1)(x-7)=0$$
$$x+1=0 \quad \text{or} \quad x-7=0$$
$$x=-1 \qquad x=7$$
The solutions are –1, 7.

23.
$$x(x-9)(x+4)=0$$
$$x=0 \quad \text{or} \quad x-9=0 \quad \text{or} \quad x+4=0$$
$$x=0 \qquad x=9 \qquad x=-4$$
The solutions are 0, 9, –4.

25.
$$(3x-2)(7x-1)=0$$
$$3x-2=0 \quad \text{or} \quad 7x-1=0$$
$$3x=2 \qquad 7x=1$$
$$x=\frac{2}{3} \qquad x=\frac{1}{7}$$
The solutions are $\frac{2}{3}, \frac{1}{7}$.

27.
$$4x^2=12x$$
$$4x^2-12x=0$$
$$4x(x-3)=0$$
$$4x=0 \quad \text{or} \quad x-3=0$$
$$x=0 \quad \text{or} \quad x=3$$
The solutions are 0, 3.

29.
$$x^2+5x=0$$
$$x(x+5)=0$$
$$x=0 \quad \text{or} \quad x+5=0$$
$$x=-5$$
The solutions are 0, –5.

31.
$$-x^2+6x=0$$
$$x(-x+6)=0$$
$$x=0 \quad \text{or} \quad -x+6=0$$
$$x=6$$
The solutions are 0, 6.

33. $3x^2 = 27x$

$3x^2 - 27x = 0$

$3x(x-9) = 0$

$3x = 0$ or $x - 9 = 0$

$x = 0$ $\quad x = 9$

The solutions are 0, 9.

35. $a^2 + 6a + 5 = 0$

$(a+5)(a+1) = 0$

$a + 5 = 0$ or $a + 1 = 0$

$a = -5$ $\quad a = -1$

The solutions are –5, –1.

37. $x^2 + x - 12 = 0$

$(x+4)(x-3) = 0$

$x + 4 = 0$ or $x - 3 = 0$

$x = -4$ $\quad x = 3$

The solutions are –4, 3.

39. $x^2 + 8x + 16 = 0$

$(x+4)(x+4) = 0$

$x + 4 = 0$ or $x + 4 = 0$

$x = -4$ $\quad x = -4$

The solution is –4.

41. $(2x+5)(x-1) = 12x$

$2x^2 - 2x + 5x - 5 = 12x$

$2x^2 - 9x - 5 = 0$

$(2x+1)(x-5) = 0$

$2x + 1 = 0$ or $x - 5 = 0$

$2x = -1$ $\quad x = 5$

$x = -\frac{1}{2}$

The solutions are $-\frac{1}{2}$, 5.

43. $2y^2 = -y + 6$

$2y^2 + y - 6 = 0$

$(2y-3)(y+2) = 0$

$2y - 3 = 0$ or $y + 2 = 0$

$2y = 3$ $\quad y = -2$

$y = \frac{3}{2}$

The solutions are $\frac{3}{2}, -2$.

45. $3x^2 - 6x - 72 = 0$

$3(x^2 - 2x - 24) = 0$

$3(x-6)(x+4) = 0$

$x - 6 = 0$ or $x + 4 = 0$

$x = 6$ $\quad x = -4$

The solutions are 6, –4.

47. $x^3 - 3x^2 = 18x$

$x^3 - 3x^2 - 18x = 0$

$x(x^2 - 3x - 18) = 0$

$x(x-6)(x+3) = 0$

$x = 0$ or $x - 6 = 0$ or $x + 3 = 0$

$x = 6$ $\quad x = -3$

The solutions are 0, 6, –3.

49. $4c^3 + 4c^2 - 48c = 0$

$4c(c^2 + c - 12) = 0$

$4c(c+4)(c-3) = 0$

$4c = 0$ or $c + 4 = 0$ or $c - 3 = 0$

$c = 0$ $\quad c = -4$ $\quad c = 3$

The solutions are 0, –4, 3.

51. $18z^3 = 15z^2 + 12z$

$18z^3 - 15z^2 - 12z = 0$

$3z(6z^2 - 5z - 4) = 0$

$3z(3z-4)(2z+1) = 0$

$3z = 0$ or $3z - 4 = 0$ or $2z + 1 = 0$

$z = 0$ $\quad 3z = 4$ $\quad 2z = -1$

$z = \frac{4}{3}$ $\quad z = -\frac{1}{2}$

The solutions are 0, $\frac{4}{3}, -\frac{1}{2}$.

53. $x^2 - 25 = 0$

$(x+5)(x-5) = 0$

$x + 5 = 0$ or $x - 5 = 0$

$x = -5$ $\quad x = 5$

The solutions are –5, 5.

55.
$$4x^2 = 9$$
$$4x^2 - 9 = 0$$
$$(2x+3)(2x-3) = 0$$
$$2x+3=0 \quad \text{or} \quad 2x-3=0$$
$$2x=-3 \qquad 2x=3$$
$$x=-\frac{3}{2} \qquad x=\frac{3}{2}$$
The solutions are $-\frac{3}{2}, \frac{3}{2}$.

57.
$$4y^3 - 36y = 0$$
$$4y(y^2-9) = 0$$
$$4y(y+3)(y-3) = 0$$
$$4y=0 \quad \text{or} \quad y+3=0 \quad \text{or} \quad y-3=0$$
$$y=0 \qquad y=-3 \qquad y=3$$
The solutions are 0, –3, 3.

59.
$$-x^2 = 2x - 99$$
$$-x^2 - 2x + 99 = 0$$
$$x^2 + 2x - 99 = 0$$
$$(x+11)(x-9) = 0$$
$$x+11=0 \quad \text{or} \quad x-9=0$$
$$x=-11 \qquad x=9$$
The solutions are –11, 9.

61.
$$(x+7)^2 - 16 = 0$$
$$[(x+7)+4][(x+7)-4] = 0$$
$$(x+11)(x+3) = 0$$
$$x+11=0 \quad \text{or} \quad x+3=0$$
$$x=-11 \qquad x=-3$$
The solutions are –11,–3.

63.
$$(2x+5)^2 - 9 = 0$$
$$[(2x+5)+3][(2x+5)-3] = 0$$
$$(2x+8)(2x+2) = 0$$
$$2(x+4)[2(x+1)] = 0$$
$$4(x+4)(x+1) = 0$$
$$x+4=0 \quad \text{or} \quad x+1=0$$
$$x=-4 \qquad x=-1$$
The solutions are –4, –1.

65.
$$6a^2 - 12 - 4a = 19a - 32$$
$$6a^2 - 23a + 20 = 0$$
$$(3a-4)(2a-5) = 0$$
$$3a-4=0 \quad \text{or} \quad 2a-5=0$$
$$3a=4 \qquad 2a=5$$
$$a=\frac{4}{3} \qquad a=\frac{5}{2}$$
The solutions are $\frac{4}{3}, \frac{5}{2}$.

67.
$$2b^3 + 16b^2 = -30b$$
$$2b^3 + 16b^2 + 30b = 0$$
$$2b(b^2 + 8b + 15) = 0$$
$$2b(b+5)(b+3) = 0$$
$$2x=0 \quad \text{or} \quad x+5=0 \quad \text{or} \quad x+3=0$$
$$x=0 \qquad x=-5 \qquad x=-3$$
The solutions are 0, –5, –3.

69. $f(x) = 3x^2 + 7x + 9$
$$7 = 3a^2 + 7a + 9$$
$$0 = 3a^2 + 7a + 2$$
$$0 = (3a+1)(a+2)$$
$$3a+1=0 \quad \text{or} \quad a+2=0$$
$$a=-\frac{1}{3} \qquad a=-2$$
The values of *a* are $-\frac{1}{3}$, –2.

71. $g(x) = 10x^2 - 31x + 16$
$$1 = 10a^2 - 31a + 16$$
$$0 = 10a^2 - 31a + 15$$
$$0 = (5a-3)(2a-5)$$
$$5a-3=0 \quad \text{or} \quad 2a-5=0$$
$$a=\frac{3}{5} \qquad a=\frac{5}{2}$$
The values of *a* are $\frac{3}{5}, \frac{5}{2}$.

73. $r(x) = x^2 - x$

$30 = a^2 - a$

$0 = a^2 - a - 30$

$0 = (a+5)(a-6)$

$a+5=0 \quad \text{or} \quad a-6=0$

$a=-5 \qquad a=6$

The values of a are –5, and 6.

75. $y = x^2 - 10x - 24$

Set $y = 0$ and solve the resulting equation.

$x^2 - 10x + 24 = 0$

$(x-4)(x-6) = 0$

$x-4=0 \quad \text{or} \quad x-6=0$

$x=4 \qquad x=6$

The x-intercepts are (4, 0) and (6, 0).

77. $y = x^2 + 16x + 64$

Set $y = 0$ and solve the resulting equation.

$x^2 + 16x + 64 = 0$

$(x+8)^2 = 0$

$x+8=0$

$x=-8$

The x-intercept is (–8, 0).

79. $y = 12x^3 - 46x^2 + 40x$

Set $y = 0$ and solve the resulting equation.

$12x^3 - 46x^2 + 40x = 0$

$2x(6x^2 - 23x + 20) = 0$

$2x(3x-4)(2x-5) = 0$

$2x=0 \quad \text{or} \quad 3x-4=0 \quad \text{or} \quad 2x-5=0$

$x=0 \qquad 3x=4 \qquad 2x=5$

$x=\frac{4}{3} \qquad x=\frac{5}{2}$

The x-intercepts are (0, 0), $\left(\frac{4}{3}, 0\right)$, and $\left(\frac{5}{2}, 0\right)$.

81. $(x+3)^2 + (x+2)^2 = (x+4)^2$

$(x^2+6x+9) + (x^2+4x+4) = x^2+8x+16$

$x^2+2x-3=0$

$(x+3)(x-1)=0$

$x+3=0 \quad \text{or} \quad x-1=0$

$x=-3 \qquad x=1$

Since $x = -3$ would result in negative length for the small side, we must reject this solution. Therefore, $x = 1$.

83. $(x)^2 + (x+7)^2 = (x+8)^2$

$(x^2) + (x^2+14x+49) = x^2+16x+64$

$x^2-2x-15=0$

$(x-5)(x+3)=0$

$x-5=0 \quad \text{or} \quad x+3=0$

$x=5 \qquad x=-3$

Since $x = -3$ would result in negative length for the small side, we must reject this solution. Therefore, $x = 5$.

85. $(x-1)^2 + (x+6)^2 = (x+8)^2$

$(x^2-2x+1) + (x^2+12x+36) = x^2+16x+64$

$x^2-6x-27=0$

$(x-9)(x+3)=0$

$x-9=0 \quad \text{or} \quad x+3=0$

$x=9 \qquad x=-3$

Since $x = -3$ would result in negative length for the small side, we must reject this solution. Therefore, $x = 9$.

87. $y = x^2 - 5x + 6$

$0 = (x-2)(x-3)$

$x-2=0 \quad \text{or} \quad x-3=0$

$x=2 \qquad x=3$

The x-intercepts are (2, 0) and (3, 0). Graph (d) matches this equation.

89. $y = x^2 + 5x + 6$

$0 = (x+2)(x+3)$

$x+2=0 \quad \text{or} \quad x+3=0$

$x=-2 \qquad x=-3$

The x-intercepts are (–3, 0) and (–2, 0). Graph (b) matches this equation.

91. $y = (x-1)(x-5)$

$y = x^2 - 6x + 5$

93. $y = (x-4)(x+2)$

$y = x^2 - 2x - 8$

95. $y = (6x+5)(x-2)$

$y = 6x^2 - 7x - 10$

97. Let w = width,
then $2w + 1$ = length
surface area = width · length

$$10 = w(2w+1)$$
$$0 = 2w^2 + w - 10$$
$$0 = (2w+5)(w-2)$$
$$2w+5=0 \quad \text{or} \quad w-2=0$$
$$w = -\frac{5}{2} \qquad w = 2$$

Since width cannot be negative, the width is 2 feet and the length is $2(2)+1=5$ feet.

99. Let b = length of base,
then $b + 6$ = height

$$\text{area} = \frac{1}{2} \cdot \text{base} \cdot \text{height}$$
$$80 = \frac{1}{2}b(b+6)$$
$$160 = b(b+6)$$
$$0 = b^2 + 6b - 160$$
$$0 = (b+16)(b-10)$$
$$b+16=0 \quad \text{or} \quad b-10=0$$
$$b=-16 \qquad b=10$$

Since the base length cannot be negative, the base is 10 feet and the height is 10 + 6 = 16 feet.

101. Let w = width of walkway.
walking area = total area – garden area

$$2w(16+w)+2w(12+w)=320-12(16)$$
$$32w+2w^2+24w+2w^2=320-192$$
$$4w^2+56w=128$$
$$4w^2+56w-128=0$$
$$4(w^2+14w-32)=0$$
$$4(w-2)(w+16)=0$$
$$w-2=0 \quad \text{or} \quad w+16=0$$
$$w=2 \qquad w=-16$$

The width cannot be negative. Therefore, the width of the walkway is 2 feet.

103. Let w = width of mulch border.
area of extra mulch = total area – garden area

$$2w(30+w)+2w(20+w)=936-20(30)$$
$$60w+2w^2+40w+2w^2=936-600$$
$$4w^2+100w=336$$
$$4w^2+100w-336=0$$
$$4(w^2+25w-84)=0$$
$$4(w-3)(w+28)=0$$
$$w-3=0 \quad \text{or} \quad w+28=0$$
$$w=3 \qquad w=-28$$

The width cannot be negative. Therefore the width is 3 feet.

105. $h(t) = -16t^2 + 32t$

$$0 = -16t^2 + 32t$$
$$0 = -16t(t-2)$$
$$-16t=0 \quad \text{or} \quad t-2=0$$
$$t=0 \qquad t=2$$

The spurt of water will return to the jet's height in 2 seconds.

107. Let x = the distance that Tim had traveled and $x + 7$ = the distance that Bob had traveled.

$$(x)^2+(x+7)^2=13^2$$
$$x^2+x^2+14x+49=169$$
$$2x^2+14x-120=0$$
$$2(x^2+7x-60)=0$$
$$2(x+12)(x-5)=0$$
$$x+12=0 \quad \text{or} \quad x-5=0$$
$$x=-12 \qquad x=5$$

Since the distance must be positive, Tim had traveled 5 miles and Bob had traveled 5 + 7 = 12 miles.

109. Let x = height from the ground where the wire is attached, then $x + 8$ = length of the wire

$$c^2=a^2+b^2$$
$$(x+8)^2=x^2+12^2$$
$$x^2+16x+64=x^2+144$$
$$16x+64=144$$
$$16x=80$$
$$x=5$$
$$x+8=13$$

The height is 5 feet, so the wire is 13 feet.

111. $R(x) = C(x)$

$$70x - x^2 = 17x + 150$$
$$0 = x^2 - 53x + 150$$
$$0 = (x-50)(x-3)$$
$$x - 50 = 0 \quad \text{or} \quad x - 3 = 0$$
$$x = 50 \quad \text{or} \quad x = 3$$

Reject 3 because it was stated in the problem that $x \geq 10$. The company must sell 50 bicycles to break even.

113. Let x = length of side of original cardboard.
Volume = length · width · height

$$162 = (x-4)(x-4) \cdot 2$$
$$162 = 2(x^2 - 8x + 16)$$
$$0 = x^2 - 8x - 65$$
$$0 = (x-13)(x+5)$$
$$x - 13 = 0 \quad \text{or} \quad x + 5 = 0$$
$$x = 13 \qquad x = -5$$

Disregard a negative length. The original cardboard measures 13 inches by 13 inches.

115. a. $V = a^3 - ab^2$

b. $V = a(a^2 - b^2) = a(a-b)(a+b)$

c. $1620 = 12(12-b)(12+b)$

$$135 = (12-b)(12+b)$$
$$135 = 144 - b^2$$
$$b^2 = 9$$
$$b = \pm 3$$

Disregard the negative. Thus, $b = 3$ in.

117. a. The x-intercepts are $x = -5$ and $x = -2$. The factors are $x + 5$ and $x + 2$. One possible representation for the function is $f(x) = (x+5)(x+2) = x^2 + 7x + 10$.

b. The quadratic equation can be $x^2 + 7x + 10 = 0$.

c. There are an infinite number. For this, express $f(x)$ as $f(x) = a(x^2 + 7x + 10)$ where a is any real number except 0.

d. There are an infinite number. For this, use $a(x^2 + 7x + 10) = 0$ where a is any real number except 0. The solution is $x = -2$ or $x = -5$.

119. a. Answers will vary. One example is:
No x-intercepts

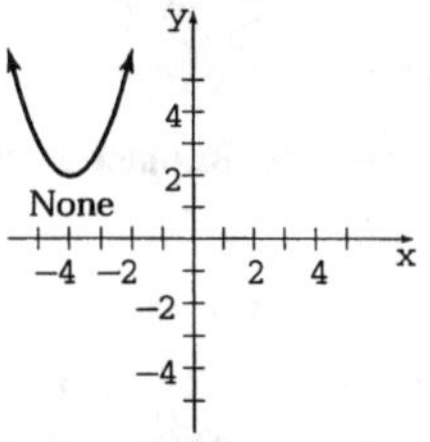

One x-intercept

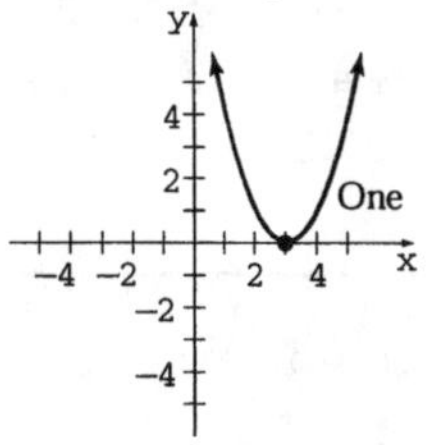

Two x-intercepts

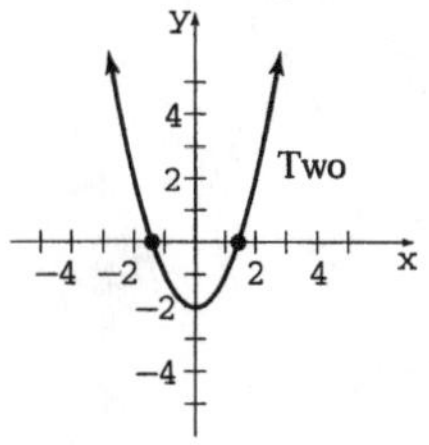

b. $ax^2 + bx + c$ could have no x-intercepts, one x-intercept, or two x-intercepts. If the graph does not cross the x-axis, there are no x-intercepts (no real solutions). If the vertex is located on the x-axis, then there is one intercept (one real solution). If the graph crosses the x-axis at two different points, then there are two x-intercepts (i.e., two real solutions).

121. $d(s) = -0.31s^2 + 59.82s - 2180.22$

$$545 = -0.31s^2 + 59.82s - 2180.22$$
$$y_1 = 545$$
$$y_2 = -0.31s^2 + 59.82s - 2180.22$$

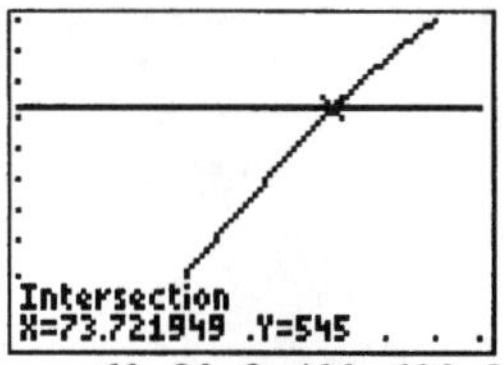

60, 80, 2, 400, 600, 20

The intersection is approximately (73.721949, 545). The car was traveling approximately 73.721949 mph.

123. $x^4 - 13x^2 = -36$

$$x^4 - 13x^2 + 36 = 0$$
$$\left(x^2\right)^2 - 13x^2 + 36 = 0$$
$$y^2 - 13y + 36 = 0 \quad \leftarrow \text{Replace } x^2 \text{ with } y$$
$$(y-9)(y-4) = 0$$
$$y - 9 = 0 \quad \text{or} \quad y - 4 = 0$$
$$x^2 - 9 = 0 \quad \text{or} \quad x^2 - 4 = 0 \quad \leftarrow \text{Replace } y \text{ with } x^2$$
$$(x-3)(x+3) = 0 \quad \text{or} \quad (x-2)(x+2) = 0$$
$$x - 3 = 0 \quad \text{or} \quad x + 3 = 0 \quad \text{or} \quad x - 2 = 0 \quad \text{or} \quad x + 2 = 0$$
$$x = 3 \qquad x = -3 \qquad x = 2 \qquad x = -2$$

The solutions are ±2, ±3.

---

128. $\left(4x^{-2}y^3\right)^{-2} = 4^{-2} \cdot x^{(-2)(-2)} \cdot y^{3(-2)}$

$$= 4^{-2} \cdot x^4 \cdot y^{-6}$$
$$= \frac{x^4}{4^2 y^6}$$
$$= \frac{x^4}{16y^6}$$

129. $-1 < \dfrac{4(3x-2)}{3} \le 5$

$$3(-1) < 3\left(\frac{4(3x-2)}{3}\right) \le 3(5)$$
$$-3 < 4(3x-2) \le 15$$
$$-3 < 12x - 8 \le 15$$
$$5 < 12x \le 23$$
$$\frac{5}{12} < x \le \frac{23}{12}$$

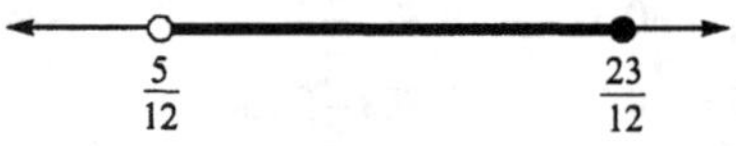

130. $3x + 4y = 2$, $2x = -5y - 1$ $\Rightarrow$ $3x + 4y = 2$ (1), $2x + 5y = -1$ (2)

Multiply equation (1) by –2 and equation (2) by 3 and then add the resulting equations.

$$-6x - 8y = -4 \quad \leftarrow -2 \text{ times equation (1)}$$
$$6x + 15y = -3 \quad \leftarrow \quad 3 \text{ times equation (2)}$$
$$7y = -7$$
$$y = -1$$

Substitute –1 for y in equation (1).

$$3x + 4(-1) = 2$$
$$3x - 4 = 2$$
$$3x = 6$$
$$x = 2$$

The solution is (2, –1).

131. $f(x) = -x^2 + 3x, \quad g(x) = x^2 + 5$

$$(f \cdot g)(x) = \left(-x^2 + 3x\right)\left(x^2 + 5\right)$$
$$(f \cdot g)(4) = \left(-(4)^2 + 3(4)\right)\left((4)^2 + 5\right)$$
$$= (-16 + 12)(16 + 5)$$
$$= (-4)(21)$$
$$= -84$$

132. $(x+1)^2 - (x+1) - 6$

$$= [(x+1) + 2][(x+1) - 3]$$
$$= (x+3)(x-2)$$

Chapter 5 Review Exercises

1. $3x^2+9$
   a. Binomial (2 terms)
   b. $3x^2+9$
   c. The degree is 2.

2. $5x+4x^3-7$
   a. Trinomial (3 terms)
   b. $4x^3+5x-7$
   c. The degree is 3.

3. $8x-x^{-1}+6$ is not a polynomial due to the negative exponent.

4. $-3-10x^2y+6xy^3+2x^4$
   a. Polynomial
   b. $2x^4-10x^2y+6xy^3-3$
   c. The degree is 4.

5. $(x^2-5x+8)+(2x+6)$
$=x^2-5x+8+2x+6$
$=x^2-3x+14$

6. $(7x^2+2x-5)-(2x^2-9x-1)$
$=7x^2-2x^2+2x+9x-5+1$
$=5x^2+11x-4$

7. $(2a-3b-2)-(-a+5b-9)$
$=2a+a-3b-5b-2+9$
$=3a-8b+7$

8. $(4x^3-4x^2-2x)+(2x^3+4x^2-7x+13)$
$=4x^3+2x^3-4x^2+4x^2-2x-7x+13$
$=6x^3-9x+13$

9. $(3x^2y+6xy-5y^2)-(4y^2+3xy)$
$=3x^2y+6xy-5y^2-4y^2-3xy$
$=3x^2y+6xy-3xy-5y^2-4y^2$
$=3x^2y+3xy-9y^2$

10. $(-8ab+2b^2-3a)+(-b^2+5ab+a)$
$=-8ab+5ab+2b^2-b^2-3a+a$
$=-3ab+b^2-2a$

11. $(x^2-3x+12)+(4x^2+10x-9)$
$=x^2+4x^2-3x+10x+12-9$
$=5x^2+7x+3$

12. $(-7a^2b-ab)-(3a^2b-2ab)$
$=-7a^2b-3a^2b-ab+2ab$
$=-10a^2b+ab$

13. $P(x)=2x^2-3x+19$
$P(2)=2(2)^2-3(2)+19$
$=8-6+19$
$=21$

14. $P(x)=x^3-3x^2+4x-10$
$P(-3)=(-3)^3-3(-3)^2+4(-3)-10$
$=-27-27-12-10$
$=-76$

15. $P=(x^2-x+7)+(x^2+1)+(x^2+x+19)$
$=x^2+x^2+x^2-x+x+7+1+19$
$=3x^2+27$

16. $P=(13x+8)+(x^2+7)+(9x+5)+(x^2+2x+3)$
$=x^2+x^2+13x+9x+2x+8+7+5+3$
$=2x^2+24x+23$

17. a. $t=2010-1997=13$
$R(t)=0.78t^2+20.28t+385.0$
$R(13)=0.78(13)^2+20.28(13)+385.0$
$R(13)=780.46$
The receipts in 2010 are estimated to be \$780.46 billion.
   b. Yes, the graph supports the answer.

18. a. $t=2010-1997=13$
$G(t)=1.74t^2+7.32t+383.91$
$G(13)=1.74(13)^2+7.32(13)+383.91$
$G(13)=773.13$
The outlays in 2010 are estimated to be \$773.13 billion.
   b. Yes, the graph supports the answer.

19. $2x(3x^2-7x+5)$
$=(2x)(3x^2)+(2x)(-7x)+(2x)(5)$
$=6x^3-14x^2+10x$

20. $-3xy^2(x^3+xy^4-4y^5)$
$=(-3xy^2)(x^3)+(-3xy^2)(xy^4)+(-3xy^2)(-4y^5)$
$=-3x^4y^2-3x^2y^6+12xy^7$

21. $(3x-5)(2x+9)$
$=(3x)(2x)+(3x)(9)+(-5)(2x)+(-5)(9)$
$=6x^2+27x-10x-45$
$=6x^2+17x-45$

22. $(5a+1)(10a-3)$
$=(5a)(10a)+(5a)(-3)+(1)(10a)+(1)(-3)$
$=50a^2-15a+10a-3$
$=50a^2-5a-3$

23. $(x+8y)^2=(x)^2+2(x)(8y)+(8y)^2$
$=x^2+16xy+64y^2$

24. $(a-11b)^2=(a)^2-2(a)(11b)+(11b)^2$
$=a^2-22ab+121b^2$

25. $(2xy-1)(5x+4y)$
$=(2xy)(5x)+(2xy)(4y)+(-1)(5x)+(-1)(4y)$
$=10x^2y+8xy^2-5x-4y$

26. $(2pq-r)(3pq+7r)$
$=(2pq)(3pq)+(2pq)(7r)+(-r)(3pq)+(-r)(7r)$
$=6p^2q^2+14pqr-3pqr-7r^2$
$=6p^2q^2+11pqr-7r^2$

27. $(2a+9b)^2=(2a)^2+2(2a)(9b)+(9b)^2$
$=4a^2+36ab+81b^2$

28. $(4x-3y)^2=(4x)^2-2(4x)(3y)+(3y)^2$
$=16x^2-24xy+9y^2$

29. $(7x+5y)(7x-5y)=(7x)^2-(5y)^2$
$=49x^2-25y^2$

30. $(2a-5b^2)(2a+5b^2)=(2a)^2-(5b^2)^2$
$=4a^2-25b^4$

31. $(4xy+6)(4xy-6)=(4xy)^2-(6)^2$
$=16x^2y^2-36$

32. $(9a^2-2b^2)(9a^2+2b^2)=(9a^2)^2-(2b^2)^2$
$=81a^4-4b^4$

33. $[(x+3y)+2]^2$
$=(x+3y)^2+2(x+3y)(2)+(2)^2$
$=(x)^2+2(x)(3y)+(3y)^2+4(x+3y)+4$
$=x^2+6xy+9y^2+4x+12y+4$

34. $[(2p-q)-5]^2$
$=(2p-q)^2-2(2p-q)(5)+(5)^2$
$=(2p)^2-2(2p)(q)+(q)^2-10(2p-q)+25$
$=4p^2-4pq+q^2-20p+10q+25$

35.
$$\begin{array}{r} 3x^2+4x-6 \\ 2x-3 \\ \hline -9x^2-12x+18 \\ 6x^3+8x^2-12x \\ \hline 6x^3-x^2-24x+18 \end{array}$$

36.
$$\begin{array}{r} 4x^3+0x^2+6x-2 \\ x+3 \\ \hline 12x^3+0x^2+18x-6 \\ 4x^4+0x^3+6x^2-5x \\ \hline 4x^4+12x^3+6x^2+13x-6 \end{array}$$

37. $A=(x)(x)+(5)(x)+(3)(x)+(5)(2)$
$=x^2+5x+3x+10$
$=x^2+8x+10$

38. $A=(x)(x)+(y)(x)+(4)(y)+(z)(x)$
$=x^2+xy+4y+xz$

39. a. $(f\cdot g)(x)=f(x)\cdot g(x)$
$=(x+1)(x-3)$
$=x^2-3x+x-3$
$=x^2-2x-3$

b. $(f\cdot g)(3)=3^2-2(3)-3$
$=9-6-3$
$=0$

40. a. $(f\cdot g)(x)=f(x)g(x)$
$=(2x-4)(x^2-3)$
$=2x^3-6x-4x^2+12$
$=2x^3-4x^2-6x+12$

b. $(f\cdot g)(3)=2(3)^3-4(3)^2-6(3)+12$
$=54-36-18+12$
$=12$

41. a. $(f\cdot g)(x)=f(x)\cdot g(x)$
$=(x^2+x-3)(x-2)$
$=x^3-2x^2+x^2-2x-3x+6$
$=x^3-x^2-5x+6$

b. $(f\cdot g)(3)=3^3-3^2-5(3)+6$
$=27-9-15+6$
$=9$

42. a. $(f\cdot g)(x)=f(x)\cdot g(x)$
$=(x^2-2)(x^2+2)$
$=x^4-4$

b. $(f\cdot g)(3)=3^4-4=81-4=77$

43. $\dfrac{4x^7y^5}{20xy^3}=\dfrac{4}{20}\cdot x^{7-1}\cdot y^{5-3}=\dfrac{1}{5}x^6y^2$

44. $\dfrac{3s^5t^8}{12s^5t^3}=\dfrac{3}{12}\cdot s^{5-5}\cdot t^{8-3}=\dfrac{1}{4}s^0t^5=\dfrac{1}{4}t^5$

45. $\dfrac{45pq-25q^2-15q}{5q}=\dfrac{45pq}{5q}-\dfrac{25q^2}{5q}-\dfrac{15q}{5q}$
$=9p-5q-3$

46. $\dfrac{7a^2-16a+32}{4}=\dfrac{7a^2}{4}-\dfrac{16a}{4}+\dfrac{32}{4}$
$=\dfrac{7}{4}a^2-4a+8$

47. $\dfrac{2x^3y^2+8x^2y^3+12xy^4}{8xy^3}=\dfrac{2x^3y^2}{8xy^3}+\dfrac{8x^2y^3}{8xy^3}+\dfrac{12xy^4}{8xy^3}$
$=\dfrac{x^2}{4y}+x+\dfrac{3y}{2}$

48.
$$\begin{array}{r} 4x-3 \\ 2x+5\overline{)8x^2+14x-15} \\ \underline{8x^2+20x}\quad \leftarrow 4x(2x+5) \\ -6x-15 \\ \underline{-6x-15}\ \leftarrow -3(2x+5) \\ 0 \end{array}$$

Thus, $\dfrac{8x^2+14x-15}{2x+5}=4x-3$

49.
$$\begin{array}{r} x^3-2x^2+\ 3x+7 \\ 2x+1\overline{)2x^4-3x^3+4x^2+17x+7} \\ \underline{2x^4+x^3\qquad\qquad\qquad} \\ -4x^3+4x^2 \\ \underline{-4x^3-2x^2} \\ 6x^2+17x \\ \underline{6x^2+\ 3x} \\ 14x+7 \\ \underline{14x+7} \\ 0 \end{array}$$

Thus,
$\dfrac{2x^4-3x^3+4x^2+17x+7}{2x+1}=x^3-2x^2+3x+7$

50. $2a-1\overline{)4a^4+0a^3-7a^2-5a+4}$ with quotient $2a^3+a^2-3a-4$

$$\begin{array}{l}
\underline{4a^4-2a^3} \\
2a^3-7a^2 \\
\underline{2a^3-a^2} \\
-6a^2-5a \\
\underline{-6a^2+3a} \\
-8a+4 \\
\underline{-8a+4} \\
0
\end{array}$$

Thus, $\dfrac{4a^4-7a^2-5a+4}{2a-1}=2a^3+a^2-3a-4$

51. $x-3\overline{)x^2+x-22}$ with quotient $x+4$

$$\begin{array}{l}
\underline{x^2-3x} \leftarrow x(x-3) \\
4x-22 \\
\underline{4x-12} \leftarrow 4(x-3) \\
-10
\end{array}$$

Thus, $\dfrac{x^2+x-22}{x-3}=x+4-\dfrac{10}{x-3}$

52. $2x+3\overline{)4x^3+12x^2+x-9}$ with quotient $2x^2+3x-4$

$$\begin{array}{l}
\underline{4x^3+6x^2} \\
6x^2+x \\
\underline{6x^2+9x} \\
-8x-9 \\
\underline{-8x-12} \\
3
\end{array}$$

Thus, $\dfrac{4x^3+12x^2+x-9}{2x+3}=2x^2+3x-4+\dfrac{3}{2x+3}$

53.

3	3	−2	0	10
		9	21	63
	3	7	21	73

Thus, $\dfrac{3x^3-2x^2+10}{x-3}=3x^2+7x+21+\dfrac{73}{x-3}$

54.

−1	2	0	−10	0	1	−2
		−2	2	8	−8	7
	2	−2	−8	8	−7	5

Thus, $\dfrac{2y^5-10y^3+y-2}{y+1}$

$=2y^4-2y^3-8y^2+8y-7+\dfrac{5}{y+1}$

55.

2	1	0	0	0	0	−18
		2	4	8	16	32
	1	2	4	8	16	14

Thus, $\dfrac{x^5-18}{x-2}=x^4+2x^3+4x^2+8x+16+\dfrac{14}{x-2}$

56.

$\frac{1}{2}$	2	1	5	−3
		1	1	3
	2	2	6	0

Thus, $\dfrac{2x^3+x^2+5x-3}{x-\frac{1}{2}}=2x^2+2x+6$

57. To find the remainder, use synthetic division:

3	1	−4	13
		3	−3
	1	−1	10

Thus, the remainder is 10.

58. To find the remainder, use synthetic division:

−4	2	−6	3	0
		−8	56	−236
	2	−14	59	−236

Thus, the remainder is −236.

59. To find the remainder, use synthetic division:

$\frac{1}{3}$	3	0	0	−6
		1	$\frac{1}{3}$	$\frac{1}{9}$
	3	1	$\frac{1}{3}$	$-\frac{53}{9}$ or $-5.\overline{8}$

Thus, the remainder is $-\dfrac{53}{9}$ or $-5.\overline{8}$.

60. To find the remainder, use synthetic division:

−2	2	0	−6	0	−8
		−4	8	−4	8
	2	−4	2	−4	0

Since the remainder is 0, then $x+2$ is a factor.

61. $4x^2+8x+32=4(x^2+2x+8)$

62. $15x^5+6x^4-12x^5y^3=3x^4(5x+2-4xy^3)$

63. $10a^3b^3-14a^2b^6=2a^2b^3(5a-7b^3)$

64. $24xy^4z^3+12x^2y^3z^2-30x^3y^2z^3$
$=6xy^2z^2(4y^2z+2xy-5x^2z)$

65. $5x^2-xy+30xy-6y^2$
$=x(5x-y)+6y(5x-y)$
$=(5x-y)(x+6y)$

66. $12a^2+8ab+15ab+10b^2$
$=4a(3a+2b)+5b(3a+2b)$
$=(3a+2b)(4a+5b)$

67. $(2x-5)(2x+1)-(2x-5)(x-8)$
$=(2x-5)[(2x+1)-(x-8)]$
$=(2x-5)(2x+1-x+8)$
$=(2x-5)(x+9)$

68. $7x(3x-7)+3(3x-7)^2$
$=(3x-7)[7x+3(3x-7)]$
$=(3x-7)(7x+9x-21)$
$=(3x-7)(16x-21)$

69. $A=13x(5x+2)-7(5x+2)$
$=(5x+2)(13x-7)$

70. $A=(14x^2+18x)-(7x+9)$
$=2x(7x+9)-1(7x+9)$
$=(7x+9)(2x-1)$

71. $V=9x(17x+3)-7(17x+3)$
$=(17x+3)(9x-7)$

72. $V=(20x^2+25x)-(8x+10)$
$=5x(4x+5)-2(4x+5)$
$=(4x+5)(5x-2)$

---

73. $x^2+9x+18=(x+6)(x+3)$ ← Use 6 and 3: (6)(3) = 18, 6 + 3 = 9

74. $x^2+3x-10=(x+5)(x-2)$ ← Use 5 and –2: (5)(–2) = –10, 5 + (–2) = 3

75. $x^2-3x-28=(x-7)(x+4)$ ← Use –7 and 4: (–7)(4) = –28, (–7) + 4 = –3

76. $x^2-10x+16=(x-8)(x-2)$ ← Use –8 and –2: (–8)(–2) = 16, (–8) + (–2) = –10

77. $-x^2+12x+45=-(x^2-12x-45)$ ← Use –15 and 3: (–15)(3) = –45, –15 + 3 = –12
$=-(x-15)(x+3)$

78. $-x^2+13x-12=-(x^2-13x+12)$ ← Use –12 and –1: (–12)(–1) = 12, –12 + (–1) = –13
$=-(x-12)(x-1)$

79. $2x^3+13x^2+6x=x(2x^2+13x+6)$
$=x(2x^2+x+12x+6)$ ← Use $x+12x=13x$
$=x[x(2x+1)+6(2x+1)]$
$=x(2x+1)(x+6)$

80. $8x^4 + 10x^3 - 25x^2 = x^2(8x^2 + 10x - 25)$

$= x^2(8x^2 + 20x - 10x - 25)$ ← Use $20x - 10x = 10x$

$= x^2[4x(2x+5) - 5(2x+5)]$

$= x^2(4x-5)(2x+5)$

81. $4a^5 - 9a^4 + 5a^3 = a^3(4a^2 - 9a + 5)$

$= a^3(4a^2 - 4a - 5a + 5)$ ← Use $-4a - 5a = -9a$

$= a^3[4a(a-1) - 5(a-1)]$

$= a^3(4a-5)(a-1)$

82. $12y^5 + 61y^4 + 5y^3 = y^3(12y^2 + 61y + 5)$

$= y^3(12y^2 + 60y + y + 5)$ ← Use $60y + y = 61y$

$= y^3[12y(y+5) + 1(y+5)]$

$= y^3(12y+1)(y+5)$

83. $x^2 - 15xy - 54y^2 = (x - 18y)(x + 3y)$ ← Use –18 and 3: $(-18)(3) = -54$, $-18 + 3 = -15$

84. $6p^2 - 19pq + 10q^2 = 6p^2 - 15pq - 4pq + 10q^2$ ← Use $-15pq - 4pq = -19pq$

$= 3p(2p - 5q) - 2q(2p - 5q)$

$= (2p - 5q)(3p - 2q)$

85. $x^4 + 10x^2 + 21 = y^2 + 10y + 21$ ← Replace x^2 by y

$= (y+3)(y+7)$ ← Use 3 and 7: $(3)(7) = 21$, $3 + 7 = 10$

$= (x^2+3)(x^2+7)$ ← Replace y by x^2

86. $x^4 + 2x^2 - 63 = y^2 + 2y - 63$ ← Replace x^2 by y

$= (y+9)(y-7)$ ← Use –7 and 9: $(-7)(9) = -63$, $(-7) + 9 = 2$

$= (x^2-7)(x^2+9)$ ← Replace y by x^2

87. $(x+3)^2 + 10(x+3) + 24 = w^2 + 10w + 24$ ← Replace $x+3$ by w

$= (w+6)(w+4)$

$= [(x+3)+6][(x+3)+4]$ ← Replace w by $x+3$

$= (x+9)(x+7)$

88. $(x-4)^2 - (x-4) - 20 = w^2 - w - 20$ ← Replace $x-4$ by w

$= (w+4)(w-5)$

$= [(x-4)+4][(x-4)-5]$ ← Replace w by $x-4$

$= x(x-9)$

89. $A=(x+9)(x+2)-(4)(2)$
$=x^2+9x+2x+18-8$
$=x^2+11x+10$
$=(x+10)(x+1)$

90. $A=(x+8)(x+4)-(4)(3)$
$=x^2+8x+4x+32-12$
$=x^2+12x+20$
$=(x+10)(x+2)$

91. $x^2-36=x^2-6^2=(x+6)(x-6)$

92. $x^2-121=x^2-11^2=(x+11)(x-11)$

93. $x^4-81=(x^2)^2-9^2$
$=(x^2+9)(x^2-9)$
$=(x^2+9)(x^2-3^2)$
$=(x^2+9)(x+3)(x-3)$

94. $x^4-16=(x^2)^2-4^2$
$=(x^2+4)(x^2-4)$
$=(x^2+4)(x^2-2^2)$
$=(x^2+4)(x+2)(x-2)$

95. $4a^2+4a+1=(2a)^2+2(2a)(1)+(1)^2$
$=(2a+1)^2$

96. $16y^2-24y+9=(4y)^2-2(4y)(3)+(3)^2$
$=(4y-3)^2$

97. $(x+2)^2-16=(x+2)^2-4^2$
$=[(x+2)+4][(x+2)-4]$
$=(x+6)(x-2)$

98. $(3y-1)^2-36=(3y-1)^2-6^2$
$=[(3y-1)+6][(3y-1)-6]$
$=(3y+5)(3y-7)$

99. $p^4+18p^2+81=(p^2)^2+2(p^2)(9)+(9)^2$
$=(p^2+9)^2$

100. $m^4-20m^2+100=(m^2)^2-2(m^2)(10)+(10)^2$
$=(m^2-10)^2$

101. $x^2+8x+16-y^2=(x+4)^2-(y)^2$
$=[(x+4)+y][(x+4)-y]$
$=(x+4+y)(x+4-y)$

102. $a^2+6ab+9b^2-36c^2$
$=(a+3b)^2-(6c)^2$
$=[(a+3b)+6c][(a+3b)-6c]$
$=(a+3b+6c)(a+3b-6c)$

103. $16x^2+8xy+y^2=(4x)^2+2(4y)(y)+(y)^2$
$=(4x+y)^2$

104. $36b^2-60bc+25c^2=(6b)^2-2(6b)(5c)+(5c)^2$
$=(6b-5c)^2$

105. $x^3-27=x^3-3^3$
$=(x-3)[x^2+x(3)+3^2]$
$=(x-3)(x^2+3x+9)$

106. $y^3+64z^3=(y)^3+(4z)^3$
$=(y+4z)[(y)^2-(y)(4z)+(4z)^2]$
$=(y+4)(y^2-4yz+16z^2)$

107. $125x^3-1=(5x)^3-1^3$
$=(5x-1)[(5x)^2+(5x)(1)+1^2]$
$=(5x-1)(25x^2+5x+1)$

108. $8a^3+27b^3=(2a)^3+(3b)^3$
$=(2a+3b)[(2a)^2-(2a)(3b)+(3b)^2]$
$=(2a+3b)(4a^2-6ab+9b^2)$

109. $y^3-64z^3=(y)^3-(4z)^3$
$=(y-4z)[(y)^2+(y)(4z)+(4z)^2]$
$=(y-4z)(y^2+4yz+16z^2)$

110. $(x-2)^3-27$
$=(x-2)^3-3^3$
$=[(x-2)-3]\left[(x-2)^2+(x-2)(3)+(3)^2\right]$
$=(x-5)\left[x^2-4x+4+3x-6+9\right]$
$=(x-5)(x^2-x+7)$

111. $(x+1)^3-8$
$=(x+1)^3-2^3$
$=[(x+1)-2]\left[(x+1)^2+(x+1)(2)+(2)^2\right]$
$=(x+1)\left[x^2+2x+1+2x+2+4\right]$
$=(x+1)(x^2+4x+7)$

112. $(a+4)^3+1$
$=(a+4)^3+1^3$
$=\left[(a+4)+1\right]\left[(a+4)^2-(a+4)(1)+(1)^2\right]$
$=(x+5)\left[a^2+8a+16-a-4+1\right]$
$=(x+5)\left(a^2+7a+13\right)$

113. The area of the large square is x^2. The area of the smaller square is $(3)(3)=9$. The area of the shaded region is x^2-9. In factored form, the area is $x^2-9=(x+3)(x-3)$.

114. The area of the large square is a^2. The sum of the areas of the four small squares is $4b^2$. The area of the shaded region is a^2-4b^2. In factored form, the area is $a^2-4b^2=(a+2b)(a-2b)$.

115. The volume of the large cube is $(2x)^3$.
The volume of the small cube is y^3.
The difference in the volumes is $(2x)^3-y^3$.
In factored form, the difference in the volumes is
$(2x)^3-y^3=(2x-y)\left[(2x)^2+(2x)(y)+y^2\right]$
$=(2x-y)\left(4x^2+2xy+y^2\right)$

116. The volume of the full figure is $(a)(a)(4a)=4a^3$.
The volume of the cut out space is
$(c)(c)(4a)=4ac^2$.
The volume of the shaded region is $4a^3-4ac^2$.
In factored form, the region is given by
$4a^3-4ac^2=4a\left(a^2-c^2\right)=4a(a+c)(a-c)$

117. $x^2y^4-2xy^4-15y^4=y^4\left(x^2-2x-15\right)$
$=y^4(x+3)(x-5)$

118. $5x^3-30x^2+40x=5x\left(x^2-6x+8\right)$
$=5x(x-4)(x-2)$

119. $3x^3y^4+18x^2y^4-6x^2y^4-36xy^4$
$=3xy^4\left(x^2+6x-2x-12\right)$
$=3xy^4\left[x(x+6)-2(x+6)\right]$
$=3xy^4(x-2)(x+6)$

120. $3y^5-75y=3y\left(y^4-25\right)$
$=3y\left[\left(y^2\right)^2-5^2\right]$
$=3y\left(y^2+5\right)\left(y^2-5\right)$

121. $4x^3y+32y=4y\left(x^3+8\right)$
$=4y\left(x^3+2^3\right)$
$=4y(x+2)\left[x^2-x(2)+2^2\right]$
$=4y(x+2)\left(x^2-2x+4\right)$

122. $5x^4y+20x^3y+20x^2y=5x^2y\left(x^2+4x+4\right)$
$=5x^2y\left[x^2+2(x)(2)+2^2\right]$
$=5x^2y(x+2)^2$

123. $6x^3-21x^2-12x=3x\left(2x^2-7x-4\right)$
$=3x\left(2x^2+x-8x-4\right)$
$=3x\left[x(2x+1)-4(2x+1)\right]$
$=3x(2x+1)(x-4)$

124. $x^2+10x+25-z^2=(x+5)^2-z^2$
$=\left[(x+5)+z\right]\left[(x+5)-z\right]$
$=(x+5+z)(x+5-z)$

125. $5x^3+40y^3=5(x^3+8y^3)$
$=5\left[x^3+(2y)^3\right]$
$=5(x+2y)\left[x^2-x(2y)+(2y)^2\right]$
$=5(x+2y)(x^2-2xy+4y^2)$

126. $x^2(x+6)-3x(x+6)-4(x+6)$
$=(x+6)(x^2+3x-4)$
$=(x+6)(x+4)(x-1)$

---

127. $4(2x+3)^2-12(2x+3)+5=4w^2-12w+5$ ← Replace $2x+3$ by w
$=4w^2-10w-2w+5$ ← Use $-10w-2w$ for $-12w$
$=2(2w-5)-1(2w-5)$
$=(2w-1)(2w-5)$
$=\left[2(2x+3)-1\right]\left[2(2x+3)-5\right]$ ← Replace w by $2x+3$
$=(4x+6-1)(4x+6-5)$
$=(4x+5)(4x+1)$

128. $4x^4+4x^2-3=4(x^2)^2+4x^2-3$
$=4w^2+4w-3$ ← Replace x^2 by w
$=4w^2+6w-2w-3$ ← Use $6w-2w$ for $4w$
$=2w(2w+3)-1(2w+3)$
$=(2w-1)(2w+3)$
$=(2x^2-1)(2x^2+3)$ ← Replace w by x^2

129. $(x+1)x^2-(x+1)x-2(x+1)=(x+1)(x^2-x-2)=(x+1)(x-2)(x+1)=(x+1)^2(x-2)$

130. $9ax-3bx+21ay-7by=3x(3a-b)+7y(3a-b)=(3a-b)(3x+7y)$

131. $6p^2q^2-5pq-6=6p^2q^2-9pq+4pq-6=3pq(2pq-3)+2(2pq-3)=(2pq-3)(3pq+2)$

132. $9x^4-12x^2+4=(3x^2)^2-2(3x^2)(2)+(2)^2=(3x^2-2)^2$

133. $16y^2-(x^2+4x+4)=(4y)^2-(x+2)^2=\left[4y+(x+2)\right]\left[4y-(x+2)\right]=(4y+x+2)(4y-x-2)$

134. $6(2a+3)^2-7(2a+3)-3=6x^2-7x-3$ ← Replace $2a+3$ by x
$=6x^2-9x+2x-3$ ← Use $-9x+2x$ for $-7x$
$=3x(2x-3)+1(2x-3)$
$=(3x+1)(2x-3)$
$=\left[3(2a+3)+1\right]\left[2(2a+3)-3\right]$ ← Replace x by $2a+3$
$=(6a+9+1)(4a+6-3)$
$=(6a+10)(4a+3)$ ← $6a+10$ has a common factor of 2
$=2(3a+5)(4a+3)$

135. $6x^4y^5+9x^3y^5-27x^2y^5$
$=3x^2y^5(2x^2+3x-9)$
$=3x^2y^5(2x^2+6x-3x-9)$
$=3x^2y^5[2x(x+3)-3(x+3)]$
$=3x^2y^5(x+3)(2x-3)$

136. $x^3-\frac{8}{27}y^6=x^3-\left(\frac{2}{3}y^2\right)^3$
$=\left(x-\frac{2}{3}y^2\right)\left[x^2+x\left(\frac{2}{3}y^2\right)+\left(\frac{2}{3}y^2\right)^2\right]$
$=\left(x-\frac{2}{3}y^2\right)\left(x^2+\frac{2}{3}xy^2+\frac{4}{9}y^4\right)$

137. The area of the large rectangle is $(x+6)(x+5)$. The area of the small cut out rectangle is $(6)(2)=12$. The area of the shaded region in factored form is
$A=(x+6)(x+5)-12$
$=x^2+5x+6x+30-12$
$=x^2+11x+18$
$=(x+9)(x+2)$

138. The area of the large rectangle is $(y+8)(y+7)$. The area of the small cut out rectangle is $(3)(2)=6$. The area of the shaded region in factored form is
$A=(y+8)(y+7)-6$
$=y^2+7y+8y+56-6$
$=y^2+15y+50$
$=(y+10)(y+5)$

139. The area of the large square is a^2. The area of the 4 small cut out squares is $4b^2$. The area of the shaded region in factored form is $A=a^2-4b^2=(a+2b)(a-2b)$.

140. The sum of the areas of the two large rectangles is $ab+ab=2ab$. The sum of the areas of the two small squares is $b^2+b^2=2b^2$. The area of the shaded region is $2ab+2b^2$. In factored form, the area is $2ab+2b^2=2b(a+b)$.

141. The sum of the areas of the three rectangles is
$a(a+3b)+a(a+3b)+b(a+3b)$
$=2a(a+3b)+b(a+3b)$. In factored form, the area is $2a(a+3b)+b(a+3b)=(2a+b)(a+3b)$

142. The area of the large square is a^2. The area of the small square is b^2. The sum of the area of the two rectangles is $ab+ab=2ab$. The area of the shaded region is $a^2+2ab+b^2$. In factored form, the area is $a^2+2ab+b^2=(a+b)^2$.

143. $(x-2)(4x+1)=0$
$x-2=0$ or $4x+1=0$
$x=2$ $\quad 4x=-1$
$x=-\frac{1}{4}$
The solutions are $2, -\frac{1}{4}$.

144. $(2x+5)(3x+10)=0$
$2x+5=0$ or $3x+10=0$
$2x=-5$ $\quad 3x=-10$
$x=-\frac{5}{2}$ $\quad x=-\frac{10}{3}$
The solutions are $-\frac{5}{2}, -\frac{10}{3}$.

145. $4x^2=8x$
$4x^2-8x=0$
$4x(x-2)=0$
$4x=0$ or $x-2=0$
$x=0$ $\quad x=2$
The solutions are 0, 2.

146. $12x^2+16x=0$
$4x(3x+4)=0$
$4x=0$ or $3x+4=0$
$x=0$ $\quad 3x=-4$
$x=-\frac{4}{3}$
The solutions are $-\frac{4}{3}$, 0.

147. $x^2+7x+12=0$
$(x+4)(x+3)=0$
$x+4=0$ or $x+3=0$
$x=-4$ $\quad x=-3$
The solutions are –4, –3.

148. $a^2+a-30=0$
$(a+6)(a-5)=0$
$a+6=0$ or $a-5=0$
$a=-6$ $\quad a=5$
The solutions are –6, 5.

149. $x^2=8x-7$
$x^2-8x+7=0$
$(x-7)(x-1)=0$
$x-7=0$ or $x-1=0$
$x=7$ $\quad x=1$
The solutions are 7, 1.

150. $c^3-6c^2+8c=0$
$c(c^2-6c+8)=0$
$c(c-2)(c-4)=0$
$c=0$ or $c-2=0$ or $c-4=0$
$c=2$ $\quad c=4$
The solutions are 0, 2, 4.

151. $5x^2=80$
$5x^2-80=0$
$5(x^2-16)=0$
$5(x+4)(x-4)=0$
$x+4=0$ or $x-4=0$
$x=-4$ $\quad x=4$
The solutions are –4, 4.

152. $x(x+3)=2(x+4)-2$
$x^2+3x=2x+8-2$
$x^2+3x=2x+6$
$x^2+x-6=0$
$(x+3)(x-2)=0$
$x+3=0$ or $x-2=0$
$x=-3$ $\quad x=2$
The solutions are –3, 2.

153. $12d^2=13d+4$
$12d^2-13d-4=0$
$(3d-4)(4d+1)=0$
$3d-4=0$ or $4d+1=0$
$3d=4$ $\quad 4d=-1$
$d=\frac{4}{3}$ $\quad d=-\frac{1}{4}$
The solutions are $-\frac{1}{4}, \frac{4}{3}$.

154. $20p^2-6=7p$
$20p^2-7p-6=0$
$(4x-3)(5x+2)=0$
$4x-3=0$ or $5x+2=0$
$4x=3$ $\quad 5x=-2$
$x=\frac{3}{4}$ $\quad x=-\frac{2}{5}$
The solutions are $\frac{3}{4}, -\frac{2}{5}$.

155. $y=2x^2-6x-36$
Set $y=0$ and solve for x.
$2x^2-6x-36=0$
$2(x^2-3x-18)=0$
$2(x+3)(x-6)=0$
$x+3=0$ or $x-6=0$
$x=-3$ $\quad x=6$
The x-intercepts are (–3, 0) and (6, 0).

156. $y=20x^2-49x+30$
Set $y=0$ and solve for x.
$20x^2-49x+30=0$
$(5x-6)(4x-5)=0$
$5x-6=0$ or $4x-5=0$
$5x=6$ $\quad 4x=5$
$x=\frac{6}{5}$ $\quad x=\frac{5}{4}$
The x-intercepts are $\left(\frac{6}{5},0\right)$ and $\left(\frac{5}{4},0\right)$.

157. $y=(x+4)(x-6)$
$=x^2-6x+4x-24$
$=x^2-2x-24$

158. $y = (2x+5)(6x+1)$
$= 12x^2 + 2x + 30x + 5$
$= 12x^2 + 32x + 5$

159. Let x be the width of the carpet. Then, $x + 3$ is the length.

$$x(x+3) = 108$$
$$x^2 + 3x = 108$$
$$x^2 + 3x - 108 = 0$$
$$(x-9)(x+12) = 0$$
$$x - 9 = 0 \quad \text{or} \quad x + 12 = 0$$
$$x = 9 \qquad x = -12$$

Reject -12 for x since width cannot be negative. Thus, the width is 9 feet and the length is $9 + 3 = 12$ feet.

160. Let x be the height. Then $2x + 5$ is the base

$$\frac{1}{2}bh = A$$
$$\frac{1}{2}(2x+5)x = 26$$
$$(2x+5)x = 52$$
$$2x^2 + 5x = 52$$
$$2x^2 + 5x - 52 = 0$$
$$(x-4)(2x+13) = 0$$
$$x - 4 = 0 \quad \text{or} \quad 2x + 13 = 0$$
$$x = 4 \qquad x = -\frac{13}{2}$$

Reject $-\frac{13}{2}$ for x. Thus, the height is 4 feet and the base is $2(4) + 5 = 8 + 5 = 13$ feet.

161. Let x be the length of a side of the smaller square. Then $x + 4$ is the length of a side of the larger square.

$$(x+4)^2 = 49$$
$$(x+4)^2 - 49 = 0$$
$$(x+4)^2 - 7^2 = 0$$
$$[(x+4) - 7][(x+4) + 7] = 0$$
$$(x-3)(x+11) = 0$$
$$x - 3 = 0 \quad \text{or} \quad x + 11 = 0$$
$$x = 3 \qquad x = -11$$

Reject -11 for x. Thus, $x = 3$ inches for the smaller square and $x + 4 = 3 + 4 = 7$ inches for the larger square.

162. $s(t) = -16t^2 + 128t + 144$
Set $s(t) = 0$

$$-16t^2 + 128t + 144 = 0$$
$$-16(t^2 - 8t - 9) = 0$$
$$-16(t-9)(t+1) = 0$$
$$t - 9 = 0 \quad \text{or} \quad t + 1 = 0$$
$$t = 9 \qquad t = -1$$

Reject $t = -1$. Thus, $t = 9$ seconds.

163. $c^2 = a^2 + b^2$

$$(x+32)^2 = x^2 + (x+31)^2$$
$$x^2 + 64x + 1024 = x^2 + x^2 + 62x + 961$$
$$0 = x^2 - 2x - 63$$
$$0 = (x-9)(x+7)$$
$$x - 9 = 0 \quad \text{or} \quad x + 7 = 0$$
$$x = 9 \qquad x = -7$$

Disregard a negative length. $x = 9$.

Chapter 5 Practice Test

1. **a.** Trinomial since it has three terms

b. $-6x^4 - 4x^2 + 3x$

c. Degree is 4

d. The leading coefficient of the polynomial is -6.

2. $(7x^2y - 5y^2 + 4x) - (3x^2y + 9y^2 - 6y)$
$= 7x^2y - 5y^2 + 4x - 3x^2y - 9y^2 + 6y$
$= 4x^2y - 14y^2 + 4x + 6y$

3. $2x^3y^2(-4x^5y + 12x^3y^2 - 6x)$
$= 2x^3y^2(-4x^5y) + 2x^3y^2(12x^3y^2) + 2x^3y^2(-6x)$
$= -8x^8y^3 + 24x^6y^4 - 12x^4y^2$

4. $(2a - 3b)(5a + b)$
$= (2a)(5a) + (2a)(b) + (-3b)(5a) + (-3b)(b)$
$= 10a^2 + 2ab - 15ab - 3b^2$
$= 10a^2 - 13ab - 3b^2$

5.
$$\begin{array}{r} 2x^2+3xy-6y^2 \\ 2x+y \\ \hline 2x^2y+3xy^2-6y^3 \\ 4x^3+6x^2y-12xy^2 \\ \hline 4x^3+8x^2y-9xy^2-6y^3 \end{array}$$

6. $\dfrac{12x^6-15x^2y+21}{3x^2}=\dfrac{12x^6}{3x^2}-\dfrac{15x^2y}{3x^2}+\dfrac{21}{3x^2}$
$=4x^4-5y+\dfrac{7}{x^2}$

7.
$$\begin{array}{r} x-5 \\ 2x+3\overline{)2x^2-7x+9} \\ \underline{2x^2+3x} \\ -10x+9 \\ \underline{-10x-15} \\ 24 \end{array}$$

Thus, $\dfrac{2x^2-7x+90}{2x+3}=x-5+\dfrac{24}{2x+3}$

8.

5	3	−12	0	−60	1
		15	15	75	75
	3	3	15	15	76

Thus, $\dfrac{3x^4-12x^3-60x+4}{x-5}$
$=3x^3+3x^2+15x+15+\dfrac{76}{x-5}$

9.

−3	2	−6	−5	8
		−6	36	−93
	−3	−12	31	−85

Thus, the remainder is −85.

10. $12x^3y+10x^2y^4-14xy^3=2xy\left(6x^2+5xy^3-7y^2\right)$

11. $x^3-2x^2-3x=x\left(x^2-2x-3\right)$
$=x\left(x-3\right)\left(x+1\right)$

12. $2a^2+4ab+3ab+6b^2=2a\left(a+2b\right)+3b\left(a+2b\right)$
$=\left(a+2b\right)\left(2a+3b\right)$

13. $2b^4+5b^2-18$
$=2\left(b^2\right)^2+5\left(b^2\right)-18$
$=2w^2+5w-18$ ← Replace b^2 by w
$=2w^2+9w-4w-18$ ← Use $9w-4w$ for $4w$
$=w\left(2w+9\right)-2\left(2w+9\right)$
$=\left(2w+9\right)\left(w-2\right)$
$=\left(2b^2+9\right)\left(b^2-2\right)$ ← Replace w by b^2

14. $4(x-5)^2+20(x-5)$
$=4w^2+20w$ ← Replace $x-5$ by w
$=4w(w+5)$
$=4(x-5)\left[(x-5)+5\right]$ ← Replace w by $x-5$
$=4(x-5)(x)$
$=4x(x-5)$

15. $(x+4)^2+2(x+4)-3$
$=w^2+2w-3$ ← Replace $x+4$ by w
$=(w+3)(w-1)$
$=[(x+4)+3][(x+4)-1]$ ← Replace w by $x+4$
$=(x+7)(x+3)$

16. $27p^3q^6-8q^6$
$=q^6\left(27p^3-8\right)$
$=q^6\left[\left(3p\right)^3-2^3\right]$
$=q^6\left(3p-2\right)\left[\left(3p\right)^2+\left(3p\right)\left(2\right)+\left(2\right)^2\right]$
$=q^6\left(3p-2\right)\left(9p^2+6p+4\right)$

17. $f\left(x\right)=3x-4,\ g\left(x\right)=x-5$

a. $\left(f\cdot g\right)\left(x\right)=f\left(x\right)\cdot g\left(x\right)$
$=\left(3x-4\right)\left(x-5\right)$
$=3x^2-19x+20$

b. $\left(f\cdot g\right)\left(2\right)=3\left(2\right)^2-19\left(2\right)+20$
$=12-38+20$
$=-6$

18. $A=(2x)(2x)-4y^2$
$=4x^2-4y^2$
$=4(x^2-y^2)$
$=4(x+y)(x-y)$

19. $A=(x+8)(x+7)-(4)(3)$
$=x^2+7x+8x+56-12$
$=x^2+15x+44$
$=(x+11)(x+4)$

20. $7x^2+25x-12=0$
$(7x-3)(x+4)=0$
$7x-3=0 \quad \text{or} \quad x+4=0$
$7x=3 \qquad x=-4$
$x=\frac{3}{7}$

The solutions are $\frac{3}{7}, -4$.

21. $x^3+3x^2-10x=0$
$x(x^2+3x-10)=0$
$x(x+5)(x-2)=0$
$x=0 \quad \text{or} \quad x+5=0 \quad \text{or} \quad x-2=0$
$x=-5 \qquad x=2$
The solutions are 0, –5, 2.

22. $y=8x^2+10x-3$
Set $y = 0$ and solve for x
$8x^2+10x-3=0$
$(4x-1)(2x+3)=0$
$4x-1=0 \quad \text{or} \quad 2x+3=0$
$4x=1 \qquad 2x=-3$
$x=\frac{1}{4} \qquad x=-\frac{3}{2}$
The x-intercepts are $\left(\frac{1}{4},0\right)$ and $\left(-\frac{3}{2},0\right)$.

23. $y=(x-2)(x-7)$
$y=x^2-7x-2x+14$
$y=x^2-9x+14$

24. Let x be the height of the triangle. Then, $2x + 3$ is the base.
$\frac{1}{2}(\text{base})(\text{height}) = A$
$\frac{1}{2}(2x+3)(x)=22$
$(2x+3)(x)=44$
$2x^2+3x=44$
$2x^2+3x-44=0$
$(x-4)(2x+11)=0$
$x-4=0 \quad \text{or} \quad 2x+11=0$
$x=4 \qquad 2x=-11$
$x=-\frac{11}{2}$

Reject $-\frac{11}{2}$. Thus, the height is 4 meters and the base is $2\cdot4+3=8+3=11$ meters.

25. $s(t)=-16t^2+48t+448$
Set $s(t) = 0$ and solve for t
$0=-16t^2+48t+448$
$0=-16(t^2-3t-28)$
$0=-16(t-7)(t+4)$
$t-7=0 \quad \text{or} \quad t+4=0$
$t=7 \qquad t=-4$
Reject $t = -4$. The baseball strikes the ground in 7 seconds.

Chapter 5 Cumulative Review Test

1. $A\cup B=\{2, 3, 4, 5, 6, 8\}$

2. 

3. $\left|\frac{3}{8}\right|\div(-4)=\frac{3}{8}\div\left(-\frac{4}{1}\right)=\frac{3}{8}\cdot\left(-\frac{1}{4}\right)=-\frac{3}{32}$

4. $(-3)^3-2^2-(-2)^2+(9-8)^2$
$=(-3)^3-2^2-(-2)^2+(1)^2$
$=-27-4-4+1$
$=-34$

5. $\left(\dfrac{2r^4s^5}{r^2}\right)^3 = \left(2r^2s^5\right)^3 = 2^3r^{2\cdot3}s^{5\cdot3} = 8r^6s^{15}$

6. $4(2x-2)-3(x+7)=-4$
$$8x-8-3x-21=-4$$
$$5x-29=-4$$
$$5x=25$$
$$x=5$$

7. $k=2(d+e)$
$$k=2d+2e$$
$$k-2d=2e$$
$$\frac{k-2d}{2}=e \text{ or } e=\frac{k-2d}{2}$$

8. Let x = the length of a side either square in meters. The total amount to be fenced is $7x$.
$$7x=91$$
$$x=13$$
Each square is 13 meters by 13 meters.

9. Let x be the number of pages.
$$0.15x+0.05(6x)=279$$
$$0.15x+0.3x=279$$
$$0.45x=279$$
$$x=\frac{279}{0.45}=620$$
The manuscript is 620 pages long.

10. $70 \le \dfrac{68+72+90+86+x}{5} < 80$
$$350 \le 316+x < 400$$
$$34 \le x < 84$$
If Santo scores at least 34 points but less than 84 points, his average is in the 70's (and he will receive a grade of C).

11. Substitute 4 for x and 1 for y to see if these satisfy the equation.
$$3x+2y=13$$
$$3(4)+2(1)\stackrel{?}{=}13$$
$$12+2\stackrel{?}{=}13$$
$$14 \ne 13$$
Thus, (4, 1) is not a solution.

12. $2=6x-3y \Rightarrow 6x-3y=2$

13. Let $(x_1,y_1)=(8,-4)$ and $(x_2,y_2)=(-1,-2)$.
$$m=\frac{y_2-y_1}{x_2-x_1}=\frac{-2-(-4)}{-1-8}=\frac{2}{-9}=-\frac{2}{9}$$

14. $f(x)=2x^3-4x^2+x+16$
$$f(-4)=2(-4)^3-4(-4)^2+(-4)+16$$
$$=2(-64)-4(16)+(-4)+16$$
$$=-128-64-4+16$$
$$=-180$$

15. $2x-y\le 6$

Graph the line $2x-y=6$ using a solid line.
$$-y=-2x+6$$
$$y=2x-6$$
For the test point, select (0, 0):
$$2x-y\le 6$$
$$2(0)-0\le 6$$
$$0-0\le 6$$
$$0\le 6$$
Since this is a true statement, shade the region which contains (0, 0).

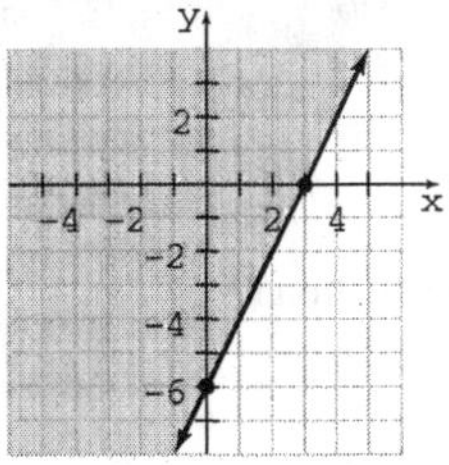

16. $\dfrac{1}{5}x+\dfrac{1}{2}y=4$
$$\frac{2}{3}x-y=\frac{8}{3}$$
To clear out the fractions, multiply the first equation by 10 and the second equation by 3.
$$10\left[\frac{1}{5}x+\frac{1}{2}y=4\right] \Rightarrow 2x+5y=40 \quad (1)$$
$$3\left[\frac{2}{3}x-y=\frac{8}{3}\right] \Rightarrow 2x-3y=8 \quad (2)$$
To eliminate x, multiply the first equation by -1 and then add.
$$-1(2x+5y=40)$$
$$2x-3y=8$$
gives

$$\begin{aligned} -2x-5y&=-40 \\ 2x-3y&=\ \ 8 \\ \hline \text{Add:}\quad -8y&=-32 \\ y&=4 \end{aligned}$$

To find x, substitute 4 for y into equation (1).

$$2x+5(4)=40$$
$$2x+20=40$$
$$2x=20$$
$$x=10$$

The solution is (10, 4).

17. $x-2y=2$ (1)

$2x+3y=11$ (2)

$-y+4z=7$ (3)

Notice that equations (1) and (2) form a system of equations in two variables. To eliminate x between these equations, multiply equation (1) by –2 and add to equation (2).

$-2[x-2y=2]$

$2x+3y=11$

gives

$$\begin{aligned} -2x+4y&=-4 \quad (4) \\ 2x+3y&=11 \quad (5) \\ \hline 7y&=7 \\ y&=1 \end{aligned}$$

Substitute 1 for y in equation (4).

$$-2x+4(1)=-4$$
$$-2x+4=-4$$
$$-2x=-8$$
$$x=4$$

Substitute 1 for y in equation (3).

$$-(1)+4z=7$$
$$4z=8$$
$$z=2$$

The solution is (4, 1, 2).

18. $\begin{vmatrix} 8 & 5 \\ -2 & 1 \end{vmatrix} = (8)(1)-(-2)(5)=8+10=18$

19.

$$\begin{array}{r} 2x^2+12x+63 \\ x-6\overline{)2x^3+0x^2-9x+15} \\ \underline{2x^3-12x^2} \\ 12x^2-\ \ 9x \\ \underline{12x^2-72x} \\ 63x+\ \ 15 \\ \underline{63x-378} \\ 393 \end{array}$$

Thus, $\dfrac{2x^3-9x+15}{x-6}=2x^2+12x+63+\dfrac{393}{x-6}$

20. $64x^3-27y^3$

$$=(4x)^3-(3y)^3$$
$$=(4x-3y)\left((4x)^2+(4x)(3y)+(3y)^2\right)$$
$$=(4x-3y)\left(16x^2+12xy+9y^2\right)$$

Chapter 6

Exercise Set 6.1

1. a. A rational expression is an expression of the form $\frac{p}{q}$, p and q polynomials, $q \neq 0$.

b. Answers will vary.

3. a. a rational function is a function of the form $f(x) = \frac{p}{q}$, p and q polynomials, $q \neq 0$.

b. Answers will vary.

5. a. The domain of a rational function is the set of values that can replace the variable.

b. $\frac{3}{x^2-25} = \frac{3}{(x+5)(x-5)}$
The domain is
$\{x|x$ is a real number, $x \neq -5, x \neq 5\}$.

7. a. Answers will vary. One possible answer is: Factor out (– 1) from the numerator and then cancel the remaining factor with the denominator.

b. $\frac{3x^2-2x-7}{-3x^2+2x+7} = \frac{-(-3x^2+2x+7)}{-3x^2+2x+7} = -1$

9. a. Answers will vary. One possible answer is: Invert the second fraction and then multiply by factoring and canceling common factors between the numerator and denominator.

b. $\frac{r+2}{r^2+9r+18} \div \frac{(r+2)^2}{r^2+5r+6}$

$= \frac{r+2}{r^2+9r+18} \cdot \frac{r^2+5r+6}{(r+2)^2}$

$= \frac{r+2}{(r+6)(r+3)} \cdot \frac{(r+3)(r+2)}{(r+2)(r+2)}$

$= \frac{1}{r+6}$

11. $\frac{4x}{5x-20} = \frac{4x}{5(x-4)}$
The excluded value is 4.

13. $\frac{4}{2x^2-15x+25} = \frac{4}{(x-5)(2x-5)}$

The excluded values are 5 and $\frac{5}{2}$.

15. $\frac{x-3}{x^2+12}$
There are no values for which $x^2+12=0$. Thus, there are no excluded values.

17. $\frac{x^2+81}{x^2-81} = \frac{x^2+81}{(x+9)(x-9)}$
The excluded values are –9 and 9.

19. $f(p) = \frac{p+1}{p-2}$

The domain is $\{p \mid p \neq 2\}$.

21. $y = \frac{5}{x^2+x-6}$
$y = \frac{5}{(x+3)(x-2)}$
The domain is $\{x| x \neq -3 \text{ and } x \neq 2\}$.

23. $f(a) = \frac{3a^2-6a+4}{2a^2+3a-2}$
$2a^2+3a-2=0$
$(2a-1)(a+2)=0$
$2a-1=0$ or $a+2=0$
$2a=1$ or $a=-2$
$a=\frac{1}{2}$

The domain is $\left\{a \,\middle|\, a \neq \frac{1}{2} \text{ and } a \neq -2\right\}$.

25. $g(x) = \frac{x^2-x+8}{x^2+4}$
There are no values for which $x^2+4=0$. Thus, there are no excluded values. The domain is $\{x|\, x$ is a real number$\}$.

27. $m(a)=\dfrac{a^2+36}{a^2-36}$

$=\dfrac{a^2+36}{(a+6)(a-6)}$

The domain is $\{a|\ a\neq -6 \text{ and } a\neq 6\}$.

29. $\dfrac{x-xy}{x}=\dfrac{x(1-y)}{x}=1-y$

31. $\dfrac{5x^2-20xy}{15x}=\dfrac{5x(x-4y)}{5x\cdot 3}=\dfrac{x-4y}{3}$

33. $\dfrac{x^3-x}{x^2-1}=\dfrac{x(x^2-1)}{x^2-1}=x$

35. $\dfrac{5r-8}{8-5r}=\dfrac{-1(8-5r)}{8-5r}=-1$

37. $\dfrac{p^2-2p-24}{6-p}=\dfrac{(p-6)(p+4)}{-1(p-6)}$

$=\dfrac{p+4}{-1}$

$=-(p+4)$ or $-p-4$

39. $\dfrac{a^2-3a-10}{a^2+5a+6}=\dfrac{(a-5)(a+2)}{(a+3)(a+2)}=\dfrac{a-5}{a+3}$

41. $\dfrac{8x^3-125y^3}{2x-5y}=\dfrac{(2x-5y)(4x^2+10xy+25y^2)}{2x-5y}$

$=4x^2+10xy+25y^2$

43. $\dfrac{(x+6)(x-3)+(x+6)(x-2)}{2(x+6)}$

$=\dfrac{(x+6)[(x-3)+(x-2)]}{2(x+6)}$

$=\dfrac{(x+6)(2x-5)}{2(x+6)}$

$=\dfrac{2x-5}{2}$

45. $\dfrac{a^2+7a-ab-7b}{a^2-ab+5a-5b}=\dfrac{a(a+7)-b(a+7)}{a(a-b)+5(a-b)}$

$=\dfrac{(a-b)(a+7)}{(a+5)(a-b)}$

$=\dfrac{a+7}{a+5}$

47. $\dfrac{x^2-x-12}{x^3+27}=\dfrac{(x+3)(x-4)}{(x+3)(x^2-3x+9)}$

$=\dfrac{x-4}{x^2-3x+9}$

49. $\dfrac{2x}{5y}\cdot\dfrac{y^3}{6}=\dfrac{x\cdot y^2}{5\cdot 3}=\dfrac{xy^2}{15}$

51. $\dfrac{9x^3}{4}\div\dfrac{3}{16y^2}=\dfrac{9x^3}{4}\cdot\dfrac{16y^2}{3}$

$=\dfrac{3x^3\cdot 4y^2}{1}$

$=\dfrac{12x^3y^2}{1}$

$=12x^3y^2$

53. $\dfrac{3-r}{r-3}\cdot\dfrac{r-9}{9-r}=\dfrac{-1(r-3)}{r-3}\cdot\dfrac{-1(9-r)}{9-r}$

$=(-1)(-1)$

$=1$

55. $\dfrac{x^2+3x-10}{4x}\cdot\dfrac{x^2-3x}{x^2-5x+6}$

$=\dfrac{(x+5)(x-2)}{4x}\cdot\dfrac{x(x-3)}{(x-3)(x-2)}$

$=\dfrac{x+5}{4}$

57. $\dfrac{r^2+10r+21}{r+7}\div\dfrac{(r^2-5r-24)}{r^3}$

$=\dfrac{r^2+10r+21}{r+7}\cdot\dfrac{r^3}{r^2-5r-24}$

$=\dfrac{(r+3)(r+7)}{r+7}\cdot\dfrac{r^3}{(r+3)(r-8)}$

$=\dfrac{r^3}{r-8}$

59. $\dfrac{x^2+12x+35}{x^2+4x-5} \div \dfrac{x^2+3x-28}{7x-7}$

$= \dfrac{x^2+12x+35}{x^2+4x-5} \cdot \dfrac{7x-7}{x^2+3x-28}$

$= \dfrac{(x+7)(x+5)}{(x+5)(x-1)} \cdot \dfrac{7(x-1)}{(x+7)(x-4)}$

$= \dfrac{7}{x-4}$

61. $\dfrac{a-b}{9a+9b} \div \dfrac{a^2-b^2}{a^2+2a+1}$

$= \dfrac{a-b}{9a+9b} \cdot \dfrac{a^2+2a+1}{a^2-b^2}$

$= \dfrac{a-b}{9(a+b)} \cdot \dfrac{(a+1)(a+1)}{(a+b)(a-b)}$

$= \dfrac{(a+1)^2}{9(a+b)^2}$

63. $\dfrac{3x^2-x-4}{4x^2+5x+1} \cdot \dfrac{2x^2-5x-12}{6x^2+x-12}$

$= \dfrac{(3x-4)(x+1)}{(4x+1)(x+1)} \cdot \dfrac{(2x+3)(x-4)}{(2x+3)(3x-4)}$

$= \dfrac{x-4}{4x+1}$

65. $\dfrac{x+2}{x^3-8} \cdot \dfrac{(x-2)^2}{x^2+4}$

$= \dfrac{x+2}{(x-2)(x^2+2x+4)} \cdot \dfrac{(x-2)(x-2)}{x^2+4}$

$= \dfrac{(x+2)(x-2)}{(x^2+2x+4)(x^2+4)}$

67. $\dfrac{x^2-y^2}{x^2-2xy+y^2} \div \dfrac{(x+y)^2}{(x-y)^2}$

$= \dfrac{x^2-y^2}{x^2-2xy+y^2} \cdot \dfrac{(x-y)^2}{(x+y)^2}$

$= \dfrac{(x+y)(x-y)}{(x-y)(x-y)} \cdot \dfrac{(x-y)(x-y)}{(x+y)(x+y)}$

$= \dfrac{x-y}{x+y}$

69. $\dfrac{2x^4+4x^2}{6x^2+14x+4} \div \dfrac{x^2+2}{3x^2+x}$

$= \dfrac{2x^4+4x^2}{6x^2+14x+4} \cdot \dfrac{3x^2+x}{x^2+2}$

$= \dfrac{2x^2(x^2+2)}{2(3x+1)(x+2)} \cdot \dfrac{x(3x+1)}{x^2+2}$

$= \dfrac{x^3}{x+2}$

71. $\dfrac{(a-b)^3}{a^3-b^3} \cdot \dfrac{a^2-b^2}{(a-b)^2}$

$= \dfrac{(a-b)^3}{(a-b)(a^2+ab+b^2)} \cdot \dfrac{(a-b)(a+b)}{(a-b)^2}$

$= \dfrac{(a-b)(a+b)}{a^2+ab+b^2}$

73. $\dfrac{4x+y}{5x+2y} \cdot \dfrac{10x^2-xy-2y^2}{8x^2-2xy-y^2}$

$= \dfrac{4x+y}{5x+2y} \cdot \dfrac{(5x+2y)(2x-y)}{(4x+y)(2x-y)}$

$= 1$

75. $\dfrac{ac-ad+bc-bd}{ac+ad+bc+bd} \cdot \dfrac{pc+pd-qc-qd}{pc-pd+qc-qd}$

$= \dfrac{a(c-d)+b(c-d)}{a(c+d)+b(c+d)} \cdot \dfrac{p(c+d)-q(c+d)}{p(c-d)+q(c-d)}$

$= \dfrac{(a+b)(c-d)}{(a+b)(c+d)} \cdot \dfrac{(p-q)(c+d)}{(p+q)(c-d)}$

$= \dfrac{p-q}{p+q}$

77. $\dfrac{3r^2+17rs+10s^2}{6r^2+13rs-5s^2} \div \dfrac{6r^2+rs-2s^2}{6r^2-5rs+s^2}$

$= \dfrac{3r^2+17rs+10s^2}{6r^2+13rs-5s^2} \cdot \dfrac{6r^2-5rs+s^2}{6r^2+rs-2s^2}$

$= \dfrac{(3r+2s)(r+5s)}{(2r+5s)(3r-s)} \cdot \dfrac{(3r-s)(2r-s)}{(2r-s)(3r+2s)}$

$= \dfrac{r+5s}{2r+5s}$

79. Use the restricted values to create factors of the denominator. One possible answer:

$\dfrac{1}{(x-2)(x+3)}$ or $\dfrac{1}{x^2+x-6}$

81. The numerator is never 0.

83. a. It is zero when the numerator is zero. That is, when $x - 4 = 0$, or $x = 4$.

b. It is undefined when the denominator is zero. That is, when

$$x^2 - 36 = 0$$
$$(x+6)(x-6) = 0$$
$$x+6=0 \quad \text{or} \quad x-6=0$$
$$x=-6 \qquad\qquad x=6$$

The function will be undefined if $x=-6$ or $x=6$.

85. Use the restricted values to create factors of the denominator. Use the value that makes the function 0 to create a factor in the numerator.

One possible answer is $f(x) = \dfrac{x-2}{(x-3)(x+1)}$.

87. $x^2 + 2x - 15 = (x-3)(x+5)$
$x + 5$ is the factor missing from the denominator of the fraction on the right side.

$$\frac{1}{x-3} \cdot \frac{x+5}{x+5} = \frac{x+5}{(x-3)(x+5)} = \frac{x+5}{x^2+2x-15}.$$

The desired numerator is $x + 5$.

89. $y^2 - y - 20 = (y-5)(y+4)$
$y - 5$ is the factor missing from the numerator of the fraction on the right side.

$$\frac{y+4}{y+1} \cdot \frac{y-5}{y-5} = \frac{(y+4)(y-5)}{(y+1)(y-5)} = \frac{y^2-y-20}{y^2-4y-5}.$$

The desired denominator is $y^2 - 4y - 5$.

91.

$$\frac{x^2-x-12}{x^2+2x-3} \cdot \frac{?}{x^2-2x-8} = 1$$
$$\frac{(x-4)(x+3)}{(x+3)(x-1)} \cdot \frac{?}{(x-4)(x+2)} = 1$$
$$\frac{?}{(x-1)(x+2)} = 1$$

The only way the left side can simplify to 1 is if the numerator is $(x-1)(x+2)$.
The corresponding polynomial is $x^2 + x - 2$.

93.

$$\frac{x^2-9}{2x^2+3x-2} \div \frac{2x^2-9x+9}{?} = \frac{x+3}{2x-1}$$
$$\frac{x^2-9}{2x^2+3x-2} \cdot \frac{?}{2x^2-9x+9} = \frac{x+3}{2x-1}$$
$$\frac{(x+3)(x-3)}{(2x-1)(x+2)} \cdot \frac{?}{(2x-3)(x-3)} = \frac{x+3}{2x-1}$$
$$\frac{x+3}{(2x-1)(x+2)} \cdot \frac{?}{(2x-3)} = \frac{x+3}{2x-1}$$

The only way the left side can simplify to the right side is if the missing numerator is $(2x-3)(x+2)$.
The corresponding polynomial is $2x^2 + x - 6$.

95. Area = (length)(width)

$$3a^2 + 7ab + 2b^2 = (2a+4b)w$$
$$\frac{3a^2+7ab+2b^2}{2a+4b} = w$$
$$\frac{(3a+b)(a+2b)}{2(a+2b)} = w$$
$$w = \frac{3a+b}{2}$$

97. Area $= \dfrac{1}{2}$(base)(height)

$$a^2 + 4ab + 3b^2 = \frac{1}{2}(a+3b)h$$
$$2(a^2+4ab+3b^2) = 2\left[\frac{1}{2}(a+3b)h\right]$$
$$2(a^2+4ab+3b^2) = (a+3b)h$$
$$\frac{2(a^2+4ab+3b^2)}{a+3b} = h$$
$$\frac{2(a+3b)(a+b)}{a+3b} = h$$
$$h = 2(a+b)$$

99. $\left(\dfrac{2x^2-3x-14}{2x^2-9x+7}\div\dfrac{6x^2+x-15}{3x^2+2x-5}\right)\cdot\dfrac{6x^2-7x-3}{2x^2-x-3}=\left(\dfrac{2x^2-3x-14}{2x^2-9x+7}\cdot\dfrac{3x^2+2x-5}{6x^2+x-15}\right)\cdot\dfrac{6x^2-7x-3}{2x^2-x-3}$

$$=\frac{(x+2)(2x-7)}{(x-1)(2x-7)}\cdot\frac{(x-1)(3x+5)}{(3x+5)(2x-3)}\cdot\frac{(3x+1)(2x-3)}{(x+1)(2x-3)}$$

$$=\frac{(x+2)(3x+1)}{(2x-3)(x+1)}$$

101. $\dfrac{5x^2(x-1)-3x(x-1)-2(x-1)}{10x^2(x-1)+9x(x-1)+2(x-1)}\cdot\dfrac{2x+1}{x+3}$

$$=\frac{(x-1)(5x^2-3x-2)}{(x-1)(10x^2+9x+2)}\cdot\frac{2x+1}{x+3}$$

$$=\frac{(x-1)(5x+2)(x-1)}{(x-1)(5x+2)(2x+1)}\cdot\frac{2x+1}{x+3}$$

$$=\frac{x-1}{x+3}$$

103. $\dfrac{(x-p)^n}{x^{-2}}\div\dfrac{(x-p)^{2n}}{x^{-6}}$

$$=x^2(x-p)^n\div x^6(x-p)^{2n}$$

$$=\frac{x^2(x-p)^n}{1}\cdot\frac{1}{x^6(x-p)^{2n}}$$

$$=\frac{x^2(x-p)^n}{1}\cdot\frac{1}{x^6(x-p)^n(x-p)^n}$$

$$=\frac{1}{x^4(x-p)^n}$$

105. $\dfrac{x^{5y}+3x^{4y}}{3x^{3y}+x^{4y}}=\dfrac{x^y(x^{4y}+3x^{3y})}{3x^{3y}+x^{4y}}=x^y$

107. a. $f(x)=\dfrac{1}{x-2}$. The denominator cannot equal zero. Therefore, $x\neq 2$. The domain is $\{x\mid x\neq 2\}$.

b.

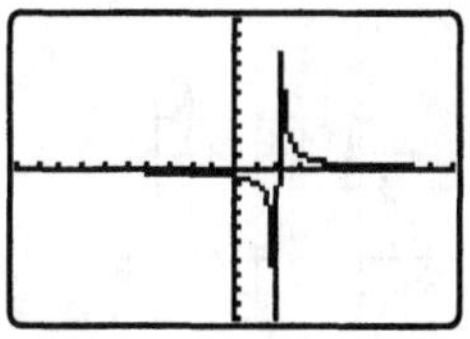

-10, 10, 1, -10, 10, 1

c. The function is decreasing as x gets closer to 2, approaching 2 from the left side.

d. The function is increasing as x gets closer to 2, approaching 2 from the right side.

109. a. $f(x)=\dfrac{x^2}{x-2}$. The denominator cannot equal zero. Therefore, $x\neq 2$. The domain is $\{x\mid x\neq 2\}$.

b.

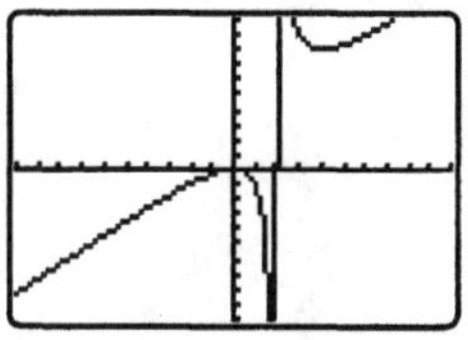

-10, 10, 1, -10, 10, 1

c. The function is decreasing as x gets closer to 2, approaching 2 from the left side.

d. The function is increasing as x gets closer to 2, approaching 2 from the right side.

111. **a.** The domain is all real numbers except where the denominator is 0. The denominator is 0 when $x = 0$ so this value must be excluded from the domain. The domain is $\{x | x \neq 0\}$.

b.

x	-10	-1	-0.5	-0.1	$-.01$
y	-0.1	-1	-2	-10	-100

x	0.01	0.1	0.5	1	10
y	100	10	2	1	0.1

c.

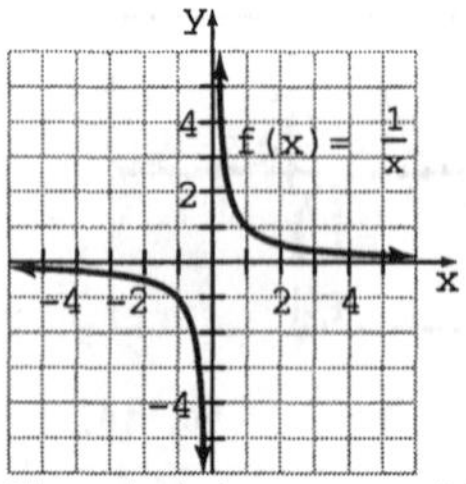

The graph increases without bound as x approaches 0 from the right and decreases without bound as x approaches 0 from the left.

d. No, the numerator can never be 0.

113. $6(x-2)+6y = 12x$

$6x - 12 + 6y = 12x$

$6y = 6x + 12$

$\frac{6y}{6} = \frac{6x+12}{6}$

$y = x + 2$

114. $4 + \frac{4x}{3} < 6$

$3\left(4 + \frac{4x}{3}\right) < 3(6)$

$12 + 4x < 18$

$4x < 6$

$x < \frac{6}{4} \Rightarrow x < \frac{3}{2}$

Interval notation: $\left(-\infty, \frac{3}{2}\right)$

115. $\left|\frac{2x-4}{12}\right| = 5$

$\frac{2x-4}{12} = -5$ or $\frac{2x-4}{12} = 5$

$2x - 4 = -60$ $\quad$ $2x - 4 = 60$

$2x = -56$ $\quad$ $2x = 64$

$x = -28$ $\quad$ $x = 32$

The solution is $\{-28, 32\}$.

116. $f(x) = |6 - 3x| - 2$

$f(1.3) = |6 - 3(1.3)| - 2$

$= |6 - 3.9| - 2$

$= |2.1| - 2$

$= 2.1 - 2$

$= 0.1$

117. $3x + 4y = 2$

$2x + 5y = -1$

To eliminate x, multiply the first equation by -2 and the second equation by 3 then add.

$-2[3x + 4y = 2]$

$3[2x + 5y = -1]$

gives

$-6x - 8y = -4$

$6x + 15y = -3$

Add: $7y = -7$

$y = -1$

To find x, substitute -1 for y into the first equation.

$3x + 4(-1) = 2$

$3x - 4 = 2$

$3x = 6$

$x = 2$

The solution is $(2, -1)$.

118. $9x^2 + 6xy + y^2 - 4$

$= (3x)^2 + 2(3x)(y) + (y)^2 - 4$

$= (3x + y)^2 - 4$

$= (3x + y)^2 - 2^2$

$= (3x + y + 2)(3x + y - 2)$

Exercise Set 6.2

1. a. Answers will vary. One possible answer is: The LCD is the 'smallest' denominator into which the individual denominators divide.

   b. Answers will vary. One possible answer is: Factor each denominator completely. Any factors that occur more than once should be expressed as powers. The LCD is the product of all of the unique factors found, each to the highest power to which it occurs in any denominator.

   c. $\frac{5}{64x^2-121}, \frac{1}{8x^2-27x+22}$

   Factor denominators:

   $64x^2-121=(8x+11)(8x-11)$

   $8x^2-27x+22=(8x-11)(x-2)$

   LCD is $(8x+11)(8x-11)(x-2)$.

3. a. The entire numerator was not subtracted.

   b. $\frac{x^2-4x}{(x+3)(x-2)}-\frac{x^2+x-2}{(x+3)(x-2)}$

   $=\frac{x^2-4x-(x^2+x-2)}{(x+3)(x-2)}$

   $=\frac{x^2-4x-x^2-x+2}{(x+3)(x-2)}$

5. $\frac{3x}{x+2}+\frac{5}{x+2}=\frac{3x+5}{x+2}$

7. $\frac{7x}{x-5}-\frac{2}{x-5}=\frac{7x-2}{x-5}$

9. $\frac{x}{x+3}+\frac{9}{x+3}-\frac{2}{x+3}=\frac{x+9-2}{x+3}=\frac{x+7}{x+3}$

11. $\frac{5x-6}{x-8}+\frac{2x-5}{x-8}=\frac{5x-6+2x-5}{x-8}=\frac{7x-11}{x-8}$

13. $\frac{x^2-2}{x^2+6x-7}-\frac{-4x+19}{x^2+6x-7}=\frac{x^2-2-(-4x)-19}{x^2+6x-7}$

   $=\frac{x^2-2+4x-19}{x^2+6x-7}$

   $=\frac{x^2+4x-21}{x^2+6x-7}$

   $=\frac{(x+7)(x-3)}{(x+7)(x-1)}$

   $=\frac{x-3}{x-1}$

15. $\frac{x^3-12x^2+45x}{x(x-8)}-\frac{x^2+5x}{x(x-8)}$

   $=\frac{x^3-12x^2+45x-(x^2+5x)}{x(x-8)}$

   $=\frac{x^3-12x^2+45x-x^2-5x}{x(x-8)}$

   $=\frac{x^3-13x^2+40x}{x(x-8)}$

   $=\frac{x(x-5)(x-8)}{x(x-8)}$

   $=x-5$

17. $\frac{3x^2-x}{2x^2-x-21}+\frac{2x-8}{2x^2-x-21}-\frac{x^2-2x+27}{2x^2-x-21}$

   $=\frac{3x^2-x+(2x-8)-(x^2-2x+27)}{2x^2-x-21}$

   $=\frac{3x^2-x+2x-8-x^2+2x-27}{2x^2-x-21}$

   $=\frac{2x^2+3x-35}{2x^2-x-21}$

   $=\frac{(2x-7)(x+5)}{(2x-7)(x+3)}$

   $=\frac{x+5}{x+3}$

19. $\frac{5}{2a^2}+\frac{9}{3a^3}$

   Factor denominators: $2a^2=2\cdot a^2$

   $3a^3=3\cdot a^3$

   The LCD is $2\cdot 3\cdot a^3=6a^3$.

21. $\dfrac{-4}{8x^2y^2}+\dfrac{7}{5x^4y^6}$

Factor denominators: $8x^2y^2 = 2^3 \cdot x^2 \cdot y^2$

$5x^4y^6 = 5 \cdot x^4 \cdot y^6$

The LCD is $2^3 \cdot 5 \cdot x^4 \cdot y^6 = 40x^4y^6$

23. $\dfrac{2}{3a^4b^2}+\dfrac{7}{2a^3b^5}$

Factor denominators: $3a^4b^2 = 3 \cdot a^4 \cdot b^2$

$2a^3b^5 = 2 \cdot a^3 \cdot b^5$

The LCD is $2 \cdot 3 \cdot a^4 \cdot b^5 = 6a^4b^5$

25. $\dfrac{4x}{x+3}+\dfrac{6}{x+9}$

The LCD is $(x + 3)(x + 9)$

27. $\dfrac{5z^2}{1}+\dfrac{9z}{z-6}$

The LCD is $1(z - 6) = z - 6$.

29. $\dfrac{x}{x^4(x-2)}-\dfrac{x+9}{x^2(x-2)^3}$

The LCD is $x^4 \cdot (x-2)^3 = x^4(x-2)^3$.

31. $\dfrac{a-2}{a^2-5a-24}+\dfrac{3}{a^2+11a+24}$

Factor denominators:

$a^2-5a-24 = (a-8)(a+3)$

$a^2+11a+24 = (a+3)(a+8)$

The LCD is $(a - 8)(a + 3)(a + 8)$

33. $\dfrac{x}{2x^2-7x+3}+\dfrac{x-3}{4x^2+4x-3}-\dfrac{x^2+1}{2x^2-3x-9}$

Factor denominators:

$2x^2-7x+3 = (2x-1)(x-3)$

$4x^2+4x-3 = (2x-1)(2x+3)$

$2x^2-3x-9 = (x-3)(2x+3)$

The LCD is $(x - 3)(2x - 1)(2x + 3)$

35. $\dfrac{2}{3r}+\dfrac{8}{r}=\dfrac{2}{3r}+\dfrac{8}{r}\cdot\dfrac{3}{3}$ ← The LCD is $3r$

$=\dfrac{2}{3r}+\dfrac{24}{3r}$

$=\dfrac{2+24}{3r}$

$=\dfrac{26}{3r}$

37. $\dfrac{5}{12x}-\dfrac{1}{4x^2}=\dfrac{5}{12x}\cdot\dfrac{x}{x}-\dfrac{1}{4x^2}\cdot\dfrac{3}{3}$ ←. The LCD is $3 \cdot 2^2 \cdot x^2 = 12x^2$

$=\dfrac{5x}{12x^2}-\dfrac{3}{12x^2}$

$=\dfrac{5x-3}{12x^2}$

39. $\dfrac{3}{8x^4y}+\dfrac{1}{5x^2y^3}=\dfrac{3}{8x^4y}\cdot\dfrac{5y^2}{5y^2}+\dfrac{1}{5x^2y^3}\cdot\dfrac{8x^2}{8x^2}$ ←. The LCD is $40x^4y^3$

$=\dfrac{15y^2}{40x^4y^3}+\dfrac{8x^2}{40x^4y^3}$

$=\dfrac{15y^2+8x^2}{40x^4y^3}$

41. $\frac{b}{a-b}-\frac{a+b}{b}=\frac{b}{a-b}\cdot\frac{b}{b}-\frac{a+b}{b}\cdot\frac{a-b}{a-b}$ ← The LCD is $b(a-b)$

$=\frac{b^2}{b(a-b)}-\frac{(a+b)(a-b)}{b(a-b)}$

$=\frac{b^2}{b(a-b)}-\frac{a^2-b^2}{b(a-b)}$

$=\frac{b^2-a^2+b^2}{b(a-b)}$

$=\frac{2b^2-a^2}{b(a-b)}$

43. $\frac{a}{a-b}-\frac{a}{b-a}=\frac{a}{a-b}-\frac{a}{b-a}\cdot\frac{(-1)}{(-1)}$ ← The LCD is $a-b$

$=\frac{a}{a-b}-\frac{-a}{a-b}$

$=\frac{2a}{a-b}$

45. $\frac{4x}{x-4}+\frac{x+3}{x+1}=\frac{4x}{x-4}\cdot\frac{x+1}{x+1}+\frac{x+3}{x+1}\cdot\frac{x-4}{x-4}$ ← The LCD is $(x-4)(x+1)$

$=\frac{4x(x+1)}{(x-4)(x+1)}+\frac{(x+3)(x-4)}{(x-4)(x+1)}$

$=\frac{4x^2+4x}{(x-4)(x+1)}+\frac{x^2-x-12}{(x-4)(x+1)}$

$=\frac{4x^2+4x+x^2-x-12}{(x-4)(x+1)}$

$=\frac{5x^2+3x-12}{(x-4)(x+1)}$

47. $\frac{3}{a+2}+\frac{3a+1}{a^2+4a+4}=\frac{3}{a+2}+\frac{3a+1}{(a+2)(a+2)}$ ← Factor denominators

$=\frac{3}{a+2}\cdot\frac{a+2}{a+2}+\frac{3a+1}{(a+2)(a+2)}$ ← The LCD is $(a+2)(a+2)$

$=\frac{3(a+2)}{(a+2)(a+2)}+\frac{3a+1}{(a+2)(a+2)}$

$=\frac{3a+6}{(a+2)(a+2)}+\frac{3a+1}{(a+2)(a+2)}$

$=\frac{3a+6+3a+1}{(a+2)(a+2)}$

$=\frac{6a+7}{(a+2)^2}$

49. $\frac{x}{x^2+2x-8}+\frac{x+1}{x^2-3x+2}=\frac{x}{(x+4)(x-2)}+\frac{x+1}{(x-2)(x-1)}$ ← Factor denominators

$=\frac{x}{(x+4)(x-2)}\cdot\frac{x-1}{x-1}+\frac{x+1}{(x-2)(x-1)}\cdot\frac{x+4}{x+4}$ ← The LCD is $(x+4)(x-2)(x-1)$

$=\frac{x(x-1)}{(x+4)(x-2)(x-1)}+\frac{(x+1)(x+4)}{(x-2)(x-1)(x+4)}$

$=\frac{x(x-1)+(x+1)(x+4)}{(x+4)(x-2)(x-1)}$

$=\frac{x^2-x+x^2+x+4x+4}{(x+4)(x-2)(x-1)}$

$=\frac{2x^2+4x+4}{(x+4)(x-2)(x-1)}$

51. Factor denominators and simplify second fraction

$\frac{5x}{x^2-9x+8}-\frac{3(x+2)}{x^2-6x-16}=\frac{5x}{(x-8)(x-1)}-\frac{3(x+2)}{(x-8)(x+2)}$

$=\frac{5x}{(x-8)(x-1)}-\frac{3}{x-8}\cdot\frac{x-1}{x-1}$ ← The LCD is $(x-8)(x-1)$

$=\frac{5x-(3x-3)}{(x-8)(x-1)}$

$=\frac{5x-3x+3}{(x-8)(x-1)}$

$=\frac{2x+3}{(x-8)(x-1)}$

53. $4-\frac{x-1}{x^2+3x-10}=\frac{4}{1}\cdot\frac{x^2+3x-10}{x^2+3x-10}-\frac{x-1}{x^2+3x-10}$ ← The LCD is $x^2+3x-10$ or $(x+5)(x-2)$

$=\frac{4x^2+12x-40}{x^2+3x-10}-\frac{x-1}{x^2+3x-10}$

$=\frac{4x^2+12x-40-(x-1)}{x^2+3x-10}$

$=\frac{4x^2+11x-39}{x^2+3x-10}$

$=\frac{4x^2+11x-39}{(x+5)(x-2)}$

55. $\dfrac{3a+2}{4a+1}-\dfrac{3a+6}{4a^2+9a+2}=\dfrac{3a+2}{4a+1}-\dfrac{3(a+2)}{(4a+1)(a+2)}$ ← . Factor the second fraction and simplify.

$$=\frac{3a+2}{4a+1}-\frac{3}{4a+1}$$

$$=\frac{3a+2-3}{4a+1}$$

$$=\frac{3a-1}{4a+1}$$

57. $\dfrac{x-y}{x^2-4xy+4y^2}+\dfrac{x-3y}{x^2-4y^2}$

$=\dfrac{x-y}{(x-2y)(x-2y)}+\dfrac{x-3y}{(x-2y)(x+2y)}$ ← . Factor denominators

$=\dfrac{x-y}{(x-2y)(x-2y)}\cdot\dfrac{x+2y}{x+2y}+\dfrac{x-3y}{(x-2y)(x+2y)}\cdot\dfrac{x-2y}{x-2y}$ ← . The LCD is $(x-2y)(x-2y)(x+2y)$

$$=\frac{(x-y)(x+2y)}{(x-2y)^2(x+2y)}+\frac{(x-3y)(x-2y)}{(x-2y)^2(x+2y)}$$

$$=\frac{x^2+xy-2y^2}{(x-2y)^2(x+2y)}+\frac{x^2-5xy+6y^2}{(x-2y)^2(x+2y)}$$

$$=\frac{x^2+xy-2y^2+x^2-5xy+6y^2}{(x-2y)^2(x+2y)}$$

$$=\frac{2x^2-4xy+4y^2}{(x-2y)^2(x+2y)}$$

59. $\dfrac{2r}{r-4}-\dfrac{2r}{r+4}+\dfrac{64}{r^2-16}=\dfrac{2r}{r-4}-\dfrac{2r}{r+4}+\dfrac{64}{(r+4)(r-4)}$ ← . Factor third denominator

$=\dfrac{2r}{r-4}\cdot\dfrac{r+4}{r+4}-\dfrac{2r}{r+4}\cdot\dfrac{r-4}{r-4}+\dfrac{64}{(r+4)(r-4)}$ ← . The LCD is $(r+4)(r-4)$

$$=\frac{2r(r+4)}{(r+4)(r-4)}-\frac{2r(r-4)}{(r+4)(r-4)}+\frac{64}{(r+4)(r-4)}$$

$$=\frac{2r^2+8r}{(r+4)(r-4)}-\frac{2r^2-8r}{(r+4)(r-4)}+\frac{64}{(r+4)(r-4)}$$

$$=\frac{2r^2+8r-(2r^2-8r)+64}{(r+4)(r-4)}$$

$$=\frac{16r+64}{(r+4)(r-4)}$$

$$=\frac{16(r+4)}{(r+4)(r-4)}=\frac{16}{r-4}$$

61.

$$\frac{-4}{x^2+2x-3}-\frac{1}{x+3}+\frac{1}{x-1}=\frac{-4}{(x+3)(x-1)}-\frac{1}{x+3}+\frac{1}{x-1} \quad \leftarrow \text{. Factor denominators}$$

$$=\frac{-4}{(x+3)(x-1)}-\frac{1}{x+3}\cdot\frac{x-1}{x-1}+\frac{1}{x-1}\cdot\frac{x+3}{x+3} \quad \leftarrow \text{. LCD is } (x+3)(x-3)$$

$$=\frac{-4}{(x+3)(x-1)}-\frac{x-1}{(x+3)(x-1)}+\frac{x+3}{(x+3)(x-1)}$$

$$=\frac{-4-(x-1)+x+3}{(x+3)(x-1)}$$

$$=\frac{-4-x+1+x+3}{(x+3)(x-1)}$$

$$=\frac{0}{(x+3)(x-1)}$$

$$=0$$

63.

$$\frac{3}{3x-2}-\frac{1}{x-4}+5=\frac{3}{3x-2}\cdot\frac{x-4}{x-4}-\frac{1}{x-4}\cdot\frac{3x-2}{3x-2}+\frac{5}{1}\cdot\frac{(x-4)(3x-2)}{(x-4)(3x-2)} \quad \leftarrow \text{ The LCD is } (3x-2)(x-4)$$

$$=\frac{3(x-4)}{(3x-2)(x-4)}-\frac{3x-2}{(3x-2)(x-4)}+\frac{5(x-4)(3x-2)}{(3x-2)(x-4)}$$

$$=\frac{3x-12}{(3x-2)(x-4)}-\frac{3x-2}{(3x-2)(x-4)}+\frac{15x^2-70x+40}{(3x-2)(x-4)}$$

$$=\frac{3x-12-(3x-2)+15x^2-70x+40}{(3x-2)(x-4)}$$

$$=\frac{15x^2-70x+30}{(3x-2)(x-4)}$$

65.

$$2-\frac{1}{8r^2+2r-15}+\frac{r+2}{4r-5}$$

$$=\frac{2}{1}-\frac{1}{(4r-5)(2r+3)}+\frac{r+2}{4r-5} \quad \leftarrow \text{. Factor second denominator}$$

$$=\frac{2}{1}\cdot\frac{(4r-5)(2r+3)}{(4r-5)(2r+3)}-\frac{1}{(4r-5)(2r+3)}+\frac{r+2}{4r-5}\cdot\frac{2r+3}{2r+3} \quad \leftarrow \text{. The LCD is } (4r-5)(2r+3)$$

$$=\frac{2(4r-5)(2r+3)}{(4r-5)(2r+3)}-\frac{1}{(4r-5)(2r+3)}+\frac{(r+2)(2r+3)}{(4r-5)(2r+3)}$$

$$=\frac{16r^2+4r-30}{(4r-5)(2r+3)}-\frac{1}{(4r-5)(2r+3)}+\frac{2r^2+7r+6}{(4r-5)(2r+3)}$$

$$=\frac{16r^2+4r-30-1+2r^2+7r+6}{(4r-5)(2r+3)}$$

$$=\frac{18r^2+11r-25}{(4r-5)(2r+3)}$$

67.

$$\frac{3}{5x+6}+\frac{x^2-x}{5x^2-4x-12}-\frac{4}{x-2}$$

$$=\frac{3}{5x+6}+\frac{x^2-x}{(5x+6)(x-2)}-\frac{4}{x-2} \quad \leftarrow \text{ Factor second denominator}$$

$$=\frac{3}{5x+6}\cdot\frac{x-2}{x-2}+\frac{x^2-x}{(5x+6)(x-2)}-\frac{4}{x-2}\cdot\frac{5x+6}{5x+6} \quad \leftarrow \text{ The LCD is } (5x+6)(x-2)$$

$$=\frac{3(x-2)}{(5x+6)(x-2)}+\frac{x^2-x}{(5x+6)(x-2)}-\frac{4(5x+6)}{(5x+6)(x-2)}$$

$$=\frac{3x-6}{(5x+6)(x-2)}+\frac{x^2-x}{(5x+6)(x-2)}-\frac{20x+24}{(5x+6)(x-2)}$$

$$=\frac{3x-6+x^2-x-(20x+24)}{(5x+6)(x-2)}$$

$$=\frac{3x-6+x^2-x-20x-24}{(5x+6)(x-2)}$$

$$=\frac{x^2-18x-30}{(5x+6)(x-2)}$$

69.

$$\frac{3m}{6m^2+13mn+6n^2}+\frac{2m}{4m^2+8mn+3n^2}$$

$$=\frac{3m}{(2m+3n)(3m+2n)}+\frac{2m}{(2m+3n)(2m+n)} \quad \leftarrow \text{ Factor denominators}$$

$$=\frac{3m}{(2m+3n)(3m+2n)}\cdot\frac{2m+n}{2m+n}+\frac{2m}{(2m+3n)(2m+n)}\cdot\frac{3m+2n}{3m+2n} \quad \leftarrow . \text{ The LCD is } (2m+3n)(3m+2n)(2m+n)$$

$$=\frac{3m(2m+n)}{(2m+3n)(3m+2n)(2m+n)}+\frac{2m(3m+2n)}{(2m+3n)(3m+2n)(2m+n)}$$

$$=\frac{6m^2+3mn}{(2m+3n)(3m+2n)(2m+n)}+\frac{6m^2+4mn}{(2m+3n)(3m+2n)(2m+n)}$$

$$=\frac{6m^2+3mn+6m^2+4mn}{(2m+3n)(3m+2n)(2m+n)}$$

$$=\frac{12m^2+7mn}{(2m+3n)(3m+2n)(2m+n)}$$

71.

$$\frac{5r-2s}{25r^2-4s^2}-\frac{2r-s}{10r^2-rs-2s^2}=\frac{5r-2s}{(5r+2s)(5r-2s)}-\frac{2r-s}{(5r+2s)(2r-s)} \quad \leftarrow . \text{ Factor denominators and simplify}$$

$$=\frac{1}{5r+2s}-\frac{1}{5r+2s}$$

$$=\frac{1-1}{5r+2s}$$

$$=\frac{0}{5r+2s}$$

$$=0$$

73. $\dfrac{2}{2x+3y}-\dfrac{4x^2-6xy+9y^2}{8x^3+27y^3}$ ← Factor second denominator and simplify

$=\dfrac{2}{2x+3y}-\dfrac{4x^2-6xy+9y^2}{(2x+3y)(4x^2-6xy+9y^2)}$

$=\dfrac{2}{2x+3y}-\dfrac{1}{2x+3y}$

$=\dfrac{2-1}{2x+3y}$

$=\dfrac{1}{2x+3y}$

75. No; the expressions should be added first, then the resulting numerator should be factored to see if simplification is possible.

77. Yes, factor –1 from the numerator and the denominator.

79. a. $f(x)=\dfrac{x+2}{x-3}$

The denominator cannot equal zero. Therefore, $x \neq 3$. The domain is $\{x \mid x \neq 3\}$.

b. $g(x)=\dfrac{x}{x+4}$

The denominator cannot equal zero. Therefore, $x \neq -4$. The domain is $\{x \mid x \neq -4\}$.

c. $(f+g)(x)=f(x)+g(x)$

$=\dfrac{x+2}{x-3}+\dfrac{x}{x+4}$

$=\dfrac{x+2}{x-3}\cdot\dfrac{x+4}{x+4}+\dfrac{x}{x+4}\cdot\dfrac{x-3}{x-3}$

$=\dfrac{(x+2)(x+4)+x(x-3)}{(x-3)(x+4)}$

$=\dfrac{x^2+6x+8+x^2-3x}{(x-3)(x+4)}$

$=\dfrac{2x^2+3x+8}{(x-3)(x+4)}$

d. The denominator cannot equal zero. Therefore, $x \neq 3$ and $x \neq -4$.
The domain is $\{x \mid x \neq 3 \text{ and } x \neq -4\}$.

81. $P(x)=R(x)-C(x)$

$P(x)=\dfrac{4x-5}{x+1}-\dfrac{2x-7}{x+2}$

$=\dfrac{4x-5}{x+1}\cdot\dfrac{x+2}{x+2}-\dfrac{2x-7}{x+2}\cdot\dfrac{x+1}{x+1}$

$=\dfrac{(4x-5)(x+2)-(2x-7)(x+1)}{(x+1)(x+2)}$

$=\dfrac{4x^2+3x-10-(2x^2-5x-7)}{(x+1)(x+2)}$

$=\dfrac{4x^2+3x-10-2x^2+5x+7}{(x+1)(x+2)}$

$=\dfrac{2x^2+8x-3}{(x+1)(x+2)}$

83. $P(x)=R(x)-C(x)$

$P(x)=\dfrac{8x-3}{x+2}-\dfrac{5x-8}{x+3}$

$=\dfrac{8x-3}{x+2}\cdot\dfrac{x+3}{x+3}-\dfrac{5x-8}{x+3}\cdot\dfrac{x+2}{x+2}$

$=\dfrac{(8x-3)(x+3)-(5x-8)(x+2)}{(x+2)(x+3)}$

$=\dfrac{8x^2+21x-9-(5x^2+2x-16)}{(x+2)(x+3)}$

$=\dfrac{8x^2+21x-9-5x^2-2x+16}{(x+2)(x+3)}$

$=\dfrac{3x^2+19x+7}{(x+2)(x+3)}$

85. The domain is $\{x \mid x \neq 2\}$.

The range is $\{y \mid y \neq 1\}$.

87. $(f+g)(x) = f(x)+g(x)$

$$= \frac{x}{x^2-4} + \frac{2}{x^2+x-6}$$
$$= \frac{x}{(x-2)(x+2)} + \frac{2}{(x-2)(x+3)}$$
$$= \frac{x}{(x-2)(x+2)} \cdot \frac{(x+3)}{(x+3)} + \frac{2}{(x-2)(x+3)} \cdot \frac{(x+2)}{(x+2)}$$
$$= \frac{x^2+3x+2x+4}{(x-2)(x+2)(x+3)}$$
$$= \frac{x^2+5x+4}{(x-2)(x+2)(x+3)}$$

89. $(f \cdot g)(x) = f(x) \cdot g(x)$

$$= \frac{x}{x^2-4} \cdot \frac{2}{x^2+x-6}$$
$$= \frac{2x}{x^4+x^3-6x^2-4x^2-4x+24}$$
$$= \frac{2x}{x^4+x^3-10x^2-4x+24}$$

91. $\frac{a}{b}+\frac{c}{d} = \frac{a}{b}\cdot\frac{d}{d}+\frac{c}{d}\cdot\frac{b}{b}$

$$= \frac{ad}{bd}+\frac{cb}{db}$$
$$= \frac{ad+cb}{bd} \text{ or } \frac{ad+bc}{bd}$$

93. a. Perimeter is

$$\frac{a+b}{a}+\frac{a+b}{a}+\frac{a-b}{a}+\frac{a-b}{a}$$
$$= \frac{a+b+a+b+a-b+a-b}{a}$$
$$= \frac{4a}{a}$$
$$= 4$$

b. Area is

$$\left(\frac{a+b}{a}\right)\left(\frac{a-b}{a}\right) = \frac{(a+b)(a-b)}{a \cdot a} = \frac{a^2-b^2}{a^2}$$

95. Let ax^2+bx+c denote the missing numerator.

$$\frac{5x^2-6}{x^2-x-1} - \frac{ax^2+bx+c}{x^2-x-1} = \frac{-2x^2+6x-12}{x^2-x-1}$$

Since the denominators are the same, the fractions on the left side can be subtracted.

$$\frac{5x^2-6-(ax^2+bx+c)}{x^2-x-1} = \frac{-2x^2+6x-12}{x^2-x-1}$$

Since the denominators are equal, the numerators must be equal for the fractions to be the same.

$$5x^2-6-ax^2-bx-c = -2x^2+6x-12$$
$$(5-a)x^2+(-b)x+(-6-c) = -2x^2+6x-12$$

Thus,

$$5-a=-2 \quad -b=6 \quad -6-c=-12$$
$$-a=-7 \quad b=-6 \quad -c=-6$$
$$a=7 \quad c=6$$

The missing numerator is $7x^2-6x+6$.

97. $\left(3+\frac{1}{x+3}\right)\left(\frac{x+3}{x-2}\right)$

$$= \left(\frac{3}{1}\cdot\frac{x+3}{x+3}+\frac{1}{x+3}\right)\left(\frac{x+3}{x-2}\right)$$
$$= \left(\frac{3x+9+1}{x+3}\right)\left(\frac{x+3}{x-2}\right)$$
$$= \left(\frac{3x+10}{x+3}\right)\left(\frac{x+3}{x-2}\right)$$
$$= \frac{3x+10}{x-2}$$

99. $\left(\frac{5}{a-5}-\frac{2}{a+3}\right) \div (3a+25)$

$$= \left(\frac{5}{a-5}-\frac{2}{a+3}\right)\cdot\frac{1}{3a+25}$$
$$= \left(\frac{5}{a-5}\cdot\frac{a+3}{a+3}-\frac{2}{a+3}\cdot\frac{a-5}{a-5}\right)\cdot\frac{1}{3a+25}$$
$$= \left(\frac{5a+15}{(a-5)(a+3)}-\frac{2a-10}{(a-5)(a+3)}\right)\cdot\frac{1}{3a+25}$$
$$= \left(\frac{5a+15-2a+10}{(a-5)(a+3)}\right)\cdot\frac{1}{3a+25}$$
$$= \frac{3a+25}{(a-5)(a+3)}\cdot\frac{1}{3a+25}$$
$$= \frac{1}{(a-5)(a+3)}$$

101. $\left(\dfrac{x+5}{x-3}-x\right)\div\dfrac{1}{x-3}=\left(\dfrac{x+5}{x-3}-x\right)(x-3)$

$=\dfrac{x+5}{x-3}\cdot(x-3)-x(x-3)$

$=x+5-x^2+3x$

$=-x^2+4x+5$

103. a. $a\left(\dfrac{x}{n}\right)+b\left(\dfrac{n-x}{n}\right)=\dfrac{ax}{n}+\dfrac{bn-bx}{n}$

$=\dfrac{ax+bn-bx}{n}$

b. $60\left(\dfrac{2}{5}\right)+92\left(\dfrac{3}{5}\right)=\dfrac{120}{5}+\dfrac{276}{5}=\dfrac{396}{5}=79.2$

105. $(a-b)^{-1}+(a-b)^{-2}=\dfrac{1}{a-b}+\dfrac{1}{(a-b)^2}$

$=\dfrac{1}{a-b}\cdot\dfrac{a-b}{a-b}+\dfrac{1}{(a-b)^2}$

$=\dfrac{a-b}{(a-b)^2}+\dfrac{1}{(a-b)^2}$

$=\dfrac{a-b+1}{(a-b)^2}$

107. $y_1=\dfrac{x-3}{x+4}+\dfrac{x}{x^2-2x-24}$

$y_2=\dfrac{x^2-10x+18}{(x+4)(x-6)}$

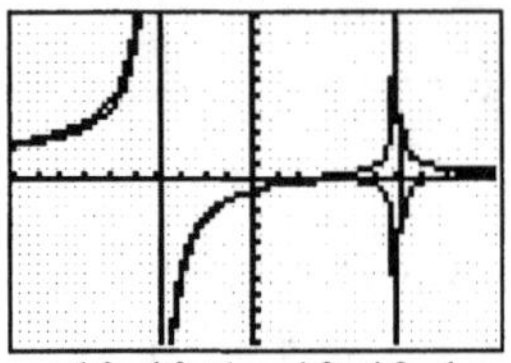

−10, 10, 1, −10, 10, 1

No, the addition is not correct because the graphs are not the same.

109. a. $1+\dfrac{1}{x}=\dfrac{x}{x}+\dfrac{1}{x}=\dfrac{x+1}{x}$

b. $1+\dfrac{1}{x}+\dfrac{1}{x^2}=1\cdot\dfrac{x^2}{x^2}+\dfrac{1}{x}\cdot\dfrac{x}{x}+\dfrac{1}{x^2}$

$=\dfrac{x^2}{x^2}+\dfrac{x}{x^2}+\dfrac{1}{x^2}$

$=\dfrac{x^2+x+1}{x^2}$

c. $1+\dfrac{1}{x}+\dfrac{1}{x^2}+\dfrac{1}{x^3}+\dfrac{1}{x^4}$

$=\dfrac{x^4+x^3+x^2+x+1}{x^4}$

d. $1+\dfrac{1}{x}+\dfrac{1}{x^2}+\cdots+\dfrac{1}{x^n}$

$=\dfrac{x^n+x^{n-1}+x^{n-2}+\cdots+1}{x^n}$

111. $g(x)=\dfrac{1}{x+1}$

$g(a+h)-g(a)=\dfrac{1}{a+h+1}-\dfrac{1}{a+1}$

$=\dfrac{1}{a+h+1}\cdot\dfrac{a+1}{a+1}-\dfrac{1}{a+1}\cdot\dfrac{a+h+1}{a+h+1}$

$=\dfrac{a+1-(a+h+1)}{(a+1)(a+h+1)}$

$=\dfrac{a+1-a-h-1}{(a+1)(a+h+1)}$

$=\dfrac{-h}{(a+1)(a+h+1)}$

112. Let t be the time at the faster speed. Then $14-t$ is the time for the slower speed.

	rate	time	amount
fast	80	t	$80t$
slow	60	$14-t$	$60(14-t)$

The two amounts are the same.

$80t=60(14-t)$

$80t=840-60t$

$140t=840$

$t=\dfrac{840}{140}=6$

a. The machine was used for 6 minutes at the faster setting.

b. While the machine was on the faster setting, $80(t) = 80(6) = 480$ boxes were filled. Since equal amounts were filled at each speed, the total number filled was $2(480) = 960$ boxes.

113.
$$|x-3|-6<-1$$
$$|x-3|<5$$
$$-5<x-3<5$$
$$-5+3<x-3+3<5+3$$
$$-2<x<8$$
$$\{x \mid -2<x<8\}$$

114. $m = \dfrac{y_2-y_1}{x_2-x_1} = \dfrac{-3-3}{7-(-2)} = \dfrac{-6}{9} = -\dfrac{2}{3}$

115. $\begin{vmatrix} -1 & 3 \\ 5 & -4 \end{vmatrix} = (-1)(-4)-(5)(3) = 4-15 = -11$

116.
$$\begin{array}{r} 3x-7 \\ 2x+3\overline{)6x^2-5x+6} \\ \underline{6x^2+9x} \quad \leftarrow 3x(2x+3) \\ -14x+6 \\ \underline{-14x-21} \leftarrow -7(2x+3) \\ 27 \leftarrow \text{Remainder} \end{array}$$

$$\frac{6x^2-5x+6}{2x+3} = 3x-7+\frac{27}{2x+3}$$

117.
$$3p^2 = 22p-7$$
$$3p^2-22p+7=0$$
$$(3p-1)(p-7)=0$$
$$3p-1=0 \quad \text{or} \quad p-7=0$$
$$p=\frac{1}{3} \qquad\qquad p=7$$

The solution set is $\left\{\dfrac{1}{3}, 7\right\}$.

Exercise Set 6.3

1. A complex fraction is one that has a fractional expression in the numerator or the denominator or both the numerator and the denominator.

3. $\dfrac{\frac{15a}{b^2}}{\frac{b^3}{5}} = \dfrac{5b^2\left(\frac{15a}{b^2}\right)}{5b^2\left(\frac{b^3}{5}\right)} = \dfrac{75a}{b^5}$ or

$\dfrac{\frac{15a}{b^2}}{\frac{b^3}{5}} = \dfrac{15a}{b^2} \div \dfrac{b^3}{5} = \dfrac{15a}{b^2} \cdot \dfrac{5}{b^3} = \dfrac{75a}{b^5}$

5. $\dfrac{\frac{36x^4}{5y^4z^5}}{\frac{9xy^2}{15z^5}} = \dfrac{15y^4z^5\left(\frac{36x^4}{5y^4z^5}\right)}{15y^4z^5\left(\frac{9xy^2}{15z^5}\right)} = \dfrac{108x^4}{9xy^6} = \dfrac{12x^3}{y^6}$

or

$\dfrac{\frac{36x^4}{5y^4z^5}}{\frac{9xy^2}{15z^5}} = \dfrac{36x^4}{5y^4z^5} \div \dfrac{9xy^2}{15z^5} = \dfrac{36x^4}{5y^4z^5} \cdot \dfrac{15z^5}{9xy^2} = \dfrac{12x^3}{y^6}$

7. $\dfrac{\frac{10x^3y^2}{9yz^4}}{\frac{40x^4y^7}{27y^2z^8}} = \dfrac{\frac{10x^3y}{9z^4}}{\frac{40x^4y^5}{27z^8}} = \dfrac{27z^8\left(\frac{10x^3y}{9z^4}\right)}{27z^8\left(\frac{40x^4y^5}{27z^8}\right)}$

$= \dfrac{30x^3yz^4}{40x^4y^5} = \dfrac{3z^4}{4xy^4}$

9. $\dfrac{1-\frac{x}{y}}{3x} = \dfrac{y\left(1-\frac{x}{y}\right)}{y(3x)} = \dfrac{y(1)-y\left(\frac{x}{y}\right)}{3xy} = \dfrac{y-x}{3xy}$

11. $\dfrac{x-\frac{x}{y}}{\frac{8+x}{y}} = \dfrac{y\left(x-\frac{x}{y}\right)}{y\left(\frac{8+x}{y}\right)}$

$= \dfrac{y(x)-y\left(\frac{x}{y}\right)}{y\left(\frac{8+x}{y}\right)}$

$= \dfrac{xy-x}{8+x}$ or $\dfrac{x(y-1)}{8+x}$

13. $\dfrac{x+\frac{5}{y}}{1+\frac{x}{y}} = \dfrac{y\left(x+\frac{5}{y}\right)}{y\left(1+\frac{x}{y}\right)} = \dfrac{y(x)+y\left(\frac{5}{y}\right)}{y+y\left(\frac{x}{y}\right)} = \dfrac{xy+5}{y+x}$

15. $\dfrac{\frac{2}{a}+\frac{1}{2a}}{a+\frac{a}{2}}=\dfrac{2a\left(\frac{2}{a}+\frac{1}{2a}\right)}{2a\left(a+\frac{a}{2}\right)}$
$=\dfrac{2a\left(\frac{2}{a}\right)+2a\left(\frac{1}{2a}\right)}{2a(a)+2a\left(\frac{a}{2}\right)}$
$=\dfrac{4+1}{2a^2+a^2}$
$=\dfrac{5}{3a^2}$

17. $\dfrac{\frac{a^2}{b}-b}{\frac{b^2}{a}-a}=\dfrac{ab\left(\frac{a^2}{b}-b\right)}{ab\left(\frac{b^2}{a}-a\right)}$
$=\dfrac{ab\left(\frac{a^2}{b}\right)-ab(b)}{ab\left(\frac{b^2}{a}\right)-ab(a)}$
$=\dfrac{a^3-ab^2}{b^3-a^2b}$
$=\dfrac{-a(-a^2+b^2)}{b(b^2-a^2)}$
$=-\dfrac{a}{b}$

19. $\dfrac{\frac{x}{y}-\frac{y}{x}}{\frac{x+y}{x}}=\dfrac{xy\left(\frac{x}{y}-\frac{y}{x}\right)}{xy\left(\frac{x+y}{x}\right)}$
$=\dfrac{xy\left(\frac{x}{y}\right)-xy\left(\frac{y}{x}\right)}{xy\left(\frac{x+y}{x}\right)}$
$=\dfrac{x^2-y^2}{y(x+y)}$
$=\dfrac{(x+y)(x-y)}{y(x+y)}$
$=\dfrac{x-y}{y}$

21. $\dfrac{\frac{a}{b}-6}{\frac{-a}{b}+6}=\dfrac{-1\left(-\frac{a}{b}+6\right)}{-\frac{a}{b}+6}=-1$

23. $\dfrac{\frac{4x+8}{3x^2}}{\frac{4x^3}{9}}=\dfrac{9x^2\left(\frac{4x+8}{3x^2}\right)}{9x^2\left(\frac{4x^3}{9}\right)}$
$=\dfrac{3(4x+8)}{4x^5}$
$=\dfrac{3\cdot 4(x+2)}{4x^5}$
$=\dfrac{3(x+2)}{x^5}$

25. $\dfrac{\frac{a}{a+1}-1}{\frac{2a+1}{a-1}}=\dfrac{(a-1)(a+1)\left(\frac{a}{a+1}-1\right)}{(a-1)(a+1)\left(\frac{2a+1}{a-1}\right)}$
$=\dfrac{(a-1)(a+1)\left(\frac{a}{a+1}\right)-(a-1)(a+1)(1)}{(a-1)(a+1)\left(\frac{2a+1}{a-1}\right)}$
$=\dfrac{a(a-1)-(a-1)(a+1)}{(a+1)(2a+1)}$
$=\dfrac{(a-1)[a-(a+1)]}{(a+1)(2a+1)}$
$=\dfrac{(a-1)(-1)}{(a+1)(2a+1)}$
$=\dfrac{-a+1}{(a+1)(2a+1)}$

27. $\dfrac{1+\frac{x}{x+1}}{\frac{2x+1}{x-1}}=\dfrac{(x-1)(x+1)\left(1+\frac{x}{x+1}\right)}{(x-1)(x+1)\left(\frac{2x+1}{x-1}\right)}$
$=\dfrac{(x-1)(x+1)(1)+(x-1)(x+1)\left(\frac{x}{x+1}\right)}{(x-1)(x+1)\left(\frac{2x+1}{x-1}\right)}$
$=\dfrac{(x-1)(x+1)+(x-1)(x)}{(x+1)(2x+1)}$
$=\dfrac{(x-1)(x+1+x)}{(x+1)(2x+1)}$
$=\dfrac{(x-1)(2x+1)}{(x+1)(2x+1)}$
$=\dfrac{x-1}{x+1}$

29. $\dfrac{\frac{a+1}{a-1}+\frac{a-1}{a+1}}{\frac{a+1}{a-1}-\frac{a-1}{a+1}}$

$$=\frac{(a-1)(a+1)\left(\frac{a+1}{a-1}+\frac{a-1}{a+1}\right)}{(a-1)(a+1)\left(\frac{a+1}{a-1}-\frac{a-1}{a+1}\right)}$$

$$=\frac{(a-1)(a+1)\left(\frac{a+1}{a-1}\right)+(a-1)(a+1)\left(\frac{a-1}{a+1}\right)}{(a-1)(a+1)\left(\frac{a+1}{a-1}\right)-(a-1)(a+1)\left(\frac{a-1}{a+1}\right)}$$

$$=\frac{(a+1)^2+(a-1)^2}{(a+1)^2-(a-1)^2}$$

$$=\frac{a^2+2a+1+a^2-2a+1}{a^2+2a+1-(a^2-2a+1)}$$

$$=\frac{2a^2+2}{4a}$$

$$=\frac{a^2+1}{2a}$$

31. $\dfrac{\frac{5}{5-x}+\frac{6}{x-5}}{\frac{3}{x}+\frac{2}{x-5}}=\dfrac{x(x-5)\left(\frac{5}{5-x}+\frac{6}{x-5}\right)}{x(x-5)\left(\frac{3}{x}+\frac{2}{x-5}\right)}$

$$=\frac{x(x-5)\left(\frac{5}{5-x}\right)+x(x-5)\left(\frac{6}{x-5}\right)}{x(x-5)\left(\frac{3}{x}\right)+x(x-5)\left(\frac{2}{x-5}\right)}$$

$$=\frac{-5x+6x}{3(x-5)+2x}$$

$$=\frac{x}{3x-15+2x}$$

$$=\frac{x}{5x-15}\text{ or }\frac{x}{5(x-3)}$$

33. $\dfrac{\frac{3}{x^2}-\frac{1}{x}+\frac{2}{x-2}}{\frac{1}{x}}=\dfrac{x^2(x-2)\left(\frac{3}{x^2}-\frac{1}{x}+\frac{2}{x-2}\right)}{x^2(x-2)\left(\frac{1}{x}\right)}$

$$=\frac{x^2(x-2)\left(\frac{3}{x^2}\right)-x^2(x-2)\left(\frac{1}{x}\right)+x^2(x-2)\left(\frac{2}{x-2}\right)}{x^2(x-2)\left(\frac{1}{x}\right)}$$

$$=\frac{3(x-2)-x(x-2)+2x^2}{x(x-2)}$$

$$=\frac{3x-6-x^2+2x+2x^2}{x(x-2)}$$

$$=\frac{x^2+5x-6}{x(x-2)}$$

35. $\dfrac{\frac{2}{a^2-3a+2}+\frac{2}{a^2-a-2}}{\frac{2}{a^2-1}+\frac{2}{a^2+4a+3}}=\dfrac{\frac{2}{(a-2)(a-1)}+\frac{2}{(a-2)(a+1)}}{\frac{2}{(a+1)(a-1)}+\frac{2}{(a+3)(a+1)}}$

$$=\frac{\frac{2(a+1)+2(a-1)}{(a-2)(a-1)(a+1)}}{\frac{2(a+3)+2(a-1)}{(a+1)(a-1)(a+3)}}$$

$$=\frac{2(a+1+a-1)}{(a-2)(a-1)(a+1)}\cdot\frac{(a+1)(a-1)(a+3)}{2(a+3+a-1)}$$

$$=\frac{2(2a)(a+3)}{2(a-2)(2a+2)}=\frac{2a(a+3)}{2(a-2)(a+1)}$$

$$=\frac{a(a+3)}{(a-2)(a+1)}$$

37. $2a^{-2}+b=\dfrac{2}{a^2}+\dfrac{b}{1}=\dfrac{2}{a^2}+\dfrac{b}{1}\cdot\dfrac{a^2}{a^2}=\dfrac{2+a^2b}{a^2}$

39. $(a^{-1}+b^{-1})^{-1}=\left(\dfrac{1}{a}+\dfrac{1}{b}\right)^{-1}$

$$=\left(\frac{1}{a}\cdot\frac{b}{b}+\frac{1}{b}\cdot\frac{a}{a}\right)^{-1}$$

$$=\left(\frac{b}{ab}+\frac{a}{ab}\right)^{-1}$$

$$=\left(\frac{b+a}{ab}\right)^{-1}$$

$$=\frac{ab}{b+a}$$

41. $\dfrac{a^{-1}+1}{b^{-1}-1}=\dfrac{\frac{1}{a}+1}{\frac{1}{b}-1}$

$$=\frac{ab\left(\frac{1}{a}+1\right)}{ab\left(\frac{1}{b}-1\right)}$$

$$=\frac{ab\left(\frac{1}{a}\right)+ab(1)}{ab\left(\frac{1}{b}\right)-ab(1)}$$

$$=\frac{b+ab}{a-ab}\text{ or }\frac{b(1+a)}{a(1-b)}$$

43. $$\frac{a^{-2}-ab^{-1}}{ab^{-2}+a^{-1}b^{-1}}=\frac{\frac{1}{a^2}-\frac{a}{b}}{\frac{a}{b^2}+\frac{1}{ab}}$$
$$=\frac{a^2b^2\left(\frac{1}{a^2}-\frac{a}{b}\right)}{a^2b^2\left(\frac{a}{b^2}+\frac{1}{ab}\right)}$$
$$=\frac{a^2b^2\left(\frac{1}{a^2}\right)-a^2b^2\left(\frac{a}{b}\right)}{a^2b^2\left(\frac{a}{b^2}\right)+a^2b^2\left(\frac{1}{ab}\right)}$$
$$=\frac{b^2-a^3b}{a^3+ab} \text{ or } \frac{b\left(b-a^3\right)}{a\left(a^2+b\right)}$$

45. $$\frac{\frac{9a}{b}+a^{-1}}{\frac{b}{a}+a^{-1}}=\frac{\frac{9a}{b}+\frac{1}{a}}{\frac{b}{a}+\frac{1}{a}}$$
$$=\frac{ab\left(\frac{9a}{b}+\frac{1}{a}\right)}{ab\left(\frac{b}{a}+\frac{1}{a}\right)}$$
$$=\frac{ab\left(\frac{9a}{b}\right)+ab\left(\frac{1}{a}\right)}{ab\left(\frac{b}{a}\right)+ab\left(\frac{1}{a}\right)}$$
$$=\frac{9a^2+b}{b^2+b} \text{ or } \frac{9a^2+b}{b(b+1)}$$

47. $$\frac{a^{-1}+b^{-1}}{(a+b)^{-1}}=\frac{\frac{1}{a}+\frac{1}{b}}{\frac{1}{a+b}}=\frac{ab(a+b)\left(\frac{1}{a}+\frac{1}{b}\right)}{ab(a+b)\left(\frac{1}{a+b}\right)}$$
$$=\frac{ab(a+b)\left(\frac{1}{a}\right)+ab(a+b)\left(\frac{1}{b}\right)}{ab(a+b)\left(\frac{1}{a+b}\right)}$$
$$=\frac{b(a+b)+a(a+b)}{ab}$$
$$=\frac{(a+b)(a+b)}{ab}$$
$$=\frac{(a+b)^2}{ab}$$

49. $$5x^{-1}-(3y)^{-1}=\frac{5}{x}-\frac{1}{3y}$$
$$=\frac{5}{x}\cdot\frac{3y}{3y}-\frac{1}{3y}\cdot\frac{x}{x}$$
$$=\frac{15y}{3xy}-\frac{x}{3xy}$$
$$=\frac{15y-x}{3xy}$$

51. $$\frac{\frac{2}{xy}-\frac{8}{y}+\frac{5}{x}}{3x^{-1}-4y^{-2}}=\frac{\frac{2}{xy}-\frac{8}{y}+\frac{5}{x}}{\frac{3}{x}-\frac{4}{y^2}}$$
$$=\frac{xy^2\left(\frac{2}{xy}-\frac{8}{y}+\frac{5}{x}\right)}{xy^2\left(\frac{3}{x}-\frac{4}{y^2}\right)}$$
$$=\frac{xy^2\left(\frac{2}{xy}\right)-xy^2\left(\frac{8}{y}\right)+xy^2\left(\frac{5}{x}\right)}{xy^2\left(\frac{3}{x}\right)-xy^2\left(\frac{4}{y^2}\right)}$$
$$=\frac{2y-8xy+5y^2}{3y^2-4x}$$

53. $$A=lw \Rightarrow l=\frac{A}{w}=A\div w$$
$$l=\frac{x^2+12x+35}{x+3}\div\frac{x^2+6x+5}{x^2+5x+6}$$
$$l=\frac{x^2+12x+35}{x+3}\cdot\frac{x^2+5x+6}{x^2+6x+5}$$
$$l=\frac{(x+7)(x+5)}{x+3}\cdot\frac{(x+3)(x+2)}{(x+1)(x+5)}$$
$$l=\frac{(x+7)(x+2)}{x+1} \text{ or } \frac{x^2+9x+14}{x+1}$$

55. $$A=lw \Rightarrow l=\frac{A}{w}=A\div w$$
$$l=\frac{x^2+11x+28}{x+5}\div\frac{x^2+8x+7}{x^2+4x-5}$$
$$l=\frac{x^2+11x+28}{x+5}\cdot\frac{x^2+4x-5}{x^2+8x+7}$$
$$l=\frac{(x+7)(x+4)}{x+5}\cdot\frac{(x+5)(x-1)}{(x+7)(x+1)}$$
$$l=\frac{(x+4)(x-1)}{x+1} \text{ or } \frac{x^2+3x-4}{x+1}$$

57. a. Substitute $\frac{2}{5}$ for h.

$$E=\frac{\frac{1}{2}h}{h+\frac{1}{2}}$$

$$=\frac{\frac{1}{2}\left(\frac{2}{5}\right)}{\frac{2}{5}+\frac{1}{2}}=\frac{\frac{1}{5}}{\frac{2}{5}+\frac{1}{2}}$$

$$=\frac{10\left(\frac{1}{5}\right)}{10\left(\frac{2}{5}+\frac{1}{2}\right)} \leftarrow \text{Multiply by LCD of 10}$$

$$=\frac{10\left(\frac{1}{5}\right)}{10\left(\frac{2}{5}\right)+10\left(\frac{1}{2}\right)}=\frac{2}{4+5}$$

$$=\frac{2}{9}$$

b. Substitute $\frac{1}{3}$ for h.

$$E=\frac{\frac{1}{2}h}{h+\frac{1}{2}}$$

$$=\frac{\frac{1}{2}\left(\frac{1}{3}\right)}{\frac{1}{3}+\frac{1}{2}}=\frac{\frac{1}{6}}{\frac{1}{3}+\frac{1}{2}}$$

$$=\frac{6\left(\frac{1}{6}\right)}{6\left(\frac{1}{3}+\frac{1}{2}\right)} \leftarrow \text{Multiply by LCD of 6}$$

$$=\frac{6\left(\frac{1}{6}\right)}{6\left(\frac{1}{3}\right)+6\left(\frac{1}{2}\right)}=\frac{1}{2+3}$$

$$=\frac{1}{5}$$

59. $$R_T=\frac{R_1R_2R_3(1)}{R_1R_2R_3\left(\frac{1}{R_1}+\frac{1}{R_2}+\frac{1}{R_3}\right)}$$

$$=\frac{R_1R_2R_3}{R_1R_2R_3\left(\frac{1}{R_1}\right)+R_1R_2R_3\left(\frac{1}{R_2}\right)+R_1R_2R_3\left(\frac{1}{R_3}\right)}$$

$$=\frac{R_1R_2R_3}{R_2R_3+R_1R_3+R_1R_2}$$

61. $f(x)=\frac{1}{x}$

$$f(a)=\frac{1}{a}$$

$$f(f(a))=\frac{1}{\left(\frac{1}{a}\right)}=a$$

63. $f(x)=\frac{1}{x}$

$$\frac{f(a+h)-f(a)}{h}=\frac{\frac{1}{a+h}-\frac{1}{a}}{h}$$

$$=\frac{\frac{a-(a+h)}{a(a+h)}}{h}$$

$$=\frac{\frac{-h}{a(a+h)}}{h}$$

$$=\frac{-1}{a(a+h)}$$

65. $f(x)=\frac{1}{x+1}$

$$\frac{f(a+h)-f(a)}{h}=\frac{\frac{1}{a+h+1}-\frac{1}{a+1}}{h}$$

$$=\frac{\frac{a+1-(a+h+1)}{(a+1)(a+h+1)}}{h}$$

$$=\frac{\frac{-h}{(a+1)(a+h+1)}}{h}$$

$$=\frac{-1}{(a+1)(a+h+1)}$$

67. $f(x)=\frac{1}{x^2}$

$$\frac{f(a+h)-f(a)}{h}=\frac{\frac{1}{(a+h)^2}-\frac{1}{a^2}}{h}$$

$$=\frac{\frac{a^2-(a+h)^2}{(a+h)^2a^2}}{h}$$

$$\frac{\frac{a^2-(a^2+2ah+h^2)}{(a+h)^2a^2}}{h}$$

$$=\frac{\frac{-2ah-h^2}{(a+h)^2a^2}}{h}$$

$$=\frac{\frac{h(-2a-h)}{(a+h)^2a^2}}{h}$$

$$=\frac{-2a-h}{a^2(a+h)^2}$$

69. $\frac{1}{2a+\frac{1}{2a+\frac{1}{2a}}}=\frac{1}{2a+\frac{1}{\frac{(2a)(2a)+1}{2a}}}$

$$=\frac{1}{2a+\frac{1}{\frac{4a^2+1}{2a}}}$$

$$=\frac{1}{2a+\frac{2a}{4a^2+1}}$$

$$=\frac{1}{\frac{2a(4a^2+1)+2a}{4a^2+1}}$$

$$=\frac{1}{\frac{8a^3+2a+2a}{4a^2+1}}$$

$$=\frac{1}{\frac{8a^3+4a}{4a^2+1}}$$

$$=\frac{4a^2+1}{8a^3+4a}$$

$$=\frac{4a^2+1}{4a(2a^2+1)}$$

71. $\frac{1}{2+\frac{1}{2+\frac{1}{2}}}=\frac{1}{2+\frac{1}{\frac{5}{2}}}=\frac{1}{2+\frac{2}{5}}=\frac{1}{\frac{12}{5}}=\frac{5}{12}$

72. $\frac{\left|-\frac{3}{9}\right|-\left(-\frac{5}{9}\right)\left|-\frac{3}{8}\right|}{\left|-5-(-3)\right|}=\frac{\frac{3}{9}+\frac{5}{9}\left(\frac{3}{8}\right)}{\left|-2\right|}=\frac{\frac{1}{3}+\frac{5}{24}}{2}=\frac{24\left(\frac{1}{3}+\frac{5}{24}\right)}{24(2)}$

$$=\frac{24\left(\frac{1}{3}\right)+24\left(\frac{5}{24}\right)}{24(2)}=\frac{8+5}{48}=\frac{13}{48}$$

73.

$$\frac{3}{5}<\frac{-x-5}{3}<6$$

$$15\left(\frac{3}{5}\right)<15\left(\frac{-x-5}{3}\right)<15(6)$$

$$9<-5x-25<90$$

$$9+25<-5x-25+25<90+25$$

$$34<-5x<115$$

$$\frac{34}{-5}>\frac{-5x}{-5}>\frac{115}{-5}$$

$$-\frac{34}{5}>x>-23 \quad\Rightarrow\quad -23<x<-\frac{34}{5}$$

In interval notation: $\left(-23,-\frac{34}{5}\right)$

74. $|x-1|=|2x-4|$

$x-1=2x-4$ or $x-1=-(2x-4)$

$x=3$ $\qquad x-1=-2x+4$

$x=\frac{5}{3}$

The solution is $\left\{3,\frac{5}{3}\right\}$.

75. $6x+2y=5$

$4x-9=-2y$

Write the equations in slope-intercept form.

$6x+2y=5$

$2y=-6x+5$

$\frac{2y}{2}=\frac{-6x+5}{2}$

$y=-3x+\frac{5}{2}$

$4x-9=-2y$

$\frac{4x-9}{-2}=\frac{-2y}{-2}$

$-2x+\frac{9}{2}=y$

$y=-2x+\frac{9}{2}$

Since the slopes are neither the same nor opposite reciprocals, the lines are neither parallel nor perpendicular.

Exercise Set 6.4

1. An extraneous root is a number obtained when solving an equation that is not a true solution to the original equation.

3. a. Multiply both sides of the equation by the LCD of 12. This removes fractions.

b. $12\left(\frac{x}{4}\right)-12\left(\frac{x}{3}\right)=12(2)$

$3x-4x=24$

$-x=24$

$x=\frac{24}{-1}=-24$

c. Write each term with the least common denominator of 12. This allows the fractions to be added or subtracted.

d. $\frac{x}{4}-\frac{x}{3}+2=\frac{x}{4}\cdot\frac{3}{3}-\frac{x}{3}\cdot\frac{4}{4}+\frac{2}{1}\cdot\frac{12}{12}$

$=\frac{3x}{12}-\frac{4x}{12}+\frac{24}{12}$

$=\frac{3x-4x+24}{12}=\frac{-x+24}{12}$

5. Similar figures are figures whose corresponding angles are the same and whose corresponding sides are in proportion.

7. Tom's solution of $x = 3$ is incorrect since this causes the denominator to be zero.

9. $\frac{5}{x}=1$

$\frac{5}{x}=\frac{1}{1}$

$(5)(1)=(x)(1)$ ← Cross multiply

$5=x$

This solution checks. The solution is 5.

11. $\frac{11}{b}=2$

$\frac{11}{b}=\frac{2}{1}$

$(11)(1)=(b)(2)$ ← Cross multiply

$11=2b$

$b=\frac{11}{2}$

This solution checks. The solution is $\frac{11}{2}$.

13. $\frac{6x+7}{5}=\frac{2x+9}{3}$

$3(6x+7)=5(2x+9)$ ← Cross multiply

$18x+21=10x+45$

$8x+21=45$

$8x=24$

$x=3$

This solution checks. The solution is 3.

15. $\frac{3x}{8}+\frac{1}{4}=\frac{2x-3}{8}$

$8\left(\frac{3x}{8}\right)+8\left(\frac{1}{4}\right)=8\left(\frac{2x-3}{8}\right)$ ← Multiply each term by the LCD of 8

$3x+2=2x-3$

$x+2=-3$

$x=-5$

This solution checks. The solution is -5.

17. $\frac{z}{3}-\frac{3z}{4}=-\frac{5z}{12}$

$12\left(\frac{z}{3}\right)-12\left(\frac{3z}{4}\right)=12\left(-\frac{5z}{12}\right)$ ← Multiply each term by the LCD of 12

$4z-9z=-5z$

$-5z=-5z$

Since this statement is true for all values of z, the solution is all real numbers.

19. $\frac{3}{4} - x = 2x$

$$\frac{3}{4} = 3x$$

$$\frac{3}{4} = \frac{3x}{1}$$

$$3(1) = 4(3x) \leftarrow \text{Cross multiply}$$

$$3 = 12x$$

$$\frac{3}{12} = x$$

$$\frac{1}{4} = x$$

This solution checks. The solution is $\frac{1}{4}$.

21. $\frac{2}{r} + \frac{5}{3r} = 1$

$$3r\left(\frac{2}{r}\right) + 3r\left(\frac{5}{3r}\right) = 3r(1) \leftarrow \text{Multiply each term by the LCD of } 3r$$

$$6 + 5 = 3r$$

$$11 = 3r$$

$$\frac{11}{3} = r$$

This solution checks. The solution is $\frac{11}{3}$.

23. $\frac{x-2}{x-5} = \frac{3}{x-5}$

Since the denominators are the same, the numerators must be equal.

$$x - 2 = 3$$

$$x = 5$$

This does not check, since both denominators are 0 when $x = 5$. There is no solution.

25. $\frac{5y-2}{7} = \frac{15y-2}{28}$

$$28(5y-2) = 7(15y-2) \leftarrow \text{Cross multiply}$$

$$140y - 56 = 105y - 14$$

$$35y = 42$$

$$y = \frac{42}{35} = \frac{6}{5}$$

This solution checks. The solution is $\frac{6}{5}$.

27. $\frac{5.6}{-p-6.2} = \frac{2}{p}$

$$5.6(p) = 2(-p-6.2) \leftarrow \text{Cross multiply}$$

$$5.6p = -2p - 12.4$$

$$7.6p = -12.4$$

$$p = \frac{-12.4}{7.6} \approx -1.63$$

This solution checks. The solution is ≈ -1.63.

29. $\frac{m+1}{m+10} = \frac{m-2}{m+4}$

$$(m+4)(m+1) = (m+10)(m-2) \leftarrow \text{Cross multiply}$$

$$m^2 + 5m + 4 = m^2 + 8m - 20$$

$$5m + 4 = 8m - 20$$

$$-3m + 4 = -20$$

$$-3m = -24$$

$$m = 8$$

This solution checks. The solution is 8.

31.

$$x - \frac{4}{3x} = -\frac{1}{3}$$

$$3x(x) - 3x\left(\frac{4}{3x}\right) = 3x\left(-\frac{1}{3}\right) \quad \leftarrow \text{ Multiply each term by the LCD of } 3x$$

$$3x^2 - 4 = -x$$

$$3x^2 + x - 4 = 0$$

$$(3x+4)(x-1) = 0$$

$$3x + 4 = 0 \quad \text{or} \quad x - 1 = 0$$

$$3x = -4 \qquad x = 1$$

$$x = -\frac{4}{3}$$

These solutions check. The solutions are $-\frac{4}{3}$ and 1.

33.

$$\frac{2x-1}{3} - \frac{x}{4} = \frac{7.4}{6}$$

$$12\left(\frac{2x-1}{3}\right) - 12\left(\frac{x}{4}\right) = 12\left(\frac{7.4}{6}\right) \leftarrow \text{ Multiply each term by the LCD of 12.}$$

$$4(2x-1) - 3x = 2(7.4)$$

$$8x - 4 - 3x = 14.8$$

$$5x - 4 = 14.8$$

$$5x = 18.8$$

$$x = \frac{18.8}{5} = 3.76$$

This solution checks. The solution is 3.76.

35.

$$x + \frac{6}{x} = -7$$

$$x(x) + x\left(\frac{6}{x}\right) = x(-7) \leftarrow \text{ Multiply each term by the LCD of } x$$

$$x^2 + 6 = -7x$$

$$x^2 + 7x + 6 = 0$$

$$(x+6)(x+1) = 0$$

$$x + 6 = 0 \text{ or } x + 1 = 0$$

$$x = -6 \qquad x = -1$$

These solutions check. The solutions are –6 and –1.

37.

$$2-\frac{5}{2b}=\frac{2b}{b+1}$$

$$2[2b(b+1)]-2b(b+1)\left(\frac{5}{2b}\right)=2b(b+1)\left(\frac{2b}{b+1}\right)$$ ← **Multiply each term by the LCD of $2b(b+1)$**

$$4b(b+1)-5(b+1)=2b(2b)$$

$$4b^2+4b-5b-5=4b^2$$

$$4b^2-b-5=4b^2$$

$$-b-5=0$$

$$b=-5$$

This solution checks. The solution is –5.

39.

$$\frac{1}{w-3}+\frac{1}{w+3}=\frac{-5}{w^2-9}$$

$$\frac{1}{w-3}+\frac{1}{w+3}=\frac{-5}{(w+3)(w-3)}$$

$$(w+3)(w-3)\left(\frac{1}{w-3}\right)+(w+3)(w-3)\left(\frac{1}{w+3}\right)=(w+3)(w-3)\left(\frac{-5}{(w+3)(w-3)}\right)$$ ← **Multiply each term by the LCD of $(w+3)(w-3)$**

$$w+3+w-3=-5$$

$$2w=-5$$

$$w=\frac{-5}{2}=-\frac{5}{2}$$

This solution checks. The solution is $-\frac{5}{2}$.

41.

$$\frac{8}{x^2-9}=\frac{2}{x-3}-\frac{4}{x+3}$$

$$\frac{8}{(x-3)(x+3)}=\frac{2}{x-3}-\frac{4}{x+3}$$

$$(x-3)(x+3)\left(\frac{8}{(x-3)(x+3)}\right)=(x-3)(x+3)\left(\frac{2}{x-3}\right)-(x-3)(x+3)\left(\frac{4}{x+3}\right)$$ ← **Multiply each term by the LCD of $(x-3)(x+3)$**

$$8=2(x+3)-4(x-3)$$

$$8=2x+6-4x+12$$

$$8=-2x+18$$

$$-10=-2x$$

$$x=\frac{-10}{-2}=5$$

This solution checks. The solution is 5.

43. $\frac{y}{2y+2}+\frac{2y-16}{4y+4}=\frac{2y-3}{y+1}$

$\frac{y}{2y+2}+\frac{2(y-8)}{4(y+1)}=\frac{2y-3}{y+1}$

$\frac{y}{2(y+1)}+\frac{y-8}{2(y+1)}=\frac{2y-3}{y+1}$

$\frac{y+y-8}{2(y+1)}=\frac{2y-3}{y+1}$

$\frac{2y-8}{2(y+1)}=\frac{2y-3}{y+1}$

$\frac{2(y-4)}{2(y+1)}=\frac{2y-3}{y+1}$

$\frac{y-4}{y+1}=\frac{2y-3}{y+1}$

Since the denominators are the same, the numerators must be equal.

$y-4=2y-3$

$-4=y-3$

$-1=y$

This does not check since all the original denominators are 0 when $y=-1$. There is no solution.

45. $\frac{x^2}{x-5}=\frac{25}{x-5}$

Since the denominators are equal, the numerators must be equal.

$x^2=25$

$x^2-25=0$

$(x-5)(x+5)=0$

$x=5$ or $x=-5$

Since $x=5$ makes a denominator 0, it must be excluded from the solution set. The only solution is $x=-5$.

47.

$$\frac{5}{x^2+4x+3}+\frac{2}{x^2+x-6}=\frac{3}{x^2-x-2}$$

$$\frac{5}{(x+3)(x+1)}+\frac{2}{(x+3)(x-2)}=\frac{3}{(x-2)(x+1)}$$

$$(x+3)(x+1)(x-2)\left(\frac{5}{(x+3)(x+1)}\right)+(x+3)(x+1)(x-2)\left(\frac{2}{(x+3)(x-2)}\right)=(x+3)(x+1)(x-2)\left(\frac{3}{(x-2)(x+1)}\right)$$

$$5(x-2)+2(x+1)=3(x+3)$$

$$5x-10+2x+2=3x+9$$

$$7x-8=3x+9$$

$$7x=3x+17$$

$$4x=17$$

$$x=\frac{17}{4}$$

This solution checks. The solution is $\frac{17}{4}$.

49. $$\frac{6x}{4}=\frac{6}{x}$$
$$x(6x)=4(6)$$
$$6x^2=24$$
$$6x^2-24=0$$
$$6(x^2-4)=0$$
$$6(x+2)(x-2)=0$$
$$x+2=0 \quad \text{or} \quad x-2=0$$
$$x=-2 \qquad x=2$$
Reject $x=-2$ since x cannot be a negative number. Thus, $x=2$ and $6x$ is $6(2)=12$. The unknown lengths are 2 and 12.

51. $$\frac{8}{2x+10}=\frac{x+3}{6}$$
$$(2x+10)(x+3)=8(6)$$
$$2x^2+16x+30=48$$
$$2x^2+16x-18=0$$
$$2(x^2+8x-9)=0$$
$$2(x+9)(x-1)=0$$
$$x+9=0 \quad \text{or} \quad x-1=0$$
$$x=-9 \qquad x=1$$
Reject $x=-9$ since x cannot be a negative number. Thus, $x=1$ so that $x+3$ is $1+3=4$ and $2x+10$ is $2\cdot 1+10=2+10=12$. The unknown lengths are 4 and 12.

53. $$f(x)=2x-\frac{4}{x}$$
$$f(a)=2a-\frac{4}{a}$$
$$-2=2a-\frac{4}{a}$$
$$a(-2)=a(2a)-a\left(\frac{4}{a}\right)$$
$$-2a=2a^2-4$$
$$0=2a^2+2a-4$$
$$0=2(a+2)(a-1)$$
$$a+2=0 \quad \text{or} \quad a-1=0$$
$$a=-2 \qquad a=1$$
$f(a)=-2$ when $a=-2$ and $a=1$.

55. $$f(x)=\frac{x-2}{x+5}$$
$$f(a)=\frac{a-2}{a+5}$$
$$\frac{3}{5}=\frac{a-2}{a+5}$$
$$3(a+5)=5(a-2)$$
$$3a+15=5a-10$$
$$25=2a$$
$$\frac{25}{2}=a$$
$f(a)=\frac{3}{5}$ when $a=\frac{25}{2}$.

57. $$f(x)=\frac{6}{x}+\frac{6}{2x}$$
$$f(a)=\frac{6}{a}+\frac{6}{2a}$$
$$=\frac{6}{a}+\frac{3}{a}$$
$$=\frac{9}{a}$$
$$6=\frac{9}{a}$$
$$a=\frac{9}{6}$$
$$a=\frac{3}{2}$$
$f(a)=6$ when $a=\frac{3}{2}$.

59. $$\frac{V_1}{V_2}=\frac{P_2}{P_1}$$
$$P_2V_2=V_1P_1$$
$$P_2=\frac{V_1P_1}{V_2}$$

61. $$\frac{V_1}{V_2}=\frac{P_2}{P_1}$$
$$P_2V_2=V_1P_1$$
$$V_2=\frac{V_1P_1}{P_2}$$

63.
$$m = \frac{y - y_1}{x - x_1}$$
$$m(x - x_1) = y - y_1$$
$$y = y_1 + m(x - x_1)$$

65.
$$z = \frac{x - \overline{x}}{s}$$
$$zs = x - \overline{x}$$
$$zs + \overline{x} = x \text{ or } x = zs + \overline{x}$$

67.
$$d = \frac{fl}{f + w}$$
$$d(f + w) = fl$$
$$df + dw = fl$$
$$dw = fl - df$$
$$w = \frac{fl - df}{d}$$

69.
$$\frac{1}{p} + \frac{1}{q} = \frac{1}{f}$$
$$pqf\left(\frac{1}{p}\right) + pqf\left(\frac{1}{q}\right) = pqf\left(\frac{1}{f}\right)$$
$$qf + pf = pq$$
$$pf = pq - qf$$
$$pf = q(p - f)$$
$$\frac{pf}{p - f} = q \text{ or } q = \frac{pf}{p - f}$$

71.
$$at_2 - at_1 + v_1 = v_2$$
$$at_2 - at_1 = v_2 - v_1$$
$$a(t_2 - t_1) = v_2 - v_1$$
$$a = \frac{v_2 - v_1}{t_2 - t_1}$$

73.
$$a_n = a_1 + nd - d$$
$$a_n - a_1 = nd - d$$
$$a_n - a_1 = d(n - 1)$$
$$\frac{a_n - a_1}{n - 1} = d \text{ or } d = \frac{a_n - a_1}{n - 1}$$

75.
$$F = \frac{Gm_1m_2}{d^2}$$
$$d^2(F) = d^2\left(\frac{Gm_1m_2}{d^2}\right)$$
$$d^2F = Gm_1m_2$$
$$\frac{d^2F}{m_1m_2} = G \text{ or } G = \frac{Fd^2}{m_1m_2}$$

77.
$$\frac{P_1V_1}{T_1} = \frac{P_2V_2}{T_2}$$
$$T_1(P_2V_2) = T_2(P_1V_1)$$
$$T_1 = \frac{T_2P_1V_1}{P_2V_2}$$

79.
$$\frac{S - S_0}{V_0 + gt} = t$$
$$S - S_0 = t(V_0 + gt)$$
$$S - S_0 = tV_0 + gt^2$$
$$S - S_0 - gt^2 = tV_0$$
$$\frac{S - S_0 - gt^2}{t} = V_0$$

81. a.
$$\frac{2}{x-2} + \frac{5}{x^2 - 4}$$
$$= \frac{2}{x-2} + \frac{5}{(x-2)(x+2)}$$
$$= \frac{2}{x-2} \cdot \frac{x+2}{x+2} + \frac{5}{(x-2)(x+2)}$$
$$= \frac{2(x+2)}{(x-2)(x+2)} + \frac{5}{(x-2)(x+2)}$$
$$= \frac{2x+4}{(x-2)(x+2)} + \frac{5}{(x-2)(x+2)}$$
$$= \frac{2x+9}{(x-2)(x+2)}$$

b.
$$\frac{2}{x-2} + \frac{5}{x^2-4} = 0$$
$$\frac{2x+9}{(x-2)(x+2)} = 0$$
$$2x + 9 = 0$$
$$2x = -9$$
$$x = -\frac{9}{2}$$

83. a.

$$\frac{b+3}{b}-\frac{b+4}{b+5}-\frac{15}{b^2+5b}$$
$$=\frac{b+3}{b}-\frac{b+4}{b+5}-\frac{15}{b(b+5)}$$
$$=\frac{b+3}{b}\cdot\frac{b+5}{b+5}-\frac{b+4}{b+5}\cdot\frac{b}{b}-\frac{15}{b(b+5)}$$
$$=\frac{(b+3)(b+5)}{b(b+5)}-\frac{b(b+4)}{b(b+5)}-\frac{15}{b(b+5)}$$
$$=\frac{b^2+8b+15}{b(b+5)}-\frac{b^2+4b}{b(b+5)}-\frac{15}{b(b+5)}$$
$$=\frac{b^2+8b+15-(b^2+4b)-15}{b(b+5)}$$
$$=\frac{4b}{b(b+5)}$$
$$=\frac{4}{b+5}$$

b.

$$\frac{b+3}{b}-\frac{b+4}{b+5}=\frac{15}{b(b+5)}$$
$$b(b+5)\left[\frac{b+3}{b}-\frac{b+4}{b+5}\right]=b(b+5)\left[\frac{15}{b(b+5)}\right]$$
$$(b+5)(b+3)-b(b+4)=15$$
$$b^2+8b+15-b^2-4b=15$$
$$4b+15=15$$
$$4b=0$$
$$b=0$$

This does not check since the denominators b and $b(b+5)$ are 0 when $b=0$.
There is no solution.

85. $c \neq 0$, since division by 0 is not defined.

87. $f(x)$ is graph **b)** and $g(x)$ is graph **a)**;
$f(x)$ is not defined for $x=3$ (hence the hole).

89. a.

$$I=\frac{AC}{0.80R}$$
$$I=\frac{50,000(10,000)}{0.80(100,000)}$$
$$I=6250$$

The insurance company will pay \$6250.

b.

$$I=\frac{AC}{0.80R}$$
$$I(0.80R)=AC$$
$$R=\frac{AC}{0.80I}$$

91. a.

$$a=\frac{v_2-v_1}{t_2-t_1}$$
$$a=\frac{60-20}{22-20}$$
$$a=20$$

The average acceleration is 20 ft/min^2.

b.

$$a=\frac{v_2-v_1}{t_2-t_1}$$
$$a(t_2-t_1)=v_2-v_1$$
$$at_2-at_1=v_2-v_1$$
$$-at_1=-at_2+v_2-v_1$$
$$at_1=at_2+v_1-v_2$$
$$t_1=\frac{at_2+v_1-v_2}{a}$$
$$t_1=t_2+\frac{v_1-v_2}{a}$$

93. a.

$$P=1-\frac{R-D}{R}$$
$$D=39.99-30.99=9$$
$$P=1-\frac{39.99-9}{39.99}$$
$$P\approx 0.225$$

The rate of discount is about 22.5%.

b.

$$P=1-\frac{R-D}{R}$$
$$1-P=\frac{R-D}{R}$$
$$R(1-P)=R-D$$
$$D=R-R(1-P)$$
$$D=R-R+PR$$
$$D=PR$$

c. $$P = 1 - \frac{R-D}{R}$$
$$P = \frac{R}{R} - \frac{R-D}{R}$$
$$P = \frac{R-R+D}{R}$$
$$P = \frac{D}{R}$$
$$R = \frac{D}{P}$$

95. $$\frac{1}{R_T} = \frac{1}{200} + \frac{1}{600}$$
$$600R_T\left(\frac{1}{R_T}\right) = 600R_T\left(\frac{1}{200}\right) + 600R_T\left(\frac{1}{600}\right)$$
$$600 = 3R_T + 1R_T$$
$$600 = 4R_T$$
$$\frac{600}{4} = R_T \text{ or } R_T = 150$$
The total resistance of the two resistors connected in parallel is 150 ohms.

97. $$\frac{1}{p} + \frac{1}{q} = \frac{1}{f}$$
Solve for p.
$$pqf\left(\frac{1}{p} + \frac{1}{q}\right) = pqf\left(\frac{1}{f}\right)$$
$$qf + pf = pq$$
$$pq - pf = qf$$
$$p(q-f) = qf$$
$$p = \frac{qf}{q-f}$$
$$p = \frac{7.5(0.1)}{7.5-0.1}$$
$$p \approx 0.101$$
The lens should be about 0.101 m from the film.

99. a. $$T_a = \frac{T_f}{1-[f+(s+c)(1-f)]}$$
$$= \frac{0.0601}{1-[0.33+(0.046+0.03)(1-0.33)]}$$
$$= \frac{0.0601}{1-[0.33+(0.076)(0.67)]}$$
$$= \frac{0.0601}{1-(0.33+0.05092)}$$
$$= \frac{0.0601}{1-0.38092}$$
$$= \frac{0.0601}{0.61908}$$
$$\approx 0.0970795$$
Thus, the taxable equivalent is $T_a \approx 9.71\%$.

b. Howard Levy should choose the Tax Free Money Market since 9.71% > 7.68%.

101. Answers will vary. One such equation is $\frac{1}{x-4} + \frac{1}{x+2} = 0$. Another one might be $\frac{1}{(x-4)(x+2)} = 0$. Make sure that the factor $(x-4)$ is in a denominator and that the factor $(x+2)$ is in a denominator (not necessarily the same).

103. Answers will vary. One possible answer is $\frac{1}{x} + \frac{1}{x} = \frac{2}{x}$. Here we have the factor x in the denominator and the equation is an identity.

105. $$-1 \le 5 - 2x < 7$$
$$-1-5 \le 5-2x-5 < 7-5$$
$$-6 \le -2x < 2$$
$$\frac{-6}{-2} \ge \frac{-2x}{-2} > \frac{2}{-2}$$
$$3 \ge x > -1$$
or
$$-1 < x \le 3$$

106. $3(y-4)=-(x-2)$

$3y-12=-x+2$

$3y=-x+14$

$\frac{3y}{3}=\frac{-x+14}{3}$

$y=-\frac{1}{3}x+\frac{14}{3}$

The slope is $-\frac{1}{3}$ and the y-intercept is $\frac{14}{3}$.

107. $3x^2y-4xy+2y^2-(3xy+6y^2+9x)$

$3x^2y-4xy+2y^2-3xy-6y^2-9x$

$3x^2y-4xy-3xy+2y^2-6y^2-9x$

$3x^2y-7xy-4y^2-9x$

108. Let x be the width of the sidewalk. The dimensions of the garden and walkway together are $(16+2x)$ feet $\times$ $(12+2x)$ feet.

Area = Length $\times$ Width

$320=(16+2x)(12+2x)$

$320=192+32x+24x+4x^2$

$0=4x^2+56x-128$

$0=4(x^2+14x-32)$

$0=4(x-2)(x+16)$

$x=2$ or $x=-16$

Disregard the negative solution. The width of the sidewalk is 2 feet.

Mid-Chapter Test: 6.1 – 6.4

1. The domain of a rational function is all real numbers except where the denominator equals 0.

$x^3-25x=0$

$x(x^2-25)=0$

$x(x-5)(x+5)=0$

$x=0$, $x=5$, and $x=-5$ will make the denominator equal 0 so they must be excluded. The domain is $\{x\,|\,x\neq 0, x\neq 5, \text{ and } x\neq -5\}$.

2. $\frac{x^2+9x+20}{2x^2+5x-12}=\frac{(x+5)(x+4)}{(2x-3)(x+4)}=\frac{x+5}{2x-3}$

3. $\frac{11a+11b}{3}\div\frac{a^3+b^3}{15b}$

$=\frac{11(a+b)}{3}\cdot\frac{15b}{(a+b)(a^2-ab+b^2)}$

$=\frac{11(a+b)}{3}\cdot\frac{3\cdot 5b}{(a+b)(a^2-ab+b^2)}$

$=\frac{55b}{a^2-ab+b^2}$

4. $\frac{x^2+4x-21}{x^2-5x-6}\cdot\frac{x^2-2x-24}{x^2+11x+28}$

$=\frac{(x+7)(x-3)}{(x-6)(x+1)}\cdot\frac{(x-6)(x+4)}{(x+7)(x+4)}$

$=\frac{\cancel{(x+7)}(x-3)}{\cancel{(x-6)}(x+1)}\cdot\frac{\cancel{(x-6)}\cancel{(x+4)}}{\cancel{(x+7)}\cancel{(x+4)}}$

$=\frac{x-3}{x+1}$

5. $\frac{4a^2+4a+1}{4a^2+6a-2a-3}\div\frac{2a^2-17a-9}{(2a+3)^2}$

$=\frac{(2a+1)^2}{2a(2a+3)-1(2a+3)}\div\frac{(2a+1)(a-9)}{(2a+3)^2}$

$=\frac{(2a+1)^2}{(2a-1)(2a+3)}\cdot\frac{(2a+3)^2}{(2a+1)(a-9)}$

$=\frac{(2a+1)(2a+3)}{(2a-1)(a-9)}$ or $\frac{4a^2+8a+3}{2a^2-19a+9}$

6. $A=lw\ \rightarrow\ w=\frac{A}{l}$

$\frac{12a^2+13ab+3b^2}{18a+6b}=\frac{(3a+b)(4a+3b)}{6(3a+b)}$

$=\frac{4a+3b}{6}$

The width of the rectangle is $\frac{4a+3b}{6}$.

7. Factor the denominators.

$x^2-x-30=(x-6)(x+5)$

$x^2-4x-12=(x-6)(x+2)$

The LCD is $(x-6)(x+5)(x+2)$.

8. $\frac{5x}{x-5}-\frac{25}{x-5}=\frac{5x-25}{x-5}=\frac{5(x-5)}{x-5}=5$

9. $\frac{10}{3x^2y}+\frac{a}{6xy^3}$

Factor the denominators.

$3x^2y=3\cdot x^2\cdot y$

$6xy^3=2\cdot 3\cdot x\cdot y^3$

The LCD is $2\cdot 3\cdot x^2\cdot y^3=6x^2y^3$

$\frac{10}{3x^2y}\cdot\frac{2y^2}{2y^2}+\frac{a}{6xy^3}\cdot\frac{x}{x}$

$=\frac{20y^2}{6x^2y^3}+\frac{ax}{6x^2y^3}=\frac{20y^2+ax}{6x^2y^3}$

10. $\frac{4}{2x^2+5x-12}-\frac{3}{x^2-16}$

Factor the denominators.

$2x^2+5x-12=(2x-3)(x+4)$

$x^2-16=(x-4)(x+4)$

The LCD is $(2x-3)(x+4)(x-4)$.

$=\frac{4}{(2x-3)(x+4)}\cdot\frac{x-4}{x-4}-\frac{3}{(x-4)(x+4)}\cdot\frac{2x-3}{2x-3}$

$=\frac{4(x-4)-3(2x-3)}{(2x-3)(x+4)(x-4)}$

$=\frac{4x-16-6x+9}{(2x-3)(x+4)(x-4)}$

$=\frac{-2x-7}{(2x-3)(x+4)(x-4)}$

11. $\frac{9+\frac{a}{b}}{\frac{3-c}{b}}=\frac{b\left(9+\frac{a}{b}\right)}{b\left(\frac{3-c}{b}\right)}=\frac{b(9)+b\left(\frac{a}{b}\right)}{3-c}=\frac{9b+a}{3-c}$

12. $\frac{\frac{5}{x}-\frac{8}{x^2}}{6-\frac{1}{x}}=\frac{x^2\left(\frac{5}{x}-\frac{8}{x^2}\right)}{x^2\left(6-\frac{1}{x}\right)}$

$=\frac{x^2\left(\frac{5}{x}\right)-x^2\left(\frac{8}{x^2}\right)}{x^2(6)-x^2\left(\frac{1}{x}\right)}$

$=\frac{5x-8}{6x^2-x}$

13. $\frac{y^{-2}+7y^{-1}}{7y^{-3}+y^{-4}}=\frac{\frac{1}{y^2}+\frac{7}{y}}{\frac{7}{y^3}+\frac{1}{y^4}}$

$=\frac{y^4\left(\frac{1}{y^2}+\frac{7}{y}\right)}{y^4\left(\frac{7}{y^3}+\frac{1}{y^4}\right)}$

$=\frac{y^4\left(\frac{1}{y^2}\right)+y^4\left(\frac{7}{y}\right)}{y^4\left(\frac{7}{y^3}\right)+y^4\left(\frac{1}{y^4}\right)}$

$=\frac{y^2+7y^3}{7y+1}$

$=\frac{y^2(7y+1)}{7y+1}=y^2$

14. An extraneous root is a number obtained when solving an equation that is not a solution to the original equation. Whenever a variable appears in the denominator, you must check the apparent solution(s).

15. $\frac{3x-1}{7}=\frac{-x+9}{2}$

$2(3x-1)=7(-x+9)$

$6x-2=-7x+63$

$13x=65$

$x=5$

16. $\frac{m-7}{m-11}=\frac{4}{m-11}$

Since the denominators are equal, the numerators must be equal.

$m-7=4$

$m=11$

This solution does not check because $m=11$ makes the denominators equal 0. There is no solution to the equation.

17.
$$x = 1 + \frac{12}{x}$$
$$x(x) = x\left(1 + \frac{12}{x}\right)$$
$$x^2 = x + 12$$
$$x^2 - x - 12 = 0$$
$$(x-4)(x+3) = 0$$
$$x - 4 = 0 \quad \text{or} \quad x + 3 = 0$$
$$x = 4 \qquad x = -3$$
These values check. The solutions are -3 and 4.

18. $\frac{1}{a} - \frac{1}{b} = \frac{1}{c}$
$$\frac{1}{a} = \frac{1}{b} + \frac{1}{c}$$
$$\frac{1}{a} = \frac{c}{bc} + \frac{b}{bc}$$
$$\frac{1}{a} = \frac{b+c}{bc}$$
$$a = \frac{bc}{b+c}$$

19.
$$x = \frac{4}{1-r}$$
$$x(1-r) = 4$$
$$1 - r = \frac{4}{x}$$
$$-r = \frac{4}{x} - 1$$
$$r = 1 - \frac{4}{x} \quad \text{or} \quad r = \frac{x-4}{x}$$

20.
$$\frac{2x}{5} = \frac{14}{x-2}$$
$$2x(x-2) = 5(14)$$
$$2x^2 - 4x = 70$$
$$2x^2 - 4x - 70 = 0$$
$$x^2 - 2x - 35 = 0$$
$$(x-7)(x+5) = 0$$
$$x - 7 = 0 \quad \text{or} \quad x + 5 = 0$$
$$x = 7 \qquad x = -5$$
Discard the negative value since the length must be positive. The unknown sides are $2(7) = 14$ units and $7 - 2 = 5$ units

Exercise Set 6.5

1. The total time needed will be equal to $\frac{1}{2}$ the time of each painting separately. In $\frac{1}{2}$ the time, each will complete $\frac{1}{2}$ the job.

3. a. Let x be the time to do the task together.

worker	Rate of Work	Time Worked	Part of Task Completed
Bill	$\frac{1}{7}$	x	$\frac{x}{7}$
Bob	$\frac{1}{9}$	x	$\frac{x}{9}$

b. $\frac{x}{7} + \frac{x}{9} = 1$

c. It will take less than 7 hours working together since Bill working alone can do the job in 7 hours.

5. Let x be the time for both working together.

worker	Rate of Work	Time Worked	Part of Task Completed
Marilyn	$\frac{1}{2}$	x	$\frac{x}{2}$
Larry	$\frac{1}{6}$	x	$\frac{x}{6}$

$$\frac{x}{2} + \frac{x}{6} = 1$$
$$6\left(\frac{x}{2} + \frac{x}{6}\right) = 6(1)$$
$$3x + x = 6$$
$$4x = 6$$
$$x = 1.5$$
It will take them 1.5 months to carve the totem pole working together.

7. Let x be the time for both working together.

worker	Rate of Work	Time Worked	Part of Task Completed
Jason	$\frac{1}{3}$	x	$\frac{x}{3}$
Tom	$\frac{1}{6}$	x	$\frac{x}{6}$

$$\frac{x}{3}+\frac{x}{6}=1$$
$$6\left(\frac{x}{3}\right)+6\left(\frac{x}{6}\right)=6(1)$$
$$2x+x=6$$
$$3x=6$$
$$x=2$$

Working together, they can shampoo the carpet in 2 hours.

9. Let x be the time for both working together.

worker	Rate of Work	Time Worked	Part of Task Completed
Jin	$\frac{1}{30}$	x	$\frac{x}{30}$
Ming	$\frac{1}{50}$	x	$\frac{x}{50}$

$$\frac{x}{30}+\frac{x}{50}=1$$
$$150\cdot\frac{x}{30}+150\cdot\frac{x}{50}=150\cdot 1$$
$$5x+3x=150$$
$$8x=150$$
$$x=\frac{150}{8}=18.75$$

Working together, the cows can be milked in 18.75 minutes.

11. Let x be the time for both working together.

worker	Rate of Work	Time Worked	Part of Task Completed
Kevin	$\frac{1}{6}$	x	$\frac{x}{6}$
Kevin's son	$\frac{1}{12}$	x	$\frac{x}{12}$

$$\frac{x}{6}+\frac{x}{12}=1$$
$$12\cdot\frac{x}{6}+12\cdot\frac{x}{12}=12\cdot 1$$
$$2x+x=12$$
$$3x=12$$
$$x=4$$

Working together, they can pick 25 bushels in 4 hours.

13. Let x be the time for both working together.

worker	Rate of Work	Time Worked	Part of Task Completed
Olga	$\frac{1}{4.5}$	x	$\frac{x}{4.5}$
Jien-Ping	$\frac{1}{5.5}$	x	$\frac{x}{5.5}$

$$\frac{x}{4.5}+\frac{x}{5.5}=1$$
$$99\cdot\frac{x}{4.5}+99\cdot\frac{x}{5.5}=99\cdot 1$$
$$22x+18x=99$$
$$40x=99$$
$$x=\frac{99}{40}=2.475\approx 2.48$$

Working together, the gutters can be cleaned in about 2.48 days.

15. Let x be the time for both working together.

worker	Rate of Work	Time Worked	Part of Task Completed
Wanda	$\frac{1}{4}$	x	$\frac{x}{4}$
Shawn	$\frac{1}{6}$	x	$\frac{x}{6}$

$$\frac{x}{4}+\frac{x}{6}=1$$

$$12\cdot\frac{x}{4}+12\cdot\frac{x}{6}=12\cdot1$$

$$3x+2x=12$$

$$5x=12$$

$$x=\frac{12}{5}=2.4$$

Working together, they can plow the field in about 2.4 hours.

17. Let x be the time for both working together.

	Rate of Work	Time Worked	Part of Task Completed
$\frac{1}{2}''$ hose	$\frac{1}{8}$	x	$\frac{x}{8}$
$\frac{4}{5}''$ hose	$\frac{1}{5}$	x	$\frac{x}{5}$

$$\frac{x}{8}+\frac{x}{5}=1$$

$$40\cdot\frac{x}{8}+40\cdot\frac{x}{5}=40\cdot1$$

$$5x+8x=40$$

$$13x=40$$

$$x=\frac{40}{13}\approx3.08$$

Working together, the pool can be filled in about 3.08 hours

19. Let x be the time for both working together. Note: emptying is 'negative filling' so the rate of work will be negative when emptying.

	Rate of Work	Time Worked	Part of Task Completed
In-valve	$\frac{1}{20}$	x	$\frac{x}{20}$
Out-valve	$-\frac{1}{25}$	x	$-\frac{x}{25}$

$$\frac{x}{20}-\frac{x}{25}=1$$

$$100\cdot\frac{x}{20}-100\cdot\frac{x}{25}=100\cdot1$$

$$5x-4x=100$$

$$x=100$$

With both valves open, the tank will fill in 100 hours.

21. Let x be the time for Henry to complete the job working by himself.

worker	Rate of Work	Time Worked	Part of Task Completed
Indiana	$\frac{1}{3.9}$	2.6	$\frac{2.6}{3.9}$
Henry	$\frac{1}{x}$	2.6	$\frac{2.6}{x}$

$$\frac{2.6}{3.9}+\frac{2.6}{x}=1$$

$$3.9x\cdot\frac{2.6}{3.9}+3.9x\cdot\frac{2.6}{x}=3.9x\cdot1$$

$$2.6x+10.14=3.9x$$

$$1.3x=10.14$$

$$x=\frac{10.14}{1.3}=7.8$$

Henry can complete the task by himself in 7.8 months.

23. Let x be the time for Shane to complete the job working by himself.

worker	Rate of Work	Time Worked	Part of Task Completed
Wade	$\frac{1}{50}$	30	$\frac{30}{50}=\frac{3}{5}$
Shane	$\frac{1}{x}$	30	$\frac{30}{x}$

$$\frac{3}{5}+\frac{30}{x}=1$$
$$5x\cdot\frac{3}{5}+5x\cdot\frac{30}{x}=5x\cdot 1$$
$$3x+150=5x$$
$$2x=150$$
$$x=75$$

Shane can complete the task by himself in 75 minutes.

25. Let x be the time to fill the tub with both valves open and the drain open.

	Rate of Work	Time Worked	Part of Task Completed
Cold	$\frac{1}{8}$	x	$\frac{x}{8}$
Hot	$\frac{1}{12}$	x	$\frac{x}{12}$
Drain	$-\frac{1}{7}$	x	$-\frac{x}{7}$

$$\frac{x}{8}+\frac{x}{12}-\frac{x}{7}=1$$
$$168\cdot\frac{x}{8}+168\cdot\frac{x}{12}-168\cdot\frac{x}{7}=168\cdot 1$$
$$21x+14x-24x=168$$
$$11x=168$$
$$x=\frac{168}{11}\approx 15.27$$

With both valves open and the drain open, the tub will fill in about 15.27 minutes.

27. Let x be the time to empty the basement if the pumps work together.

	Rate of Work	Time Worked	Part of Task Completed
Pump 1	$\frac{1}{6}$	x	$\frac{x}{6}$
Pump 2	$\frac{1}{5}$	x	$\frac{x}{5}$
Pump 3	$\frac{1}{4}$	x	$\frac{x}{4}$

$$\frac{x}{6}+\frac{x}{5}+\frac{x}{4}=1$$
$$60\cdot\frac{x}{6}+60\cdot\frac{x}{5}+60\cdot\frac{x}{4}=60\cdot 1$$
$$10x+12x+15x=60$$
$$37x=60$$
$$x=\frac{60}{37}\approx 1.62$$

With all three pumps, the basement can be emptied in about 1.62 hours.

29. Let x be the time for Anna to complete the job.

worker	Rate of Work	Time Worked	Part of Task Completed
Gary	$\frac{1}{15}$	6	$\frac{6}{15}$
Anna	$\frac{1}{20}$	x	$\frac{x}{20}$

$$\frac{6}{15}+\frac{x}{20}=1$$
$$60\cdot\frac{6}{15}+60\cdot\frac{x}{20}=60\cdot 1$$
$$24+3x=60$$
$$3x=36$$
$$x=12$$

Anna can finish the job in 12 hours.

31. Let x be the unknown number.

$$\frac{4x}{3+x}=\frac{5}{2}$$
$$5(3+x)=4x(2)$$
$$15+5x=8x$$
$$15=3x$$
$$\frac{15}{3}=x$$
$$5=x$$

33. Let x and $2x$ be the two numbers. Their reciprocals are $\frac{1}{x}$ and $\frac{1}{2x}$.

$$\frac{1}{x}+\frac{1}{2x}=\frac{3}{4}$$
$$4x\left(\frac{1}{x}\right)+4x\left(\frac{1}{2x}\right)=4x\left(\frac{3}{4}\right)$$
$$4+2=3x$$
$$6=3x$$
$$\frac{6}{3}=x$$
$$2=x$$

Thus, $x=2$ and $2x=2\cdot 2=4$. The two numbers are 2 and 4.

35. Let x and $x+2$ be the two consecutive even integers. Their reciprocals are $\frac{1}{x}$ and $\frac{1}{x+2}$.

$$\frac{1}{x}+\frac{1}{x+2}=\frac{5}{12}$$
$$12x(x+2)\left[\frac{1}{x}+\frac{1}{x+2}\right]=12x(x+2)\left(\frac{5}{12}\right)$$
$$12(x+2)+12x(1)=5x(x+2)$$
$$12x+24+12x=5x^2+10x$$
$$24x+24=5x^2+10x$$
$$0=5x^2-14x-24$$
$$0=(5x+6)(x-4)$$
$$5x+6=0 \quad \text{or} \quad x-4=0$$
$$5x=-6 \quad \text{or} \quad x=4$$
$$x=-\frac{6}{5}$$

Reject $x=-\frac{6}{5}$ since x must be an integer. Thus, $x=4$ and $x+2=4+2=6$. The two integers are 4 and 6.

37. Let x be the unknown number.

$$\frac{2}{x}+3=\frac{31}{10}$$
$$10x\left(\frac{2}{x}\right)+10x(3)=10x\left(\frac{31}{10}\right)$$
$$20+30x=31x$$
$$20=x$$

39. Let x be the unknown number.

$$3x+\frac{2}{x}=5$$
$$x(3x)+x\left(\frac{2}{x}\right)=5(x)$$
$$3x^2+2=5x$$
$$3x^2-5x+2=0$$
$$(3x-2)(x-1)=0$$
$$3x-2=0 \quad \text{or} \quad x-1=0$$
$$3x=2 \qquad x=1$$
$$x=\frac{2}{3}$$

The two numbers that work are $\frac{2}{3}$ and 1.

41. Let x be the rate of the current.

	d	r	$t=\frac{d}{r}$
upstream	2.3	$3-x$	$\frac{2.3}{3-x}$
downstream	2.4	$3+x$	$\frac{2.4}{3+x}$

$$\frac{2.3}{3-x}=\frac{2.4}{3+x}$$
$$2.3(3+x)=2.4(3-x)$$
$$6.9+2.3x=7.2-2.4x$$
$$4.7x=0.3$$
$$x=\frac{0.3}{4.7}\approx 0.064$$

The rate of the current is about 0.064 mph.

43. Let r be the rate of Nancy walking. Then $r + 2$ is the rate of Nancy walking on the moving sidewalk.

	d	r	$t=\frac{d}{r}$
On sidewalk	120	$r+2$	$\frac{120}{r+2}$
Off sidewalk	52	r	$\frac{52}{r}$

The time is the same for both.

$$\frac{120}{r+2}=\frac{52}{r}$$
$$120(r)=52(r+2)$$
$$120r=52r+104$$
$$68r=104$$
$$r\approx 1.53$$

Nancy walks at a rate of about 1.53 ft / sec.

45. Let x be the length of the trail.

	d	r	$t=\frac{d}{r}$
Bonnie	x	6	$\frac{x}{6}$
Clide	x	10	$\frac{x}{10}$

Bonnie's time is a half hour more than Clide's time.

$$\frac{x}{10}+\frac{1}{2}=\frac{x}{6}$$
$$30\left(\frac{x}{10}+\frac{1}{2}\right)=30\left(\frac{x}{6}\right)$$
$$3x+15=5x$$
$$2x=15$$
$$x=7.5$$

The trail is 7.5 miles long.

47. Let x be Phil's average speed (in mph) driving to Yosemite National Park.

	d	r	$t=\frac{d}{r}$
driving	60	x	$\frac{60}{x}$

He spent twice as much time visiting as he did driving. So he spent $2\left(\frac{60}{x}\right)$ hours visiting. The total time was 5 hours.

$$\frac{60}{x}+2\left(\frac{60}{x}\right)=5$$
$$\frac{60}{x}+\frac{120}{x}=5$$
$$\frac{180}{x}=5$$
$$5x=180$$
$$x=36$$

Phil's average speed was 36 mph.

49. Let x be the distance the ball traveled at 14.7 yards per second. Then $80 - x$ is the distance Steve traveled at 5.8 yards per second.

	d	r	$t=\frac{d}{r}$
Ball	x	14.7	$\frac{x}{14.7}$
Steve	$80-x$	5.8	$\frac{80-x}{5.8}$

The total time of the play is 10.6 seconds.

$$\frac{x}{14.7}+\frac{80-x}{5.8}=10.6$$
$$85.26\left(\frac{x}{14.7}\right)+85.26\left(\frac{80-x}{5.8}\right)=85.26(10.6)$$
$$5.8x+14.7(80-x)=903.756$$
$$5.8x+1176-14.7x=903.756$$
$$-8.9x=-272.244$$
$$x\approx 30.59$$

The ball traveled about 30.59 yards before Steve caught it.

51. Let x be the speed of the local train.

	d	r	$t=\frac{d}{r}$
local	$24.2-7.8=16.4$	x	$\frac{16.4}{x}$
express	24.2	$x+5.2$	$\frac{24.2}{x+5.2}$

The time is the same.

$$\frac{16.4}{x}=\frac{24.2}{x+5.2}$$
$$16.4(x+5.2)=24.2x$$
$$16.4x+85.28=24.2x$$
$$85.28=7.8x$$
$$10.93\approx x$$
$$16.13\approx x+5.2$$

The local train's speed is about 10.93 mph and the express train's speed is about 16.13 mph.

53. Let x be the speed of the train. Then $2x$ is the speed of the car.

	d	r	$t=\frac{d}{r}$
car	390	$2x$	$\frac{390}{2x}$
train	390	x	$\frac{390}{x}$

The car arrives 6.5 hours before the train.

$$\frac{390}{2x}+6.5=\frac{390}{x}$$
$$2x\left(\frac{390}{2x}\right)+2x(6.5)=2x\left(\frac{390}{x}\right)$$
$$390+13x=780$$
$$13x=390$$
$$x=\frac{390}{13}=30$$

The speed of the train is 30 mph and the speed of the car is 2(30) = 60 mph.

55. Let r be the speed of Mary Ann's car.

	d	r	$t=\frac{d}{r}$
Mary Ann	600	r	$\frac{600}{r}$
Carla	600	$r-10$	$\frac{600}{r-10}$

$$\frac{600}{r}+2=\frac{600}{r-10}$$
$$r(r-10)\left(\frac{600}{r}\right)+2r(r-10)=r(r-10)\left(\frac{600}{r-10}\right)$$
$$600(r-10)+2r(r-10)=600r$$
$$600r-6000+2r^2-20r=600r$$
$$2r^2-20r-6000=0$$
$$2(r-60)(r+50)=0$$
$$r-60=0 \quad \text{or} \quad r+50=0$$
$$r=60 \qquad r=-50$$

Since a speed cannot be negative, $r=60$. Mary Ann's car travels at 60 mph.

57. Let x be the speed of the helicopter going to the glacier.

	d	r	$t=\dfrac{d}{r}$
to glacier	60	x	$\dfrac{60}{x}$
to Te Anu	140	$x + 20$	$\dfrac{140}{x+20}$

The entire trip took 2 hours.

$$\frac{60}{x}+\frac{1}{2}+\frac{140}{x+20}=2$$

$$2x(x+20)\left(\frac{60}{x}+\frac{1}{2}+\frac{140}{x+20}\right)=2x(x+20)(2)$$

$$120(x+20)+x(x+20)+2x(140)=4x(x+20)$$

$$120x+2400+x^2+20x+280x=4x^2+80x$$

$$0=3x^2-340x-2400$$

$$0=(3x+20)(x-120)$$

$$3x+20=0 \quad \text{or} \quad x-120=0$$

$$x=-\frac{20}{3} \qquad x=120$$

Since a speed cannot be negative, $x = 120$. The average speed of the helicopter going to the glacier is 120 kph.

59. Let t be the time spent pedaling at 6 mph.

	r	t	$d=r\cdot t$
Slow part	6	t	$6t$
Fast part	10	$2.5-t$	$10(2.5-t)$

The total distance is 17 miles.

$$6t+10(2.5-t)=17$$

$$6t+25-10t=17$$

$$-4t=-8$$

$$t=2$$

Robert pedals for 2 hours at 6 mph and he pedals for $\frac{1}{2}$ hour at 10 mph.

61. Let x be Phil's average speed.

	d	r	$t=\frac{d}{r}$
Phil	450	x	$\frac{450}{x}$
Heim	450	$x+2$	$\frac{450}{x+2}$

Heim's time plus 2.5 minutes is Phil's time.

$$\frac{450}{x+2}+\frac{5}{2}=\frac{450}{x}$$

$$2x(x+2)\left(\frac{450}{x+2}\right)+2x(x+2)\left(\frac{5}{2}\right)=2x(x+2)\left(\frac{450}{x}\right)$$

$$900x+5x(x+2)=900(x+2)$$

$$900x+5x^2+10x=900x+1800$$

$$5x^2+10x-1800=0$$

$$(5x+100)(x-18)=0$$

$$5x+100=0 \quad \text{or} \quad x-18=0$$

$$x=-20 \quad \text{or} \quad x=18$$

Since a speed cannot be negative, $x = 18$.
Phil's average speed is 18 feet per minute.

63. Let x be the distance between the space station and NASA headquarters.

	d	r	$t=\frac{d}{r}$
First rocket	x	20,000	$\frac{x}{20{,}000}$
Second rocket	x	18,000	$\frac{x}{18{,}000}$

$$\frac{x}{20{,}000}+0.6=\frac{x}{18{,}000}$$

$$180{,}000\left(\frac{x}{20{,}000}+0.6\right)=180{,}000\left(\frac{x}{18{,}000}\right)$$

$$9x+108{,}000=10x$$

$$108{,}000=x$$

The space station is 108,000 miles from NASA headquarters.

65. Answers will vary.

67. a. $\text{time}=\frac{\text{distance}}{\text{rate}}$

$$=\frac{10}{(150-90)}$$

$$=\frac{10}{60}=\frac{1}{6}\text{ hour}=10\text{ minutes}$$

The aircraft will reach the car in 10 minutes.

b. $d=r\cdot t$

$$d=90\cdot\frac{1}{6}=15$$

The car will have traveled 15 miles before the plane catches up.

c. Let r be the rate of the plane.

	r	t	$d=r\cdot t$
Car	90	$\frac{8}{60}$	$90\left(\frac{8}{60}\right)$
Aircraft	r	$\frac{8}{60}$	$r\left(\frac{8}{60}\right)$

The aircraft must fly 10 more miles than the car.

$$90\left(\frac{8}{60}\right)+10=r\left(\frac{8}{60}\right)$$

$$12+10=\frac{2r}{15}$$

$$22=\frac{2r}{15}$$

$$r=165$$

The plane must travel at a speed of 165 mph to catch the car in 8 minutes.

68. $\dfrac{\left(2x^{-2}y^{-2}\right)^{-3}}{\left(3x^{-1}y^{3}\right)^{2}} = \dfrac{2^{-3}x^{(-2)(-3)}y^{(-2)(-3)}}{3^{2}x^{(-1)(2)}y^{(3)(2)}}$

$= \dfrac{2^{-3}x^{6}y^{6}}{3^{2}x^{-2}y^{6}}$

$= \dfrac{x^{6}x^{2}y^{6}}{3^{2}2^{3}y^{6}}$ ← the *y*'s cancel out

$= \dfrac{x^{6+2}}{9\cdot 8}$

$= \dfrac{x^{8}}{72}$

69. $9,260,000,000 = 9.26\times 10^{9}$

70. Let *x* be Sandy's total dollar volume of all sales. Sandy's weekly salary is $0.12x + 240$.

$0.12x + 240 = 540$

$0.12x = 300$

$x = 2500$

Sandy must have \$2500 in total sales volume per week in order to earn \$540.

71. Make a table of values to graph the equation.

$y = |x| - 2$

$x = -3 \Rightarrow y = |-3| - 2 = 3 - 2 = 1$

$x = -2 \Rightarrow y = |-2| - 2 = 2 - 2 = 0$

$x = -1 \Rightarrow y = |-1| - 2 = 1 - 2 = -1$

$x = 0 \Rightarrow y = |0| - 2 = 0 - 2 = -2$

$x = 1 \Rightarrow y = |1| - 2 = 1 - 2 = -1$

$x = 2 \Rightarrow y = |2| - 2 = 2 - 2 = 0$

$x = 3 \Rightarrow y = |3| - 2 = 3 - 2 = 1$

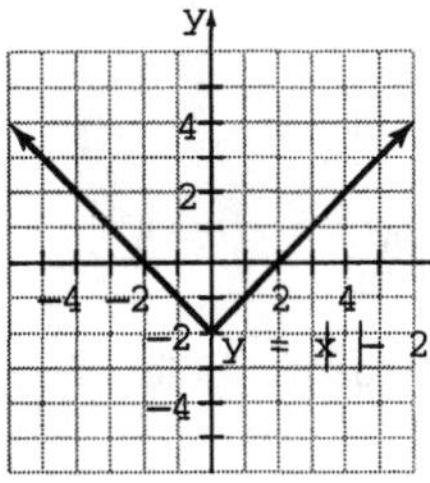

72. $2a^{4} - 2a^{3} - 5a^{2} + 5a$ ← GCF is *a*

$a\left(2a^{3} - 2a^{2} - 5a + 5\right)$

$a\left[2a^{2}(a-1) - 5(a-1)\right]$

$a(a-1)\left(2a^{2} - 5\right)$

Exercise Set 6.6

1. a. As one quantity increases, the other increases.

b. Answers will vary.

c. Answers will vary.

3. One quantity varies as a product of two or more quantities.

5. a. *y* decreases; as *x* increases, the denominator increases which makes the value of the whole fraction smaller.

b. inverse variation; as one quantity increases, the other decreases.

7. Direct

9. Inverse

11. Direct

13. Direct

15. Direct

17. Inverse

19. Direct

21. Inverse

23. Inverse

25. a. $x = ky$

b. Substitute 12 for *y* and 6 for *k*.

$x = 6(12)$

$x = 72$

27. a. $y = kR$

b. Substitute 180 for *R* and 1.7 for *k*.

$y = 1.7(180)$

$y = 306$

29. a. $R = \frac{k}{W}$

b. Substitute 160 for W and 8 for k.

$R = \frac{8}{160}$

$R = \frac{1}{20} = 0.05$

31. a. $A = \frac{kB}{C}$

b. Substitute 12 for B, 4 for C, and 3 for k.

$A = \frac{3(12)}{4} = \frac{36}{4} = 9$

33. a. $x = ky$

b. To find k, substitute 12 for x and 3 for y.

$12 = k(3)$

$\frac{12}{3} = k$

$4 = k$

Thus, $x = 4y$.

Now substitute 5 for y.

$x = 4(5) = 20$

35. a. $y = kR^2$

b. To find k, substitute 5 for y and 5 for R.

$5 = k(5)^2$

$5 = k(25)$

$\frac{5}{25} = k$

$\frac{1}{5} = k$

Thus $y = \frac{1}{5}R^2$.

Now substitute 10 for R.

$y = \frac{1}{5}(10)^2 = \frac{1}{5}(100) = 20$

37. a. $S = \frac{k}{G}$

b. To find k, substitute 12 for S and 0.4 for G.

$12 = \frac{k}{0.4}$

$12(0.4) = k$

$4.8 = k$

Thus $S = \frac{4.8}{G}$.

Now substitute 5 for G.

$S = \frac{4.8}{5} = 0.96$

39. a. $x = \frac{k}{P^2}$

b. To find k, substitute 5 for P and 4 for x.

$4 = \frac{k}{5^2}$

$4 = \frac{k}{25}$

$100 = k$

Thus $x = \frac{100}{P^2}$.

Now substitute 2 for P.

$x = \frac{100}{(2)^2} = \frac{100}{4} = 25$

41. a. $F = \frac{kM_1M_2}{d}$

b. To find k, substitute 5 for M_1, 10 for M_2, 0.2 for d, and 20 for F.

$20 = \frac{k(5)(10)}{0.2}$

$20 = k(250)$

$k = \frac{20}{250} = 0.08$

Thus $F = \frac{0.08M_1M_2}{d}$.

Now substitute 10 for M_1, 20 for M_2, and 0.4 for d.

$F = \frac{0.08(10)(20)}{0.4} = \frac{16}{0.4} = 40$

43. $a = kb$

$k(2b) = 2(kb) = 2a$

If b is doubled, a is doubled.

45. $y = \frac{k}{x}$

$\frac{k}{2x} = \frac{1}{2}\left(\frac{k}{x}\right) = \frac{1}{2}y$

If x is doubled, y is halved.

47. $F = \frac{km_1m_2}{d^2}$

$\frac{k(2m_1)m_2}{d^2} = \frac{2km_1m_2}{d^2} = 2 \cdot \frac{km_1m_2}{d^2} = 2F$

If m_1 is doubled, F is doubled.

49. $F = \frac{km_1m_2}{d^2}$

$\frac{k(2m_1)\left(\frac{1}{2}m_2\right)}{d^2} = \frac{2 \cdot \frac{1}{2}km_1m_2}{d^2} = \frac{1 \cdot km_1m_2}{d^2} = F$

If m_1 is doubled and m_2 is halved, F is unchanged.

51. $F = \frac{km_1m_2}{d^2}$

$\frac{k\left(\frac{1}{2}m_1\right)(4m_2)}{d^2} = \frac{\frac{1}{2} \cdot 4km_1m_2}{d^2} =$

$\frac{2 \cdot km_1m_2}{d^2} = 2 \cdot \frac{km_1m_2}{d^2} = 2F$

If m_1 is halved and m_2 is quadrupled, F is doubled.

53. Notice that as x gets bigger, y gets smaller. This suggests that the variation is inverse rather than direct. Therefore use the equation $y = \frac{k}{x}$. To determine the value of k, choose one of the ordered pairs and substitute the values into the equation $y = \frac{k}{x}$ and solve for k. We'll use the ordered pair (5, 1).

$y = \frac{k}{x}$

$1 = \frac{k}{5} \Rightarrow k = 5$

55. The equation is $p = kl$ To find k substitute 150 for l and 2542.50 for p.

$2542.50 = k(150)$

$k = \frac{2542.50}{150}$

$k = 16.95$

Thus $p = 16.95l$.

Now substitute 520 for l.

$p = 16.95(520)$

$p = 8814$

The profit would be \$8814.

57. The equation is $d = kw$. To find k, substitute 2376 for d and 132 for w.

$2376 = k132$

$k = \frac{2376}{132}$

$k = 18$

Thus $d = 18w$. Now substitute 172 for w.

$d = 18(172)$

$d = 3096$

The recommended dosage for Nathan is 3096 mg.

59. The equation is $S = kF$. To find k, substitute 1.4 for S and 20 for F.

$1.4 = k(20)$

$\frac{1.4}{20} = k$

$0.07 = k$

Thus, $S = 0.07F$.

Now substitute 15 for F.

$S = 0.07(15) = 1.05$

The spring will stretch 1.05 inches.

61. The equation is $V = \frac{k}{P}$. To find *k*, substitute 800 for *V* and 200 for *P*.

$800 = \frac{k}{200}$

$800(200) = k$

$160,000 = k$

Thus $V = \frac{160,000}{P}$.

Now substitute 25 for *P*.

$V = \frac{160,000}{25} = 6400$

The volume is 6400 cc.

63. The equation is $t = \frac{k}{s}$. To find *k*, substitute 6 for *s* and 2.6 for *t*.

$2.6 = \frac{k}{6}$

$k = 6(2.6)$

$k = 15.6$

Thus $t = \frac{15.6}{s}$.

Now substitute 5 for *s*.

$t = \frac{15.6}{5}$

$t = 3.12$

Jackie will take 3.12 hours.

65. The equation is $I = \frac{k}{d^2}$. To find *k*, substitute 20 for *I* and 15 for *d*.

$20 = \frac{k}{(15)^2}$

$20 = \frac{k}{225}$

$k = 20(225) = 4500$

Thus $I = \frac{4500}{d^2}$.

Now substitute 10 for *d*.

$I = \frac{4500}{(10)^2}$

$I = \frac{4500}{100}$

$I = 45$

The intensity is 45 foot-candles.

67. The equation is $d = ks^2$. To find *k*, substitute 40 for *s* and 60 for *d*.

$d = ks^2$

$60 = k(40)^2$

$60 = 1600k$

$k = \frac{60}{1600} = 0.0375$

Thus $d = 0.0375s^2$. Now substitute 56 for *s*.

$d = 0.0375s^2$

$d = 0.0375(56)^2$

$d = 117.6$

The stopping distance is 117.6 feet.

69. The equation is $V = kBh$. To find *k*, substitute 160 for *V*, 48 for *B*, and 10 for *h*.

$160 = k(48)(10)$

$160 = 480k$

$\frac{160}{480} = k$

$\frac{1}{3} = k$

Thus, $V = \frac{1}{3}Bh$. Now substitute 42 for *B* and 9 for *h*.

$V = \frac{1}{3}(42)(9) = 14(9) = 126$

The volume of the pyramid would be 126 m^3.

71. The equation is $R = \frac{kA}{P}$. To find *k*, substitute 400 for *A*, 2 for *P*, and 4600 for *R*.

$4600 = \frac{k(400)}{2}$

$4600 = 200k$

$k = \frac{4600}{200} = 23$

Thus $R = \frac{23A}{P}$. Now substitute 500 for *A* and 2.50 for *P*.

$R = \frac{23(500)}{2.50}$

$R = \frac{11,500}{2.50}$

$R = 4600$ DVDs

They would still rent 4600 DVDs per week.

73. The equation is $w = \dfrac{k}{d^2}$. To find *k*, substitute 140 for *w* and 4000 for *d*.

$$140 = \frac{k}{(4000)^2}$$

$$140 = \frac{k}{16,000,000}$$

$$140(16,000,000) = k$$

$$2,240,000,000 = k$$

Thus $w = \dfrac{2,240,000,000}{d^2}$. Now substitute 4100 for *d*.

$$w = \frac{2,240,000,000}{(4100)^2}$$

$$w = \frac{2,240,000,000}{16,810,000}$$

$$w \approx 133.25$$

The weight is about 133.25 pounds.

75. The equation is $N = \dfrac{kp_1p_2}{d}$. To find *k*, substitute 100,000 for *N*, 300 for *d*, 60,000 for p_1, and 200,000 for p_2.

$$100,000 = \frac{k(60,000)(200,000)}{300}$$

$$100,000 = 40,000,000k$$

$$\frac{100,000}{40,000,000} = k$$

$$0.0025 = k$$

Thus $N = \dfrac{0.0025p_1p_2}{d}$.

Now substitute 450 for *d*, 125,000 for p_1, and 175,000 for p_2.

$$N = \frac{0.0025(125,000)(175,000)}{450}$$

$$N \approx 121,528$$

About 121,528 calls are made.

77. Let *I* be the intensity of the illumination and *d* be the distance the subject is from the flash.

The equation is $I = \dfrac{k}{d^2}$. To find *k*, substitute $\dfrac{1}{16}$ for *I* and 4 for *d*.

$$\frac{1}{16} = \frac{k}{4^2}$$

$$\frac{1}{16} = \frac{k}{16}$$

$$1 = k$$

Thus $I = \dfrac{1}{d^2}$.

Now, substitute 7 for *d*.

$$I = \frac{1}{7^2} = \frac{1}{49}$$

The illumination is $\dfrac{1}{49}$ of the light of the flash.

79. a. The equation is $P = 14.7 + kx$

b. The find *k*, substitute 40.5 for *P* and 60 for *x*.

$$40.5 = 14.7 + 60k$$

$$25.8 = 60k$$

$$\frac{25.8}{60} = k$$

$$0.43 = k$$

c. The equation is $P = 14.7 + 0.43x$. Substitute 160 for *P*.

$$160 = 14.7 + 0.43x$$

$$145.3 = 0.43x$$

$$\frac{145.3}{0.43} = x$$

$$337.9 \approx x$$

The submarine can go about 337.9 feet deep.

80. $V = \dfrac{4}{3}\pi r^2 h$

$$3(V) = 3\left(\frac{4}{3}\pi r^2 h\right)$$

$$3V = 4\pi r^2 h$$

$$\frac{3V}{4\pi r^2} = \frac{4\pi r^2 h}{4\pi r^2}$$

$$\frac{3V}{4\pi r^2} = h \Rightarrow h = \frac{3V}{4\pi r^2}$$

81. $f(x)=x^2-4,\ \text{and } g(x)=-5x+1$

$$\begin{aligned}f(-4)\cdot g(-2)&=\left[(-4)^2-4\right]\cdot\left[-5(-2)+1\right]\\&=\left[16-4\right]\cdot\left[10+1\right]\\&=12\cdot 11\\&=132\end{aligned}$$

82.

$$\begin{array}{r}-2x^2-4x+5\\ \times\qquad\qquad 7x-3\\ \hline 6x^2+12x-15\\ -14x^3-28x^2+35x\qquad\\ \hline -14x^3-22x^2+47x-15\end{array}$$

83.

$$\begin{aligned}(x+1)^2-(x+1)-6&=w^2-w-6 \leftarrow \text{let } w=x+1\\&=(w-3)(w+2)\\&=((x+1)-3)((x+1)+2)\\&=(x-2)(x+3)\end{aligned}$$

Chapter 6 Review Exercises

1. $\dfrac{3}{x-5}$

The excluded value is 5.

2. $\dfrac{x}{x+1}$

The excluded value is -1.

3. $\dfrac{-2x}{x^2+9}$

There are no values for which $x^2+9=0$

Thus, there are no excluded values.

4. $y=\dfrac{5}{(x+3)^2}$

$x+3=0$

$x=-3$

The domain is $\{x\mid x\neq -3\}$.

5. $f(x)=\dfrac{x+6}{x^2}$

$x^2=0$

$x=0$

The domain is $\{x\mid x\neq 0\}$.

6. $f(x)=\dfrac{x^2-2}{x^2+4x-12}$

$x^2+4x-12=0$

$(x+6)(x-2)=0$

$x+6=0 \text{ or } x-2=0$

$x=-6 \qquad x=2$

The domain is $\{x\mid x\neq -6 \text{ and } x\neq 2\}$.

7. $\dfrac{x^2+xy}{x+y}=\dfrac{x(x+y)}{x+y}=\dfrac{x}{1}=x$

8. $\dfrac{x^2-36}{x+6}=\dfrac{(x+6)(x-6)}{x+6}=x-6$

9. $\dfrac{7-5x}{5x-7}=\dfrac{-(5x-7)}{5x-7}=-1$

10. $\dfrac{x^2+5x-6}{x^2+4x-12}=\dfrac{(x+6)(x-1)}{(x+6)(x-2)}=\dfrac{x-1}{x-2}$

11.

$$\begin{aligned}\frac{2x^2-6x+5x-15}{2x^2+7x+5}&=\frac{2x(x-3)+5(x-3)}{(2x+5)(x+1)}\\&=\frac{(2x+5)(x-3)}{(2x+5)(x+1)}\\&=\frac{x-3}{x+1}\end{aligned}$$

12.

$$\begin{aligned}&\frac{a^3-8b^3}{a^2-4b^2}\\&=\frac{(a-2b)(a^2+2ab+4b^2)}{(a-2b)(a+2b)}\\&=\frac{a^2+2ab+4b^2}{a+2b}\end{aligned}$$

13. $\frac{27x^3+y^3}{9x^2-y^2}=\frac{(3x)^3+(y)^3}{(3x)^2-(y)^2}$

$=\frac{(3x+y)\left((3x)^2-(3x)(y)+(y)^2\right)}{(3x+y)(3x-y)}$

$=\frac{(3x+y)(9x^2-3xy+y^2)}{(3x+y)(3x-y)}$

$=\frac{(9x^2-3xy+y^2)}{(3x-y)}$

14. $\frac{2x^2+x-6}{x^3+8}=\frac{(2x-3)(x+2)}{(x+2)(x^2-2x+4)}$

$=\frac{(2x-3)}{(x^2-2x+4)}$

15. Factor denominators:
$x+4$ is $x+4$
x is x
The LCD is $x(x+4)$.

16. Factor denominators: $x+2y$ is $x+2y$
$x^2-4y^2=(x+2y)(x-2y)$
The LCD is $(x+2y)(x-2y)$.

17. Factor denominators:
$x^2+2x-35=(x+7)(x-5)$
$x^2-3x-10=(x-5)(x+2)$
The LCD is $(x+7)(x-5)(x+2)$.

18. Factor denominators:
$(x+2)^2=(x+2)(x+2)$
$x^2-4=(x+2)(x-2)$
$x+3=x+3$
The LCD is $(x+2)^2(x-2)(x+3)$.

19. $\frac{30x^2y^3}{3z}\cdot\frac{6z^3}{5xy^3}=\frac{12xz^2}{1}=12xz^2$

20. $\frac{x}{x-9}\cdot\frac{9-x}{6}=\frac{x}{x-9}\cdot\frac{-(x-9)}{6}=\frac{-x}{6}$ or $-\frac{x}{6}$

21. $\frac{18x^2y^4}{xz^5}\div\frac{2x^2y^4}{x^4z^{10}}=\frac{18x^2y^4}{xz^5}\cdot\frac{x^4z^{10}}{2x^2y^4}=9x^3z^5$

22. $\frac{11}{3x}+\frac{2}{x^2}=\frac{11\cdot x}{3x\cdot x}+\frac{2\cdot 3}{x^2\cdot 3}$

$=\frac{11x}{3x^2}+\frac{6}{3x^2}$

$=\frac{11x+6}{3x^2}$

23. $\frac{4x-4y}{x^2y}\cdot\frac{y^3}{16x}=\frac{4(x-y)}{x^2y}\cdot\frac{y^3}{16x}=\frac{(x-y)y^2}{4x^3}$

24. $\frac{4x^2-11x+4}{x-3}-\frac{x^2-4x+10}{x-3}$

$=\frac{4x^2-11x+4-(x^2-4x+10)}{x-3}$

$=\frac{4x^2-11x+4-x^2+4x-10}{x-3}$

$=\frac{3x^2-7x-6}{x-3}$

$=\frac{(3x+2)(x-3)}{x-3}$

$=3x+2$

25. $\frac{6}{xy}+\frac{3y}{5x^2}=\frac{6\cdot 5x}{xy\cdot 5x}+\frac{3y\cdot y}{5x^2\cdot y}$

$=\frac{30x}{5x^2y}+\frac{3y^2}{5x^2y}$

$=\frac{30x+3y^2}{5x^2y}$

26. $\frac{x+2}{x-1}\cdot\frac{x^2+3x-4}{x^2+6x+8}=\frac{x+2}{x-1}\cdot\frac{(x+4)(x-1)}{(x+4)(x+2)}=1$

27. $\frac{3x^2-7x+4}{3x^2-14x-5}-\frac{x^2+2x+9}{3x^2-14x-5}$

$$=\frac{\left(3x^2-7x+4\right)-\left(x^2+2x+9\right)}{3x^2-14x-5}$$
$$=\frac{3x^2-7x+4-x^2-2x-9}{3x^2-14x-5}$$
$$=\frac{2x^2-9x-5}{3x^2-14x-5}$$
$$=\frac{(2x+1)(x-5)}{(3x+1)(x-5)}$$
$$=\frac{2x+1}{3x+1}$$

28. $5+\frac{a+2}{a+1}=\frac{5}{1}\cdot\frac{a+1}{a+1}+\frac{a+2}{a+1}$

$$=\frac{5(a+1)}{a+1}+\frac{a+2}{a+1}$$
$$=\frac{5a+5}{a+1}+\frac{a+2}{a+1}$$
$$=\frac{5a+5+a+2}{a+1}$$
$$=\frac{6a+7}{a+1}$$

29. $7-\frac{b+1}{b-1}=\frac{7}{1}\cdot\frac{b-1}{b-1}-\frac{b+1}{b-1}$

$$=\frac{7(b-1)}{b-1}-\frac{b+1}{b-1}$$
$$=\frac{7b-7}{b-1}-\frac{b+1}{b-1}$$
$$=\frac{7b-7-b-1}{b-1}$$
$$=\frac{6b-8}{b-1}$$

30. $\frac{a^2-b^2}{a+b}\cdot\frac{a^2+2ab+b^2}{a^3+a^2b}$

$$=\frac{(a+b)(a-b)}{a+b}\cdot\frac{(a+b)^2}{a^2(a+b)}$$
$$=\frac{(a-b)(a+b)}{a^2}$$
$$=\frac{a^2-b^2}{a^2}$$

31. $\frac{1}{a^2+8a+15}\div\frac{3}{a+5}=\frac{1}{a^2+8a+15}\cdot\frac{a+5}{3}$

$$=\frac{1}{(a+5)(a+3)}\cdot\frac{a+5}{3}$$
$$=\frac{1}{3(a+3)}$$

32. $\frac{a+c}{c}-\frac{a-c}{a}=\frac{a+c}{c}\cdot\frac{a}{a}-\frac{a-c}{a}\cdot\frac{c}{c}$

$$=\frac{a(a+c)}{ac}-\frac{c(a-c)}{ac}$$
$$=\frac{a^2+ac}{ac}-\frac{ac-c^2}{ac}$$
$$=\frac{a^2+ac-(ac-c^2)}{ac}$$
$$=\frac{a^2+ac-ac+c^2}{ac}$$
$$=\frac{a^2+c^2}{ac}$$

33. $\frac{4x^2+8x-5}{2x+5}\cdot\frac{x+1}{4x^2-4x+1}$

$$=\frac{(2x-1)(2x+5)}{2x+5}\cdot\frac{x+1}{(2x-1)(2x-1)}$$
$$=\frac{x+1}{2x-1}$$

34. $(a+b)\div\frac{a^2-2ab-3b^2}{a-3b}$

$$=\frac{a+b}{1}\cdot\frac{a-3b}{a^2-2ab-3b^2}$$
$$=\frac{a+b}{1}\cdot\frac{a-3b}{(a-3b)(a+b)}$$
$$=1$$

35. $\frac{x^2-3xy-10y^2}{6x}\div\frac{x+2y}{24x^2}$

$$=\frac{x^2-3xy-10y^2}{6x}\cdot\frac{24x^2}{x+2y}$$
$$=\frac{(x+2y)(x-5y)}{6x}\cdot\frac{24x^2}{x+2y}$$
$$=4x(x-5y)$$

36. $\frac{a+1}{2a}+\frac{3}{4a+8}=\frac{a+1}{2a}+\frac{3}{4(a+2)}$

$=\frac{a+1}{2a}\cdot\frac{2(a+2)}{2(a+2)}+\frac{3}{4(a+2)}\cdot\frac{a}{a}$

$=\frac{2(a+1)(a+2)}{4a(a+2)}+\frac{3a}{4a(a+2)}$

$=\frac{2a^2+6a+4}{4a(a+2)}+\frac{3a}{4a(a+2)}$

$=\frac{2a^2+6a+4+3a}{4a(a+2)}$

$=\frac{2a^2+9a+4}{4a(a+2)}$

37. $\frac{x-2}{x-5}-\frac{3}{x+5}=\frac{x-2}{x-5}\cdot\frac{x+5}{x+5}-\frac{3}{x+5}\cdot\frac{x-5}{x-5}$

$=\frac{(x-2)(x+5)}{(x-5)(x+5)}-\frac{3(x-5)}{(x-5)(x+5)}$

$=\frac{x^2+3x-10}{(x-5)(x+5)}-\frac{3x-15}{(x-5)(x+5)}$

$=\frac{x^2+3x-10-(3x-15)}{(x-5)(x+5)}$

$=\frac{x^2+5}{(x-5)(x+5)}$

38. $\frac{x+4}{x^2-4}-\frac{3}{x-2}=\frac{x+4}{(x+2)(x-2)}-\frac{3}{x-2}$

$=\frac{x+4}{(x+2)(x-2)}-\frac{3}{x-2}\cdot\frac{x+2}{x+2}$

$=\frac{x+4}{(x+2)(x-2)}-\frac{3(x+2)}{(x-2)(x+2)}$

$=\frac{x+4}{(x+2)(x-2)}-\frac{3x+6}{(x-2)(x+2)}$

$=\frac{x+4-3x-6}{(x+2)(x-2)}$

$=\frac{-2x-2}{(x+2)(x-2)}$ or $-\frac{2(x+1)}{x^2-4}$

39. $\frac{x+1}{x-3}\cdot\frac{x^2+2x-15}{x^2+7x+6}=\frac{x+1}{x-3}\cdot\frac{(x+5)(x-3)}{(x+6)(x+1)}=\frac{x+5}{x+6}$

40. $\frac{2}{x^2-x-6}-\frac{3}{x^2-4}$

$=\frac{2}{(x+2)(x-3)}-\frac{3}{(x-2)(x+2)}$

$=\frac{2}{(x+2)(x-3)}\cdot\frac{x-2}{x-2}-\frac{3}{(x+2)(x-2)}\cdot\frac{x-3}{x-3}$

$=\frac{2(x-2)}{(x+2)(x-3)(x-2)}-\frac{3(x-3)}{(x+2)(x-3)(x-2)}$

$=\frac{2x-4-3x+9}{(x+2)(x-3)(x-2)}$

$=\frac{-x+5}{(x+2)(x-3)(x-2)}$

41. $\frac{4x^2-16y^2}{9}\div\frac{(x+2y)^2}{12}$

$=\frac{4x^2-16y^2}{9}\cdot\frac{12}{(x+2y)^2}$

$=\frac{4(x+2y)(x-2y)}{9}\cdot\frac{12}{(x+2y)(x+2y)}$

$=\frac{16(x-2y)}{3(x+2y)}$

42. $\frac{a^2+5a+6}{a^2+4a+4}\cdot\frac{3a+6}{a^4+3a^3}$

$=\frac{(a+3)(a+2)}{(a+2)^2}\cdot\frac{3(a+2)}{a^3(a+3)}$

$=\frac{3}{a^3}$

43. $\dfrac{x+5}{x^2-15x+50}-\dfrac{x-2}{x^2-25}$

$=\dfrac{x+5}{(x-5)(x-10)}-\dfrac{x-2}{(x-5)(x+5)}$

$=\dfrac{x+5}{(x-5)(x-10)}\cdot\dfrac{x+5}{x+5}-\dfrac{x-2}{(x-5)(x+5)}\cdot\dfrac{x-10}{x-10}$

$=\dfrac{(x+5)(x+5)}{(x-5)(x-10)(x+5)}-\dfrac{(x-2)(x-10)}{(x-5)(x-10)(x+5)}$

$=\dfrac{x^2+10x+25}{(x-5)(x-10)(x+5)}-\dfrac{x^2-12x+20}{(x-5)(x-10)(x+5)}$

$=\dfrac{x^2+10x+25-(x^2-12x+20)}{(x-5)(x-10)(x+5)}$

$=\dfrac{x^2+10x+25-x^2+12x-20}{(x-5)(x-10)(x+5)}$

$=\dfrac{22x+5}{(x-5)(x-10)(x+5)}$

44. $\dfrac{x+2}{x^2-x-6}+\dfrac{x-3}{x^2-8x+15}$

$=\dfrac{x+2}{(x+2)(x-3)}+\dfrac{x-3}{(x-3)(x-5)}$

$=\dfrac{1}{x-3}+\dfrac{1}{x-5}$

$=\dfrac{1}{x-3}\cdot\dfrac{x-5}{x-5}+\dfrac{1}{x-5}\cdot\dfrac{x-3}{x-3}$

$=\dfrac{x-5}{(x-3)(x-5)}+\dfrac{x-3}{(x-3)(x-5)}$

$=\dfrac{x-5+x-3}{(x-3)(x-5)}$

$=\dfrac{2x-8}{(x-3)(x-5)}$ or $\dfrac{2(x-4)}{(x-3)(x-5)}$

45. $\dfrac{1}{x+3}-\dfrac{2}{x-3}+\dfrac{6}{x^2-9}$

$=\dfrac{1}{x+3}-\dfrac{2}{x-3}+\dfrac{6}{(x+3)(x-3)}$

$=\dfrac{1}{x+3}\cdot\dfrac{x-3}{x-3}-\dfrac{2}{x-3}\cdot\dfrac{x+3}{x+3}+\dfrac{6}{(x+3)(x-3)}$

$=\dfrac{x-3}{(x+3)(x-3)}-\dfrac{2(x+3)}{(x+3)(x-3)}+\dfrac{6}{(x+3)(x-3)}$

$=\dfrac{x-3}{(x+3)(x-3)}-\dfrac{2x+6}{(x+3)(x-3)}+\dfrac{6}{(x+3)(x-3)}$

$=\dfrac{x-3-(2x+6)+6}{(x+3)(x-3)}$

$=\dfrac{x-3-2x-6+6}{(x+3)(x-3)}$

$=\dfrac{-x-3}{(x+3)(x-3)}$

$=\dfrac{-(x+3)}{(x+3)(x-3)}$

$=\dfrac{-1}{x-3}$ or $-\dfrac{1}{x-3}$

46. $\dfrac{a-4}{a-5}-\dfrac{3}{a+5}-\dfrac{10}{a^2-25}$

$=\dfrac{a-4}{a-5}-\dfrac{3}{a+5}-\dfrac{10}{(a+5)(a-5)}$

$=\dfrac{a-4}{a-5}\cdot\dfrac{a+5}{a+5}-\dfrac{3}{a+5}\cdot\dfrac{a-5}{a-5}-\dfrac{10}{(a+5)(a-5)}$

$=\dfrac{(a-4)(a+5)}{(a+5)(a-5)}-\dfrac{3(a-5)}{(a+5)(a-5)}-\dfrac{10}{(a+5)(a-5)}$

$=\dfrac{a^2+a-20}{(a+5)(a-5)}-\dfrac{3a-15}{(a+5)(a-5)}-\dfrac{10}{(a+5)(a-5)}$

$=\dfrac{a^2+a-20-(3a-15)-10}{(a+5)(a-5)}$

$=\dfrac{a^2+a-20-3a+15-10}{(a+5)(a-5)}$

$=\dfrac{a^2-2a-15}{(a+5)(a-5)}$

$=\dfrac{(a+3)(a-5)}{(a+5)(a-5)}$

$=\dfrac{a+3}{a+5}$

47. $\dfrac{x^3+64}{2x^2-32}\div\dfrac{x^2-4x+16}{2x+12}$

$=\dfrac{x^3+64}{2x^2-32}\cdot\dfrac{2x+12}{x^2-4x+16}$

$=\dfrac{(x+4)(x^2-4x+16)}{2(x-4)(x+4)}\cdot\dfrac{2(x+6)}{x^2-4x+16}$

$=\dfrac{x+6}{x-4}$

48. $\dfrac{a^2-b^4}{a^2+2ab^2+b^4}\div\dfrac{3a-3b^2}{a^2+3ab^2+2b^4}$

$=\dfrac{\left(a+b^2\right)\left(a-b^2\right)}{\left(a+b^2\right)^2}\cdot\dfrac{\left(a+2b^2\right)\left(a+b^2\right)}{3\left(a-b^2\right)}$

$=\dfrac{a+2b^2}{3}$

49. $\left(\dfrac{x^2-x-56}{x^2+14x+49}\cdot\dfrac{x^2+4x-21}{x^2-9x+8}\right)+\dfrac{3}{x^2+8x-9}$

$=\left(\dfrac{(x-8)(x+7)}{(x+7)(x+7)}\cdot\dfrac{(x+7)(x-3)}{(x-8)(x-1)}\right)+\dfrac{3}{(x-1)(x+9)}$

$=\dfrac{x-3}{x-1}+\dfrac{3}{(x-1)(x+9)}$

$=\dfrac{x-3}{x-1}\cdot\dfrac{x+9}{x+9}+\dfrac{3}{(x-1)(x+9)}$

$=\dfrac{(x-3)(x+9)}{(x-1)(x+9)}+\dfrac{3}{(x-1)(x+9)}$

$=\dfrac{x^2+6x-27}{(x-1)(x+9)}+\dfrac{3}{(x-1)(x+9)}$

$=\dfrac{x^2+6x-27+3}{(x-1)(x+9)}$

$=\dfrac{x^2+6x-24}{(x-1)(x+9)}$

50. $\left(\dfrac{x^2-8x+16}{2x^2-x-6}\cdot\dfrac{2x^2-7x-15}{x^2-2x-24}\right)\div\dfrac{x^2-9x+20}{x^2+2x-8}$

$=\left(\dfrac{x^2-8x+16}{2x^2-x-6}\cdot\dfrac{2x^2-7x-15}{x^2-2x-24}\right)\cdot\dfrac{x^2+2x-8}{x^2-9x+20}$

$=\dfrac{(x-4)(x-4)}{(2x+3)(x-2)}\cdot\dfrac{(2x+3)(x-5)}{(x-6)(x+4)}\cdot\dfrac{(x+4)(x-2)}{(x-5)(x-4)}$

$=\dfrac{x-4}{x-6}$

51. a. $f(x)=\dfrac{x+1}{x+2}$

The denominator cannot equal zero.
Therefore $x\neq-2$.
The domain is $\{x|x\neq-2\}$

b. $g(x)=\dfrac{x}{x+4}$

The denominator cannot equal zero.
Therefore, $x\neq-4$.
The domain is $\{x|x\neq-4\}$

c. $(f+g)(x)=f(x)+g(x)$

$=\dfrac{x+1}{x+2}+\dfrac{x}{x+4}$

$=\dfrac{x+1}{x+2}\cdot\dfrac{x+4}{x+4}+\dfrac{x}{x+4}\cdot\dfrac{x+2}{x+2}$

$=\dfrac{(x+1)(x+4)+x(x+2)}{(x+2)(x+4)}$

$=\dfrac{x^2+5x+4+x^2+2x}{(x+2)(x+4)}$

$=\dfrac{2x^2+7x+4}{(x+2)(x+4)}$

d. The denominator cannot equal zero.
Therefore, $x\neq-2, x\neq-4$.
The domain is $\{x|x\neq-2 \text{ and } x\neq-4\}$.

52. a. $f(x)=\dfrac{x}{x^2-9}$ or $f(x)=\dfrac{x}{(x+3)(x-3)}$

The denominator cannot equal zero.
Therefore, $x\neq-3, x\neq3$.
The domain is $\{x|x\neq-3 \text{ and } x\neq3\}$.

b. $g(x)=\dfrac{x+4}{x-3}$

The denominator cannot equal zero.
Therefore, $x \neq 3$.
The domain is $\{x|x \neq 3\}$.

c. $(f+g)(x)=f(x)+g(x)$

$$=\frac{x}{x^2-9}+\frac{x+4}{x-3}$$
$$=\frac{x}{(x-3)(x+3)}+\frac{(x+4)}{(x-3)}\cdot\frac{(x+3)}{(x+3)}$$
$$=\frac{x+(x+4)(x+3)}{(x-3)(x+3)}$$
$$=\frac{x+x^2+7x+12}{(x-3)(x+3)}$$
$$=\frac{x^2+8x+12}{(x-3)(x+3)}$$

d. The denominator cannot equal zero.
Therefore, $x \neq -3$ and $x \neq 3$.
The domain is $\{x|x \neq -3 \text{ and } x \neq 3\}$.

53. $$\frac{\frac{9a^2b}{2c}}{\frac{6ab^4}{4c^3}}=\frac{9a^2b}{2c}\cdot\frac{4c^3}{6ab^4}=\frac{3ac^2}{b^3}$$

54. $$\frac{\frac{2}{x}+\frac{4}{y}}{\frac{x}{y}+y^2}=\frac{xy\left(\frac{2}{x}+\frac{4}{y}\right)}{xy\left(\frac{x}{y}+y^2\right)}=\frac{4x+2y}{x^2+xy^3}$$

55. $$\frac{\frac{3}{y}-\frac{1}{y^2}}{7+\frac{1}{y^2}}=\frac{y^2\left(\frac{3}{y}-\frac{1}{y^2}\right)}{y^2\left(7+\frac{1}{y^2}\right)}$$
$$=\frac{y^2\left(\frac{3}{y}\right)-y^2\left(\frac{1}{y^2}\right)}{y^2(7)+y^2\left(\frac{1}{y^2}\right)}$$
$$=\frac{3y-1}{7y^2+1}$$

56. $$\frac{a^{-1}+5}{a^{-1}+\frac{1}{a}}=\frac{\frac{1}{a}+5}{\frac{1}{a}+\frac{1}{a}}=\frac{\frac{1}{a}+5}{\frac{2}{a}}$$
$$=\left(\frac{1}{a}+5\right)\left(\frac{a}{2}\right)=\frac{\left(\frac{1}{a}+5\right)a}{2}$$
$$=\frac{1+5a}{2}=\frac{5a+1}{2}$$

57. $$\frac{x^{-2}+\frac{3}{x}}{\frac{1}{x^2}-\frac{1}{x}}=\frac{\frac{1}{x^2}+\frac{3}{x}}{\frac{1}{x^2}-\frac{1}{x}}=\frac{x^2\left(\frac{1}{x^2}+\frac{3}{x}\right)}{x^2\left(\frac{1}{x^2}-\frac{1}{x}\right)}$$
$$=\frac{x^2\left(\frac{1}{x^2}\right)+x^2\left(\frac{3}{x}\right)}{x^2\left(\frac{1}{x^2}\right)-x^2\left(\frac{1}{x}\right)}$$
$$=\frac{1+3x}{1-x} \text{ or } \frac{3x+1}{-x+1}$$

58. $$\frac{\frac{1}{x^2-3x-18}+\frac{2}{x^2-2x-15}}{\frac{3}{x^2-11x+30}+\frac{1}{x^2-9x+20}}$$
$$=\frac{\frac{1}{(x-6)(x+3)}+\frac{2}{(x+3)(x-5)}}{\frac{3}{(x-5)(x-6)}+\frac{1}{(x-5)(x-4)}}$$
$$=\frac{\frac{(x-5)+2(x-6)}{(x-6)(x+3)(x-5)}}{\frac{3(x-4)+1(x-6)}{(x-5)(x-6)(x-4)}}$$
$$=\frac{x-5+2x-12}{(x-6)(x+3)(x-5)}\cdot\frac{(x-5)(x-6)(x-4)}{3x-12+x-6}$$
$$=\frac{(3x-17)(x-4)}{(4x-18)(x+3)}$$
$$=\frac{3x^2-29x+68}{4x^2-6x-54}$$

59. $l = A \div w$

$$l = \frac{x^2+5x+6}{x+4} \div \frac{x^2+8x+15}{x^2+5x+4}$$
$$= \frac{x^2+5x+6}{x+4} \cdot \frac{x^2+5x+4}{x^2+8x+15}$$
$$= \frac{(x+3)(x+2)}{x+4} \cdot \frac{(x+4)(x+1)}{(x+5)(x+3)}$$
$$= \frac{(x+2)(x+1)}{x+5}$$
$$= \frac{x^2+3x+2}{x+5}$$

60. $l = A \div w$

$$l = \frac{x^2+10x+24}{x+5} \div \frac{x^2+9x+18}{x^2+7x+10}$$
$$= \frac{x^2+10x+24}{x+5} \cdot \frac{x^2+7x+10}{x^2+9x+18}$$
$$= \frac{(x+6)(x+4)}{x+5} \cdot \frac{(x+5)(x+2)}{(x+6)(x+3)}$$
$$= \frac{(x+4)(x+2)}{x+3}$$
$$= \frac{x^2+6x+8}{x+3}$$

61. $\frac{2}{x} = \frac{5}{9}$

$$5x = 2 \cdot 9$$
$$5x = 18$$
$$x = \frac{18}{5} \text{ or } 3\frac{3}{5}$$

This solution checks. The solution is $\frac{18}{5}$ or $3\frac{3}{5}$.

62. $\frac{x}{1.5} = \frac{x-4}{4.5}$

$$4.5x = 1.5(x-4)$$
$$4.5x = 1.5x - 6$$
$$3x = -6$$
$$x = -2$$

This solution checks. The solution is –2.

63. $\frac{3x+4}{5} = \frac{2x-8}{3}$

$$3(3x+4) = 5(2x-8)$$
$$9x+12 = 10x-40$$
$$9x+52 = 10x$$
$$52 = x$$

This solution checks. The solution is 52.

64. $\frac{x}{4.8} + \frac{x}{2} = 1.7$

$$9.6\left(\frac{x}{4.8}\right) + 9.6\left(\frac{x}{2}\right) = 9.6(1.7)$$
$$2x + 4.8x = 16.32$$
$$6.8x = 16.32$$
$$x = \frac{16.32}{6.8} = 2.4$$

This solution checks. The solution is 2.4.

65. $\frac{2}{y} + \frac{1}{5} = \frac{3}{y}$

$$5y\left(\frac{2}{y}\right) + 5y\left(\frac{1}{5}\right) = 5y\left(\frac{3}{y}\right)$$
$$10 + y = 15$$
$$y = 5$$

This solution checks. The solution is 5.

66.

$$\frac{2}{x+4}-\frac{3}{x-4}=\frac{-11}{x^2-16}$$

$$\frac{2}{x+4}-\frac{3}{x-4}=\frac{-11}{(x+4)(x-4)}$$

$$(x+4)(x-4)\left(\frac{2}{x+4}\right)-(x+4)(x-4)\left(\frac{3}{x-4}\right)=(x+4)(x-4)\left(\frac{-11}{(x+4)(x-4)}\right)$$

$$2(x-4)-3(x+4)=-11$$

$$2x-8-3x-12=-11$$

$$-x-20=-11$$

$$-x=9$$

$$x=-9$$

This solution checks. The solution is –9.

67.

$$\frac{x}{x^2-9}+\frac{2}{x+3}=\frac{4}{x-3}$$

$$\frac{x}{(x+3)(x-3)}+\frac{2}{x+3}=\frac{4}{x-3}$$

$$(x+3)(x-3)\left(\frac{x}{(x+3)(x-3)}\right)+(x+3)(x-3)\left(\frac{2}{x+3}\right)=(x+3)(x-3)\left(\frac{4}{x-3}\right)$$

$$x+2(x-3)=4(x+3)$$

$$x+2x-6=4x+12$$

$$3x-6=4x+12$$

$$-6=x+12$$

$$-18=x$$

This solution checks. The solution is –18.

68.

$$\frac{7}{x^2-25}+\frac{3}{x+5}=\frac{4}{x-5}$$

$$\frac{7}{(x+5)(x-5)}+\frac{3}{x+5}=\frac{4}{x-5}$$

$$(x+5)(x-5)\left(\frac{7}{(x+5)(x-5)}\right)+(x+5)(x-5)\left(\frac{3}{x+5}\right)=(x+5)(x-5)\left(\frac{4}{x-5}\right)$$

$$7+3(x-5)=4(x+5)$$

$$7+3x-15=4x+20$$

$$3x-8=4x+20$$

$$-8=x+20$$

$$-28=x$$

This solution checks. The solution is –28.

69.

$$\frac{x-3}{x-2}+\frac{x+1}{x+3}=\frac{2x^2+x+1}{x^2+x-6}$$

$$\frac{x-3}{x-2}+\frac{x+1}{x+3}=\frac{2x^2+x+1}{(x+3)(x-2)}$$

$$(x+3)(x-2)\left(\frac{x-3}{x-2}\right)+(x+3)(x-2)\left(\frac{x+1}{x+3}\right)=(x+3)(x-2)\left(\frac{2x^2+x+1}{(x+3)(x-2)}\right)$$

$$(x+3)(x-3)+(x-2)(x+1)=2x^2+x+1$$

$$x^2-9+x^2-x-2=2x^2+x+1$$

$$2x^2-x-11=2x^2+x+1$$

$$-x-11=x+1$$

$$-11=2x+1$$

$$-12=2x$$

$$-6=x$$

This solution checks. The solution is –6.

70.

$$\frac{x+1}{x+3}+\frac{x+2}{x-4}=\frac{2x^2-18}{x^2-x-12}$$

$$\frac{x+1}{x+3}+\frac{x+2}{x-4}=\frac{2x^2-18}{(x+3)(x-4)}$$

$$(x+3)(x-4)\left(\frac{x+1}{x+3}\right)+(x+3)(x-4)\left(\frac{x+2}{x-4}\right)=(x+3)(x-4)\left(\frac{2x^2-18}{(x+3)(x-4)}\right)$$

$$(x-4)(x+1)+(x+3)(x+2)=2x^2-18$$

$$x^2-3x-4+x^2+5x+6=2x^2-18$$

$$2x^2+2x+2=2x^2-18$$

$$2x+2=-18$$

$$2x=-20$$

$$x=-10$$

This solution checks. The solution is –10.

71.

$$\frac{1}{a}+\frac{1}{b}=\frac{1}{c}$$

$$abc\cdot\frac{1}{a}+abc\cdot\frac{1}{b}=abc\cdot\frac{1}{c}$$

$$bc+ac=ab$$

$$ac=ab-bc$$

$$ac=b(a-c)$$

$$\frac{ac}{a-c}=\frac{b(a-c)}{a-c}$$

$$\frac{ac}{a-c}=b \text{ or } b=\frac{ac}{a-c}$$

72. $z=\frac{x-\overline{x}}{s}$

$$zs=x-\overline{x}$$

$$zs-x=-\overline{x}$$

$$-zs+x=\overline{x} \text{ or } \overline{x}=x-sz$$

73. $\frac{1}{R_T}=\frac{1}{R_1}+\frac{1}{R_2}+\frac{1}{R_3}$

Substitute 100 for R_1, 200 for R_2, and 600 for R_3.

$$\frac{1}{R_T}=\frac{1}{100}+\frac{1}{200}+\frac{1}{600}$$

$$600R_T\left(\frac{1}{R_T}\right)=600R_T\left(\frac{1}{100}+\frac{1}{200}+\frac{1}{600}\right)$$

$$600=6R_T+3R_T+R_T$$

$$600=10R_T$$

$$\frac{600}{10}=R_T$$

$$R_T=60$$

The total resistance is 60 ohms.

74. $\frac{1}{p}+\frac{1}{q}=\frac{1}{f}$

$$\frac{1}{6}+\frac{1}{3}=\frac{1}{f}$$

$$\frac{1}{6}+\frac{2}{6}=\frac{1}{f}$$

$$\frac{3}{6}=\frac{1}{f}$$

$$\frac{1}{2}=\frac{1}{f}$$

$$f=2$$

The focal length is 2 centimeters.

75. $\frac{2x}{5}=\frac{4}{x-3}$

$$(2x)(x-3)=(5)(4)$$

$$2x^2-6x=20$$

$$2x^2-6x-20=0$$

$$2(x^2-3x-10)=0$$

$$2(x-5)(x+2)=0$$

$$x-5=0 \quad \text{or} \quad x+2=0$$

$$x=5 \qquad\qquad x=-2$$

The negative value of x produces negative lengths so we may disregard this value of x. Therefore, the values for the missing sides using $x=5$ are $2(5)=10$ and $(5)-3=2$.

76. $\frac{2x+1}{9}=\frac{7}{\frac{x}{2}-2}$

$$(2x+1)\left(\frac{x}{2}-2\right)=(9)(7)$$

$$x^2-4x+\frac{1}{2}x-2=63 \quad \leftarrow \text{ multiply by 2}$$

$$2x^2-8x+x-4=126$$

$$2x^2-7x-130=0$$

$$(2x+13)(x-10)=0$$

$$2x+13=0 \quad \text{or} \quad x-10=0$$

$$x=-\frac{13}{2} \qquad\qquad x=10$$

The negative value of x produces negative lengths so we may disregard this value of x. Therefore, the values for the missing sides using $x=10$ are $2(10)+1=21$ and $\frac{10}{2}-2=5-2=3$.

77. Let x be time needed for both of them to pick the string beans.

	Rate	Time Worked	Part of job completed
Jerome	$\frac{1}{30}$	x	$\frac{x}{30}$
Sanford	$\frac{1}{40}$	x	$\frac{x}{40}$

$$\frac{x}{30}+\frac{x}{40}=1$$

$$120\left(\frac{x}{30}\right)+12\left(\frac{x}{40}\right)=120(1)$$

$$4x+3x=120$$

$$7x=120$$

$$x=\frac{120}{7}\approx 17.14$$

Working together, they can pick a basket of string beans in about 17.14 minutes.

78. Let x be the time needed for Fran to plant the garden by herself.

	Rate	Time Worked	Part of job completed
Fran	$\frac{1}{x}$	4.2	$\frac{4.2}{x}$
Sam	$\frac{1}{6}$	4.2	$\frac{4.2}{6}$

$$\frac{4.2}{6}+\frac{4.2}{x}=1$$
$$6x\left(\frac{4.2}{6}\right)+6x\left(\frac{4.2}{x}\right)=6x(1)$$
$$4.2x+25.2=6x$$
$$25.2=1.8x$$
$$\frac{25.2}{1.8}=x$$
$$14=x$$

Working alone, it takes Fran 14 hours to plant the garden.

79. Let x be the unknown number.

$$\frac{1+x}{11-x}=\frac{1}{2}$$
$$2(1+x)=1(11-x)$$
$$2+2x=11-x$$
$$3x=9$$
$$x=3$$

The desired number is 3.

80. Let x be the number.

$$1-\frac{1}{2x}=\frac{1}{3x}$$
$$6x(1)-6x\left(\frac{1}{2x}\right)=6x\left(\frac{1}{3x}\right)$$
$$6x-3=2$$
$$6x=5$$
$$x=\frac{5}{6}$$

The desired number is $\frac{5}{6}$.

81. Let x be the speed of the current.

	d	r	$t=\frac{d}{r}$
With current	20	$15+x$	$\frac{20}{15+x}$
Against current	10	$15-x$	$\frac{10}{15-x}$

The times are the same.

$$\frac{20}{15+x}=\frac{10}{15-x}$$
$$20(15-x)=10(15+x)$$
$$300-20x=150+10x$$
$$300=150+30x$$
$$150=30x$$
$$5=x$$

The speed of the current is 5 mph.

82. Let x be the speed of the car.

	d	r	$t=\frac{d}{r}$
Car	450	x	$\frac{450}{x}$
Plane	450	$3x$	$\frac{450}{3x}$

$$\frac{450}{3x}+6=\frac{450}{x}$$
$$\frac{150}{x}+6=\frac{450}{x}$$
$$x\left(\frac{150}{x}\right)+x(6)=x\left(\frac{450}{x}\right)$$
$$150+6x=450$$
$$6x=300$$
$$x=\frac{300}{6}=50$$

The speed of the car is 50 mph and the speed of the plane is 3(50) = 150 mph.

83. The equation is $x = ky^2$. To find *k*, substitute 45 for *x* and 3 for *y*.

$45 = k(3)^2$

$45 = 9k$

$\frac{45}{9} = k$

$5 = k$

Thus $x = 5y^2$. Now substitute 2 for *y*.

$x = 5(2)^2 = 5(4) = 20$

84. The equation is $W = \frac{kL^2}{A}$. To find *k*, substitute 4 for *W*, 2 for *L*, and 10 for *A*.

$4 = \frac{k(2)^2}{10}$

$4 = \frac{4k}{10}$

$40 = 4k$

$10 = k$

Thus $W = \frac{10L^2}{A}$. Now substitute 5 for *L* and 20 for *A*.

$W = \frac{10 \cdot (5)^2}{20} = \frac{250}{20} = \frac{25}{2}$

85. The equation is $z = \frac{kxy}{r^2}$. To find *k*, substitute 12 for *z*, 20 for *x*, 8 for *y*, and 8 for *r*.

$12 = \frac{k(20)(8)}{(8)^2}$

$12 = \frac{160k}{64}$

$12(64) = 160k$

$768 = 160k$

$\frac{768}{160} = k$

$k = 4.8$

Thus $z = \frac{4.8xy}{r^2}$. Now substitute 10 for *x*, 80 for *y*, and 3 for *r*.

$z = \frac{4.8(10)(80)}{3^2} \approx 426.7$

86. Let *s* represent the surcharge and let *E* represent the energy used in kilowatt-hours.
The equation is $s = kE$.
To find *k*, substitute 7.20 for *s* and 3600 for *E*.

$s = kE$

$7.20 = k(3600) \Rightarrow k = \frac{7.20}{3600} = 0.002$

Thus $s = 0.002E$. Now substitute 4200 for *E*.

$s = 0.002E$

$s = 0.002(4200)$

$s = 8.40$

The surcharge is \$8.40.

87. The equation is $d = kt^2$. To find *k*, substitute 16 for *d* and 1 for *t*.

$16 = k(1)^2$

$16 = k(1)$

$16 = k$

Thus $d = 16t^2$. Now substitute 10 for *t*.

$d = 16(10)^2 = 16(100) = 1600$

The person will fall 1600 feet in 10 seconds.

88. The equation is $A = kr^2$. To find *k*, substitute 78.5 for *A* and 5 for *r*.

$78.5 = k(5)^2$

$78.5 = k(25)$

$3.14 = k$

Thus $A = 3.14r^2$. Now substitute 8 for *r*.

$A = 3.14(8)^2$

$A = 3.14(64)$

$A = 200.96$

The area is 200.96 square units.

89. The equation is $t = \frac{k}{w}$ where *t* is time and *w* is water temperature. To find *k*, substitute 1.7 for *t* and 70 for *w*.

$1.7 = \frac{k}{70}$

$(1.7)(70) = k$

$119 = k$

Thus $t = \frac{119}{w}$. Now substitute 50 for *w*.

$t = \frac{119}{50} = 2.38$

It takes the ice cube 2.38 minutes to melt.

Chapter 6 Practice Test

1. $\dfrac{x+4}{x^2+3x-28}=\dfrac{x+4}{(x-4)(x+7)}$
The denominator cannot equal zero. Therefore, the excluded values are 4 and –7.

2. $f(x)=\dfrac{x^2+7}{2x^2+7x-4}=\dfrac{x^2+7}{(2x-1)(x+4)}$
The denominator cannot equal zero. Therefore,
$2x-1\neq 0$ or $x+4\neq 0$
$x\neq\dfrac{1}{2}$ $\quad x\neq -4$
The domain is $\left\{x\middle|\ x\neq\dfrac{1}{2}\text{ and } x\neq -4\right\}$.

3. $\dfrac{10x^7y^2+16x^2y+22x^3y^3}{2x^2y}$
$=\dfrac{10x^7y^2}{2x^2y}+\dfrac{16x^2y}{2x^2y}+\dfrac{22x^3y^3}{2x^2y}$
$=5x^5y+8+11xy^2$

4. $\dfrac{x^2-4xy-12y^2}{x^2+3xy+2y^2}=\dfrac{(x-6y)(x+2y)}{(x+2y)(x+y)}$
$=\dfrac{x-6y}{x+y}$

5. $\dfrac{3xy^4}{6x^2y^3}\cdot\dfrac{2x^2y^4}{x^5y^7}=\dfrac{3\cdot 2\cdot x^{1+2}y^{4+4}}{6\cdot 1\cdot x^{2+5}y^{7+3}}$
$=\dfrac{6x^3y^8}{6x^7y^{10}}$
$=\dfrac{1}{x^4y^2}$

6. $\dfrac{x+1}{x^2-7x-8}\cdot\dfrac{x^2-x-56}{x^2+9x+14}$
$=\dfrac{x+1}{(x-8)(x+1)}\cdot\dfrac{(x-8)(x+7)}{(x+7)(x+2)}$
$=\dfrac{1}{x+2}$

7. $\dfrac{7a+14b}{a^2-4b^2}\div\dfrac{a^3+a^2b}{a^2-2ab}$
$=\dfrac{7a+14b}{a^2-4b^2}\cdot\dfrac{a^2-2ab}{a^3+a^2b}$
$=\dfrac{7(a+2b)}{(a+2b)(a-2b)}\cdot\dfrac{a(a-2b)}{a^2(a+b)}$
$=\dfrac{7}{a(a+b)}$

8. $\dfrac{x^3+y^3}{x+y}\div\dfrac{x^2-xy+y^2}{x^2+y^2}$
$=\dfrac{x^3+y^3}{x+y}\cdot\dfrac{x^2+y^2}{x^2-xy+y^2}$
$=\dfrac{(x+y)(x^2-xy+y^2)}{x+y}\cdot\dfrac{x^2+y^2}{x^2-xy+y^2}$
$=x^2+y^2$

9. $\dfrac{5}{x+1}+\dfrac{2}{x^2}=\dfrac{5}{x+1}\cdot\dfrac{x^2}{x^2}+\dfrac{2}{x^2}\cdot\dfrac{(x+1)}{(x+1)}$
$=\dfrac{5x^2}{x^2(x+1)}+\dfrac{2x+2}{x^2(x+1)}$
$=\dfrac{5x^2+2x+2}{x^2(x+1)}$

10. $\dfrac{x-1}{x^2-9}-\dfrac{x}{x^2-2x-3}$
$=\dfrac{x-1}{(x+3)(x-3)}-\dfrac{x}{(x-3)(x+1)}$
$=\dfrac{x-1}{(x+3)(x-3)}\cdot\dfrac{x+1}{x+1}-\dfrac{x}{(x-3)(x+1)}\cdot\dfrac{x+3}{x+3}$
$=\dfrac{x^2-1}{(x+3)(x-3)(x+1)}-\dfrac{x^2+3x}{(x+3)(x-3)(x+1)}$
$=\dfrac{x^2-1-x^2-3x}{(x+3)(x-3)(x+1)}$
$=\dfrac{-3x-1}{(x+3)(x-3)(x+1)}$

11. $\dfrac{m}{12m^2+4mn-5n^2}+\dfrac{2m}{12m^2+28mn+15n^2}$

$=\dfrac{m}{(6m+5n)(2m-n)}+\dfrac{2m}{(6m+5n)(2m+3n)}$

$=\dfrac{m}{(6m+5n)(2m-n)}\cdot\dfrac{2m+3n}{2m+3n}+\dfrac{2m}{(6m+5n)(2m+3n)}\cdot\dfrac{2m-n}{2m-n}$

$=\dfrac{m(2m+3n)+2m(2m-n)}{(6m+5n)(2m-n)(2m+3n)}$

$=\dfrac{2m^2+3mn+4m^2-2mn}{(6m+5n)(2m-n)(2m+3n)}$

$=\dfrac{6m^2+mn}{(6m+5n)(2m-n)(2m+3n)}$ or $\dfrac{m(6m+n)}{(6m+5n)(2m-n)(2m+3n)}$

12. $\dfrac{x+1}{4x^2-4x+1}+\dfrac{3}{2x^2+5x-3}=\dfrac{x+1}{(2x-1)(2x-1)}+\dfrac{3}{(2x-1)(x+3)}$

$=\dfrac{x+1}{(2x-1)(2x-1)}\cdot\dfrac{x+3}{x+3}+\dfrac{3}{(2x-1)(x+3)}\cdot\dfrac{2x-1}{2x-1}$

$=\dfrac{(x+1)(x+3)}{(2x-1)(2x-1)(x+3)}+\dfrac{3(2x-1)}{(2x-1)(2x-1)(x+3)}$

$=\dfrac{x^2+4x+3}{(2x-1)(2x-1)(x+3)}+\dfrac{6x-3}{(2x-1)(2x-1)(x+3)}$

$=\dfrac{x^2+4x+3+6x-3}{(2x-1)(2x-1)(x+3)}$

$=\dfrac{x^2+10x}{(2x-1)(2x-1)(x+3)}$ or $\dfrac{x(x+10)}{(2x-1)^2(x+3)}$

13. $\dfrac{x^3-8}{x^2+5x-14}\div\dfrac{x^2+2x+4}{x^2+10x+21}$

$=\dfrac{x^3-8}{x^2+5x-14}\cdot\dfrac{x^2+10x+21}{x^2+2x+4}$

$=\dfrac{(x-2)(x^2+2x+4)}{(x+7)(x-2)}\cdot\dfrac{(x+7)(x+3)}{x^2+2x+4}$

$=x+3$

14. a. $(f+g)(x)=f(x)+g(x)$

$=\dfrac{x-3}{x+5}+\dfrac{x}{2x+3}$

$=\dfrac{x-3}{x+5}\cdot\dfrac{2x+3}{2x+3}+\dfrac{x}{2x+3}\cdot\dfrac{x+5}{x+5}$

$=\dfrac{(x-3)(2x+3)+x(x+5)}{(x+5)(2x+3)}$

$=\dfrac{2x^2-3x-9+x^2+5x}{(x+5)(2x+3)}$

$=\dfrac{3x^2+2x-9}{(x+5)(2x+3)}$

b. The denominator of $(f+g)(x)$ cannot be zero.

Therefore $x+5\neq 0$ or $2x+3\neq 0$

$x\neq -5$ $\qquad$ $x\neq -\frac{3}{2}$

The domain is $\left\{x\middle|x\neq -5 \text{ and } x\neq -\frac{3}{2}\right\}$.

15. area = length × width

width = area ÷ length

$$\text{width} = \frac{x^2+11x+30}{x+2} \div \frac{x^2+9x+18}{x+3}$$
$$= \frac{x^2+11x+30}{x+2} \cdot \frac{x+3}{x^2+9x+18}$$
$$= \frac{(x+6)(x+5)}{x+2} \cdot \frac{x+3}{(x+6)(x+3)}$$
$$= \frac{x+5}{x+2}$$

16. $$\frac{\frac{1}{x}+\frac{2}{y}}{\frac{1}{x}-\frac{3}{y}} = \frac{xy\left(\frac{1}{x}+\frac{2}{y}\right)}{xy\left(\frac{1}{x}-\frac{3}{y}\right)} = \frac{xy\left(\frac{1}{x}\right)+xy\left(\frac{2}{y}\right)}{xy\left(\frac{1}{x}\right)-xy\left(\frac{3}{y}\right)} = \frac{y+2x}{y-3x}$$

17. $$\frac{\frac{a^2-b^2}{ab}}{\frac{a+b}{b^2}} = \frac{\frac{a^2}{ab}-\frac{b^2}{ab}}{\frac{a}{b^2}+\frac{b}{b^2}} = \frac{ab^2\left(\frac{a^2}{ab}-\frac{b^2}{ab}\right)}{ab^2\left(\frac{a}{b^2}+\frac{b}{b^2}\right)}$$
$$= \frac{a^2b-b^3}{a^2+ab} = \frac{b\left(a^2-b^2\right)}{a(a+b)}$$
$$= \frac{b(a+b)(a-b)}{a(a+b)}$$
$$= \frac{b(a-b)}{a}$$

18. $$\frac{\frac{7}{x}-\frac{6}{x^2}}{4-\frac{1}{x}} = \frac{x^2\left(\frac{7}{x}-\frac{6}{x^2}\right)}{x^2\left(4-\frac{1}{x}\right)} = \frac{7x-6}{4x^2-x}$$

19. $$\frac{x}{5}-\frac{x}{4} = -1$$
$$20\left(\frac{x}{5}\right)-20\left(\frac{x}{4}\right) = 20(-1)$$
$$4x-5x = -20$$
$$-x = -20$$
$$x = 20$$

20. $$\frac{x}{x-8}+\frac{6}{x-2} = \frac{x^2}{x^2-10x+16}$$
$$\frac{x}{x-8}+\frac{6}{x-2} = \frac{x^2}{(x-8)(x-2)}$$
$$(x-8)(x-2)\left(\frac{x}{x-8}\right)+(x-8)(x-2)\left(\frac{6}{x-2}\right) = (x-8)(x-2)\left(\frac{x^2}{(x-8)(x-2)}\right)$$
$$x(x-2)+6(x-8) = x^2$$
$$x^2-2x+6x-48 = x^2$$
$$x^2+4x-48 = x^2$$
$$4x-48 = 0$$
$$4x = 48$$
$$x = 12$$

This solution checks. The solution is 12.

21. $A = \frac{2b}{C-d}$

$A(C-d) = 2b$

$AC - Ad = 2b$

$AC = 2b + Ad$

$\frac{AC}{A} = \frac{2b + Ad}{A}$

$C = \frac{2b + Ad}{A}$

22. The equation is $W = kI^2R$.
To find k, substitute 10 for W, 1 for I, and 1000 for R.

$10 = k \cdot 1^2 \cdot 1000$

$10 = 1000k$

$0.01 = k$

Thus $W = 0.01I^2R$. Now substitute 0.5 for I and 300 for R.

$W = 0.01(0.5)^2(300)$

$W = 0.75$

The wattage is 0.75 watt.

23. The equation is $R = \frac{kP}{T^2}$.

To find k, substitute 30 for R, 40 for P, and 2 for T.

$30 = \frac{k(40)}{2^2}$

$30 = \frac{40k}{4}$

$30 = 10k$

$3 = k$

Thus $R = \frac{3P}{T^2}$.

Now substitute 50 for P, and 5 for T.

$R = \frac{3(50)}{5^2} = \frac{150}{25} = 6$

24. Let x be the amount of time needed for both of them to wash the windows.

	Rate	Time worked	Part of job
Paul	$\frac{1}{10}$	x	$\frac{x}{10}$
Nancy	$\frac{1}{8}$	x	$\frac{x}{8}$

$\frac{x}{10} + \frac{x}{8} = 1$

$80\left(\frac{x}{10}\right) + 80\left(\frac{x}{8}\right) = 80(1)$

$8x + 10x = 80$

$18x = 80$

$x = \frac{80}{18} \approx 4.44$

Working together, they can wash the windows in about 4.44 hrs.

25. Let x be the length of the trail.

	d	r	$t = \frac{d}{r}$
Cameron	x	8	$\frac{x}{8}$
Ashley	x	5	$\frac{x}{5}$

$\frac{x}{8} = \frac{x}{5} - \frac{1}{2}$

$40\left(\frac{x}{8}\right) = 40\left(\frac{x}{5}\right) - 40\left(\frac{1}{2}\right)$

$5x = 8x - 20$

$20 = 3x$

$\frac{20}{3} = x$

$x = 6\frac{2}{3}$

The trail is $6\frac{2}{3}$ miles long.

Chapter 6 Cumulative Review Test

1. $\left\{x \middle| -\frac{5}{3} < x \le \frac{19}{4}\right\}$

2. $-3x^3 - 2x^2y + \frac{1}{2}xy^2$

$= -3(2)^3 - 2(2)^2\left(\frac{1}{2}\right) + \frac{1}{2}(2)\left(\frac{1}{2}\right)^2$

$= -3(8) - 2(4)\left(\frac{1}{2}\right) + \frac{1}{2}(2)\left(\frac{1}{4}\right)$

$= -24 - 4 + \frac{1}{4}$

$= -28 + \frac{1}{4}$

$= -27\frac{3}{4}$

3. $2(x+1) = \frac{1}{2}(x-5)$

$2x + 2 = \frac{1}{2}x - \frac{5}{2}$

$2(2x+2) = 2\left(\frac{1}{2}x - \frac{5}{2}\right)$

$4x + 4 = x - 5$

$3x = -9$

$x = -3$

4. a. other

$= 100\% - (20\% + 15\% + 10\% + 7\% + 7\% + 13\%)$

$= 100\% - (72\%)$

$= 28\%$

The "other" category makes up 28%.

b. $20\% \text{ of } 220{,}000 = (0.20)(220{,}000)$

$= 44{,}000$

About 44,000 business degrees were awarded.

5. $4x^2 - 3y - 8$

$4(4)^2 - 3(-2) - 8$

$4(16) - 3(-2) - 8$

$64 - 3(-2) - 8$

$64 + 6 - 8$

62

6. $\left(\frac{6x^5y^6}{12x^4y^7}\right)^3 = \left(\frac{x}{2y}\right)^3 = \frac{x^3}{8y^3}$

7. $F = \frac{mv^2}{r}$

$rF = r\left(\frac{mv^2}{r}\right)$

$rF = mv^2$

$\frac{rF}{v^2} = \frac{mv^2}{v^2}$

$\frac{rF}{v^2} = m$ or $m = \frac{rF}{v^2}$

8. Let r be the intersest rate as a decimal.

$3180 = 3000 + 3000r$

$180 = 3000r$

$r = \frac{180}{3000} = 0.06$

The simple interest rate was 6%.

9. Let x be the time, in hours, until they meet.

	distance	rate	Time
Dawn	$60t$	60	t
Paula	$50t$	50	t

The total distance is 330 miles.

$60t + 50t = 330$

$110t = 330$

$t = 3$

It will take them 3 hours to meet. Since they started at 8 am, they will meet at 11am.

10. $\left|\frac{3x+5}{3}\right|-3=6$

$\left|\frac{3x+5}{3}\right|=9$

$\frac{3x+5}{3}=9$ or $\frac{3x+5}{3}=-9$

$3x+5=27$ $\quad$ $3x+5=-27$

$3x=22$ $\quad$ $3x=-32$

$x=\frac{22}{3}$ $\quad$ $x=-\frac{32}{3}$

The solution is $\left\{-\frac{32}{3}, \frac{22}{3}\right\}$.

11. $y=x^2-2$

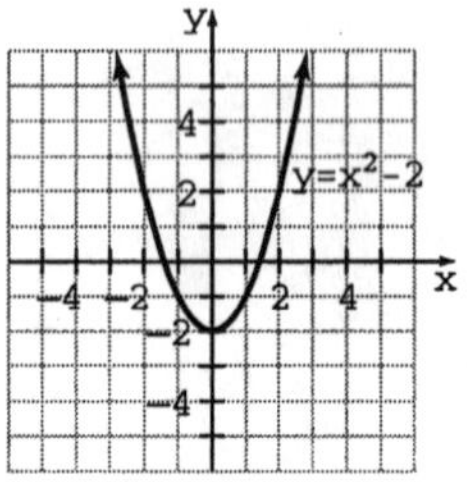

12. $f(x)=\sqrt{2x+7}$

$f(9)=\sqrt{2(9)+7}$

$=\sqrt{18+7}$

$=\sqrt{25}$

$=5$

13. $m=\frac{y_2-y_1}{x_2-x_1}$

$=\frac{-3-(-4)}{-5-2}$

$=\frac{-3+4}{-7}$

$=\frac{1}{-7}$ or $-\frac{1}{7}$

14. First find the slope of the given line.

$2x+3y-9=0$

$3y=-2x+9$

$y=-\frac{2}{3}x+3 \Rightarrow m=-\frac{2}{3}$

Now use the point-slope equation using the slope $-\frac{2}{3}$ and the given point $\left(\frac{1}{2}, 1\right)$ to find the equation.

$y-y_1=m(x-x_1)$

$y-1=-\frac{2}{3}\left(x-\frac{1}{2}\right)$

$y-1=-\frac{2}{3}x+\frac{1}{3}$

$3(y-1)=3\left(-\frac{2}{3}x+\frac{1}{3}\right)$

$3y-3=-2x+1$

$2x+3y=4$

15. $10x-y=2$

$4x+3y=11$

To eliminate y, multiply the first equation by 3 then add.

$3[10x-y=2]$

$4x+3y=11$

gives

$30x-3y=6$

Add: $4x+3y=11$

$34x \quad =17$

$x=\frac{17}{34}=\frac{1}{2}$

Substitute $\frac{1}{2}$ for x in the second equation.

$4\left(\frac{1}{2}\right)+3y=11$

$2+3y=11$

$3y=9 \Rightarrow y=3$

The solution is $\left(\frac{1}{2}, 3\right)$.

16. $(3x^2-5y)(3x^2+5y)$

$=9x^4+15x^2y-15x^2y-25y^2$

$=9x^4-25y^2$

17. $3x^2-30x+75=3(x^2-10x+25)$

$=3(x-5)(x-5)$

$=3(x-5)^2$

18. $y=|x|+2$

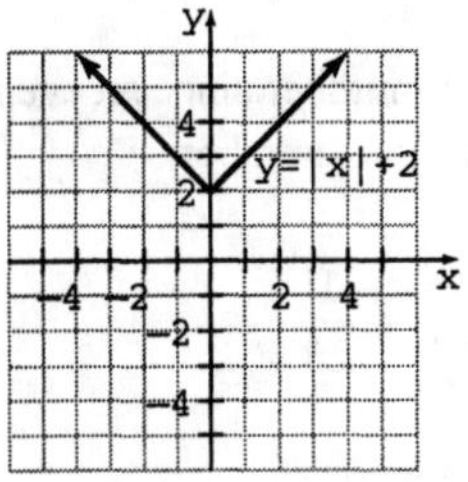

19. $\frac{7}{3x^2+x-4}+\frac{9x+2}{3x^2-2x-8}$

$$=\frac{7}{(3x+4)(x-1)}+\frac{9x+2}{(3x+4)(x-2)}$$

$$=\frac{7}{(3x+4)(x-1)}\cdot\frac{x-2}{x-2}+\frac{9x+2}{(3x+4)(x-2)}\cdot\frac{x-1}{x-1}$$

$$=\frac{7(x-2)}{(3x+4)(x-1)(x-2)}+\frac{(9x+2)(x-1)}{(3x+4)(x-2)(x-1)}$$

$$=\frac{7(x-2)+(9x+2)(x-1)}{(3x+4)(x-1)(x-2)}$$

$$=\frac{7x-14+9x^2-9x+2x-2}{(3x+4)(x-1)(x-2)}$$

$$=\frac{9x^2-16}{(3x+4)(x-1)(x-2)}$$

$$=\frac{(3x+4)(3x-4)}{(3x+4)(x-1)(x-2)}$$

$$=\frac{3x-4}{(x-1)(x-2)}$$

20.

$$\frac{3y-2}{y+1}=4-\frac{y+2}{y-1}$$

$$(y+1)(y-1)\left(\frac{3y-2}{y+1}=4-\frac{y+2}{y-1}\right)$$

$$(y-1)(3y-2)=4(y+1)(y-1)-(y+1)(y+2)$$

$$3y^2-5y+2=4(y^2-1)-(y^2+3y+2)$$

$$3y^2-5y+2=4y^2-4-y^2-3y-2$$

$$-2y=-8$$

$$y=4$$

Chapter 7

Exercise Set 7.1

1. a. Every real number has two square roots; a positive, or principal, square root and a negative square root.

   b. The square roots of 49 are 7 and −7.

   c. When we say square root, we are referring to the principal square root.

   d. $\sqrt{49} = 7$

3. There is no real number which, when squared, results in −81.

5. No. If the number under the radical is negative, the answer is not a real number.

7. a. $\sqrt{(1.3)^2} = \sqrt{1.69} = 1.3$

   b. $\sqrt{(-1.3)^2} = \sqrt{1.69} = 1.3$

9. a. $\sqrt[3]{27} = 3$ since $3^3 = 27$

   b. $-\sqrt[3]{27} = -3$

   c. $\sqrt[3]{-27} = -3$ since $(-3)^3 = -27$

11. $\sqrt{36} = 6$ since $6^2 = 36$

13. $\sqrt[3]{-64} = -4$ since $(-4)^3 = -64$

15. $\sqrt[3]{-125} = -5$ since $(-5)^3 = -125$

17. $\sqrt[5]{-1} = -1$ since $(-1)^5 = -1$

19. $\sqrt[5]{1} = 1$ since $1^5 = 1$

21. $\sqrt[6]{-64}$ is not a real number since an even root of a negative is not a real number.

23. $\sqrt[3]{-343} = -7$ since $(-7)^3 = -343$

25. $\sqrt{-36}$ is not a real number since an even root of a negative is not a real number.

27. $\sqrt{-45.3}$ is not a real number since an even root of a negative is not a real number.

29. $\sqrt{\frac{1}{25}} = \frac{1}{5}$ since $\left(\frac{1}{5}\right)^2 = \frac{1}{25}$

31. $\sqrt[3]{\frac{1}{8}} = \frac{1}{2}$ since $\left(\frac{1}{2}\right)^3 = \frac{1}{8}$

33. $\sqrt{\frac{4}{49}} = \frac{2}{7}$ since $\left(\frac{2}{7}\right)^2 = \frac{4}{49}$

35. $\sqrt[3]{-\frac{8}{27}} = -\frac{2}{3}$ since $\left(-\frac{2}{3}\right)^3 = -\frac{8}{27}$

37. $-\sqrt[4]{18.2} \approx -2.07$

39. $\sqrt{7^2} = |7| = 7$

41. $\sqrt{(19)^2} = |19| = 19$

43. $\sqrt{119^2} = |119| = 119$

45. $\sqrt{(235.23)^2} = |235.23| = 235.23$

47. $\sqrt{(0.06)^2} = |0.06| = 0.06$

49. $\sqrt{\left(\frac{12}{13}\right)^2} = \left|\frac{12}{13}\right| = \frac{12}{13}$

51. $\sqrt{(x-4)^2} = |x-4|$

53. $\sqrt{(x-3)^2} = |x-3|$

55. $\sqrt{(3x^2-1)^2} = |3x^2-1|$

57. $\sqrt{(6a^3-5b^4)^2} = |6a^3-5b^4|$

59. $\sqrt{a^{14}} = \sqrt{(a^7)^2} = |a^7|$

61. $\sqrt{z^{32}} = \sqrt{(z^{16})^2} = \left|z^{16}\right| = z^{16}$

63. $\sqrt{a^2 - 8a + 16} = \sqrt{(a-4)^2} = |a-4|$

65. $\sqrt{9a^2 + 12ab + 4b^2} = \sqrt{(3a+2b)^2} = |3a+2b|$

67. $\sqrt{49x^2} = \sqrt{(7x)^2} = 7x$

69. $\sqrt{16c^6} = \sqrt{\left(4c^3\right)^2} = 4c^3$

71. $\sqrt{x^2 + 4x + 4} = \sqrt{(x+2)^2} = x + 2$

73. $\sqrt{4x^2 + 4xy + y^2} = \sqrt{(2x+y)^2} = 2x + y$

75.
$$\begin{aligned} f(x) &= \sqrt{5x-6} \\ f(2) &= \sqrt{5\cdot 2-6} \\ &= \sqrt{10-6} \\ &= \sqrt{4} \\ &= 2 \end{aligned}$$

77.
$$\begin{aligned} q(x) &= \sqrt{76-3x} \\ q(4) &= \sqrt{76-3\cdot 4} \\ &= \sqrt{76-12} \\ &= \sqrt{64} \\ &= 8 \end{aligned}$$

79.
$$\begin{aligned} t(a) &= \sqrt{-15a-9} \\ t(-6) &= \sqrt{-15(-6)-9} \\ &= \sqrt{90-9} \\ &= \sqrt{81} \\ &= 9 \end{aligned}$$

81.
$$\begin{aligned} g(x) &= \sqrt{64-8x} \\ g(-3) &= \sqrt{64-8(-3)} \\ &= \sqrt{64+24} \\ &= \sqrt{88} \\ &\approx 9.381 \end{aligned}$$

83.
$$\begin{aligned} h(x) &= \sqrt[3]{9x^2+4} \\ h(4) &= \sqrt[3]{9(4)^2+4} \\ &= \sqrt[3]{144+4} \\ &= \sqrt[3]{148} \\ &\approx 5.290 \end{aligned}$$

85.
$$\begin{aligned} f(x) &= \sqrt[3]{-2x^2+x-6} \\ f(-3) &= \sqrt[3]{-2(-3)^2+(-3)-6} \\ &= \sqrt[3]{-18-3-6} \\ &= \sqrt[3]{-27} \\ &= -3 \end{aligned}$$

87.
$$\begin{aligned} f(x) &= x+\sqrt{x}+7 \\ f(81) &= 81+\sqrt{81}+7 \\ &= 81+9+7 \\ &= 97 \end{aligned}$$

89.
$$\begin{aligned} t(x) &= \frac{x}{2}+\sqrt{2x}-4 \\ t(18) &= \frac{18}{2}+\sqrt{2(18)}-4 \\ &= 9+\sqrt{36}-4 \\ &= 9+6-4 \\ &= 11 \end{aligned}$$

91.
$$\begin{aligned} k(x) &= x^2+\sqrt{\frac{x}{2}}-21 \\ k(8) &= (8)^2+\sqrt{\frac{8}{2}}-21 \\ &= 64+\sqrt{4}-21 \\ &= 64+2-21 \\ &= 45 \end{aligned}$$

93. Choose a value for x that will make the expression $2x+1$ a negative number. For example, select $x = -1$.

$\sqrt{(2(-1)+1)^2} \neq 2(-1)+1$

$\sqrt{(-1)^2} \neq -1$

$\sqrt{1} \neq -1$

$1 \neq -1$

Choosing any value for x less than $-\frac{1}{2}$ will show this inequality.

95. $\sqrt{(x-1)^2}=x-1$ for all $x \geq 1$. The expression $\sqrt{(x-1)^2}=x-1$, when $(x-1)$ is equal to zero or a positive number. Therefore, solving for x,
$x-1 \geq 0$
$x \geq 1$.

97. $\sqrt{(2x-6)^2}=2x-6$ for all $x \geq 3$. The expression $\sqrt{(2x-6)^2}=2x-6$, when $(2x-6)$ is positive or equal to 0. Therefore, solving for x,
$2x-6 \geq 0$
$x \geq 3$

99. **a.** $\sqrt{a^2}=|a|$ for all real values

b. $\sqrt{a^2}=a$ when $a \geq 0$

c. $\sqrt[3]{a^3}=a$ for all real values

101. If n is even, we are finding the even root of a positive number. If n is odd, the expression is a real number even if the radicand is negative.

103. $\dfrac{\sqrt{x+5}}{\sqrt[3]{x+5}}$ The denominator cannot equal zero.
$\sqrt[3]{x+5} \neq 0$
$x \neq -5$
The radicand in the numerator must be greater than or equal to zero.
$x+5 \geq 0$
$x \geq -5$
Therefore the domain is $\{x \mid x > -5\}$.

105. $f(x)=\sqrt{x}$ matches graph d).
The x-intercept is 0 and the domain is $x \geq 0$.

107. $f(x)=\sqrt{x-5}$ matches graph a). The x-intercept is 5 and the domain is $x-5 \geq 0$, or $x \geq 5$.

109. Answers may vary. One answer is
$f(x)=\sqrt{x-8}$

111. $f(x)=-\sqrt{x}$

a. No; since $\sqrt{x} \geq 0$, $-\sqrt{x}$ must be ≤ 0.

b. Yes, if $x=0$.

c. Yes; since $\sqrt{x} \geq 0$, $-\sqrt{x}$ must be ≤ 0.

113. $V=\sqrt{64.4h}$

a. $V=\sqrt{64.4(20)}$
$=\sqrt{1288}$
≈ 35.89
The velocity will be about 35.89 ft/sec.

b. $V=\sqrt{64.4(40)}$
$=\sqrt{2576}$
≈ 50.75
The velocity will be about 50.75 ft/sec.

115. $f(x)=\sqrt{x+1}$

x	$f(x)$
−1	0
0	1
3	2
8	3

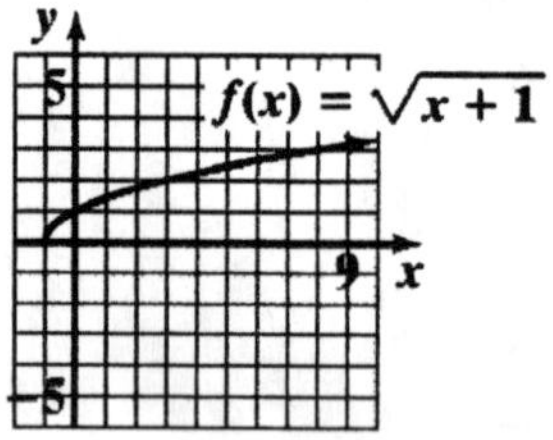

117. $g(x)=\sqrt{x}+1$

x	$g(x)$
0	1
4	3
9	4

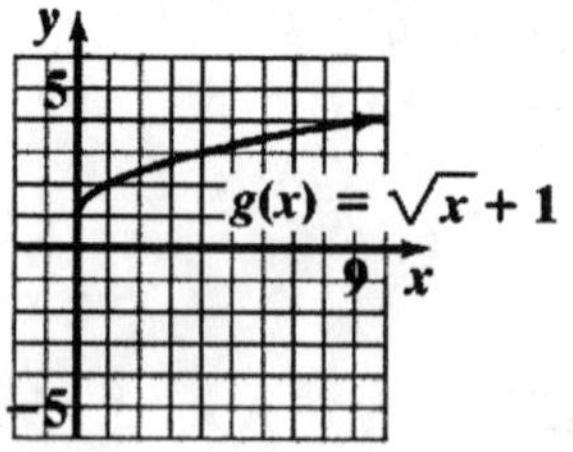

119. $y_1 = \sqrt{x+1}$

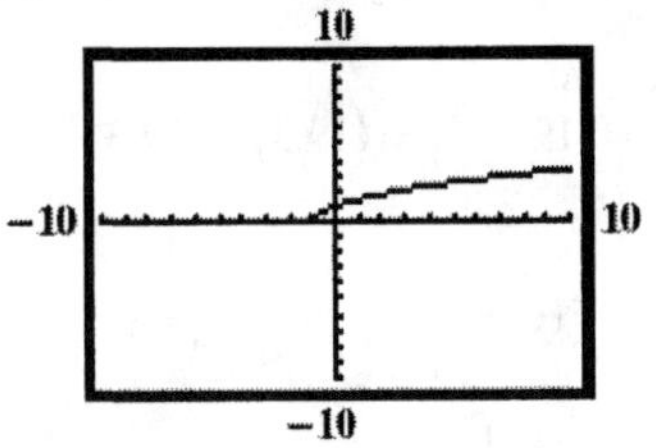

121. $y_1 = \dfrac{\sqrt{x+5}}{\sqrt[3]{x+5}}$

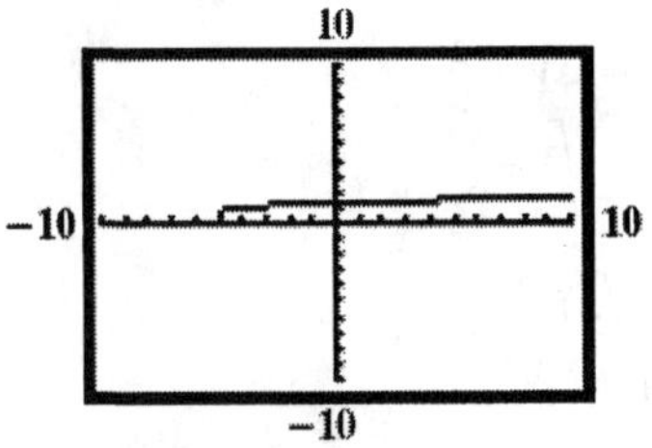

Yes, the domain is $x > -5$.

123. $y = \sqrt[3]{x+4}$

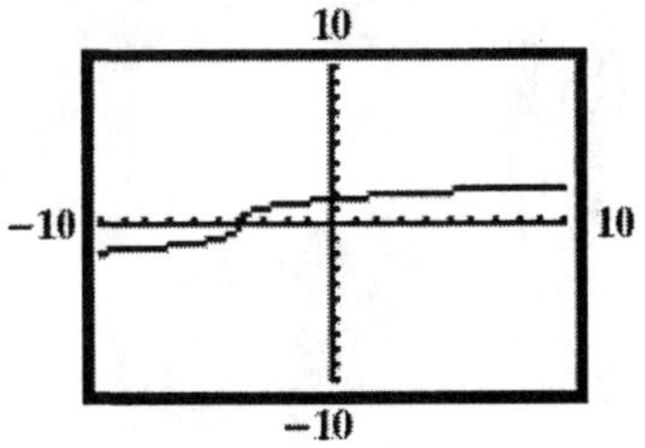

125. $a \geq 0$ and $b \geq 0$

127. $9ax - 3bx + 12ay - 4by$
$= 3x(3a-b) + 4y(3a-b)$
$= (3a-b)(3x+4y)$

128. $3x^3 - 18x^2 + 24x = 3x(x^2 - 6x + 8)$
$= 3x(x-4)(x-2)$

129. $8x^4 + 10x^2 - 3$
$= 8(x^2)^2 + 10(x^2) - 3 \leftarrow$ use y for x^2
$= 8y^2 + 10y - 3$
$= (4y-1)(2y+3) \quad \leftarrow$ use x^2 for y
$= (4x^2-1)(2x^2+3)$
$= ((2x)^2 - 1^2)(2x^2+3)$
$= (2x-1)(2x+1)(2x^2+3)$

130. $x^3 - \dfrac{8}{27}y^3 = x^3 - \left(\dfrac{2}{3}y\right)^3$

$$= \left(x - \frac{2}{3}y\right)\left(x^2 + \frac{2}{3}xy + \left(\frac{2}{3}y\right)^2\right)$$

$$= \left(x - \frac{2}{3}y\right)\left(x^2 + \frac{2}{3}xy + \frac{4}{9}y^2\right)$$

Exercise Set 7.2

1. a. $\sqrt[n]{a}$ is a real number when n is even and $a \geq 0$, or n is odd.

b. $\sqrt[n]{a}$ can be expressed with rational exponents as $a^{1/n}$.

3. a. $\sqrt[n]{a^n}$ is always real

b. $\sqrt[n]{a^n} = a$ when $a \geq 0$ and n is even

c. $\sqrt[n]{a^n} = a$ when n is odd

d. $\sqrt[n]{a^n} = |a|$ when n is even and a is any real number

5. a. No, $(xy)^{1/2} = x^{1/2}y^{1/2} \neq xy^{1/2}$

b. No, since $(xy)^{-1/2} = \dfrac{1}{(xy)^{1/2}} = \dfrac{1}{x^{1/2}y^{1/2}}$

but $\dfrac{x^{1/2}}{y^{-1/2}} = x^{1/2}y^{1/2}$

7. $\sqrt{a^3} = a^{3/2}$

9. $\sqrt{9^5} = 9^{5/2}$

11. $\sqrt[3]{z^5} = z^{5/3}$

13. $\sqrt[3]{7^{10}} = 7^{10/3}$

15. $\sqrt[4]{9^7} = 9^{7/4}$

17. $\left(\sqrt[3]{y}\right)^{14} = y^{14/3}$

19. $\sqrt[4]{a^3b} = (a^3b)^{1/4}$

21. $\sqrt[4]{x^9z^5} = (x^9z^5)^{1/4}$

23. $\sqrt[6]{3a+8b} = (3a+8b)^{1/6}$

25. $\sqrt[5]{\dfrac{2x^6}{11y^7}} = \left(\dfrac{2x^6}{11y^7}\right)^{1/5}$

27. $a^{1/2} = \sqrt{a}$

29. $c^{5/2} = \sqrt{c^5}$

31. $18^{5/3} = \sqrt[3]{18^5}$

33. $(24x^3)^{1/2} = \sqrt{24x^3}$

35. $(11b^2c)^{3/5} = \left(\sqrt[5]{11b^2c}\right)^3$

37. $(6a+5b)^{1/5} = \sqrt[5]{6a+5b}$

39. $(b^3-d)^{-1/3} = \dfrac{1}{\sqrt[3]{b^3-d}}$

41. $\sqrt{a^6} = a^{6/2} = a^3$

43. $\sqrt[3]{x^9} = x^{9/3} = x^3$

45. $\sqrt[6]{y^2} = y^{2/6} = y^{1/3} = \sqrt[3]{y}$

47. $\sqrt[6]{y^3} = y^{3/6} = y^{1/2} = \sqrt{y}$

49. $\left(\sqrt{19.3}\right)^2 = (19.3)^{2/2} = (19.3)^1 = 19.3$

51. $\left(\sqrt[3]{xy^2}\right)^{15} = (xy^2)^{15/3}$
$= (xy^2)^5$
$= x^5y^{10}$

53. $(\sqrt[8]{xyz})^4 = (xyz)^{4/8}$
$= (xyz)^{1/2}$
$= \sqrt{xyz}$

55. $\sqrt{\sqrt{x}} = (\sqrt{x})^{1/2}$
$= (x^{1/2})^{1/2}$
$= x^{1/4}$
$= \sqrt[4]{x}$

57. $\sqrt{\sqrt[4]{y}} = \left(\sqrt[4]{y}\right)^{1/2}$
$= \left(y^{1/4}\right)^{1/2}$
$= y^{1/8}$
$= \sqrt[8]{y}$

59. $\sqrt[3]{\sqrt[3]{x^2y}} = (\sqrt[3]{x^2y})^{1/3}$
$= (x^2y)^{1/3 \cdot 1/3}$
$= (x^2y)^{1/9}$
$= \sqrt[9]{x^2y}$

61. $\sqrt{\sqrt[5]{a^9}} = \left(\sqrt[5]{a^9}\right)^{1/2}$
$= \left(a^{9/5}\right)^{1/2}$
$= a^{9/10}$
$= \sqrt[10]{a^9}$

63. $25^{1/2} = \sqrt{25} = 5$

65. $64^{1/3} = \sqrt[3]{64} = 4$

67. $64^{2/3} = \left(\sqrt[3]{64}\right)^2 = 4^2 = 16$

69. $(-49)^{1/2} = \sqrt{-49}$ is not a real number.

71. $\left(\frac{25}{9}\right)^{1/2} = \sqrt{\frac{25}{9}} = \frac{5}{3}$

73. $\left(\frac{1}{8}\right)^{1/3} = \sqrt[3]{\frac{1}{8}} = \frac{1}{2}$

75. $-81^{1/2} = -\sqrt{81} = -9$

77. $-64^{1/3} = -\sqrt[3]{64} = -4$

79. $64^{-1/3} = \frac{1}{64^{1/3}} = \frac{1}{\sqrt[3]{64}} = \frac{1}{4}$

81. $16^{-3/2} = \frac{1}{16^{3/2}} = \frac{1}{\left(\sqrt{16}\right)^3} = \frac{1}{4^3} = \frac{1}{64}$

83. $\left(\frac{64}{27}\right)^{-1/3} = \left(\frac{27}{64}\right)^{1/3} = \sqrt[3]{\frac{27}{64}} = \frac{3}{4}$

85. $(-100)^{3/2} = \left(\sqrt{-100}\right)^3$ is not a real number.

87. $121^{1/2} + 169^{1/2} = \sqrt{121} + \sqrt{169} = 11 + 13 = 24$

89.
$$\begin{aligned} 343^{-1/3} + 16^{-1/2} &= \frac{1}{343^{1/3}} + \frac{1}{16^{1/2}} \\ &= \frac{1}{\sqrt[3]{343}} + \frac{1}{\sqrt{16}} \\ &= \frac{1}{7} + \frac{1}{4} \\ &= \frac{4}{28} + \frac{7}{28} \\ &= \frac{11}{28} \end{aligned}$$

91. $x^4 \cdot x^{1/2} = x^{4+1/2} = x^{9/2}$

93. $\frac{x^{1/2}}{x^{1/3}} = x^{1/2-1/3} = x^{3/6-2/6} = x^{1/6}$

95. $(x^{1/2})^{-2} = x^{1/2(-2)} = x^{-1} = \frac{1}{x}$

97. $(9^{-1/3})^0 = 9^{-1/3(0)} = 9^0 = 1$

99. $\frac{5y^{-1/3}}{60y^{-2}} = \frac{1}{12}y^{-1/3-(-2)} = \frac{1}{12}y^{5/3} = \frac{y^{5/3}}{12}$

101.
$$\begin{aligned} 4x^{5/3} \cdot 3x^{-7/2} &= 4 \cdot 3 \cdot x^{5/3} \cdot x^{-7/2} = 12x^{5/3-7/2} \\ &= 12x^{10/6-21/6} = 12x^{-11/6} = \frac{12}{x^{11/6}} \end{aligned}$$

103. $\left(\frac{3}{24x}\right)^{1/3} = \left(\frac{1}{8x}\right)^{1/3} \frac{1^{1/3}}{8^{1/3}x^{1/3}} = \frac{1}{2x^{1/3}}$

105.
$$\begin{aligned} \left(\frac{22x^{3/7}}{2x^{1/2}}\right)^2 &= (11x^{3/7-1/2})^2 = (11x^{6/14-7/14})^2 \\ &= (11x^{-1/14})^2 = (11)^2(x^{-1/14})^2 = 121x^{-1/7} \\ &= \frac{121}{x^{1/7}} \end{aligned}$$

107.
$$\begin{aligned} \left(\frac{a^4}{4a^{-2/5}}\right)^{-3} &= \frac{a^{-12}}{4^{-3}a^{6/5}} \\ &= 4^3a^{-12-6/5} \\ &= 64a^{-66/5} \\ &= \frac{64}{a^{66/5}} \end{aligned}$$

109.
$$\begin{aligned} \left(\frac{x^{3/4}y^{-3}}{x^{1/2}y^2}\right)^4 &= \left(x^{3/4-1/2}y^{-3-2}\right)^4 \\ &= (x^{1/4}y^{-5})^4 \\ &= (x^{1/4})^4(y^{-5})^4 \\ &= xy^{-20} \\ &= \frac{x}{y^{20}} \end{aligned}$$

111.
$$\begin{aligned} 4z^{-1/2}(2z^4 - z^{1/2}) &= 4z^{-1/2} \cdot 2z^4 - 4z^{-1/2}z^{1/2} \\ &= 8z^{-1/2+4} - 4z^{-1/2+1/2} \\ &= 8z^{7/2} - 4z^0 \\ &= 8z^{7/2} - 4 \end{aligned}$$

113. $5x^{-1}(x^{-4}+4x^{-1/2}) = 5x^{-1}\cdot x^{-4}+5x^{-1}\cdot 4x^{-1/2}$
$= 5x^{-1-4}+20x^{-1-1/2}$
$= 5x^{-5}+20x^{-3/2}$
$= \dfrac{5}{x^5}+\dfrac{20}{x^{3/2}}$

115. $-6x^{5/3}(-2x^{1/2}+3x^{1/3})$
$= (-6x^{5/3})(-2x^{1/2})+(-6x^{5/3})(3x^{1/3})$
$= 12x^{5/3+1/2}-18x^{5/3+1/3}$
$= 12x^{13/6}-18x^{6/3}$
$= 12x^{13/6}-18x^{2}$

117. $\sqrt{180}\approx 13.42$

119. $\sqrt[5]{402.83}\approx 3.32$

121. $93^{2/3}\approx 20.53$

123. $1000^{-1/2}\approx 0.03$

125. $\sqrt[n]{a^n}=\left(\sqrt[n]{a}\right)^n = a$ when n is odd, or n is even with $a\ge 0$.

127. To show $(a^{1/2}+b^{1/2})^2\ne a+b$, use $a=9$ and $b=16$. Then $(a^{1/2}+b^{1/2})^2$ becomes $(9^{1/2}+16^{1/2})^2=(3+4)^2=7^2=49$ whereas $a+b$ becomes $9+16=25$. Since $49\ne 25$, then $(a^{1/2}+b^{1/2})^2\ne a+b$. Answers will vary.

129. To show $(a^{1/3}+b^{1/3})^3\ne a+b$, use $a=1$ and $b=1$. Then $(a^{1/3}+b^{1/3})^3$ becomes $(1^{1/3}+1^{1/3})^3=(\sqrt[3]{1}+\sqrt[3]{1})^3=(1+1)^3=2^3=8$ whereas $a+b$ becomes $1+1=2$. Since $8\ne 2$, then $(a^{1/3}+b^{1/3})^3\ne a+b$. Answers will vary.

131. $x^{3/2}+x^{1/2}=x^{1/2}\cdot x^1+x^{1/2}$
$= x^{1/2}(x+1)$

133. $y^{1/3}-y^{7/3}=y^{1/3}-y^{1/3}y^2$
$= y^{1/3}(1-y^2)$
$= y^{1/3}(1-y)(1+y)$

135. $y^{-2/5}+y^{8/5}=y^{-2/5}+y^{-2/5}y^2$
$= y^{-2/5}(1+y^2)$
$= \dfrac{1+y^2}{y^{2/5}}$

137. a. $B(t)=2^{10}\cdot 2^t$
$B(0)=2^{10}\cdot 2^0$
$= 2^{10}\cdot 1$
$= 2^{10}$
Initially, there are 2^{10} or 1024 bacteria.

b. $B\left(\dfrac{1}{2}\right)=2^{10}\cdot 2^{1/2}$
$= 2^{10}\sqrt{2}$
≈ 1448.15
After $\dfrac{1}{2}$ hour there are about 1448 bacteria.

139. $A(t)=2.69t^{3/2}$

a. $t=200-1993=7$
$A(7)=2.69(7)^{3/2}$
≈ 49.82
In 2000, there was about $49.82 billion in total assets in the U.S. in 401(k) plans.

b. $t=2009-1993=16$
$A(16)=2.69(16)^{3/2}$
$=172.16$
In 2009, there will be about $172.16 billion in total assets in the U.S. in 401(k) plans.

141. $(3^{\sqrt{2}})^{\sqrt{2}}=3^{\sqrt{2}\cdot\sqrt{2}}=3^2=9$

143. $f(x) = (x-7)^{1/2}(x+3)^{-1/2}$

$$= \frac{(x-7)^{1/2}}{(x+3)^{1/2}}$$

$$= \frac{\sqrt{x-7}}{\sqrt{x+3}}$$

The denominator must be greater than zero.

$\sqrt{x+3} > 0$

$x > -3$

The numerator must be greater than or equal to zero.

$\sqrt{x-7} \geq 0$

$x \geq 7$

Therefore, the domain is $\{x \mid x \geq 7\}$.

145. a. If n is even: $\sqrt[n]{(x-6)^{2n}} = \sqrt[n]{\left((x-6)^2\right)^n}$

$$= \left|(x-6)^2\right|$$

$$= (x-6)^2$$

b. If n is odd: $\sqrt[n]{(x-6)^{2n}} = \sqrt[n]{\left((x-6)^2\right)^n}$

$$= (x-6)^2$$

147. Let a be the unknown index in the shaded area.

$$\sqrt[4]{\sqrt[5]{\sqrt[a]{\sqrt[3]{z}}}} = z^{1/120}$$

$$\left(\left(\left(z^{1/3}\right)^{1/a}\right)^{1/5}\right)^{1/4} = z^{1/120}$$

$$z^{1/60a} = z^{1/120}$$

$$\frac{1}{60a} = \frac{1}{120} \leftarrow \text{Equate exponents}$$

$$60a = 120$$

$$a = 2$$

149. a. The graph is a relation but not a function because it fails the vertical line test.

b. The graph is a relation but not a function because it fails the vertical line test.

c. The graph is both a relation and a function. It passes the vertical line test.

150. $\dfrac{a^{-2}+ab^{-1}}{ab^{-2}-a^{-2}b^{-1}} = \dfrac{\frac{1}{a^2}+\frac{a}{b}}{\frac{a}{b^2}-\frac{1}{a^2b}}$

$$= \frac{a^2b^2\left(\frac{1}{a^2}\right)+a^2b^2\left(\frac{a}{b}\right)}{a^2b^2\left(\frac{a}{b^2}\right)-a^2b^2\left(\frac{1}{a^2b}\right)}$$

$$= \frac{b^2+a^3b}{a^3-b}$$

151. $\dfrac{3x-2}{x+4} = \dfrac{2x+1}{3x-2}$

$(3x-2)(3x-2) = (2x+1)(x+4)$

$9x^2-12x+4 = 2x^2+9x+4$

$7x^2-21x = 0$

$7x(x-3) = 0$

$7x = 0$ or $x-3 = 0$

$x = 0$ $\quad$ $x = 3$

The solutions are 0 and 3.

152. Let y be the speed of the plane in still air. The table is

	d	r	$t=\frac{d}{r}$
With wind	560	$y+25$	$\frac{560}{y+25}$
Against wind	500	$y-25$	$\frac{500}{y-25}$

Since the time is the same for both parts of the trip. The equation is

$$\frac{560}{y+25} = \frac{500}{y-25}$$

$$560(y-25) = 500(y+25)$$

$$560y - 14{,}000 = 500y + 12{,}500$$

$$560y = 500y + 26{,}500$$

$$60y = 26{,}500$$

$$y = \frac{26{,}500}{60}$$

$$\approx 441.67 \text{ mph}$$

The speed of the plane in still air is about 441.67 mph.

Exercise Set 7.3

1. a. Square the natural numbers.

b. $1^2 = 1, 2^2 = 4, 3^2 = 9,$
$4^2 = 16, 5^2 = 25, 6^2 = 36$

3. a. Raise natural numbers to the fifth power.

b. $1^5 = 1, 2^5 = 32, 3^5 = 243,$
$4^5 = 1024, 5^5 = 3125$

5. If n is even and a is negative, $\sqrt[n]{a}$ is not a real number and the rule does not apply. Similarly for the case when b is negative.

7. If n is even and a is negative, $\sqrt[n]{a}$ is not a real number and the rule does not apply. Similarly for the case when b is negative.

9. $\sqrt{8} = \sqrt{4 \cdot 2} = \sqrt{4}\sqrt{2} = 2\sqrt{2}$

11. $\sqrt{24} = \sqrt{4 \cdot 6} = 2\sqrt{6}$

13. $\sqrt{32} = \sqrt{16 \cdot 2} = \sqrt{16}\sqrt{2} = 4\sqrt{2}$

15. $\sqrt{50} = \sqrt{25 \cdot 2} = \sqrt{25}\sqrt{2} = 5\sqrt{2}$

17. $\sqrt{75} = \sqrt{25 \cdot 3} = \sqrt{25}\sqrt{3} = 5\sqrt{3}$

19. $\sqrt{40} = \sqrt{4 \cdot 10} = \sqrt{4}\sqrt{10} = 2\sqrt{10}$

21. $\sqrt[3]{16} = \sqrt[3]{8 \cdot 2} = \sqrt[3]{8}\sqrt[3]{2} = 2\sqrt[3]{2}$

23. $\sqrt[3]{54} = \sqrt[3]{27 \cdot 2} = \sqrt[3]{27}\sqrt[3]{2} = 3\sqrt[3]{2}$

25. $\sqrt[3]{32} = \sqrt[3]{8 \cdot 4} = \sqrt[3]{8}\sqrt[3]{4} = 2\sqrt[3]{4}$

27. $\sqrt[3]{40} = \sqrt[3]{8 \cdot 5} = \sqrt[3]{8}\sqrt[3]{5} = 2\sqrt[3]{5}$

29. $\sqrt[4]{48} = \sqrt[4]{16 \cdot 3} = \sqrt[4]{16}\,\sqrt[4]{3} = 2\sqrt[4]{3}$

31. $-\sqrt[5]{64} = -\sqrt[5]{32 \cdot 2} = -\sqrt[5]{32}\,\sqrt[5]{2} = -2\sqrt[5]{2}$

33. $\sqrt[3]{b^9} = b^3$

35. $\sqrt[3]{x^6} = x^2$

37. $\sqrt{x^3} = \sqrt{x^2 \cdot x} = \sqrt{x^2}\sqrt{x} = x\sqrt{x}$

39. $\sqrt{a^{11}} = \sqrt{a^{10} \cdot a} = \sqrt{a^{10}}\sqrt{a} = a^5\sqrt{a}$

41. $8\sqrt[3]{z^{32}} = 8\sqrt[3]{z^{30} \cdot z^2} = 8\sqrt[3]{z^{30}}\,\sqrt[3]{z^2} = 8z^{10}\sqrt[3]{z^2}$

43. $\sqrt[4]{b^{23}} = \sqrt[4]{b^{20} \cdot b^3} = \sqrt[4]{b^{20}}\,\sqrt[4]{b^3} = b^5\sqrt[4]{b^3}$

45. $\sqrt[6]{x^9} = \sqrt[6]{x^6 \cdot x^3} = \sqrt[6]{x^6}\,\sqrt[6]{x^3} = x\,\sqrt[6]{x^3}$ or $x\sqrt{x}$

47. $3\sqrt[5]{y^{23}} = 3\sqrt[5]{y^{20} \cdot y^3} = 3\sqrt[5]{y^{20}}\,\sqrt[5]{y^3} = 3y^4\,\sqrt[5]{y^3}$

49.
$$\begin{aligned} 2\sqrt{50y^9} &= 2\sqrt{25 \cdot 2 \cdot y^8 \cdot y} \\ &= 2\sqrt{25y^8 \cdot 2y} \\ &= 2\sqrt{25y^8}\sqrt{2y} \\ &= 10y^4\sqrt{2y} \end{aligned}$$

51.
$$\begin{aligned} \sqrt[3]{x^3y^7} &= \sqrt[3]{x^3 \cdot y^6 \cdot y} \\ &= \sqrt[3]{x^3y^6 \cdot y} \\ &= \sqrt[3]{x^3y^6}\,\sqrt[3]{y} \\ &= xy^2\sqrt[3]{y} \end{aligned}$$

53.
$$\begin{aligned} \sqrt[5]{a^6b^{23}} &= \sqrt[5]{a^5b^{20} \cdot ab^3} \\ &= \sqrt[5]{a^5b^{20}} \cdot \sqrt[5]{ab^3} \\ &= ab^4\,\sqrt[5]{ab^3} \end{aligned}$$

55.
$$\begin{aligned} \sqrt{24x^{15}y^{20}z^{27}} &= \sqrt{4 \cdot 6 \cdot x^{14} \cdot x \cdot y^{20} \cdot z^{26} \cdot z} \\ &= \sqrt{4x^{14}y^{20}z^{26} \cdot 6xz} \\ &= \sqrt{4x^{14}y^{20}z^{26}}\,\sqrt{6xz} \\ &= 2x^7y^{10}z^{13}\sqrt{6xz} \end{aligned}$$

57.
$$\begin{aligned} \sqrt[3]{81a^6b^8} &= \sqrt[3]{27 \cdot 3 \cdot a^6 \cdot b^6 \cdot b^2} \\ &= \sqrt[3]{27a^6b^6 \cdot 3b^2} \\ &= \sqrt[3]{27a^6b^6}\,\sqrt[3]{3b^2} \\ &= 3a^2b^2\sqrt[3]{3b^2} \end{aligned}$$

59.
$$\begin{aligned} \sqrt[4]{32x^8y^9z^{19}} &= \sqrt[4]{16 \cdot 2 \cdot x^8 \cdot y^8 \cdot y \cdot z^{16} \cdot z^3} \\ &= \sqrt[4]{16x^8y^8z^{16} \cdot 2yz^3} \\ &= \sqrt[4]{16x^8y^8z^{16}}\,\sqrt[4]{2yz^3} \\ &= 2x^2y^2z^4\sqrt[4]{2yz^3} \end{aligned}$$

61. $\sqrt[4]{81a^8b^9} = \sqrt[4]{81 \cdot a^8 \cdot b^8 \cdot b}$
$= \sqrt[4]{81a^8b^8 \cdot b}$
$= \sqrt[4]{81a^8b^8}\sqrt[4]{b}$
$= 3a^2b^2\sqrt[4]{b}$

63. $\sqrt[5]{32a^{10}b^{12}} = \sqrt[5]{32 \cdot a^{10} \cdot b^{10} \cdot b^2}$
$= \sqrt[5]{32a^{10}b^{10} \cdot b^2}$
$= \sqrt[5]{32a^{10}b^{10}}\sqrt[5]{b^2}$
$= 2a^2b^2\sqrt[5]{b^2}$

65. $\sqrt{\frac{75}{3}} = \sqrt{25} = 5$

67. $\sqrt{\frac{81}{100}} = \frac{\sqrt{81}}{\sqrt{100}} = \frac{9}{10}$

69. $\frac{\sqrt{27}}{\sqrt{3}} = \sqrt{\frac{27}{3}} = \sqrt{9} = 3$

71. $\frac{\sqrt{3}}{\sqrt{48}} = \sqrt{\frac{3}{48}} = \sqrt{\frac{1}{16}} = \frac{1}{4}$

73. $\sqrt[3]{\frac{3}{24}} = \sqrt[3]{\frac{1}{8}} = \frac{\sqrt[3]{1}}{\sqrt[3]{8}} = \frac{1}{2}$

75. $\frac{\sqrt[3]{3}}{\sqrt[3]{81}} = \sqrt[3]{\frac{3}{81}} = \sqrt[3]{\frac{1}{27}} = \frac{1}{3}$

77. $\sqrt[4]{\frac{3}{48}} = \sqrt[4]{\frac{1}{16}} = \frac{1}{2}$

79. $\sqrt[5]{\frac{96}{3}} = \sqrt[5]{32} = 2$

81. $\sqrt{\frac{r^4}{4}} = \frac{\sqrt{r^4}}{\sqrt{4}} = \frac{r^2}{2}$

83. $\sqrt{\frac{16x^4}{25y^{10}}} = \frac{\sqrt{16x^4}}{\sqrt{25y^{10}}} = \frac{4x^2}{5y^5}$

85. $\sqrt[3]{\frac{c^6}{64}} = \frac{\sqrt[3]{c^6}}{\sqrt[3]{64}} = \frac{c^2}{4}$

87. $\sqrt[3]{\frac{a^8b^{12}}{b^{-8}}} = \sqrt[3]{a^8b^{12+8}}$
$= \sqrt[3]{a^8b^{20}}$
$= \sqrt[3]{a^6b^{18} \cdot a^2b^2}$
$= \sqrt[3]{a^6b^{18}} \cdot \sqrt[3]{a^2b^2}$
$= a^2b^6\sqrt[3]{a^2b^2}$

89. $\frac{\sqrt{24}}{\sqrt{3}} = \sqrt{\frac{24}{3}} = \sqrt{8} = \sqrt{4 \cdot 2} = 2\sqrt{2}$

91. $\frac{\sqrt{27x^6}}{\sqrt{3x^2}} = \sqrt{\frac{27x^6}{3x^2}} = \sqrt{9x^4} = 3x^2$

93. $\frac{\sqrt{48x^6y^9}}{\sqrt{6x^2y^6}} = \sqrt{\frac{40x^6y^9}{5x^2y^6}}$
$= \sqrt{8x^4y^3}$
$= \sqrt{4x^4y^2 \cdot 2y}$
$= 2x^2y\sqrt{2y}$

95. $\sqrt[3]{\frac{5xy}{8x^{13}}} = \sqrt[3]{\frac{5y}{8x^{12}}} = \frac{\sqrt[3]{5y}}{\sqrt[3]{8x^{12}}} = \frac{\sqrt[3]{5y}}{2x^4}$

97. $\sqrt[3]{\frac{25x^2y^9}{5x^8y^2}} = \sqrt[3]{\frac{5y^7}{x^6}}$
$= \frac{\sqrt[3]{5y^7}}{\sqrt[3]{x^6}}$
$= \frac{\sqrt[3]{y^6 \cdot 5y}}{x^2}$
$= \frac{y^2\sqrt[3]{5y}}{x^2}$

99. $\sqrt[4]{\frac{10x^4y}{81x^{-8}}} = \sqrt[4]{\frac{10x^{12}y}{81}}$
$= \frac{\sqrt[4]{10x^{12}y}}{\sqrt[4]{81}}$
$= \frac{\sqrt[4]{x^{12}}\sqrt[4]{10y}}{\sqrt[4]{81}}$
$= \frac{x^3\sqrt[4]{10y}}{3}$

101. $\sqrt{a\cdot b}=(a\cdot b)^{1/2}=a^{1/2}\cdot b^{1/2}=\sqrt{a}\cdot\sqrt{b}$

103. No, for example $\frac{\sqrt{18}}{\sqrt{2}}=\sqrt{\frac{18}{2}}=\sqrt{9}=3$.

105. **a.** no

b. $\frac{\sqrt[n]{x}}{\sqrt[n]{x}}$ is equal to 1 when $\sqrt[n]{x}$ is a real number and not equal to 0.

106.
$$F=\frac{9}{5}C+32$$
$$F-32=\frac{9}{5}C$$
$$\frac{5}{9}(F-32)=\frac{5}{9}\left(\frac{9}{5}C\right)$$
$$\frac{5}{9}(F-32)=C \text{ or } C=\frac{5}{9}(F-32)$$

107.
$$\left|\frac{2x-4}{5}\right|=12$$
$$\frac{2x-4}{5}=-12 \quad\text{or}\quad \frac{2x-4}{5}=12$$
$$2x-4=-60 \qquad 2x-4=60$$
$$2x=-56 \qquad 2x=64$$
$$x=-28 \qquad x=32$$
The solution is {−28, 32}.

108.
$$\frac{15x^{12}-5x^9+20x^6}{5x^6}=\frac{15x^{12}}{5x^6}-\frac{5x^9}{5x^6}+\frac{20x^6}{5x^6}$$
$$=3x^6-x^3+4$$

109.
$$(x-3)^3+8$$
$$=(x-3)^3+(2)^3$$
$$=((x-3)+2)\left((x-3)^2-(x-3)(2)+(2)^2\right)$$
$$=(x-1)\left(x^2-6x+9-2x+6+4\right)$$
$$=(x-1)\left(x^2-8x+19\right)$$

Exercise Set 7.4

1. Like radicals are radicals with the same radicands and index.

3.
$$\sqrt{3}+3\sqrt{2}\approx 1.732+3(1.414)$$
$$\approx 1.732+4.242$$
$$\approx 5.974 \text{ or } 5.97$$

5. No. To see this, let $a=16$ and $b=9$. Then, the left side is $\sqrt{a}+\sqrt{b}=\sqrt{16}+\sqrt{9}=4+3=7$ whereas the right side is $\sqrt{a+b}=\sqrt{16+9}=\sqrt{25}=5$.

7. $\sqrt{3}-\sqrt{3}=0$

9. $6\sqrt{5}-2\sqrt{5}=4\sqrt{5}$

11. $2\sqrt{3}-2\sqrt{3}-4\sqrt{3}+5=-4\sqrt{3}+5$

13. $2\sqrt[4]{y}-9\sqrt[4]{y}=-7\sqrt[4]{y}$

15. $3\sqrt{5}-\sqrt[3]{x}+6\sqrt{5}+3\sqrt[3]{x}=2\sqrt[3]{x}+9\sqrt{5}$

17. $5\sqrt{x}-8\sqrt{y}+3\sqrt{x}+2\sqrt{y}-\sqrt{x}=7\sqrt{x}-6\sqrt{y}$

19.
$$\sqrt{5}+\sqrt{20}=\sqrt{5}+\sqrt{4}\cdot\sqrt{5}$$
$$=\sqrt{5}+2\sqrt{5}$$
$$=3\sqrt{5}$$

21.
$$-6\sqrt{75}+5\sqrt{125}=-6\sqrt{25}\cdot\sqrt{3}+5\sqrt{25}\cdot\sqrt{5}$$
$$=-6\left(5\sqrt{3}\right)+5\left(5\sqrt{5}\right)$$
$$=-30\sqrt{3}+25\sqrt{5}$$

23.
$$-4\sqrt{90}+3\sqrt{40}+2\sqrt{10}$$
$$=-4\sqrt{9}\cdot\sqrt{10}+3\sqrt{4}\cdot\sqrt{10}+2\sqrt{10}$$
$$=-4\left(3\sqrt{10}\right)+3\left(2\sqrt{10}\right)+2\left(\sqrt{10}\right)$$
$$=-12\sqrt{10}+6\sqrt{10}+2\sqrt{10}$$
$$=-4\sqrt{10}$$

25. $\sqrt{500xy^2} + y\sqrt{320x}$
$= \sqrt{100y^2} \cdot \sqrt{5x} + y\sqrt{64}\sqrt{5x}$
$= 10y\sqrt{5x} + 8y\sqrt{5x}$
$= 18y\sqrt{5x}$

27. $2\sqrt{5x} - 3\sqrt{20x} - 4\sqrt{45x}$
$= 2\sqrt{5x} - 3\sqrt{4} \cdot \sqrt{5x} - 4\sqrt{9} \cdot \sqrt{5x}$
$= 2\sqrt{5x} - 3\left(2\sqrt{5x}\right) - 4\left(3\sqrt{5x}\right)$
$= 2\sqrt{5x} - 6\sqrt{5x} - 12\sqrt{5x}$
$= -16\sqrt{5x}$

29. $3\sqrt{50a^2} - 3\sqrt{72a^2} - 8a\sqrt{18}$
$= 3\sqrt{25a^2} \cdot \sqrt{2} - 3\sqrt{36a^2} \cdot \sqrt{2} - 8a\sqrt{9} \cdot \sqrt{2}$
$= 3\left(5a\sqrt{2}\right) - 3\left(6a\sqrt{2}\right) - 8a\left(3\sqrt{2}\right)$
$= 15a\sqrt{2} - 18a\sqrt{2} - 24a\sqrt{2}$
$= -27a\sqrt{2}$

31. $\sqrt[3]{108} + \sqrt[3]{32} = \sqrt[3]{27} \cdot \sqrt[3]{4} + \sqrt[3]{8} \cdot \sqrt[3]{4}$
$= 3\sqrt[3]{4} + 2\sqrt[3]{4}$
$= 5\sqrt[3]{4}$

33. $\sqrt[3]{27} - 5\sqrt[3]{8} = 3 - 5 \cdot 2 = 3 - 10 = -7$

35. $2\sqrt[3]{a^4b^2} + 4a\sqrt[3]{ab^2} = 2\sqrt[3]{a^3} \cdot \sqrt[3]{ab^2} + 4a\sqrt[3]{ab^2}$
$= 2a\sqrt[3]{ab^2} + 4a\sqrt[3]{ab^2}$
$= 6a\sqrt[3]{ab^2}$

37. $\sqrt{4r^7s^5} + 3r^2\sqrt{r^3s^5} - 2rs\sqrt{r^5s^3}$
$= \sqrt{4r^6s^4} \cdot \sqrt{rs} + 3r^2\sqrt{r^2s^4} \cdot \sqrt{rs} - 2rs\sqrt{r^4s^2} \cdot \sqrt{rs}$
$= 2r^3s^2\sqrt{rs} + 3r^2(rs^2\sqrt{rs}) - 2rs(r^2s\sqrt{rs})$
$= 2r^3s^2\sqrt{rs} + 3r^3s^2\sqrt{rs} - 2r^3s^2\sqrt{rs}$
$= 3r^3s^2\sqrt{rs}$

39. $\sqrt[3]{128x^8y^{10}} - 2x^2y\sqrt[3]{16x^2y^7}$
$= \sqrt[3]{64x^8y^9}\sqrt[3]{2y} - 2x^2y\sqrt[3]{8x^2y^6}\sqrt[3]{2y}$
$= 4x^2y^3\sqrt[3]{2x^2y} - 2x^2y\left(2y^2\sqrt[3]{2x^2y}\right)$
$= 4x^2y^3\sqrt[3]{2x^2y} - 4x^2y^3\sqrt[3]{2x^2y}$
$= 0$

41. $\sqrt{3}\sqrt{27} = \sqrt{3 \cdot 27} = \sqrt{81} = 9$

43. $\sqrt[3]{4}\sqrt[3]{14} = \sqrt[3]{56} = \sqrt[3]{8 \cdot 7} = 2\sqrt[3]{7}$

45. $\sqrt{9m^3n^7}\sqrt{3mn^4} = \sqrt{9m^3n^7 \cdot 3mn^4}$
$= \sqrt{27m^4n^{11}}$
$= \sqrt{9 \cdot 3 \cdot m^4 \cdot n^{10} \cdot n}$
$= \sqrt{9m^4n^{10} \cdot 3n}$
$= \sqrt{9m^4n^{10}}\sqrt{3n}$
$= 3m^2n^5\sqrt{3n}$

47. $\sqrt[3]{9x^7y^{10}}\sqrt[3]{6x^4y^3} = \sqrt[3]{9x^7y^{10} \cdot 6x^4y^3}$
$= \sqrt[3]{54x^{11}y^{13}}$
$= \sqrt[3]{27 \cdot 2 \cdot x^9 \cdot x^2 \cdot y^{12} \cdot y}$
$= \sqrt[3]{27x^9y^{12} \cdot 2x^2y}$
$= \sqrt[3]{27x^9y^{12}}\sqrt[3]{2x^2y}$
$= 3x^3y^4\sqrt[3]{2x^2y}$

49. $\sqrt[5]{x^{24}y^{30}z^9}\sqrt[5]{x^{13}y^8z^7}$
$= \sqrt[4]{x^{24}y^{30}z^9 \cdot x^{13}y^8z^7}$
$= \sqrt[5]{x^{37}y^{38}z^{16}}$
$= \sqrt[5]{x^{35} \cdot x^2 \cdot y^{35} \cdot y^3 \cdot z^{15} \cdot z}$
$= \sqrt[5]{x^{35}y^{35}z^{15} \cdot x^2y^3z}$
$= \sqrt[5]{x^{35}y^{35}z^{15}}\sqrt[5]{x^2y^3z}$
$= x^7y^7z^3\sqrt[5]{x^2y^3z}$

51. $\left(\sqrt[3]{2x^3y^4}\right)^2 = \sqrt[3]{(2x^3y^4)^2}$
$= \sqrt[3]{4x^6y^8}$
$= \sqrt[3]{4 \cdot x^6 \cdot y^6 \cdot y^2}$
$= \sqrt[3]{x^6y^6 \cdot 4y^2}$
$= \sqrt[3]{x^6y^6}\sqrt[3]{4y^2}$
$= x^2y^2\sqrt[3]{4y^2}$

53. $\sqrt{5}\left(\sqrt{5}-\sqrt{3}\right)=\left(\sqrt{5}\right)\left(\sqrt{5}\right)-\left(\sqrt{5}\right)\left(\sqrt{3}\right)$
$=\sqrt{25}-\sqrt{15}$
$=5-\sqrt{15}$

55. $\sqrt[3]{y}\left(2\sqrt[3]{y}-\sqrt[3]{y^8}\right)$
$=\left(\sqrt[3]{y}\right)\left(2\sqrt[3]{y}\right)-\left(\sqrt[3]{y}\right)\left(\sqrt[3]{y^8}\right)$
$=2\sqrt[3]{y^2}-\sqrt[3]{y^9}$
$=2\sqrt[3]{y^2}-y^3$

57. $2\sqrt[3]{x^4y^5}\left(\sqrt[3]{8x^{12}y^4}+\sqrt[3]{16xy^9}\right)$
$=\left(2\sqrt[3]{x^4y^5}\right)\left(\sqrt[3]{8x^{12}y^4}\right)+\left(2\sqrt[3]{x^4y^5}\right)\left(\sqrt[3]{16xy^9}\right)$
$=2\sqrt[3]{8x^{16}y^9}+2\sqrt[3]{16x^5y^{14}}$
$=2\sqrt[3]{8x^{15}y^9}\sqrt[3]{x}+2\sqrt[3]{8x^3y^{12}}\sqrt[3]{2x^2y^2}$
$=2\cdot 2x^5y^3\sqrt[3]{x}+2\cdot 2xy^4\sqrt[3]{2x^2y^2}$
$=4x^5y^3\sqrt[3]{x}+4xy^4\sqrt[3]{2x^2y^2}$

59. $(8+\sqrt{5})(8-\sqrt{5})=8^2-(\sqrt{5})^2$
$=64-5$
$=59$

61. $\left(\sqrt{6}+x\right)\left(\sqrt{6}-x\right)=\left(\sqrt{6}\right)^2-x^2$
$=6-x^2$

63. $\left(\sqrt{7}-\sqrt{z}\right)\left(\sqrt{7}+\sqrt{z}\right)=\left(\sqrt{7}\right)^2-\left(\sqrt{z}\right)^2$
$=7-z$

65. $\left(\sqrt{3}+4\right)\left(\sqrt{3}+5\right)=\sqrt{9}+5\sqrt{3}+4\sqrt{3}+20$
$=3+9\sqrt{3}+20$
$=23+9\sqrt{3}$

67. $\left(3-\sqrt{2}\right)\left(4-\sqrt{8}\right)=12-3\sqrt{8}-4\sqrt{2}+\sqrt{16}$
$=12-3\cdot 2\sqrt{2}-4\sqrt{2}+4$
$=12-6\sqrt{2}-4\sqrt{2}+4$
$=16-10\sqrt{2}$

69. $\left(4\sqrt{3}+\sqrt{2}\right)\left(\sqrt{3}-\sqrt{2}\right)$
$=4\sqrt{9}-4\sqrt{6}+\sqrt{6}-\sqrt{4}$
$=4\cdot 3-3\sqrt{6}-2$
$=12-3\sqrt{6}-2$
$=10-3\sqrt{6}$

71. $\left(2\sqrt{5}-3\right)^2=\left(2\sqrt{5}-3\right)\left(2\sqrt{5}-3\right)$
$=4\sqrt{25}-6\sqrt{5}-6\sqrt{5}+9$
$=4\cdot 5-12\sqrt{5}+9$
$=20+9-12\sqrt{5}$
$=29-12\sqrt{5}$

73. $\left(2\sqrt{3x}-\sqrt{y}\right)\left(3\sqrt{3x}+\sqrt{y}\right)$
$=6\left(\sqrt{3x}\right)^2+2\sqrt{3x}\sqrt{y}-3\sqrt{3x}\sqrt{y}-\left(\sqrt{y}\right)^2$
$=6(3x)+2\sqrt{3xy}-3\sqrt{3xy}-y$
$=18x-\sqrt{3xy}-y$

75. $\left(\sqrt[3]{4}-\sqrt[3]{6}\right)\left(\sqrt[3]{2}-\sqrt[3]{36}\right)$
$=\sqrt[3]{4}\sqrt[3]{2}-\sqrt[3]{4}\sqrt[3]{36}-\sqrt[3]{6}\sqrt[3]{2}+\sqrt[3]{6}\sqrt[3]{36}$
$=\sqrt[3]{8}-\sqrt[3]{144}-\sqrt[3]{12}+\sqrt[3]{216}$
$=2-2\sqrt[3]{18}-\sqrt[3]{12}+6$
$=8-2\sqrt[3]{18}-\sqrt[3]{12}$

77. $(f\cdot g)(x)=f(x)\cdot g(x)$
$=\sqrt{2x}(\sqrt{8x}-\sqrt{32})$
$=\sqrt{2x}\cdot\sqrt{8x}-\sqrt{2x}\cdot\sqrt{32}$
$=\sqrt{16x^2}-\sqrt{64x}$
$=4x-8\sqrt{x}$

79. $(f\cdot g)(x)=f(x)\cdot g(x)$
$=\sqrt[3]{x}\left(\sqrt[3]{x^5}+\sqrt[3]{x^4}\right)$
$=\sqrt[3]{x}\,\sqrt[3]{x^5}+\sqrt[3]{x}\,\sqrt[3]{x^4}$
$=\sqrt[3]{x^6}+\sqrt[3]{x^5}$
$=\sqrt[3]{x^6}+\sqrt[3]{x^3\cdot x^2}$
$=x^2+x\sqrt[3]{x^2}$

81. $(f \cdot g)(x) = f(x) \cdot g(x)$
$= \sqrt[4]{3x^2}\left(\sqrt[4]{9x^4} - \sqrt[4]{x^7}\right)$
$= \sqrt[4]{3x^2}\,\sqrt[4]{9x^4} - \sqrt[4]{3x^2}\,\sqrt[4]{x^7}$
$= \sqrt[4]{27x^6} - \sqrt[4]{3x^9}$
$= \sqrt[4]{x^4}\,\sqrt[4]{27x^2} - \sqrt[4]{x^8}\,\sqrt[4]{3x}$
$= x\sqrt[4]{27x^2} - x^2\sqrt[4]{3x}$

83. $\sqrt{24} = \sqrt{4 \cdot 6} = 2\sqrt{6}$

85. $\sqrt{125} - \sqrt{20} = \sqrt{25 \cdot 5} - \sqrt{4 \cdot 5}$
$= 5\sqrt{5} - 2\sqrt{5}$
$= 3\sqrt{5}$

87. $\left(3\sqrt{2} - 4\right)\left(\sqrt{2} + 5\right)$
$= 3\left(\sqrt{2}\right)^2 + 15\sqrt{2} - 4\sqrt{2} - 20$
$= 6 + 11\sqrt{2} - 20$
$= -14 + 11\sqrt{2}$

89. $\sqrt{6}\left(5 - \sqrt{2}\right) = \sqrt{6} \cdot 5 - \sqrt{6} \cdot \sqrt{2}$
$= 5\sqrt{6} - \sqrt{12}$
$= 5\sqrt{6} - 2\sqrt{3}$

91. $\sqrt{150}\sqrt{3} = \sqrt{450} = \sqrt{225 \cdot 2} = 15\sqrt{2}$

93. $\sqrt[3]{80x^{11}} = \sqrt[3]{8x^9 \cdot 10x^2} = 2x^3\sqrt[3]{10x^2}$

95. $\sqrt[6]{128ab^{17}c^9} = \sqrt[6]{64b^{12}c^6 \cdot 2ab^5c^3}$
$= 2b^2c\sqrt[6]{2ab^5c^3}$

97. $2b\sqrt[4]{a^4b} + ab\sqrt[4]{16b} = 2b\sqrt[4]{a^4} \cdot \sqrt[4]{b} + ab\sqrt[4]{16} \cdot \sqrt[4]{b}$
$= 2b\left(a\sqrt[4]{b}\right) + ab\left(2\sqrt[4]{b}\right)$
$= 2ab\sqrt[4]{b} + 2ab\sqrt[4]{b}$
$= 4ab\sqrt[4]{b}$

99. $\left(\sqrt[3]{x^2} - \sqrt[3]{y}\right)\left(\sqrt[3]{x} - 2\sqrt[3]{y^2}\right)$
$= \sqrt[3]{x^2}\,\sqrt[3]{x} - 2\sqrt[3]{x^2}\,\sqrt[3]{y^2} - \sqrt[3]{y}\,\sqrt[3]{x} + 2\sqrt[3]{y}\,\sqrt[3]{y^2}$
$= \sqrt[3]{x^3} - 2\sqrt[3]{x^2y^2} - \sqrt[3]{xy} + 2\sqrt[3]{y^3}$
$= x - 2\sqrt[3]{x^2y^2} - \sqrt[3]{xy} + 2y$

101. $\sqrt[3]{3ab^2}\left(\sqrt[3]{4a^4b^3} - \sqrt[3]{8a^5b^4}\right)$
$= \left(\sqrt[3]{3ab^2}\right)\left(\sqrt[3]{4a^4b^3}\right) - \left(\sqrt[3]{3ab^2}\right)\left(\sqrt[3]{8a^5b^4}\right)$
$= \sqrt[3]{12a^5b^5} - \sqrt[3]{24a^6b^6}$
$= \sqrt[3]{a^3b^3 \cdot 12a^2b^2} - \sqrt[3]{8a^6b^6 \cdot 3}$
$= ab\sqrt[3]{12a^2b^2} - 2a^2b^2\sqrt[3]{3}$

103. $f\left(x\right) = \sqrt{2x - 5}\sqrt{2x - 5} = 2x - 5$

No absolute value needed since $x \geq \dfrac{5}{2}$.

105. $h\left(r\right) = \sqrt{4r^2 - 32r + 64}$
$= \sqrt{4\left(r^2 - 8r + 16\right)}$
$= \sqrt{4\left(r - 4\right)^2}$
$= 2\left|r - 4\right|$

107. Perimeter $= \sqrt{45} + \sqrt{45} + \sqrt{80} + \sqrt{80}$
$= 2\sqrt{45} + 2\sqrt{80}$
$= 2\sqrt{9}\sqrt{5} + 2\sqrt{16}\sqrt{5}$
$= 2\left(3\sqrt{5}\right) + 2\left(4\sqrt{5}\right)$
$= 6\sqrt{5} + 8\sqrt{5}$
$= 14\sqrt{5}$

Area $= \sqrt{45}\sqrt{80}$
$= 3\sqrt{5} \cdot 4\sqrt{5}$
$= 12\left(\sqrt{5}\right)^2$
$= 12 \cdot 5$
$= 60$

109. Perimeter $= \sqrt{245} + \sqrt{180} + \sqrt{80}$
$= \sqrt{49}\sqrt{5} + \sqrt{36}\sqrt{5} + \sqrt{16}\sqrt{5}$
$= 7\sqrt{5} + 6\sqrt{5} + 4\sqrt{5}$
$= 17\sqrt{5}$

Area $= \frac{1}{2}\sqrt{245}\sqrt{45}$
$= \frac{1}{2}\sqrt{49}\sqrt{5}\sqrt{9}\sqrt{5}$
$= \frac{1}{2}\cdot 7\cdot 3\left(\sqrt{5}\right)^2$
$= \frac{21}{2}\cdot 5$
$= 52.5$

111. No, for example $-\sqrt{2} + \sqrt{2} = 0$

113. **a.** $s = \sqrt{30FB}$
$s = \sqrt{30(0.85)(80)} \approx 45.17$
The car's speed was about 45.17 mph.

b. $s = \sqrt{30FB}$
$s = \sqrt{30(0.52)(80)} \approx 35.33$
The car's speed was about 35.33 mph.

115. $f(t) = 3\sqrt{t} + 19$

a. $f(36) = 3\sqrt{36} + 19$
$= 3(6) + 19$
$= 18 + 19$
$= 37$
The length at 36 months is 37 inches.

b. $f(40) = 3\sqrt{40} + 19$
$= 3\sqrt{4}\sqrt{10} + 19$
$= 3\cdot 2\sqrt{10} + 19$
$= 6\sqrt{10} + 19$
≈ 37.97
The length at 40 months is about 37.97 inches.

117. **a.** $f(x) = \sqrt{x}$
$g(x) = 2$
$(f+g)(x) = f(x) + g(x) = \sqrt{x} + 2$

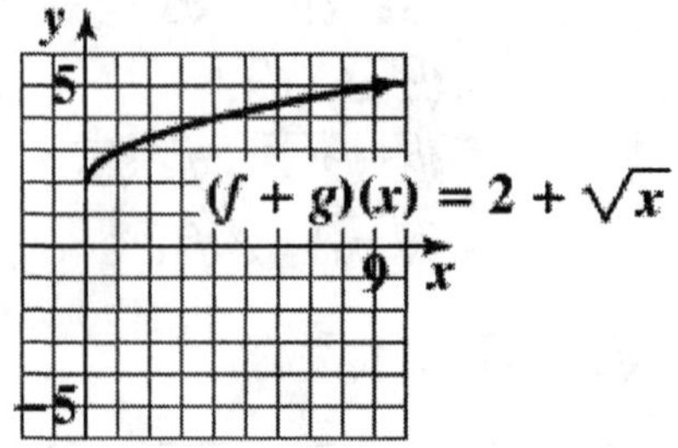

b. It raises the graph 2 units.

119. **a.** $(f-g)(x) = f(x) - g(x)$
$= \sqrt{x} - \left(\sqrt{x} - 2\right)$
$= \sqrt{x} - \sqrt{x} + 2$
$= 2$

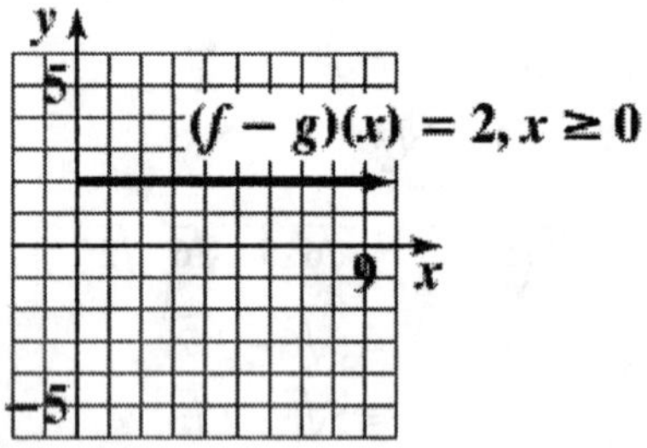

b. $\sqrt{x} \geq 0$, so $x \geq 0$
The domain is $\{x | x \geq 0\}$.

121. $f(x) = \sqrt{x^2}$

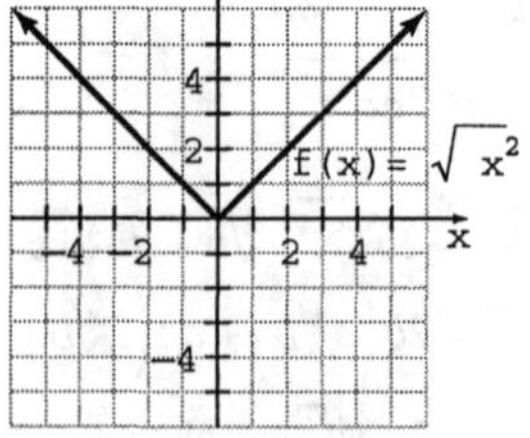

123. A rational number is a quotient of two integers with a nonzero denominator.

124. A real number is a number that can represented on a real number line.

125. An irrational number is a real number that cannot be expressed as the quotient of two integers.

126. $|a| = \begin{cases} a, & a \geq 0 \\ -a, & a < 0 \end{cases}$

127. $E = \frac{1}{2}mv^2$

$2E = 2\left(\frac{1}{2}mv^2\right)$

$2E = mv^2$

$\frac{2E}{v^2} = m$ or $m = \frac{2E}{v^2}$

128. **a.** $-4 < 2x - 3 \leq 7$

$-4 + 3 < 2x - 3 + 3 \leq 7 + 3$

$-1 < 2x \leq 10$

$-\frac{1}{2} < x \leq 5$

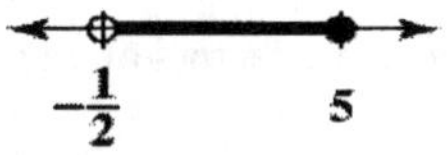

b. $\left(-\frac{1}{2}, 5\right]$

c. $\left\{x \middle| -\frac{1}{2} < x \leq 5\right\}$

Mid-Chapter Test: 7.1 – 7.4

1. $\sqrt{121} = \sqrt{(11)^2} = 11$

2. $\sqrt[3]{-\frac{27}{64}} = \sqrt[3]{-\left(\frac{3}{4}\right)^3} = -\frac{3}{4}$

3. $\sqrt{(-16.3)^2} = |-16.3| = 16.3$

4. $\sqrt{(3a^2 - 4b^3)^2} = |3a^2 - 4b^3|$

5. $g(x) = \frac{x}{8} + \sqrt{4x} - 7$

$g(16) = \frac{16}{8} + \sqrt{4(16)} - 7$

$= 2 + \sqrt{64} - 7$

$= 2 + 8 - 7$

$= 3$

6. $\sqrt[5]{7a^4b^3} = (7a^4b^3)^{1/5}$

7. $-49^{1/2} + 81^{3/4} = -\sqrt{49} + \left(\sqrt[4]{81}\right)^3$

$= -7 + (3)^3$

$= -7 + 27$

$= 20$

8. $\left(\sqrt[4]{a^2b^3c}\right)^{20} = \left[\left(a^2b^3c\right)^{1/4}\right]^{20} = \left(a^2b^3c\right)^{(1/4)\cdot 20}$

$= \left(a^2b^3c\right)^5 = a^{2\cdot 5}b^{3\cdot 5}c^{1\cdot 5}$

$= a^{10}b^{15}c^5$

9. $7x^{-5/2} \cdot 2x^{3/2} = 7 \cdot 2 \cdot x^{-5/2} \cdot x^{3/2}$

$= 14x^{-5/2 + 3/2}$

$= 14x^{-2/2}$

$= 14x^{-1}$

$= \frac{14}{x}$

10. $8x^{-2}\left(x^3 + 2x^{-1/2}\right)$

$= 8x^{-2}\left(x^3\right) + 8x^{-2}\left(2x^{-1/2}\right)$

$= 8x^{-2+3} + 16x^{-2+(-1/2)}$

$= 8x^1 + 16x^{-5/2}$

$= 8x + \frac{16}{x^{5/2}}$

11. $\sqrt{32x^4y^9} = \sqrt{16x^4y^8} \cdot \sqrt{2y}$

$= 4x^2y^4\sqrt{2y}$

12. $\sqrt[6]{64a^{13}b^{23}c^{15}} = \sqrt[6]{64a^{12}b^{18}c^{12}} \cdot \sqrt[6]{ab^5c^3}$

$= 2a^2b^3c^2\sqrt[6]{ab^5c^3}$

13. $\frac{\sqrt[3]{3}}{\sqrt[3]{81}} = \sqrt[3]{\frac{3}{81}} = \sqrt[3]{\frac{1}{27}} = \frac{1}{3}$

14. $\frac{\sqrt{20x^5y^{12}}}{\sqrt{180x^{15}y^7}} = \sqrt{\frac{20x^5y^{12}}{180x^{15}y^7}} = \sqrt{\frac{y^5}{9x^{10}}}$

$= \frac{\sqrt{y^4} \cdot \sqrt{y}}{\sqrt{9x^{10}}} = \frac{y^2\sqrt{y}}{3x^5}$

15. $2\sqrt{x}-3\sqrt{y}+9\sqrt{x}+15\sqrt{y}$
$=2\sqrt{x}+9\sqrt{x}-3\sqrt{y}+15\sqrt{y}$
$=11\sqrt{x}+12\sqrt{y}$

16. $2\sqrt{90x^2y}+3x\sqrt{490y}$
$=2\sqrt{9x^2}\cdot\sqrt{10y}+3x\sqrt{49}\cdot\sqrt{10y}$
$=2(3x)\sqrt{10y}+3x(7)\sqrt{10y}$
$=6x\sqrt{10y}+21x\sqrt{10y}$
$=27x\sqrt{10y}$ (assuming $x\geq 0$)

17. $(x+\sqrt{5})(2x-3\sqrt{5})$
$=x(2x)-x(3\sqrt{5})+\sqrt{5}(2x)-(\sqrt{5})(3\sqrt{5})$
$=2x^2-3x\sqrt{5}+2x\sqrt{5}-3(5)$
$=2x^2-x\sqrt{5}-15$

18. $2\sqrt{3a}\left(\sqrt{27a^2}-5\sqrt{4a}\right)$
$=2\sqrt{3a}\cdot\sqrt{27a^2}-2\sqrt{3a}\cdot 5\sqrt{4a}$
$=2\sqrt{81a^3}-10\sqrt{12a^2}$
$=2\sqrt{81a^2}\cdot\sqrt{a}-10\sqrt{4a^2}\cdot\sqrt{3}$
$=2(9a)\sqrt{a}-10(2a)\sqrt{3}$
$=18a\sqrt{a}-20a\sqrt{3}$

19. $3b\sqrt[4]{a^5b}+2ab\sqrt[4]{16ab}$
$=3b\sqrt[4]{a^4}\cdot\sqrt[4]{ab}+2ab\sqrt[4]{16}\cdot\sqrt[4]{ab}$
$=3b(a)\sqrt[4]{ab}+2ab(2)\sqrt[4]{ab}$
$=3ab\sqrt[4]{ab}+4ab\sqrt[4]{ab}$
$=7ab\sqrt[4]{ab}$ (assuming $a\geq 0$)

20. a. $\sqrt{(x-3)^2}=|x-3|$
The absolute value is needed here because we do not know the domain for *x*. Since the expression $x-3$ is negative for $x<3$, the absolute value is required.

b. $\sqrt{64x^2}=\sqrt{(8x)^2}=8x$
Since $x\geq 0$, we know that $8x\geq 0$. Since $8x$ is never negative, the absolute value is not needed.

Exercise Set 7.5

1. a. The conjugate of a binomial is a binomial with the same two terms as the original but the sign of the second term is changed. The conjugate of $a+b$ is $a-b$. Also, the conjugate of $a-b$ is $a+b$.

b. The conjugate of $x-\sqrt{3}$ is $x+\sqrt{3}$.

3. a. Answers will vary. Possible answer: Multiply the numerator and denominator by a quantity that will result in no radicals in the denominator.

b. $\dfrac{4}{\sqrt{3y}}=\dfrac{4}{\sqrt{3y}}\cdot\dfrac{\sqrt{3y}}{\sqrt{3y}}=\dfrac{4\sqrt{3y}}{\sqrt{9y^2}}=\dfrac{4\sqrt{3y}}{3y}$

5. (1) No perfect powers are factors of any radicand.
(2) No radicand contains fractions.
(3) No radicals are in any denominator.

7. $\dfrac{1}{\sqrt{3}}=\dfrac{1}{\sqrt{3}}\cdot\dfrac{\sqrt{3}}{\sqrt{3}}=\dfrac{\sqrt{3}}{3}$

9. $\dfrac{4}{\sqrt{5}}=\dfrac{4}{\sqrt{5}}\cdot\dfrac{\sqrt{5}}{\sqrt{5}}=\dfrac{4\sqrt{5}}{5}$

11. $\dfrac{6}{\sqrt{6}}=\dfrac{6}{\sqrt{6}}\cdot\dfrac{\sqrt{6}}{\sqrt{6}}=\dfrac{6\sqrt{6}}{6}=\sqrt{6}$

13. $\dfrac{1}{\sqrt{z}}=\dfrac{1}{\sqrt{z}}\cdot\dfrac{\sqrt{z}}{\sqrt{z}}=\dfrac{\sqrt{z}}{z}$

15. $\dfrac{p}{\sqrt{2}}=\dfrac{p}{\sqrt{2}}\cdot\dfrac{\sqrt{2}}{\sqrt{2}}=\dfrac{p\sqrt{2}}{2}$

17. $\dfrac{\sqrt{y}}{\sqrt{7}}=\dfrac{\sqrt{y}}{\sqrt{7}}\cdot\dfrac{\sqrt{7}}{\sqrt{7}}=\dfrac{\sqrt{7y}}{7}$

19. $\dfrac{6\sqrt{3}}{\sqrt{6}}=\dfrac{6\sqrt{3}}{\sqrt{6}}\cdot\dfrac{\sqrt{6}}{\sqrt{6}}=\dfrac{6\sqrt{18}}{6}=\dfrac{6\cdot 3\sqrt{2}}{6}=3\sqrt{2}$

21. $\dfrac{\sqrt{x}}{\sqrt{y}}=\dfrac{\sqrt{x}}{\sqrt{y}}\cdot\dfrac{\sqrt{y}}{\sqrt{y}}=\dfrac{\sqrt{xy}}{y}$

23. $\sqrt{\frac{5m}{8}}=\frac{\sqrt{5m}}{\sqrt{8}}$
$=\frac{\sqrt{5m}}{2\sqrt{2}}$
$=\frac{\sqrt{5m}}{2\sqrt{2}}\cdot\frac{\sqrt{2}}{\sqrt{2}}$
$=\frac{\sqrt{10m}}{2\cdot 2}=\frac{\sqrt{10m}}{4}$

25. $\frac{2n}{\sqrt{18n}}=\frac{2n}{\sqrt{9}\sqrt{2n}}$
$=\frac{2n}{3\sqrt{2n}}$
$=\frac{2n}{3\sqrt{2n}}\cdot\frac{\sqrt{2n}}{\sqrt{2n}}$
$=\frac{2n\sqrt{2n}}{3\cdot 2n}=\frac{\sqrt{2n}}{3}$

27. $\sqrt{\frac{18x^4y^3}{2z^3}}=\sqrt{\frac{9x^4y^3}{z^3}}=\frac{\sqrt{9x^4y^3}}{\sqrt{z^2\cdot z}}$
$=\frac{\sqrt{9x^4y^2}\sqrt{y}}{\sqrt{z^2}\sqrt{z}}=\frac{3x^2y\sqrt{y}}{z\sqrt{z}}$
$=\frac{3x^2y\sqrt{y}}{z\sqrt{z}}\cdot\frac{\sqrt{z}}{\sqrt{z}}$
$=\frac{3x^2y\sqrt{yz}}{z^2}$

29. $\sqrt{\frac{20y^4z^3}{3xy^{-4}}}=\sqrt{\frac{20y^8z^3}{3x}}=\frac{\sqrt{20y^8z^3}}{\sqrt{3x}}$
$=\frac{\sqrt{4y^8z^2}\sqrt{5z}}{\sqrt{3x}}=\frac{2y^4z\sqrt{5z}}{\sqrt{3x}}$
$=\frac{2y^4z\sqrt{5z}}{\sqrt{3x}}\cdot\frac{\sqrt{3x}}{\sqrt{3x}}$
$=\frac{2y^4z\sqrt{15xz}}{3x}$

31. $\sqrt{\frac{48x^6y^5}{3z^3}}=\sqrt{\frac{16x^6y^5}{z^3}}=\frac{\sqrt{16x^6y^5}}{\sqrt{z^2\cdot z}}$
$=\frac{\sqrt{16x^6y^4}\sqrt{y}}{\sqrt{z^2}\sqrt{z}}=\frac{4x^3y^2\sqrt{y}}{z\sqrt{z}}$
$=\frac{4x^3y^2\sqrt{y}}{z\sqrt{z}}\cdot\frac{\sqrt{z}}{\sqrt{z}}$
$=\frac{4x^3y^2\sqrt{yz}}{z^2}$

33. $\frac{1}{\sqrt[3]{2}}=\frac{1}{\sqrt[3]{2}}\cdot\frac{\sqrt[3]{4}}{\sqrt[3]{4}}=\frac{\sqrt[3]{4}}{\sqrt[3]{8}}=\frac{\sqrt[3]{4}}{2}$

35. $\frac{8}{\sqrt[3]{y}}=\frac{8}{\sqrt[3]{y}}\cdot\frac{\sqrt[3]{y^2}}{\sqrt[3]{y^2}}=\frac{8\sqrt[3]{y^2}}{y}$

37. $\frac{1}{\sqrt[4]{3}}=\frac{1}{\sqrt[4]{3}}\cdot\frac{\sqrt[4]{27}}{\sqrt[4]{27}}=\frac{\sqrt[4]{27}}{\sqrt[4]{27}}=\frac{\sqrt[4]{27}}{3}$

39. $\frac{a}{\sqrt[4]{8}}=\frac{a}{\sqrt[4]{8}}\cdot\frac{\sqrt[4]{2}}{\sqrt[4]{2}}=\frac{a\sqrt[4]{2}}{\sqrt[4]{16}}=\frac{a\sqrt[4]{2}}{2}$

41. $\frac{5}{\sqrt[4]{z^2}}=\frac{5}{\sqrt[4]{z^2}}\cdot\frac{\sqrt[4]{z^2}}{\sqrt[4]{z^2}}=\frac{5\sqrt[4]{z^2}}{\sqrt[4]{z^4}}=\frac{5\sqrt[4]{z^2}}{z}$

43. $\frac{10}{\sqrt[5]{y^3}}=\frac{10}{\sqrt[5]{y^3}}\cdot\frac{\sqrt[5]{y^2}}{\sqrt[5]{y^2}}=\frac{10\sqrt[5]{y^2}}{\sqrt[5]{y^5}}=\frac{10\sqrt[5]{y^2}}{y}$

45. $\frac{2}{\sqrt[7]{a^4}}=\frac{2}{\sqrt[7]{a^4}}\cdot\frac{\sqrt[7]{a^3}}{\sqrt[7]{a^3}}=\frac{2\sqrt[7]{a^3}}{\sqrt[7]{a^7}}=\frac{2\sqrt[7]{a^3}}{a}$

47. $\sqrt[3]{\frac{1}{2x}}=\frac{\sqrt[3]{1}}{\sqrt[3]{2x}}=\frac{1}{\sqrt[3]{2x}}\cdot\frac{\sqrt[3]{4x^2}}{\sqrt[3]{4x^2}}=\frac{\sqrt[3]{4x^2}}{2x}$

49. $\frac{5m}{\sqrt[4]{2}}=\frac{5m}{\sqrt[4]{2}}\cdot\frac{\sqrt[4]{2^3}}{\sqrt[4]{2^3}}=\frac{5m\sqrt[4]{8}}{2}$

51. $\sqrt[4]{\frac{5}{3x^3}}=\frac{\sqrt[4]{5}}{\sqrt[4]{3x^3}}\cdot\frac{\sqrt[4]{3^3x}}{\sqrt[4]{3^3x}}=\frac{\sqrt[4]{135x}}{3x}$

53. $\sqrt[3]{\frac{3x^2}{2y^2}}=\frac{\sqrt[3]{3x^2}}{\sqrt[3]{2y^2}}\cdot\frac{\sqrt[3]{4y}}{\sqrt[3]{4y}}=\frac{\sqrt[3]{12x^2y}}{2y}$

55. $\sqrt[3]{\frac{14xy^2}{2z^2}}=\sqrt[3]{\frac{7xy^2}{z^2}}$
$=\frac{\sqrt[3]{7xy^2}}{\sqrt[3]{z^2}}$
$=\frac{\sqrt[3]{7xy^2}}{\sqrt[3]{z^2}}\cdot\frac{\sqrt[3]{z}}{\sqrt[3]{z}}$
$=\frac{\sqrt[3]{7xy^2z}}{z}$

57. $(5-\sqrt{6})(5+\sqrt{6})=5^2-(\sqrt{6})^2$
$=25-6$
$=19$

59. $(8+\sqrt{2})(8-\sqrt{2})=8^2-(\sqrt{2})^2=64-2=62$

61. $(2-\sqrt{10})(2+\sqrt{10})=2^2-(\sqrt{10})^2=4-10=-6$

63. $(\sqrt{a}-\sqrt{b})(\sqrt{a}+\sqrt{b})=(\sqrt{a})^2-(\sqrt{b})^2=a-b$

65. $(2\sqrt{x}-3\sqrt{y})(2\sqrt{x}+3\sqrt{y})=(2\sqrt{x})^2-(3\sqrt{y})^2$
$=4x-9y$

67. $\frac{2}{\sqrt{3}+1}=\frac{2}{\sqrt{3}+1}\cdot\frac{(\sqrt{3}-1)}{(\sqrt{3}-1)}$
$=\frac{2(\sqrt{3}-1)}{(\sqrt{3})^2-1^2}$
$=\frac{2(\sqrt{3}-1)}{3-1}$
$=\frac{2(\sqrt{3}-1)}{2}$
$=\sqrt{3}-1$

69. $\frac{1}{2+\sqrt{3}}=\frac{1}{2+\sqrt{3}}\cdot\frac{(2-\sqrt{3})}{(2-\sqrt{3})}$
$=\frac{2-\sqrt{3}}{2^2-(\sqrt{3})^2}$
$=\frac{2-\sqrt{3}}{4-3}$
$=\frac{2-\sqrt{3}}{1}$
$=2-\sqrt{3}$

71. $\frac{5}{\sqrt{2}-7}=\frac{5}{\sqrt{2}-7}\cdot\frac{(\sqrt{2}+7)}{(\sqrt{2}+7)}$
$=\frac{5\sqrt{2}+35}{2-49}$
$=\frac{5\sqrt{2}+35}{-47}$
$=\frac{-5\sqrt{2}-35}{47}$

73. $\frac{\sqrt{5}}{2\sqrt{5}-\sqrt{6}}=\frac{\sqrt{5}}{2\sqrt{5}-\sqrt{6}}\cdot\frac{(2\sqrt{5}+\sqrt{6})}{(2\sqrt{5}+\sqrt{6})}$
$=\frac{10+\sqrt{30}}{20-6}$
$=\frac{10+\sqrt{30}}{14}$

75. $\frac{3}{6+\sqrt{x}}=\frac{3}{6+\sqrt{x}}\cdot\frac{(6-\sqrt{x})}{(6-\sqrt{x})}=\frac{18-3\sqrt{x}}{36-x}$

77. $\frac{4\sqrt{x}}{\sqrt{x}-y}=\frac{4\sqrt{x}}{\sqrt{x}-y}\cdot\frac{(\sqrt{x}+y)}{(\sqrt{x}+y)}=\frac{4x+4y\sqrt{x}}{x-y^2}$

79. $\frac{\sqrt{2}-2\sqrt{3}}{\sqrt{2}+4\sqrt{3}}=\frac{\sqrt{2}-2\sqrt{3}}{\sqrt{2}+4\sqrt{3}}\cdot\frac{(\sqrt{2}-4\sqrt{3})}{(\sqrt{2}-4\sqrt{3})}$

$=\frac{2-4\sqrt{6}-2\sqrt{6}+8\cdot 3}{2-16\cdot 3}$

$=\frac{26-6\sqrt{6}}{-46}$

$=\frac{-13+3\sqrt{6}}{23}$

81. $\frac{\sqrt{a^3}+\sqrt{a^7}}{\sqrt{a}}=\frac{\sqrt{a^3}+\sqrt{a^7}}{\sqrt{a}}\cdot\frac{(\sqrt{a})}{(\sqrt{a})}$

$=\frac{\sqrt{a^4}+\sqrt{a^8}}{a}$

$=\frac{a^2+a^4}{a}$

$=a+a^3$

83. $\frac{4}{\sqrt{x+2}-3}=\frac{4}{\sqrt{x+2}-3}\cdot\frac{(\sqrt{x+2}+3)}{(\sqrt{x+2}+3)}$

$=\frac{4\sqrt{x+2}+12}{(\sqrt{x+2})^2-3^2}$

$=\frac{4\sqrt{x+2}+12}{x+2-9}$

$=\frac{4\sqrt{x+2}+12}{x-7}$

85. $\sqrt{\frac{x}{16}}=\frac{\sqrt{x}}{\sqrt{16}}=\frac{\sqrt{x}}{4}$

87. $\sqrt{\frac{2}{9}}=\frac{\sqrt{2}}{\sqrt{9}}=\frac{\sqrt{2}}{3}$

89. $(\sqrt{7}+\sqrt{6})(\sqrt{7}-\sqrt{6})=(\sqrt{7})^2-(\sqrt{6})^2$

$=7-6$

$=1$

91. $\sqrt{\frac{24x^3y^6}{5z}}=\frac{\sqrt{24x^3y^6}}{\sqrt{5z}}$

$=\frac{\sqrt{4x^2y^6}\sqrt{6x}}{\sqrt{5z}}$

$=\frac{2xy^3\sqrt{6x}}{\sqrt{5z}}\cdot\frac{\sqrt{5z}}{\sqrt{5z}}$

$=\frac{2xy^3\sqrt{30xz}}{5z}$

93. $\sqrt{\frac{28xy^4}{2x^3y^4}}=\sqrt{\frac{14}{x^2}}=\frac{\sqrt{14}}{\sqrt{x^2}}=\frac{\sqrt{14}}{x}$

95. $\frac{1}{\sqrt{a}+7}=\frac{1}{\sqrt{a}+7}\cdot\frac{\sqrt{a}-7}{\sqrt{a}-7}$

$=\frac{\sqrt{a}-7}{(\sqrt{a}+7)(\sqrt{a}-7)}$

$=\frac{\sqrt{a}-7}{(\sqrt{a})^2-(7)^2}$

$=\frac{\sqrt{a}-7}{a-49}$

97. $-\frac{7\sqrt{x}}{\sqrt{98}}=-\frac{7\sqrt{x}}{7\sqrt{2}}$

$=-\frac{\sqrt{x}}{\sqrt{2}}\cdot\frac{\sqrt{2}}{\sqrt{2}}$

$=-\frac{\sqrt{2x}}{\sqrt{4}}$

$=-\frac{\sqrt{2x}}{2}$

99. $\sqrt[4]{\frac{3y^2}{2x}}=\frac{\sqrt[4]{3y^2}}{\sqrt[4]{2x}}\cdot\frac{\sqrt[4]{8x^3}}{\sqrt[4]{8x^3}}=\frac{\sqrt[4]{24x^3y^2}}{\sqrt[4]{16x^4}}=\frac{\sqrt[4]{24x^3y^2}}{2x}$

101. $\sqrt[3]{\dfrac{32y^{12}z^{10}}{2x}}=\sqrt[3]{\dfrac{16y^{12}z^{10}}{x}}$

$=\dfrac{\sqrt[3]{16y^{12}z^{10}}}{\sqrt[3]{x}}$

$=\dfrac{\sqrt[3]{8y^{12}z^{9}}\sqrt[3]{2z}}{\sqrt[3]{x}}$

$=\dfrac{2y^4z^3\sqrt[3]{2z}}{\sqrt[3]{x}}\cdot\dfrac{\sqrt[3]{x^2}}{\sqrt[3]{x^2}}$

$=\dfrac{2y^4z^3\sqrt[3]{2x^2z}}{\sqrt[3]{x^3}}$

$=\dfrac{2y^4z^3\sqrt[3]{2x^2z}}{x}$

103. $\dfrac{\sqrt{ar}}{\sqrt{a}-2\sqrt{r}}\cdot\dfrac{\left(\sqrt{a}+2\sqrt{r}\right)}{\left(\sqrt{a}+2\sqrt{r}\right)}=\dfrac{\sqrt{ar}\left(\sqrt{a}+2\sqrt{r}\right)}{\left(\sqrt{a}\right)^2-\left(2\sqrt{r}\right)^2}$

$=\dfrac{a\sqrt{r}+2r\sqrt{a}}{a-4r}$

105. $\dfrac{\sqrt[3]{6x}}{\sqrt[3]{5xy}}=\sqrt[3]{\dfrac{6x}{5xy}}$

$=\sqrt[3]{\dfrac{6}{5y}}$

$=\dfrac{\sqrt[3]{6}}{\sqrt[3]{5y}}\cdot\dfrac{\sqrt[3]{25y^2}}{\sqrt[3]{25y^2}}$

$=\dfrac{\sqrt[3]{150y^2}}{5y}$

107. $\sqrt[4]{\dfrac{2x^7y^{12}z^4}{3x^9}}=\sqrt[4]{\dfrac{2y^{12}z^4}{3x^2}}$

$=\dfrac{\sqrt[4]{2y^{12}z^4}}{\sqrt[4]{3x^2}}$

$=\dfrac{\sqrt[4]{y^{12}z^4}\sqrt[4]{2}}{\sqrt[4]{3x^2}}$

$=\dfrac{y^3z\sqrt[4]{2}}{\sqrt[4]{3x^2}}\cdot\dfrac{\sqrt[4]{27x^2}}{\sqrt[4]{27x^2}}$

$=\dfrac{y^3z\sqrt[4]{54x^2}}{\sqrt[4]{81x^4}}$

$=\dfrac{y^3z\sqrt[4]{54x^2}}{3x}$

109. $\dfrac{1}{\sqrt{2}}+\dfrac{\sqrt{2}}{2}=\dfrac{1}{\sqrt{2}}\cdot\dfrac{\sqrt{2}}{\sqrt{2}}+\dfrac{\sqrt{2}}{2}$

$=\dfrac{\sqrt{2}}{\sqrt{4}}+\dfrac{\sqrt{2}}{2}$

$=\dfrac{\sqrt{2}}{2}+\dfrac{\sqrt{2}}{2}$

$=\dfrac{2\sqrt{2}}{2}=\sqrt{2}$

111. $\sqrt{5}-\dfrac{2}{\sqrt{5}}=\sqrt{5}-\dfrac{2}{\sqrt{5}}\cdot\dfrac{\sqrt{5}}{\sqrt{5}}$

$=\sqrt{5}-\dfrac{2\sqrt{5}}{5}$

$=\dfrac{5\sqrt{5}}{5}-\dfrac{2\sqrt{5}}{5}$

$=\dfrac{5\sqrt{5}-2\sqrt{5}}{5}$

$=\dfrac{3\sqrt{5}}{5}$

113. $4\sqrt{\frac{1}{6}}+\sqrt{24}=\frac{4}{\sqrt{6}}+2\sqrt{6}$

$$=\frac{4}{\sqrt{6}}\cdot\frac{\sqrt{6}}{\sqrt{6}}+2\sqrt{6}$$
$$=\frac{4\sqrt{6}}{6}+2\sqrt{6}$$
$$=\frac{2\sqrt{6}}{3}+\frac{(2\sqrt{6})3}{3}$$
$$=\frac{2\sqrt{6}}{3}+\frac{6\sqrt{6}}{3}$$
$$=\frac{8\sqrt{6}}{3}$$

115. $5\sqrt{2}-\frac{2}{\sqrt{8}}+\sqrt{50}=5\sqrt{2}-\frac{2}{\sqrt{8}}\cdot\frac{\sqrt{8}}{\sqrt{8}}+\sqrt{25}\sqrt{2}$

$$=3\sqrt{2}-\frac{2\sqrt{4}\sqrt{2}}{\sqrt{64}}+5\sqrt{2}$$
$$=\frac{5\sqrt{2}}{1}-\frac{4\sqrt{2}}{8}+\frac{5\sqrt{2}}{1}$$
$$=\frac{10\sqrt{2}}{2}-\frac{\sqrt{2}}{2}+\frac{10\sqrt{2}}{2}$$
$$=\frac{(10-1+10)\sqrt{2}}{2}$$
$$=\frac{19\sqrt{2}}{2}$$

117. $\sqrt{\frac{1}{2}}+7\sqrt{2}+\sqrt{18}=\frac{\sqrt{1}}{\sqrt{2}}\cdot\frac{\sqrt{2}}{\sqrt{2}}+7\sqrt{2}+\sqrt{9}\cdot\sqrt{2}$

$$=\frac{\sqrt{2}}{\sqrt{4}}+7\sqrt{2}+3\sqrt{2}$$
$$=\frac{\sqrt{2}}{2}+10\sqrt{2}$$
$$=\frac{\sqrt{2}}{2}+\frac{20\sqrt{2}}{2}$$
$$=\frac{21\sqrt{2}}{2}$$

119. $\frac{2}{\sqrt{50}}-3\sqrt{50}-\frac{1}{\sqrt{8}}$

$$=\frac{2}{5\sqrt{2}}-3(5\sqrt{2})-\frac{1}{2\sqrt{2}}$$
$$=\frac{2}{5\sqrt{2}}\cdot\frac{\sqrt{2}}{\sqrt{2}}-15\sqrt{2}-\frac{1}{2\sqrt{2}}\cdot\frac{\sqrt{2}}{\sqrt{2}}$$
$$=\frac{2\sqrt{2}}{10}-15\sqrt{2}-\frac{\sqrt{2}}{4}$$
$$=\frac{\sqrt{2}}{5}-15\sqrt{2}-\frac{\sqrt{2}}{4}$$
$$=\frac{4\sqrt{2}}{20}-\frac{300\sqrt{2}}{20}-\frac{5\sqrt{2}}{20}$$
$$=-\frac{301\sqrt{2}}{20}$$

121. $\sqrt{\frac{3}{8}}+\sqrt{\frac{3}{2}}=\frac{\sqrt{3}}{\sqrt{8}}+\frac{\sqrt{3}}{\sqrt{2}}$

$$=\frac{\sqrt{3}}{2\sqrt{2}}+\frac{\sqrt{3}}{\sqrt{2}}$$
$$=\frac{\sqrt{3}}{2\sqrt{2}}\cdot\frac{\sqrt{2}}{\sqrt{2}}+\frac{\sqrt{3}}{\sqrt{2}}\cdot\frac{2\sqrt{2}}{2\sqrt{2}}$$
$$=\frac{\sqrt{6}}{4}+\frac{2\sqrt{6}}{4}$$
$$=\frac{3\sqrt{6}}{4}$$

123. $-2\sqrt{\frac{x}{y}}+3\sqrt{\frac{y}{x}}=-2\frac{\sqrt{x}}{\sqrt{y}}+3\frac{\sqrt{y}}{\sqrt{x}}$

$$=-2\frac{\sqrt{x}}{\sqrt{y}}\cdot\frac{\sqrt{y}}{\sqrt{y}}+3\frac{\sqrt{y}}{\sqrt{x}}\cdot\frac{\sqrt{x}}{\sqrt{x}}$$
$$=-2\frac{\sqrt{xy}}{y}+3\frac{\sqrt{xy}}{x}$$
$$=\left(-\frac{2}{y}+\frac{3}{x}\right)\sqrt{xy}$$

125. $\frac{3}{\sqrt{a}}-\sqrt{\frac{9}{a}}+2\sqrt{a}=\frac{3}{\sqrt{a}}-\frac{\sqrt{9}}{\sqrt{a}}+2\sqrt{a}$

$$=\frac{3}{\sqrt{a}}\left(\frac{\sqrt{a}}{\sqrt{a}}\right)-\frac{\sqrt{9}}{\sqrt{a}}\left(\frac{\sqrt{a}}{\sqrt{a}}\right)+2\sqrt{a}$$
$$=\frac{3\sqrt{a}}{a}-\frac{3\sqrt{a}}{a}+2\sqrt{a}$$
$$=2\sqrt{a}$$

127. $\dfrac{\sqrt{(a+b)^4}}{\sqrt[3]{a+b}}=\dfrac{(a+b)^{4/2}}{(a+b)^{1/3}}$

$=(a+b)^{6/3-1/3}$

$=(a+b)^{5/3}$

$=\sqrt[3]{(a+b)^5}$

129. $\dfrac{\sqrt[5]{(a+2b)^4}}{\sqrt[3]{(a+2b)^2}}=\dfrac{(a+2b)^{4/5}}{(a+2b)^{2/3}}$

$=(a+2b)^{4/5-2/3}$

$=(a+2b)^{2/15}$

$=\sqrt[15]{(a+2b)^2}$

131. $\dfrac{\sqrt[3]{r^2s^4}}{\sqrt{rs}}=\dfrac{(r^2s^4)^{1/3}}{(rs)^{1/2}}$

$=\dfrac{r^{2/3}s^{4/3}}{r^{1/2}s^{1/2}}$

$=r^{2/3-1/2}s^{4/3-1/2}$

$=r^{1/6}s^{5/6}$

$=(rs^5)^{1/6}$

$=\sqrt[6]{rs^5}$

133. $\dfrac{\sqrt[5]{x^4y^6}}{\sqrt[3]{(xy)^2}}=\dfrac{(x^4y^6)^{1/5}}{(xy)^{2/3}}$

$=\dfrac{x^{4/5}y^{6/5}}{x^{2/3}y^{2/3}}$

$=x^{4/5-2/3}y^{6/5-2/3}$

$=x^{2/15}y^{8/15}$

$=(x^2y^8)^{1/15}$

$=\sqrt[15]{x^2y^8}$

135. $d=\sqrt{\dfrac{72}{I}}$

$d=\sqrt{\dfrac{72}{5.3}}\approx 3.69$

The person is about 3.69 m from the light source.

137. $r=\sqrt[3]{\dfrac{3V}{4\pi}}$

$r=\sqrt[3]{\dfrac{3(7238.23)}{4\pi}}=12$

The radius of the tank is 12 inches.

139. $N(t)=\dfrac{6.21}{\sqrt[4]{t}}$

a. $t=1960-1959=1$

$N(1)=\dfrac{6.21}{\sqrt[4]{1}}=6.21$

The number of farms in 1960 was 6.21 million.

b. $t=2008-1959=49$

$N(49)=\dfrac{6.21}{\sqrt[4]{49}}\approx 2.35$

The number of farms in 2008 will be about 2.35 million.

141. $\dfrac{2}{\sqrt{2}}=\dfrac{2}{\sqrt{2}}\cdot\dfrac{\sqrt{2}}{\sqrt{2}}=\dfrac{2\sqrt{2}}{\sqrt{4}}=\dfrac{2\sqrt{2}}{2}=\sqrt{2}$

$\dfrac{3}{\sqrt{3}}=\dfrac{3}{\sqrt{3}}\cdot\dfrac{\sqrt{3}}{\sqrt{3}}=\dfrac{3\sqrt{3}}{\sqrt{9}}=\dfrac{3\sqrt{3}}{3}=\sqrt{3}$

Since $3>2$, then $\sqrt{3}>\sqrt{2}$ and we conclude that $\dfrac{3}{\sqrt{3}}>\dfrac{2}{\sqrt{2}}$.

143. $\dfrac{1}{\sqrt{3}+2}=\dfrac{1}{\sqrt{3}+2}\cdot\dfrac{\sqrt{3}-2}{\sqrt{3}-2}$

$=\dfrac{\sqrt{3}-2}{\left(\sqrt{3}\right)^2-2^2}$

$=\dfrac{\sqrt{3}-2}{3-4}$

$=\dfrac{\sqrt{3}-2}{-1}$

$=-\sqrt{3}+2$

$=2-\sqrt{3}$

$2+\sqrt{3}>2-\sqrt{3}$

Therefore, $2+\sqrt{3}>\dfrac{1}{\sqrt{3}+2}$.

145. $f(x) = x^{a/2}$, $g(x) = x^{b/3}$

a. $x^{4/2} = x^2$

$x^{12/2} = x^6$

$x^{8/2} = x^4$

Therefore $x^{a/2}$ is a perfect square when $a = 4, 8, 12$.

b. $x^{9/3} = x^3$

$x^{18/3} = x^6$

$x^{27/3} = x^9$

Therefore, $x^{b/3}$ is a perfect cube when $b = 9, 18, 27$.

c.
$$\begin{aligned}(f \cdot g)(x) &= f(x) \cdot g(x)\\ &= x^{a/2} \cdot x^{b/3}\\ &= x^{a/2+b/3}\\ &= x^{3a/6+2b/6}\\ &= x^{(3a+2b)/6}\end{aligned}$$

d.
$$\begin{aligned}\left(\frac{f}{g}\right)(x) &= \frac{f(x)}{g(x)}\\ &= \frac{x^{a/2}}{x^{b/3}}\\ &= x^{a/2-b/3}\\ &= x^{3a/6-2b/6}\\ &= x^{(3a-2b)/6}\end{aligned}$$

147. $$\frac{3}{\sqrt{2a-3b}} = \frac{3}{\sqrt{2a-3b}} \cdot \frac{\sqrt{2a-3b}}{\sqrt{2a-3b}} = \frac{3\sqrt{2a-3b}}{2a-3b}$$

149.
$$\begin{aligned}\frac{5-\sqrt{5}}{6} &= \frac{5-\sqrt{5}}{6} \cdot \frac{5+\sqrt{5}}{5+\sqrt{5}}\\ &= \frac{25-5}{30+6\sqrt{5}} = \frac{20}{2(15+3\sqrt{5})}\\ &= \frac{10}{15+3\sqrt{5}}\end{aligned}$$

151.
$$\begin{aligned}\frac{\sqrt{x+h}-\sqrt{x}}{h} &= \frac{\sqrt{x+h}-\sqrt{x}}{h} \cdot \frac{\sqrt{x+h}+\sqrt{x}}{\sqrt{x+h}+\sqrt{x}}\\ &= \frac{x+h-x}{h\left(\sqrt{x+h}+\sqrt{x}\right)}\\ &= \frac{h}{h\left(\sqrt{x+h}+\sqrt{x}\right)}\\ &= \frac{1}{\sqrt{x+h}+\sqrt{x}}\end{aligned}$$

154.
$$\begin{aligned}A &= \frac{1}{2}h(b_1+b_2)\\ 2A &= h(b_1+b_2)\\ \frac{2A}{h} &= b_1+b_2\\ \frac{2A}{h} - b_1 &= b_2\\ b_2 &= \frac{2A}{h} - b_1\end{aligned}$$

155. Let r be the rate of the slower car and $r + 10$ be the rate of the faster.
Distance the first traveled plus distance the second traveled is 270 miles.

$$\begin{aligned}3r + 3(r+10) &= 270\\ 3r + 3r + 30 &= 270\\ 6r + 30 &= 270\\ 6r &= 240\\ r &= 40\end{aligned}$$

The rate of the slower car is 40 mph and the rate of the faster car is $r + 10 = 50$ mph.

156.
$$\begin{array}{r} 4x^2 + 9x - 2\\ x - 2\\ \hline -8x^2 - 18x + 4\\ 4x^3 + 9x^2 - 2x\\ \hline 4x^3 + x^2 - 20x + 4 \end{array}$$

157.
$$\begin{aligned}\frac{x}{2} - \frac{4}{x} &= -\frac{7}{2}\\ 2x\left(\frac{x}{2} - \frac{4}{x}\right) &= 2x\left(-\frac{7}{2}\right)\\ x^2 - 8 &= -7x\\ x^2 + 7x - 8 &= 0\\ (x+8)(x-1) &= 0\\ x &= -8 \text{ or } 1\end{aligned}$$

Exercise Set 7.6

1. a. Answers will vary.

b.
$$\sqrt{2x+26}-2=4$$
$$\sqrt{2x+26}-2+2=4+2$$
$$\sqrt{2x+26}=6$$
$$\left(\sqrt{2x+26}\right)^2=6^2$$
$$2x+26=36$$
$$2x=10$$
$$x=\frac{10}{2}$$
$$x=5$$

3. 0 is the only solution to the equation. For all other values, the left side of the equation is negative and the right side is positive.

5. Answers will vary. Possible answer:
The equation has no solution since the left side is a positive number whereas the right side is 0. A positive number is never equal to 0.
Also, the equation can be written as $\sqrt{x-3}=-4$ for which the left side is positive and the right side is negative. It is impossible for $\sqrt{x-3}$ to equal a negative number.

7. One solution, $x = 25$. Note that the domain of x is all nonnegative numbers. Therefore we have
$$\left(\sqrt{x}\right)^2=5^2$$
$$x=25$$

9.
$$\sqrt{x}=4$$
$$\left(\sqrt{x}\right)^2=4^2$$
$$x=16$$

Check: $\sqrt{16}=4$
$4=4$ True

11. $\sqrt{x}=-9$ No real solution. The principal square root is never negative.

13.
$$\sqrt[3]{x}=-4$$
$$\left(\sqrt[3]{x}\right)^3=(-4)^3$$
$$x=-64$$

Check: $\sqrt[3]{-64}=-4$
$\sqrt[3]{(-4)^3}=-4$
$-4=-4$ True

15.
$$\sqrt{2x+3}=5$$
$$\left(\sqrt{2x+3}\right)^2=(5)^2$$
$$2x+3=25$$
$$2x=22$$
$$x=11$$

Check: $\sqrt{2x+3}=5$
$\sqrt{2(11)+3}=5$
$\sqrt{22+3}=5$
$\sqrt{25}=5$
$5=5$ True

17.
$$\sqrt[3]{3x}+4=7$$
$$\sqrt[3]{3x}=3$$
$$\left(\sqrt[3]{3x}\right)^3=(3)^3$$
$$3x=27$$
$$x=9$$

Check: $\sqrt[3]{3(9)}+4=7$
$\sqrt[3]{27}+4=7$
$3+4=7$
$7=7$ True

19.
$$\sqrt[3]{2x+29}=3$$
$$\left(\sqrt[3]{2x+29}\right)^3=3^3$$
$$2x+29=27$$
$$2x=-2$$
$$x=-1$$

Check: $\sqrt[3]{2(-1)+29}=3$
$\sqrt[3]{-2+29}=3$
$\sqrt[3]{27}=3$
$3=3$ True

21. $\sqrt[4]{x}=3$

$\left(\sqrt[4]{x}\right)^4=3^4$

$x=81$

Check: $\sqrt[4]{81}=3$

$\sqrt[4]{3^4}=3$

$3=3$ True

23. $\sqrt[4]{x+10}=3$

$\left(\sqrt[4]{x+10}\right)^4=3^4$

$x+10=81$

$x=71$

Check: $\sqrt[4]{71+10}=3$

$\sqrt[4]{81}=3$

$\sqrt[4]{3^4}=3$

$3=3$ True

25. $\sqrt[4]{2x+1}+6=2$

$\sqrt[4]{2x+1}=-4$

No real solution exists because an even root of a real number is never negative.

27. $\sqrt{x+8}=\sqrt{x-8}$

$\left(\sqrt{x+8}\right)^2=\left(\sqrt{x-8}\right)^2$

$x+8=x-8$

$8=-8$ False

A contradiction results so the equation has no solution.

29. $2\sqrt[3]{x-1}=\sqrt[3]{x^2+2x}$

$\left(2\sqrt[3]{x-1}\right)^3=\left(\sqrt[3]{x^2+2x}\right)^3$

$8(x-1)=x^2+2x$

$8x-8=x^2+2x$

$x^2-6x+8=0$

$(x-4)(x-2)=0$

$x=4$ or $x=2$

A check will show that both 2 and 4 are solutions to the equation.

31. $\sqrt[4]{x+8}=\sqrt[4]{2x}$

$\left(\sqrt[4]{x+8}\right)^4=\left(\sqrt[4]{2x}\right)^4$

$x+8=2x$

$x=8$

A check will show that the solution is 8.

33. $\sqrt{5x+1}-6=0$

$\sqrt{5x+1}=6$

$\left(\sqrt[2]{5x+1}\right)^2=(6)^2$

$5x+1=36$

$5x=35$

$x=7$

A check will show that the solution is 7.

35. $\sqrt{m^2+6m-4}=m$

$\left(\sqrt{m^2+6m-4}\right)^2=(m)^2$

$m^2+6m-4=m^2$

$6m-4=0$

$6m=4$

$m=\frac{2}{3}$

Check: $\sqrt{\left(\frac{2}{3}\right)^2+6\left(\frac{2}{3}\right)-4}=\frac{2}{3}$

$\sqrt{\frac{4}{9}+4-4}=\frac{2}{3}$

$\sqrt{\frac{4}{9}}=\frac{2}{3}$

$\frac{2}{3}=\frac{2}{3}$ True

37. $\sqrt{5c+1}-9=0$

$\sqrt{5c+1}=9$

$\left(\sqrt{5c+1}\right)^2=9^2$

$5c+1=81$

$5c=80$

$c=16$

A check will show that the solution is 16.

39. $\sqrt{z^2+5}=z+1$

$$\left(\sqrt{z^2+5}\right)^2=(z+1)^2$$
$$z^2+5=z^2+2z+1$$
$$5=2z+1$$
$$4=2z$$
$$2=z$$

A check will show that the solution is 2.

41. $\sqrt{2y+5}+5-y=0$

$$\sqrt{2y+5}=y-5$$
$$\left(\sqrt{2y+5}\right)^2=(y-5)^2$$
$$2y+5=y^2-10y+25$$
$$0=y^2-12y+20$$
$$0=(y-10)(y-2)$$
$$y=10 \text{ or } y=2$$

Check: $\sqrt{2y+5}+5-y=0$

$$\sqrt{2(10)+5}+5-(10)=0$$
$$\sqrt{25}-5=0$$
$$0=0 \text{ True}$$

$$\sqrt{2y+5}+5-y=0$$
$$\sqrt{2(2)+5}+5-(2)=0$$
$$\sqrt{9}+3=0$$
$$6=0 \text{ False}$$

This check shows that 10 is the only solution to this equation.

43. $\sqrt{5x+6}=2x-6$

$$\left(\sqrt{5x+6}\right)^2=(2x-6)^2$$
$$5x+6=4x^2-24x+36$$
$$0=4x^2-29x+30$$
$$0=(4x-5)(x-6)$$
$$x=\frac{5}{4} \text{ or } x=6$$

Check:

$$\sqrt{5x+6}=2x-6$$
$$\sqrt{5\left(\frac{5}{4}\right)+6}=2\left(\frac{5}{4}\right)-6$$
$$\sqrt{\frac{49}{4}}=-\frac{14}{4}$$
$$\frac{7}{2}=-\frac{7}{2} \text{ False}$$

$$\sqrt{5x+6}=2x-6$$
$$\sqrt{5(6)+6}=2(6)-6$$
$$\sqrt{36}=12-6$$
$$6=6 \text{ True}$$

This check shows that 6 is the only solution to this equation.

45. $(2a+9)^{1/2}-a+3=0$

$$(2a+9)^{1/2}=a-3$$
$$[(2a+9)^{1/2}]^2=(a-3)^2$$
$$2a+9=a^2-6a+9$$
$$0=a^2-8a$$
$$0=a(a-8)$$
$$a=0 \text{ or } a=8$$

Check: $(2a+9)^{1/2}-a+3=0$

$$(2\cdot 0+9)^{1/2}-0+3=0$$
$$3+3=0$$
$$6=0 \text{ False}$$

Check: $(2a+9)^{1/2}-a+3=0$

$$(2\cdot 8+9)^{1/2}-8+3=0$$
$$5-8+3=0$$
$$0=0 \text{ True}$$

This check shows that 8 is the only solution to the equation.

47. $(2x^2+4x+9)^{1/2}=\sqrt{2x^2+9}$

$$[(2x^2+4x+9)^{1/2}]^2=\left(\sqrt{2x^2+9}\right)^2$$
$$2x^2+4x+9=2x^2+9$$
$$4x=0$$
$$x=0$$

Check: $\left(2(0)^2+4(0)+9\right)^{1/2}=\sqrt{2(0)^2+9}$

$$(9)^{1/2}=\sqrt{9}$$
$$3=3 \text{ True}$$

49. $(r+4)^{1/3}=(3r+10)^{1/3}$

$$[(r+4)^{1/3}]^3=[(3r+10)^{1/3}]^3$$
$$r+4=3r+10$$
$$4=2r+10$$
$$-6=2r$$
$$-3=r$$

Check: $(-3+4)^{1/3}=[3(-3)+10]^{1/3}$

$$(1)^{1/3}=(1)^{1/3}$$
$$1=1 \quad \text{True}$$

51. $(5x+7)^{1/4}=(9x+1)^{1/4}$

$$[(5x+7)^{1/4}]^4=[(9x+1)^{1/4}]^4$$
$$5x+7=9x+1$$
$$7=4x+1$$
$$6=4x$$
$$\frac{3}{2}=x$$

Check: $(5x+7)^{1/4}=(9x+1)^{1/4}$

$$\left(5\cdot\frac{3}{2}+7\right)^{1/4}=\left(9\cdot\frac{3}{2}+1\right)^{1/4}$$
$$\left(\frac{15}{2}+7\right)^{1/4}=\left(\frac{27}{2}+1\right)^{1/4}$$
$$\left(\frac{29}{2}\right)^{1/4}=\left(\frac{29}{2}\right)^{1/4} \quad \text{True}$$

53. $\sqrt[4]{x+5}=-2$

$$\left(\sqrt[4]{x+5}\right)^4=(-2)^4$$
$$x+5=16$$
$$x=11$$

Check: $\sqrt[4]{x+5}=-2$

$$\sqrt[4]{11+5}=-2$$
$$\sqrt[4]{16}=-2$$
$$2=-2 \quad \text{False}$$

Thus, 11 is not a solution to this equation and we conclude that there is no real solution.
Note: We could have determined that there was no real solution immediately by noting that an even root of a real number is never negative.

55. $\sqrt{4x+1}=\sqrt{2x}+1$

$$\left(\sqrt{4x+1}\right)^2=\left(\sqrt{2x}+1\right)^2$$
$$4x+1=\left(\sqrt{2x}\right)^2+2\cdot1\cdot\sqrt{2x}+1^2$$
$$4x+1=2x+2\sqrt{2x}+1$$
$$2x=2\sqrt{2x}$$
$$x=\sqrt{2x}$$
$$x^2=\left(\sqrt{2x}\right)^2$$
$$x^2=2x$$
$$x^2-2x=0$$
$$x(x-2)=0$$
$$x=0 \quad \text{or} \quad x-2=0$$
$$x=2$$

A check will show that both 0 and 2 are solutions to the equation.

57. $\sqrt{3a+1}=\sqrt{a-4}+3$

$$\left(\sqrt{3a+1}\right)^2=\left(\sqrt{a-4}+3\right)^2$$
$$3a+1=\left(\sqrt{a-4}\right)^2+2\cdot3\sqrt{a-4}+3^2$$
$$3a+1=a-4+6\sqrt{a-4}+9$$
$$2a-4=6\sqrt{a-4}$$
$$(2a-4)^2=\left(6\sqrt{a-4}\right)^2$$
$$4a^2-16a+16=36(a-4)$$
$$4a^2-52a+160=0$$
$$4\left(a^2-13a+40\right)=0$$
$$4(a-8)(a-5)=0$$
$$a=8 \text{ or } a=5$$

A check will show that both 5 and 8 are solutions to the equation.

59. $\sqrt{x+3}=\sqrt{x}-3$

$$\left(\sqrt{x+3}\right)^2=\left(\sqrt{x}-3\right)^2$$
$$x+3=x-6\sqrt{x}+9$$
$$-6=-6\sqrt{x}$$
$$1=\sqrt{x}$$
$$(1)^2=\left(\sqrt{x}\right)^2$$
$$1=x$$

Check: $\sqrt{x+3}=\sqrt{x}-3$
$$\sqrt{1+3}=\sqrt{1}-3$$
$$\sqrt{4}=1-3$$
$$2=-2 \quad \text{False}$$

Thus, 1 is not a solution to this equation and we conclude that there is no real solution.

61. $\sqrt{x+7}=6-\sqrt{x-5}$

$$\left(\sqrt{x+7}\right)^2=\left(6-\sqrt{x-5}\right)^2$$
$$x+7=36-12\sqrt{x-5}+x-5$$
$$12\sqrt{x-5}=24$$
$$\sqrt{x-5}=2$$
$$x-5=4$$
$$x=9$$

A check will show that the solution is 9.

63. $\sqrt{4x-3}=2+\sqrt{2x-5}$

$$\left(\sqrt{4x-3}\right)^2=\left(2+\sqrt{2x-5}\right)^2$$
$$4x-3=4+4\sqrt{2x-5}+2x-5$$
$$2x-2=4\sqrt{2x-5}$$
$$x-1=2\sqrt{2x-5}$$
$$(x-1)^2=\left(2\sqrt{2x-5}\right)^2$$
$$x^2-2x+1=4(2x-5)$$
$$x^2-2x+1=8x-20$$
$$x^2-10x+21=0$$
$$(x-7)(x-3)=0$$
$$x-7=0 \quad \text{or} \quad x-3=0$$
$$x=7 \qquad\qquad x=3$$

A check will show that both 3 and 7 are solutions to the equation.

65. $\sqrt{y+1}=\sqrt{y+10}-3$

$$\left(\sqrt{y+1}\right)^2=\left(\sqrt{y+10}-3\right)^2$$
$$y+1=\left(\sqrt{y+10}\right)^2-6\sqrt{y+10}+9$$
$$y+1=y+10-6\sqrt{y+10}+9$$
$$6\sqrt{y+10}=18$$
$$\sqrt{y+10}=3$$
$$\left(\sqrt{y+10}\right)^2=3^2$$
$$y+10=9$$
$$y=-1$$

A check will show that the solution is -1.

67. $f(x)=g(x)$

$$\sqrt{x+8}=\sqrt{2x+1}$$
$$\left(\sqrt{x+8}\right)^2=\left(\sqrt{2x+1}\right)^2$$
$$x+8=2x+1$$
$$7=x$$

A check will show that $x=7$ is the solution to the equation $f(x)=g(x)$.

69. $f(x)=g(x)$

$$\sqrt[3]{5x-19}=\sqrt[3]{6x-23}$$
$$\left(\sqrt[3]{5x-19}\right)^3=\left(\sqrt[3]{6x-23}\right)^3$$
$$5x-19=6x-23$$
$$x=4$$

A check will show that $x=4$ is the solution to the equation $f(x)=g(x)$.

71. $f(x)=g(x)$

$$2(8x+24)^{1/3}=4(2x-2)^{1/3}$$
$$[2(8x+24)^{1/3}]^3=[4(2x-2)^{1/3}]^3$$
$$8(8x+24)=64(2x-2)$$
$$64x+192=128x-128$$
$$64x=320$$
$$x=5$$

A check will show that $x=5$ is the solution to the equation $f(x)=g(x)$.

73. $p=\sqrt{2v}$

$p^2=\left(\sqrt{2v}\right)^2$

$p^2=2v$

$\frac{p^2}{2}=v$ or $v=\frac{p^2}{2}$

75. $v=\sqrt{2gh}$

$v^2=\left(\sqrt{2gh}\right)^2$

$v^2=2gh$

$g=\frac{v^2}{2h}$

77. $v=\sqrt{\frac{FR}{M}}$

$v^2=\left(\sqrt{\frac{FR}{M}}\right)^2$

$v^2=\frac{FR}{M}$

$Mv^2=FR$

$F=\frac{Mv^2}{R}$

79. $x=\sqrt{\frac{m}{k}V_0}$

$x^2=\left(\sqrt{\frac{m}{k}V_0}\right)^2$

$x^2=\frac{mV_0^2}{k}$

$x^2k=mV_0^2$

$m=\frac{x^2k}{V_0^2}$

81. $r=\sqrt{\frac{A}{\pi}}$

$r^2=\left(\sqrt{\frac{A}{\pi}}\right)^2$

$r^2=\frac{A}{\pi}$

$\pi r^2=A$ or $A=\pi r^2$

83. $a^2+b^2=c^2$

$\left(\sqrt{6}\right)^2+9^2=x^2$

$6+81=x^2$

$87=x^2 \Rightarrow x=\sqrt{87}$

85. $a^2+b^2=c^2$

$x^2+5^2=\left(\sqrt{65}\right)^2$

$x^2+25=65$

$x^2=40$

$x=\sqrt{40} \Rightarrow x=2\sqrt{10}$

87. $\sqrt{x+5}-\sqrt{x}=\sqrt{x-3}$

$\left(\sqrt{x+5}-\sqrt{x}\right)^2=\left(\sqrt{x-3}\right)^2$

$x+5-2\sqrt{x(x+5)}+x=x-3$

$x+8=2\sqrt{x^2+5x}$

$(x+8)^2=\left(2\sqrt{x^2+5x}\right)^2$

$x^2+16x+64=4\left(x^2+5x\right)$

$x^2+16x+64=4x^2+20x$

$3x^2+4x-64=0$

$(3x+16)(x-4)=0 \Rightarrow x=-\frac{16}{3}$ or $x=4$

A check will show that $-\frac{16}{3}$ is an extraneous solution. The solution to the equation is 4.

89. $\sqrt{4y+6}+\sqrt{y+5}=\sqrt{y+1}$

$\left(\sqrt{4y+6}+\sqrt{y+5}\right)^2=\left(\sqrt{y+1}\right)^2$

$4y+6+2\sqrt{(4y+6)(y+5)}+y+5=y+1$

$2\sqrt{(4y+6)(y+5)}=-4y-10$

$\left(2\sqrt{4y^2+26y+30}\right)^2=(-4y-10)^2$

$4\left(4y^2+26y+30\right)=16y^2+80y+100$

$16y^2+104y+120=16y^2+80y+100$

$24y=-20 \Rightarrow y=-\frac{5}{6}$

Upon checking, this value does not satisfy the equation. There is no solution.

91. $\sqrt{c+1}+\sqrt{c-2}=\sqrt{3c}$

$\left(\sqrt{c+1}+\sqrt{c-2}\right)^2=\left(\sqrt{3c}\right)^2$

$c+1+2\sqrt{(c+1)(c-2)}+c-2=3c$

$2\sqrt{(c+1)(c-2)}=c+1$

$\left(2\sqrt{c^2-c-2}\right)^2=(c+1)^2$

$4\left(c^2-c-2\right)=c^2+2c+1$

$4c^2-4c-8=c^2+2c+1$

$3c^2-6c-9=0$

$3(c+1)(c-3)=0 \Rightarrow c=-1 \text{ or } c=3$

Upon checking, only $c=3$ satisfies the equation.

93. $\sqrt{a+2}-\sqrt{a-3}=\sqrt{a-6}$

$\left(\sqrt{a+2}-\sqrt{a-3}\right)^2=\left(\sqrt{a-6}\right)^2$

$a+2-2\sqrt{(a+2)(a-3)}+a-3=a-6$

$a+5=2\sqrt{(a+2)(a-3)}$

$(a+5)^2=\left(2\sqrt{a^2-a-6}\right)^2$

$a^2+10a+25=4\left(a^2-a-6\right)$

$a^2+10a+25=4a^2-4a-24$

$3a^2-14a-49=0$

$(3a+7)(a-7)=0 \Rightarrow a=-\frac{7}{3} \text{ or } a=7$

Upon checking, only $a=7$ satisfies the equation.

95. $\sqrt{2-\sqrt{x}}=\sqrt{x}$

$\left(\sqrt{2-\sqrt{x}}\right)^2=\left(\sqrt{x}\right)^2$

$2-\sqrt{x}=x$

$2-x=\sqrt{x}$

$(2-x)^2=\left(\sqrt{x}\right)^2$

$4-4x+x^2=x$

$x^2-5x+4=0$

$(x-4)(x-1)=0 \Rightarrow x=4 \text{ or } x=1$

Upon checking, only $x=1$ satisfies the equation.

97. $\sqrt{2+\sqrt{x+1}}=\sqrt{7-x}$

$\left(\sqrt{2+\sqrt{x+1}}\right)^2=\left(\sqrt{7-x}\right)^2$

$2+\sqrt{x+1}=7-x$

$\sqrt{x+1}=5-x$

$\left(\sqrt{x+1}\right)^2=(5-x)^2$

$x+1=25-10x+x^2$

$x^2-11x+24=0$

$(x-8)(x-3)=0 \Rightarrow x=8 \text{ or } x=3$

Upon checking, only $x=3$ satisfies the equation.

99. $c=\sqrt{90^2+90^2}=\sqrt{2(90)^2}=90\sqrt{2}\approx 127.28$

Second base is about 127.28 feet from home plate.

101. $s=\sqrt{A}$

$s=\sqrt{169}$

$=13$ feet

103. $T = 2\pi\sqrt{\frac{l}{32}}$

a. Let $l = 8$

$$T = 2\pi\sqrt{\frac{l}{32}}$$
$$= 2\pi\sqrt{\frac{8}{32}} = 2\pi\sqrt{\frac{1}{4}}$$
$$= 2\pi\frac{1}{2} = \pi$$
$$\approx 3.14 \text{ seconds}$$

b. Replace l with $2l$:

$$T_D = 2\pi\sqrt{\frac{2l}{32}}$$
$$= 2\pi\sqrt{2}\sqrt{\frac{l}{32}}$$
$$= \sqrt{2}\left(2\pi\sqrt{\frac{l}{32}}\right)$$
$$= \sqrt{2}\cdot T$$

c. This part must be solved in two phases. First, we need to find the length of the pendulum:

$$T = 2\pi\sqrt{\frac{l}{g}}$$
$$2 = 2\pi\sqrt{\frac{l}{32}}$$
$$\frac{1}{\pi} = \sqrt{\frac{l}{32}}$$
$$\left(\frac{1}{\pi}\right)^2 = \frac{l}{32}$$
$$l = \frac{32}{\pi^2}$$

Now, find T using $g = \frac{32}{6}$ and $l = \frac{32}{\pi^2}$

$$T = 2\pi\sqrt{\frac{l}{g}}$$
$$= 2\pi\sqrt{\frac{\frac{32}{\pi^2}}{\frac{32}{6}}} = 2\pi\sqrt{\frac{6}{\pi^2}}$$
$$= 2\pi\frac{\sqrt{6}}{\pi} = 2\sqrt{6}$$
$$\approx 4.90 \text{ seconds}$$

105.
$$r = \sqrt[4]{\frac{8\mu l}{\pi R}}$$
$$r^4 = \left(\sqrt[4]{\frac{8\mu l}{\pi R}}\right)^4$$
$$r^4 = \frac{8\mu l}{\pi R^4}$$
$$\pi R r^4 = 8\mu l$$
$$R = \frac{8\mu l}{\pi r^4}$$

107. $N = 0.2\left(\sqrt{R}\right)^3$

$$N = 0.2\left(\sqrt{149.4}\right)^3$$
$$= 0.2(12.223)^3$$
$$= 0.2(1826.106)$$
$$\approx 365.2 \text{ days}$$

109. $R = \sqrt{F_1^2 + F_2^2}$

$$R = \sqrt{60^2 + 80^2}$$
$$= \sqrt{10{,}000}$$
$$= 100 \text{ lb}$$

111. $c = \sqrt{gH} = \sqrt{32\cdot 10} = \sqrt{320} \approx 17.89$ ft/sec

113. The diagonal and the two given sides form a right triangle. Use the Pythagorean formula to solve for the diagonal.

$$a^2 + b^2 = c^2$$
$$25^2 + 32^2 = c^2$$
$$625 + 1024 = c^2$$
$$1649 = c^2 \;\Rightarrow\; c = \sqrt{1649} \approx 40.61$$

The diagonal is about 40.61 meters in length.

115. $x = \frac{-b \pm \sqrt{b^2 - 4ac}}{2a}$

$$x = \frac{-0 \pm \sqrt{0^2 - 4(1)(-4)}}{2(1)} = \frac{\pm\sqrt{16}}{2} = \frac{\pm 4}{2} = \pm 2$$

Now, $x = 2$ or $x = -2$.

117. $x = \dfrac{-b \pm \sqrt{b^2 - 4ac}}{2a}$

$x = \dfrac{-4 \pm \sqrt{4^2 - 4(-1)(5)}}{2(-1)}$

$= \dfrac{-4 \pm \sqrt{36}}{-2}$

$= \dfrac{-4 \pm 6}{-2}$

Now, $x = \dfrac{-4+6}{-2} = \dfrac{2}{-2} = -1$ or

$x = \dfrac{-4-6}{-2} = \dfrac{-10}{-2} = 5$

119. $f(x) = \sqrt{x-5}$

$5 = \sqrt{x-5}$

$5^2 = \left(\sqrt{x-5}\right)^2$

$25 = x - 5$

$30 = x$

121. $f(x) = \sqrt{3x^2 - 11} + 7$

$15 = \sqrt{3x^2 - 11} + 7$

$8 = \sqrt{3x^2 - 11}$

$8^2 = \left(\sqrt{3x^2 - 11}\right)^2$

$64 = 3x^2 - 11$

$75 = 3x^2$

$25 = x^2$

$\pm 5 = x$

123. a. $y = \sqrt{4x-12}$, $y = x - 3$

The points of intersection are (3, 0) and (7, 4). The x-values are 3 and 7.

b. $\sqrt{4x-12} = x - 3$

For $x = 3$:	For $x = 7$:
$\sqrt{4 \cdot 3 - 12} = 3 - 3$	$\sqrt{4x-12} = x - 3$
$\sqrt{12-12} = 0$	$\sqrt{4 \cdot 7 - 12} = 7 - 3$
$\sqrt{0} = 0$	$\sqrt{16} = 4$
$0 = 0$ True	$4 = 4$ True

Both values are solutions to the equation.

c. $\sqrt{4x-12} = x - 3$

$\left(\sqrt{4x-12}\right)^2 = (x-3)^2$

$4x - 12 = x^2 - 6x + 9$

$0 = x^2 - 10x + 21$

$0 = (x-3)(x-7)$

$x - 3 = 0$ or $x - 7 = 0$

$x = 3$ $\quad$ $x = 7$

The answers agree.

125. At $x = 4$, $g(x) = 0$ or $y = 0$.

Therefore, the graph must have an x-intercept at 4.

127. $L_1 = p - 1.96\sqrt{\dfrac{p(1-p)}{n}}$

$L_1 = 0.60 - 1.96\sqrt{\dfrac{0.60(1-0.60)}{36}}$

$\approx 0.60 - 0.16$

≈ 0.44

$L_2 = p + 1.96\sqrt{\dfrac{p(1-p)}{n}}$

$L_2 = 0.60 + 1.96\sqrt{\dfrac{0.60(1-0.60)}{36}}$

$= 0.60 + 0.16$

≈ 0.76

129. $\sqrt{x^2 + 49} = (x^2 + 49)^{1/2}$

$\left(\sqrt{x^2 + 49}\right)^2 = [(x^2 + 49)^{1/2}]^2$

$x^2 + 49 = x^2 + 49$

$49 = 49$

All real numbers, x, satisfy this equation.

131. Graph:

$y_1 = \sqrt{x+8}$

$y_2 = \sqrt{3x+5}$

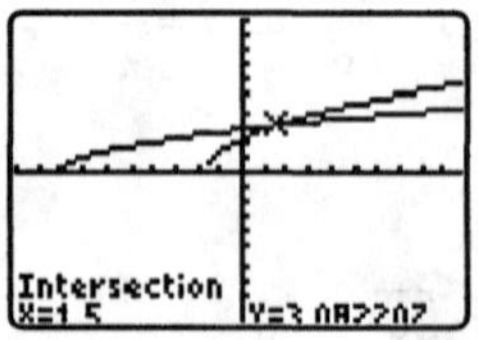

-10, 10, 1, -10, 10, 1

The graphs of the equations intersect when $x = 1.5$.

133. Graph:

$y = \sqrt[3]{5x^2 - 6} - 4$

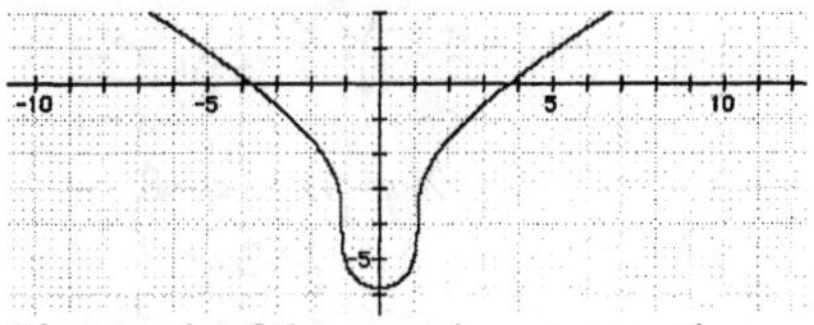

The graph of the equation crosses the x-axis at $x \approx -3.7$ and $x \approx 3.7$.

135.
$$\sqrt{\sqrt{x+25} - \sqrt{x}} = 5$$
$$\left(\sqrt{\sqrt{x+25} - \sqrt{x}}\right)^2 = 5^2$$
$$\sqrt{x+25} - \sqrt{x} = 25$$
$$\sqrt{x+25} = 25 + \sqrt{x}$$
$$\left(\sqrt{x+25}\right)^2 = \left(25 + \sqrt{x}\right)^2$$
$$x + 25 = 625 + 50\sqrt{x} + x$$
$$-600 = 50\sqrt{x}$$
$$-12 = \sqrt{x}$$
$$(-12)^2 = \left(\sqrt{x}\right)^2$$
$$144 = x$$

Check:
$$\sqrt{\sqrt{x+25} - \sqrt{x}} = 5$$
$$\sqrt{\sqrt{144+25} - \sqrt{144}} = 5$$
$$\sqrt{\sqrt{169} - \sqrt{144}} = 5$$
$$\sqrt{13-12} = 5$$
$$\sqrt{1} = 5$$
$$1 = 5 \text{ False}$$

Thus, 144 is not a solution and we conclude that there is no solution.

137.
$$z = \frac{\overline{x} - \mu}{\dfrac{\sigma}{\sqrt{n}}}$$
$$z\left(\frac{\sigma}{\sqrt{n}}\right) = \overline{x} - \mu$$
$$\left(z\frac{\sigma}{\sqrt{n}}\right)^2 = (\overline{x} - \mu)^2$$
$$\frac{z^2\sigma^2}{n} = (\overline{x} - \mu)^2$$
$$z^2\sigma^2 = n(\overline{x} - \mu)^2$$
$$\frac{z^2\sigma^2}{(\overline{x} - \mu)^2} = n \text{ or } n = \frac{z^2\sigma^2}{(\overline{x} - \mu)^2}$$

139. a.
$$S = \frac{3+4+5}{2} = \frac{12}{2} = 6$$
$$A = \sqrt{6(6-3)(6-4)(6-5)}$$
$$= \sqrt{6(3)(2)(1)}$$
$$= \sqrt{36}$$
$$= 6$$

The area is 6 square inches.

b. Answers will vary.

c. Answers will vary.

140.
$$P_1P_2 - P_1P_3 = P_2P_3 \quad \text{Solve for } P_2.$$
$$P_1P_2 - P_1P_2 - P_1P_3 = P_2P_3 - P_1P_2$$
$$-P_1P_3 = P_2P_3 - P_1P_2$$
$$-P_1P_3 = P_2(P_3 - P_1)$$
$$\frac{-P_1P_3}{P_3 - P_1} = \frac{P_2(P_3 - P_1)}{P_3 - P_1}$$
$$\frac{P_1P_3}{P_1 - P_3} = P_2 \text{ or } P_2 = \frac{P_1P_3}{P_1 - P_3}$$

141.
$$\frac{x(x-5) + x(x-2)}{2x-7} = \frac{x^2 - 5x + x^2 - 2x}{2x-7}$$
$$= \frac{2x^2 - 7x}{2x-7}$$
$$= \frac{x(2x-7)}{2x-7}$$
$$= x$$

142. $\frac{4a^2-9b^2}{4a^2+12ab+9b^2}\cdot\frac{6a^2b}{8a^2b^2-12ab^3}$
$=\frac{(2a-3b)(2a+3b)}{(2a+3b)(2a+3b)}\cdot\frac{6a^2b}{4ab^2(2a-3b)}$
$=\frac{3a}{2b(2a+3b)}$

143. $(t^2-2t-15)\div\frac{t^2-9}{t^2-3t}$
$=(t^2-2t-15)\cdot\frac{t^2-3t}{t^2-9}$
$=(t-5)(t+3)\cdot\frac{t(t-3)}{(t+3)(t-3)}$
$=\frac{t(t-3)(t-5)(t+3)}{(t+3)(t-3)}$
$=\frac{t\cancel{(t-3)}(t-5)\cancel{(t+3)}}{\cancel{(t+3)}\cancel{(t-3)}}$
$=t(t-5)$

144. $\frac{2}{x+3}-\frac{1}{x-3}+\frac{2x}{x^2-9}$
$=\frac{2}{x+3}-\frac{1}{x-3}+\frac{2x}{(x+3)(x-3)}$
$=\frac{2}{x+3}\cdot\frac{x-3}{x-3}-\frac{1}{x-3}\cdot\frac{x+3}{x+3}+\frac{2x}{(x+3)(x-3)}$
$=\frac{2(x-3)}{(x+3)(x-3)}-\frac{x+3}{(x+3)(x-3)}+\frac{2x}{(x+3)(x-3)}$
$=\frac{2x-6}{(x+3)(x-3)}-\frac{x+3}{(x+3)(x-3)}+\frac{2x}{(x+3)(x-3)}$
$=\frac{2x-6-(x+3)+2x}{(x+3)(x-3)}$
$=\frac{2x-6-x-3+2x}{(x+3)(x-3)}$
$=\frac{3x-9}{(x+3)(x-3)}$
$=\frac{3(x-3)}{(x+3)(x-3)}$
$=\frac{3}{x+3}$

145. $2+\frac{3x}{x-1}=\frac{8}{x-1}$
$(x-1)(2)+(x-1)\left(\frac{3x}{x-1}\right)=(x-1)\left(\frac{8}{x-1}\right)$
$2(x-1)+3x=8$
$2x-2+3x=8$
$5x-2=8$
$5x=10$
$x=2$

Exercise Set 7.7

1. a. $i=\sqrt{-1}$

b. $i^2=-1$

3. Yes, all the numbers listed are complex numbers.

5. Yes

7. The conjugate of $a+bi$ is $a-bi$.

9. Answers will vary. Possible answers:
a. $\sqrt{2}$ **b.** 1 **c.** $\sqrt{-3}$ or $2i$ **d.** 6
e. Every number we have studied is a complex number.

11. $7=7+0i$

13. $\sqrt{25}=5=5+0i$

15. $21-\sqrt{-36}=21-\sqrt{36}\sqrt{-1}$
$=21-6i$

17. $\sqrt{-24}=\sqrt{4}\sqrt{-1}\sqrt{6}=2i\sqrt{6}=0+2i\sqrt{6}$

19. $8-\sqrt{-12}=8-\sqrt{12}\sqrt{-1}$
$=8-\sqrt{4}\sqrt{3}\sqrt{-1}$
$=8-2i\sqrt{3}$

21. $3+\sqrt{-98}=3+\sqrt{98}\sqrt{-1}$
$=3+\sqrt{49}\sqrt{2}\sqrt{-1}$
$=3+7i\sqrt{2}$

23. $12-\sqrt{-25}=12-\sqrt{25}\sqrt{-1}$
$=12-5i$

25. $7i-\sqrt{-45}=0+7i-\sqrt{9}\sqrt{5}\sqrt{-1}$
$=0+7i-3i\sqrt{5}$
$=0+\left(7-3\sqrt{5}\right)i$

27. $(19-i)+(2+9i)=19-i+2+9i$
$=21+8i$

29. $(8-3i)+(-8+3i)=8-3i-8+3i$
$=0$

31. $\left(1+\sqrt{-1}\right)+\left(-18-\sqrt{-169}\right)$
$=1+\sqrt{-1}-18-\sqrt{-169}$
$=1+i-18-13i$
$=-17-12i$

33. $\left(\sqrt{3}+\sqrt{2}\right)+\left(3\sqrt{2}-\sqrt{-8}\right)$
$=\sqrt{3}+\sqrt{2}+3\sqrt{2}-\sqrt{-8}$
$=\sqrt{3}+\sqrt{2}+3\sqrt{2}-\sqrt{-4\cdot 2}$
$=\sqrt{3}+\sqrt{2}+3\sqrt{2}-2i\sqrt{2}$
$=\sqrt{3}+4\sqrt{2}-2i\sqrt{2}$
$=\left(\sqrt{3}+4\sqrt{2}\right)-2i\sqrt{2}$

35. $\left(5-\sqrt{-72}\right)+\left(6+\sqrt{-8}\right)$
$=5-\sqrt{-72}+6+\sqrt{-8}$
$=5-\sqrt{36}\sqrt{2}\sqrt{-1}+6+\sqrt{4}\sqrt{2}\sqrt{-1}$
$=5-6i\sqrt{2}+6+2i\sqrt{2}$
$=11-4i\sqrt{2}$

37. $\left(\sqrt{4}-\sqrt{-45}\right)+\left(-\sqrt{25}+\sqrt{-5}\right)$
$=\sqrt{4}-\sqrt{-45}-\sqrt{25}+\sqrt{-5}$
$=\sqrt{4}-\sqrt{9}\sqrt{5}\sqrt{-1}-\sqrt{25}+\sqrt{-1}\sqrt{5}$
$=2-3i\sqrt{5}-5+i\sqrt{5}$
$=-3-2i\sqrt{5}$

39. $2(3-i)=6-2i$

41. $i(4+9i)=4i+9i^2=4i+9(-1)=-9+4i$

43. $\sqrt{-9}(6+11i)=3i(6+11i)$
$=18i+33i^2$
$=18i+33(-1)$
$=-33+18i$

45. $\sqrt{-16}\left(\sqrt{3}-7i\right)=4i\left(\sqrt{3}-7i\right)$
$=4i\sqrt{3}-28i^2$
$=4i\sqrt{3}-28(-1)$
$=4i\sqrt{3}+28$
$=28+4i\sqrt{3}$

47. $\sqrt{-27}\left(\sqrt{3}-\sqrt{-3}\right)=\sqrt{-9}\sqrt{3}\left(\sqrt{3}-\sqrt{-3}\right)$
$=3i\sqrt{3}\left(\sqrt{3}-i\sqrt{3}\right)$
$=3i\cdot 3-3i^2\cdot 3$
$=9i-3(-1)\cdot 3$
$=9+9i$

49. $(3+2i)(1+i)$
$=3(1)+3(i)+2i(1)+2i(i)$
$=3+3i+2i+2i^2$
$=3+3i+2i+2(-1)$
$=3+3i+2i-2$
$=1+5i$

51. $(10-3i)(10+3i)=100+30i-30i-9i^2$
$=100+30i-30i-9(-1)$
$=100+9$
$=109$

53. $\left(7+\sqrt{-2}\right)\left(5-\sqrt{-8}\right)$
$=\left(7+i\sqrt{2}\right)\left(5-2i\sqrt{2}\right)$
$=35-14i\sqrt{2}+5i\sqrt{2}-2i^2\cdot 2$
$=35-14i\sqrt{2}+5i\sqrt{2}-2(-1)\cdot 2$
$=35-14i\sqrt{2}+5i\sqrt{2}+4$
$=39-9i\sqrt{2}$

55. $\left(\frac{1}{2}-\frac{1}{3}i\right)\left(\frac{1}{4}+\frac{2}{3}i\right)$

$=\frac{1}{8}+\frac{1}{3}i-\frac{1}{12}i-\frac{2}{9}i^2$

$=\frac{1}{8}+\frac{1}{3}i-\frac{1}{12}i-\frac{2}{9}(-1)$

$=\frac{1}{8}+\frac{1}{3}i-\frac{1}{12}i+\frac{2}{9}$

$=\frac{1}{8}+\frac{2}{9}+\left(\frac{1}{3}-\frac{1}{12}\right)i$

$=\frac{25}{72}+\frac{1}{4}i$

57. $\frac{8}{3i}=\frac{8}{3i}\cdot\frac{-i}{-i}=\frac{-8i}{-3i^2}=\frac{-8i}{-3(-1)}=-\frac{8i}{3}$

59. $\frac{2+3i}{2i}=\frac{2+3i}{2i}\cdot\frac{-i}{-i}$

$=\frac{(2+3i)(-i)}{-2i^2}$

$=\frac{-2i-3i^2}{-2i^2}$

$=\frac{-2i-3(-1)}{-2(-1)}$

$=\frac{-2i+3}{2}$

$=\frac{3-2i}{2}$

61. $\frac{6}{2-i}=\frac{6}{2-i}\cdot\frac{2+i}{2+i}$

$=\frac{6(2+i)}{(2-i)(2+i)}$

$=\frac{12+6i}{4+2i-2i-i^2}$

$=\frac{12+6i}{4+2i-2i-(-1)}$

$=\frac{12+6i}{5}$

63. $\frac{3}{1-2i}=\frac{3}{1-2i}\cdot\frac{1+2i}{1+2i}$

$=\frac{3(1+2i)}{(1-2i)(1+2i)}$

$=\frac{3+6i}{1+2i-2i-4i^2}$

$=\frac{3+6i}{1+2i-2i-4(-1)}$

$=\frac{3+6i}{5}$

65. $\frac{6-3i}{4+2i}=\frac{6-3i}{4+2i}\cdot\frac{4-2i}{4-2i}$

$=\frac{(6-3i)(4-2i)}{16-4i^2}$

$=\frac{24-12i-12i+6i^2}{16-4i^2}$

$=\frac{24-12i-12i-6}{16+4}$

$=\frac{18-24i}{20}$

$=\frac{2(9-12i)}{20}$

$=\frac{9-12i}{10}$

67. $\frac{4}{6-\sqrt{-4}}=\frac{4}{6-\sqrt{4}\sqrt{-1}}$

$=\frac{4}{6-2i}\cdot\frac{6+2i}{6+2i}$

$=\frac{4(6+2i)}{36-4i^2}$

$=\frac{24+8i}{36-4(-1)}$

$=\frac{24+8i}{36+4}$

$=\frac{8(3+i)}{40}$

$=\frac{3+i}{5}$

69. $\dfrac{\sqrt{2}}{5+\sqrt{-12}} = \dfrac{\sqrt{2}}{5+\sqrt{4}\sqrt{3}\sqrt{-1}}$

$= \dfrac{\sqrt{2}}{5+2i\sqrt{3}} \cdot \dfrac{5-2i\sqrt{3}}{5-2i\sqrt{3}}$

$= \dfrac{\sqrt{2}\left(5-2i\sqrt{3}\right)}{25-4i^2\sqrt{3}^2}$

$= \dfrac{5\sqrt{2}-2i\sqrt{6}}{25-4(-1)(3)}$

$= \dfrac{5\sqrt{2}-2i\sqrt{6}}{25+12}$

$= \dfrac{5\sqrt{2}-2i\sqrt{6}}{37}$

71. $\dfrac{\sqrt{10}+\sqrt{-3}}{5-\sqrt{-20}}$

$= \dfrac{\sqrt{10}+\sqrt{3}\sqrt{-1}}{5-\sqrt{4}\sqrt{5}\sqrt{-1}}$

$= \dfrac{\sqrt{10}+i\sqrt{3}}{5-2i\sqrt{5}} \cdot \dfrac{5+2i\sqrt{5}}{5+2i\sqrt{5}}$

$= \dfrac{(\sqrt{10}+i\sqrt{3})(5+2i\sqrt{5})}{5^2-4i^2\sqrt{5}^2}$

$= \dfrac{5\sqrt{10}+2i\sqrt{50}+5i\sqrt{3}+2i^2\sqrt{15}}{5^2-4(-1)(5)}$

$= \dfrac{5\sqrt{10}+2i\sqrt{25}\sqrt{2}+5i\sqrt{3}+2(-1)\sqrt{15}}{25+20}$

$= \dfrac{(5\sqrt{10}-2\sqrt{15})+(10\sqrt{2}+5\sqrt{3})i}{45}$

73. $\dfrac{\sqrt{-75}}{\sqrt{-3}} = \dfrac{\sqrt{25}\sqrt{3}\sqrt{-1}}{\sqrt{3}\sqrt{-1}} = \dfrac{5i\sqrt{3}}{i\sqrt{3}} = 5$

75. $\dfrac{\sqrt{-32}}{\sqrt{-18}\sqrt{8}} = \dfrac{\sqrt{32}\sqrt{-1}}{\sqrt{144}\sqrt{-1}} = \dfrac{i\sqrt{16}\sqrt{2}}{i\sqrt{144}} = \dfrac{4i\sqrt{2}}{12i} = \dfrac{\sqrt{2}}{3}$

77. $(9-2i)+(3-5i) = 9+3-2i-5i = 12-7i$

79. $\left(\sqrt{50}-\sqrt{2}\right)-\left(\sqrt{-12}-\sqrt{-48}\right)$

$= \left(5\sqrt{2}-\sqrt{2}\right)-\left(2i\sqrt{3}-4i\sqrt{3}\right)$

$= 4\sqrt{2}-\left(-2i\sqrt{3}\right)$

$= 4\sqrt{2}+2i\sqrt{3}$

81. $5.2(4-3.2i) = 5.2(4)-5.2(3.2i) = 20.8-16.64i$

83. $(9+2i)(3-5i)$

$= 27-45i+6i-10i^2$

$= 27-39i-10(-1)$

$= 27+10-39i$

$= 37-39i$

85. $\dfrac{11+4i}{2i} = \dfrac{11+4i}{2i} \cdot \dfrac{-2i}{-2i}$

$= \dfrac{(11+4i)(-2i)}{-4i^2}$

$= \dfrac{-22i-8i^2}{-4i^2}$

$= \dfrac{-22i-8(-1)}{-4(-1)}$

$= \dfrac{8-22i}{4}$

$= \dfrac{4-11i}{2}$

87. $\dfrac{6}{\sqrt{3}-\sqrt{-4}} = \dfrac{6}{\sqrt{3}-2i}$

$= \dfrac{6}{\sqrt{3}-2i} \cdot \dfrac{\sqrt{3}+2i}{\sqrt{3}+2i}$

$= \dfrac{6(\sqrt{3}+2i)}{(\sqrt{3})^2-(2i)^2}$

$= \dfrac{6\sqrt{3}+12i}{3-4i^2}$

$= \dfrac{6\sqrt{3}+12i}{3-4(-1)}$

$= \dfrac{6\sqrt{3}+12i}{7}$

89. $\left(11-\dfrac{5}{9}i\right)-\left(4-\dfrac{3}{5}i\right) = 11-\dfrac{5}{9}i-4+\dfrac{3}{5}i$

$= 7-\dfrac{5}{9}i+\dfrac{3}{5}i$

$= 7-\dfrac{25}{45}i+\dfrac{27}{45}i$

$= 7+\dfrac{2}{45}i$

91. $\left(\frac{2}{3}-\frac{1}{5}i\right)\left(\frac{3}{5}-\frac{3}{4}i\right)$

$=\left(\frac{2}{3}\right)\left(\frac{3}{5}\right)-\frac{2}{3}\left(\frac{3}{4}i\right)-\left(\frac{1}{5}i\right)\left(\frac{3}{5}\right)+\left(\frac{1}{5}i\right)\left(\frac{3}{4}i\right)$

$=\frac{2}{5}-\frac{1}{2}i-\frac{3}{25}i+\frac{3}{20}i^2$

$=\frac{2}{5}-\frac{1}{2}i-\frac{3}{25}i+\frac{3}{20}(-1)$

$=\left(\frac{2}{5}-\frac{3}{20}\right)+\left(-\frac{1}{2}-\frac{3}{25}\right)i$

$=\left(\frac{8}{20}-\frac{3}{20}\right)+\left(-\frac{25}{50}-\frac{6}{50}\right)i$

$=\frac{5}{20}-\frac{31}{50}i$

$=\frac{1}{4}-\frac{31}{50}i$

93. $\frac{\sqrt{-48}}{\sqrt{-12}}=\frac{\sqrt{48}\sqrt{-1}}{\sqrt{12}\sqrt{-1}}$

$=\frac{i\sqrt{48}}{i\sqrt{12}}$

$=\frac{\sqrt{48}}{\sqrt{12}}$

$=\sqrt{\frac{48}{12}}$

$=\sqrt{4}$

$=2$

95. $(5.23-6.41i)-(9.56+4.5i)$
$=5.23-6.41i-9.56-4.5i$
$=-4.33-10.91i$

97. $i^6=i^4\cdot i^2=1\cdot(-1)=-1$

99. $i^{160}=(i^4)^{40}=1^{40}=1$

101. $i^{93}=i^{92}\cdot i^1=(i^4)^{23}i=1^{23}\cdot i=1(i)=i$

103. $i^{811}=i^{808}\cdot i^3$

$=(i^4)^{202}\cdot i^3$

$=1^{202}\cdot i^3$

$=1\cdot(i^3)$

$=i^3$

$=-i$

105. a. The additive inverse of $2+3i$ is its opposite, $-2-3i$. Note that $(2+3i)+(-2-3i)=0$.

b. The multiplicative inverse of $2+3i$ is its reciprocal, $\frac{1}{2+3i}$. To simplify this, multiply the numerator and denominator by the conjugate of the denominator.

$\frac{1}{2+3i}=\frac{1}{2+3i}\cdot\frac{2-3i}{2-3i}$

$=\frac{2-3i}{(2+3i)(2-3i)}$

$=\frac{2-3i}{4-6i+6i-9i^2}$

$=\frac{2-3i}{13}$

107. True. The product of two pure imaginary numbers is always a real number. Consider two pure imaginary numbers *bi* and *di* where *b*, *d* are non-zero real numbers whose product

$(bi)(di)=bdi^2$

$=bd(-1)$

$=-bd$

which is a real number.

109. False. The product of two complex numbers is not always a real number. For example,

$(1+i)(1+i)=1+i+i+i^2$

$=1+2i+(-1)$

$=0+2i$

which is not a real number.

111. Even values of *n* will result in i^n being a real number. $i^2=-1, i^{2n}=(i^2)^n=(-1)^n$

113. $f(x)=x^2$

$f(2i)=(2i)^2=4i^2=4(-1)=-4$

115. $f(x)=x^4-2x$

$f(2i)=(2i)^4-2(2i)$

$=2^4i^4-4i$

$=16(1)-4i$

$=16-4i$

117. $f(x)=x^2+2x$

$f(3+i)=(3+i)^2+2(3+i)$

$=9+6i+i^2+6+2i$

$=9+6i-1+6+2i$

$=14+8i$

119. x^2-2x+5

$=(1+2i)^2-2(1+2i)+5$

$=1^2+2(1)(2i)+(2i)^2-2-4i+5$

$=1+4i-4-2-4i+5$

$=0$

121. x^2+2x+7

$=(-1+i\sqrt{5})^2+2(-1+i\sqrt{5})+7$

$=(-1)^2-2(1)(i\sqrt{5})+(i\sqrt{5})^2-2+2i\sqrt{5}+7$

$=1-2i\sqrt{5}-5-2+2i\sqrt{5}+7$

$=1+0i$

$=1$

123.

$$x^2-4x+5=0$$
$$(2-i)^2-4(2-i)+5=0$$
$$2^2-2(2)(i)+(i)^2-8+4i+5=0$$
$$4-4i-1-8+4i+5=0$$
$$0+0i=0$$
$$0=0 \quad \text{True}$$

$2-i$ is a solution .

125.

$$x^2-6x+11=0$$
$$(-3+i\sqrt{3})^2-6(-3+i\sqrt{3})+11=0$$
$$(-3)^2-2(3)(i\sqrt{3})+(i\sqrt{3})^2+18-6i\sqrt{3}+11=0$$
$$9-6i\sqrt{3}-3+18-6i\sqrt{3}+11=0$$
$$35-12i\sqrt{3}=0$$

False, $-3+i\sqrt{3}$ is not a solution.

127. $Z=\dfrac{V}{I}$

$$Z=\frac{1.8+0.5i}{0.6i}$$
$$=\frac{1.8+0.5i}{0.6i}\cdot\frac{-0.6i}{-0.6i}$$
$$=\frac{-1.08i-0.3i^2}{-0.36i^2}$$
$$=\frac{-1.08i-0.3(-1)}{-0.36(-1)}$$
$$=\frac{0.3-1.08i}{0.36}$$
$$\approx 0.83-3i$$

129. $Z_T=\dfrac{Z_1Z_2}{Z_1+Z_2}$

$$=\frac{(2-i)(4+i)}{(2-i)+(4+i)}$$
$$=\frac{8+2i-4i-i^2}{6}$$
$$=\frac{8-2i-(-1)}{6}$$
$$=\frac{9-2i}{6}$$
$$\approx 1.5-0.33i$$

131. $i^{-1}=\dfrac{1}{i}=\dfrac{1}{i}\cdot\dfrac{i}{i}=\dfrac{i}{i^2}=\dfrac{i}{-1}=-i$

133. $x^2-2x+6=0$

$a=1,\ b=-2,\ c=6$

$$x=\frac{-(-2)\pm\sqrt{(-2)^2-4(1)(6)}}{2(1)}$$
$$=\frac{2\pm\sqrt{4-24}}{2}$$
$$=\frac{2\pm\sqrt{-20}}{2}$$
$$=\frac{2\pm 2i\sqrt{5}}{2}$$
$$=\frac{2\left(1\pm i\sqrt{5}\right)}{2}$$
$$=1\pm i\sqrt{5}$$

135. $a+b=5+2i\sqrt{3}+1+i\sqrt{3}$
$=5+1+2i\sqrt{3}+i\sqrt{3}$
$=6+3i\sqrt{3}$

137. $ab=(5+2i\sqrt{3})(1+i\sqrt{3})$
$=5(1)+(5)(i\sqrt{3})+(2i\sqrt{3})(1)+(2i\sqrt{3})(i\sqrt{3})$
$=5+5i\sqrt{3}+2i\sqrt{3}+2i^2(\sqrt{3})^2$
$=5+5i\sqrt{3}+2i\sqrt{3}-6$
$=-1+7i\sqrt{3}$

139. This problem can be solved using a single variable. To do this, let x be the amount that is \$5.50 per pound. Then $40-x$ is the amount that is \$6.30 per pound and the equation is

$5.50(x)+6.30(40-x)=6(40)$
$5.5x+252-6.3x=240$
$252-0.8x=240$
$-0.8x=-12$
$x=15$

Thus, combine 15 lb of the \$5.50 per pound coffee with 40 – 15 = 25 lb of the \$6.30 per pound coffee to obtain 40 lb of \$6.00 per pound coffee.

140. $4c+9\overline{)8c^2+6c-35}$ with quotient $2c-3$

$8c^2+18c \quad \leftarrow 2c(4c+9)$
$-12c-35$
$-12c-27 \quad \leftarrow -3(4c+9)$
$-8 \quad \leftarrow$ Remainder

Thus, $\dfrac{8c^2+6c-35}{4c+9}=2c-3-\dfrac{8}{4c+9}$

141. $\dfrac{b}{a-b}+\dfrac{a+b}{b}=\dfrac{b}{a-b}\cdot\dfrac{b}{b}+\dfrac{a+b}{b}\cdot\dfrac{a-b}{a-b}$
$=\dfrac{b^2}{b(a-b)}+\dfrac{(a+b)(a-b)}{b(a-b)}$
$=\dfrac{b^2+(a+b)(a-b)}{b(a-b)}$
$=\dfrac{b^2+a^2-b^2}{b(a-b)}$
$=\dfrac{a^2}{b(a-b)}$

142. $\dfrac{x}{4}+\dfrac{1}{2}=\dfrac{x-1}{2}$
$4\left(\dfrac{x}{4}+\dfrac{1}{2}\right)=4\left(\dfrac{x-1}{2}\right)$
$x+2=2(x-1)$
$x+2=2x-2$
$x=4$

Chapter 7 Review Exercises

1. $\sqrt{100}=\sqrt{10^2}=10$

2. $\sqrt[3]{-27}=\sqrt[3]{(-3)^3}=-3$

3. $\sqrt[3]{-125}=\sqrt[3]{(-5)^3}=-5$

4. $\sqrt[4]{256}=\sqrt[4]{4^4}=4$

5. $\sqrt{(-8)^2}=|-8|=8$

6. $\sqrt{(38.2)^2}=38.2$

7. $\sqrt{x^2}=|x|$

8. $\sqrt{(x-3)^2}=|x-3|$

9. $\sqrt{(x-y)^2}=|x-y|$

10. $\sqrt{\left(x^2-4x+12\right)^2}=|x^2-4x+12|$

(Note: absolute value would not actually be required because $x^2-4x+12>0$ for all x.)

11. $f(x)=\sqrt{10x+9}$
$f(4)=\sqrt{10(4)+9}$
$=\sqrt{40+9}$
$=\sqrt{49}$
$=7$

12. $k(x) = 2x + \sqrt{\frac{x}{3}}$

$k(27) = 2(27) + \sqrt{\frac{27}{3}}$
$= 54 + \sqrt{9}$
$= 54 + 3$
$= 57$

13. $g(x) = \sqrt[3]{2x+3}$

$g(4) = \sqrt[3]{2(4)+3}$
$= \sqrt[3]{11}$
≈ 2.2

14. Area $= (\text{side})^2$

$144 = s^2$

$\pm 12 = s$

Disregard the negative value since lengths must be positive. The length of each side is 12 m.

15. $\sqrt{x^7} = x^{7/2}$

16. $\sqrt[3]{x^5} = x^{5/3}$

17. $\left(\sqrt[4]{y}\right)^{13} = y^{13/4}$

18. $\sqrt[7]{6^{-2}} = 6^{-2/7}$

19. $x^{1/2} = \sqrt{x}$

20. $a^{4/5} = \sqrt[5]{a^4}$

21. $(8m^2n)^{7/4} = \left(\sqrt[4]{8m^2n}\right)^7$

22. $(x+y)^{-5/3} = \frac{1}{\left(\sqrt[3]{x+y}\right)^5}$

23. $\sqrt[3]{4^6} = 4^{6/3} = 4^2 = 16$

24. $\sqrt{x^{12}} = x^{12/2} = x^6$

25. $\left(\sqrt[4]{9}\right)^8 = 9^{8/4} = 9^2 = 81$

26. $\sqrt[20]{a^5} = a^{5/20} = a^{1/4} = \sqrt[4]{a}$

27. $-36^{1/2} = -\sqrt{36} = -6$

28. $(-36)^{1/2}$ is not a real number.

29. $\left(\frac{64}{27}\right)^{-1/3} = \left(\frac{27}{64}\right)^{1/3} = \sqrt[3]{\frac{27}{64}} = \frac{\sqrt[3]{27}}{\sqrt[3]{64}} = \frac{3}{4}$

30. $64^{-1/2} + 8^{-2/3} = \frac{1}{64^{1/2}} + \frac{1}{8^{2/3}}$
$= \frac{1}{\sqrt{64}} + \frac{1}{\left(\sqrt[3]{8}\right)^2}$
$= \frac{1}{8} + \frac{1}{2^2}$
$= \frac{1}{8} + \frac{1}{4}$
$= \frac{3}{8}$

31. $x^{3/5}x^{-1/3} = x^{3/5-1/3} = x^{9/15-5/15} = x^{4/15}$

32. $\left(\frac{64}{y^9}\right)^{1/3} = \sqrt[3]{\frac{64}{y^9}} = \frac{\sqrt[3]{64}}{\sqrt[3]{y^9}} = \frac{4}{y^3}$

33. $\left(\frac{a^{-6/5}}{a^{2/5}}\right)^{2/3} = \frac{a^{-6/5\cdot 2/3}}{a^{2/5\cdot 2/3}}$
$= \frac{a^{-12/15}}{a^{4/15}}$
$= a^{-12/15-(4/15)}$
$= a^{-16/15}$
$= \frac{1}{a^{16/15}}$

34. $\left(\frac{20x^5y^{-3}}{4y^{1/2}}\right)^2 = \left(\frac{5x^5}{y^{7/2}}\right)^2$
$= \frac{5^2x^{5\cdot 2}}{y^{(7/2)\cdot 2}}$
$= \frac{25x^{10}}{y^7}$

35. $a^{1/2}\left(5a^{3/2}-3a^2\right)=a^{1/2}\left(5a^{3/2}\right)-a^{1/2}\left(3a^2\right)$
$=5a^{1/2+3/2}-3a^{1/2+2}$
$=5a^{4/2}-3a^{1/2+4/2}$
$=5a^2-3a^{5/2}$

36. $4x^{-2/3}\left(x^{-1/2}+\dfrac{11}{4}x^{2/3}\right)$
$=4x^{-2/3}\left(x^{-1/2}\right)+4x^{-2/3}\left(\dfrac{11}{4}x^{2/3}\right)$
$=4x^{-2/3+(-1/2)}+11x^{-2/3+(2/3)}$
$=4x^{-4/6+(-3/6)}+11x^{0/3}$
$=4x^{-7/6}+11x^0$
$=\dfrac{4}{x^{7/6}}+11$

37. $x^{2/5}+x^{7/5}=x^{2/5}+x^{2/5}\cdot x^1$
$=x^{2/5}(1+x)$

38. $a^{-1/2}+a^{3/2}=a^{-1/2}+a^{-1/2}\cdot a^{4/2}$
$=a^{-1/2}(1+a^2)$
$=\dfrac{1+a^2}{a^{1/2}}$

39. $f(x)=\sqrt{6x-11}$
$f(6)=\sqrt{6(6)-11}$
$=\sqrt{36-11}$
$=\sqrt{25}$
$=5$

40. $g(x)=\sqrt[3]{9x-17}$
$g(4)=\sqrt[3]{9(4)-17}$
$=\sqrt[3]{36-17}$
$=\sqrt[3]{19}$
≈ 2.668

41. $f(x)=\sqrt{x}$

x	$f(x)$
0	0
1	1
4	2
9	3

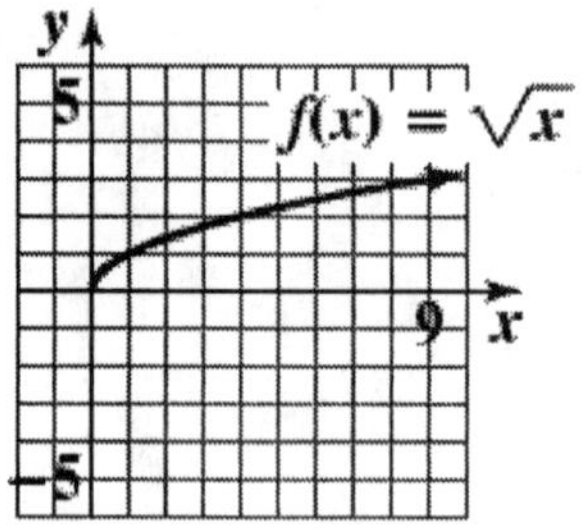

42. $f(x)=\sqrt{x}-4$

x	$f(x)$
0	−4
1	−3
4	−2
9	−1
1	0

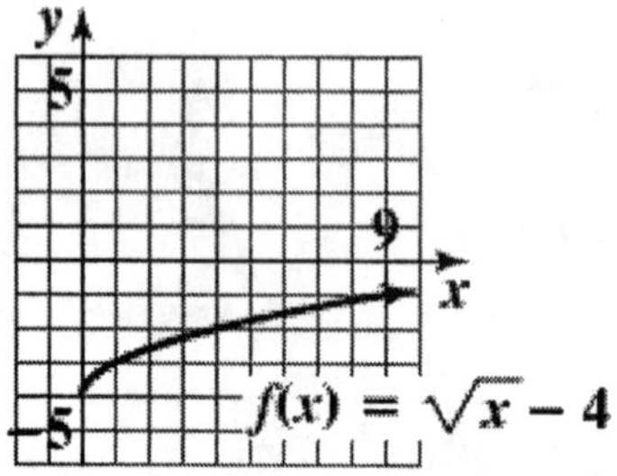

43. $\sqrt{48}=\sqrt{16}\sqrt{3}=4\sqrt{3}$

44. $\sqrt[3]{128}=\sqrt[3]{64}\sqrt[3]{2}=4\sqrt[3]{2}$

45. $\sqrt{\dfrac{49}{9}}=\dfrac{\sqrt{49}}{\sqrt{9}}=\dfrac{7}{3}$

46. $\sqrt[3]{\dfrac{8}{125}}=\dfrac{\sqrt[3]{8}}{\sqrt[3]{125}}=\dfrac{2}{5}$

47. $-\sqrt{\frac{81}{49}} = -\frac{\sqrt{81}}{\sqrt{49}} = -\frac{9}{7}$

48. $\sqrt[3]{-\frac{27}{125}} = \frac{\sqrt[3]{-27}}{\sqrt[3]{125}} = \frac{-3}{5} = -\frac{3}{5}$

49. $\sqrt{32}\sqrt{2} = \sqrt{32 \cdot 2} = \sqrt{64} = 8$

50. $\sqrt[3]{32} \cdot \sqrt[3]{2} = \sqrt[3]{32 \cdot 2} = \sqrt[3]{64} = 4$

51. $\sqrt{18x^2y^3z^4} = \sqrt{9x^2y^2z^4}\sqrt{2y} = 3xyz^2\sqrt{2y}$

52. $\sqrt{75x^3y^7} = \sqrt{25x^2y^6}\sqrt{3xy} = 5xy^3\sqrt{3xy}$

53. $\sqrt[3]{54a^7b^{10}} = \sqrt[3]{27a^6b^9}\sqrt[3]{2ab} = 3a^2b^3\sqrt[3]{2ab}$

54. $\sqrt[3]{125x^8y^9z^{16}} = \sqrt[3]{125x^6y^9z^{15}}\sqrt[3]{x^2z}$
$= 5x^2y^3z^5\sqrt[3]{x^2z}$

55. $\left(\sqrt[6]{x^2y^3z^5}\right)^{42} = (x^2y^3z^5)^{42/6}$
$= (x^2y^3z^5)^7$
$= x^{14}y^{21}z^{35}$

56. $\left(\sqrt[5]{2ab^4c^6}\right)^{15} = \left(2ab^4c^6\right)^{15/5}$
$= \left(2ab^4c^6\right)^3$
$= 8a^3b^{12}c^{18}$

57. $\sqrt{5x}\sqrt{8x^5} = \sqrt{40x^6} = \sqrt{4x^6}\sqrt{10} = 2x^3\sqrt{10}$

58. $\sqrt[3]{2x^2y}\,\sqrt[3]{4x^9y^4} = \sqrt[3]{8x^{11}y^5} = 2x^3y\,\sqrt[3]{x^2y^2}$

59. $\sqrt[3]{2x^4y^5}\,\sqrt[3]{16x^4y^4} = \sqrt[3]{32x^8y^9}$
$= \sqrt[3]{8x^6y^9}\,\sqrt[3]{4x^2}$
$= 2x^2y^3\sqrt[3]{4x^2}$

60. $\sqrt[4]{4x^4y^7}\,\sqrt[4]{4x^5y^9} = \sqrt[4]{16x^9y^{16}}$
$= \sqrt[4]{16x^8y^{16}}\,\sqrt[4]{x}$
$= 2x^2y^4\sqrt[4]{x}$

61. $\sqrt{3x}(\sqrt{12x} - \sqrt{20}) = \sqrt{36x^2} - \sqrt{60x}$
$= \sqrt{36x^2} - \sqrt{4}\sqrt{15x}$
$= 6x - 2\sqrt{15x}$

62. $\sqrt[3]{2x^2y}\left(\sqrt[3]{4x^4y^7} + \sqrt[3]{9x}\right)$
$= \sqrt[3]{8x^6y^8} + \sqrt[3]{18x^3y}$
$= \sqrt[3]{8x^6y^6}\,\sqrt[3]{y^2} + \sqrt[3]{x^3}\,\sqrt[3]{18y}$
$= 2x^2y^2\sqrt[3]{y^2} + x\sqrt[3]{18y}$

63. $\sqrt{\sqrt{a^3b^2}} = \left(\sqrt{a^3b^2}\right)^{1/2}$
$= \left[\left(a^3b^2\right)^{1/2}\right]^{1/2}$
$= \left(a^3b^2\right)^{1/4}$
$= \sqrt[4]{a^3b^2}$

64. $\sqrt{\sqrt[3]{x^5y^2}} = \left(\sqrt[3]{x^5y^2}\right)^{1/2}$
$= \left[\left(x^5y^2\right)^{1/3}\right]^{1/2}$
$= \left(x^5y^2\right)^{1/6}$
$= \sqrt[6]{\left(x^5y^2\right)}$

65. $\left(\frac{4r^2p^{1/3}}{r^{1/2}p^{4/3}}\right)^3 = (4r^{2-1/2}p^{1/3-4/3})^3$
$= (4r^{3/2}p^{-1})^3$
$= 4^3(r^{3/2})^3(p^{-1})^3$
$= 64r^{9/2}p^{-3}$
$= \frac{64r^{9/2}}{p^3}$

66. $\left(\dfrac{6y^{2/5}z^{1/3}}{x^{-1}y^{3/5}}\right)^{-1}=\left(\dfrac{6y^{2/5-3/5}z^{1/3}}{x^{-1}}\right)^{-1}$

$=\left(\dfrac{6y^{-1/5}z^{1/3}}{x^{-1}}\right)^{-1}$

$=\left(\dfrac{6xz^{1/3}}{y^{1/5}}\right)^{-1}$

$=\dfrac{y^{1/5}}{6xz^{1/3}}$

67. $\sqrt{\dfrac{3}{5}}=\dfrac{\sqrt{3}}{\sqrt{5}}\cdot\dfrac{\sqrt{5}}{\sqrt{5}}=\dfrac{\sqrt{15}}{5}$

68. $\sqrt[3]{\dfrac{7}{9}}=\dfrac{\sqrt[3]{7}}{\sqrt[3]{9}}=\dfrac{\sqrt[3]{7}}{\sqrt[3]{9}}\cdot\dfrac{\sqrt[3]{3}}{\sqrt[3]{3}}=\dfrac{\sqrt[3]{21}}{\sqrt[3]{27}}=\dfrac{\sqrt[3]{21}}{3}$

69. $\sqrt[4]{\dfrac{5}{4}}=\dfrac{\sqrt[4]{5}}{\sqrt[4]{4}}=\dfrac{\sqrt[4]{5}}{\sqrt[4]{4}}\cdot\dfrac{\sqrt[4]{2^2}}{\sqrt[4]{2^2}}=\dfrac{\sqrt[4]{20}}{\sqrt[4]{2^4}}=\dfrac{\sqrt[4]{20}}{2}$

70. $\dfrac{x}{\sqrt{10}}=\dfrac{x}{\sqrt{10}}\cdot\dfrac{\sqrt{10}}{\sqrt{10}}=\dfrac{x\sqrt{10}}{10}$

71. $\dfrac{8}{\sqrt{x}}=\dfrac{8}{\sqrt{x}}\cdot\dfrac{\sqrt{x}}{\sqrt{x}}=\dfrac{8\sqrt{x}}{x}$

72. $\dfrac{m}{\sqrt[3]{25}}=\dfrac{m}{\sqrt[3]{5^2}}\cdot\dfrac{\sqrt[3]{5}}{\sqrt[3]{5}}=\dfrac{m\sqrt[3]{5}}{\sqrt[3]{5^3}}=\dfrac{m\sqrt[3]{5}}{5}$

73. $\dfrac{10}{\sqrt[3]{y^2}}=\dfrac{10}{\sqrt[3]{y^2}}\cdot\dfrac{\sqrt[3]{y}}{\sqrt[3]{y}}=\dfrac{10\sqrt[3]{y}}{\sqrt[3]{y^3}}=\dfrac{10\sqrt[3]{y}}{y}$

74. $\dfrac{9}{\sqrt[4]{z}}=\dfrac{9}{\sqrt[4]{z}}\cdot\dfrac{\sqrt[4]{z^3}}{\sqrt[4]{z^3}}=\dfrac{9\sqrt[4]{z^3}}{\sqrt[4]{z^4}}=\dfrac{9\sqrt[4]{z^3}}{z}$

75. $\sqrt[3]{\dfrac{x^3}{27}}=\dfrac{\sqrt[3]{x^3}}{\sqrt[3]{27}}=\dfrac{x}{3}$

76. $\dfrac{\sqrt[3]{2x^{10}}}{\sqrt[3]{16x^7}}=\sqrt[3]{\dfrac{2x^{10}}{16x^7}}=\sqrt[3]{\dfrac{x^3}{8}}=\dfrac{\sqrt[3]{x^3}}{\sqrt[3]{8}}=\dfrac{x}{2}$

77. $\sqrt{\dfrac{32x^2y^5}{2x^8y}}=\sqrt{\dfrac{16y^4}{x^6}}=\dfrac{\sqrt{16y^4}}{\sqrt{x^6}}=\dfrac{4y^2}{x^3}$

78. $\sqrt[4]{\dfrac{48x^9y^{15}}{3xy^3}}=\sqrt[4]{16x^8y^{12}}=2x^2y^3$

79. $\sqrt{\dfrac{6x^4}{y}}=\dfrac{\sqrt{6x^4}}{\sqrt{y}}$

$=\dfrac{\sqrt{x^4}\sqrt{6}}{\sqrt{y}}$

$=\dfrac{x^2\sqrt{6}}{\sqrt{y}}\cdot\dfrac{\sqrt{y}}{\sqrt{y}}$

$=\dfrac{x^2\sqrt{6y}}{y}$

80. $\sqrt{\dfrac{12a}{7b}}=\dfrac{\sqrt{12a}}{\sqrt{7b}}$

$=\dfrac{2\sqrt{3a}}{\sqrt{7b}}$

$=\dfrac{2\sqrt{3a}}{\sqrt{7b}}\cdot\dfrac{\sqrt{7b}}{\sqrt{7b}}$

$=\dfrac{2\sqrt{21ab}}{7b}$

81. $\sqrt{\dfrac{18x^4y^5}{3z}}=\dfrac{\sqrt{18x^4y^5}}{\sqrt{3z}}$

$=\dfrac{\sqrt{9x^4y^4}\sqrt{2y}}{\sqrt{3z}}$

$=\dfrac{3x^2y^2\sqrt{2y}}{\sqrt{3z}}$

$=\dfrac{3x^2y^2\sqrt{2y}}{\sqrt{3z}}\cdot\dfrac{\sqrt{3z}}{\sqrt{3z}}$

$=\dfrac{3x^2y^2\sqrt{6yz}}{3z}$

$=\dfrac{x^2y^2\sqrt{6yz}}{z}$

82. $\sqrt{\dfrac{125x^2y^5}{3z}}=\dfrac{\sqrt{125x^2y^5}}{\sqrt{3z}}$
$=\dfrac{\sqrt{25x^2y^4}\sqrt{5y}}{\sqrt{3z}}$
$=\dfrac{5xy^2\sqrt{5y}}{\sqrt{3z}}\cdot\dfrac{\sqrt{3z}}{\sqrt{3z}}$
$=\dfrac{5xy^2\sqrt{15yz}}{3z}$

83. $\sqrt[3]{\dfrac{108x^3y^7}{2y^3}}=\sqrt[3]{54x^3y^4}=3xy\sqrt[3]{2y}$

84. $\sqrt[3]{\dfrac{3x}{5y}}=\dfrac{\sqrt[3]{3x}}{\sqrt[3]{5y}}\cdot\dfrac{\sqrt[3]{25y^2}}{\sqrt[3]{25y^2}}=\dfrac{\sqrt[3]{75xy^2}}{5y}$

85. $\sqrt[3]{\dfrac{9x^5y^3}{x^6}}=\sqrt[3]{\dfrac{9y^3}{x}}$
$=\dfrac{\sqrt[3]{9y^3}}{\sqrt[3]{x}}$
$=\dfrac{\sqrt[3]{y^3}\sqrt[3]{9}}{\sqrt[3]{x}}$
$=\dfrac{y\sqrt[3]{9}}{\sqrt[3]{x}}\cdot\dfrac{\sqrt[3]{x^2}}{\sqrt[3]{x^2}}$
$=\dfrac{y\sqrt[3]{9x^2}}{x}$

86. $\sqrt[3]{\dfrac{y^6}{5x^2}}=\dfrac{\sqrt[3]{y^6}}{\sqrt[3]{5x^2}}=\dfrac{y^2}{\sqrt[3]{5x^2}}\cdot\dfrac{\sqrt[3]{25x}}{\sqrt[3]{25x}}=\dfrac{y^2\sqrt[3]{25x}}{5x}$

87. $\sqrt[4]{\dfrac{2a^2b^{11}}{a^5b}}=\sqrt[4]{\dfrac{2b^{10}}{a^3}}$
$=\dfrac{\sqrt[4]{2b^{10}}}{\sqrt[4]{a^3}}\cdot\dfrac{\sqrt[4]{a}}{\sqrt[4]{a}}$
$=\dfrac{\sqrt[4]{2ab^{10}}}{\sqrt[4]{a^4}}$
$=\dfrac{b^2\sqrt[4]{2ab^2}}{a}$

88. $\sqrt[4]{\dfrac{3x^2y^6}{8x^3}}=\sqrt[4]{\dfrac{3y^6}{8x}}$
$=\dfrac{\sqrt[4]{3y^6}}{\sqrt[4]{8x}}$
$=\dfrac{y\sqrt[4]{3y^2}}{\sqrt[4]{8x}}$
$=\dfrac{y\sqrt[4]{3y^2}}{\sqrt[4]{8x}}\cdot\dfrac{\sqrt[4]{2x^3}}{\sqrt[4]{2x^3}}$
$=\dfrac{y\sqrt[4]{6x^3y^2}}{2x}$

89. $\left(3-\sqrt{2}\right)\left(3+\sqrt{2}\right)=3^2-\left(\sqrt{2}\right)^2=9-2=7$

90. $\left(\sqrt{x}+y\right)\left(\sqrt{x}-y\right)=\left(\sqrt{x}\right)^2-y^2=x-y^2$

91. $\left(x-\sqrt{y}\right)\left(x+\sqrt{y}\right)=x^2-\left(\sqrt{y}\right)^2=x^2-y$

92. $\left(\sqrt{3}+2\right)^2=\left(\sqrt{3}\right)^2+2(2)\left(\sqrt{3}\right)+2^2$
$=3+4\sqrt{3}+4$
$=7+4\sqrt{3}$

93. $\left(\sqrt{x}-\sqrt{3y}\right)\left(\sqrt{x}+\sqrt{5y}\right)$
$=\left(\sqrt{x}\right)^2+\sqrt{x}\sqrt{5y}-\sqrt{x}\sqrt{3y}-\sqrt{3y}\sqrt{5y}$
$=x+\sqrt{5xy}-\sqrt{3xy}-\sqrt{15y^2}$
$=x+\sqrt{5xy}-\sqrt{3xy}-y\sqrt{15}$

94. $\left(\sqrt[3]{2x}-\sqrt[3]{3y}\right)\left(\sqrt[3]{3x}-\sqrt[3]{2y}\right)$
$=\sqrt[3]{2x}\left(\sqrt[3]{3x}\right)-\left(\sqrt[3]{2x}\right)\left(\sqrt[3]{2y}\right)$
$-\sqrt[3]{3y}\left(\sqrt[3]{3x}\right)+\sqrt[3]{3y}\sqrt[3]{2y}$
$=\sqrt[3]{6x^2}-\sqrt[3]{4xy}-\sqrt[3]{9xy}+\sqrt[3]{6y^2}$

95. $\dfrac{6}{2+\sqrt{5}}=\dfrac{6}{2+\sqrt{5}}\cdot\dfrac{2-\sqrt{5}}{2-\sqrt{5}}$
$=\dfrac{6\left(2-\sqrt{5}\right)}{2^2-\left(\sqrt{5}\right)^2}$
$=\dfrac{12-6\sqrt{5}}{4-5}$
$=\dfrac{12-6\sqrt{5}}{-1}$
$=-12+6\sqrt{5}$

96. $\dfrac{x}{4+\sqrt{x}}=\dfrac{x}{4+\sqrt{x}}\cdot\dfrac{4-\sqrt{x}}{4-\sqrt{x}}$
$=\dfrac{x\left(4-\sqrt{x}\right)}{4^2-\left(\sqrt{x}\right)^2}$
$=\dfrac{4x-x\sqrt{x}}{16-x}$

97. $\dfrac{a}{4-\sqrt{b}}=\dfrac{a}{4-\sqrt{b}}\cdot\dfrac{4+\sqrt{b}}{4+\sqrt{b}}$
$=\dfrac{a\left(4+\sqrt{b}\right)}{\left(4-\sqrt{b}\right)\left(4+\sqrt{b}\right)}$
$=\dfrac{4a+a\sqrt{b}}{16+4\sqrt{b}-4\sqrt{b}-b}$
$=\dfrac{4a+a\sqrt{b}}{16-b}$

98. $\dfrac{x}{\sqrt{y}-7}=\dfrac{x}{\sqrt{y}-7}\cdot\dfrac{\sqrt{y}+7}{\sqrt{y}+7}$
$=\dfrac{x\left(\sqrt{y}+7\right)}{\left(\sqrt{y}-7\right)\left(\sqrt{y}+7\right)}$
$=\dfrac{x\sqrt{y}+7x}{y+7\sqrt{y}-7\sqrt{y}-49}$
$=\dfrac{x\sqrt{y}+7x}{y-49}$

99. $\dfrac{\sqrt{x}}{\sqrt{x}+\sqrt{y}}=\dfrac{\sqrt{x}}{\sqrt{x}+\sqrt{y}}\cdot\dfrac{\sqrt{x}-\sqrt{y}}{\sqrt{x}-\sqrt{y}}$
$=\dfrac{\sqrt{x}\left(\sqrt{x}-\sqrt{y}\right)}{\left(\sqrt{x}\right)^2-\left(\sqrt{y}\right)^2}$
$=\dfrac{\sqrt{x^2}-\sqrt{xy}}{x-y}$
$=\dfrac{x-\sqrt{xy}}{x-y}$

100. $\dfrac{\sqrt{x}-3\sqrt{y}}{\sqrt{x}-\sqrt{y}}=\dfrac{\sqrt{x}-3\sqrt{y}}{\sqrt{x}-\sqrt{y}}\cdot\dfrac{\sqrt{x}+\sqrt{y}}{\sqrt{x}+\sqrt{y}}$
$=\dfrac{x+\sqrt{xy}-3\sqrt{xy}-3y}{\left(\sqrt{x}\right)^2-\left(\sqrt{y}\right)^2}$
$=\dfrac{x-2\sqrt{xy}-3y}{x-y}$

101. $\dfrac{2}{\sqrt{a-1}-2}=\dfrac{2}{\sqrt{a-1}-2}\cdot\dfrac{\sqrt{a-1}+2}{\sqrt{a-1}+2}$
$=\dfrac{2\left(\sqrt{a-1}+2\right)}{\left(\sqrt{a-1}-2\right)\left(\sqrt{a-1}+2\right)}$
$=\dfrac{2\sqrt{a-1}+4}{a-1+2\sqrt{a-1}-2\sqrt{a-1}-4}$
$=\dfrac{2\sqrt{a-1}+4}{a-5}$

102. $\dfrac{5}{\sqrt{y+2}-3}=\dfrac{5}{\sqrt{y+2}-3}\cdot\dfrac{\sqrt{y+2}+3}{\sqrt{y+2}+3}$
$=\dfrac{5\sqrt{y+2}+15}{\left(\sqrt{y+2}\right)^2-3^2}$
$=\dfrac{5\sqrt{y+2}+15}{y+2-9}$
$=\dfrac{5\sqrt{y+2}+15}{y-7}$

103. $\sqrt[3]{x}+10\sqrt[3]{x}-2\sqrt[3]{x}=9\sqrt[3]{x}$

104. $\sqrt{3}+\sqrt{27}-\sqrt{192}=\sqrt{3}+3\sqrt{3}-\sqrt{64}\sqrt{3}$
$=\sqrt{3}+3\sqrt{3}-8\sqrt{3}$
$=-4\sqrt{3}$

105. $\sqrt[3]{16}-5\sqrt[3]{54}+3\sqrt[3]{64}$
$=\sqrt[3]{8}\sqrt[3]{2}-5\sqrt[3]{27}\sqrt[3]{2}+3\sqrt[3]{64}$
$=2\sqrt[3]{2}-5\left(3\sqrt[3]{2}\right)+3(4)$
$=2\sqrt[3]{2}-15\sqrt[3]{2}+12$
$=12-13\sqrt[3]{2}$

106. $\sqrt{2}-\frac{3}{\sqrt{32}}+\sqrt{50}$
$=\sqrt{2}-\frac{3}{4\sqrt{2}}+5\sqrt{2}$
$=\sqrt{2}-\frac{3}{4\sqrt{2}}\cdot\frac{\sqrt{2}}{\sqrt{2}}+5\sqrt{2}$
$=\sqrt{2}-\frac{3\sqrt{2}}{8}+5\sqrt{2}$
$=\frac{8}{8}\left(\sqrt{2}\right)-\frac{3\sqrt{2}}{8}+\left(\frac{8}{8}\right)5\sqrt{2}$
$=\frac{8\sqrt{2}}{8}-\frac{3\sqrt{2}}{8}+\frac{40\sqrt{2}}{8}$
$=\frac{45\sqrt{2}}{8}$

107. $9\sqrt{x^5y^6}-\sqrt{16x^7y^8}$
$=9\sqrt{x^4y^6}\sqrt{x}-\sqrt{16x^6y^8}\sqrt{x}$
$=9x^2y^3\sqrt{x}-4x^3y^4\sqrt{x}$
$=\left(9x^2y^3-4x^3y^4\right)\sqrt{x}$

108. $8\sqrt[3]{x^7y^8}-\sqrt[3]{x^4y^2}+3\sqrt[3]{x^{10}y^2}$
$=8\sqrt[3]{x^6y^6}\sqrt[3]{xy^2}-\sqrt[3]{x^3}\sqrt[3]{xy^2}+3\sqrt[3]{x^9}\sqrt[3]{xy^2}$
$=8x^2y^2\sqrt[3]{xy^2}-x\sqrt[3]{xy^2}+3x^3\sqrt[3]{xy^2}$
$=(8x^2y^2-x+3x^3)\sqrt[3]{xy^2}$

109. $(f\cdot g)(x)=f(x)\cdot g(x)$
$=\sqrt{3x}\cdot\left(\sqrt{6x}-\sqrt{15}\right)$
$=\sqrt{3x}\sqrt{6x}-\sqrt{3x}\sqrt{15}$
$=\sqrt{18x^2}-\sqrt{45x}$
$=\sqrt{9x^2}\sqrt{2}-\sqrt{9}\sqrt{5x}$
$=3x\sqrt{2}-3\sqrt{5x}$

110. $(f\cdot g)(x)=f(x)\cdot g(x)$
$=\sqrt[3]{2x^2}\left(\sqrt[3]{4x^4}+\sqrt[3]{16x^5}\right)$
$=\sqrt[3]{2x^2}\sqrt[3]{4x^4}+\sqrt[3]{2x^2}\sqrt[3]{16x^5}$
$=\sqrt[3]{8x^6}+\sqrt[3]{32x^7}$
$=\sqrt[3]{8x^6}+\sqrt[3]{8x^6}\sqrt[3]{4x}$
$=2x^2+2x^2\sqrt[3]{4x}$

111. $f(x)=\sqrt{2x+7}\sqrt{2x+7},\ x\ge-\frac{7}{2}$
$=\sqrt{(2x+7)^2}$
$=|2x+7|$
$=2x+7\quad\text{since } x\ge-\frac{7}{2}$

112. $g(a)=\sqrt{20a^2+100a+125}$
$=\sqrt{5(4a^2+20a+25)}$
$=\sqrt{5(2a+5)^2}$
$=\sqrt{5}\,|2a+5|$

113. $\frac{\sqrt[3]{(x+5)^5}}{\sqrt{(x+5)^3}}=\frac{(x+5)^{5/3}}{(x+5)^{3/2}}$
$=(x+5)^{5/3-3/2}$
$=(x+5)^{1/6}$
$=\sqrt[6]{x+5}$

114. $\frac{\sqrt[3]{a^3b^2}}{\sqrt[4]{a^4b}} = \frac{a\sqrt[3]{b^2}}{a\sqrt[4]{b}}$

$= \frac{\sqrt[3]{b^2}}{\sqrt[4]{b}}$

$= \frac{b^{2/3}}{b^{1/4}}$

$= b^{2/3-1/4}$

$= b^{5/12}$

$= \sqrt[12]{b^5}$

115. a. $P = 2l + 2w$

$P = 2\sqrt{48} + 2\sqrt{12}$

$= 2\sqrt{16\cdot 3} + 2\sqrt{4\cdot 3}$

$= 8\sqrt{3} + 4\sqrt{3}$

$= 12\sqrt{3}$

b. $A = lw$

$A = \left(\sqrt{48}\right)\left(\sqrt{12}\right)$

$= \sqrt{576}$

$= 24$

116. a. $P = s_1 + s_2 + s_3$

$P = \sqrt{125} + \sqrt{45} + \sqrt{130}$

$= 5\sqrt{5} + 3\sqrt{5} + \sqrt{130}$

$= 8\sqrt{5} + \sqrt{130}$

b. $A = \frac{1}{2}bh$

$A = \frac{1}{2}\left(\sqrt{130}\right)\left(\sqrt{20}\right)$

$= \frac{1}{2}\sqrt{5200}$

$= \frac{1}{2}\sqrt{400\cdot 13}$

$= \frac{20}{2}\sqrt{13}$

$= 10\sqrt{13}$

117. a. $f(x) = \sqrt{x} + 2$

$g(x) = -3$

$(f+g)(x) = f(x) + g(x)$

$= \sqrt{x} + 2 - 3$

$= \sqrt{x} - 1$

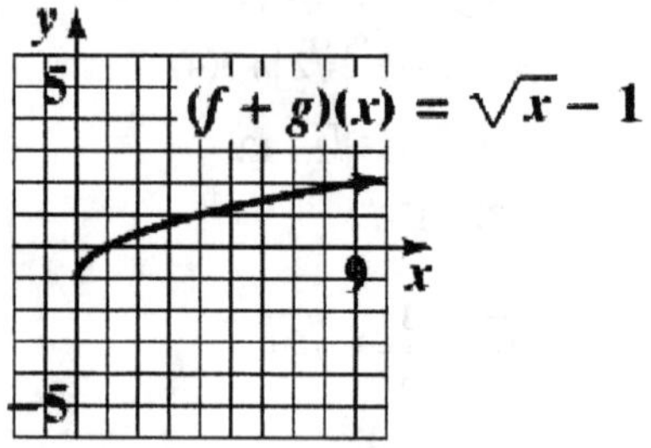

b. $\sqrt{x} \geq 0, x \geq 0$

The domain is $\{x|\ x \geq 0\}$.

118. a. $f(x) = -\sqrt{x}$

$g(x) = \sqrt{x} + 2$

$(f+g)(x) = f(x) + g(x)$

$= -\sqrt{x} + \sqrt{x} + 2$

$= 2$

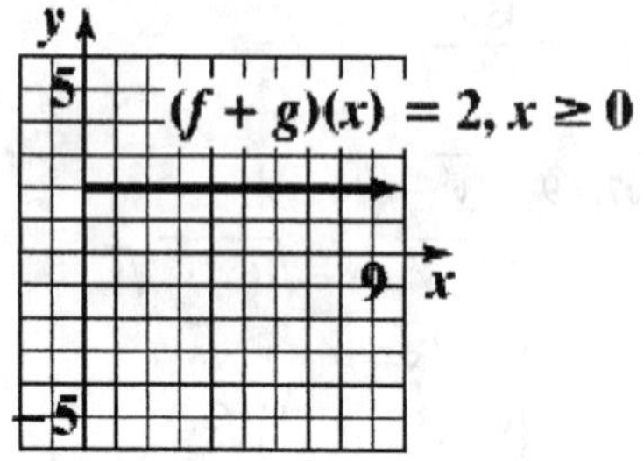

b. $\sqrt{x} \geq 0, x \geq 0$

The domain is $\{x|\ x \geq 0\}$.

119. $\sqrt{x} = 9$

$\left(\sqrt{x}\right)^2 = 9^2$

$x = 81$

Check: $\sqrt{81} = 9$ True

81 is the solution.

120. $\sqrt{x}=-4$

$\left(\sqrt{x}\right)^2=(-4)^2$

$x=16$

Check: $\sqrt{16}=-4$

$4=-4$ False

no solution

121. $\sqrt[3]{x}=4$

$\left(\sqrt[3]{x}\right)^3=4^3$

$x=64$

Check: $\sqrt[3]{x}=4$

$\sqrt[3]{64}=4$

$4=4$ True

The solution is 64.

122. $\sqrt[3]{x}=-5$

$\left(\sqrt[3]{x}\right)^3=(-5)^3$

$x=-125$

Check: $\sqrt[3]{x}=-5$

$\sqrt[3]{-125}=-5$

$-5=-5$ True

-125 is the solution.

123. $7+\sqrt{x}=10$

$\sqrt{x}=3$

$\left(\sqrt{x}\right)^2=3^2$

$x=9$

Check: $7+\sqrt{x}=10$

$7+\sqrt{9}=10$

$7+3=10$

$10=10$ True

9 is the solution.

124. $7+\sqrt[3]{x}=12$

$\sqrt[3]{x}=5$

$\left(\sqrt[3]{x}\right)^3=5^3$

$x=125$

Check: $7+\sqrt[3]{x}=12$

$7+\sqrt[3]{125}=12$

$7+5=12$

$12=12$ True

125 is the solution.

125 $\sqrt{3x+4}=\sqrt{5x+14}$

$\left(\sqrt{3x+4}\right)^2=\left(\sqrt{5x+14}\right)^2$

$3x+4=5x+14$

$-10=2x$

$-5=x$

Check: $\sqrt{3x+4}$ becomes

$\sqrt{3(-5)+4}=\sqrt{-15+4}=\sqrt{-11}$

which is not a real number. -5 is not a solution so there is no real solution.

126. $\sqrt{x^2+2x-8}=x$

$\left(\sqrt{x^2+2x-8}\right)^2=(x)^2$

$x^2+2x-8=x^2$

$2x-8=0$

$2x=8$

$x=4$

Check: $\sqrt{2^2+2\cdot 4-8}=2$

$\sqrt{4+8-8}=2$

$\sqrt{4}=2$

$2=2$ True

4 is the solution.

127. $\sqrt[3]{x-9}=\sqrt[3]{5x+3}$

$$\left(\sqrt[3]{x-9}\right)^3=\left(\sqrt[3]{5x+3}\right)^3$$
$$x-9=5x+3$$
$$-4x=12$$
$$x=-3$$

Check: $\sqrt[3]{-3-9}=\sqrt[3]{5(-3)+3}$

$$\sqrt[3]{-12}=\sqrt[3]{-15+3}$$
$$\sqrt[3]{-12}=\sqrt[3]{-12}\text{ True}$$

−3 is the solution.

128. $(x^2+7)^{1/2}=x+1$

$$\left[(x^2+7)^{1/2}\right]^2=(x+1)^2$$
$$x^2+7=x^2+2x+1$$
$$7=2x+1$$
$$6=2x$$
$$x=3$$

Check: $(3^2+7)^{1/2}=3+1$

$$(9+7)^{1/2}=4$$
$$16^{1/2}=4$$
$$4=4\text{ True}$$

3 is the solution.

129. $\sqrt{x}+3=\sqrt{3x+9}$

$$\left(\sqrt{x}+3\right)^2=\left(\sqrt{3x+9}\right)^2$$
$$\left(\sqrt{x}+3\right)\left(\sqrt{x}+3\right)=3x+9$$
$$x+6\sqrt{x}+9=3x+9$$
$$6\sqrt{x}=2x$$
$$\left(6\sqrt{x}\right)^2=(2x)^2$$
$$36x=4x^2$$
$$4x^2-36x=0$$
$$4x(x-9)=0$$
$$4x=0\quad\text{or}\quad x-9=0$$
$$x=0\qquad\qquad x=9$$

A check shows that 0 and 9 are both solutions.

130. $\sqrt{6x-5}-\sqrt{2x+6}-1=0$

$$\sqrt{6x-5}=\sqrt{2x+6}+1$$
$$\left(\sqrt{6x-5}\right)^2=\left(\sqrt{2x+6}+1\right)^2$$
$$6x-5=2x+6+2\sqrt{2x+6}+1$$
$$6x-5=2x+7+2\sqrt{2x+6}$$
$$4x-12=2\sqrt{2x+6}$$
$$\frac{4x}{2}-\frac{12}{2}=\frac{2}{2}\sqrt{2x+6}$$
$$2x-6=\sqrt{2x+6}$$
$$(2x-6)^2=\left(\sqrt{2x+6}\right)^2$$
$$4x^2-24x+36=2x+6$$
$$4x^2-26x+30=0$$
$$2x^2-13x+15=0$$

Express the middle term, $-13x$, as $-10x-3x$.

$$2x^2-10x-3x+15=0$$
$$2x(x-5)-3(x-5)=0$$
$$(2x-3)(x-5)=0$$
$$2x-3=0\quad\text{or}\quad x-5=0$$
$$x=\frac{3}{2}\qquad\qquad x=5$$

The solution $x=5$ checks in the original equation but $x=\frac{3}{2}$ does not check. Therefore the only solution is $x=5$.

131. $f(x)=g(x)$

$$\sqrt{3x+4}=2\sqrt{2x-4}$$
$$\left(\sqrt{3x+4}\right)^2=\left(2\sqrt{2x-4}\right)^2$$
$$3x+4=4(2x-4)$$
$$3x+4=8x-16$$
$$20=5x$$
$$4=x$$

132. $f(x)=g(x)$

$$(4x+5)^{1/3}=(6x-7)^{1/3}$$
$$[(4x+5)^{1/3}]^3=[(6x-7)^{1/3}]^3$$
$$4x+5=6x-7$$
$$12=2x$$
$$6=x$$

133. $V=\sqrt{\frac{2L}{w}}$ Solve for L.

$V^2=\frac{2L}{w}$

$V^2w=2L$

$\frac{V^2w}{2}=L$ or $L=\frac{V^2w}{2}$

134. $r=\sqrt{\frac{A}{\pi}}$ Solve for A.

$r^2=\left(\sqrt{\frac{A}{\pi}}\right)^2$

$r^2=\frac{A}{\pi}$

$\pi r^2=A$ or $A=\pi r^2$

135. Pythagorean Theorem: $a^2+b^2=c^2$

$\left(\sqrt{20}\right)^2+6^2=x^2$

$20+36=x^2$

$56=x^2$

$\sqrt{56}=x$ or $x=2\sqrt{14}$

136. Pythagorean Theorem: $a^2+b^2=c^2$

$\left(\sqrt{26}\right)^2+x^2=\left(\sqrt{101}\right)^2$

$26+x^2=101$

$x^2=75$

$x=\sqrt{75}$ or $x=5\sqrt{3}$

137. $l=\sqrt{a^2+b^2}$

$=\sqrt{5^2+2^2}$

$=\sqrt{29}$

≈ 5.39 m

138. $v=\sqrt{2gh}$

$=\sqrt{2(32)(20)}$

$=\sqrt{1280}\approx 35.78$ ft/sec

139. $T=2\pi\sqrt{\frac{L}{32}}$

$=2\pi\sqrt{\frac{64}{32}}$

$=2\pi\sqrt{2}$

≈ 8.89 sec

140. $V=\sqrt{\frac{2K}{m}}$

$=\sqrt{\frac{2(45)}{0.145}}$

≈ 24.91 meters per second

141. $m=\frac{m_0}{\sqrt{1-\frac{v^2}{c^2}}}$

$=\frac{m_0}{\sqrt{1-\frac{(0.98c)^2}{c^2}}}$

$=\frac{m_0}{\sqrt{1-\frac{0.9604c^2}{c^2}}}$

$=\frac{m_0}{\sqrt{1-0.9604}}$

$=\frac{m_0}{\sqrt{0.0396}}$

$\approx 5m_0$

It is about 5 times its original mass.

142. $5=5+0i$

143. $-8=-8+0i$

144. $7-\sqrt{-256}=7-\sqrt{-1}\sqrt{256}$

$=7-16i$

145. $9+\sqrt{-16}=9+\sqrt{16}\sqrt{-1}$

$=9+4i$

146. $(3+2i)+(10-i)=3+2i+10-i=13+i$

147. $(9-6i)-(3-4i)=9-6i-3+4i=6-2i$

148. $\left(\sqrt{3}+\sqrt{-5}\right)+\left(11\sqrt{3}-\sqrt{-7}\right)$
$=\sqrt{3}+\sqrt{5}\sqrt{-1}+11\sqrt{3}-\sqrt{7}\sqrt{-1}$
$=\sqrt{3}+i\sqrt{5}+11\sqrt{3}-i\sqrt{7}$
$=12\sqrt{3}+\left(\sqrt{5}-\sqrt{7}\right)i$

149. $\sqrt{-6}\left(\sqrt{6}+\sqrt{-6}\right)=\sqrt{6}\sqrt{-1}\left(\sqrt{6}+\sqrt{6}\sqrt{-1}\right)$
$=i\sqrt{6}\left(\sqrt{6}+i\sqrt{6}\right)$
$=i\sqrt{36}+i^2\sqrt{36}$
$=6i+6(-1)=-6+6i$

150. $(4+3i)(2-3i)=8-12i+6i-9i^2$
$=8-6i-9(-1)$
$=8-6i+9$
$=17-6i$

151. $\left(6+\sqrt{-3}\right)\left(4-\sqrt{-15}\right)$
$=\left(6+\sqrt{3}\sqrt{-1}\right)\left(4-\sqrt{15}\sqrt{-1}\right)$
$=\left(6+i\sqrt{3}\right)\left(4-i\sqrt{15}\right)$
$=24-6i\sqrt{15}+4i\sqrt{3}-i^2\sqrt{45}$
$=24-6i\sqrt{15}+4i\sqrt{3}-(-1)\sqrt{9}\sqrt{5}$
$=\left(24+3\sqrt{5}\right)+\left(4\sqrt{3}-6\sqrt{15}\right)i$

152. $\frac{8}{3i}=\frac{8}{3i}\cdot\frac{-3i}{-3i}$
$=\frac{-24i}{-9i^2}$
$=\frac{-24i}{-9(-1)}$
$=-\frac{24i}{9}$
$=-\frac{8i}{3}$

153. $\frac{2+\sqrt{3}}{2i}=\frac{2+\sqrt{3}}{2i}\cdot\frac{-2i}{-2i}$
$=\frac{-4i-2i\sqrt{3}}{-4i^2}$
$=\frac{-4i-2i\sqrt{3}}{-4(-1)}$
$=\frac{2\left(-2i-i\sqrt{3}\right)}{4}$
$=\frac{-2i-i\sqrt{3}}{2}$
$=\frac{\left(-2-\sqrt{3}\right)i}{2}$

154. $\frac{4}{3+2i}=\frac{4}{3+2i}\cdot\frac{3-2i}{3-2i}$
$=\frac{4(3-2i)}{9-4i^2}$
$=\frac{12-8i}{9-4(-1)}$
$=\frac{12-8i}{9+4}$
$=\frac{12-8i}{13}$

155. $\frac{\sqrt{3}}{5-\sqrt{-6}}=\frac{\sqrt{3}}{5-i\sqrt{6}}$
$=\frac{\sqrt{3}}{\left(5-i\sqrt{6}\right)}\cdot\frac{5+i\sqrt{6}}{5+i\sqrt{6}}$
$=\frac{5\sqrt{3}+i\sqrt{18}}{(5)^2-\left(i\sqrt{6}\right)^2}$
$=\frac{5\sqrt{3}+3i\sqrt{2}}{(5)^2+\left(\sqrt{6}\right)^2}$
$=\frac{5\sqrt{3}+3i\sqrt{2}}{25+6}$
$=\frac{5\sqrt{3}+3i\sqrt{2}}{31}$

156. $x^2 - 2x + 9$

$= \left(1 + 2i\sqrt{2}\right)^2 - 2\left(1 + 2i\sqrt{2}\right) + 9$

$= 1^2 + 2(1)\left(2i\sqrt{2}\right) + \left(2i\sqrt{2}\right)^2 - 2 - 4i\sqrt{2} + 9$

$= 1 + 4i\sqrt{2} - 8 - 2 - 4i\sqrt{2} + 9$

$= 0 + 0i$

$= 0$

157. $x^2 - 2x + 12$

$= (1 - 2i)^2 - 2(1 - 2i) + 12$

$= 1^2 - 2(1)(2i) + (2i)^2 - 2 + 4i + 12$

$= 1 - 4i - 4 - 2 + 4i + 12$

$= 7 + 0i$

$= 7$

158. $i^{33} = i^{32}i = (i^4)^8 = 1^8 \cdot i = i$

159. $i^{59} = i^{56}i^3 = (i^4)^{14}i^3 = 1^{14} \cdot i^3 = 1(i^3) = i^3 = -i$

160. $i^{404} = (i^4)^{101} = 1^{101} = 1$

161. $i^{802} = i^{800}i^2$

$= (i^4)^{200}i^2$

$= 1^{200} \cdot i^2$

$= 1(i^2)$

$= i^2$

$= -1$

Chapter 7 Practice Test

1. $\sqrt{(5x-3)^2} = |5x - 3|$

2. $\left(\dfrac{x^{2/5} \cdot x^{-1}}{x^{3/5}}\right)^2 = \left(x^{2/5-3/5-1}\right)^2$

$= \left(x^{2/5-3/5-5/5}\right)^2$

$= \left(x^{-6/5}\right)^2$

$= x^{-12/5}$

$= \dfrac{1}{x^{12/5}}$

3. $x^{-2/3} + x^{4/3} = x^{-2/3}(1) + x^{-2/3}\left(x^{6/3}\right)$

$= x^{-2/3}\left(1 + x^{6/3}\right)$

$= x^{-2/3}\left(1 + x^2\right)$

$= \dfrac{1 + x^2}{x^{2/3}}$

4. $g(x) = \sqrt{x} + 1$

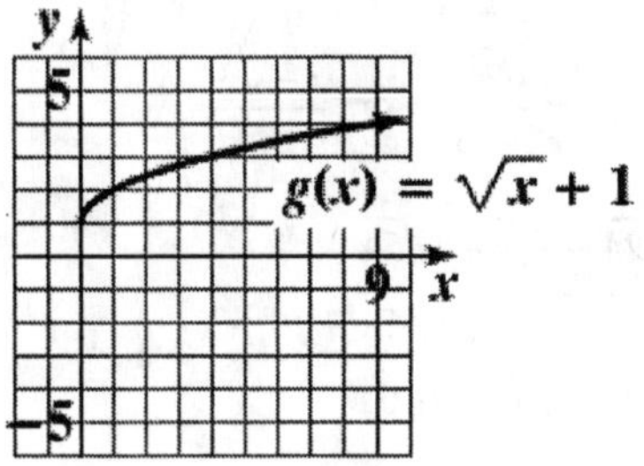

5. $\sqrt{54x^7y^{10}} = \sqrt{9x^6y^{10}}\sqrt{6x}$

$= 3x^3y^5\sqrt{6x}$

6. $\sqrt[3]{25x^5y^2}\sqrt[3]{10x^6y^8} = \sqrt[3]{250x^{11}y^{10}}$

$= \sqrt[3]{125x^9y^9} \cdot \sqrt[3]{2x^2y}$

$= 5x^3y^3\sqrt[3]{2x^2y}$

7. $\sqrt{\dfrac{7x^6y^3}{8z}} = \dfrac{\sqrt{7x^6y^3}}{\sqrt{8z}}$

$= \dfrac{\sqrt{x^6y^2}\sqrt{7y}}{\sqrt{4}\sqrt{2z}}$

$= \dfrac{x^3y\sqrt{7y}}{2\sqrt{2z}} \cdot \dfrac{\sqrt{2z}}{\sqrt{2z}}$

$= \dfrac{x^3y\sqrt{14yz}}{2(2z)}$

$= \dfrac{x^3y\sqrt{14yz}}{4z}$

8. $\dfrac{9}{\sqrt[3]{x}} = \dfrac{9}{\sqrt[3]{x}} \cdot \dfrac{\sqrt[3]{x^2}}{\sqrt[3]{x^2}} = \dfrac{9\sqrt[3]{x^2}}{x}$

9. $\dfrac{\sqrt{3}}{3+\sqrt{27}} = \dfrac{\sqrt{3}}{3+\sqrt{27}} \cdot \dfrac{3-\sqrt{27}}{3-\sqrt{27}}$

$= \dfrac{\sqrt{3}\left(3-\sqrt{27}\right)}{\left(3+\sqrt{27}\right)\left(3-\sqrt{27}\right)}$

$= \dfrac{3\sqrt{3}-\sqrt{81}}{9-3\sqrt{27}+3\sqrt{27}-27}$

$= \dfrac{3\sqrt{3}-9}{-18}$

$= \dfrac{\sqrt{3}-3}{-6}$ or $\dfrac{3-\sqrt{3}}{6}$

10. $2\sqrt{24}-6\sqrt{6}+3\sqrt{54}$

$= 2\sqrt{4}\sqrt{6}-6\sqrt{6}+3\sqrt{9}\sqrt{6}$

$= 4\sqrt{6}-6\sqrt{6}+9\sqrt{6}$

$= 7\sqrt{6}$

11. $\sqrt[3]{8x^3y^5}+4\sqrt[3]{x^6y^8}$

$= \sqrt[3]{8x^3y^3}\sqrt[3]{y^2}+4\sqrt[3]{x^6y^6}\sqrt[3]{y^2}$

$= 2xy\sqrt[3]{y^2}+4x^2y^2\sqrt[3]{y^2}$

$= (2xy+4x^2y^2)\sqrt[3]{y^2}$

12. $\left(\sqrt{3}-2\right)\left(6-\sqrt{8}\right) = \sqrt{3}(6)-\sqrt{3}\sqrt{8}-2(6)+2\sqrt{8}$

$= 6\sqrt{3}-\sqrt{24}-12+2\sqrt{8}$

$= 6\sqrt{3}-\sqrt{4\cdot 6}-12+2\sqrt{4\cdot 2}$

$= 6\sqrt{3}-2\sqrt{6}-12+4\sqrt{2}$

13. $\sqrt[4]{\sqrt{x^5y^3}} = \sqrt[4]{\left(x^5y^3\right)^{1/2}}$

$= \left[\left(x^5y^3\right)^{1/2}\right]^{1/4}$

$= \left(x^5y^3\right)^{1/8}$

$= \sqrt[8]{x^5y^3}$

14. $\dfrac{\sqrt[4]{(7x+2)^5}}{\sqrt[3]{(7x+2)^2}} = \dfrac{(7x+2)^{5/4}}{(7x+2)^{2/3}}$

$= (7x+2)^{5/4-2/3}$

$= (7x+2)^{7/12}$

$= \sqrt[12]{(7x+2)^7}$

15. $\sqrt{2x+19} = 3$

$\left(\sqrt{2x+19}\right)^2 = 3^2$

$2x+19 = 9$

$2x = -10$

$x = -5$

Check: $\sqrt{2x+19} = 3$

$\sqrt{2(-5)+19} = 3$

$\sqrt{9} = 3$

$3 = 3$ True

This check confirms that -5 is the solution to the equation.

16. $\sqrt{x^2-x-12} = x+3$

$\left(\sqrt{x^2-x-12}\right)^2 = (x+3)^2$

$x^2-x-12 = x^2+6x+9$

$-x-12 = 6x+9$

$-12 = 7x+9$

$-21 = 7x$

$x = -3$

Check:

$\sqrt{(-3)^2-(-3)-12} = -3+3$

$\sqrt{9+3-12} = -3+3$

$\sqrt{0} = 0$

$0 = 0$

This check confirms that -3 is the solution to the equation.

17. $\sqrt{a-8}=\sqrt{a}-2$

$$\left(\sqrt{a-8}\right)^2=\left(\sqrt{a}-2\right)^2$$
$$a-8=a-4\sqrt{a}+4$$
$$-12=-4\sqrt{a}$$
$$\sqrt{a}=3$$
$$a=3^2=9$$

Check:

$$\sqrt{a-8}=\sqrt{a}-2$$
$$\sqrt{9-8}=\sqrt{9}-2$$
$$\sqrt{1}=3-2$$
$$1=1 \quad \text{True}$$

This check confirms that 9 is the solution.

18. $f(x)=g(x)$

$$(9x+37)^{1/3}=2(2x+2)^{1/3}$$
$$[(9x+37)^{1/3}]^3=[2(2x+2)^{1/3}]^3$$
$$9x+37=8(2x+2)$$
$$9x+37=16x+16$$
$$21=7x$$
$$3=x$$

19. $w=\dfrac{\sqrt{2gh}}{4}$ Solve for g.

$$4w=\sqrt{2gh}$$
$$(4w)^2=2gh$$
$$\frac{16w^2}{2h}=\frac{2gh}{2h}$$
$$\frac{8w^2}{h}=g$$

20. $V=\sqrt{64.4h}$

$$V=\sqrt{64.4(200)}$$
$$=\sqrt{12{,}880}$$
$$\approx 113.49 \text{ ft/sec}$$

21. Let x be the length of the ladder.

$$x=\sqrt{12^2+5^2}$$
$$=169$$
$$=13 \text{ feet}$$

22. $T=2\pi\sqrt{\dfrac{m}{k}}$

$$T=2\pi\sqrt{\frac{1400}{65{,}000}}$$
$$\approx 0.92 \text{ sec}$$

23. $\left(6-\sqrt{-4}\right)\left(2+\sqrt{-16}\right)=(6-2i)(2+4i)$

$$=12+24i-4i-8i^2$$
$$=12+20i-8(-1)$$
$$=12+20i+8$$
$$=20+20i$$

24. $\dfrac{5-i}{7+2i}=\dfrac{5-i}{7+2i}\cdot\dfrac{7-2i}{7-2i}$

$$=\frac{(5-i)(7-2i)}{(7+2i)(7-2i)}$$
$$=\frac{35-10i-7i+2i^2}{49-14i+14i-4i^2}$$
$$=\frac{35-17i+2(-1)}{49-4(-1)}$$
$$=\frac{35-17i-2}{49+4}$$
$$=\frac{33-17i}{53}$$

25. $x^2+6x+12$

$$=(-3+i)^2+6(-3+i)+12$$
$$=(-3)^2-2(3)(i)+(i)^2-18+6i+12$$
$$=9-6i-1-18+6i+12$$
$$=2+0i$$
$$=2$$

Chapter 7 Cumulative Review Test

1. $\frac{1}{5}(x-3)=\frac{3}{4}(x+3)-x$

$20\left[\frac{1}{5}(x-3)\right]=20\left[\frac{3}{4}(x+3)\right]-20x$

$4(x-3)=5(3)(x+3)-20x$

$4x-12=15x+45-20x$

$4x-12=-5x+45$

$9x-12=45$

$9x=57$

$x=\frac{57}{9}$

2. $3(x-4)=6x-(4-5x)$

$3x-12=6x-4+5x$

$3x-12=11x-4$

$-8x=8$

$x=-1$

3. Let x be the original price of the sweater.

$x-60\%x=16$

$x-0.60x=16$

$0.40x=16$

$x=\frac{16}{0.40}$

$x=40$

The original price of the sweater is \$40.

4. $|3-2x|<5$

$-5<3-2x<5$

$-5-3<3-2x-3<5-3$

$-8<-2x<2$

$\frac{-8}{-2}>\frac{-2x}{-2}>\frac{2}{-2}$

$4>x>-1$ or $-1<x<4$

The solution set is $\{x|-1<x<4\}$

5. $y=\frac{3}{2}x-3$

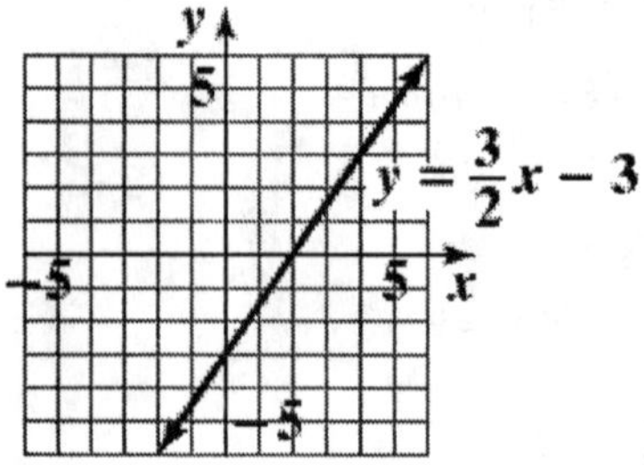

6. $y=3x-8 \Rightarrow m_1=3$

$6y=18x+12 \Rightarrow y=3x+2 \Rightarrow m_2=3$

The slope of both lines is 3. Since the slopes of the two lines are the same, and the y-intercepts are different, the lines are parallel.

7. $f(x)=x^2-3x+4$ and $g(x)=2x-9$

$(g-f)(x)=(2x-9)-(x^2-3x+4)$

$=2x-9-x^2+3x-4$

$=-x^2+5x-13$

8. First find the slope of the given line.

$3x-2y=6$

$-2y=-3x+6$

$y=\frac{3}{2}x-3 \Rightarrow$ The slope is $\frac{3}{2}$.

The slope of any line perpendicular to this line must have an opposite reciprocal slope. Therefore, the slope of the line perpendicular to the given line is $-\frac{2}{3}$. Finally, use the point-slope formula to find the equation of the line.

$y-y_1=m(x-x_1)$ with $m=-\frac{2}{3}$ and $(1,-4)$

$y-(-4)=-\frac{2}{3}(x-1)$

$y+4=-\frac{2}{3}x+\frac{2}{3}$

$y=-\frac{2}{3}x-\frac{10}{3}$

9. $x+2y=12 \quad (1)$

$4x=8 \quad (2)$

$3x-4y+5z=20 \quad (3)$

Using equation (2), solve for x.

$4x=8 \Rightarrow x=\frac{8}{4} \Rightarrow x=2$

Substitute 2 for x in equation (1) in order to solve for y.

$2+2y=12 \Rightarrow 2y=10 \Rightarrow y=5$

Substitute 2 for x and 5 for y in equation (3) in order to solve for z.

$$3(2)-4(5)+5z=20$$
$$6-20+5z=20$$
$$-14+5z=20$$
$$5z=34$$
$$z=\frac{34}{5}$$

The solution is $\left(2, 5, \frac{34}{5}\right)$.

10. $\begin{vmatrix} 3 & -6 & -1 \\ 2 & 1 & -2 \\ 1 & 3 & 1 \end{vmatrix}$

$$=3\begin{vmatrix} 1 & -2 \\ 3 & 1 \end{vmatrix}-(-6)\begin{vmatrix} 2 & -2 \\ 1 & 1 \end{vmatrix}+(-1)\begin{vmatrix} 2 & 1 \\ 1 & 3 \end{vmatrix}$$
$$=3(1-(-6))+6(2-(-2))-(6-1)$$
$$=3(1+6)+6(2+2)-(6-1)$$
$$=3(7)+6(4)-(5)$$
$$=21+24-5$$
$$=40$$

11. $V=lwh$

$$6r^3+5r^2+r=(3r+1)(w)(r)$$
$$\frac{6r^3+5r^2+r}{(3r+1)(r)}=\frac{(3r+1)(w)(r)}{(3r+1)(r)}$$
$$\frac{r(6r^2+5r+1)}{r(3r+1)}=w$$
$$\frac{r(3r+1)(2r+1)}{r(3r+1)}=w$$
$$w=2r+1$$

12. $(5xy-3)(5xy+3)$

$$=25x^2y^2+15xy-15xy-9$$
$$=25x^2y^2-9$$

13. $\sqrt{2x^2+7}+3=8$

$$\sqrt{2x^2+7}=5$$
$$\left(\sqrt{2x^2+7}\right)^2=5^2$$
$$2x^2+7=25$$
$$2x^2=18$$
$$x^2=9$$
$$x=3 \text{ or } -3$$

Check:

$\sqrt{2(3)^2+7}+3=8$ or $\sqrt{2(-3)^2+7}+3=8$

$\sqrt{25}=5 \qquad \sqrt{25}=5$

$5=5 \qquad 5=5$

Both values check. The solutions are 3 and –3.

14. $4x^3-9x^2+5x=x(4x^2-9x+5)$

$$=x(4x-5)(x-1)$$

15. $(x+1)^3-27=(x+1)^3-3^3$

$$=(x+1-3)\left((x+1)^2+3(x+1)+3^2\right)$$
$$=(x-2)(x^2+2x+1+3x+3+9)$$
$$=(x-2)(x^2+5x+13)$$

16. $8x^2-3=-10x$

$$8x^2+10x-3=0$$
$$(4x-1)(2x+3)=0$$

$4x-1=0$ or $2x+3=0$

$4x=1 \qquad 2x=-3$

$x=\frac{1}{4} \qquad x=-\frac{3}{2}$

17. $\frac{4x+4y}{x^2y}\cdot\frac{y^3}{12x}=\frac{4(x+y)}{x^2y}\cdot\frac{y^3}{12x}$

$=\frac{x+y}{x^2}\cdot\frac{y^2}{3x}$

$=\frac{(x+y)y^2}{3x^3}$

18. $\frac{x-4}{x-5}-\frac{3}{x+5}-\frac{10}{x^2-25}$

$=\frac{x-4}{x-5}-\frac{3}{x+5}-\frac{10}{(x+5)(x-5)}$

$=\frac{x-4}{x-5}\cdot\frac{x+5}{x+5}-\frac{3}{x+5}\cdot\frac{x-5}{x-5}-\frac{10}{(x+5)(x-5)}$

$=\frac{(x-4)(x+5)}{(x+5)(x-5)}-\frac{3(x-5)}{(x+5)(x-5)}-\frac{10}{(x+5)(x-5)}$

$=\frac{(x-4)(x+5)-3(x-5)-10}{(x+5)(x-5)}$

$=\frac{x^2+x-20-3x+15-10}{(x+5)(x-5)}$

$=\frac{x^2-2x-15}{(x+5)(x-5)}$

$=\frac{(x-5)(x+3)}{(x+5)(x-5)}$

$=\frac{x+3}{x+5}$

19. $\frac{4}{x}-\frac{1}{6}=\frac{1}{x}$

$6x\left(\frac{4}{x}-\frac{1}{6}=\frac{1}{x}\right)$

$24-x=6$

$-x=-18$

$x=18$

Upon checking, this value satisfies the equation.
The solution is 18.

20. $d=kt^2$

$16=k(1)^2 \Rightarrow k=16$

$d=16(5)^2$

$=16\cdot 25$

$=400$

The object will fall 400 feet in 5 seconds.

Chapter 8

Exercise Set 8.1

1. The two square roots of 36 are $\pm\sqrt{36} = \pm 6$.

3. The square root property is: If $x^2 = a$, where a is a real number, then $x = \pm\sqrt{a}$.

5. A trinomial, $x^2 + bx + c$, is a perfect square trinomial if $\left(\frac{b}{2}\right)^2 = c$.

7. **a.** Yes, $x = 4$ is the solution to the equation. It is the only real number that satisfies the equation.

b. No, $x = 2$ is not the solution. Both -2 and 2 satisfy the equation.

9. Multiply the equation by $\frac{1}{2}$ to obtain a leading coefficient of 1.

11. You should add the square of half the coefficient of the first degree term: $\left(\frac{-6}{2}\right)^2 = (-3)^2 = 9$.

13.
$$\begin{aligned} x^2 - 25 &= 0 \\ x^2 &= 25 \\ x &= \pm\sqrt{25} = \pm 5 \end{aligned}$$

15.
$$\begin{aligned} x^2 + 49 &= 0 \\ x^2 &= -49 \\ x &= \pm\sqrt{-49} = \pm 7i \end{aligned}$$

17.
$$\begin{aligned} y^2 + 24 &= 0 \\ y^2 &= -24 \\ y &= \pm\sqrt{-24} = \pm 2i\sqrt{6} \end{aligned}$$

19.
$$\begin{aligned} y^2 + 10 &= -51 \\ y^2 &= -61 \\ y &= \pm\sqrt{-61} = \pm i\sqrt{61} \end{aligned}$$

21.
$$\begin{aligned} (p-4)^2 &= 16 \\ p-4 &= \pm\sqrt{16} \\ p-4 &= \pm 4 \\ p &= 4 \pm 4 \end{aligned}$$
$$p = 4+4 \quad \text{or} \quad p = 4-4$$
$$p = 8 \qquad\qquad p = 0$$

23.
$$\begin{aligned} (x+3)^2 + 25 &= 0 \\ (x+3)^2 &= -25 \\ x+3 &= \pm\sqrt{-25} \\ x+3 &= \pm 5i \\ x &= -3 \pm 5i \end{aligned}$$

25.
$$\begin{aligned} (a-2)^2 + 45 &= 0 \\ (a-2)^2 &= -45 \\ a-2 &= \pm\sqrt{-45} \\ a-2 &= \pm 3i\sqrt{5} \\ a &= 2 \pm 3i\sqrt{5} \end{aligned}$$

27.
$$\begin{aligned} \left(b+\frac{1}{3}\right)^2 &= \frac{4}{9} \\ b+\frac{1}{3} &= \pm\sqrt{\frac{4}{9}} \\ b+\frac{1}{3} &= \pm\frac{2}{3} \\ b &= -\frac{1}{3} \pm \frac{2}{3} \end{aligned}$$
$$b = -\frac{1}{3} + \frac{2}{3} \quad \text{or} \quad b = -\frac{1}{3} - \frac{2}{3}$$
$$b = \frac{1}{3} \qquad\qquad b = -\frac{3}{3}$$
$$b = -1$$

29. $\left(b-\frac{2}{3}\right)^2+\frac{4}{9}=0$

$$\left(b-\frac{2}{3}\right)^2=-\frac{4}{9}$$
$$b-\frac{2}{3}=\pm\sqrt{-\frac{4}{9}}$$
$$b-\frac{2}{3}=\pm\frac{2}{3}i$$
$$b=\frac{2}{3}\pm\frac{2}{3}i \text{ or } b=\frac{2\pm 2i}{3}$$

31. $(x+0.8)^2=0.81$

$$x+0.8=\pm\sqrt{0.81}$$
$$x+0.8=\pm 0.9$$
$$x=-0.8\pm 0.9$$
$$x=-0.8+0.9 \quad \text{or} \quad x=-0.8-0.9$$
$$x=0.1 \qquad\qquad x=-1.7$$

33. $(2a-5)^2=18$

$$2a-5=\pm\sqrt{18}$$
$$2a-5=\pm 3\sqrt{2}$$
$$2a=5\pm 3\sqrt{2}$$
$$a=\frac{5\pm 3\sqrt{3}}{2}$$

35. $\left(2y+\frac{1}{2}\right)^2=\frac{4}{25}$

$$2y+\frac{1}{2}=\pm\sqrt{\frac{4}{25}}$$
$$2y+\frac{1}{2}=\pm\frac{2}{5}$$
$$2y+\frac{1}{2}=\frac{2}{5} \quad \text{or} \quad 2y+\frac{1}{2}=-\frac{2}{5}$$
$$2y=-\frac{1}{2}+\frac{2}{5} \qquad 2y=-\frac{1}{2}-\frac{2}{5}$$
$$2y=-\frac{1}{10} \qquad 2y=-\frac{9}{10}$$
$$y=-\frac{1}{20} \qquad y=-\frac{9}{20}$$

37. $x^2+3x-4=0$

$$x^2+3x=4$$
$$x^2+3x+\frac{9}{4}=4+\frac{9}{4}$$
$$x^2+3x+\frac{9}{4}=\frac{16}{4}+\frac{9}{4}$$
$$\left(x+\frac{3}{2}\right)^2=\frac{25}{4}$$
$$x+\frac{3}{2}=\pm\sqrt{\frac{25}{4}}$$
$$x+\frac{3}{2}=\pm\frac{5}{2}$$
$$x=-\frac{3}{2}\pm\frac{5}{2}$$
$$x=-\frac{3}{2}+\frac{5}{2} \quad \text{or} \quad x=-\frac{3}{2}-\frac{5}{2}$$
$$x=\frac{2}{2} \qquad x=\frac{-8}{2}$$
$$x=1 \qquad x=-4$$

39. $x^2+2x-15=0$

$$x^2+2x=15$$
$$x^2+2x+1=15+1$$
$$(x+1)^2=16$$
$$x+1=\pm\sqrt{16}$$
$$x+1=\pm 4$$
$$x=-1\pm 4$$
$$x=-1+4 \quad \text{or} \quad x=-1-4$$
$$x=3 \qquad x=-5$$

41. $x^2+6x+8=0$

$$x^2+6x=-8$$
$$x^2+6x+9=-8+9$$
$$x^2+6x+9=1$$
$$(x+3)^2=1$$
$$x+3=\pm\sqrt{1}$$
$$x+3=\pm 1$$
$$x=-3\pm 1$$
$$x=-3+1 \quad \text{or} \quad x=-3-1$$
$$x=-2 \qquad x=-4$$

43.
$$x^2-7x+6=0$$
$$x^2-7x=-6$$
$$x^2-7x+\frac{49}{4}=-6+\frac{49}{4}$$
$$x^2-7x+\frac{49}{4}=-\frac{24}{4}+\frac{49}{4}$$
$$x^2-7x+\frac{49}{4}=\frac{25}{4}$$
$$\left(x-\frac{7}{2}\right)^2=\frac{25}{4}$$
$$x-\frac{7}{2}=\pm\sqrt{\frac{25}{4}}$$
$$x-\frac{7}{2}=\pm\frac{5}{2}$$
$$x=\frac{7}{2}\pm\frac{5}{2}$$
$$x=\frac{7}{2}+\frac{5}{2} \quad \text{or} \quad x=\frac{7}{2}-\frac{5}{2}$$
$$x=\frac{12}{2} \qquad x=\frac{2}{2}$$
$$x=6 \qquad x=1$$

45.
$$2x^2+x-1=0$$
$$\frac{1}{2}\left(2x^2+x-1\right)=\frac{1}{2}(0)$$
$$x^2+\frac{1}{2}x-\frac{1}{2}=0$$
$$x^2+\frac{1}{2}x=\frac{1}{2}$$
$$x^2+\frac{1}{2}x+\frac{1}{16}=\frac{1}{2}+\frac{1}{16}$$
$$\left(x+\frac{1}{4}\right)^2=\frac{9}{16}$$
$$x+\frac{1}{4}=\pm\sqrt{\frac{9}{16}}$$
$$x+\frac{1}{4}=\pm\frac{3}{4}$$
$$x=-\frac{1}{4}\pm\frac{3}{4}$$
$$x=-\frac{1}{4}+\frac{3}{4} \quad \text{or} \quad x=-\frac{1}{4}-\frac{3}{4}$$
$$x=\frac{2}{4} \qquad x=-\frac{4}{4}$$
$$x=\frac{1}{2} \qquad x=-1$$

47.
$$2z^2-7z-4=0$$
$$\frac{1}{2}\left(2z^2-7z-4\right)=\frac{1}{2}(0)$$
$$z^2-\frac{7}{2}z-2=0$$
$$z^2-\frac{7}{2}z=2$$
$$z^2-\frac{7}{2}z+\frac{49}{16}=2+\frac{49}{16}$$
$$\left(z-\frac{7}{4}\right)^2=\frac{81}{16}$$
$$z-\frac{7}{4}=\pm\sqrt{\frac{81}{16}}$$
$$z-\frac{7}{4}=\pm\frac{9}{4}$$
$$z=\frac{7}{4}\pm\frac{9}{4}$$
$$z=\frac{7}{4}+\frac{9}{4} \quad \text{or} \quad z=\frac{7}{4}-\frac{9}{4}$$
$$z=\frac{16}{4} \qquad z=-\frac{2}{4}$$
$$z=4 \qquad z=-\frac{1}{2}$$

49.
$$x^2-13x+40=0$$
$$x^2-13x=-40$$
$$x^2-13x+\frac{169}{4}=-40+\frac{169}{4}$$
$$\left(x-\frac{13}{2}\right)^2=\frac{9}{4}$$
$$x-\frac{13}{2}=\pm\sqrt{\frac{9}{4}}$$
$$x-\frac{13}{2}=\pm\frac{3}{2}$$
$$x=\frac{13}{2}\pm\frac{3}{2}$$
$$x=\frac{13}{2}+\frac{3}{2} \quad \text{or} \quad x=\frac{13}{2}-\frac{3}{2}$$
$$x=\frac{16}{2} \qquad x=\frac{10}{2}$$
$$x=8 \qquad x=5$$

51.
$$-x^2+6x+7=0$$
$$-1\left(-x^2+6x+7\right)=-1(0)$$
$$x^2-6x-7=0$$
$$x^2-6x=7$$
$$x^2-6x+9=7+9$$
$$(x-3)^2=16$$
$$x-3=\pm\sqrt{16}$$
$$x-3=\pm4$$
$$x=3\pm4$$
$$x=3+4 \quad \text{or} \quad x=3-4$$
$$x=7 \qquad\qquad x=-1$$

53.
$$-z^2+9z-20=0$$
$$-1(-z^2+9z-20)=-1(0)$$
$$z^2-9z+20=0$$
$$z^2-9z=-20$$
$$z^2-9z+\frac{81}{4}=-20+\frac{81}{4}$$
$$\left(z-\frac{9}{2}\right)^2=\frac{1}{4}$$
$$z-\frac{9}{2}=\pm\sqrt{\frac{1}{4}}$$
$$z-\frac{9}{2}=\pm\frac{1}{2}$$
$$z=\frac{9}{2}\pm\frac{1}{2}$$
$$z=\frac{9}{2}+\frac{1}{2} \quad \text{or} \quad z=\frac{9}{2}-\frac{1}{2}$$
$$z=\frac{10}{2} \qquad\qquad z=\frac{8}{2}$$
$$z=5 \qquad\qquad z=4$$

55.
$$b^2=3b+28$$
$$b^2-3b=28$$
$$b^2-3b+\frac{9}{4}=\frac{112}{4}+\frac{9}{4}$$
$$\left(b-\frac{3}{2}\right)^2=\frac{121}{4}$$
$$b-\frac{3}{2}=\pm\sqrt{\frac{121}{4}}$$
$$b-\frac{3}{2}=\pm\frac{11}{2}$$
$$b=\frac{3}{2}\pm\frac{11}{2}$$
$$b=\frac{3}{2}+\frac{11}{2} \quad \text{or} \quad b=\frac{3}{2}-\frac{11}{2}$$
$$b=\frac{14}{2} \qquad\qquad b=-\frac{8}{2}$$
$$b=7 \qquad\qquad b=-4$$

57.
$$x^2+10x=11$$
$$x^2+10x+25=11+25$$
$$(x+5)^2=36$$
$$x+5=\pm\sqrt{36}$$
$$x+5=\pm6$$
$$x=-5\pm6$$
$$x=\frac{11}{2}-\frac{9}{2} \quad \text{or} \quad x=-\frac{11}{2}-\frac{9}{2}$$
$$x=\frac{2}{2} \qquad\qquad x=-\frac{20}{2}$$
$$x=1 \qquad\qquad x=-10$$

59.
$$x^2-4x-10=0$$
$$x^2-4x=10$$
$$x^2-4x+4=10+4$$
$$(x-2)^2=14$$
$$x-2=\pm\sqrt{14}$$
$$x=2\pm\sqrt{14}$$

61.
$$r^2+8r+5=0$$
$$r^2+8r=-5$$
$$r^2+8r+16=-5+16$$
$$(r+4)^2=11$$
$$r+4=\pm\sqrt{11}$$
$$r=-4\pm\sqrt{11}$$

63.
$$c^2-c-3=0$$
$$c^2-c=3$$
$$c^2-c+\frac{1}{4}=3+\frac{1}{4}$$
$$\left(c-\frac{1}{2}\right)^2=\frac{13}{4}$$
$$c-\frac{1}{2}=\pm\sqrt{\frac{13}{4}}$$
$$c=\frac{1}{2}\pm\frac{\sqrt{13}}{2}=\frac{1\pm\sqrt{13}}{2}$$

65. $x^2+3x+6=0$

$$x^2+3x=-6$$

$$x^2+3x+\frac{9}{4}=-6+\frac{9}{4}$$

$$\left(x+\frac{3}{2}\right)^2=\frac{-15}{4}$$

$$x+\frac{3}{2}=\pm\sqrt{\frac{-15}{4}}$$

$$x+\frac{3}{2}=\pm\frac{i\sqrt{15}}{2}$$

$$x=-\frac{3}{2}\pm\frac{i\sqrt{15}}{2}=\frac{-3\pm i\sqrt{15}}{2}$$

67. $9x^2-9x=0$

$$\frac{1}{9}(9x^2-9x)=\frac{1}{9}(0)$$

$$x^2-x=0$$

$$x^2-x+\frac{1}{4}=\frac{1}{4}$$

$$\left(x-\frac{1}{2}\right)^2=\frac{1}{4}$$

$$x-\frac{1}{2}=\pm\frac{1}{2}$$

$$x=\frac{1}{2}\pm\frac{1}{2}$$

$$x=\frac{1}{2}+\frac{1}{2} \quad \text{or} \quad x=\frac{1}{2}-\frac{1}{2}$$

$$x=1 \qquad x=0$$

69. $-\frac{3}{4}b^2-\frac{1}{2}b=0$

$$-\frac{4}{3}\left(-\frac{3}{4}b^2-\frac{1}{2}b\right)=-\frac{4}{3}(0)$$

$$b^2+\frac{2}{3}b=0$$

$$b^2+\frac{2}{3}b+\frac{1}{9}=0+\frac{1}{9}$$

$$\left(b+\frac{1}{3}\right)^2=\frac{1}{9}$$

$$b+\frac{1}{3}=\pm\sqrt{\frac{1}{9}}$$

$$b+\frac{1}{3}=\pm\frac{1}{3}$$

$$b=-\frac{1}{3}\pm\frac{1}{3}$$

$$b=-\frac{1}{3}+\frac{1}{3} \quad \text{or} \quad b=-\frac{1}{3}-\frac{1}{3}$$

$$b=0 \qquad b=-\frac{2}{3}$$

71. $36z^2-6z=0$

$$\frac{1}{36}\left(36z^2-6z\right)=\frac{1}{36}(0)$$

$$z^2-\frac{1}{6}z=0$$

$$z^2-\frac{1}{6}z+\frac{1}{144}=0+\frac{1}{144}$$

$$\left(z-\frac{1}{12}\right)^2=\frac{1}{144}$$

$$z-\frac{1}{12}=\pm\sqrt{\frac{1}{144}}$$

$$z-\frac{1}{12}=\pm\frac{1}{12}$$

$$z=\frac{1}{12}\pm\frac{1}{12}$$

$$z=\frac{1}{12}+\frac{1}{12} \quad \text{or} \quad z=\frac{1}{12}-\frac{1}{12}$$

$$z=\frac{2}{12} \qquad z=0$$

$$z=\frac{1}{6}$$

73. $-\frac{1}{2}p^2-p+\frac{3}{2}=0$

$$-2\left(-\frac{1}{2}p^2-p+\frac{3}{2}\right)=-2(0)$$

$$p^2+2p-3=0$$

$$p^2+2p=3$$

$$p^2+2p+1=3+1$$

$$(p+1)^2=4$$

$$p+1=\pm\sqrt{4}$$

$$p+1=\pm2$$

$$p=-1\pm2$$

$$p=-1+2 \quad \text{or} \quad p=-1-2$$

$$p=1 \qquad p=-3$$

75.
$$2x^2 = 8x+64$$
$$\frac{1}{2}\left(2x^2\right) = \frac{1}{2}(8x+64)$$
$$x^2 = 4x+32$$
$$x^2 - 4x = 32$$
$$x^2 - 4x + 4 = 32 + 4$$
$$(x-2)^2 = 36$$
$$x-2 = \pm\sqrt{36}$$
$$x-2 = \pm 6$$
$$x = 2 \pm 6$$
$$x = 2+6 \quad \text{or} \quad x = 2-6$$
$$x = 8 \quad \text{or} \quad x = -4$$

77.
$$2x^2 + 18x + 4 = 0$$
$$\frac{1}{2}\left(2x^2 + 18x + 4\right) = \frac{1}{2}(0)$$
$$x^2 + 9x + 2 = 0$$
$$x^2 + 9x = -2$$
$$x^2 + 9x + \frac{81}{4} = -2 + \frac{81}{4}$$
$$\left(x+\frac{9}{2}\right)^2 = -\frac{8}{4} + \frac{81}{4}$$
$$\left(x+\frac{9}{2}\right)^2 = \frac{73}{4}$$
$$x + \frac{9}{2} = \pm\frac{\sqrt{73}}{2}$$
$$x = -\frac{9}{2} \pm \frac{\sqrt{73}}{2}$$
$$x = \frac{-9 \pm \sqrt{73}}{2}$$

79.
$$\frac{3}{4}w^2 + \frac{1}{2}w - \frac{1}{4} = 0$$
$$\frac{4}{3}\left(\frac{3}{4}w^2 + \frac{1}{2}w - \frac{1}{4}\right) = \frac{4}{3}(0)$$
$$w^2 + \frac{2}{3}w - \frac{1}{3} = 0$$
$$w^2 + \frac{2}{3}w = \frac{1}{3}$$
$$w^2 + \frac{2}{3}w + \frac{1}{9} = \frac{1}{3} + \frac{1}{9}$$
$$\left(w + \frac{1}{3}\right)^2 = \frac{4}{9}$$
$$w + \frac{1}{3} = \pm\sqrt{\frac{4}{9}}$$
$$w + \frac{1}{3} = \pm\frac{2}{3}$$
$$w = -\frac{1}{3} \pm \frac{2}{3}$$
$$w = -\frac{1}{3} + \frac{2}{3} \quad \text{or} \quad w = -\frac{1}{3} - \frac{2}{3}$$
$$w = \frac{1}{3} \qquad\qquad w = -1$$

81.
$$2x^2 - x = -5$$
$$\frac{1}{2}\left(2x^2 - x\right) = \frac{1}{2}(-5)$$
$$x^2 - \frac{1}{2}x = -\frac{5}{2}$$
$$x^2 - \frac{1}{2}x + \frac{1}{16} = -\frac{40}{16} + \frac{1}{16}$$
$$\left(x - \frac{1}{4}\right)^2 = -\frac{39}{16}$$
$$x - \frac{1}{4} = \pm\frac{i\sqrt{39}}{4}$$
$$x = \frac{1}{4} \pm \frac{i\sqrt{39}}{4}$$
$$x = \frac{1 \pm i\sqrt{39}}{4}$$

83.
$$-3x^2 + 6x = 6$$
$$-\frac{1}{3}\left(-3x^2 + 6x\right) = -\frac{1}{3}(6)$$
$$x^2 - 2x = -2$$
$$x^2 - 2x + 1 = -2 + 1$$
$$(x-1)^2 = -1$$
$$x - 1 = \pm\sqrt{-1}$$
$$x - 1 = \pm i$$
$$x = 1 \pm i$$

85. **a.** $21 = (x+2)(x-2)$

b.
$$21 = (x+2)(x-2)$$
$$21 = x^2 - 2x + 2x - 4$$
$$0 = x^2 - 25$$
$$0 = (x+5)(x-5)$$
$$x+5=0 \quad \text{or} \quad x-5=0$$
$$x=-5 \qquad\qquad x=5$$

Disregard the negative answer since x represents a distance. $x = 5$.

87. **a.** $18 = (x+4)(x+2)$

b.
$$18 = (x+4)(x+2)$$
$$18 = x^2 + 2x + 4x + 8$$
$$0 = x^2 + 6x - 10$$
$$x^2 + 6x = 10$$
$$x^2 + 6x + 9 = 10 + 9$$
$$(x+3)^2 = 19$$
$$x+3 = \pm\sqrt{19}$$
$$x = -3 \pm \sqrt{19}$$

Disregard the negative answer since x represents a distance. $x = -3+\sqrt{19}$.

89.
$$d = \frac{1}{6}x^2$$
$$150 = \frac{1}{6}x^2$$
$$6 \cdot 150 = x^2$$
$$900 = x^2$$
$$x = \pm\sqrt{900} = \pm 30$$

Disregard the negative answer since speed must be positive. The car's speed was about 30 mph.

91. Let x be the first integer. Then $x + 2$ is the next consecutive odd integer.
$$x(x+2) = 35$$
$$x^2 + 2x = 35$$
$$x^2 + 2x + 1 = 35 + 1$$
$$(x+1)^2 = 36$$
$$x+1 = \pm\sqrt{36}$$
$$x+1 = \pm 6$$
$$x = -1 \pm 6$$
$$x = -1+6 \quad \text{or} \quad x = -1-6$$
$$x = 5 \qquad\qquad x = -7$$

Since it was given that the integers are positive, one integer is 5 and the other is 5 + 2 = 7.

93. Let x be the width of the rectangle. Then $2x+2$ is the length. Use length · width = area.
$$(2x+2)x = 60$$
$$2x^2 + 2x = 60$$
$$x^2 + x = 30$$
$$x^2 + x + \frac{1}{4} = 30 + \frac{1}{4}$$
$$\left(x+\frac{1}{2}\right)^2 = \frac{120}{4} + \frac{1}{4}$$
$$\left(x+\frac{1}{2}\right)^2 = \frac{121}{4}$$
$$x + \frac{1}{2} = \pm\sqrt{\frac{121}{4}}$$
$$x + \frac{1}{2} = \pm\frac{11}{2}$$
$$x = -\frac{1}{2} \pm \frac{11}{2}$$
$$x = -\frac{1}{2} + \frac{11}{2} \quad \text{or} \quad x = -\frac{1}{2} - \frac{11}{2}$$
$$x = \frac{10}{2} \qquad\qquad x = -\frac{12}{2}$$
$$x = 5 \qquad\qquad x = -6$$

Since the width cannot be negative, the width is 5 ft.
Length $= 2(5) + 2 = 10 + 2 = 12$ ft.
The rectangle is 5 ft by 12 ft.

95. Let s be the length of the side. Then $s + 6$ is the length of the diagonal (d). Use $s^2 + s^2 = d^2$.
$$2s^2 = (s+6)^2$$
$$2s^2 = s^2 + 12s + 36$$
$$s^2 = 12s + 36$$
$$s^2 - 12s = 36$$
$$s^2 - 12s + 36 = 36 + 36$$
$$(s-6)^2 = 72$$
$$s - 6 = \pm\sqrt{72}$$
$$s - 6 = \pm 6\sqrt{2}$$
$$s = 6 \pm 6\sqrt{2}$$

Length is never negative. Thus, $s = 6 + 6\sqrt{2} \approx 14.49$.
The patio is about 14.49 ft by 14.49 ft.

97. Since the radius is 10 inches, the diameter (d) is 20 inches. Use the formula $s^2+s^2=d^2$ to find the length (s) of the other two sides.

$$s^2+s^2=d^2$$
$$s^2+s^2=20^2$$
$$2s^2=400$$
$$s^2=200$$
$$s=\pm\sqrt{200}=\pm 10\sqrt{2}$$

Length is never negative.
Thus, $s=10\sqrt{2}\approx 14.14$ inches.

99. $A=\pi r^2$

$$24\pi=\pi r^2$$
$$24=r^2$$
$$r=\pm\sqrt{24}=\pm 2\sqrt{6}$$

Length is never negative.
Thus, $r=2\sqrt{6}\approx 4.90$ feet.

101. $A=P\left(1+\frac{r}{n}\right)^{nt}$

$$540.80=500\left(1+\frac{r}{1}\right)^{1(2)}$$
$$540.80=500(1+r)^2$$
$$1.0816=(1+r)^2$$
$$1+r=\pm\sqrt{1.0816}$$
$$1+r=\pm 1.04$$
$$r=-1\pm 1.04$$
$$r=-1+1.04 \quad\text{or}\quad r=-1-1.04$$
$$r=0.04 \qquad\text{or}\quad r=-2.04$$

An interest rate is never negative. Thus $r=0.04=4\%$.

103. $A=P\left(1+\frac{r}{n}\right)^{nt}$

$$1432.86=1200\left(1+\frac{r}{2}\right)^{2(3)}$$
$$1432.86=1200\left(1+\frac{r}{2}\right)^{6}$$
$$1.19405=\left(1+\frac{r}{2}\right)^{6}$$
$$1+\frac{r}{2}\approx\pm 1.03$$
$$\frac{r}{2}\approx -1\pm 1.03$$
$$r\approx -2\pm 2.06$$
$$r\approx -2+2.06 \quad\text{or}\quad r\approx -2-2.06$$
$$r\approx 0.06 \qquad\text{or}\quad r\approx -4.06$$

An interest rate is never negative. Thus, Steve Rodi's annual interest rate is about 6%.

105. a. To find the surface area, we must first determine the radius. Use $V=\pi r^2h$ with $V=160$ and $h=10$ to get

$$160=\pi r^2(10)$$
$$16=\pi r^2$$
$$\frac{16}{\pi}=r^2$$
$$r=\pm\frac{4}{\sqrt{\pi}}$$

The length cannot be negative, so $r=\frac{4}{\sqrt{\pi}}$.

Use the formula $S=2\pi r^2+2\pi rh$ to calculate the surface area.

$$S=2\pi\left(\frac{4}{\sqrt{\pi}}\right)^2+2\pi\left(\frac{4}{\sqrt{\pi}}\right)(10)$$
$$=2\pi\left(\frac{16}{\pi}\right)+\frac{80\pi}{\sqrt{\pi}}$$
$$=32+80\sqrt{\pi}\approx 173.80$$

The surface area is about 173.80 square inches.

b. Use $V=\pi r^2h$ with $V=160$ and $h=10$ to obtain $160=\pi r^2(10)$. In part (a) this was solved for r to get

$$r=\frac{4}{\sqrt{\pi}}=\frac{4}{\sqrt{\pi}}\cdot\frac{\sqrt{\pi}}{\sqrt{\pi}}=\frac{4\sqrt{\pi}}{\pi}\approx 2.26$$

The radius is about 2.26 inches.

c. Use $S = 2\pi r^2 + 2\pi rh$ with $S = 160$ and $h = 10$.

$$160 = 2\pi r^2 + 2\pi r(10)$$
$$160 = 2\pi r^2 + 20\pi r$$
$$\frac{160}{2\pi} = \frac{2\pi r^2}{2\pi} + \frac{20\pi r}{2\pi}$$
$$\frac{80}{\pi} = r^2 + 10r$$
$$\frac{80}{\pi} + 25 = r^2 + 10r + 25$$
$$\frac{80+25\pi}{\pi} = (r+5)^2$$
$$r + 5 = \pm\sqrt{\frac{80+25\pi}{\pi}}$$
$$r = -5 \pm \sqrt{\frac{80+25\pi}{\pi}}$$

The radius is never negative so $r \approx 2.1$ in.

107. $-4(2z-6) = -3(z-4) + z$
$$-8z + 24 = -3z + 12 + z$$
$$-8z + 24 = -2z + 12$$
$$-6z = -12$$
$$z = 2$$

108. Let x = the amount invested at 7%. Then the amount invested at $6\frac{1}{4}\%$ will be $10{,}000 - x$.
The interest earned at 7% will be $0.07x$. and the amount of interest earned at 6.25% will be $.0625(10{,}000 - x)$. The total interest earned is \$656.50.
$$0.07x + 0.0625(10{,}000 - x) = 656.50$$
$$0.07x + 625 - 0.0625x = 656.50$$
$$0.0075x = 31.5$$
$$x = 4200$$
Thus, \$4200 was invested at 7% and \$10,000 – \$4200 = \$5800 was invested at $6\frac{1}{4}\%$

109. $|x+3| = |2x-7|$

$x + 3 = 2x - 7$ or $x + 3 = -(2x-7)$
$-x = -10$ $\qquad x + 3 = -2x + 7$
$x = 10$ $\qquad 3x = 4$
$\qquad x = \frac{4}{3}$

110. $m = \frac{y_2 - y_1}{x_2 - x_1} = \frac{5-5}{0-(-2)} = \frac{0}{0+2} = \frac{0}{2} = 0$

111.
$$\begin{array}{r} 4x^2 + 9x - 3 \\ x - 2 \\ \hline -8x^2 - 18x + 6 \\ 4x^3 + 9x^2 - 3x \\ \hline 4x^3 + x^2 - 21x + 6 \end{array}$$

Exercise Set 8.2

1. For a quadratic equation in standard form, $ax^2 + bx + c = 0$, the quadratic formula is $x = \frac{-b \pm \sqrt{b^2 - 4ac}}{2a}$.

3. $6x - 3x^2 + 8 = 0$
$-3x^2 + 6x + 8 = 0$
$a = -3$, $b = 6$, and $c = 8$

5. Yes, multiply both sides of the equation $-6x^2 + \frac{1}{2}x - 5 = 0$ by -1 to obtain $6x^2 - \frac{1}{2}x + 5 = 0$. The equations are equivalent so they will have the same solutions.

7. a. For a quadratic equation in standard form, $ax^2 + bx + c = 0$, the discriminant is the expression under the square root symbol in the quadratic formula, $b^2 - 4ac$.

b. $3x^2 - 6x + 10 = 0$, $a = 3$, $b = -6$, and $c = 10$.
$b^2 - 4ac = (-6)^2 - 4(3)(10) = 36 - 120 = -84$

c. If $b^2 - 4ac > 0$, then the quadratic equation will have two distinct real solutions. Since there is a positive number under the radical sign in the quadratic formula, the value of the radical will be real and there will be two real solutions. If $b^2 - 4ac = 0$, then the equation has the single real solution $\frac{-b}{2a}$. If $b^2 - 4ac < 0$, then the expression under the radical sign in the quadratic formula is negative. Thus, the equation will have no real solution.

9. $x^2+3x+1=0$

$b^2-4ac=(3)^2-4(1)(1)=9-4=5$

Since $5>0$, there are two real solutions.

11. $4z^2+6z+5=0$

$b^2-4ac=6^2-4(4)(5)=36-80=-44$

Since $-44<0$, there is no real solution.

13. $5p^2+3p-7=0$

$b^2-4ac=3^2-4(5)(-7)=9+140=149$

Since $149>0$, there are two real solutions.

15. $-5x^2+5x-8=0$

$b^2-4ac=5^2-4(-5)(-8)=25-160=-135$

Since $-135<0$, there is no real solution.

17. $x^2+10.2x+26.01=0$

$$\begin{aligned}b^2-4ac&=(10.2)^2-4(1)(26.01)\\&=104.04-104.04\\&=0\end{aligned}$$

Since the discriminant is 0, there is one real solution.

19. $b^2=-3b-\frac{9}{4}$

$b^2+3b+\frac{9}{4}=0$

$b^2-4ac=3^2-4(1)\left(\frac{9}{4}\right)=9-9=0$

Since the discriminant is 0, there is one real solution.

21. $x^2-9x+18=0$

$$\begin{aligned}x&=\frac{-(-9)\pm\sqrt{(-9)^2-4(1)(18)}}{2(1)}\\&=\frac{9\pm\sqrt{81-72}}{2}\\&=\frac{9\pm\sqrt{9}}{2}\\&=\frac{9\pm3}{2}\end{aligned}$$

$$x=\frac{9+3}{2}=\frac{12}{2}=6 \quad\text{or}\quad x=\frac{9-3}{2}=\frac{6}{2}=3$$

The solutions are 6 and 3.

23. $a^2-6a+8=0$

$$\begin{aligned}a&=\frac{-b\pm\sqrt{b^2-4ac}}{2a}\\&=\frac{6\pm\sqrt{(-6)^2-4(1)(8)}}{2(1)}\\&=\frac{6\pm\sqrt{36-32}}{2}\\&=\frac{6\pm\sqrt{4}}{2}\\&=\frac{6\pm2}{2}\end{aligned}$$

$$a=\frac{6-2}{2}=\frac{4}{2}=2 \quad\text{or}\quad a=\frac{6+2}{2}=\frac{8}{2}=4$$

The solutions are 2 and 4.

25. $x^2=-6x+7$

$x^2+6x-7=0$

$$\begin{aligned}x&=\frac{-6\pm\sqrt{6^2-4(1)(-7)}}{2(1)}\\&=\frac{-6\pm\sqrt{36+28}}{2}\\&=\frac{-6\pm\sqrt{64}}{2}\\&=\frac{-6\pm8}{2}\end{aligned}$$

$$x=\frac{-6+8}{2}=\frac{2}{2}=1 \quad\text{or}\quad x=\frac{-6-8}{2}=\frac{-14}{2}=-7$$

The solutions are 1 and −7.

27. $-b^2 = 4b - 20$

$b^2 + 4b - 20 = 0$

$b = \frac{-4 \pm \sqrt{(4)^2 - 4(1)(-20)}}{2(1)}$

$= \frac{-4 \pm \sqrt{16 + 80}}{2}$

$= \frac{-4 \pm \sqrt{96}}{2}$

$= \frac{-4 \pm 4\sqrt{6}}{2}$

$= -2 \pm \sqrt{6}$

The solutions are $-2 + 2\sqrt{6}$ and $-2 - 2\sqrt{6}$.

29. $b^2 - 64 = 0$

$b = \frac{0 \pm \sqrt{0^2 - 4(1)(-64)}}{2(1)}$

$= \frac{\pm\sqrt{256}}{2}$

$= \frac{\pm 16}{2}$

$= \pm 8$

The solutions are 8 and –8.

31. $3w^2 - 4w + 5 = 0$

$w = \frac{-(-4) \pm \sqrt{(-4)^2 - 4(3)(5)}}{2(3)}$

$= \frac{4 \pm \sqrt{16 - 60}}{6}$

$= \frac{4 \pm \sqrt{-44}}{6}$

$= \frac{4 \pm 2i\sqrt{11}}{6}$

$= \frac{2\left(2 \pm i\sqrt{11}\right)}{6}$

$= \frac{2 \pm i\sqrt{11}}{3}$

The solutions are $\frac{2 - i\sqrt{11}}{3}$ and $\frac{2 + i\sqrt{11}}{3}$.

33. $c^2 - 5c = 0$

$c = \frac{-(-5) \pm \sqrt{(-5)^2 - 4(1)(0)}}{2(1)}$

$= \frac{5 \pm \sqrt{25}}{2}$

$= \frac{5 \pm 5}{2}$

$c = \frac{5+5}{2}$ or $c = \frac{5-5}{2}$

$= \frac{10}{2}$ $\quad = \frac{0}{2}$

$= 5$ $\quad = 0$

The solutions are 5 and 0.

35. $4s^2 - 8s + 6 = 0$

$\frac{1}{2}(4s^2 - 8s + 6) = 0$

$2s^2 - 4s + 3 = 0$

$s = \frac{-(-4) \pm \sqrt{(-4)^2 - 4(2)(3)}}{2(2)}$

$= \frac{4 \pm \sqrt{16 - 24}}{4}$

$= \frac{4 \pm \sqrt{-8}}{4}$

$= \frac{4 \pm 2i\sqrt{2}}{4}$

$= \frac{2 \pm i\sqrt{2}}{2}$

The solutions are $\frac{2 - i\sqrt{2}}{2}$ and $\frac{2 + i\sqrt{2}}{2}$.

37. $a^2 + 2a + 1 = 0$

$a = \frac{-2 \pm \sqrt{2^2 - 4(1)(1)}}{2(1)}$

$= \frac{-2 \pm \sqrt{4 - 4}}{2}$

$= \frac{-2 \pm \sqrt{0}}{2}$

$= \frac{-2}{2}$

$= -1$

The solution is –1.

39. $16x^2 - 8x + 1 = 0$

$$x = \frac{-(-8) \pm \sqrt{(-8)^2 - 4(16)(1)}}{2(16)} = \frac{8 \pm \sqrt{64 - 64}}{32} = \frac{8 \pm \sqrt{0}}{32} = \frac{8}{32} = \frac{1}{4}$$

The solution is $\frac{1}{4}$.

41. $x^2 - 2x - 1 = 0$

$$x = \frac{-(-2) \pm \sqrt{(-2)^2 - 4(1)(-1)}}{2(1)} = \frac{2 \pm \sqrt{4+4}}{2} = \frac{2 \pm \sqrt{8}}{2} = \frac{2 \pm 2\sqrt{2}}{2} = 1 \pm \sqrt{2}$$

The solutions are $1 - \sqrt{2}$ and $1 + \sqrt{2}$.

43. $-n^2 = 3n + 6$

$$0 = n^2 + 3n + 6$$

$$n = \frac{-3 \pm \sqrt{3^2 - 4(1)(6)}}{2(1)} = \frac{-3 \pm \sqrt{9 - 24}}{2} = \frac{-3 \pm \sqrt{-15}}{2} = \frac{-3 \pm i\sqrt{15}}{2}$$

The solutions are $\frac{-3 + i\sqrt{15}}{2}$ and $\frac{-3 - i\sqrt{15}}{2}$.

45. $2x^2 + 5x - 3 = 0$

$$x = \frac{-5 \pm \sqrt{5^2 - 4(2)(-3)}}{2(2)} = \frac{-5 \pm \sqrt{25 + 24}}{4} = \frac{-5 \pm \sqrt{49}}{4} = \frac{-5 \pm 7}{4}$$

$$x = \frac{-5+7}{4} = \frac{2}{4} = \frac{1}{2} \quad \text{or} \quad x = \frac{-5-7}{4} = \frac{-12}{4} = -3$$

The solutions are $\frac{1}{2}$ and -3.

47. $(2a+3)(3a-1) = 2$

$$6a^2 + 7a - 3 = 2$$

$$6a^2 + 7a - 5 = 0$$

$$a = \frac{-(7) \pm \sqrt{(7)^2 - 4(6)(-5)}}{2(6)} = \frac{-7 \pm \sqrt{49 + 120}}{12} = \frac{-7 \pm \sqrt{169}}{12} = \frac{-7 \pm 13}{12}$$

$$a = \frac{-7+13}{12} = \frac{6}{12} = \frac{1}{2} \quad \text{or} \quad a = \frac{-7-13}{12} = \frac{-20}{12} = -\frac{5}{3}$$

The solutions are $\frac{1}{2}$ and $-\frac{5}{3}$.

49. $\frac{1}{2}t^2+t-12=0$

$2\left(\frac{1}{2}t^2+t-12\right)=2(0)$

$t^2+2t-24=0$

$$t=\frac{-(2)\pm\sqrt{(2)^2-4(1)(-24)}}{2(1)}=\frac{-2\pm\sqrt{4+96}}{2}=\frac{-2\pm\sqrt{100}}{2}=\frac{-2\pm10}{2}$$

$t=\frac{-2+10}{2}=\frac{8}{2}=4$ or $t=\frac{-2-10}{2}=\frac{-12}{2}=-6$

The solutions are 4 and -6.

51. $9r^2+3r-2=0$

$$r=\frac{-3\pm\sqrt{3^2-4(9)(-2)}}{2(9)}=\frac{-3\pm\sqrt{9+72}}{18}=\frac{-3\pm\sqrt{81}}{18}=\frac{-3\pm9}{18}$$

$r=\frac{-3+9}{18}=\frac{6}{18}=\frac{1}{3}$ or $r=\frac{-3-9}{18}=\frac{-12}{18}=-\frac{2}{3}$

The solutions are $\frac{1}{3}$ and $-\frac{2}{3}$.

53. $\frac{1}{2}x^2+2x+\frac{2}{3}=0$

$6\left(\frac{1}{2}x^2+2x+\frac{2}{3}\right)=6(0)$

$3x^2+12x+4=0$

$$x=\frac{-12\pm\sqrt{(12)^2-4(3)(4)}}{2(3)}=\frac{-12\pm\sqrt{144-48}}{6}=\frac{-12\pm\sqrt{96}}{6}=\frac{-12\pm4\sqrt{6}}{6}=\frac{2(-6\pm2\sqrt{6})}{2(3)}=\frac{-6\pm2\sqrt{6}}{3}$$

The solutions are $\frac{-6+2\sqrt{6}}{3}$ and $\frac{-6-2\sqrt{6}}{3}$.

55. $a^2-\frac{a}{5}-\frac{1}{3}=0$

$15\left(a^2-\frac{a}{5}-\frac{1}{3}\right)=15(0)$

$15a^2-3a-5=0$

$$a=\frac{-(-3)\pm\sqrt{(-3)^2-4(15)(-5)}}{2(15)}=\frac{3\pm\sqrt{9+300}}{30}=\frac{3\pm\sqrt{309}}{30}$$

The solutions are $\frac{3-\sqrt{309}}{30}$ and $\frac{3+\sqrt{309}}{30}$.

57.
$$c = \frac{c-6}{4-c}$$
$$c(4-c) = c-6$$
$$4c - c^2 = c-6$$
$$0 = c^2 - 3c - 6$$
$$c = \frac{-(-3) \pm \sqrt{(-3)^2 - 4(1)(-6)}}{2(1)}$$
$$= \frac{3 \pm \sqrt{9+24}}{2}$$
$$= \frac{3 \pm \sqrt{33}}{2}$$
The solutions are $\frac{3+\sqrt{33}}{2}$ and $\frac{3-\sqrt{33}}{2}$.

59. $2x^2 - 4x + 5 = 0$
$$x = \frac{-(-4) \pm \sqrt{(-4)^2 - 4(2)(5)}}{2(2)}$$
$$= \frac{4 \pm \sqrt{16-40}}{4}$$
$$= \frac{4 \pm \sqrt{-24}}{4}$$
$$= \frac{4 \pm 2i\sqrt{6}}{4}$$
$$= \frac{2 \pm i\sqrt{6}}{2}$$
The solutions are $\frac{2+i\sqrt{6}}{2}$ and $\frac{2-i\sqrt{6}}{2}$.

61.
$$y^2 + \frac{y}{2} = -\frac{3}{2}$$
$$2\left(y^2 + \frac{y}{2}\right) = 2\left(-\frac{3}{2}\right)$$
$$2y^2 + y = -3$$
$$2y^2 + y + 3 = 0$$
$$y = \frac{-1 \pm \sqrt{(1)^2 - 4(2)(3)}}{2(2)}$$
$$= \frac{-1 \pm \sqrt{1-24}}{4}$$
$$= \frac{-1 \pm \sqrt{-23}}{4}$$
$$= \frac{-1 \pm i\sqrt{23}}{4}$$
The solutions are $\frac{-1+i\sqrt{23}}{4}$ and $\frac{-1-i\sqrt{23}}{4}$.

63.
$$0.1x^2 + 0.6x - 1.2 = 0$$
$$10(0.1x^2 + 0.6x - 1.2) = 10(0)$$
$$x^2 + 6x - 12 = 0$$
$$x = \frac{-6 \pm \sqrt{6^2 - 4(1)(-12)}}{2(1)}$$
$$= \frac{-6 \pm \sqrt{36+48}}{2}$$
$$= \frac{-6 \pm \sqrt{84}}{2}$$
$$= \frac{-6 \pm 2\sqrt{21}}{2}$$
$$= -3 \pm \sqrt{21}$$
The solutions are $-3+\sqrt{21}$ or $-3-\sqrt{21}$.

65. $f(x) = x^2 - 2x + 5$, $f(x) = 5$
$$x^2 - 2x + 5 = 5$$
$$x^2 - 2x = 0$$
$$x = \frac{2 \pm \sqrt{(-2)^2 - 4(1)(0)}}{2(1)}$$
$$= \frac{2 \pm \sqrt{4}}{2}$$
$$= \frac{2 \pm 2}{2}$$
$$x = \frac{2+2}{2} = 2 \quad \text{or} \quad x = \frac{2-2}{2} = 0$$
The values of x are 2 and 0.

67. $k(x) = x^2 - x - 15$, $k(x) = 15$
$$x^2 - x - 15 = 15$$
$$x^2 - x - 30 = 0$$
$$x = \frac{1 \pm \sqrt{(-1)^2 - 4(1)(-30)}}{2(1)}$$
$$= \frac{1 \pm \sqrt{1+120}}{2}$$
$$= \frac{1 \pm \sqrt{121}}{2}$$
$$= \frac{1 \pm 11}{2}$$
$$x = \frac{1+11}{2} = \frac{12}{2} = 6 \quad \text{or} \quad x = \frac{1-11}{2} = \frac{-10}{2} = -5$$
The values of x are 6 and -5.

69. $h(t) = 2t^2 - 7t + 6$, $h(t) = 2$

$$2t^2 - 7t + 6 = 2$$
$$2t^2 - 7t + 4 = 0$$
$$t = \frac{7 \pm \sqrt{(-7)^2 - 4(2)(4)}}{2(2)}$$
$$= \frac{7 \pm \sqrt{49 - 32}}{4}$$
$$= \frac{7 \pm \sqrt{17}}{4}$$

The values of t are $\frac{7+\sqrt{17}}{4}$ and $\frac{7-\sqrt{17}}{4}$.

71. $g(a) = 2a^2 - 3a + 16$, $g(a) = 14$

$$2a^2 - 3a + 16 = 14$$
$$2a^2 - 3a + 2 = 0$$
$$a = \frac{3 \pm \sqrt{(-3)^2 - 4(2)(2)}}{2(2)}$$
$$= \frac{3 \pm \sqrt{9 - 16}}{4}$$
$$= \frac{3 \pm \sqrt{-7}}{4}$$

There are no real values of a for which $g(a) = 14$.

73. $x = 2$ and $x = 5$

$x - 2 = 0$ and $x - 5 = 0$

$$(x-2)(x-5) = 0$$
$$x^2 - 5x - 2x + 10 = 0$$
$$x^2 - 7x + 10 = 0$$

75. $x = 1$ and $x = -9$

$x - 1 = 0$ and $x + 9 = 0$

$$(x-1)(x+9) = 0$$
$$x^2 + 9x - x - 9 = 0$$
$$x^2 + 8x - 9 = 0$$

77. $x = -\frac{3}{5}$ and $x = \frac{2}{3}$

$5x = -3$ and $3x = 2$

$5x + 3 = 0$ and $3x - 2 = 0$

$$(5x+3)(3x-2) = 0$$
$$15x^2 - 10x + 9x - 6 = 0$$
$$15x^2 - x - 6 = 0$$

79. $x = \sqrt{2}$ and $x = -\sqrt{2}$

$x - \sqrt{2} = 0$ and $x + \sqrt{2} = 0$.

$$\left(x-\sqrt{2}\right)\left(x+\sqrt{2}\right) = 0$$
$$x^2 + x\sqrt{2} - x\sqrt{2} - 2 = 0$$
$$x^2 - 2 = 0$$

81. $x = 3i$ and $x = -3i$

$x - 3i = 0$ and $x + 3i = 0$.

$$(x-3i)(x+3i) = 0$$
$$x^2 + 3ix - 3ix - 9i^2 = 0$$
$$x^2 - 9i^2 = 0$$
$$x^2 - 9(-1) = 0$$
$$x^2 + 9 = 0$$

83. $x = 3 + \sqrt{2}$ and $x = 3 - \sqrt{2}$

$x - 3 - \sqrt{2} = 0$ and $x - 3 + \sqrt{2} = 0$

$$\left(x-3-\sqrt{2}\right)\left(x-3+\sqrt{2}\right) = 0$$
$$\left[(x-3)-\sqrt{2}\right]\left[(x-3)+\sqrt{2}\right] = 0$$
$$(x-3)^2 - \left(\sqrt{2}\right)^2 = 0$$
$$x^2 - 6x + 9 - 2 = 0$$
$$x^2 - 6x + 7 = 0$$

85. $x = 2 + 3i$ and $x = 2 - 3i$

$x - 2 - 3i = 0$ and $x - 2 + 3i = 0$

$$(x-2-3i)(x-2+3i) = 0$$
$$[(x-2)-3i][(x-2)+3i] = 0$$
$$(x-2)^2 - (3i)^2 = 0$$
$$x^2 - 4x + 4 - 9i^2 = 0$$
$$x^2 - 4x + 4 + 9 = 0$$
$$x^2 - 4x + 13 = 0$$

87. **a.** $n(10-0.02n)=450$

b.
$$n(10-0.02n)=450$$
$$10n-0.02n^2=450$$
$$0.02n^2-10n+450=0$$
$$n=\frac{-(-10)\pm\sqrt{(-10)^2-4(0.02)(450)}}{2(0.02)}$$
$$n=\frac{10\pm\sqrt{100-36}}{0.04}$$
$$n=\frac{10\pm\sqrt{64}}{0.04}$$
$$n=\frac{10\pm 8}{0.04}$$
$$n=450 \text{ or } n=50$$
Since $n\le 65$, the number of lamps that must be sold is 50.

89. **a.** $n(50-0.2n)=1680$

b.
$$n(50-0.4n)=660$$
$$50n-0.4n^2=660$$
$$0.4n^2-50n+660=0$$
$$n=\frac{-(-50)\pm\sqrt{(-50)^2-4(0.4)(660)}}{2(0.4)}$$
$$n=\frac{50\pm\sqrt{2500-1056}}{0.8}$$
$$n=\frac{50\pm\sqrt{1444}}{0.8}$$
$$n=\frac{50\pm 38}{0.8}$$
$$n=110 \text{ or } n=15$$
Since $n\le 50$, the number of chairs that must be sold is 15.

91. Any quadratic equation for which the discriminant is a non-negative perfect square can be solved by factoring. Any quadratic equation for which the discriminant is a positive number but not a perfect square can be solved by the quadratic formula but not by factoring over the set of integers.

93. Yes. If the discriminant is a perfect square, the simplified expression will not contain a radical and the quadratic equation can be solved by factoring.

95. Let x be the number.
$$2x^2+3x=27$$
$$2x^2+3x-27=0$$
$$x=\frac{-3\pm\sqrt{3^2-4(2)(-27)}}{2(2)}$$
$$x=\frac{-3\pm\sqrt{9+112}}{4}$$
$$x=\frac{-3\pm\sqrt{121}}{4}$$
$$x=\frac{-3\pm 11}{4}$$
$$x=\frac{8}{4} \text{ or } x=\frac{-14}{4}$$
$$x=2 \text{ or } x=-\frac{7}{2}$$
Since the number must be positive, it is 3.

97. Let x be the width. Then $3x-1$ is the length. Use A = (length)(width).
$$24=(3x-1)(x)$$
$$24=3x^2-x$$
$$0=3x^2-x-24$$
$$x=\frac{-(-1)\pm\sqrt{(-1)^2-4(3)(-24)}}{2(3)}$$
$$=\frac{1\pm\sqrt{1+288}}{6}$$
$$=\frac{1\pm\sqrt{289}}{6}$$
$$=\frac{1\pm 17}{6}$$
Since width is positive, use
$$x=\frac{1+17}{6}=\frac{18}{6}=3$$
$3x-1=3(3)-1=9-1=8$
The width is 3 feet and the length is 8 feet.

99. Let x be the amount by which each side is to be reduced.
Then $6-x$ is the new width
and $8-x$ is the new length

new area $= \frac{1}{2}$ (old area) $= \frac{1}{2}(6 \cdot 8) = 24$

new area = (new width)(new length)

$$24 = (6-x)(8-x)$$
$$0 = 48 - 14x + x^2 - 24$$
$$0 = x^2 - 14x + 24$$
$$x = \frac{-(-14) \pm \sqrt{(-14)^2 - 4(1)(24)}}{2(1)}$$
$$= \frac{14 \pm \sqrt{196-96}}{2}$$
$$= \frac{14 \pm \sqrt{100}}{2}$$
$$= \frac{14 \pm 10}{2}$$
$$x = \frac{14+10}{2} \quad \text{or} \quad x = \frac{14-10}{2}$$
$$= \frac{24}{2} \qquad\qquad = \frac{4}{2}$$
$$= 12 \qquad\qquad = 2$$

We reject $x = 12$, since this would give negative values for width and length.
The only meaningful value is $x = 2$ inches since this gives positive values for the new width and length.

101. Substitute $h = 0$ in the formula $h = -16t^2 + 308$.

$$0 = -16t^2 + 308$$
$$16t^2 = 308$$
$$t^2 = \frac{308}{16}$$
$$t^2 = \frac{77}{4}$$
$$t = \pm\sqrt{\frac{77}{7}} = \pm\frac{\sqrt{77}}{2}$$

Time must be positive, so $t = \frac{\sqrt{77}}{2} \approx 4.39$. It take approximately 4.39 seconds for the water to reach the bottom of the falls.

103. a. $h = \frac{1}{2}at^2 + v_0 t + h_0$

$$20 = \frac{1}{2}(-32)t^2 + 60t + 80$$
$$0 = -16t^2 + 60t + 60$$
$$0 = 16t^2 - 60t - 60$$
$$t = \frac{-(-60) \pm \sqrt{(-60)^2 - 4(16)(-60)}}{2(16)}$$
$$t = \frac{60 \pm \sqrt{7440}}{32}$$

Since time must be positive, use

$$t = \frac{60 + \sqrt{7440}}{32} \approx 4.57$$

The horseshoe is 20 feet from the ground after about 4.57 seconds.

b. $0 = \frac{1}{2}(-32)t^2 + 60t + 80$

$$0 = -16t^2 + 60t + 80$$
$$0 = 16t^2 - 60t - 80$$
$$t = \frac{-(-60) \pm \sqrt{(-60)^2 - 4(16)(-80)}}{2(16)}$$
$$= \frac{60 \pm \sqrt{8720}}{32}$$

Since time must be positive, use

$$t = \frac{60 + \sqrt{8720}}{32} \approx 4.79$$

The horseshoes strike the ground after about 4.79 seconds.

105. $x^2 - \sqrt{5}x - 10 = 0,\ a = 1, b = -\sqrt{5}, c = -10$

$$x = \frac{-(-\sqrt{5}) \pm \sqrt{\left(-\sqrt{5}\right)^2 - 4(1)(-10)}}{2(1)}$$
$$= \frac{\sqrt{5} \pm \sqrt{5+40}}{2}$$
$$= \frac{\sqrt{5} \pm \sqrt{45}}{2}$$
$$= \frac{\sqrt{5} \pm 3\sqrt{5}}{2}$$
$$x = \frac{\sqrt{5} + 3\sqrt{5}}{2} \quad \text{or} \quad x = \frac{\sqrt{5} - 3\sqrt{5}}{2}$$
$$= \frac{4\sqrt{5}}{2} \qquad\qquad = \frac{-2\sqrt{5}}{2}$$
$$= 2\sqrt{5} \qquad\qquad = -\sqrt{5}$$

The solutions are $2\sqrt{5}$ and $-\sqrt{5}$.

107. Let s be the length of the side of the original cube. Then $s + 0.2$ is the length of the side of the expanded cube.

$$(s+0.2)^3 = s^3 + 6$$
$$s^3 + 0.6s^2 + 0.12s + 0.008 = s^3 + 6$$
$$0.6s^2 + 0.12s + 0.008 = 6$$
$$0.6s^2 + 0.12s - 5.992 = 0$$
$$s = \frac{-0.12 \pm \sqrt{(0.12)^2 - 4(0.6)(-5.992)}}{2(0.6)}$$
$$= \frac{-0.12 \pm \sqrt{0.0144 + 14.3803}}{1.2}$$
$$= \frac{-0.12 \pm \sqrt{14.3952}}{1.2}$$

Use the positive value since a length cannot be negative.

$$s = \frac{-0.12 + \sqrt{14.3952}}{1.2}$$
$$\approx \frac{-0.12 + 3.7941}{1.2}$$
$$\approx 3.0618$$

The original side was about 3.0618 mm long.

109. a. $h = \frac{1}{2}at^2 + v_0 t + h_0$

$$0 = \frac{1}{2}(-32)t^2 + 0t + 60$$
$$0 = -16t^2 + 0t + 60$$
$$t = \frac{-(0) \pm \sqrt{0^2 - 4(-16)(60)}}{2(-16)}$$
$$t = \frac{0 \pm \sqrt{3840}}{-32}$$

$t \approx -1.94$ or $t \approx 1.94$

Since time must be positive, use 1.94 sec.

b. $h = \frac{1}{2}at^2 + v_0 t + h_0$

$$0 = \frac{1}{2}(-32)t^2 + 0t + 120$$
$$0 = -16t^2 + 0t + 120$$
$$t = \frac{-(0) \pm \sqrt{0^2 - 4(-16)(120)}}{2(-16)}$$
$$t = \frac{0 \pm \sqrt{7680}}{-32}$$

$t \approx -2.74$ or $t \approx 2.74$

Since time must be positive, use 2.74 sec.

c. The height of Travis' rock is given by $h(t) = -16t^2 + 100t + 60$. Set the height equal to 0 to determine the how long it will take Travis' rock to strike the ground.

$$-16t^2 + 100t + 60 = 0$$
$$4t^2 - 25t - 15 = 0$$
$$t = \frac{-(-25) \pm \sqrt{(-25)^2 - 4(4)(-15)}}{2(4)}$$
$$= \frac{25 \pm \sqrt{625 + 240}}{8}$$
$$= \frac{25 \pm \sqrt{865}}{8}$$

$t \approx 6.80$ or $t \approx -0.55$

Time must be positive, so it will take about 6.80 seconds for Travis' rock to strike the ground.

The height of Courtney's rock is given by $h(t) = -16t^2 + 60t + 120$. Set the height equal to 0 to determine the how long it will take Courtney's rock to strike the ground.

$$-16t^2 + 60t + 120 = 0$$
$$4t^2 - 15t - 30 = 0$$
$$t = \frac{-(-15) \pm \sqrt{(-15)^2 - 4(4)(-30)}}{2(4)}$$
$$= \frac{15 \pm \sqrt{225 + 480}}{8}$$
$$= \frac{15 \pm \sqrt{705}}{8}$$

$t \approx 5.19$ or $t \approx -1.44$

Time must be positive, so it will take about 5.19 seconds for Travis' rock to strike the ground.

Thus, Courtney's rock will strike the ground sooner than Travis'.

d. The height of Travis' rock is given by $h(t) = -16t^2 + 100t + 60$. The height of Courtney's rock is given by $h(t) = -16t^2 + 60t + 120$. We want to know when these will be equal.

$$-16t^2 + 100t + 60 = -16t^2 + 60t + 120$$
$$100t + 60 = 60t + 120$$
$$40t = 60$$
$$t = \frac{60}{40} = 1.5$$

The rocks will be the same distance above the ground after 1.5 seconds.

110. $\dfrac{5.55\times10^3}{1.11\times10^1}=\dfrac{5.55}{1.11}\times\dfrac{10^3}{10^1}=5\times10^2$ or 500

111. $f(x)=x^2+2x-8$

$$\begin{aligned}f(3)&=(3)^2+2(3)-8\\&=9+6-8\\&=7\end{aligned}$$

112. $3x+4y=2$

$2x=-5y-1$

Rewrite the system in standard form.

$3x+4y=2$

$2x+5y=-1$

To eliminate the x variable, multiply the first equation by 2 and the second equation by –3, and then add.

$$\begin{aligned}6x+8y&=4\\-6x-15y&=3\\\hline -7y&=7\\y&=-1\end{aligned}$$

Substitute –1 for y in the first equation to find x.

$$\begin{aligned}3x+4(-1)&=2\\3x-4&=2\\3x&=6\\x&=2\end{aligned}$$

The solution is (2, –1).

113. $$\begin{aligned}2x^{-1}-(3y)^{-1}&=\frac{2}{x}-\frac{1}{3y}\\&=\frac{2}{x}\cdot\frac{3y}{3y}-\frac{1}{3y}\cdot\frac{x}{x}\\&=\frac{6y-x}{3xy}\end{aligned}$$

114. $$\begin{aligned}\sqrt{x^2-6x-4}&=x\\x^2-6x-4&=x^2\\-6x-4&=0\\-6x&=4\\x&=\frac{4}{-6}=-\frac{2}{3}\end{aligned}$$

Upon checking, this value does not satisfy the equation. There is no real solution.

Exercise Set 8.3

1. Answers will vary.

3. $A=s^2$, for s.

$\sqrt{A}=s$

5. $d=4.9t^2$, for t

$$\frac{d}{4.9}=t^2$$

$$\sqrt{\frac{d}{4.9}}=t$$

7. $E=i^2r$, for i

$$\frac{E}{r}=i^2$$

$$\sqrt{\frac{E}{r}}=i$$

9. $d=16t^2$, for t

$$\frac{d}{16}=t^2$$

$$\sqrt{\frac{d}{16}}=t \Rightarrow t=\frac{\sqrt{d}}{4}$$

11. $E=mc^2$, for c

$$\frac{E}{m}=c^2$$

$$\sqrt{\frac{E}{m}}=c$$

13. $V=\frac{1}{3}\pi r^2h$, for r

$$3V=\pi r^2h$$

$$\frac{3V}{\pi h}=r^2$$

$$\sqrt{\frac{3V}{\pi h}}=r$$

15. $d=\sqrt{L^2+W^2}$, for W

$$\begin{aligned}d^2&=L^2+W^2\\d^2-L^2&=W^2\\\sqrt{d^2-L^2}&=W\end{aligned}$$

17. $a^2+b^2=c^2$, for b

$$\begin{aligned}b^2&=c^2-a^2\\b&=\sqrt{c^2-a^2}\end{aligned}$$

19. $d=\sqrt{L^2+W^2+H^2}$, for H

$$d^2=L^2+W^2+H^2$$
$$d^2-L^2-W^2=H^2$$
$$\sqrt{d^2-L^2-W^2}=H$$

21. $h=-16t^2+s_0$, for t

$$16t^2=s_0-h$$
$$t^2=\frac{s_0-h}{16}$$
$$t=\sqrt{\frac{s_0-h}{16}}$$
$$t=\frac{\sqrt{s_0-h}}{4}$$

23. $E=\frac{1}{2}mv^2$, for v

$$2E=mv^2$$
$$\frac{2E}{m}=v^2$$
$$\sqrt{\frac{2E}{m}}=v$$

25. $a=\frac{v_2^2-v_1^2}{2d}$, for v_1

$$2ad=v_2^2-v_1^2$$
$$2ad+v_1^2=v_2^2$$
$$v_1^2=v_2^2-2ad$$
$$v_1=\sqrt{v_2^2-2ad}$$

27. $v'=\sqrt{c^2-v^2}$, for c

$$(v')^2=c^2-v^2$$
$$(v')^2+v^2=c^2$$
$$\sqrt{(v')^2+v^2}=c$$

29. a. $P(n)=2.7n^2+9n-3$

$P(5)=2.7(5)^2+9(5)-3=109.5$

The profit would be \$10,950.

b. $P(n)=2.7n^2+9n-3$

$$200=2.7n^2+9n-3$$
$$0=2.7n^2+9n-203$$
$$x=\frac{-9\pm\sqrt{9^2-4(2.7)(-203)}}{2(2.7)}$$
$$x=\frac{-9\pm\sqrt{2273.4}}{5.4}$$
$$x\approx 7 \text{ or } x\approx -10$$

We disregard the negative answer. Thus 7 tractors must be sold.

31. $T=6.2t^2+12t+32$

a. When the car is turned on, $t=0$.

$T=6.2(0)^2+12(0)+32=32$

The temperature is 32°F.

b. $T=6.2(2)^2+12(2)+32=80.8$

The temperature after 2 minutes is 80.8°F.

c. $120=6.2t^2+12t+32$

$$0=6.2t^2+12t-88$$
$$t=\frac{-12\pm\sqrt{12^2-4(6.2)(-88)}}{2(6.2)}$$
$$t\approx\frac{-12\pm 48.23}{12.4}$$
$$t\approx 2.92 \text{ or } t\approx -4.86$$

The radiator temperature will reach 120°F about 2.92 min. after the engine is started.

33. a. In 2006, $t=4$.

$D(4)=0.04(4)^2-0.03(4)+0.01=0.53$

The number of downloads in 2006 was about 0.53 billion.

b. $D(t)=0.04t^2-0.03t+0.01$, $D(t)=1$

$$1=0.04t^2-0.03t+0.01$$
$$0=0.04t^2-0.03t-0.99$$
$$t=\frac{-(-0.03)\pm\sqrt{(-0.03)^2-4(0.04)(-0.99)}}{2(0.04)}$$
$$=\frac{0.03\pm\sqrt{0.1593}}{0.08}$$
$$\approx\frac{0.03\pm 0.3991}{0.08}$$
$$t\approx\frac{0.03+0.3991}{0.08} \quad\text{or}\quad t\approx\frac{0.03-0.3991}{0.08}$$
$$\approx\frac{0.4291}{0.08} \qquad\qquad \approx\frac{-0.3691}{0.08}$$
$$\approx 5.36 \qquad\qquad \approx -4.61$$

We disregard the negative answer. The number of downloads will be 1 billion 5.36 years after 2002, which is the year 2007.

35. a. In 2003, $t = 3$.

$Y(3) = 0.66(3)^2 - 2.49(3) + 12.93 = 11.4$

In 2003, the yield was about 11.4 tons per acre.

b. $0.66t^2 - 2.49t + 12.93 = 13$

$0.66t^2 - 2.49t - 0.07 = 0$

$$t = \frac{-(-2.49) \pm \sqrt{(-2.49)^2 - 4(0.66)(-0.07)}}{2(0.66)}$$

$$= \frac{2.49 \pm \sqrt{6.3849}}{2(0.66)}$$

$$= \frac{2.49 \pm 2.5268}{1.32}$$

$$t = \frac{2.49 + 2.5268}{1.32} \quad \text{or} \quad t = \frac{2.49 - 2.5268}{1.32}$$

$$= \frac{5.0168}{1.32} \qquad\qquad = \frac{-0.0368}{1.32}$$

$$\approx 3.80 \qquad\qquad \approx -0.03$$

We disregard the negative answer. Thus the yield will be 13 tons per acre about 3.8 years after 2000, which would be in the year 2003.

37. a. $M = -0.00434t^2 + 0.142t + 0.315$

In 2007, $t = 10$.

$M(1) = -0.00434(10)^2 + 0.142(10) + 0.315$

$= 1.301$

In 2007, about 1.301 million motorcycles was sold in the United States.

b. $M = -0.00434t^2 + 0.142t + 0.315$, $M = 1.4$

$1.4 = -0.00434t^2 + 0.142t + 0.315$

$0.00434t^2 - 0.142t + 1.085 = 0$

$$t = \frac{-(-0.142) \pm \sqrt{(-0.142)^2 - 4(0.00434)(1.085)}}{2(0.00434)}$$

$$= \frac{0.142 \pm \sqrt{0.0013284}}{0.00868}$$

$$\approx \frac{0.142 \pm 0.036447}{0.00868}$$

$$t \approx \frac{0.142 + 0.036447}{0.00868} = \frac{0.178447}{0.00868} \approx 20.56$$

or

$$t \approx \frac{0.142 - 0.036447}{0.00868} = \frac{0.105553}{0.00868} \approx 12.16$$

The number of motorcycles sold in the U.S. will be 1.4 million 12.16 years after 1997, which is the year 2009, and 20.56 years after 1997, which is the year 2017.

39. Let x be the width of the playground. Then the length is given by $x + 10$.

Area = length × width

$600 = (x + 10)x$

$0 = x^2 + 10x - 600$

$0 = (x - 20)(x + 30)$

$x - 20 = 0 \quad \text{or} \quad x + 30 = 0$

$x = 20 \quad \text{or} \quad x = -30$

Disregard the negative value. The width of the playground is 20 meters and the length is 20 + 10 = 30 meters.

41. Let r be the rate at which the present equipment drills.

	d	r	$t = \frac{d}{r}$
Present equipment	64	r	$\frac{64}{r}$
New equipment	64	$r + 1$	$\frac{64}{r+1}$

They would have hit water in 3.2 hours less time with the new equipment.

$$\frac{64}{r+1} = \frac{64}{r} - 3.2$$

$$r(r+1)\left(\frac{64}{r+1}\right) = r(r+1)\left(\frac{64}{r}\right) - r(r+1)(3.2)$$

$$64r = 64(r+1) - 3.2r(r+1)$$

$$64r = 64r + 64 - 3.2r^2 - 3.2r$$

$3.2r^2 + 3.2r - 64 = 0$

$r^2 + r - 20 = 0$

$(r + 5)(r - 4) = 0$

$r + 5 = 0 \quad \text{or} \quad r - 4 = 0$

$r = -5 \qquad r = 4$

Disregard the negative answer. Thus the present equipment drills at a rate of 4 ft/hr.

43. Let x be Latoya's rate going uphill so $x+2$ is her rate going downhill. Using $\frac{d}{r}=t$ gives

$$t_{\text{uphill}}+t_{\text{downhill}}=1.75$$
$$\frac{6}{x}+\frac{6}{x+2}=1.75$$
$$x(x+2)\left(\frac{6}{x}\right)+x(x+2)\left(\frac{6}{x+2}\right)=x(x+2)(1.75)$$
$$6(x+2)+6x=1.75x(x+2)$$
$$6x+12+6x=1.75x^2+3.5x$$
$$0=1.75x^2-8.5x-12$$
$$x=\frac{-(-8.5)\pm\sqrt{(-8.5)^2-4(1.75)(-12)}}{2(1.75)}$$
$$=\frac{8.5\pm\sqrt{156.25}}{3.5}=\frac{8.5\pm12.5}{3.5}$$
$$x=6 \quad \text{or} \quad x\approx-1.14$$

Since the time must be positive, Latoya's uphill rate is 6 mph and her downhill rate is $x+2=8$ mph.

45. Let x be the Bonita's time, then $x + 1$ is the Pamela's time.

$$\frac{6}{x}+\frac{6}{x+1}=1$$
$$x(x+1)\left(\frac{6}{x}\right)+x(x+1)\left(\frac{6}{x+1}\right)=x(x+1)(1)$$
$$6(x+1)+6x=x^2+x$$
$$6x+6+6x=x^2+x$$
$$0=x^2-11x-6$$
$$x=\frac{-(-11)\pm\sqrt{(-11)^2-4(1)(-6)}}{2(1)}$$
$$=\frac{11\pm\sqrt{121+24}}{2}$$
$$=\frac{11\pm\sqrt{145}}{2}$$
$$x=\frac{11+\sqrt{145}}{2} \quad \text{or} \quad x=\frac{11-\sqrt{145}}{2}$$
$$\approx11.52 \qquad \approx-0.52$$

Since the time must be positive, it takes Bonita about 11.52 hours and Pamela about 12.52 hours to rebuild the engine.

47. Let r be the speed of the plane in still air.

	d	r	$t=\frac{d}{r}$
With wind	80	$r+30$	$\frac{80}{r+30}$
Against wind	80	$r-30$	$\frac{80}{r-30}$

The total time is 1.3 hours

$$\frac{80}{r+30}+\frac{80}{r-30}=1.3$$
$$(r+30)(r-30)\left(\frac{80}{r+30}+\frac{80}{r-30}=1.3\right)$$
$$80(r-3)+80(r+30)=1.3(r^2-900)$$
$$80r-240+80r+240=1.3r^2-1170$$
$$160r=1.3r^2-1170$$
$$0=1.3r^2-160r-1170$$
$$r=\frac{-(-160)\pm\sqrt{(-160)^2-4(1.3)(-1170)}}{2(1.3)}$$
$$=\frac{160\pm\sqrt{25,600+6084}}{2.6}$$
$$=\frac{160\pm\sqrt{31,684}}{2.6}$$
$$=\frac{160\pm178}{2.6}$$
$$r=\frac{160+178}{2.6} \quad \text{or} \quad r=\frac{160-178}{2.6}$$
$$=\frac{338}{2.6} \qquad =\frac{-18}{2.6}$$
$$=130 \qquad \approx-6.92$$

Since speed must be positive, the speed of the plane in still air is 130 mph.

49. Let t be the number of hours for Chris to clean alone. Then $t + 0.5$ is the number of hours for John to clean alone.

	Rate of work	Time worked	Part of Task completed
Chris	$\frac{1}{t}$	6	$\frac{6}{t}$
John	$\frac{1}{t+0.5}$	6	$\frac{6}{t+0.5}$

$$1 = \frac{6}{t} + \frac{6}{t+0.5}$$

$$t(t+0.5)(1) = t(t+0.5)\left(\frac{6}{t}\right) + t(t+0.5)\left(\frac{6}{t+0.5}\right)$$

$$t(t+0.5) = 6(t+0.5) + 6t$$

$$t^2 + 0.5t = 6t + 3 + 6t$$

$$t^2 + 0.5t = 12t + 3$$

$$t^2 - 11.5t - 3 = 0$$

$$t = \frac{11.5 \pm \sqrt{(-11.5)^2 - 4(1)(-3)}}{2(1)}$$

$$= \frac{11.5 \pm \sqrt{132.25 + 12}}{2}$$

$$= \frac{11.5 \pm \sqrt{144.25}}{2}$$

$$t = \frac{11.5 + \sqrt{144.25}}{2} \quad \text{or} \quad t = \frac{11.5 - \sqrt{144.25}}{2}$$

$$\approx 11.76 \qquad\qquad \approx -0.26$$

Since the time must be positive, it takes Chris about 11.76 hours and John about $11.76 + 0.5 = 12.26$ hours to clean alone

51. Let x be the speed of the trip from Lubbock to Plainview. Then $x - 10$ is the speed from Plainview to Amarillo.

	d	r	t
First part	75	x	$\frac{75}{x}$
Second part	195	$x - 10$	$\frac{195}{x-10}$

Including the 2 hours she spent in Plainview, the entire trip took Lisa 6 hours.

$$\frac{75}{x} + 2 + \frac{195}{x-10} = 6$$

$$\frac{75}{x} + \frac{195}{x-10} = 4$$

$$x(x-10)\left(\frac{75}{x} + \frac{195}{x-10}\right) = x(x-10)4$$

$$75(x-10) + 195x = 4x^2 - 40x$$

$$75x - 750 + 195x = 4x^2 - 40x$$

$$270x - 750 = 4x^2 - 40x$$

$$0 = 4x^2 - 310x + 750$$

$$0 = 2(2x^2 - 155x + 375)$$

$$0 = 2(2x-5)(x-75)$$

$$2x - 5 = 0 \quad \text{or} \quad x - 75 = 0$$

$$x = \frac{5}{2} \qquad\qquad x = 75$$

We disregard $x = \frac{5}{2}$ since it would result in a negative speed for the second part of the trip. Shywanda's average speed from San Antonio to Austin was 75 mph.

53. From the figure, $16x$ = the length and $9x$ = the height.

$$(16x)^2 + (9x)^2 = 40^2$$

$$256x^2 + 81x^2 = 1600$$

$$337x^2 = 1600$$

$$x^2 = \frac{1600}{337}$$

$$x = \pm\sqrt{\frac{1600}{337}} \approx \pm 2.179$$

Disregard the negative answer.
The length is $16(2.179) \approx 34.86$ inches and the height is $9(2.179) \approx 19.61$ inches.

55. Answers will vary.

57. Let l = original length and w = original width. A system of equations that describes this situation is

$$l \cdot w = 18$$
$$(l+2)(w+3) = 48$$

Solve the first equation for l to obtain $l = \frac{18}{w}$.

Substitute $\frac{18}{w}$ for l in the second equation. The result is an equation in only one variable which can be solved.

$$(l+2)(w+3) = 48$$
$$\left(\frac{18}{w}+2\right)(w+3) = 48$$
$$18 + \frac{54}{w} + 2w + 6 = 48$$
$$2w - 24 + \frac{54}{w} = 0$$
$$w\left(2w - 24 + \frac{54}{w}\right) = w(0)$$
$$2w^2 - 24w + 54 = 0$$
$$2(w-3)(w-9) = 0$$
$$w = 3 \text{ or } w = 9$$

If $w = 3$, then $l = \frac{18}{3} = 6$. One possible set of dimensions for the original rectangle is 6 meters by 3 meters. If $w = 9$, then $l = \frac{18}{9} = 2$. Another possible set of dimensions for the original rectangle is 2 meters by 9 meters.

59.
$$-\left[4(5-3)^3\right] + 2^4 = -\left[4(2)^3\right] + 2^4$$
$$= -\left[4(8)\right] + 16$$
$$= -32 + 16$$
$$= -16$$

60. $IR + Ir = E$, for R

$$IR = E - Ir$$
$$R = \frac{E - Ir}{I}$$

61.
$$\frac{r}{r-4} - \frac{r}{r+4} + \frac{32}{r^2-16}$$
$$= \frac{r}{r-4}\cdot\frac{r+4}{r+4} - \frac{r}{r+4}\cdot\frac{r-4}{r-4} + \frac{32}{(r+4)(r-4)}$$
$$= \frac{r(r+4) - r(r-4) + 32}{(r+4)(r-4)} = \frac{r^2 + 4r - r^2 + 4r + 32}{(r+4)(r-4)}$$
$$= \frac{8r+32}{(r+4)(r-4)} = \frac{8(r+4)}{(r+4)(r-4)} = \frac{8}{r-4}$$

62.
$$\left(\frac{x^{3/4}y^{-2}}{x^{1/2}y^2}\right)^8 = \left(x^{(3/4)-(1/2)}y^{-2-2}\right)^8$$
$$= \left(x^{1/4}y^{-4}\right)^8$$
$$= x^2y^{-32}$$
$$= \frac{x^2}{y^{32}}$$

63.
$$\sqrt{x^2+3x+12} = x$$
$$x^2 + 3x + 12 = x^2$$
$$3x + 12 = 0$$
$$3x = -12$$
$$x = -4$$

Upon checking, this value does not satisfy the equation. There is no real solution.

Mid-Chapter Test: 8.1-8.3

1.
$$x^2 - 12 = 86$$
$$x^2 = 98$$
$$x = \pm\sqrt{98} = \pm 7\sqrt{2}$$

2.
$$(a-3)^2 + 20 = 0$$
$$(a-3)^2 = -20$$
$$a - 3 = \pm\sqrt{-20}$$
$$a - 3 = \pm 2i\sqrt{5}$$
$$a = 3 \pm 2i\sqrt{5}$$

3.
$$(2m+7)^2 = 36$$
$$2m + 7 = \pm\sqrt{36}$$
$$2m + 7 = \pm 6$$
$$2m = -7 \pm 6$$
$$m = \frac{-7 \pm 6}{2}$$
$$m = \frac{-7+6}{2} \quad \text{or} \quad m = \frac{-7-6}{2}$$
$$m = -\frac{1}{2} \qquad m = -\frac{13}{2}$$

4. $y^2+4y-12=0$

$$y^2+4y=12$$
$$y^2+4y+4=12+4$$
$$(y+2)^2=16$$
$$y+2=\pm\sqrt{16}$$
$$y+2=\pm 4$$
$$y=-2\pm 4$$
$$y=-2+4 \quad \text{or} \quad y=-2-4$$
$$y=2 \qquad\qquad y=-6$$

5. $3a^2-12a-30=0$

$$\frac{1}{3}(3a^2-12a-30)=\frac{1}{3}(0)$$
$$a^2-4a-10=0$$
$$a^2-4a=10$$
$$a^2-4a+4=10+4$$
$$(a-2)^2=14$$
$$a-2=\pm\sqrt{14}$$
$$y=2\pm\sqrt{14}$$

6. $4c^2+c=-9$

$$\frac{1}{4}\left(4c^2+c\right)=\frac{1}{4}(-9)$$
$$c^2+\frac{1}{4}c=-\frac{9}{4}$$
$$c^2+\frac{1}{4}c+\frac{1}{64}=-\frac{9}{4}+\frac{1}{64}$$
$$\left(c+\frac{1}{8}\right)^2=-\frac{143}{64}$$
$$\left(c+\frac{1}{8}\right)^2=-\frac{143}{64}$$
$$c+\frac{1}{8}=\sqrt{-\frac{143}{64}}$$
$$c+\frac{1}{8}=\pm\frac{i\sqrt{143}}{8}$$
$$c=-\frac{1}{8}\pm\frac{i\sqrt{143}}{8}$$
$$c=\frac{-1\pm i\sqrt{143}}{8}$$

7. Let x = the length of one side of the patio. Then $x+6$ is the length of the diagonal.

$$x^2+x^2=(x+6)^2$$
$$2x^2=x^2+12x+36$$
$$x^2-12x=36$$
$$x^2-12x+36=36+36$$
$$(x-6)^2=72$$
$$x-6=\pm\sqrt{72}$$
$$x-6=\pm 6\sqrt{2}$$
$$x=6\pm 6\sqrt{2}$$

The length of one side is $\left(6\pm 6\sqrt{2}\right)$ meters.

8. a. b^2-4ac

b. If $b^2-4ac>0$, then the quadratic equation will have two distinct real solutions. If $b^2-4ac=0$, then the equation has the single real solution $\frac{-b}{2a}$. If $b^2-4ac<0$, then the expression under the radical sign in the quadratic formula is negative. Thus, the equation will have no real solution.

9. $2b^2-6b-11=0$

$b^2-4ac=(-6)^2-4(2)(-11)=36+88=124$

Since the discriminant is positive, the equation will have two distinct real solutions.

10. $6n^2+n=15$

$$6n^2+n-15=0$$
$$n=\frac{-b\pm\sqrt{b^2-4ac}}{2a}$$
$$=\frac{-1\pm\sqrt{1^2-4(6)(-15)}}{2(6)}$$
$$=\frac{-1\pm\sqrt{1+360}}{12}$$
$$=\frac{-1\pm\sqrt{361}}{12}$$
$$=\frac{-1\pm 19}{12}$$
$$n=\frac{-1+19}{12} \quad \text{or} \quad n=\frac{-1-19}{12}$$
$$=\frac{18}{12} \qquad\qquad =\frac{-20}{12}$$
$$=\frac{3}{2} \qquad\qquad =-\frac{5}{3}$$

11.
$$p^2 = -4p + 8$$
$$p^2 + 4p - 8 = 0$$
$$p = \frac{-b \pm \sqrt{b^2 - 4ac}}{2a}$$
$$= \frac{-4 \pm \sqrt{4^2 - 4(1)(-8)}}{2(1)}$$
$$= \frac{-4 \pm \sqrt{16 + 32}}{2}$$
$$= \frac{-4 \pm \sqrt{48}}{2}$$
$$= \frac{-4 \pm 4\sqrt{3}}{2}$$
$$= -2 \pm 2\sqrt{3}$$

12. $3d^2 - 2d + 5 = 0$
$$d = \frac{-b \pm \sqrt{b^2 - 4ac}}{2a}$$
$$= \frac{-(-2) \pm \sqrt{(-2)^2 - 4(3)(5)}}{2(3)}$$
$$= \frac{2 \pm \sqrt{4 - 60}}{6}$$
$$= \frac{2 \pm \sqrt{-56}}{6}$$
$$= \frac{2 \pm 2i\sqrt{14}}{6}$$
$$= \frac{1 \pm i\sqrt{14}}{3}$$

13. $x = 7$ and $x = -2$
$x - 7 = 0$ and $x + 2 = 0$
$$(x - 7)(x + 2) = 0$$
$$x^2 + 2x - 7x - 14 = 0$$
$$x^2 - 5x - 14 = 0$$

14. $x = 2 + \sqrt{5}$ and $x = 2 - \sqrt{5}$
$x - 2 - \sqrt{5} = 0$ and $x - 2 + \sqrt{5} = 0$
$$\left(x - 2 - \sqrt{5}\right)\left(x - 2 + \sqrt{5}\right) = 0$$
$$\left[(x - 2) - \sqrt{5}\right]\left[(x - 2) + \sqrt{5}\right] = 0$$
$$(x - 2)^2 - \left(\sqrt{5}\right)^2 = 0$$
$$x^2 - 4x + 4 - 5 = 0$$
$$x^2 - 4x - 1 = 0$$

15. The revenue function is
$R(n) = n(60 - 0.5n) = 60n - 0.5n^2$
Let $R(n) = 550$.
$$550 = 60n - 0.5n^2$$
$$0.5n^2 - 60n + 550 = 0$$
$$2(0.5n^2 - 60n + 550) = 2(0)$$
$$n^2 - 120n + 1100 = 0$$
$$n = \frac{-b \pm \sqrt{b^2 - 4ac}}{2a}$$
$$= \frac{-(-120) \pm \sqrt{(-120)^2 - 4(1)(1100)}}{2(1)}$$
$$= \frac{120 \pm \sqrt{14,400 - 4400}}{2}$$
$$= \frac{120 \pm \sqrt{10,000}}{2}$$
$$= \frac{120 \pm 100}{2}$$
$$n = \frac{120 + 100}{2} \quad \text{or} \quad n = \frac{120 - 100}{2}$$
$$= \frac{220}{2} \qquad = \frac{20}{2}$$
$$= 110 \qquad = 10$$
We have a restriction of $n \le 20$, so disregard 110. Thus 10 lamps must be sold.

16. $y = x^2 - r^2$, for r
$$r^2 = x^2 - y$$
$$r = \sqrt{x^2 - y}$$

17. $A = \frac{1}{3}kx^2$, for r
$$3(A) = 3\left(\frac{1}{3}kx^2\right)$$
$$3A = kx^2$$
$$\frac{3A}{k} = x^2$$
$$\sqrt{\frac{3A}{k}} = x$$

18. $d = \sqrt{x^2 + y^2}$, for y
$$d^2 = x^2 + y^2$$
$$d^2 - x^2 = y^2$$
$$\sqrt{d^2 - x^2} = y$$

19. Let w = the width of the rectangle. Then $2w+2$ = the length.

$$w(2w+2)=60$$
$$2w^2+2w-60=0$$
$$\frac{1}{2}(2w^2+2w-60)=\frac{1}{2}(0)$$
$$w^2+w-30=0$$
$$w=\frac{-1\pm\sqrt{1^2-4(1)(-30)}}{2(1)}$$
$$=\frac{-1\pm\sqrt{1+120}}{2}$$
$$=\frac{-1\pm\sqrt{121}}{2}$$
$$=\frac{-1\pm 11}{2}$$
$$w=\frac{-1+11}{2} \quad\text{or}\quad w=\frac{-1-11}{2}$$
$$=\frac{10}{2} \qquad\qquad =\frac{-12}{2}$$
$$=5 \qquad\qquad =-6$$

The width must be positive, so disregard the negative answer. The width is 5 feet and the length is $2(5)+2=12$ feet.

20. Note: 2000 = 20 hundreds

$$p(n)=2n^2+n-35;\ p(n)=20$$
$$20=2n^2+n-35$$
$$0=2n^2+n-55$$
$$0=2n^2+n-55$$
$$n=\frac{-1\pm\sqrt{1^2-4(2)(-55)}}{2(2)}$$
$$=\frac{-1\pm\sqrt{1+440}}{4}$$
$$=\frac{-1\pm\sqrt{441}}{4}$$
$$=\frac{-1\pm 21}{4}$$
$$n=\frac{-1+21}{4} \quad\text{or}\quad n=\frac{-1-21}{4}$$
$$=\frac{20}{4} \qquad\qquad =\frac{-22}{4}$$
$$=5 \qquad\qquad =-\frac{11}{2}$$

Disregard the negative answer. Thus 5 clocks must be sold.

Exercise Set 8.4

1. A given equation can be expressed as an equation in quadratic form if the equation can be written in the form $au^2+bu+c=0$.

3. Let $u=x^2$. Then

$$3x^4-5x^2+1=0$$
$$3\left(x^2\right)^2-5x^2+1=0$$
$$3u^2-5u+1=0$$

5. Let $u=z^{-1}$. Then

$$z^{-2}-z^{-1}=56$$
$$\left(z^{-1}\right)^2-z^{-1}=56$$
$$u^2-u=56$$

7.

$$x^4-10x^2+9=0$$
$$\left(x^2\right)^2-10x^2+9=0$$

Let $u=x^2$.

$$u^2-10u+9=0$$
$$(u-9)(u-1)=0$$
$$u-9=0 \quad\text{or}\quad u-1=0$$
$$u=9 \qquad\qquad u=1$$

Substitute x^2 for u.

$$x^2=9 \quad\text{or}\quad x^2=1$$
$$x=\pm\sqrt{9}=\pm 3 \qquad x=\pm\sqrt{1}=\pm 1$$

The solutions are 3, –3, 1, and –1.

9.

$$x^4+17x^2+16=0$$
$$(x^2)^2+17x^2+16=0$$

Let $u=x^2$.

$$u^2+17u+16=0$$
$$(u+16)(u+1)=0$$
$$u+16=0 \quad\text{or}\quad u+1=0$$
$$u=-16 \qquad\qquad u=-1$$

Substitute x^2 for u.

$$x^2=-16 \quad\text{or}\quad x^2=-1$$
$$x=\pm\sqrt{-16}=\pm 4i \qquad x=\pm\sqrt{-1}=\pm i$$

The solutions are $4i$, $-4i$, i, and $-i$.

11. $x^4 - 13x^2 + 36 = 0$

$(x^2)^2 - 13x^2 + 36 = 0$

Let $u = x^2$.

$u^2 - 13u + 36 = 0$

$(u-9)(u-4) = 0$

$u - 9 = 0$ or $u - 4 = 0$

$u = 9$ $\quad u = 4$

Substitute x^2 for u.

$x^2 = 9$ or $x^2 = 4$

$x = \pm\sqrt{9} = \pm 3$ $\quad x = \pm\sqrt{4} = \pm 2$

The solutions are 3, –3, 2, and –2.

13. $a^4 - 7a^2 + 12 = 0$

$\left(a^2\right)^2 - 7a^2 + 12 = 0$

Let $u = a^2$.

$u^2 - 7u + 12 = 0$

$(u-4)(u-3) = 0$

$u - 4 = 0$ or $u - 3 = 0$

$u = 4$ $\quad u = 3$

Substitute a^2 for u.

$a^2 = 4$ or $a^2 = 3$

$a = \pm\sqrt{4} = \pm 2$ $\quad a = \pm\sqrt{3}$

The solutions are 2, –2, $\sqrt{3}$, and $-\sqrt{3}$.

15. $4x^4 - 17x^2 + 4 = 0$

$4(x^2)^2 - 17x^2 + 4 = 0$

Let $u = x^2$.

$4u^2 - 17u + 4 = 0$

$(4u-1)(u-4) = 0$

$4u - 1 = 0$ or $u - 4 = 0$

$u = \frac{1}{4}$ $\quad u = 4$

Substitute x^2 for u.

$x^2 = \frac{1}{4}$ or $x^2 = 4$

$x = \pm\sqrt{\frac{1}{4}} = \pm\frac{1}{2}$ $\quad x = \pm\sqrt{4} = \pm 4$

The solutions are $\frac{1}{2}, -\frac{1}{2}$, 2, and –2.

17. $r^4 - 8r^2 = -15$

$r^4 - 8r^2 + 15 = 0$

$(r^2)^2 - 8r^2 + 15 = 0$

Let $u = r^2$.

$u^2 - 8u + 15 = 0$

$(u-3)(u-5) = 0$

$u - 3 = 0$ or $u - 5 = 0$

$u = 3$ $\quad u = 5$

Substitute r^2 for u.

$r^2 = 3$ or $r^2 = 5$

$r = \pm\sqrt{3}$ $\quad r = \pm\sqrt{5}$

The solutions are $\sqrt{3}, -\sqrt{3}, \sqrt{5}$, and $-\sqrt{5}$.

19. $z^4 - 7z^2 = 18$

$z^4 - 7z^2 - 18 = 0$

$\left(z^2\right)^2 - 7z^2 - 18 = 0$

Let $u = z$.

$u^2 - 7u - 18 = 0$

$(u-9)(u+2) = 0$

$u - 9 = 0$ or $u + 2 = 0$

$u = 9$ $\quad u = -2$

Substitute z^2 for u.

$z^2 = 9$ or $z^2 = -2$

$z = \pm 3$ $\quad z = \pm\sqrt{-2} = \pm i\sqrt{2}$

The solutions are $3, -3, i\sqrt{2}$, and $-i\sqrt{2}$.

21. $-c^4 = 4c^2 - 5$

$0 = c^4 + 4c^2 - 5$

$0 = (c^2)^2 + 4c^2 - 5$

Let $u = c$.

$u^2 + 4u - 5 = 0$

$(u-1)(u+5) = 0$

$u - 1 = 0$ or $u + 5 = 0$

$u = 1$ $\quad u = -5$

Substitute c^2 for u.

$c^2 = 1$ or $c^2 = -5$

$c = \pm 1$ $\quad c = \pm\sqrt{-5} = \pm i\sqrt{5}$

The solutions are $1, -1, i\sqrt{5}$, and $-i\sqrt{5}$.

23. $\sqrt{x} = 2x - 6$

$$0 = 2x - \sqrt{x} - 6$$
$$0 = 2\left(x^{1/2}\right)^2 - x^{1/2} - 6$$

Let $u = x^{1/2}$.

$$2u^2 - u - 6 = 0$$
$$(2u+3)(u-2) = 0$$
$$2u + 3 = 0 \quad \text{or} \quad u - 2 = 0$$
$$u = -\frac{3}{2} \qquad u = 2$$

Substitute $x^{1/2}$ for u.

$$x^{1/2} = -\frac{3}{2} \quad \text{or} \quad x^{1/2} = 2$$

Not real $\qquad x = 2^2 = 4$

The solution is 4.

25. $x - \sqrt{x} = 6$

$$x - \sqrt{x} - 6 = 0$$
$$\left(x^{1/2}\right)^2 - x^{1/2} - 6 = 0$$

Let $u = x^{1/2}$.

$$u^2 - u - 6 = 0$$
$$(u+2)(u-3) = 0$$
$$u + 2 = 0 \quad \text{or} \quad u - 3 = 0$$
$$u = -2 \quad \text{or} \quad u = 3$$

Substitute $x^{1/2}$ for u.

$$x^{1/2} = -2 \quad \text{or} \quad x^{1/2} = 3$$

Not real $\qquad x = 3^2 = 9$

The solution is 9.

27. $9x + 3\sqrt{x} = 2$

$$9x + 3\sqrt{x} - 2 = 0$$
$$9\left(x^{1/2}\right)^2 + 3x^{1/2} - 2 = 0$$

Let $u = x^{1/2}$.

$$9u^2 + 3u - 2 = 0$$
$$(3u+2)(3u-1) = 0$$
$$3u + 2 = 0 \quad \text{or} \quad 3u - 1 = 0$$
$$u = -\frac{2}{3} \qquad u = \frac{1}{3}$$

Substitute $x^{1/2}$ for u.

$$x^{1/2} = -\frac{2}{3} \quad \text{or} \quad x^{1/2} = \frac{1}{3}$$

Not real $\qquad x = \left(\frac{1}{3}\right)^2 = \frac{1}{9}$

The solution is $\frac{1}{9}$.

29. $(x+3)^2 + 2(x+3) = 24$

$$(x+3)^2 + 2(x+3) - 24 = 0$$

Let $u = x + 3$.

$$u^2 + 2u - 24 = 0$$
$$(u-4)(u+6) = 0$$
$$u - 4 = 0 \quad \text{or} \quad u + 6 = 0$$
$$u = 4 \qquad u = -6$$

Substitute $x+3$ for u.

$$x + 3 = 4 \quad \text{or} \quad x + 3 = -6$$
$$x = 1 \qquad x = -9$$

The solutions are 1 and –9.

31. $6(a-2)^2 = -19(a-2) - 10$

$$6(a-2)^2 + 19(a-2) + 10 = 0$$

Let $u = a - 2$.

$$6u^2 + 19u + 10 = 0$$
$$(3u+2)(2u+5) = 0$$
$$3u + 2 = 0 \quad \text{or} \quad 2u + 5 = 0$$
$$u = -\frac{2}{3} \qquad u = -\frac{5}{2}$$

Substitute $a-2$ for u.

$$a - 2 = -\frac{2}{3} \quad \text{or} \quad a - 2 = -\frac{5}{2}$$
$$a = \frac{4}{3} \qquad a = -\frac{1}{2}$$

The solutions are $\frac{4}{3}$ and $-\frac{1}{2}$.

33. $(x^2-3)^2 - (x^2-3) - 6 = 0$

Let $u = x^2 - 3$.

$$u^2 - u - 6 = 0$$
$$(u+2)(u-3) = 0$$
$$u + 2 = 0 \quad \text{or} \quad u - 3 = 0$$
$$u = -2 \qquad u = 3$$

Substitute $x^2 - 1$ for u.

$$x^2 - 3 = -2 \quad \text{or} \quad x^2 - 3 = 3$$
$$x^2 = 1 \qquad x^2 = 6$$
$$x = \pm\sqrt{1} = \pm 1 \qquad x = \pm\sqrt{6}$$

The solutions are $1, -1, \sqrt{6}$, and $-\sqrt{6}$.

35. $2(b+3)^2+5(b+3)-3=0$

Let $u=b+3$.

$2u^2+5u-3=0$

$(u+3)(2u-1)=0$

$u+3=0$ or $2u-1=0$

$u=-3$ $\quad u=\frac{1}{2}$

Substitute $b+3$ for u.

$b+3=-3$ or $b+3=\frac{1}{2}$

$b=-6$ $\quad b=-\frac{5}{2}$

The solutions are –6 and $-\frac{5}{2}$.

37. $18(x^2-5)^2+27(x^2-5)+10=0$

Let $u=x^2-5$.

$18u^2+27u+10=0$

$(3u+2)(6u+5)=0$

$3u+2=0$ or $6u+5=0$

$u=-\frac{2}{3}$ $\quad u=-\frac{5}{6}$

Substitute x^2-5 for u.

or $x^2-5=-\frac{5}{6}$

$x^2-5=-\frac{2}{3}$ or $x^2-5=-\frac{5}{6}$

$x^2=\frac{13}{3}$ $\quad x^2=\frac{25}{6}$

$x=\pm\sqrt{\frac{13}{3}}$ $\quad x=\pm\sqrt{\frac{25}{6}}$

$=\pm\frac{\sqrt{13}}{\sqrt{3}}\cdot\frac{\sqrt{3}}{\sqrt{3}}$ $\quad =\pm\frac{5}{\sqrt{6}}\cdot\frac{\sqrt{6}}{\sqrt{6}}$

$=\pm\frac{\sqrt{39}}{3}$ $\quad =\pm\frac{5\sqrt{6}}{6}$

The solutions are $\frac{\sqrt{39}}{3}, -\frac{\sqrt{39}}{3}, \frac{5\sqrt{6}}{6}$, and $-\frac{5\sqrt{6}}{6}$.

39. $a^{-2}+4a^{-1}+4=0$

$(a^{-1})^2+4(a^{-1})+4=0$

Let $u=a^{-1}$.

$u^2+4u+4=0$

$(u+2)^2=0$

$u+2=0$

$u=-2$

Substitute a^{-1} for u.

$a^{-1}=-2$

$a=-\frac{1}{2}$

The solution is $-\frac{1}{2}$.

41. $12b^{-2}-7b^{-1}+1=0$

$12(b^{-1})^2-7(b^{-1})+1=0$

Let $u=b^{-1}$.

$12u^2-7u+1=0$

$(4u-1)(3u-1)=0$

$4u-1=0$ or $3u-1=0$

$u=\frac{1}{4}$ $\quad u=\frac{1}{3}$

Substitute b^{-1} for u.

$b^{-1}=\frac{1}{4}$ or $b^{-1}=\frac{1}{3}$

$b=4$ $\quad b=3$

The solutions are 4 and 3.

43. $2b^{-2}=7b^{-1}-3$

$2b^{-2}-7b^{-1}+3=0$

$2(b^{-1})^2-7(b^{-1})+3=0$

Let $u=b^{-1}$.

$2u^2-7u+3=0$

$(2u-1)(u-3)=0$

$2u-1=0$ or $u-3=0$

$u=\frac{1}{2}$ $\quad u=3$

Substitute b^{-1} for u.

$b^{-1}=\frac{1}{2}$ or $b^{-1}=3$

$b=2$ $\quad b=\frac{1}{3}$

The solutions are 2 and $\frac{1}{3}$.

45. $x^{-2}+9x^{-1}=10$

$x^{-2}+9x^{-1}-10=0$

$(x^{-1})^2+9(x^{-1})-10=0$

Let $u=x^{-1}$.

$u^2+9u-10=0$

$(u+10)(u-1)=0$

$u+10=0$ or $u-1=0$

$u=-10$ $\quad u=1$

Substitute x^{-1} for u.

$x^{-1}=-10$ or $x^{-1}=1$

$x=-\frac{1}{10}$ $\quad x=1$

The solutions are $-\frac{1}{10}$ and 1.

47. $x^{-2}=4x^{-1}+12$

$x^{-2}-4x^{-1}-12=0$

$(x^{-1})^2-4(x^{-1})-12=0$

Let $u=x^{-1}$.

$u^2-4u-12=0$

$(u+2)(u-6)=0$

$u+2=0$ or $u-6=0$

$u=-2$ $\quad u=6$

Substitute x^{-1} for u.

$x^{-1}=-2$ or $x^{-1}=6$

$x=-\frac{1}{2}$ $\quad x=\frac{1}{6}$

The solutions are $-\frac{1}{2}$ and $\frac{1}{6}$.

49. $x^{2/3}-4x^{1/3}=-3$

$(x^{1/3})^2-4x^{1/3}+3=0$

Let $u=x^{1/3}$.

$u^2-4u+3=0$

$(u-1)(u-3)=0$

$u-1=0$ or $u-3=0$

$u=1$ $\quad u=3$

Substitute $x^{1/3}$ for u.

$x^{1/3}=1$ or $x^{1/3}=3$

$x=1^3=1$ $\quad x=3^3=27$

The solutions are 1 and 27.

51. $b^{2/3}-9b^{1/3}+18=0$

$(b^{1/3})^2-9b^{1/3}+18=0$

Let $u=b^{1/3}$.

$u^2-9u+18=0$

$(u-6)(u-3)=0$

$u-6=0$ or $u-3=0$

$u=6$ $\quad u=3$

Substitute $b^{1/3}$ for u.

$b^{1/3}=6$ or $b^{1/3}=3$

$b=6^3=216$ $\quad b=3^3=27$

The solutions are 216 and 27.

53. $-2a-5a^{1/2}+3=0$

$-2(a^{1/2})^2-5a^{1/2}+3=0$

Let $u=a^{1/2}$.

$-2u^2-5u+3=0$

$2u^2+5u-3=0$

$(2u-1)(u+3)=0$

$2u-1=0$ or $u+3=0$

$u=\frac{1}{2}$ $\quad u=-3$

Substitute $a^{1/2}$ for u.

$a^{1/2}=2$ or $a^{1/2}=-3$

$a=\left(\frac{1}{2}\right)^2=\frac{1}{4}$ $\quad$ not real

The solution is $\frac{1}{4}$.

55. $c^{2/5}+3c^{1/5}+2=0$

$(c^{1/5})^2+3c^{1/5}+2=0$

Let $u=c^{1/5}$.

$u^2+3u+2=0$

$(u+2)(u+1)=0$

$u+2=0$ or $u+1=0$

$u=-2$ $\quad u=-1$

Substitute $c^{1/5}$ for u.

$c^{1/5}=-2$ or $c^{1/5}=-1$

$c=(-2)^5=-32$ $\quad c=(-1)^5=-1$

The solutions are -32 and -1.

57. $f(x) = x - 5\sqrt{x} + 4$, $f(x) = 0$

$0 = (x^{1/2})^2 - 5x^{1/2} + 4$

Let $u = x^{1/2}$.

$0 = u^2 - 5u + 4$

$0 = (u-1)(u-4)$

$u - 1 = 0$ or $u - 4 = 0$

$u = 1$ $\quad u = 4$

Substitute $x^{1/2}$ for u.

$x^{1/2} = 1$ or $x^{1/2} = 4$

$x = 1^2 = 1$ $\quad x = 4^2 = 16$

The x-intercepts are (1, 0) and (16, 0).

59. $h(x) = x + 14\sqrt{x} + 45$, $h(x) = 0$

$0 = (x^{1/2})^2 + 14x^{1/2} + 45$

Let $u = x^{1/2}$.

$0 = u^2 + 14u + 45$

$0 = (u+9)(u+5)$

$u + 9 = 0$ or $u + 5 = 0$

$u = -9$ $\quad u = -5$

Substitute $x^{1/2}$ for u.

$x^{1/2} = -9$ or $x^{1/2} = -5$

There are no values of x for which $x^{1/2} = -9$ or $x^{1/2} = -5$. There are no x-intercepts.

61. $p(x) = 4x^{-2} - 19x^{-1} - 5$, $p(x) = 0$

$0 = 4(x^{-1})^2 - 19x^{-1} - 5$

Let $u = x^{-1}$.

$0 = 4u^2 - 19u - 5$

$0 = (4u+1)(u-5)$

$4u + 1 = 0$ or $u - 5 = 0$

$u = -\dfrac{1}{4}$ $\quad u = 5$

Substitute x^{-1} for u.

$x^{-1} = -\dfrac{1}{4}$ or $x^{-1} = 5$

$x = -4$ $\quad x = \dfrac{1}{5}$

The x-intercepts are (–4, 0) and $\left(\dfrac{1}{5}, 0\right)$.

63. $f(x) = x^{2/3} - x^{1/3} - 6$, $f(x) = 0$

$0 = (x^{1/3})^2 - x^{1/3} - 6$

Let $u = x^{1/3}$.

$0 = u^2 - u - 6$

$0 = (u-3)(u+2)$

$u - 3 = 0$ or $u + 2 = 0$

$u = 3$ $\quad u = -2$

Substitute $x^{1/3}$ for u.

$x^{1/3} = 3^3 = 27$ or $x^{1/3} = (-2)^3 = -8$

The x-intercepts are (27, 0) and (–8, 0).

65. $g(x) = (x^2 - 3x)^2 + 2(x^2 - 3x) - 24$, $g(x) = 0$

Let $u = x^2 - 3x$.

$0 = u^2 + 2u - 24$

$0 = (u+6)(u-4)$

$u + 6 = 0$ or $u - 4 = 0$

$u = -6$ $\quad u = 4$

Substitute $(x^2 - 3x)$ for u.

$x^2 - 3x = -6$ or $x^2 - 3x = 4$

$x^2 - 3x + 6 = 0$ $\quad x^2 - 3x - 4 = 0$

$(x-4)(x+1) = 0$

$x = 4$ or $x = -1$

There are no x-intercepts for $x^2 - 3x + 6 = 0$ since $b^2 - 4ac = (-3)^2 - 4(1)(6) = -15$ (that is, since the discriminant is negative). The x-intercepts are (4, 0) or (–1, 0).

67. $f(x) = x^4 - 29x + 100$, $f(x) = 0$

$0 = (x^2)^2 - 29x^2 + 100$

Let $u = x^2$.

$0 = u^2 - 29u + 100$

$0 = (u-25)(u-4)$

$u - 25 = 0$ or $u - 4 = 0$

$u = 25$ $\quad u = 4$

Substitute x^2 for u.

$x^2 = 25$ or $x^2 = 4$

$x = \pm\sqrt{25} = \pm 5$ $\quad x = \pm\sqrt{4} = \pm 2$

The x-intercepts are (5, 0), (–5, 0), (2, 0), and (–2, 0).

69. When solving an equation of the form $ax^4 + bx^2 + c = 0$, let $u = x^2$.

71. When solving an equation of the form $ax^{-2} + bx^{-1} + c = 0$, let $u = x^{-1}$.

73. If the solutions are ± 2 and ± 1, the factors must be $(x-2), (x+2), (x-1)$, and $(x+1)$.

$$(x-2)(x+2)(x-1)(x+1)=0$$
$$(x^2-4)(x^2-1)=0$$
$$x^4-5x^2+4=0$$

75. If the solutions are $\pm\sqrt{2}$ and $\pm\sqrt{5}$, the factors must be $(x+\sqrt{2}), (x-\sqrt{2}), (x+\sqrt{5}), (x-\sqrt{5})$.

$$(x+\sqrt{2})(x-\sqrt{2})(x+\sqrt{5})(x-\sqrt{5})=0$$
$$(x^2-2)(x^2-5)=0$$
$$x^4-7x^2+10=0$$

77. No. An equation of the form $ax^4+bx^2+c=0$ can have no imaginary solutions, two imaginary solutions, or four imaginary solutions.

79. a.

$$\frac{3}{x^2}-\frac{3}{x}=60 \quad \text{The LCD is } x^2$$
$$x^2\left(\frac{3}{x^2}\right)-x^2\left(\frac{3}{x}\right)=x^2(60)$$
$$3-3x=60x^2$$
$$0=60x^2+3x-3$$
$$0=3(20x^2+x-1)$$
$$0=3(5x-1)(4x+1)$$

$5x-1=0$ or $4x+1=0$

$x=\frac{1}{5}$ $\quad$ $x=-\frac{1}{4}$

The solutions are $\frac{1}{5}$ and $-\frac{1}{4}$.

b.

$$\frac{3}{x^2}-\frac{3}{x}=60$$
$$3x^{-2}-3x^{-1}=60$$
$$3(x^{-1})^2-3x^{-1}-60=0$$

Let $u=x^{-1}$.

$$3u^2-3u-60=0$$
$$3(u^2-u-20)=0$$
$$3(u-5)(u+4)=0$$

$u-5=0$ or $u+4=0$

$u=5$ $\quad$ $u=-4$

Substitute x^{-1} for u.

$x^{-1}=5$ or $x^{-1}=-4$

$x=\frac{1}{5}$ $\quad$ $x=-\frac{1}{4}$

The solutions are $\frac{1}{5}$ and $-\frac{1}{4}$.

81.

$$15(r+2)+22=-\frac{8}{r+2}$$
$$15(r+2)(r+2)+22(r+2)=-\frac{8}{r+2}(r+2)$$
$$15(r+2)^2+22(r+2)=-8$$
$$15(r+2)^2+22(r+2)+8=0$$

Let $u=r+2$.

$$15u^2+22u+8=0$$
$$(5u+4)(3u+2)=0$$

$5u+4=0$ or $3u+2=0$

$u=-\frac{4}{5}$ $\quad$ $u=-\frac{2}{3}$

Substitute $r+2$ for u.

$r+2=-\frac{4}{5}$ or $r+2=-\frac{2}{3}$

$r=-\frac{14}{5}$ $\quad$ $r=-\frac{8}{3}$

The solutions are $-\frac{14}{5}$ and $-\frac{8}{3}$.

83.

$$4-(x-1)^{-1}=3(x-1)^{-2}$$
$$3(x-1)^{-2}+(x-1)^{-1}-4=0$$
$$3\left[(x-1)^{-1}\right]^2+(x-1)^{-1}-4=0$$

Let $u=(x-1)^{-1}$

$$3u^2+u-4=0$$
$$(3u+4)(u-1)=0$$

$3u+4=0$ or $u-1=0$

$u=-\frac{4}{3}$ $\quad$ $u=1$

Substitute $(x-1)^{-1}$ for u.

$(x-1)^{-1}=-\frac{4}{3}$ or $(x-1)^{-1}=1$

$x-1=-\frac{3}{4}$ $\quad$ $x-1=1$

$x=\frac{1}{4}$ $\quad$ $x=2$

The solutions are $\frac{1}{4}$ and 2.

85. $x^6 - 9x^3 + 8 = 0$

$(x^3)^2 - 9x^3 + 8 = 0$

Let $u = x^3$.

$u^2 - 9u + 8 = 0$

$(u-8)(u-1) = 0$

$u - 8 = 0$ or $u - 1 = 0$

$u = 8 \qquad u = 1$

Substitute x^3 for u.

$x^3 = 8$ or $x^3 = 1$

$x = \sqrt[3]{8} = 2 \qquad x = \sqrt[3]{1} = 1$

The solutions are 2 and 1.

87. $(x^2 + 2x - 2)^2 - 7(x^2 + 2x - 2) + 6 = 0$

Let $u = x^2 + 2x - 2$.

$u^2 - 7u + 6 = 0$

$(u-6)(u-1) = 0$

$u = 6$ or $u = 1$

Substitute $x^2 + 2x - 2$ for u.

$x^2 + 2x - 2 = 6$ or $x^2 + 2x - 2 = 1$

$x^2 + 2x - 8 = 0 \qquad x^2 + 2x - 3 = 0$

$(x+4)(x-2) = 0 \qquad (x+3)(x-1) = 0$

$x = -4$ or $x = 2 \qquad x = -3$ or $x = 1$

The solutions are −4, 2, −3, and 1.

89. $2n^4 - 6n^2 - 3 = 0$

$2(n^2)^2 - 6n^2 - 3 = 0$

Let $u = n^2$.

$2u^2 - 6u - 3 = 0$

$$u = \frac{6 \pm \sqrt{(-6)^2 - 4(2)(-3)}}{2(2)}$$

$$= \frac{6 \pm \sqrt{60}}{4} = \frac{6 \pm 2\sqrt{15}}{4} = \frac{3 \pm \sqrt{15}}{2}$$

Substitute n^2 for u.

$$n^2 = \frac{3 \pm \sqrt{15}}{2}$$

$$n = \pm\sqrt{\frac{3 \pm \sqrt{15}}{2}}$$

91. $$\frac{4}{5} - \left(\frac{3}{4} - \frac{2}{3}\right) = \frac{4}{5} - \left(\frac{9}{12} - \frac{8}{12}\right)$$

$$= \frac{4}{5} - \frac{1}{12}$$

$$= \frac{48}{60} - \frac{5}{60}$$

$$= \frac{43}{60}$$

92. $3(x+2) - 2(3x+3) = -3$

$3x + 6 - 6x - 6 = -3$

$-3x = -3$

$x = 1$

93. $y = (x-3)^2$

Domain: $\mathbb{R}$

Range: $\{y \mid y \geq 0\}$

94. $\sqrt[3]{16x^3y^6} = \sqrt[3]{8 \cdot 2 \cdot x^3 (y^2)^3} = 2xy^2\sqrt[3]{2}$

95. $\sqrt{75} + \sqrt{48} = \sqrt{25 \cdot 3} + \sqrt{16 \cdot 3}$

$= 5\sqrt{3} + 4\sqrt{3}$

$= 9\sqrt{3}$

Exercise Set 8.5

1. The graph of a quadratic equation is called a parabola.

3. The axis of symmetry of a parabola is the line where, if the graph is folded, the two sides overlap.

5. For $f(x) = ax^2 + bx + c$, the vertex of the graph is $\left(-\frac{b}{2a}, \frac{4ac - b^2}{4}\right)$.

7. a. For $f(x) = ax^2 + bx + c$, $f(x)$ will have a minimum if $a > 0$ since the graph opens upward.

b. For $f(x) = ax^2 + bx + c$, $f(x)$ will have a maximum if $a < 0$ since the graph opens downward.

9. To find the y-intercepts of the graph of a quadratic function, set $x = 0$ and solve for y.

11. a. For $f(x) = ax^2$, the general shape of $f(x)$ if $a > 0$ is

b. For $f(x) = ax^2$, the general shape of $f(x)$ if $a < 0$ is

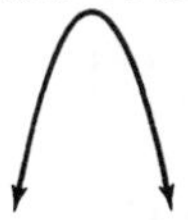

13. Since $a = 3$ is greater than 0, the graph opens upward, and therefore has a minimum value.

15. $f(x) = x^2 + 8x + 15$

a. Since $a = 1$, the parabola opens upward.

b. $y = (0)^2 + 8(0) + 15 = 15$
The y-intercept is (0, 15).

c. $x = -\dfrac{b}{2a} = -\dfrac{8}{2(1)} = -\dfrac{8}{2} = -4$

$$y = \frac{4ac - b^2}{4a} = \frac{4(1)(15) - 8^2}{4(1)} = \frac{60 - 64}{4} = \frac{-4}{4} = -1$$

The vertex is (–4, –1).

d.
$$x^2 + 8x + 15 = 0$$
$$(x+5)(x+3) = 0$$
$$x + 5 = 0 \quad \text{or} \quad x + 3 = 0$$
$$x = -5 \qquad\qquad x = -3$$
The x-intercepts are (–5, 0) and (–3, 0).

e.

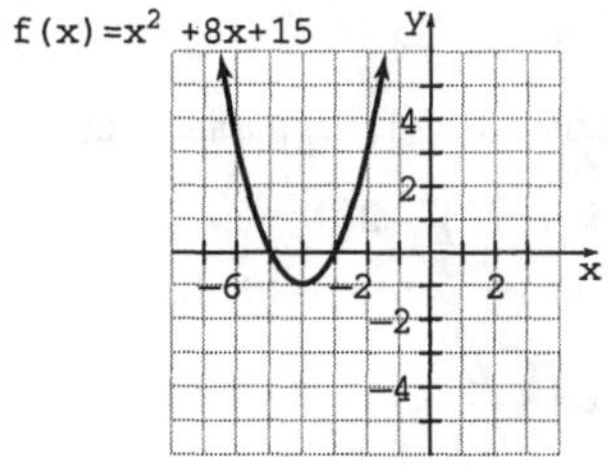

17. $f(x) = x^2 - 4x + 3$

a. Since $a = 1$, the parabola opens upward.

b. $y = 0^2 - 4(0) + 3 = 3$
The y-intercept is (0, 3).

c. $x = -\dfrac{b}{2a} = -\dfrac{-4}{2(1)} = \dfrac{4}{2} = 2$

$$y = \frac{4ac - b^2}{4a} = \frac{4(1)(3) - (-4)^2}{4(1)} = \frac{12 - 16}{4} = \frac{-4}{4} = -1$$

The vertex is (2, –1).

d.
$$x^2 - 4x + 3 = 0$$
$$(x-3)(x-1) = 0$$
$$x - 3 = 0 \quad \text{or} \quad x - 1 = 0$$
$$x = 3 \qquad\qquad x = 1$$
The x-intercepts are (3, 0) and (1, 0).

e.

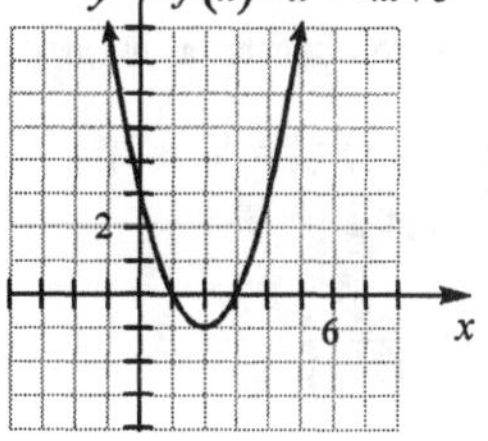

19. $f(x) = -x^2 - 2x + 8$

a. Since $a = -1$, the parabola opens downward.

b. $y = -(0)^2 - 2(0) + 8 = 8$
The y-intercept is (0, 8).

c. $x = -\frac{b}{2a} = -\frac{-2}{2(-1)} = -\frac{2}{2} = -1$

$$y = \frac{4ac - b^2}{4a}$$
$$= \frac{4(-1)(8) - (-2)^2}{4(-1)}$$
$$= \frac{-32 - 4}{-4} = \frac{-36}{-4} = 9$$

The vertex is $(-1, 9)$.

d.
$$-x^2 - 2x + 8 = 0$$
$$x^2 + 2x - 8 = 0$$
$$(x+4)(x-2) = 0$$
$$x + 4 = 0 \quad \text{or} \quad x - 2 = 0$$
$$x = -4 \qquad\qquad x = 2$$

The x-intercepts are (–4, 0) and (2, 0).

e.

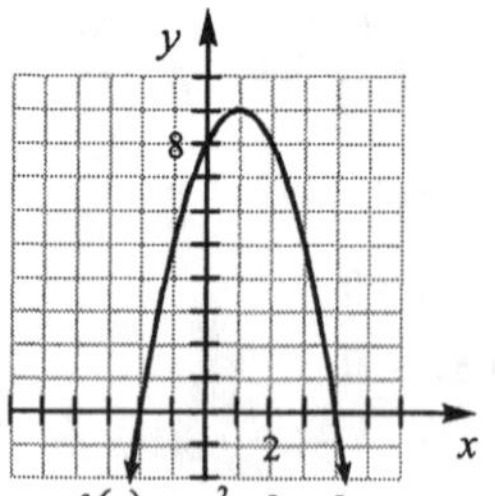

$f(x) = -x^2 - 2x + 8$

21. $g(x) = -x^2 + 4x + 5$

a. Since $a = -1$, the parabola opens downward.

b. $y = -(0)^2 + 4(0) + 5 = 8$
The y-intercept is (0, 5).

c. $x = -\frac{b}{2a} = -\frac{4}{2(-1)} = \frac{4}{2} = 2$

$$y = \frac{4ac - b^2}{4a}$$
$$= \frac{4(-1)(5) - (4)^2}{4(-1)}$$
$$= \frac{-20 - 16}{-4} = \frac{-36}{-4} = 9$$

The vertex is (2, 9).

d.
$$-x^2 + 4x + 5 = 0$$
$$x^2 - 4x - 5 = 0$$
$$(x-5)(x+1) = 0$$
$$x - 5 = 0 \quad \text{or} \quad x + 1 = 0$$
$$x = 5 \qquad\qquad x = -1$$

The x-intercepts are (5, 0) and (–1, 0).

e.

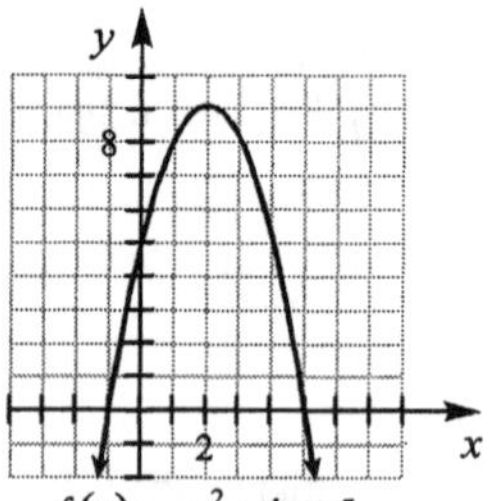

$f(x) = -x^2 + 4x + 5$

23. $t(x) = -x^2 + 4x - 5$

a. Since $a = -1$, the parabola opens downward.

b. $y = -0^2 + 4(0) - 5 = -5$; The y-intercept is (0, –5).

c. $x = -\frac{b}{2a} = -\frac{4}{2(-1)} = -\frac{4}{-2} = 2$

$$y = \frac{4ac - b^2}{4a} = \frac{4(-1)(-5) - (4)^2}{4(-1)}$$
$$= \frac{20 - 16}{-4} = \frac{4}{-4} = -1$$

The vertex is (2, –1)

d. $0 = -x^2 + 4x - 5$
Since this is not factorable, check the discriminant.
$b^2 - 4ac = 4^2 - 4(-1)(-5) = 16 - 20 = -4$
Since $-4 < 0$ there are no real roots. Thus, there are no x-intercepts.

e.

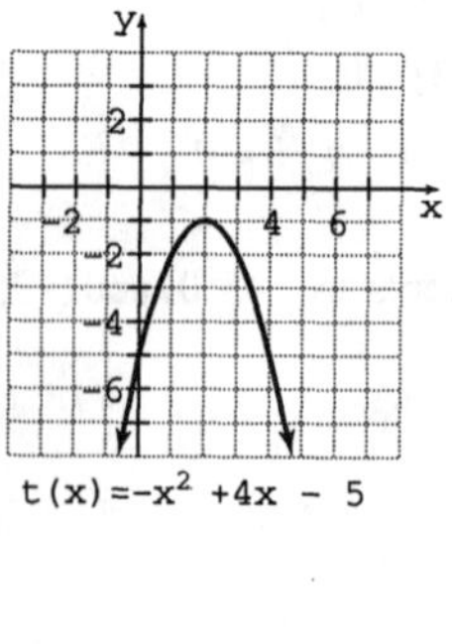

$t(x) = -x^2 + 4x - 5$

25. $f(x) = x^2 - 4x + 4$

a. Since $a = 1$, the parabola opens upward.

b. $y = 0^2 - 4(0) + 4 = 4$
The y-intercept is (0, 4).

c. $x = -\frac{b}{2a} = -\frac{-4}{2(1)} = \frac{4}{2} = 2$

$y = \frac{4ac - b^2}{4a} = \frac{4(1)(4) - (-4)^2}{4(1)}$

$= \frac{16 - 16}{4} = \frac{0}{4} = 0$

The vertex is (2, 0).

d. $x^2 - 4x + 4 = 0$

$(x - 2)^2 = 0$

$x - 2 = 0$

$x = 2$

The x-intercept (2, 0).

e.

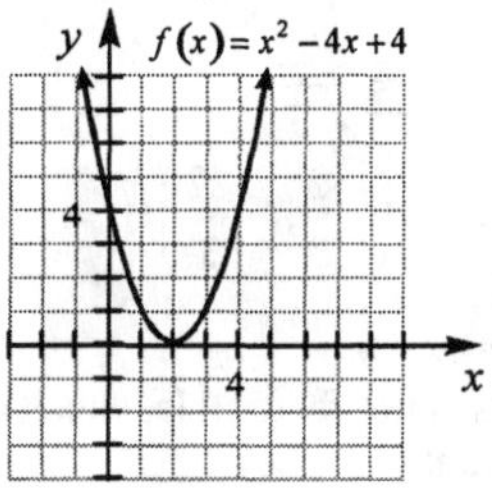

27. $r(x) = x^2 + 2$

a. Since $a = 1$ the parabola opens upward.

b. $y = 0^2 + 2 = 2$
The y-intercept is (0, 2).

c. $x = -\frac{b}{2a} = -\frac{0}{2(1)} = \frac{0}{2} = 0$

$y = \frac{4ac - b^2}{4a} = \frac{4(1)(2) - 0^2}{4(1)} = \frac{8}{4} = 2$

The vertex is (0, 2).

d. $0 = x^2 + 2$
Since this is not factorable, check the discriminant.
$b^2 - 4ac = 0 - 4(1)2 = -8$

There are no real roots. Thus, there are no x-intercepts.

e.

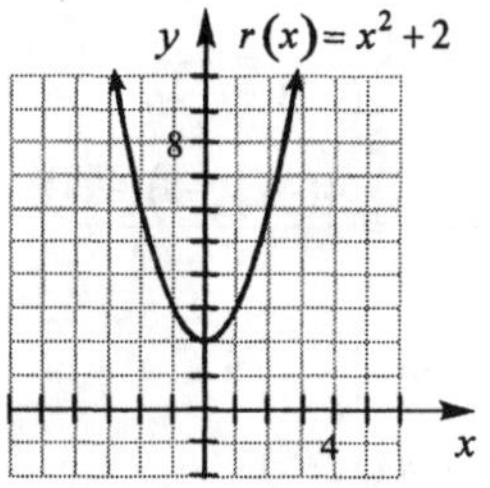

29. $l(x) = -x^2 + 5$

a. Since $a = -1$, the parabola opens downward.

b. $y = -(0)^2 + 5 = 5$
The y-intercept is (0, 5).

c. $x = -\frac{b}{2a} = -\frac{0}{2(-1)} = \frac{0}{2} = 0$

$y = \frac{4ac - b^2}{4a} = \frac{4(-1)(5) - (0)^2}{4(-1)}$

$= \frac{-20 - 0}{-4} = \frac{-20}{-4} = 5$

The vertex is (0, 5).

d. $-x^2 + 5 = 0$

$-x^2 = -5$

$x^2 = 5$

$x = \pm\sqrt{5}$

The x-intercepts are $(\sqrt{5}, 0)$ and $(-\sqrt{5}, 0)$.

e.

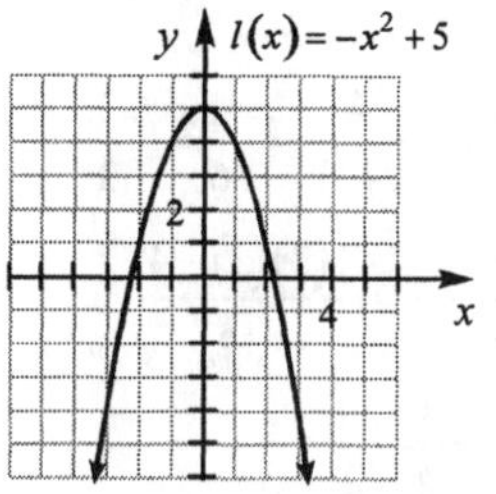

31. $y = -2x^2 + 4x - 8$

a. Since $a = -2$ the parabola opens downward.

b. $y = -2(0)^2 + 4(0) - 8 = -8$
The y-intercept is (0, –8).

c. $x=-\frac{b}{2a}=-\frac{4}{2(-2)}=-\frac{4}{-4}=1$

$y=\frac{4ac-b^2}{4a}=\frac{4(-2)(-8)-(4)^2}{4(-2)}$

$=\frac{64-16}{-8}=\frac{48}{-8}=-6$

The vertex is (1, –6).

d. $-2x^2+4x-8=0$

$-2(x^2-2x+4)=0$

Since this is not factorable, check the discriminant.

$b^2-4ac=4^2-4(-2)(-8)=16-64=-48$

Since –48 < 0, there are no real roots. Thus, there are no x-intercepts.

e. 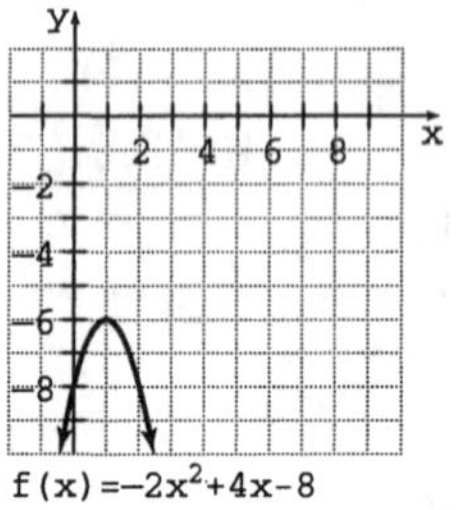

33. $m(x)=3x^2+4x+3$

a. Since $a=3$ the parabola opens upward.

b. $y=3(0)+4(0)+3=3$

The y-intercept is (0, 3).

c. $x=-\frac{b}{2a}=-\frac{4}{2(3)}=-\frac{4}{6}=-\frac{2}{3}$

$y=\frac{4ac-b^2}{4a}=\frac{4(3)(3)-4^2}{4(3)}$

$=\frac{36-16}{12}=\frac{20}{12}=\frac{5}{3}$

The vertex is $\left(-\frac{2}{3},\frac{5}{3}\right)$.

d. $0=3x^2+4x+3$

Since this is not factorable, check the discriminant.

$b^2-4ac=4^2-4(3)(3)=16-36=-20$

Since –20 < 0 there are no real roots. Thus, there are no x-intercepts.

e. 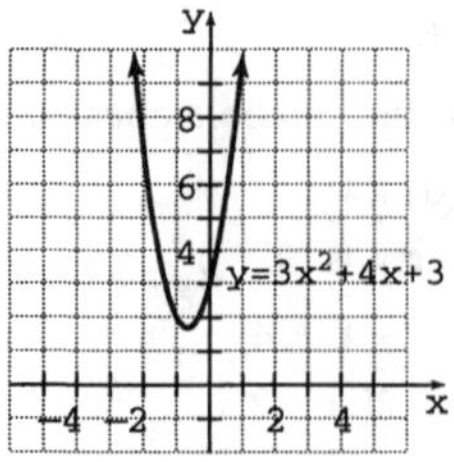

35. $y=3x^2+4x-6$

a. Since $a=3$ the parabola opens upward.

b. $y=3(0)^2+4(0)-6=-6$

The y-intercept is (0, –6).

c. $x=-\frac{b}{2a}=-\frac{4}{2(3)}=-\frac{4}{6}=-\frac{2}{3}$

$y=\frac{4ac-b^2}{4a}=\frac{4(3)(-6)-4^2}{4(3)}$

$=\frac{-72-16}{12}=-\frac{22}{3}$

The vertex is $\left(-\frac{2}{3},-\frac{22}{3}\right)$.

d. $0=3x^2+4x-6$

Since this is not factorable, check the discriminant.

$b^2-4ac=4^2-4(3)(-6)=88$

Since 88 > 0 there are two real roots.

$x=\frac{-b\pm\sqrt{b^2-4ac}}{2a}=\frac{-4\pm\sqrt{88}}{2(3)}$

$=\frac{-4\pm2\sqrt{22}}{6}=\frac{-2\pm\sqrt{22}}{3}$

The x-intercepts are

$\left(\frac{-2+\sqrt{22}}{3},0\right)$ and $\left(\frac{-2-\sqrt{22}}{3},0\right)$.

e. 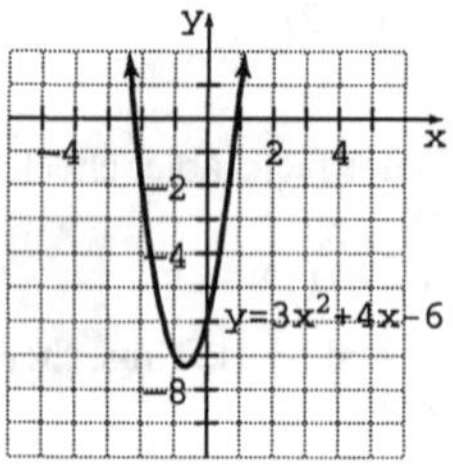

37. $y = 2x^2 - x - 6$

a. Since $a = 2$ the parabola opens upward.

b. $y = 2(0)^2 - 0 - 6 = -6$
The y-intercept is $(0, -6)$.

c. $x = -\frac{b}{2a} = -\frac{-1}{2(2)} = \frac{1}{4}$

$$y = \frac{4ac - b^2}{4a} = \frac{4(2)(-6) - (-1)^2}{4(2)}$$
$$= \frac{-48 - 1}{8} = -\frac{49}{8}$$

The vertex is $\left(\frac{1}{4}, -\frac{49}{8}\right)$.

d.
$$2x^2 - x - 6 = 0$$
$$(2x + 3)(x - 2) = 0$$
$$2x + 3 = 0 \quad \text{or} \quad x - 2 = 0$$
$$x = -\frac{3}{2} \qquad x = 2$$

The x-intercepts are $\left(-\frac{3}{2}, 0\right)$ and $(2, 0)$.

e.

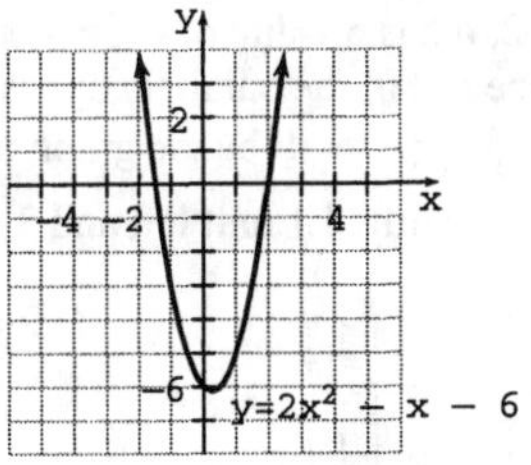

39. $f(x) = -x^2 + 3x - 5$

a. Since $a = -1$ the parabola opens downward.

b. $y = -0^2 + 3(0) - 5 = -5$
The y-intercept is $(0, -5)$.

c. $x = -\frac{b}{2a} = -\frac{3}{2(-1)} = -\frac{3}{-2} = \frac{3}{2}$

$$y = \frac{4ac - b^2}{4a} = \frac{4(-1)(-5) - 3^2}{4(-1)} = -\frac{11}{4}$$

The vertex is $\left(\frac{3}{2}, -\frac{11}{4}\right)$.

d. $0 = -x^2 + 3x - 5$
Since this is not factorable, check the discriminant.
$b^2 - 4ac = 3^2 - 4(-1)(-5) = 9 - 20 = -11$
Since $-11 < 0$ there are no real roots. Thus, there are no x-intercepts.

e.

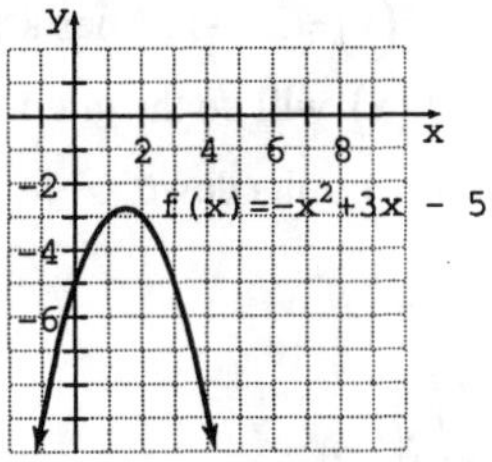

41. In the function $f(x) = (x-3)^2$, h has a value of 3. The graph of $f(x)$ is the graph of $g(x) = x^2$ shifted 3 units to the right.

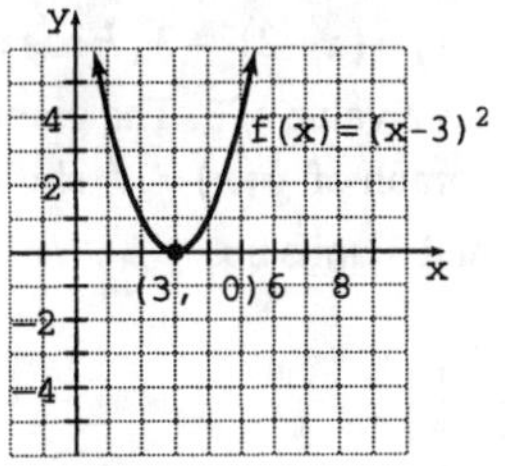

43. In the function $f(x) = (x+1)^2$, h has a value of −1. The graph of $f(x)$ is the graph of $g(x) = x^2$ shifted 1 unit to the left.

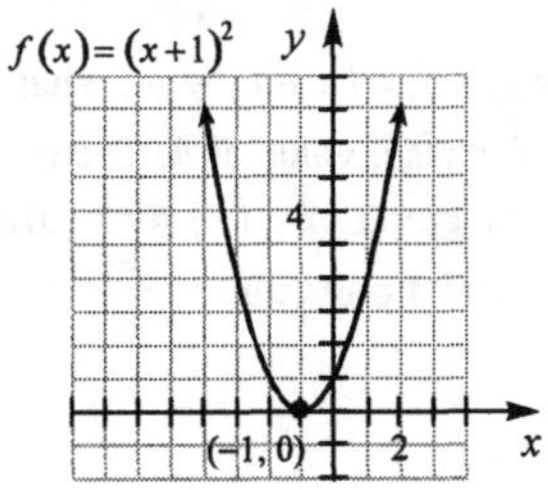

45. In the function $f(x) = x^2 + 3$, k has a value of 3. The graph $f(x)$ will be the graph of $g(x) = x^2$ shifted 3 units up.

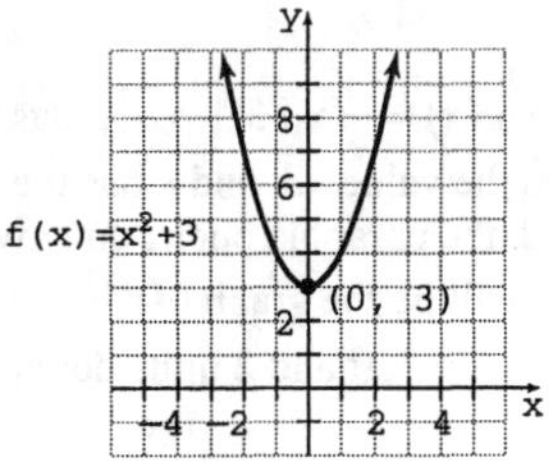

47. In the function $f(x) = x^2 - 1$, k has a value of -1. The graph $f(x)$ will be the graph of $g(x) = x^2$ shifted 1 units down.

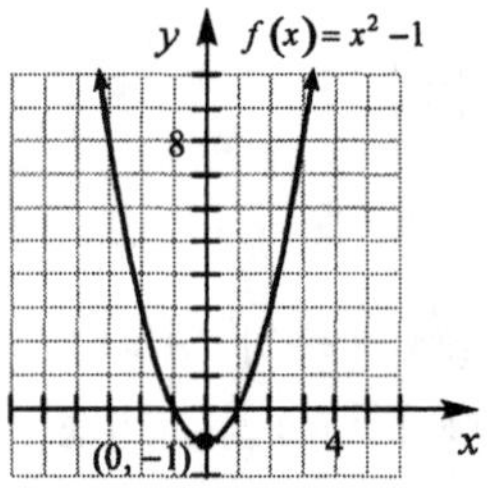

49. In the function $f(x) = (x-2)^2 + 3$, h has a value of 2 and k has a value of 3. The graph of $f(x)$ will be the graph of $g(x) = x^2$ shifted 2 units to the right and 3 units up.

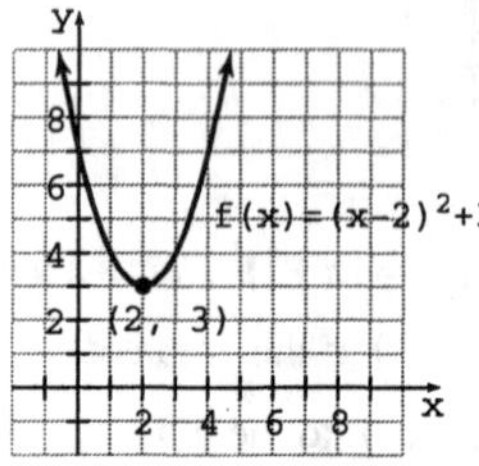

51. In the function $f(x) = (x+4)^2 + 4$, h has a value of -4 and k has a value of 4. The graph of $f(x)$ will be the graph of $g(x) = x^2$ shifted 4 units to the left and 4 units up.

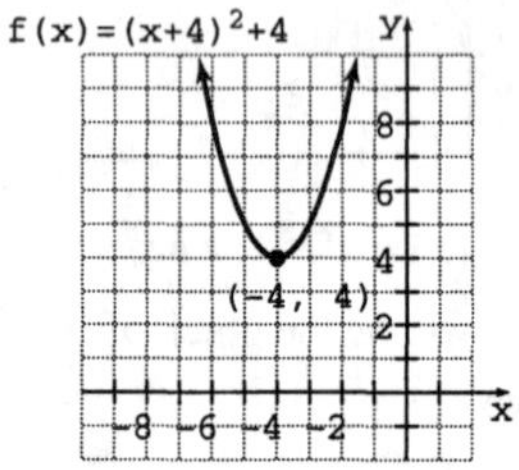

53. In the function $g(x) = -(x+3)^2 - 2$, a has the value -1, h has the value -3, and k has the value -2. Since $a < 0$, the parabola opens downward. The graph of $g(x)$ will be the graph of $f(x) = -x^2$ shifted 3 units to the left and 2 units down.

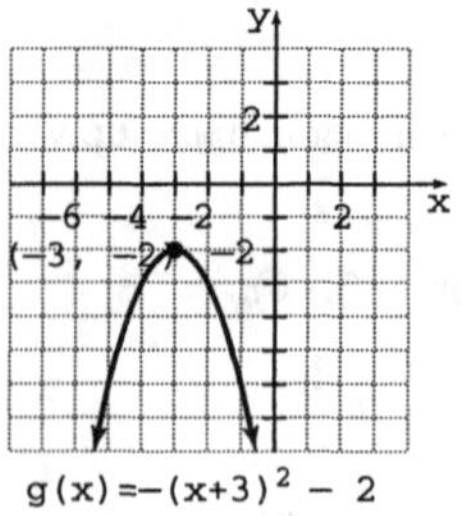

55. In the function $y = -2(x-2)^2 + 2$, a has a value of -2, h has a value of 2, and k has a value of 2. The graph of y will be the graph of $g(x) = -2x^2$ shifted 2 units to the right and 2 units up.

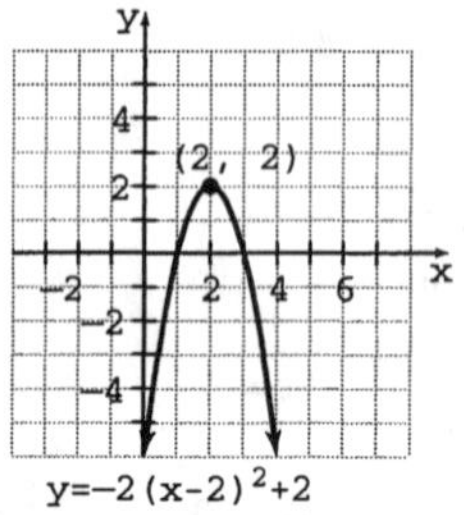

57. In the function $h(x) = -2(x+1)^2 - 3$, a has a value of -2, h has a value of -1, and k has a value of -3. Since $a < 0$, the parabola opens downward. The graph of $h(x)$ will be the graph of $f(x) = -2x^2$ shifted 1 unit left and 3 units down.

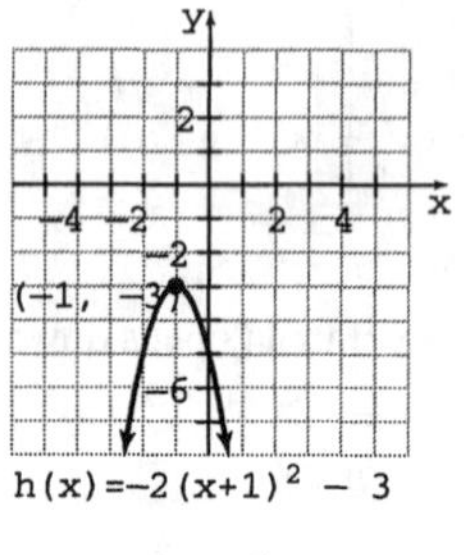

59. a.

$$f(x) = x^2 - 6x + 8$$
$$f(x) = (x^2 - 6x + 9) + 8 - 9$$
$$f(x) = (x-3)^2 - 1$$

b. Since $h = 3$ and $k = -1$, the vertex is $(3, -1)$

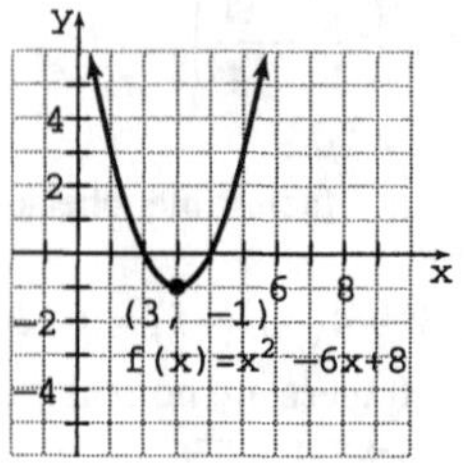

61. a. $g(x)=x^2-x-3$

$$g(x)=\left(x^2-x+\frac{1}{4}\right)-3-\frac{1}{4}$$

$$g(x)=\left(x-\frac{1}{2}\right)^2-\frac{13}{4}$$

b. Since $h=\frac{1}{2}$ and $k=-\frac{13}{4}$, the vertex is $\left(\frac{1}{2},-\frac{13}{4}\right)$.

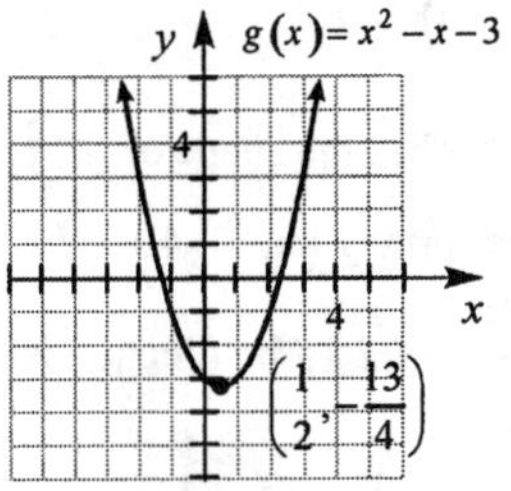

63. a. $f(x)=-x^2-4x-6$

$$f(x)=-(x^2+4x)-6$$

$$f(x)=-(x^2+4x+4)-6-(-4)$$

$$f(x)=-(x+2)^2-2$$

b. Since $h=-2$ and $k=-2$, the vertex is $(-2,-2)$. Since $a<0$, the parabola opens downward.

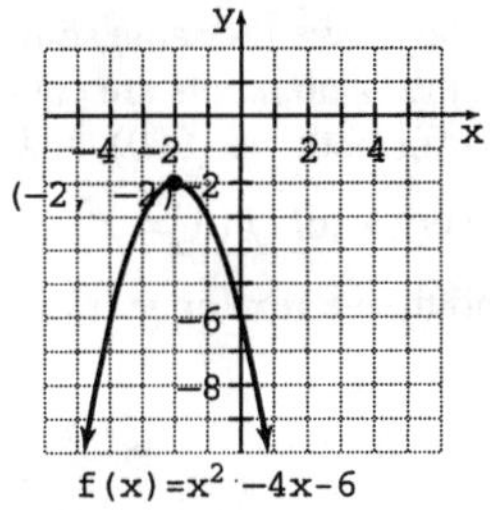

65. a. $g(x)=x^2-4x-1$

$$g(x)=(x^2-4x+4)-1-4$$

$$g(x)=(x-2)^2-5$$

b. Since $h=2$ and $k=-5$, the vertex is $(2,-5)$.

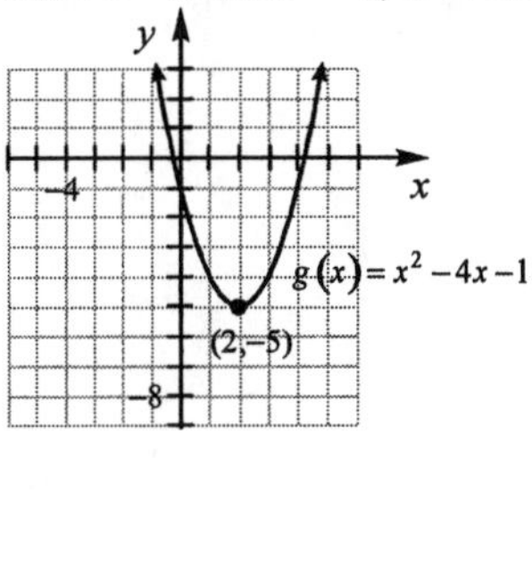

67. a. $f(x)=2x^2+5x-3$

$$f(x)=2\left(x^2+\frac{5}{2}x\right)-3$$

$$f(x)=2\left(x^2+\frac{5}{2}x+\frac{25}{16}\right)-3-2\left(\frac{25}{16}\right)$$

$$f(x)=2\left(x^2+\frac{5}{2}x+\frac{25}{16}\right)-3-\frac{25}{8}$$

$$f(x)=2\left(x+\frac{5}{4}\right)^2-\frac{49}{8}$$

b. Since $h=-\frac{5}{4}$ and $k=-\frac{49}{8}$, the vertex is $\left(-\frac{5}{4},-\frac{49}{8}\right)$. The graph of $f(x)$ will be the graph of $g(x)=2x^2$ shifted $\frac{5}{4}$ units left and $\frac{49}{8}$ units down.

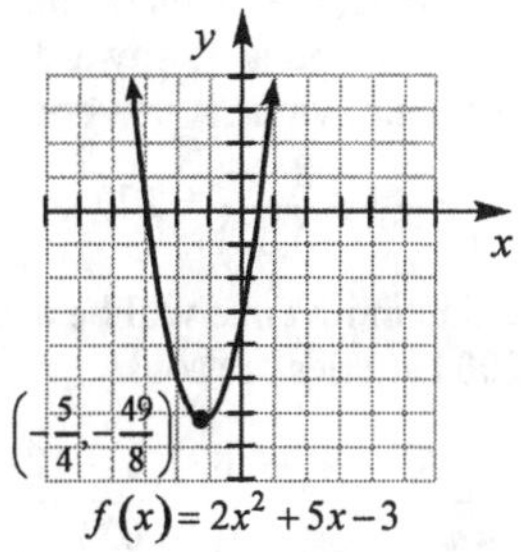

69. $f(x)=2(x+3)^2-1$. The vertex is $(h, k)=(-3,-1)$. Since $a>0$, the parabola opens up. The graph is (d).

71. $f(x)=2(x-1)^2+3$ The vertex is $(h, k)=(1, 3)$. Since $a>0$, the parabola opens up. The graph is (b).

73. a.

$$\begin{aligned} f(x) &= (x+4)(18-x) \\ &= 18x-x^2+72-4x \\ &= -x^2+14x+72 \end{aligned}$$

Since $a=-1$, the graph of this function is a parabola that opens downward and thus has a maximum value at its vertex.

$$x=-\frac{b}{2a}=-\frac{14}{2(-1)}=7$$

b.

$$y=\frac{4ac-b^2}{4a}=\frac{4(-1)(72)-(14)^2}{4(-1)}$$

$$=\frac{-288-196}{-4}=\frac{-484}{-4}=121$$

The maximum area is 121 square units.

75. **a.** $f(x)=(x+5)(26-x)$

$=26x-x^2+130-5x$

$=-x^2+21x+130$

Since $a=-1$, the graph of this function is a parabola that opens downward and thus has a maximum value at its vertex.

$x=-\frac{b}{2a}=-\frac{21}{2(-1)}=10.5$

b. $y=\frac{4ac-b^2}{4a}=\frac{4(-1)(130)-(21)^2}{4(-1)}$

$=\frac{-520-441}{-4}=240.25$

The maximum area is 240.25 square units.

77. **a.** $R(n)=-0.02n^2+8n$

Since $a=-0.02$, the graph of this function is a parabola that opens downward and thus has a maximum value at its vertex.

$x=-\frac{b}{2a}=-\frac{8}{2(-0.02)}=200$

The maximum revenue will be achieved when 200 batteries are sold.

b. $y=\frac{4ac-b^2}{4a}=\frac{4(-0.02)(0)-(8)^2}{4(-0.02)}$

$=\frac{0-64}{-0.08}=\frac{-64}{-0.08}=800$

The maximum revenue is \$800.

79. $N(t)=-0.043t^2+1.82t+46.0$

Since $a=-0.043$, the graph of this function is a parabola that opens downward and thus has a maximum value at its vertex.

$x=-\frac{b}{2a}=-\frac{1.82}{2(-0.043)}\approx 21.2$

The maximum enrollment will be obtained about 21 years after 1989 which is the year 2010.

81. For $f(x)=(x-2)^2+\frac{5}{2}$, the vertex is $\left(2, \frac{5}{2}\right)$.

For $g(x)=(x-2)^2-\frac{3}{2}$, the vertex is $\left(2, -\frac{3}{2}\right)$.

These points are on the vertical line $x=2$. The distance between the two points is

$\frac{5}{2}-\left(-\frac{3}{2}\right)=\frac{5}{2}+\frac{3}{2}=\frac{8}{2}=4$ units.

83. For $f(x)=2(x+4)^2-3$, the vertex is (–4, –3).

For $g(x)=-(x+1)^2-3$, the vertex is (–1, –3).

These points are on the horizontal line $y=-3$. The distance between the two points is $-1-(-4)=-1+4=3$ units.

85. A function that has the shape of $f(x)=2x^2$ will have the form $f(x)=2(x-h)^2+k$. If $(h, k)=(3, -2)$, the function is $f(x)=2(x-3)^2-2$.

87. A function that has the shape of $f(x)=-4x^2$ will have the form $f(x)=-4(x-h)^2+k$. If $(h, k)=\left(-\frac{3}{5}, -\sqrt{2}\right)$, the function is

$f(x)=-4\left(x+\frac{3}{5}\right)^2-\sqrt{2}$.

89. **a.** The graphs will have the same x-intercepts but $f(x)=x^2-8x+12$ will open up and $g(x)=-x^2+8x-12$ will open down.

b. Yes, because the x-intercepts are located by setting $x^2-8x+12$ and $-x^2+8x-12$ equal to zero. They have the same solution set, therefore the x-intercepts are equal. The x-intercepts for both are (6, 0), and (2, 0).

c. No. The vertex for $f(x)=x^2-8x+12$ is (4,–4) and the vertex for $g(x)=-x^2+8x-12$ is (4, 4).

d.

91. a. The vertex $x=-\frac{b}{2a}=-\frac{24}{2(-1)}=12$

$$\begin{aligned} I &= -(12)^2+24(12)-44 \\ &= -144+288-44 \\ &= 100 \end{aligned}$$

The vertex is at (12, 100). To find the roots set $I = 0$.

$$\begin{aligned} 0 &= -x^2+24x-44 \\ &= -1(x^2-24x+44) \\ &= -1(x-2)(x-22) \end{aligned}$$

$$x-2=0 \quad \text{or} \quad x-22=0$$
$$x=2 \qquad\qquad x=22$$

The roots are 2 and 22.

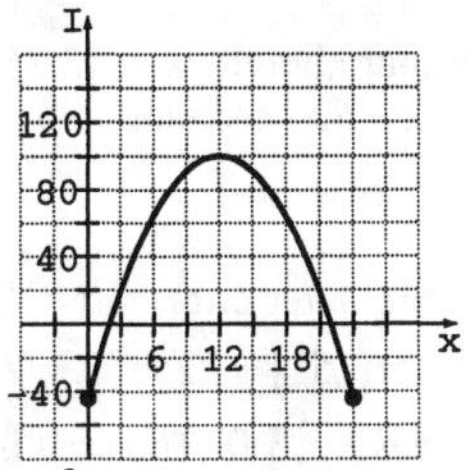

$I=-x^2+24x-44, \quad 0 \le x \le 24$

b. The minimum cost will be $2 since the smaller root is 2.

c. The maximum cost is $22 since the larger root is 22.

d. The maximum value will occur at the vertex of the parabola, (12, 100). Therefore, they should charge $12.

e. The maximum value will occur at the vertex of the parabola, (12, 100). Since I is in hundreds of dollars the maximum income is 100($100) = $10,000.

93. a. The number of bird feeders sold for the maximum profit will be the x-coordinate of the vertex.

$$f(x)=-0.4x^2+80x-200$$

$$x=-\frac{b}{2a}=-\frac{80}{2(-0.4)}=-\frac{80}{-0.8}=100$$

The company must sell 100 bird feeders for maximum profit.

b. The maximum profit will be the y-coordinate of the vertex, $y=f(100)$.

$$\begin{aligned} f(100) &= -0.4(100)^2+80(100)-200 \\ &= -0.4(10{,}000)+8000-200 \\ &= -4000+8000-200 \\ &= 3800 \end{aligned}$$

The maximum profit will be $3800.

95. a. The maximum height is h, the y-coordinate of the vertex.

$$h(t)=-4.9t^2+24.5t+9.8$$

$$\begin{aligned} h &= \frac{4ac-b^2}{4a}=\frac{4(-4.9)(9.8)-(24.5)^2}{4(-4.9)} \\ &= \frac{-192.08-600.25}{-19.6}=\frac{-792.33}{-19.6}=40.425 \end{aligned}$$

The maximum height obtained by the cannonball is 40.425 meters.

b. The time the cannonball reaches the maximum height is t, the x-coordinate of the vertex,. $t=-\frac{b}{2a}=-\frac{24.5}{2(-4.9)}=2.5$

The cannonball will reach the maximum height after 2.5 seconds.

c. $f(x) = 0$ when the cannonball hits the ground.

$$h(t)=-4.9t^2+24.5t+9.8$$

$$0=-4.9t^2+24.5t+9.8$$

$$\begin{aligned} t &= \frac{-24.5\pm\sqrt{(24.5)^2-4(-4.9)(9.8)}}{2(-4.9)} \\ &= \frac{-24.5\pm\sqrt{792.33}}{-9.8} \\ &\approx \frac{-24.5\pm 28.14836}{-9.8} \end{aligned}$$

$t\approx-0.37$ or $t\approx 5.37$

Disregard the negative value. The cannonball will hit the ground after about 5.37 seconds.

97. a. The year 2007 is represented by $t=13$.

$$r(t)=-2.723t^2+35.273t+579$$

$$r(13)=-2.723(13)^2+35.273(13)+579\approx 577$$

The average monthly rent should be about $577.

b. The function will reach a maximum at it's vertex: $t=-\frac{b}{2a}=-\frac{35.273}{2(-2.723)}\approx 6.5$

The monthly rent for an apartment reached a maximum about 6.5 years after 1994, which was during the year 2000.

99. If the perimeter of the room is 8 ft., then $80 = 2l + 2w$, where l is the length and w is the width. Then $80 = 2(l + w)$ and $40 = l + w$.
Therefore, $l = 40 - w$. The area of the room is $A = lw = (40 - w)w = 40w - w^2 = -w^2 + 40w$.
The maximum area is the y-coordinate of the vertex.

$$A = \frac{4ac - b^2}{4a} = \frac{4(-1)(0) - 40^2}{4(-1)}$$
$$= \frac{0 - 1600}{-4} = \frac{-1600}{-4} = 400$$

The maximum area is 400 square feet.

101. If two numbers differ by 8 and x is one of the numbers, then $x + 8$ is the other number. The product is $f(x) = x(x + 8) = x^2 + 8x$. The maximum product is the y-coordinate of the vertex.

$$x = -\frac{b}{2a} = -\frac{8}{2(1)} = -4$$
$$y = f(-4) = (-4)^2 + 8(-4) = -16$$

The maximum product is –16. The numbers are –4 and –4 + 8 = 4.

103. If two numbers add to 60 and x is one of the numbers, then $60 - x$ is the other number. The product is
$f(x) = x(60 - x) = 60x - x^2 = -x^2 + 60x$.
The maximum product is the y-coordinate of the vertex.

$$x = -\frac{b}{2a} = -\frac{60}{2(-1)} = -\frac{60}{-2} = 30$$
$$y = f(30) = -30^2 + 60(30) = -900 + 1800 = 900$$

The maximum product is 900. The numbers are 30 and 60 – 30 = 30.

105. $C(x) = 2000 + 40x,\ R(x) = 800x - x^2$

$$P(x) = R(x) - C(x)$$
$$P(x) = (800x - x^2) - (2000 + 40x)$$
$$= 800x - x^2 - 2000 - 40x$$
$$= -x^2 + 760x - 2000$$

The maximum profit is $P(x)$, the y-coordinate of the vertex. The number of items that must be produced and sold to obtain maximum profit is the x coordinate of the vertex.

$$x = -\frac{b}{2a} = -\frac{760}{2(-1)} = 380$$

a.
$$P(380) = -(380)^2 + 760(380) - 2000$$
$$= -144{,}400 + 288800 - 2000$$
$$= 142{,}400$$

The maximum profit is \$142,400.

b. The number of items that must be produced and sold to obtain maximum profit is 380.

107. a.
$$f(t) = -16t^2 + 52t + 3$$
$$= -16(t^2 - 3.25t) + 3$$
$$= -16\left(t^2 - 3.25t + \left(\frac{3.25}{2}\right)^2\right) + 3 + (16)\left(\left(\frac{3.25}{2}\right)^2\right)$$
$$= -16(t - 1.625)^2 + 3 + 42.25$$
$$= -16(t - 1.625)^2 + 45.25$$

b. The maximum height was 45.25 feet obtained at 1.625 seconds.

c. It is the same answer.

109. The radius of the outer circle is $r = 15$ ft. The area $A = \pi r^2 = \pi(15^2) = 225\pi\ \text{ft}^2$. The radius of the inner circle is $r = 5$ ft. The area is $A = \pi r^2 = \pi(5^2) = 25\pi\ \text{ft}^2$. The blue shaded area is $225\pi\ \text{ft}^2 - 25\pi\ \text{ft}^2 = 200\pi\ \text{ft}^2$.

110. $y \le \frac{2}{3}x + 3$

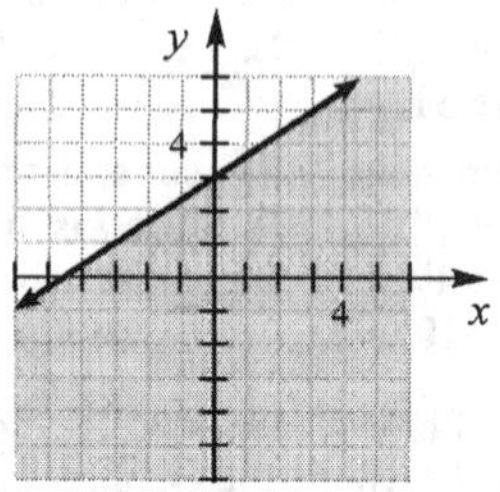

111.
$$\begin{aligned} x - y \quad &= -5 \quad (1)\\ 2x + 2y - z &= 0 \quad (2)\\ x + y + z &= 3 \quad (3) \end{aligned}$$

First eliminate the variable z from equations (2) and (3) by adding these equations.

$$\begin{aligned} 2x + 2y - z &= 0\\ x + y + z &= 3\\ \hline 3x + 3y \quad &= 3 \quad (4) \end{aligned}$$

Equations (1) and (4) form a system of equations in two variables. Multiply equation (1) by 3 and add the result to equation (4) to eliminate the variable y.

$$\begin{aligned} 3(x-y)=3(-5) \;&\Rightarrow\; 3x-3y=-15 \\ 3x+3y=3 \;&\Rightarrow\; \underline{3x+3y=\;\;3} \\ &\qquad 6x \quad\;\; =-12 \\ &\qquad\quad x=-2 \end{aligned}$$

Substitute –2 for x in equation (1) to find y.

$$\begin{aligned} (-2)-y&=-5 \\ -y&=-3 \\ y&=3 \end{aligned}$$

Substitute –2 for x and 3 for y in equation (3) to find z.

$$\begin{aligned} (-2)+(3)+z&=3 \\ 1+z&=3 \\ z&=2 \end{aligned}$$

The solution is (–2, 3, 2).

112. $\begin{vmatrix} \frac{1}{2} & 3 \\ 2 & -4 \end{vmatrix} = \left(\frac{1}{2}\right)(-4)-(2)(3)=-2-6=-8$

113. $$\begin{aligned} (x-3)\div\frac{x^2+3x-18}{x} &= \frac{x-3}{1}\cdot\frac{x}{x^2+3x-18} \\ &= \frac{x-3}{1}\cdot\frac{x}{(x+6)(x-3)} \\ &= \frac{x}{x+6} \end{aligned}$$

Exercise Set 8.6

1. **a.** For $f(x)=x^2-7x+10$, $f(x)>0$ when the graph is above the x-axis. The solution is $x<2$ or $x>5$.

b. For $f(x)=x^2-7x+10$ $f(x)<0$ when the graph is below the x-axis. The solution is $2<x<5$.

3. Yes. The boundary values 5 and –3 are included in the solution set since this is a greater than *or equal to* inequality. These values make the expression equal to zero.

5. The boundary values –2 and 1 are included in the solution set since this is a less than *or equal to* inequality. These values make the expression equal to zero. However, the boundary value –1 is not included in the solution set since it would result in a zero in the denominator, which is undefined.

7. $x^2-2x-8\ge 0$

$(x+2)(x-4)\ge 0$

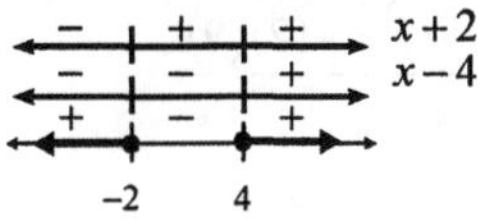

9. $x^2+7x+6>0$

$(x+6)(x+1)>0$

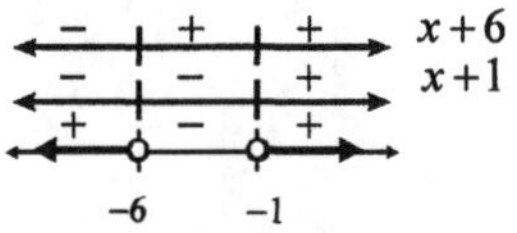

11. $n^2-6n+9\ge 0$

$(n-3)^2\ge 0$

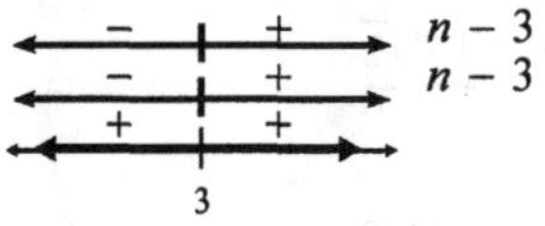

13. $x^2-16<0$

$(x+4)(x-4)<0$

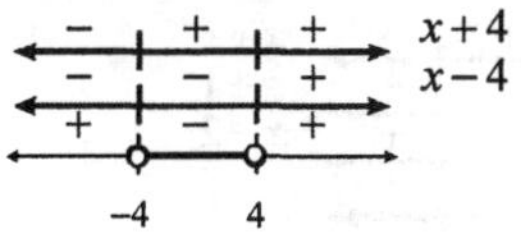

15. $2x^2+5x-3\ge 0$

$(2x-1)(x+3)\ge 0$

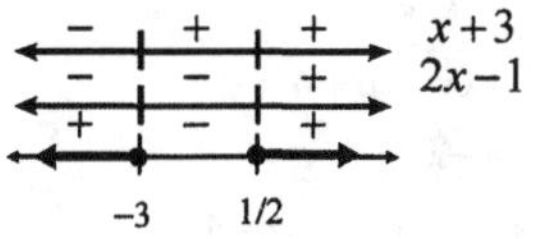

17. $$\begin{aligned} 5x^2+6x&\le 8 \\ 5x^2+6x-8&\le 0 \\ (x+2)(5x-4)&\le 0 \end{aligned}$$

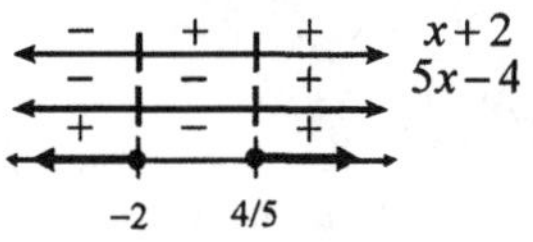

19. $2x^2-12x+9\le 0$

$2x^2-12x+9=0$

$x=\dfrac{-(-12)\pm\sqrt{(-12)^2-4(2)(9)}}{2(2)}$

$=\dfrac{12\pm\sqrt{144-72}}{4}$

$=\dfrac{12\pm\sqrt{72}}{4}$

$=\dfrac{12\pm6\sqrt{2}}{4}$

$=\dfrac{6\pm3\sqrt{2}}{2}$

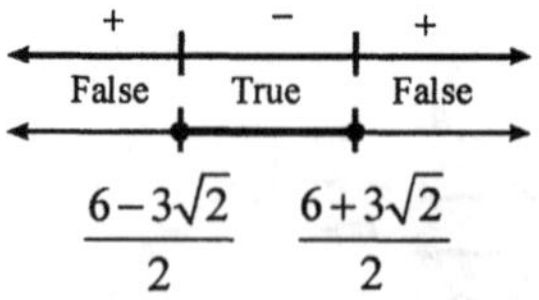

21. $(x-2)(x+1)(x+5)\ge 0$

$x-2=0 \quad x+1=0 \quad x+5=0$

$x=2 \quad x=-1 \quad x=-5$

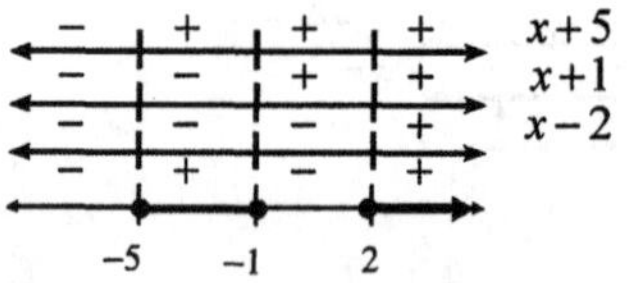

$[-5,-1]\cup[2,\infty)$

23. $(a-3)(a+2)(a+4)<0$

$a-3=0 \quad a+2=0 \quad a+4=0$

$a=3 \quad a=-2 \quad a=-4$

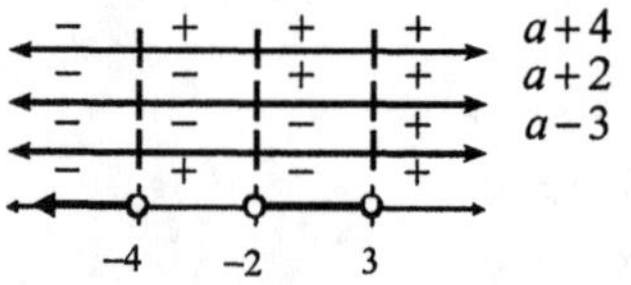

$(-\infty,-4)\cup(-2,3)$

25. $(2c+5)(3c-6)(c+6)>0$

$2c+5=0 \quad 3c-6=0 \quad c+6=0$

$2c=-5 \quad 3c=6 \quad c=-6$

$c=-\dfrac{5}{2} \quad c=2$

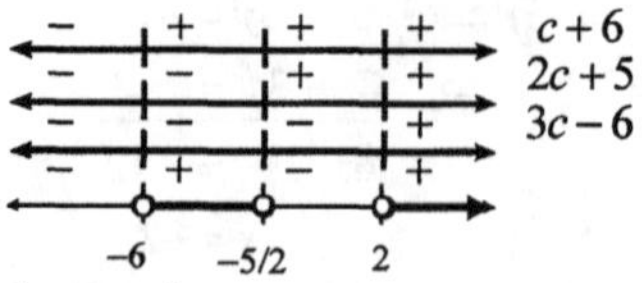

$\left(-6,-\dfrac{5}{2}\right)\cup(2,\infty)$

27. $(3x+5)(x-3)(x+1)>0$

$3x+5=0 \quad x-3=0 \quad x+1=0$

$3x=-5 \quad x=3 \quad x=-1$

$x=-\dfrac{5}{3}$

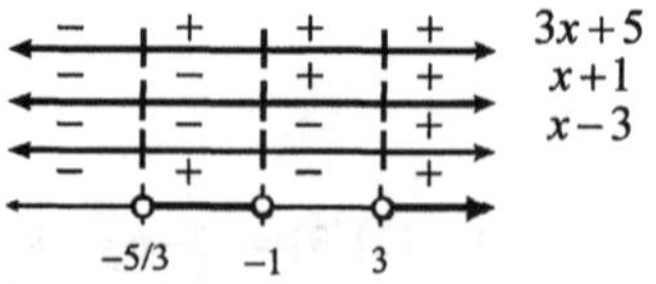

$\left(-\dfrac{5}{3},-1\right)\cup(3,\infty)$

29. $(x+2)(x+2)(3x-8)\ge 0$

$x+2=0 \quad 3x-8=0$

$x=-2 \quad x=\dfrac{8}{3}$

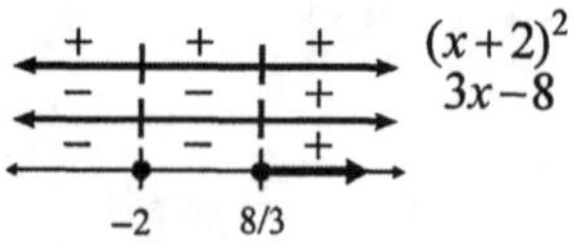

$\{-2\}\cup\left[\dfrac{8}{3},\infty\right)$

31. $x^3-6x^2+9x<0$

$x(x^2-6x+9)<0$

$x(x-3)^2<0$

$x=0 \quad x-3=0$

$x=3$

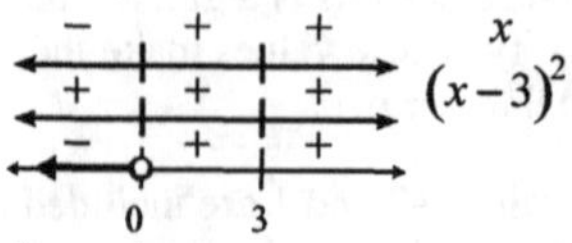

$(-\infty,0)$

33. $f(x) = x^2 - 6x,\ f(x) \ge 0$

$x^2 - 6x \ge 0$

$x(x-6) \ge 0$

$x = 0 \quad x - 6 = 0$

$x = 6$

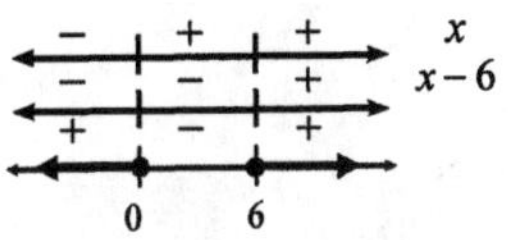

35. $f(x) = x^2 + 4x,\ f(x) > 0$

$x^2 + 4x > 0$

$x(x+4) > 0$

$x = 0 \quad x + 4 = 0$

$x = -4$

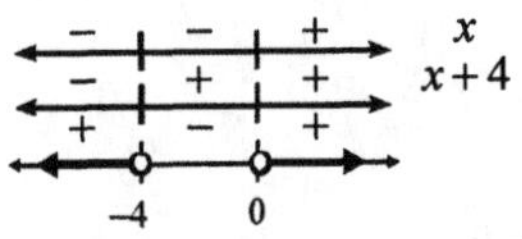

37. $f(x) = x^2 - 14x + 48,\ f(x) < 0$

$x^2 - 14x + 48 < 0$

$(x-6)(x-8) < 0$

$x - 6 = 0 \quad x - 8 = 0$

$x = 6 \qquad x = 8$

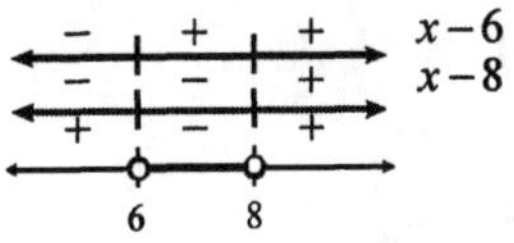

39. $f(x) = 2x^2 + 9x - 1,\ f(x) \le 5$

$2x^2 + 9x - 1 \le 5$

$2x^2 + 9x - 6 \le 0$

$$x = \frac{-9 \pm \sqrt{9^2 - 4(2)(-6)}}{2(2)}$$

$$= \frac{-9 \pm \sqrt{129}}{4}$$

$$x = \frac{-9 - \sqrt{129}}{4} \quad x = \frac{-98 + \sqrt{129}}{4}$$

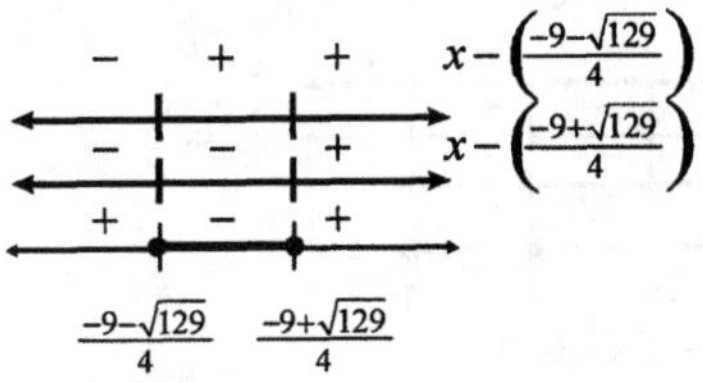

41. $f(x) = 2x^3 + 9x^2 - 35x,\ f(x) \ge 0$

$2x^3 + 9x^2 - 35x \ge 0$

$x(2x^2 + 9x - 35) \ge 0$

$x(2x-5)(x+7) \ge 0$

$x = 0 \quad 2x - 5 = 0 \quad x + 7 = 0$

$x = \frac{5}{2} \qquad x = -7$

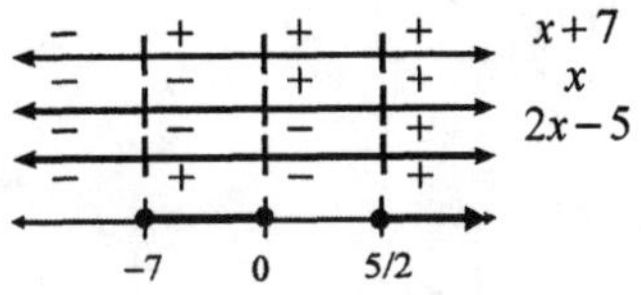

43. $\frac{x+2}{x-4} > 0$

$x \ne 4$

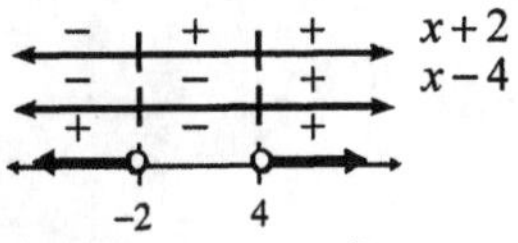

$\{x | x < -2 \text{ or } x > 4\}$

45. $\frac{x-1}{x+5} < 0$

$x \ne -5$

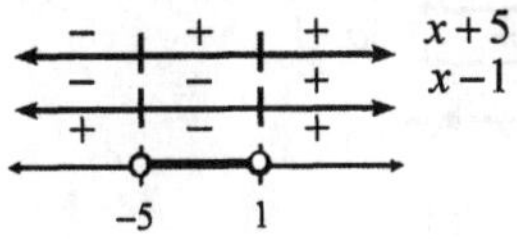

$\{x | -5 < x < 1\}$

47. $\frac{x+3}{x-2} \ge 0$

$x \ne 2$

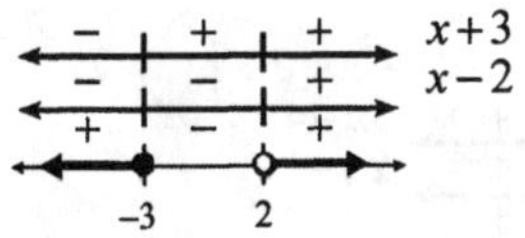

$\{x | \ x \le -3 \text{ or } x > 2\}$

49. $\frac{a-9}{a+5} < 0$

$a \ne -5$

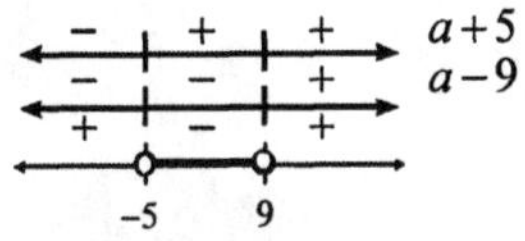

$\{a | -5 < a < 9\}$

51. $\dfrac{c-10}{c-4}>0$

$c\neq 4$

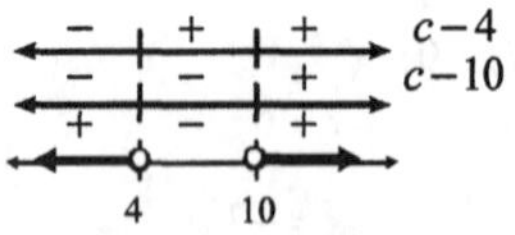

$\{c \mid c<4 \text{ or } c>10\}$

53. $\dfrac{3y+6}{y+4}\leq 0$

$y\neq -4,$

$3y+6=0 \;\Rightarrow\; y=-2$

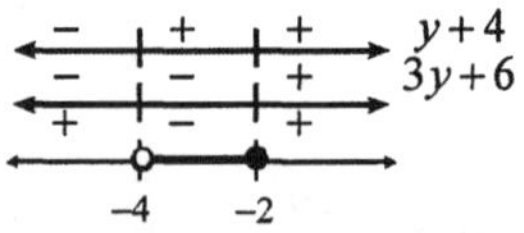

$\{y \mid -4<y\leq -2\}$

55. $\dfrac{5a+10}{3a-1}\geq 0$

$a\neq \dfrac{1}{3},$

$5a+10=0 \;\Rightarrow\; a=-2$

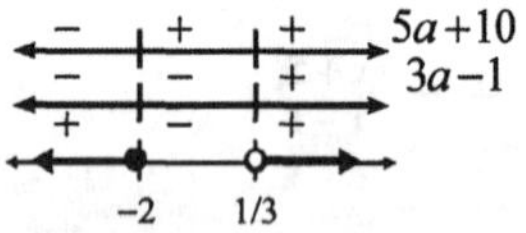

$\left\{a \,\middle|\, a\leq -2 \text{ or } a>\dfrac{1}{3}\right\}$

57. $\dfrac{3x+4}{2x-1}<0$

$x\neq \dfrac{1}{2}$

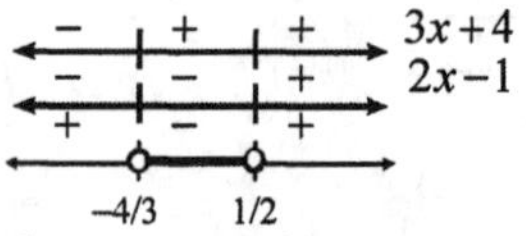

$\left\{x \,\middle|\, -\dfrac{4}{3}<x<\dfrac{1}{2}\right\}$

59. $\dfrac{3x+8}{x-2}\leq 0$

$x\neq 2$

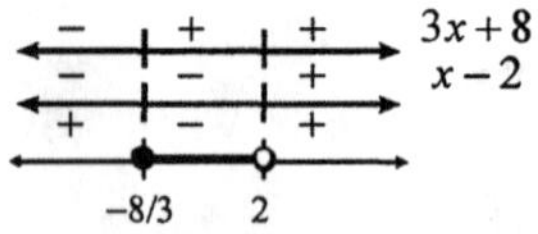

$\left\{x \,\middle|\, -\dfrac{8}{3}\leq x<2\right\}$

61. $\dfrac{(x+1)(x-6)}{x+3}<0$

$x\neq -3$

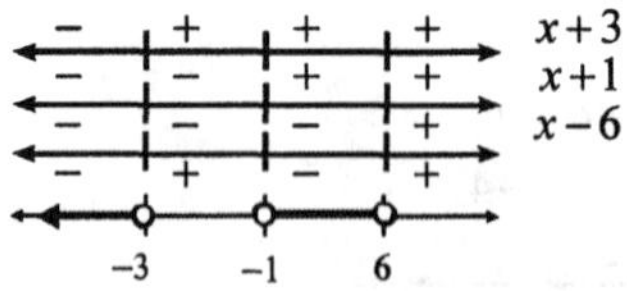

$(-\infty,-3)\cup(-1,\,6)$

63. $\dfrac{(x-2)(x+3)}{x-5}>0$

$x\neq 5$

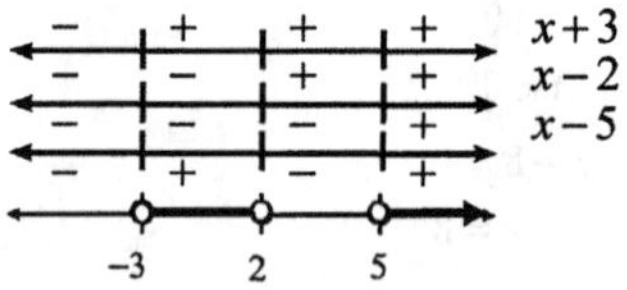

$(-3,\,2)\cup(5\;\infty)$

65. $\dfrac{(a-1)(a-7)}{a+2}\geq 0$

$a\neq -2$

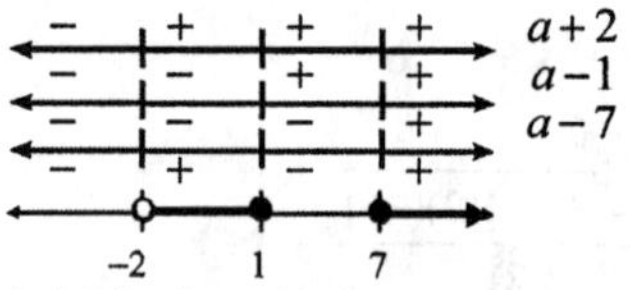

$(-2,\,1]\cup[7,\,\infty)$

67. $\dfrac{c}{(c-3)(c+8)}\leq 0$

$c\neq 3,\; c\neq -8$

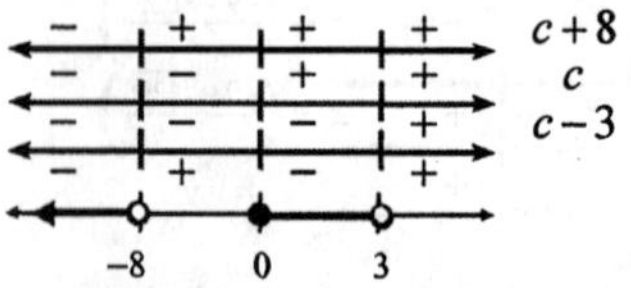

$(-\infty,-8)\cup[0,\,3)$

69. $\dfrac{x-6}{(x+4)(x-1)} \le 0$

$x \ne -4,\ x \ne 1$

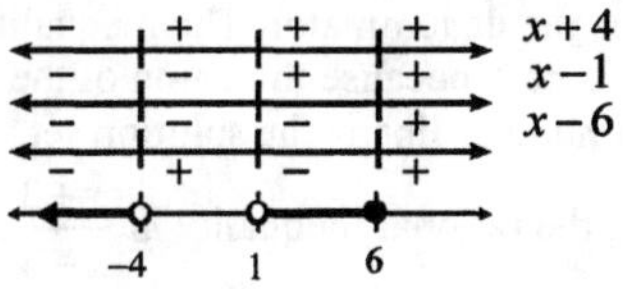

$(-\infty, -4) \cup)(1, 6]$

71. $\dfrac{(x-3)(2x+5)}{x-4} \ge 0$

$x \ne 4$

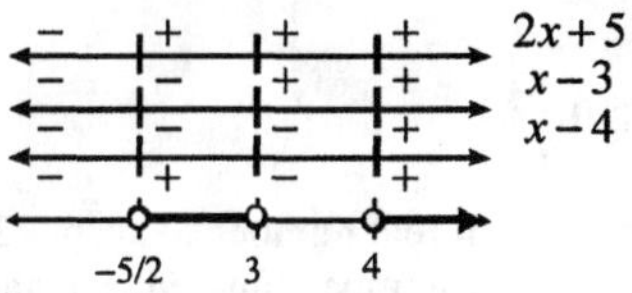

$\left[-\dfrac{5}{2}, 3\right] \cup (4, \infty)$

73.
$$\frac{2}{x-4} \ge 1$$
$$\frac{2}{x-4} - 1 \ge 0$$
$$\frac{2}{x-4} + \frac{-1(x-4)}{x-4} \ge 0$$
$$\frac{2-x+4}{x-4} \ge 0$$
$$\frac{6-x}{x-4} \ge 0, \quad x \ne 4$$

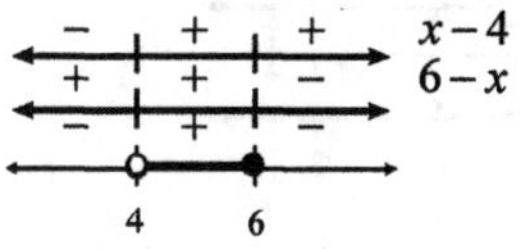

75.
$$\frac{3}{x-1} > -1$$
$$\frac{3}{x-1} + 1 > 0$$
$$\frac{3}{x-1} + \frac{1(x-1)}{x-1} > 0$$
$$\frac{3+x-1}{x-1} > 0$$
$$\frac{x+2}{x-1} > 0, \quad x \ne 1$$

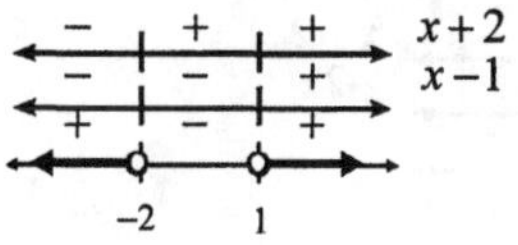

77.
$$\frac{5}{x+2} \le 1$$
$$\frac{5}{x+2} - 1 \le 0$$
$$\frac{5}{x+2} + \frac{-1(x+2)}{x+2} \le 0$$
$$\frac{5-x-2}{x+2} \le 0$$
$$\frac{3-x}{x+2} \le 0$$
$$x \ne -2$$

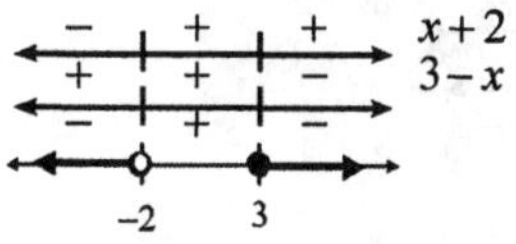

79.
$$\frac{2p-5}{p-4} \le 1$$
$$\frac{2p-5}{p-4} - 1 \le 0$$
$$\frac{2p-5}{p-4} - \frac{1(p-4)}{p-4} \le 0$$
$$\frac{2p-5-p+4}{p-4} \le 0$$
$$\frac{p-1}{p-4} \le 0, \quad p \ne 4$$

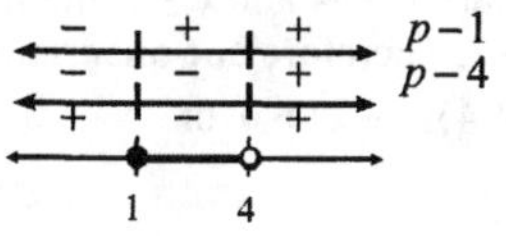

81.
$$\frac{4}{x+2} \geq 2$$
$$\frac{4}{x+2} - 2 \geq 0$$
$$\frac{4}{x+2} - \frac{2(x+2)}{x+2} \geq 0$$
$$\frac{4-2x-4}{x+2} \geq 0$$
$$\frac{-2x}{x+2} \geq 0, \quad x \neq -2$$

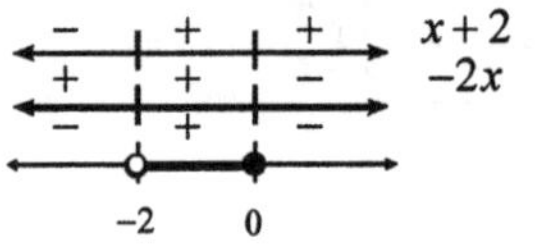

83.
$$\frac{w}{3w-2} > -2$$
$$\frac{w}{3w-2} + 2 > 0$$
$$\frac{w}{3w-2} + \frac{2(3w-2)}{3w-2} > 0$$
$$\frac{w+6w-4}{3w-2} > 0$$
$$\frac{7w-4}{3w-2} > 0, \quad w \neq \frac{2}{3}$$

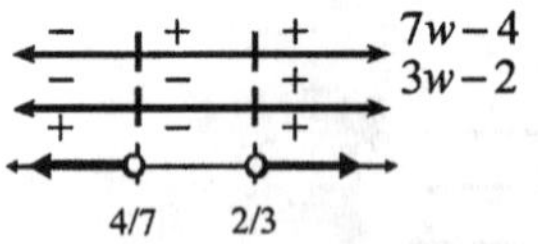

85. a. $y = \dfrac{x^2-4x+4}{x-4} > 0$ where the graph of *y* is above the *x*-axis, on the interval (4, ∞).

b. $y = \dfrac{x^2-4x+4}{x-4} < 0$ where the graph of *y* is below the *x*-axis, on the interval $(-\infty, 2) \cup (2, 4)$.

87. A quadratic inequality with the union of the two outer regions of the number line as its solution, not including the boundary values, will be of the form $ax^2+bx+c>0$ with $a > 0$. Since the boundary values are $x = -4$ and $x = 2$, the factors are $x + 4$ and $x - 2$. Therefore one quadratic inequality is $(x+4)(x-2)>0$ or $x^2+2x-8>0$.

89. Since the solution set is $x \leq -3$ and $x > 4$, the factors are $x + 3$ and $x - 4$. Because -3 is included in the solution set, $x + 3$ is the numerator. Since 4 is not included in the solution set, $x - 4$ is the denominator. The inequality symbol will be ≥ because the union of the outer regions of number line is the solution set.

Therefore, the rational inequality is $\dfrac{x+3}{x-4} \geq 0$.

91. $(x+3)^2(x-1)^2 \geq 0$

The solution is all real numbers since any nonzero number squared is positive and zero squared is zero.

93. $\dfrac{x^2}{(x+2)^2} \geq 0$

This solution is all real numbers, except −2, since any nonzero number squared is negative. It is undefined when $x = -2$. Therefore, −2 is not a solution.

95. If $f(x) = ax^2+bx+c$ and $a > 0$, the graph of $f(x)$ opens upward. If the discriminant is negative, the graph of $f(x)$ has no *x*-intercepts. Therefore, the graph lies above the *x*-axis and $f(x) < 0$ has no solution.

97. $(x+1)(x-3)(x+5)(x+8) \geq 0$

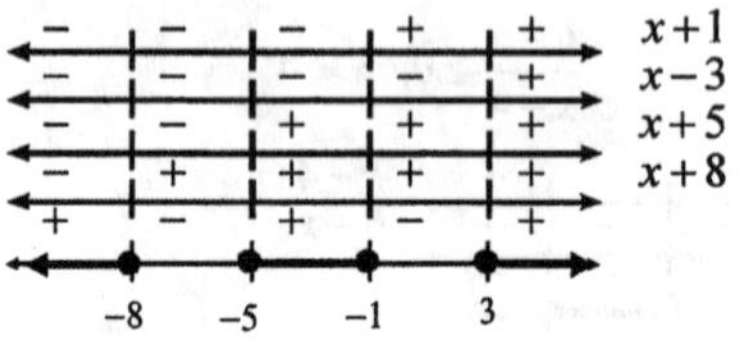

99. One possible answer is: Use a parabola that opens upward and has *x*-intercepts of (0, 0) and (3, 0). The *x*-values for which the parabola lies above the *x*-axis are $(-\infty, 0) \cup (3, \infty)$.

$x^2 - 3x > 0$

101. One possible answer is: Use a parabola that opens upward and has its vertex on or above the *x*-axis. Then there are no *x*-values for which the parabola lies below the *x*-axis.

$x^2 < 0$

103. $x^4 - 10x^2 + 9 > 0$

$(x^2)^2 - 10x^2 + 9 > 0$

Let $u = x^2$.

$u^2 - 10u + 9 > 0$

$(u-9)(u-1) > 0$

Substitute x^2 for u.

$\left(x^2 - 9\right)\left(x^2 - 1\right) > 0$

$(x+3)(x-3)(x+1)(x-1) > 0$

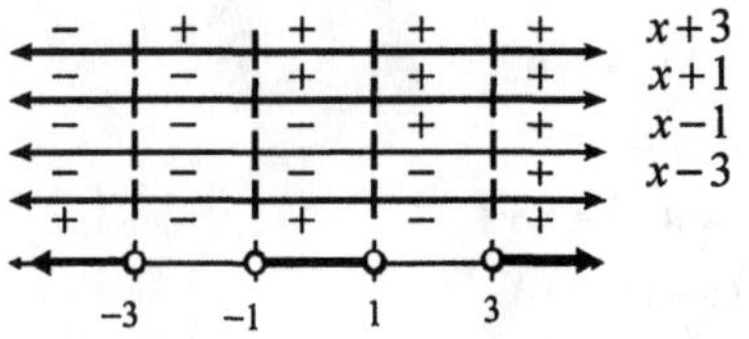

$(-\infty, -3) \cup (-1, 1) \cup (3, \infty)$

105. $x^3 + x^2 - 4x - 4 \geq 0$

$x^2(x+1) - 4(x+1) \geq 0$

$(x^2 - 4)(x+1) \geq 0$

$(x+2)(x-2)(x+1) \geq 0$

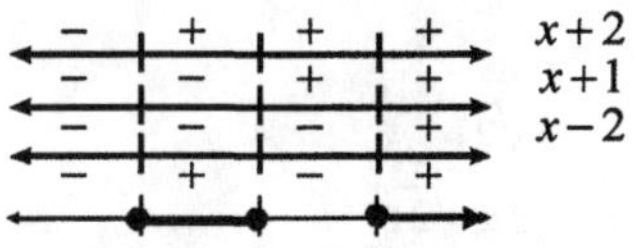

$[-2, -1] \cup [2, \infty)$

109. Let x = the number of quarts of 100% antifreeze added. The equation below describes the situation.

$100\%(x) + 20\%(10) = 50\%(x+10)$

$x + 0.2(10) = 0.5(x+10)$

$x + 2 = 0.5x + 5$

$0.5x = 3$

$x = 6$

Paul should add 6 quarts of 100% antifreeze.

110. $h(x) = \dfrac{x^2 + 4x}{x+9}$

$h(-3) = \dfrac{(-3)^2 + 4(-3)}{(-3)+9} = \dfrac{9-12}{6} = \dfrac{-3}{6} = -\dfrac{1}{2}$

111. $(6r + 5s - t) + (-3r - 2s - 8t)$

$= 6r + 5s - t - 3r - 2s - 8t$

$= 3r + 3s - 9t$

112. $\dfrac{\dfrac{1 + \dfrac{x}{x+1}}{2x+1}}{x-3} = \dfrac{\dfrac{\dfrac{1}{1} \cdot \dfrac{x+1}{x+1} + \dfrac{x}{x+1}}{2x+1}}{x-3}$

$= \dfrac{\dfrac{2x+1}{x+1}}{\dfrac{2x+1}{x-3}}$

$= \dfrac{2x+1}{x+1} \cdot \dfrac{x-3}{2x+1}$

$= \dfrac{x-3}{x+1}$

113. $(3-4i)(6+5i) = 18 + 15i - 24i - 20i^2$

$= 18 - 9i - 20(-1)$

$= 18 - 9i + 20$

$= 38 - 9i$

Chapter 8 Review Exercises

1. $(x-5)^2 = 24$

$x - 5 = \pm\sqrt{24}$

$x - 5 = \pm 2\sqrt{6}$

$x = 5 \pm 2\sqrt{6}$

$x = 5 + 2\sqrt{6}$ or $x = 5 - 2\sqrt{6}$

2. $(2x+1)^2 = 60$

$2x + 1 = \pm\sqrt{60}$

$2x + 1 = \pm 2\sqrt{15}$

$2x = -1 \pm 2\sqrt{15}$

$x = \dfrac{-1 \pm 2\sqrt{15}}{2}$

$x = \dfrac{-1 + 2\sqrt{15}}{2}$ or $x = \dfrac{-1 - 2\sqrt{15}}{2}$

3. $\left(x-\frac{1}{3}\right)^2=\frac{4}{9}$

$$x-\frac{1}{3}=\pm\sqrt{\frac{4}{9}}$$

$$x-\frac{1}{3}=\pm\frac{2}{3}$$

$$x=\frac{1}{3}\pm\frac{2}{3}$$

$$x=\frac{1}{3}+\frac{2}{3} \quad \text{or} \quad x=\frac{1}{3}-\frac{2}{3}$$

$$=1 \qquad\qquad =-\frac{1}{3}$$

4. $\left(2x-\frac{1}{2}\right)^2=4$

$$2x-\frac{1}{2}=\pm\sqrt{4}$$

$$2x-\frac{1}{2}=\pm 2$$

$$2x=\frac{1}{2}\pm 2$$

$$2x=\frac{1\pm 4}{2}$$

$$x=\frac{1\pm 4}{4}$$

$$x=\frac{1+4}{4} \quad \text{or} \quad x=\frac{1-4}{4}$$

$$=\frac{5}{4} \qquad\qquad =-\frac{3}{4}$$

5. $x^2-7x+12=0$

$$x^2-7x=-12$$

$$x^2-7x+\frac{49}{4}=-\frac{48}{4}+\frac{49}{4}$$

$$x^2-7x+\frac{49}{4}=\frac{1}{4}$$

$$\left(x-\frac{7}{2}\right)^2=\frac{1}{4}$$

$$x-\frac{7}{2}=\pm\frac{1}{2}$$

$$x=\frac{7}{2}\pm\frac{1}{2}$$

$$x=\frac{7}{2}+\frac{1}{2} \quad \text{or} \quad x=\frac{7}{2}-\frac{1}{2}$$

$$=4 \qquad\qquad =3$$

6. $x^2+4x-32=0$

$$x^2+4x=32$$

$$x^2+4x+4=32+4$$

$$(x+2)^2=36$$

$$x+2=\pm\sqrt{36}$$

$$x+2=\pm 6$$

$$x=-2\pm 6$$

$$x=4 \quad \text{or} \quad x=-8$$

7. $a^2+2a-9=0$

$$a^2+2a=9$$

$$a^2+2a+1=9+1$$

$$(a+1)^2=10$$

$$a+1=\pm\sqrt{10}$$

$$a=-1\pm\sqrt{10}$$

$$a=-1+\sqrt{10} \quad \text{or} \quad a=-1-\sqrt{10}$$

8. $z^2+6z=12$

$$z^2+6z+9=12+9$$

$$(z+3)^2=21$$

$$z+3=\pm\sqrt{21}$$

$$z=-3\pm\sqrt{21}$$

$$z=-3+\sqrt{21} \quad \text{or} \quad z=-3-\sqrt{21}$$

9. $x^2-2x+10=0$

$$x^2-2x=-10$$

$$x^2-2x+1=-10+1$$

$$(x-1)^2=-9$$

$$(x-1)^2=\sqrt{-9}$$

$$x-1=\pm 3i$$

$$x=1\pm 3i$$

$$x=1+3i \quad \text{or} \quad x=1-3i$$

10. $2r^2 - 8r = -64$

$r^2 - 4r = -32$

$r^2 - 4r + 4 = -32 + 4$

$(r-2)^2 = -28$

$r - 2 = \pm\sqrt{-28}$

$r = 2 \pm \sqrt{4}\sqrt{7}\sqrt{-1}$

$r = 2 \pm 2i\sqrt{7}$

$r = 2 + 2i\sqrt{7}$ or $r = 2 - 2i\sqrt{7}$

11. a. Area = length × width

$32 = (x+5)(x+1)$

b. $32 = (x+5)(x+1)$

$32 = x^2 + x + 5x + 5$

$0 = x^2 + 6x - 27$

$0 = (x-3)(x+9)$

$x = 3$ or $x = -9$

Disregard the negative value. $x = 3$.

12. a. Area = length × width

$63 = (x+2)(x+4)$

b. $63 = (x+2)(x+4)$

$63 = x^2 + 4x + 2x + 8$

$0 = x^2 + 6x - 55$

$0 = (x-5)(x+11)$

$x = 5$ or $x = -11$

Disregard the negative value. $x = 5$.

13. Let x = the smaller integer. The larger will then be $x + 1$. Their product is 42.

$x(x+1) = 42$

$x^2 + x = 42$

$x^2 + x - 42 = 0$

$(x+7)(x-6) = 0$

$x + 7 = 0$ or $x - 6 = 0$

$x = -7$ $\quad$ $x = 6$

Since the integers must be positive, disregard the negative value. The smaller integer is 6 and the larger is 7.

14. Let x = the length of side of the square room. The diagonal can then be described by $x + 7$. Two of the adjacent sides and the diagonal make up a right triangle. Use the Pythagorean theorem to solve the problem.

$a^2 + b^2 = c^2$

$x^2 + x^2 = (x+7)^2$

$2x^2 = x^2 + 14x + 49$

$x^2 - 14x - 49 = 0$

$$x = \frac{-(-14) \pm \sqrt{(-14)^2 - 4(1)(-49)}}{2(1)}$$

$$x = \frac{14 \pm \sqrt{392}}{2}$$

$x \approx 16.90$ or $x \approx -2.90$

Disregard the negative value. The room is about 16.90 feet by 16.90 feet.

15. $2x^2 - 5x - 1 = 0$

$a = 2, b = -5, c = -1$

$b^2 - 4ac = (-5)^2 - 4(2)(-1) = 25 + 8 = 33$

Since the discriminant is positive, this equation has two distinct real solutions.

16. $3x^2 + 2x = -6$

$3x^2 + 2x + 6 = 0$

$a = 3, b = 2, c = 6$

$b^2 - 4ac = (2)^2 - 4(3)(6) = 4 - 72 = -68$

Since the discriminant is negative, this equation has no real solutions.

17. $r^2 + 16r = -64$

$r^2 + 16r + 64 = 0$

$a = 1, b = 16, c = 64$

$b^2 - 4ac = (16)^2 - 4(1)(64) = 256 - 256 = 0$

Since the discriminant is 0, the equation has one real solution.

18. $5x^2 - x + 2 = 0$

$a = 5, b = -1, c = 2$

$b^2 - 4ac = (-1)^2 - 4(5)(2) = 1 - 40 = -39$

Since the discriminant is negative, this equation has no real solutions.

19. $a^2-14n=-49$

$a^2-14n+49=0$
$a=1, b=-14, c=49$
$b^2-4ac=(-14)^2-4(1)(49)=196-196=0$

Since the discriminant is 0, the equation has one real solution.

20. $\frac{1}{2}x^2-3x=8$

$\frac{1}{2}x^2-3x-8=0$

$a=\frac{1}{2}, b=-3, c=-8$

$b^2-4ac=(-3)^2-4\left(\frac{1}{2}\right)(-8)=9+16=25$

Since the discriminant is positive, this equation has two real solutions.

21. $3x^2+4x=0$
$a=3, b=4, c=0$

$$x=\frac{-b\pm\sqrt{b^2-4ac}}{2a}$$

$$x=\frac{-4\pm\sqrt{4^2-4(3)(0)}}{2(3)}$$

$$=\frac{-4\pm\sqrt{16-0}}{6}$$

$$=\frac{-4\pm\sqrt{16}}{6}$$

$$=\frac{-4\pm4}{6}$$

$x=\frac{-4+4}{6}$ or $x=\frac{-4-4}{6}$

$x=\frac{0}{6}$ $\qquad$ $x=\frac{-8}{6}$

$x=0$ $\qquad$ $x=-\frac{4}{3}$

22. $x^2-11x=-18$

$x^2-11x+18=0$
$a=1, b=-11, c=18$

$$x=\frac{-b\pm\sqrt{b^2-4ac}}{2a}$$

$$x=\frac{-(-11)\pm\sqrt{(-11)^2-4(1)(18)}}{2(1)}$$

$$=\frac{11\pm\sqrt{121-72}}{2}$$

$$=\frac{11\pm\sqrt{49}}{2}$$

$$=\frac{11\pm7}{2}$$

$x=\frac{11+7}{2}$ or $x=\frac{11-7}{2}$

$=\frac{18}{2}$ $\qquad$ $=\frac{4}{2}$

$=9$ $\qquad$ $=2$

23. $r^2=3r+40$

$r^2-3r-40=0$
$a=1, b=-3, c=-40$

$$r=\frac{-b\pm\sqrt{b^2-4ac}}{2a}$$

$$r=\frac{-(-3)\pm\sqrt{(-3)^2-4(1)(-40)}}{2(1)}$$

$$=\frac{3\pm\sqrt{9+160}}{2}$$

$$=\frac{3\pm\sqrt{169}}{2}$$

$$=\frac{3\pm13}{2}$$

$r=\frac{3+13}{2}$ or $r=\frac{3-13}{2}$

$=\frac{16}{2}$ $\qquad$ $=\frac{-10}{2}$

$=8$ $\qquad$ $=-5$

24. $7x^2 = 9x$

$$7x^2 - 9x = 0$$
$$a = 7,\ b = -9,\ c = 0$$
$$x = \frac{-b \pm \sqrt{b^2 - 4ac}}{2a}$$
$$x = \frac{-(-9) \pm \sqrt{(-9)^2 - 4(7)(0)}}{2(7)}$$
$$= \frac{9 \pm \sqrt{81 - 0}}{14}$$
$$= \frac{9 \pm \sqrt{81}}{14}$$
$$= \frac{9 \pm 9}{14}$$
$$x = \frac{9+9}{14} \quad \text{or} \quad x = \frac{9-9}{14}$$
$$= \frac{18}{14} \qquad = \frac{0}{14}$$
$$= \frac{9}{7} \qquad = 0$$

25. $6a^2 + a - 15 = 0$

$$a = 6,\ b = 1,\ c = -15$$
$$a = \frac{-b \pm \sqrt{b^2 - 4ac}}{2a}$$
$$a = \frac{-1 \pm \sqrt{1^2 - 4(6)(-15)}}{2(6)}$$
$$= \frac{-1 \pm \sqrt{1 + 360}}{12}$$
$$= \frac{-1 \pm \sqrt{361}}{12}$$
$$= \frac{-1 \pm 19}{12}$$
$$a = \frac{-1+19}{12} \quad \text{or} \quad a = \frac{-1-19}{12}$$
$$= \frac{18}{12} \qquad = \frac{-20}{12}$$
$$= \frac{3}{2} \qquad = -\frac{5}{3}$$

26. $4x^2 + 11x = 3$

$$4x^2 + 11x - 3 = 0$$
$$a = 4,\ b = 11,\ c = -3$$
$$x = \frac{-b \pm \sqrt{b^2 - 4ac}}{2a}$$
$$x = \frac{-(11) \pm \sqrt{(11)^2 - 4(4)(-3)}}{2(4)}$$
$$= \frac{-11 \pm \sqrt{121 + 48}}{8}$$
$$= \frac{-11 \pm \sqrt{169}}{8}$$
$$= \frac{-11 \pm 13}{8}$$
$$x = \frac{-11+13}{8} \quad \text{or} \quad x = \frac{-11-13}{8}$$
$$= \frac{2}{8} \qquad = \frac{-24}{8}$$
$$= \frac{1}{4} \qquad = -3$$

27. $x^2 + 8x + 5 = 0$

$$a = 1,\ b = 8,\ c = 5$$
$$x = \frac{-b \pm \sqrt{b^2 - 4ac}}{2a}$$
$$x = \frac{-8 \pm \sqrt{8^2 - 4(1)(5)}}{2(1)}$$
$$= \frac{-8 \pm \sqrt{64 - 20}}{2}$$
$$= \frac{-8 \pm \sqrt{44}}{2}$$
$$= \frac{-8 \pm 2\sqrt{11}}{2}$$
$$= -4 \pm \sqrt{11}$$
$$x = -4 + \sqrt{11} \quad \text{or} \quad x = -4 - \sqrt{11}$$

28. $b^2 + 4b = 8$

$b^2 + 4b - 8 = 0$

$a = 1, b = 4, c = -8$

$$b = \frac{-b \pm \sqrt{b^2 - 4ac}}{2a}$$

$$b = \frac{-(4) \pm \sqrt{(4)^2 - 4(1)(-8)}}{2(1)}$$

$$= \frac{-4 \pm \sqrt{16 + 32}}{2}$$

$$= \frac{-4 \pm \sqrt{48}}{2}$$

$$= \frac{-4 \pm 4\sqrt{3}}{2}$$

$$= -2 \pm 2\sqrt{3}$$

$x = -2 + 2\sqrt{3}$ or $x = -2 - 2\sqrt{3}$

29. $2x^2 + 4x - 3 = 0$

$a = 2, b = 4, c = -3$

$$x = \frac{-b \pm \sqrt{b^2 - 4ac}}{2a}$$

$$x = \frac{-4 \pm \sqrt{4^2 - 4(2)(-3)}}{2(2)}$$

$$= \frac{-4 \pm \sqrt{16 + 24}}{4}$$

$$= \frac{-4 \pm \sqrt{40}}{4}$$

$$= \frac{-4 \pm 2\sqrt{10}}{4}$$

$$= \frac{2\left(-2 \pm \sqrt{10}\right)}{2(2)}$$

$$= \frac{-2 \pm \sqrt{10}}{2}$$

$x = \frac{-2 + \sqrt{10}}{2}$ or $x = \frac{-2 - \sqrt{10}}{2}$

30. $3y^2 - 6y = 8$

$3y^2 - 6y - 8 = 0$

$a = 3, b = -6, c = -8$

$$y = \frac{-b \pm \sqrt{b^2 - 4ac}}{2a}$$

$$y = \frac{-(-6) \pm \sqrt{(-6)^2 - 4(3)(-8)}}{2(3)}$$

$$= \frac{6 \pm \sqrt{36 + 96}}{6}$$

$$= \frac{6 \pm \sqrt{132}}{6}$$

$$= \frac{6 \pm 2\sqrt{33}}{6}$$

$$= \frac{2\left(3 \pm \sqrt{33}\right)}{6}$$

$$= \frac{3 \pm \sqrt{33}}{3}$$

$y = \frac{3 + \sqrt{33}}{3}$ or $y = \frac{3 - \sqrt{33}}{3}$

31. $x^2 - x + 13 = 0$

$a = 1, b = -1, c = 13$

$$x = \frac{-b \pm \sqrt{b^2 - 4ac}}{2a}$$

$$x = \frac{-(-1) \pm \sqrt{(-1)^2 - 4(1)(13)}}{2(1)}$$

$$= \frac{1 \pm \sqrt{1 - 52}}{2}$$

$$= \frac{1 \pm \sqrt{-51}}{2}$$

$$= \frac{1 \pm i\sqrt{51}}{2}$$

$x = \frac{1 + i\sqrt{51}}{2}$ or $x = \frac{1 - i\sqrt{51}}{2}$

32. $x^2 - 2x + 11 = 0$

$a = 1, b = -2, c = 11$

$$x = \frac{-b \pm \sqrt{b^2 - 4ac}}{2a}$$

$$x = \frac{-(-2) \pm \sqrt{(-2)^2 - 4(1)(11)}}{2(1)}$$

$$= \frac{2 \pm \sqrt{4 - 44}}{2}$$

$$= \frac{2 \pm \sqrt{-40}}{2}$$

$$= \frac{2 \pm 2i\sqrt{10}}{2}$$

$$= 1 \pm i\sqrt{10}$$

$x = 1 + i\sqrt{10}$ or $x = 1 - i\sqrt{10}$

33. $2x^2 - \frac{5}{3}x = \frac{25}{3}$

$$3\left(2x^2 - \frac{5}{3}x\right) = 3\left(\frac{25}{3}\right)$$

$$6x^2 - 5x = 25$$

$$6x^2 - 5x - 25 = 0$$

$a = 6, b = -5, c = -25$

$$x = \frac{-b \pm \sqrt{b^2 - 4ac}}{2a}$$

$$x = \frac{-(-5) \pm \sqrt{(-5)^2 - 4(6)(-25)}}{2(6)}$$

$$x = \frac{5 \pm \sqrt{25 + 600}}{12}$$

$$= \frac{5 \pm \sqrt{625}}{12}$$

$$= \frac{5 \pm 25}{12}$$

$x = \frac{5 + 25}{12}$ or $x = \frac{5 - 25}{12}$

$= \frac{30}{12}$ $\qquad$ $= \frac{-20}{12}$

$= \frac{5}{2}$ $\qquad$ $= -\frac{5}{3}$

34. $4x^2 + 5x - \frac{3}{2} = 0$

$$2\left(4x^2 + 5x - \frac{3}{2}\right) = 2(0)$$

$$8x^2 + 10x - 3 = 0$$

$a = 8, b = 10, c = -3$

$$x = \frac{-b \pm \sqrt{b^2 - 4ac}}{2a}$$

$$= \frac{-10 \pm \sqrt{10^2 - 4(8)(-3)}}{2(8)}$$

$$= \frac{-10 \pm \sqrt{100 + 96}}{16}$$

$$= \frac{-10 \pm \sqrt{196}}{16}$$

$$= \frac{-10 \pm 14}{16}$$

$x = \frac{-10 + 14}{16}$ or $x = \frac{-10 - 14}{16}$

$= \frac{1}{4}$ $\qquad$ $= -\frac{3}{2}$

35. $f(x) = x^2 - 4x - 35$, $f(x) = 25$

$$x^2 - 4x - 35 = 25$$

$$x^2 - 4x - 60 = 0$$

$$(x - 10)(x + 6) = 0$$

$x - 10 = 0$ or $x + 6 = 0$

$x = 10$ $\qquad$ $x = -6$

The solutions are 10 and –6.

36. $g(x) = 6x^2 + 5x$, $g(x) = 6$

$$6x^2 + 5x = 6$$

$$6x^2 + 5x - 6 = 0$$

$$(2x + 3)(3x - 2) = 0$$

$2x + 3 = 0$ or $3x - 2 = 0$

$x = -\frac{3}{2}$ $\qquad$ $x = \frac{2}{3}$

The solutions are $-\frac{3}{2}$ and $\frac{2}{3}$.

37. $h(r) = 5r^2 - 7r - 10$, $h(r) = -8$

$$5r^2 - 7r - 10 = -8$$
$$5r^2 - 7r - 2 = 0$$
$$r = \frac{7 \pm \sqrt{(-7)^2 - 4(5)(-2)}}{2(5)}$$
$$= \frac{7 \pm \sqrt{49 + 40}}{10}$$
$$= \frac{7 \pm \sqrt{89}}{10}$$

The solutions are $\frac{7+\sqrt{89}}{10}$ and $\frac{7-\sqrt{89}}{10}$.

38. $f(x) = -2x^2 + 6x + 7$, $f(x) = -2$

$$-2x^2 + 6x + 7 = -2$$
$$-2x^2 + 6x + 9 = 0$$
$$x = \frac{-6 \pm \sqrt{6^2 - 4(-2)(9)}}{2(-2)}$$
$$= \frac{-6 \pm \sqrt{36 + 72}}{-4}$$
$$= \frac{-6 \pm \sqrt{108}}{-4}$$
$$= \frac{-6 \pm 6\sqrt{3}}{-4}$$
$$= \frac{3 \pm 3\sqrt{3}}{2}$$

The solutions are $\frac{3+3\sqrt{3}}{2}$ and $\frac{3-3\sqrt{3}}{2}$.

39. Solutions are $x = 3$ and $x = -1$.

$x - 3 = 0$ and $x + 1 = 0$.

$$(x-3)(x+1) = 0$$
$$x^2 - 2x - 3 = 0$$

40. Solutions are $x = \frac{2}{3}$ and $x = -2$.

$3x = 2$ and $x + 2 = 0$

$$3x - 2 = 0$$
$$(3x-2)(x+2) = 0$$
$$3x^2 + 4x - 4 = 0$$

41. Solutions are $x = -\sqrt{11}$ and $x = \sqrt{11}$.

$x + \sqrt{11} = 0$ and $x - \sqrt{11} = 0$.

$$(x+\sqrt{11})(x-\sqrt{11}) = 0$$
$$x^2 - 11 = 0$$

42. Solutions are $x = 3 - 2i$ and $x = 3 + 2i$.

$x - (3-2i) = 0$ and $x - (3+2i) = 0$

$x - 3 + 2i = 0$ $\quad x - 3 - 2i = 0$

$$(x-3+2i)(x-3-2i) = 0$$
$$[(x-3)+2i][(x-3)-2i] = 0$$
$$(x-3)^2 - (2i)^2 = 0$$
$$x^2 - 6x + 9 - 4i^2 = 0$$
$$x^2 - 6x + 9 + 4 = 0$$
$$x^2 - 6x + 13 = 0$$

43. Let x = the width of the garden. Then the length is $x + 4$.

Area = length × width

$$96 = (x+4)x$$
$$96 = x^2 + 4x$$
$$0 = x^2 + 4x - 96$$
$$0 = (x+12)(x-8)$$

$x + 12 = 0$ or $x - 8 = 0$

$x = -12$ $\quad x = 8$

Disregard the negative value. The width is 8 feet and the length is 8 + 4 = 12 feet.

44. Using the Pythagorean Theorem, $a^2 + b^2 = c^2$

$$8^2 + 8^2 = x^2$$
$$64 + 64 = x^2$$
$$128 = x^2$$
$$\sqrt{128} = x$$
$$x = 8\sqrt{2} \approx 11.31$$

45. $A = P\left(1 + \frac{r}{n}\right)^{nt}$

$$1081.60 = 1000\left(1 + \frac{r}{1}\right)^{1(2)}$$
$$1081.60 = 1000(1+r)^2$$
$$1.08160 = (1+r)^2$$
$$\pm 1.04 = 1 + r$$
$$r = -1 \pm 1.04$$

The rate must be positive, so $r = -1 + 1.04 = 0.04$.
The annual interest is 4%.

46. Let x be the smaller positive number. Then $x + 4$ is the larger positive number.

$$x(x+4)=77$$
$$x^2+4x-77=0$$
$$(x+11)(x-7)=0$$
$$x=-11 \quad \text{or} \quad x=7$$

Since x must be positive, $x = 7$ and $7+2=9$.

47. Let x be the width. Then $2x - 4$ is the length and the equation is $A = lw$.

$$96=(2x-4)(x)$$
$$96=2x^2-4x$$
$$0=2x^2-4x-96$$
$$0=(2x+12)(x-8)$$
$$0=2x+12 \quad \text{or} \quad 0=x-8$$
$$-12=2x \qquad 8=x$$
$$-6=x$$

Since the width must be positive, $x = 8$.
The width is 8 inches and the length $2x - 4$ is $2(8)-4=16-4=12$ inches.

48. $V=12d-0.05d^2$, $d = 60$

$V=12(60)-0.05(60)^2=720-180=540$

The value is $540.

49. a. Note that 2004 is represented by $t = 3$.

$$E(t)=7t^2-7.8t+82.2$$
$$E(3)=7(3)^2-7.8(3)+82.2=121.8$$

The expenditure of oil companies in 2004 was about $121.8 billion.

b. $E(t)=579$

$$579=7t^2-7.8t+82.2$$
$$0=7t^2-7.8t-496.8$$
$$0=7t^2-7.8t-496.8$$
$$t=\frac{-(-7.8)\pm\sqrt{(-7.8)^2-4(7)(-496.8)}}{2(7)}$$
$$=\frac{7.8\pm\sqrt{13,971.24}}{14}=\frac{7.8\pm118.2}{14}$$
$$t=\frac{7.8+118.2}{14} \quad \text{or} \quad t=\frac{7.8-118.2}{14}$$
$$=\frac{126}{14}=9 \qquad =-\frac{110.4}{14}$$

Since the time must be positive, $t = 9$, which represents the year 2010. If the trend continues, the expenditure by oil companies will be $579 billion in the year 2010.

50. $d=-16t^2+784$

a. $d=-16(2)^2+784=-64+784=720$

The object is 720 feet from the ground 2 seconds after being dropped.

b.
$$0=-16t^2+784$$
$$16t^2=784$$
$$t^2=49$$
$$t=\pm\sqrt{49}=\pm7$$

Since the time must be positive, $t = 7$ seconds.

51. a. $L(t)=0.0004t^2+0.16t+20$

$L(100)=0.0004(100)^2+0.16(100)+20=40$

40 milliliters will leak out at 100°C.

b.
$$53=0.0004t^2+0.16t+20$$
$$0=0.0004t^2+0.16t-33$$
$$t=\frac{-0.16\pm\sqrt{(0.16)^2-4(0.0004)(-33)}}{2(0.0004)}$$
$$=\frac{-0.16\pm\sqrt{0.0784}}{0.0008}$$
$$=\frac{-0.16\pm0.28}{0.0008}$$
$$t=\frac{-0.16+0.28}{0.0008} \quad \text{or} \quad t=\frac{-0.16-0.28}{0.0008}$$
$$=150 \qquad =-550$$

Since the temperature must be positive, $t = 150$. The operating temperature is 150°C.

52. Let x be the time ror the smaller machine to do the job then $x - 1$ is the time for the larger machine.

$$\frac{12}{x}+\frac{12}{x-1}=1$$
$$x(x-1)\left(\frac{12}{x}\right)+x(x-1)\left(\frac{12}{x-1}\right)=x(x-1)$$
$$12(x-1)+12x=x^2-x$$
$$12x-12+12x=x^2-x$$
$$-12+24x=x^2-x$$
$$0=x^2-25x+12$$
$$x=\frac{-(-25)\pm\sqrt{(-25)^2-4(1)(12)}}{2(1)}=\frac{25\pm\sqrt{577}}{2}$$
$$x=\frac{25+\sqrt{577}}{2}\approx24.5 \text{ or } x=\frac{25-\sqrt{577}}{2}\approx0.49$$

x cannot equal 0.49 since this would mean the smaller machine could do the work in 0.49 hours and the larger can do the work in $x - 1$ or -0.51 hours. Therefore, the smaller machine does the work in 24.51 hours and the larger machine does the work in 23.51 hours.

53. Let x be the speed (in miles per hour) for the first 25 miles. Then, the speed for the next 65 miles is $x + 15$. The time for the first 20 miles is $\frac{d}{r} = \frac{25}{x}$ and the time for the next 30 miles is $\frac{d}{r} = \frac{65}{x+15}$.

The total time is 1.5 hours.

$$\frac{25}{x} + \frac{65}{x+15} = 1.5$$

$$x(x+15)\left(\frac{25}{x} + \frac{65}{x+15}\right) = x(x+15)(1.5)$$

$$25(x+15) + x(65) = (x^2 + 15x)(1.5)$$

$$90x + 375 = 1.5x^2 + 22.5x$$

$$0 = 1.5x^2 - 67.5x - 375$$

$$0 = 3x^2 - 135x - 750$$

$$0 = 3(x-50)(x+5)$$

$$x - 50 = 0 \quad \text{or} \quad x + 5 = 0$$

$$x = 50 \quad \text{or} \quad x = -5$$

Since the speed must be positive, $x = 50$. Thus, the speed was 50 mph.

54. Let r be the speed (in miles per hour) of the canoe in still water. For the trip downstream, the rate is $r + 0.4$ and the distance 3 miles so that the time is $t = \frac{3}{r+0.4}$. For the trip upstream the rate is $r - 0.4$ and the distance is 3 miles so that the time is $t = \frac{3}{r-0.4}$. The total time is 4 hours.

$$\frac{3}{r+0.4} + \frac{3}{r-0.4} = 4$$

$$(r+0.4)(r-0.4)\left[\frac{3}{r+0.4} + \frac{3}{r-0.4} = 4\right]$$

$$3(r-0.4) + 3(r+0.4) = 4(r+0.4)(r-0.4)$$

$$3r - 1.2 + 3r + 1.2 = 4(r^2 - 0.16)$$

$$6r = 4r^2 - 0.64$$

$$0 = 4r^2 - 6r - 0.64$$

$$0 = 2r^2 - 3r - 0.32$$

$$r = \frac{-(-3) \pm \sqrt{(-3)^2 - 4(2)(-0.32)}}{2(2)}$$

$$= \frac{3 \pm \sqrt{9 + 2.56}}{4}$$

$$= \frac{3 \pm \sqrt{11.56}}{4}$$

$$= \frac{3 \pm 3.4}{4}$$

$$r = \frac{3+3.4}{4} \quad \text{or} \quad r = \frac{3-3.4}{4}$$

$$= \frac{6.4}{4} \qquad = \frac{-0.4}{4}$$

$$= 1.6 \qquad = -0.1$$

Since the rate must be positive, $r = 1.6$. Rachel canoes 1.6 miles per hour in still water.

55. Let x be the length. The width is $x - 2$.
Area = length × width

$$80 = x(x-2)$$

$$80 = x^2 - 2x$$

$$0 = x^2 - 2x - 80$$

$$x = \frac{-(-2) \pm \sqrt{(-2)^2 - 4(1)(-80)}}{2(1)}$$

$$= \frac{2 \pm \sqrt{324}}{2} = \frac{2 \pm 18}{2}$$

$$x = \frac{20}{2} = 10 \quad \text{or} \quad x = \frac{-16}{2} = -8$$

Since the width must be positive, $x = 10$. The length is 10 units and the width is $10 - 2 = 8$ units.

56. If the business sells n tables at a price of $(60 - 0.3n)$ dollars per table, then the revenue is given by $R(n) = n(60 - 0.3n)$ with $n \le 40$. Set this equal to 1080 and solve for n.

$$R(n) = n(60 - 0.3n)$$

$$1080 = n(60 - 0.3n)$$

$$1080 = 60n - 0.3n^2$$

$$0.3n^2 - 60n + 1080 = 0$$

$$n^2 - 200n + 3600 = 0$$

$$(n-20)(n-180) = 0$$

$$n = 20 \quad \text{or} \quad n = 180$$

Disregard $n = 180$ since $n \le 40$. The business must sell 20 tables.

57. $a^2 + b^2 = c^2$

$a^2 = c^2 - b^2$

$\sqrt{a^2} = \pm\sqrt{c^2 - b^2}$

$a = \pm\sqrt{c^2 - b^2}$

Since a refers to a length, it cannot be negative. Therefore disregard the negative sign.

$a = \sqrt{c^2 - b^2}$

58. $h = -4.9t^2 + c$

$h - c = -4.9t^2$

$\frac{h-c}{-4.9} = \frac{-4.9t^2}{-4.9}$

$\frac{h-c}{-4.9} = t^2$

$\pm\sqrt{\frac{h-c}{-4.9}} = t$ or $t = \pm\sqrt{\frac{c-h}{4.9}}$

Disregard the negative value for time:

$t = \sqrt{\frac{c-h}{4.9}}$

59. $v_x^2 + v_y^2 = v^2$ for v_y

$v_y^2 = v^2 - v_x^2$

$v_y = \sqrt{v^2 - v_x^2}$

60. $a = \frac{v_2^2 - v_1^2}{2d}$ for v_2

$2ad = v_2^2 - v_1^2$

$2ad + v_1^2 = v_2^2$

$v_2 = \sqrt{v_1^2 + 2ad}$

61. $x^4 - 13x^2 + 36 = 0$

$(x^2)^2 - 13x^2 + 36 = 0$

Let $u = x^2$

$u^2 - 13u + 36 = 0$

$(u-9)(u-4) = 0$

$u - 9 = 0$ or $u - 4 = 0$

$u = 9$ $\quad$ $u = 4$

Substitute x^2 for u.

$x^2 = 9$ or $x^2 = 4$

$x = \pm 3$ $\quad$ $x = \pm 2$

The solutions are ± 3 and ± 2.

62. $x^4 - 21x^2 + 80 = 0$

$(x^2)^2 - 21x^2 + 80 = 0$

Let $u = x^2$

$u^2 - 21u + 80 = 0$

$(u-16)(u-5) = 0$

$u - 16 = 0$ or $u - 5 = 0$

$u = 16$ $\quad$ $u = 5$

Substitute x^2 for u.

$x^2 = 16$ or $x^2 = 5$

$x = \pm 4$ $\quad$ $x = \pm\sqrt{5}$

The solutions are ± 4 and $\pm\sqrt{5}$.

63. $a^4 = 5a^2 + 24$

$a^4 - 5a^2 - 24 = 0$

$(a^2)^2 - 5a^2 - 24 = 0$

Let $u = a^2$

$u^2 - 5u - 24 = 0$

$(u-8)(u+3) = 0$

$u - 8 = 0$ or $u + 3 = 0$

$u = 8$ $\quad$ $u = -3$

Substitute a^2 for u.

$a^2 = 8$ or $a^2 = -3$

$a = \pm\sqrt{8}$ $\quad$ $a = \pm\sqrt{-3}$

$= \pm 2\sqrt{2}$ $\quad$ $= \pm i\sqrt{3}$

The solutions are $\pm 2\sqrt{2}$ and $\pm i\sqrt{3}$.

64. $3y^{-2} + 16y^{-1} = 12$

$3(y^{-1})^2 + 16(y^{-1}) - 12 = 0$

Let $u = y^{-1}$.

$3u^2 + 16u - 12 = 0$

$(3u - 2)(u + 6) = 0$

$3u - 2 = 0$ or $u + 6 = 0$

$u = \frac{2}{3}$ $\quad$ $u = -6$

Substitute y^{-1} for u.

$y^{-1} = \frac{2}{3}$ or $y^{-1} = -6$

$y = \frac{3}{2}$ $\quad$ $y = -\frac{1}{6}$

The solutions are $\frac{3}{2}$ and $-\frac{1}{6}$.

65. $3r + 11\sqrt{r} - 4 = 0$

$3\left(r^{1/2}\right)^2 + 11r^{1/2} - 4 = 0$

Let $u = r^{1/2}$

$3u^2 + 11u - 4 = 0$

$(3u - 1)(u + 4) = 0$

$3u - 1 = 0$ or $u + 4 = 0$

$u = \frac{1}{3}$ $\quad u = -4$

Substitute $r^{1/2}$ for u.

$r^{1/2} = \frac{1}{3}$ or $r^{1/2} = -6$

not real

$r = \left(\frac{1}{3}\right)^2 = \frac{1}{9}$

There are no solutions for $r^{1/2} = -6$ since there is no real number x for which $r^{1/2} = -6$.

The solution is $\frac{1}{9}$.

66. $2p^{2/3} - 7p^{1/3} + 6 = 0$

$2\left(p^{1/3}\right)^2 - 7p^{1/3} + 6 = 0$

Let $u = p^{1/3}$

$2u^2 - 7u + 6 = 0$

$(2u - 3)(u - 2) = 0$

$2u - 3 = 0$ or $u - 2 = 0$

$u = \frac{3}{2}$ $\quad u = 2$

Substitute $p^{1/3}$ for u.

$p^{1/3} = \frac{3}{2}$ $\quad p^{1/3} = 2$

$p = \left(\frac{3}{2}\right)^3$ $\quad p = 2^3$

$= \frac{27}{8}$ $\quad = 8$

The solutions are $\frac{27}{8}$ and 8.

67. $6(x-2)^{-2} = -13(x-2)^{-1} + 8$

$6\left[(x-2)^{-1}\right]^2 + 13(x-2)^{-1} - 8 = 0$

Let $u = (x-2)^{-1}$

$6u^2 + 13u - 8 = 0$

$(2u - 1)(3u + 8) = 0$

$2u - 1 = 0$ or $3u + 8 = 0$

$u = \frac{1}{2}$ $\quad u = -\frac{8}{3}$

Substitute $(x-2)^{-1}$ for u.

$(x-2)^{-1} = \frac{1}{2}$ or $(x-2)^{-1} = -\frac{8}{3}$

$x - 2 = 2$ $\quad x - 2 = -\frac{3}{8}$

$x = 4$ $\quad x = \frac{13}{8}$

The solutions are 4 and $\frac{13}{8}$.

68. $10(r+1) = \frac{12}{r+1} - 7$

$(r+1)\left[10(r+1)\right] = (r+1)\left[\frac{12}{r+1} - 7\right]$

$10(r+1)^2 + 7(r+1) = 12$

$10(r+1)^2 + 7(r+1) - 12 = 0$

Let $u = r + 1$.

$10u^2 + 7u - 12 = 0$

$(5u - 4)(2u + 3) = 0$

$5u - 4 = 0$ or $2u + 3 = 0$

$u = \frac{4}{5}$ $\quad u = -\frac{3}{2}$

Substitute $(r+1)$ for u.

$r + 1 = \frac{4}{5}$ or $r + 1 = -\frac{3}{2}$

$r = -\frac{1}{5}$ $\quad r = -\frac{5}{2}$

The solutions are $-\frac{1}{5}$ and $-\frac{5}{2}$.

69. $f(x) = x^4 - 82x^2 + 81$

To find the x-intercepts, set $f(x) = 0$.

$0 = x^4 - 82x^2 + 81$

$0 = (x^2)^2 - 82x^2 + 81$

Let $u = x^2$.

$0 = u^2 - 82u + 81$

$0 = (u - 81)(u - 1)$

$u - 81 = 0$ or $u - 1 = 0$

$u = 81$ $\quad$ $u = 1$

Substitute x^2 for u.

$x^2 = 81$ or $x^2 = 1$

$x = \pm 9$ $\quad$ $x = \pm 1$

The x-intercepts are (9, 0), (–9, 0), (1, 0), and (–1, 0).

70. $f(x) = 30x + 13\sqrt{x} - 10$

To find the x-intercepts, set $f(x) = 0$.

$0 = 30x + 13\sqrt{x} - 10$

$0 = 30\left(\sqrt{x}\right)^2 + 13\sqrt{x} - 10$

Let $u = \sqrt{x}$.

$0 = 30u^2 + 13u - 10$

$0 = (6u + 5)(5u - 2)$

$6u + 5 = 0$ or $5u - 2 = 0$

$u = -\dfrac{5}{6}$ $\quad$ $u = \dfrac{2}{5}$

Substitute $\sqrt{x}$ for u.

$\sqrt{x} = -\dfrac{5}{6}$ or $\sqrt{x} = \dfrac{2}{5}$

not real $\quad$ $x = \dfrac{4}{25}$

Since $\sqrt{x}$ cannot be negative, the solution is $\dfrac{4}{25}$. The only x-intercept is $\left(\dfrac{4}{25}, 0\right)$.

71. $f(x) = x - 6\sqrt{x} + 12$

To find the x-intercepts, set $f(x) = 0$.

$0 = x - 6\sqrt{x} + 12$

$0 = \left(\sqrt{x}\right)^2 - 6\sqrt{x} + 12$

Let $u = \sqrt{x}$.

$0 = u^2 - 6u + 12$

$$u = \frac{-(-6) \pm \sqrt{(-6)^2 - 4(1)(10)}}{2(1)}$$

$$= \frac{6 \pm \sqrt{-4}}{2} = \frac{6 \pm 2i}{2} = 3 \pm i$$

Substitute $\sqrt{x}$ for u.

$\sqrt{x} = 3 \pm i$

Since x-intercepts must be real numbers, this function has no x-intercepts.

72. $g(x) = (x^2 - 6x)^2 - 5(x^2 - 6x) - 24$

To find the x-intercepts, set $g(x) = 0$.

$0 = (x^2 - 6x)^2 - 5(x^2 - 6x) - 24$

Let $u = x^2 - 6x$.

$0 = u^2 - 5u - 24$

$0 = (u + 3)(u - 8)$

$u + 3 = 0$ or $u - 8 = 0$

$u = -3$ $\quad$ $u = 8$

Substitute $(x^2 - 6x)$ for u.

$x^2 - 6x = -3$ or $x^2 - 6x = 8$

$x^2 - 6x + 3 = 0$ $\quad$ $x^2 - 6x - 8 = 0$

$\dfrac{6 \pm \sqrt{(-6)^2 - 4(1)(3)}}{2(1)}$ $\quad$ $\dfrac{6 \pm \sqrt{(-6)^2 - 4(1)(-8)}}{2(1)}$

$\dfrac{6 \pm \sqrt{24}}{2}$ $\quad$ $\dfrac{6 \pm \sqrt{68}}{2}$

$\dfrac{6 \pm 2\sqrt{6}}{2}$ $\quad$ $\dfrac{6 \pm 2\sqrt{17}}{2}$

$3 \pm \sqrt{6}$ $\quad$ $3 \pm \sqrt{17}$

The x-intercepts are $\left(3 + \sqrt{6}, 0\right), \left(3 - \sqrt{6}, 0\right)$, $\left(3 + \sqrt{17}, 0\right)$ and $\left(3 - \sqrt{17}, 0\right)$.

73. $f(x) = x^2 + 5x$

a. Since $a = 1$ the parabola opens upward.

b. $y = f(0) = 0^2 + 5(0) = 0$
The y-intercept is (0, 0).

c. $x = -\frac{b}{2a} = -\frac{5}{2(1)} = -\frac{5}{2}$

$$y = \frac{4ac - b^2}{4a} = \frac{4(1)(0) - 5^2}{4(1)} = -\frac{25}{4}$$

The vertex is $\left(-\frac{5}{2}, -\frac{25}{4}\right)$.

d. $0 = x^2 + 5x$
$0 = x(x+5)$
$0 = x$ or $0 = x + 5$
$x = 0$ $\quad x = -5$
The x-intercepts are (0, 0) and (–5, 0).

e.
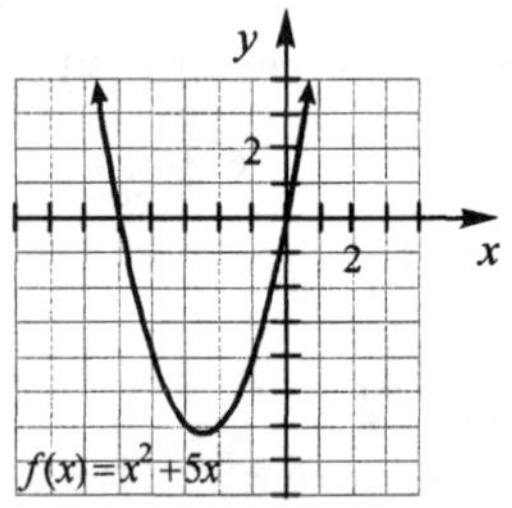

74. $f(x) = x^2 - 2x - 8$

a. Since $a = 1$ the parabola opens upward.

b. $y = f(0) = (0)^2 - 2(0) - 8 = -8$
The y-intercept is (0, –8).

c. $x = -\frac{b}{2a} = -\frac{-2}{2(1)} = \frac{2}{2} = 1$

$$y = \frac{4ac - b^2}{4a} = \frac{4(1)(-8) - (-2)^2}{4(1)} = \frac{-36}{4} = -9$$

The vertex is (1, –9).

d. $0 = x^2 - 2x - 8$
$0 = (x-4)(x+2)$
$0 = x - 4$ or $0 = x + 2$
$4 = x$ or $-2 = x$
The x-intercepts are (4, 0) and (–2, 0).

e.
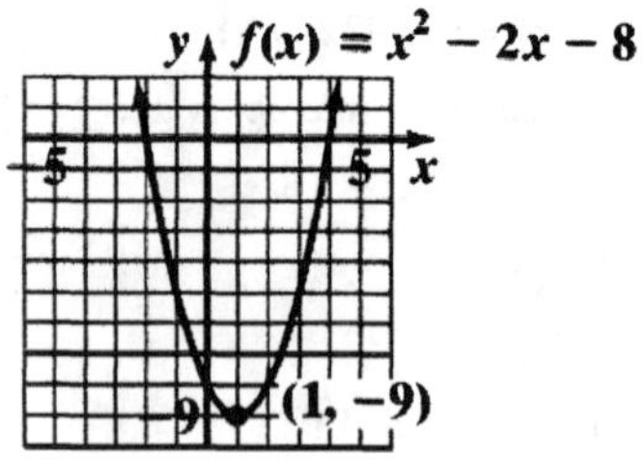

75. $g(x) = -x^2 - 2$

a. Since $a = -1$ the parabola opens downward.

b. $y = g(0) = -(0)^2 - 2 = -2$
The y-intercept is (0, –2).

c. $x = -\frac{b}{2a} = -\frac{0}{2(-1)} = -\frac{0}{-2} = 0$

$$y = \frac{4ac - b^2}{4a} = \frac{4(-1)(-2) - 0^2}{4(-1)} = \frac{8}{-4} = -2$$

The vertex is (0, –2).

d. $0 = -x^2 - 2$
$x^2 = -2$
$x = \pm\sqrt{-2} = \pm i\sqrt{2}$
There are no real roots. Thus, there are no x-intercepts.

e.
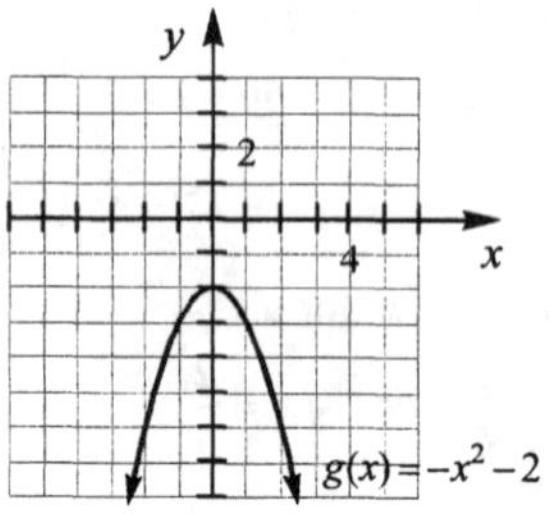

76. $g(x) = -2x^2 - x + 15$

a. Since $a = -2$ the parabola opens downward.

b. $y = -2(0)^2 - 0 + 15 = 15$
The y-intercept is (0, 15).

c. $x = -\frac{b}{2a} = -\frac{-1}{2(-2)} = \frac{1}{-4} = -\frac{1}{4}$

$$y = \frac{4ac - b^2}{4a} = \frac{4(-2)(15) - (-1)^2}{4(-2)} = \frac{121}{8}$$

The vertex is $\left(-\frac{1}{4}, \frac{121}{8}\right)$.

d. $0=-1\left(2x^2+x-15\right)$

$0=-1(2x-5)(x+3)$

$0=(2x-5)$ or $0=x+3$

$5=2x$

$\frac{5}{2}=x$ $\quad -3=x$

The x-intercepts are (–3, 0) and $\left(\frac{5}{2}, 0\right)$.

e.

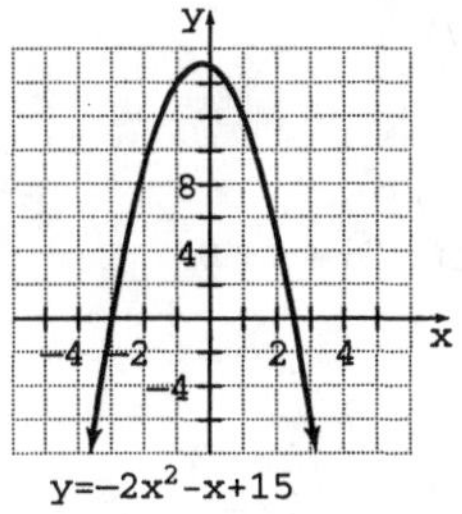

77. a. $I=-x^2+22x-45$, $2 \le x \le 20$

The x-coordinate of the vertex will be the cost per ticket to maximize profit.

$x=-\frac{b}{2a}=-\frac{22}{2(-1)}=11$

They should charge \$11 per ticket.

b. The maximum profit in hundreds is the y-coordinate of the vertex.

$I(11)=-11^2+22(11)-45$

$=-121+242-45$

$=76$

The maximum profit is \$76 hundred or \$7600.

78. a. $s(t)=-16t^2+80t+75$

The ball will attain maximum height at the x-coordinate of the vertex.

$t=-\frac{b}{2a}=-\frac{80}{2(-16)}=-\frac{80}{-32}=2.5$

The ball will attain maximum height 2.5 seconds after being thrown.

b. The maximum height is the y-coordinate of the vertex.

$s(2.5)=-16(2.5)^2+80(2.5)+74$

$=-100+200+75$

$=175$

The maximum height is 175 feet.

79. The graph of $f(x)=(x-3)^2$ has vertex (3, 0). The graph will be $g(x)=x^2$ shifted right 3 units.

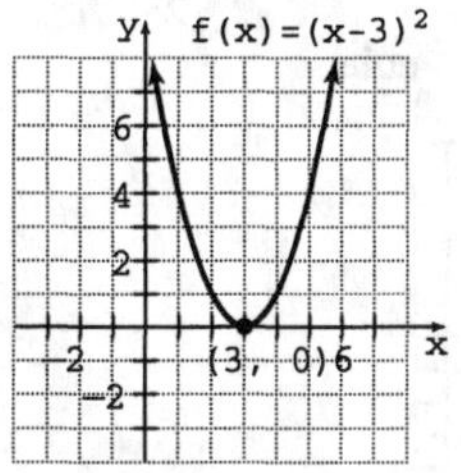

80. The graph of $f(x)=-(x+2)^2-3$ has vertex (–2, –3). Since $a<0$, the parabola opens downward. The graph will be $g(x)=-x^2$ shifted left 2 units and down 3 units.

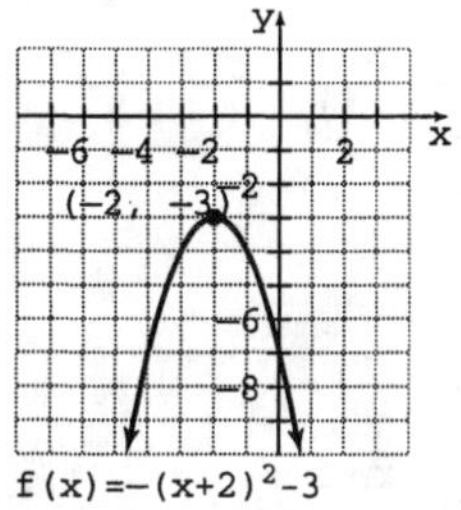

81. The graph of $g(x)=-2(x+4)^2-1$ has vertex $(-4, -1)$. Since $a<0$, the parabola opens downward. The graph will be $f(x)=-2x^2$ shifted left 4 units and down 1 unit.

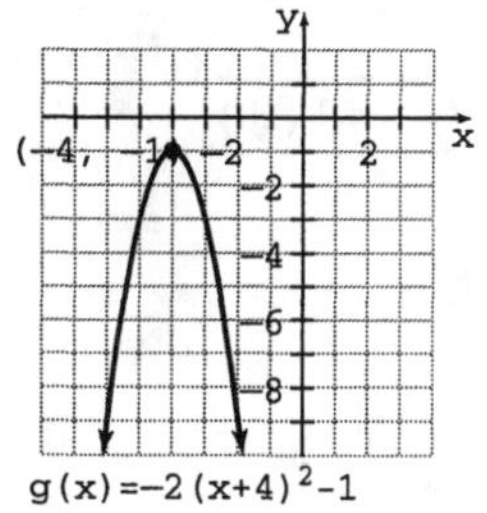

82. The graph of $h(x)=\frac{1}{2}(x-1)^2+3$ has vertex (1, 3). The graph will be $f(x)=\frac{1}{2}x^2$ shifted right 1 unit and up 3 units.

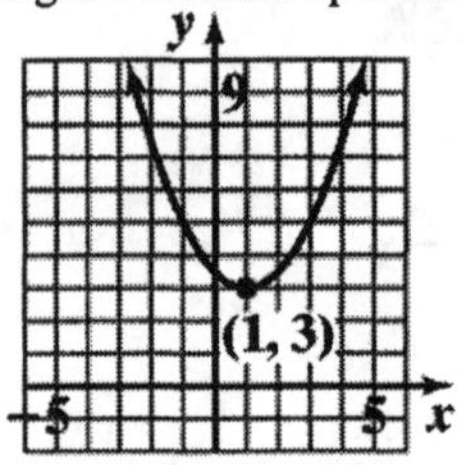

$h(x)=\frac{1}{2}(x-1)^2+3$

83. $x^2+4x+3\ge 0$

$(x+1)(x+3)\ge 0$

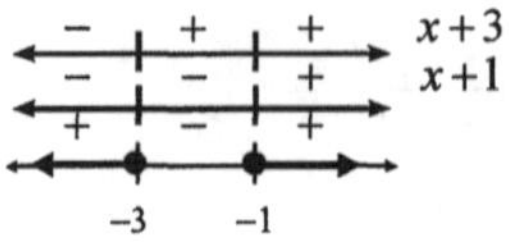

84. $x^2+3x-10\le 0$

$(x+5)(x-2)\le 0$

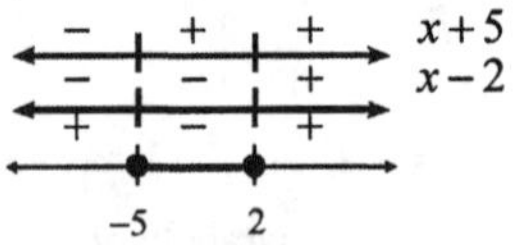

85. $x^2\le 11x-20$

$x^2-11x+20\le 0$

$x^2-11x+20=0$

$x=\frac{-(-11)\pm\sqrt{(-11)^2-4(1)(20)}}{2(1)}$

$=\frac{11\pm\sqrt{121-80}}{2}$

$=\frac{11\pm\sqrt{41}}{2}$

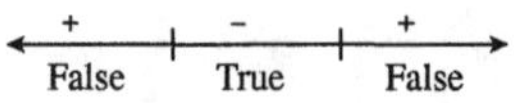

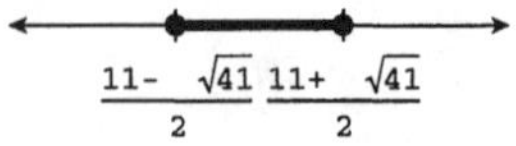

86. $3x^2+8x>16$

$3x^2+8x-16>0$

$(3x-4)(x+4)>0$

$3x-4=0$ or $x+4=0$

$x=\frac{4}{3}$ $\quad x=-4$

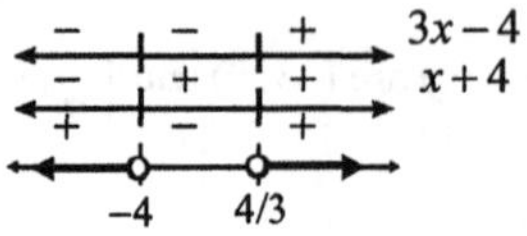

87. $4x^2-9\le 0$

$(2x-3)(2x+3)\le 0$

$2x-3=0$ or $2x+3=0$

$x=\frac{3}{2}$ $\quad x=-\frac{3}{2}$

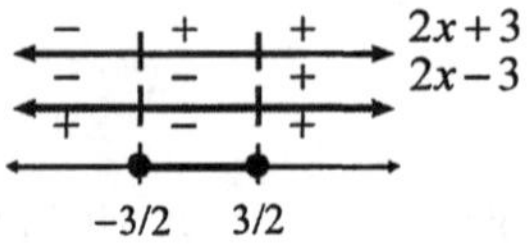

88. $6x^2-30>0$

$6(x^2-5)>0$

$6(x+\sqrt{5})(x-\sqrt{5})>0$

$x+\sqrt{5}=0$ or $x-\sqrt{5}=0$

$x=-\sqrt{5}$ $\quad x=\sqrt{5}$

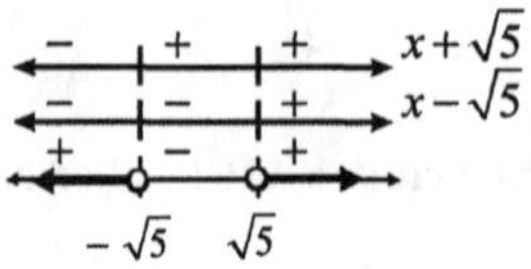

89. $\frac{x+1}{x-5}>0$

$x\ne 5$

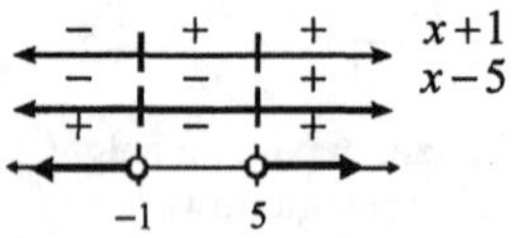

$\{x|x<-1 \text{ or } x>5\}$

90. $\dfrac{x-3}{x+2} \le 0$

$x = -2$

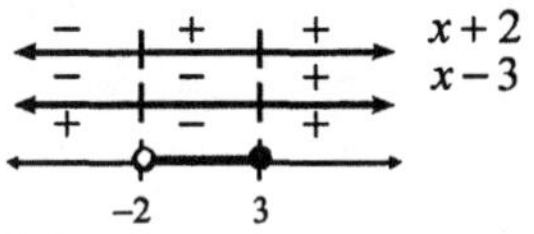

$\{x \mid -2 < x \le 3\}$

91. $\dfrac{2x-4}{x+3} \ge 0$

$\dfrac{2(x-2)}{x+3} \ge 0$

$x \ne -3$

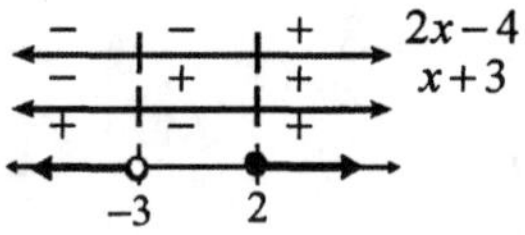

$\{x \mid x < -3 \text{ or } x \ge 2\}$

92. $\dfrac{3x+5}{x-6} < 0$

$x \ne 6$

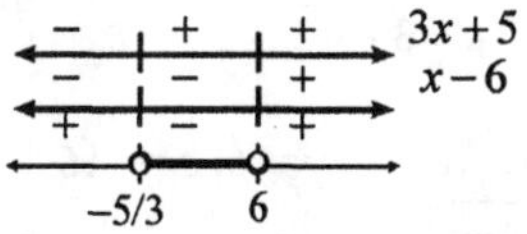

$\left\{x \,\middle|\, -\dfrac{5}{3} < x < 6\right\}$

93. $(x+4)(x+1)(x-2) > 0$

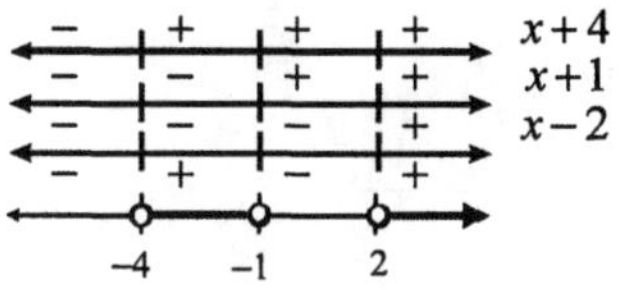

$\{x \mid -4 < x < -1 \text{ or } x > 2\}$

94. $x(x-3)(x-6) \le 0$

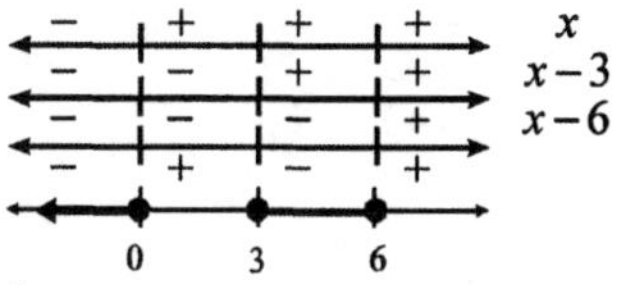

$\{x \mid x \le 0 \text{ or } 3 \le x \le 6\}$

95. $(3x+4)(x-1)(x-3) \ge 0$

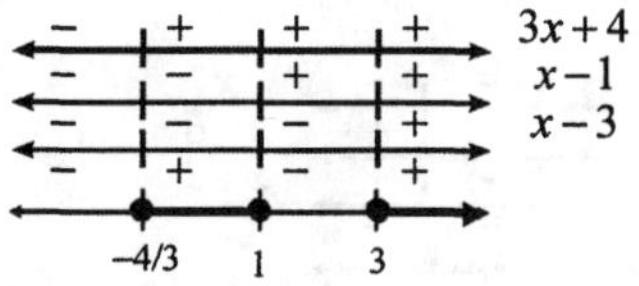

$\left[-\dfrac{4}{3}, 1\right] \cup [3, \infty)$

96. $2x(x+2)(x+4) < 0$

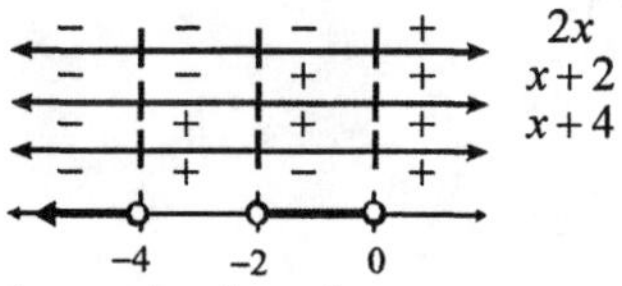

$(-\infty, -4) \cup (-2, 0)$

97. $\dfrac{x(x-4)}{x+2} > 0$

$x \ne -2$

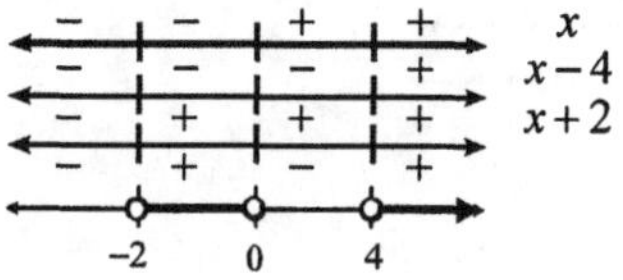

$(-2, 0) \cup (4, \infty)$

98. $\dfrac{(x-2)(x-8)}{x+3} < 0$

$x \ne -3$

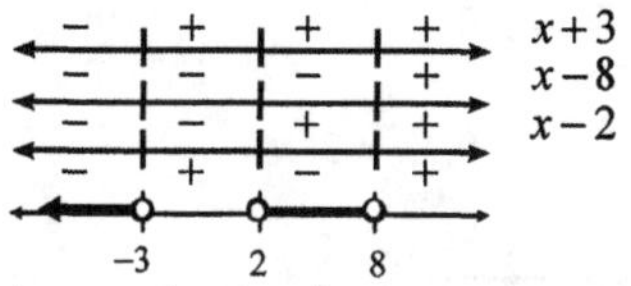

$(-\infty, -3) \cup (2, 8)$

99. $\dfrac{x-3}{(x+2)(x-7)} \ge 0$

$x \ne -2, x \ne 7$

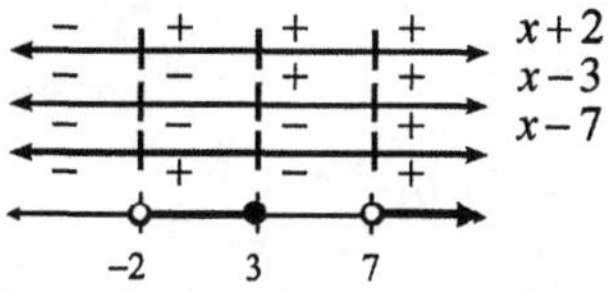

$(-2, 3] \cup (7, \infty)$

100. $\dfrac{x(x-6)}{x+3} \le 0, \; x \ne -3$

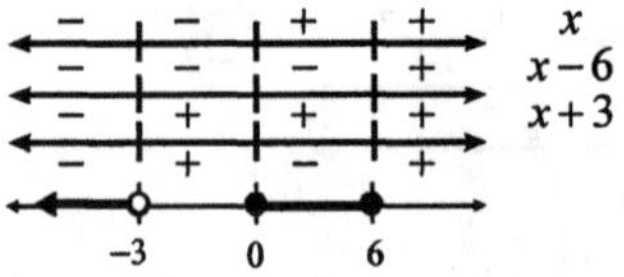

$(-\infty, -3) \cup [0, 6]$

101.

$$\frac{5}{x+4} \ge -1$$
$$\frac{5}{x+4} + 1 \ge 0$$
$$\frac{5+1(x+4)}{x+4} \ge 0$$
$$\frac{5+x+4}{x+4} \ge 0$$
$$\frac{x+9}{x+4} \ge 0, \; x \ne -4$$

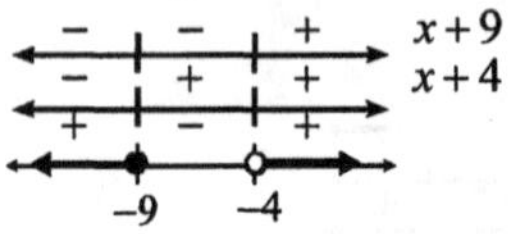

102.

$$\frac{2x}{x-2} \le 1$$
$$\frac{2x}{x-2} - 1 \le 0$$
$$\frac{2x}{x-2} - \frac{1(x-2)}{x-2} \le 0$$
$$\frac{2x-x+2}{x-2} \le 0$$
$$\frac{x+2}{x-2} \le 0, \; x \ne 2$$

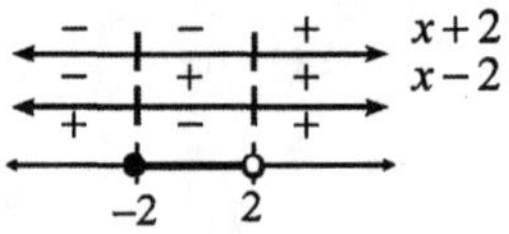

103.

$$\frac{2x+3}{3x-5} < 4$$
$$\frac{2x+3}{3x-5} - 4 < 0$$
$$\frac{2x+3-4(3x-5)}{3x-5} < 0$$
$$\frac{2x+3-12x+20}{3x-5} < 0$$
$$\frac{-10x+23}{3x-5} < 0$$

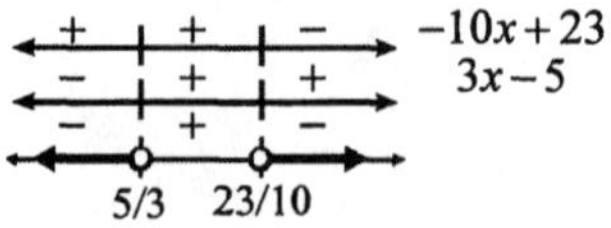

Chapter 8 Practice Test

1.
$$x^2+2x-15=0$$
$$x^2+2x=15$$
$$x^2+2x+1=15+1$$
$$(x+1)^2=16$$
$$x+1=\pm 4$$
$$x=-1\pm 4$$
$$x=3 \text{ or } x=-5$$

2.
$$a^2+7=6a$$
$$a^2-6a=-7$$
$$a^2-6a+9=-7+9$$
$$(a-3)^2=2$$
$$a-3=\pm\sqrt{2}$$
$$a=3\pm\sqrt{2}$$
$$a=3+\sqrt{2} \text{ or } a=3-\sqrt{2}$$

3. $x^2-6x-16=0, \; a=1, b=-6, c=-16$

$$x=\frac{-b\pm\sqrt{b^2-4ac}}{2a}$$
$$x=\frac{-(-6)\pm\sqrt{(-6)^2-4(1)(-16)}}{2(1)}$$
$$=\frac{6\pm\sqrt{36+64}}{2}$$
$$=\frac{6\pm\sqrt{100}}{2}$$
$$=\frac{6\pm 10}{2}$$
$$x=\frac{6+10}{2} \text{ or } x=\frac{6-10}{2}$$
$$=\frac{16}{2} \qquad =\frac{-4}{2}$$
$$=8 \qquad =-2$$

4. $x^2-4x=-11$

$x^2-4x+11=0$

$a=1,\ b=-4,\ c=11$

$$x=\frac{-b\pm\sqrt{b^2-4ac}}{2a}$$

$$x=\frac{-(-4)\pm\sqrt{(-4)^2-4(1)(11)}}{2(1)}$$

$$=\frac{4\pm\sqrt{16-44}}{2}=\frac{4\pm\sqrt{-28}}{2}$$

$$=\frac{4\pm 2i\sqrt{7}}{2}=2\pm i\sqrt{7}$$

$x=2+i\sqrt{7}$ or $x=2-i\sqrt{7}$

5. $3r^2+r=2$

$3r^2+r-2=0$

$(3r-2)(r+1)=0$

$3r-2=0$ or $r+1=0$

$3r=2$ $\quad r=-1$

$r=\frac{2}{3}$

6. $p^2+4=-7p$

$p^2+7p+4=0$

$$p=\frac{-7\pm\sqrt{(7)^2-4(1)(4)}}{2(1)}$$

$$=\frac{-7\pm\sqrt{49-16}}{2}$$

$$=\frac{-7\pm\sqrt{33}}{2}$$

$p=\frac{-7+\sqrt{33}}{2}$ or $p=\frac{-7-\sqrt{33}}{2}$

7. $x=4$ and $x=-\frac{2}{5}$

$x-4=0$ and $5x+2=0$

$(x-4)(5x+2)=0$

$5x^2-18x-8=0$

8. $K=\frac{1}{2}mv^2$ for v

$2K=mv^2$

$\frac{2K}{m}=v^2$

$v=\sqrt{\frac{2K}{m}}$

9. a. $c(s)=-0.01s^2+78s+22{,}000$

$c(1600)=-0.01(1600)^2+78(1600)+22{,}000$

$=-25{,}600+124{,}800+22{,}000$

$=121{,}200$

The cost is about \$121,200.

b. $160{,}000=-0.01s^2+78s+22{,}000$

$0=-0.01s^2+78s-138{,}000$

$$s=\frac{-78\pm\sqrt{78^2-4(-0.01)(-138{,}000)}}{2(-0.01)}$$

$$=\frac{-78\pm\sqrt{564}}{-0.02}$$

$$s=\frac{-78+\sqrt{564}}{-0.02}\approx 2712.57$$

$$s=\frac{-78-\sqrt{564}}{-0.02}\approx 5087.43$$

Since $1300\le s\le 3900$, the house should have about 2712.57 square feet.

10. The formula $d=rt$ can be written $t=\frac{d}{r}$.

Let r = his actual rate.

	distance	rate	time $=\frac{d}{r}$
Actual trip	520	r	$\frac{520}{r}$
Faster trip	520	$r+15$	$\frac{520}{r+15}$

The faster trip would have taken 2.4 hours less time than the actual trip.

$$\frac{520}{r+15}=\frac{520}{r}-2.4$$

$$r(r+15)\left(\frac{520}{r+15}\right)=r(r+15)\left(\frac{520}{r}-2.4\right)$$

$$520r=520(r+15)-2.4r(r+15)$$

$$520r=520r+7800-2.4r^2-36r$$

$0=-2.4r^2-36r+7800$

$0=r^2+15r-3250$

$0=(r+65)(r-50)$

$r+65=0$ $\quad r-50=0$

$r=-65$ $\quad r=50$

Since speed is never negative, Tom drove an average speed of 50 mph.

11. $2x^4+15x^2-50=0$

$2(x^2)^2+15x^2-50=0$

Let $u=x^2$.

$2u^2+15u-50=0$

$(u+10)(2u-5)=0$

$u+10=0$ or $2u-5=0$

$u=-10$ $\quad u=\frac{5}{2}$

Substitute x^2 for u.

$x^2=-10$ or $x^2=\frac{5}{2}$

$x=\pm\sqrt{-10}$ $\quad x=\pm\sqrt{\frac{5}{2}}$

$=\pm i\sqrt{10}$ $\quad =\pm\frac{\sqrt{5}}{\sqrt{2}}\cdot\frac{\sqrt{2}}{\sqrt{2}}$

$=\pm\frac{\sqrt{10}}{2}$

12. $3r^{2/3}+11r^{1/3}-42=0$

$3(r^{1/3})^2+11r^{1/3}-42=0$

Let $u=r^{1/3}$.

$3u^2+11u-42=0$

$(3u-7)(u+6)=0$

$3u-7=0$ or $u+6=0$

$u=\frac{7}{3}$ $\quad u=-6$

$r^{1/3}=\frac{7}{3}$ or $r^{1/3}=-6$

$r=\left(\frac{7}{3}\right)^3$ $\quad r=(-6)^3$

$=\frac{343}{27}$ $\quad r=-216$

13. $f(x)=16x-24\sqrt{x}+9$

$0=16(\sqrt{x})^2-24\sqrt{x}+9$

Let $u=\sqrt{x}$.

$0=16u^2-24u+9$

$0=(4u-3)^2$

$4u-3=0$

$u=\frac{3}{4}$

Substitute $\sqrt{x}$ for u.

$\sqrt{x}=\frac{3}{4}$

$x=\left(\frac{3}{4}\right)^2=\frac{9}{16}$

The x-intercept is $\left(\frac{9}{16}, 0\right)$.

14. $f(x)=(x-3)^2+2$

The vertex is (3, 2). The graph will be the graph of $g(x)=x^2$ shifted right 3 units and up 2 units.

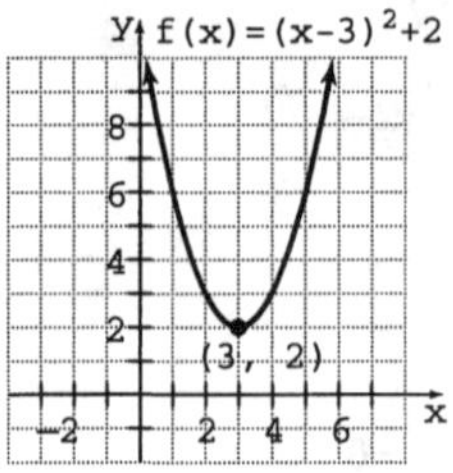

15. $h(x)=-\frac{1}{2}(x-2)^2-2$

The vertex is (2, –2). The graph will be the graph of $g(x)=-\frac{1}{2}x^2$ shifted right 2 units and down 2 units.

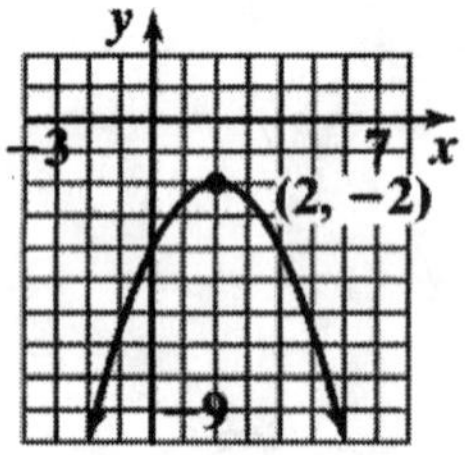

$h(x)=-\frac{1}{2}(x-2)^2-2$

16. $6x^2=2x+3$

$6x^2-2x-3=0$

$a=6, b=-2, c=-3$

The discriminant is

$b^2-4ac=(-2)^2-4(6)(-3)=4+72=76$

Since the discriminant is greater than 0, the quadratic equation has two distinct real solutions.

17. $y=x^2+2x-8$

a. Since $a=1$ the parabola opens upward.

b. Let $x=0$: $y=0^2+2(0)-8=-8$

The y-intercept is (0, –8).

c. $x = -\frac{b}{2a} = -\frac{2}{2(1)} = -1$

$y = (-1)^2 + 2(-1) - 8 = -9$

The vertex is (–1, –9).

d. The x-intercepts occur when $y = 0$.

$0 = x^2 + 2x - 8$

$0 = (x+4)(x-2)$

$x + 4 = 0$ or $x - 2 = 0$

$x = -4$ $\quad$ $x = 2$

The x-intercepts are (2, 0) and (–4, 0).

e.

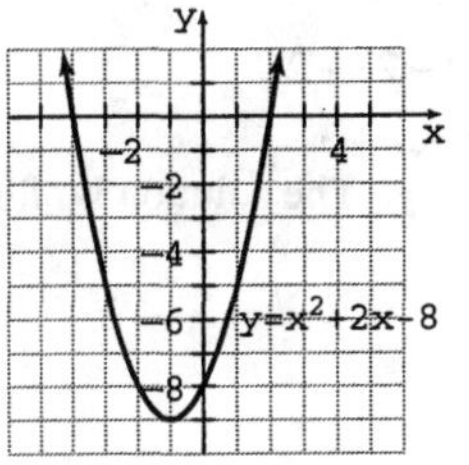

18. $x = -7$ and $x = \frac{1}{2}$

$x + 7 = 0$ and $2x - 1 = 0$

$(x+7)(2x-1) = 0$

$2x^2 + 13x - 7 = 0$

19. $x^2 - x \geq 42$

$x^2 - x - 42 \geq 0$

$(x-7)(x+6) \geq 0$

$x - 7 = 0$ or $x + 6 = 0$

$x = 7$ $\quad$ $x = -6$

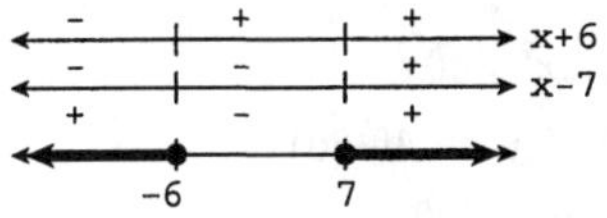

20. $\frac{(x+5)(x-4)}{x+1} \geq 0$

$x \neq -1$

$x + 5 = 0$ or $x - 4 = 0$ or $x + 1 = 0$

$x = -5$ $\quad$ $x = 4$ $\quad$ $x = -1$

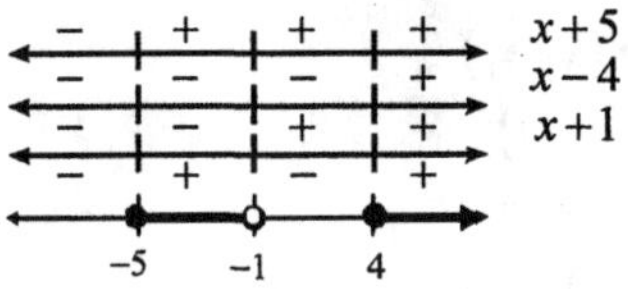

21. $\frac{x+3}{x+2} \leq -1$

$\frac{x+3}{x+2} + 1 \leq 0$

$\frac{x+3}{x+2} + \frac{x+2}{x+2} \leq 0$

$\frac{2x+5}{x+2} \leq 0$

$x \neq -2$

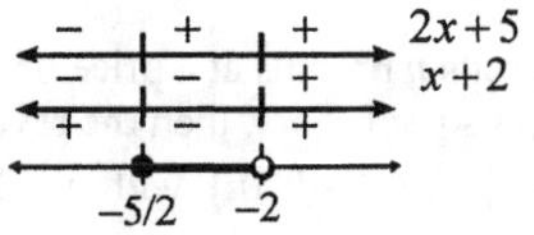

a. $\left[-\frac{5}{2}, -2\right)$

b. $\left\{x \middle| -\frac{5}{2} \leq x < -2\right\}$

22. Let x be the width of the carpet. Then $2x + 3$ is the length $A = lw$.

$65 = x(2x+3)$

$65 = 2x^2 + 3x$

$0 = 2x^2 + 3x - 65$

$0 = (2x+13)(x-5)$

$2x + 13 = 0$ or $x - 5 = 0$

$x = -\frac{13}{2}$ $\quad$ $x = 5$

Disregard the negative answer. The width is 5 feet and the length is $2 \cdot 5 + 3 = 10 + 3 = 13$ feet.

23. $d = -16t^2 + 80t + 96$

$d = 0$ when the ball strikes the ground.

$0 = -16t^2 + 80t + 96$

$0 = -16(t^2 - 5t - 6)$

$0 = -16(t-6)(t+1)$

$t - 6 = 0$ or $t + 1 = 0$

$t = 6$ $\quad$ $t = -1$

The time must be positive, so $t = 6$.

Thus, the ball strikes the ground in 6 seconds.

24. a. $f(x) = -1.4x^2 + 56x - 70$

$$x = -\frac{b}{2a} = -\frac{56}{2(-1.4)} = -\frac{56}{-2.8} = 20$$

The company must sell 20 carvings.

b. $f(20) = -1.4(20)^2 + 56(20) - 70$

$$= -560 + 1120 - 70$$
$$= 490$$

The maximum weekly profit is \$490.

25. If the business sells n brooms at a price of $(10 - 0.1n)$ dollars per broom, then the revenue is given by $R(n) = n(10 - 0.1n)$ with $n \le 32$. Set this equal to 160 and solve for n.

$$R(n) = n(10 - 0.1n)$$
$$210 = n(10 - 0.1n)$$
$$210 = 10n - 0.1n^2$$
$$0.1n^2 - 10n + 210 = 0$$
$$n^2 - 100n + 2100 = 0$$
$$(n-30)(n-70) = 0$$
$$n = 30 \text{ or } n = 70$$

Disregard $n = 70$ since $n \le 32$. The business must sell 30 brooms.

Chapter 8 Cumulative Review Test

1. $-4 \div (-2) + 18 - \sqrt{49} = -4 \div (-2) + 18 - 7$

$$= 2 + 18 - 7$$
$$= 20 - 7$$
$$= 13$$

2. Evaluate $2x^2 + 3x + 4$ when $x = 2$.

$2(2)^2 + 3(2) + 4 = 8 + 6 + 4 = 18$

3. $2{,}540{,}000 = 2.54 \times 10^6$

4. $|4 - 2x| = 5$

$4 - 2x = 5$ or $4 - 2x = -5$

$-2x = 1$ $\quad$ $-2x = -9$

$x = -\frac{1}{2}$ $\quad$ $x = \frac{9}{2}$

The solution set is $\left\{-\frac{1}{2}, \frac{9}{2}\right\}$

5. $6x - \{3 - [2(x-2) - 5x]\}$

$$= 6x - \{3 - [2x - 4 - 5x]\}$$
$$= 6x - \{3 - [-4 - 3x]\}$$
$$= 6x - \{3 + 4 + 3x\}$$
$$= 6x - \{7 + 3x\}$$
$$= 6x - 7 - 3x$$
$$= 3x - 7$$

6. $-\frac{1}{2}(4x - 6) = \frac{1}{3}(3 - 6x) + 2$

$$-2x + 3 = 1 - 2x + 2$$
$$-2x + 3 = -2x + 3$$

This is an identity. The solution is all real numbers.

7. $-4 < \frac{x+4}{2} < 6$

$$-8 < x + 4 < 12$$
$$-12 < x < 8$$

In interval notation the solution is $(-12, 8)$.

8. $9x + 7y = 15$

$$7y = -9x + 15$$
$$y = \frac{-9x + 15}{7}$$
$$y = -\frac{9}{7}x + \frac{15}{7}$$

slope $= -\frac{9}{7}$; y-intercept is $\left(0, \frac{15}{7}\right)$

9. $N(x) = -0.2x^2 + 40x$

$$N(50) = -0.2(50)^2 + 40(50)$$
$$= -500 + 2000$$
$$= 1500$$

50 trees would produce about 1500 baskets of apples.

10. $m = \frac{y_2 - y_1}{x_2 - x_1} = \frac{3-5}{4-6} = \frac{-2}{-2} = 1$

$$y - y_1 = m(x - x_1)$$
$$y - 3 = 1(x - 4)$$
$$y - 3 = x - 4$$
$$y = x - 1$$

11. a. No, the graph is not a function since each x-value does not have a unique y-value.

b. The domain is the set of x-values, Domain: $\{x|x \ge -2\}$.
The range is the set of y-values, Range: $\mathbb{R}$.

12. a. $x = -4$ is a vertical line.

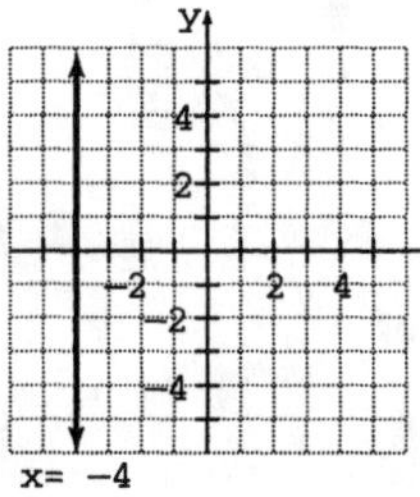

b. $y = 2$ is a horizontal line.

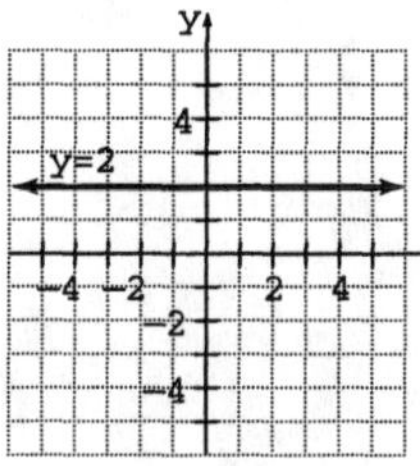

13. $\begin{vmatrix} 4 & 0 & -2 \\ 3 & 5 & 1 \\ 1 & -1 & 7 \end{vmatrix} = 4\begin{vmatrix} 5 & 1 \\ -1 & 7 \end{vmatrix} - 0\begin{vmatrix} 3 & 1 \\ 1 & 7 \end{vmatrix} + (-2)\begin{vmatrix} 3 & 5 \\ 1 & -1 \end{vmatrix}$

$$= 4(35+1) - 0(21-1) - 2(-3-5)$$
$$= 4(36) - 0(20) - 2(-8)$$
$$= 144 - 0 + 16$$
$$= 160$$

14. $4x - 3y = 10$ (1)
$2x + y = 5$ (2)

To eliminate the y variable, multiply equation (2) by 3 and add the result to equation (1).

$$\begin{array}{lcl} 4x - 3y = 10 & & 4x - 3y = 10 \\ 3(2x + y = 5) & \Rightarrow & \underline{6x + 3y = 15} \\ & & 10x = 25 \\ & & x = \dfrac{5}{2} \end{array}$$

Substitute $\frac{5}{2}$ for x in equation (2) and then solve for y.

$$2\left(\frac{5}{2}\right) + y = 5$$
$$5 + y = 5$$
$$y = 0$$

The solution is $\left(\frac{5}{2}, 0\right)$.

15. $(x+3)^2 + 10(x+3) + 24$
$= [(x+3)+4][(x+3)+6]$
$= (x+7)(x+9)$

16. a. $a \cdot a + a \cdot b + a \cdot b + b \cdot b$
$= a^2 + 2ab + b^2$

b. $(a+b)^2$

17. $\dfrac{x+2}{x^2 - x - 6} + \dfrac{x-3}{x^2 - 8x + 15}$

$$= \frac{x+2}{(x-3)(x+2)} + \frac{x-3}{(x-5)(x-3)}$$
$$= \frac{1}{(x-3)} + \frac{1}{(x-5)}$$
$$= \frac{1}{x-3} \cdot \frac{x-5}{x-5} + \frac{1}{x-5} \cdot \frac{x-3}{x-3}$$
$$= \frac{x-5+x-3}{(x-3)(x-5)}$$
$$= \frac{2x-8}{(x-3)(x-5)} \text{ or } \frac{2(x-4)}{(x-3)(x-5)}$$

18.
$$\frac{1}{a-2} = \frac{4a-1}{a^2+5a-14} + \frac{2}{a+7}$$
$$\frac{1}{a-2} = \frac{4a-1}{(a+7)(a-2)} + \frac{2}{a+7}$$
$$(a+7)(a-2)\left[\frac{1}{a-2} = \frac{4a-1}{(a+7)(a-2)} + \frac{2}{a+7}\right]$$
$$a + 7 = 4a - 1 + 2(a-2)$$
$$a + 7 = 4a - 1 + 2a - 4$$
$$a + 7 = 6a - 5$$
$$-5a + 7 = -5$$
$$-5a = -12$$
$$a = \frac{12}{5}$$

19. $w = kI^2R,\ w = 12,\ I = 2,\ R = 100$

$$12 = k(2^2)(100)$$

$$12 = 400k$$

$$\frac{12}{400} = k$$

$$k = \frac{3}{100}$$

$$w = \frac{3}{100}I^2R,\ I = 0.8,\ R = 600$$

$$w = \frac{3}{100}(0.8)^2(600)$$

$$w = 11.52$$

The wattage is 11.52 watts.

20. $$\frac{3-4i}{2+5i} = \left(\frac{3-4i}{2+5i}\right)\left(\frac{2-5i}{2-5i}\right)$$

$$= \frac{6-23i+20i^2}{4-25i^2}$$

$$= \frac{6-23i-20}{4+25}$$

$$= \frac{-14-23i}{29}$$

Chapter 9

Exercise Set 9.1

1. To find $(f \circ g)(x)$, substitute $g(x)$ for x in $f(x)$.

3. **a.** Each y has a unique x in a one-to-one function.

   b. Use the horizontal line test to determine whether a graph is one-to-one.

5. **a.** Yes; each first coordinate is paired with only one second coordinate.

   b. Yes; each second coordinate is paired with only one first coordinate.

   c. {(5, 3), (2, 4), (3, −1), (−2, 0)}
   Reverse each ordered pair.

7. The domain of f is the range of f^{-1} and the range of f is the domain of f^{-1}.

9. $f(x)=x^2+1$, $g(x)=x+2$

   a. $(f \circ g)(x)=(x+2)^2+1$
   $=x^2+4x+4+1$
   $=x^2+4x+5$

   b. $(f \circ g)(4)=4^2+4(4)+5=37$

   c. $(g \circ f)(x)=(x^2+1)+2=x^2+3$

   d. $(g \circ f)(4)=4^2+3=19$

11. $f(x)=x+3,\ g(x)=x^2+x-4$

   a. $(f \circ g)(x)=(x^2+x-4)+3=x^2+x-1$

   b. $(f \circ g)(4)=4^2+4-1=19$

   c. $(g \circ f)(x)=(x+3)^2+(x+3)-4$
   $=x^2+6x+9+x+3-4$
   $=x^2+7x+8$

   d. $(g \circ f)(4)=4^2+7(4)+8=52$

13. $f(x)=\dfrac{1}{x},\ g(x)=2x+3$

   a. $(f \circ g)(x)=\dfrac{1}{2x+3}$

   b. $(f \circ g)(4)=\dfrac{1}{2(4)+3}=\dfrac{1}{11}$

   c. $(g \circ f)(x)=2\left(\dfrac{1}{x}\right)+3=\dfrac{2}{x}+3$

   d. $(g \circ f)(4)=\dfrac{2}{4}+3=3\dfrac{1}{2}$

15. $f(x)=3x+1,\ g(x)=\dfrac{3}{x}$

   a. $(f \circ g)(x)=3\left(\dfrac{3}{x}\right)+1=\dfrac{9}{x}+1$

   b. $(f \circ g)(4)=\dfrac{9}{4}+1=3\dfrac{1}{4}$

   c. $(g \circ f)(x)=\dfrac{3}{3x+1}$

   d. $(g \circ f)(4)=\dfrac{3}{3(4)+1}=\dfrac{3}{13}$

17. $f(x)=x^2+1, g(x)=x^2+5$

   a. $(f \circ g)(x)=(x^2+5)^2+1$
   $=x^4+10x^2+25+1$
   $=x^4+10x^2+26$

   b. $(f \circ g)(4)=4^4+10(4)^2+26=442$

   c. $(g \circ f)(x)=(x^2+1)^2+5$
   $=x^4+2x^2+1+5$
   $=x^4+2x^2+6$

   d. $(g \circ f)(4)=4^4+2(4)^2+6=294$

19. $f(x) = x - 4,\ g(x) = \sqrt{x+5}, x \geq -5$

a. $(f \circ g)(x) = \sqrt{x+5} - 4$

b. $(f \circ g)(4) = \sqrt{4+5} - 4$
$= \sqrt{9} - 4$
$= 3 - 4$
$= -1$

c. $(g \circ f)(x) = \sqrt{(x-4)+5} = \sqrt{x+1}$

d. $(g \circ f)(4) = \sqrt{4+1} = \sqrt{5}$

21. This function is not a one-to-one function since it does not pass the horizontal line test.

23. This function is a one-to-one function since it passes the horizontal line test.

25. Yes, the ordered pairs represent a one-to-one function. For each value of x there is a unique value for y and each y-value has a unique x-value.

27. No, the ordered pairs do not represent a one-to-one function. For each value of x there is a unique y, but for each y-value there is not a unique x since (–4, 2) and (0, 2) are ordered pairs in the given set.

29. $y = 2x + 5$ is a line with a slope of 2 and having a y-intercept of 5. It is a one-to-one function since it passes both the vertical line test and the horizontal line test.

31. $y = x^2 - 1$ is a parabola with vertex at (0, –1). It is not a one-to-one function since it does not pass the horizontal line test. Horizontal lines above $y = -1$ intersect the graph at 2 different points.

33. $y = x^2 - 2x + 5$ is a parabola with vertex at (1, 4). It is not a one-to-one function since it does not pass the horizontal line test. Horizontal lines above $y = 4$ intersect the graph at two different points.

35. $y = x^2 - 9, x \geq 0$ is the right side of a parabola. It is a one-to-one function since it passes both the vertical line test and the horizontal line test.

37. $y = \sqrt{x}$ is a one-to-one function since it passes both the vertical line test and the horizontal line test.

39. $y = |x|$ is not a one-to-one function since it does not pass the vertical line test and the horizontal line test. Horizontal lines above $y = 0$ intersect the graph at two different points.

41. $y = \sqrt[3]{x}$ is a one-to-one function since it passes both the vertical line test and the horizontal line test.

43. For $f(x)$: Domain: {–2, –1, 2, 4, 8}
Range: {0, 4, 6, 7, 9}
For $f^{-1}(x)$: Domain: {0, 4, 6, 7, 9}
Range: {–2, –1, 2, 4, 8}

45. For $f(x)$: Domain: { –1, 1, 2, 4}
Range: {–3, –1, 0, 2}
For $f^{-1}(x)$: Domain: {–3, –1, 0, 2}
Range: {–1, 1, 2, 4}

47. For $f(x)$: Domain: $\{x | x \geq 2\}$
Range: $\{y | y \geq 0\}$
For $f^{-1}(x)$: Domain: $\{x | x \geq 0\}$
Range: $\{y | y \geq 2\}$

49. **a.** Yes, $f(x) = x - 2$ is a one-to-one function.

b.
$$y = x - 2$$
$$x = y - 2$$
$$x + 2 = y$$
$$y = x + 2$$
$$f^{-1}(x) = x + 2$$

51. **a.** Yes, $h(x) = 4x$ is a one-to-one function.

b.
$$y = 4x$$
$$x = 4y$$
$$y = \frac{x}{4}$$
$$h^{-1}(x) = \frac{x}{4}$$

53. a. No, $p(x) = 3x^2$ is not a one-to-one function.

b. Does not exist

55. a. No; $t(x) = x^2 + 3$ is not a one-to-one function.

b. Does not exist

57. a. Yes; $g(x) = \frac{1}{x}$ is a one-to-one function.

b.
$$y = \frac{1}{x}$$
$$x = \frac{1}{y}$$
$$y = \frac{1}{x}$$
$$g^{-1}(x) = \frac{1}{x}$$

59. a. No; $f(x) = x^2 + 10$ is not a one-to-one function.

b. Does not exist

61. a. Yes, $g(x) = x^3 - 6$ is a one-to-one function.

b.
$$y = x^3 - 6$$
$$x = y^3 - 6$$
$$x + 6 = y^3$$
$$\sqrt[3]{x+6} = y$$
$$g^{-1}(x) = \sqrt[3]{x+6}$$

63. a. Yes, $g(x) = \sqrt{x+2}$, $x \geq -2$ is a one-to-one function.

b.
$$y = \sqrt{x+2}$$
$$x = \sqrt{y+2}$$
$$x^2 = y + 2$$
$$x^2 - 2 = y$$
$$g^{-1}(x) = x^2 - 2, x \geq 0$$

65. a. Yes, $h(x) = x^2 - 4, x \geq 0$ is one-to-one.

b.
$$y = x^2 - 4$$
$$x = y^2 - 4$$
$$x + 4 = y^2$$
$$y = \sqrt{x+4}$$
$$h^{-1}(x) = \sqrt{x+4}, x \geq -4$$

67. $f(x) = 2x + 8$

a.
$$y = 2x + 8$$
$$x = 2y + 8$$
$$x - 8 = 2y$$
$$\frac{x-8}{2} = y$$
$$f^{-1}(x) = \frac{x-8}{2}$$

b.

x	$f(x)$
0	8
−4	0

x	$f^{-1}(x)$
0	−4
8	0

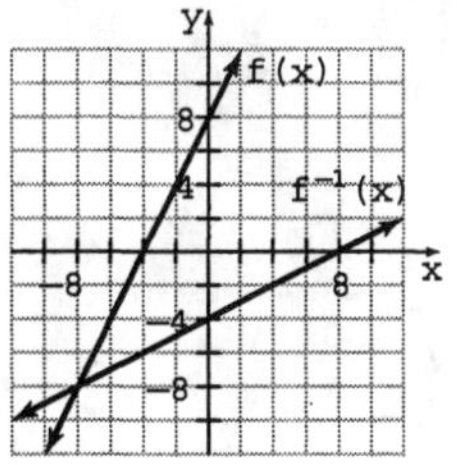

69. $f(x) = \sqrt{x}, x \geq 0$

a.
$$y = \sqrt{x}$$
$$x = \sqrt{y}$$
$$x^2 = \left(\sqrt{y}\right)^2$$
$$x^2 = y$$
$$f^{-1}(x) = x^2 \text{ for } x \geq 0$$

b.

x	$f(x)$
0	0
1	1
4	2

x	$f^{-1}(x)$
0	0
1	1
2	4

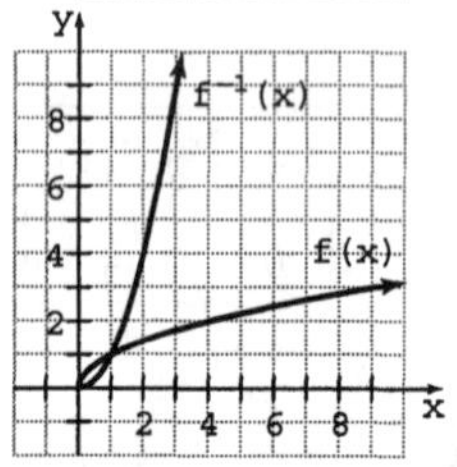

71. $f(x) = \sqrt{x-1}, x \geq 1$

a.
$$y = \sqrt{x-1}$$
$$x = \sqrt{y-1}$$
$$x^2 = \left(\sqrt{y-1}\right)^2$$
$$x^2 = y-1$$
$$x^2 + 1 = y$$
$$f^{-1}(x) = x^2 + 1 \text{ for } x \geq 0$$

b.

x	$f(x)$
1	0
2	1
5	2

x	$f^{-1}(x)$
0	1
1	2
2	5

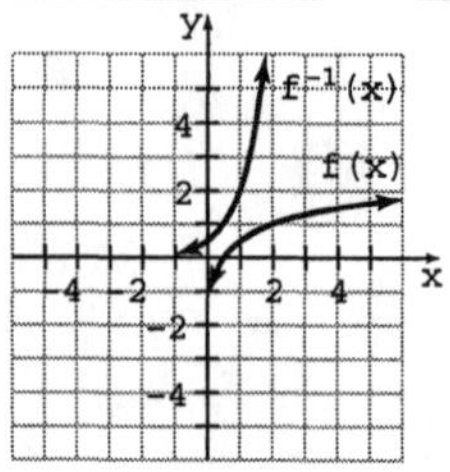

73. $f(x) = \sqrt[3]{x}$

a.
$$y = \sqrt[3]{x}$$
$$x = \sqrt[3]{y}$$
$$x^3 = \left(\sqrt[3]{y}\right)^3$$
$$x^3 = y$$
$$f^{-1}(x) = x^3$$

b.

x	$f(x)$
−8	−2
−1	−1
0	0
1	1
8	2

x	$f^{-1}(x)$
−2	−8
−1	−1
0	0
1	1
2	8

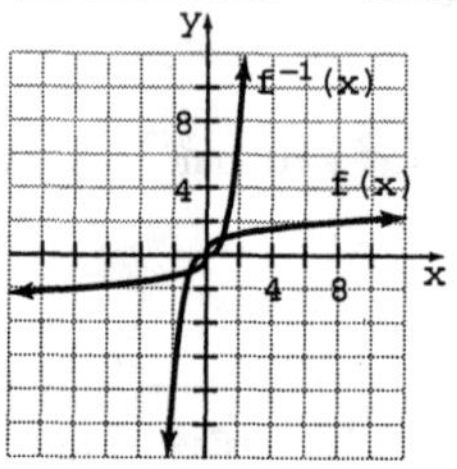

75. $f(x) = \dfrac{1}{x}, x > 0$

a.
$$y = \frac{1}{x}$$
$$x = \frac{1}{y}$$
$$xy = 1$$
$$y = \frac{1}{x}$$
$$f^{-1}(x) = \frac{1}{x}, x > 0$$

b.

x	$f(x)$
$\frac{1}{2}$	2
1	1
3	$\frac{1}{3}$

x	$f^{-1}(x)$
2	$\frac{1}{2}$
1	1
$\frac{1}{3}$	3

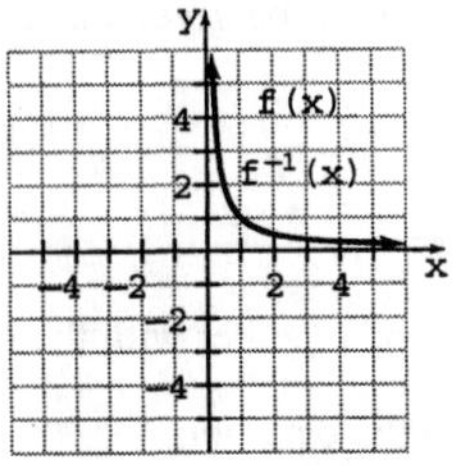

77. $(f \circ f^{-1})(x) = (x+8) - 8 = x$

$(f^{-1} \circ f)(x) = (x-8) + 8 = x$

79. $(f \circ f^{-1})(x) = \frac{1}{2}(2x-6) + 3$

$= x - 3 + 3$

$= x$

$(f^{-1} \circ f)(x) = 2\left(\frac{1}{2}x + 3\right) - 6$

$= x + 6 - 6$

$= x$

81. $(f \circ f^{-1})(x) = \sqrt[3]{(x^3+2) - 2}$

$= \sqrt[3]{x^3}$

$= x$

$(f^{-1} \circ f)(x) = \left(\sqrt[3]{x-2}\right)^3 + 2$

$= x - 2 + 2$

$= x$

83. $(f \circ f^{-1})(x) = \frac{3}{\frac{3}{x}} = 3 \cdot \frac{x}{3} = x$

$(f^{-1} \circ f)(x) = \frac{3}{\frac{3}{x}} = 3 \cdot \frac{x}{3} = x$

85. No, composition of functions is not commutative.
Let $f(x) = x^2$ and $g(x) = x + 1$.
Then $(f \circ g)(x) = (x+1)^2 = x^2 + 2x + 1$ while $(g \circ f)(x) = x^2 + 1$.

87. a. $(f \circ g)(x) = f[g(x)]$

$= \left(\sqrt[3]{x-2}\right)^3 + 2$

$= x - 2 + 2$

$= x$

$(g \circ f)(x) = g[f(x)]$

$= \sqrt[3]{\left(x^3+2\right) - 2}$

$= \sqrt[3]{x^3}$

$= x$

b. The domain of *f* is all real numbers and the domain of *g* is all real numbers. The domains of $(f \circ g)(x)$ and $(g \circ f)(x)$ are also all real numbers.

89. The range of $f^{-1}(x)$ is the domain of $f(x)$.

91. $f(x) = 3x$ converts yards, x, into feet, y.

$y = 3x$

$x = 3y$

$\frac{x}{3} = y$

$f^{-1}(x) = \frac{x}{3}$

Here, x is feet and $f^{-1}(x)$ is yards. The inverse function converts feet to yards.

93. $f(x) = \frac{5}{9}(x - 32)$ where x is degrees Fahrenheit and $f(x)$ is degrees Celsius.

$y = \frac{5}{9}(x-32)$

$x = \frac{5}{9}(y-32)$

$\frac{9}{5}x = \frac{9}{5}\left[\frac{5}{9}(y-32)\right]$

$\frac{9}{5}x = y - 32$

$\frac{9}{5}x + 32 = y$

$f^{-1}(x) = \frac{9}{5}x + 32$

Here, x is degrees Celsius and $f^{-1}(x)$ is degrees Fahrenheit. The inverse function converts Celsius to Fahrenheit.

95. $f(x)=16x$; $g(x)=28.35x$

$(f \circ g)(x)=16(28.35x)=453.6x$

In this composition, x represents pounds and $(f \circ g)(x)$ represents grams. The composition converts pounds to grams.

97. $f(x)=3x$; $g(x)=0.305x$

$(f \circ g)(x)=3(0.305x)=0.915x$

In this composition, x represents yards and $(f \circ g)(x)$ represents meters. The composition converts yards to meters.

99.

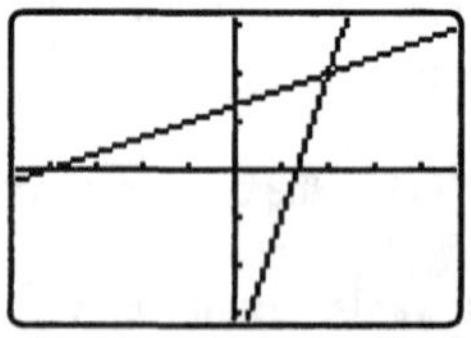

-4.7, 4.7, 1, -3.1, 3.1, 1

Yes, the functions are inverses.

101.

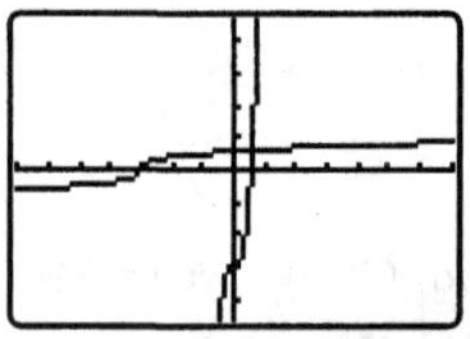

-28.2, 28.2, 4, -18.6, 18.6, 4

Yes, the functions are inverses.

103. **a.** $r(3)=2(3)=6$

The radius is 6 feet.

b. $A=\pi r^2$

$A=\pi(6)^2$

$A=36\pi \approx 113.10$

The surface area is $36\pi \approx 113.10$ square feet.

c. $(A \circ r)(t)=\pi(2t)^2=\pi(4t^2)=4\pi t^2$

d. $4\pi(3)^2=4\pi(9)=36\pi$

e. The answers should agree.

106. $\left|\frac{-9}{4}\right| \div \left|\frac{-4}{9}\right| = \left|-\frac{9}{4}\right| \div \left|-\frac{4}{9}\right|$

$=\frac{9}{4} \div \frac{4}{9} = \frac{9}{4} \cdot \frac{9}{4}$

$=\frac{81}{16}$

107. First find the slope of the given line.

$2x+3y-9=0$

$3y=-2x+9$

$y=-\frac{2}{3}x+3 \Rightarrow m=-\frac{2}{3}$

Now use this slope together with the given point $\left(\frac{1}{2}, 3\right)$ to find the equation.

point-slope form:

$y-y_1=m(x-x_1)$

$y-3=-\frac{2}{3}\left(x-\frac{1}{2}\right)$

$y-3=-\frac{2}{3}x+\frac{1}{3}$

$3\left(y-3=-\frac{2}{3}x+\frac{1}{3}\right)$

$3y-9=-2x+1$

$2x+3y=10$

108. $\dfrac{\frac{3}{x^2}-\frac{2}{x}}{\frac{x}{6}} = \dfrac{6x^2\left(\frac{3}{x^2}-\frac{2}{x}\right)}{6x^2\left(\frac{x}{6}\right)} = \dfrac{18-12x}{x^3}$

109. $\frac{1}{f}=\frac{1}{p}+\frac{1}{q}$ for p

$fpq\left(\frac{1}{f}=\frac{1}{p}+\frac{1}{q}\right)$

$pq=fq+fp$

$pq-fp=fq$

$p(q-f)=fq$

$p=\frac{fq}{q-f}$

110. $x^2+2x-10=0$

$x^2+2x=10$

$x^2+2x+1=10+1$

$(x+1)^2=11$

$x+1=\pm\sqrt{11}$

$x=-1\pm\sqrt{11}$

Exercise Set 9.2

1. Exponential functions are functions of the form $f(x)=a^x, a>0, a\neq 1$.

3. a. $y=\left(\frac{1}{2}\right)^x$; as x increases, y decreases.

b. No, y can never be zero because $\left(\frac{1}{2}\right)^x$ can never be 0.

c. No, y can never be negative because $\left(\frac{1}{2}\right)^x$ is never negative.

5. $y=2^x$ and $y=3^x$

a. Let $x=0$
$y=2^0 \quad y=3^0$
$y=1 \quad y=1$
They have the same y-intercept at (0, 1).

b. $y=3^x$ will be steeper than $y=2^x$ for $x>0$ because it has a larger base.

7. $y=2^x$

x	−2	−1	0	1	2
y	$\frac{1}{4}$	$\frac{1}{2}$	1	2	4

Domain: $\mathbb{R}$; Range: $\{y\mid y>0\}$

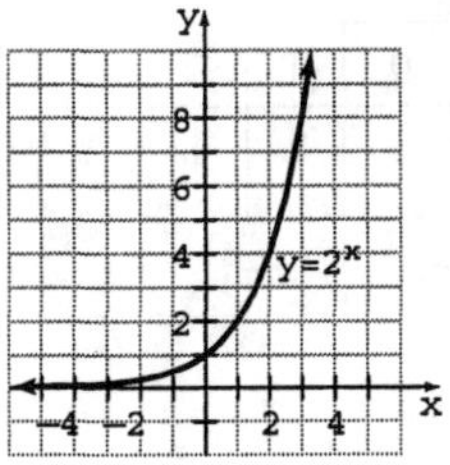

9. $y=\left(\frac{1}{2}\right)^x$

x	−2	−1	0	1	2
y	4	2	1	$\frac{1}{2}$	$\frac{1}{4}$

Domain: $\mathbb{R}$; Range: $\{y\mid y>0\}$

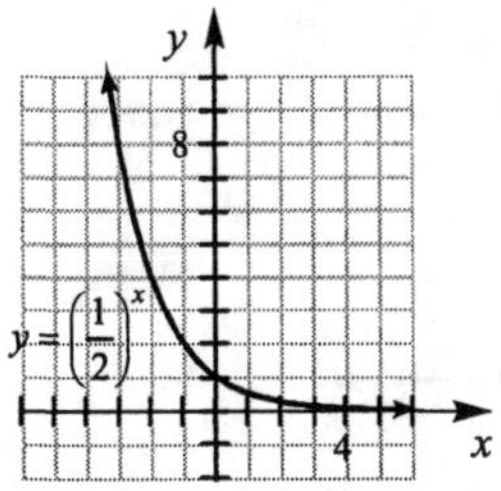

11. $y=4^x$

x	−2	−1	0	1	2
y	$\frac{1}{16}$	$\frac{1}{4}$	1	4	16

Domain: $\mathbb{R}$; Range: $\{y\mid y>0\}$

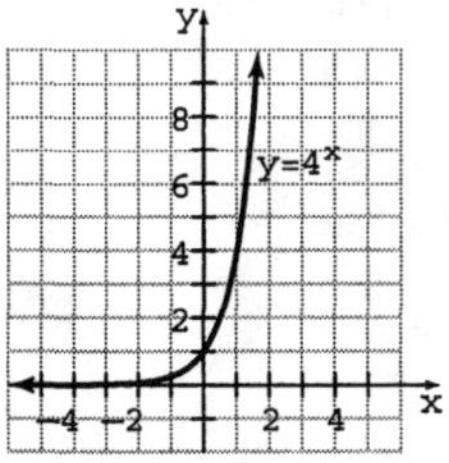

13. $y=\left(\frac{1}{4}\right)^x$

x	−2	−1	0	1	2
y	16	4	1	$\frac{1}{4}$	$\frac{1}{16}$

Domain: $\mathbb{R}$; Range: $\{y\mid y>0\}$

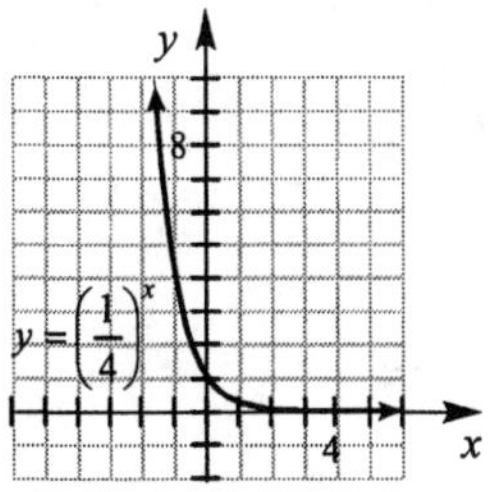

15. $y = 3^{-x} = \dfrac{1}{3^x} = \left(\dfrac{1}{3}\right)^x$

x	-2	-1	0	1	2
y	9	3	1	$\frac{1}{3}$	$\frac{1}{9}$

Domain: $\mathbb{R}$; Range: $\{y \mid y > 0\}$

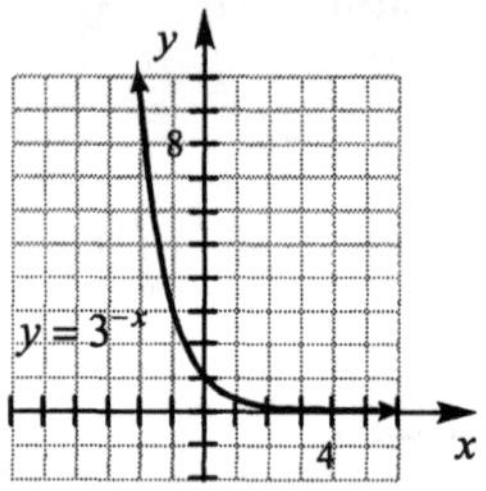

21. $y = \left(\dfrac{1}{3}\right)^{x+1}$

x	-3	-2	-1	0	1
y	9	3	1	$\frac{1}{3}$	$\frac{1}{9}$

Domain: $\mathbb{R}$; Range: $\{y \mid y > 0\}$

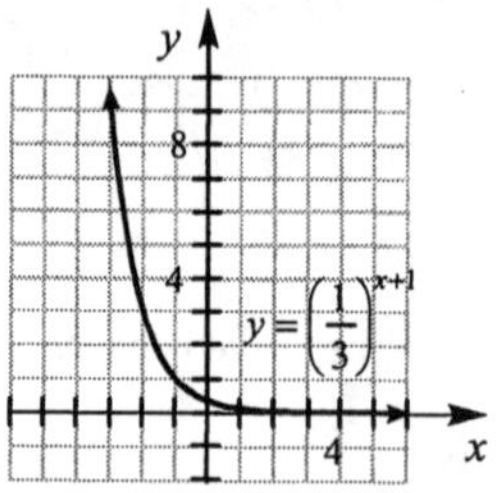

17. $y = \left(\dfrac{1}{3}\right)^{-x} = 3^x$

x	-2	-1	0	1	2
y	$\frac{1}{9}$	$\frac{1}{3}$	1	3	9

Domain: $\mathbb{R}$; Range: $\{y \mid y > 0\}$

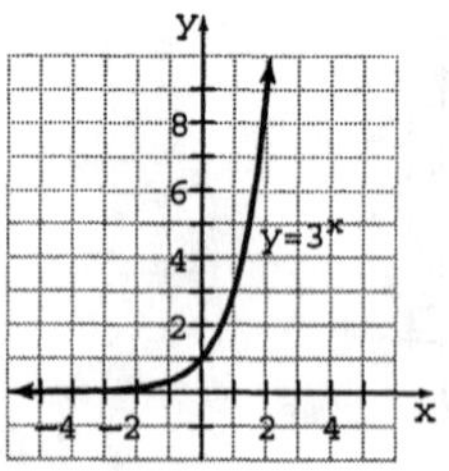

23. $y = 2^x + 1$

x	-2	-1	0	1	2	3
y	$\frac{5}{4}$	$\frac{3}{2}$	2	3	5	9

Domain: $\mathbb{R}$; Range: $\{y \mid y > 1\}$

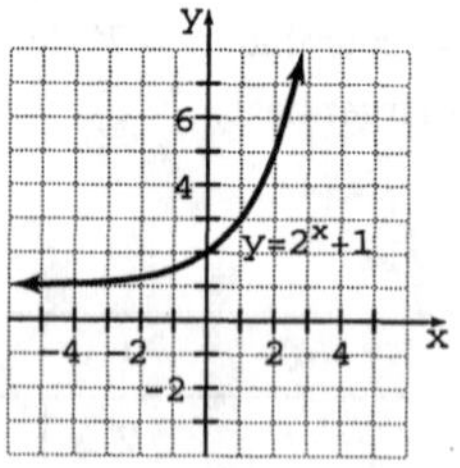

19. $y = 2^{x-1}$

x	-2	0	2	4	6
y	$\frac{1}{8}$	$\frac{1}{2}$	2	8	32

Domain: $\mathbb{R}$; Range: $\{y \mid y > 0\}$

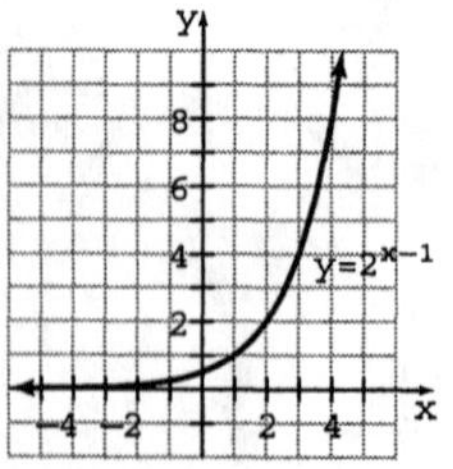

25. $y = 3^x - 1$

x	-2	-1	0	1	2
y	$-\frac{8}{9}$	$-\frac{2}{3}$	0	2	8

Domain: $\mathbb{R}$; Range: $\{y \mid y > -1\}$

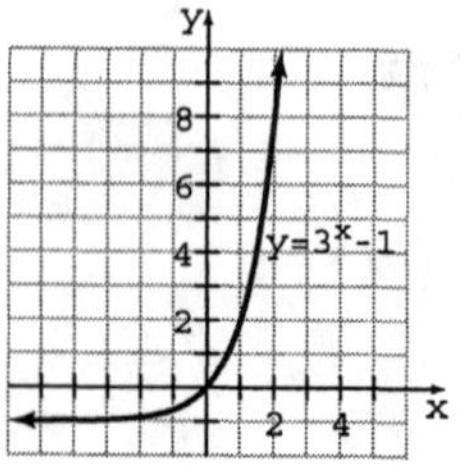

27. a. The graph is the horizontal line through $y = 1$.

b. Yes. A horizontal line will pass the vertical line test.

c. No. $f(x)$ is not one-to-one and therefore does not have an inverse function.

29. The graph of $y = a^x - k$ will have the same basic shape as the graph of $y = a^x$. However, the graph of $y = a^x - k$ will be k units lower than that of $y = a^x$.

31. The graph of $y = a^{x+2}$ is the graph of $y = a^x$ shifted 2 units to the left.

33. a. For 2060 we have $t = 2060 - 1960 = 100$.

$f(100) = 0.592(1.042)^{100} \approx 36.232$

The function estimates that in 2060 there will be 36.232 million people in the U.S. aged 85 or older.

b. For 2100 we have $t = 2100 - 1960 = 140$.

$f(140) = 0.592(1.042)^{140} \approx 187.846$

The function estimates that in 2100 there will be 187.846 million people in the U.S. aged 85 or older.

35. The amount each day is given by the function $A(d) = 2^d$ where d is the number of days.

$A(9) = 2^9 = 512$

After 9 days, the amount would be \$512.

37. a. About 14 years

b. About 10 years

c. From the graph, the difference is about \$25. Using the formulas given, we get:

$A = 100(1.07)^{10} = 196.72$ (exponential)

$A = 100 + 100(.07)(10) = 170$ (linear)

The difference is \$26.72.

d. For daily compounding we would get

$$A = 100\left(1 + \frac{0.07}{365}\right)^{(365 \cdot 10)} = 201.36$$

This is about a \$5 increase over compounding annually. Daily compounding increases the amount.

39. $N(t) = 5(3)^t, t = 2$

$N(2) = 5(3)^2 = 5 \cdot 9 = 45$

There will be 45 bacteria in the petri dish after two days.

41. $A = p\left(1 + \frac{r}{n}\right)^{nt}$.

Use $p = 5000$, $r = 6\% = 0.06$ and $n = 4$ and $t = 4$.

$$A = 5000\left(1 + \frac{0.06}{4}\right)^{4 \cdot 4}$$

$A = 5000(1 + 0.015)^{16}$

$A = 5000(1.015)^{16}$

$A \approx 5000(1.2689855)$

$A \approx 6344.93$

He has \$6344.93 after 4 years.

43. $A = A_0 2^{-t/5600}$

Use $A_0 = 12$ and $t = 1000$.

$A = 12(2^{-1000/5600})$

$A \approx 12(2^{-0.18})$

$A \approx 12(0.88)$

$A \approx 10.6$ grams

There are about 10.6 grams left.

45. $y = 80(2)^{-0.4t}$

a. $t = 10$

$y = 80(2)^{-0.4(10)}$

$$y = 80(2)^{-4} = 80\left(\frac{1}{16}\right) = 5$$

After 10 years, 5 grams remain.

b. $t = 100$

$y = 80(2)^{-0.4(100)}$

$y = 80(2)^{-40}$

$y \approx 80(9.094947 \times 10^{-13})$

$y \approx 7.28 \times 10^{-11}$

After 100 years, about 7.28×10^{-11} grams are left.

47. $y = 2000(1.2)^{0.1t}$

a. $t = 10$

$y = 2000(1.2)^{0.1(10)}$

$y = 2000(1.2)^1$

$y = 2400$

In 10 years, the population is expected to be 2400.

b. $t = 50$

$y = 2000(1.2)^{0.1(50)}$

$y = 2000(1.2)^5$

$y = 2000\left(2.48832\right)$

$y \approx 4977$

In 50 years, the population is expected to be about 4,977.

49. $V\left(t\right) = 24,000(0.82)^t$, $t = 4$

$V\left(t\right) = 24,000(0.82)^4 \approx 10,850.92$

The SUV will be worth about $10,850.92 in 4 years.

51. a. Answers will vary. One possible answer is: Since the amount is reduced by 5%, the consumption is 95% of the previous year, or 0.95. Thus, $A(t) = 580,000(0.95)^t$.

b. $t = 2009 - 2005 = 4$

$A(4) = 580,000(0.95)^4$

$\approx 472,414$

The expected average use in 2009 is about 472,414 gallons.

53. $A = 41.97(0.996)^x$

$A(389) = 41.97(0.996)^{389}$

$A \approx 8.83$

The altitude at the top of Mt. Everest is about 8.83 kilometers.

55. a. $A = p\left(1 + \dfrac{r}{n}\right)^{nt}$

$A = 100\left(1 + \dfrac{0.07}{365}\right)^{365 \cdot 10}$

$A \approx 100(1.0001918)^{3650}$

$A = 201.36$

The amount is $201.36.

b. For simple interest,

$A = 100 + 100(0.07)t$

$A = 100 + 100(0.07)(10)$

$A = 100 + 70$

$A = 170$

$\$201.36 - \$170 = \$31.36$

57. a. $y_1 = 3^{x-5}$

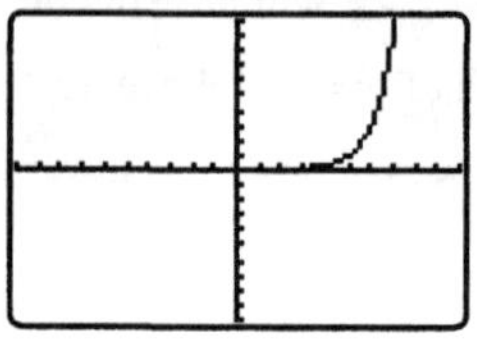

-10, 10, 1, -10, 10, 1

b. $4 = 3^{x-5}$ when $x \approx 6.26$.

59. a. Day 15: $2^{15-1} = 2^{14} = \$16,384$

b. Day 20: $2^{20-1} = 2^{19} = \$524,288$

c. nth Day: 2^{n-1}

d. Day 30: $2^{30-1} = \$2^{29} = \$536,870,912$

e. $2^0 + 2^1 + 2^2 + \cdots + 2^{29}$

61. a. $2.3x^4y - 6.2x^6y^2 + 9.2x^5y^2$

$= -6.2x^6y^2 + 9.2x^5y^2 + 2.3x^4y$

b. $-6.2x^6y^2$ is the leading term.

$6 + 2 = 8$ is the degree of the polynomial.

c. $-6.2x^6y^2$ is the leading term, so -6.2 is the leading coefficient.

62. $(f \cdot g)(x) = f(x) \cdot g(x)$

$= (x+5)(x^2 - 2x + 4)$

$= x^3 - 2x^2 + 4x + 5x^2 - 10x + 20$

$= x^3 + 3x^2 - 6x + 20$

63. $\sqrt{a^2 - 8a + 16} = \sqrt{(a-4)^2} = |a-4|$

64. $\sqrt[4]{\frac{32x^5y^9}{2y^3z}} = \sqrt[4]{\frac{16x^5y^6}{z}}$

$= \frac{\sqrt[4]{16x^5y^6}}{\sqrt[4]{z}}$

$= \frac{\sqrt[4]{16x^4y^4 \cdot xy^2}}{\sqrt[4]{z}}$

$= \frac{2xy\sqrt[4]{xy^2}}{\sqrt[4]{z}} \cdot \frac{\sqrt[4]{z^3}}{\sqrt[4]{z^3}}$

$= \frac{2xy\sqrt[4]{xy^2z^3}}{\sqrt[4]{z^4}}$

$= \frac{2xy\sqrt[4]{xy^2z^3}}{z}$

Exercise Set 9.3

1. $y = \log_a x$

a. The base a must be positive and must not be equal to one.

b. The argument x represents a number that is greater than 0. Thus, the domain is $\{x | x > 0\}$.

c. $\mathbb{R}$

3. The functions $f(x) = a^x$ and $g(x) = \log_a x$ are inverse functions. Therefore, some of the points on the function are $g(x) = \log_a x$ are $\left(\frac{1}{27}, -3\right)\left(\frac{1}{9}, -2\right), \left(\frac{1}{3}, -1\right)$(1, 0), (3, 1), (9, 2), and (27, 3). These points were obtained by switching the coordinates of the given points for the graph of $f(x) = a^x$.

5. The functions $y = a^x$ and $y = \log_a x$ for $a \neq 1$ are inverses of each other, thus the graphs are symmetric with respect to the line $y = x$. For each ordered pair (x, y) on the graph of $y = a^x$, the ordered pair (y, x) is on the graph of $y = \log_a x$.

7. $y = \log_2 x$

Convert to exponential form.

$2^y = x$

x	$\frac{1}{4}$	$\frac{1}{2}$	1	2	4
y	−2	−1	0	1	2

Domain: $\{x | x > 0\}$

Range: $\mathbb{R}$

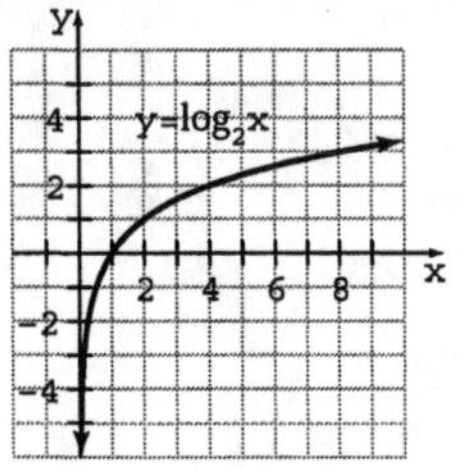

9. $y = \log_{1/2} x$

Convert to exponential form.

$x = \left(\frac{1}{2}\right)^y$

x	4	2	1	$\frac{1}{2}$	$\frac{1}{4}$
y	−2	−1	0	1	2

Domain: $\{x | x > 0\}$

Range: $\mathbb{R}$

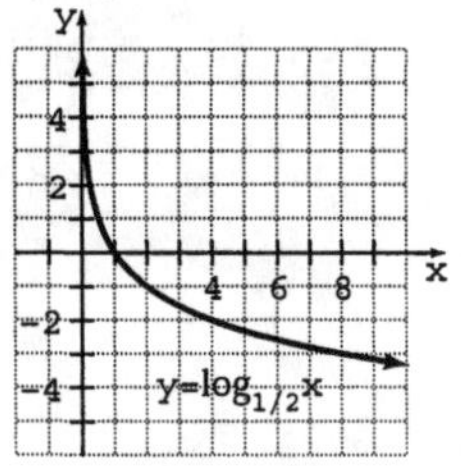

11. $y = \log_5 x$

Convert to the exponential form.

$x = 5^y$

x	$\frac{1}{25}$	$\frac{1}{5}$	1	5	25
y	-2	-1	0	1	2

Domain: $\{x \mid x > 0\}$

Range: $\mathbb{R}$

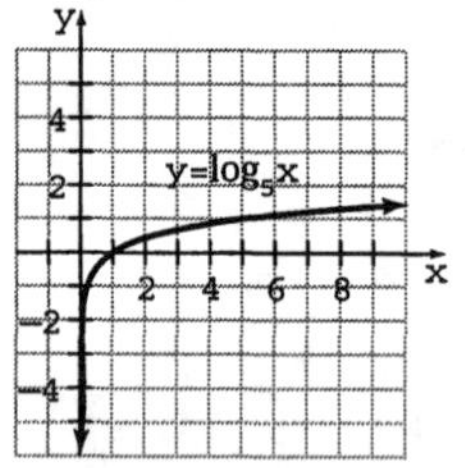

13. $y = \log_{1/5} x$ Convert to exponential form.

$$x = \left(\frac{1}{5}\right)^y$$

x	25	5	1	$\frac{1}{5}$	$\frac{1}{25}$
y	-2	-1	0	1	2

Domain: $\{x \mid x > 0\}$

Range: $\mathbb{R}$

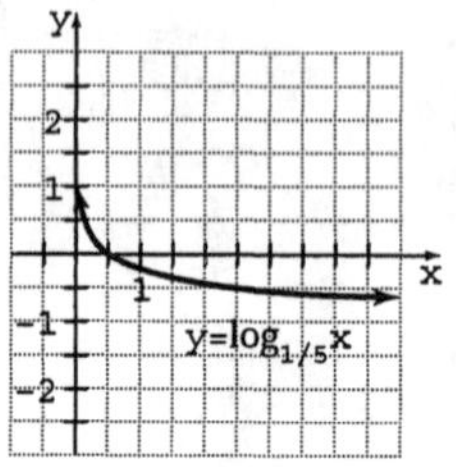

15. $y = 2^x$

x	-2	-1	0	1	2
y	$\frac{1}{4}$	$\frac{1}{2}$	1	2	4

$y = \log_{1/2} x$

Convert to exponential form.

$$x = \left(\frac{1}{2}\right)^y$$

x	4	2	1	$\frac{1}{2}$	$\frac{1}{4}$
y	-2	-1	0	1	2

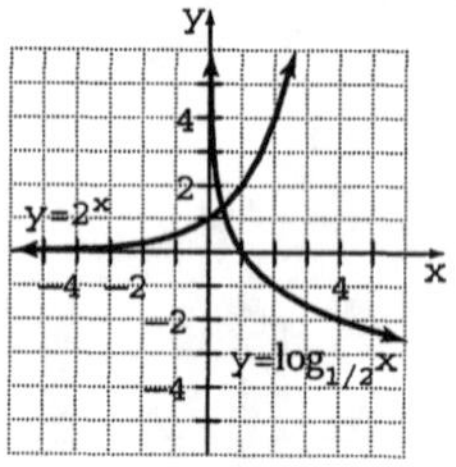

17. $y = 2^x$

x	-2	-1	0	1	2
y	$\frac{1}{4}$	$\frac{1}{2}$	1	2	4

$y = \log_2 x$

Convert to exponential form.

$x = 2^y$

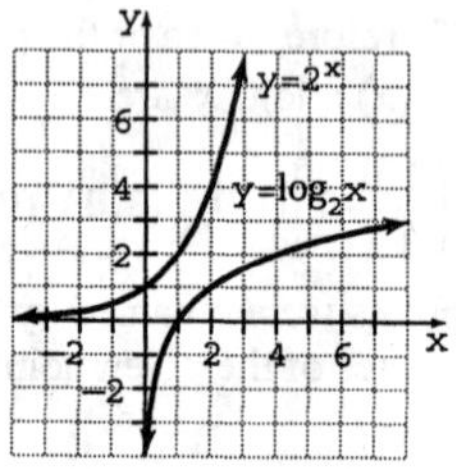

19. $2^3 = 8$

$\log_2 8 = 3$

21. $3^2 = 9$

$\log_3 9 = 2$

23. $16^{1/2} = 4$

$\log_{16} 4 = \frac{1}{2}$

25. $8^{1/3} = 2$

$\log_8 2 = \frac{1}{3}$

27. $\left(\frac{1}{2}\right)^5 = \frac{1}{32}$

$\log_{1/2}\left(\frac{1}{32}\right) = 5$

29. $2^{-3} = \frac{1}{8}$

$\log_2 \frac{1}{8} = -3$

31. $4^{-3} = \frac{1}{64}$

$\log_4 \frac{1}{64} = -3$

33. $64^{1/3} = 4$

$\log_{64} 4 = \frac{1}{3}$

35. $8^{-1/3} = \frac{1}{2}$

$\log_8 \frac{1}{2} = -\frac{1}{3}$

37. $81^{-1/4} = \frac{1}{3}$

$\log_{81} \frac{1}{3} = -\frac{1}{4}$

39. $10^{0.8451} = 7$

$\log_{10} 7 = 0.8451$

41. $e^2 = 7.3891$

$\log_e 7.3891 = 2$

43. $a^n = b$

$\log_a b = n$

45. $\log_2 8 = 3$

$2^3 = 8$

47. $\log_{1/3} \frac{1}{27} = 3$

$\left(\frac{1}{3}\right)^3 = \frac{1}{27}$

49. $\log_5 \frac{1}{25} = -2$

$5^{-2} = \frac{1}{25}$

51. $\log_{49} 7 = \frac{1}{2}$

$49^{1/2} = 7$

53. $\log_9 \frac{1}{81} = -2$

$9^{-2} = \frac{1}{81}$

55. $\log_{10} \frac{1}{1000} = -3$

$10^{-3} = \frac{1}{1000}$

57. $\log_6 216 = 3$

$6^3 = 216$

59. $\log_{10} 0.62 = -0.2076$

$10^{-0.2076} = 0.62$

61. $\log_e 6.52 = 1.8749$

$e^{1.8749} = 6.52$

63. $\log_w s = -p$

$w^{-p} = s$

65. $\log_4 64 = y$

$4^y = 64$

$4^y = 4^3$

$y = 3$

67. $\log_a 125 = 3$

$a^3 = 125$

$a^3 = 5^3$

$a = 5$

69. $\log_3 x = 3$

$3^3 = x$

$27 = x$

71. $\log_2 \frac{1}{16} = y$

$2^y = \frac{1}{16}$

$2^y = 2^{-4}$

$y = -4$

73. $\log_{1/2} x = 6$

$\left(\frac{1}{2}\right)^6 = x$

$\frac{1}{64} = x$

75. $\log_a \frac{1}{27} = -3$

$a^{-3} = \frac{1}{27}$

$a^{-3} = 3^{-3}$

$a = 3$

77. $\log_{10} 1 = 0$ because $10^0 = 1$

79. $\log_{10} 100 = 2$ because $10^2 = 100$

81. $\log_{10} \frac{1}{100} = -2$ because $10^{-2} = \frac{1}{10^2} = \frac{1}{100}$

83. $\log_{10} 10{,}000 = 4$ because $10^4 = 10{,}000$

85. $\log_4 256 = 4$ because $4^4 = 256$

87. $\log_3 \frac{1}{81} = -4$ because $3^{-4} = \frac{1}{3^4} = \frac{1}{81}$

89. $\log_8 \frac{1}{64} = -2$ because $8^{-2} = \frac{1}{8^2} = \frac{1}{64}$

91. $\log_9 1 = 0$ because $9^0 = 1$

93. $\log_9 9 = 1$ because $9^1 = 9$

95. $\log_4 1024 = 5$ because $4^5 = 1024$

97. If $f(x) = 5^x$, then $f^{-1}(x) = \log_5 x$.

99. $\log_3 62$ lies between 3 and 4 since 62 lies between $3^3 = 27$ and $3^4 = 81$.

101. $\log_{10} 425$ lies between 2 and 3 since 425 lies between $10^2 = 100$ and $10^3 = 1000$.

103. For $x > 1$, 2^x will grow faster than $\log_{10} x$. Note that when $x = 10$, $2^x = 1024$ while $\log_{10} x = 1$.

105. $x = \log_{10} 10^6$

$10^x = 10^6$

$x = 6$

107. $x = \log_b b^8$

$b^x = b^8$

$x = 8$

109. $x = 10^{\log_{10} 3}$

$\log_{10} x = \log_{10} 3$

$x = 3$

111. $x = b^{\log_b 9}$

$\log_b x = \log_b 9$

$x = 9$

113. $R = \log_{10} I$

$7 = \log_{10} I$

$10^7 = I$

$I = 10{,}000{,}000$

The earthquake is 10,000,000 times more intense than the smallest measurable activity.

115.

$$\begin{aligned} R &= \log_{10} I & R &= \log_{10} I \\ 6 &= \log_{10} I & 2 &= \log_{10} I \\ 10^6 &= I & 10^2 &= I \\ 1{,}000{,}000 &= I & 100 &= I \end{aligned}$$

$$\frac{1{,}000{,}000}{100} = 10{,}000$$

An earthquake that measures 6 is 10,000 times more intense than one that measures 2.

117. $y = \log_2(x-1)$ or $2^y = x - 1$

x	$1\frac{1}{4}$	$1\frac{1}{2}$	2	3	5
y	−2	−1	0	1	2

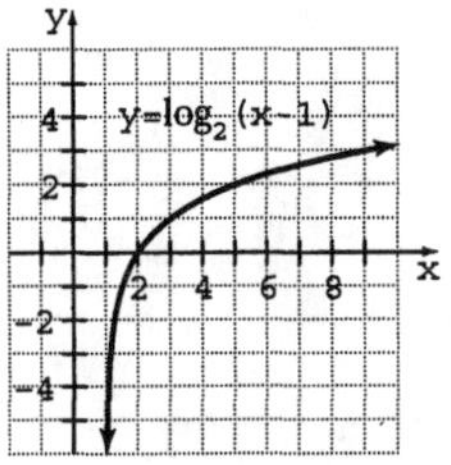

119. $2x^3 - 6x^2 - 36x = 2x\left(x^2 - 3x - 18\right)$
$= 2x(x+3)(x-6)$

120. $x^4 - 16 = \left(x^2 - 4\right)\left(x^2 + 4\right)$
$= (x-2)(x+2)\left(x^2+4\right)$

121. $40x^2 + 52x - 12 = 4\left(10x^2 + 13x - 3\right)$
$= 4(2x+3)(5x-1)$

122. $6r^2s^2 + rs - 1 = (3rs - 1)(2rs + 1)$

Exercise Set 9.4

1. Answers will vary.

3. Answers will vary.

5. Yes. This is true because of the product rule for logarithms.

7. $\log_4(3 \cdot 10) = \log_4 3 + \log_4 10$

9. $\log_8 7(x+3) = \log_8 7 + \log_8(x+3)$

11. $\log_2 \frac{27}{11} = \log_2 27 - \log_2 11$

13.
$$\begin{aligned} \log_{10} \frac{\sqrt{x}}{x-9} &= \log_{10} \frac{x^{1/2}}{x-9} \\ &= \log_{10} x^{1/2} - \log_{10}(x-9) \\ &= \frac{1}{2}\log_{10} x - \log_{10}(x-9) \end{aligned}$$

15. $\log_6 x^7 = 7\log_6 x$

17. $\log_4(r+7)^5 = 5\log_4(r+7)$

19.
$$\begin{aligned} \log_4 \sqrt{\frac{a^3}{a+2}} &= \log_4 \left(\frac{a^3}{a+2}\right)^{1/2} \\ &= \frac{1}{2}\log_4 \frac{a^3}{a+2} \\ &= \frac{1}{2}[\log_4 a^3 - \log_4(a+2)] \\ &= \frac{1}{2}[3\log_4 a - \log_4(a+2)] \\ &= \frac{3}{2}\log_4 a - \frac{1}{2}\log_4(a+2) \end{aligned}$$

21.
$$\begin{aligned} \log_3 \frac{d^6}{(a-8)^4} &= \log_3 d^6 - \log_3(a-8)^4 \\ &= 6\log_3 d - 4\log_3(a-8) \end{aligned}$$

23.
$$\begin{aligned} \log_8 \frac{y(y+4)}{y^3} &= \log_8 y + \log_8(y+4) - \log_8 y^3 \\ &= \log_8 y + \log_8(y+4) - 3\log_8 y \\ &= \log_8(y+4) - 2\log_8 y \end{aligned}$$

25.
$$\begin{aligned} \log_{10} \frac{9m}{8n} &= \log_{10} 9m - \log_{10} 8n \\ &= \log_{10} 9 + \log_{10} m - (\log_{10} 8 + \log_{10} n) \\ &= \log_{10} 9 + \log_{10} m - \log_{10} 8 - \log_{10} n \end{aligned}$$

27. $\log_5 2 + \log_5 8 = \log_5(2 \cdot 8) = \log_5 16$

29. $\log_2 9 - \log_2 5 = \log_2 \frac{9}{5}$

31. $6\log_4 2 = \log_4 2^6 = \log_4 64$

33. $\log_{10} x + \log_{10}(x+3) = \log_{10} x(x+3)$

35. $2\log_9 z - \log_9(z-2) = \log_9 z^2 - \log_9(z-2)$
$= \log_9 \frac{z^2}{z-2}$

37. $4(\log_5 p - \log_5 3) = 4\log_5 \frac{p}{3}$
$= \log_5 \left(\frac{p}{3}\right)^4$

39. $\log_2 n + \log_2(n+4) - \log_2(n-3)$
$\log_2 n(n+4) - \log_2(n-3)$
$\log_2 \frac{n(n+4)}{n-3}$

41. $\frac{1}{2}\left[\log_5(x-8) - \log_5 x\right] = \frac{1}{2}\log_5 \frac{x-8}{x}$
$= \log_5 \left[\frac{x-8}{x}\right]^{1/2}$
$= \log_5 \sqrt{\frac{x-8}{x}}$

43. $2\log_9 4 + \frac{1}{3}\log_9(r-6) - \frac{1}{2}\log_9 r$
$= \log_9 4^2 + \log_9(r-6)^{1/3} - \log_9 r^{1/2}$
$= \log_9 16 + \log_9 \sqrt[3]{r-6} - \log_9 \sqrt{r}$
$= \log_9 16\sqrt[3]{r-6} - \log_9 \sqrt{r}$
$= \log_9 \frac{16\sqrt[3]{r-6}}{\sqrt{r}}$

45. $4\log_6 3 - [2\log_6(x+3) + 4\log_6 x]$
$= \log_6 3^4 - [\log_6(x+3)^2 + \log_6 x^4]$
$= \log_6 81 - \log_6(x+3)^2 x^4$
$= \log_6 \frac{81}{(x+3)^2 x^4}$

47. $\log_a 10 = \log_a(2)(5)$
$= \log_a 2 + \log_a(5)$
$= 0.3010 + 0.6990$
$= 1$

49. $\log_a 0.4 = \log_a \frac{2}{5}$
$= \log_a 2 - \log_a 5$
$= 0.3010 - 0.6990$
$= -0.3980$

51. $\log_a 25 = \log_a 5^2$
$= 2(\log_a 5)$
$= 2(0.6990)$
$= 1.3980$

53. $5^{\log_5 10} = 10$

55. $(2^3)^{\log_8 7} = 8^{\log_8 7} = 7$

57. $\log_3 27 = \log_3 3^3 = 3$

59. $5\left(\sqrt[3]{27}\right)^{\log_3 5} = 5(3)^{\log_3 5}$
$= 5(5)$
$= 25$

61. Yes

63. $\log_a \frac{x}{y} = \log_a xy^{-1}$
$= \log_a x + \log_a y^{-1}$
$= \log_a x + \log_a \frac{1}{y}$

65. $\log_a(x^2-4) - \log_a(x+2) = \log_a \frac{x^2-4}{x+2}$
$= \log_a \frac{(x+2)(x-2)}{x+2}$
$= \log_a(x-2)$

67. Yes, assuming $x+4>0$.
$\log_a(x^2+8x+16) = \log_a(x+4)^2$
$= 2\log_a(x+4)$

69. $\log_{10} x^2 = 2\log_{10} x$
$= 2(0.4320)$
$= 0.8640$

71. $\log_{10} \sqrt[4]{x} = \log_{10} x^{1/4}$
$= \frac{1}{4}\log_{10} x$
$= \frac{1}{4}(0.4320) = 0.1080$

73. $\log_{10} xy = \log_{10} x + \log_{10} y$
$= 0.5000 + 0.2000$
$= 0.7000$

75. No; answers will vary. There is no simplification rule for the log of a sum.

77. $\log_2 \frac{\sqrt[4]{xy}\,\sqrt[3]{a}}{\sqrt[5]{a-b}} = \log_2 \sqrt[4]{xy}\,\sqrt[3]{a} - \log_2 \sqrt[5]{a-b}$
$= \log_2 (xy)^{1/4} + \log_2 a^{1/3} - \log_2 (a-b)^{1/5}$
$= \frac{1}{4}\log_2 xy + \frac{1}{3}\log_2 a - \frac{1}{5}\log_2 (a-b)$
$= \frac{1}{4}\log_2 x + \frac{1}{4}\log_2 y + \frac{1}{3}\log_2 a - \frac{1}{5}\log_2 (a-b)$

79. Let $\log_a x = m$ and $\log_a y = n$. Then

$a^m = x$ and $a^n = y$, so $\frac{x}{y} = \frac{a^m}{a^n} = a^{m-n}$.

Thus, $\log_a \frac{x}{y} = m - n = \log_a x - \log_a y$.

82. $\frac{x-4}{2} - \frac{2x-5}{5} > 3$
$10\left(\frac{x-4}{2} - \frac{2x-5}{5}\right) > 3 \cdot 10$
$5(x-4) - 2(2x-5) > 30$
$5x - 20 - 4x + 10 > 30$
$x - 10 > 30$
$x > 40$

a. $\{x \mid x > 40\}$

b. $(40, \infty)$

83. a. $a^2 - 4c^2$

b. $(a+2c)(a-2c)$

84. $\frac{15}{x} + \frac{9x-7}{x+2} = 9$
$x(x+2)\left(\frac{15}{x} + \frac{9x-7}{x+2}\right) = 9x(x+2)$
$15(x+2) + x(9x-7) = 9x(x+2)$
$15x + 30 + 9x^2 - 7x = 9x^2 + 18x$
$15x + 30 - 7x = 18x$
$8x + 30 = 18x$
$30 = 10x$
$3 = x$

85. $(3i+4)(2i-5) = 6i^2 - 15i + 8i - 20$
$= 6(-1) - 7i - 20$
$= -6 - 7i - 20$
$= -26 - 7i$

86. $a - 6\sqrt{a} = 7$
$a - 6\sqrt{a} - 7 = 0$
$u^2 - 6u - 7 = 0$
$(u-7)(u+1) = 0$
$u - 7 = 0$ or $u + 1 = 0$
$u = 7$ $\quad u = -1$
$\sqrt{a} = 7$ $\quad \sqrt{a} = -1$
$a = 49$ $\quad$ not possible
The solution is $a = 49$.

Mid-Chapter Test: 9.1 – 9.4

1. a. In $f(x)$, replace x by $g(x)$.

b. $f(x)=3x+3;\ g(x)=2x+5$

$$(f\circ g)(x)=3(2x+5)+3$$
$$=6x+15+3$$
$$=6x+18$$

2. a. $(f\circ g)(x)=\left(\frac{6}{x}\right)^2+5$

$$=\frac{36}{x^2}+5$$

b. $(f\circ g)(3)=\frac{36}{3^2}+5=\frac{36}{9}+5=4+5=9$

c. $(g\circ f)(x)=\frac{6}{x^2+5}$

d. $(g\circ f)(3)=\frac{6}{3^2+5}=\frac{6}{9+5}=\frac{6}{14}=\frac{3}{7}$

3. a. Answers will vary. A function is one-to-one if each input corresponds to exactly one output and each output corresponds to exactly one input.

b. No, the function is not one-to-one because it fails the horizontal line test.

4. a. The function is one-to-one. Each value in the range corresponds to exactly one value in the domain.

b. $\{(2,-3),(3,2),(1,5),(8,6)\}$

5. a. The function is one-to-one. Each value in the range corresponds to exactly one value in the domain.

b.
$$y=\frac{1}{3}x-5$$
$$x=\frac{1}{3}y-5$$
$$x+5=\frac{1}{3}y$$
$$3(x+5)=y$$
$$3x+15=y \text{ or } p^{-1}(x)=3x+15$$

6. a. The function is one-to-one. Each value in the range corresponds to exactly one value in the domain.

b.
$$y=\sqrt{x-4}$$
$$x=\sqrt{y-4}$$
$$x^2=y-4$$
$$x^2+4=y \text{ or } k^{-1}(x)=x^2+4,\ x\ge 0$$

7. $m(x)=-2x+4$
$$y=-2x+4$$
$$x=-2y+4$$
$$x-4=-2y$$
$$\frac{x-4}{-2}=y$$
$$2-\frac{1}{2}x=y \text{ or } m^{-1}(x)=-\frac{1}{2}x+2$$

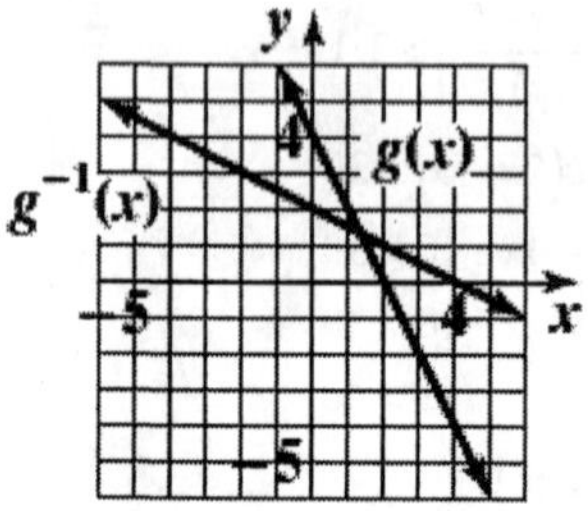

8. $y=2^x$

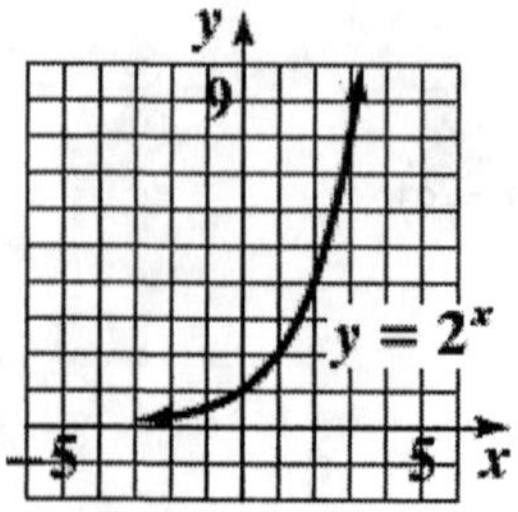

9. $y = 3^{-x}$

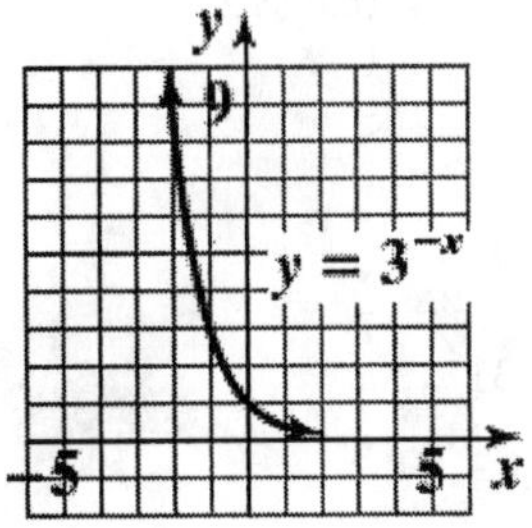

10. $y = \log_2 x$

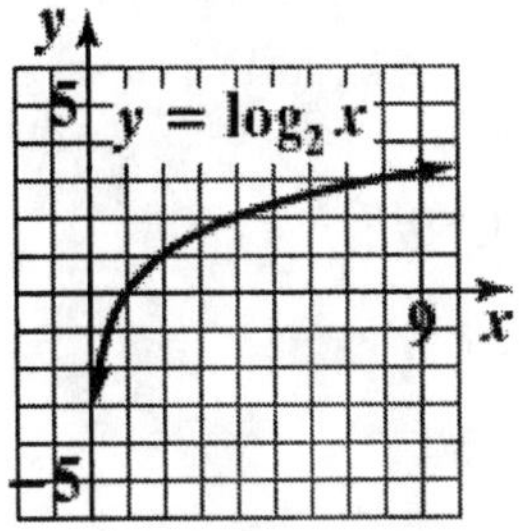

11. a. $N(t) = 5(2)^t$

$N(1) = 5(2)^1 = 5(2) = 10$

After 1 hour there are 10 bacteria in the dish.

b. $N(t) = 5(2)^t$

$N(6) = 5(2)^6 = 5(64) = 320$

After 6 hours there are 320 bacteria in the dish.

12. $27^{2/3} = 9 \quad \Leftrightarrow \quad \log_{27} 9 = \dfrac{2}{3}$

13. $\log_2 \dfrac{1}{64} = -6 \quad \Leftrightarrow \quad 2^{-6} = \dfrac{1}{64}$

14. $\log_5 125 = \log_5 5^3 = 3\log_5 5 = 3 \cdot 1 = 3$

15.

$$\log_{1/4} \frac{1}{16} = x$$

$$\log_{1/4} \left(\frac{1}{4}\right)^2 = x$$

$$2\log_{1/4} \frac{1}{4} = x$$

$$2 \cdot 1 = x$$

$$2 = x$$

16. $\log_x 64 = 3$

Write the equivalent exponential equation and solve for x.

$x^3 = 64$

$x^3 = 4^3 \quad \Rightarrow \quad x = 4$

17.

$$\log_9 x^2(x-5) = \log_9 x^2 + \log_9 (x-5)$$
$$= 2\log_9 x + \log_9 (x-5)$$

18.

$$\log_5 \frac{7m}{\sqrt{n}} = \log_5 (7m) - \log_5 \sqrt{n}$$
$$= \log_5 (7m) - \log_5 n^{1/2}$$
$$= \log_5 7 + \log_5 m - \frac{1}{2}\log_5 n$$

19.

$$3\log_2 x + \log_2 (x+7) - 4\log_2 (x+1)$$
$$= \log_2 x^3 + \log_2 (x+7) - \log_2 (x+1)^4$$
$$= \log_2 x^3 (x+7) - \log_2 (x+1)^4$$
$$= \log_2 \frac{x^3 (x+7)}{(x+1)^4}$$

20.

$$\frac{1}{2}\left[\log_7 (x+2) - \log_7 x\right]$$
$$= \frac{1}{2}\left[\log_7 \frac{(x+2)}{x}\right]$$
$$= \log_7 \left[\frac{(x+2)}{x}\right]^{1/2}$$
$$= \log_7 \sqrt{\frac{x+2}{x}}$$

Exercise Set 9.5

1. Common logarithms are logarithms with base 10.

3. Antilogarithms are numbers obtained by taking the base of the logarithm and raising it to the power that the logarithm is equal to. They are the numbers *inside* the logarithm.

5. $\log 86 = 1.9345$

7. $\log 19{,}200 = 4.2833$

9. $\log 0.0613 = -1.2125$

11. $\log 100 = 2.0000$

13. $\log 3.75 = 0.5740$

15. $\log 0.0173 = -1.7620$

17. $\text{antilog } 0.2137 = 1.64$

19. $\text{antilog } 4.6283 = 42{,}500$

21. $\text{antilog } (-1.7086) = 0.0196$

23. $\text{antilog } 0.0000 = 1.00$

25. $\text{antilog } 2.7625 = 579$

27. $\text{antilog } (-4.1390) = 0.0000726$

29. $\log N = 2.0000$
$N = \text{antilog } 2.000$
$N = 100$

31. $\log N = 3.3817$
$N = \text{antilog } 3.3817$
$N = 2410$

33. $\log N = 4.1409$
$N = \text{antilog } 4.1409$
$N = 13{,}800$

35. $\log N = -1.06$
$N = \text{antilog } (-1.06)$
$N = 0.0871$

37. $\log N = -0.6218$
$N = \text{antilog } (-0.6218)$
$N = 0.239$

39. $\log N = -0.1256$
$N = \text{antilog } (-0.1256)$
$N = 0.749$

41. $\log 3560 = 3.5514$
Therefore, $10^{3.5514} \approx 3560$.

43. $\log 0.0727 = -1.1385$
Therefore, $10^{-1.1385} \approx 0.0727$.

45. $\log 243 = 2.3856$
Therefore, $10^{2.3856} \approx 243$.

47. $\log 0.00592 = -2.2277$
Therefore, $10^{-2.2277} \approx 0.00592$.

49. $10^{2.8316} = 679$

51. $10^{-0.5186} = 0.303$

53. $10^{-1.4802} = 0.0331$

55. $10^{1.3503} = 22.4$

57. $\log 1 = x$
$10^x = 1$
$10^x = 10^0$
$x = 0$
Therefore, $\log 1 = 0$.

59. $\log 0.1 = x$
$10^x = 0.1$
$10^x = \frac{1}{10}$
$10^x = 10^{-1}$
$x = -1$
Therefore, $\log 0.1 = -1$.

61. $\log 0.01 = x$
$10^x = 0.01$
$10^x = \frac{1}{100}$
$10^x = 10^{-2}$
$x = -2$
Therefore, $\log 0.01 = -2$.

63. $\log 0.001 = x$
$10^x = 0.001$
$10^x = \frac{1}{1000}$
$10^x = 10^{-3}$
$x = -3$
Therefore, $\log 0.001 = -3$.

65. $\log 10^7 = 7$

67. $10^{\log 7} = 7$

69. $4 \log 10^{5.2} = 4(5.2) = 20.8$

71. $5(10^{\log 8.3}) = 5(8.3) = 41.5$

73. No; $10^2 = 100$ and since $462 > 100$, $\log 462$ must be greater than 2.

75. No; $10^0 = 1$ and $10^{-1} = 0.1$ and since $0.1 < 0.163 < 1$, $\log 0.163$ must be between 0 and -1.

77. No;

$$\begin{aligned}\log\frac{y}{4x} &= \log y - \log 4x \\ &= \log y - (\log 4 + \log x) \\ &= \log y - \log 4 - \log x\end{aligned}$$

79.
$$\begin{aligned}\log 125 &= \log(25 \cdot 5) \\ &= \log 25 + \log 5 \\ &= 1.3979 + 0.6990 \\ &= 2.0969\end{aligned}$$

81.
$$\begin{aligned}\log\frac{1}{5} &= \log 5^{-1} \\ &= -\log 5 \\ &= -1(0.6990) \\ &= -0.6990\end{aligned}$$

83.
$$\begin{aligned}\log 625 &= \log 25^2 \\ &= 2\log 25 \\ &= 2(1.3979) \\ &= 2.7958\end{aligned}$$

85. $R = \log I$, $R = 3.4$

$3.4 = \log I$

$I = \text{antilog}(3.4)$

$I \approx 2510$

This earthquake is about 2,510 times more intense than the smallest measurable activity.

87. $R = \log I$, $R = 5.7$

$5.7 = \log I$

$I = \text{antilog}(5.7)$

$I \approx 501{,}000$

This earthquake is about 501,000 times more intense than the smallest measurable activity.

89. $\log d = 3.7 - 0.2g$

a. $g = 11$

$$\begin{aligned}\log d &= 3.7 - 0.2(11) \\ &= 3.7 - 2.2 \\ &= 1.5\end{aligned}$$

$d = \text{antilog } 1.5 = 31.62$

A planet with absolute magnitude of 11 has a diameter of 31.62 kilometers.

b. $g = 20$

$$\begin{aligned}\log d &= 3.7 - 0.2(20) \\ &= 3.7 - 4 \\ &= -0.3\end{aligned}$$

$d = \text{antilog }(-0.3) = 0.50$

A planet with absolute magnitude of 20 has a diameter of 0.50 kilometers.

c. $d = 5.8$

$$\begin{aligned}\log 5.8 &= 3.7 - 0.2g \\ \log 5.8 - 3.7 &= -0.2g \\ 0.76343 - 3.7 &= -0.2g \\ -2.93657 &= -0.2g \\ \frac{-2.93657}{-0.2} &= g \\ 14.68 &= g\end{aligned}$$

A planet with diameter 5.8 kilometers has an absolute magnitude of 14.68.

91. $R(t) = 94 - 46.8\log(t+1)$

a.
$$\begin{aligned}R(2) &= 94 - 46.8\log(2+1) \\ &= 94 - 46.8\log(3) \\ &\approx 72\end{aligned}$$

After two months, Sammy will remember about 72% of the course material.

b.
$$\begin{aligned}R(48) &= 94 - 46.8\log(2+48) \\ &= 94 - 46.8\log(50) \\ &\approx 15\end{aligned}$$

After forty-eight months, Sammy will remember about 15% of the course material.

93. $R = \log I,\ R = 3.8$

$3.8 = \log I$

$I = \text{anti}\log(3.8)$

$I \approx 6310$

This earthquake is about 6310 times more intense than the smallest measurable activity.

95. $\log E = 11.8 + 1.5m_s$

a. $\log E = 11.8 + 1.5(6)$

$\log E = 20.8$

$10^{20.8} = E$

$E = 6.31 \times 10^{20}$

The energy released is 6.31×10^{20}.

b. $\log(1.2 \times 10^{15}) = 11.8 + 1.5m_s$

$15.07918125 = 11.8 + 1.5m_s$

$3.27918125 = 1.5m_s$

$m_s \approx 2.19$

The surface wave has magnitude 2.19.

97. $M = \dfrac{\log E - 11.8}{1.5}$

$M = \dfrac{\log(1.259 \times 10^{21}) - 11.8}{1.5}$

$= \dfrac{\log 1.259 + \log 10^{21} - 11.8}{1.5}$

$= \dfrac{\log 1.259 + 21 - 11.8}{1.5}$

$= \dfrac{\log 1.259 + 9.2}{1.5}$

$\approx \dfrac{0.1000 + 9.2}{1.5}$

≈ 6.2

The magnitude is about 6.2.

99. $R = \log I$

$\text{antilog}(R) = \text{antilog}(\log I)$

$\text{antilog}(R) = I$

101. $R = 26 - 41.9\log(t+1)$

$R - 26 = -41.9\log(t+1)$

$\dfrac{R-26}{-41.9} = \dfrac{-41.9\log(t+1)}{-41.9}$

$\dfrac{26-R}{41.9} = \log(t+1)$

$\text{antilog}\left(\dfrac{26-R}{41.9}\right) = \text{antilog}(\log(t+1))$

$\text{antilog}\left(\dfrac{26-R}{41.9}\right) = t+1$

$\text{antilog}\left(\dfrac{26-R}{41.9}\right) - 1 = t$

104. Let r equal the rate of car 2.

	d	r	t
Car 1	$4(r+5)$	$r+5$	4
Car 2	$4r$	r	4

The total distance was 420 miles.

$4(r+5) + 4r = 420$

$4r + 20 + 4r = 420$

$8r + 20 = 420$

$8r = 400$

$r = 50$

The rate of car 2 is 50 mph and the rate of car 1 is 50 + 5 = 55 mph.

105. $3r = -4s - 6 \Rightarrow 3r + 4s = -6$ (1)

$3s = -5r + 1 \Rightarrow 5r + 3s = 1$ (2)

To eliminate the variable r, multiply equation (1) by –5 and equation (2) by 3 then add.

$-5(3r + 4s = -6) \Rightarrow -15r - 20s = 30$

$3(5r + 3s = 1) \Rightarrow \underline{15r + 9s = 3}$

$-11s = 33$

$s = -3$

Substitute –3 for s in equation (1) and solve for r.

$3r + 4(-3) = -6$

$3r - 12 = -6$

$3r = 6$

$r = 2$

The solution is (2, –3).

106. $3x^3+3x^2-36x=0$

$3x\left(x^2+x-12\right)=0$

$3x(x+4)(x-3)=0$

$3x=0 \quad \text{or} \quad x+4=0 \quad \text{or} \quad x-3=0$

$x=0 \qquad x=-4 \qquad x=3$

The solution set is $\{0,-4,3\}$.

107. $\sqrt{\left(3x^2-y\right)^2}=\left|3x^2-y\right|$

108. $(x-5)(x+4)(x-2)\le 0$

$x-5=0 \qquad x+4=0 \qquad x-2=0$

$x=5 \qquad x=-4 \qquad x=2$

–	–	–	+	$(x-5)$
–	+	+	+	$(x+4)$
–	–	+	+	$(x-2)$
–	+	–	+	$(x-5)(x+4)(x-2)$
	–4	2	5	

$(-\infty,-4]\cup[2,\ 5]$

Exercise Set 9.6

1. $c=d$

3. Check for extraneous solutions.

5. log (–2) is not a real number

7. $5^x=125$

$5^x=5^3$

$x=3$

9. $3^x=81$

$3^x=3^4$

$x=4$

11. $64^x=8$

$(8^2)^x=8^1$

$8^{2x}=8^1$

$2x=1$

$x=\frac{1}{2}$

13. $7^{-x}=\frac{1}{49}$

$7^{-x}=7^{-2}$

$-x=-2$

$x=2$

15. $27^x=\frac{1}{3}$

$\left(3^3\right)^x=3^{-1}$

$3^{3x}=3^{-1}$

$3x=-1$

$x=-\frac{1}{3}$

17. $2^{x+2}=64$

$2^{x+2}=2^6$

$x+2=6$

$x=4$

19. $2^{3x-2}=128$

$2^{3x-2}=2^7$

$3x-2=7$

$3x=9$

$x=3$

21. $27^x=3^{2x+3}$

$3^{3x}=3^{2x+3}$

$3x=2x+3$

$x=3$

23. $7^x=50$

$\log 7^x=\log 50$

$x\log 7=\log 50$

$x=\frac{\log 50}{\log 7}$

$x\approx 2.01$

25.
$$4^{x-1} = 35$$
$$\log 4^{x-1} = \log 35$$
$$(x-1)\log 4 = \log 35$$
$$x-1 = \frac{\log 35}{\log 4}$$
$$x = \frac{\log 35}{\log 4} + 1$$
$$x \approx 3.56$$

27.
$$1.63^{x+1} = 25$$
$$\log 1.63^{x+1} = \log 25$$
$$(x+1)\log 1.63 = \log 25$$
$$x+1 = \frac{\log 25}{\log 1.63}$$
$$x+1 \approx 6.59$$
$$x \approx 5.59$$

29.
$$3^{x+4} = 6^x$$
$$\log 3^{x+4} = \log 6^x$$
$$(x+4)\log 3 = x\log 6$$
$$x\log 3 + 4\log 3 = x\log 6$$
$$4\log 3 = x\log 6 - x\log 3$$
$$4\log 3 = x(\log 6 - \log 3)$$
$$\frac{4\log 3}{\log 6 - \log 3} = x$$
$$6.34 \approx x$$

31. $\log_{36} x = \frac{1}{2}$
$$36^{1/2} = x$$
$$\sqrt{36} = x$$
$$6 = x$$

33. $\log_{125} x = \frac{1}{3}$
$$125^{1/3} = x$$
$$\sqrt[3]{125} = x$$
$$5 = x$$

35. $\log_2 x = -4$
$$2^{-4} = x$$
$$\frac{1}{2^4} = x$$
$$\frac{1}{16} = x$$

37. $\log x = 2$
$$\log_{10} x = 2$$
$$10^2 = x$$
$$100 = x$$

39. $\log_2(5-3x) = 3$
$$2^3 = 5-3x$$
$$8 = 5-3x$$
$$3x = -3$$
$$x = -1$$

41. $\log_5(x+1)^2 = 2$
$$(x+1)^2 = 5^2$$
$$(x+1)^2 = 25$$
$$x+1 = \pm\sqrt{25}$$
$$x+1 = \pm 5$$
$$x+1 = 5 \quad \text{or} \quad x+1 = -5$$
$$x = 4 \qquad\qquad x = -6$$
Both values check. The solutions are 4 and –6.

43. $\log_2(r+4)^2 = 4$
$$(r+4)^2 = 2^4$$
$$r^2 + 8r + 16 = 16$$
$$r^2 + 8r = 0$$
$$r(r+8) = 0$$
$$r = 0 \quad \text{or} \quad r+8 = 0$$
$$r = -8$$
Both values check. The solutions are 0 and –8.

45. $\log(x+8) = 2$
$$\log_{10}(x+8) = 2$$
$$10^2 = x+8$$
$$100 = x+8$$
$$x = 92$$

47. $\log_2 x + \log_2 5 = 2$

$$\log_2 5x = 2$$
$$5x = 2^2$$
$$x = \frac{4}{5}$$

49. $\log(r+2) = \log(3r-1)$

$$r + 2 = 3r - 1$$
$$3 = 2r$$
$$\frac{3}{2} = r$$

51. $\log(2x+1) + \log 4 = \log(7x+8)$

$$\log(8x+4) = \log(7x+8)$$
$$8x + 4 = 7x + 8$$
$$x = 4$$

53. $\log n + \log(3n-5) = \log 2$

$$\log(3n^2 - 5n) = \log 2$$
$$3n^2 - 5n = 2$$
$$3n^2 - 5n - 2 = 0$$
$$(3n+1)(n-2) = 0$$
$$3n + 1 = 0 \quad \text{or} \quad n - 2 = 0$$
$$3n = -1 \qquad n = 2$$
$$n = -\tfrac{1}{3}$$

Check: $n = -\dfrac{1}{3}$

$$\log n + \log(3n-5) = \log 2$$
$$\log\left(-\frac{1}{3}\right) + \log\left[3\left(\frac{-1}{3}\right) - 5\right] = \log 2$$

Logarithms of negative numbers are not real numbers so $-\dfrac{1}{3}$ is an extraneous solution.

Check: $n = 2$

$$\log n + \log(3n-5) = \log 2$$
$$\log 2 + \log[3(2) - 5] = \log 2$$
$$\log 2 + \log 1 = \log 2$$
$$\log(2 \cdot 1) = \log 2$$
$$\log 2 = \log 2$$

2 is the only solution.

55. $\log 6 + \log y = 0.72$

$$\log 6y = 0.72$$
$$6y = \text{antilog}\, 0.72$$
$$y = \frac{\text{antilog}\, 0.72}{6}$$
$$y \approx 0.87$$

57. $2\log x - \log 9 = 2$

$$\log x^2 - \log 9 = 2$$
$$\log \frac{x^2}{9} = 2$$
$$\frac{x^2}{9} = \text{antilog}\ 2$$
$$\frac{x^2}{9} = 100$$
$$x^2 = 900$$
$$x^2 - 900 = 0$$
$$(x+30)(x-30) = 0$$
$$x + 30 = 0 \quad \text{or} \quad x - 30 = 0$$
$$x = -30 \qquad x = 30$$

Check: $x = -30$

$$2\log x - \log 9 = 2$$
$$2\log(-30) - \log 9 = 2$$

Logarithms of negative numbers are not real numbers so -30 is an extraneous solution.

Check: $x = 30$

$$2\log x - \log 9 = 2$$
$$2\log 30 - \log 9 = 2$$
$$\log \frac{900}{9} = 2$$
$$\log 100 = 2$$
$$100 = \text{antilog}\ 2$$
$$100 = 100$$

Thus, 30 is the only solution.

59. $\log x + \log(x-3) = 1$

$\log(x^2 - 3x) = 1$

$x^2 - 3x = \text{antilog } 1$

$x^2 - 3x = 10$

$x^2 - 3x - 10 = 0$

$(x-5)(x+2) = 0$

$x - 5 = 0$ or $x + 2 = 0$

$x = 5 \qquad x = -2$

A check shows that –2 is an extraneous solution so 5 is the only solution.

61. $\log x = \frac{1}{3}\log 64$

$\log x = \log 64^{1/3}$

$\log x = \log 4$

$x = 4$

63. $\log_8 x = 4\log_8 2 - \log_8 8$

$\log_8 x = \log_8 2^4 - \log_8 8$

$\log_8 x = \log_8 \frac{16}{8}$

$\log_8 x = \log_8 2$

$x = 2$

65. $\log_5(x+3) + \log_5(x-2) = \log_5 6$

$\log_5(x+3)(x-2) = \log_5 6$

$\log_5(x^2 + x - 6) = \log_5 6$

$x^2 + x - 6 = 6$

$x^2 + x - 12 = 0$

$(x+4)(x-3) = 0$

$x + 4 = 0$ or $x - 3 = 0$

$x = -4 \qquad x = 3$

Check $x = -4$:

$\log_5(-4+3) + \log_5(-4-2) = \log_5 6$

$\log_5(-1) + \log_5(-6) = \log_5 6$

Since we cannot take the logarithm of a negative number, –4 is an extraneous solution.

Check $x = 3$:

$\log_5(3+3) + \log_5(3-2) = \log_5 6$

$\log_5(6) + \log_5(1) = \log_5 6$

$\log_5 6 = \log_5 6$

Therefore, $x = 3$ is the only solution.

67. $\log_2(x+3) - \log_2(x-6) = \log_2 4$

$\log_2 \frac{x+3}{x-6} = \log_2 4$

$\frac{x+3}{x-6} = 4$

$x + 3 = 4x - 24$

$27 = 3x$

$9 = x$

69. $50{,}000 = 4500(2^t)$

$\frac{50{,}000}{4500} = 2^t$

$\log \frac{50{,}000}{4500} = \log 2^t$

$\log \frac{50{,}000}{4500} = t \log 2$

$\log 50{,}000 - \log 4500 = t \log 2$

$\frac{\log 50{,}000 - \log 4500}{\log 2} = t$

$3.47 \approx t$

There are 50,000 bacteria after about 3.47 hours.

71. $80 = 200(0.75)^t$

$0.4 = (0.75)^t$

$\log 0.4 = \log(0.75)^t$

$\log 0.4 = t \log 0.75$

$\frac{\log 0.4}{\log 0.75} = t$

$3.19 \approx t$

80 grams remain after about 3.19 years.

73. $A = P\left(1 + \frac{r}{n}\right)^{nt}$

$4600 = 2000\left(1 + \frac{0.05}{1}\right)^{1 \cdot t}$

$4600 = 2000(1.05)^t$

$\frac{4600}{2000} = 1.05^t$

$2.3 = 1.05^t$

$\log 2.3 = \log 1.05^t$

$\log 2.3 = t \log 1.05$

$\frac{\log 2.3}{\log 1.05} = t \quad \Rightarrow \quad t \approx 17.07$ years

The \$2000 will grow to \$4600 in about 17.07 years.

75. $f(t)=26-12.1\cdot\log(t+1)$

a. $x=1990-1960=30$
$f(30)=26-12.1\log(30+1)=7.95$
In 1990, the rate was about 7.95 deaths per 1000 live births.

b. $x=2005-1960=45$
$f(45)=26-12.1\log(45+1)\approx 5.88$
In 2005, the rate was about 5.88 deaths per 1000 live births.

77. $c=50{,}000$, $n=12$, $r=0.15$.

$S=c(1-r)^n$

$S=50{,}000(1-0.15)^{12}$

$S=50{,}000(0.85)^{12}$

$S\approx 7112.09$

The scrap value is about \$7112.09.

79. $P_{\text{out}}=12.6$ and $P_{\text{in}}=0.146$

$P=10\log\left(\frac{12.6}{0.146}\right)$

$P\approx 10\log 86.30137$

$P\approx 10(1.936)$

$P\approx 19.36$

The power gain is about 19.36 watts.

81. a. $d=120$

$d=10\log I$

$120=10\log I$

$12=\log I$

$I=\text{antilog }12$

$I=10^{12}$

$I=1{,}000{,}000{,}000{,}000$

The intensity is 1,000,000,000,000 times the minimum intensity of audible sound.

b. $d=50$

$d=10\log I$

$50=10\log I$

$5=\log I$

$I=\text{ antilog }5$

$I=10^5$

$I=100{,}000$

$$\frac{1{,}000{,}000{,}000{,}000}{100{,}000}=10{,}000{,}000$$

The sound of an airplane engine is 10,000,000 times more intense than the noise in a busy city street.

83. $8^x=16^{x-2}$

$2^{3x}=2^{4(x-2)}$

$3x=4(x-2)$

$3x=4x-8$

$8=x$

85. $2^{2x}-6(2^x)+8=0$

$(2^x)^2-6(2^x)+8=0$

$y^2-6y+8=0\leftarrow$ Replace 2^x with y

$(y-4)(y-2)=0$

$y-4=0$ or $y-2=0$

$y=4 \qquad y=2$

$2^x=4 \qquad 2^x=2\leftarrow$ Replace y with 2^x

$2^x=2^2 \qquad 2^x=2^1$

$x=2 \qquad x=1$

The solutions are $x=2$ and $x=1$.

87. $2^x=8^y$

$x+y=4$

The first equation simplifies to

$2^x=(2^3)^y$

$2^x=2^{3y}$

$x=3y$

The system becomes

$x=3y$

$x+y=4$

Substitute $3y$ for x in the second equation.

$x+y=4$

$3y+y=4$

$4y=4$

$y=1$

Now, substitute 1 for y in the first equation.

$x=3y$

$x=3(1)=3$

The solution is (3, 1).

89. $\log(x+y)=2$

$x-y=8$

The first equation can be written as

$x+y=10^2$

$x+y=100$

The system becomes

$$\begin{aligned} x+y&=100\\ x-y&=\ 8\\ \hline \text{Add: } 2x\quad &=108\\ x\quad &=54 \end{aligned}$$

Substitute 54 for x in the first equation.

$54+y=100$

$y=46$

The solution is (54, 46).

91.

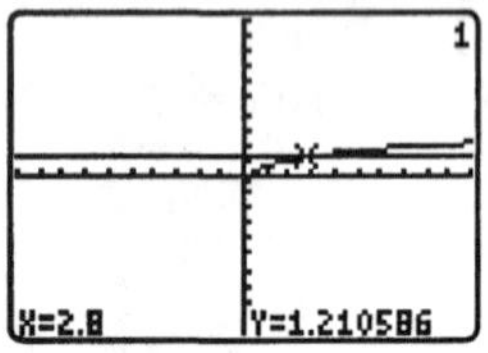

-10, 30, 5, -10, 10

The solution is $x \approx 2.8$.

93.

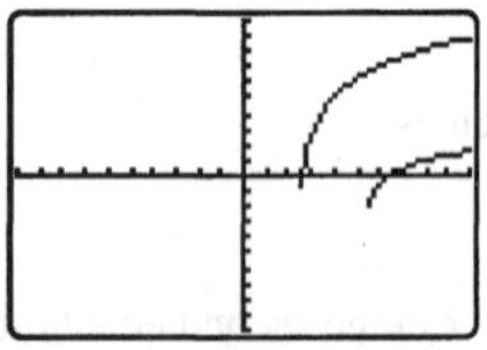

-10, 10, 1, -10, 10

There is no real-number solution.

95. Volume of cylinder:

$$V_1=\pi r^2 h=\pi\left(\frac{3}{2}\right)^2\cdot 4\approx 28.2743 \text{ cubic feet}$$

Volume of box:

$$V_2=l\cdot w\cdot h=(3)(3)(4)=36 \text{ cubic feet}$$

The box has the greater volume.

Difference in volumes:

$V_2-V_1\approx 7.73$ cubic feet

96. $f(x)=x^2-x,\ g(x)=x-1$

$$\begin{aligned}(g-f)(x)&=(x-1)-(x^2-x)\\ &=x-1-x^2+x\\ &=-x^2+2x-1\end{aligned}$$

$$\begin{aligned}(g-f)(3)&=-(3)^2+2(3)-1\\ &=-9+6-1\\ &=-4\end{aligned}$$

97. $3x-4y\le 6$

$y>-x+4$

For $3x-4y\le 6$, graph the line $3x-4y=6$ using a solid line. For the check point, select (0, 0):

$3x-4y\le 6$

$3(0)-4(0)\le 6$

$0\le 6$ True

Since this is a true statement, shade the region which contains the point (0, 0). This is the region "above" the line.

For $y>-x+4$, graph the line $y=-x+4$ using a dashed line. For the check point, select (0, 0):

$y>-x+4$

$(0)>-(0)+4$

$0>4$ False

Since this is a false statement, shade the region which does not contain the point (0, 0). This is the region "above" the line. To obtain the final region, take the intersection of the above two regions.

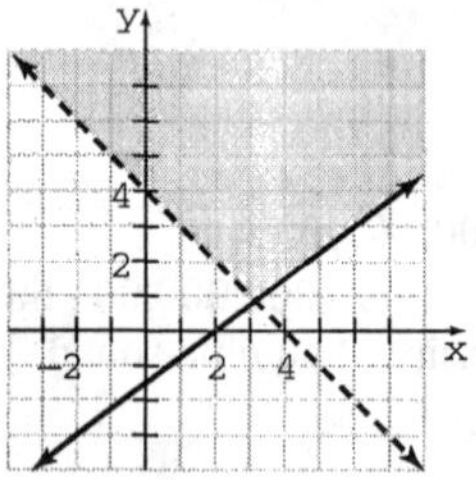

98. $$\frac{2\sqrt{xy}-\sqrt{xy}}{\sqrt{x}+\sqrt{y}}=\frac{2\sqrt{xy}-\sqrt{xy}}{\sqrt{x}+\sqrt{y}}\cdot\frac{\sqrt{x}-\sqrt{y}}{\sqrt{x}-\sqrt{y}}$$
$$=\frac{\left(2\sqrt{xy}-\sqrt{xy}\right)\cdot\left(\sqrt{x}-\sqrt{y}\right)}{\left(\sqrt{x}+\sqrt{y}\right)\cdot\left(\sqrt{x}-\sqrt{y}\right)}$$
$$=\frac{2\sqrt{x^2y}-2\sqrt{xy^2}-\sqrt{x^2y}+\sqrt{xy^2}}{\sqrt{x^2}-\sqrt{xy}+\sqrt{xy}-\sqrt{y^2}}$$
$$=\frac{\sqrt{x^2y}-\sqrt{xy^2}}{\sqrt{x^2}-\sqrt{y^2}}$$
$$=\frac{x\sqrt{y}-y\sqrt{x}}{x-y}$$

99. $E=mc^2$, for c
$$\frac{E}{m}=\frac{mc^2}{m}$$
$$\frac{E}{m}=c^2$$
$$\sqrt{\frac{E}{m}}=c$$
(Recall that when using the square root property to solve for a variable in a formula we only use the principal square root because we are generally solving for a variable that cannot be negative – see section 8.3 in the text.)

100. Use $f(x)=a(x-h)^2+k$, where $a=2$ and $(h, k)=(3,-5)$.
$$f(x)=2(x-3)^2-5$$

Exercise Set 9.7

1. a. The base in the natural exponential function is e.

b. The approximate value of e is 2.7183.

3. The domain of $\ln x$ is $\{x\,|\,x>0\}$.

5. $\log_a x=\dfrac{\log_b x}{\log_b a}$

7. $\ln e^x=x$

9. The inverse of $\ln x$ is e^x.

11. P decreases when t increases for $k<0$.

13. $\ln 62\approx 4.1271$

15. $\ln 0.813\approx -0.2070$

17. $\ln N=1.6$
$$e^{\ln N}=e^{1.6}$$
$$N=e^{1.6}\approx 4.95$$

19. $\ln N=-2.85$
$$e^{\ln N}=e^{-2.85}$$
$$N=e^{-2.85}\approx 0.0578$$

21. $\ln N=-0.0287$
$$e^{\ln N}=e^{-0.0287}$$
$$N=e^{-0.0287}\approx 0.972$$

23. $\log_3 56=\dfrac{\log 56}{\log 3}\approx 3.6640$

25. $\log_2 21=\dfrac{\log 21}{\log 2}\approx 4.3923$

27. $\log_4 11=\dfrac{\log 11}{\log 4}\approx 1.7297$

29. $\log_5 82=\dfrac{\log 82}{\log 5}\approx 2.7380$

31. $\log_6 185=\dfrac{\log 185}{\log 6}\approx 2.9135$

33. $\ln 51=\dfrac{\log 51}{\log e}\approx 3.9318$

35. $\log_5 0.463=\dfrac{\log 0.463}{\log 5}\approx -0.4784$

37. $\ln x + \ln(x-1) = \ln 12$

$\ln x(x-1) = \ln 12$

$e^{\ln[x(x-1)]} = e^{\ln 12}$

$x(x-1) = 12$

$x^2 - x - 12 = 0$

$(x-4)(x+3) = 0$

$x-4=0 \qquad x+3=0$

$x=4 \qquad x=-3$

Only $x=4$ checks since $x=-3$ is an extraneous solution. Note that when $x=-3$ we would get $\ln(-3)$ and $\ln(-4)$ which are not real numbers.

39. $\ln x + \ln(x+4) = \ln 5$

$\ln x(x+4) = \ln 5$

$e^{\ln(x^2+4x)} = e^{\ln 5}$

$x^2 + 4x = 5$

$x^2 + 4x - 5 = 0$

$(x+5)(x-1) = 0$

$x+5=0$ or $x-1=0$

$x=-5 \qquad x=1$

Only $x=1$ checks since $x=-5$ is an extraneous solution. Note that when $x=-5$ we would get $\ln(-5)$ and $\ln(-1)$ which are not real numbers.

41. $\ln x = 5\ln 2 - \ln 8$

$\ln x = \ln 2^5 - \ln 8$

$\ln x = \ln \dfrac{32}{8}$

$\ln x = \ln 4$

$e^{\ln x} = e^{\ln 4}$

$x = 4$

$x=4$ checks.

43. $\ln(x^2-4) - \ln(x+2) = \ln 4$

$\ln(x^2-4) - \ln(x+2) = \ln 4$

$\ln\left(\dfrac{x^2-4}{x+2}\right) = \ln 4$

$\ln(x-2) = \ln 4$

$e^{\ln(x-2)} = e^{\ln(4)}$

$x-2=4$

$x=6$

$x=6$ checks.

45. $P = 120e^{(2.3)(1.6)}$

$P = 120e^{3.68}$

$P \approx 4757.5673$

47. $50 = P_0 e^{-0.5(3)}$

$50 = P_0 e^{-1.5}$

$\dfrac{50}{e^{-1.5}} = P_0$

$P_0 \approx 224.0845$

49. $60 = 20e^{1.4t}$

$3 = e^{1.4t}$

$\ln 3 = \ln e^{1.4t}$

$\ln 3 = 1.4t$

$t = \dfrac{\ln 3}{1.4} \approx 0.7847$

51. $86 = 43e^{k(3)}$

$2 = e^{3k}$

$\ln 2 = \ln e^{3k}$

$\ln 2 = 3k$

$k = \dfrac{\ln 2}{3} \approx 0.2310$

53. $20 = 40e^{k(2.4)}$

$0.5 = e^{2.4k}$

$\ln 0.5 = \ln e^{2.4k}$

$\ln 0.5 = 2.4k$

$k = \dfrac{\ln 0.5}{2.4} \approx -0.2888$

55. $A = 6000e^{-0.08(3)}$

$A = 6000e^{-0.24}$

$A \approx 4719.7672$

57. $V = V_0 e^{kt}$

$\dfrac{V}{e^{kt}} = V_0$ or $V_0 = \dfrac{V}{e^{kt}}$

59.
$$P = 150e^{7t}$$
$$\frac{P}{150} = e^{7t}$$
$$\ln\frac{P}{150} = \ln e^{7t}$$
$$\ln\frac{P}{150} = 7t$$
$$\frac{\ln P - \ln 150}{7} = t \text{ or } t = \frac{\ln P - \ln 150}{7}$$

61.
$$A = A_0 e^{kt}$$
$$\frac{A}{A_0} = e^{kt}$$
$$\ln\frac{A}{A_0} = \ln e^{kt}$$
$$\ln A - \ln A_0 = kt$$
$$\frac{\ln A - \ln A_0}{t} = k \text{ or } k = \frac{\ln A - \ln A_0}{t}$$

63. $\ln y - \ln x = 2.3$
$$\ln\frac{y}{x} = 2.3$$
$$e^{\ln(y/x)} = e^{2.3}$$
$$\frac{y}{x} = e^{2.3}$$
$$y = xe^{2.3}$$

65. $\ln y - \ln(x+6) = 5$
$$\ln\frac{y}{x+6} = 5$$
$$e^{\ln\frac{y}{x+6}} = e^5$$
$$\frac{y}{x+6} = e^5$$
$$y = (x+6)e^5$$

67. $e^x = 12.183$

Take the natural logarithm of both sides of the equation.

$\ln e^x = \ln 12.183$

$x = \ln 12.183 \approx 2.5000$

69. $P = P_0 e^{kt}$

a. $P = 5000e^{0.06(2)}$
$$= 5000e^{0.12}$$
$$\approx 5637.48$$
The amount will be $5637.48.

b. If the amount in the account is to double, then $P = 2(5000) = 10{,}000$.
$$10{,}000 = 5000e^{0.06t}$$
$$2 = e^{0.06t}$$
$$\ln 2 = \ln e^{0.06t}$$
$$\ln 2 = 0.06t$$
$$\frac{\ln 2}{0.06} = t$$
$$11.55 \approx t$$
It would take about 11.55 years for the value to double.

71. $P = P_0 e^{-0.028t}$
$$P = 70e^{-0.028(20)}$$
$$P = 70e^{-0.56}$$
$$P \approx 39.98$$
After 20 years, about 39.98 grams remain.

73. $f(t) = 1 - e^{-0.04t}$

a. $f(t) = 1 - e^{-0.04(50)} = 1 - e^{-2} \approx 0.8647$

About 86.47% of the target market buys the drink after 50 days of advertising.

b.
$$0.75 = 1 - e^{-0.04t}$$
$$-0.25 = -e^{-0.04t}$$
$$0.25 = e^{-0.04t}$$
$$\ln 0.25 = \ln e^{-0.04t}$$
$$\ln 0.25 = -0.04t$$
$$t = \frac{\ln 0.25}{-0.04}$$
$$t \approx 34.66$$
About 34.66 days of advertising are needed if 75% of the target market is to buy the soft drink.

75. $f(P) = 0.37 \ln P + 0.05$

a. $f(972{,}000) = 0.37\ln(972{,}000) + 0.05$
$\approx 5.1012311 + 0.05$
≈ 5.15
The average walking speed in Nashville, Tennessee is 5.15 feet per second.

b. $f(8{,}567{,}000) = 0.37\ln(8{,}567{,}000) + 0.05$
$\approx -5.906 + 0.05$
≈ 5.96
The average walking speed in New York City is 5.96 feet per second.

c. $5 = 0.37\ln P + 0.05$
$4.95 = 0.37\ln P$
$13.378378 = \ln P$
$e^{13.378378} = e^{\ln P}$
$P = e^{13.378378}$
$P \approx 646{,}000$
The population is about 646,000.

77. $V(t) = 24e^{0.08t}, \quad t = 2008 - 1626 = 382$
$V(377) = 24e^{0.08(382)}$
$\approx 449{,}004{,}412{,}200{,}000$
The value of Manhattan in 2003 is about $449,004,412,200,000.

79. $P(t) = 6.30e^{0.013t}$

a. $t = 2010 - 2003 = 7$
$P(7) = 6.30e^{0.013(7)}$
$= 6.30e^{0.091}$
≈ 6.9
The world's population in 2010 is expected to be about 6.9 billion.

b. 2(6.30 billion) $\Rightarrow$ 12.60 billion
$12.60 = 6.30e^{0.013t}$
$\frac{12.60}{6.30} = \frac{6.30e^{0.013t}}{6.30}$
$2 = e^{0.013t}$
$\ln 2 = \ln e^{0.013t}$
$\ln 2 = 0.013t$
$t = \frac{\ln 2}{0.013} \approx 53$
The world's population will double in about 53 years.

81. $d(t) = 2.19e^{0.0164t}$

a. $t = 2025 - 2005 = 20$
$d(20) = 2.19e^{0.0164(20)} = 2.19e^{0.328} \approx 3.04$
According to the trend, the demand for nurses in 2025 will be approximately 3.04 million.

b. $t = 2040 - 2005 = 35$
$d(35) = 2.19e^{0.0164(35)} = 2.19e^{0.574} \approx 3.89$
According to the trend, the demand for nurses in 2040 will be approximately 3.89 million.

83. $r(t) = 1182.3e^{0.0715t}$

a. $t = 2006 - 1994 = 12$
$r(12) = 1182.3e^{0.0715(12)}$
$= 1182.3e^{0.858} \approx 2788.38$
According to the trend, the average annual tax refund was about $2788.38 in 2006.

b. $t = 2010 - 1994 = 16$
$r(16) = 1182.3e^{0.0715(16)}$
$= 1182.3e^{1.144} \approx 3711.59$
According to the trend, the average annual tax refund will be about $3711.59 in 2010.

85. $y = 15.29 + 5.93\ln x$

a. $y(18) = 15.29 + 5.93\ln(18) \approx 32.43$ in.

b. $y(30) = 15.29 + 5.93\ln(30) \approx 35.46$ in.

87. $f(t) = v_0 e^{-0.0001205t}$

a. Use $f(t) = 9$ and $v_0 = 20$.

$$9 = 20e^{-0.0001205t}$$
$$0.45 = e^{-0.0001205t}$$
$$\ln 0.45 = -0.0001205t$$
$$\frac{\ln 0.45}{-0.0001205} = t$$
$$t \approx 6626.62$$

The bone is about 6626.62 years old.

b. Let x equal the original amount of carbon 14 then $0.5x$ equals the remaining amount.

$$0.5x = xe^{-0.0001205t}$$
$$\frac{0.5x}{x} = \frac{xe^{-0.0001205t}}{x}$$
$$0.5 = e^{-0.0001205t}$$
$$\ln 0.5 = \ln e^{-0.0001205t}$$
$$\ln 0.5 = -0.0001205t$$
$$\frac{\ln 0.5}{-0.0001205} = t$$
$$t \approx 5752.26$$

If 50% of the carbon 14 remains, the item is about 5752.26 years old.

89. Let P_0 be the initial investment, then $P = 20{,}000$, $k = 0.06$, and $t = 18$.

$$P = P_0 e^{kt}$$
$$20{,}000 = P_0 e^{0.06(18)}$$
$$20{,}000 = P_0 e^{1.08}$$
$$\frac{20{,}000}{e^{1.08}} = P_0$$
$$6791.91 \approx P_0$$

The initial investment should be \$6791.91.

91. a. Strontium 90 has a higher decay rate so it will decompose more quickly.

b. $P = P_0 e^{-kt}$

$$P = P_0 e^{-0.023(50)}$$
$$= P_0 e^{-1.15}$$
$$\approx P_0(0.3166)$$

About 31.66% of the original amount will remain.

93. Answers will vary.

95. $e^{x-4} = 12\ln(x+2)$

$$y_1 = e^{x-4}$$
$$y_2 = 12\ln(x+2).$$

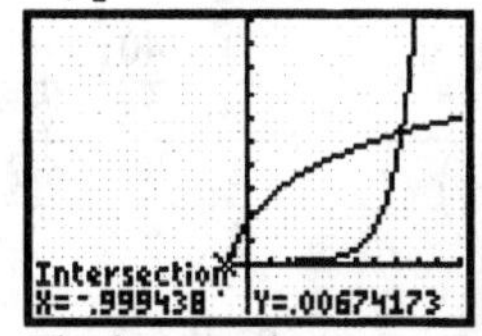

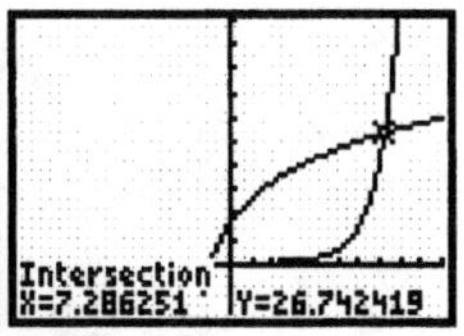

The intersections are approximately $(-0.999, 0.007)$ and $(7.286, 26.742)$. Therefore, $x \approx -0.999$ or $x \approx 7.286$.

97. $3x - 6 = 2e^{0.2x} - 12$

$$y_1 = 3x - 6$$
$$y_2 = 2e^{0.2x} - 12$$

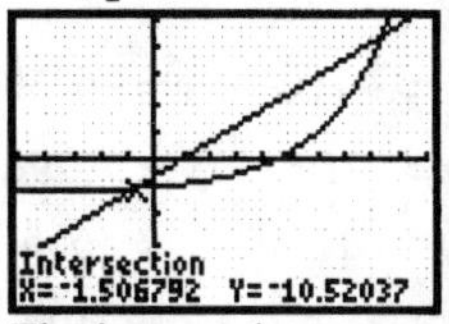

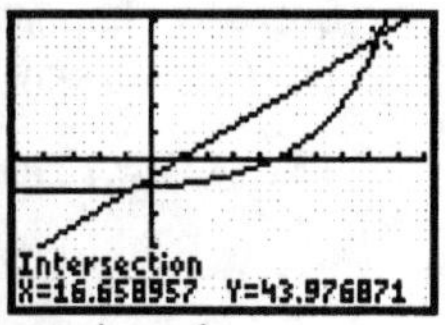

The intersections are approximately $(-1.507, -10.520)$ and $(16.659, 43.977)$. Therefore, $x \approx -1.507$ and 16.659.

99. $x = \frac{1}{k}\ln(kv_0 t + 1)$

$$xk = \ln(kv_0 t + 1)$$
$$e^{xk} = e^{\ln(kv_0 t + 1)}$$
$$e^{xk} = kv_0 t + 1$$
$$e^{xk} - 1 = kv_0 t$$
$$\frac{e^{xk} - 1}{kt} = v_0 \text{ or } v_0 = \frac{e^{xk} - 1}{kt}$$

101. $\ln i - \ln I = \dfrac{-t}{RC}$

$$\ln\frac{i}{I} = \frac{-t}{RC}$$
$$e^{\ln(i/I)} = e^{-t/RC}$$
$$\frac{i}{I} = e^{-t/RC}$$
$$i = Ie^{-t/RC}$$

102. $h(x) = \dfrac{x^2+4x}{x+6}$

a. $h(-4) = \dfrac{(-4)^2+4(-4)}{(-4)+6} = \dfrac{16-16}{2} = \dfrac{0}{2} = 0$

b. $h\left(\dfrac{2}{5}\right) = \dfrac{\left(\frac{2}{5}\right)^2+4\left(\frac{2}{5}\right)}{\left(\frac{2}{5}\right)+6} = \dfrac{\frac{4}{25}+\frac{40}{25}}{\frac{2}{5}+\frac{30}{5}} = \dfrac{\frac{44}{25}}{\frac{32}{5}}$

$= \dfrac{44}{25}\cdot\dfrac{5}{32} = \dfrac{11}{40}$ or 0.275

103. Let x be the number of adult tickets sold and y be the number of children's tickets sold. The following system describes the situation.

$$x + \quad y = 550$$
$$15x + 11y = 7290$$

Solve by elimination.

$$-11(x+y=550) \Rightarrow -11x - 11y = -6050$$
$$15x+11y = 7290 \Rightarrow \underline{15x+11y = \ \ 7290}$$
$$4x \qquad = 1240$$
$$x = 310$$

Substitute $x = 310$ into the first equation to find y.

$310 + y = 550 \quad \Rightarrow \quad y = 240$

Thus, 310 adult tickets and 240 children's tickets must be sold.

104. $(3xy^2+y)(4x-3xy)$

$= 12x^2y^2 - 9x^2y^3 + 4xy - 3xy^2$

$= -9x^2y^3 + 12x^2y^2 - 3xy^2 + 4xy$

105. $4x^2 + bx + 25 = (2x)^2 + bx + (5)^2$ will be a perfect square trinomial if

$bx = \pm 2(2x)(5) \quad \Rightarrow \quad b = \pm 20$

106. $\sqrt[3]{x}\left(\sqrt[3]{x^2}+\sqrt[3]{x^5}\right) = \sqrt[3]{x}\cdot\sqrt[3]{x^2}+\sqrt[3]{x}\cdot\sqrt[3]{x^5}$

$= \sqrt[3]{x^3}+\sqrt[3]{x^6}$

$= x + x^2$

Chapter 9 Review Exercises

1. $(f \circ g)(x) = (2x-5)^2 - 3(2x-5) + 4$

$= 4x^2 - 20x + 25 - 6x + 15 + 4$

$= 4x^2 - 26x + 44$

2. $(f \circ g)(x) = 4x^2 - 26x + 44$

$(f \circ g)(3) = 4(3)^2 - 26(3) + 44$

$= 36 - 78 + 44$

$= 2$

3. $(g \circ f)(x) = 2(x^2 - 3x + 4) - 5$

$= 2x^2 - 6x + 8 - 5$

$= 2x^2 - 6x + 3$

4. $(g \circ f)(x) = 2x^2 - 6x + 3$

$(g \circ f)(-3) = 2(-3)^2 - 6(-3) + 3$

$= 18 + 18 + 3$

$= 39$

5. $(f \circ g)(x) = 6\sqrt{x-3}+7,\ x \ge 3$

6. $(g \circ f)(x) = \sqrt{(6x+7)-3} = \sqrt{6x+4},\ x \ge -\dfrac{2}{3}$

7. This function is one-to-one since it passes the horizontal line test.

8. The function is not one-to-one since the graph does not pass the horizontal line test.

9. Yes, the ordered pairs represent a one-to-one function. For each value of x, there is a unique value for y and each y-value has a unique x-value.

10. No, the ordered pairs do not represent a one-to-one function since the pairs (0, –2) and (3, –2) have different x-values but the same y-value.

11. Yes, the function $y = \sqrt{x+8}, x \ge -8$, is a one-to-one function since each output, y, corresponds to exactly one intput, x. The graph passes the horizontal line test.

12. No, $y = x^2 - 9$ is a parabola with vertex at (0, –9). It is not a one-to-one function since it does not pass the horizontal line test. Horizontal lines above $y = -9$ intersect the graph at two points.

13. $f(x)$: Domain: $\{-4, -1, 5, 6\}$
Range: $\{-3, 2, 3, 8\}$
$f^{-1}(x)$: Domain: $\{-3, 2, 3, 8\}$
Range: $\{-4, -1, 5, 6\}$

14. $f(x)$: Domain: $\{x \mid x \geq 0\}$
Range: $\{y \mid y \geq 4\}$
$f^{-1}(x)$: Domain: $\{x \mid x \geq 4\}$
Range: $\{y \mid y \geq 0\}$

15. $y = f(x) = 4x - 2$

$$x = 4y - 2$$
$$x + 2 = 4y$$
$$\frac{x+2}{4} = y \text{ or } y = f^{-1}(x) = \frac{x+2}{4}$$

x	$f(x)$
0	–2
$\frac{1}{2}$	0

x	$f^{-1}(x)$
0	$\frac{1}{2}$
–2	0

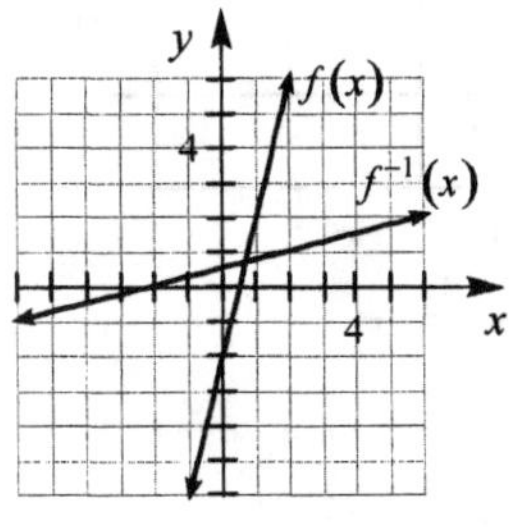

16. $y = f(x) = \sqrt[3]{x-1} = (x-1)^{1/3}$

$$x = (y-1)^{1/3}$$
$$x^3 = [(y-1)^{1/3}]^3$$
$$x^3 = y - 1$$
$$x^3 + 1 = y$$
$$f^{-1}(x) = x^3 + 1$$

x	$f(x)$
–7	–2
0	–1
1	0
2	1
9	2

x	$f^{-1}(x)$
–2	–7
–1	0
0	1
1	2
2	9

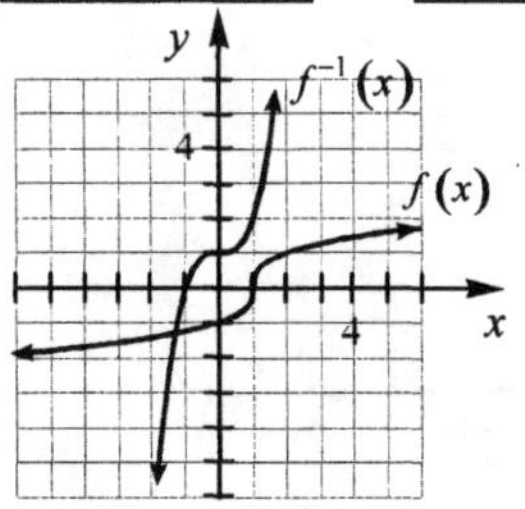

17. $f(x) = 36x \Rightarrow y = 36x$

$$x = 36y \Rightarrow y = \frac{x}{36} \Rightarrow f^{-1}(x) = \frac{x}{36}$$

$f^{-1}(x)$ represents yards and x represents inches.

18. $f(x) = 4x \Rightarrow y = 4x$

$$x = 4y \Rightarrow y = \frac{x}{4} \Rightarrow f^{-1}(x) = \frac{x}{4}$$

$f^{-1}(x)$ represents gallons and x represents quarts.

19. $y = 2^x$

x	–2	–1	0	1	2	3
y	$\frac{1}{4}$	$\frac{1}{2}$	1	2	4	8

Domain: $\mathbb{R}$
Range: $\{y \mid y > 0\}$

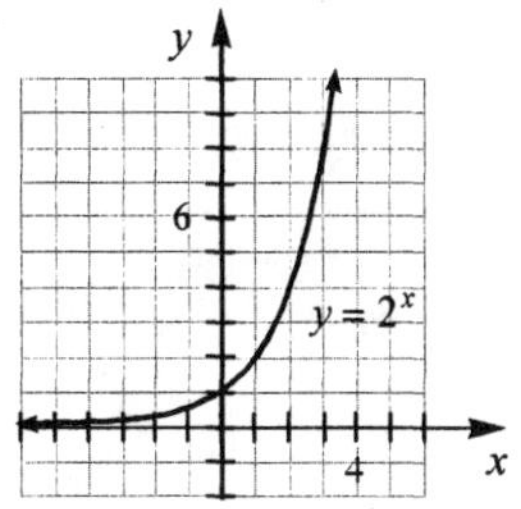

20. $y = \left(\frac{1}{2}\right)^x$

x	-2	-1	0	1	2
y	4	2	1	$\frac{1}{2}$	$\frac{1}{4}$

Domain: $\mathbb{R}$
Range: $\{y \mid y > 0\}$

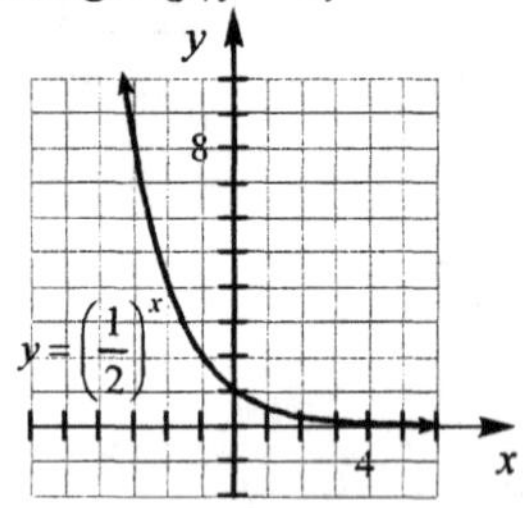

21. $f(t) = 7.02e^{0.365t}$

a. $t = 2003 - 1999 = 4$

$f(4) = 7.02e^{0.365(4)}$
≈ 30.23

The worldwide shipment in 2003 is about 30.23 million.

b. $t = 2005 - 1999 = 6$

$f(6) = 7.02e^{0.365(6)}$
≈ 62.73

The worldwide shipment in 2005 will be about 62.73 million.

c. $t = 2008 - 1999 = 9$

$f(9) = 7.02e^{0.365(9)}$
≈ 187.50

The worldwide shipment in 2008 will be about 187.50 million.

22. $8^2 = 64$

$\log_8 64 = 2$

23. $81^{1/4} = 3$

$\log_{81} 3 = \frac{1}{4}$

24. $5^{-3} = \frac{1}{125}$

$\log_5 \frac{1}{125} = -3$

25. $\log_2 32 = 5$

$2^5 = 32$

26. $\log_{1/4} \frac{1}{16} = 2$

$\left(\frac{1}{4}\right)^2 = \frac{1}{16}$

27. $\log_6 \frac{1}{36} = -2$

$6^{-2} = \frac{1}{36}$

28. $3 = \log_4 x$

$x = 4^3$

$x = 64$

29. $4 = \log_a 81$

$a^4 = 81$

$a^4 = 3^4$

$a = 3$

30. $-3 = \log_{1/5} x$

$x = \left(\frac{1}{5}\right)^{-3}$

$x = \frac{1}{\left(\frac{1}{5}\right)^3}$

$x = \frac{1}{\frac{1}{125}}$

$x = 125$

31. $y = \log_3 x$

$x = 3^y$

x	$\frac{1}{9}$	$\frac{1}{3}$	1	3	9	27
y	-2	-1	0	1	2	3

Domain: $\{x \mid x > 0\}$

Range: $\mathbb{R}$

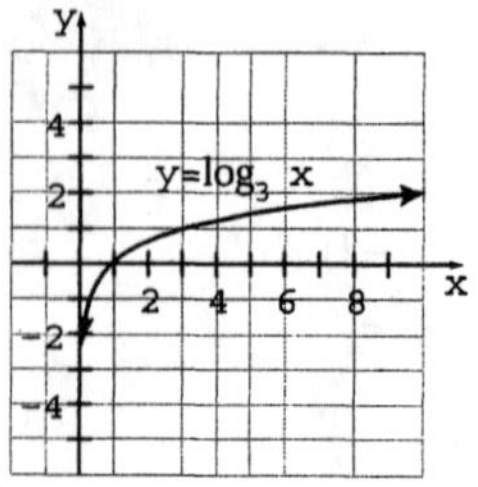

32. $y = \log_{1/2} x$

$x = \left(\frac{1}{2}\right)^y$

x	4	2	1	$\frac{1}{2}$	$\frac{1}{4}$
y	−2	−1	0	1	2

Domain: $\{x \mid x > 0\}$

Range: $\mathbb{R}$

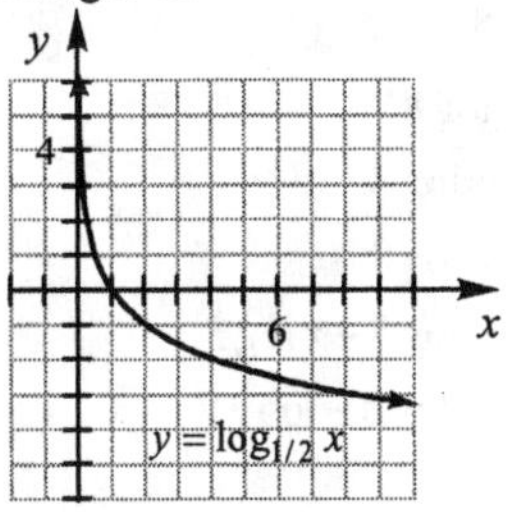

33. $\log_5 17^8 = 8\log_5 17$

34. $\log_3 \sqrt{x-9} = \log_3 (x-9)^{1/2} = \frac{1}{2}\log_3(x-9)$

35. $\log \frac{6(a+1)}{19} = \log\left[6(a+1)\right] - \log 19$

$= \log 6 + \log(a+1) - \log 19$

36. $\log \frac{x^4}{7(2x+3)^5} = \log x^4 - \log\left(7(2x+3)^5\right)$

$= 4\log x - [\log 7 + \log(2x+3)^5]$

$= 4\log x - [\log 7 + 5\log(2x+3)]$

$= 4\log x - \log 7 - 5\log(2x+3)$

37. $5\log x - 3\log(x+1) = \log x^5 - \log(x+1)^3$

$= \log \frac{x^5}{(x+1)^3}$

38. $4\left(\log 2 + \log x\right) - \log y = 4\left(\log\left(2x\right)\right) - \log y$

$= \log\left(2x\right)^4 - \log y$

$= \log \frac{\left(2x\right)^4}{y}$ or $\log \frac{16x^4}{y}$

39. $\frac{1}{3}[\ln x - \ln(x+2)] - \ln 2$

$= \frac{1}{3}\left(\ln \frac{x}{x+2}\right) - \ln 2$

$= \ln\left(\frac{x}{x+2}\right)^{1/3} - \ln 2$

$= \ln \frac{\sqrt[3]{\frac{x}{x+2}}}{2}$

40. $3\ln x + \frac{1}{2}\ln(x+1) - 6\ln(x+4)$

$= \ln x^3 + \ln(x+1)^{1/2} - \ln(x+4)^6$

$= \ln \frac{x^3\sqrt{x+1}}{(x+4)^6}$

41. $8^{\log_8 10} = 10$

42. $\log_4 4^5 = 5$

43. $11^{\log_9 81} = 11^{\log_9 9^2}$

$= 11^2$

$= 121$

44. $9^{\log_8 \sqrt{8}} = 9^{\log_8 8^{1/2}}$

$= 9^{1/2}$

$= \sqrt{9}$

$= 3$

45. log 819 = 2.9133

46. ln 0.0281 = −3..5720

47. antilog 3.159 = 1440

48. antilog(−3.157) = 0.000697

49. $\log N = 4.063$

$N = \text{antilog } 4.063$

$N = 11{,}600$

50. $\log N = -1.2262$

$N = \text{antilog } (-1.2262)$

$N = 0.0594$

51. $\log 10^5 = 5$

52. $10^{\log 9} = 9$

53. $7\log 10^{3.2} = 7(3.2) = 22.4$

54. $2\left(10^{\log 4.7}\right) = 2(4.7) = 9.4$

55.
$$\begin{aligned} 625 &= 5^x \\ 5^4 &= 5^x \\ 4 &= x \end{aligned}$$

56.
$$\begin{aligned} 49^x &= \frac{1}{7} \\ (7^2)^x &= 7^{-1} \\ 7^{2x} &= 7^{-1} \\ 2x &= -1 \\ x &= -\frac{1}{2} \end{aligned}$$

57.
$$\begin{aligned} 2^{3x-1} &= 32 \\ 2^{3x-1} &= 2^5 \\ 3x-1 &= 5 \\ 3x &= 6 \\ x &= 2 \end{aligned}$$

58.
$$\begin{aligned} 27^x &= 3^{2x+5} \\ (3^3)^x &= 3^{2x+5} \\ 3^{3x} &= 3^{2x+5} \\ 3x &= 2x+5 \\ x &= 5 \end{aligned}$$

59.
$$\begin{aligned} 7^x &= 152 \\ \log 7^x &= \log 152 \\ x\log 7 &= \log 152 \\ x &= \frac{\log 152}{\log 7} \\ x &\approx 2.582 \end{aligned}$$

60.
$$\begin{aligned} 3.1^x &= 856 \\ \log 3.1^x &= \log 856 \\ x\log 3.1 &= \log 856 \\ x &= \frac{\log 856}{\log 3.1} \\ x &\approx 5.968 \end{aligned}$$

61.
$$\begin{aligned} 12.5^{x+1} &= 381 \\ \log 12.5^{x+1} &= \log 381 \\ (x+1)\log 12.5 &= \log 381 \\ x+1 &= \frac{\log 381}{\log 12.5} \\ x &= \frac{\log 381}{\log 12.5} - 1 \\ x &\approx 1.353 \end{aligned}$$

62.
$$\begin{aligned} 3^{x+2} &= 8^x \\ \log 3^{x+2} &= \log 8^x \\ (x+2)\log 3 &= x\log 8 \\ x\log 3 + 2\log 3 &= x\log 8 \\ 2\log 3 &= x\log 8 - x\log 3 \\ 2\log 3 &= x(\log 8 - \log 3) \\ \frac{2\log 3}{\log 8 - \log 3} &= x \\ 2.240 &\approx x \end{aligned}$$

63.
$$\begin{aligned} \log_7(2x-3) &= 2 \\ 7^2 &= 2x-3 \\ 49 &= 2x-3 \\ 52 &= 2x \\ 26 &= x \end{aligned}$$

64.
$$\begin{aligned} \log x + \log(4x-19) &= \log 5 \\ \log\left(x(4x-19)\right) &= \log 5 \\ x(4x-19) &= 5 \\ 4x^2 - 19x - 5 &= 0 \\ (4x+1)(x-5) &= 0 \end{aligned}$$

$$4x+1=0 \quad \text{or} \quad x-5=0$$
$$x = -\frac{1}{4} \qquad\qquad x = 5$$

Only $x = 5$ checks. $x = -\frac{1}{4}$ is an extraneous solution since $\log(4x-19)$ becomes $\log(-20)$ which is not a real number.

65. $\log_3 x + \log_3(2x+1) = 1$

$$\log_3 x(2x+1) = 1$$
$$3^1 = x(2x+1)$$
$$0 = 2x^2 + x - 3$$
$$0 = (2x+3)(x-1)$$

$2x+3=0$ or $x-1=0$

$x = -\frac{3}{2}$ $\quad x = 1$

Only $x = 1$ checks. $x = -\dfrac{3}{2}$ is an extraneous solution since $\log_3 x$ becomes $\log_3\left(-\dfrac{3}{2}\right)$ which is not a real number.

66. $\ln(x+1) - \ln(x-2) = \ln 4$

$$\ln\frac{x+1}{x-2} = \ln 4$$
$$\frac{x+1}{x-2} = 4$$
$$x+1 = 4(x-2)$$
$$x+1 = 4x-8$$
$$1 = 3x-8$$
$$9 = 3x$$
$$x = 3$$

67.

$$50 = 25e^{0.6t}$$
$$2 = e^{0.6t}$$
$$\ln 2 = \ln e^{0.6t}$$
$$\ln 2 = 0.6t$$
$$\frac{\ln 2}{0.6} = t$$
$$1.155 \approx t$$

68.

$$100 = A_0 e^{-0.42(3)}$$
$$100 = A_0 e^{-1.26}$$
$$\frac{100}{e^{-1.26}} = A_0$$
$$352.542 \approx A_0$$

69.

$$A = A_0 e^{kt}$$
$$\frac{A}{A_0} = e^{kt}$$
$$\ln\frac{A}{A_0} = \ln e^{kt}$$
$$\ln\frac{A}{A_0} = kt$$
$$\frac{\ln\frac{A}{A_0}}{k} = t$$
$$\frac{\ln A - \ln A_0}{k} = t \text{ or } t = \frac{\ln A - \ln A_0}{k}$$

70.

$$200 = 800e^{kt}$$
$$\frac{200}{800} = e^{kt}$$
$$0.25 = e^{kt}$$
$$\ln 0.25 = \ln e^{kt}$$
$$\ln 0.25 = kt$$
$$\frac{\ln 0.25}{t} = k \text{ or } k = \frac{\ln 0.25}{t}$$

71.

$$\ln y - \ln x = 6$$
$$\ln\frac{y}{x} = 6$$
$$e^{\ln\frac{y}{x}} = e^6$$
$$\frac{y}{x} = e^6$$
$$y = xe^6$$

72. $\ln(y+1) - \ln(x+8) = \ln 3$

$$\ln\frac{y+1}{x+8} = \ln 3$$
$$\frac{y+1}{x+8} = 3$$
$$y+1 = 3(x+8)$$
$$y = 3(x+8) - 1$$
$$y = 3x + 24 - 1$$
$$y = 3x + 23$$

73. $\log_2 196 = \dfrac{\log 196}{\log 2} \approx 7.6147$

74. $\log_3 47 = \dfrac{\log 47}{\log 3} \approx 3.5046$

75. $A = P(1+r)^n$
$= 12,000(1+0.06)^8$
$= 12,000(1.06)^8$
$= 19,126.18$
The amount is \$19,126.18.

76. $P = P_0 e^{kt}$
$P_0 = 6,000, k = 0.04,$ and $P = 12,000$
$12,000 = 12,000e^{(0.04)t}$
$2 = e^{0.04t}$
$\ln 2 = 0.04t$
$t = \frac{\ln 2}{0.04}$
$t \approx 17.3$
It will take about 17.3 years for the \$6,000 to double.

77. $N(t) = 2000(2)^{0.05t}$

a. Let $N(t) = 50,000$.
$50,000 = 2000(2)^{0.05t}$
$\frac{50,000}{2000} = 2^{0.05t}$
$25 = 2^{0.05t}$
$\log 25 = \log 2^{0.05t}$
$\log 25 = 0.05t \log 2$
$\frac{\log 25}{0.05 \log 2} = t$
$92.88 \approx t$
The time is 92.88 minutes.

b. Let $N(t) = 120,000$.
$120,000 = 2000(2)^{0.05t}$
$\frac{120,000}{2000} = 2^{0.05t}$
$60 = 2^{0.05t}$
$\log 60 = \log 2^{0.05t}$
$\log 60 = 0.05t \log 2$
$\frac{\log 60}{0.05 \log 2} = t$
$118.14 \approx t$
The time is 118.14 minutes.

78. $P = 14.7e^{-0.00004x}$
$P = 14.7e^{-0.00004(8842)}$
$P = 14.7e^{-0.35368}$
$P \approx 14.7(0.7021)$
$P \approx 10.32$
The atmospheric pressure is 10.32 pounds per square inch at 8,842 feet above sea level.

79. $A(n) = 72 - 18\log(n+1)$

a. $A(0) = 72 - 18\log(0+1)$
$= 72 - 18\log(1)$
$= 72 - 18(0)$
$= 72$
The original class average was 72.

b. $A(3) = 72 - 18\log(3+1)$
$= 72 - 18\log(4)$
$\approx 72 - 10.8$
$= 61.2$
After 3 months, the class average was 61.2.

c. Let $A(n) = 58.0$.
$58.0 = 72 - 18\log(n+1)$
$-14 = -18\log(n+1)$
$\frac{7}{9} = \log(n+1)$
$10^{7/9} = 10^{\log(n+1)}$
$10^{7/9} = n+1$
$10^{7/9} - 1 = n$
$5.0 \approx n$
It takes about 5 months.

Chapter 9 Practice Test

1. a. Yes, {(4, 2), (–3, 8), (–1, 3), (6, –7)} is one-to-one.

b. {(2, 4), (8, –3), (3, –1), (–7, 6)} is the inverse function.

2. a. $(f \circ g)(x) = f[g(x)]$
$= f(x+2)$
$= (x+2)^2 - 3$
$= x^2 + 4x + 4 - 3$
$= x^2 + 4x + 1$

b. $(f \circ g)(6) = 6^2 + 4(6) + 1$
$= 36 + 24 + 1$
$= 61$

3. a. $(g \circ f)(x) = g[f(x)]$
$= g(x^2 + 8)$
$= \sqrt{x^2 + 8 - 5}$
$= \sqrt{x^2 + 3}$

b. $(g \circ f)(7) = \sqrt{7^2 + 3}$
$= \sqrt{49 + 3}$
$= \sqrt{52}$
$= 2\sqrt{13}$

4. a. $y = f(x) = -3x - 5$
$x = -3y - 5$
$x + 5 = -3y$
$\frac{x+5}{-3} = y$
$-\frac{1}{3}(x+5) = y$
$f^{-1}(x) = -\frac{1}{3}(x+5)$

b.

x	$f(x)$
0	−5
$-\frac{5}{3}$	0

x	$f^{-1}(x)$
0	$-\frac{5}{3}$
−5	0

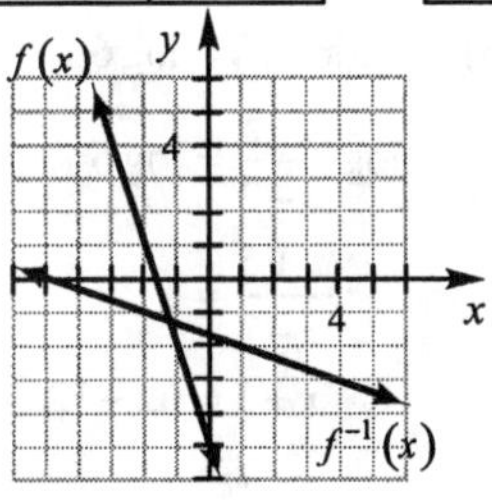

5. a. $y = f(x) = \sqrt{x-1}, x \geq 1$
$x = (y-1)^{1/2}$
$x^2 = [(y-1)^{1/2}]^2$
$x^2 = y - 1$
$x^2 + 1 = y$
$f^{-1}(x) = x^2 + 1,\ x \geq 0$

b.

x	$f(x)$
1	0
2	1
5	2

x	$f^{-1}(x)$
0	1
1	2
2	5

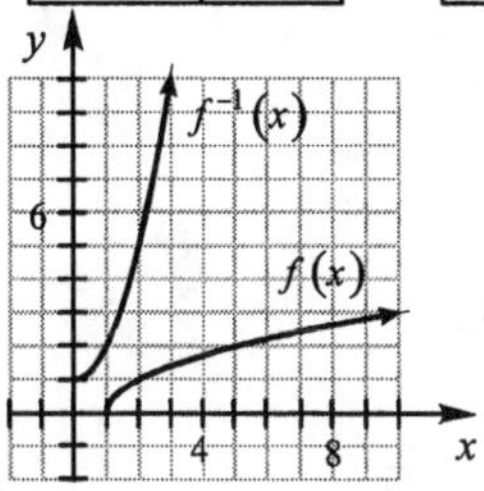

6. The domain of $y = \log_5(x)$ is $\{x \mid x > 0\}$.

7. $\log_4 \frac{1}{256} = \log_4 4^{-4} = -4$

8. $y = 3^x$

x	−2	−1	0	2	3
y	$\frac{1}{9}$	$\frac{1}{3}$	1	9	27

Domain: $\mathbb{R}$
Range: $\{y \mid y > 0\}$

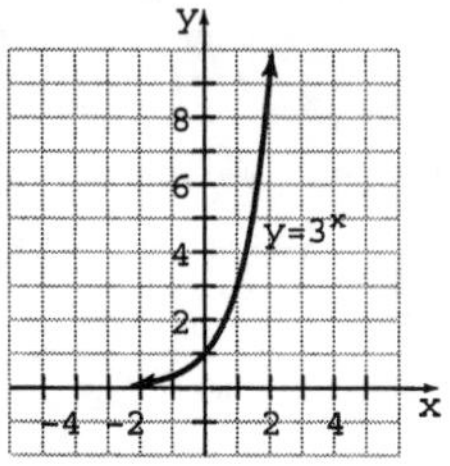

9. $y = \log_2 x$

$x = 2^y$

x	$\frac{1}{4}$	$\frac{1}{2}$	1	2	4
y	−2	−1	0	1	2

Domain: $\{x \mid x > 0\}$

Range: $\mathbb{R}$

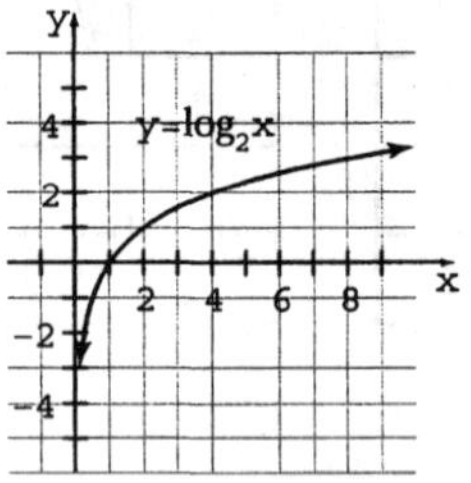

10. $2^{-5} = \frac{1}{32}$

$\log_2 \frac{1}{32} = -5$

11. $\log_5 125 = 3$

$5^3 = 125$

12. $4 = \log_2(x+3)$

$2^4 = x+3$

$16 = x+3$

$13 = x$

13. $y = \log_{64} 16$

$64^y = 16$

$\left(4^3\right)^y = 4^2$

$4^{3y} = 4^2$

$3y = 2$

$y = \frac{2}{3}$

14. $\log_2 \frac{x^3(x-4)}{x+2}$

$= \log_2 x^3(x-4) - \log_2(x+2)$

$= \log_2 x^3 + \log_2(x-4) - \log_2(x+2)$

$= 3\log_2 x + \log_2(x-4) - \log_2(x+2)$

15. $7\log_6(x-4) + 2\log_6(x+3) - \frac{1}{2}\log_6 x$

$= \log_6(x-4)^7 + \log_6(x+3)^2 - \log_6 x^{1/2}$

$= \log_6 \frac{(x-4)^7(x+3)^2}{\sqrt{x}}$

16. $10\log_9 \sqrt{9} = 10\log_9 9^{1/2}$

$= 10 \cdot \frac{1}{2}$

$= 5$

17. a. $\log 4620 \approx 3.6646$

b. $\ln 0.0692 \approx -2.6708$

18. $3^x = 19$

$\log 3^x = \log 19$

$x \log 3 = \log 19$

$x = \frac{\log 19}{\log 3}$

$x \approx 2.68$

19. $\log 4x = \log(x+3) + \log 2$

$\log 4x = \log 2(x+3)$

$4x = 2(x+3)$

$4x = 2x+6$

$2x = 6$

$x = 3$

20. $\log(x+5) - \log(x-2) = \log 6$

$\log \frac{x+5}{x-2} = \log 6$

$\frac{x+5}{x-2} = 6$

$x+5 = 6x-12$

$17 = 5x$

$\frac{17}{5} = x$

21. $\ln N = 2.79$

$e^{2.79} = N$

$16.2810 \approx N$

22. $\log_6 40 = \dfrac{\log 40}{\log 6} \approx 2.0588$

23.
$$100 = 250e^{-0.03t}$$
$$\frac{100}{250} = e^{-0.03t}$$
$$\ln\frac{100}{250} = -0.03t$$
$$\frac{\ln\frac{100}{250}}{-0.03} = t$$
$$t \approx 30.5430$$

24. $A = p\left(1+\dfrac{r}{n}\right)^{nt}$

Use $p = 3500$, $r = 0.04$ and $n = 4$

$t = 10$

$$A = 3500\left(1+\frac{0.04}{4}\right)^{4\cdot 10}$$
$$= 3500(1.01)^{40}$$
$$= 5211.02$$

The amount in the account is \$5211.02.

25. $v = v_0 e^{-0.0001205t}$

Use $v = 40$, and $v_0 = 60$.

$$40 = 60e^{-0.0001205t}$$
$$\frac{40}{60} = e^{-0.0001205t}$$
$$\frac{2}{3} = e^{-0.0001205t}$$
$$\ln\frac{2}{3} = \ln e^{-0.0001205t}$$
$$\ln\frac{2}{3} = -0.0001205t$$
$$\frac{\ln\frac{2}{3}}{-0.0001205} = t$$
$$3364.86 \approx t$$

The fossil is approximately 3364.86 years old.

Chapter 9 Cumulative Review Test

1.
$$\frac{\left(2xy^2z^{-3}\right)^2}{\left(3x^{-1}yz^2\right)^{-1}} = \frac{2^2x^2y^4z^{-6}}{3^{-1}x^1y^{-1}z^{-2}}$$
$$= \frac{2^2\cdot 3xy^5}{z^4}$$
$$= \frac{12xy^5}{z^4}$$

2.
$$5^2-\left(2-3^2\right)^2+4^3$$
$$5^2-(2-9)^2+4^3$$
$$5^2-(-7)^2+4^3$$
$$25-49+64$$
$$40$$

3. Let r be the tax rate.
$$92+92r = 98.90$$
$$92r = 6.90$$
$$r = \frac{6.90}{92}$$
$$\approx 0.075$$

The tax rate is 7.5%.

4.
$$-3 \le 2x-7 < 8$$
$$-3+7 \le 2x < 8+7$$
$$4 \le 2x < 15$$
$$2 \le x < \frac{15}{2}$$
$$\left\{x \,\middle|\, 2 \le x < \frac{15}{2}\right\},\ \left[2, \frac{15}{2}\right)$$

5.
$$2x-3y = 8$$
$$-3y = 8-2x$$
$$y = \frac{8-2x}{-3} \text{ or } y = \frac{2x-8}{3}$$

6.
$$h(x) = \frac{x^2+4x}{x+6}$$
$$h(-4) = \frac{(-4)^2+4(-4)}{(-4)+6}$$
$$= \frac{16-16}{2} = \frac{0}{2} = 0$$

7. Two points on the line are (0, 3) and (–2,–1).

$m=\frac{y_2-y_1}{x_2-x_1}=\frac{-1-3}{-2-0}=\frac{-4}{-2}=2$

Use point-slope form with $(0,3)$ and $m=2$.

$y-3=2(x-0)$

$y-3=2x$

$y=2x+3$

8. $4x=3y-3 \Rightarrow y=\frac{4}{3}x+3$

Plot the y-intercept (0, 3) and use the slope to plot additional points.

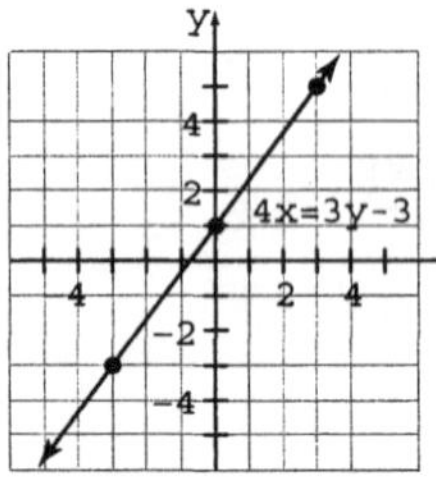

9. $y\le\frac{1}{3}x+6$

Plot a solid line at $y=\frac{1}{3}x+6$.

Use the point (0, 0) as the check point.

$y\le\frac{1}{3}x+6$

$0\le\frac{1}{3}(0)+6$

$0\le 6$ True

Therefore, shade the half-plane containing $(0,0)$.

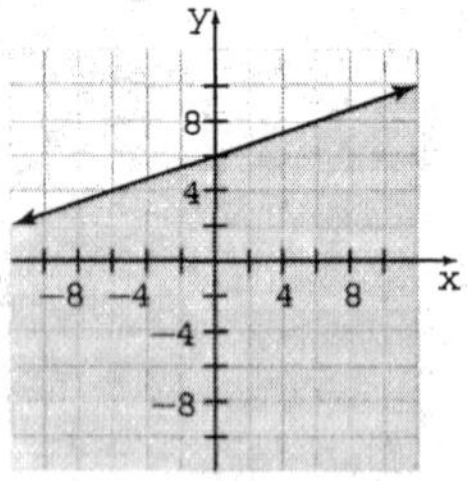

10. $\frac{1}{2}x+\frac{1}{3}y=13$

$\frac{1}{5}x+\frac{1}{8}y=5$

Clear out the fractions by multiplying through by its least common denominator.

$6\left(\frac{1}{2}x+\frac{1}{3}y=13\right) \Rightarrow 3x+2y=78 \quad (1)$

$40\left(\frac{1}{5}x+\frac{1}{8}y=5\right) \Rightarrow 8x+5y=200 \quad (2)$

Multiply equation (1) by –5, multiply equation (2) by 2 and add the results.

$$\begin{aligned} -5(3x+2y=78) &\Rightarrow & -15x-10y &= -390 \\ 2(8x+5y=200) &\Rightarrow & 16x+10y &= 400 \\ \hline & & x &= 10 \end{aligned}$$

Substitute 10 for x in equation (1).

$3(10)+2y=78$

$30+2y=78$

$2y=48$

$y=24$

The solution is (10, 24).

11. $\frac{x^3+3x^2+5x+9}{x+1}$

Using synthetic division with $c=-1$:

$$\begin{array}{r|rrrr} -1 & 1 & 3 & 5 & 9 \\ & & -1 & -2 & -3 \\ \hline & 1 & 2 & 3 & 6 \end{array}$$

$\frac{x^3+3x^2+5x+9}{x+1}=x^2+2x+3+\frac{6}{x+1}$

12. $x^2-2xy+y^2-64$

$(x^2-2xy+y^2)-64$

$(x-y)^2-64$

$(x-y+8)(x-y-8)$

13. $(2x+1)^2-9=0$

$(2x+1)^2=9$

$2x+1=\pm\sqrt{9}$

$2x+1=\pm 3$

$x=\dfrac{-1\pm 3}{2}$

$x=\dfrac{-1+3}{2}$ or $x=\dfrac{-1-3}{2}$

$=\dfrac{2}{2}$ $\quad=\dfrac{-4}{2}$

$=1$ $\quad=-2$

The solutions are 1 and -2.

14. $\dfrac{2x+3}{x+1}=\dfrac{3}{2}$

$(2)(2x+3)=(3)(x+1)$

$4x+6=3x+3$

$x=-3$

15. $a_n=a_1+nd-d$, for d

$a_n-a_1=nd-d$

$a_n-a_1=d(n-1)$

$\dfrac{a_n-a_1}{n-1}=d$

16. The equation of variation is

$L=\dfrac{k}{P^2}$, with $P=4$ and $k=100$.

$L=\dfrac{100}{4^2}=\dfrac{100}{16}=6.25$

17. $4\sqrt{45x^3}+\sqrt{5x}$

$4\sqrt{9x^2\cdot 5x}+\sqrt{5x}$

$12x\sqrt{5x}+\sqrt{5x}$

$(12x+1)\sqrt{5x}$

18. $\sqrt{2a+9}-a+3=0$

$\sqrt{2a+9}=a-3$

$\left(\sqrt{2a+9}\right)^2=(a-3)^2$

$2a+9=a^2-6a+9$

$a^2-8a=0$

$a(a-8)=0$

$a=0$ or $a-8=0$

$a=8$

Upon checking, $a=0$ is an extraneous solution.
The only solution is $a=8$.

19. $(x^2-5)^2+3(x^2-5)-10=0$

$u^2+3u-10=0$

$(u+5)(u-2)=0$

$u+5=0$ or $u-2=0$

$x^2-5+5=0$ $\quad x^2-5-2=0$

$x^2=0$ $\quad x^2=7$

$x=0$ $\quad x=\pm\sqrt{7}$

The solutions are 0, $-\sqrt{7}$, and $\sqrt{7}$.

20. $g(x)=x^2-4x-5$

a. $g(x)=x^2-4x-5$

$g(x)=(x^2-4x+4)-5-4$

$g(x)=(x-2)^2-9$

b.

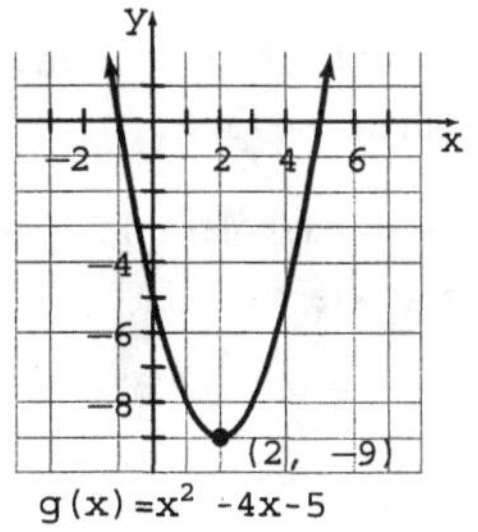

Chapter 10

Exercise Set 10.1

1.

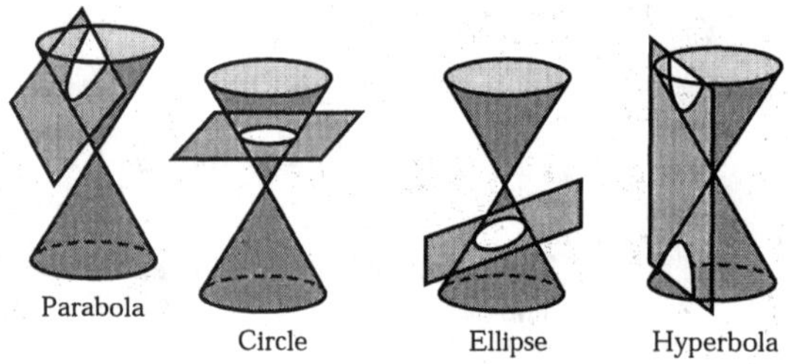

3. Yes, any parabola in the form $y = a(x-h)^2 + k$ is a function because each value of x corresponds to only one value of y. The domain is $\mathbb{R}$, the set of all real numbers. Since the vertex is at (h, k) and $a > 0$, the range is $\{y \mid y \geq k\}$.

5. The graphs have the same vertex, (3, 4). The first graph opens upward, and the second one opens downward.

7. The distance is always a positive number because both distances are squared and we use the principal square root.

9. A circle is the set of all points in a plane that are the same distance from a fixed point.

11. No, the coefficients of the y^2 - term and the x^2 - term must both be the same.

13. No, the coefficients of the y^2 - term and the x^2 - term must both be the same.

15. No, equations of parabolas do not include both x^2 - and y^2 - terms.

17. $y = (x-2)^2 + 3$

This is a parabola in the form $y = a(x-h)^2 + k$ with $a = 1$, $h = 2$ and $k = 3$. Since $a > 0$, the parabola opens upward. The vertex is (2, 3). The y-intercept is (0, 7). There are no x-intercepts.

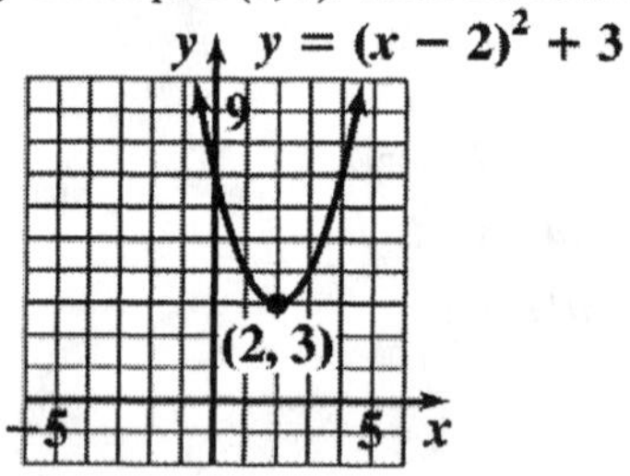

19. $y = (x+3)^2 + 2$

This is a parabola in the form $y = a(x-h)^2 + k$ with $a = 1$, $h = -3$ and $k = 2$. Since $a > 0$, the parabola opens upward. The vertex is (–3, 2). The y-intercept is (0, 11). There are no x-intercepts.

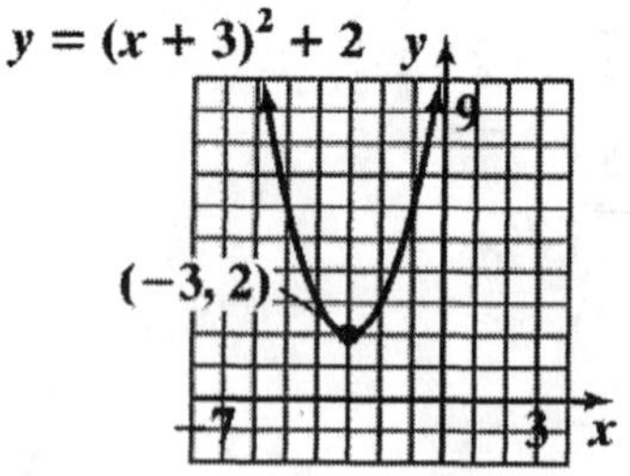

21. $y=(x-2)^2-1$

This is a parabola in the form $y=a(x-h)^2+k$ with $a=1$, $h=2$ and $k=-1$. Since $a>0$, the parabola opens upward. The vertex is $(2, -1)$. The y-intercept is $(0, 3)$. The x-intercepts are $(1, 0)$ and $(3, 0)$.

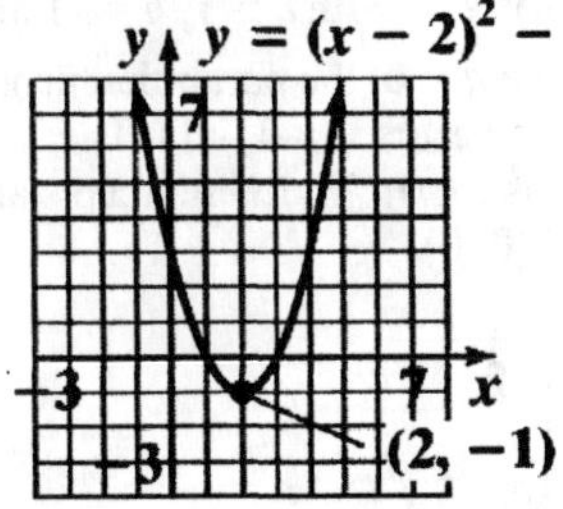

23. $y=-(x-1)^2+1$

This is a parabola in the form $y=a(x-h)^2+k$ with $a=-1$, $h=1$ and $k=1$. Since $a<0$, the parabola opens downward. The vertex is $(1, 1)$. The y-intercept is $(0, 0)$. The x-intercepts are $(0, 0)$ and $(2, 0)$.

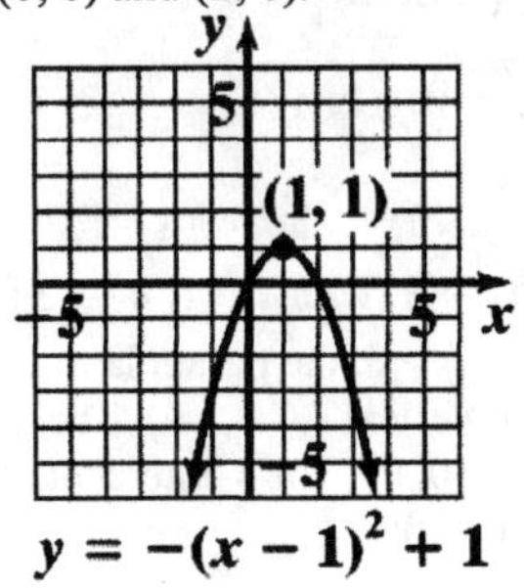

25. $y=-(x+3)^2+4$

This is a parabola in the form $y=a(x-h)^2+k$ with $a=-1$, $h=-3$ and $k=4$. Since $a<0$, the parabola opens downward. The vertex is $(-3, 4)$. The y-intercept is $(0, -5)$. The x-intercepts are $(-5, 0)$ and $(-1, 0)$.

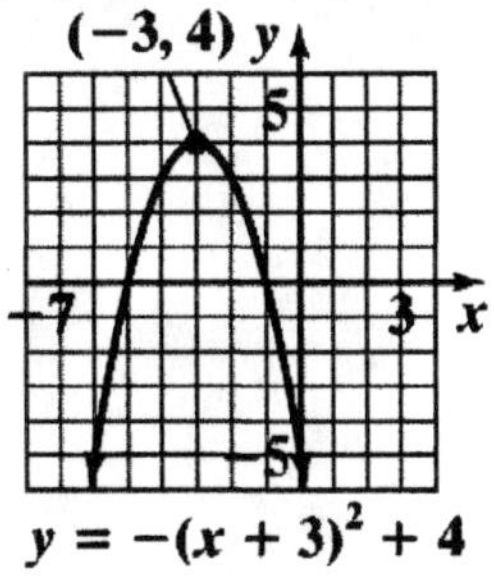

27. $y=-3(x-5)^2+3$

This is a parabola in the form $y=a(x-h)^2+k$ with $a=-3$, $h=5$ and $k=3$. Since $a<0$, the parabola opens downward. The vertex is $(5, 3)$. The y-intercept is $(0, -72)$. The x-intercepts are $(4, 0)$ and $(6, 0)$.

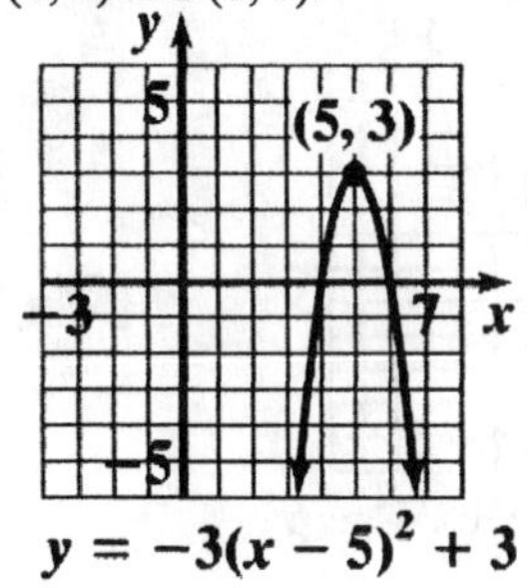

29. $x=(y-4)^2-3$

This is a parabola in the form $x=a(y-k)^2+h$ with $a=1$, $h=-3$ and $k=4$. Since $a>0$, the parabola opens to the right. The vertex is $(-3, 4)$. The y-intercepts are about $(0, 2.27)$ and $(0, 5.73)$. The x-intercept is $(13, 0)$.

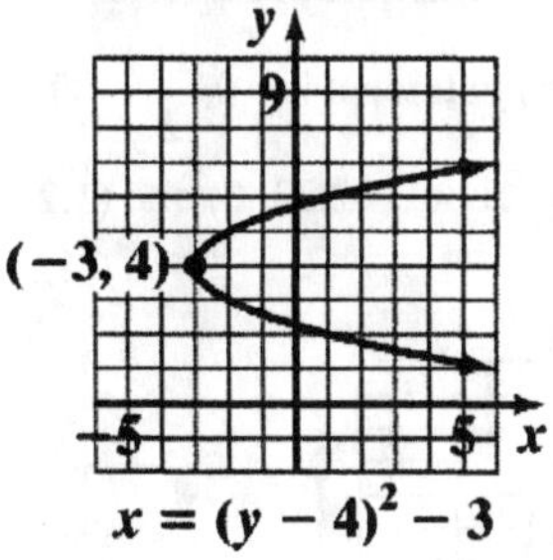

31. $x=-(y-5)^2+4$

This is a parabola in the form $x=a(y-k)^2+h$ with $a=-1$, $h=4$ and $k=5$. Since $a<0$, the parabola opens to the left. The vertex is $(4, 5)$. The y-intercepts are $(0, 3)$ and $(0, 7)$. The x-intercept is $(-21, 0)$.

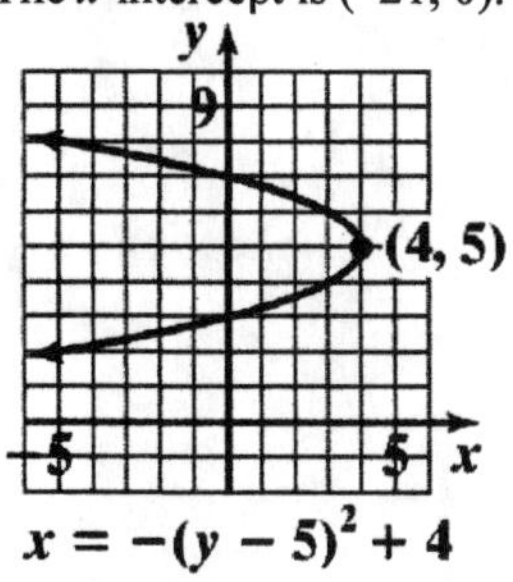

33. $x = -5(y+3)^2 - 6$

This is a parabola in the form $x = a(y-k)^2 + h$ with $a = -5$, $h = -6$ and $k = -3$. Since $a < 0$, the parabola opens to the left. The vertex is $(-6, -3)$. There are no y-intercepts. The x-intercept is $(-51, 0)$

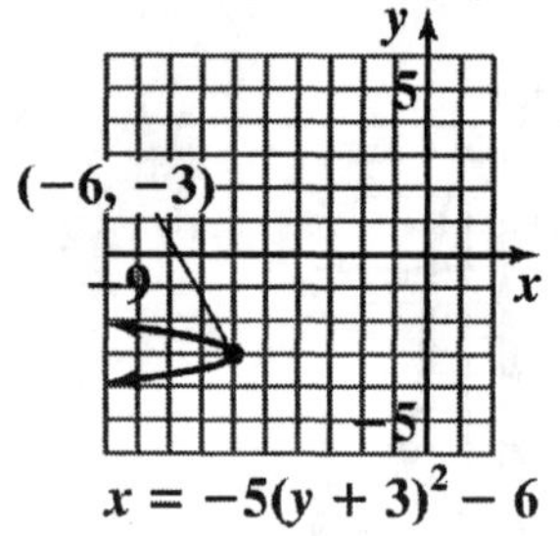

35. $y = -2\left(x+\frac{1}{2}\right)^2 + 6$

This is a parabola in the form $y = a(x-h)^2 + k$ with $a = -2$, $h = -\frac{1}{2}$ and $k = 6$. Since $a < 0$, the parabola opens downward. The vertex is $\left(-\frac{1}{2}, 6\right)$. The y-intercept is $\left(0, \frac{11}{2}\right)$. The x-intercepts are about $(-2.23, 0)$ and $(1.23, 0)$.

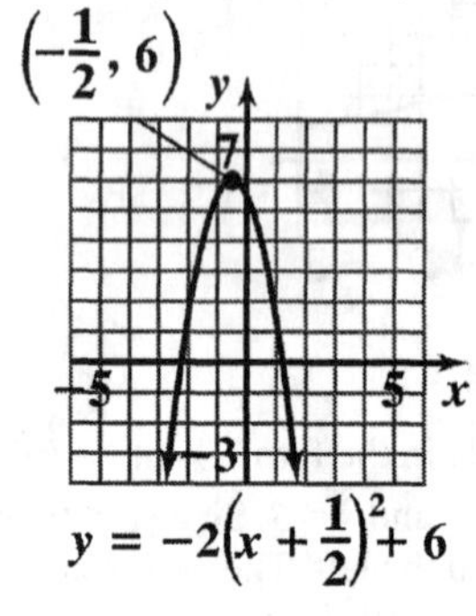

37. a. $y = x^2 + 2x$

$y = (x^2 + 2x + 1) - 1$

$y = (x+1)^2 - 1$

b. This is a parabola in the form $y = a(x-h)^2 + k$ with $a = 1$, $h = -1$ and $k = -1$. Since $a > 0$, the parabola opens upward. The vertex is $(-1, -1)$. The y-intercept is $(0, 0)$. The x-intercepts are $(-2, 0)$ and $(0, 0)$.

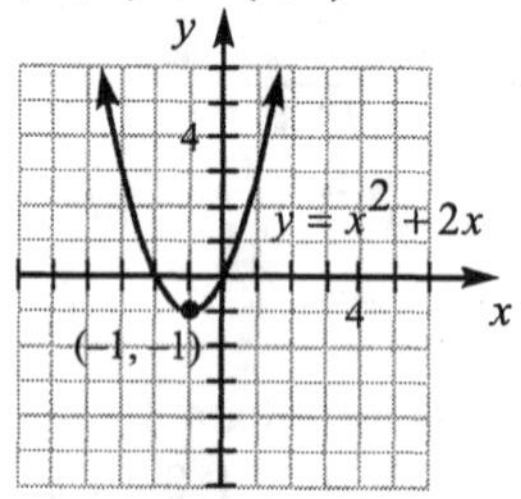

39. a. $y = x^2 + 6x$

$y = (x^2 + 6x + 9) - 9$

$y = (x+3)^2 - 9$

b. This is a parabola in the form $y = a(x-h)^2 + k$ with $a = 1$, $h = -3$ and $k = -9$. Since $a > 0$, the parabola opens upward. The vertex is $(-3, -9)$. The y-intercept is $(0, 0)$. The x-intercepts are $(-6, 0)$ and $(0, 0)$.

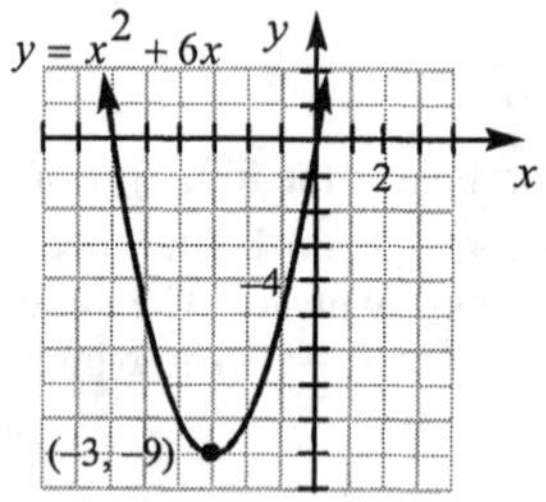

41. a. $x = y^2 + 4y$

$x = (y^2 + 4y + 4) - 4$

$x = (y + 2)^2 - 4$

b. This is a parabola in the form $x = a(y-k)^2 + h$ with $a = 1$, $h = -4$ and $k = -2$. Since $a > 0$, the parabola opens to the right. The vertex is (–4, –2). The y-intercepts are (0, –4) and (0, 0). The x-intercept is (0, 0).

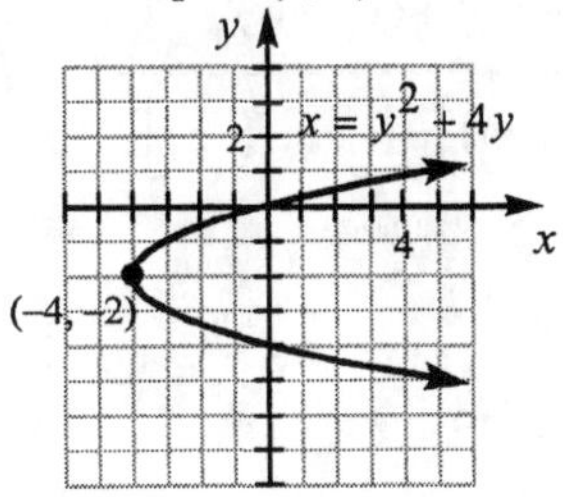

43. a. $y = x^2 + 7x + 10$

$y = \left(x^2 + 7x + \frac{49}{4}\right) - \frac{49}{4} + 10$

$y = \left(x + \frac{7}{2}\right)^2 - \frac{9}{4}$

b. This is a parabola in the form $y = a(x-h)^2 + k$ with $a = 1$, $h = -\frac{7}{2}$ and $k = -\frac{9}{4}$. Since $a > 0$, the parabola opens upward. The vertex is $\left(-\frac{7}{2}, -\frac{9}{4}\right)$. The y-intercept is (0, 10). The x-intercepts are (–5, 0) and (–2, 0).

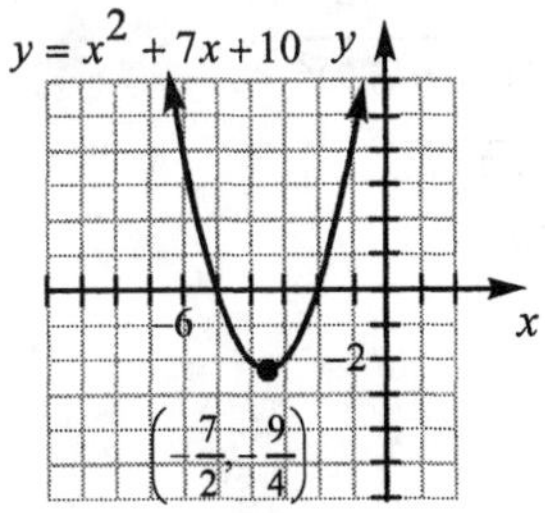

45. a. $x = -y^2 + 6y - 9$

$x = -(y^2 - 6y) - 9$

$x = -(y^2 - 6y + 9) + 9 - 9$

$x = -(y - 3)^2$

b. This is a parabola in the form $x = a(y-k)^2 + h$ with $a = -1$, $h = 0$ and $k = 3$. Since $a < 0$, the parabola opens to the left. The vertex is (0, 3). The y-intercept is (0, 3). The x-intercept is (–9, 0).

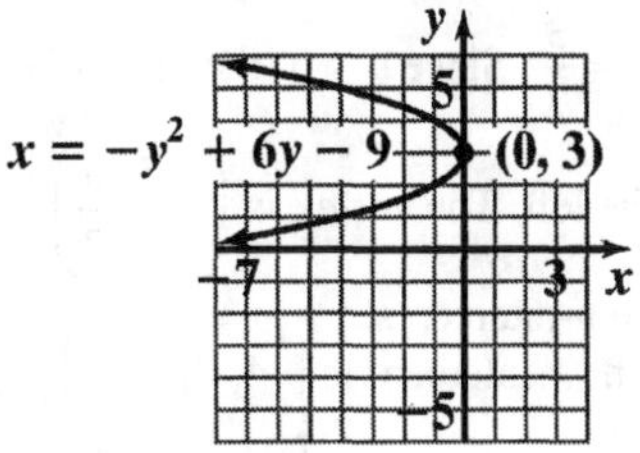

47. a. $y = -x^2 + 4x - 4$

$y = -(x^2 - 4x) - 4$

$y = -(x^2 - 4x + 4) + 4 - 4$

$y = -(x - 2)^2$

b. This is a parabola in the form $y = a(x-h)^2 + k$ with $a = -1$, $h = 2$ and $k = 0$. Since $a < 0$, the parabola opens downward. The vertex is (2, 0). The y-intercept is (0, –4). The x-intercept is (2, 0).

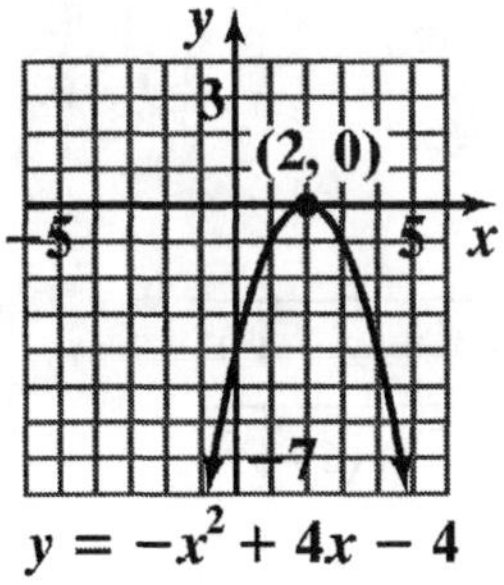

49. a. $x = -y^2 + 3y - 4$

$x = -\left(y^2 - 3y\right) - 4$

$x = -\left(y^2 - 3y + \frac{9}{4}\right) + \frac{9}{4} - 4$

$x = -\left(y - \frac{3}{2}\right)^2 - \frac{7}{4}$

b. This is a parabola in the form $x = a(y-k)^2 + h$ with $a = -1$, $h = -\frac{7}{4}$ and $k = \frac{3}{2}$. Since $a < 0$, the parabola opens to the left. The vertex is $\left(-\frac{7}{4}, \frac{3}{2}\right)$. There are no y-intercepts.
The x-intercept is $(-4, 0)$.

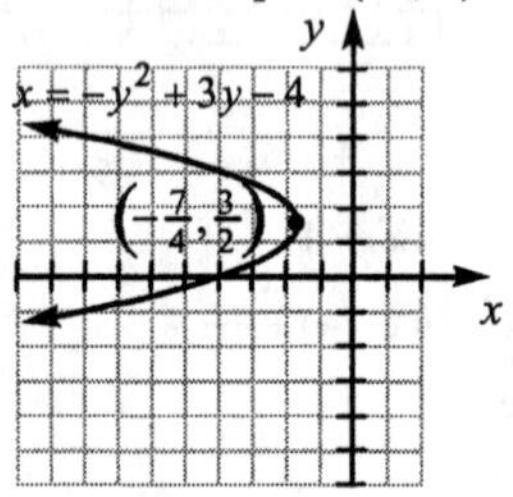

51. $d = \sqrt{(x_2 - x_1)^2 + (y_2 - y_1)^2}$
$= \sqrt{(5-5)^2 + [-6-(-1)]^2}$
$= \sqrt{0^2 + (-5)^2}$
$= \sqrt{0 + 25}$
$= \sqrt{25}$
$= 5$

53. $d = \sqrt{(x_2 - x_1)^2 + (y_2 - y_1)^2}$
$= \sqrt{[8-(-1)]^2 + (6-6)^2}$
$= \sqrt{9^2 + 0^2}$
$= \sqrt{81 + 0}$
$= \sqrt{81}$
$= 9$

55. $d = \sqrt{(x_2 - x_1)^2 + (y_2 - y_1)^2}$
$= \sqrt{[4-(-1)]^2 + [9-(-3)]^2}$
$= \sqrt{5^2 + 12^2}$
$= \sqrt{25 + 144}$
$= \sqrt{169}$
$= 13$

57. $d = \sqrt{(x_2 - x_1)^2 + (y_2 - y_1)^2}$
$= \sqrt{[5-(-4)]^2 + [-2-(-5)]^2}$
$= \sqrt{9^2 + 3^2}$
$= \sqrt{81 + 9}$
$= \sqrt{90} \approx 9.49$

59. $d = \sqrt{(x_2 - x_1)^2 + (y_2 - y_1)^2}$
$= \sqrt{\left(\frac{1}{2} - 3\right)^2 + [4-(-1)]^2}$
$= \sqrt{\left(-\frac{5}{2}\right)^2 + 5^2}$
$= \sqrt{\frac{25}{4} + 25}$
$= \sqrt{\frac{125}{4}}$
≈ 5.59

61. $d = \sqrt{(x_2 - x_1)^2 + (y_2 - y_1)^2}$
$= \sqrt{[-4.3-(-1.6)]^2 + (-1.7-3.5)^2}$
$= \sqrt{(-2.7)^2 + (-5.2)^2}$
$= \sqrt{7.29 + 27.04}$
$= \sqrt{34.33}$
≈ 5.86

63. $d = \sqrt{(x_2 - x_1)^2 + (y_2 - y_1)^2}$
$= \sqrt{\left(0 - \sqrt{7}\right)^2 + \left[0 - \sqrt{3}\right]^2}$
$= \sqrt{\left(-\sqrt{7}\right)^2 + \left(\sqrt{3}\right)^2}$
$= \sqrt{7 + 3}$
$= \sqrt{10}$
≈ 3.16

65. $\text{Midpoint} = \left(\frac{x_1 + x_2}{2}, \frac{y_1 + y_2}{2}\right)$
$= \left(\frac{1+5}{2}, \frac{9+3}{2}\right)$
$= (3, 6)$

67. $\text{Midpoint} = \left(\frac{x_1 + x_2}{2}, \frac{y_1 + y_2}{2}\right)$
$= \left(\frac{-7+7}{2}, \frac{2+(-2)}{2}\right)$
$= (0, 0)$

69. $\text{Midpoint} = \left(\frac{x_1 + x_2}{2}, \frac{y_1 + y_2}{2}\right)$
$= \left(\frac{-1+4}{2}, \frac{4+6}{2}\right)$
$= \left(\frac{3}{2}, 5\right)$

71. $\text{Midpoint} = \left(\frac{x_1 + x_2}{2}, \frac{y_1 + y_2}{2}\right)$
$= \left(\frac{3+2}{2}, \frac{\frac{1}{2}+(-4)}{2}\right)$
$= \left(\frac{5}{2}, -\frac{7}{4}\right)$

73. $\text{Midpoint} = \left(\frac{x_1 + x_2}{2}, \frac{y_1 + y_2}{2}\right)$
$= \left(\frac{\sqrt{3}+\sqrt{2}}{2}, \frac{2+7}{2}\right)$
$= \left(\frac{\sqrt{3}+\sqrt{2}}{2}, \frac{9}{2}\right)$

75. $(x-h)^2 + (y-k)^2 = r^2$
$(x-0)^2 + (y-0)^2 = 4^2$
$x^2 + y^2 = 16$

77. $(x-h)^2 + (y-k)^2 = r^2$
$(x-2)^2 + (y-0)^2 = 5^2$
$(x-2)^2 + y^2 = 25$

79. $(x-h)^2 + (y-k)^2 = r^2$
$(x-0)^2 + \left(y-5\right)^2 = 1^2$
$x^2 + (y-5)^2 = 1$

81. $(x-h)^2 + (y-k)^2 = r^2$
$(x-3)^2 + (y-4)^2 = \left(8\right)^2$
$(x-3)^2 + (y-4)^2 = 64$

83. $(x-h)^2 + (y-k)^2 = r^2$
$\left(x-7\right)^2 + \left[y-(-6)\right]^2 = 10^2$
$(x-7)^2 + (y+6)^2 = 100$

85. $(x-h)^2 + (y-k)^2 = r^2$
$(x-1)^2 + (y-2)^2 = \left(\sqrt{5}\right)^2$
$(x-1)^2 + (y-2)^2 = 5$

87. The center is (0, 0) and the radius is 4.
$(x-h)^2 + (y-k)^2 = r^2$
$(x-0)^2 + (y-0)^2 = 4^2$
$x^2 + y^2 = 16$

89. The center is (3, –2) and the radius is 3.
$(x-h)^2 + (y-k)^2 = r^2$
$(x-3)^2 + [y-(-2)]^2 = 3^2$
$(x-3)^2 + (y+2)^2 = 9$

91. $x^2 + y^2 = 16$

$x^2 + y^2 = 4^2$

The graph is a circle with its center at the origin and radius 4.

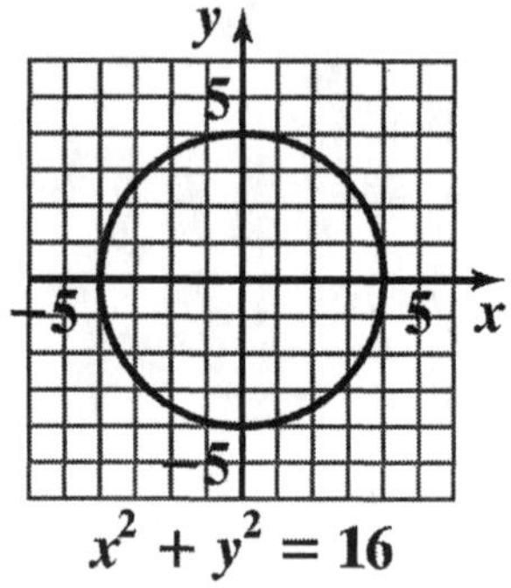

93. $x^2 + y^2 = 10$

$x^2 + y^2 = \left(\sqrt{10}\right)^2$

The graph is a circle with its center at the origin and radius $\sqrt{10}$.

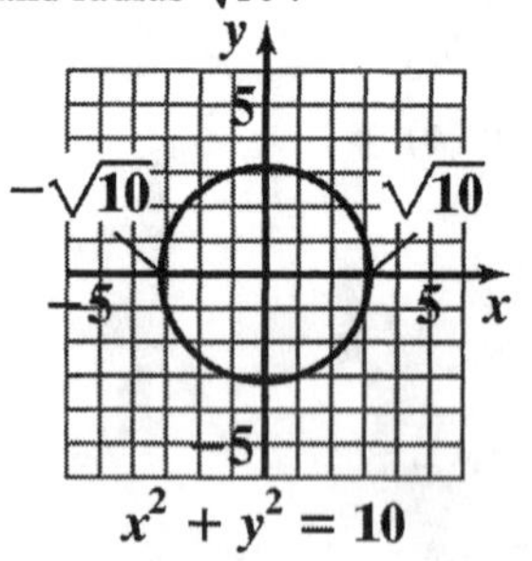

95. $(x+4)^2 + y^2 = 25$

$(x+4)^2 + (y-0)^2 = 5^2$

The graph is a circle with its center at (−4, 0) and radius 5.

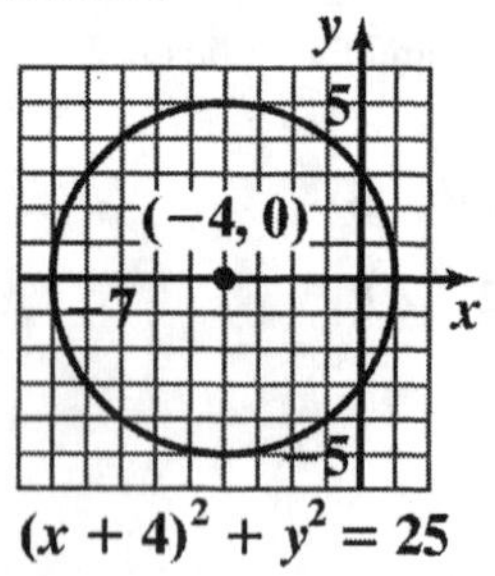

97. $x^2 + (y-3)^2 = 4$

$(x-0)^2 + (y-3)^2 = (2)^2$

The graph is a circle with its center at (0, 3) and radius 2.

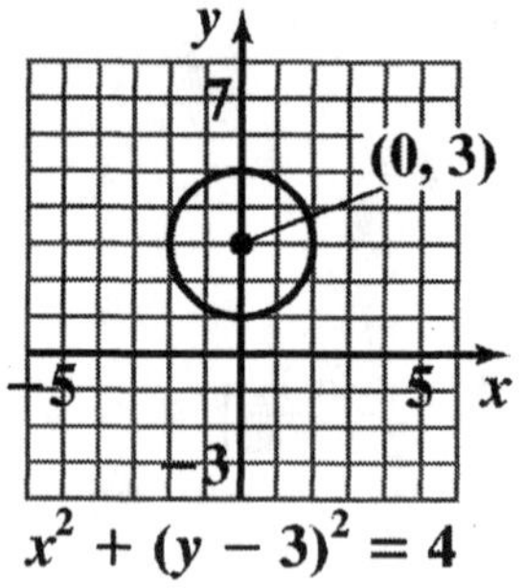

99. $(x+8)^2 + (y+2)^2 = 9$

$(x+8)^2 + (y+2)^2 = 3^2$

The graph is a circle with its center at (−8, −2) and radius 3.

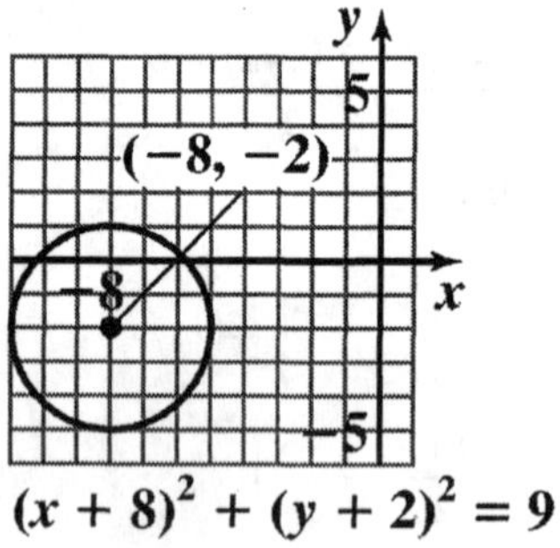

101. $y = \sqrt{25 - x^2}$

If we solve $x^2 + y^2 = 25$ for y, we obtain $y = \pm\sqrt{25 - x^2}$. Therefore, the graph of $y = \sqrt{25 - x^2}$ is the upper half ($y \geq 0$) of a circle with its center at the origin and radius 5.

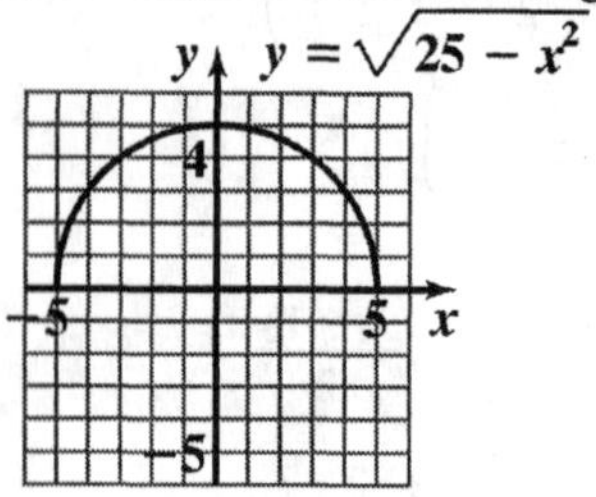

103. $y=-\sqrt{4-x^2}$

If we solve $x^2+y^2=4$ for *y*, we obtain $y=\pm\sqrt{4-x^2}$. Therefore, the graph of $y=-\sqrt{4-x^2}$ is the lower half ($y \le 0$) of a circle with its center at the origin and radius 2.

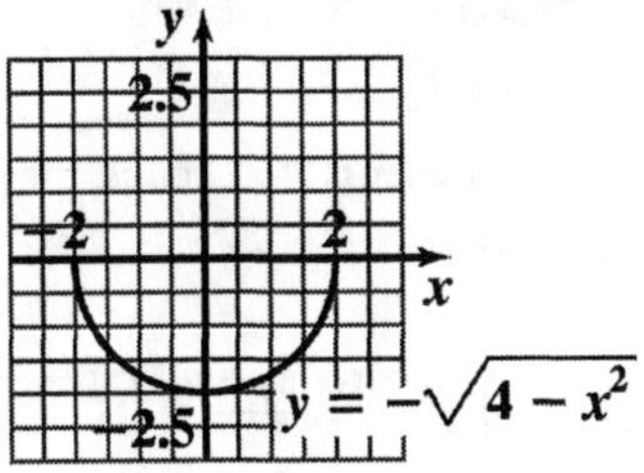

105. a.

$$\begin{aligned} x^2+y^2+8x+15&=0\\ x^2+8x+y^2&=-15\\ (x^2+8x+16)+y^2&=-15+16\\ (x+4)^2+y^2&=1\\ (x+4)^2+y^2&=1^2 \end{aligned}$$

b. The graph is a circle with center (–4, 0) and radius 1.

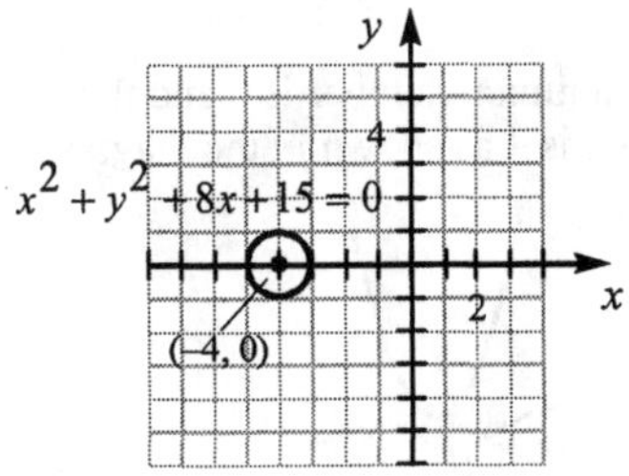

107. a.

$$\begin{aligned} x^2+y^2+6x-4y+4&=0\\ x^2+6x+y^2-4y&=-4\\ (x^2+6x+9)+(y^2-4y+4)&=-4+9+4\\ (x+3)^2+(y-2)^2&=9\\ (x+3)^2+(y-2)^2&=3^2 \end{aligned}$$

b. The graph is a circle with center (–3, 2) and radius 3.

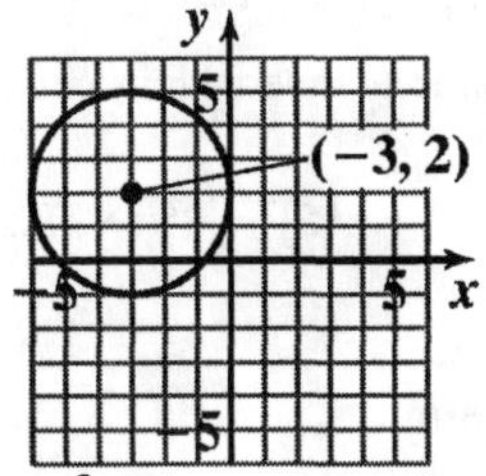

$x^2+y^2+6x-4y+4=0$

109. a.

$$\begin{aligned} x^2+y^2+6x-2y+6&=0\\ x^2+6x+y^2-2y&=-6\\ (x^2+6x+9)+(y^2-2y+1)&=-6+9+1\\ (x+3)^2+(y-1)^2&=4\\ (x+3)^2+(y-1)^2&=2^2 \end{aligned}$$

b. The graph is a circle with center (–3, 1) and radius 2.

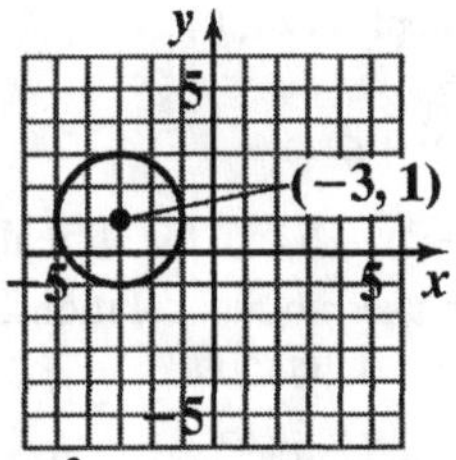

$x^2+y^2+6x-2y+6=0$

111. a.

$$\begin{aligned} x^2+y^2-8x+2y+13&=0\\ x^2-8x+y^2+2y&=-13\\ (x^2-8x+16)+(y^2+2y+1)&=-13+16+1\\ (x-4)^2+(y+1)^2&=4\\ (x-4)^2+(y+1)^2&=2^2 \end{aligned}$$

b. The graph is a circle with center (4, –1) and radius 2.

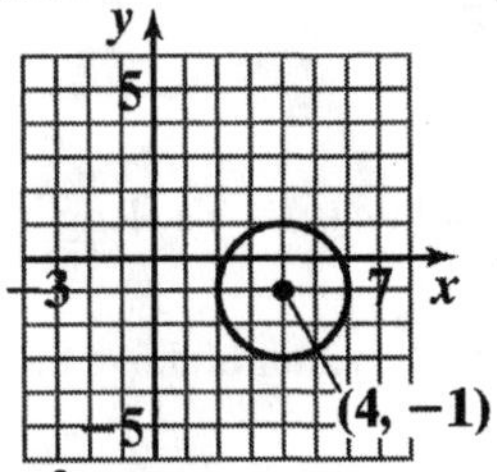

$x^2+y^2-8x+2y+13=0$

113. $(x+4)^2 + y^2 = 25$

$(x+4)^2 + (y-0)^2 = 5^2$

The graph is a circle with its center at (–4, 0) and radius 5.

Area $= \pi r^2 = \pi(5)^2 = 25\pi \approx 78.5$ sq. units

115. x-intercept:

$x = 0^2 - 6(0) - 7$

$x = -7$

The x-intercept is (–7, 0)

y-intercepts:

$0 = y^2 - 6y - 7$

$0 = (y+1)(y-7)$

$y = -1$ or $y = 7$

The y-intercepts are (0, –1) and (0, 7).

117. x-intercept:

$x = 2(0-3)^2 + 6$

$x = 24$

The x-intercept is (24, 0).

y-intercepts:

$0 = 2(y-3)^2 + 6$

Since $2(y-3)^2 + 6 \geq 6$ for all real values of y, this equation has no real solutions.

There are no y-intercepts.

119. No. For example, the origin is the midpoint of both the segment from (1, 1) to (–1, –1) and the segment from (2, 2) to (–2, –2), but these segments have different lengths.

121. The distance from the midpoint (4, –6) to the endpoint (7, –2) is half the length of the line segment.

$$\frac{d}{2} = \sqrt{(7-4)^2 + [-2-(-6)]^2}$$
$$= \sqrt{3^2 + 4^2}$$
$$= \sqrt{25}$$
$$= 5$$

Since $\frac{d}{2} = 5$, $d = 10$. The length is 10 units.

123. Since (–6, 2) is 2 units above the x-axis, the radius is 2.

$(x-h)^2 + (y-k)^2 = r^2$

$(x+6)^2 + (y-2)^2 = 2^2$

$(x+6)^2 + (y-2)^2 = 4$

125. a. Diameter $= \sqrt{(x_2 - x_1)^2 + (y_2 - y_1)^2}$

$= \sqrt{(9-5)^2 + (8-4)^2}$

$= \sqrt{4^2 + 4^2}$

$= \sqrt{16+16}$

$= \sqrt{32}$

$= 4\sqrt{2}$

Since the diameter is $4\sqrt{2}$ units, the radius is $2\sqrt{2}$ units.

b. Midpoint $= \left(\frac{x_1 + x_2}{2}, \frac{y_1 + y_2}{2}\right)$

$= \left(\frac{5+9}{2}, \frac{4+8}{2}\right)$

$= (7, 6)$

The center is (7, 6).

c. $(x-h)^2 + (y-k)^2 = r^2$

$(x-7)^2 + (y-6)^2 = \left(2\sqrt{2}\right)^2$

$(x-7)^2 + (y-6)^2 = 8$

127. The minimum number is 0 and the maximum number is 4 as shown in the diagrams.

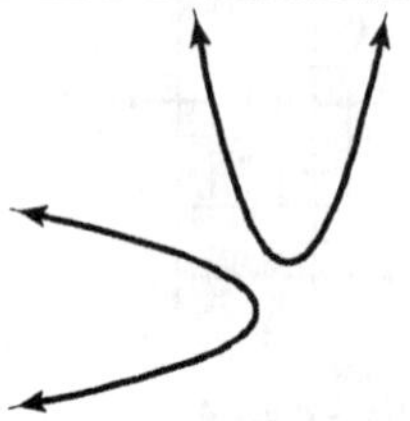

No points of intersection

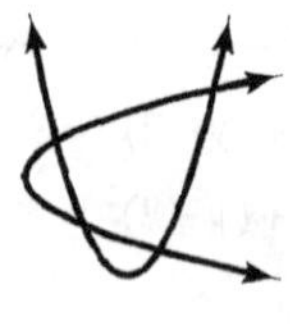

Four points of intersection

129. a. Since $150 - 2(68.2) = 13.6$, the clearance is 13.6 feet.

b. Since $150 - 68.2 = 81.8$, the center of the wheel is 81.8 feet above the ground.

c. $$(x-h)^2+(y-k)^2=r^2$$
$$(x-0)^2+(y-81.8)^2=68.2^2$$
$$x^2+(y-81.8)^2=68.2^2$$
$$x^2+(y-81.8)^2=4651.24$$

131. a. The center of the blue circle is the origin, and the radius is 4.
$$x^2+y^2=r^2$$
$$x^2+y^2=4^2$$
$$x^2+y^2=16$$

b. The center of the red circle is (2, 0), and the radius is 2.
$$(x-h)^2+(y-k)^2=r^2$$
$$(x-2)^2+(y-0)^2=2^2$$
$$(x-2)^2+y^2=4$$

c. The center of the green circle is (–2, 0), and the radius is 2.
$$(x-h)^2+(y-k)^2=r^2$$
$$[x-(-2)]^2+(y-0)^2=2^2$$
$$(x+2)^2+y^2=4$$

d. Shaded area = (blue circle area) – (red circle area) – (green circle area)
$$=\pi(4^2)-\pi(2^2)-\pi(2^2)$$
$$=16\pi-4\pi-4\pi$$
$$=8\pi$$

133. The radii are 4 and 8, respectively. So, the area between the circles is
$\pi(8)^2-\pi(4)^2=64\pi-16\pi=48\pi$ square units.

136. $$\frac{6x^{-3}y^4}{18x^{-2}y^3}=\frac{6}{18}x^{-3-(-2)}y^{4-3}$$
$$=\frac{1}{3}x^{-3+2}y^1$$
$$=\frac{1}{3}x^{-1}y \text{ or } \frac{y}{3x}$$

137. $$-4<3x-4<17$$
$$-4+4<3x-4+4<17+4$$
$$0<3x<21$$
$$\frac{0}{3}<\frac{3x}{3}<\frac{21}{3}$$
$$0<x<7$$
In interval notation: $(0, 7)$

138. $$\begin{vmatrix}4&0&3\\5&2&-1\\3&6&4\end{vmatrix}=4\begin{vmatrix}2&-1\\6&4\end{vmatrix}-5\begin{vmatrix}0&3\\6&4\end{vmatrix}+3\begin{vmatrix}0&3\\2&-1\end{vmatrix}$$
$$=4(8+6)-5(0-18)+3(0-6)$$
$$=4(14)-5(-18)+3(-6)$$
$$=56+90-18$$
$$=128$$

139. a. area 1: a^2 area 2: ab
area 3: ab area 4: b^2

b. $$a^2+ab+ab+b^2=a^2+2ab+b^2$$
$$=(a+b)^2$$

140. This is a parabola in the form $y=a(x-h)^2+k$ with $a=1$, $h=4$ and $k=1$. Since $a>0$, the parabola opens upward. The vertex is (4, 1). The y-intercept is (0, 17). There are no x-intercepts.

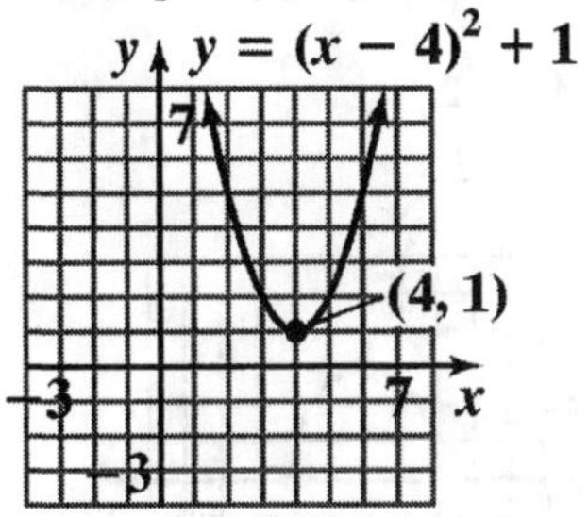

Exercise Set 10.2

1. An ellipse is a set of points in a plane, the sum of whose distances from two fixed points is constant.

3. $\frac{(x-h)^2}{a^2}+\frac{(y-k)^2}{b^2}=1$

5. If $a=b$, the formula for a circle is obtained.

7. First divide both sides by 180.

9. No, the sign in front of the y^2 is negative.

11. $\frac{x^2}{4}+\frac{y^2}{1}=1$

Since $a^2=4, a=2$.

Since $b^2=1, b=1$.

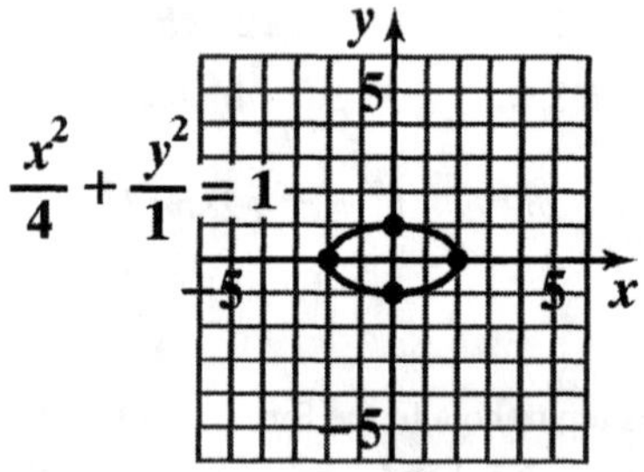

13. $\frac{x^2}{4}+\frac{y^2}{9}=1$

Since $a^2=4, a=2$.

Since $b^2=9, b=3$.

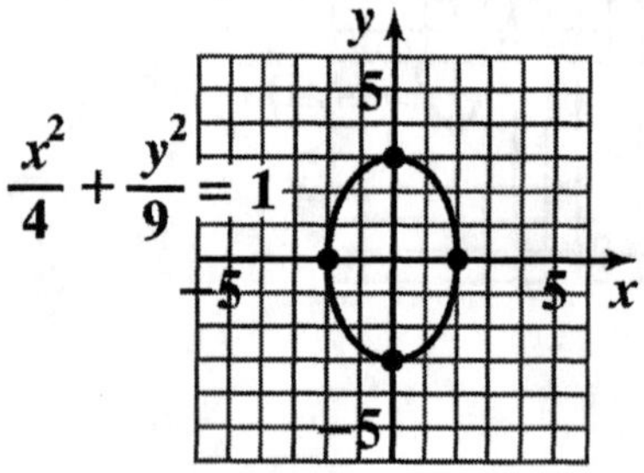

15. $\frac{x^2}{25}+\frac{y^2}{9}=1$

Since $a^2=25, a=5$.

Since $b^2=9, b=3$.

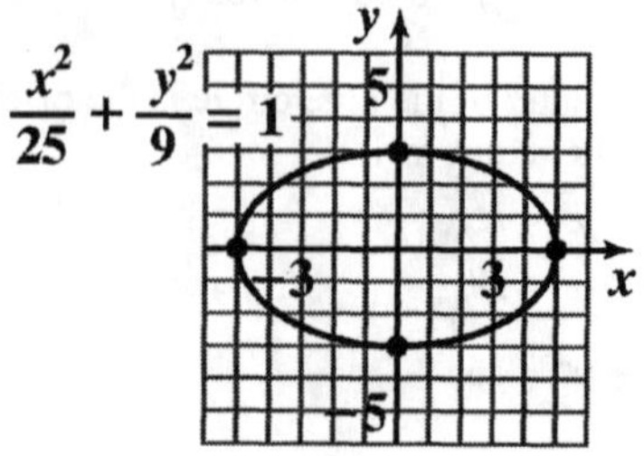

17. $\frac{x^2}{16}+\frac{y^2}{25}=1$

Since $a^2=16, a=4$.

Since $b^2=25, b=5$.

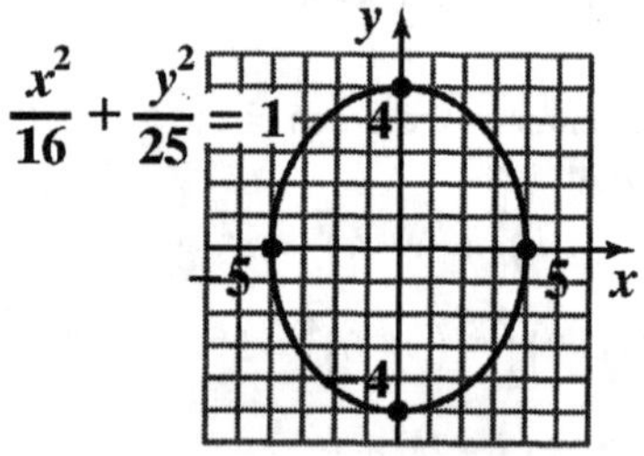

19. $x^2+16y^2=16$

$\frac{x^2}{16}+\frac{16y^2}{16}=1$

$\frac{x^2}{16}+\frac{y^2}{1}=1$

Since $a^2=16, a=4w$.

Since $b^2=1, b=1$.

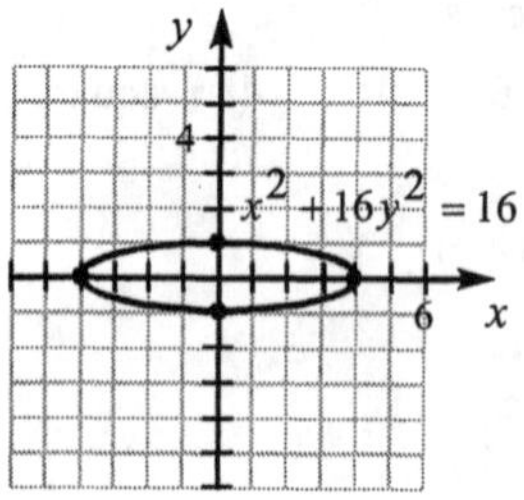

21. $49x^2 + y^2 = 49$

$$\frac{49x^2}{49} + \frac{y^2}{49} = 1$$

$$\frac{x^2}{1} + \frac{y^2}{49} = 1$$

Since $a^2 = 1, a = 1.$

Since $b^2 = 49, b = 7.$

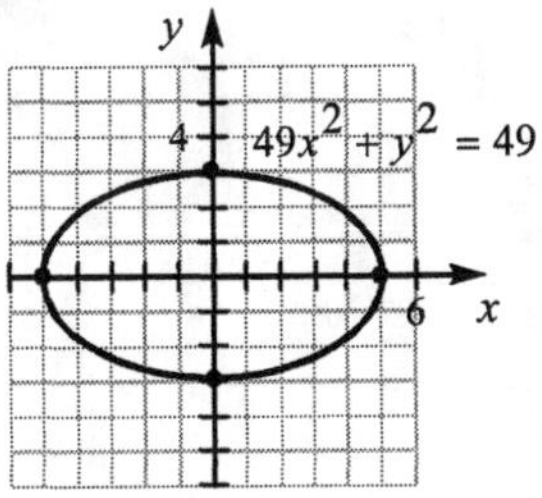

23. $9x^2 + 16y^2 = 144$

$$\frac{9x^2}{144} + \frac{16y^2}{144} = 1$$

$$\frac{x^2}{16} + \frac{y^2}{9} = 1$$

Since $a^2 = 16, a = 4$.

Since $b^2 = 9, b = 3.$

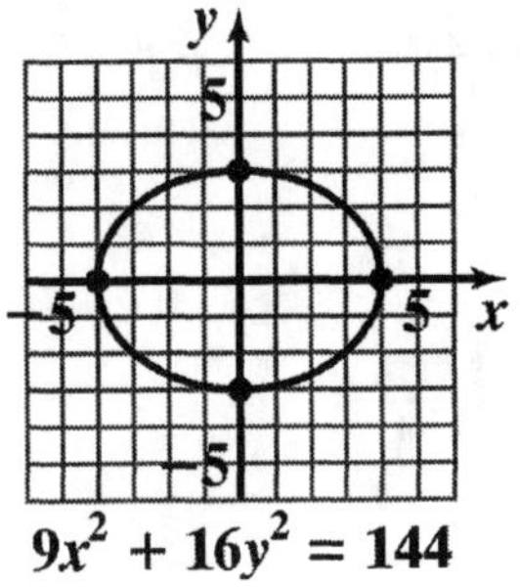

25. $25x^2 + 100y^2 = 400$

$$x^2 + 4y^2 = 16$$

$$\frac{x^2}{16} + \frac{4y^2}{16} = 1$$

$$\frac{x^2}{16} + \frac{y^2}{4} = 1$$

Since $a^2 = 16, a = 4.$

Since $b^2 = 4, b = 2.$

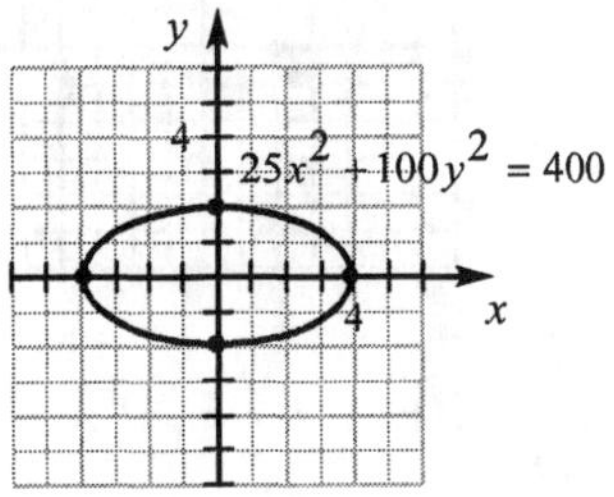

27. $x^2 + 2y^2 = 8$

$$\frac{x^2}{8} + \frac{2y^2}{8} = 1$$

$$\frac{x^2}{8} + \frac{y^2}{4} = 1$$

Since $a^2 = 8, a = \sqrt{8} = 2\sqrt{2} \approx 2.83.$

Since $b^2 = 4, b = 2.$

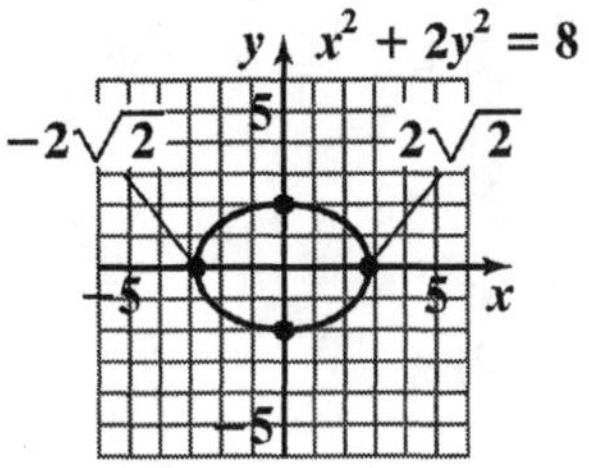

29. $\frac{x^2}{16}+\frac{(y-2)^2}{9}=1$

The center is (0, 2).

Since $a^2=16, a=4$.

Since $b^2=9, b=3$.

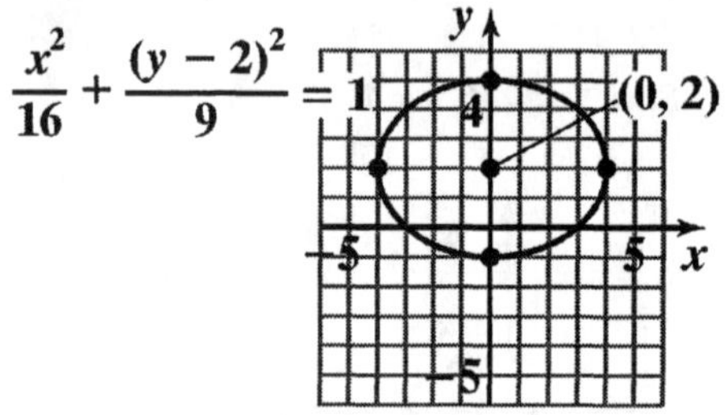

31. $\frac{(x-4)^2}{9}+\frac{(y+3)^2}{25}=1$

The center is (4, −3).

Since $a^2=9, a=3$.

Since $b^2=25, b=5$.

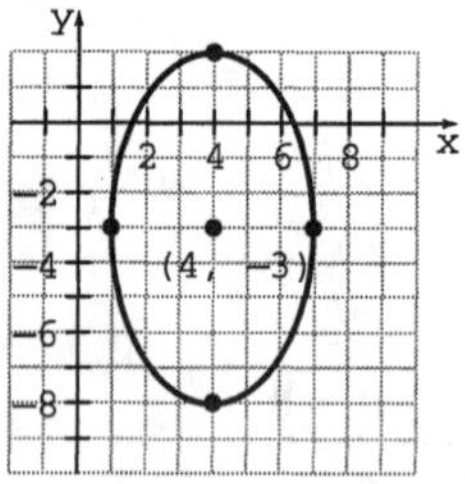

33. $\frac{(x+1)^2}{9}+\frac{(y-2)^2}{4}=1$

The center is (−1, 2).

Since $a^2=9, a=3$.

Since $b^2=4, b=2$.

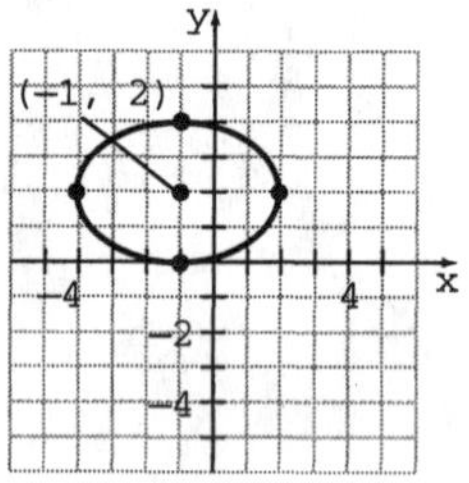

35. $(x+3)^2+9(y+1)^2=81$

$\frac{(x+3)^2}{81}+\frac{9(y+1)^2}{81}=1$

$\frac{(x+3)^2}{81}+\frac{(y+1)^2}{9}=1$

The center is (−3, −1).

Since $a^2=81, a=9$.

Since $b^2=9, b=3$.

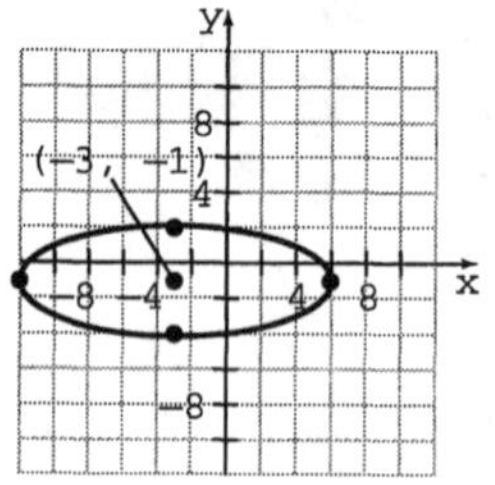

37. $(x-5)^2+4(y+4)^2=4$

$\frac{(x-5)^2}{4}+\frac{4(y+4)^2}{4}=1$

$\frac{(x-5)^2}{4}+\frac{(y+4)^2}{1}=1$

The center is (5, −4).

Since $a^2=4, a=2$.

Since $b^2=1, b=1$.

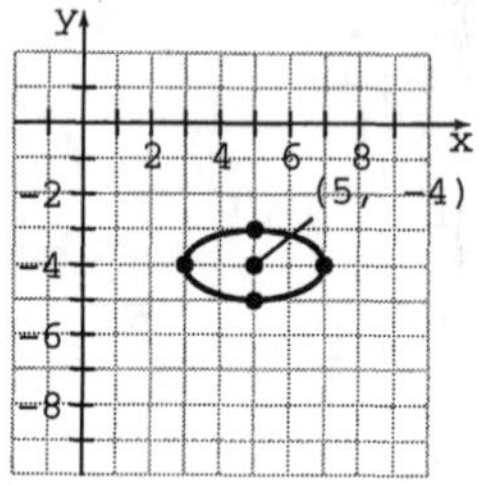

39. $12(x+4)^2 + 3(y-1)^2 = 48$

$$\frac{12(x+4)^2}{48} + \frac{3(y-1)^2}{48} = 1$$

$$\frac{(x+4)^2}{4} + \frac{(y-1)^2}{16} = 1$$

The center is (–4, 1).

Since $a^2 = 4, a = 2$.

Since $b^2 = 16, b = 4$.

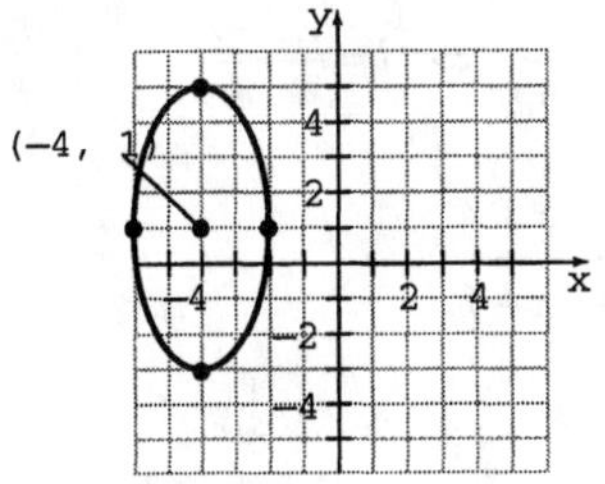

41. $\frac{x^2}{4} + \frac{y^2}{1} = 1$

Since $a^2 = 4, a = 2$; Since $b^2 = 1, b = 1$.

$$\text{Area} = \pi ab$$
$$= \pi(2)(1)$$
$$= 2\pi \approx 6.3 \text{ square units}$$

43. There is only one point, at (0, 0). Two non-negative numbers can only sum to 0 if they are both 0.

45. The center is the origin, $a = 3$, and $b = 4$.

$$\frac{x^2}{a^2} + \frac{y^2}{b^2} = 1$$

$$\frac{x^2}{3^2} + \frac{y^2}{4^2} = 1$$

$$\frac{x^2}{9} + \frac{y^2}{16} = 1$$

47. The center is the origin, $a = 2$, and $b = 3$.

$$\frac{x^2}{a^2} + \frac{y^2}{b^2} = 1$$

$$\frac{x^2}{2^2} + \frac{y^2}{3^2} = 1$$

$$\frac{x^2}{4} + \frac{y^2}{9} = 1$$

49. There are no points of intersection, because the ellipse with $a = 4$ and $b = 5$ is completely inside the circle of radius 7.

51.

$$x^2 + 4y^2 + 6x + 16y - 11 = 0$$
$$x^2 + 6x + 4y^2 + 16y = 11$$
$$(x^2 + 6x + 9) + 4(y^2 + 4y + 4) = 11 + 9 + 16$$
$$(x+3)^2 + 4(y+2)^2 = 36$$
$$\frac{(x+3)^2}{36} + \frac{(y+2)^2}{9} = 1$$

The center is (–3, –2).

53. Since 90.2 – 20.7 = 69.5, the distance between the foci is 69.5 feet.

55. **a.** Consider the ellipse to be centered at the origin, (0, 0). Here, $a = 10$ and $b = 24$. The equation is

$$\frac{x^2}{10^2} + \frac{y^2}{24^2} = 1 \Rightarrow \frac{x^2}{100} + \frac{y^2}{576} = 1.$$

b.
$$\text{Area} = \pi ab$$
$$= \pi(10)(24)$$
$$= 240\pi \approx 753.98 \text{ square feet}$$

c. Area of opening is half the area of ellipse

$$= \frac{\pi ab}{2}$$
$$= \frac{\pi(10)(24)}{2}$$
$$= 120\pi \approx 376.99 \text{ square feet}$$

57. Using $a = 3$ and $b = 2$, we may assume that the ellipse has the equation $\frac{x^2}{9}+\frac{y^2}{4}=1$ and that the foci are located at $(\pm c, 0)$. Apply the definition of an ellipse using the points (3, 0) and (0, 2). That is, the distance from (3, 0) to (–c, 0) plus the distance from (3, 0) to (c, 0) is the same as the sum of the distance from (0, 2) to (–c, 0) and the distance from (0, 2) to (c, 0).

$$\sqrt{[3-(-c)]^2+(0-0)^2}+\sqrt{(3-c)^2+(0-0)^2}$$
$$=\sqrt{[0-(-c)]^2+(2-0)^2}+\sqrt{(0-c)^2+(2-0)^2}$$
$$|3+c|+|3-c|=\sqrt{c^2+4}+\sqrt{(-c)^2+4}$$

Note that the foci are inside the ellipse, so $3+c>0$ and $3-c>0$. So, $|3+c|=3+c$ and $|3-c|=3-c$.

$$(3+c)+(3-c)=2\sqrt{c^2+4}$$
$$6=2\sqrt{c^2+4}$$
$$3=\sqrt{c^2+4}$$
$$9=c^2+4$$
$$5=c^2$$
$$c=\pm\sqrt{5}$$

The foci are located at $(\pm\sqrt{5}, 0)$. That is, the foci are $\sqrt{5}\approx 2.24$ feet, in both directions, from the center of the ellipse, along the major axis.

59. Answers will vary.

61. Answers will vary.

63. $h=\frac{-3+11}{2}=\frac{8}{2}=4$; $k=\frac{5+(-1)}{2}=\frac{4}{2}=2$

Thus, the center is (4, 2).

$$a=\frac{11-(-3)}{2}=\frac{14}{2}=7;\quad b=\frac{5-(-1)}{2}=\frac{6}{2}=3$$
$$\frac{(x-h)^2}{a^2}+\frac{(y-k)^2}{b^2}=1$$
$$\frac{(x-4)^2}{7^2}+\frac{(y-2)^2}{3^2}=1$$
$$\frac{(x-4)^2}{49}+\frac{(y-2)^2}{9}=1$$

65. $S=\frac{n}{2}(f+l)$, for l

$$S=\frac{nf}{2}+\frac{nl}{2}$$
$$S-\frac{nf}{2}=\frac{nl}{2}$$
$$2\left(S-\frac{nf}{2}\right)=2\left(\frac{nl}{2}\right)$$
$$2S-nf=nl$$
$$\frac{2S-nf}{n}=\frac{nl}{n}\Rightarrow l=\frac{2S-nf}{n}$$

66.

$$\begin{array}{r} x+\frac{5}{2} \\ 2x-3\,\overline{\big)\,2x^2+2x-7} \\ \underline{2x^2-3x} \\ 5x-7 \\ \underline{5x-\frac{15}{2}} \\ \frac{1}{2} \end{array}$$

$$\frac{2x^2+2x-7}{2x-3}=x+\frac{5}{2}+\frac{1}{2(2x-3)}$$

67. $\sqrt{3b-2}=10-b$

$$3b-2=(10-b)^2$$
$$3b-2=100-20b+b^2$$
$$b^2-23b+102=0$$
$$(b-6)(b-17)=0$$
$$b-6=0 \quad\text{or}\quad b-17=0$$
$$b=6 \qquad\qquad b=17$$

Upon checking $b = 17$ is extraneous. The solution is $b = 6$.

68. $\frac{3x+5}{x-4}\le 0$

$$3x+5=0\Rightarrow x=-\frac{5}{3}$$
$$x-4=0\Rightarrow x=4$$
$$\frac{3x+5}{x-4}\le 0\Rightarrow -\frac{5}{3}\le x<4 \text{ or } \left[-\frac{5}{3}, 4\right)$$

69. $\log_8 321=\frac{\ln 321}{\ln 8}\approx 2.7755$

Mid-Chapter Test: 10.1 – 10.2

1. This is a parabola in the form $y = a(x-h)^2 + k$ with $a = 1$, $h = 2$, and $k = -1$. Since $a > 0$, the parabola opens upward. The vertex is $(2,8)$. The y-intercept is $(0,3)$. The x-intercepts are $(1,0)$ and $(3,0)$.

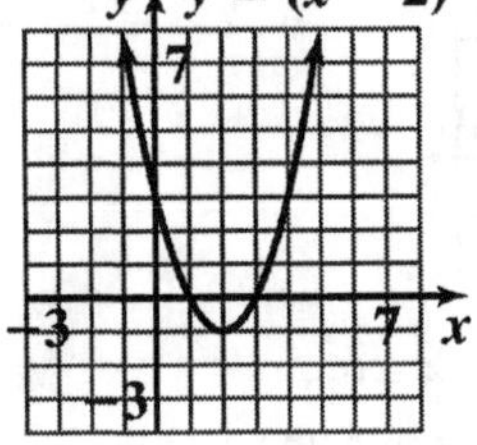

$y = (x-2)^2 - 1$

2. This is a parabola in the form $y = a(x-h)^2 + k$ with $a = -1$, $h = -1$, and $k = 3$. Since $a < 0$, the parabola opens downward. The vertex is $(-1,3)$. The y-intercept is $(0,2)$. The x-intercepts are about $(-2.732,0)$ and $(0.732,0)$.

$y = -(x+1)^2 + 3$

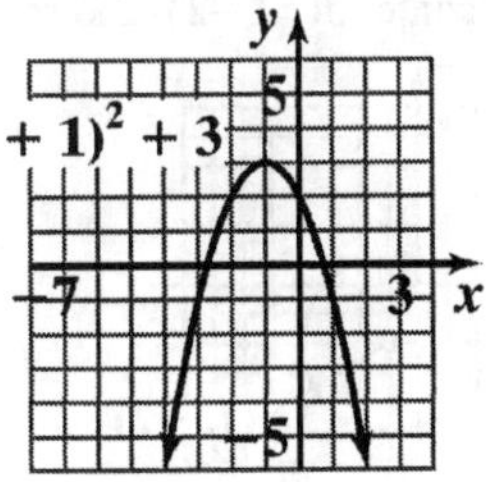

3. This is a parabola in the form $x = a(y-k)^2 + h$ with $a = -1$, $h = 1$, and $k = 4$. Since $a < 0$, the parabola opens to the left. The vertex is $(1,4)$. The x-intercept is $(-15,0)$. The y-intercepts are $(0,5)$ and $(0,3)$.

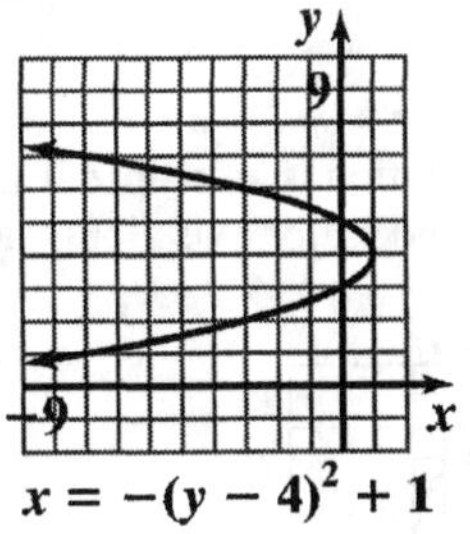

$x = -(y-4)^2 + 1$

4. This is a parabola in the form $x = a(y-k)^2 + h$ with $a = 2$, $h = -2$, and $k = -3$. Since $a > 0$, the parabola opens to the right. The vertex is $(-2,-3)$. The x-intercept is $(16,0)$. The y-intercepts are $(0,-4)$ and $(0,-2)$.

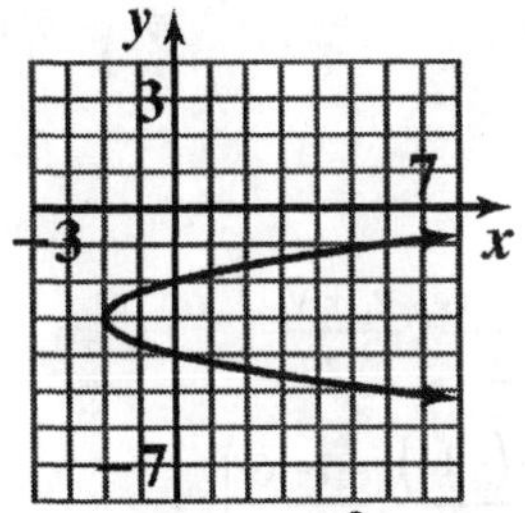

$x = 2(y+3)^2 - 2$

5. $y = x^2 + 6x + 10$

$= (x^2 + 6x) + 10$

$= (x^2 + 6x + 9) + 10 - 9$

$= (x+3)^2 + 1$

This is a parabola in the form $y = a(x-h)^2 + k$ with $a = 1$, $h = -3$, and $k = 1$. Since $a > 0$, the parabola opens upward. The vertex is $(-3,1)$. The y-intercept is $(0,10)$. There are no x-intercepts.

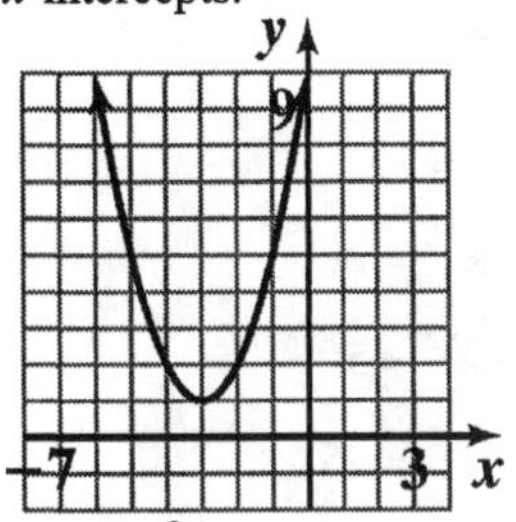

$y = x^2 + 6x + 10$

6. $d = \sqrt{(x_2 - x_1)^2 + (y_2 - y_1)^2}$

$= \sqrt{(-2-(-7))^2 + (-8-4)^2}$

$= \sqrt{(5)^2 + (-12)^2}$

$= \sqrt{25 + 144}$

$= \sqrt{169}$

$= 13$

7. $d=\sqrt{(x_2-x_1)^2+(y_2-y_1)^2}$

$=\sqrt{(2-5)^2+(9-(-3))^2}$

$=\sqrt{(-3)^2+(12)^2}$

$=\sqrt{9+144}$

$=\sqrt{153}$

≈ 12.37

8. $\text{Midpoint}=\left(\frac{x_1+x_2}{2},\frac{y_1+y_2}{2}\right)$

$=\left(\frac{9+(-11)}{2},\frac{-1+6}{2}\right)$

$=\left(\frac{-2}{2},\frac{5}{2}\right)$

$=\left(-1,\frac{5}{2}\right)$

9. $\text{Midpoint}=\left(\frac{x_1+x_2}{2},\frac{y_1+y_2}{2}\right)$

$=\left(\frac{-\frac{5}{2}+8}{2},\frac{7+\frac{1}{2}}{2}\right)$

$=\left(\frac{11}{4},\frac{15}{4}\right)$

10. $(h,k)=(-3,2);\ r=5$

$(x-h)^2+(y-k)^2=r^2$

$(x-(-3))^2+(y-2)^2=5^2$

$(x+3)^2+(y-2)^2=25$

11. $x^2+(y-1)^2=16$

$x^2+(y-1)^2=4^2$

The graph is a circle with its center at $(0,1)$ and radius 4.

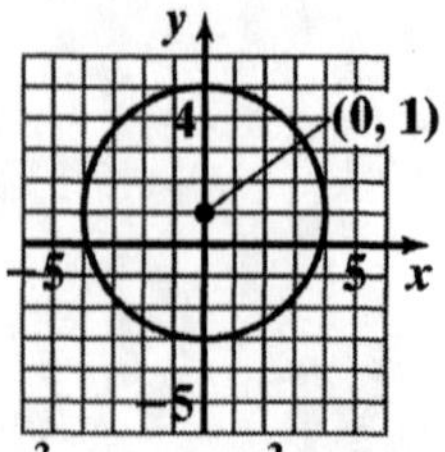

$x^2+(y-1)^2=16$

12. $y=\sqrt{36-x^2}$

If we solve $x^2+y^2=36$ for *y*, we obtain $y=\pm\sqrt{36-x^2}$. Therefore, the graph of $y=\sqrt{36-x^2}$ is the upper half ($y\ge 0$) of a circle with its center at the origin and radius 6.

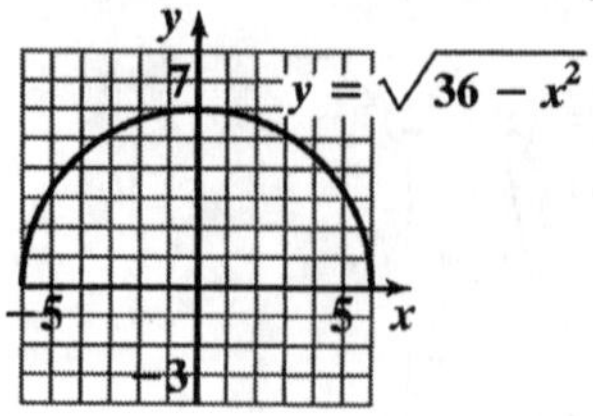

13. $x^2+y^2-2x+4y-4=0$

Complete the square in both *x* and *y*.

$(x^2-2x)+(y^2+4y)=4$

$(x^2-2x+1)+(y^2+4y+4)=4+1+4$

$(x-1)^2+(y+2)^2=9$

The graph of $x^2+y^2-2x+4y-4=0$ is a circle with its center at $(1,-2)$ and radius 3.

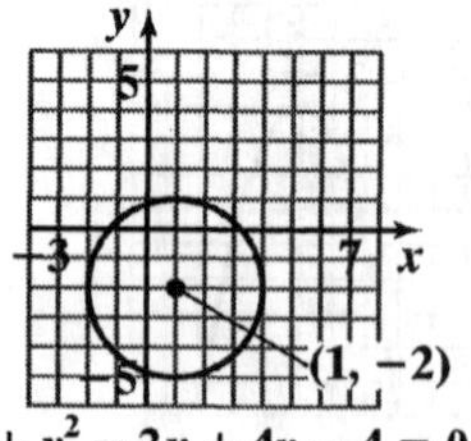

$x^2+y^2-2x+4y-4=0$

14. A circle is defined to be the set of points that are equidistant from a fixed point. This distance is called the radius and the fixed point is called the center.

15. $\frac{x^2}{4}+\frac{y^2}{9}=1$

$a^2=4\rightarrow a=2;\ b^2=9\rightarrow b=3$

The graph of this equation is an ellipse with center at the origin and major axis along the *y*-axis.

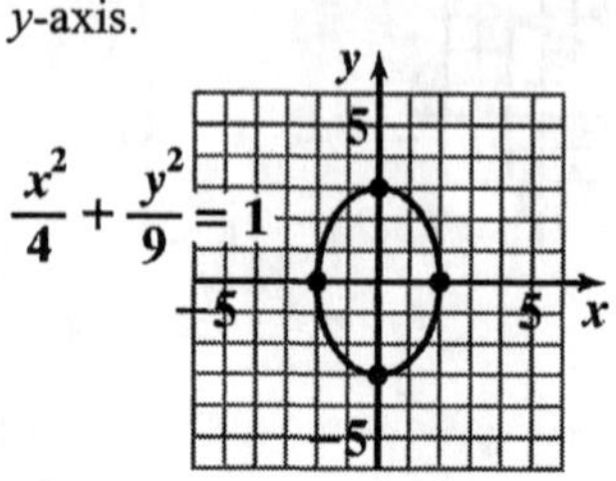

16. $\frac{x^2}{81}+\frac{y^2}{25}=1$

$a^2=81 \to a=9$; $b^2=25 \to b=5$

The graph of this equation is an ellipse with center at the origin and major axis along the x-axis.

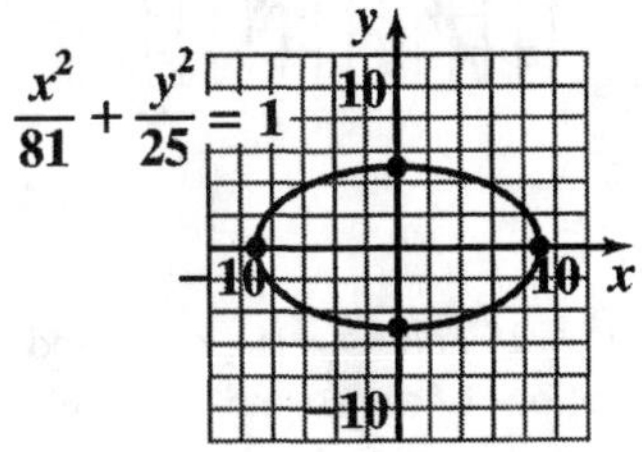

17. $\frac{(x-1)^2}{49}+\frac{(y+2)^2}{4}=1$

$a^2=49 \to a=7$; $b^2=4 \to b=2$

The graph of this equation is an ellipse with center at $(1,-2)$ and major axis parallel to the x-axis.

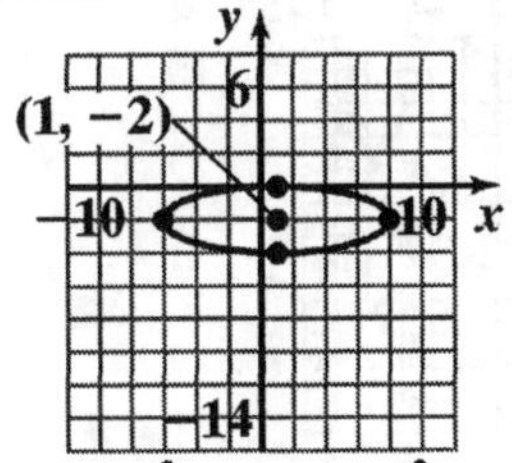

$\frac{(x-1)^2}{49}+\frac{(y+2)^2}{4}=1$

18. $36(x+3)^2+(y-4)^2=36$

$$\frac{36(x+3)^2}{36}+\frac{(y-4)^2}{36}=\frac{36}{36}$$

$$\frac{(x+3)^2}{1}+\frac{(y-4)^2}{36}=1$$

$a^2=1 \to a=1$; $b^2=36 \to b=6$

The graph of this equation is an ellipse with center at $(-3,4)$ and major axis parallel to the y-axis.

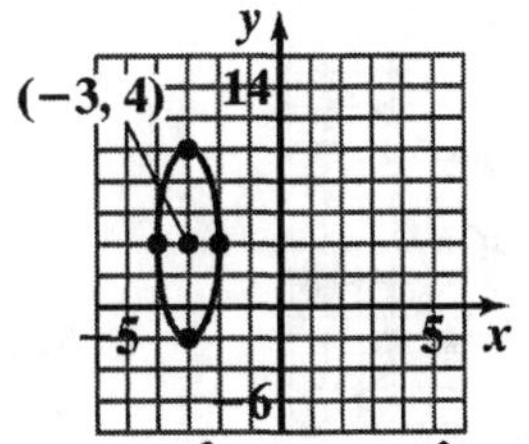

$36(x+3)^2+(y-4)^2=36$

19. $\frac{x^2}{4}+\frac{y^2}{9}=1$

Since $a^2=4, a=2$; Since $b^2=9, b=3$.

Area $=\pi ab=\pi(2)(3)$

$=6\pi \approx 18.85$ square units

20. The center of the ellipse is the origin, $(0,0)$. We have $a=8$ and $b=5$ so $a^2=64$ and $b^2=25$. Thus, the equation of the ellipse is

$$\frac{(x-0)^2}{8^2}+\frac{(y-0)^2}{5^2}=1$$

$$\frac{x^2}{64}+\frac{y^2}{25}=1$$

Exercise Set 10.3

1. A hyperbola is the set of points in a plane, the difference of whose distances from two fixed points (called foci) is a constant.

3. The graph of $\frac{x^2}{a^2}-\frac{y^2}{b^2}=1$ is a hyperbola with vertices at $(a, 0)$ and $(-a, 0)$. Its transverse axis lies along the x-axis. The asymptotes are $y=\pm\frac{b}{a}x$.

5. No, equations of hyperbolas have one positive square term and one negative square term. This equation has two positive square terms.

7. Yes, equations of hyperbolas have one positive square term and one negative square term. This equation satisfies that condition.

9. The first step is to divide both sides by 81 in order to make the right side equal to 1.

11. a. $\frac{x^2}{9}-\frac{y^2}{4}=1$

Since $a^2=9$ and $b^2=4$, $a=3$ and $b=2$. The equations of the asymptotes are $y=\pm\frac{b}{a}x$, or $y=\pm\frac{2}{3}x$.

b. To graph the asymptotes, plot the points (3, 2), (–3, 2), (3, –2), and (–3, –2). The graph intersects the *x*-axis at (–3, 0) and (3, 0).

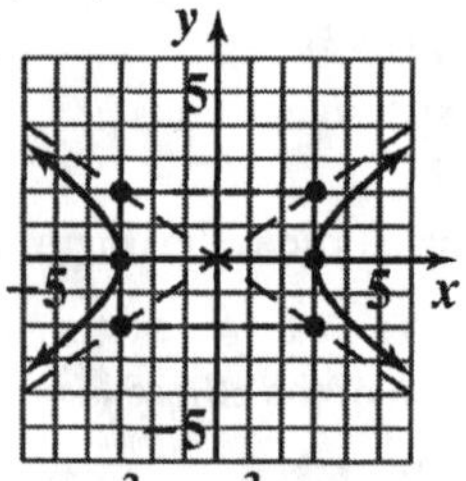

$$\frac{x^2}{9} - \frac{y^2}{4} = 1$$

13. a. $\frac{x^2}{4} - \frac{y^2}{1} = 1$

Since $a^2 = 4$ and $b^2 = 1$, $a = 2$ and $b = 1$. The equations of the asymptotes are $y = \pm\frac{b}{a}x$, or $y = \pm\frac{1}{2}x$.

b. To graph the asymptotes, plot the points (2, 1), (–2, 1), (2, –1), and (–2, –1). The graph intersects the *x*-axis at (–2, 0) and (2, 0).

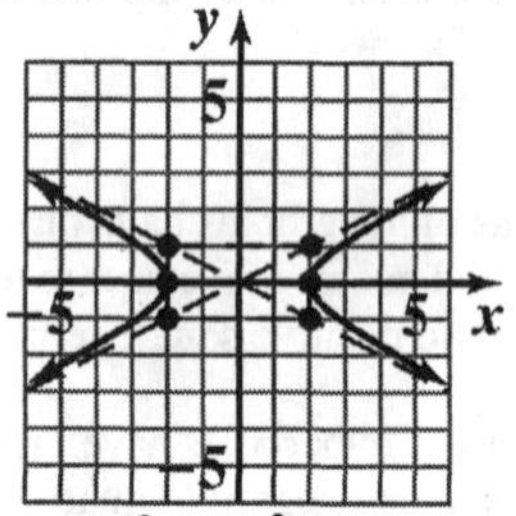

$$\frac{x^2}{4} - \frac{y^2}{1} = 1$$

15. a. $\frac{x^2}{9} - \frac{y^2}{25} = 1$

Since $a^2 = 9$ and $b^2 = 25$, $a = 3$ and $b = 5$. The equations of the asymptotes are $y = \pm\frac{b}{a}x$, or $y = \pm\frac{5}{3}x$.

b. To graph the asymptotes, plot the points (3, 5), (–3, 5), (3, –5), and (–3, –5). The graph intersects the *x*-axis at (–3, 0) and (3, 0).

$$\frac{x^2}{9} - \frac{y^2}{25} = 1$$

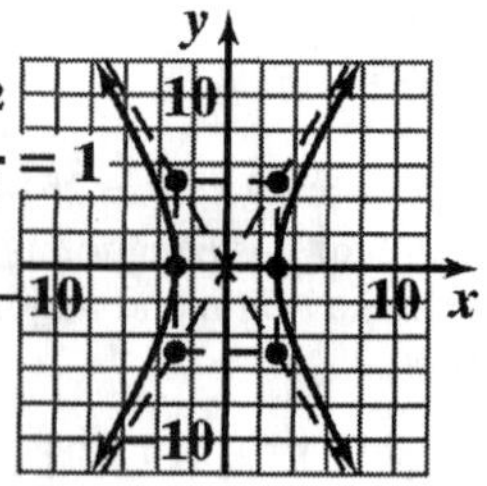

17. a. $\frac{x^2}{25} - \frac{y^2}{16} = 1$

Since $a^2 = 25$ and $b^2 = 16$, $a = 5$ and $b = 4$. The equations of the asymptotes are $y = \pm\frac{b}{a}x$, or $y = \pm\frac{4}{5}x$.

b. To graph the asymptotes, plot the points (5, 4), (–5, 4), (5, –4), and (–5, –4). The graph intersects the *x*-axis at (–5, 0) and (5, 0).

$$\frac{x^2}{25} - \frac{y^2}{16} = 1$$

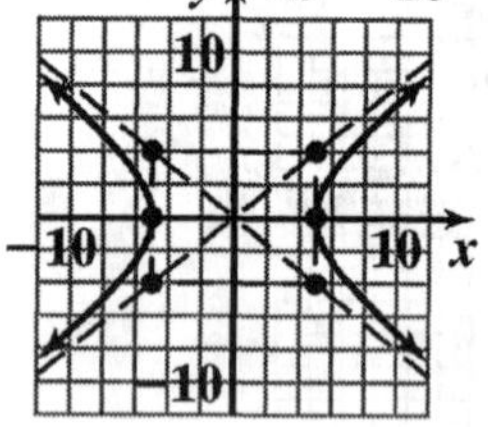

19. a. $\frac{y^2}{25} - \frac{x^2}{36} = 1$

Since $a^2 = 36$ and $b^2 = 25$, $a = 6$ and $b = 5$. The equations of the asymptotes are $y = \pm\frac{b}{a}x$, or $y = \pm\frac{5}{6}x$.

b. To graph the asymptotes, plot the points (6, 5), (–6, 5), (6, –5), and (–6, –5). The graph intersects the *y*-axis at (0, –5) and (0, 5).

$$\frac{y^2}{25} - \frac{x^2}{36} = 1$$

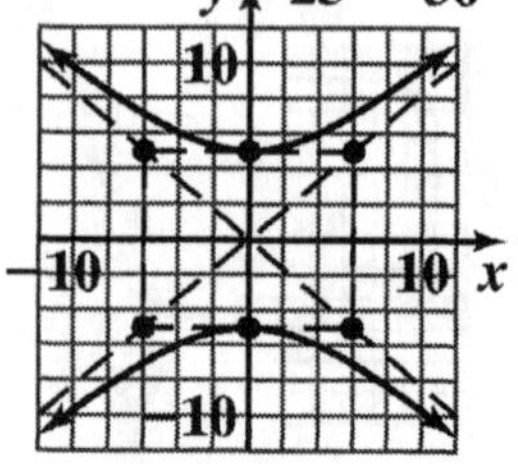

21. a. $\frac{y^2}{9}-\frac{x^2}{16}=1$

Since $a^2=16$ and $b^2=9$, $a=4$ and $b=3$. The equations of the asymptotes are $y=\pm\frac{b}{a}x$, or $=y\pm\frac{3}{4}x$.

b. To graph the asymptotes, plot the points (4, 3), (–4, 3), (4, –3) and (–4, –3). The graph intersects the *y*-axis at (0, –3) and (0, 3).

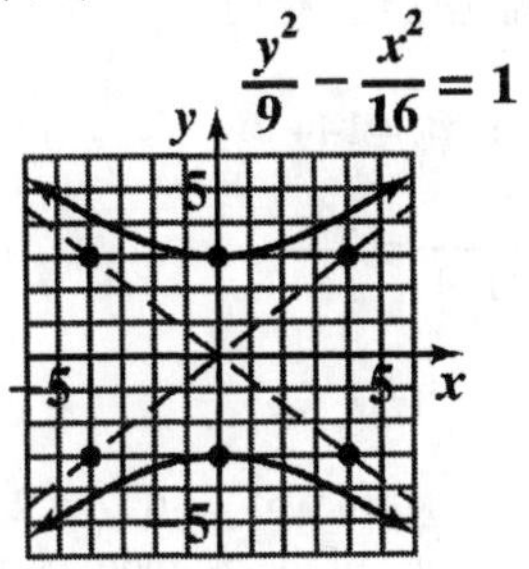

23. a. $\frac{y^2}{25}-\frac{x^2}{4}=1$

Since $a^2=4$ and $b^2=25$, $a=2$ and $b=5$. The equations of the asymptotes are $y=\pm\frac{b}{a}x$ or $y=\pm\frac{5}{2}x$.

b. To graph the asymptotes, plot the points (2, 5), (–2, 5), (2, –5) and (–2, –5). The graph intersects the *y*-axis at (0, –5) and (0, 5).

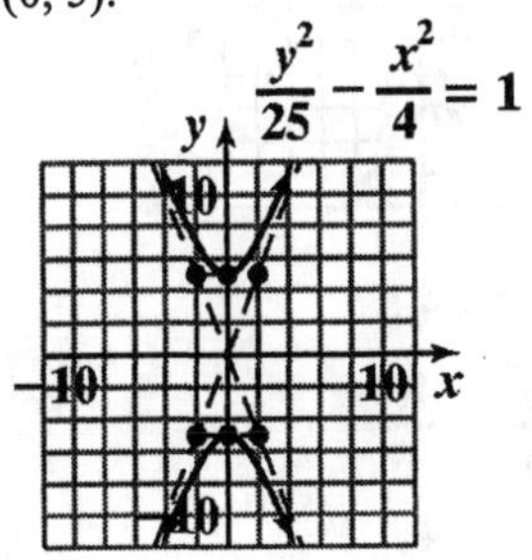

25. a. $\frac{x^2}{81}-\frac{y^2}{16}=1$

Since $a^2=81$ and $b^2=16$, $a=9$ and $b=4$. The equations of the asymptotes are $y=\pm\frac{b}{a}x$, or $y=\pm\frac{4}{9}x$.

b. To graph the asymptotes, plot the points (9, 4), (–9, 4), (9, –4), and (–9, –4). The graph intersects the *x*-axis at (–9, 0) and (9, 0).

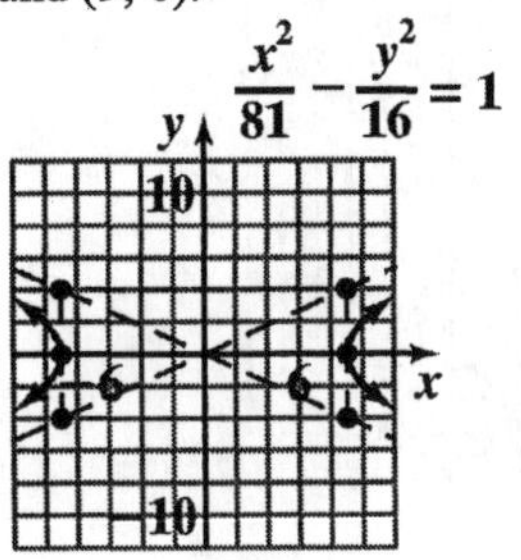

27. a. $x^2-25y^2=25$

$\frac{x^2}{25}-\frac{25y^2}{25}=1$

$\frac{x^2}{25}-\frac{y^2}{1}=1$

Since $a^2=25$ and $b^2=1$, $a=5$ and $b=1$. The equations of the asymptotes are $y=\pm\frac{b}{a}x$, or $y=\pm\frac{1}{5}x$.

b. To graph the asymptotes, plot the points (5, 1), (–5, 1), (5, –1), and (–5, –1). The graph intersects the *x*-axis at (–5, 0), and (5, 0).

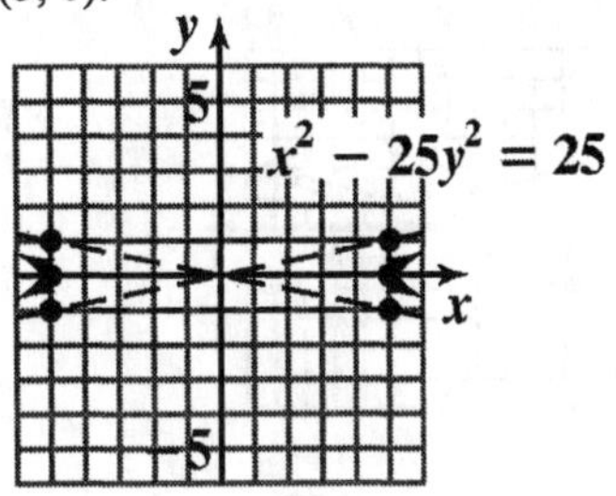

29. a. $4y^2-16x^2=64$

$\frac{4y^2}{64}-\frac{16x^2}{64}=1$

$\frac{y^2}{16}-\frac{x^2}{4}=1$

Since $a^2=4$ and $b^2=16$, $a=2$ and $b=4$. The equations of the asymptotes are $y=\pm\frac{b}{a}x$, or $y=\pm 2x$.

b. To graph the asymptotes, plot (2, 4), (–2, 4), (2 –4), and (–2, –4). The graph intersects the *y*-axis at (0, –4) and (0, 4).

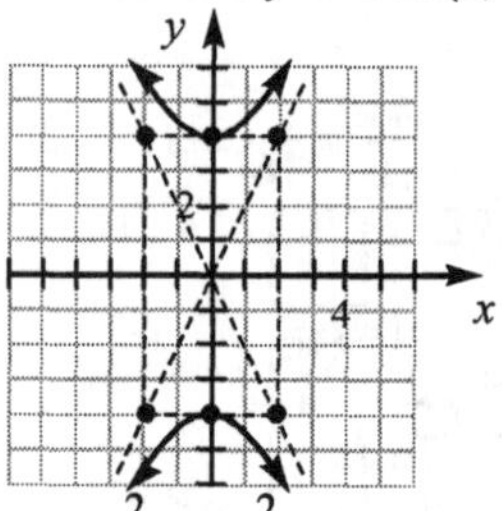

31. a. $9y^2 - x^2 = 9$

$$\frac{9y^2}{9} - \frac{x^2}{9} = 1$$

$$\frac{y^2}{1} - \frac{x^2}{9} = 1$$

Since $a^2 = 9$ and $b^2 = 1$, $a = 3$ and $b = 1$. The equations of the asymptotes are $y = \pm\frac{b}{a}x$, or $\pm\frac{1}{3}x$.

b. To graph the asymptotes, plot the points (3, 1), (–3, 1), (3, –1), and (–3, –1). The graph intersects the *y*-axis at (0, –1) and (0, 1).

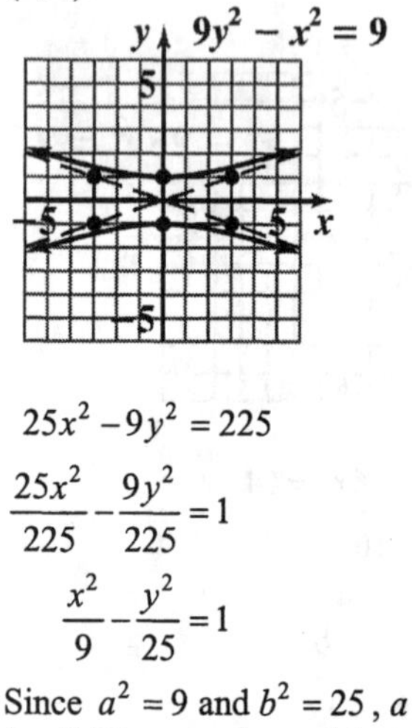

33. a. $25x^2 - 9y^2 = 225$

$$\frac{25x^2}{225} - \frac{9y^2}{225} = 1$$

$$\frac{x^2}{9} - \frac{y^2}{25} = 1$$

Since $a^2 = 9$ and $b^2 = 25$, $a = 3$ and $b = 5$. The equations of the asymptotes are $y = \pm\frac{b}{a}x$, or $y = \pm\frac{5}{3}x$.

b. To graph the asymptotes, plot the points (3, 5), (–3, 5), (3, –5), and (–3, –5). The graph intersects the *x*-axis at (–3, 0) and (3, 0).

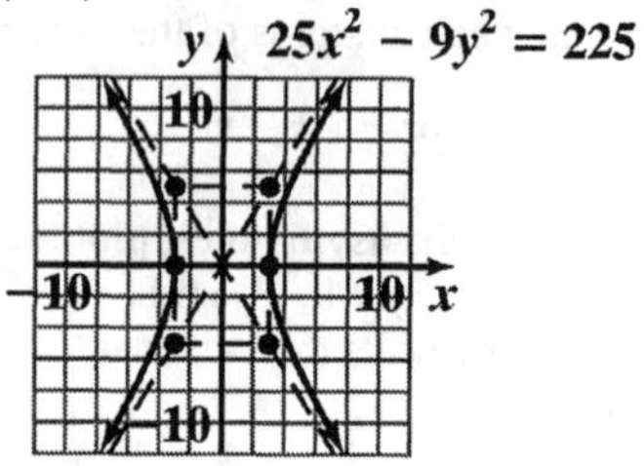

35. a. $4y^2 - 36x^2 = 144$

$$\frac{4y^2}{144} - \frac{36x^2}{144} = \frac{144}{144}$$

$$\frac{y^2}{36} - \frac{x^2}{4} = 1$$

Since $a^2 = 4$ and $b^2 = 36$, $a = 2$ and $b = 6$. The equations of the asymptotes are $y = \pm\frac{b}{a}x$, or $y = \pm 3x$.

b. To graph the asymptotes, plot the points (2, 6), (–2, 6), (2, –6), and (–2, –6). The graph intersects the *y*-axis at (0, –6) and (0, 6).

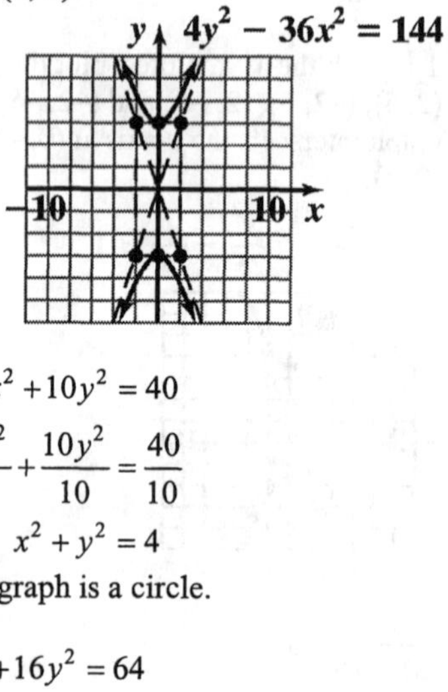

37. $10x^2 + 10y^2 = 40$

$$\frac{10x^2}{10} + \frac{10y^2}{10} = \frac{40}{10}$$

$$x^2 + y^2 = 4$$

The graph is a circle.

39. $x^2 + 16y^2 = 64$

$$\frac{x^2}{64} + \frac{16y^2}{64} = \frac{64}{64}$$

$$\frac{x^2}{64} + \frac{y^2}{4} = 1$$

The graph is an ellipse.

41. $4x^2 - 4y^2 = 29$

$$\frac{4x^2}{29} - \frac{4y^2}{29} = \frac{29}{29}$$

$$\frac{x^2}{\frac{29}{4}} - \frac{y^2}{\frac{29}{4}} = 1$$

The graph is a hyperbola.

43. $2y = 12x^2 - 8x + 16$

$$y = 6x^2 - 4x + 8$$

The graph is a parabola.

45. $6x^2 + 9y^2 = 54$

$$\frac{6x^2}{54} + \frac{9y^2}{54} = \frac{54}{54}$$

$$\frac{x^2}{9} + \frac{y^2}{6} = 1$$

The graph is an ellipse.

47. $3x = -2y^2 + 9y - 15$

$$x = -\frac{2}{3}y^2 + 3y - 5$$

The graph is a parabola.

49. $6x^2 + 6y^2 = 36$

$$\frac{6x^2}{6} + \frac{6y^2}{6} = \frac{36}{6}$$

$$x^2 + y^2 = 6$$

The graph is a circle.

51. $14y^2 = 7x^2 + 35$

$$14y^2 - 7x^2 = 35$$

$$\frac{14y^2}{35} - \frac{7x^2}{35} = \frac{35}{35}$$

$$\frac{y^2}{\frac{5}{2}} - \frac{x^2}{5} = 1$$

The graph is a hyperbola.

53. $x + y = 2y^2 + 6$

$$x = 2y^2 - y + 6$$

The graph is a parabola.

55. $12x^2 = 4y^2 + 48$

$$12x^2 - 4y^2 = 48$$

$$\frac{12x^2}{48} - \frac{4y^2}{48} = \frac{48}{48}$$

$$\frac{x^2}{4} - \frac{y^2}{12} = 1$$

The graph is a hyperbola.

57. $y - x + 4 = x^2$

$$y = x^2 + x - 4$$

The graph is a parabola.

59. $-3x^2 - 3y^2 = -27$

$$\frac{-3x^2}{-3} + \frac{-3y^2}{-3} = \frac{-27}{-3}$$

$$x^2 + y^2 = 9$$

The graph is a circle.

61. Since the vertices are (0, ±2), the hyperbola is of the form $\frac{y^2}{b^2} - \frac{x^2}{a^2} = 1$ with $b = 2$. Since the asymptotes are $y = \pm\frac{1}{2}x$, we have $\frac{b}{a} = \frac{1}{2}$. Therefore, $\frac{2}{a} = \frac{1}{2}$, so $a = 4$. The equation of the hyperbola is $\frac{y^2}{2^2} - \frac{x^2}{4^2} = 1$, or $\frac{y^2}{4} - \frac{x^2}{16} = 1$.

63. Since the vertices are $(0, \pm 3)$, the hyperbola is of the form $\frac{x^2}{a^2} - \frac{y^2}{b^2} = 1$ with $a = 3$. Since the asymptotes are $y = \pm 2x$, we have $\frac{b}{a} = 2$. Therefore, $\frac{b}{3} = 2$, so $b = 6$. The equation of the hyperbola is $\frac{x^2}{3^2} - \frac{y^2}{6^2} = 1$, or $\frac{x^2}{9} - \frac{y^2}{36} = 1$.

65. Since the transverse axis is along the x-axis, the equation is of the form $\frac{x^2}{a^2}-\frac{y^2}{b^2}=1$ Since the asymptotes are $y=\pm\frac{5}{3}x$, we require $\frac{b}{a}=\frac{5}{3}$.
Using $a=3$ and $b=5$, the equation of the hyperbola is $\frac{x^2}{3^2}-\frac{y^2}{5^2}=1$, or $\frac{x^2}{9}-\frac{y^2}{25}=1$.
No, this is not the only possible answer, because a and b are not uniquely determined.
$\frac{x^2}{18}-\frac{y^2}{50}=1$ and others will also work.

67. No, for each value of x with $|x|>a$, there are 2 possible values of y.

69. $\frac{x^2}{25}-\frac{y^2}{4}=1$. This hyperbola has its transverse axis along the x-axis with vertices at $(\pm5,0)$.
Domain: $(-\infty,-5]\cup[5,\infty)$
Range: $\mathbb{R}$

71. The equation is changed from $\frac{x^2}{a^2}-\frac{y^2}{b^2}=1$ to $\frac{x^2}{b^2}-\frac{y^2}{a^2}=1$. Both graphs have a transverse axis along the x-axis. The vertices of the second graph will be closer to the origin, at $(\pm b,0)$ instead of $(\pm a,0)$. The second graph will open wider.

73. Answers will vary.

75. The points are $(-6,4)$ and $(-2,2)$.

$$m=\frac{y_2-y_1}{x_2-x_1}=\frac{2-4}{-2-(-6)}=\frac{-2}{4}=-\frac{1}{2}$$

Use $y-y_1=m(x-x_1)$, with $m=-\frac{1}{2}$, $(-2,2)$

$$y-2=-\frac{1}{2}(x-(-2))$$
$$y-2=-\frac{1}{2}x-1$$
$$y=-\frac{1}{2}x+1$$

76. $f(x)=3x^2-x+5,\ g(x)=6-4x^2$

$$(f+g)(x)=(3x^2-x+5)+(6-4x^2)$$
$$=-x^2-x+11$$

77.

$$5(-4x+9y=7)\ \Rightarrow\ -20x+45y=35$$
$$4(5x+6y=-3)\ \Rightarrow\ 20x+24y=-12$$
$$69y=23$$
$$y=\frac{1}{3}$$

$$5x+6\left(\frac{1}{3}\right)=-3\ \Rightarrow\ 5x+2=-3\ \Rightarrow\ x=-1$$

The solution is $\left(-1,\ \frac{1}{3}\right)$.

78.

$$\frac{3x}{2x-3}+\frac{2x+4}{2x^2+x-6}$$
$$=\frac{3x}{2x-3}+\frac{2x+4}{(2x-3)(x+2)}$$
$$=\frac{3x}{2x-3}\cdot\frac{x+2}{x+2}+\frac{2x+4}{(2x-3)(x+2)}$$
$$=\frac{3x^2+6x+2x+4}{(2x-3)(x+2)}$$
$$=\frac{3x^2+8x+4}{(2x-3)(x+2)}$$
$$=\frac{(3x+2)(x+2)}{(2x-3)(x+2)}$$
$$=\frac{3x+2}{2x-3}$$

79. $E=\frac{1}{2}mv^2$, for v

$$2E=mv^2$$
$$\frac{2E}{m}=v^2$$
$$\sqrt{\frac{2E}{m}}=v\ \text{ or }\ v=\sqrt{\frac{2E}{m}}$$

80. $\log(x+4) = \log 5 - \log x$

$\log(x+4) = \log\frac{5}{x}$

$x+4 = \frac{5}{x}$

$x(x+4) = 5$

$x^2 + 4x - 5 = 0$

$(x+5)(x-1) = 0$

$x+5 = 0$ or $x-1 = 0$

$x = -5$ $\quad x = 1$

Upon checking, $x = -5$ does not satisfy the equation. The solution is $x = 1$.

Exercise Set 10.4

1. A nonlinear system of equations is a system in which at least one equation is nonlinear.

3. Yes

5. Yes

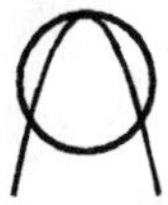

7. $x^2 + y^2 = 18$

$x + y = 0$

Solve $x + y = 0$ for x: $x = -y$.

Substitute $x = -y$ for x in $x^2 + y^2 = 18$.

$x^2 + y^2 = 18$

$(-y)^2 + y^2 = 18$

$y^2 + y^2 = 18$

$2y^2 = 18$

$y^2 = 9$

$y = \pm 3$

$y = 3$ or $y = -3$

$x = -3$ $\quad x = 3$

The solutions are (3, –3) and (–3, 3).

9. $x^2 + y^2 = 9$

$x + 2y = 3$

Solve $x + 2y = 3$ for x: $x = 3 - 2y$.

Substitute $3 - 2y$ for x in $x^2 + y^2 = 9$.

$x^2 + y^2 = 9$

$(3-2y)^2 + y^2 = 9$

$9 - 12y + 4y^2 + y^2 = 9$

$5y^2 - 12y = 0$

$y(5y - 12) = 0$

$y = 0$ or $y = \frac{12}{5}$

$x = 3 - 2y$

$x = 3 - 2(0)$

$x = 3$

$x = 3 - 2y$

$x = 3 - 2\left(\frac{12}{5}\right)$

$x = -\frac{9}{5}$

The solutions are (3, 0) and $\left(-\frac{9}{5}, \frac{12}{5}\right)$.

11. $y = x^2 - 5$

$3x + 2y = 10$

Substitute $x^2 - 5$ for y in $3x + 2y = 10$

$3x + 2y = 10$

$3x + 2(x^2 - 5) = 10$

$3x + 2x^2 - 10 = 10$

$2x^2 + 3x - 20 = 0$

$(x+4)(2x-5) = 0$

$x = -4$ or $x = \frac{5}{2}$

$y = x^2 - 5$

$y = (-4)^2 - 5$

$y = 11$

$y = x^2 - 5$

$y = \left(\frac{5}{2}\right)^2 - 5$

$y = \frac{5}{4}$

The solutions are $(-4, 11)$ and $\left(\frac{5}{2}, \frac{5}{4}\right)$.

13. $x^2 + y = 6$

$y = x^2 + 4$

Substitute $x^2 + 4$ for y in $x^2 + y = 6$

$x^2 + x^2 + 4 = 6$

$2x^2 = 2$

$x^2 = 1$

$x = \pm 1$

$y = (1)^2 + 4$ or $y = (-1)^2 + 4$

$y = 1 + 4$ $\qquad$ $y = 1 + 4$

$y = 5$ $\qquad$ $y = 5$

The solutions are (1, 5) and (–1, 5).

15. $2x^2 + y^2 = 16$

$x^2 - y^2 = -4$

Solve $x^2 - y^2 = -4$ for y^2: $y^2 = x^2 + 4$

Substitute $x^2 + 4$ for y^2 in $2x^2 + y^2 = 16$.

$2x^2 + y^2 = 16$

$2x^2 + (x^2 + 4) = 16$

$3x^2 + 4 = 16$

$3x^2 = 12$

$x^2 = 4$

$x = 2$ or $x = -2$

$y^2 = x^2 + 4$ $\qquad$ $y^2 = x^2 + 4$

$y^2 = 2^2 + 4$ $\qquad$ $y = (-2)^2 + 4$

$y^2 = 8$ $\qquad$ $y = 8$

$y = \pm\sqrt{8} = \pm 2\sqrt{2}$ $\qquad$ $y = \pm\sqrt{8} = \pm 2\sqrt{2}$

The solutions are $(2, 2\sqrt{2}), (2, -2\sqrt{2})$, $(-2, 2\sqrt{2})$, and $(-2, -2\sqrt{2})$.

17. $x^2 + y^2 = 4$

$y = x^2 - 6 \Rightarrow x^2 = y + 6$

Substitute $y + 6$ for x^2 in $x^2 + y^2 = 4$.

$x^2 + y^2 = 4$

$(y + 6) + y^2 = 4$

$y^2 + y + 2 = 0$

$$y = \frac{-1 \pm \sqrt{1^2 - 4(1)(2)}}{2(1)}$$

$$= \frac{-1 \pm \sqrt{-7}}{2}$$

$$= \frac{-1 \pm i\sqrt{31}}{2}$$

There is no real solution.

19. $x^2 + y^2 = 9$

$y = x^2 - 3$

Solve $y = x^2 - 3$ for x^2: $x^2 = y + 3$.

Substitute $y + 3$ for x^2 in $x^2 + y^2 = 9$.

$x^2 + y^2 = 9$

$(y + 3) + y^2 = 9$

$y^2 + y - 6 = 0$

$(y - 2)(y + 3) = 0$

$y = 2$ or $y = -3$

$x^2 = y + 3$ $\qquad$ $x^2 = y + 3$

$x^2 = 2 + 3$ $\qquad$ $x^2 = -3 + 3$

$x^2 = 5$ $\qquad$ $x^2 = 0$

$x = \pm\sqrt{5}$ $\qquad$ $x = 0$

The solutions are $(0, -3)$, $(\sqrt{5}, 2)$, and $(-\sqrt{5}, 2)$.

21. $2x^2 - y^2 = -8$

$x - y = 6$

Solve the second equation for *y*: $y = x - 6$.

Substitute $x - 6$ for *y* in $2x^2 - y^2 = -8$.

$$2x^2 - y^2 = -8$$
$$2x^2 - (x-6)^2 = -8$$
$$2x^2 - (x^2 - 12x + 36) = -8$$
$$2x^2 - x^2 + 12x - 36 = -8$$
$$x^2 + 12x - 28 = 0$$
$$(x-2)(x+14) = 0$$

or

$x = 2$ $\quad$ $x = -14$

$y = x - 6$ $\quad$ $y = x - 6$

$y = 2 - 6$ $\quad$ $y = -14 - 6$

$y = -4$ $\quad$ $y = -20$

The solutions are (2, –4) and (–14, –20).

23. $x^2 - y^2 = 4$

$\underline{2x^2 + y^2 = 8}$

$3x^2 \quad = 12$

$x^2 = 4$

$x = 2$ or $x = -2$

$x^2 - y^2 = 4$ $\quad$ $x^2 - y^2 = 4$

$2^2 - y^2 = 4$ $\quad$ $(-2)^2 - y^2 = 4$

$y^2 = 0$ $\quad$ $y^2 = 0$

$y = 0$ $\quad$ $y = 0$

The solutions are (2, 0) and (–2, 0).

25. $x^2 + y^2 = 16$ (1)

$2x^2 - 5y^2 = 25$ (2)

$-2x^2 - 2y^2 = -32$ (1) multiplied by –2

$\underline{2x^2 - 5y^2 = 25}$ (2)

$-7y^2 = -7$

$y^2 = 1$

$y = 1$ or $y = -1$

$x^2 + y^2 = 16$ $\quad$ $x^2 + y^2 = 16$

$x^2 + 1^2 = 16$ $\quad$ $x^2 + (-1)^2 = 16$

$x^2 = 15$ $\quad$ $x^2 = 15$

$x = \pm\sqrt{15}$ $\quad$ $x = \pm\sqrt{15}$

The solutions are $(-\sqrt{15}, 1)$, $(-\sqrt{15}, -1)$, $(\sqrt{15}, -1)$, and $(\sqrt{15}, 1)$.

27. $3x^2 - y^2 = 4$ (1)

$x^2 + 4y^2 = 10$ (2)

$12x^2 - 4y^2 = 16$ (1) multiplied by 4

$\underline{x^2 + 4y^2 = 10}$ (2)

$13x^2 = 26$

$x^2 = 2$

$x = \sqrt{2}$ or $x = -\sqrt{2}$

$3x^2 - y^2 = 4$ $\quad$ $3x^2 - y^2 = 4$

$3(\sqrt{2})^2 - y^2 = 4$ $\quad$ $3(-\sqrt{2})^2 - y^2 = 4$

$6 - y^2 = 4$ $\quad$ $6 - y^2 = 4$

$y^2 = 2$ $\quad$ $y^2 = 2$

$y = \pm\sqrt{2}$ $\quad$ $y = \pm\sqrt{2}$

The solutions are $(\sqrt{2}, \sqrt{2})$, $(\sqrt{2}, -\sqrt{2})$, $(-\sqrt{2}, \sqrt{2})$, and $(-\sqrt{2}, -\sqrt{2})$.

29. $4x^2 + 9y^2 = 36$

$2x^2 - 9y^2 = 18$

$6x^2 = 54$

$x^2 = 9$

$x = 3$ or $x = -3$

$4x^2 + 9y^2 = 36$ $\quad$ $4x^2 + 9y^2 = 36$

$4(3)^2 + 9y^2 = 36$ $\quad$ $4(-3)^2 + 9y^2 = 36$

$9y^2 = 0$ $\quad$ $9y^2 = 0$

$y^2 = 0$ $\quad$ $y^2 = 0$

$y = 0$ $\quad$ $y = 0$

The solutions are (3, 0) and (–3, 0).

31. $2x^2 - y^2 = 7$ (1)

$x^2 + 2y^2 = 6$ (2)

$4x^2 - 2y^2 = 14$ (1) multiplied by 2

$x^2 + 2y^2 = 6$ (2)

$5x^2 = 20$

$x^2 = 4$

$x = 2$ or $x = -2$

$2(2)^2 - y^2 = 7$ $\quad$ $2(-2)^2 - y^2 = 7$

$8 - y^2 = 7$ $\quad$ $8 - y^2 = 7$

$y^2 = 1$ $\quad$ $y^2 = 1$

$x = \pm 1$ $\quad$ $x = \pm 1$

The solutions are (2, 1), (2, –1), (–2, 1), and (–2, –1).

33. $x^2 + y^2 = 25$ (1)

$2x^2 - 3y^2 = -30$ (2)

$3x^2 + 3y^2 = 75$ (1) multiplied by 3

$2x^2 - 3y^2 = -30$ (2)

$5x^2 = 45$

$x^2 = 9$

$x = 3$ or $x = -3$

$x^2 + y^2 = 25$ $\quad$ $x^2 + y^2 = 25$

$(3)^2 + y^2 = 25$ $\quad$ $(-3)^2 + y^2 = 25$

$y^2 = 16$ $\quad$ $y^2 = 16$

$y = \pm 4$ $\quad$ $y = \pm 4$

The solutions are(3, 4), (3, –4), (–3, 4), and (–3, –4).

35. $x^2 + y^2 = 9$ (1)

$16x^2 - 4y^2 = 64$ (2)

$4x^2 + 4y^2 = 36$ (1) multiplied by 4

$16x^2 - 4y^2 = 64$ (2)

$20x^2 = 100$

$x^2 = 5$

$x = \sqrt{5}$ or $x = -\sqrt{5}$

$x^2 + y^2 = 9$ $\quad$ $x^2 + y^2 = 9$

$(\sqrt{5})^2 + y^2 = 9$ $\quad$ $(-\sqrt{5})^2 + y^2 = 9$

$y^2 = 4$ $\quad$ $y^2 = 4$

$y = \pm 2$ $\quad$ $y = \pm 2$

The solutions are $(\sqrt{5}, 2)$, $(\sqrt{5}, -2)$, $(-\sqrt{5}, 2)$, and $(-\sqrt{5}, -2)$.

37. $x^2+y^2=4$ (1)

$16x^2+9y^2=144$ (2)

Multiply the first equation by -16 and add.

$$\begin{aligned} -16x^2-16y^2 &= -164 \\ 16x^2+9y^2 &= 144 \\ \hline 7y^2 &= -20 \\ y^2 &= -\frac{20}{7} \end{aligned}$$

This equation has no real solutions so the system has no real solutions.

39. $x^2+4y^2=4$

$10y^2-9x^2=90$

Write the equations in standard form.

$$x^2+4y^2=4$$
$$-9x^2+10y^2=90$$

Multiply the first equation by -2 and the second equation by 2, then add.

$$\begin{aligned} -5x^2-20y^2 &= -20 \\ -18x^2+20y^2 &= 180 \\ \hline -23x^2 &= 160 \\ x^2 &= -\frac{160}{23} \end{aligned}$$

This equation has no real solutions so the system has no real solutions.

41. Answers will vary.

43. Let x = length; y = width

$xy=440$

$2x+2y=84$

Solve $2x+2y=84$ for y: $y=42-x$.

Substitute $42-x$ for y in $xy=440$.

$$xy=440$$
$$x(42-x)=440$$
$$42x-x^2=440$$
$$x^2-42x+440=0$$
$$(x-20)(x-22)=0$$

$x-20=0$ or $x-22=0$

$x=20$ $\quad$ $x=22$

$y=42-x$ or $y=42-22$

$y=42-20$ $\quad$ $y=42-22$

$y=22$ $\quad$ $y=20$

The solutions are (20, 22) and (22, 20).
The dimensions of the floor are 20 m by 22 m.

45. Let x = length

y = width

$xy=270$

$2x+2y=78$

Solve $2x+2y=78$ for y: $y=39-x$.

Substitute $39-x$ for y in $xy=270$.

$$xy=270$$
$$x(39-x)=270$$
$$39x-x^2=270$$
$$x^2-39x+270=0$$
$$(x-9)(x-30)=0$$

$x-9=0$ or $x-30=0$

$x=9$ $\quad$ $x=30$

$y=39-x$ $\quad$ $y=39-x$

$y=39-9$ $\quad$ $y=39-30$

$y=30$ $\quad$ $y=9$

The solutions are (9, 30) and (30, 9). The dimensions of the garden are 9 ft by 30 ft.

47. Let x = length; y = width

$xy=112$

$x^2+y^2=\left(\sqrt{260}\right)^2$

Solve $xy=112$ for y: $y=\dfrac{112}{x}$.

$$x^2+y^2=260$$
$$x^2+\left(\frac{112}{x}\right)^2=260$$
$$x^2+\frac{12,544}{x^2}=260$$
$$x^4+12,544=260x^2$$
$$x^4-260x^2+12,544=0$$
$$(x^2-64)(x^2-196)=0$$

$x^2-64=0$ or $x^2-196=0$

$x^2=64$ $\quad$ $x^2=196$

$x=\pm 8$ $\quad$ $x=\pm 14$

Since x must be positive, $x=8$ or $x=14$.

If $x=8$, then $y=\dfrac{112}{8}=14$.

If $x=14$, then $y=\dfrac{112}{14}=8$.

The dimensions of the new bill are 8 cm by 14 cm.

49. Let x = length

y = width

$$x^2 + y^2 = 34^2$$

$$x + y + 34 = 80$$

Solve $x + y + 34 = 80$ for y: $y = 46 - x$.

Substitute $46 - x$ for y in $x^2 + y^2 = 34^2$.

$$x^2 + y^2 = 34^2$$

$$x^2 + (46 - x)^2 = 34^2$$

$$x^2 + (2116 - 92x + x^2) = 1156$$

$$2x^2 - 92 + 2116 = 1156$$

$$2x^2 - 92x + 960 = 0$$

$$x^2 - 46x + 480 = 0$$

$$(x - 16)(x - 30) = 0$$

$x - 16 = 0$ or $x - 30 = 0$

$x = 16$ $\qquad$ $x = 30$

$y = 46 - x$ $\qquad$ $y = 46 - x$

$y = 46 - 16$ $\qquad$ $y = 46 - 30$

$y = 30$ $\qquad$ $y = 16$

The solutions are (16, 30) and (30, 16).
The dimensions of the piece of wood are 16 in. by 30 in.

51. $d = -16t^2 + 64t$

$d = -16t^2 + 16t + 80$

Substitute $-16t^2 + 64t$ for d in $d = -16t^2 + 16t + 80$.

$$d = -16t^2 + 16t + 80$$

$$-16t^2 + 64t = -16t^2 + 16t + 80$$

$$64t = 16t + 80$$

$$48t = 80$$

$$t = \frac{80}{48} = \frac{5}{3} \approx 1.67$$

The balls are the same height above the ground at $t \approx 1.67$ sec.

53. Since $t = 1$ year, we may write the formula as $i = pr$.

$7.50 = pr$

$7.50 = (p + 25)(r - 0.01)$

Rewrite the second equation by multiplying the binomials. Then substitute 7.50 for pr and solve for r.

$$7.50 = (p + 25)(r - 0.01)$$

$$7.50 = pr - 0.01p + 25r - 0.25$$

$$7.50 = 7.50 - 0.01p + 25r - 0.25$$

$$0 = -0.01p + 25r - 0.25$$

$$0.01p + 0.25 = 25r$$

$$\frac{0.01p}{25} + \frac{0.25}{25} = \frac{25r}{25}$$

$$r = 0.0004p + 0.01$$

Substitute $0.0004p + 0.01$ for r in $7.50 = pr$.

$7.50 = pr$

$7.50 = p(0.0004p + 0.01)$

$7.50 = 0.0004p^2 + 0.01p$

$0 = 0.0004p^2 + 0.01p - 7.50$

$0 = p^2 + 25p - 18,750$

$0 = (p - 125)(p + 150)$

$p - 125 = 0$ or $p + 150 = 0$

$p = 125$ $\qquad$ $p = -150$

Since the principal must be positive, use $p = 125$.

$r = 0.0004p + 0.01$

$r = 0.0004(125) + 0.01$

$r = 0.06$

The principal is \$125 and the interest rate is 6%.

55. $C = 10x + 300$

$R = 30x - 0.1x^2$

$$C = R$$

$$10x + 300 = 30x - 0.1x^2$$

$$0.1x^2 - 20x + 300 = 0$$

$$x = \frac{-b \pm \sqrt{b^2 - 4ac}}{2a}$$

$$= \frac{-(-20) \pm \sqrt{(-20)^2 - 4(0.1)(300)}}{2(0.1)}$$

$$= \frac{20 \pm \sqrt{280}}{0.2}$$

$$x = \frac{20 + \sqrt{280}}{0.2} \approx 183.7$$

or

$$x = \frac{20 - \sqrt{280}}{0.2} \approx 16.3$$

The break-even points are ≈ 16 and ≈ 184.

57. $C = 12.6x + 150$

$R = 42.8x - 0.3x^2$

$$C = R$$

$$12.6x + 150 = 42.8x - 0.3x^2$$

$$0.3x^2 - 30.2x + 150 = 0$$

$$x = \frac{-b \pm \sqrt{b^2 - 4ac}}{2a}$$

$$= \frac{-(-30.2) \pm \sqrt{(-30.2)^2 - 4(0.3)(150)}}{2(0.3)}$$

$$= \frac{30.2 \pm \sqrt{732.04}}{0.6}$$

$$x = \frac{30.2 + \sqrt{732.04}}{0.6} \approx 95.4$$

or

$$x = \frac{30.2 - \sqrt{732.04}}{0.6} \approx 5.2$$

The break-even points are ≈ 5 and ≈ 95.

59. Solve each equation for y.

$$3x - 5y = 12 \qquad x^2 + y^2 = 10$$

$$-5y = -3x + 12 \qquad y^2 = 10 - x^2$$

$$y = \frac{3}{5}x - \frac{12}{5} \qquad y = \pm\sqrt{10 - x^2}$$

Use $y_1 = \frac{3}{5}x - \frac{12}{5}$, $y_2 = \sqrt{10 - x^2}$, and $y_2 = -\sqrt{10 - x^2}$.

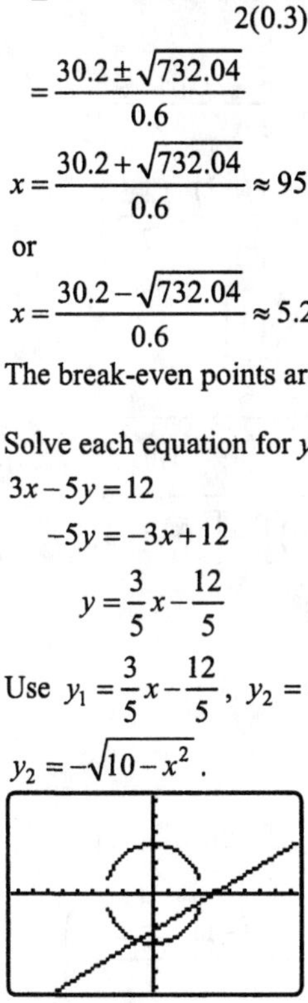

-9.4, 9.4, 1, -6.2, 6.2, 1

Approximate solutions: (–1, –3), (3.12, –0.53)

61. Let x = length of one leg

y = length of other leg

$$x^2 + y^2 = 26^2$$

$$\frac{1}{2}xy = 120$$

Solve $\frac{1}{2}xy = 120$ for y: $y = \frac{240}{x}$.

Substitute $\frac{240}{x}$ for y in $x^2 + y^2 = 26^2$

$$x^2 + y^2 = 26^2$$

$$x^2 + \left(\frac{240}{x}\right)^2 = 676$$

$$x^2 + \frac{57,600}{x^2} = 676$$

$$x^4 + 57,600 = 676x^2$$

$$x^4 - 676x^2 + 57,600 = 0$$

$$(x^2 - 100)(x^2 - 576) = 0$$

$$x^2 - 100 = 0 \quad \text{or} \quad x^2 - 576 = 0$$

$$x^2 = 100 \qquad x^2 = 576$$

$$x = \pm 10 \qquad x = \pm 24$$

Since x is a length, x must be positive.

If $x = 10$, then $y = \frac{240}{10} = 24$. If $x = 24$, then $y = \frac{240}{24} = 10$. The legs have lengths 10 yards and 24 yards.

63. The operations are evaluated in the following order: parentheses, exponents, multiplication or division, addition or subtraction.

64. $(x+1)^3 + 1$

$$= (x+1)^3 + (1)^3$$

$$= (x+1+1)\left((x+1)^2 - (x+1)(1) + (1)^2\right)$$

$$= (x+2)(x^2 + 2x + 1 - x - 1 + 1)$$

$$= (x+2)(x^2 + x + 1)$$

65. $x = \frac{k}{P^2}$

$10 = \frac{k}{6^2} \Rightarrow k = 360 \Rightarrow x = \frac{360}{P^2}$

$x = \frac{360}{20^2} = \frac{360}{400} = \frac{9}{10}$ or 0.9

66. $\frac{5}{\sqrt{x+2}-3} = \frac{5}{\sqrt{x+2}-3} \cdot \frac{\sqrt{x+2}+3}{\sqrt{x+2}+3}$

$= \frac{5\sqrt{x+2}+15}{x+2+3\sqrt{x+2}-3\sqrt{x+2}-9}$

$= \frac{5\sqrt{x+2}+15}{x-7}$

67. $A = A_0 e^{kt}$, for k

$\frac{A}{A_0} = e^{kt}$

$\ln \frac{A}{A_0} = \ln e^{kt}$

$\ln A - \ln A_0 = kt$

$\frac{\ln A - \ln A_0}{t} = k$

Chapter 10 Review Exercises

1. $d = \sqrt{(x_2 - x_1)^2 + (y_2 - y_1)^2}$

$= \sqrt{(5-0)^2 + (-12-0)^2}$

$= \sqrt{5^2 + (-12)^2}$

$= \sqrt{25+144}$

$= \sqrt{169}$

$= 13$

$\text{Midpoint} = \left(\frac{x_1+x_2}{2}, \frac{y_1+y_2}{2}\right)$

$= \left(\frac{0+5}{2}, \frac{0+(-12)}{2}\right)$

$= \left(\frac{5}{2}, -6\right)$

2. $d = \sqrt{(x_2 - x_1)^2 + (y_2 - y_1)^2}$

$= \sqrt{(-1-(-4))^2 + (5-1)^2}$

$= \sqrt{3^2 + 4^2}$

$= \sqrt{9+16}$

$= \sqrt{25}$

$= 5$

$\text{Midpoint} = \left(\frac{x_1+x_2}{2}, \frac{y_1+y_2}{2}\right)$

$= \left(\frac{-4+(-1)}{2}, \frac{1+5}{2}\right)$

$= \left(-\frac{5}{2}, 3\right)$

3. $d = \sqrt{(x_2 - x_1)^2 + (y_2 - y_1)^2}$

$= \sqrt{[-1-(-9)]^2 + [10-(-5)]^2}$

$= \sqrt{(8)^2 + 15^2}$

$= \sqrt{64+225}$

$= \sqrt{289}$

$= 17$

$\text{Midpoint} = \left(\frac{x_1+x_2}{2}, \frac{y_1+y_2}{2}\right)$

$= \left(\frac{-9+(-1)}{2}, \frac{-5+10}{2}\right)$

$= \left(-5, \frac{5}{2}\right)$

4. $d = \sqrt{(x_2 - x_1)^2 + (y_2 - y_1)^2}$

$= \sqrt{[-2-(-4)]^2 + (5-3)^2}$

$= \sqrt{2^2 + 2^2}$

$= \sqrt{4+4}$

$= \sqrt{8}$

≈ 2.83

$\text{Midpoint} = \left(\dfrac{x_1 + x_2}{2}, \dfrac{y_1 + y_2}{2}\right)$

$= \left(\dfrac{-4+(-2)}{2}, \dfrac{3+5}{2}\right)$

$= (-3, 4)$

5. $y = (x-2)^2 + 1$

This is a parabola in the form $y = a(x-h)^2 + k$ with $a = 1$, $h = 2$, and $k = 1$. Since $a > 0$, the parabola opens upward. The vertex is (2, 1). The y-intercept is (0, 5).
There are no x-intercepts.

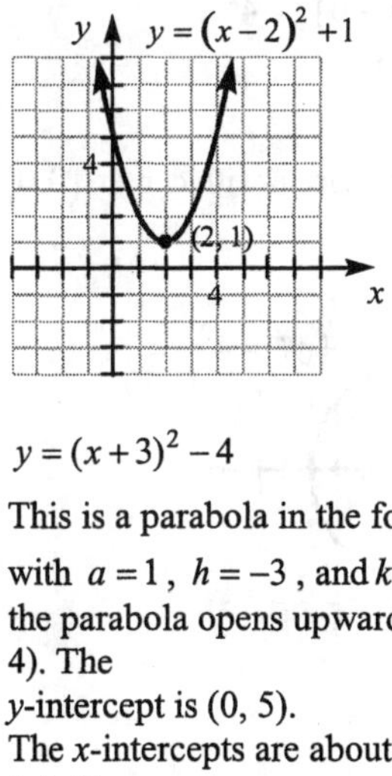

6. $y = (x+3)^2 - 4$

This is a parabola in the form $y = a(x-h)^2 + k$ with $a = 1$, $h = -3$, and $k = -4$. Since $a > 0$, the parabola opens upward. The vertex is (–3, –4). The
y-intercept is (0, 5).
The x-intercepts are about (–5, 0) and (–1, 0).

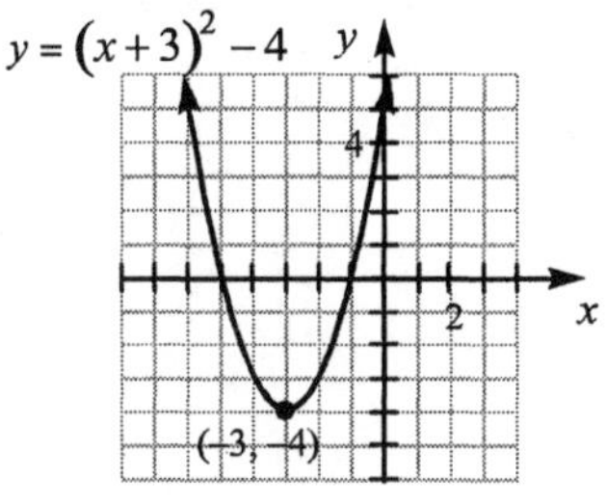

7. $x = (y-1)^2 + 4$

This is a parabola in the form $x = a(y-k)^2 + h$ with $a = 1$, $h = 4$, and $k = 1$. Since $a > 0$, the parabola opens to the right. The vertex is (4, 1). There are no
y-intercepts.
The x-intercept is (5, 0).

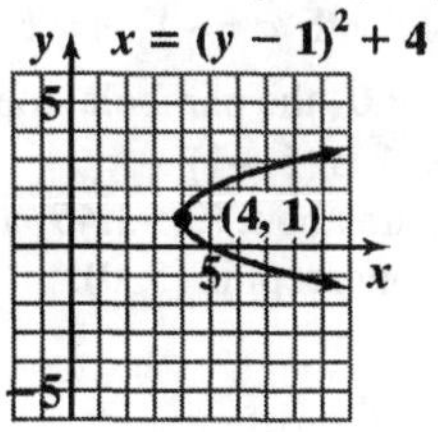

8. $x = -2(y+4)^2 - 3$

This is a parabola in the form $x = a(y-k)^2 + h$ with $a = -2$, $h = -3$, and $k = -4$. Since $a < 0$, the parabola opens to the left. The vertex is (–3, –4). There are no y-intercepts.
The x-intercept is (–35, 0).

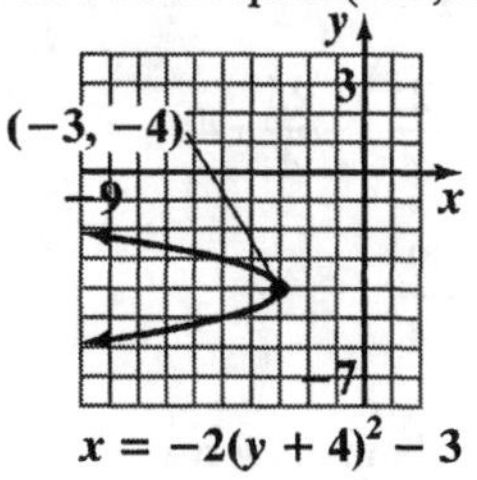

9. a. $y = x^2 - 8x + 22$

$y = (x^2 - 8x + 16) + 22 - 16$

$y = (x-4)^2 + 6$

b. This is a parabola in the form $y = a(x-h)^2 + k$ with $a = 1$, $h = 4$, and $k = 6$. Since $a > 0$, the parabola opens upward. The vertex is (4, 6). The y-intercept is (0, 22). There are no y-intercepts.

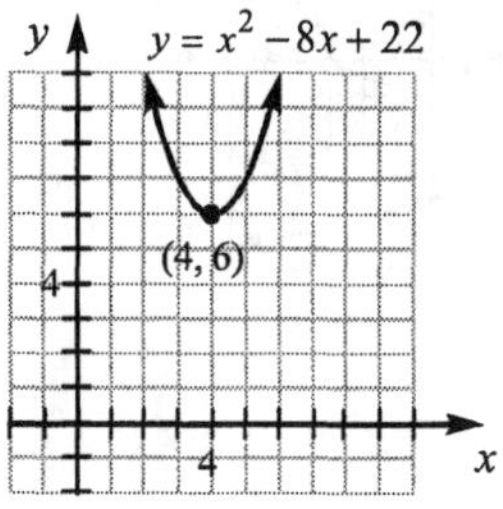

10. a.
$$x = -y^2 - 2y + 5$$
$$x = -(y^2 + 2y) + 5$$
$$x = -(y^2 + 2y + 1) + 1 + 5$$
$$x = -(y+1)^2 + 6$$

b. This is a parabola in the form $x = a(y-k)^2 + h$ with $a = -1$, $h = 6$, and $k = -1$. Since $a < 0$, the parabola opens to the left. The vertex is (6, –1).
The *y*-intercepts are about (0, –3.45) and (0, 1.45). The *x*-intercept is (5, 0).

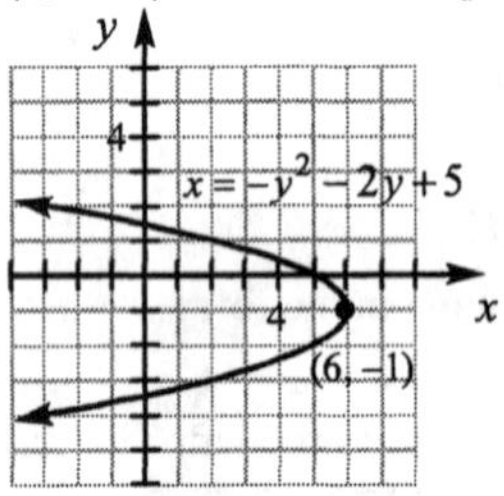

11. a.
$$x = y^2 + 5y + 4$$
$$x = \left(y^2 + 5y + \frac{25}{4}\right) - \frac{25}{4} + 4$$
$$x = \left(y + \frac{5}{2}\right)^2 - \frac{9}{4}$$

b. This is a parabola in the form $x = a(y-k)^2 + h$ with $a = 1$, $h = -\frac{9}{4}$, and $k = -\frac{5}{2}$. Since $a > 0$, the parabola opens to the right. The vertex is $\left(-\frac{9}{4}, -\frac{5}{2}\right)$.
The *y*-intercepts are (0, –4) and (0, –1).
The *x*-intercept is (4, 0).

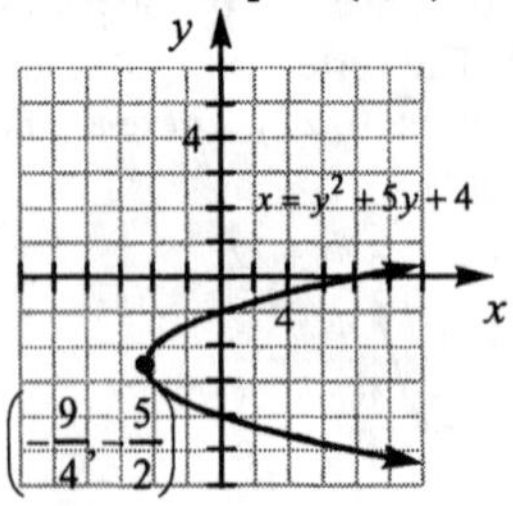

12. a.
$$y = 2x^2 - 8x - 24$$
$$y = 2(x^2 - 4x) - 24$$
$$y = 2(x^2 - 4x + 4) - 8 - 24$$
$$y = 2(x-2)^2 - 32$$

b. This is a parabola in the form $y = a(x-h)^2 + k$ with $a = 2$, $h = 2$, and $k = -32$. Since $a > 0$, the parabola opens upward. The vertex is (2, –32). The *y*-intercept is (0, –24). The *x*-intercepts are (–2, 0) and (6, 0).

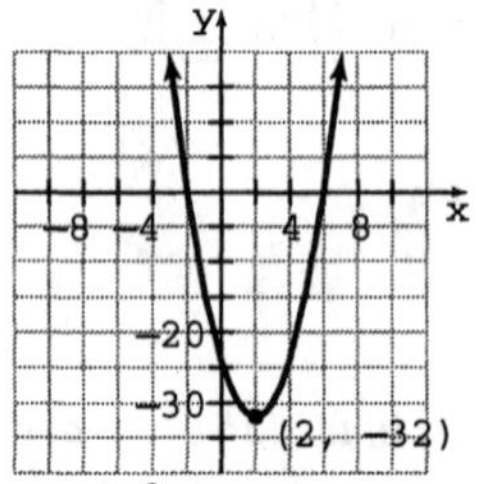

y=2x²-8x-24

13. a.
$$(x-h)^2 + (y-k)^2 = r^2$$
$$(x-0)^2 + (y-0)^2 = 4^2$$
$$x^2 + y^2 = 4^2$$

b. The graph is a circle with center $(0,0)$ and radius 4.

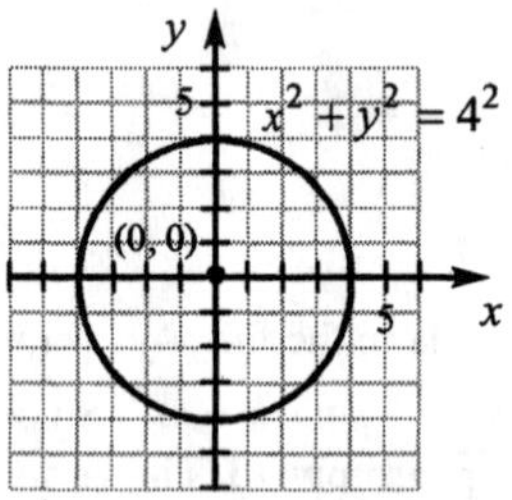

14. a.
$$(x-h)^2+(y-k)^2=r^2$$
$$\left[x-(-3)\right]^2+(y-4)^2=1^2$$
$$(x+3)^2+(y-4)^2=1^2$$

b. The graph is a circle with center $(-3,4)$ and radius 1.

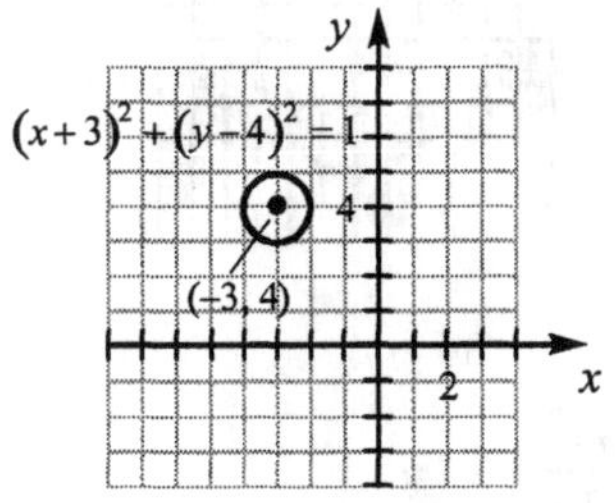

15. a.
$$x^2+y^2-4y=0$$
$$x^2+(y^2-4y+4)=4$$
$$x^2+(y-2)^2=2^2$$

b. The graph is a circle with center (0, 2) and radius 2.

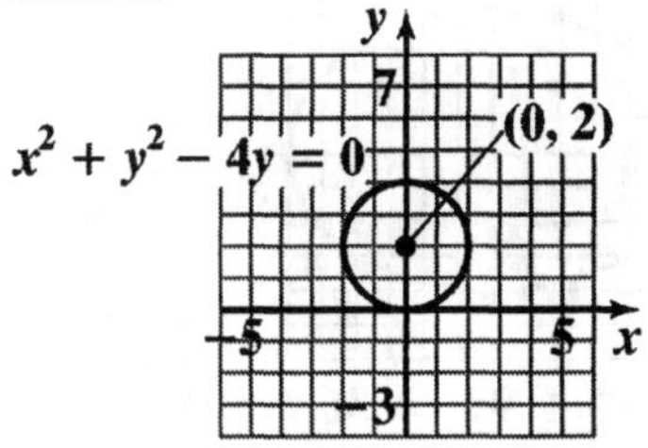

16. a.
$$x^2+y^2-2x+6y+1=0$$
$$x^2-2x+y^2+6y=-1$$
$$(x^2-2x+1)+(y^2+6y+9)=-1+1+9$$
$$(x-1)^2+(y+3)^2=9$$
$$(x-1)^2+(y+3)^2=3^2$$

b. The graph is a circle with center (1, –3) and radius 3.

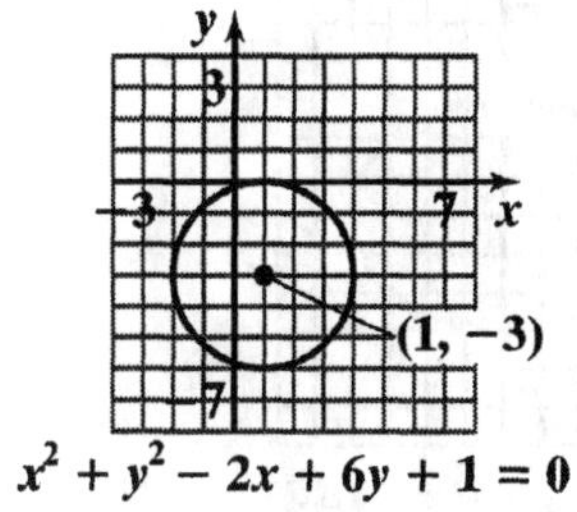

17. a.
$$x^2-8x+y^2-10y+40=0$$
$$(x^2-8x+16)+(y^2-10y+25)=-40+16+25$$
$$(x-4)^2+(y-5)^2=1$$
$$(x-4)^2+(y-5)^2=1^2$$

b. The graph is a circle with center (4, 5) and radius 1.

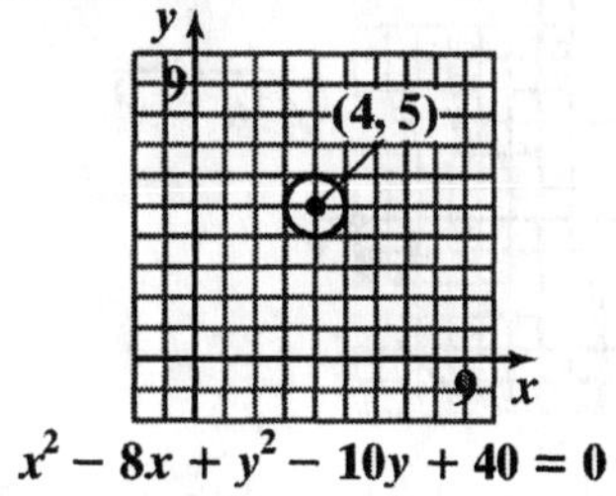

18. a.
$$x^2+y^2-4x+10y+17=0$$
$$x^2-4x+y^2+10y=-17$$
$$(x^2-4x+4)+(y^2+10y+25)=-17+4+25$$
$$(x-2)^2+(y+5)^2=12$$
$$(x-2)^2+(y+5)^2=\left(\sqrt{12}\right)^2$$

b. The graph is a circle with center (2, –5) and radius $\sqrt{12}\approx 3.46$.

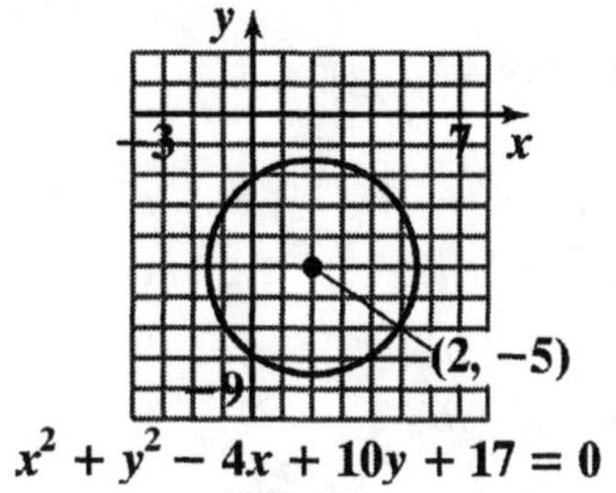

19. $y=\sqrt{9-x^2}$

If we solve $x^2+y^2=9$ for y, we obtain $y=\pm\sqrt{9-x^2}$. Therefore, the graph of $y=\sqrt{9-x^2}$ is the upper half $(y\ge 0)$ of a circle with its center at the origin and radius 4.

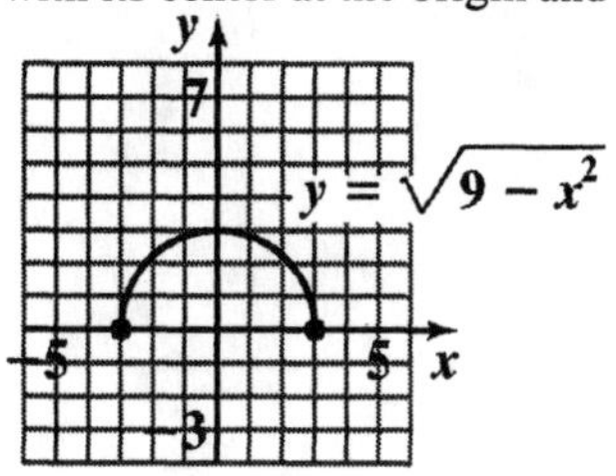

20. $y=-\sqrt{36-x^2}$

If we solve $x^2+y^2=36$ for *y*, we obtain $y=\pm\sqrt{36-x^2}$. Therefore, the graph of $y=-\sqrt{36-x^2}$ is the lower half $(y\le 0)$ of a circle with its center at the origin and radius 6.

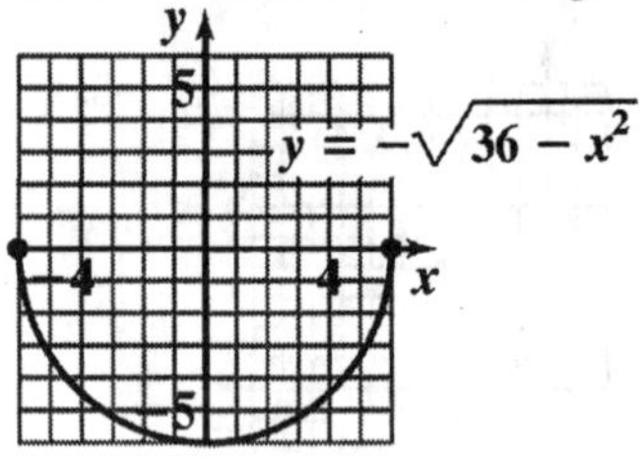

21. The center is (–1, 1) and the radius is 2.

$$(x-h)^2+(y-k)^2=r^2$$
$$[x-(-1)]^2+(y-1)^2=2^2$$
$$(x+1)^2+(y-1)^2=4$$

22. The center is (5, –3) and the radius is 3.

$$(x-h)^2+(y-k)^2=r^2$$
$$(x-5)^2+[y-(-3)]^2=3^2$$
$$(x-5)^2+(y+3)^2=9$$

23. $\frac{x^2}{4}+\frac{y^2}{9}=1$

Since $a^2=4, a=2$.

Since $b^2=9, b=3$.

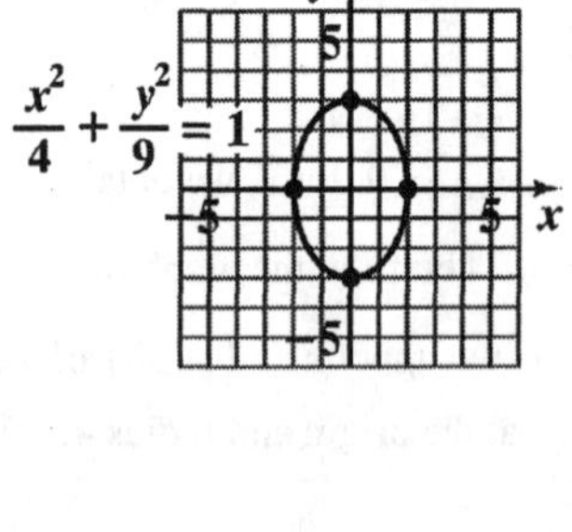

24. $\frac{x^2}{36}+\frac{y^2}{64}=1$

Since $a^2=36, a=6$.

Since $b^2=64$, $b=8$.

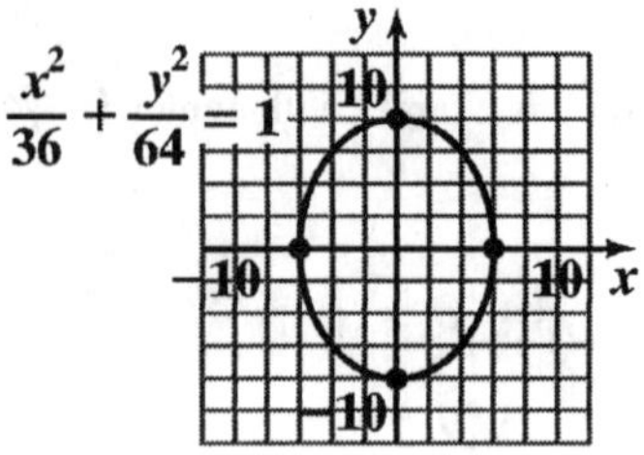

25. $4x^2+9y^2=36$

$$\frac{4x^2}{36}+\frac{9y^2}{36}=1$$
$$\frac{x^2}{9}+\frac{y^2}{4}=1$$

Since $a^2=9$, $a=3$.

Since $b^2=4$, $b=2$.

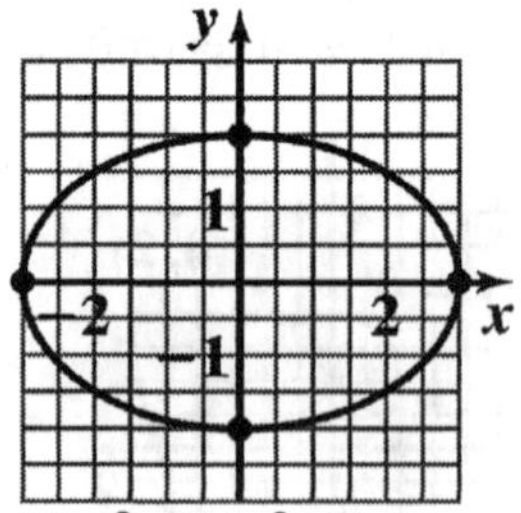

$4x^2+9y^2=36$

26. $9x^2+16y^2=144$

$$\frac{9x^2}{144}+\frac{16y^2}{144}=1$$
$$\frac{x^2}{16}+\frac{y^2}{9}=1$$

Since $a^2=16$, $a=4$.

Since $b^2=9$, $b=3$.

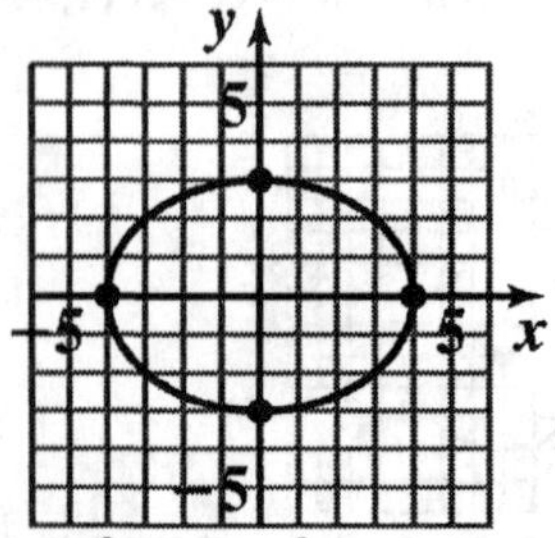

$9x^2+16y^2=144$

27. $\dfrac{(x-3)^2}{16}+\dfrac{(y+2)^2}{4}=1$

The center is (3, –2).

Since $a^2=16$, $a=4$.

Since $b^2=4$, $b=2$.

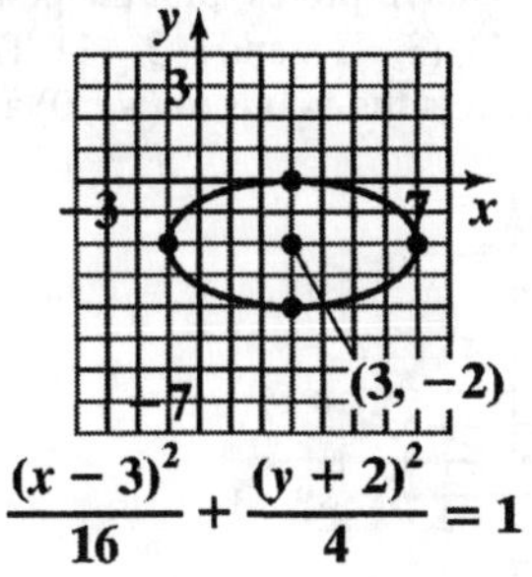

28. $\dfrac{(x+3)^2}{9}+\dfrac{y^2}{25}=1$

The center is (–3, 0).

Since $a^2=9$, $a=3$.

Since $b^2=25$, $b=5$.

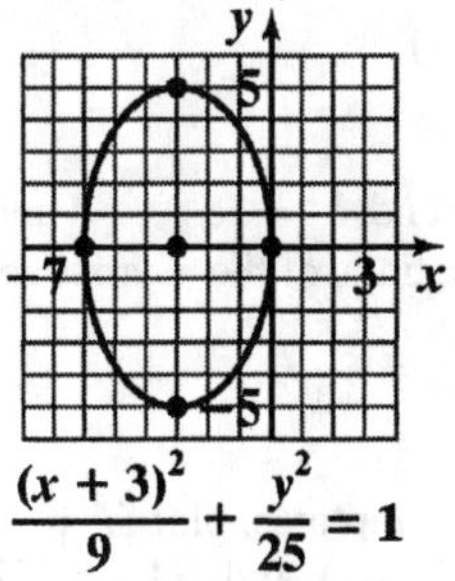

29. $25(x-2)^2+9(y-1)^2=225$

$$\frac{25(x-2)^2}{225}+\frac{9(y-1)^2}{225}=1$$

$$\frac{(x-2)^2}{9}+\frac{(y-1)^2}{25}=1$$

The center is (2, 1).

Since $a^2=9$, $a=3$.

Since $b^2=25$, $b=5$.

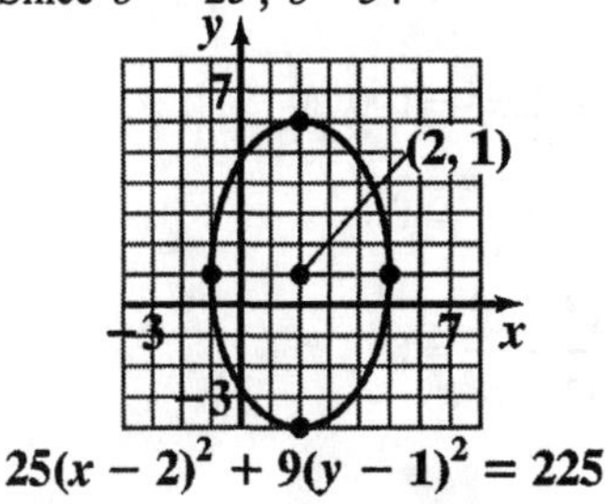

30. $\dfrac{x^2}{4}+\dfrac{y^2}{9}=1$

Since $a^2=4$, $a=2$.

Since $b^2=9$, $b=3$.

Area $=\pi ab=\pi(2)(3)=6\pi\approx 18.85$ sq. units

31. a. $\dfrac{x^2}{4}-\dfrac{y^2}{16}=1$

Since $a^2=4$ and $b^2=16$, $a=2$ and $b=4$. The equations of the asymptotes are $y=\pm\dfrac{b}{a}x$, or $y=\pm 2x$.

b. To graph the asymptotes, plot the points (2, 4), (–2, 4), (2, –4), and (–2, –4). The graph intersects the x-axis at (–2, 0) and (2, 0).

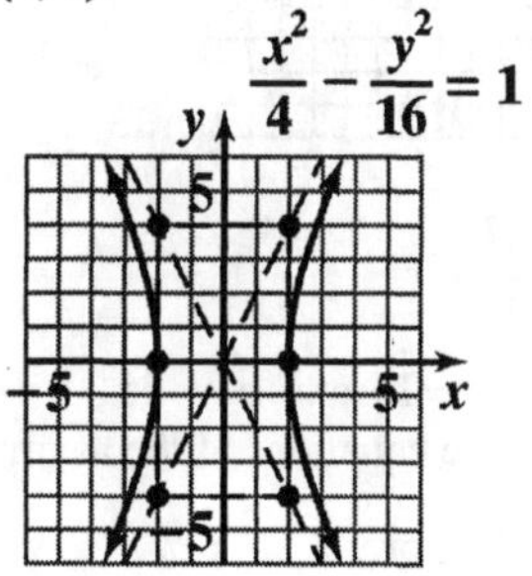

32. a. $\dfrac{x^2}{4}-\dfrac{y^2}{4}=1$

Since $a^2=4$ and $b^2=4$, $a=2$ and $b=2$. The equations of the asymptotes are $y=\pm\dfrac{b}{a}x$, or $y=\pm\dfrac{2}{2}x=\pm x$.

b. To graph the asymptotes, plot the points (2, 2), (–2, 2), (2, –2), and (–2, –2). The graph intersects the x-axis at (–2, 0) and (2, 0).

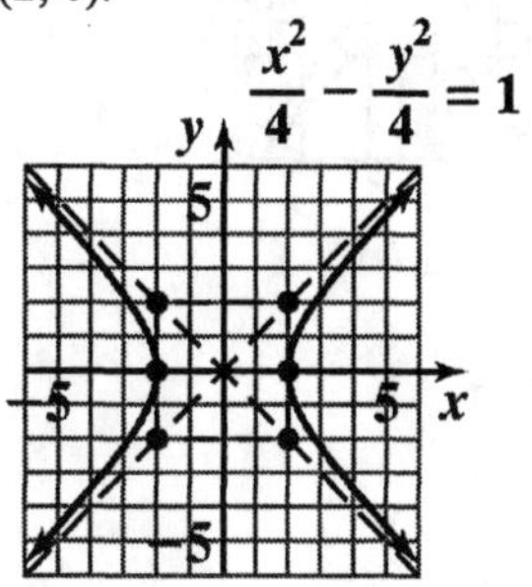

33. a. $\frac{y^2}{4}-\frac{x^2}{36}=1$

Since $a^2=36$ and $b^2=4$, $a=6$ and $b=2$. The equations of the asymptotes are $y=\pm\frac{b}{a}x$, or $y=\pm\frac{1}{3}$.

b. To graph the asymptotes, plot the points (6, 2), (6, –2), (–6, 2), and (–6, –2). The graph intersects the *y*-axis at (0,–2) and (0, 2).

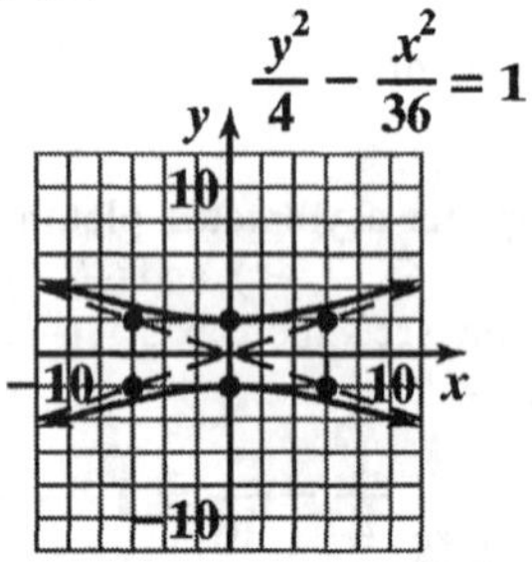

34. a. $\frac{y^2}{25}-\frac{x^2}{16}=1$

Since $a^2=16$ and $b^2=25$, $a=4$ and $b=5$. The equations of the asymptotes are $y=\pm\frac{b}{a}x$, or $y=\pm\frac{5}{4}x$.

b. To graph the asymptotes, plot the points (4, 5), (4, –5), (–4, 5), and (–4, –5). The graph intersects the *y*-axis at (0, –5) and (0, 5).

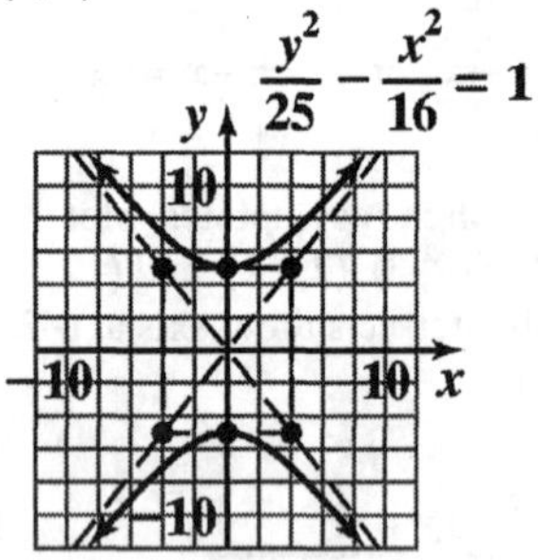

35. a. $x^2-9y^2=9$

$$\frac{x^2}{9}-\frac{9y^2}{9}=1$$

$$\frac{x^2}{9}-\frac{y^2}{1}=1$$

b. Since $a^2=9$ and $b^2=1$, $a=3$ and $b=1$. The equations of the asymptotes are $y=\pm\frac{b}{a}x$, or $y=\pm\frac{1}{3}x$.

c. To graph the asymptotes, plot the points (3, 1), (–3, 1), (3, –1), and (–3, –1). The graph intersects the *x*-axis at (–3, 0) and (3, 0).

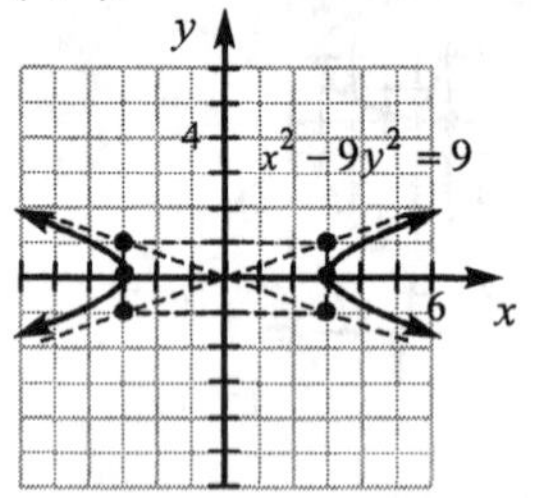

36. a. $25x^2-16y^2=400$

$$\frac{25x^2}{400}-\frac{16y^2}{400}=1$$

$$\frac{x^2}{16}-\frac{y^2}{25}=1$$

b. Since $a^2=16$ and $b^2=25$, $a=4$ and $b=5$. The equations of the asymptotes are $y=\pm\frac{b}{a}x$, or $y=\pm\frac{5}{4}x$.

c. To graph the asymptotes, plot the points (4, 5), (–4, 5), (4, –5), and (–4, –5). The graph intersects the *x*-axis at (–4, 0) and (4, 0).

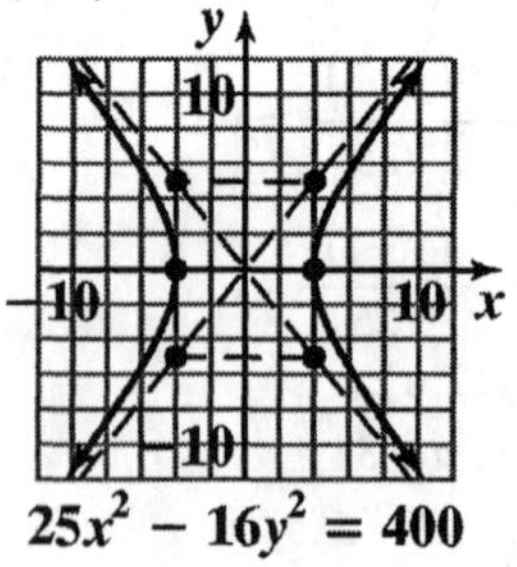

37. a. $4y^2-25x^2=100$

$$\frac{4y^2}{100}-\frac{25x^2}{100}=1$$

$$\frac{y^2}{25}-\frac{x^2}{4}=1$$

b. Since $a^2 = 4$ and $b^2 = 25$, $a = 2$ and $b = 5$. The equations of the asymptotes are $y = \pm\frac{b}{a}x$, or $y = \pm\frac{5}{2}x$.

c. To graph the asymptotes, plot the points (2, 5), (2, –5), (–2, 5), and (–2, –5). The graph intersects the y-axis at (0, –5) and (0, 5).

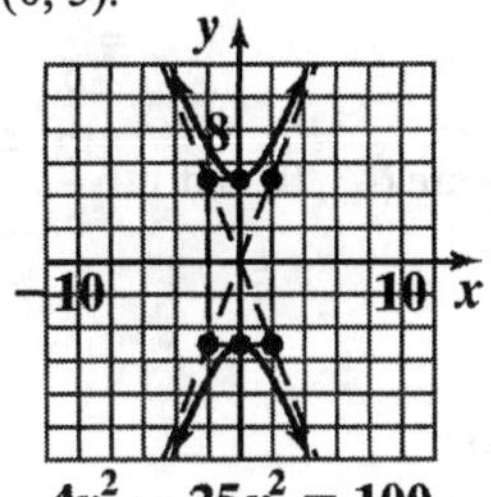

$4y^2 - 25x^2 = 100$

38. a. $49y^2 - 9x^2 = 441$

$$\frac{49y^2}{441} - \frac{9x^2}{441} = 1$$

$$\frac{y^2}{9} - \frac{x^2}{49} = 1$$

b. Since $a^2 = 49$ and $b^2 = 9$, $a = 7$ and $b = 3$. The equations of the asymptotes are $y = \pm\frac{b}{a}x$, or $y = \pm\frac{3}{7}x$.

c. To graph the asymptotes, plot the points (7, 3), (–7, 3), (7, –3), and (–7, –3). The graph intersects the y-axis at (0, –3) and (0, 3).

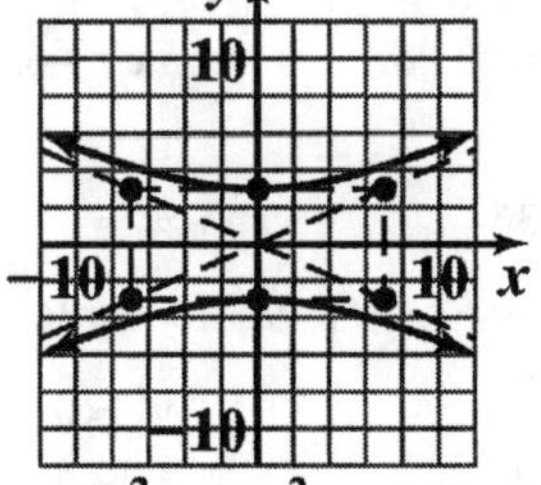

$49y^2 - 9x^2 = 441$

39. $\frac{x^2}{49} - \frac{y^2}{16} = 1$

The graph is a hyperbola.

40. $4x^2 + 8y^2 = 32$

$$\frac{4x^2}{32} + \frac{8y^2}{32} = \frac{32}{32}$$

$$\frac{x^2}{8} + \frac{y^2}{4} = 1$$

The graph is an ellipse.

41. $5x^2 + 5y^2 = 125$

$$\frac{5x^2}{5} + \frac{5y^2}{5} = \frac{125}{5}$$

$$x^2 + y^2 = 25$$

The graph is a circle.

42. $4x^2 - 25y^2 = 25$

$$\frac{4x^2}{25} - \frac{25y^2}{25} = \frac{25}{25}$$

$$\frac{x^2}{6.25} - \frac{y^2}{1} = 1$$

The graph is a hyperbola.

43. $\frac{x^2}{18} + \frac{y^2}{9} = 1$

The graph is an ellipse.

44. $y = (x-2)^2 + 1$

The graph is a parabola.

45. $12x^2 + 9y^2 = 108$

$$\frac{12x^2}{108} + \frac{9y^2}{108} = \frac{108}{108}$$

$$\frac{x^2}{9} + \frac{y^2}{12} = 1$$

The graph is an ellipse.

46. $x = -y^2 + 8y - 9$

The graph is a parabola.

47. $x^2+2y^2=25$

$x^2-3y^2=25 \Rightarrow x^2=3y^2+25$

Substitute $3y^2+25$ for x^2 in $x^2+2y^2=25$.

$x^2+2y^2=25$

$3y^2+25+2y^2=25$

$5y^2=0$

$y^2=0$

Substitute 0 for y^2 in $x^2-3y^2=25$

$x^2-0=25 \Rightarrow x=\pm 5$

The solutions are (5, 0) and (–5, 0).

48. $x^2=y^2+4$

$x+y=4$

Solve $x+y=4$ for y: $y=4-x$.

Substitute $4-x$ for y in $x^2=y^2+4$.

$x^2=y^2+4$

$x^2=(4-x)^2+4$

$x^2=(16-8x+x^2)+4$

$8x-16=4$

$8x=20$

$x=\frac{5}{2}$

$y=4-x$

$y=4-\frac{5}{2}$

$y=\frac{3}{2}$

The solution is $\left(\frac{5}{2},\frac{3}{2}\right)$.

49. $x^2+y^2=9$

$y=3x+9$

Substitute $3x+9$ for y in $x^2+y^2=9$.

$x^2+y^2=9$

$x^2+(3x+9)^2=9$

$x^2+9x^2+54x+81=9$

$10x^2+54x+72=0$

$5x^2+27x+36=0$

$(x+3)(5x+12)=0$

$x+3=0$ or $5x+12=0$

$x=-3$ $\qquad x=-\frac{12}{5}$

$y=3x+9$

$y=3(-3)+9$

$y=0$

$y=3x+9$

$y=3\left(-\frac{12}{5}\right)+9$

$y=\frac{9}{5}$

The solutions are (–3, 0) and $\left(-\frac{12}{5},\frac{9}{5}\right)$.

50. $x^2+2y^2=9$

$x^2-6y^2=36$

Solve $x^2+2y^2=9$ for x^2: $x^2=9-2y^2$.

Substitute $9-2y^2$ for x^2 in $x^2-6y^2=36$.

$x^2-6y^2=36$

$(9-2y^2)-6y^2=36$

$9-8y^2=36$

$-8y^2=27$

$y^2=-\frac{27}{8}$

There is no real solution to this equation so the system has no real solution.

51. $x^2+y^2=36$

$x^2-y^2=36$

$2x^2=72$

$x^2=36$

$x=6$ or $x=-6$

$x^2+y^2=36$

$6^2+y^2=36$

$y^2=0$

$y=0$

$x^2+y^2=36$

$(-6)^2+y^2=36$

$y^2=0$

$y=0$

The solutions are (6, 0) and (–6, 0).

52. $x^2+y^2=25$ (1)

$x^2-2y^2=-2$ (2)

$2x^2+2y^2=50$ (1) multiplied by 2

$x^2-2y^2=-2$ (2)

$3x^2=48$

$x^2=16$

$x=4$ or $x=-4$

$x^2+y^2=25$ $\quad$ $x^2+y^2=25$

$4^2+y^2=25$ $\quad$ $(-4)^2+y^2=25$

$y^2=9$ $\quad$ $y^2=9$

$y=\pm3$ $\quad$ $y=\pm3$

The solutions are (4, 3), (4, –3), (–4, 3) and (–4, –3).

53. $-4x^2+y^2=-15$ (1)

$8x^2+3y^2=-5$ (2)

$-8x^2+2y^2=-30$ (1) multiplied by 2

$8x^2+3y^2=-5$ (2)

$5y^2=-35$

$y^2=-7$

This equation has no real solution so there is no real solution to the system.

54. $3x^2+2y^2=6$ (1)

$4x^2+5y^2=15$ (2)

$-12x^2-8y^2=-24$ (1) multiplied by -4

$12x^2+15y^2=45$ (2) multiplied by 3

$7y^2=21$

$y^2=3$

$y=\pm\sqrt{3}$

$3x^2+2y^2=6$

$3x^2+2(3)=6$

$3x^2=0$

$x^2=0$

The solutions are $\left(0,\sqrt{3}\right)$ and $\left(0,-\sqrt{3}\right)$.

55. Let x = length

y = width

$xy=45$

$2x+2y=28$

Solve $2x+2y=28$ for y: $y=14-x$.

Substitute $14-x$ for y in $xy=45$.

$xy=45$

$x(14-x)=45$

$14x-x^2=45$

$x^2-14x+45=0$

$(x-5)(x-9)=0$

$x-5=0$ or $x-9=0$

$x=5$ $\quad$ $x=9$

$y=14-x$ $\quad$ $y=14-x$

$y=14-5$ $\quad$ $y=14-9$

$y=9$ $\quad$ $y=5$

The solutions are (5, 9) and (9, 5).

The dimensions of the pool table are 5 feet by 9 feet.

56. $C=20.3x+120$

$R=50.2x-0.2x^2$

$C=R$

$20.3x+120=50.2x-0.2x^2$

$0.2x^2-29.9+120=0$

$$x=\frac{29.9\pm\sqrt{(-29.9)^2-4(0.2)(120)}}{2(0.2)}$$

$$x=\frac{29.9\pm\sqrt{798.01}}{0.4}$$

$x\approx145$ or 4

The company must sell either 4 bottles or 145 bottles to break even.

57. Since $t = 1$ year, we may rewrite the formula $i = prt$ as $i = pr$.

$120 = pr$

$120 = (p+2000)(r-0.01)$

Rewrite the second equation by multiplying the binomials. Then substitute 120 for pr and solve for r.

$$120 = pr - 0.01p + 2000r - 20$$
$$120 = 120 - 0.01p + 2000r - 20$$
$$0 = -0.01p + 2000r - 20$$
$$0.01p + 20 = 2000r$$
$$\frac{0.01p}{2000} + \frac{20}{2000} = r$$
$$r = 0.000005p + 0.01$$

Substitute $0.000005p + 0.01$ for r in $120 = pr$.

$$120 = pr$$
$$120 = p(0.000005p + 0.01)$$
$$120 = 0.000005p^2 + 0.01p$$
$$0 = 0.000005p^2 + 0.01p - 120$$
$$0 = p^2 + 2000p - 24{,}000{,}000$$
$$0 = (p-4000)(p+6000)$$

$p - 4000 = 0$ or $p + 6000 = 0$

$p = 4000$ $\quad$ $p = -6000$

The principal must be positive, so use $p = 4000$.

$$r = 0.000005p + 0.01$$
$$r = 0.000005(4000) + 0.01$$
$$r = 0.03$$

The principal is $4000 and the rate is 3%.

Chapter 10 Practice Test

1. They are formed by cutting a cone or pair of cones.

2. $d = \sqrt{(x_2 - x_1)^2 + (y_2 - y_1)^2}$

$= \sqrt{[6-(-1)]^2 + (7-8)^2}$

$= \sqrt{7^2 + (-1)^2}$

$= \sqrt{49+1}$

$= \sqrt{50}$

The length is $\sqrt{50} \approx 7.07$ units.

3. Midpoint $= \left(\dfrac{x_1 + x_2}{2}, \dfrac{y_1 + y_2}{2}\right)$

$= \left(\dfrac{-9+7}{2}, \dfrac{4+(-1)}{2}\right)$

$= \left(-1, \dfrac{3}{2}\right)$

4. $y = -2(x+3)^2 + 1$

This is a parabola in the form $y = a(x-h)^2 + k$ with $a = -2$, $h = -3$, and $k = 1$. Since $a < 0$, the parabola opens downward. The vertex is (–3, 1). The y-intercept is (0, –17). The x-intercepts are about (–3.71, 0) and (–2.29, 0).

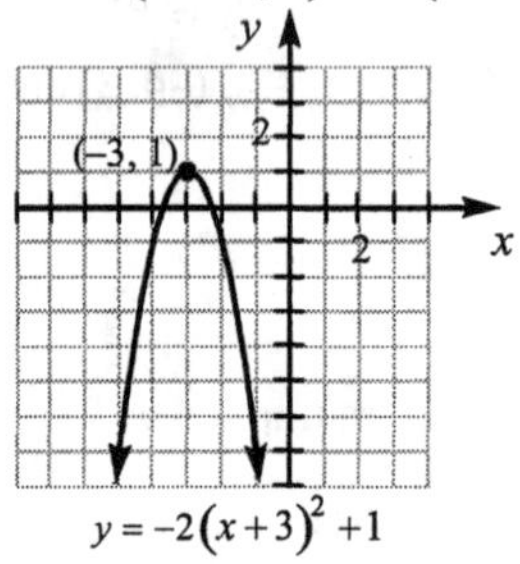

5. $x = y^2 - 2y + 4$

$x = (y^2 - 2y + 1) - 1 + 4$

$x = (y-1)^2 + 3$

This is a parabola in the form $x = a(y-k)^2 + h$ with $a = 1$, $h = 3$ and $k = 1$. Since $a > 0$, the parabola opens to the right. The vertex is (3, 1). There is no y-intercept. The x-intercept is (4, 0).

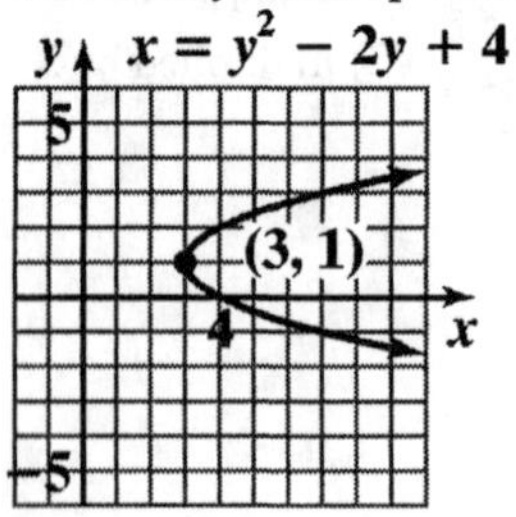

6. $x = -y^2 - 4y - 5$

$x = -(y^2 + 4y) - 5$

$x = -(y^2 + 4y + 4) + 4 - 5$

$x = -(y+2)^2 - 1$

This is a parabola in the form $x = a(y-k)^2 + h$ with $a = -1$, $h = -1$, and $k = -2$. Since $a < 0$, the parabola opens to the left. The vertex is $(-1, -2)$. There are no y-intercepts. The x-intercept is $(-5,0)$.

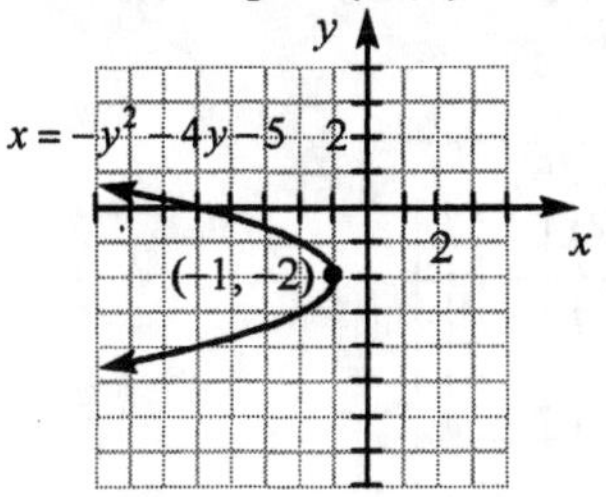

7. $(x-h)^2 + (y-k)^2 = r^2$

$[x-2]^2 + [y-4]^2 = 3^2$

$(x-2)^2 + (y-4)^2 = 9$

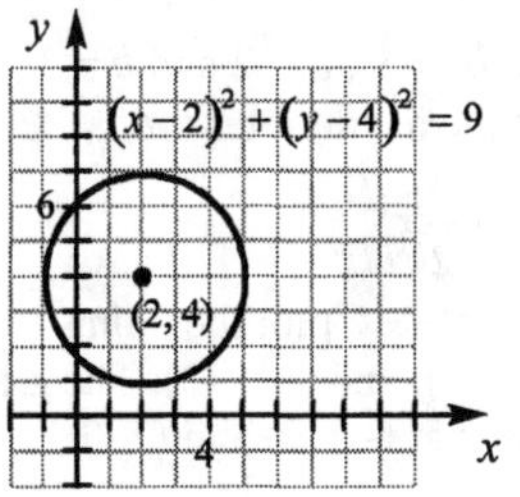

8. $(x+2)^2 + (y-8)^2 = 9$. The graph of this equation is a circle with center $(-2, 8)$ and radius 3.

Area $= \pi r^2$

$= \pi 3^2 = 9\pi \approx 28.27$ sq. units

9. The center is $(3, -1)$ and the radius is 4.

$(x-h)^2 + (y-k)^2 = r^2$

$(x-3)^2 + [y-(-1)]^2 = 4^2$

$(x-3)^2 + (y+1)^2 = 16$

10. $y = -\sqrt{16 - x^2}$

If we solve $x^2 + y^2 = 16$ for y, we obtain $y = \pm\sqrt{16 - x^2}$. Therefore, the graph of $y = -\sqrt{16 - x^2}$ is the lower half ($y \le 0$) of a circle with its center at the origin and radius 4.

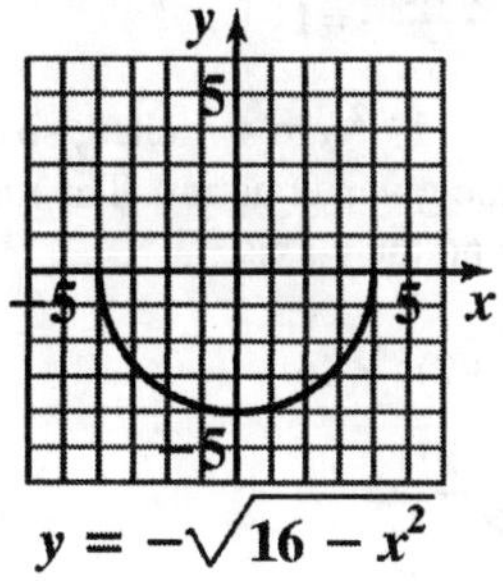

11.

$x^2 + y^2 + 2x - 6y + 1 = 0$

$x^2 + 2x + y^2 - 6y = -1$

$(x^2 + 2x + 1) + (y^2 - 6y + 9) = -1 + 1 + 9$

$(x+1)^2 + (y-3)^2 = 9$

The graph is a circle with center (1, 3) and radius 3.

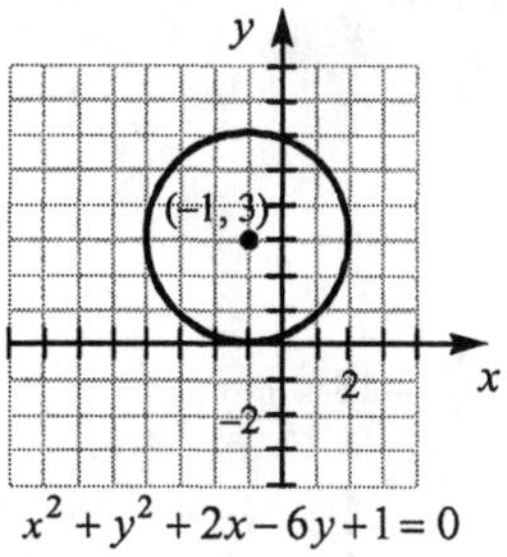

12. $4x^2 + 25y^2 = 100$

$\frac{4x^2}{100} + \frac{25y^2}{100} = 1$

$\frac{x^2}{25} + \frac{y^2}{4} = 1$

Since $a^2 = 25$, $a = 5$

Since $b^2 = 4$, $b = 2$.

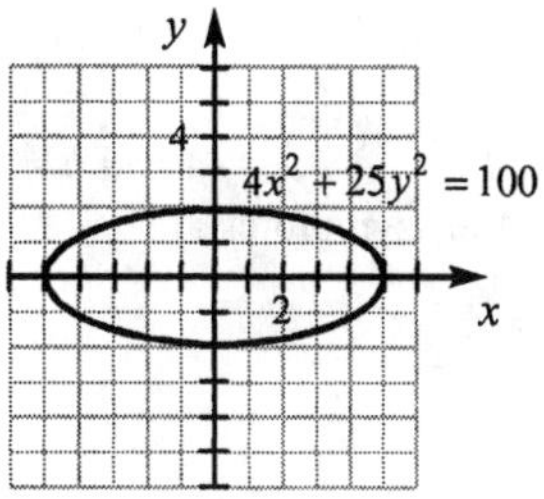

13. The center is (–2, –1), $a = 4$, and $b = 2$.

$$\frac{(x-h)^2}{a^2}+\frac{(y-k)^2}{b^2}=1$$

$$\frac{[x-(-2)]^2}{4^2}+\frac{[y-(-1)]^2}{2^2}=1$$

$$\frac{(x+2)^2}{16}+\frac{(y+1)^2}{4}=1$$

The values of a^2 and b^2 are switched, so this is not the graph of the given equation. The major axis should be along the y-axis.

14. $4(x-4)^2+36(y+2)^2=36$

$$\frac{4(x-4)^2}{36}+\frac{36(y+2)^2}{36}=1$$

$$\frac{(x-4)^2}{9}+\frac{(y+2)^2}{1}=1$$

The center is (4, –2). Since $a^2=9$, $a=3$

Since $b^2=1$, $b=1$

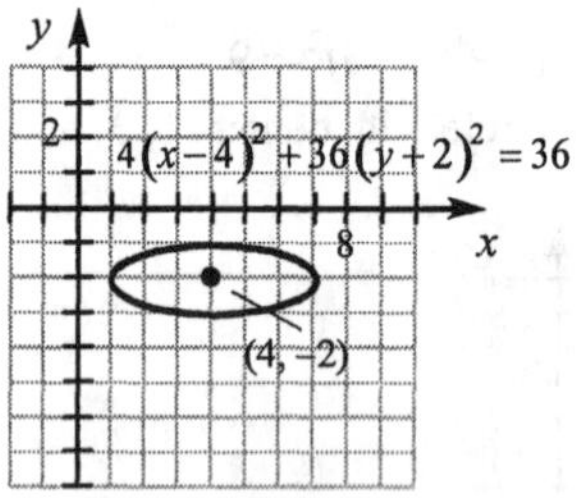

15. $3(x-8)^2+6(y+7)^2=18$

$$\frac{3(x-8)^2}{18}+\frac{6(y+7)^2}{18}=\frac{18}{18}$$

$$\frac{(x-8)^2}{6}+\frac{(y+7)^2}{3}=1$$

The center is $(8, -7)$.

16. The transverse axis lies along the axis corresponding to the positive term of the equation in standard form.

17. $\frac{x^2}{16}-\frac{y^2}{49}=1$

Since $a^2=16$ and $b^2=49$, $a=4$ and $b=7$. The equations of the asymptotes are

$y=\pm\frac{b}{a}x$, or $y=\pm\frac{7}{4}x$.

18. $\frac{y^2}{25}-\frac{x^2}{1}=1$

Since $a^2=1$ and $b^2=25$, $a=1$ and $b=5$. The equations of the asymptotes are

$y=\pm\frac{b}{a}x$, or $y=\pm 5x$.

To graph the asymptotes, plot the points
(1, 5),
(–1, 5), (1, –5), and (–1, –5).
The graph intersects the y-axis at (0, –5) and (0, 5).

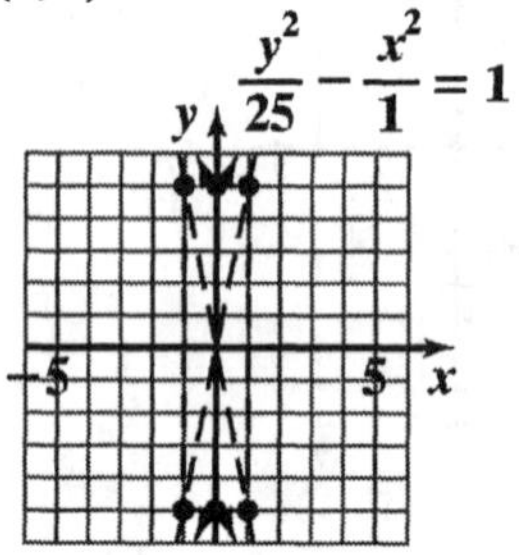

19. $\frac{x^2}{4}-\frac{y^2}{9}=1$

Since $a^2=4$ and $b^2=9$, $a=2$ and $b=3$. The equations of the asymptotes are

$y=\pm\frac{b}{a}x$, or $y=\pm\frac{3}{2}x$.

To graph the asymptotes, plot the points (2, 3), (–2, 3), (2, –3), and (–2, –3). The graph intersects the x-axis at (–2, 0) and (2, 0).

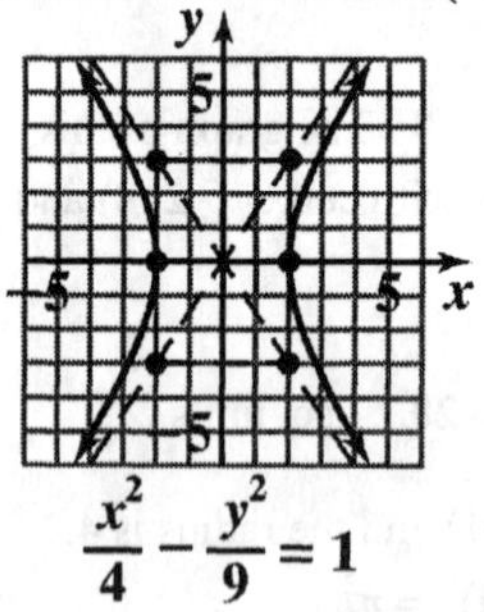

20. $4x^2-15y^2=30$

$$\frac{4x^2}{30}-\frac{15y^2}{30}=\frac{30}{30}$$

$$\frac{x^2}{\frac{15}{2}}-\frac{y^2}{2}=1$$

Since the equation is of the form

$\frac{x^2}{a^2}-\frac{y^2}{b^2}=1$, the graph is a hyperbola.

21. $25x^2 + 4y^2 = 100$

$$\frac{25x^2}{100} + \frac{4y^2}{100} = \frac{100}{100}$$

$$\frac{x^2}{4} + \frac{y^2}{25} = 1$$

Since the equation is of the form

$\frac{x^2}{a^2} + \frac{y^2}{b^2} = 1$, the graph is an ellipse.

22. $x^2 + y^2 = 7 \overset{\times 3}{\Rightarrow} 3x^2 + 3y^2 = 21$

$2x^2 - 3y^2 = -1 \qquad \underline{2x^2 - 3y^2 = -1}$

$$5x^2 = 20$$

$$x^2 = 4$$

$x = 2$ or $x = -2$

$x^2 + y^2 = 7 \qquad x^2 + y^2 = 7$

$(2)^2 + y^2 = 7 \qquad (-2)^2 + y^2 = 7$

$4 + y^2 = 7 \qquad 4 + y^2 = 7$

$y^2 = 3 \qquad y^2 = 3$

$y = \pm\sqrt{3} \qquad y = \pm\sqrt{3}$

The solutions are $(2, \sqrt{3}), (2, -\sqrt{3})$, $(-2, \sqrt{3})$, and $(-2, -\sqrt{3})$.

23. $x + y = 8$

$x^2 + y^2 = 4$

Solve $x + y = 8$ for y: $y = 8 - x$.

Substitute $8 - x$ for y in $x^2 + y^2 = 4$.

$$x^2 + y^2 = 4$$

$$x^2 + (8 - x)^2 = 4$$

$$x^2 + 64 - 16x + x^2 = 4$$

$$2x^2 - 16x + 60 = 0$$

$$x^2 - 8x + 30 = 0$$

$$x = \frac{-(-8) \pm \sqrt{(-8)^2 - 4(1)(30)}}{2(1)}$$

$$= \frac{8 \pm \sqrt{-56}}{2} = 4 \pm i\sqrt{14}$$

There is no real solution.

24. Let x = length, y = width.

$xy = 1500$

$2x + 2y = 160$

Solve $2x + 2y = 160$ for y: $y = 80 - x$.

Substitute $80 - x$ for y in $xy = 1500$.

$$xy = 1500$$

$$x(80 - x) = 1500$$

$$80x - x^2 = 1500$$

$$x^2 - 80x + 1500 = 0$$

$$(x - 30)(x - 50) = 0$$

$x - 30 = 0$ or $x - 50 = 0$

$x = 30 \qquad x = 50$

$y = 80 - 30 \qquad y = 80 - 50$

$y = 50 \qquad y = 30$

The solutions are (30, 50) and (50, 30).

The dimensions are 30 m by 50 m.

25. Let x = length, y = width

$xy = 60$

$x^2 + y^2 = 13^2$

Solve $xy = 60$ for y: $y = \frac{60}{x}$.

Substitute $\frac{60}{x}$ for y in $x^2 + y^2 = 13^2$.

$$x^2 + y^2 = 13^2$$

$$x^2 + \left(\frac{60}{x}\right)^2 = 169$$

$$x^2 + \frac{3600}{x^2} = 169$$

$$x^4 + 3600 = 169x^2$$

$$x^4 - 169x^2 + 3600 = 0$$

$$(x^2 - 25)(x^2 - 144) = 0$$

$x^2 - 25 = 0$ or $x^2 - 144 = 0$

$x^2 = 25 \qquad x^2 = 144$

$x = \pm 5 \qquad x = \pm 12$

Since x must be positive, $x = 5$ or $x = 12$.

If $x = 5$, then $y = \frac{60}{5} = 12$. If $x = 12$, then $y = \frac{60}{12} = 5$. The dimensions of the bed of the truck are 5 feet by 12 feet

Chapter 10 Cumulative Review Test

1. $(9x^2y^5)(-3xy^4) = (9)(-3)x^{2+1}y^{5+4}$
$= -27x^3y^9$

2. $4x - 2(3x - 7) = 2x - 5$
$4x - 6x + 14 = 2x - 5$
$-2x + 14 = 2x - 5$
$-4x = -19$
$x = \frac{-19}{-4} = \frac{19}{4}$

3. $2(x - 5) + 2x = 4x - 7$
$2x - 10 + 2x = 4x - 7$
$4x - 10 = 4x - 7$
$-10 = -7$
This is a contradiction so the solution set is $\varnothing$.

4. $|3x + 1| > 4$
$3x + 1 > 4 \qquad 3x + 1 < -4$
$3x > 3 \qquad 3x < -5$
$x > 1 \qquad x < -\frac{5}{3}$
The solution set is $\left\{x \,\middle|\, x < -\frac{5}{3} \text{ or } x > 1\right\}$

5. $y = -2x + 2$
The slope is –2 and the y-intercept is $(0, 2)$.

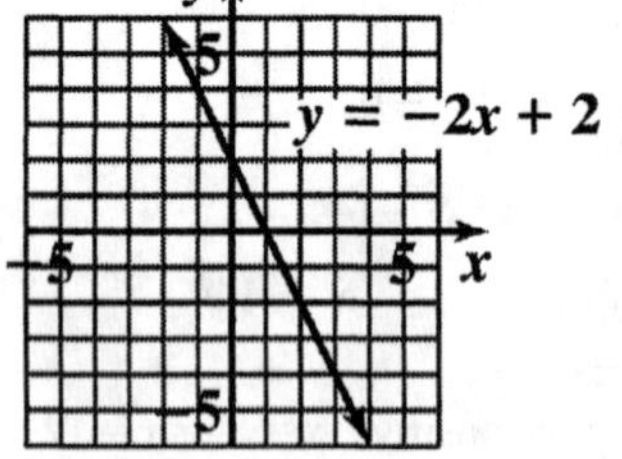

6. $f(x) = x^2 + 3x + 9$
$f(10) = (10)^2 + 3(10) + 9$
$= 100 + 30 + 9$
$= 139$

7. $\frac{1}{2}x - \frac{1}{3}y = 2 \overset{\times(-6)}{\Rightarrow} -3x + 2y = -12$
$\frac{1}{4}x + \frac{2}{3}y = 6 \overset{\times 12}{\Rightarrow} \underline{3x + 8y = 72}$
$10y = 60$
$y = 6$
Substitute $y = 6$ into $-3x + 2y = -12$
$-3x + 2(6) = -12$
$-3x + 12 = -12$
$-3x = -24 \Rightarrow x = 8$
The solution is (8, 6).

8. $x^4 - x^2 - 42 \qquad$ let $u = x^2$
$u^2 - u - 42$
$(u - 7)(u + 6)$
$(x^2 - 7)(x^2 + 6)$

9. Let x be the base of the sign. Then the height can be expressed as $x - 6$.
Area $= \frac{1}{2}$ (base $\times$ height)
$56 = \frac{1}{2}(x - 6)(x)$
$112 = x^2 - 6x$
$0 = x^2 - 6x - 112$
$0 = (x - 14)(x + 8)$
$x - 14 = 0$ or $x + 8 = 0$
$x = 14 \qquad x = -8$
Disregard the negative value. The base of the sign is 14 feet and the height is 14 – 6 = 8 ft.

10. $\frac{3x^2 - x - 4}{4x^2 + 7x + 3} \cdot \frac{2x^2 - 5x - 12}{6x^2 + x - 12}$
$\frac{(3x - 4)(x + 1)}{(4x + 3)(x + 1)} \cdot \frac{(2x + 3)(x - 4)}{(3x - 4)(2x + 3)}$
$\frac{(3x - 4)(x + 1)(2x + 3)(x - 4)}{(4x + 3)(x + 1)(3x - 4)(2x + 3)}$
$\frac{x - 4}{4x + 3}$

11. $\frac{x}{x+3}-\frac{x+5}{2x^2-2x-24}$

$=\frac{x}{x+3}-\frac{x+5}{2(x+3)(x-4)}$

$=\frac{2x(x-4)}{2(x+3)(x-4)}-\frac{x+5}{2(x+3)(x-4)}$

$=\frac{2x^2-8x-(x+5)}{2(x+3)(x-4)}$

$=\frac{2x^2-9x-5}{2(x+3)(x-4)}$

12. $\frac{3}{x+3}+\frac{5}{x+4}=\frac{12x+19}{x^2+7x+12}$

$\left(\frac{3}{x+3}+\frac{5}{x+4}=\frac{12x+19}{(x+3)(x+4)}\right)(x+3)(x+4)$

$3(x+4)+5(x+3)=12x+19$

$3x+12+5x+15=12x+19$

$8x+27=12x+19$

$-4x=-8$

$x=2$

13. $\left(\frac{18x^{1/2}y^3}{2x^{3/2}}\right)^{1/2}=\left(\frac{18}{2}x^{1/2-3/2}y^3\right)^{1/2}$

$=\left(9x^{-1}y^3\right)^{1/2}$

$=9^{1/2}x^{-1/2}y^{3/2}$

$=\sqrt{9}\cdot x^{-1/2}y^{3/2}$

$=\frac{3y^{3/2}}{x^{1/2}}$

14. $\frac{6\sqrt{x}}{\sqrt{x}-y}=\frac{6\sqrt{x}}{\sqrt{x}-y}\cdot\frac{\sqrt{x}+y}{\sqrt{x}+y}$

$=\frac{6x+6y\sqrt{x}}{x+y\sqrt{x}-y\sqrt{x}-y^2}$

$=\frac{6x+6y\sqrt{x}}{x-y^2}$

15. $3\sqrt[3]{2x+2}=\sqrt[3]{80x-24}$

$\left(3\sqrt[3]{2x+2}\right)^3=\left(\sqrt[3]{80x-24}\right)^3$

$27(2x+2)=(80x-24)$

$54x+54=80x-24$

$54x=80x-78$

$-26x=-78$

$x=3$

Check: $3\sqrt[3]{2(3)+2}\stackrel{?}{=}\sqrt[3]{80(3)-24}$

$3\sqrt[3]{8}\stackrel{?}{=}\sqrt[3]{216}$

$3\cdot 2=6$ True

The solution is 3.

16. $3x^2-4x+5=0$

$x=\frac{-b\pm\sqrt{b^2-4ac}}{2a}$

$=\frac{-(-4)\pm\sqrt{(-4)^2-4(3)(5)}}{2(3)}$

$=\frac{4\pm\sqrt{-44}}{6}$

$=\frac{4\pm 2i\sqrt{11}}{6}$

$=\frac{2\pm i\sqrt{11}}{3}$

17. $\log(3x-4)+\log(4)=\log(x+6)$

$\log 4(3x-4)=\log(x+6)$

$\log(12x-16)=\log(x+6)$

$12x-16=x+6$

$12x=x+22$

$11x=22$

$x=2$

Check: $\log[3(2)-4]+\log 4\stackrel{?}{=}\log(2+6)$

$\log 2+\log 4\stackrel{?}{=}\log 8$

$\log(2\cdot 4)=\log(8)$ True

The solution is 2.

18.
$$35 = 70e^{-0.3t}$$
$$\frac{1}{2} = e^{-0.3t}$$
$$\ln\frac{1}{2} = \ln e^{-0.3t}$$
$$\ln\frac{1}{2} = -0.3t$$
$$-\frac{1}{0.3}\ln\frac{1}{2} = t$$
$$t = -\frac{1}{0.3}\ln\frac{1}{2} \approx 2.31$$

19. $9x^2 + 4y^2 = 36$
$$\frac{9x^2}{36} + \frac{4y^2}{36} = 1$$
$$\frac{x^2}{4} + \frac{y^2}{9} = 1$$
Since $a^2 = 4$, $a = 2$.
Since $b^2 = 9$, $b = 3$.

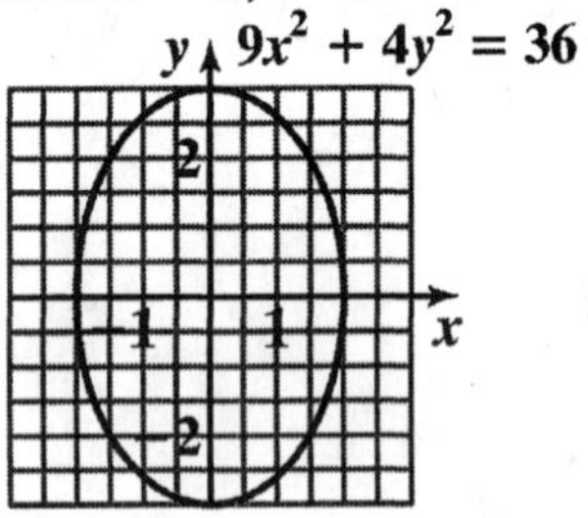

20. $\frac{y^2}{25} - \frac{x^2}{16} = 1$

Since $a^2 = 25$ and $b^2 = 16$, $a = 5$ and $b = 4$

The equations of the asymptotes are $y = \pm\frac{b}{a}x$,

or $y = \pm\frac{5}{4}x$. To graph the asymptotes, plot the points (4, 5), (4, –5), (–4, 5), and (–4, –5). The graph intersects the y-axis at (0, –5) and (0, 5).

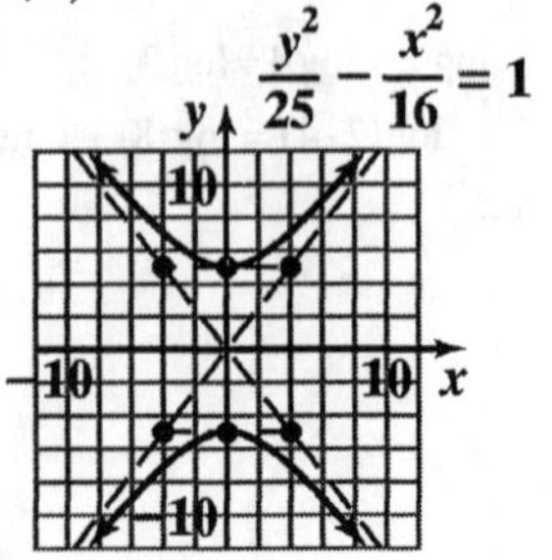

Chapter 11

Exercise Set 11.1

1. A sequence is a list of numbers arranged in a specific order.

3. A finite sequence is a function whose domain includes only the first n natural numbers.

5. In a decreasing sequence, the terms decrease.

7. A series is the sum of the terms of a sequence.

9. $\sum_{i=1}^{5}(i+4)$
The sum as i goes from 1 to 5 of $i + 4$.

11. The sequence $a_n = 2n-1$ is an increasing sequence since the coefficient of n is positive.

13. Yes, $a_n = 1+(-2)^n$ is an alternating sequence since $(-2)^n$ alternates between positive and negative values as n alternates from odd to even.

15. $a_n = 6n$
$a_1 = 6(1) = 6$
$a_2 = 6(2) = 12$
$a_3 = 6(3) = 18$
$a_4 = 6(4) = 24$
$a_5 = 6(5) = 30$
The terms are 6, 12, 18, 24, 30.

17. $a_n = 4n-1$
$a_1 = 4(1)-1 = 3$
$a_2 = 4(2)-1 = 7$
$a_3 = 4(3)-1 = 11$
$a_4 = 4(4)-1 = 15$
$a_5 = 4(5)-1 = 19$
The terms are 3, 7, 11, 15, 19.

19. $a_n = \frac{7}{n}$
$a_1 = \frac{7}{1} = 7$
$a_2 = \frac{7}{2}$
$a_3 = \frac{7}{3}$
$a_4 = \frac{7}{4}$
$a_5 = \frac{7}{5}$
The terms are 7, $\frac{7}{2}$, $\frac{7}{3}$, $\frac{7}{4}$, and $\frac{7}{5}$.

21. $a_n = \frac{n+2}{n+1}$
$a_1 = \frac{1+2}{1+1} = \frac{3}{2}$
$a_2 = \frac{2+2}{2+1} = \frac{4}{3}$
$a_3 = \frac{3+2}{3+1} = \frac{5}{4}$
$a_4 = \frac{4+2}{4+1} = \frac{6}{5}$
$a_5 = \frac{5+2}{5+1} = \frac{7}{6}$
The terms are $\frac{3}{2}$, $\frac{4}{3}$, $\frac{5}{4}$, $\frac{6}{5}$, and $\frac{7}{6}$.

23. $a_n = (-1)^n$
$a_1 = (-1)^1 = -1$
$a_2 = (-1)^2 = 1$
$a_3 = (-1)^3 = -1$
$a_4 = (-1)^4 = 1$
$a_5 = (-1)^5 = -1$
The terms are -1, 1, -1, 1, -1.

25. $a_n = (-2)^{n+1}$
$a_1 = (-2)^{1+1} = (-2)^2 = 4$
$a_2 = (-2)^{2+1} = (-2)^3 = -8$
$a_3 = (-2)^{3+1} = (-2)^4 = 16$
$a_4 = (-2)^{4+1} = (-2)^5 = -32$
$a_5 = (-2)^{5+1} = (-2)^6 = 64$
The terms are 4, -8, 16, -32, 64.

27. $a_n = 2n+7$
$a_{12} = 2(12)+7 = 24+7 = 31$

29. $a_n = \frac{n}{4}+8$
$a_{16} = \frac{16}{4}+8 = 4+8 = 12$

31. $a_n = (-1)^n$
$a_8 = (-1)^8 = 1$

33. $a_n = n(n+2)$
$a_9 = 9(9+2) = 9(11) = 99$

35. $a_n = \frac{n^2}{2n+7}$
$a_9 = \frac{9^2}{2(9)+7} = \frac{81}{18+7} = \frac{81}{25}$

37. $a_n = 3n-1$
$a_1 = 3(1)-1 = 3-1 = 2$
$a_2 = 3(2)-1 = 6-1 = 5$
$a_3 = 3(3)-1 = 9-1 = 8$

$s_1 = a_1 = 2$
$s_3 = a_1+a_2+a_3 = 2+5+8 = 15$

39. $a_n = 2^n+1$
$a_1 = 2^1+1 = 2+1 = 3$
$a_2 = 2^2+1 = 4+1 = 5$
$a_3 = 2^3+1 = 8+1 = 9$

$s_1 = a_1 = 3$
$s_3 = a_1+a_2+a_3 = 3+5+9 = 17$

41. $a_n = \frac{n-1}{n+2}$
$a_1 = \frac{1-1}{1+2} = \frac{0}{3} = 0$
$a_2 = \frac{2-1}{2+2} = \frac{1}{4}$
$a_3 = \frac{3-1}{3+2} = \frac{2}{5}$

$s_1 = 0$
$s_3 = 0+\frac{1}{4}+\frac{2}{5} = \frac{5}{20}+\frac{8}{20} = \frac{13}{20}$

43. $a_n = (-1)^n$
$a_1 = (-1)^1 = -1$
$a_2 = (-1)^2 = 1$
$a_3 = (-1)^3 = -1$

$s_1 = a_1 = -1$
$s_3 = a_1+a_2+a_3 = -1+1+-1 = -1$

45. $a_n = \frac{n^2}{2}$
$a_1 = \frac{1^2}{2} = \frac{1}{2}$
$a_2 = \frac{2^2}{2} = \frac{4}{2} = 2$
$a_3 = \frac{3^2}{2} = \frac{9}{2}$

$s_1 = a_1 = \frac{1}{2}$
$s_3 = a_1+a_2+a_3 = \frac{1}{2}+\frac{4}{2}+\frac{9}{2} = \frac{14}{2} = 7$

47. Each term is twice the preceding term. The next three terms are 64, 128, 256.

49. Each term is two more than the preceding term. The next three terms are 17, 19, 21.

51. Each denominator is one more than the preceding one while each numerator is one. The next three terms are $\frac{1}{6}, \frac{1}{7}, \frac{1}{8}$.

53. Each term is –1 times the previous term. The next three terms are 1, –1, 1.

55. Each denominator is three times the previous one while each numerator is one. The next three terms are $\frac{1}{81}, \frac{1}{243}, \frac{1}{729}$.

57. Each term is $-\frac{1}{2}$ times the preceding term. The next three terms are $\frac{1}{16}, -\frac{1}{32}, \frac{1}{64}$.

59. Each term is 5 less than the preceding term. The next three terms are 17, 12, 7.

61. $\sum_{i=1}^{5}(3i-1)=[3(1)-1]+[3(2)-1]+[3(3)-1]+[3(4)-1]+[3(5)-1]$

$=2+5+8+11+14$

$=40$

63. $\sum_{i=1}^{6}(i^2+1)=(1^2+1)+(2^2+1)+(3^2+1)+(4^2+1)+(5^2+1)+(6^2+1)$

$=(1+1)+(4+1)+(9+1)+(16+1)+(25+1)+(36+1)$

$=2+5+10+17+26+37$

$=97$

65. $\sum_{i=1}^{4}\frac{i^2}{2}=\frac{1^2}{2}+\frac{2^2}{2}+\frac{3^2}{2}+\frac{4^2}{2}=\frac{1}{2}+\frac{4}{2}+\frac{9}{2}+\frac{16}{2}=\frac{1}{2}+2+\frac{9}{2}+8=\frac{30}{2}=15$

67. $\sum_{i=4}^{9}\frac{i^2+i}{i+1}=\frac{4^2+4}{4+1}+\frac{5^2+5}{5+1}+\frac{6^2+6}{6+1}+\frac{7^2+7}{7+1}+\frac{8^2+8}{8+1}+\frac{9^2+9}{9+1}$

$=\frac{20}{5}+\frac{30}{6}+\frac{42}{7}+\frac{56}{8}+\frac{72}{9}+\frac{90}{10}$

$=4+5+6+7+8+9$

$=39$

---

69. $a_n=n+8$

The fifth partial sum is $\sum_{i=1}^{5}(i+8)$.

71. $a_n=\frac{n^2}{4}$

The third partial sum is $\sum_{i=1}^{3}\frac{i^2}{4}$.

73. $\sum_{i=1}^{5}x_i=x_1+x_2+x_3+x_4+x_5$

$=2+3+5+(-1)+4$

$=13$

75. $\left(\sum_{i=1}^{5}x_i\right)^2=(x_1+x_2+x_3+x_4+x_5)^2$

$=(2+3+5+(-1)+4)^2$

$=13^2$

$=169$

77. $\sum_{i=1}^{5}x_i^2=x_1^2+x_2^2+x_3^2+x_4^2+x_5^2$

$=2^2+3^2+5^2+(-1)^2+4^2$

$=55$

79. $\bar{x}=\frac{15+20+25+30+35}{5}=\frac{125}{5}=25$

81. $\bar{x}=\frac{72+83+4+60+18+20}{6}=\frac{257}{6}\approx 42.83$

83. **a.** Perimeter of rectangle: $p=2l+2w$

$p_1=2(1)+2(2\cdot 1)=2+4=6$

$p_2=2(2)+2(2\cdot 2)=4+8=12$

$p_3=2(3)+2(2\cdot 3)=6+12=18$

$p_4=2(4)+2(2\cdot 4)=8+16=24$

b. $p_n=2n+2(2n)=2n+4n=6n$

85 – 87. Answers will vary.

89. $\bar{x}=\frac{\sum x}{n}$

$n\bar{x}=n\cdot\frac{\sum x}{n}$

$n\bar{x}=\sum x$ or $\sum x=n\bar{x}$

91. Yes, $\sum_{i=1}^{n} 4x_i = 4\sum_{i=1}^{n} x_i$. Examples will vary.

93. a. $\sum x = x_1 + x_2 + x_3 = 3+5+2 = 10$

b. $\sum y = y_1 + y_2 + y_3 = 4+1+6 = 11$

c. $\sum x \cdot \sum y = 10 \cdot 11 = 110$

d.
$$\begin{aligned}\sum xy &= x_1y_1 + x_2y_2 + x_3y_3\\ &= 3(4)+5(1)+2(6)\\ &= 12+5+12\\ &= 29\end{aligned}$$

e. No, $\sum xy \neq \sum x \cdot \sum y$.

94.
$$\left|\frac{1}{2}x+\frac{3}{5}\right| = \left|\frac{1}{2}x-1\right|$$

$$\frac{1}{2}x+\frac{3}{5} = \frac{1}{2}x-1 \quad \text{or} \quad \frac{1}{2}x+\frac{3}{5} = -\left(\frac{1}{2}x-1\right)$$

$$\frac{3}{5} = -1 \qquad\qquad \frac{1}{2}x+\frac{3}{5} = -\frac{1}{2}x+1$$

$$\varnothing \qquad\qquad x = 1-\frac{3}{5}$$

$$x = \frac{2}{5}$$

The solution is $x = \frac{2}{5}$.

95.
$$\begin{aligned}8y^3-64x^6 &= 8\left(y^3-8x^6\right)\\ &= 8\left[(y)^3-(2x^2)^3\right]\\ &= 8\left[\left(y-2x^2\right)\left(y^2+2x^2y+(2x^2)^2\right)\right]\\ &= 8\left[\left(y-2x^2\right)\left(y^2+2x^2y+4x^4\right)\right]\end{aligned}$$

96.
$$\begin{aligned}\sqrt{x+5}-1 &= \sqrt{x-2}\\ \left(\sqrt{x+5}-1\right)^2 &= x-2\\ \left(\sqrt{x+5}\right)^2-2\left(\sqrt{x+5}\right)(1)+(1)^2 &= x-2\\ x+5-2\sqrt{x+5}+1 &= x-2\\ -2\sqrt{x+5} &= -8\\ \sqrt{x+5} &= 4\\ x+5 &= 4^2\\ x+5 &= 16\\ x &= 11\end{aligned}$$

Since we have raised both sides of the equation to an even-numbered power, we must check for extraneous solutions:

$$\begin{aligned}\sqrt{11+5}-1 &= \sqrt{11-2}\,?\\ \sqrt{16}-1 &= \sqrt{9}\,?\\ 4-1 &= 3\,?\\ 3 &= 3 \text{ True}\end{aligned}$$

Thus, $x = 11$ is the solution to the equation.

97.
$$\begin{aligned}V &= \pi r^2h, \text{ for } r\\ \frac{V}{\pi h} &= \frac{\pi r^2 h}{\pi h}\\ \frac{V}{\pi h} &= r^2\\ \sqrt{\frac{V}{\pi h}} &= r\end{aligned}$$

Exercise Set 11.2

1. In an arithmetic sequence, each term differs by a constant amount.

3. It is called the common difference.

5. The common difference, d, must be a positive number.

7. Yes. For example, $-1, -2, -3, -4, \ldots$ is an arithmetic sequence with $a_1 = -1$ and $d = -1$.

9. Yes. For example, $2, 4, 6, 8, \ldots$ is an arithmetic sequence with $a_1 = 2$ and $d = 2$.

11.
$$\begin{aligned}a_1 &= 4\\ a_2 &= 4+(2-1)(3) = 4+3 = 7\\ a_3 &= 4+(3-1)(3) = 4+2(3) = 4+6 = 10\\ a_4 &= 4+(4-1)(3) = 4+3(3) = 4+9 = 13\\ a_5 &= 4+(5-1)(3) = 4+4(3) = 4+12 = 16\end{aligned}$$

The terms are 4, 7, 10, 13, 16. The general term is $a_n = 4+(n-1)3$ or $a_n = 3n+1$.

13.
$$\begin{aligned}a_1 &= 7\\ a_2 &= 7+(2-1)(-2) = 7-2 = 5\\ a_3 &= 7+(3-1)(-2) = 7+2(-2) = 7-4 = 3\\ a_4 &= 7+(4-1)(-2) = 7+3(-2) = 7-6 = 1\\ a_5 &= 7+(5-1)(-2) = 7+4(-2) = 7-8 = -1\end{aligned}$$

The terms are 7, 5, 3, 1, -1. The general term is $a_n = 7+(n-1)(-2)$ or $a_n = -2n+9$.

15. $a_1 = \frac{1}{2}$

$a_2 = \frac{1}{2} + (2-1)\left(\frac{3}{2}\right) = \frac{1}{2} + \frac{3}{2} = \frac{4}{2} = 2$

$a_3 = \frac{1}{2} + (3-1)\left(\frac{3}{2}\right) = \frac{1}{2} + 2\left(\frac{3}{2}\right) = \frac{1}{2} + \frac{6}{2} = \frac{7}{2}$

$a_4 = \frac{1}{2} + (4-1)\left(\frac{3}{2}\right) = \frac{1}{2} + 3\left(\frac{3}{2}\right) = \frac{1}{2} + \frac{9}{2} = \frac{10}{2} = 5$

$a_5 = \frac{1}{2} + (5-1)\left(\frac{3}{2}\right) = \frac{1}{2} + 4\left(\frac{3}{2}\right) = \frac{1}{2} + \frac{12}{2} = \frac{13}{2}$

The terms are $\frac{1}{2}, 2, \frac{7}{2}, 5, \frac{13}{2}$. The general term is

$a_n = \frac{1}{2} + (n-1)\frac{3}{2}$ or $a_n = \frac{3}{2}n - 1$.

17. $a_1 = 100$

$a_2 = 100 + (2-1)(-5)$
$= 100 + (-5) = 100 - 5 = 95$

$a_3 = 100 + (3-1)(-5)$
$= 100 + 2(-5) = 100 - 10 = 90$

$a_4 = 100 + (4-1)(-5)$
$= 100 + 3(-5) = 100 - 15 = 85$

$a_5 = 100 + (5-1)(-5)$
$= 100 + 4(-5) = 100 - 20 = 80$

The terms are 100, 95, 90, 85, 80. The general term is $a_n = 100 + (n-1)(-5)$ or $a_n = -5n + 105$.

19. $a_n = a_1 + (n-1)d$

$a_4 = 5 + (4-1)3 = 5 + 3 \cdot 3 = 5 + 9 = 14$

21. $a_n = a_1 + (n-1)d$

$a_{10} = -9 + (10-1)(4) = -9 + 9(4) = -9 + 36 = 27$

23. $a_n = a_1 + (n-1)d$

$a_{13} = -8 + (13-1)\left(\frac{5}{3}\right)$
$= -8 + 12\left(\frac{5}{3}\right) = -8 + 20 = 12$

25. $a_n = a_1 + (n-1)d$

$27 = 11 + (9-1)d$
$27 = 11 + 8d$
$16 = 8d$
$2 = d$

27. $a_n = a_1 + (n-1)d$

$28 = 4 + (n-1)(3)$
$28 = 4 + 3n - 3$
$28 = 1 + 3n$
$27 = 3n$
$9 = n$

29. $a_n = a_1 + (n-1)d$

$42 = 82 + (n-1)(-8)$
$42 = 82 - 8n + 8$
$42 = 90 - 8n$
$-48 = -8n$
$6 = n$

31. $s_{10} = \frac{10(a_1 + a_{10})}{2} = \frac{10(1+19)}{2} = 5(20) = 100$

$a_{10} = a_1 + (10-1)d$
$a_{10} = a_1 + 9d$
$19 = 1 + 9d$
$18 = 9d$
$2 = d$

33. $s_8 = \frac{8(a_1 + a_8)}{2} = \frac{8\left(\frac{3}{5}+2\right)}{2} = 4\left(\frac{3}{5}+2\right)$

$= 4\left(\frac{3}{5} + \frac{10}{5}\right) = 4\left(\frac{13}{5}\right) = \frac{52}{5}$

$a_8 = a_1 + (8-1)d$
$a_8 = a_1 + 7d$
$2 = \frac{3}{5} + 7d$
$\frac{7}{5} = 7d$
$d = \frac{1}{7} \cdot \frac{7}{5} = \frac{1}{5}$

35. $s_6 = \frac{6(a_1 + a_6)}{2} = \frac{6(-5+13.5)}{2} = \frac{6(8.5)}{2} = 25.5$

$a_6 = a_1 + (6-1)d$
$a_6 = a_1 + 5d$
$13.5 = -5 + 5d$
$18.5 = 5d$
$3.7 = d$

37. $s_{11} = \frac{11(a_1 + a_{11})}{2} = \frac{11(7+67)}{2} = \frac{11(74)}{2} = 407$

$a_{11} = a_1 + (11-1)d$
$a_{11} = a_1 + 10d$
$67 = 7 + 10d$
$60 = 10d$
$6 = d$

39. $a_1 = 4$

$a_2 = 4 + (2-1)(3) = 4 + 3 = 7$
$a_3 = 4 + (3-1)(3) = 4 + 2(3) = 4 + 6 = 10$
$a_4 = 4 + (4-1)(3) = 4 + 3(3) = 4 + 9 = 13$

The terms are 4, 7, 10, 13.

$a_{10} = 4 + (10-1)(3) = 4 + 9(3) = 4 + 27 = 31$

$s_{10} = \frac{10(4+31)}{2} = \frac{10(35)}{2} = 175$

41. $a_1 = -6$

$a_2 = -6 + (2-1)(2) = -6 + 1(2) = -6 + 2 = -4$
$a_3 = -6 + (3-1)(2) = -6 + 2(2) = -6 + 4 = -2$
$a_4 = -6 + (4-1)(2) = -6 + 3(2) = -6 + 6 = 0$

The terms are –6, –4, –2, 0.

$a_{10} = -6 + (10-1)(2) = -6 + 9(2) = -6 + 18 = 12$

$s_{10} = \frac{10(-6+12)}{2} = \frac{10(6)}{2} = \frac{60}{2} = 30$

43. $a_1 = -8$

$a_2 = -8 + (2-1)(-5) = -8 - 5 = -13$
$a_3 = -8 + (3-1)(-5) = -8 + 2(-5) = -8 - 10 = -18$
$a_4 = -8 + (4-1)(-5) = -8 + 3(-5) = -8 - 15 = -23$

The terms are –8, –13, –18, –23.

$a_{10} = -8 + (10-1)(-5)$
$= -8 + 9(-5) = -8 - 45 = -53$

$s_{10} = \frac{10[-8+(-53)]}{2} = \frac{10(-61)}{2} = \frac{-610}{2} = -305$

45. $a_1 = \frac{7}{2}$

$a_2 = \frac{7}{2} + (2-1)\left(\frac{5}{2}\right) = \frac{7}{2} + 1\left(\frac{5}{2}\right) = \frac{7}{2} + \frac{5}{2} = \frac{12}{2} = 6$

$a_3 = \frac{7}{2} + (3-1)\left(\frac{5}{2}\right) = \frac{7}{2} + 2\left(\frac{5}{2}\right) = \frac{7}{2} + \frac{10}{2} = \frac{17}{2}$

$a_4 = \frac{7}{2} + (4-1)\left(\frac{5}{2}\right) = \frac{7}{2} + 3\left(\frac{5}{2}\right) = \frac{7}{2} + \frac{15}{2} = \frac{22}{2} = 11$

The terms are $\frac{7}{2}$, 6, $\frac{17}{2}$, 11.

$a_{10} = \frac{7}{2} + (10-1)\left(\frac{5}{2}\right)$

$= \frac{7}{2} + 9\left(\frac{5}{2}\right) = \frac{7}{2} + \frac{45}{2} = \frac{52}{2} = 26$

$s_{10} = \frac{10(3.5+26)}{2} = \frac{10(29.5)}{2} = 147.5$

47. $a_1 = 100$

$a_2 = 100 + (2-1)(-7)$
$= 100 + 1(-7) = 100 - 7 = 93$
$a_3 = 100 + (3-1)(-7)$
$= 100 + 2(-7) = 100 - 14 = 86$
$a_4 = 100 + (4-1)(-7)$
$= 100 + 3(-7) = 100 - 21 = 79$

The terms are 100, 93, 86, 79.

$a_{10} = 100 + (10-1)(-7)$
$= 100 + 9(-7) = 100 - 63 = 37$

$s_{10} = \frac{10(100+37)}{2} = \frac{10(137)}{2} = 685$

49. $d = 4 - 1 = 3$

$a_n = a_1 + (n-1)d$
$43 = 1 + (n-1)(3)$
$43 = 1 + 3n - 3$
$43 = -2 + 3n$
$45 = 3n$
$15 = n$

$s_{15} = \frac{15(a_1 + a_{15})}{2} = \frac{15(1+43)}{2} = \frac{15(44)}{2} = 330$

51. $d = -5 - (-9) = -5 + 9 = 4$

$a_n = a_1 + (n-1)d$
$31 = -9 + (n-1)(4)$
$31 = -9 + 4n - 4$
$31 = -13 + 4n$
$44 = 4n$
$11 = n$

$s_{10} = \frac{11(a_1 + a_{10})}{2} = \frac{11(-9+31)}{2} = \frac{11(22)}{2} = 121$

53. $d = \frac{2}{2} - \frac{1}{2} = \frac{1}{2}$

$a_n = \frac{1}{2} + (n-1)\left(\frac{1}{2}\right)$

$\frac{17}{2} = \frac{1}{2} + \frac{1}{2}n - \frac{1}{2}$

$\frac{17}{2} = \frac{1}{2}n$

$17 = n$

$s_{17} = \frac{17(a_1 + a_{17})}{2}$

$= \frac{17\left(\frac{1}{2} + \frac{17}{2}\right)}{2} = \frac{17\left(\frac{18}{2}\right)}{2} = \frac{17(9)}{2} = \frac{153}{2}$

55. $d = 10 - 7 = 3$

$a_n = a_1 + (n-1)d$

$91 = 7 + (n-1)(3)$

$91 = 7 + 3n - 3$

$91 = 4 + 3n$

$87 = 3n$

$29 = n$

$s_{29} = \frac{29(a_1 + a_{29})}{2} = \frac{29(7+91)}{2} = \frac{29(98)}{2} = 1421$

57. $s_n = \frac{n(a_1 + a_n)}{2}$

$s_{50} = \frac{50(1+50)}{2} = \frac{50(51)}{2} = 1275$

59. $s_n = \frac{n(a_1 + a_n)}{2}$

$s_{50} = \frac{50(1+99)}{2} = \frac{50(100)}{2} = 2500$

61. $s_n = \frac{n(a_1 + a_n)}{2}$

$s_{30} = \frac{30(3+90)}{2} = \frac{30(93)}{2} = 1395$

63. The smallest number greater than 7 that is divisible by 6 is 12. The largest number less than 1610 that is divisible by 6 is 1608. Now find n in the equation $a_n = a_1 + (n-1)d$.

$1608 = 12 + (n-1)6$

$1596 = 6(n-1)$

$266 = n - 1$

$267 = n$

There are 267 numbers between 7 and 1610 that are divisible by 6.

65. $a_1 = 20,\ d = 2,\ n = 12$

$a_n = a_1 + (n-1)d$

$a_{12} = 20 + (12-1)(2) = 20 + 11(2) = 20 + 22 = 42$

$s_n = \frac{n(a_1 + a_n)}{2}$

$s_{12} = \frac{12(20+42)}{2} = \frac{12(62)}{2} = \frac{744}{2} = 372$

There are 42 seats in the twelfth row and 372 seats in the first twelve rows.

67. $26 + 25 + 24 + \cdots + 1$ or $1 + 2 + 3 + \cdots + 26$

$s_n = \frac{n(a_1 + a_n)}{2}$

$s_{26} = \frac{26(1+26)}{2} = \frac{26(27)}{2} = \frac{702}{2} = 351$

There are 351 logs in the pile.

69. $a_1 = 1,\ d = 2,\ n = 14$

$a_n = a_1 + (n-1)d$

$a_{14} = 1 + (14-1)(2) = 1 + 13(2) = 1 + 26 = 27$

$s_n = \frac{n(a_1 + a_n)}{2}$

$s_{14} = \frac{14(1+27)}{2} = \frac{14(28)}{2} = \frac{392}{2} = 196$

There are 27 glasses in the 14th row and 196 glasses in all.

71. $1 + 2 + 3 + \cdots + 100$

$= (1+100) + (2+99) + \cdots + (50+51)$

$= 101 + 101 + \cdots + 101$

$= 50(101)$

$= 5050$

73. $s_n = \frac{n(a_1 + a_n)}{2}$

$= \frac{n[1 + (2n-1)]}{2} = \frac{n(2n)}{2} = \frac{2n^2}{2} = n^2$

75. a. $a_1 = 22,\ d = -\frac{1}{2},\ n = 7$

$a_n = a_1 + (n-1)d$

$a_7 = 22 + (7-1)\left(-\frac{1}{2}\right) = 22 - 3 = 19$

Her seventh swing is 19 feet.

b. $s_n = \frac{n(a_1 + a_n)}{2}$

$s_7 = \frac{7(22+19)}{2} = 143.5$

She travels 143.5 feet during the seven swings.

77. $d = -6 \text{ in.} = -\frac{1}{2} \text{ ft}, a_1 = 6$

$a_n = a_1 + (n-1)d$

$a_9 = 6 + (9-1)\left(-\frac{1}{2}\right) = 6 + 8\left(-\frac{1}{2}\right) = 6 - 4 = 2$

The ball bounces 2 feet on the ninth bounce.

79. a. Note that if March17th is day 1, then March 22nd is day 6.

$a_1 = 105, d = 10, n = 6$

$a_n = a_1 + (n-1)d$

$a_6 = 105 + (6-1)(10)$

$= 105 + 5(10) = 105 + 50 = 155$

He can prepare 155 packages for shipment on March 22nd.

b. $s_n = \frac{n(a_1 + a_n)}{2}$

$s_5 = \frac{5(105+155)}{2} = \frac{5(260)}{2} = \frac{1300}{2} = 650$

He can prepare 650 packages for shipment from March 17th through March 22nd.

81. $s_n = \frac{n(a_1 + a_n)}{2}$

$s_{31} = \frac{31(1+31)}{2} = \frac{31(32)}{2} = 496$

On day 31, Craig will have saved \$496.

83. a. $a_{10} = 42{,}000 + (10-1)(400) = 45{,}600$

She will receive \$45,600 in her tenth year of retirement.

b. $s_{10} = \frac{10(42{,}000 + 45{,}600)}{2}$

$= \frac{10(87{,}600)}{2}$

$= 438{,}000$

In her first 10 years, she will receive a total of \$438,000.

85. $360 - 180 = 180$

$540 - 360 = 180$

$720 - 540 = 180$

The terms form an arithmetic sequence with $d = 180$ and $a_3 = 180$.

$a_n = a_1 + (n-1)d$

$180 = a_1 + (3-1)180$

$180 = a_1 + 360$

$-180 = a_1$

$a_n = a_1 + (n-1)d$

$a_n = -180 + (n-1)(180)$

$= -180 + 180n - 180$

$= 180n - 360$

$= 180(n-2)$

93. $A = P + Prt$

$A - P = Prt$

$\frac{A-P}{Pt} = r$ or $r = \frac{A-P}{Pt}$

94. $y = 2x + 1$

$3x - 2y = 1$

Substitute $2x + 1$ for y in the second equation.

$3x - 2y = 1$

$3x - 2(2x+1) = 1$

$3x - 4x - 2 = 1$

$-x - 2 = 1$

$-x = 3$

$x = -3$

Substitute -3 for x in the first equation.

$y = 2(-3) + 1 = -6 + 1 = -5$

The solution is $(-3, -5)$.

95. $12n^2 - 6n - 30n + 15$

$= 3(4n^2 - 2n - 10n + 5)$

$= 3[(4n^2 - 2n) - (10n - 5)]$

$= 3[2n(2n-1) - 5(2n-1)]$

$= 3[(2n-1)(2n-5)]$

$= 3(2n-1)(2n-5)$

96. $(x+4)^2 + y^2 = 25$

$(x+4)^2 + y^2 = 5^2$

The center is $(-4, 0)$ and the radius is 5.

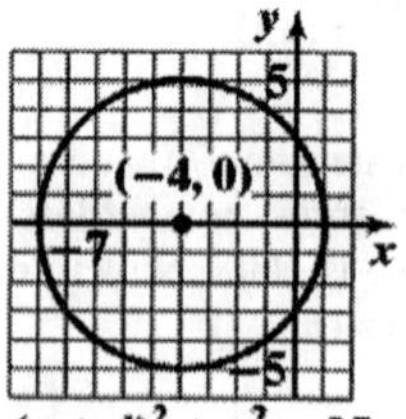

$(x+4)^2 + y^2 = 25$

Exercise Set 11.3

1. A geometric sequence is a sequence in which each term after the first is the same multiple of the preceding term.

3. To find the common ratio, take any term except the first and divide by the term that precedes it.

5. r^n approaches 0 as n gets larger and larger when $|r|<1$.

7. Yes

9. Yes, s_∞ exists. $s_\infty = \frac{a_1}{1-r} = \frac{6}{1-\frac{1}{4}} = \frac{6}{\frac{3}{4}} = 6\cdot\frac{4}{3} = 8$

   This is true since $|r|<1$.

11. $a_1 = 2$
   $a_2 = 2(3)^{2-1} = 2(3) = 6$
   $a_3 = 2(3)^{3-1} = 2(3)^2 = 2(9) = 18$
   $a_4 = 2(3)^{4-1} = 2(3)^3 = 2(27) = 54$
   $a_5 = 2(3)^{5-1} = 2(3)^4 = 2(81) = 162$
   The terms are 2, 6, 18, 54, 162.

13. $a_1 = 6$
   $a_2 = 6\left(-\frac{1}{2}\right)^{2-1} = 6\left(-\frac{1}{2}\right) = -3$
   $a_3 = 6\left(-\frac{1}{2}\right)^{3-1} = 6\left(-\frac{1}{2}\right)^2 = 6\left(\frac{1}{4}\right) = \frac{3}{2}$
   $a_4 = 6\left(-\frac{1}{2}\right)^{4-1} = 6\left(-\frac{1}{2}\right)^3 = 6\left(-\frac{1}{8}\right) = -\frac{3}{4}$
   $a_5 = 6\left(-\frac{1}{2}\right)^{5-1} = 6\left(-\frac{1}{2}\right)^4 = 6\left(\frac{1}{16}\right) = \frac{3}{8}$
   The terms are $6, -3, \frac{3}{2}, -\frac{3}{4}, \frac{3}{8}$.

15. $a_1 = 72$
   $a_2 = 72\left(\frac{1}{3}\right)^{2-1} = 72\left(\frac{1}{3}\right) = 24$
   $a_3 = 72\left(\frac{1}{3}\right)^{3-1} = 72\left(\frac{1}{3}\right)^2 = 72\left(\frac{1}{9}\right) = 8$
   $a_4 = 72\left(\frac{1}{3}\right)^{4-1} = 72\left(\frac{1}{3}\right)^3 = 72\left(\frac{1}{27}\right) = \frac{8}{3}$
   $a_5 = 72\left(\frac{1}{3}\right)^{5-1} = 72\left(\frac{1}{3}\right)^4 = 72\left(\frac{1}{81}\right) = \frac{8}{9}$
   The terms are 72, 24, 8, $\frac{8}{3}, \frac{8}{9}$.

17. $a_1 = 90$
   $a_2 = 90\left(-\frac{1}{3}\right)^{2-1} = 90\left(-\frac{1}{3}\right) = -30$
   $a_3 = 90\left(-\frac{1}{3}\right)^{3-1} = 90\left(-\frac{1}{3}\right)^2 = 90\left(\frac{1}{9}\right) = 10$
   $a_4 = 90\left(-\frac{1}{3}\right)^{4-1} = 90\left(-\frac{1}{3}\right)^3 = 90\left(-\frac{1}{27}\right) = -\frac{10}{3}$
   $a_5 = 90\left(-\frac{1}{3}\right)^{5-1} = 90\left(-\frac{1}{3}\right)^4 = 90\left(\frac{1}{81}\right) = \frac{10}{9}$
   The terms are 90, –30, 10, $-\frac{10}{3}, \frac{10}{9}$.

19. $a_1 = -1$
   $a_2 = -1(3)^{2-1} = -1(3) = -3$
   $a_3 = -1(3)^{3-1} = -1(3)^2 = -1(9) = -9$
   $a_4 = -1(3)^{4-1} = -1(3)^3 = -1(27) = -27$
   $a_5 = -1(3)^{5-1} = -1(3)^4 = -1(81) = -81$
   The terms are –1, –3, –9, –27, –81.

21. $a_1 = 5$
   $a_2 = 5(-2)^{2-1} = 5(-2)^1 = 5(-2) = -10$
   $a_3 = 5(-2)^{3-1} = 5(-2)^2 = 5(4) = 20$
   $a_4 = 5(-2)^{4-1} = 5(-2)^3 = 5(-8) = -40$
   $a_5 = 5(-2)^{5-1} = 5(-2)^4 = 5(16) = 80$
   The terms are 5, –10, 20, –40, 80.

23. $a_1 = \frac{1}{3}$
   $a_2 = \frac{1}{3}\left(\frac{1}{2}\right)^{2-1} = \frac{1}{3}\left(\frac{1}{2}\right) = \frac{1}{6}$
   $a_3 = \frac{1}{3}\left(\frac{1}{2}\right)^{3-1} = \frac{1}{3}\left(\frac{1}{2}\right)^2 = \frac{1}{3}\left(\frac{1}{4}\right) = \frac{1}{12}$
   $a_4 = \frac{1}{3}\left(\frac{1}{2}\right)^{4-1} = \frac{1}{3}\left(\frac{1}{2}\right)^3 = \frac{1}{3}\left(\frac{1}{8}\right) = \frac{1}{24}$
   $a_5 = \frac{1}{3}\left(\frac{1}{2}\right)^{5-1} = \frac{1}{3}\left(\frac{1}{2}\right)^4 = \frac{1}{3}\left(\frac{1}{16}\right) = \frac{1}{48}$
   The terms are $\frac{1}{3}, \frac{1}{6}, \frac{1}{12}, \frac{1}{24}, \frac{1}{48}$.

25. $a_1 = 3$

$a_2 = 3\left(\frac{3}{2}\right)^{2-1} = 3\left(\frac{3}{2}\right) = \frac{9}{2}$

$a_3 = 3\left(\frac{3}{2}\right)^{3-1} = 3\left(\frac{3}{2}\right)^2 = 3\left(\frac{9}{4}\right) = \frac{27}{4}$

$a_4 = 3\left(\frac{3}{2}\right)^{4-1} = 3\left(\frac{3}{2}\right)^3 = 3\left(\frac{27}{8}\right) = \frac{81}{8}$

$a_5 = 3\left(\frac{3}{2}\right)^{5-1} = 3\left(\frac{3}{2}\right)^4 = 3\left(\frac{81}{16}\right) = \frac{243}{16}$

The terms are 3, $\frac{9}{2}, \frac{27}{4}, \frac{81}{8}, \frac{243}{16}$.

27. $a_6 = a_1 r^{6-1}$

$a_6 = 4(2)^{6-1} = 4(2)^5 = 4(32) = 128$

29. $a_9 = a_1 r^{9-1}$

$a_9 = -12\left(\frac{1}{2}\right)^{9-1} = -12\left(\frac{1}{2}\right)^8 = -12\left(\frac{1}{256}\right) = -\frac{3}{64}$

31. $a_{10} = a_1 r^{10-1}$

$a_{10} = \frac{1}{4}(2)^{10-1} = \frac{1}{4}(2)^9 = \frac{1}{4}(512) = 128$

33. $a_{12} = a_1 r^{12-1}$

$a_{12} = -3(-2)^{12-1} = -3(-2)^{11} = -3(-2048) = 6144$

35. $a_8 = a_1 r^{8-1}$

$a_8 = 2\left(\frac{1}{2}\right)^{8-1} = 2\left(\frac{1}{2}\right)^7 = 2\left(\frac{1}{128}\right) = \frac{1}{64}$

37. $a_7 = a_1 r^{7-1}$

$a_7 = 50\left(\frac{1}{3}\right)^{7-1} = 50\left(\frac{1}{3}\right)^6 = 50\left(\frac{1}{729}\right) = \frac{50}{729}$

39. $s_5 = \frac{a_1(1-r^5)}{1-r}$

$s_5 = \frac{5(1-2^5)}{1-2} = \frac{5(1-32)}{-1} = \frac{5(-31)}{-1} = \frac{-155}{-1} = 155$

41. $s_6 = \frac{a_1(1-r^6)}{1-r}$

$s_6 = \frac{2(1-5^6)}{1-5} = \frac{2(1-15{,}625)}{-4} = \frac{2(-15{,}624)}{-4} = 7812$

43. $s_7 = \frac{a_1(1-r^7)}{1-r}$

$s_7 = \frac{80(1-2^7)}{1-2}$

$= \frac{80(1-128)}{-1} = \frac{80(-127)}{-1} = \frac{-10{,}160}{-1} = 10{,}160$

45. $s_9 = \frac{a_1(1-r^9)}{1-r}$

$s_9 = \frac{-15\left[1-\left(-\frac{1}{2}\right)^9\right]}{1-\left(-\frac{1}{2}\right)}$

$= \frac{-15\left[1-\left(-\frac{1}{512}\right)\right]}{\frac{3}{2}}$

$= \frac{-15\left(1+\frac{1}{512}\right)}{\frac{3}{2}}$

$= \frac{-15\left(\frac{513}{512}\right)}{\frac{3}{2}}$

$= -15\left(\frac{513}{512}\right)\left(\frac{2}{3}\right)$

$= -\frac{2565}{256}$

47. $s_5 = \frac{a_1(1-r^5)}{1-r}$

$s_5 = \frac{-9\left[1-\left(\frac{2}{5}\right)^5\right]}{1-\frac{2}{5}} = \frac{-9\left(1-\frac{32}{3125}\right)}{\frac{3}{5}}$

$= \frac{-9\left(\frac{3093}{3125}\right)}{\frac{3}{5}} = -9\left(\frac{3093}{3125}\right)\left(\frac{5}{3}\right) = -\frac{9279}{625}$

49. $r = \frac{3}{2} \div 3 = \frac{3}{2} \cdot \frac{1}{3} = \frac{1}{2}$

$a_n = 3\left(\frac{1}{2}\right)^{n-1}$

51. $r = 18 \div 9 = 2$

$a_n = 9(2)^{n-1}$

53. $r = -6 \div 2 = -3$

$a_n = 2(-3)^{n-1}$

55. $r = \frac{1}{2} \div \frac{3}{4} = \frac{1}{2} \cdot \frac{4}{3} = \frac{2}{3}$

$a_n = \frac{3}{4}\left(\frac{2}{3}\right)^{n-1}$

57. $r = \frac{1}{2} \div 1 = \frac{1}{2}$; $s_\infty = \frac{1}{1-\frac{1}{2}} = \frac{1}{\frac{1}{2}} = 1\left(\frac{2}{1}\right) = 2$

59. $r = \frac{1}{5} \div 1 = \frac{1}{5}$; $s_\infty = \frac{1}{1-\frac{1}{5}} = \frac{1}{\frac{4}{5}} = 1\left(\frac{5}{4}\right) = \frac{5}{4}$

61. $r = 3 \div 6 = \frac{1}{2}$; $s_\infty = \frac{6}{1-\frac{1}{2}} = \frac{6}{\frac{1}{2}} = 6\left(\frac{2}{1}\right) = 12$

63 $r = 2 \div 5 = \frac{2}{5}$; $s_\infty = \frac{5}{1-\frac{2}{5}} = \frac{5}{\frac{3}{5}} = 5\left(\frac{5}{3}\right) = \frac{25}{3}$

65.
$$\begin{aligned} s_n &= \frac{a_1(1-r^n)}{1-r} \\ 93 &= \frac{3(1-2^n)}{1-2} \\ 93 &= \frac{3(1-2^n)}{-1} \\ 93 &= -3(1-2^n) \\ -31 &= 1-2^n \\ -32 &= -2^n \\ 32 &= 2^n \\ n &= 5 \text{ since } 2^5 = 32 \end{aligned}$$

67.
$$\begin{aligned} s_n &= \frac{a_1(1-r^n)}{1-r} \\ \frac{189}{32} &= \frac{3\left[1-\left(\frac{1}{2}\right)^n\right]}{1-\frac{1}{2}} \\ \frac{189}{32} &= \frac{3\left[1-\left(\frac{1}{2}\right)^n\right]}{\frac{1}{2}} \\ \frac{1}{2}\cdot\frac{1}{3}\cdot\frac{189}{32} &= 1-\left(\frac{1}{2}\right)^n \\ \frac{63}{64} &= 1-\left(\frac{1}{2}\right)^n \\ -\frac{1}{64} &= -\left(\frac{1}{2}\right)^n \\ \frac{1}{64} &= \left(\frac{1}{2}\right)^n \\ n &= 6 \text{ since } \left(\frac{1}{2}\right)^6 = \frac{1}{64} \end{aligned}$$

69. $r = 1 \div 2 = \frac{1}{2}$

$s_\infty = \frac{2}{1-\frac{1}{2}} = \frac{2}{\frac{1}{2}} = 2\left(\frac{2}{1}\right) = 4$

71. $r = \frac{16}{3} \div 8 = \frac{16}{3}\left(\frac{1}{8}\right) = \frac{2}{3}$

$s_\infty = \frac{8}{1-\frac{2}{3}} = \frac{8}{\frac{1}{3}} = 8\left(\frac{3}{1}\right) = 24$

73. $r = 20 \div -60 = \frac{20}{-60} = -\frac{1}{3}$

$s_\infty = \frac{-60}{1-\left(-\frac{1}{3}\right)} = \frac{-60}{1+\frac{1}{3}} = \frac{-60}{\frac{4}{3}} = -60\left(\frac{3}{4}\right) = -45$

75. $r = -\frac{12}{5} \div -12 = -\frac{12}{5}\left(-\frac{1}{12}\right) = \frac{1}{5}$

$s_\infty = \frac{-12}{1-\frac{1}{5}} = \frac{-12}{\frac{4}{5}} = -12\left(\frac{5}{4}\right) = -15$

77.
$$\begin{aligned} 0.242424\ldots &= 0.24 + 0.0024 + 0.000024 + \cdots \\ &= 0.24 + 0.24(0.01) + 0.24(0.01)^2 + \cdots \end{aligned}$$

$r = 0.01$ and $a_1 = 0.24$

$s_\infty = \frac{0.24}{1-0.01} = \frac{0.24}{0.99} = \frac{24}{99} = \frac{8}{33}$

79.
$$\begin{aligned} 0.8888\ldots &= 0.8 + 0.08 + 0.008 + \cdots \\ &= 0.8 + 0.8(0.1) + 0.8(0.1)^2 + \cdots \end{aligned}$$

$r = 0.1$ and $a_1 = 0.8$

$s_\infty = \frac{0.8}{1-0.1} = \frac{0.8}{0.9} = \frac{8}{9}$

81.
$$\begin{aligned} 0.515151\cdots &= 0.51 + 0.0051 + 0.000051 + \cdots \\ &= 0.51 + 0.51(0.01) + 0.51(0.01)^2 + \cdots \end{aligned}$$

$r = 0.01$ and $a_1 = 0.51$

$s_\infty = \frac{0.51}{1-0.01} = \frac{0.51}{0.99} = \frac{51}{99} = \frac{17}{33}$

83. Consider a new series $b_1, b_2, b_3, \ldots$ where $b_1 = 15$ and $b_4 = 405$. Now $b_4 = b_1 r^{4-1}$ becomes

$$\begin{aligned} 405 &= 15r^3 \\ \frac{405}{15} &= r^3 \\ 27 &= r^3 \\ \sqrt[3]{27} &= r \\ 3 &= r \end{aligned}$$

From the original series, $a_1 = \frac{a_2}{r} = \frac{15}{3} = 5$.

85. Consider a new series $b_1, b_2, b_3, \ldots$ where $b_1 = 28$ and $b_3 = 112$. Now $b_3 = b_1 r^{3-1}$ becomes

$112 = 28r^2$

$\frac{112}{28} = r^2$

$4 = r^2$

so that $r = 2$ or $r = -2$. From the original series

$a_1 = \frac{a_3}{r^2} = \frac{28}{4} = 7$.

87. $a_1 = 1.40$, $n = 9$, $r = 1.03$

$a_n = a_1 r^{n-1}$

$a_9 = 1.4(1.03)^{9-1} = 1.4(1.03)^8 \approx 1.77$

In 8 years, a loaf of bread would cost $1.77.

89. $r = \frac{1}{2}$. Let a_n be the amount left after the nth day. After 1 day there are $600\left(\frac{1}{2}\right) = 300$ grams left, so $a_1 = 300$.

a. $37.5 = 300\left(\frac{1}{2}\right)^{n-1}$

$\frac{37.5}{300} = \left(\frac{1}{2}\right)^{n-1}$

$\frac{1}{8} = \left(\frac{1}{2}\right)^{n-1}$

$\left(\frac{1}{2}\right)^3 = \left(\frac{1}{2}\right)^{n-1}$

$n - 1 = 3$

$n = 4$

37.5 grams are left after 4 days.

b. $a_9 = 300\left(\frac{1}{2}\right)^{9-1} = 300\left(\frac{1}{256}\right) \approx 1.172$

After 9 days, about 1.172 grams of the substance remain.

91. After ten years will be at the beginning of the eleventh year. If the population increases by 1.1% each year, then the population at the beginning of a year will be 1.011 times the population at the beginning of the previous year.

a. Use $a_1 = 296.5$, $r = 1.011$, and $n = 11$.

$a_{11} = a_1 r^{11-1} = 296.5(1.011)^{10} \approx 330.78$

After ten years (the beginning of the eleventh year), the population is about 330.78 million people.

b. $a_n = 2(296.5) = 593$

Now, use $a_n = a_1 r^{n-1}$

$593 = 296.5(1.011)^{n-1}$

$\frac{593}{296.5} = (1.011)^{n-1}$

$2 = (1.011)^{n-1}$

Now, use logarithms.

$\log 2 = \log(1.011)^{n-1}$

$\log 2 = (n-1)\log 1.011$

$\frac{\log 2}{\log 1.011} = n - 1$

$63.4 \approx n - 1$

$64.4 \approx n$

The population will be double at the beginning of the 64.4 year, which is the end of the 63.4 year.

93. a. After 1 meter there is $\frac{1}{2}$ of the original light remaining, so $a_1 = \frac{1}{2}$.

$a_1 = \frac{1}{2}$

$a_2 = \frac{1}{2}\left(\frac{1}{2}\right)^{2-1} = \left(\frac{1}{2}\right)\left(\frac{1}{2}\right) = \frac{1}{4}$

$a_3 = \frac{1}{2}\left(\frac{1}{2}\right)^{3-1} = \left(\frac{1}{2}\right)\left(\frac{1}{4}\right) = \frac{1}{8}$

$a_4 = \frac{1}{2}\left(\frac{1}{2}\right)^{4-1} = \left(\frac{1}{2}\right)\left(\frac{1}{8}\right) = \frac{1}{16}$

$a_5 = \frac{1}{2}\left(\frac{1}{2}\right)^{5-1} = \left(\frac{1}{2}\right)\left(\frac{1}{16}\right) = \frac{1}{32}$

b. $a_n = \frac{1}{2}\left(\frac{1}{2}\right)^{n-1} = \left(\frac{1}{2}\right)^n$

c. $a_7 = \left(\frac{1}{2}\right)^7 = \frac{1}{128} \approx 0.0078$ or 0.78%

95. After 8 years will be the beginning of the ninth year. In any given year (except the first) there will be 106% of the previous amount in the account. Use $a_1 = 10{,}000$, $r = 1.06$, and $n = 9$

$a_9 = 10{,}000(1.06)^{9-1}$

$\approx 10{,}000(1.5938481)$

$\approx 15{,}938.48$

At the end of 8 years, there is $15,938.48 in the account.

97. a. $a_1 = 0.6(220) = 132$, $r = 0.6$

$a_n = a_1 r^{n-1}$

$a_4 = 132(0.6)^{4-1} = 132(0.6)^3 = 28.512$

The height of the fourth bounce is 28.512 feet.

b. $s_\infty = \frac{a_1}{1-r} = \frac{220}{1-0.6} = \frac{220}{0.4} = 550$

She travels a total of 550 feet in the downward direction.

99. a. $a_1 = 30(0.7) = 21$, $r = 0.7$

$a_n = a_1 r^{n-1}$

$a_3 = 21(0.7)^{3-1} = 21(0.7)^2 = 10.29$

The ball will bounce 10.29 inches on the third bounce.

b. $s_\infty = \frac{a_1}{1-r} = \frac{30}{1-0.7} = \frac{30}{0.3} = 100$

The ball travels a total of 100 inches in the downward direction.

101. Blue: $a_1 = 1$, $r = 2$

Red: $a_1 = 1$, $r = 3$

$a_6 = a_1 r^{6-1} = a_1 r^5$

Blue: $a_6 = 1(2)^5 = 32$

Red: $a_6 = 1(3)^5 = 243$

$243 - 32 = 211$

There are 211 more chips in the sixth stack of red chips.

103. Let a_n = value left after the *n*th year. After the first year there is $15000\left(\frac{4}{5}\right) = 12000$ of the value left, so $a_1 = 12000$, $r = \frac{4}{5}$.

$a_2 = 12000\left(\frac{4}{5}\right)^{2-1} = 12000\left(\frac{4}{5}\right) = 9600$

$a_3 = 12000\left(\frac{4}{5}\right)^{3-1} = 12000\left(\frac{16}{25}\right) = 7680$

$a_4 = 12000\left(\frac{4}{5}\right)^{4-1} = 12000\left(\frac{64}{125}\right) = 6144$

a. \$12,000, \$9,600, \$7,680, \$6,144

b. $a_n = 12000\left(\frac{4}{5}\right)^{n-1}$

c. $a_5 = 12000\left(\frac{4}{5}\right)^{5-1} = 12000\left(\frac{256}{625}\right) = 4915.20$

After 5 years, the value of the car is \$4,915.20.

105. Each time the ball bounces it goes up and then comes down the same distance. Therefore, the total vertical distance will be twice the height it rises after each bounce plus the initial 10 feet. The heights after each bounce form an infinite geometric sequence with $r = 0.9$ and $a_1 = 9$.

$s_\infty = \frac{9}{1-0.9} = \frac{9}{0.1} = 90$

Total distance: 6

$2(s_\infty) + 10 = 2(90) + 10 = 190$

The total vertical distance is 190 feet.

107. a. y_2 goes up more steeply.

b.

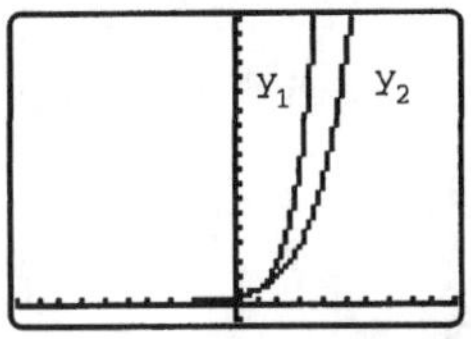

-10, 10, 1, -1, 19, 1

From the graph y_2 goes up more steeply, which agrees with our answer to part (a).

109. This is a geometric sequence with $r = \frac{2}{1} = 2$.

Also, $a_n = 1{,}048{,}576 = 2^{20}$. Using $a_n = a_1 r^{n-1}$ gives

$2^{20} = 1(2)^{n-1}$

$2^{20} = 2^{n-1}$

$20 = n - 1$

$21 = n$

Thus, there are 21 terms in the sequence.

$s_{21} = \frac{a_1(1-r^{21})}{1-r} = \frac{1(1-2^{21})}{1-2}$

$= \frac{1-2{,}097{,}152}{-1} = \frac{-2{,}097{,}151}{-1} = 2{,}097{,}151$

110. $f(x) = x^2 - 4$, $g(x) = x - 3$

$(f \cdot g)(4) = f(4) \cdot g(4)$

$= (4^2 - 4) \cdot (4 - 3)$

$= (16 - 4)(1)$

$= 12 \cdot 1$

$= 12$

111.
$$\begin{array}{r} 3x^2+4xy-2y^2 \\ \times \qquad\qquad 2x-3y \\ \hline -9x^2y-12xy^2+6y^3 \\ 6x^3+8x^2y-4xy^2 \qquad\qquad \\ \hline 6x^3-x^2y-16xy^2+6y^3 \end{array}$$

112. $S=\dfrac{2a}{1-r}$, for r

$$S(1-r)=2a$$
$$1-r=\frac{2a}{S}$$
$$-r=\frac{2a}{S}-1$$
$$r=1-\frac{2a}{S} \text{ or } r=\frac{S-2a}{S}$$

113.
$$g(x)=x^3+9$$
$$y=x^3+9$$
$$x=y^3+9$$
$$x-9=y^3$$
$$\sqrt[3]{x-9}=y$$
$$g^{-1}(x)=\sqrt[3]{x-9}$$

114.
$$\log x+\log(x-1)=\log 20$$
$$\log[x(x-1)]=\log 20$$
$$x(x-1)=20$$
$$x^2-x-20=0$$
$$(x+4)(x-5)=0$$
$$x+4=0 \quad\text{or}\quad x-5=0$$
$$x=-4 \qquad\qquad x=5$$

$x=-4$ is an extraneous solution. The solution is $x=5$.

115. Let x and y be the lengths of the legs. Then

$$x^2+y^2=15^2$$
$$x+y+15=36 \Rightarrow y=21-x$$
$$x^2+(21-x)^2=225$$
$$x^2+441-42x+x^2=225$$
$$2x^2-42x+216=0$$
$$2(x^2-21x+108)=0$$
$$2(x-9)(x-12)=0$$
$$x-9=0 \quad\text{or}\quad x-12=0$$
$$x=9 \qquad\qquad x=12$$
$$y=12 \qquad\qquad y=9$$

The solutions are (9, 12) and (12, 9). The legs are 9 meters and 12 meters.

Mid-Chapter Test: 11.1-11.3

1. $a_n=-3n+5$

$$a_1=-3(1)+5=-3+5=2$$
$$a_2=-3(2)+5=-6+5=-1$$
$$a_3=-3(3)+5=-9+5=-4$$
$$a_4=-3(4)+5=-12+5=-7$$
$$a_5=-3(5)+5=-15+5=-10$$

The terms are 2, –1, –4, –7, –10.

2. $a_n=n(n+6)$

$$a_7=7(7+6)=7(13)=91$$

3 $a_n=2^n-1$

$$a_1=2^1-1=2-1=1$$
$$a_2=2^2-1=4-1=3$$
$$a_3=2^3-1=8-1=7$$

$$s_1=a_1=1$$
$$s_3=a_1+a_2+a_3=1+3+7=11$$

4. Each term is 4 less than the previous term. The next three terms are $-15, -19, -23$.

5. $\sum_{i=1}^{5}(4i-3)$

$$=[4(1)-3]+[4(2)-3]+[4(3)-3]+[4(4)-3]+[4(5)-3]$$
$$=[4-3]+[8-3]+[12-3]+[16-3]+[20-3]$$
$$=1+5+9+13+17$$
$$=45$$

6. $a_n=\dfrac{1}{3}n+7$

The fifth partial sum is $\sum_{i=1}^{5}\left(\dfrac{1}{3}i+7\right)$.

7. $a_1=-6$

$$a_2=-6+(2-1)(5)=-6+1(5)=-6+5=-1$$
$$a_3=-6+(3-1)(5)=-6+2(5)=-6+10=4$$
$$a_4=-6+(4-1)(5)=-6+3(5)=-6+15=9$$
$$a_5=-6+(5-1)(5)=-6+4(5)=-6+20=14$$

The terms are –6, –1, 4, 9, 14. The general term is $a_n=-6+(n-1)5$ or $a_n=5n-11$.

8. $a_n = a_1 + (n-1)d$

$-\frac{1}{2} = \frac{11}{2} + (7-1)d$

$-\frac{1}{2} = \frac{11}{2} + 6d$

$-\frac{12}{2} = 6d$

$-6 = 6d$

$-1 = d$

9. $a_n = a_1 + (n-1)d$

$-3 = 22 + (n-1)(-5)$

$-3 = 22 - 5n + 5$

$-3 = 27 - 5n$

$-30 = -5n$

$6 = n$

10. $a_n = a_1 + (n-1)d$

$7 = -8 + (6-1)d$

$7 = -8 + 5d$

$15 = 5d$

$3 = d$

$$s_6 = \frac{6(a_1 + a_6)}{2} = \frac{6(-8+7)}{2} = \frac{6(-1)}{2} = \frac{-6}{2} = -3$$

11. $a_n = a_1 + (n-1)d$

$$a_{10} = \frac{5}{2} + (10-1)\left(\frac{1}{2}\right) = \frac{5}{2} + 9\left(\frac{1}{2}\right) = \frac{5}{2} + \frac{9}{2} = \frac{14}{2} = 7$$

$$s_{10} = \frac{10(a_1 + a_{10})}{2}$$

$$= \frac{10(\frac{5}{2}+7)}{2} = \frac{25+70}{2} = \frac{95}{2} = 47\frac{1}{2}$$

12. $d = 0 - (-7) = 0 + 7 = 7$

$a_n = a_1 + (n-1)d$

$63 = -7 + (n-1)(7)$

$63 = -7 + 7n - 7$

$63 = 7n - 14$

$77 = 7n$

$11 = n$

There are 11 terms in the sequence.

13. $16+15+14+\cdots+1$ or $1+2+3+\cdots+16$

$$s_n = \frac{n(a_1 + a_n)}{2}$$

$$s_{16} = \frac{16(1+16)}{2} = \frac{16(17)}{2} = \frac{272}{2} = 136$$

There are 136 logs in the pile.

14. $a_1 = 80$

$$a_2 = 80\left(-\frac{1}{2}\right)^{2-1} = 80\left(-\frac{1}{2}\right) = -40$$

$$a_3 = 80\left(-\frac{1}{2}\right)^{3-1} = 80\left(-\frac{1}{2}\right)^2 = 80\left(\frac{1}{4}\right) = 20$$

$$a_4 = 80\left(-\frac{1}{2}\right)^{4-1} = 80\left(-\frac{1}{2}\right)^3 = 80\left(-\frac{1}{8}\right) = -10$$

$$a_5 = 80\left(-\frac{1}{2}\right)^{5-1} = 80\left(-\frac{1}{2}\right)^4 = 80\left(\frac{1}{16}\right) = 5$$

The terms are 80, –40, 20, –10, 5.

15. $a_7 = a_1 r^{7-1}$

$$a_7 = 81\left(\frac{1}{3}\right)^{7-1} = 81\left(\frac{1}{3}\right)^6 = 81\left(\frac{1}{729}\right) = \frac{1}{9}$$

16. $s_6 = \frac{a_1(1-r^6)}{1-r}$

$$s_6 = \frac{5(1-2^6)}{1-2} = \frac{5(1-64)}{-1} = \frac{5(-63)}{-1} = 315$$

17. $r = -\frac{16}{3} \div 8 = -\frac{16}{3} \cdot \frac{1}{8} = -\frac{2}{3}$

18. $r = 4 \div 12 = \frac{4}{12} = \frac{1}{3}$

$$s_\infty = \frac{a_1}{1-r} = \frac{12}{1-\frac{1}{3}} = \frac{12}{\frac{2}{3}} = 12\left(\frac{3}{2}\right) = 18$$

19. $0.878787\ldots = 0.87 + 0.0087 + 0.000087 + \cdots$

$= 0.87 + 0.87(0.01) + 0.87(0.01)^2 + \cdots$

$r = 0.01$ and $a_1 = 0.87$

$$s_\infty = \frac{0.87}{1-0.01} = \frac{0.87}{0.99} = \frac{87}{99} = \frac{29}{33}$$

20. a. A sequence is a list of numbers arranged in a specific order.

b. An arithmetic sequence is a sequence where each term differs by a constant amount.

c. A geometric sequence is a sequence in which each term after the first is the same multiple of the preceding term.

d. A series is the sum of the terms of a sequence.

Exercise Set 11.4

1. Answers will vary. One possibility follows. The first and last numbers in each row are 1 and the inner numbers are obtained by adding the two numbers in the row above (to the right and left).

$$\begin{array}{ccccccccc} & & & & 1 & & & & \\ & & & 1 & & 1 & & & \\ & & 1 & & 2 & & 1 & & \\ & 1 & & 3 & & 3 & & 1 & \\ 1 & & 4 & & 6 & & 4 & & 1 \end{array}$$

3. $1! = 1$

5. No. Factorials are only defined for nonnegative integers.

7. The expansion of $(a+b)^{13}$ has 14 terms, one more than the power to which the binomial is raised.

9. $\binom{5}{2} = \frac{5!}{2!\cdot(5-2)!}$

$= \frac{5!}{2!\cdot 3!}$

$= \frac{5\cdot 4\cdot \cancel{3\cdot 2\cdot 1}}{(2\cdot 1)(\cancel{3\cdot 2\cdot 1})}$

$= \frac{20}{2}$

$= 10$

11. $\binom{5}{5} = \frac{5!}{5!(5-5)!} = \frac{\cancel{5!}}{\cancel{5!}\cdot 0!} = \frac{1}{0!} = \frac{1}{1} = 1$

13. $\binom{7}{0} = \frac{7!}{0!(7-0)!} = \frac{7!}{0!\cdot 7!} = \frac{\cancel{7!}}{0!\cdot \cancel{7!}} = \frac{1}{0!} = \frac{1}{1} = 1$

15. $\binom{8}{4} = \frac{8!}{4!\cdot(8-4)!}$

$= \frac{8!}{4!\cdot 4!}$

$= \frac{8\cdot 7\cdot 6\cdot 5\cdot \cancel{4\cdot 3\cdot 2\cdot 1}}{(4\cdot 3\cdot 2\cdot 1)(\cancel{4\cdot 3\cdot 2\cdot 1})}$

$= \frac{1680}{24}$

$= 70$

17. $\binom{8}{2} = \frac{8!}{2!\cdot(8-2)!}$

$= \frac{8!}{2!\cdot 6!}$

$= \frac{8\cdot 7\cdot \cancel{6\cdot 5\cdot 4\cdot 3\cdot 2\cdot 1}}{(2\cdot 1)\cdot(\cancel{6\cdot 5\cdot 4\cdot 3\cdot 2\cdot 1})}$

$= \frac{56}{2}$

$= 28$

19. $(x+4)^3$

$= \binom{3}{0}x^3 4^0 + \binom{3}{1}x^2 4^1 + \binom{3}{2}x^1 4^2 + \binom{3}{3}x^0 4^3$

$= 1x^3(1) + 3x^2(4) + 3x(16) + 1(1)64$

$= x^3 + 12x^2 + 48x + 64$

---

21. $(2x-3)^3 = \binom{3}{0}(2x)^3(-3)^0 + \binom{3}{1}(2x)^2(-3)^1 + \binom{3}{2}(2x)^1(-3)^2 + \binom{3}{3}(2x)^0(-3)^3$

$= 1(8x^3)(1) + 3(4x^2)(-3) + 3(2x)(9) + 1(1)(-27)$

$= 8x^3 - 36x^2 + 54x - 27$

23. $(a-b)^4 = \binom{4}{0}a^4(-b)^0 + \binom{4}{1}a^3(-b)^1 + \binom{4}{2}a^2(-b)^2 + \binom{4}{3}a^1(-b)^3 + \binom{4}{4}a^0(-b)^4$
$= 1a^4(1) + 4a^3(-b) + 6a^2b^2 + 4a(-b^3) + 1(1)b^4$
$= a^4 - 4a^3b + 6a^2b^2 - 4ab^3 + b^4$

25. $(3a-b)^5 = \binom{5}{0}(3a)^5(-b)^0 + \binom{5}{1}(3a)^4(-b)^1 + \binom{5}{2}(3a)^3(-b)^2 + \binom{5}{3}(3a)^2(-b)^3 + \binom{5}{4}(3a)^1(-b)^4 + \binom{5}{5}(3a)^0(-b)^5$
$= 1(243a^5)(1) + 5(81a^4)(-b) + 10(27a^3)b^2 + 10(9a^2)(-b^3) + 5(3a)b^4 + 1(1)(-b^5)$
$= 243a^5 - 405a^4b + 270a^3b^2 - 90a^2b^3 + 15ab^4 - b^5$

27. $\left(2x+\frac{1}{2}\right)^4 = \binom{4}{0}(2x)^4\left(\frac{1}{2}\right)^0 + \binom{4}{1}(2x)^3\left(\frac{1}{2}\right)^1 + \binom{4}{2}(2x)^2\left(\frac{1}{2}\right)^2 + \binom{4}{3}(2x)^1\left(\frac{1}{2}\right)^3 + \binom{4}{4}(2x)^0\left(\frac{1}{2}\right)^4$
$= 1(16x^4)(1) + 4(8x^3)\left(\frac{1}{2}\right) + 6(4x^2)\left(\frac{1}{4}\right) + 4(2x)\left(\frac{1}{8}\right) + 1(1)\left(\frac{1}{16}\right)$
$= 16x^4 + 16x^3 + 6x^2 + x + \frac{1}{16}$

29. $\left(\frac{x}{2}-3\right)^4 = \binom{4}{0}\left(\frac{x}{2}\right)^4(-3)^0 + \binom{4}{1}\left(\frac{x}{2}\right)^3(-3)^1 + \binom{4}{2}\left(\frac{x}{2}\right)^2(-3)^2 + \binom{4}{3}\left(\frac{x}{2}\right)^1(-3)^3 + \binom{4}{4}\left(\frac{x}{2}\right)^0(-3)^4$
$= 1\left(\frac{x^4}{16}\right)(1) + 4\left(\frac{x^3}{8}\right)(-3) + 6\left(\frac{x^2}{4}\right)(9) + 4\left(\frac{x}{2}\right)(-27) + 1(1)(81)$
$= \frac{x^4}{16} - \frac{3x^3}{2} + \frac{27x^2}{2} - 54x + 81$

31. $(x+10)^{10} = \binom{10}{0}x^{10}(10)^0 + \binom{10}{1}x^9(10)^1 + \binom{10}{2}x^8(10)^2 + \binom{10}{3}x^7(10)^3 + \cdots$
$= 1x^{10}(1) + \frac{10}{1}x^9(10) + \frac{10\cdot 9}{2\cdot 1}x^8(100) + \frac{10\cdot 9\cdot 8}{3\cdot 2\cdot 1}x^7(1000) + \cdots$
$= x^{10} + 100x^9 + 4{,}500x^8 + 120{,}000x^7 + \cdots$

33. $(3x-y)^7 = \binom{7}{0}(3x)^7(-y)^0 + \binom{7}{1}(3x)^6(-y)^1 + \binom{7}{2}(3x)^5(-y)^2 + \binom{7}{3}(3x)^4(-y)^3 + \cdots$
$= 1(2187x^7)(1) + \frac{7}{1}(729x^6)(-y) + \frac{7\cdot 6}{2\cdot 1}(243x^5)(y^2) + \frac{7\cdot 6\cdot 5}{3\cdot 2\cdot 1}(81x^4)(-y^3) + \cdots$
$= 2187x^7 + 7(729x^6)(-y) + 21(243x^5)(y^2) + 35(81x^4)(-y^3) + \cdots$
$= 2187x^7 - 5103x^6y + 5103x^5y^2 - 2835x^4y^3 + \cdots$

35. $(x^2-3y)^8 = \binom{8}{0}(x^2)^8(-3y)^0 + \binom{8}{1}(x^2)^7(-3y)^1 + \binom{8}{2}(x^2)^6(-3y)^2 + \binom{8}{3}(x^2)^5(-3y)^3 + \cdots$
$= 1(x^{16})(1) + \frac{8}{1}(x^{14})(-3y) + \frac{8\cdot 7}{2\cdot 1}(x^{12})(9y^2) + \frac{8\cdot 7\cdot 6}{3\cdot 2\cdot 1}(x^{10})(-27y^3) + \cdots$
$= x^{16} + 8(x^{14})(-3y) + 28(x^{12})(9y^2) + 56(x^{10})(-27y^3) + \cdots$
$= x^{16} - 24x^{14}y + 252x^{12}y^2 - 1512x^{10}y^3 + \cdots$

37. Yes, $n! = n \cdot (n-1)!$

$$4! = 4 \cdot 3 \cdot 2 \cdot 1 = 4 \cdot (3 \cdot 2 \cdot 1) = 4 \cdot (3)! = 4 \cdot (4-1)!$$

39. Yes, $(n-3)! = (n-3)(n-4)(n-5)!$ for $n \geq 5$.
Let $n = 7$:
$(7-3)! = (7-3)(7-4)(7-5)!$ or
$4! = 4 \cdot 3 \cdot 2! = 4 \cdot 3 \cdot 2 \cdot 1 = 4!$

41. $\binom{n}{m} = 1$ when either $n = m$ or $m = 0$.

43. $(x+3)^8$

First term is $\binom{8}{0}(x)^8(3)^0 = 1(x^8)(1) = x^8$.

Second term is $\binom{8}{1}(x)^7(3)^1 = 8(x^7)(3) = 24x^7$.

Next to last term is
$\binom{8}{7}(x)^1(3)^7 = 8(x)(2187) = 17{,}496x$.

Last term is $\binom{8}{8}(x)^0(3)^8 = 1(1)(6561) = 6561$.

45. $(a+b)^n = \sum_{i=0}^{n} \binom{n}{i} a^{n-i} b^i$

47. Let $x = 0$.

$$2x + y = 10$$
$$2(0) + y = 6$$
$$y = 6$$

The y-intercept is (0, 10).

48. $\frac{1}{5}x + \frac{1}{2}y = 4 \overset{\times(-10)}{\Rightarrow} -2x - 5y = -40 \quad (1)$

$\frac{2}{3}x - y = \frac{8}{3} \overset{\times 3}{\Rightarrow} \underline{2x - 3y = \ 8} \quad (2)$

$$-8y = -32$$
$$y = 4$$

Substitute $y = 4$ into equation (2):

$$2x - 3(4) = 8$$
$$2x = 20$$
$$x = 10$$

The solution is $(10, 4)$.

49.

$$x(x-11) = -18$$
$$x^2 - 11x + 18 = 0$$
$$(x-9)(x-2) = 0$$
$$x - 9 = 0 \quad \text{or} \quad x - 2 = 0$$
$$x = 9 \qquad\qquad x = 2$$

50. $\sqrt{20xy^4}\sqrt{6x^5y^7} = \sqrt{120x^6y^{11}} = \sqrt{4x^6y^{10} \cdot 30y} = 2x^3y^5\sqrt{30y}$

51.

$$f(x) = 3x + 8$$
$$y = 3x + 8$$
$$x = 3y + 8$$
$$3y = x - 8$$
$$y = \frac{x-8}{3}$$
$$f^{-1}(x) = \frac{x-8}{3}$$

Chapter 11 Review Exercises

1. $a_n = n + 5$

$$a_1 = 1 + 5 = 6$$
$$a_2 = 2 + 5 = 7$$
$$a_3 = 3 + 5 = 8$$
$$a_4 = 4 + 5 = 9$$
$$a_5 = 5 + 5 = 10$$

The terms are 6, 7, 8, 9, 10.

2. $a_n = n^2 + n - 3$

$$a_1 = 1^2 + 1 - 3 = -1$$
$$a_2 = 2^2 + 2 - 3 = 3$$
$$a_3 = 3^2 + 3 - 3 = 9$$
$$a_4 = 4^2 + 4 - 3 = 17$$
$$a_5 = 5^2 + 5 - 3 = 27$$

The terms are -1, 3, 9, 17, 27.

3. $a_n = \frac{6}{n}$

$a_1 = \frac{6}{1} = 6$

$a_2 = \frac{6}{2} = 3$

$a_3 = \frac{6}{3} = 2$

$a_4 = \frac{6}{4} = \frac{3}{2}$

$a_5 = \frac{6}{5}$

The terms are 6, 3, 2, $\frac{3}{2}$, $\frac{6}{5}$.

4. $a_n = \frac{n^2}{n+4}$

$a_1 = \frac{1^2}{1+4} = \frac{1}{5}$

$a_2 = \frac{2^2}{2+4} = \frac{4}{6} = \frac{2}{3}$

$a_3 = \frac{3^2}{3+4} = \frac{9}{7}$

$a_4 = \frac{4^2}{4+4} = \frac{16}{8} = 2$

$a_5 = \frac{5^2}{5+4} = \frac{25}{9}$

The terms are $\frac{1}{5}, \frac{2}{3}, \frac{9}{7}, 2, \frac{25}{9}$.

5. $a_n = 3n - 10$

$a_7 = 3(7) - 10 = 21 - 10 = 11$

6. $a_n = (-1)^n + 5$

$a_7 = (-1)^7 + 5 = -1 + 5 = 4$

7. $a_n = \frac{n+17}{n^2}$

$a_9 = \frac{9+17}{9^2} = \frac{26}{81}$

8. $a_n = (n)(n-3)$

$a_{11} = (11)(11-3) = (11)(8) = 88$

9. $a_n = 2n + 5$

$a_1 = 2(1) + 5 = 2 + 5 = 7$

$a_2 = 2(2) + 5 = 4 + 5 = 9$

$a_3 = 2(3) + 5 = 6 + 5 = 11$

$s_1 = a_1 = 7$

$s_3 = a_1 + a_2 + a_3 = 7 + 9 + 11 = 27$

10. $a_n = n^2 + 8$

$a_1 = (1)^2 + 8 = 1 + 8 = 9$

$a_2 = (2)^2 + 8 = 4 + 8 = 12$

$a_3 = (3)^2 + 8 = 9 + 8 = 17$

$s_1 = a_1 = 9$

$s_3 = a_1 + a_2 + a_3 = 9 + 12 + 17 = 38$

11. $a_n = \frac{n+3}{n+2}$

$a_1 = \frac{1+3}{1+2} = \frac{4}{3}$

$a_2 = \frac{2+3}{2+2} = \frac{5}{4}$

$a_3 = \frac{3+3}{3+2} = \frac{6}{5}$

$s_1 = a_1 = \frac{4}{3}$

$$\begin{aligned} s_3 &= a_1 + a_2 + a_3 \\ &= \frac{4}{3} + \frac{5}{4} + \frac{6}{5} \\ &= \frac{80}{60} + \frac{75}{60} + \frac{72}{60} \\ &= \frac{227}{60} \end{aligned}$$

12. $a_n = (-1)^n (n+8)$

$a_1 = (-1)^1 (1+8) = (-1)(9) = -9$

$a_2 = (-1)^2 (2+8) = 1(10) = 10$

$a_3 = (-1)^3 (3+8) = -1(11) = -11$

$s_1 = a_1 = -9$

$s_3 = a_1 + a_2 + a_3 = -9 + 10 - 11 = -10$

13. This is a geometric sequence with $r = 4 \div 2 = 2$ and $a_1 = 2$.
$a_5 = 2(2)^{5-1} = 2 \cdot 2^4 = 32$
$a_6 = 2(2)^{6-1} = 2 \cdot 2^5 = 64$
$a_7 = 2(2)^{7-1} = 2 \cdot 2^6 = 128$
The terms are 32, 64, 128.
$a_n = 2(2)^{n-1} = 2^1 2^{n-1} = 2^n$

14. This is a geometric sequence with $r = 9 \div (-27) = \frac{9}{-27} = -\frac{1}{3}$ and $a_1 = -27$.
$a_5 = -27\left(-\frac{1}{3}\right)^4 = -27\left(\frac{1}{81}\right) = -\frac{1}{3}$
$a_6 = -27\left(-\frac{1}{3}\right)^5 = -27\left(-\frac{1}{243}\right) = \frac{1}{9}$
$a_5 = -27\left(-\frac{1}{3}\right)^6 = -27\left(\frac{1}{729}\right) = -\frac{1}{27}$
The terms are $-\frac{1}{3}, \frac{1}{9}, -\frac{1}{27}$.
$a_n = -27\left(-\frac{1}{3}\right)^{n-1}$ or $a_n = (-1)^n (3)^{4-n}$

15. This is a geometric sequence with $r = \frac{2}{7} \div \frac{1}{7} = \frac{2}{7} \cdot \frac{7}{1} = 2$ and $a_1 = \frac{1}{7}$.
$a_5 = \frac{1}{7}(2)^{5-1} = \frac{2^4}{7} = \frac{16}{7}$
$a_6 = \frac{1}{7}(2)^{6-1} = \frac{2^5}{7} = \frac{32}{7}$
$a_7 = \frac{1}{7}(2)^{7-1} = \frac{2^6}{7} = \frac{64}{7}$
The terms are $\frac{16}{7}, \frac{32}{7}, \frac{64}{7}$.
$a_n = \frac{1}{7}(2)^{n-1} = \frac{2^{n-1}}{7}$

16. This is an arithmetic sequence with $d = 9 - 13 = -4$ and $a_1 = 13$.
$a_5 = 13 + (5-1)(-4) = 13 - 16 = -3$
$a_6 = 13 + (6-1)(-4) = 13 - 20 = -7$
$a_7 = 13 + (7-1)(-4) = 13 - 24 = -11$
The terms are $-3, -7, -11$.
$a_n = a_1 + (n-1)d$
$= 13 + (n-1)(-4)$
$= 13 - 4n + 4$
$= 17 - 4n$

17. $\sum_{i=1}^{3}(i^2+9) = (1^2+9)+(2^2+9)+(3^2+9)$
$= (1+9)+(4+9)+(9+9)$
$= 10+13+18$
$= 41$

18. $\sum_{i=1}^{4} i(i+5)$
$= 1(1+5)+2(2+5)+3(3+5)+4(4+5)$
$= 1(6)+2(7)+3(8)+4(9)$
$= 6+14+24+36$
$= 80$

19. $\sum_{i=1}^{5} \frac{i^2}{6} = \frac{1^2}{6}+\frac{2^2}{6}+\frac{3^2}{6}+\frac{4^2}{6}+\frac{5^2}{6}$
$= \frac{1}{6}+\frac{4}{6}+\frac{9}{6}+\frac{16}{6}+\frac{25}{6}$
$= \frac{55}{6}$

20. $\sum_{i=1}^{4} \frac{i}{i+1} = \frac{1}{1+1}+\frac{2}{2+1}+\frac{3}{3+1}+\frac{4}{4+1}$
$= \frac{1}{2}+\frac{2}{3}+\frac{3}{4}+\frac{4}{5}$
$= \frac{163}{60}$

21. $\sum_{i=1}^{4} x_i = x_1+x_2+x_3+x_4 = 3+9+7+10 = 29$

22. $\sum_{i=1}^{4} (x_i)^2 = x_1^2+x_2^2+x_3^2+x_4^2$
$= 3^2+9^2+7^2+10^2$
$= 9+81+49+100$
$= 239$

23. $\sum_{i=2}^{3} (x_i^2+1) = (x_2^2+1)+(x_3^2+1)$
$= (9^2+1)+(7^2+1)$
$= (81+1)+(49+1)$
$= 82+50$
$= 132$

24. $\left(\sum_{i=1}^{4} x_i\right)^2 = (x_1+x_2+x_3+x_4)^2$
$= (3+9+7+10)^2$
$= (29)^2$
$= 841$

25. a. perimeter of rectangle: $p = 2l + 2w$

$p_1 = 2(1) + 2(1+3) = 2 + 8 = 10$
$p_2 = 2(2) + 2(2+3) = 4 + 10 = 14$
$p_3 = 2(3) + 2(3+3) = 6 + 12 = 18$
$p_4 = 2(4) + 2(4+3) = 8 + 14 = 22$

b. $p_n = 2n + 2(n+3) = 2n + 2n + 6 = 4n + 6$

26. a. area of rectangle: $a = l \cdot w$

$a_1 = 1(1+3) = 4$
$a_2 = 2(2+3) = 10$
$a_3 = 3(3+3) = 18$
$a_4 = 4(4+3) = 28$

b. $a_n = n(n+3) = n^2 + 3n$

27. $a_1 = 5$

$a_2 = 5 + (2-1)(3) = 5 + 1(3) = 5 + 3 = 8$
$a_3 = 5 + (3-1)(3) = 5 + 2(3) = 5 + 6 = 11$
$a_4 = 5 + (4-1)(3) = 5 + 3(3) = 5 + 9 = 14$
$a_5 = 5 + (5-1)(3) = 5 + 4(3) = 5 + 12 = 17$

The terms are 5, 7, 9, 11, 13.

28. $a_1 = 5$

$a_2 = 5 + (2-1)\left(-\frac{1}{3}\right) = 5 + 1\left(-\frac{1}{3}\right) = \frac{15}{3} - \frac{1}{3} = \frac{14}{3}$

$a_3 = 5 + (3-1)\left(-\frac{1}{3}\right) = 5 + 2\left(-\frac{1}{3}\right) = \frac{15}{3} - \frac{2}{3} = \frac{13}{3}$

$a_4 = 5 + (4-1)\left(-\frac{1}{3}\right) = 5 + 3\left(-\frac{1}{3}\right) = 5 - 1 = 4$

$a_5 = 5 + (5-1)\left(-\frac{1}{3}\right) = 5 + 4\left(-\frac{1}{3}\right) = \frac{15}{3} - \frac{4}{3} = \frac{11}{3}$

The terms are 5, $\frac{14}{3}$, $\frac{13}{3}$, 4, $\frac{11}{3}$.

29. $a_1 = \frac{1}{2}$

$a_2 = \frac{1}{2} + (2-1)(-2) = \frac{1}{2} - 2 = -\frac{3}{2}$

$a_3 = \frac{1}{2} + (3-1)(-2) = \frac{1}{2} - 4 = -\frac{7}{2}$

$a_4 = \frac{1}{2} + (4-1)(-2) = \frac{1}{2} - 6 = -\frac{11}{2}$

$a_5 = \frac{1}{2} + (5-1)(-2) = \frac{1}{2} - 8 = -\frac{15}{2}$

The terms are $\frac{1}{2}, -\frac{3}{2}, -\frac{7}{2}, -\frac{11}{2}, -\frac{15}{2}$.

30. $a_1 = -100$

$a_2 = -100 + (2-1)\left(\frac{1}{5}\right) = -100 + \frac{1}{5} = -\frac{499}{5}$

$a_3 = -100 + (3-1)\left(\frac{1}{5}\right) = -100 + \frac{2}{5} = -\frac{498}{5}$

$a_4 = -100 + (4-1)\left(\frac{1}{5}\right) = -100 + \frac{3}{5} = -\frac{497}{5}$

$a_5 = -100 + (5-1)\left(\frac{1}{5}\right) = -100 + \frac{4}{5} = -\frac{496}{5}$

The terms are $-100, -\frac{499}{5}, -\frac{498}{5}, -\frac{497}{5}, -\frac{496}{5}$.

31. $a_9 = a_1 + (9-1)d$

$a_9 = 6 + (9-1)(3) = 6 + 8(3) = 6 + 24 = 30$

32. $a_7 = a_1 + (7-1)d$

$-14 = 10 + 6d$
$-24 = 6d$
$-4 = d$

33. $a_{11} = a_1 + (11-1)d$

$2 = -3 + 10d$
$5 = 10d$
$d = \frac{5}{10} = \frac{1}{2}$

34. $a_n = a_1 + (n-1)d$

$-3 = 22 + (n-1)(-5)$
$-3 = 22 - 5n + 5$
$-3 = 27 - 5n$
$-30 = -5n$
$6 = n$

35. $a_8 = a_1 + (8-1)d$

$21 = 7 + 7d$
$14 = 7d$
$2 = d$

$s_8 = \frac{8(a_1 + a_8)}{2} = \frac{8(7+21)}{2} = \frac{8(28)}{2} = \frac{224}{2} = 112$

36. $a_7 = a_1 + (7-1)d$

$-48 = -12 + 6d$
$-36 = 6d$
$-6 = d$

$s_7 = \frac{7(a_1 + a_7)}{2} = \frac{7(-12-48)}{2} = \frac{7(-60)}{2} = -210$

37.
$$a_6 = a_1 + (6-1)d$$
$$\frac{13}{5} = \frac{3}{5} + 5d$$
$$\frac{13}{5} - \frac{3}{5} = 5d$$
$$\frac{10}{5} = 5d$$
$$2 = 5d$$
$$\frac{2}{5} = d$$
$$s_6 = \frac{6(a_1 + a_6)}{2}$$
$$= \frac{6\left(\frac{3}{5} + \frac{13}{5}\right)}{2} = \frac{6\left(\frac{16}{5}\right)}{2} = 6\left(\frac{16}{5}\right)\left(\frac{1}{2}\right) = \frac{48}{5}$$

38.
$$a_9 = a_1 + (9-1)d$$
$$-6 = -\frac{10}{3} + 8d$$
$$-6 + \frac{10}{3} = 8d$$
$$-\frac{8}{3} = 8d$$
$$d = \frac{1}{8}\left(-\frac{8}{3}\right) = -\frac{1}{3}$$
$$s_9 = \frac{9(a_1 + a_n)}{2}$$
$$= \frac{9\left(-\frac{10}{3} - 6\right)}{2} = \frac{9\left(-\frac{28}{3}\right)}{2} = 9\left(-\frac{28}{3}\right)\left(\frac{1}{2}\right) = -42$$

39. $a_1 = -7$

$a_2 = -7 + (2-1)(4) = -7 + 1(4) = -7 + 4 = -3$

$a_3 = -7 + (3-1)(4) = -7 + 2(4) = -7 + 8 = 1$

$a_4 = -7 + (4-1)(4) = -7 + 3(4) = -7 + 12 = 5$

The terms are –7, –3, 1, 5.

$a_{10} = -7 + (10-1)(4) = -7 + 9(4) = -7 + 36 = 29$

$$s_{10} = \frac{10(-7+29)}{2} = \frac{10(22)}{2} = \frac{220}{2} = 110$$

40. $a_1 = 4$

$a_2 = 4 + (2-1)(-3) = 4 + 1(-3) = 4 - 3 = 1$

$a_3 = 4 + (3-1)(-3) = 4 + 2(-3) = 4 - 6 = -2$

$a_4 = 4 + (4-1)(-3) = 4 + 3(-3) = 4 - 9 = -5$

The terms are 4, 1, –2, –5.

$a_{10} = 4 + (10-1)(-3) = 4 + 9(-3) = 4 - 27 = -23$

$$s_{10} = \frac{10[4 + (-23)]}{2} = \frac{10(-19)}{2} = \frac{-190}{2} = -95$$

41. $a_1 = \frac{5}{6}$

$$a_2 = \frac{5}{6} + (2-1)\left(\frac{2}{3}\right) = \frac{5}{6} + \frac{2}{3} = \frac{9}{6} = \frac{3}{2}$$
$$a_3 = \frac{5}{6} + (3-1)\left(\frac{2}{3}\right) = \frac{5}{6} + 2\left(\frac{2}{3}\right) = \frac{5}{6} + \frac{4}{3} = \frac{13}{6}$$
$$a_4 = \frac{5}{6} + (4-1)\left(\frac{2}{3}\right) = \frac{5}{6} + 3\left(\frac{2}{3}\right) = \frac{5}{6} + \frac{6}{3} = \frac{17}{6}$$

The terms are $\frac{5}{6}, \frac{3}{2}, \frac{13}{6}, \frac{17}{6}$.

$$a_{10} = \frac{5}{6} + (10-1)\left(\frac{2}{3}\right) = \frac{5}{6} + 9\left(\frac{2}{3}\right) = \frac{5}{6} + 6 = \frac{41}{6}$$
$$s_{10} = \frac{10\left(\frac{5}{6} + \frac{41}{6}\right)}{2} = \frac{10\left(\frac{46}{6}\right)}{2} = 5\left(\frac{46}{6}\right) = 5\left(\frac{23}{3}\right) = \frac{115}{3}$$

42. $a_1 = -60$

$a_2 = -60 + (2-1)(5) = -60 + 1(5) = -60 + 5 = -55$

$a_3 = -60 + (3-1)(5) = -60 + 2(5) = -60 + 10 = -50$

$a_4 = -60 + (4-1)(5) = -60 + 3(5) = -60 + 15 = -45$

The terms are –60, –55, –50, –45.

$a_{10} = -60 + (10-1)(5) = -60 + 45 = -15$

$$s_{10} = \frac{10(-60-15)}{2} = 5(-75) = -375$$

43. $d = 9 - 4 = 5$

$a_n = a_1 + (n-1)d$

$64 = 4 + (n-1)5$

$64 = 4 + 5n - 5$

$64 = 5n - 1$

$65 = 5n$

$13 = n$

$$s_{13} = \frac{13(a_1 + a_{13})}{2} = \frac{13(4+64)}{2} = \frac{13(68)}{2} = 442$$

44. $d = -4 - (-7) = -4 + 7 = 3$

$a_n = a_1 + (n-1)d$

$11 = -7 + (n-1)3$

$11 = -7 + 3n - 3$

$11 = 3n - 10$

$21 = 3n$

$7 = n$

$$s_7 = \frac{7(a_1 + a_7)}{2} = \frac{7(-7+11)}{2} = \frac{7(4)}{2} = \frac{28}{2} = 14$$

45. $d = \frac{9}{10} - \frac{6}{10} = \frac{3}{10}$

$a_n = a_1 + (n-1)d$

$\frac{36}{10} = \frac{6}{10} + (n-1)\frac{3}{10}$

$\frac{36}{10} = \frac{6}{10} + \frac{3}{10}n - \frac{3}{10}$

$\frac{36}{10} = \frac{3}{10} + \frac{3}{10}n$

$\frac{33}{10} = \frac{3}{10}n$

$n = \frac{10}{3}\left(\frac{33}{10}\right) = 11$

$s_{11} = \frac{11(a_1 + a_{11})}{2}$

$= \frac{11\left(\frac{6}{10} + \frac{36}{10}\right)}{2}$

$= \frac{11\left(\frac{42}{10}\right)}{2}$

$= 11\left(\frac{42}{10}\right)\left(\frac{1}{2}\right)$

$= \frac{231}{10}$

46. $d = -3 - (-9) = -3 + 9 = 6$

$a_n = a_1 + (n-1)d$

$45 = -9 + (n-1)6$

$45 = -9 + 6n - 6$

$45 = -15 + 6n$

$60 = 6n$

$10 = n$

$s_{10} = \frac{10(a_1 + a_{10})}{2}$

$= \frac{10(-9+45)}{2}$

$= \frac{10(36)}{2}$

$= 180$

47. $a_1 = 6$

$a_2 = 6(2)^{2-1} = 6(2) = 12$

$a_3 = 6(2)^{3-1} = 6(2)^2 = 6(4) = 24$

$a_4 = 6(2)^{4-1} = 6(2)^3 = 6(8) = 48$

$a_5 = 6(2)^{5-1} = 6(2)^4 = 6(16) = 96$

The terms are 6, 12, 24, 48, 96.

48. $a_1 = -12$

$a_2 = -12\left(\frac{1}{2}\right)^{2-1} = -12\left(\frac{1}{2}\right) = -6$

$a_3 = -12\left(\frac{1}{2}\right)^{3-1} = -12\left(\frac{1}{4}\right) = -3$

$a_4 = -12\left(\frac{1}{2}\right)^{4-1} = -12\left(\frac{1}{8}\right) = -\frac{3}{2}$

$a_5 = -12\left(\frac{1}{2}\right)^{5-1} = -12\left(\frac{1}{16}\right) = -\frac{3}{4}$

The terms are –12, –6, –3, $-\frac{3}{2}, -\frac{3}{4}$.

49. $a_1 = 20$

$a_2 = 20\left(-\frac{2}{3}\right)^{2-1} = 20\left(-\frac{2}{3}\right) = -\frac{40}{3}$

$a_3 = 20\left(-\frac{2}{3}\right)^{3-1} = 20\left(-\frac{2}{3}\right)^2 = 20\left(\frac{4}{9}\right) = \frac{80}{9}$

$a_4 = 20\left(-\frac{2}{3}\right)^{4-1} = 20\left(-\frac{2}{3}\right)^3 = 20\left(-\frac{8}{27}\right) = -\frac{160}{27}$

$a_5 = 20\left(-\frac{2}{3}\right)^{5-1} = 20\left(-\frac{2}{3}\right)^4 = 20\left(\frac{16}{81}\right) = \frac{320}{81}$

The terms are $20, -\frac{40}{3}, \frac{80}{9}, -\frac{160}{27}, \frac{320}{81}$.

50. $a_1 = -20$

$a_2 = -20\left(\frac{1}{5}\right)^{2-1} = -20\left(\frac{1}{5}\right) = -4$

$a_3 = -20\left(\frac{1}{5}\right)^{3-1} = -20\left(\frac{1}{25}\right) = -\frac{4}{5}$

$a_4 = -20\left(\frac{1}{5}\right)^{4-1} = -20\left(\frac{1}{125}\right) = -\frac{4}{25}$

$a_5 = -20\left(\frac{1}{5}\right)^{5-1} = -20\left(\frac{1}{625}\right) = -\frac{4}{125}$

The terms are –20, –4, $-\frac{4}{5}, -\frac{4}{25}, -\frac{4}{125}$.

51. $a_5 = a_1 \cdot r^{5-1} = 6\left(\frac{1}{3}\right)^{5-1} = 6\left(\frac{1}{3}\right)^4 = 6\left(\frac{1}{81}\right) = \frac{2}{27}$

52. $a_6 = a_1 \cdot r^{6-1} = 15(2)^{6-1} = 15(2)^5 = 15(32) = 480$

53. $a_4 = -8(-3)^{4-1} = -8(-3)^3 = -8(-27) = 216$

54. $a_5 = \frac{1}{12}\left(\frac{2}{3}\right)^{5-1}$
$= \frac{1}{12}\left(\frac{2}{3}\right)^4$
$= \frac{1}{12}\left(\frac{16}{81}\right)$
$= \frac{4}{243}$

55. $s_6 = \frac{6(1-2^6)}{1-2}$
$= \frac{6(1-64)}{-1}$
$= \frac{6(-63)}{-1}$
$= 378$

56. $s_5 = \frac{-84\left[1-\left(-\frac{1}{4}\right)^5\right]}{1-\left(-\frac{1}{4}\right)}$
$= \frac{-84\left[1-\left(-\frac{1}{1024}\right)\right]}{1+\frac{1}{4}}$
$= \frac{-84\left(1+\frac{1}{1024}\right)}{\frac{5}{4}}$
$= -84 \cdot \left(\frac{1025}{1024}\right) \cdot \frac{4}{5}$
$= -\frac{4305}{64}$

57. $s_4 = \frac{9\left[1-\left(\frac{3}{2}\right)^4\right]}{1-\frac{3}{2}}$
$= \frac{9\left(1-\frac{81}{16}\right)}{-\frac{1}{2}}$
$= 9\left(-\frac{65}{16}\right)\left(-\frac{2}{1}\right)$
$= \frac{585}{8}$

58. $s_7 = \frac{8\left[1-\left(\frac{1}{2}\right)^7\right]}{1-\frac{1}{2}}$
$= \frac{8\left(1-\frac{1}{128}\right)}{\frac{1}{2}}$
$= 8\left(\frac{127}{128}\right)\left(\frac{2}{1}\right)$
$= \frac{127}{8}$

59. $r = 12 \div 6 = 2$
$a_n = 6(2)^{n-1}$

60. $r = -20 \div (-4) = 5$
$a_n = -4(5)^{n-1}$

61. $r = \frac{10}{3} \div 10 = \frac{10}{3} \cdot \frac{1}{10} = \frac{1}{3}$
$a_n = 10\left(\frac{1}{3}\right)^{n-1}$

62. $r = \frac{18}{15} \div \frac{9}{5} = \frac{18}{15} \cdot \frac{5}{9} = \frac{2}{3}$
$a_n = \frac{9}{5}\left(\frac{2}{3}\right)^{n-1}$

63. $r = \frac{5}{2} \div 5 = \frac{5}{2} \cdot \frac{1}{5} = \frac{1}{2}$
$s_\infty = \frac{5}{1-\frac{1}{2}} = \frac{5}{\frac{1}{2}} = 5\left(\frac{2}{1}\right) = 10$

64. $r = 1 \div \frac{5}{2} = 1\left(\frac{2}{5}\right) = \frac{2}{5}$
$s_\infty = \frac{\frac{5}{2}}{1-\frac{2}{5}} = \frac{\frac{5}{2}}{\frac{3}{5}} = \frac{5}{2}\left(\frac{5}{3}\right) = \frac{25}{6}$

65. $r = \frac{8}{3} \div (-8) = \frac{8}{3}\left(-\frac{1}{8}\right) = -\frac{1}{3}$
$s_\infty = \frac{-8}{1-\left(-\frac{1}{3}\right)} = \frac{-8}{\frac{4}{3}} = -8\left(\frac{3}{4}\right) = -6$

66. $r = -4 \div -6 = \dfrac{-4}{-6} = \dfrac{2}{3}$

$S_\infty = \dfrac{-6}{1-\frac{2}{3}} = \dfrac{-6}{\frac{1}{3}} = -6\left(\dfrac{3}{1}\right) = -18$

67. $r = 8 \div 16 = \dfrac{8}{16} = \dfrac{1}{2}$

$S_\infty = \dfrac{16}{1-\frac{1}{2}} = \dfrac{16}{\frac{1}{2}} = 16\left(\dfrac{2}{1}\right) = 32$

68. $r = \dfrac{9}{3} \div 9 = \dfrac{9}{3} \cdot \dfrac{1}{9} = \dfrac{1}{3}$

$S_\infty = \dfrac{9}{1-\frac{1}{3}} = \dfrac{9}{\frac{2}{3}} = 9\left(\dfrac{3}{2}\right) = \dfrac{27}{2}$

69. $r = -1 \div 5 = -\dfrac{1}{5}$

$S_\infty = \dfrac{5}{1-(-\frac{1}{5})} = \dfrac{5}{\frac{6}{5}} = 5\left(\dfrac{5}{6}\right) = \dfrac{25}{6}$

70. $r = -\dfrac{8}{3} \div -4 = -\dfrac{8}{3}\left(-\dfrac{1}{4}\right) = \dfrac{2}{3}$

$S_\infty = \dfrac{-4}{1-\frac{2}{3}} = \dfrac{-4}{\frac{1}{3}} = -4\left(\dfrac{3}{1}\right) = -12$

71. $0.363636... = 0.36 + 0.0036 + 0.000036 + ...$

$= 0.36 + 0.36(0.01) + 0.36(0.01)^2 + \cdots$

$a_1 = 0.36$ and $r = 0.01$

$S_\infty = \dfrac{0.36}{1-0.01} = \dfrac{0.36}{0.99} = \dfrac{36}{99} = \dfrac{4}{11}$

72. $0.621621...$

$= 0.621 + 0.000621 + 0.000000621 + \cdots$

$= 0.621 + 0.621(0.001) + 0.621(0.001)^2 + \cdots$

$a_1 = 0.621$ and $r = 0.001$

$S_\infty = \dfrac{0.621}{1-0.001} = \dfrac{0.621}{0.999} = \dfrac{621}{999} = \dfrac{23}{37}$

---

73. $(3x+y)^4 = \binom{4}{0}(3x)^4(y)^0 + \binom{4}{1}(3x)^3(y)^1 + \binom{4}{2}(3x)^2(y)^2 + \binom{4}{3}(3x)^1(y)^3 + \binom{4}{4}(3x)^0(y)^4$

$= 1(81x^4)(1) + 4(27x^3)(y) + 6(9x^2)(y^2) + 4(3x)(y^3) + 1(1)(y^4)$

$= 81x^4 + 108x^3y + 54x^2y^2 + 12xy^3 + y^4$

74. $(2x-3y^2)^3 = \binom{3}{0}(2x)^3(-3y^2)^0 + \binom{3}{1}(2x)^2(-3y^2)^1 + \binom{3}{2}(2x)^1(-3y^2)^2 + \binom{3}{3}(2x)^0(-3y^2)^3$

$= 1(8x^3)(1) + 3(4x^2)(-3y^2) + 3(2x)(9y^4) + 1(1)(-27y^6)$

$= 8x^3 - 36x^2y^2 + 54xy^4 - 27y^6$

75. $(x-2y)^9 = \binom{9}{0}(x)^9(-2y)^0 + \binom{9}{1}(x)^8(-2y)^1 + \binom{9}{2}(x)^7(-2y)^2 + \binom{9}{3}(x)^6(-2y)^3 + \cdots$

$= 1(x^9)(1) + 9(x^8)(-2y) + 36(x^7)(4y^2) + 84(x^6)(-8y^3) + \cdots$

$= x^9 - 18x^8y + 144x^7y^2 - 672x^6y^3 + \cdots$

76. $(2a^2+3b)^8 = \binom{8}{0}(2a^2)^8(3b)^0 + \binom{8}{1}(2a^2)^7(3b)^1 + \binom{8}{2}(2a^2)^6(3b)^2 + \binom{8}{3}(2a^2)^5(3b)^3 + \cdots$

$= 1(256a^{16})(1) + 8(128a^{14})(3b) + 28(64a^{12})(9b^2) + 56(32a^{10})(27b^3) + \cdots$

$= 256a^{16} + 3072a^{14}b + 16{,}128a^{12}b^2 + 48{,}384a^{10}b^3 + \cdots$

77. This is an arithmetic series with $d = 1$, $a_1 = 101$, and $a_n = 200$.

$$a_n = a_1 + (n-1)d$$
$$200 = 101 + (n-1)(1)$$
$$200 = 101 + n - 1$$
$$200 = n + 100$$
$$100 = n$$

The sum is

$$s_{100} = \frac{n(a_1 + a_{100})}{2} = \frac{100(101+200)}{2} = \frac{100(301)}{2} = 15{,}050$$

78. $21 + 20 + 19 + \cdots + 1$ or $1 + 2 + 3 + \cdots + 21$

$$s_n = \frac{n(a_1 + a_n)}{2}$$
$$s_{21} = \frac{21(1+21)}{2} = \frac{21(22)}{2} = \frac{462}{2} = 231$$

There are 231 barrels in the stack.

79. This is an arithmetic sequence with $d = 1000$

a. $a_1 = 36{,}000$

$$a_2 = 36{,}000 + (2-1)(1000) = 36{,}000 + 1000 = 37{,}000$$
$$a_3 = 36{,}000 + (3-1)(1000) = 36{,}000 + 2000 = 38{,}000$$
$$a_4 = 36{,}000 + (4-1)(1000) = 36{,}000 + 3000 = 39{,}000$$

Ahmed's salaries for the first four years are \$36,000, \$37,000, \$38,000, and \$39,000.

b. $a_n = 36{,}000 + (n-1)(1000) = 36{,}000 + 1000n - 1000 = 35{,}000 + 1000n$

c. $a_6 = 35{,}000 + 1000(6) = 35{,}000 + 6{,}000 = 41{,}000$

In the 6th year, his salary would be \$41,000.

d. $a_{11} = 36{,}000 + (11-1)1000 = 46{,}000$

$$s_n = \frac{n(a_1 + a_n)}{2}$$
$$s_{11} = \frac{11(36{,}000 + 46{,}000)}{2} = \frac{11(82{,}000)}{2} = 451{,}000$$

Ahmed will make a total of \$451,000 in his first 11 years.

80. This is a geometric series with $r = 2$, $a_1 = 100$, and $n = 11$. (Note: There are 11 terms here since 200 represents the first doubling.)

$$a_{11} = a_1 \cdot r^{11-1}$$
$$= 100(2)^{11-1} = 100(2)^{10} = 100(1024) = 102{,}400$$

You would have \$102,400.

81. $a_1 = 1600$, $r = 1.04$, $a_n = a_1 r^{n-1}$

a. July is the 7th month, so $n = 7$.

$a_7 = 1600(1.04)^{7-1} = 1600(1.04)^6 \approx 2024.51$

Gertrude's salary will be \$2,024.51 in July.

b. December is the 12th month, so $n = 12$.

$a_{12} = 1600(1.04)^{12-1} = 1600(1.04)^{11} \approx 2463.13$

Gertrude's salary will be \$2,463.13 in December.

c. $s_n = \dfrac{a_1(1-r^n)}{1-r}$, $n = 12$

$$s_{12} = \frac{1600\left[1-(1.04)^{12}\right]}{1-1.04} \approx 24{,}041.29$$

Gertrude will make \$24,041.29 in 2006.

82. Each year, the cost of the object will be 1.08 times greater than the previous year. After 12 years will be the 13th year. Therefore,

$a_{13} = 200(1.08)^{13-1} \approx 200(2.51817) \approx 503.63$

The item would cost \$503.63.

83. This is an infinite geometric series with $r = 0.92$ and $a_1 = 12$.

$$s_\infty = \frac{12}{1-0.92} = \frac{12}{0.08} = 150$$

The pendulum travels a total distance of 100 feet.

Chapter 11 Practice Test

1. A series is the sum of the terms of a sequence

2. a. An arithmetic sequence is one whose terms differ by a constant amount.

   b. A geometric sequence is one whose terms differ by a common multiple.

3. $a_n = \frac{n-2}{3n}$

$a_1 = \frac{1-2}{3(1)} = \frac{-1}{3} = -\frac{1}{3}$

$a_2 = \frac{2-2}{3(2)} = \frac{0}{6} = 0$

$a_3 = \frac{3-2}{3(3)} = \frac{1}{9}$

$a_4 = \frac{4-2}{3(4)} = \frac{2}{12} = \frac{1}{6}$

$a_5 = \frac{5-2}{3(5)} = \frac{3}{15} = \frac{1}{5}$

The terms are $-\frac{1}{3}$, 0, $\frac{1}{9}$, $\frac{1}{6}$, $\frac{1}{5}$.

4. $a_n = \frac{2n+1}{n^2}$

$a_1 = \frac{2(1)+1}{1^2} = \frac{2+1}{1} = 3$

$a_2 = \frac{2(2)+1}{2^2} = \frac{4+1}{4} = \frac{5}{4}$

$a_3 = \frac{2(3)+1}{3^2} = \frac{6+1}{9} = \frac{7}{9}$

$s_1 = a_1 = 3$

$s_3 = a_1 + a_2 + a_3$

$= 3 + \frac{5}{4} + \frac{7}{9}$

$= \frac{181}{36}$

5. $\sum_{i=1}^{5}(2i^2+3)$

$= [2(1)^2+3]+[2(2)^2+3]+[2(3)^2+3]$
$\quad +[2(4)^2+3]+[2(5)^2+3]$

$= (2+3)+(8+3)+(18+3)+(32+3)+(50+3)$

$= 5+11+21+35+53$

$= 125$

6. $\sum_{i=1}^{4}(x_i)^2 = x_1^2 + x_2^2 + x_3^2 + x_4^2$

$= 4^2 + 2^2 + 8^2 + 10^2$

$= 16+4+64+100$

$= 184$

7. $d = \frac{2}{3} - \frac{1}{3} = \frac{1}{3}$

$a_n = a_1 + (n-1)d$

$= \frac{1}{3} + (n-1)\left(\frac{1}{3}\right)$

$= \frac{1}{3} + \frac{1}{3}n - \frac{1}{3}$

$= \frac{1}{3}n$

8. $r = 10 \div 5 = \frac{10}{5} = 2$

$a_n = a_1 r^{n-1} = 5(2)^{n-1}$

9. $a_1 = 15$

$a_2 = 15+(2-1)(-6) = 15+1(-6) = 15-6 = 9$

$a_3 = 15+(3-1)(-6) = 15+2(-6) = 15-12 = 3$

$a_4 = 15+(4-1)(-6) = 15+3(-6) = 15-18 = -3$

The terms are 15, 9, 3, −3.

10. $a_1 = \frac{5}{12}$

$a_2 = a_1 r^1 = \frac{5}{12}\left(\frac{2}{3}\right)^1 = \frac{5}{12}\left(\frac{2}{3}\right) = \frac{5}{18}$

$a_3 = a_1 r^2 = \frac{5}{12}\left(\frac{2}{3}\right)^2 = \frac{5}{12}\left(\frac{4}{9}\right) = \frac{5}{27}$

$a_4 = a_1 r^3 = \frac{5}{12}\left(\frac{2}{3}\right)^3 = \frac{5}{12}\left(\frac{8}{27}\right) = \frac{10}{81}$

The terms are $\frac{5}{12}$, $\frac{5}{18}$, $\frac{5}{27}$, $\frac{10}{81}$.

11. $a_{11} = a_1 + 10d$

$= 40 + (10)(-8)$

$= 40 - 80$

$= -40$

12. $s_8 = \frac{8(a_1 + a_8)}{2}$
$= \frac{8[7+(-12)]}{2}$
$= \frac{8(-5)}{2}$
$= -20$

13. $d = -16-(-4) = -16+4 = -12$
$a_n = a_1 + (n-1)d$
$-136 = -4 + (n-1)(-12)$
$-136 = -4 - 12n + 12$
$-136 = 8 - 12n$
$-144 = -12n$
$12 = n$

14. $a_6 = a_1 r^5$
$= 8\left(\frac{2}{3}\right)^5$
$= 8\left(\frac{32}{243}\right)$
$= \frac{256}{243}$

15. $s_7 = \frac{a_1\left(1-r^7\right)}{1-r}$
$= \frac{\frac{3}{5}\left[1-(-5)^7\right]}{1-(-5)}$
$= \frac{\frac{3}{5}[1-(-78,125)]}{1-(-5)}$
$= \frac{\frac{3}{5}(1+78,125)}{1+5}$
$= \frac{\frac{3}{5}(78,126)}{6}$
$= \frac{3(78,126)}{5\cdot 6}$
$= \frac{78,126}{5\cdot 2}$
$= \frac{39,063}{5}$

16. $r = 5 \div 15 = \frac{5}{15} = \frac{1}{3}$
$a_n = a_1 r^{n-1} = 15\left(\frac{1}{3}\right)^{n-1}$

17. $r = \frac{8}{3} \div 4 = \frac{8}{3}\cdot\frac{1}{4} = \frac{2}{3}$
$s_\infty = \frac{a_1}{1-r}$
$= \frac{4}{1-\frac{2}{3}}$
$= \frac{4}{\frac{1}{3}}$
$= 4\cdot\frac{3}{1}$
$= 12$

18. $0.3939\cdots = 0.39 + 0.0039 + 0.000039 + \cdots$
$= 0.39 + 0.39(0.01) + 0.39(0.01)^2 + \cdots$
$r = 0.01$ and $a_1 = 0.39$
$s_\infty = \frac{0.39}{1-0.01}$
$= \frac{0.39}{0.99}$
$= \frac{39}{99}$
$= \frac{13}{33}$

19. $\binom{8}{3} = \frac{8!}{3!(8-3)!}$
$= \frac{8!}{3!5!}$
$= \frac{8\cdot 7\cdot 6\cdot \cancel{5\cdot 4\cdot 3\cdot 2\cdot 1}}{(3\cdot 2\cdot 1)(\cancel{5\cdot 4\cdot 3\cdot 2\cdot 1})}$
$= \frac{336}{6}$
$= 56$

20. $(x+2y)^4 = \binom{4}{0}(x)^4(2y)^0 + \binom{4}{1}(x)^3(2y)^1 + \binom{4}{2}(x)^2(2y)^2 + \binom{4}{3}(x)^1(2y)^3 + \binom{4}{4}(x)^0(2y)^4$
$= 1(x^4)(1) + 4(x^3)(2y) + 6(x^2)(4y^2) + 4(x)(8y^3) + 1(1)(16y^4)$
$= x^4 + 8x^3y + 24x^2y^2 + 32xy^3 + 16y^4$

---

21. $\bar{x} = \frac{\sum x}{n}$
$= \frac{76+93+83+87+71}{5}$
$= \frac{410}{5}$
$= 82$

22. $13+12+11+\cdots+1$ or $1+2+3+\cdots+13$
$s_n = \frac{n(a_1+a_n)}{2}$
$s_{13} = \frac{13(1+13)}{2} = \frac{13(14)}{2} = \frac{182}{2} = 91$
There are 91 logs in the pile.

23. $a_1 = 1000$, $n = 20$
$a_n = a_1 + (n-1)d$
$a_{20} = 1000 + (20-1)(1000) = 20{,}000$
$s_{20} = \frac{20(1000+20{,}000)}{2}$
$= \frac{20(21{,}000)}{2}$
$= 210{,}000$
After 20 years, she will have saved \$210,000.

24. $a_1 = 700,\ r = 1.04,\ a_n = a_1r^{n-1}$
$a_6 = 700(1.04)^{6-1} = 700(1.04)^5 \approx 851.66$
She will be making about \$851.66 in the sixth week.

25. $r = 3,\ a_1 = 500(3) = 1500$
$a_6 = 1500(3)^{6-1} = 1500(3)^5 = 364{,}500$
At the end of the sixth hour, there will be 364,500 bacteria in the culture.

Chapter 11 Cumulative Review Test

1. $A = \frac{1}{2}bh$, for b
$2A = bh$
$\frac{2A}{h} = \frac{bh}{h}$
$\frac{2A}{h} = b$

2. $m = \frac{y_2 - y_1}{x_2 - x_1} = \frac{9-(-2)}{1-4} = \frac{11}{-3} = -\frac{11}{3}$

Use the point-slope equation with $m = -\frac{11}{3}$ and $(4,-2)$.
$y - y_1 = m(x - x_1)$
$y - (-2) = -\frac{11}{3}(x-4)$
$y + 2 = -\frac{11}{3}x + \frac{44}{3}$
$y = -\frac{11}{3}x + \frac{44}{3} - 2$
$y = -\frac{11}{3}x + \frac{38}{3}$

3. $x + y + z = 1$ (1)
$2x + 2y + 2z = 2$ (2)
$3x + 3y + 3z = 3$ (3)
Notice that if you multiply equation (1) by 2, the result is exactly the same as equation (2). This implies that this is a dependent system and therefore has infinitely many solutions.

4.
$$\begin{array}{r} 5x^3+4x^2-6x+2 \\ x+5 \\ \hline 5x^4+4x^3-\ \ 6x^2+\ \ 2x\quad\ \ \\ 25x^3+20x^2-30x+10 \\ \hline 5x^4+29x^3+14x^2-28x+10 \end{array}$$

5.
$$\begin{aligned} x^3+2x-5x^2-10 &= \left(x^3+2x\right)-\left(6x^2+12\right) \\ &= x\left(x^2+2\right)-6\left(x^2+2\right) \\ &= \left(x^2+2\right)\left(x-6\right) \end{aligned}$$

6. $(a+b)^2+8(a+b)+16$ [Let $u=a+b$.]

$=u^2+8u+16$

$=(u+4)(u+4)$

$=(u+4)^2$ [Back substitute $a+b$ for u.]

$=(a+b+4)^2$

7.
$$\begin{aligned} &5-\frac{x-1}{x^2+3x-10} \\ &=\frac{5(x^2+3x-10)}{x^2+3x-10}-\frac{x-1}{x^2+3x-10} \\ &=\frac{5x^2+15x-50}{x^2+3x-10}-\frac{x-1}{x^2+3x-10} \\ &=\frac{5x^2+15x-50-(x-1)}{x^2+3x-10} \\ &=\frac{5x^2+15x-50-x+1}{x^2+3x-10} \\ &=\frac{5x^2+14x-49}{x^2+3x-10} \end{aligned}$$

8. $y=kz^2$

$80=k(20)^2$

$80=400k$

$k=\frac{80}{400}=0.2$

$y=0.2z^2$

Let $z=50$: $y=0.2(50)^2=0.2(2500)=500$

9. $f(x)=3\sqrt[3]{x-2},\ g(x)=\sqrt[3]{7x-14}$
$$\begin{aligned} f(x)&=g(x) \\ 2\sqrt[3]{x-3}&=\sqrt[3]{5x-15} \\ \left(2\sqrt[3]{x-3}\right)^3&=\left(\sqrt[3]{5x-15}\right)^3 \\ 8(x-3)&=5x-15 \\ 8x-24&=5x-15 \\ 3x&=9 \\ x&=3 \end{aligned}$$

10.
$$\begin{aligned} \sqrt{6x-5}-\sqrt{2x+6}-1&=0 \\ \sqrt{6x-5}&=1+\sqrt{2x+6} \\ \left(\sqrt{6x-5}\right)^2&=\left(1+\sqrt{2x+6}\right)^2 \\ 6x-5&=1+2\sqrt{2x+6}+2x+6 \\ 4x-12&=2\sqrt{2x+6} \\ 2x-6&=\sqrt{2x+6} \\ (2x-6)^2&=\left(\sqrt{2x+6}\right)^2 \\ 4x^2-24x+36&=2x+6 \\ 4x^2-26x+30&=0 \\ 2(2x^2-13x+15)&=0 \\ 2(2x-3)(x-5)&=0 \end{aligned}$$
$$2x-3=0 \quad\text{or}\quad x-5=0$$
$$x=\frac{3}{2} \qquad\qquad x=5$$

Upon checking, $x=\frac{3}{2}$ is an extraneous solution.

The solution is $x=5$.

11.
$$\begin{aligned} x^2+2x+15&=0 \\ x^2+2x\qquad&=-15 \\ x^2+2x+1&=-15+1 \\ (x+1)^2&=-14 \\ x+1&=\pm\sqrt{-14} \\ x+1&=\pm i\sqrt{14} \\ x&=-1\pm i\sqrt{14} \end{aligned}$$

12. $x^2 - \frac{x}{5} - \frac{1}{3} = 0$

$15\left(x^2 - \frac{x}{5} - \frac{1}{3}\right) = 15(0)$

$15x^2 - 3x - 5 = 0$

$x = \frac{-b \pm \sqrt{b^2 - 4ac}}{2a}$

$= \frac{-(-3) \pm \sqrt{(-3)^2 - 4(15)(-5)}}{2(15)}$

$= \frac{3 \pm \sqrt{9 + 300}}{30}$

$= \frac{3 \pm \sqrt{309}}{30}$

13. Let x be the number. Then

$2x^2 - 9x = 5$

$2x^2 - 9x - 5 = 0$

$(2x + 1)(x - 5) = 0$

$2x + 1 = 0$ or $x - 5 = 0$

$x = -\frac{1}{2}$ $\quad x = 5$

Since the number must be positive, disregard $x = -\frac{1}{2}$. The number is 5.

14. $y = x^2 - 4x$

$y = \left(x^2 - 4x + 4\right) - 4$

$y = (x - 2)^2 - 4$

The vertex is $(2, -4)$.

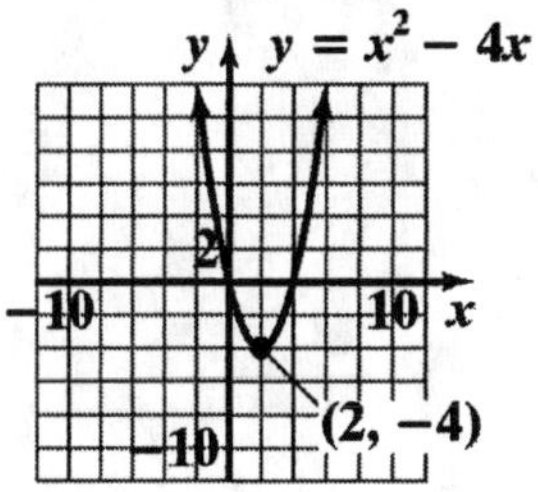

15. $\log_a \frac{1}{64} = 6$

$a^6 = \frac{1}{64}$

$a^6 = \left(\frac{1}{2}\right)^6$

$a = \frac{1}{2}$

16. $y = 2^x - 1$

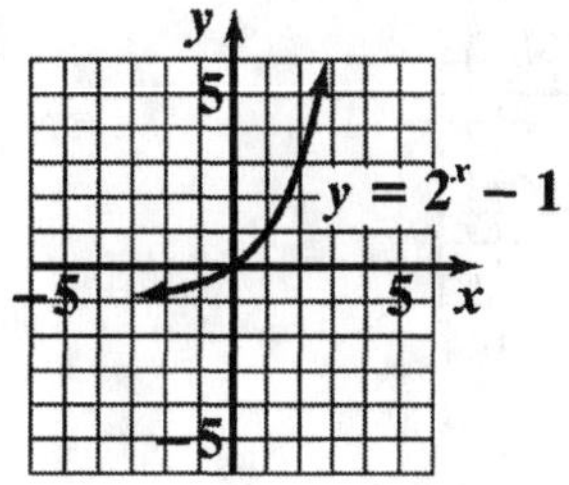

17. $(h, k) = (-6, 2),\ r = 7$

$(x - h)^2 + (y - k)^2 = r^2$

$\left(x - (-6)\right)^2 + \left(y - 2\right)^2 = 7^2$

$(x + 6)^2 + (y - 2)^2 = 49$

18. $(x + 3)^2 + (y + 1)^2 = 16$

$(x + 3)^2 + (y + 1)^2 = 4^2$

center: $(-3, -1)$ $\quad$ radius: 4

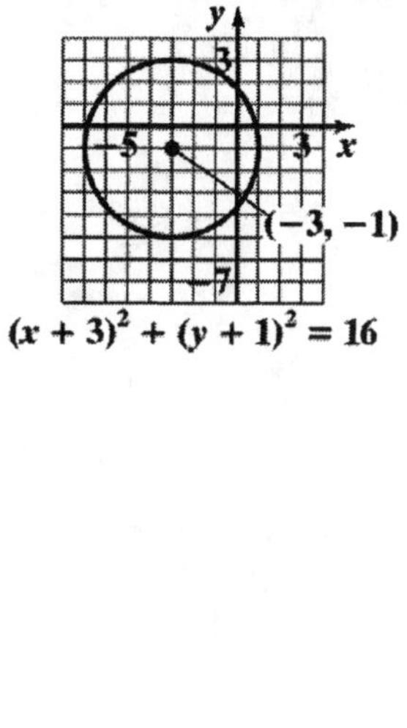

19. $9x^2 + 16y^2 = 144$

$$\frac{9x^2}{144} + \frac{16y^2}{144} = 1$$
$$\frac{x^2}{16} + \frac{y^2}{9} = 1$$
$$\frac{x^2}{4^2} + \frac{y^2}{3^2} = 1$$

$a = 4$ and $b = 3$

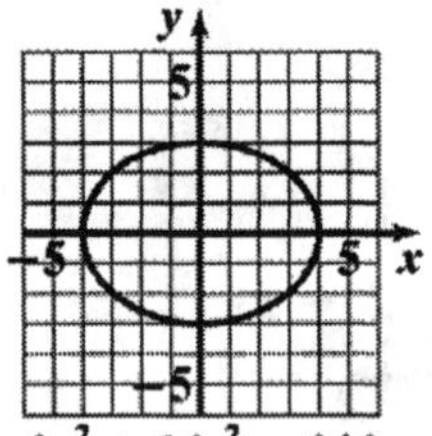

$9x^2 + 16y^2 = 144$

20. $r = 4 \div 6 = \frac{4}{6} = \frac{2}{3}$

$$\begin{aligned} S_\infty &= \frac{a_1}{1-r} \\ &= \frac{6}{1-\frac{2}{3}} \\ &= \frac{6}{\frac{1}{3}} \\ &= 6 \cdot \frac{3}{1} \\ &= 18 \end{aligned}$$

Chapter 10

Exercise Set 10.1

1. No. A function will only have a local maximum or local minimum if the function has an interval where it changes either from increasing to decreasing or from decreasing to increasing. Functions that are always increasing or always decreasing will not have a local maximum or a local minimum.

3. Yes. Since $f(x)=x^2-4$ is a polynomial function, it is continuous.

5. No. Since $2(-5)-1\neq 2(-5)+6$, f is discontinuous. That is, since the pieces give different values at $x=5$, f is discontinuous.

7. Discontinuous.

9. Continuous.

11. Continuous.

13. Discontinuous.

15. Continuous.

17. Continuous.

19. Continuous.

21. Discontinuous. Point of discontinuity at $x=-5$.

23. Discontinuous. Points of discontinuity at $x=\ldots,-\frac{3}{2},-1,-\frac{1}{2},0,\frac{1}{2},1,\frac{3}{2},\ldots$

25. Continuous.

27. Discontinuous. Point of discontinuity at $x=3$.

29. Continuous.

31. For $x<-2$, graph $f(x)=x+5$.
    $f(-5)=-5+5=0$
    $f(-2)=-2+5=3$
    For $x\geq -2$, graph $f(x)=-x+5$.
    $f(-2)=-(-2)+5=7$
    $f(5)=-5+5=0$

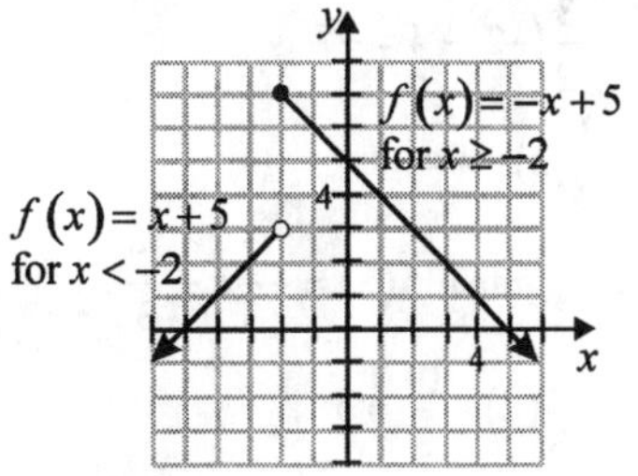

33. For $x\leq 0$, graph $g(x)=-x+4$.
    $g(-3)=-(-3)+4=7$
    $g(-3)=-(0)+4=4$
    For $x>0$, graph $g(x)=4$.
    $g(0)=4$
    $g(4)=4$

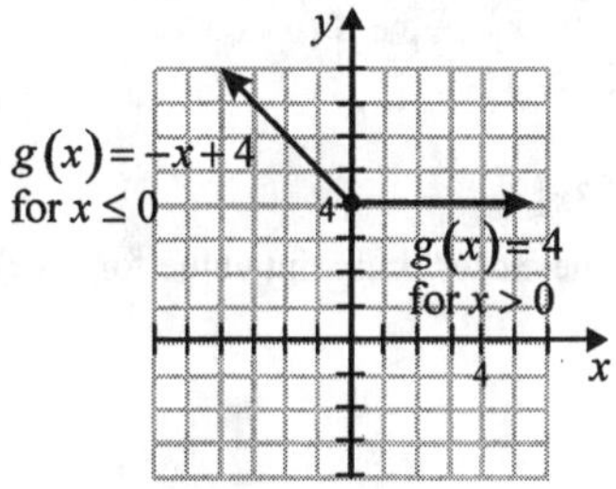

35. For $x<\frac{1}{2}$, graph $h(x)=-2x+1$.
    $h(-3)=-2(-3)+1=7$
    $h\left(\frac{1}{2}\right)=-2\left(\frac{1}{2}\right)+1=0$
    For $x\geq\frac{1}{2}$, graph $h(x)=2x-1$.
    $h\left(\frac{1}{2}\right)=2\left(\frac{1}{2}\right)-1=0$
    $h(4)=2(4)-1=7$

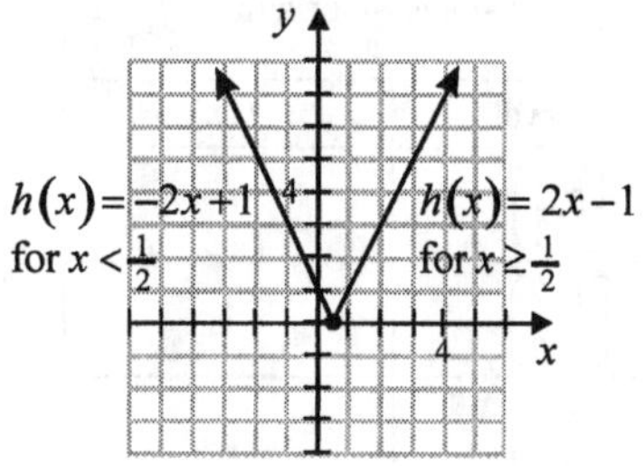

37. For $x \le 2$, graph $p(x) = \frac{3}{2}x - 4$.

$p(-2) = \frac{3}{2}(-2) - 4 = -7$

$p(2) = \frac{3}{2}(2) - 4 = -1$

For $x > 2$, graph $p(x) = -\frac{2}{3}x + \frac{19}{3}$.

$p(2) = -\frac{2}{3}(2) + \frac{19}{3} = 5$

$p(8) = -\frac{2}{3}(8) + \frac{19}{3} = 1$

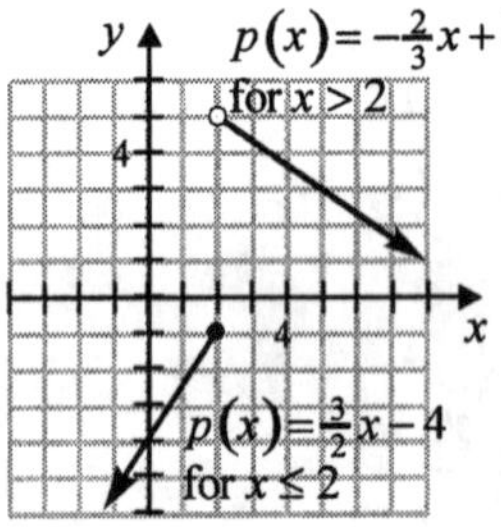

39. $f(x) = [\![3x]\!]$

To find the closed endpoint value for $f(x) = 0$, solve

$0 = 3x$

$x = 0$

To find the closed endpoint value for $f(x) = 1$ solve

$1 = 3x$

$x = \frac{1}{3}$

To find the closed endpoint value for $f(x) = 2$ solve

$2 = 3x$

$x = \frac{2}{3}$

From these points, we know the interval for each line segment spans $\frac{1}{3}$ of a unit.

Interval	$f(x)$
$\left[0, \frac{1}{3}\right)$	0
$\left[\frac{1}{3}, \frac{2}{3}\right)$	1
$\left[\frac{2}{3}, 1\right)$	2

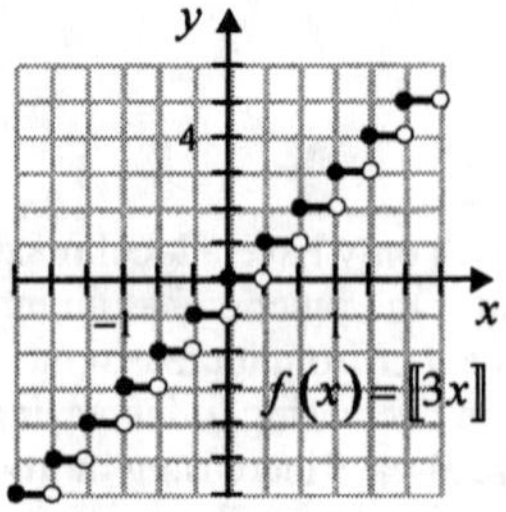

41. $h(x) = [\![-2x]\!]$

To find the closed endpoint value for $h(x) = 0$, solve

$0 = -2x$

$x = 0$

To find the closed endpoint value for $h(x) = 1$ solve

$1 = -2x$

$x = -\frac{1}{2}$

To find the closed endpoint value for $h(x) = 2$ solve

$2 = -2x$

$x = -1$

From these points, we know the interval for each line segment spans $\frac{1}{2}$ of a unit.

Interval	$h(x)$
$\left(-\frac{1}{2}, 0\right]$	0
$\left(-1, -\frac{1}{2}\right]$	1
$\left(-\frac{3}{2}, -1\right]$	2

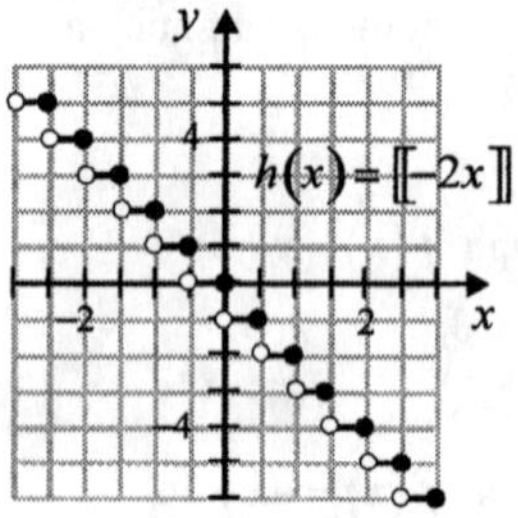

43. $g(x) = \left[\!\left[-\frac{1}{2}x + 1 \right]\!\right]$

To find the closed endpoint value for $g(x) = 0$, solve

$0 = -\frac{1}{2}x + 1$

$\frac{1}{2}x = 1$

$x = 2$

To find the closed endpoint value for $g(x) = 1$ solve

$1 = -\frac{1}{2}x + 1$

$\frac{1}{2}x = 0$

$x = 0$

To find the closed endpoint value for $g(x) = 2$ solve

$2 = -\frac{1}{2}x + 1$

$\frac{1}{2}x = -1$

$x = -2$

From these points, we know the interval for each line segment spans 2 units.

Interval	$g(x)$
$(0,2]$	0
$(-2,0]$	1
$(-4,-2]$	2

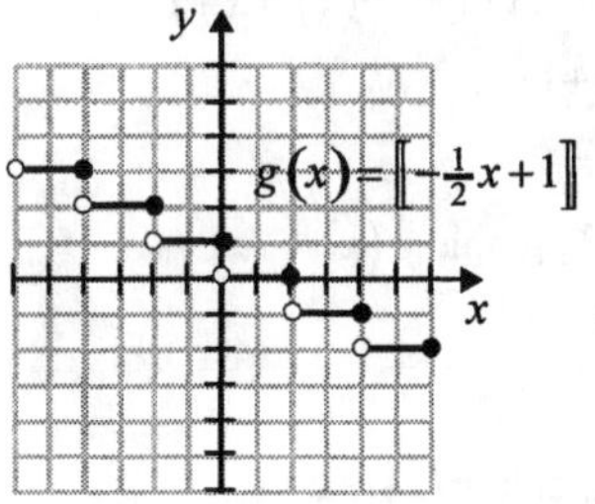

45. $f(x) = \left[\!\left[\frac{2x+7}{10} \right]\!\right]$

To find the closed endpoint value for $f(x) = 0$, solve

$0 = \frac{2x+7}{10}$

$0 = 2x + 7$

$-2x = 7$

$x = -3.5$

To find the closed endpoint value for $f(x) = 1$ solve

$1 = \frac{2x+7}{10}$

$10 = 2x + 7$

$-2x = -3$

$x = 1.5$

To find the closed endpoint value for $f(x) = 2$ solve

$2 = \frac{2x+7}{10}$

$20 = 2x + 7$

$-2x = -13$

$x = 6.5$

From these points, we know the interval for each line segment spans 5 units.

Interval	$f(x)$
$[-3.5,1.5)$	0
$[1.5,6.5)$	1
$[6.5,11.5)$	2

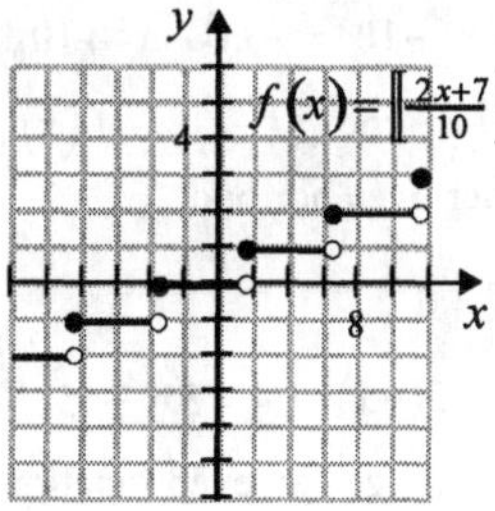

47. Kelly can purchase 1 greeting card for \$1.25, 2 greeting cards for \$2.50, 3 greeting cards for \$3.75, and so on. To determine the maximum number of greeting cards that Kelly can buy, divide the given dollar amount by \$1.25 and round down. Thus, the greatest integer function is $f(x)=\left[\!\left[\frac{x}{1.25}\right]\!\right]$.

49. The office supply manager can purchase 1 calendar for \$3.99, 2 calendars for \$7.98, 3 calendars for \$11.97, and so on. To determine the maximum number of calendars she can buy, divide the given dollar amount by \$3.99 and round down. Thus, the greatest integer function is $f(x)=\left[\!\left[\frac{x}{3.99}\right]\!\right]$.

51. If $x \le 3$ hours, the cost for renting the bike will be $5x$. If $x > 3$ hours, the cost will be the cost of the first 3 hours, \$15, plus \$3 per hour for the number of hours over 3, $x-3$. That is, if $x>3$ hours, the cost will be $15+3(x-3)$. Thus, the piecewise function is

$$f(x)=\begin{cases}5x & \text{if } x\le 3\\ 15+3(x-3) & \text{if } x>3\end{cases}.$$

53. $f(x)=x^2-15$

$f(-x)=(-x)^2-15=x^2-15$.

Since $f(-x)=f(x)$, f is an even function.

55. $f(x)=3x^3+x^2-10$

$f(-x)=3(-x)^3+(-x)^2-10=-3x^3+x^2-10$

$-f(x)=-(3x^3+x^2-10)=-3x^3-x^2+10$

Since $f(-x)\ne f(x)$ and $f(-x)\ne -f(x)$, the function f is neither even nor odd.

57. $g(x)=3x^6-15x^2+4$

$g(-x)=3(-x)^6-15(-x)^2+4=3x^6-15x^2+4$

Since $g(-x)=g(x)$, g is an even function.

59. $g(x)=e^{x^2}+5$

$g(-x)=e^{(-x)^2}+5=e^{x^2}+5$

Since $g(-x)=g(x)$, g is an even function.

61. $h(x)=x^4-x^3+x^2$

$h(-x)=(-x)^4-(-x)^3+(-x)^2=x^4+x^3+x^2$

$-h(x)=-(x^4-x^3+x^2)=-x^4+x^3-x^2$

Since $h(-x)\ne h(x)$ and $h(-x)\ne -h(x)$, the function h is neither even nor odd.

63. For $x<-1$, graph $f(x)=2x$.

$f(-3)=2(-3)=-6$

$f(-1)=2(-1)=-2$

For $-1\le x<2$, graph $f(x)=-x+5$.

$f(-1)=-(-1)+5=6$

$f(2)=-2+5=3$

For $x\ge 2$, graph $f(x)=x+1$.

$f(2)=2+1=3$

$f(4)=4+1=5$

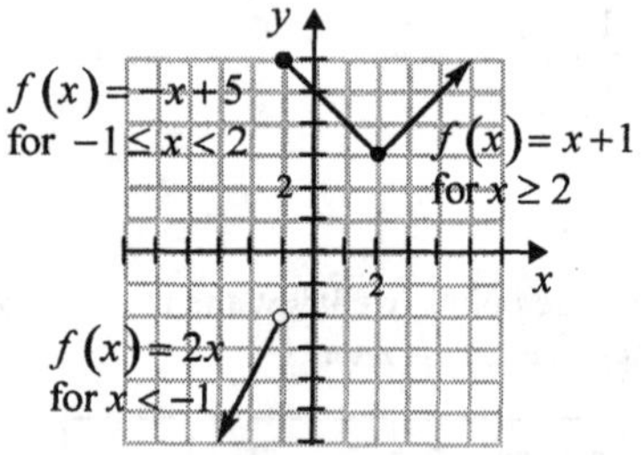

65. For $x<-1$, graph $p(x)=2x+3$.

$p(-3)=2(-3)+3=-6+3=-3$

$p(-1)=2(-1)+3=-2+3=1$

For $-1\le x<1$, graph $p(x)=x+1$.

$p(-1)=-1+1=0$

$p(1)=1+1=2$

For $1\le x<3$, graph $p(x)=2x$.

$p(1)=2(1)=2$

$p(3)=2(3)=6$

For $x\ge 3$, graph $p(x)=x-1$.

$p(3)=3-1=2$

$p(5)=5-1=4$

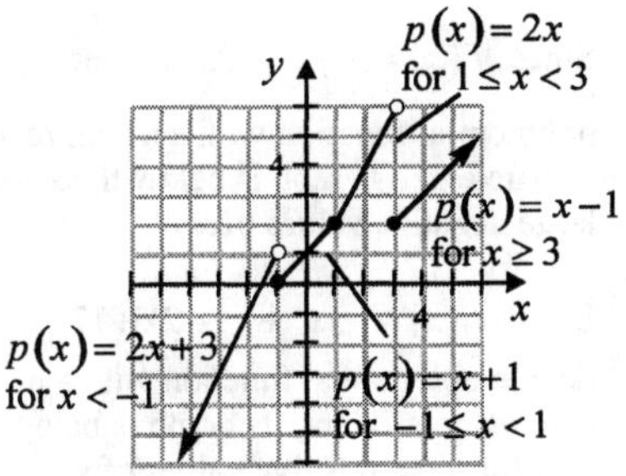

69. $y = -x^3 - 1$

x	y
−2	7
−1	0
0	−1
1	−2
2	−9

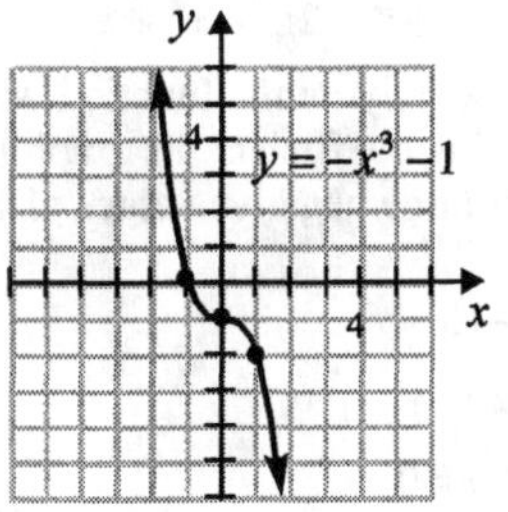

70. $y = \frac{1}{x-2}$

x	y
−1	$-\frac{1}{3}$
0	$-\frac{1}{2}$
1	−1
1.5	−2
2	undefined
2.5	2
3	1
4	$\frac{1}{2}$

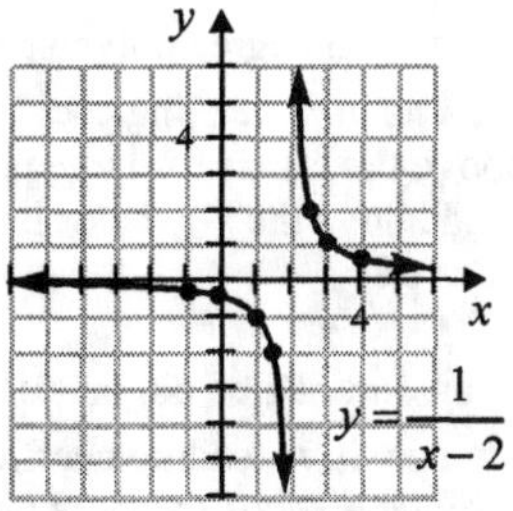

71. a. Since the highest degree term of the polynomial $4x^5 - 7x^4 - 6x^3 + x^2 - 9$ is degree 5, the degree of the polynomial is 5.

b. Since the coefficient of the highest degree term of the polynomial is 4, the leading coefficient is 4.

72. $g(x) = \frac{x-7}{x^2 - 4x - 21} = \frac{x-7}{(x+3)(x-7)}$

Since −3 and 7 cause zero in the denominator, the domain is all real numbers except −3 and 7. That is, the domain is $\{x | x \neq -3 \text{ and } x \neq 7\}$.

Exercise Set 10.2

1. a. The constant term is negative because the function crosses the y-axis below the origin.

b. The leading coefficient is negative because the value of the function is negative for large positive values of x.

c. The degree of the polynomial function is even because the end behavior for large values of $|x|$ is below the x-axis.

d. Yes, the degree of the polynomial function can be 6 because the function has even degree and has fewer than 5 turning points.

e. No, the degree of the polynomial function cannot be 7 because the degree of the polynomial is even.

f. The polynomial has 4 real zeros because it crosses the x-axis 4 times.

3. a. $f(x)=x^8$ is an even degree polynomial function with a positive leading coefficient. Thus, it is above the x-axis for large positive values of x and above the x-axis for large negative values of x.

b. $f(x)=x^9$ is an odd degree polynomial function with a positive leading coefficient. Thus, it is above the x-axis for large positive values of x and below the x-axis for large negative values of x.

5. $P(x)=-2x^3+5x-8$ is a third degree polynomial function with a negative leading coefficient. Thus, for large positive values of x the graph will be below the x-axis and for large negative values of x will be above the x-axis. Thus, the answer is *b*.

7. $P(x)=3x^5+5x^3-25x^2+20$ is a fifth degree polynomial function with a positive leading coefficient. Thus, for large values of x the graph will be above the x-axis and for large negative values of x it will be below the x-axis. Thus, the answer is *a*.

9. a. Since $P(x)=x^7$ is a seventh degree polynomial function with a positive leading coefficient, its graph is above the x-axis for large positive values of x.

b. Since $P(x)=x^7$ is a seventh degree polynomial function with a positive leading coefficient, its graph is below the x-axis for large negative values of x.

11. a. Since $f(x)=-3x^7$ is a seventh degree polynomial function with a negative leading coefficient, its graph is below the x-axis for large positive values of x.

b. Since $f(x)=-3x^7$ is a seventh degree polynomial function with a negative leading coefficient, its graph is above the x-axis for large negative values of x.

13. a. Since $f(x)=-\frac{1}{4}x^2+2x$ is a second degree polynomial function with a negative leading coefficient, its graph is below the x-axis for large positive values of x.

b. Since $f(x)=-\frac{1}{4}x^2+2x$ is a second degree polynomial function with a negative leading coefficient, its graph is below the x-axis for large negative values of x.

15. a. Since $f(x)=-3x^4+x^3-2x+17$ is a fourth degree polynomial function with a negative leading coefficient, its graph is below the x-axis for large positive values of x.

b. Since $f(x)=-3x^4+x^3-2x+17$ is a fourth degree polynomial function with a negative leading coefficient, its graph is below the x-axis for large negative values of x.

17. a. Since $f(x)=-5x^7+2x^5-x^4+x^2-x+1$ is a seventh degree polynomial function with a negative leading coefficient, its graph is below the x-axis for large positive values of x.

b. Since $f(x)=-5x^7+2x^5-x^4+x^2-x+1$ is a seventh degree polynomial function with a negative leading coefficient, its graph is above the x-axis for large negative values of x.

19. $f(x)=x^2-2x-15$

$$0=(x-5)(x+3)$$
$$x-5=0 \text{ or } x+3=0$$
$$x=5 \text{ or } x=-3$$

The zeros are 5 and -3.

21. $f(x)=x^3+6x^2+5x$

$$0=x(x^2+6x+5)$$
$$0=x(x+5)(x+1)$$
$$x=0 \text{ or } x+5=0 \text{ or } x+1=0$$
$$x=0 \text{ or } x=-5 \text{ or } x=-1$$

The zeros are $0, -5$, and -1.

23. $f(x)=x^2-25$

$$0=(x+5)(x-5)$$
$$x+5=0 \text{ or } x-5=0$$
$$x=-5 \text{ or } x=5$$

The zeros are -5 and 5.

25. $P(x)=(x+7)(x-2)^3$

$0=(x+7)(x-2)^3$

$x+7=0$ or $x-2=0$

$x=-7$ or $x=2$

The zeros are -7 and 2 with a multiplicity of 3.

27. $g(x)=(x+3)(2x+1)(3x-1)$

$0=(x+3)(2x+1)(3x-1)$

$x+3=0$ or $2x+1=0$ or $3x-1=0$

$x=-3$ or $2x=-1$ or $3x=1$

$x=-3$ or $x=-\frac{1}{2}$ or $x=\frac{1}{3}$

The zeros are $-3, -\frac{1}{2}$, and $\frac{1}{3}$.

29. $h(x)=(x+2)^3(x-4)^2$

$0=(x+2)^3(x-4)^2$

$x+2=0$ or $x-4=0$

$x=-2$ or $x=4$

The zeros are -2 of multiplicity 3, and 4 of multiplicity 2.

31. $P(x)=x(x+1)^2(x-1)^3(x-6)^4$

$0=x(x+1)^2(x-1)^3(x-6)^4$

$x=0$ or $x+1=0$ or $x-1=0$ or $x-6=0$

$x=0$ or $x=-1$ or $x=1$ or $x=6$

The zeros are $0, -1$ of multiplicity 2, 1 of multiplicity 3, and 6 of multiplicity 4.

33. Since the degree of $f(x)=4x^5+x^3-8x^2+64$ is 5, the maximum number of turning points is $n-1=5-1=4$.

35. Since the degree of $h(x)=\frac{1}{2}x-7$ is 1, the maximum number of turning points is $n-1=1-1=0$.

37. Since the degree of $g(x)=3x^6+x^5-x^4+6x^2-5x+12$ is 6, the maximum number of turning points is $n-1=6-1=5$.

39. The degree of $f(x)=x(x-5)(3x+4)$ is the degree of the product of the x-terms of each factor: $x(x)(3x)=3x^3$. Thus, the degree is 3 and the maximum number of turning points is $n-1=3-1=2$.

41. The degree of $h(x)=(x-1)(x+2)$ is the degree of the product of the x-terms of each factor: $x(x)=x^2$. Thus, the degree is 2 and the maximum number of turning points is $n-1=2-1=1$.

43. The degree of $g(x)=(2x-3)^3(x+8)(x-1)$ is the degree of the product of the x-terms of each factor (don't forget that exponents indicate how many times a factor occurs): $(2x)^3(x)(x)=8x^5$. Thus, the degree is 5 and the maximum number of turning points is $n-1=5-1=4$.

45. $f(x)=x(x+3)(x-1)$

Find the x-intercepts by solving the equation $f(x)=0$.

$0=x(x+3)(x-1)$

$x=0$ or $x=-3$ or $x=1$

The x-intercepts are $(-3,0),(0,0)$, and $(1,0)$. These break up the graph into four regions. Find a point in each region, then graph the points along with the *x*-intercepts. Connect all the points with a smooth curve. Since all the multiplicities are 1, the graph will cross the *x*-axis at each *x*-intercept.

Interval	x	$f(x)=x(x+3)(x-1)$
$(-\infty,-3)$	-4	$f(-4)=-20$
$(-3,0)$	-2	$f(-2)=6$
$(0,1)$	0.5	$f(0.5)=-0.875$
$(1,\infty)$	2	$f(2)=10$

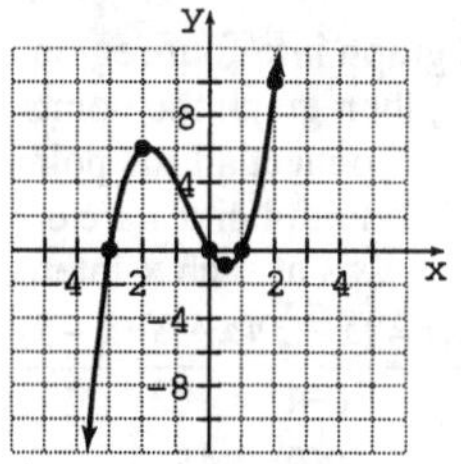

f (x) =x (x+3) (x-1)

47. $P(x)=(x+3)^2(x-1)$
Find the x-intercepts by solving the equation $P(x)=0$.

$0=(x+3)^2(x-1)$

$(x+3)^2=0$ or $x-1=0$

$x=-3$ or $x=1$

The x-intercepts are $(-3,0)$ and $(1,0)$. These break up the graph into three regions. Find a point in each region, then graph the points along with the x-intercepts. Connect all the points with a smooth curve. Since the multiplicity of the zero $x=-3$ is 2 (even), the graph will touch, but not cross, the x-axis at $x=-3$. The graph will cross the x-axis at $x=1$ since this zero has an odd multiplicity.

Interval	x	$P(x)=(x+3)^2(x-1)$
$(-\infty,-3)$	-4	$P(-4)=-5$
$(-3,1)$	0	$P(0)=-9$
$(1,\infty)$	2	$P(2)=25$

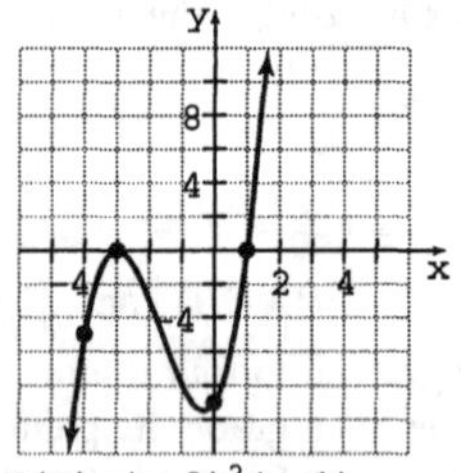

P(x) = (x+3)² (x-1)

49. $P(x)=(x+3)(x-3)(x-5)$
Find the x-intercepts by solving the equation $P(x)=0$.

$0=(x+3)(x-3)(x-5)$

$x+3=0$ or $x-3=0$ or $x-5=0$

$x=-3$ or $x=3$ or $x=5$

The x-intercepts are $(-3,0)$, $(3,0)$, and $(5,0)$. These break up the graph into four regions. Find a point in each region, then graph the points along with the x-intercepts. Connect all the points with a smooth curve. All the multiplicities are odd so the graph will cross the x-axis at each x-intercept.

Interval	x	$P(x)=(x+3)(x-3)(x-5)$
$(-\infty,-3)$	-4	$P(-4)=-63$
$(-3,3)$	0	$P(0)=45$
$(3,5)$	4	$P(4)=-7$
$(5,\infty)$	6	$P(6)=27$

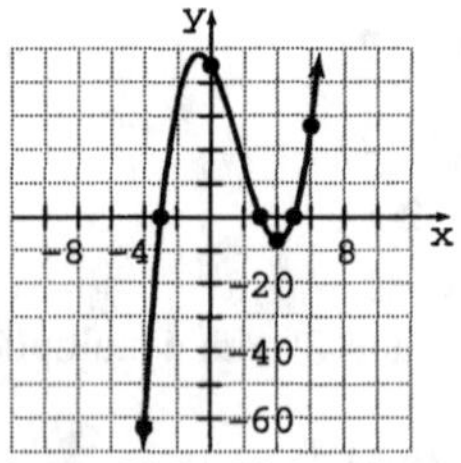

P(x) = (x+3) (x-3) (x-5)

51. $g(x)=(x+1)(x^2-36)$
Find the x-intercepts by solving the equation $g(x)=0$.

$0=(x+1)(x^2-36)$

$0=(x+1)(x-6)(x+6)$

$x+1=0$ or $x-6=0$ or $x+6=0$

$x=-1$ or $x=6$ or $x=-6$

The x-intercepts are $(-1,0)$, $(6,0)$, and $(-6,0)$. These break up the graph into four regions. Find a point in each region, then graph the points along with the x-intercepts. Connect all the points with a smooth curve. All the multiplicities are odd so the graph will cross the x-axis at each x-intercept.

Interval	x	$g(x)=(x+1)(x^2-36)$
$(-\infty,-6)$	-7	$g(-7)=-78$
$(-6,-1)$	-3	$g(-3)=54$
$(-1,6)$	3	$g(3)=-108$
$(6,\infty)$	7	$g(7)=104$

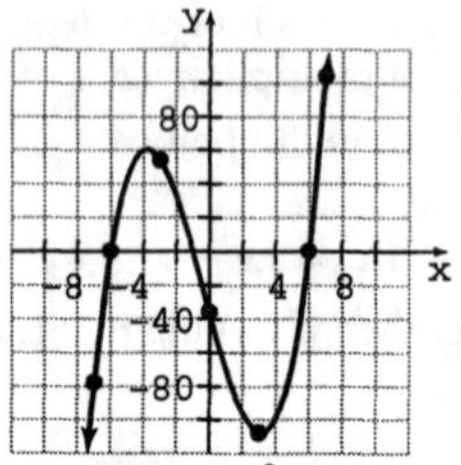

g(x) = (x+1) (x²-36)

53. Since the graph has 1 turning point, the degree of the polynomial function is at least 2.

55. Since the graph has 2 turning points, the degree of the polynomial function is at least 3.

57. Since the graph has 2 turning points, the degree of the polynomial function is at least 3.

59. Since the graph has 3 turning points, the degree of the polynomial function is at least 4.

61. Since the graph has 3 turning points, the degree of the polynomial function is at least 4.

63. $f(x)=x^3-2x^2-3x$

Find the x-intercepts by solving the equation $f(x)=0$.

$0=x^3-2x^2-3x$

$0=x(x^2-2x-3)$

$0=x(x-3)(x+1)$

$x=0$ or $x-3=0$ or $x+1=0$

$x=0$ or $x=3$ or $x=-1$

The x-intercepts are $(0,0)$, $(3,0)$, and $(-1,0)$. These break up the graph into four regions. Find a point in each region, then graph the points along with the x-intercepts. Connect all the points with a smooth curve. All the multiplicities are odd so the graph will cross the x-axis at each x-intercept.

Interval	x	$f(x)=x^3-2x^2-3x$
$(-\infty,-1)$	-2	$f(-2)=-10$
$(-1,0)$	-0.5	$f(-0.5)=0.875$
$(0,3)$	2	$f(2)=-6$
$(3,\infty)$	4	$f(4)=20$

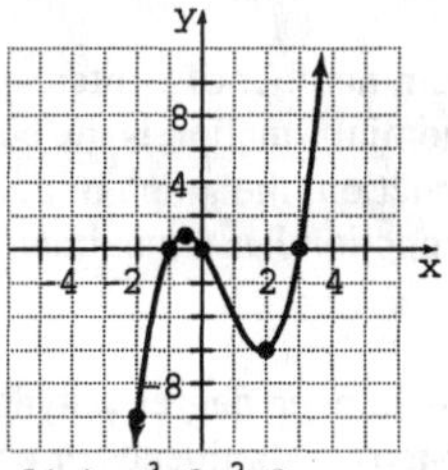

f (x) =x³-2x²-3x

65. $f(x)=5x^3+4x^2-x$

Find the x-intercepts by solving the equation $f(x)=0$.

$0=5x^3+4x^2-x$

$0=x(5x^2+4x-1)$

$0=x(5x-1)(x+1)$

$x=0$ or $5x-1=0$ or $x+1=0$

$x=0$ or $x=\frac{1}{5}$ or $x=-1$

The x-intercepts are $(0,0)$, $\left(\frac{1}{5},0\right)$, and $(-1,0)$.

These break up the graph into four regions. Find a point in each region, then graph the points along with the x-intercepts. Connect all the points with a smooth curve. All the multiplicities are odd so the graph will cross the x-axis at each x-intercept.

Interval	x	$f(x)=5x^3+4x^2-x$
$(-\infty,-1)$	-2	$f(-2)=-22$
$(-1,0)$	-0.5	$f(-0.5)=0.875$
$\left(0,\frac{1}{5}\right)$	$\frac{1}{10}$	$f\left(\frac{1}{10}\right)=-0.1$
$\left(\frac{1}{5},\infty\right)$	1	$f(1)=8$

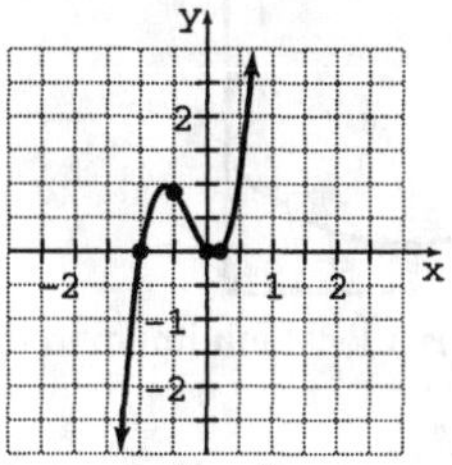

f (x) =5x³+4x²-x

67. $f(x)=x^3-4x^2+4x$

Find the x-intercepts by solving the equation $f(x)=0$.

$0=x^3-4x^2+4x$

$0=x(x^2-4x+4)$

$0=x(x-2)^2$

$x=0$ or $(x-2)^2=0$

$x=0$ or $x-2=0$

$x=0$ or $x=2$

The x-intercepts are $(0,0)$ and $(2,0)$. These break up the graph into three regions. Find a point in each region, then graph the points along with the x-intercepts. Connect all the points with a smooth curve. The zero $x=2$ has an even multiplicity so the graph will touch, but not cross, the x-axis at $x=2$. The other zero has an odd multiplicity so the graph will cross the x-axis at $x=0$.

Interval	x	$f(x)=x^3-4x^2+4x$
$(-\infty,0)$	-1	$f(-1)=-9$
$(0,2)$	1	$f(1)=1$
$(2,\infty)$	3	$f(3)=3$

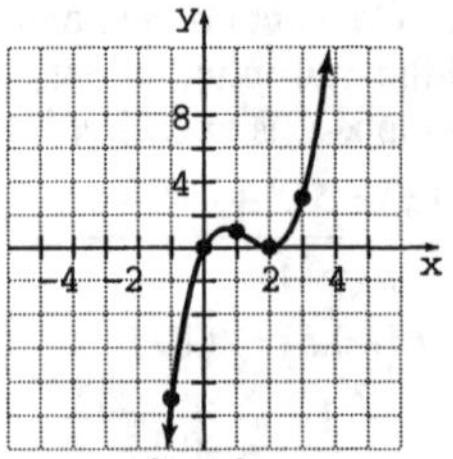

f(x) = x^3-4x^2+4x

69. **a.** Window: $[-1,1,1,2,8,1]$

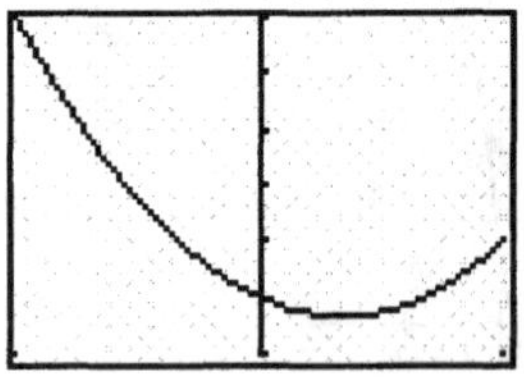

b. The function has a local minimum in the interval $(-1,1)$.

c. The turning point occurs at the point $(0.33, 2.67)$.

71. **a.** Window: $[-2,0,1,-5,1,1]$

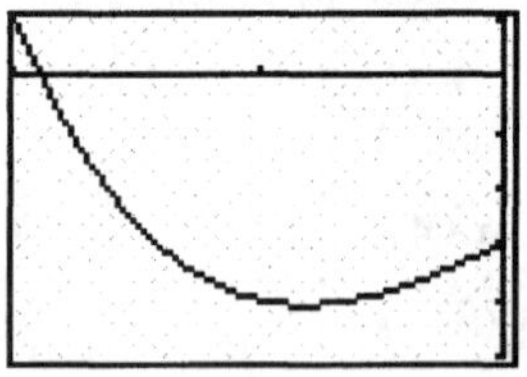

b. The function has a local minimum in the interval $(-2,0)$.

c. The turning point occurs at the point $(-0.82,-4.09)$.

74.

3	4	−3	2	−1	−250
		12	27	87	258
	4	9	29	86	8

$$\left(4x^4-3x^3+2x^2-x-250\right)\div(x-3)$$
$$=4x^3+9x^2+29x+86+\frac{8}{x-3}$$

75. $\left(5x^3-7x^2+9x-11\right)\div(x+4)$

$$=\left(5x^3-7x^2+9x-11\right)\div(x-(-4))$$

$$P(x)=5x^3-7x^2+9x-11$$
$$P(-4)=5(-4)^3-7(-4)^2+9(-4)-11$$
$$=5(-64)-7(16)-36-11$$
$$=-320-112-36-11$$
$$=-479$$

Since $P(-4)=-479$, the remainder, when $5x^3-7x^2+9x-11$ is divided by $x+4$, is -479.

76. $\frac{3}{5x}+\frac{x}{4}=\frac{3\cdot4}{5x\cdot4}+\frac{x\cdot5x}{4\cdot5x}=\frac{12}{20x}+\frac{5x^2}{20x}=\frac{5x^2+12}{20x}$

77. $\frac{3x-1}{y-4}-\frac{2x+5}{4-y}=\frac{3x-1}{y-4}-\frac{2x+5}{-(y-4)}$

$$=\frac{3x-1}{y-4}+\frac{2x+5}{y-4}$$
$$=\frac{5x+4}{y-4}$$

Exercise Set 10.3

1. Because the maximum number of x-intercepts of the graph of a polynomial function is the degree of the polynomial function, the graph of a fourth degree polynomial function has a maximum of 4 x-intercepts.

3. Because the graph of an even degree polynomial is either above the x-axis on both ends or below the x-axis on both ends, it is possible for the graph of the polynomial to be completely above the x-axis or completely below the x-axis. Therefore, the minimum number of x-intercepts of the graph of a sixth degree polynomial function is 0.

5. If the coefficients of a polynomial $P(x)$ are all positive, then the coefficients have no variation in signs. Therefore, by Descartes' Rule of Signs, the graph of the polynomial has no positive x-intercepts.

7. Since *f* is a polynomial function of degree 2, the maximum number of x-intercepts is 2. Since the degree is an even number, and its leading coefficient and constant term have the same sign, the minimum number of x-intercepts is 0.

9. Since g is a polynomial function of degree 3, the maximum number of x-intercepts is 3. Since the degree is an odd number, the minimum number of x-intercepts is 1.

11. Since r is a polynomial function of degree 4, the maximum number of x-intercepts is 4. Since the degree is an even number, and its leading coefficient and constant term have the same sign, the minimum number of x-intercepts is 0.

13. Since m is a polynomial function of degree 5, the maximum number of x-intercepts is 5. Since the degree is an odd number, the minimum number of x-intercepts is 1.

15. Since r is a polynomial function of degree 6, the maximum number of x-intercepts is 6. Since the degree is an even number, and its leading coefficient and constant term have the same sign, the minimum number of x-intercepts is 0.

17. Since $f(x)=x^3+2x^2+4x+7$ has no variation in signs, the number of positive zeros is 0. Since

$$f(-x)=(-x)^3+2(-x)^2+4(-x)+7$$
$$=-x^3+2x^2-4x+7$$

has three variations in signs, the number of negative zeros is either 3 or 1.

19. Since $f(x)=-6x^3+2x^2+3x-1$ has two variations in signs, the number of positive zeros is either 2 or 0. Since

$$f(-x)=-6(-x)^3+2(-x)^2+3(-x)-1$$
$$=6x^3+2x^2-3x-1$$

has one variation in signs, the number of negative zeros is 1.

21. Since $g(x)=5x^4-2x^3+x^2-9x-14$ has three variations in signs, the number of positive zeros is either 3 or 1. Since

$$g(-x)=5(-x)^4-2(-x)^3+(-x)^2-9(-x)-14$$
$$=5x^4+2x^3+x^2+9x-14$$

has one variation in signs, the number of negative zeros is 1.

23. Since $g(x)=-3x^4-x^3-6x^2-2x-3$ has no variations in signs, the number of positive zeros is 0. Since

$$g(-x)=-3(-x)^4-(-x)^3-6(-x)^2-2(-x)-3$$
$$=-3x^4+x^3-6x^2+2x-3$$

has four variations in signs, the number of negative zeros is either 4 or 2 or 0.

25. Since $r(x)=x^5-x^4+x^3+2x^2+7x-3$ has three variations in signs, the number of positive zeros is either 3 or 1. Since $r(-x)=$

$$(-x)^5-(-x)^4+(-x)^3+2(-x)^2+7(-x)-3$$
$$=-x^5-x^4-x^3+2x^2-7x-3$$

has two variations in signs, the number of negative zeros is either 2 or 0.

27. Since $r(x)=7x^5+2x^4+x^3+5x^2+9x+20$ has no variations in signs, the number of positive zeros is 0. Since $r(-x)=$

$$7(-x)^5+2(-x)^4+(-x)^3+5(-x)^2+9(-x)+20$$
$$=-7x^5+2x^4-x^3+5x^2-9x+20$$

has five variations in signs, the number of negative zeros is either 5 or 3 or 1.

29. The possible rational zeros of $f(x)=x^2+3x+2$ are of the form $\frac{p}{q}$ where p is a factor of 2 and q is a factor of 1.

Factors of 2: $\pm1,\ \pm2$

Factors of 1: ±1

Possible rational zeros: $\pm1,\ \pm2$.

31. The possible rational zeros of $f(x)=3x^2+4x+3$ are of the form $\frac{p}{q}$ where p is a factor of 3 and q is a factor of 3.

Factors of 3: $\pm1,\ \pm3$

Factors of 1: $\pm1,\ \pm3$

Possible rational zeros: $\pm1,\ \pm3,\ \pm\frac{1}{3}$.

33. The possible rational zeros of $g(x)=x^3-6x^2+3x+12$ are of the form $\frac{p}{q}$ where p is a factor of 12 and q is a factor of 1.

Factors of 12: $\pm1,\ \pm2,\ \pm3,\ \pm4,\ \pm6,\ \pm12$

Factors of 1: ±1

Possible rational zeros:

$\pm1,\ \pm2,\ \pm3,\ \pm4,\ \pm6,\ \pm12$.

35. The possible rational zeros of $g(x)=3x^3-7x^2+17x-4$ are of the form $\frac{p}{q}$ where p is a factor of -4 and q is a factor of 3.

Factors of -4: $\pm 1,\ \pm 2,\ \pm 4$

Factors of 3: $\pm 1,\ \pm 3$

Possible rational zeros:

$$\pm 1,\ \pm 2,\ \pm 4,\ \pm\frac{1}{3},\ \pm\frac{2}{3},\ \pm\frac{4}{3}.$$

37. The possible rational zeros of $h(x)=6x^4+5x^3-x^2+9x+4$ are of the form $\frac{p}{q}$ where p is a factor of 4 and q is a factor of 6.

Factors of 4: $\pm 1,\ \pm 2,\ \pm 4$

Factors of 6: $\pm 1,\ \pm 2,\ \pm 3,\ \pm 6$

Possible rational zeros:

$$\pm 1,\ \pm 2,\ \pm 4,\ \pm\frac{1}{2},\ \pm\frac{1}{3},\ \pm\frac{2}{3},\ \pm\frac{4}{3},\ \pm\frac{1}{6}.$$

39. The possible rational zeros of $r(x)=6x^5+x^4-x^2+11x+3$ are of the form $\frac{p}{q}$ where p is a factor of 4 and q is a factor of 6.

Factors of 3: $\pm 1,\ \pm 3$

Factors of 6: $\pm 1,\ \pm 2,\ \pm 3,\ \pm 6$

Possible rational zeros:

$$\pm 1,\ \pm 3,\ \pm\frac{1}{2},\ \pm\frac{3}{2},\ \pm\frac{1}{3},\ \pm\frac{1}{6}.$$

41. The possible rational zeros of $r(x)=7x^5-x^4+15x^3-18x^2+4$ are of the form $\frac{p}{q}$ where p is a factor of 4 and q is a factor of 7.

Factors of 4: $\pm 1,\ \pm 2,\ \pm 4$

Factors of 7: $\pm 1,\ \pm 7$

Possible rational zeros:

$$\pm 1,\ \pm 2,\ \pm 4,\ \pm\frac{1}{7},\ \pm\frac{2}{7},\ \pm\frac{4}{7}.$$

43. Letting $f(x)=0$, and using factoring and the zero factor property, we obtain:

$$x^2-6x+8=0$$
$$(x-2)(x-4)=0$$
$$x-2=0 \text{ or } x-4=0$$
$$x=2 \text{ or } x=4$$

Thus, the zeros of f are 2 and 4.

45. Letting $f(x)=0$, and using factoring and the zero factor property, we obtain:

$$3x^2+14x-5=0$$
$$(3x-1)(x+5)=0$$
$$3x-1=0 \text{ or } x+5=0$$
$$x=\frac{1}{3} \text{ or } x=-5$$

Thus, the zeros of f are -5 and $\frac{1}{3}$.

47. The possible rational zeros of $g(x)=x^3+7x^2-x-7$ are of the form $\frac{p}{q}$ where p is a factor of -7 and q is a factor of 1.

Factors of -7: $\pm 1,\ \pm 7$

Factors of 1: ± 1

Possible rational zeros: $\pm 1,\ \pm 7$.

By Descartes' Rule of Signs, since $g(x)$ has one variation in signs, it has 1 positive zero. Since $g(-x)=-x^3+7x^2+x-7$ has two variations in signs, $g(x)$ has either 2 or 0 negative zeros. Using synthetic division to find the 1 positive zero and checking the possibility of 1 being that zero, we get

1	1	7	−1	−7
		1	8	7
	1	8	7	0

Since the remainder is 0, 1 is a zero of g and $x-1$ is a factor of g. Writing g in factored form, we get

$$g(x)=x^3+7x^2-x-7$$
$$=(x-1)(x^2+8x+7)$$
$$=(x-1)(x+1)(x+7)$$

Thus, g has three rational zeros: 1, -1, and -7.

49. The possible rational zeros of $g(x)=2x^3+x^2-8x-4$ are of the form $\frac{p}{q}$ where p is a factor of -4 and q is a factor of 2.

Factors of -4: $\pm 1,\ \pm 2,\ \pm 4$

Factors of 2: $\pm 1,\ \pm 2$

Possible rational zeros: $\pm 1,\ \pm 2,\ \pm 4,\ \pm\frac{1}{2}$.

By Descartes' Rule of Signs, since $g(x)$ has one variation in signs, it has 1 positive zero.

Since $g(-x) = -2x^3 + x^2 + 8x - 4$ has two variations in signs, $g(x)$ has either 2 or 0 negative zeros. Using synthetic division to find the 1 positive zero and checking the possibility of 2 being that zero, we get

$$\begin{array}{r|rrrr} 2 & 2 & 1 & -8 & -4 \\ & & 4 & 10 & 4 \\ \hline & 2 & 5 & 2 & 0 \end{array}$$

Since the remainder is 0, 2 is the one positive zero of *g*. Writing *g* in factored form, we get

$$\begin{aligned} g(x) &= 2x^3 + x^2 - 8x - 4 \\ &= (x-2)(2x^2 + 5x + 2) \\ &= (x-2)(2x+1)(x+2) \end{aligned}$$

Thus, *g* has three rational zeros: 2, $-\frac{1}{2}$, and –2.

51. The possible rational zeros of $r(x) = x^4 - 12x^2 - 64$ are of the form $\frac{p}{q}$ where *p* is a factor of –64 and *q* is a factor of 1.

Factors of –64:

$\pm 1, \pm 2, \pm 4, \pm 8, \pm 16, \pm 32, \pm 64$

Factors of 1: ± 1

Possible rational zeros:

$\pm 1, \pm 2, \pm 4, \pm 8, \pm 16, \pm 32, \pm 64$.

By Descartes' Rule of Signs, since $r(x)$ has one variation in signs, it has 1 positive zero. Since $r(-x) = x^4 - 12x^2 - 64$ has one variation in signs, it has 1 negative zero. Note then, that *r* has only 2 real zeros, 1 positive and 1 negative. Using synthetic division to find the one positive zero and checking the possibility of 4 being that zero, we get

$$\begin{array}{r|rrrrr} 4 & 1 & 0 & -12 & 0 & -64 \\ & & 4 & 16 & 16 & 64 \\ \hline & 1 & 4 & 4 & 16 & 0 \end{array}$$

Since the remainder is 0, 4 is the one positive zero of *r*. Thus, *r* can be written as follows:

$r(x) = x^4 - 12x^2 - 64 = (x-4)(x^3 + 4x^2 + 4x + 16)$

Now, we know that the one positive real zero is 4, we now need to find the one negative real zero. To do this, focus on the polynomial $h(x) = x^3 + 4x^2 + 4x + 16$. Using synthetic division and checking the possibility of –4 being that zero, we get

$$\begin{array}{r|rrrr} -4 & 1 & 4 & 4 & 16 \\ & & -4 & 0 & -16 \\ \hline & 1 & 0 & 4 & 0 \end{array}$$

Since the remainder is 0, –4 is the negative real zero of *r*. In summary, the two rational zeros of *r* are 4 and –4.

Note: *r* can be written in the form: $r(x) = (x-4)(x+4)(x^2+4)$. Also note that $x^2 + 4$ will have no rational zeros.

53. The possible rational zeros of $r(x) = x^4 - 8x^2 + 16$ are of the form $\frac{p}{q}$ where *p* is a factor of 16 and *q* is a factor of 1.

Factors of 16: $\pm 1, \pm 2, \pm 4, \pm 8, \pm 16$

Factors of 1: ± 1

Possible rational zeros: $\pm 1, \pm 2, \pm 4, \pm 8, \pm 16$

By Descartes' Rule of Signs, since $r(x)$ has two variations in signs, it has either 2 or 0 positive zeros. Since $r(-x) = x^4 - 8x^2 + 16$ has two variations in signs, it has either 2 or 0 negative zeros. Using synthetic division to find one of the positive zeros and checking the possibility of 2 being that zero, we get

$$\begin{array}{r|rrrrr} 2 & 1 & 0 & -8 & 0 & 16 \\ & & 2 & 4 & -8 & -16 \\ \hline & 1 & 2 & -4 & -8 & 0 \end{array}$$

Since the remainder is 0, 2 is a positive zero of *r*. Thus, *r* can be written as follows:

$$\begin{aligned} r(x) &= x^4 - 8x^2 + 16 \\ &= (x-2)(x^3 + 2x^2 - 4x - 8) \end{aligned}$$

Now, we know that *r* has 2 as a positive real zero. Thus, since the number of positive real zeros is either 0 or 2, *r* must have another positive real zero. To find it, focus on the polynomial $h(x) = x^3 + 2x^2 - 4x - 8$. Using synthetic division and checking the possibility of 2 being a positive zero for a second time (i.e., multiplicity 2), we get

$$\begin{array}{r|rrrr} 2 & 1 & 2 & -4 & -8 \\ & & 2 & 8 & 8 \\ \hline & 1 & 4 & 4 & 0 \end{array}$$

Since the remainder is 0, 2 is a real zero of *r* with multiplicity 2. Writing *r* in factored form, we get

$$\begin{aligned} r(x) &= x^4 - 8x^2 + 16 \\ &= (x-2)(x-2)(x^2 + 4x + 4) \\ &= (x-2)(x-2)(x+2)(x+2) \end{aligned}$$

Thus, *r* has four rational zeros: 2 and –2, both with multiplicity 2.

55. The possible rational zeros of $r(x)=x^4+x^3-7x^2-13x-6$ are of the form $\frac{p}{q}$ where p is a factor of –6 and q is a factor of 1.

Factors of –6: ±1, ±2, ±3, ±6

Factors of 1: ±1

Possible rational zeros: ±1, ±2, ±3, ±6

By Descartes' Rule of Signs, since $r(x)$ has one variation in signs, it has 1 positive zero. Since $r(-x)=x^4-x^3-7x^2+13x-6$ has three variations in signs, it has either 3 or 1 negative zero. Using synthetic division to find the positive zero and checking the possibility of 3 being that zero, we get

3	1	1	–7	–13	–6
		3	12	15	6
	1	4	5	2	0

Since the remainder is 0, 3 is the one positive zero of r. Thus, r can be written as follows:

$$r(x)=x^4+x^3-7x^2-13x-6$$
$$=(x-3)(x^3+4x^2+5x+2)$$

We now need to determine any negative real zeros. To do this, focus on the polynomial $h(x)=x^3+4x^2+5x+2$. Using synthetic division and checking the possibility of –1 being a zero, we get

–1	1	4	5	2
		–1	–3	–2
	1	3	2	0

Since the remainder is 0, –1 is a negative real zero of r. Thus, r can be written as follows:

$$r(x)=x^4+x^3-7x^2-13x-6$$
$$=(x-3)(x^3+4x^2+5x+2)$$
$$=(x-3)(x+1)(x^2+3x+2)$$
$$=(x-3)(x+1)(x+1)(x+2)$$

Thus, r has four rational zeros: 3, –1 with multiplicity 2, and –2.

57. The possible rational zeros of $r(x)=x^4-8x^3+22x^2-24x+9$ are of the form $\frac{p}{q}$ where p is a factor of 9 and q is a factor of 1.

Factors of 9: ±1, ±3, ±9

Factors of 1: ±1

Possible rational zeros: ±1, ±3, ±9

By Descartes' Rule of Signs, since $r(x)$ has four variations in signs, it has either 4 or 2 or 0 positive zeros. Since $r(-x)=x^4+8x^3+22x^2+24x+9$ has no variations in signs, it has 0 negative zeros. Using synthetic division and checking the possibility of 1 being a zero , we get

1	1	–8	22	–24	9
		1	–7	15	–9
	1	–7	15	–9	0

Since the remainder is 0, 1 is a positive zero of r. Thus, r can be written as follows:

$$r(x)=x^4-8x^3+22x^2-24x+9$$
$$=(x-1)(x^3-7x^2+15x-9)$$

We have found one positive real zero. We now see if there are others. To do this, focus on the polynomial $h(x)=x^3-7x^2+15x-9$. Using synthetic division and checking the possibility of 3 being a zero , we get

3	1	–7	15	–9
		3	–12	9
	1	–4	3	0

Since the remainder is 0, 3 is another positive real zero of r. Thus, r can be written as follows:

$$r(x)=x^4-8x^3+22x^2-24x+9$$
$$=(x-1)(x^3-7x^2+15x-9)$$
$$=(x-1)(x-3)(x^2-4x+3)$$
$$=(x-1)(x-3)(x-1)(x-3)$$

Thus, r has four rational zeros: 1 and 3, both with multiplicity 2.

59. a. $f(2)=(2)^3+(2)^2-10(2)+8$
$=8+4-20+8$
$=0$

Yes, 2 is a zero of f.

b. $f(-4)=(-4)^3+(-4)^2-10(-4)+8$
$=-64+16+40+8$
$=0$

Yes, –4 is a zero of f.

61. a. $g(-6)=(-6)^4-3(-6)^3-3(-6)^2+11(-6)-6$
$=1296-3(216)-3(36)+11(-6)-6$
$=1296-648-108-66-6$
$=468$

No, –6 is not a zero of g.

b. $g(-1) = (-1)^4 - 3(-1)^3 - 3(-1)^2 + 11(-1) - 6$
$= 1 - 3(-1) - 3(1) + 11(-1) - 6$
$= 1 + 3 - 3 - 11 - 6$
$= -16$
No, -1 is not a zero of g.

63.
$$\begin{array}{r|rrrr} 1 & 1 & 2 & -5 & 11 \\ & & 1 & 3 & -2 \\ \hline & 1 & 3 & -2 & 9 \end{array}$$

The remainder when $x^3 + 2x^2 - 5x + 11$ is divided by $x - 1$ is 9.

65.
$$\begin{array}{r|rrrrr} -3 & 2 & 5 & -4 & -15 & 8 \\ & & -6 & 3 & 3 & 36 \\ \hline & 2 & -1 & -1 & -12 & 44 \end{array}$$

The remainder when $2x^4 + 5x^3 - 4x^2 - 15x + 8$ is divided by $x + 3$ is 44.

67. $f(x) = x^2 + 1$
$x^2 + 1 = 0$
$x^2 = -1$
$x = \pm\sqrt{-1}$
$x = \pm i$
Thus, the two zeros of f are i and $-i$.

69. The possible rational zeros of $f(x) = x^4 - 3x^2 - 4$ are of the form $\frac{p}{q}$ where p is a factor of -4 and q is a factor of 1.
Factors of -4: $\pm 1, \pm 2, \pm 4$
Factors of 1: ± 1
Possible rational zeros: $\pm 1, \pm 2, \pm 4$
By Descartes' Rule of Signs, since $f(x)$ has one variation in signs, it has 1 positive real zero. Since $f(-x) = x^4 - 3x^2 - 4$ has one variation in signs, it has 1 negative real zero. Using synthetic division to find the one positive real zero and checking the possibility of 2 being that zero, we get

$$\begin{array}{r|rrrrr} 2 & 1 & 0 & -3 & 0 & -4 \\ & & 2 & 4 & 2 & 4 \\ \hline & 1 & 2 & 1 & 2 & 0 \end{array}$$

Since the remainder is 0, 2 is the one positive real zero of f. Thus, f can be written as follows:
$f(x) = x^4 - 3x^2 - 4$
$= (x-2)(x^3 + 2x^2 + x + 2)$

Now, we know that the one positive real zero is 2, so we need to find the one negative real zero. To do this, focus on the polynomial $h(x) = x^3 + 2x^2 + x + 2$. Using synthetic division and checking the possibility of -2 being that zero, we get

$$\begin{array}{r|rrrr} -2 & 1 & 2 & 1 & 2 \\ & & -2 & 0 & -2 \\ \hline & 1 & 0 & 1 & 0 \end{array}$$

Since the remainder is 0, -2 is the negative real zero of f. Thus, f can be written as follows:
$f(x) = x^4 - 3x^2 - 4$
$= (x-2)(x^3 + 2x^2 + x + 2)$
$= (x-2)(x+2)(x^2+1)$
Now consider the polynomial $g(x) = x^2 + 1$. The zeros of g are
$x^2 + 1 = 0$
$x^2 = -1$
$x = \pm\sqrt{-1}$
$x = \pm i$
Thus, the four zeros of f are 2, -2, i, and $-i$.

71. The possible rational zeros of $h(x) = x^4 + x^3 - x - 1$ are of the form $\frac{p}{q}$ where p is a factor of -1 and q is a factor of 1.
Factors of -1: ± 1
Factors of 1: ± 1
Possible rational zeros: ± 1
By Descartes' Rule of Signs, since $h(x)$ has one variation in signs, it has 1 positive real zero. Since $h(-x) = x^4 - x^3 + x - 1$ has three variations in signs, it has either 3 or 1 negative real zero. Using synthetic division to find the one positive real zero and checking the possibility of 1 being that zero, we get

$$\begin{array}{r|rrrrr} 1 & 1 & 1 & 0 & -1 & -1 \\ & & 1 & 2 & 2 & 1 \\ \hline & 1 & 2 & 2 & 1 & 0 \end{array}$$

Since the remainder is 0, 1 is the one positive real zero of h. Thus, h can be written as follows:
$h(x) = x^4 + x^3 - x - 1 = (x-1)(x^3 + 2x^2 + 2x + 1)$
Now, we know that the one positive real zero is 1, so we need to find any negative real zeros. To do this, focus on the polynomial $r(x) = x^3 + 2x^2 + 2x + 1$. Using synthetic division and checking the possibility of -1 being that zero, we get

$$\begin{array}{r|rrrr} -1 & 1 & 2 & 2 & 1 \\ & & -1 & -1 & -1 \\ \hline & 1 & 1 & 1 & 0 \end{array}$$

Since the remainder is 0, –1 is a negative real zero of h. Thus, h can be written as follows:

$$h(x) = x^4 + x^3 - x - 1$$
$$= (x-1)(x^3 + 2x^2 + 2x + 1)$$
$$= (x-1)(x+1)(x^2 + x + 1)$$

Now consider the polynomial $t(x) = x^2 + x + 1$.

The zeros of t are

$$x^2 + x + 1 = 0$$
$$x = \frac{-1 \pm \sqrt{1^2 - 4(1)(1)}}{2(1)}$$
$$x = \frac{-1 \pm \sqrt{-3}}{2}$$
$$x = \frac{-1 \pm i\sqrt{3}}{2}$$

Thus, the four zeros of h are 1, –1, $\frac{-1+i\sqrt{3}}{2}$, and $\frac{-1-i\sqrt{3}}{2}$.

73. The possible rational zeros of $g(x) = x^4 + 3x^3 - x^2 - 12x - 12$ are of the form $\frac{p}{q}$ where p is a factor of –12 and q is a factor of 1.

Factors of –12: ±1, ±2, ±3, ±4, ±6, ±12
Factors of 1: ±1

Possible rational zeros:
±1, ±2, ±3, ±4, ±6, ±12

By Descartes' Rule of Signs, since $g(x)$ has one variation in signs, it has 1 positive real zero. Since $g(-x) = x^4 - 3x^3 - x^2 + 12x - 12$ has three variations in signs, it has either 3 or 1 negative real zero. Using synthetic division to find the one positive real zero and checking the possibility of 2 being that zero, we get

$$\begin{array}{r|rrrrr} 2 & 1 & 3 & -1 & -12 & -12 \\ & & 2 & 10 & 18 & 12 \\ \hline & 1 & 5 & 9 & 6 & 0 \end{array}$$

Since the remainder is 0, 2 is the one positive real zero of g. Thus, g can be written as follows:

$$g(x) = x^4 + 3x^3 - x^2 - 12x - 12$$
$$= (x-2)(x^3 + 5x^2 + 9x + 6)$$

Now, we know that the one positive real zero is 2, so we need to find any negative real zeros. To do this, focus on the polynomial $r(x) = x^3 + 5x^2 + 9x + 6$. Using synthetic division and checking the possibility of –2 being that zero, we get

$$\begin{array}{r|rrrr} -2 & 1 & 5 & 9 & 6 \\ & & -2 & -6 & -6 \\ \hline & 1 & 3 & 3 & 0 \end{array}$$

Since the remainder is 0, –2 is a negative real zero of g. Thus, g can be written as follows:

$$g(x) = x^4 + 3x^3 - x^2 - 12x - 12$$
$$= (x-2)(x^3 + 5x^2 + 9x + 6)$$
$$= (x-2)(x+2)(x^2 + 3x + 3)$$

Now consider the polynomial $t(x) = x^2 + 3x + 3$.

The zeros of t are

$$x^2 + 3x + 3 = 0$$
$$x = \frac{-3 \pm \sqrt{3^2 - 4(1)(3)}}{2(1)} = \frac{-3 \pm \sqrt{-3}}{2} = \frac{-3 \pm i\sqrt{3}}{2}$$

Thus, the four zeros of g are 2, –2, $\frac{-3+i\sqrt{3}}{2}$, and $\frac{-3-i\sqrt{3}}{2}$.

75. The possible rational zeros of $f(x) = x^5 - x^4 + 2x^3 - 2x^2 + x - 1$ are of the form $\frac{p}{q}$ where p is a factor of –1 and q is a factor of 1.

Factors of –1: ±1
Factors of 1: ±1

Possible rational zeros: ±1

By Descartes' Rule of Signs, since $f(x)$ has five variations in signs, it has either 5 or 3 or 1 positive real zeros. Since $f(-x) = -x^5 - x^4 - 2x^3 - 2x^2 - x - 1$ has no variations in signs, it has either 0 negative zeros. Thus, the only possible value for real zeros is 1. Using synthetic division to verify and factor, we get

$$\begin{array}{r|rrrrrr} 1 & 1 & -1 & 2 & -2 & 1 & -1 \\ & & 1 & 0 & 2 & 0 & 1 \\ \hline & 1 & 0 & 2 & 0 & 1 & 0 \end{array}$$

Since the remainder is 0, 1 is the positive zero of *f*. Thus, *f* can be written as follows:

$$f(x)=x^5-x^4+2x^3-2x^2+x-1$$
$$=(x-1)(x^4+2x^2+1)$$

We have found one real zero. The remaining zeros are imaginary. To find them, focus on the polynomial $h(x)=x^4+2x^2+1$.

The zeros of *h* are

$$x^4+2x^2+1=0$$
$$(x^2+1)(x^2+1)=0$$
$$x^2+1=0 \text{ or } x^2+1=0$$
$$x^2=-1 \text{ or } x^2=-1$$
$$x=\pm\sqrt{-1} \text{ or } x=\pm\sqrt{-1}$$
$$x=\pm i \text{ or } x=\pm i$$

Thus, the five zeros of *f* are 1, *i* with multiplicity 2, and $-i$ with multiplicity 2.

77. a. $P(x)=(x^3+5x^2+x)-(x^2+6)$
$$=x^3+4x^2+x-6$$

b. $P(0)=0^3+4(0)^2+0-6=-6$

If no items are produced, the manufacturer will lose $6000.

$R(0)=0^3+5(0)^2+0=0$

If no items are produced, the manufacturer will bring in $0 in revenue.

$C(0)=0^2+6=6$

It will cost the manufacturer $6000 if no items are produced.

c. The possible rational zeros of $P(x)=x^3+4x^2+x-6$ are of the form $\frac{p}{q}$ where *p* is a factor of –6 and *q* is a factor of 1.

Factors of –6: $\pm1,\ \pm2,\ \pm3,\ \pm6$

Factors of 1: ±1

Possible rational zeros: $\pm1,\ \pm2,\ \pm3,\ \pm6$

By Descartes' Rule of Signs, since $P(x)$ has one variation in signs, it has 1 positive real zero. Using synthetic division and checking the possibility of 1 being the zero, we get

$$\begin{array}{r|rrrr} 1 & 1 & 4 & 1 & -6 \\ & & 1 & 5 & 6 \\ \hline & 1 & 5 & 6 & 0 \end{array}$$

Since the remainder is 0, 1 is the positive zero of *P*. Thus, *P* can be written as follows:

$$P(x)=x^3+4x^2+x-6$$
$$=(x-1)(x^2+5x+6)$$
$$=(x-1)(x+2)(x+3)$$

Thus, the three zeros of *P* are 1, –2, and –3. Now since 1 is the only zero within the given domain of $0\le x\le 3$, the break even point is $x=1$ (when 100 items are produced).

d. $P(x)>0$ when $1<x\le 3$.

e. $P(x)<0$ when $0\le x<1$.

f. If more than 100 items are produced, then the manufacturer will make a profit. If less than 100 items are produced, then the manufacturer will lose money.

79. a. $P(x)=(x^3+x^2+x)-(3x^2+14x+10)$
$$=x^3-2x^2-13x-10$$

b. $P(0)=0^3-2(0)^2-13(0)-10=-10$

If no items are produced, the manufacturer will lose $10,000.

$R(0)=0^3+0^2+0=0$

If no items are produced, the manufacturer will bring in $0 in revenue.

$C(0)=3(0)^2+14(0)+10=10$

It will cost the manufacturer $10,000 if no items are produced.

c. The possible rational zeros of $P(x)=x^3-2x^2-13x-10$ are of the form $\frac{p}{q}$ where *p* is a factor of –10 and *q* is a factor of 1.

Factors of –10: $\pm1,\ \pm2,\ \pm5,\ \pm10$

Factors of 1: ±1

Possible rational zeros: $\pm1,\ \pm2,\ \pm5,\ \pm10$

By Descartes' Rule of Signs, since $P(x)$ has one variation in signs, it has 1 positive real zero. Using synthetic division and checking the possibility of 5 being the zero, we get

$$\begin{array}{r|rrrr} 5 & 1 & -2 & -13 & -10 \\ & & 5 & 15 & 10 \\ \hline & 1 & 3 & 2 & 0 \end{array}$$

Since the remainder is 0, 5 is the positive zero of *P*. Thus, *P* can be written as follows:

$$P(x) = x^3 - 2x^2 - 13x - 10$$
$$= (x-5)(x^2+3x+2)$$
$$= (x-5)(x+1)(x+2)$$

Thus, the three zeros of P are 5, –1, and –2. Now since 5 is the only zero within the given domain of $0 \le x \le 8$, the break even point is $x = 5$ (when 500 items are produced).

d. $P(x) > 0$ when $5 < x \le 8$.

e. $P(x) < 0$ when $0 \le x < 5$.

f. If more than 500 items are produced, then the manufacturer will make a profit. If less than 500 items are produced, then the manufacturer will lose money.

81. a. $P(x) = (2x^3 + x^2 + 2x) - (2x^2 + 7x + 2)$
$$= 2x^3 - x^2 - 5x - 2$$

b. $P(0) = 2(0)^3 - 0^2 - 5(0) - 2 = -2$

If no items are produced, the manufacturer will lose \$2000.

$R(0) = 2(0)^3 + 0^2 + 2(0) = 0$

If no items are produced, the manufacturer will bring in \$0 in revenue.

$C(0) = 2(0)^2 + 7(0) + 2 = 2$

It will cost the manufacturer \$2000 if no items are produced.

c. The possible rational zeros of $P(x) = 2x^3 - x^2 - 5x - 2$ are of the form $\frac{p}{q}$ where p is a factor of –2 and q is a factor of 2.

Factors of –2: $\pm 1, \pm 2$

Factors of 2: $\pm 1, \pm 2$

Possible rational zeros: $\pm 1, \pm 2, \pm\frac{1}{2}$

By Descartes' Rule of Signs, since $P(x)$ has one variation in signs, it has 1 positive real zero. Using synthetic division and checking the possibility of 2 being the zero, we get

2	2	–1	–5	–2
		4	6	2
	2	3	1	0

Since the remainder is 0, 2 is the positive zero of P. Thus, P can be written as follows:

$$P(x) = 2x^3 - x^2 - 5x - 2$$
$$= (x-2)(2x^2+3x+1)$$
$$= (x-2)(x+1)(2x+1)$$

Thus, the three zeros of P are 2, –1, and $-\frac{1}{2}$. Now since 2 is the only zero within the given domain of $0 \le x \le 5$, the break even point is $x = 2$ (when 200 items are produced).

d. $P(x) > 0$ when $2 < x \le 5$.

e. $P(x) < 0$ when $0 \le x < 2$.

f. If more than 200 items are produced, then the manufacturer will make a profit. If less than 200 items are produced, then the manufacturer will lose money.

83. $$\left(\frac{y^{-2/3}}{y^{-6}}\right)^{1/2} = \frac{\left(y^{-2/3}\right)^{1/2}}{\left(y^{-6}\right)^{1/2}} = \frac{y^{(-2/3)(1/2)}}{y^{(-6)(1/2)}} = \frac{y^{-1/3}}{y^{-3}} = y^{-1/3-(-3)} = y^{-1/3+3} = y^{8/3}$$

84. $3\sqrt{72x^3y} = 3\sqrt{36x^2 \cdot 2xy} = 3 \cdot 6x\sqrt{2xy} = 18x\sqrt{2xy}$

85. $$\frac{5}{\sqrt{5}-10} = \frac{5}{\sqrt{5}-10} \cdot \frac{\sqrt{5}+10}{\sqrt{5}+10} = \frac{5(\sqrt{5}+10)}{5-100} = \frac{5(\sqrt{5}+10)}{-95} = \frac{\sqrt{5}+10}{-19} \text{ or } -\frac{10+\sqrt{5}}{19}$$

86.
$$\sqrt{5x+6} = x$$
$$\left(\sqrt{5x+6}\right)^2 = (x)^2$$
$$5x+6 = x^2$$
$$x^2 - 5x - 6 = 0$$
$$(x+1)(x-6) = 0$$
$$x+1 = 0 \text{ or } x-6 = 0$$
$$x = -1 \text{ or } x = 6$$

Now, –1 is extraneous, so 6 is the only solution.

Mid-Chapter Test: 10.1-10.3

1. Continuous

2. Continuous

3. Discontinuous. . Points of discontinuity at $x=-3$ and $x=3$.

4. For $x \le -1$, graph $m(x)=2x$.
   $m(-3)=2(-3)=-6$
   $m(-1)=2(-1)=-2$
   For $x > -1$, graph $m(x)=3x$.
   $m(-1)=3(-1)=-3$
   $m(2)=3(2)=6$

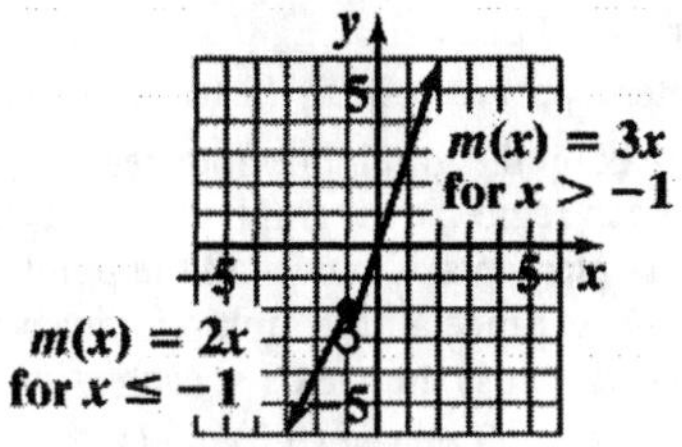

5. For $x < -3$, graph $t(x)=-x-3$.
   $t(-5)=-(-5)-3=5-3=2$
   $t(-3)=-(-3)-3=3-3=0$
   For $x \ge -3$, graph $t(x)=2$.
   $t(-3)=2$
   $t(5)=2$

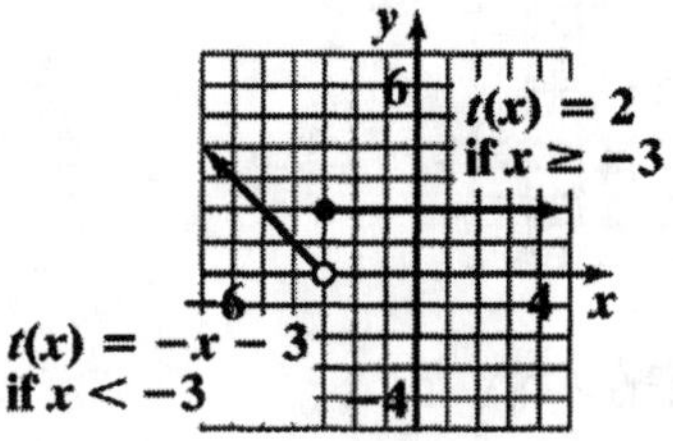

6. $n(x)=[\![2x]\!]$
   To find the closed endpoint value for $n(x)=0$, solve
   $0=2x$
   $x=0$
   To find the closed endpoint value for $n(x)=1$ solve
   $1=2x$
   $x=\frac{1}{2}$

To find the closed endpoint value for $n(x)=2$ solve
$2=2x$
$x=1$
From these points, we know the interval for each line segment spans $\frac{1}{2}$ of a unit.

Interval	$n(x)$
$\left[0,\frac{1}{2}\right)$	0
$\left[\frac{1}{2},1\right)$	1
$\left[1,\frac{3}{2}\right)$	2

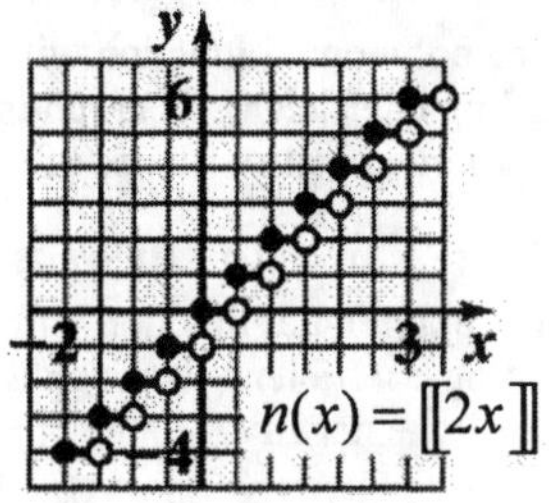

7. $q(x)=\left[\!\!\left[\frac{x+1}{3}\right]\!\!\right]$
   To find the closed endpoint value for $q(x)=0$, solve
   $0=\frac{x+1}{3}$
   $0=x+1$
   $x=-1$
   To find the closed endpoint value for $q(x)=1$ solve
   $1=\frac{x+1}{3}$
   $3=x+1$
   $x=2$
   To find the closed endpoint value for $q(x)=2$ solve
   $2=\frac{x+1}{3}$
   $6=x+1$
   $x=5$
   From these points, we know the interval for each line segment spans 3 units.

Interval	$f(x)$
$[-1,2)$	0
$[2,5)$	1
$[5,8)$	2

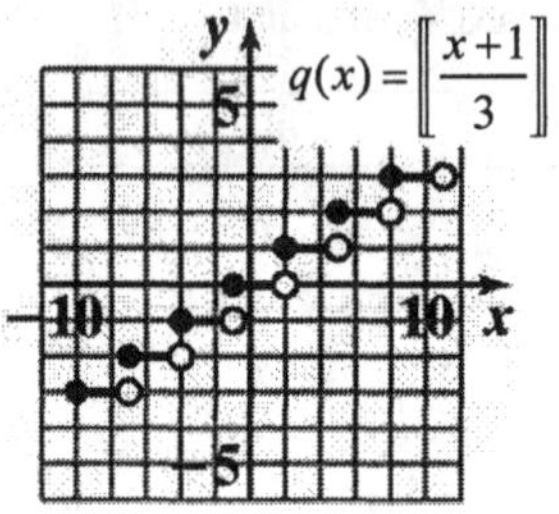

8. a. Since $f(x)=x^4-2x^3+10x^2+x-9$ is a fourth degree polynomial function with a positive leading coefficient, its graph is above the x-axis for large positive values of x.

b. Since $f(x)=x^4-2x^3+10x^2+x-9$ is a fourth degree polynomial function with a positive leading coefficient, its graph is above the x-axis for large negative values of x.

9. a. Since $f(x)=-3x^5+x^4+2x^3-7x+6$ is a fifth degree polynomial function with a negative leading coefficient, its graph is below the x-axis for large positive values of x.

b. Since $f(x)=-3x^5+x^4+2x^3-7x+6$ is a fifth degree polynomial function with a negative leading coefficient, its graph is above the x-axis for large negative values of x.

10. $f(x)=x^3+2x^2-24x$

$0=x^3+2x^2-24x$

$0=x(x^2+2x-24)$

$0=x(x+6)(x-4)$

$x=0$ or $x+6=0$ or $x-4=0$

$x=0$ or $x=-6$ or $x=4$

The zeros are −6, 0, and 4.

11. $h(x)=(x+5)(x+1)^2(x-3)^4$

$0=(x+5)(x+1)^2(x-3)^4$

$x+5=0$ or $x+1=0$ or $x-3=0$

$x=-5$ or $x=-1$ or $x=3$

The zeros are −5, −1 of multiplicity 2, 3 of multiplicity 4.

12. Since the degree of $g(x)=2x^3-x^2+7x-19$ is 3, the maximum number of turning points is $n-1=3-1=2$.

13. The degree of $k(x)=(x-5)(x-3)(x^2+1)(x^2-10)$ is the degree of the product of the x-terms of each factor: $x(x)(x^2)(x^2)=x^6$. Thus, the degree is 6 and the maximum number of turning points is $n-1=6-1=5$.

14. $f(x)=x(x+3)(x-1)$

Find the x-intercepts by solving the equation $f(x)=0$.

$0=x(x+3)(x-1)$

$x=0$ or $x=-3$ or $x=1$

The x-intercepts are $(-3,0)$, $(0,0)$, and $(1,0)$.

These break up the graph into four regions. Find a point in each region, then graph the points along with the x-intercepts. Connect all the points with a smooth curve. Since all the multiplicities are 1, the graph will cross the x-axis at each x-intercept.

Interval	x	$f(x)=x(x+3)(x-1)$
$(-\infty,-3)$	−4	$f(-4)=-20$
$(-3,0)$	−2	$f(-2)=6$
$(0,1)$	0.5	$f(0.5)=-0.875$
$(1,\infty)$	2	$f(2)=10$

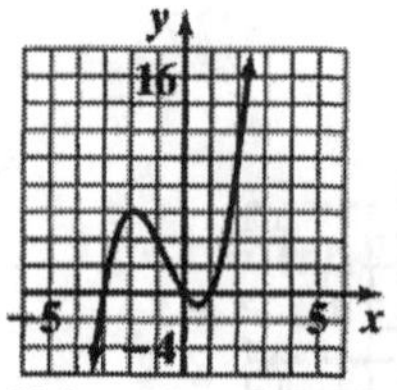

$f(x)=x(x+3)(x-1)$

15. $g(x)=x^2(x+4)(x-2)$

Find the x-intercepts by solving the equation $g(x)=0$.

$0=x^2(x+4)(x-2)$

$x=0$ or $x=-4$ or $x=2$

The x-intercepts are $(0,0)$, $(-4,0)$, and $(2,0)$.

They break the graph into four regions. Find a point in each region, then graph the points along with the x-intercepts. Connect all the points with a smooth curve. Since the zero at $x=0$ has multiplicity 2, the graph will touch the x-axis, but not cross, at $x=0$. The graph will cross the x-axis at the other x-intercepts since they have odd multiplicities.

Interval	x	$g(x)=x^2(x+4)(x-2)$
$(-\infty,-4)$	-5	$g(-5)=175$
$(-4,0)$	-2	$g(-2)=-32$
$(0,2)$	1	$g(1)=-5$
$(2,\infty)$	3	$g(3)=63$

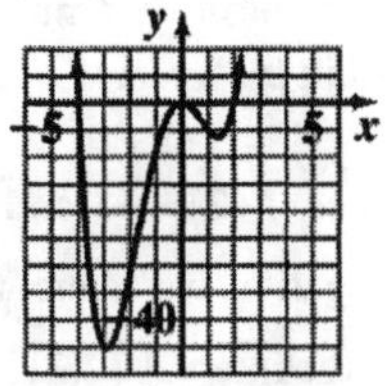

$g(x)=x^2(x+4)(x-2)$

16. Since $f(x)=5x^3+2x^2-6x-7$ is a polynomial function of degree 3, the maximum number of *x*-intercepts is 3. Since the degree is an odd number, the minimum number of *x*-intercepts is 1.

17. Since $t(x)=x^5+3x^4-x^3-12x^2+5x+11$ has two variations in signs, the number of positive zeros is either 2 or 0. Since $t(-x)=$

$$(-x)^5+3(-x)^4-(-x)^3-12(-x)^2+5(-x)+11$$
$$=-x^5+3x^4+x^3-12x^2-5x+11$$

has three variations in signs, the number of negative zeros is either 3 or 1.

18. The possible rational zeros of $q(x)=3x^4-8x^3+7x^2-x+4$ are of the form $\frac{p}{q}$ where *p* is a factor of 4 and *q* is a factor of 3.

Factors of 4: $\pm1,\ \pm2,\ \pm4$

Factors of 3: $\pm1,\ \pm3$

Possible rational zeros: $\pm1,\pm2,\pm4,\pm\frac{1}{3},\pm\frac{2}{3},\pm\frac{4}{3}$.

19. The possible rational zeros of $f(x)=x^4-x^3-19x^2-x-20$ are of the form $\frac{p}{q}$ where *p* is a factor of -20 and *q* is a factor of 1.

Factors of -20: $\pm1,\pm2,\pm4,\pm5,\pm10,\pm20$

Factors of 1: ±1

Possible rational zeros: $\pm1,\pm2,\pm4,\pm5,\pm10,\pm20$

By Descartes' Rule of Signs, since $f(x)$ has one variation in signs, it has 1 positive zero. Since $f(-x)=x^4+x^3-19x^2+x-20$ has three variations in signs, it has either 3 or 1 negative zero. Using synthetic division to find the positive zero and checking the possibility of 5 being that zero, we get

5	1	−1	−19	−1	−20
		5	20	5	20
	1	4	1	4	0

Since the remainder is 0, 5 is the one positive zero of *f*. Thus, *f* can be written as follows:

$$f(x)=x^4-x^3-19x^2-x-20$$
$$=(x-5)(x^3+4x^2+x+4)$$

We now need to determine any negative real zeros. To do this, focus on the polynomial $h(x)=x^3+4x^2+x+4$. Using synthetic division and checking the possibility of -4 being a zero, we get

−4	1	4	1	4
		−4	0	−4
	1	0	1	0

Since the remainder is 0, -4 is a negative real zero of *f*. Thus, *f* can be written as follows:

$$f(x)=x^4-x^3-19x^2-x-20$$
$$=(x-5)(x^3+4x^2+x+4)$$
$$=(x-5)(x+4)(x^2+1)$$

Now focus on the polynomial $g(x)=x^2+1$. The zeros of *g* are

$$x^2+1=0$$
$$x^2=-1$$
$$x=\pm\sqrt{-1}$$
$$x=\pm i$$

Thus, the four zeros of *f* are 5, -4, *i*, and $-i$.

20. The possible rational zeros of $g(x)=10x^3-29x^2-5x+6$ are of the form $\frac{p}{q}$ where *p* is a factor of 6 and *q* is a factor of 10.

Factors of 6: $\pm1,\ \pm2,\ \pm3,\ \pm6$

Factors of 10: $\pm1,\ \pm2,\ \pm5,\ \pm10$

Possible rational zeros:

$\pm1,\pm2,\pm3,\pm6,\pm\frac{1}{2},\pm\frac{3}{2},\pm\frac{1}{5},\pm\frac{2}{5},\pm\frac{3}{5},\pm\frac{6}{5},\pm\frac{1}{10},\pm\frac{3}{10}$.

By Descartes' Rule of Signs, since $g(x)$ has two variations in signs, it has either 0 or 2 positive zeros. Since $g(-x)=-10x^3-29x^2+5x+6$ has one variation in signs, $g(x)$ has 1 negative zero. Using synthetic division to find the 1 negative zero and checking the possibility of $-\frac{1}{2}$ being that zero, we get

$$\begin{array}{r|rrrr} -\frac{1}{2} & 10 & -29 & -5 & 6 \\ & & -5 & 17 & -6 \\ \hline & 10 & -34 & 12 & 0 \end{array}$$

Since the remainder is 0, $-\frac{1}{2}$ is the one positive zero of g. Writing g in factored form, we get

$$\begin{aligned} g(x) &= 10x^3 - 29x^2 - 5x + 6 \\ &= \left(x+\frac{1}{2}\right)\left(10x^2 - 34x + 12\right) \\ &= 2\left(x+\frac{1}{2}\right)\left(5x^2 - 17x + 6\right) \\ &= 2\left(x+\frac{1}{2}\right)(5x-2)(x-3) \end{aligned}$$

Thus, g has three rational zeros: $-\frac{1}{2}, \frac{2}{5}$, and 3.

Exercise Set 10.4

1. a. Yes. Using synthetic division with $x = 5$, we obtain

$$\begin{array}{r|rrrr} 5 & 1 & -2 & -5 & 2 \\ & & 5 & 15 & 50 \\ \hline & 1 & 3 & 10 & 52 \end{array}$$

By the Boundedness Theorem, since the bottom row in the synthetic division contains all positive values, 5 must be an upper bound for the real zeros of the function $f(x) = x^3 - 2x^2 - 5x + 2$.

b. Yes. Using synthetic division with $x = 4$, we obtain

$$\begin{array}{r|rrrr} 4 & 1 & -2 & -5 & 2 \\ & & 4 & 8 & 6 \\ \hline & 1 & 2 & 3 & 8 \end{array}$$

By the Boundedness Theorem, since the bottom row in the synthetic division contains all positive values, 4 must be an upper bound for the real zeros of *f*.

c. Cannot determine. Using synthetic division with $x = 3$, we obtain

$$\begin{array}{r|rrrr} 3 & 1 & -2 & -5 & 2 \\ & & 3 & 3 & -6 \\ \hline & 1 & 1 & -2 & -4 \end{array}$$

Since the bottom row in the synthetic division does not contain all positive values, we cannot tell if 3 is an upper bound for the real zeros of *f*.

3. a. No. Since $f(1) = -4$ and $f(2) = -8$ are both negative, we cannot tell if *f* contain a zero within the interval $(1, 2)$.

b. Yes. Since $f(-2) = -4$ and $f(-1) = 4$, one negative and one positive, *f* must contain a zero within the interval $(-2, -1)$.

5. Cannot determine. Since $f(2) = -5$ and $f(3) = -5$ are both negative, we cannot tell if *f* contains a zero within the interval $(2, 3)$.

7. Yes. Since $f(1) = -3$ and $f(2) = 4$, one negative and one positive, *f* must contain a zero within the interval $(1, 2)$.

9. Yes. Since $f(-3) = -8$ and $f(-2) = 4$, one negative and one positive, *f* must contain a zero within the interval $(-3, -2)$.

11. Yes. Since $f(-2) = -4$ and $f(-1) = 4$, one negative and one positive, *f* must contain a zero within the interval $(-2, -1)$.

13. Yes. Since $f(1) = -8$ and $f(3) = 60$, one negative and one positive, *f* must contain a zero within the interval $(1, 3)$.

15. Cannot determine. Since $f(-2) = -7$ and $f(-1) = -22$ are both negative, we cannot tell if *f* contains a zero within the interval $(-2, -1)$.

17. Yes. Since $f(-1) = 10$ and $f(0) = -12$, one positive and one negative, *f* must contain a zero within the interval $(-1, 0)$.

19. Cannot determine. Since $f(0) = -17$ and $f(1) = -13$ are both negative, we cannot tell if *f* contains a zero within the interval $(0, 1)$.

21. Yes. Using synthetic division with $x = -1$, we obtain

$-1\rfloor$	1	−6	3
		−1	7
	1	−7	10

With $x = 6$, we obtain

$6\rfloor$	1	−6	3
		6	0
	1	0	3

By the Boundedness Theorem, since the bottom row of numbers in the first synthetic division has alternating signs, −1 must be a lower bound for the real zeros of $f(x) = x^2 - 6x + 3$. Since the bottom row of numbers in the second synthetic division contains all positive values, 6 must be an upper bound for the real zeros of *f*.

23. Yes. Using synthetic division with $x = -2$, we obtain

$-2\rfloor$	5	−2	7
		−10	24
	5	−12	31

With $x = 2$, we obtain

$2\rfloor$	5	−2	7
		10	16
	5	8	23

By the Boundedness Theorem, since the bottom row of numbers in the first synthetic division has alternating signs, −2 must be a lower bound for the real zeros of $f(x) = 5x^2 - 2x + 7$. Since the bottom row of numbers in the second synthetic division contains all positive values, 2 must be an upper bound for the real zeros of *f*.

25. Cannot determine. Using synthetic division with $x = -2$, we obtain

$-2\rfloor$	1	2	−5	−6
		−2	0	10
	1	0	−5	4

With $x = 2$, we obtain

$4\rfloor$	1	2	−5	−6
		4	24	76
	1	6	19	70

By the Boundedness Theorem, since the bottom row of numbers in the second synthetic division contains all positive values, 4 must be an upper bound for the real zeros of $g(x) = x^3 + 2x^2 - 5x - 6$. However, since the bottom row of numbers in the first synthetic division does not have alternating signs, we cannot tell if −2 is a lower bound for the real zeros of *g*.

27. Yes. Using synthetic division with $x = -5$, we obtain

$-5\rfloor$	1	−5	9	−53
		−5	50	−295
	1	−10	59	−348

With $x = 6$, we obtain

$6\rfloor$	1	−5	9	−53
		6	6	80
	1	1	15	27

By the Boundedness Theorem, since the bottom row of numbers in the first synthetic division has alternating signs, −5 must be a lower bound for the real zeros of $g(x) = x^3 - 5x^2 + 9x - 53$. Since the bottom row of numbers in the second synthetic division contains all positive values, 6 must be an upper bound for the real zeros of *g*.

29. Cannot determine. Using synthetic division with $x = -2$, we obtain

$-2\rfloor$	1	1	−7	−13	−6
		−2	2	10	6
	1	−1	−5	−3	0

With $x = 2$, we obtain

$2\rfloor$	1	1	−7	−13	−6
		2	6	−2	−30
	1	3	−1	−15	−36

Since the bottom row of numbers in the first synthetic division does not have alternating signs, we cannot tell if −2 is a lower bound for the real zeros of $h(x) = x^4 + x^3 - 7x^2 - 13x - 6$. Likewise, since the bottom row of numbers in the second synthetic division does not contain all positive values, we cannot tell if 2 is an upper bound for the real zeros of *h*.

31. Yes. Using synthetic division with $x=-1$, we obtain

$$\begin{array}{r|rrrrr} -1 & 6 & -1 & 5 & -1 & 10 \\ & & -6 & 7 & -12 & 13 \\ \hline & 6 & -7 & 12 & -13 & 23 \end{array}$$

With $x=2$, we obtain

$$\begin{array}{r|rrrrr} 2 & 6 & -1 & 5 & -1 & 10 \\ & & 12 & 22 & 54 & 106 \\ \hline & 6 & 11 & 27 & 53 & 116 \end{array}$$

By the Boundedness Theorem, since the bottom row of numbers in the first synthetic division has alternating signs, –1 must be a lower bound for the real zeros of $h(x)=6x^4-x^3+5x^2-x+10$. Since the bottom row of numbers in the second synthetic division contains all positive values, 2 must be an upper bound for the real zeros of *h*.

33. Cannot determine. Using synthetic division with $x=-5$, we obtain

$$\begin{array}{r|rrrrr} -5 & 1 & 5 & -12 & -13 & 2 \\ & & -5 & 0 & 60 & -235 \\ \hline & 1 & 0 & -12 & 47 & -233 \end{array}$$

With $x=3$, we obtain

$$\begin{array}{r|rrrrr} 3 & 1 & 5 & -12 & -13 & 2 \\ & & 3 & 24 & 36 & 69 \\ \hline & 1 & 8 & 12 & 23 & 71 \end{array}$$

By the Boundedness Theorem, since the bottom row of numbers in the second synthetic division contains all positive values, 3 must be an upper bound for the real zeros of $h(x)=x^4+5x^3-12x^2-13x+2$. However, since the bottom row of numbers in the first synthetic division does not have alternating signs, we cannot tell if –5 is a lower bound for the real zeros of *h*.

35. $f(x)=x^2-x-20$

To find the *y*-intercept, substitute 0 for *x*: $f(0)=-20$. The *y*-intercept is $(0,-20)$.

$f(x)=x^2-x-20=(x+4)(x-5)$, so the two zeros are –4 and 5. These two *x*-intercepts divide the graph into three regions. We will find a point in each region:

Interval	*x*	Value of *f*(*x*)
$(-\infty,-4)$	–5	$f(-5)=10$
$(-4,5)$	0	$f(0)=-20$
$(5,\infty)$	6	$f(6)=10$

Graph the intercepts and points from each region. Then connect them with a smooth curve.

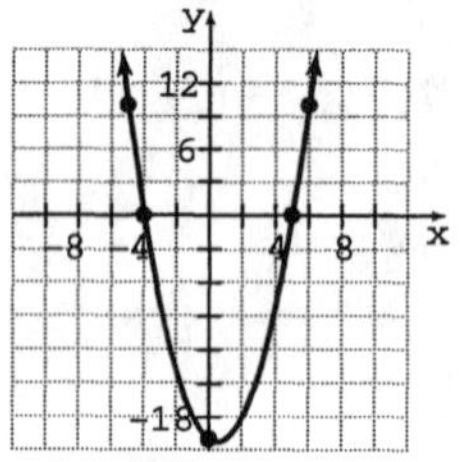

f(x)=x²-x-20

37. $g(x)=x^3+2x^2-5x-6$

To find the *y*-intercept, substitute 0 for *x*: $g(0)=-6$. The *y*-intercept is $(0,-6)$.

Since the coefficients of $g(x)$ have 1 variation in signs, it has 1 positive real zero. Since the coefficients of $g(-x)=-x^3+2x^2+5x-6$ have 2 variations in signs, *g* has either 2 or 0 negative real zeros. The possible rational zeros of *g* are of the form $\frac{p}{q}$ where *p* is a factor of –6 and *q* is a factor of 1.

Factors of –6: $\pm1, \pm2, \pm3, \pm6$

Factors of 1: ±1

Possible rational zeros: $\pm1, \pm2, \pm3, \pm6$.

Using synthetic division to find the 1 positive zero and checking the possibility of 2 being that zero, we get

$$\begin{array}{r|rrrr} 2 & 1 & 2 & -5 & -6 \\ & & 2 & 8 & 6 \\ \hline & 1 & 4 & 3 & 0 \end{array}$$

Since the remainder is 0, 2 is a zero of *g* and $x-2$ is a factor of *g*. Writing *g* in factored form, we get

$$\begin{aligned} g(x) &= x^3+2x^2-5x-6 \\ &= (x-2)(x^2+4x+3) \\ &= (x-2)(x+1)(x+3) \end{aligned}$$

Thus, *g* has three rational zeros: 2, –1, and –3. These three *x*-intercepts divide the graph into four regions. We will find a point in each region:

Interval	*x*	Value of *g*(*x*)
$(-\infty,-3)$	–4	$g(-4)=-18$
$(-3,-1)$	–2	$g(-2)=4$
$(-1,1)$	1	$g(1)=-8$
$(1,\infty)$	3	$g(3)=24$

Graph the intercepts and points from each region. Then connect them with a smooth curve.

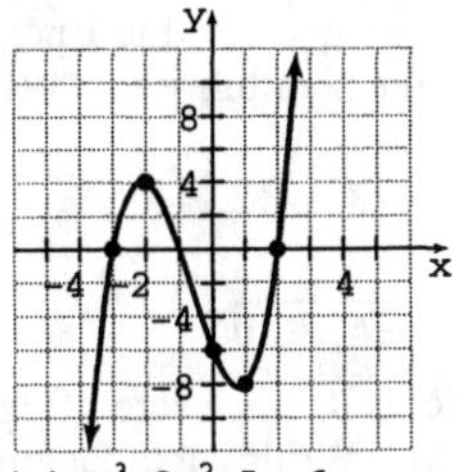

39. $g(x) = 2x^3 + x^2 - 8x - 4$

To find the y-intercept, substitute 0 for x: $g(0) = -4$. The y-intercept is $(0,-4)$.

Since the coefficients of $g(x)$ have 1 variation in signs, it has 1 positive real zero. Since the coefficients of $g(-x) = -2x^3 + x^2 + 8x - 4$ have 2 variations in signs, g has either 2 or 0 negative real zeros. The possible rational zeros of g are of the form $\frac{p}{q}$ where p is a factor of -4 and q is a factor of 2.

Factors of -4: $\pm1,\ \pm2,\ \pm4$

Factors of 2: $\pm1,\ \pm2$

Possible rational zeros: $\pm1,\ \pm2,\ \pm4,\ \pm\frac{1}{2}$.

Using synthetic division to find the 1 positive zero and checking the possibility of 2 being that zero, we get

$$\begin{array}{r|rrrr} 2 & 2 & 1 & -8 & -4 \\ & & 4 & 10 & 4 \\ \hline & 2 & 5 & 2 & 0 \end{array}$$

Since the remainder is 0, 2 is a zero of g and $x-2$ is a factor of g. Writing g in factored form, we get

$$\begin{aligned} g(x) &= 2x^3 + x^2 - 8x - 4 \\ &= (x-2)(2x^2+5x+2) \\ &= (x-2)(2x+1)(x+2) \end{aligned}$$

Thus, g has three rational zeros: 2, $-\frac{1}{2}$, and -2.

These three x-intercepts divide the graph into four regions. We will find a point in each region:

Interval	x	Value of $g(x)$
$(-\infty,-2)$	-3	$g(-3) = -25$
$\left(-2,-\frac{1}{2}\right)$	-1	$g(-1) = 3$
$\left(-\frac{1}{2},2\right)$	1	$g(1) = -9$
$(2,\infty)$	3	$g(3) = 35$

Graph the intercepts and points from each region. Then connect them with a smooth curve.

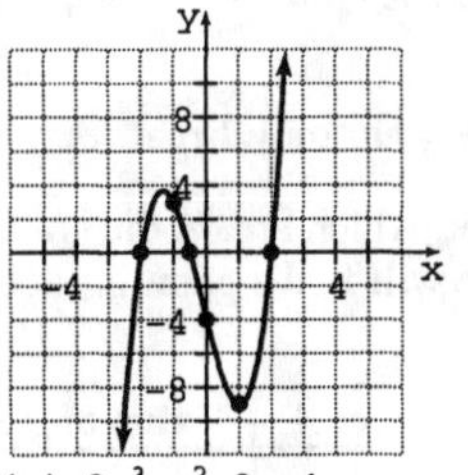

41. $g(x) = x^3 - x^2 - x - 2$

To find the y-intercept, substitute 0 for x: $g(0) = -2$. The y-intercept is $(0,-2)$.

Since the coefficients of $g(x)$ have 1 variation in signs, it has 1 positive real zeros. Since the coefficients of $g(-x) = -x^3 - x^2 + x - 2$ have two variations in signs, g has either 2 or 0 negative real zeros. The possible rational zeros of g are of the form $\frac{p}{q}$ where p is a factor of -2 and q is a factor of 1.

Factors of -2: $\pm1,\ \pm2$

Factors of 1: ±1

Possible rational zeros: $\pm1,\ \pm2$.

Using synthetic division to find the 1 positive zero and checking the possibility of 2 being that zero, we get

$$\begin{array}{r|rrrr} 2 & 1 & -1 & -1 & -2 \\ & & 2 & 2 & 2 \\ \hline & 1 & 1 & 1 & 0 \end{array}$$

Since the remainder is 0, 2 is a zero of g and $x-2$ is a factor of g. Writing g in factored form, we get

$$\begin{aligned} g(x) &= x^3 - x^2 - x - 2 \\ &= (x-2)(x^2+x+1) \end{aligned}$$

$$x-2=0 \qquad x^2+x+1=0$$
$$x=2 \qquad x=\frac{-1\pm\sqrt{1^2-4(1)(1)}}{2(1)}$$
$$=\frac{-1\pm\sqrt{-3}}{2}$$
$$=\frac{-1\pm i\sqrt{3}}{2}$$

Thus, the three zeros are 2, $\frac{-1-i\sqrt{3}}{2}$, and $\frac{-1+i\sqrt{3}}{2}$. Since g has only 1 real zero, it has only 1 x-intercept, which breaks the graph into two regions. We will find a point in each region:

Interval	x	Value of $g(x)$
$(-\infty,2)$	-1	$g(-1)=-3$
$(2,\infty)$	3	$g(3)=13$

Since we currently only have a few points to plot, it is wise to find a couple of other points: $g(1)=-3$ and $g(-2)=-12$. Graph the intercepts, the points from each region, and the additional points. Then connect them with a smooth curve.

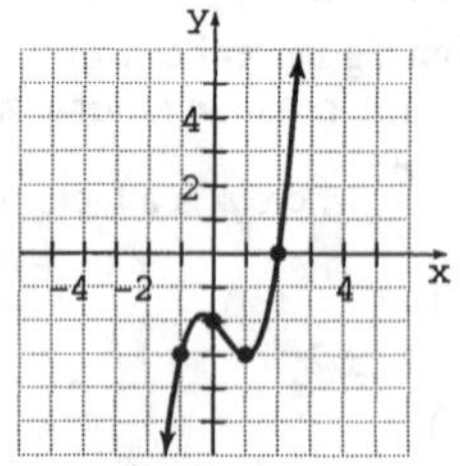

$g(x)=x^3-x^2-x-2$

43. $g(x)=x^3+x^2-8x-12$

To find the y-intercept, substitute 0 for x: $g(0)=-12$. The y-intercept is $(0,-12)$.

Since the coefficients of $g(x)$ have one variation in signs, it has 1 positive real zeros. Since the coefficients of $g(-x)=-x^3+x^2+8x-12$ have two variations in signs, g has either 2 or 0 negative real zeros. The possible rational zeros of g are of the form $\frac{p}{q}$ where p is a factor of -12 and q is a factor of 1.

Factors of -12:
$\pm1, \pm2, \pm3, \pm4, \pm6, \pm12$
Factors of 1: ±1

Possible rational zeros:
$\pm1, \pm2, \pm3, \pm4, \pm6, \pm12$.

Using synthetic division to find the 1 positive zero and checking the possibility of 3 being that zero, we get

$$\begin{array}{r|rrrr} 3 & 1 & 1 & -8 & -12 \\ & & 3 & 12 & 12 \\ \hline & 1 & 4 & 4 & 0 \end{array}$$

Since the remainder is 0, 3 is a zero of g and $x-3$ is a factor of g. Writing g in factored form, we get

$$g(x)=x^3+x^2-8x-12$$
$$=(x-3)(x^2+4x+4)$$
$$=(x-3)(x+2)(x+2)$$

Thus, the three zeros are 3 and -2 with multiplicity 2. The two distinct x-intercepts divide the graph into three regions. We will find a point in each region:

Interval	x	Value of $g(x)$
$(-\infty,-2)$	-3	$g(-3)=-6$
$(-2,3)$	-1	$g(-1)=-4$
$(3,\infty)$	4	$g(4)=36$

Graph the intercepts and points from each region. Then connect them with a smooth curve.

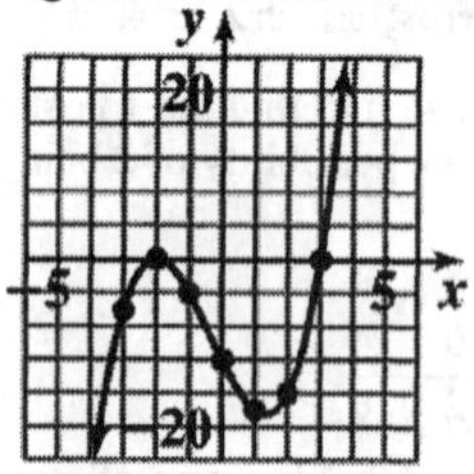

$g(x)=x^3+x^2-8x-12$

45. $g(x)=x^3-10x^2+17x+28$

To find the y-intercept, substitute 0 for x: $g(0)=28$. The y-intercept is $(0,28)$.

Since the coefficients of $g(x)$ have two variations in signs, it has either 2 or 0 positive real zeros. Since the coefficients of $g(-x)=-x^3-10x^2-17x+28$ have one variation in signs, g has 1 negative real zero. The possible rational zeros of g are of the form $\frac{p}{q}$ where p is a factor of 28 and q is a factor of 1.

Factors of 28:
$\pm1, \pm2, \pm4, \pm7, \pm14, \pm28$

Factors of 1: ± 1
Possible rational zeros:
$\pm 1, \pm 2, \pm 4, \pm 7, \pm 14, \pm 28$.
Using synthetic division to find the 1 negative zero and checking the possibility of -1 being that zero, we get

$$\begin{array}{r|rrrr} -1 & 1 & -10 & 17 & 28 \\ & & -1 & 11 & -28 \\ \hline & 1 & -11 & 28 & 0 \end{array}$$

Since the remainder is 0, -1 is a zero of g and $x+1$ is a factor of g. Writing g in factored form, we get

$$\begin{aligned} g(x) &= x^3 - 10x^2 + 17x + 28 \\ &= (x+1)(x^2 - 11x + 28) \\ &= (x+1)(x-4)(x-7) \end{aligned}$$

Thus, the three zeros are -1, 4, and 7. The three x-intercepts divide the graph into four regions. We will find a point in each region:

Interval	x	Value of $g(x)$
$(-\infty,-1)$	-2	$g(-2) = -54$
$(-1,4)$	2	$g(2) = 30$
$(4,7)$	5	$g(5) = -12$
$(7,\infty)$	8	$g(8) = 36$

Graph the intercepts and points from each region. Then connect them with a smooth curve.

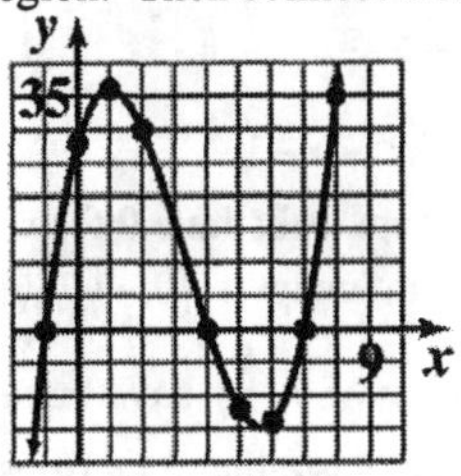

$g(x) = x^3 - 10x^2 + 17x + 28$

47. $h(x) = x^4 + 2x^3 - 16x^2 - 2x + 15$
To find the y-intercept, substitute 0 for x: $h(0) = 15$. The y-intercept is $(0,15)$.

Since the coefficients of $h(x)$ have two variations in signs, it has either 2 or 0 positive real zeros. Since the coefficients of $h(-x) = x^4 - 2x^3 - 16x^2 + 2x + 15$ have two variation in signs, h has either 2 or 0 negative real zeros. The possible rational zeros of h are of the form $\frac{p}{q}$ where p is a factor of 15 and q is a factor of 1.

Factors of 15: $\pm 1, \pm 3, \pm 5, \pm 15$
Factors of 1: ± 1
Possible rational zeros: $\pm 1, \pm 3, \pm 5, \pm 15$
Using synthetic division to find a real zero and checking the possibility of 1 being that zero, we get

$$\begin{array}{r|rrrrr} 1 & 1 & 2 & -16 & -2 & 15 \\ & & 1 & 3 & -13 & -15 \\ \hline & 1 & 3 & -13 & -15 & 0 \end{array}$$

Since the remainder is 0, 1 is a positive zero of h and $x-1$ is a factor of h.. Thus, h can be written as follows:

$$\begin{aligned} h(x) &= x^4 + 2x^3 - 16x^2 - 2x + 15 \\ &= (x-1)(x^3 + 3x^2 - 13x - 15) \end{aligned}$$

Now, we know that h has 1 as a positive real zero. Thus, since the number of positive real zeros is either 0 or 2, h must have another positive real zero. To find it, focus on the polynomial $r(x) = x^3 + 3x^2 - 13x - 15$. Using synthetic division and checking the possibility of 3 being a second positive zero, we get

$$\begin{array}{r|rrrr} 3 & 1 & 3 & -13 & -15 \\ & & 3 & 18 & 15 \\ \hline & 1 & 6 & 5 & 0 \end{array}$$

Since the remainder is 0, 3 is a real zero of h and $x-3$ is a factor of h.. Writing h in factored form, we get

$$\begin{aligned} h(x) &= x^4 + 2x^3 - 16x^2 - 2x + 15 \\ &= (x-1)(x^3 + 3x^2 - 13x - 15) \\ &= (x-1)(x-3)(x^2 + 6x + 5) \\ &= (x-1)(x-3)(x+1)(x+5) \end{aligned}$$

Thus, the four zeros are 1, 3, -1 and -5. These four x-intercepts divide the graph into five regions. We will find a point in each region:

Interval	x	Value of $h(x)$
$(-\infty,-5)$	-6	$h(-6) = 315$
$(-5,-1)$	-3	$h(-3) = -96$
$(-1,1)$	0	$h(0) = 15$
$(1,3)$	2	$h(2) = -21$
$(3,\infty)$	4	$h(4) = 135$

Graph the intercepts and points from each region. Then connect them with a smooth curve.

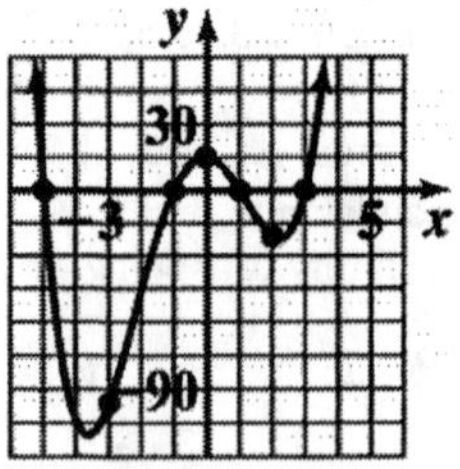

$h(x)=x^4+2x^3-16x^2-2x+15$

49. $h(x)=x^4+x^3-7x^2-13x-6$

To find the y-intercept, substitute 0 for x:
$h(0)=-6$. The y-intercept is $(0,-6)$.

Since the coefficients of $h(x)$ have one variation in signs, it has 1 positive real zero. Since the coefficients of $h(-x)=x^4-x^3-7x^2+13x-6$ have three variations in signs, h has either 3 or 1 negative real zeros. The possible rational zeros of h are of the form $\frac{p}{q}$ where p is a factor of –6 and q is a factor of 1.

Factors of –6: ±1, ±2, ±3, ±6

Factors of 1: ±1

Possible rational zeros: ±1, ±2, ±3, ±6

Using synthetic division to find the positive real zero and checking the possibility of 3 being that zero, we get

3⌋	1	1	–7	–13	–6
		3	12	15	6
	1	4	5	2	0

Since the remainder is 0, 3 is the one positive zero of h and $x-3$ is a factor of h.. Thus, h can be written as follows:

$$h(x)=x^4+x^3-7x^2-13x-6$$
$$=(x-3)(x^3+4x^2+5x+2)$$

We now need to determine any negative real zeros. To do this, focus on the polynomial. $r(x)=x^3+4x^2+5x+2$. Using synthetic division and checking the possibility of –1 being a zero, we get

–1⌋	1	4	5	2
		–1	–3	–2
	1	3	2	0

Since the remainder is 0, –1 is a real zero of h and $x+1$ is a factor of h.. Writing h in factored form, we get

$$h(x)=x^4+x^3-7x^2-13x-6$$
$$=(x-3)(x^3+4x^2+5x+2)$$
$$=(x-3)(x+1)(x^2+3x+2)$$
$$=(x-3)(x+1)(x+1)(x+2)$$

Thus, the four zeros are 3, –1 with multiplicity 2, and –2. The three distinct x-intercepts divide the graph into four regions. We will find a point in each region:

Interval	x	Value of $h(x)$
$(-\infty,-2)$	–3	$h(-3)=24$
$(-2,-1)$	–1.5	$h(-1.5)=-0.5625$
$(-1,3)$	2	$h(2)=-36$
$(3,\infty)$	4	$h(4)=150$

Graph the intercepts and points from each region. Then connect them with a smooth curve.

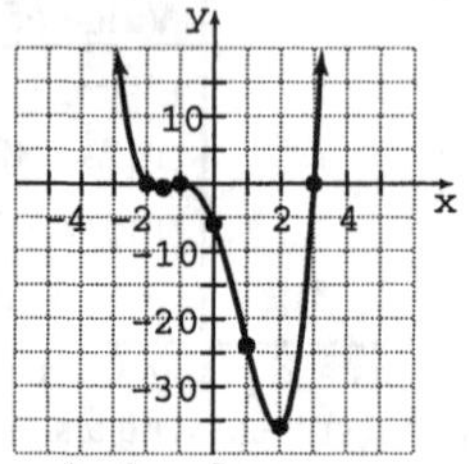

h(x) =x⁴+x³-7x²-13x-6

51. $h(x)=6x^4-x^3+5x^2-x-1$

To find the y-intercept, substitute 0 for x:
$h(0)=-1$. The y-intercept is $(0,-1)$.

Since the coefficients of $h(x)$ have three variations in signs, it has either 3 or 1 positive real zeros. Since the coefficients of $h(-x)=6x^4+x^3+5x^2+x-1$ have one variation in signs, h has 1 negative real zero. The possible rational zeros of h are of the form $\frac{p}{q}$ where p is a factor of –1 and q is a factor of 6.

Factors of –1: ±1

Factors of 6: ±1, ±2, ±3, ±6

Possible rational zeros: $\pm1,\ \pm\frac{1}{2},\ \pm\frac{1}{3},\ \pm\frac{1}{6}$

Using synthetic division to find the one negative real zero and checking the possibility of $-\frac{1}{3}$ being that zero, we get

$$\begin{array}{r|rrrrr} -\frac{1}{3} & 6 & -1 & 5 & -1 & -1 \\ & & -2 & 1 & -2 & 1 \\ \hline & 6 & -3 & 6 & -3 & 0 \end{array}$$

Since the remainder is 0, $-\frac{1}{3}$ is a zero of h and $x+\frac{1}{3}$ is a factor of h.. Thus, h can be written as follows:

$$h(x) = 6x^4 - x^3 + 5x^2 - x - 1$$
$$= \left(x+\frac{1}{3}\right)\left(6x^3 - 3x^2 + 6x - 3\right)$$

We now need to determine any positive real zeros. To do this, focus on the polynomial. $r(x) = 6x^3 - 3x^2 + 6x - 3$. Using synthetic division and checking the possibility of $\frac{1}{2}$ being a zero, we get

$$\begin{array}{r|rrrr} \frac{1}{2} & 6 & -3 & 6 & -3 \\ & & 3 & 0 & 3 \\ \hline & 6 & 0 & 6 & 0 \end{array}$$

Since the remainder is 0, $\frac{1}{2}$ is a real zero of h and $x-\frac{1}{2}$ is a factor of h.. Writing h in factored form, we get

$$h(x) = 6x^4 - x^3 + 5x^2 - x - 1$$
$$= \left(x+\frac{1}{3}\right)\left(6x^3 - 3x^2 + 6x - 3\right)$$
$$= \left(x+\frac{1}{3}\right)\left(x-\frac{1}{2}\right)\left(6x^2 + 6\right)$$
$$= 6\left(x+\frac{1}{3}\right)\left(x-\frac{1}{2}\right)\left(x^2 + 1\right)$$

$$x+\frac{1}{3} = 0 \quad \text{or} \quad x-\frac{1}{2} = 0 \quad \text{or} \quad x^2 + 1 = 0$$
$$x = -\frac{1}{3} \qquad\qquad x = \frac{1}{2} \qquad\qquad x^2 = -1$$
$$x = \pm\sqrt{-1}$$
$$x = \pm i$$

Thus, the four zeros are $-\frac{1}{3}$, $\frac{1}{2}$, $-i$, and i. Since h only has two real zeros, it has only two x-intercepts which divide the graph into three regions. We will find a point in each region:

Interval	x	Value of $h(x)$
$\left(-\infty, -\frac{1}{3}\right)$	-1	$h(-1) = 12$
$\left(-\frac{1}{3}, \frac{1}{2}\right)$	0	$h(0) = -1$
$\left(\frac{1}{2}, \infty\right)$	1	$h(1) = 8$

Graph the intercepts and points from each region. Then connect them with a smooth curve.

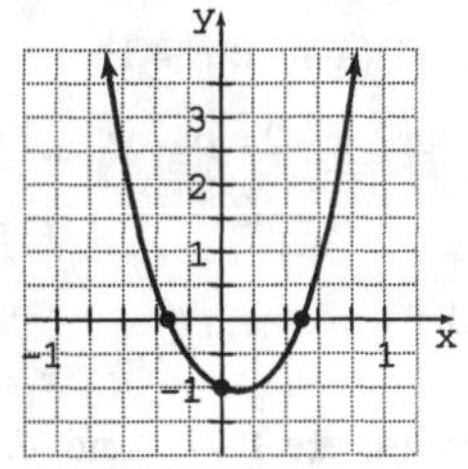

$h(x) = 6x^4 - x^3 + 5x^2 - x - 1$

53. $h(x) = x^4 - x^3 - 5x^2 - x - 6$

To find the y-intercept, substitute 0 for x: $h(0) = -6$. The y-intercept is $(0, -6)$.

Since the coefficients of $h(x)$ have one variation in signs, it has 1 positive real zero. Since the coefficients of $h(-x) = x^4 + x^3 - 5x^2 + x - 6$ have three variations in signs, h has either 3 or 1 negative real zero. The possible rational zeros of h are of the form $\frac{p}{q}$ where p is a factor of -6 and q is a factor of 1.

Factors of -6: ± 1, ± 2, ± 3, ± 6

Factors of 1: ± 1

Possible rational zeros: ± 1, ± 2, ± 3, ± 6

Using synthetic division to find the one positive real zero and checking the possibility of 3 being that zero, we get

$$\begin{array}{r|rrrrr} 3 & 1 & -1 & -5 & -1 & -6 \\ & & 3 & 6 & 3 & 6 \\ \hline & 1 & 2 & 1 & 2 & 0 \end{array}$$

Since the remainder is 0, 3 is a zero of h and $x-3$ is a factor of h.. Thus, h can be written as follows:

$$h(x) = x^4 - x^3 - 5x^2 - x - 6$$
$$= (x-3)\left(x^3 + 2x^2 + x + 2\right)$$

We now need to determine any negative real zeros. To do this, focus on the polynomial. $r(x) = x^3 + 2x^2 + x + 2$. Using synthetic division and checking the possibility of -2 being a zero, we get

$$\begin{array}{r|rrrr} -2 & 1 & 2 & 1 & 2 \\ & & -2 & 0 & -2 \\ \hline & 1 & 0 & 1 & 0 \end{array}$$

Since the remainder is 0, -2 is a real zero of h and $x+2$ is a factor of h.. Writing h in factored form, we get

$$\begin{aligned} h(x) &= x^4 - x^3 - 5x^2 - x - 6 \\ &= (x-3)(x^3+2x^2+x+2) \\ &= (x-3)(x+2)(x^2+1) \end{aligned}$$

$x-3=0$ or $x+2=0$ or $x^2+1=0$

$x=3$ $\quad x=-2$ $\quad x^2=-1$

$x=\pm\sqrt{-1}$

$x=\pm i$

Thus, the four zeros are 3, –2, *i* and –*i*. Since *h* only has two real zeros, it has only two *x*-intercepts which divide the graph into three regions. We will find a point in each region:

Interval	x	Value of $h(x)$
$(-\infty,-2)$	–3	$h(-3)=60$
$(-2,3)$	2	$h(2)=-20$
$(3,\infty)$	4	$h(4)=102$

Since we currently only have a few points to plot, it is wise to find a couple of extra points: $h(-1)=-8$ and $h(1)=-12$. Graph the intercepts, the points from each region, and the additional points. Then connect them with a smooth curve.

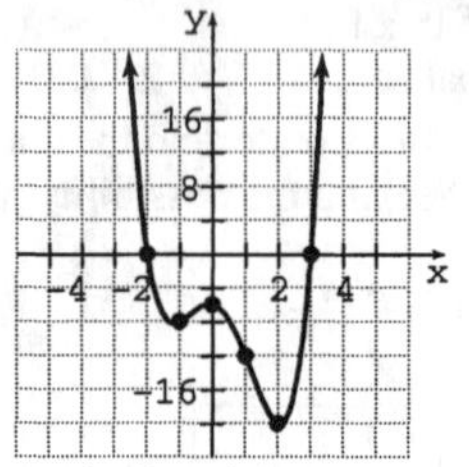

$h(x)=x^4-x^3-5x^2-x-6$

55. Since the coefficients of $f(x)=x^3+x^2+x-4$ have one variation in signs, it has 1 positive real zeros. Since the coefficients of $f(-x)=-x^3+x^2-x-4$ have two variations in signs, *f* has either 2 or 0 negative real zeros. Since the possible rational zeros of *f* are of the form $\frac{p}{q}$ where *p* is a factor of –4 and *q* is a factor of 1, the possible rational zeros are ±1, ±2, and ±4. Using synthetic division to find the one positive real zero and checking the possibility of 1 being that zero, we get

$$\begin{array}{r|rrrr} 1 & 1 & 1 & 1 & -4 \\ & & 1 & 2 & 3 \\ \hline & 1 & 2 & 3 & -1 \end{array}$$

Since the remainder is not 0, 1 is *not* a zero of *f*. Still, we know that $f(1)=-1$ which is below the *x*-axis.

Using synthetic division to check the possibility of 2 being the one positive zero, we get

$$\begin{array}{r|rrrr} 2 & 1 & 1 & 1 & -4 \\ & & 2 & 6 & 14 \\ \hline & 1 & 3 & 7 & 10 \end{array}$$

Since the remainder is not 0, 2 is *not* a zero of *f*. Still, we know that $f(2)=10$ which is above the *x*-axis.

Now since the graph of *f* is below the *x*-axis at $x=1$ and above the *x*-axis at $x=2$, an irrational zero exists in the interval $(1,2)$.

x	$f(x)$	Above or below *x*-axis	Interval containing the zero
1.5	3.125	above	$(0,1.5)$
1.2	0.368	above	$(0,1.2)$
1.1	–0.359	below	$(1.1,1.2)$
1.15	–0.007	below	$(1.15,1.2)$
1.16	0.067	above	$(1.15,1.16)$
1.155	0.030	above	$(1.15,1.155)$

The one positive real zero is $x\approx1.15$.

Using synthetic division to find a negative real zero and checking the possibility of –1 being that zero, we get

$$\begin{array}{r|rrrr} -1 & 1 & 1 & 1 & -4 \\ & & -1 & 0 & -1 \\ \hline & 1 & 0 & 1 & -5 \end{array}$$

Since the remainder is not 0, –1 is *not* a zero of *f*. Still, we know that $f(-1)=-5$ which is below the *x*-axis. Further, by the Boundedness Theorem, since the numbers in the bottom row of the synthetic division are alternatively positive and negative, –1 is a lower bound for all of the zeros of *f*. This means that if *f* has negative zeros, they must be in the interval $(-1,0)$. Using a graphing calculator to graph f below, we see that f has no negative real zeros.

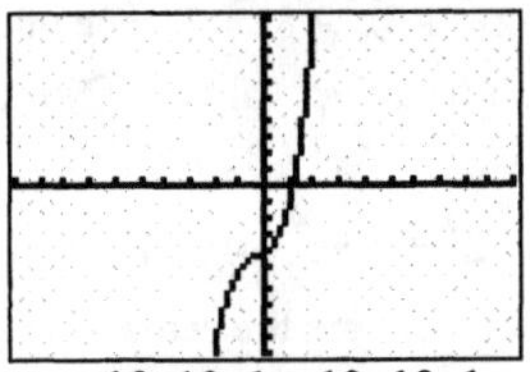

−10, 10, 1, −10, 10, 1

In summary, the only real zero of f is $x \approx 1.15$.

57. Since the coefficients of $h(x) = 2x^4 - 3x^3 + x^2 - 12$ have three variations in signs, it has either 3 or 1 positive real zeros. Since the coefficients of $h(-x) = 2x^4 + 3x^3 + x^2 - 12$ have one variation in signs, h has 1 negative real zero. Since the possible rational zeros of h are of the form $\frac{p}{q}$ where p is a factor of −12 and q is a factor of 2, the possible rational zeros are ±1, ±2, ±3, ±4, ±6, ±12, $\pm\frac{1}{2}$, and $\pm\frac{3}{2}$.

Using synthetic division to find a positive real zero and checking the possibility of 1 being that zero, we get

1	2	−3	1	0	−12
		2	−1	0	0
	2	−1	0	0	−12

Since the remainder is not 0, 1 is *not* a zero of h. Still, we know that $h(1) = -12$ which is below the x-axis.

Using synthetic division to check the possibility of 2 being a zero, we get

2	2	−3	1	0	−12
		5	2	6	12
	2	1	3	6	0

Since the remainder is 0, 2 is a zero of h. Further, h can be written as follows:

$$\begin{aligned} h(x) &= 2x^4 - 3x^3 + x^2 - 12 \\ &= (x-2)(2x^3 + x^2 + 3x + 6) \end{aligned}$$

The other zeros of h are the zeros of $r(x) = 2x^3 + x^2 + 3x + 6$. Now since the coefficients of r have no variations in signs, r has no positive zeros. Thus, 2 is the only positive zero of h.

Now, the one negative real zero of h will also be a zero of r. Checking the possibility of −1 being that zero, we get

−1	2	1	3	6
		−2	1	−4
	2	−1	4	2

Since the remainder is not 0, −1 is *not* a zero of r. Still, we know that $r(-1) = 2$ which is above the x-axis.

Checking the possibility of −2 being the one negative zero, we get

−2	2	1	3	6
		−4	6	−18
	2	−3	9	−12

Since the remainder is not 0, −2 is *not* a zero of r. Still, we know that $r(-2) = -12$ which is above the x-axis.

Now since the graph of r is above the x-axis at $x = -1$ and below the x-axis at $x = -2$, an irrational zero exists in the interval $(-2,-1)$.

x	$r(x)$	**Above or below x-axis**	**Interval containing the zero**
−1.5	−3	below	(−1.5,−1)
−1.2	0.384	above	(−1.5,−1.2)
−1.3	−0.604	below	(−1.3,−1.2)
−1.25	−0.094	below	(−1.3,−1.25)
−1.24	0.004	above	(−1.25,−1.24)
−1.245	−0.045	below	(−1.245,−1.24)

The one negative real zero is $x \approx -1.24$.

In summary, the two real zeros of h are $x \approx -1.24$ and $x = 2$.

59. Since the coefficients of $g(x) = 8x^4 - 2x^2 + 5x - 1$ have three variations in signs, it has either 3 or 1 positive real zero. Since the coefficients of $g(-x) = 8x^4 - 2x^2 - 5x - 1$ have one variation in signs, g has 1 negative real zero. Since the possible rational zeros of g are of the form $\frac{p}{q}$ where p is a factor of −1 and q is a factor of 8, the possible rational zeros are ±1, $\pm\frac{1}{2}$, $\pm\frac{1}{4}$, and $\pm\frac{1}{8}$.

Using synthetic division to find the one negative real zero and checking the possibility of −1 being that zero, we get

$$\begin{array}{r|rrrrr} -1 & 8 & 0 & -2 & 5 & -1 \\ & & -8 & 8 & -6 & 1 \\ \hline & 8 & -8 & 6 & -1 & 0 \end{array}$$

Since the remainder is 0, –1 is the one negative zero of *g*. Further, *g* can be written as follows:

$$g(x) = 8x^4 - 2x^2 + 5x - 1$$
$$= (x+1)(8x^3 - 8x^2 + 6x - 1)$$

The other zeros of *g* are the zeros of $r(x) = 8x^3 - 8x^2 + 6x - 1$.

Using synthetic division to find a positive real zero and checking the possibility of 1 being a zero, we get

$$\begin{array}{r|rrrr} 1 & 8 & -8 & 6 & -1 \\ & & 8 & 0 & 6 \\ \hline & 8 & 0 & 6 & 5 \end{array}$$

Since the remainder is not 0, 1 is *not* a zero of *r*. Still, we know that $r(1) = 5$ which is above the *x*-axis. Further, by the Boundedness Theorem, since the numbers in the bottom row of the synthetic division are all positive, 1 is an upper bound for all of the zeros of *r*. This means that the positive zeros of *r* must be in the interval $(0,1)$.

x	*r*(*x*)	Above or below *x*-axis	Interval containing the zero
0.5	1	above	$(0, 0.5)$
0.2	–0.056	below	$(0.2, 0.5)$
0.3	0.296	above	$(0.2, 0.3)$
0.22	0.018	above	$(0.2, 0.22)$
0.21	–0.019	below	$(0.21, 0.22)$
0.215	–0.000	below	$(0.215, 0.22)$

A positive real zero occurs at $x \approx 0.22$.

We have found one positive zero. Since it is possible for *g* to have three positive zeros, two zeros might exist in the interval $(0.5, 1)$. Since we cannot determine if zeros exist in this interval, we use a graphing calculator to graph *g*.

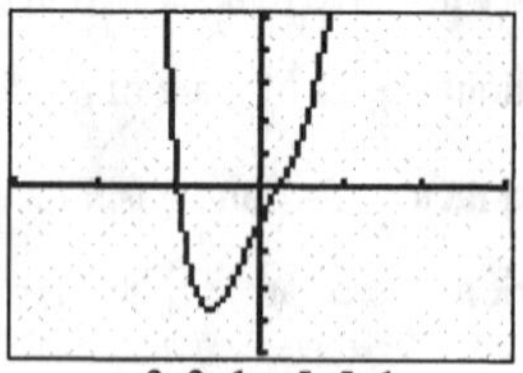

–3, 3, 1, –5, 5, 1

Thus, *g* has only 1 positive zero.

The two real zeros of *g* are $x = -1$ and $x \approx 0.22$.

61. Since the coefficients of $f(x) = 2x^3 + 6x^2 - 8x + 2$ have two variations in signs, it has either 2 or 0 positive real zeros. Since the coefficients of $f(-x) = -2x^3 + 6x^2 + 8x + 2$ have one variation in signs, *f* has 1 negative real zero. Now,

$$f(x) = 2x^3 + 6x^2 - 8x + 2$$
$$= 2(x^3 + 3x^2 - 4x + 1)$$

Then the zeros of *f* are the zeros of $r(x) = x^3 + 3x^2 - 4x + 1$. Since the rational zero of *r* are of the form $\frac{p}{q}$ where *p* is a factor of 1 and *q* is a factor of 1, the possible rational zeros are ± 1. Using synthetic division to check if –1 is the one negative zero of *r*, we get

$$\begin{array}{r|rrrr} -1 & 1 & 3 & -4 & 1 \\ & & -1 & -2 & 6 \\ \hline & 1 & 2 & -6 & 7 \end{array}$$

Since the remainder is not 0, –1 is *not* a zero of *r*. Still, we know that $r(-1) = 7$ which is above the *x*-axis. Since *f* is a third degree function with a positive leading coefficient, its graph must eventually fall below the *x*-axis.

x	*r*(*x*)	Above or below *x*-axis	Interval containing the zero
–5	–29	below	$(-5, -1)$
–4	1	above	$(-5, -4)$
–4.1	–1.091	below	$(-4.1, -4)$
–4.05	–0.023	below	$(-4.05, -4)$
–4.04	0.186	above	$(-4.05, -4.04)$
–4.045	0.082	above	$(-4.05, -4.045)$

The one negative real zero occurs at $x \approx -4.05$.

Using synthetic division to check if 1 is a positive zero of *r*, we get

$$\begin{array}{r|rrrr} 1 & 1 & 3 & -4 & 1 \\ & & 1 & 4 & 0 \\ \hline & 1 & 4 & 0 & 1 \end{array}$$

Since the remainder is not 0, 1 is *not* a zero of *r*. Still, we know that $r(1) = 1$ which is above the *x*-axis. . Further, by the Boundedness Theorem, since the numbers in the bottom row of the synthetic division are all positive, 1 is an upper bound for all of the zeros of *r*. This means that any positive zeros of *r* must be in the interval

$(0,1)$. However, since $r(0)=1$ is also above the x-axis, the Intermediate Value Theorem does not help us to determine if a zero exist in the interval. Thus, we use a graphing calculator to graph r.

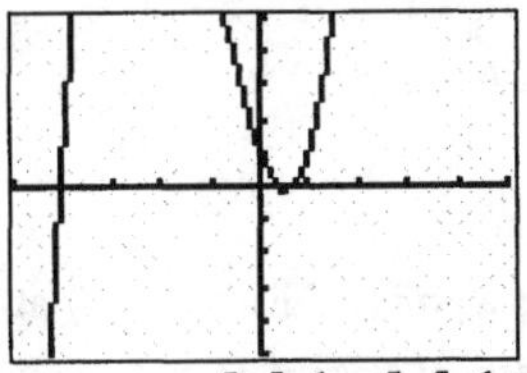

$-5, 5, 1, -5, 5, 1$

From the graph, we know that two positive irrational zero exist in the interval $(0,1)$.

x	$r(x)$	Above or below x-axis	Interval containing the zero
0.3	0.097	above	$(0.3,1)$
0.4	−0.056	below	$(0.3,0.4)$
0.35	0.010	above	$(0.35,0.4)$
0.36	−0.005	below	$(0.35,0.36)$
0.355	0.003	above	$(0.355,0.36)$

A positive zero occurs at $x \approx 0.36$.

x	$r(x)$	Above or below x-axis	Interval containing the zero
0.7	0.013	above	$(0.4,0.7)$
0.65	−0.058	below	$(0.65,0.7)$
0.68	−0.018	below	$(0.68,0.7)$
0.69	−0.003	below	$(0.69,0.7)$
0.695	0.005	above	$(0.69,0.695)$

A positive zero occurs at $x \approx 0.69$.

In summary, the three real zeros of f are $x \approx -4.05$, $x \approx 0.36$, and $x \approx 0.69$.

63. $$\frac{3x^2+9x^3-12x^4y}{6x^4y} = \frac{3x^2\left(1+3x-4x^2y\right)}{3x^2\cdot 2x^2y} = \frac{1+3x-4x^2y}{2x^2y}$$

64. $$\frac{4}{x+2}+\frac{5}{x+3} = \frac{4(x+3)+5(x+2)}{(x+2)(x+3)} = \frac{4x+12+5x+10}{(x+2)(x+3)} = \frac{9x+22}{(x+2)(x+3)}$$

65. $$\frac{3}{4x}-\frac{5}{6y} = \frac{3(6y)-5(4x)}{(4x)(6y)} = \frac{18y-20x}{24xy} = \frac{2(9y-10x)}{2\cdot 12xy} = \frac{9y-10x}{12xy}$$

66. $$\frac{3}{y}+\frac{1}{3}=\frac{2}{3y}$$
$$3y\left(\frac{3}{y}+\frac{1}{3}\right)=3y\left(\frac{2}{3y}\right)$$
$$3y\left(\frac{3}{y}\right)+3y\left(\frac{1}{3}\right)=2$$
$$9+y=2$$
$$y=-7$$

Exercise Set 10.5

1. No; if a rational function has no variables in the denominator when it is written in lowest terms, it will not have any asymptotes.

3. No; consider the three cases:
   1) If $f(x)$ is a proper rational function, the values of $f(x)$ approach 0 as $x \to \infty$ and as $x \to -\infty$.
   2) If the degrees of the numerator and denominator are the same, the values of $f(x)$ approach the ratio of the leading coefficients of the numerator and denominator as $x \to \infty$ and $x \to -\infty$.
   3) If the degree of the numerator is greater than the degree of the denominator, the graph of $f(x)$ does not have a horizontal asymptote.

5. The point of discontinuity can be found by finding the values of x that make the denominator of the function equal to 0.

7. A graph of a rational function can have an oblique asymptote when the degree of the numerator is one greater than the degree of the denominator.

9. Since $f(x)=\frac{3-x}{x}$ is in lowest terms, each real zero of the denominator results in a point of discontinuity with a vertical asymptote.
$x=0$
Thus, $f(x)$ has a point of discontinuity at $x=0$ resulting from a vertical asymptote.

11. Since $h(x)=\frac{x+2}{x+5}$ is in lowest terms, each real zero of the denominator will yield a point of discontinuity with a vertical asymptote.
$x+5=0$
$x=-5$
Thus, $h(x)$ has a point of discontinuity at $x=-5$ resulting from a vertical asymptote.

13. Since $g(x)=\frac{x^2-25}{x+5}=\frac{(x-5)(x+5)}{x+5}$ is not in lowest terms, we first note the values which make the denominator equal 0 and then simplify.
$x+5=0$
$x=-5$
$g(x)=\frac{(x-5)(x+5)}{x+5}=x-5$
Notice that $x=-5$ no longer makes the denominator equal to 0. Thus, $g(x)$ has a point of discontinuity at $x=-5$ resulting from a hole in the graph.

15. Since $f(x)=\frac{x^2+8x-9}{x-1}=\frac{(x+9)(x-1)}{x-1}$ is not in lowest terms, we first note the values that make the denominator equal 0 and then simplify.
$x-1=0$
$x=1$
$f(x)=\frac{(x+9)(x-1)}{x-1}=x+9$
Notice that $x=1$ no longer makes the denominator equal 0. Thus, $f(x)$ has a point of discontinuity at $x=1$ resulting from a hole in the graph.

17. Since $g(x)=\frac{4x-7}{x^2-16}=\frac{4x-7}{(x-4)(x+4)}$ is in lowest terms, each real zero of the denominator will yield a point of discontinuity with a vertical asymptote.
$(x-4)(x+4)=0$
$x-4=0$ or $x+4=0$
$x=4$ or $x=-4$
Thus, $g(x)$ has a point of discontinuity at $x=4$ resulting in a vertical asymptote and a point of discontinuity at $x=-4$ resulting in a vertical asymptote.

19. Since $f(x)=\frac{x-1}{x^2-7x+6}=\frac{x-1}{(x-6)(x-1)}$ is not in lowest terms, we first note the values which make the denominator equal 0 and then simplify.
$(x-6)(x-1)=0$
$x-6=0$ or $x-1=0$
$x=6$ or $x=1$
$f(x)=\frac{x-1}{(x-6)(x-1)}=\frac{1}{x-6}$
Notice that $x=6$ still makes the denominator equal 0 while $x=1$ does not. Thus, $f(x)$ has a point of discontinuity at $x=6$ resulting from a vertical asymptote and a point of discontinuity at $x=1$ resulting from a hole in the graph.

21. Since $h(x)=\frac{x+7}{x^2+8x+15}=\frac{x+7}{(x+5)(x+3)}$ is in lowest terms, each real zero of the denominator will result in a point of discontinuity with a vertical asymptote.
$(x+5)(x+3)=0$
$x+5=0$ or $x+3=0$
$x=-5$ or $x=-3$
Thus, $h(x)$ has a point of discontinuity at $x=-5$ resulting from a vertical asymptote and a point of discontinuity at $x=-3$ resulting from a vertical asymptote.

23. Since $g(x)=\frac{x+5}{x^2-25}=\frac{x+5}{(x-5)(x+5)}$ is not in lowest terms, we first note the values which make the denominator equal 0 and then simplify.

$(x+5)(x-5)=0$

$x+5=0$ or $x-5=0$

$x=-5$ or $x=5$

$f(x)=\frac{x+5}{(x-5)(x+5)}=\frac{1}{x-5}$

Notice that $x=5$ still makes the denominator equal 0 while $x=-5$ does not. Thus, $f(x)$ has a point of discontinuity at $x=5$ resulting from a vertical asymptote and a point of discontinuity at $x=-5$ resulting from a hole in the graph.

25. Since $f(x)=\frac{x^2+4x+3}{x^2-4x+3}=\frac{(x+3)(x+1)}{(x-3)(x-1)}$ is in lowest terms, each real zero of the denominator will result in a point of discontinuity with a vertical asymptote.

$(x-3)(x-1)=0$

$x-3=0$ or $x-1=0$

$x=3$ or $x=1$

Thus, $f(x)$ has a point of discontinuity at $x=3$ resulting from a vertical asymptote and a point of discontinuity at $x=1$ resulting from a vertical asymptote.

27. Since $h(x)=\frac{x+5}{x^2+5}$ is in lowest terms, each real zero of the denominator will result in a point of discontinuity with a vertical asymptote.

$x^2+5=0$

$x^2=-5$

$x=\pm\sqrt{-5}$

$x=\pm i\sqrt{5}$

Since the denominator has no real zeros, the function has no points of discontinuity.

29. $f(x)=\frac{7}{x}$

Since the degree of the denominator is larger than the degree of the numerator, there is a horizontal asymptote of $y=0$.

31. $h(x)=\frac{6x-1}{2x+3}$

Since the degree of the numerator is the same as the degree of the denominator, the graph will have a horizontal asymptote at the ratio of the leading coefficients. The horizontal asymptote is $y=\frac{6}{2}=3$.

33. $g(x)=\frac{12x+5}{3x^2}$

Since the degree of the denominator is larger than the degree of the numerator, the graph will have a horizontal asymptote of $y=0$.

35. $f(x)=\frac{4x^2-1}{x^2-9}$

Since the degree of the numerator is the same as the degree of the denominator, the graph will have a horizontal asymptote at the ratio of the leading coefficients. The horizontal asymptote is $y=\frac{4}{1}=4$.

37. $f(x)=\frac{x^2-8x+15}{x+2}$

Since the degree of the numerator is exactly one more than the degree of the denominator, the graph may have an oblique asymptote.

$$\begin{array}{r} x-10 \\ x+2\overline{)x^2-8x+15} \\ -(x^2+2x) \\ \hline -10x+15 \\ -(-10x-20) \\ \hline 35 \end{array}$$

$y=x-10+\frac{35}{x+2}$

Since the remainder is nonzero, the graph has an oblique asymptote of $y=x-10$.

39. $f(x)=\dfrac{2x^2+8x+12}{x-2}$

Since the degree of the numerator is exactly one more than the degree of the denominator, the graph may have an oblique asymptote.

$$\begin{array}{r} 2x+12 \\ x-2\overline{)2x^2+8x+12} \\ \underline{-(2x^2-4x)} \\ 12x+12 \\ \underline{-(12x-24)} \\ 36 \end{array}$$

$f(x)=2x+12+\dfrac{36}{x-2}$

Since the remainder is nonzero, the graph has an oblique asymptote of $y=2x+12$.

41. $f(x)=\dfrac{4x^2-6}{8x^3-1}$

Since the degree of the denominator is larger than the degree of the numerator, the graph has a horizontal asymptote of $y=0$.

43. $f(x)=\dfrac{x^3+2}{x^2+2x}$

Since the degree of the numerator is exactly one more than the degree of the denominator, the graph may have an oblique asymptote.

$$\begin{array}{r} x-2 \\ x^2+2x\overline{)x^3 \qquad +2} \\ \underline{-(x^3+2x^2)} \\ -2x^2 \qquad +2 \\ \underline{-(-2x^2-4x)} \\ 4x+2 \end{array}$$

$f(x)=x-2+\dfrac{4x+2}{x^2+2x}$

Since the remainder is nonzero, the graph has an oblique asymptote of $y=x-2$.

45. $f(x)=\dfrac{1}{x+2}$

Find the y-intercept by evaluating $f(0)$.

$f(0)=\dfrac{1}{0+2}=\dfrac{1}{2}$

The y-intercept is $\left(0,\dfrac{1}{2}\right)$.

Since f is a proper rational fraction, there is a horizontal asymptote at $y=0$.

The function is in lowest terms so we can find the points of discontinuity by finding the real zeros of the denominator.

$x+2=0$

$x=-2$

The function has a vertical asymptote at $x=-2$.

Evaluate $f(x)$ as $x\to -2$ from the right.

x	$f(x)$
−1	1
−1.5	2
−1.9	10

As $x\to -2$ from the right, $f(x)$ increases.

Evaluate $f(x)$ as $x\to -2$ from the left.

x	$f(x)$
−3	−1
−2.5	−2
−2.1	−10

As $x\to -2$ from the left, $f(x)$ decreases.

Graph the asymptotes with a dashed line and plot the points. Connect the points on either side of the vertical asymptote with a smooth curve that approaches the asymptotes.

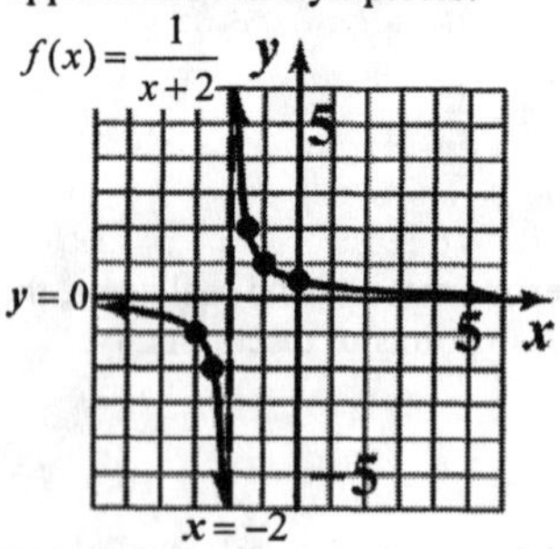

47. $h(x)=\dfrac{5}{x-1}$

Find the y-intercept by evaluating $h(0)$.

$h(0)=\dfrac{5}{0-1}=-5$

The y-intercept is $(0,-5)$.

Since h is a proper rational function, there is a horizontal asymptote at $y=0$.

The function is in lowest terms so we can find the points of discontinuity by finding the real zeros of the denominator.

$x-1=0$

$x=1$

The function has a vertical asymptote at $x=1$.

Evaluate $h(x)$ as $x\to 1$ from the right.

x	$h(x)$
2	5
1.5	10
1.1	50

As $x\to 1$ from the right, $h(x)$ increases.

Evaluate $h(x)$ as $x\to 1$ from the left.

x	$h(x)$
0	−5
.5	−10
.9	−50

As $x\to 1$ from the left, $h(x)$ decreases.

Graph the asymptotes with dashed lines and plot the points. Connect the points on either side of the vertical asymptote with a smooth curve that approaches the asymptotes.

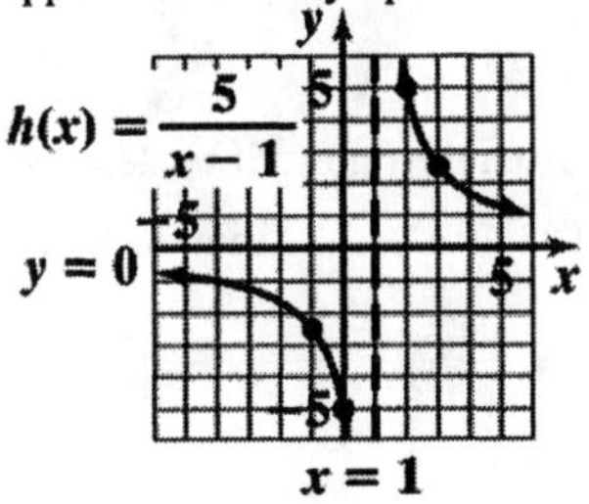

49. $h(x)=\dfrac{x+1}{x(x+4)}$

Find the y-intercept by evaluating $h(0)$.

$h(0)=\dfrac{0+1}{0(0+4)}=\dfrac{1}{0}$ which is undefined

The function has no y-intercept.

Since h is a proper rational fraction, there is a horizontal asymptote at $y=0$.

The function is in lowest terms so we can find the points of discontinuity by finding the real zeros of the denominator.

$x(x+4)=0$

$x=0$ or $x+4=0$

$x=0$ or $x=-4$

The function has vertical asymptotes at $x=0$ and $x=-4$.

Evaluate $h(x)$ as $x\to 0$ from the right.

x	$h(x)$
1	0.4
0.5	0.67
0.1	2.68

As $x\to 0$ from the right, $h(x)$ increases.

Evaluate $h(x)$ as $x\to 0$ from the left.

x	$h(x)$
−1	0
−0.5	−0.29
−0.1	−2.31

As $x\to 0$ from the left, $h(x)$ decreases.

Evaluate $h(x)$ as $x\to -4$ from the right.

x	$h(x)$
−3	0.67
−3.5	1.43
−3.9	7.44

As $x\to -4$ from the right, $h(x)$ increases.

Evaluate $h(x)$ as $x\to -4$ from the left.

x	$h(x)$
−5	−0.8
−4.5	−1.56
−4.1	−7.56

As $x\to -4$ from the left, $h(x)$ decreases.

Graph the asymptotes with dashed lines and plot the points. Connect the points within each region defined by the vertical asymptotes with a smooth curve that approaches the asymptotes.

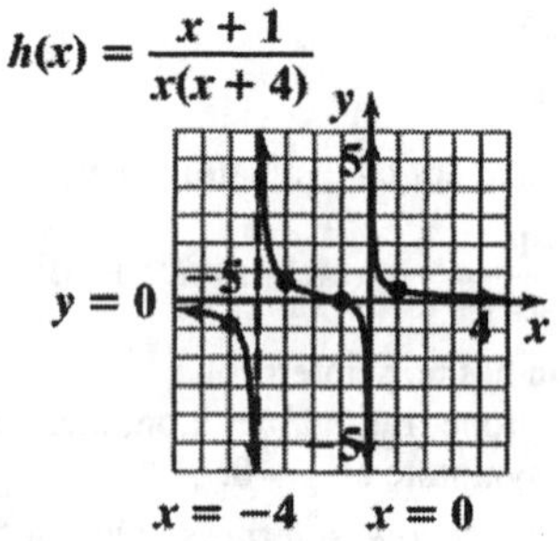

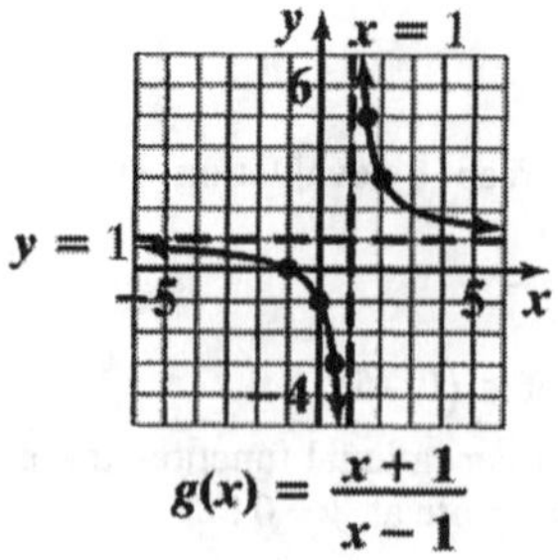

51. $g(x)=\dfrac{x+1}{x-1}$

Find the y-intercept by evaluating $g(0)$.

$g(0)=\dfrac{0+1}{0-1}=-1$

The y-intercept is $(0,-1)$.

Since the numerator and denominator have the same degree, there is a horizontal asymptote at the ratio of their leading coefficients, $y=\dfrac{1}{1}=1$.

The function is in lowest terms so we can find the points of discontinuity by finding the real zeros of the denominator.

$x-1=0$

$x=1$

The function has a vertical asymptote at $x=1$.

Evaluate $g(x)$ as $x\to 1$ from the right.

x	$g(x)$
2	3
1.5	5
1.1	21

As $x\to 1$ from the right, $g(x)$ increases.

Evaluate $g(x)$ as $x\to 1$ from the left.

x	$g(x)$
0	−1
0.5	−3
0.9	−19

As $x\to 1$ from the left, $g(x)$ decreases.

Graph the asymptotes with dashed lines and plot the points. Connect the points on either side of the vertical asymptote with a smooth curve that approaches the asymptotes.

53. $f(x)=\dfrac{3x+3}{2x+4}$

Find the y-intercept by evaluating $f(0)$.

$f(0)=\dfrac{3(0)+3}{2(0)+4}=\dfrac{3}{4}$

The y-intercept is $\left(0,\dfrac{3}{4}\right)$.

Since the numerator and denominator have the same degree, there is a horizontal asymptote at the ratio of their leading coefficients, $y=\dfrac{3}{2}$.

The function is in lowest terms so we can find the points of discontinuity by finding the real zeros of the denominator.

$2x+4=0$

$2x=-4$

$x=-2$

The function has a vertical asymptote at $x=-2$.

Evaluate $f(x)$ as $x\to -2$ from the right.

x	$f(x)$
−1	0
−1.5	−1.5
−1.9	−13.5

As $x\to -2$ from the right, $f(x)$ decreases.

Evaluate $f(x)$ as $x\to -2$ from the left.

x	$f(x)$
−3	3
−2.5	4.5
−2.1	16.5

As $x\to -2$ from the left, $f(x)$ increases.

Graph the asymptotes as dashed lines and plot the points. Connect the points on either side of the vertical asymptote with a smooth curve that approaches the asymptotes.

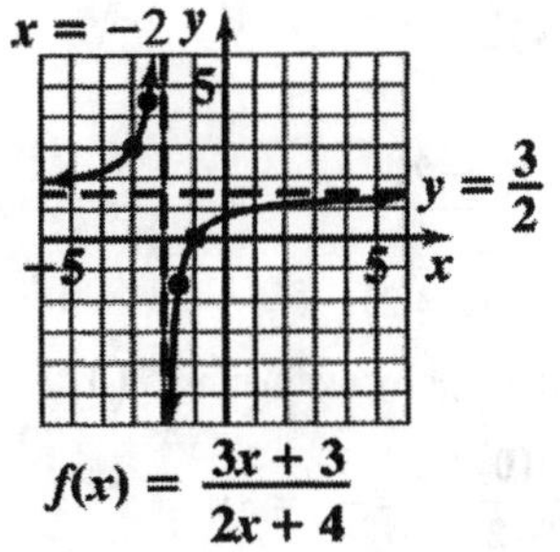

55. $h(x)=\dfrac{15x^2+10}{3x^2}$

Find the y-intercept by evaluating $h(0)$.

$h(0)=\dfrac{15(0)^2+10}{3(0)^2}=\dfrac{10}{0}$ which is undefined

The function has no y-intercept.

Since the numerator and denominator have the same degree, there is a horizontal asymptote at the ratio of their leading coefficients,

$y=\dfrac{15}{3}=5$.

The function is in lowest terms so we can find the points of discontinuity by finding the real zeros of the denominator.

$3x^2=0$

$x=0$

The function has a vertical asymptote at $x=0$.

Evaluate $h(x)$ as $x\to 0$ from the right.

x	$h(x)$
2	5.8
1	8.3
0.5	18.3

As $x\to 0$ from the right, $h(x)$ increases.

Evaluate $h(x)$ as $x\to 0$ from the left.

x	$h(x)$
−2	5.8
−1	8.3
−0.5	18.3

As $x\to 0$ from the left, $h(x)$ increases.

Graph the asymptotes as dashed lines and plot the points. Connect the points on either side of the vertical asymptote with a smooth curve that approaches the asymptotes.

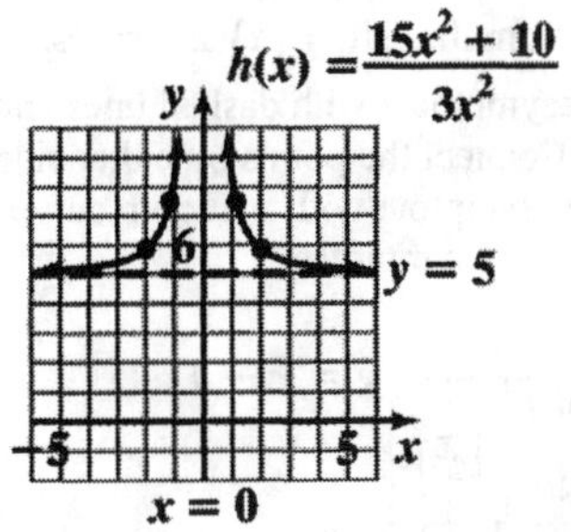

57. $g(x)=\dfrac{x^2-3x-4}{x-2}$

Find the y-intercept by evaluating $g(0)$.

$g(0)=\dfrac{0^2-3(0)-4}{0-2}=\dfrac{-4}{-2}=2$

The y-intercept is $(0,2)$.

Since the degree of the numerator is exactly one more than the degree of the denominator, the function may have an oblique asymptote,

$$\begin{array}{r} x-1 \\ x-2\overline{)x^2-3x-4} \\ \underline{-(x^2-2x)} \\ -x-4 \\ \underline{-(-x+2)} \\ -6 \end{array}$$

The remainder is nonzero so the function has an oblique asymptote of $y=x-1$.

The function is in lowest terms so we can find the points of discontinuity by finding the real zeros of the denominator.

$x-2=0$

$x=2$

The function has a vertical asymptote at $x=2$.

Evaluate $g(x)$ as $x\to 2$ from the right.

x	$g(x)$
5	2
4	0
3	−4

As $x\to 2$ from the right, $g(x)$ decreases.

Evaluate $g(x)$ as $x\to 2$ from the left.

x	$g(x)$
−1	0
0	2
1	6

As $x \to 2$ from the left, $g(x)$ increases.

Graph the asymptotes with dashed lines and plot the points. Connect the points on either side of the vertical asymptote with a smooth curve that approaches the asymptotes.

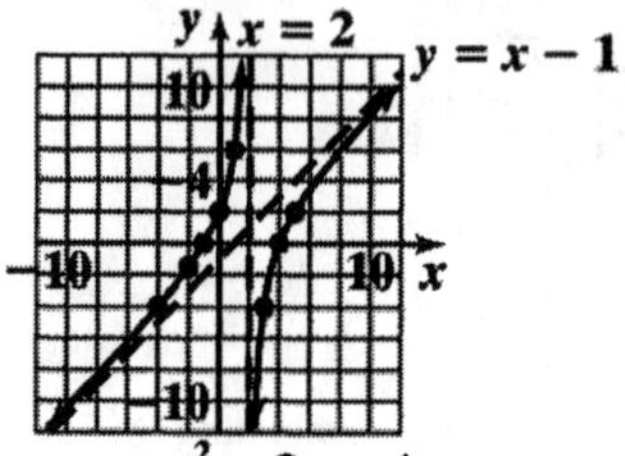

$g(x) = \dfrac{x^2 - 3x - 4}{x - 2}$

59. $g(x) = \dfrac{x^2 + 2x - 8}{x + 4}$
$= \dfrac{(x+4)(x-2)}{x+4}$
$= x - 2, \quad (x \neq -4)$

Find the y-intercept by evaluating $g(0)$.

$$g(0) = \frac{0^2 + 2(0) - 8}{0 + 4} = \frac{-8}{4} = -2$$

The y-intercept is $(0, -2)$.

At $x = -4$, both the numerator and denominator equal 0. Since $x = -4$ does not make the denominator of the simplified function 0, there is a hole in the graph at $x = -4$.

Graph the line $y = x - 2$, making a hole in the graph at $x = -4$.

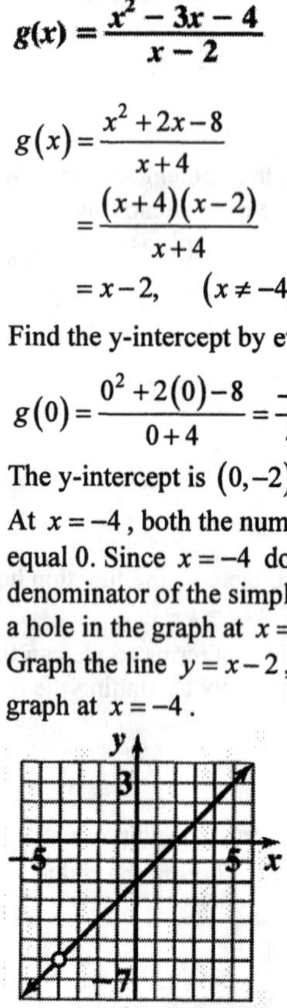

$g(x) = \dfrac{x^2 + 2x - 8}{x + 4}$

61. $r(x) = \dfrac{x^2 - 5x + 6}{x - 2}$
$= \dfrac{(x-3)(x-2)}{x-2}$
$= x - 3, \quad (x \neq 2)$

Find the y-intercept by evaluating $r(0)$.

$$r(0) = \frac{0^2 - 5(0) + 6}{0 - 2} = \frac{6}{-2} = -3$$

The y-intercept is $(0, -3)$.

At $x = 2$, both the numerator and denominator have a value of 0. Since $x = 2$ does not make the denominator of the simplified function 0, there is a hole in the graph at $x = 2$.

Graph the line $y = x - 3$, making a hole in the graph at $x = 2$.

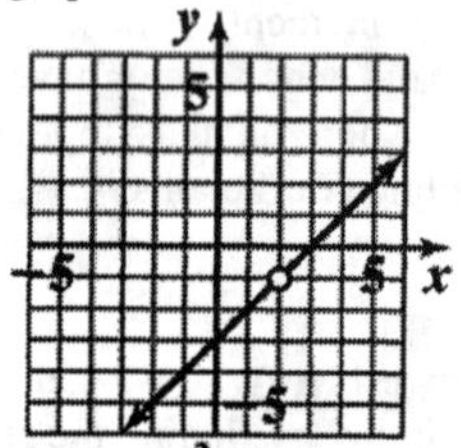

$r(x) = \dfrac{x^2 - 5x + 6}{x - 2}$

63. $g(x) = \dfrac{2x^3 - x^2 + x - 8}{x^2 - 2x + 1} = \dfrac{2x^3 - x^2 + x - 8}{(x-1)^2}$

Find the y-intercept by evaluating $g(0)$.

$$g(0) = \frac{2(0)^3 - (0)^2 + 0 - 8}{(0)^2 - 2(0) + 1} = \frac{-8}{1} = -8$$

The y-intercept is $(0, -8)$.

Since the degree of the numerator is exactly one more than the degree of the denominator, the function may have an oblique asymptote.

$$\begin{array}{r} 2x+3 \\ x^2 - 2x + 1 \overline{\big) 2x^3 - x^2 + x - 8} \\ \underline{-(2x^3 - 4x^2 + 2x)} \\ 3x^2 - x - 8 \\ \underline{-(3x^2 - 6x + 3)} \\ 5x - 11 \end{array}$$

Since the remainder is nonzero, the function has an oblique asymptote of $y = 2x + 3$.

The function is in lowest terms so we can find the points of discontinuity by finding the real zeros of the denominator.

$$(x-1)^2 = 0$$
$$x-1=0$$
$$x=1$$

The function has a vertical asymptote at $x=1$.

Evaluate $g(x)$ as $x \to 1$ from the right.

x	$g(x)$
4	12
2	6
1.5	–8

As $x \to 1$ from the right, $g(x)$ decreases.

Evaluate $g(x)$ as $x \to 1$ from the left.

x	$g(x)$
–2	–3.3
0	–8
0.5	–30

As $x \to 1$ from the left, $g(x)$ decreases.

Graph the asymptotes with dashed lines and plot the points. Connect the points on either side of the vertical asymptote with a smooth curve that approaches the asymptotes.

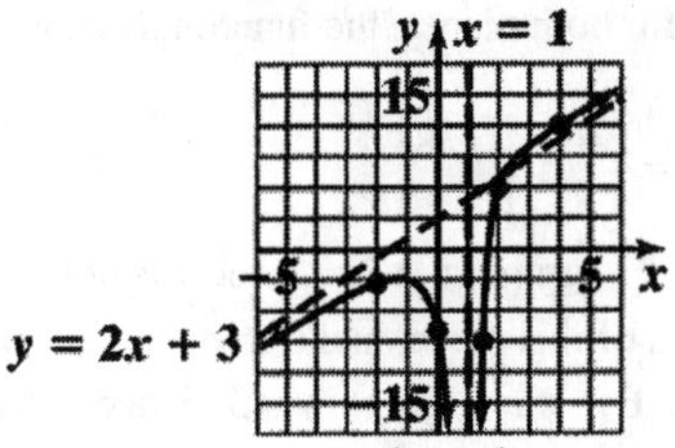

$$g(x) = \frac{2x^3 - x^2 + x - 8}{x^2 - 2x + 1}$$

65. Answers will vary.
To have a horizontal asymptote at $y=0$, the degree of the denominator must be larger than the degree of the numerator. To have a vertical asymptote at $x=-4$, the factor $x+4$ must be in the denominator. One example of such a function is $f(x)=\dfrac{1}{x+4}$.

67. Answers will vary.
To have a hole in the graph at $(4,3)$, the factor $x-4$ must be in both the numerator and denominator the same number of times. Since the function, in simplest form, is $f(x)=2x-5$, the factor $2x-5$ must also be in the numerator. One example of such a function is

$$g(x)=\frac{(2x-5)(x-4)}{x-4}.$$

69. Answers will vary.
To have a horizontal asymptote at $y=0$, the degree of the numerator must be the same as the degree of the denominator. To have no vertical asymptotes, the denominator (when the function is in simplest form) must never equal 0. One example of such a function is $f(x)=\dfrac{1}{x^2+1}$.

72. $f(x)=\dfrac{1}{2}x+3$

$$y=\frac{1}{2}x+3$$

Start by interchanging *x* and *y*.

$$x=\frac{1}{2}y+3$$

Now solve for *y*.

$$x-3=\frac{1}{2}y$$
$$2x-6=y$$
$$f^{-1}(x)=2x-6$$

73. $f(x)=\dfrac{5}{x}$

$$y=\frac{5}{x}$$

Start by interchanging *x* and *y*.

$$x=\frac{5}{y}$$

Now solve for *y*.

$$y=\frac{5}{x}$$
$$f^{-1}(x)=\frac{5}{x}$$

74. $\log_a \frac{1}{144} = -2$
Change to the equivalent exponential form.
$a^{-2} = \frac{1}{144}$
Invert both sides.
$a^2 = 144$
Write both sides as perfect squares.
$a^2 = 12^2$
$a = 12$

75. $\log_2 x = -5$
Change to the equivalent exponential expression.
$2^{-5} = x$
$x = 2^{-5}$
$x = \frac{1}{2^5}$
$x = \frac{1}{32}$

Chapter 10 Review Exercises

1. Since the graph of the function can be traced without lifting the pencil, the function is continuous.

2. Since the graph of the function cannot be traced without lifting the pencil, the function is not continuous.

3. Since $f(x) = -4x^3 + 8x^2 - 12x + 19$ is a polynomial function, the function is continuous.

4. Since $g(x) = \frac{x+8}{x^2+2}$ is a rational function, its points of discontinuity, if any, will occur at the real zeros of the denominator.
$x^2 + 2 = 0$
$x^2 = -2$
$x = \pm\sqrt{-2}$
$x = \pm i\sqrt{2}$
Since the denominator has no real zeros, the function is continuous.

5. Since $h(x) = \left[\!\left[\frac{x+5}{4} \right]\!\right]$ is a greatest integer function, it is discontinuous.
The points of discontinuity can be found by solving $y = \frac{x+5}{4}$ for x.
$x + 5 = 4y$
$x = 4y - 5$
For every integer value of y, we get a point of discontinuity. For example, when $y = 1$ we get a point of discontinuity at $x = 4(1) - 5 = -1$, when $y = 2$ we get a point of discontinuity at $x = 4(2) - 5 = 3$, etc.

6. Since $f(x) = \begin{cases} x-2 & \text{if } x < 2 \\ 2-x & \text{if } x \geq 2 \end{cases}$ is a piecewise function, it may be discontinuous at $x = 2$, the point where the definition of the function changes. The individual pieces are continuous so to determine whether the function is continuous, we need to evaluate both pieces for $x = 2$.
$y = x - 2 \qquad y = 2 - x$
$= 2 - 2 \qquad = 2 - 2$
$= 0 \qquad = 0$
Since the value of the function at $x = 2$ is the same for both pieces, the function is continuous.

7. $f(x) = \begin{cases} 2x & \text{if } x < 1 \\ 3 & \text{if } x \geq 1 \end{cases}$
For $x < 1$, graph $y = 2x$. Since 1 is not in the domain of this piece, make an open circle at $(1,2)$. For $x \geq 1$, graph $y = 3$. Since 1 is in the domain of this piece, make a solid dot at $(1,3)$.

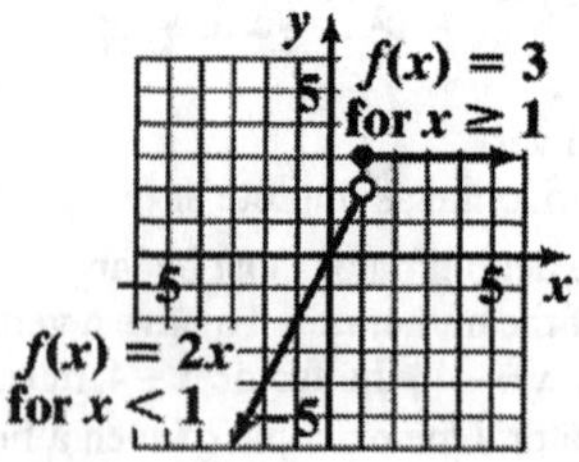

8. $g(x)=\begin{cases} 4-x & \text{if } x\le -2 \\ \frac{1}{4}x & \text{if } x>-2 \end{cases}$

For $x\le -2$, graph $y=4-x$. Since $x=-2$ is in the domain of this piece, make a solid dot at $(-2,6)$. For $x>-2$, graph $y=\frac{1}{4}x$. Since $x=-2$ is not in the domain of this piece, make an open circle at $\left(-2,-\frac{1}{2}\right)$.

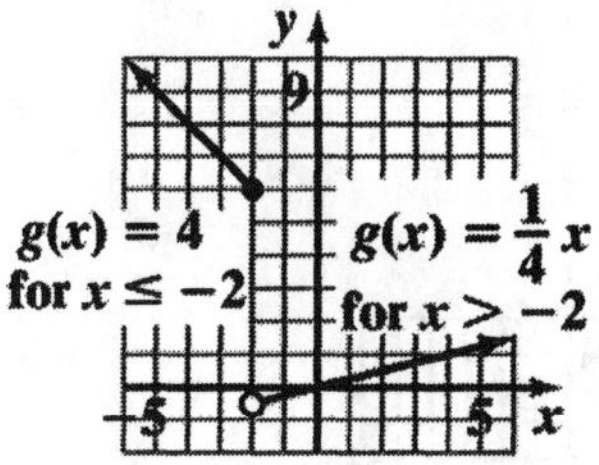

9. $h(x)=\begin{cases} -4 & \text{if } x<0 \\ 4 & \text{if } x\ge 0 \end{cases}$

For $x<0$, graph $y=-4$. Since $x=0$ is not in the domain of this piece, make an open circle at $(0,-4)$. For $x\ge 0$, graph $y=4$. Since $x=0$ is in the domain of this piece, make a solid dot at $(0,4)$.

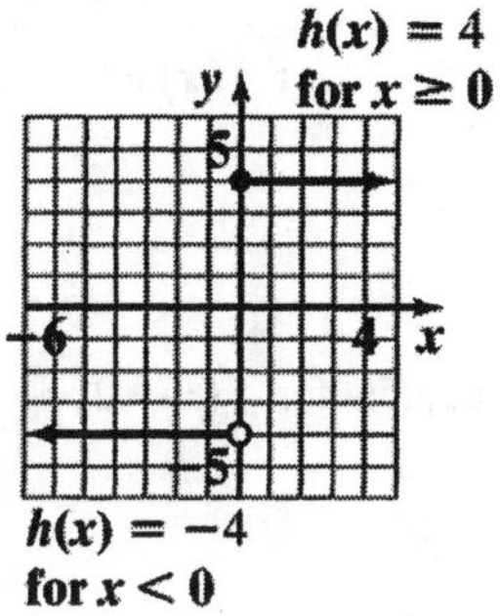

10. $f(x)=\begin{cases} 3 & \text{if } x\le 3 \\ x & \text{if } x>3 \end{cases}$

For $x\le 3$, graph $y=3$. Since $x=3$ is in the domain of this piece, make a solid dot at $(3,3)$. For $x>3$, graph $y=x$. Even though $x=3$ is not in the domain of this piece, we do not need an open circle at $(3,3)$ because the same point is obtained using the other piece.

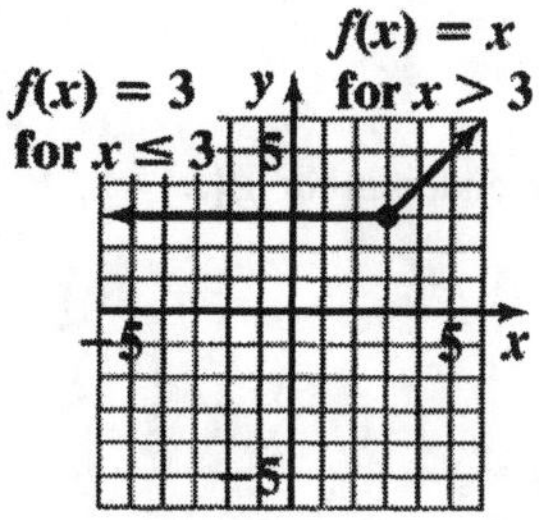

11. $f(x)=[\![2x-3]\!]$

Find the closed endpoint for $f(x)=1$.

$1=2x-3$

$4=2x$

$2=x$

Find the closed endpoint for $f(x)=2$.

$2=2x-3$

$5=2x$

$2.5=x$

Find the closed endpoint for $f(x)=3$.

$3=2x-3$

$6=2x$

$3=x$

From these points, we know the interval for each line segment spans $\frac{1}{2}$ of a unit.

Interval	$f(x)$
$[2,2.5)$	1
$[2.5,3)$	2
$[3,3.5)$	3

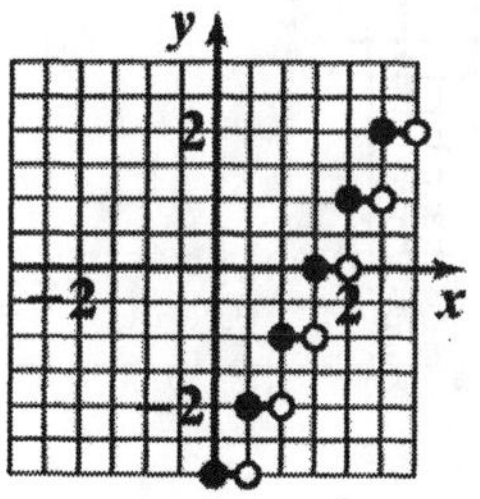

$f(x)=[\![2x-3]\!]$

12. $g(x)=\left[\!\left[\frac{1}{2}x+1\right]\!\right]$

Find the closed endpoint for $g(x)=1$.

$1=\frac{1}{2}x+1$

$0=\frac{1}{2}x$

$0=x$

Find the closed endpoint for $g(x)=2$.

$2=\frac{1}{2}x+1$

$1=\frac{1}{2}x$

$2=x$

Find the closed endpoint for $g(x)=3$.

$3=\frac{1}{2}x+1$

$2=\frac{1}{2}x$

$4=x$

From these points, we know the interval for each line segment spans 2 units.

Interval	$g(x)$
$[0,2)$	1
$[2,4)$	2
$[4,6)$	3

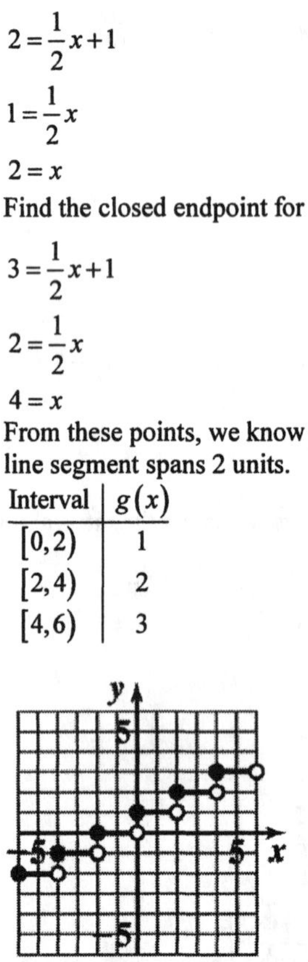

$g(x)=\left[\!\left[\frac{1}{2}x+1\right]\!\right]$

13. $h(x)=\left[\!\left[\frac{x}{12}\right]\!\right]$

Find the closed endpoint for $h(x)=1$.

$1=\frac{x}{12}$

$12=x$

Find the closed endpoint for $h(x)=2$.

$2=\frac{x}{12}$

$24=x$

Find the closed endpoint for $h(x)=3$.

$3=\frac{x}{12}$

$36=x$

From these points, we know the interval for each line segment spans 12 units.

Interval	$h(x)$
$[12,24)$	1
$[24,36)$	2
$[36,48)$	3

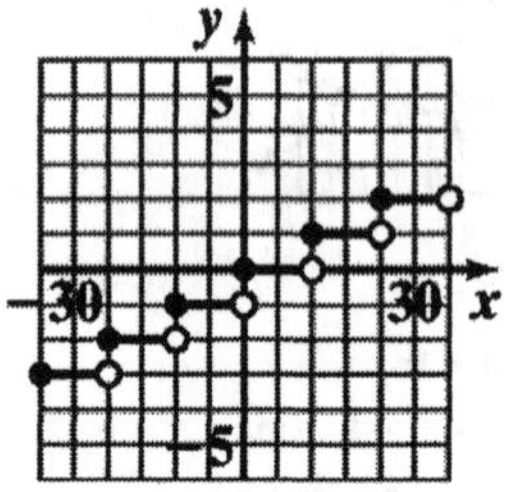

$h(x)=\left[\!\left[\frac{x}{12}\right]\!\right]$

14. $f(x)=\left[\!\left[\sqrt{x}\right]\!\right]$

Find the closed endpoint for $f(x)=1$.

$1=\sqrt{x}$

$1^2=\left(\sqrt{x}\right)^2$

$1=x$

Find the closed endpoint for $f(x)=2$.

$2=\sqrt{x}$

$2^2=\left(\sqrt{x}\right)^2$

$4=x$

Find the closed endpoint for $f(x)=3$.

$3=\sqrt{x}$

$3^2=\left(\sqrt{x}\right)^2$

$9=x$

From these points, we know the interval lengths are not constant. Moving from left to right, their lengths increase 2 units to the next odd number. The lengths are 1, 3, 5, 7, …
The function is not defined for $x<0$.

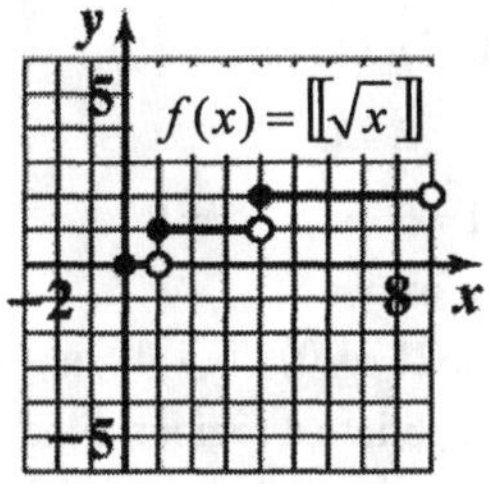

15. Since the pencils cost \$0.37 each, the maximum number of pencilss that she can buy for x dollars is the greatest integer less than or equal to $\frac{x}{0.37}$.

The corresponding function is $f(x)=\left[\!\left[\frac{x}{0.37}\right]\!\right]$.

16. Since the videos cost \$2.25 each, the maximum number of videos he can rent for x dollars is the greatest integer less than or equal to $\frac{x}{2.25}$. The corresponding function is $f(x)=\left[\!\left[\frac{x}{2.25}\right]\!\right]$.

17. $f(x)=-2x^3-6x^2+7x-3$

a. Since f is a polynomial of odd degree with a negative leading coefficient, large positive values of x will be graphed below the x-axis.

b. Since f is a polynomial of odd degree with a negative leading coefficient, large negative values of x will be graphed above the x-axis.

18. $f(x)=2x^4-x^3+7x^2+26x$

a. Since f is a polynomial of even degree with a positive leading coefficient, large positive values of x will be graphed above the x-axis.

b. Since f is a polynomial of even degree with a positive leading coefficient, large negative values of x will be graphed above the x-axis.

19. $f(x)=-x^6+3x^4-5x^2+11$

a. Since f is an even degree polynomial with a negative leading coefficient, large positive values of x will be graphed below the x-axis.

b. Since f is an even degree polynomial with a negative leading coefficient, large negative values of x will be graphed below the x-axis.

20. $f(x)=7x^5-2x^4-5x^3+7x-1$

a. Since f is an odd degree polynomial with a positive leading coefficient, large positive values of x will be graphed above the x-axis.

b. Since f is an odd degree polynomial with a positive leading coefficient, large negative values of x will be graphed below the x-axis.

21. $f(x)=x^2+11x+24$

$x^2+11x+24=0$

$(x+8)(x+3)=0$

$x+8=0$ or $x+3=0$

$x=-8$ or $x=-3$

The zeros are $x=-3$ and $x=-8$.

22. $g(x)=x^3+5x^2-36x$

$x^3+5x^2-36x=0$

$x(x^2+5x-36)=0$

$x(x+9)(x-4)=0$

$x=0$ or $x+9=0$ or $x-4=0$

$x=0$ or $x=-9$ or $x=4$

The zeros are $x=0$, $x=-9$, and $x=4$.

23. $h(x)=(x+2)^3(2x-1)$

$(x+2)^3(2x-1)=0$

$x+2=0$ or $2x-1=0$

$x=-2$ or $2x=1$

$x=-2$ or $x=\frac{1}{2}$

The zeros are $x=-2$, with multiplicity 3, and $x=\frac{1}{2}$.

24. $f(x)=x^2-7$

$x^2-7=0$

$x^2=7$

$x=\pm\sqrt{7}$

The zeros are $x=-\sqrt{7}$ and $x=\sqrt{7}$.

25. $f(x)=3x^2+6x-16$

Since the degree of the function is 2, it has a maximum of $2-1=1$ turning points.

26. $g(x)=-2x^4-5x^3+6x+11$

Since the degree of the function is 4, it has a maximum of $4-1=3$ turning points.

27. $h(x)=2x^3+7x^2-x+9$

Since the degree of the function is 3, it has a maximum of $3-1=2$ turning points.

28. $f(x)=-5x^6+6x^5+13x-2$

Since the degree of the function is 6, it has a maximum of $6-1=5$ turning points.

29. Since there are 4 turning points, the polynomial must be at least of degree $4+1=5$.

30. Since there are 5 turning points, the polynomial must be at least of degree $5+1=6$.

31. $f(x)=(x+4)(x+1)(x-5)$

$(x+4)(x+1)(x-5)=0$

$x+4=0$ or $x+1=0$ or $x-5=0$

$x=-4$ or $x=-1$ or $x=5$

The x-intercepts are $(-4,0)$, $(-1,0)$, and $(5,0)$. They break the graph into 4 regions. Find a point in each region.

Interval	x	$f(x)$
$(-\infty,-4)$	-5	$f(-5)=-40$
$(-4,-1)$	-2.5	$f(-2.5)=16.875$
$(-1,5)$	2	$f(2)=-54$
$(5,\infty)$	6	$f(6)=70$

Find the y-intercept by evaluating $f(0)$. Since $f(0)=-20$, the y-intercept is $(0,-20)$. Graph the intercepts and the points in each region. Then connect the points with a smooth curve.

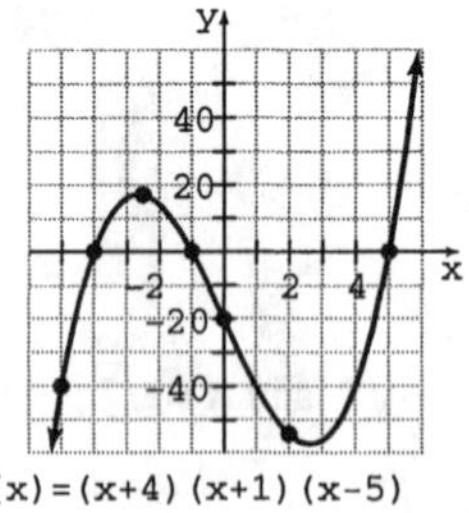

f(x)=(x+4)(x+1)(x-5)

32. $g(x)=(x+4)(x+1)^2(x-5)$

$(x+4)(x+1)^2(x-5)=0$

$x+4=0$ or $x+1=0$ or $x-5=0$

$x=-4$ or $x=-1$ or $x=5$

The x-intercepts are $(-4,0)$, $(-1,0)$, and $(5,0)$. They break the graph into 4 regions. Find a point in each region.

Interval	x	$g(x)$
$(-\infty,-4)$	-5	$g(-5)=160$
$(-4,-1)$	-2.5	$g(-2.5)=-25.3125$
$(-1,5)$	2	$g(2)=-162$
$(5,\infty)$	6	$g(6)=490$

To find the y-intercept, evaluate $g(0)$. Since $g(0)=-20$, the y-intercept is $(0,-20)$. Graph the intercepts and the points from each region. Then connect the points with a smooth curve.

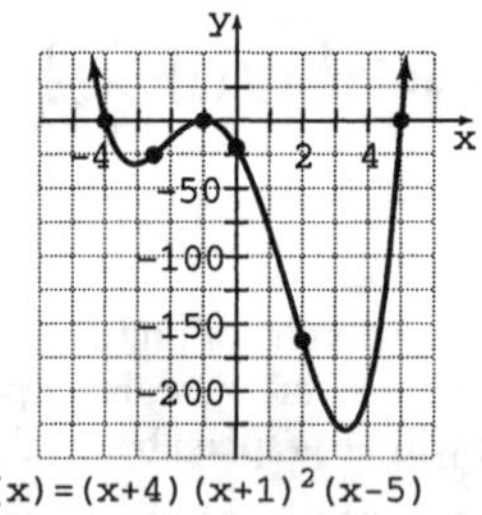

g(x)=(x+4)(x+1)²(x-5)

33. $h(x)=x^4-7x^3+6x^2$

$x^4-7x^3+6x^2=0$

$x^2(x^2-7x+6)=0$

$x^2(x-6)(x-1)=0$

$x=0$ or $x-6=0$ or $x-1=0$

$x=0$ or $x=6$ or $x=1$

The x-intercepts are $(0,0)$, $(6,0)$, and $(1,0)$. They break the graph into 4 regions. Find a point in each region.

Interval	x	$h(x)$
$(-\infty,0)$	-2	$h(-2)=96$
$(0,1)$	0.5	$h(0.5)=0.6875$
$(1,6)$	4	$h(4)=-96$
$(6,\infty)$	7	$h(7)=294$

Graph the intercepts and the points from each region. Then connect the points with a smooth curve.

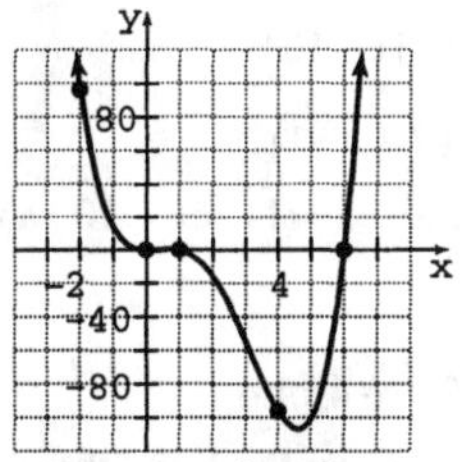

$h(x)=x^4-7x^3+6x^2$

34. $f(x)=x^5-7x^4+6x^3$

$x^5-7x^4+6x^3=0$

$x^3(x^2-7x+6)=0$

$x^3(x-6)(x-1)=0$

$x^3=0$ or $x-6=0$ or $x-1=0$

$x=0$ or $x=6$ or $x=1$

The x-intercepts are $(0,0)$, $(6,0)$, and $(1,0)$.

They break the graph into 4 regions. Find a point in each region.

Interval	x	$f(x)$
$(-\infty,0)$	-2	$f(-2)=192$
$(0,1)$	0.5	$f(0.5)=0.34375$
$(1,6)$	4	$f(4)=-384$
$(6,\infty)$	7	$f(7)=2058$

Graph the intercepts and the points from each region. Then connect the points with a smooth curve.

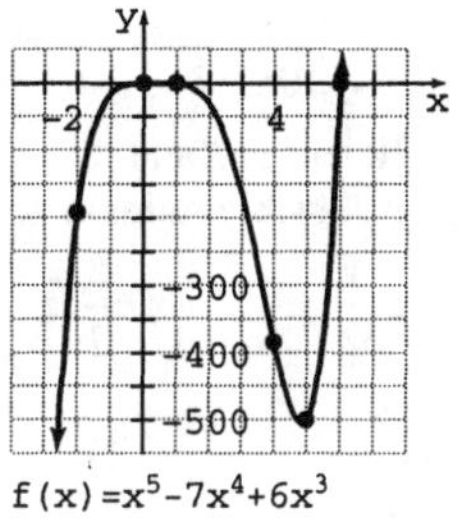

$f(x)=x^5-7x^4+6x^3$

35. $f(x)=-x^3+5x^2-7x-9$

Since the degree is 3, it will have a maximum of 3 x-intercepts. Since its degree is odd, it will have at least 1 x-intercept.

36. $g(x)=x^4-x^3-x^2+x$

Since the degree is 4, it will have a maximum of 4 x-intercepts. Since its degree is even and its leading coefficient and constant term have opposite signs, it has a minimum of 2 x-intercepts.

37. $h(x)=x^5-x^4-2x^3+20$

Since the degree is 5, it will have a maximum of 5 x-intercepts. Since its degree is odd, it will have at least 1 x-intercept.

38. $f(x)=-x^6+5x^4+3x^2-18$

Since the degree is 6, it will have a maximum of 6 x-intercepts. Since its degree is even and its leading coefficient has the same sign as the constant term, it has a minimum of 0 x-intercepts.

39. a. $f(x)=x^2-12x+36$

There are two sign changes in $f(x)$ so there are 2 or 0 positive real zeros.

b. $f(-x)=(-x)^2-12(-x)+36$

$=x^2+12x+36$

Since there are no sign changes in $f(-x)$, there are no negative real zeros.

40. a. $g(x)=x^2+3x-21$

Since there is one sign change in $g(x)$, there is exactly 1 positive real zero.

b. $g(-x)=(-x)^2+3(-x)-21$

$=x^2-3x-21$

Since there is one sign change in $g(-x)$, there is exactly 1 negative real zero.

41. a. $h(x)=x^3+5x^2+3x+9$

Since there are no sign changes in $h(x)$, there are no positive real zeros.

b. $h(-x)=(-x)^3+5(-x)^2+3(-x)+9$

$=-x^3+5x^2-3x+9$

Since there are 3 sign changes in $h(-x)$, there are 3 or 1 negative real zeros.

42. a. $f(x)=x^4-x^3-6x^2+2x+3$

Since there are two sign changes in $f(x)$, there are 2 or 0 positive real zeros.

b. $f(-x)=(-x)^4-(-x)^3-6(-x)^2+2(-x)+3$

$=x^4+x^3-6x^2-2x+3$

Since there are two sign changes in $f(-x)$, there are 2 or 0 negative real zeros.

43. $f(x) = x^3 + 3x^2 - 9x + 12$

The possible rational zeros are of the form $\frac{p}{q}$ where p is a factor of 12 and q is a factor of 1.
Factors of 12: $\pm1, \pm2, \pm3, \pm4, \pm6, \pm12$
Factors of 1: ±1
Possible rational zeros: $\pm1, \pm2, \pm3, \pm4, \pm6, \pm12$

44. $g(x) = 2x^5 - 9x^2 - 16x - 5$

The possible rational zeros are of the form $\frac{p}{q}$ where p is a factor of -5 and q is a factor of 2.
Factors of -5: $\pm1, \pm5$
Factors of 2: $\pm1, \pm2$
Possible rational zeros: $\pm1, \pm5, \pm\frac{1}{2}, \pm\frac{5}{2}$

45. $h(x) = 3x^4 - 12x^3 - 8x^2 + 4x - 9$

The possible rational zeros are of the form $\frac{p}{q}$ where p is a factor of -9 and q is a factor of 3.
Factors of -9: $\pm1, \pm3, \pm9$
Factors of 3: $\pm1, \pm3$
Possible rational zeros: $\pm1, \pm\frac{1}{3}, \pm3, \pm9$

46. $f(x) = x^6 - 5x^4 + 6x - 64$

The possible rational zeros are of the form $\frac{p}{q}$ where p is a factor of -64 and q is a factor of 1.
Factors of -64: $\pm1, \pm2, \pm4, \pm8, \pm16, \pm32, \pm64$
Factors of 1: ±1
Possible rational zeros:
$\pm1, \pm2, \pm4, \pm8, \pm16, \pm32, \pm64$

47. $f(x) = x^3 + 10x^2 - x - 10$
Since the leading coefficient is 1 and the constant term is -10, the possible rational zeros are $\pm1, \pm2, \pm5, \pm10$.

1	1	10	−1	−10
		1	11	10
	1	11	10	0

Since 1 is a zero of *f*, we know $(x-1)$ is a factor. We can write the function as

$$f(x) = (x-1)(x^2 + 11x + 10)$$
$$= (x-1)(x+10)(x+1)$$

The 3 rational zeros are $x = 1$, $x = -10$, and $x = -1$.

48. $g(x) = x^4 - 4x^3 + 5x^2 - 4x + 4$
Since the leading coefficient is 1 and the constant term is 4, the possible rational zeros are $\pm1, \pm2, \pm4$.

1	1	−4	5	−4	4
		1	−3	2	−2
	1	−3	2	−2	2

1 is not a zero since the remainder is not 0.

2	1	−4	5	−4	4
		2	−4	2	−4
	1	−2	1	−2	0

2 is a zero, so $(x-2)$ is a factor. We can write the function as

$$g(x) = (x-2)(x^3 - 2x^2 + x - 2)$$
$$= (x-2)(x^2(x-2) + (x-2))$$
$$= (x-2)(x^2+1)(x-2)$$
$$= (x-2)^2(x^2+1)$$

Since $x^2 + 1$ is never 0, the rational zero is $x = 2$ with a multiplicity of 2.

49. $h(x) = 3x^4 + 10x^3 - 5x^2 + 10x - 8$
Since the leading coefficient is 3 and the constant term is -8, the possible rational zeros are $\pm1, \pm2, \pm4, \pm8, \pm\frac{1}{3}, \pm\frac{1}{2}, \pm\frac{2}{3}, \pm\frac{4}{3}, \pm\frac{8}{3}$.

1	3	10	−5	10	−8
		3	13	8	18
	3	13	8	18	10

1 is not a zero since the remainder is not 0. Since all the numbers in the last row of the synthetic division are positive, 1 is an upper bound for the zeros. The only remaining positive rational zeros that are possible are $\frac{1}{3}$ and $\frac{2}{3}$.

$\frac{1}{3}$	3	10	−5	10	−8
		1	$\frac{11}{3}$	$-\frac{4}{9}$	$\frac{86}{27}$
	3	11	$-\frac{4}{3}$	$\frac{86}{9}$	$\frac{-130}{27}$

$\frac{1}{3}$ is not a zero because the remainder is not 0.

$\frac{2}{3}\rfloor$	3	10	−5	10	−8
		2	8	2	8
	3	12	3	12	0

Since $\frac{2}{3}$ is a zero, we know $\left(x-\frac{2}{3}\right)$ is a factor.

$$\begin{aligned} h(x) &= \left(x-\frac{2}{3}\right)\left(3x^3+12x^2+3x+12\right) \\ &= 3\left(x-\frac{2}{3}\right)\left(x^3+4x^2+x+4\right) \\ &= 3\left(x-\frac{2}{3}\right)\left(x^2+1\right)(x+4) \end{aligned}$$

Since x^2+1 is never 0, the only rational zeros are $x=\frac{2}{3}$ and $x=-4$.

50. $f(x)=x^4+x^3-35x^2-57x+90$

Since the leading coefficient it 1 and the constant term is 90, the possible rational zeros are $\pm1, \pm2, \pm3, \pm5, \pm6, \pm9, \pm10, \pm15, \pm18, \pm30, \pm45, \pm90$

$1\rfloor$	1	1	−35	−57	90
		1	2	−33	−90
	1	2	−33	−90	0

1 is a zero so we know $x-1$ is a factor.

$f(x)=(x-1)\left(x^3+2x^2-33x-90\right)$

Now consider the reduced polynomial:

$Q(x)=\left(x^3+2x^2-33x-90\right)$

This polynomial has the same potential rational zeros as the original polynomial.

$3\rfloor$	1	2	−33	−90
		3	15	−54
	1	5	−18	−144

3 is not a zero since the remainder is not zero.

$5\rfloor$	1	2	−33	−90
		5	35	10
	1	7	2	−80

5 is not a zero since the remainder is not zero.

$6\rfloor$	1	2	−33	−90
		6	48	90
	1	8	15	0

6 is a zero so we know that $x-6$ is a factor. The original function can be written as:

$$\begin{aligned} f(x) &= (x-1)(x-6)\left(x^2+8x+15\right) \\ &= (x-1)(x-6)(x+5)(x+3) \end{aligned}$$

The rational zeros are $x=1$, $x=6$, $x=-5$, and $x=-3$.

51. $f(x)=x^4-4x^3+11x^2-64x-80$

There are 3 sign changes so there are 3 or 1 positive real zeros.

$f(-x)=x^4+4x^3+11x^2+64x-80$ has 1 sign change so there is exactly 1 negative real zero.

The leading coefficient is 1 and the constant term is −80, so the potential rational zeros are $\pm1, \pm2, \pm4, \pm5, \pm8, \pm10, \pm16, \pm20, \pm40, \pm80$

$1\rfloor$	1	−4	11	−64	−80
		1	−3	8	−56
	1	−3	8	−56	−136

1 is not a zero since the remainder is not 0.

$2\rfloor$	1	−4	11	−64	−80
		2	−4	14	−100
	1	−2	7	−50	−180

2 is not a zero since the remainder is not 0.

$4\rfloor$	1	−4	11	−64	−80
		4	0	44	−80
	1	0	11	−20	−160

4 is not a zero since the remainder is not 0.

$5\rfloor$	1	−4	11	−64	−80
		5	5	80	80
	1	1	16	16	0

Since 5 is a zero, we know that $x-5$ is a factor.

$$\begin{aligned} f(x) &= (x-5)\left(x^3+x^2+16x+16\right) \\ &= (x-5)\left(x^2(x+1)+16(x+1)\right) \\ &= (x-5)\left(x^2+16\right)(x+1) \end{aligned}$$

$x^2+16=0$

$x^2=-16$

$x=\pm\sqrt{-16}$

$x=\pm4i$

The zeros are $x=5$, $x=-1$, $x=4i$, and $x=-4i$.

52. $g(x)=x^4+x^3+19x^2+25x-150$
There is one sign change so there is exactly one positive real zero.
$g(-x)=x^4-x^3+19x^2-25x-150$ has 3 sign changes so there are 3 or 1 negative real zeros.
Since the leading coefficient is 1 and the constant term is –150, the potential rational zeros are:
$\pm1,\pm2,\pm3,\pm5,\pm6,\pm10,\pm15,\pm25,\pm50,\pm75,\pm150$

$$\begin{array}{r|rrrrr} 1 & 1 & 1 & 19 & 25 & -150 \\ & & 1 & 2 & 21 & 46 \\ \hline & 1 & 2 & 21 & 46 & -104 \end{array}$$

1 is not a zero since the remainder is not 0.

$$\begin{array}{r|rrrrr} 2 & 1 & 1 & 19 & 25 & -150 \\ & & 2 & 6 & 50 & 150 \\ \hline & 1 & 3 & 25 & 75 & 0 \end{array}$$

Since 2 is a zero, we know that $x-2$ is a factor.

$$\begin{aligned} g(x) &= (x-2)\left(x^3+3x^2+25x+75\right) \\ &= (x-2)\left(x^2(x+3)+25(x+3)\right) \\ &= (x-2)(x+3)\left(x^2+25\right) \end{aligned}$$

$x^2+25=0$

$x^2=-25$

$x=\pm\sqrt{-25}$

$x=\pm5i$

The zeros are $x=2$, $x=-3$, $x=5i$, and $x=-5i$.

53. $g(x)=x^3-4x^2-5x+13$

$$\begin{aligned} g(-3) &= (-3)^3-4(-3)^2-5(-3)+13 \\ &= -27-36+15+13 \\ &= -35 \end{aligned}$$

$$\begin{aligned} g(-1) &= (-1)^3-4(-1)^2-5(-1)+13 \\ &= -1-4+5+13 \\ &= 13 \end{aligned}$$

Since $g(-3)<0$ and $g(-1)>0$, a zero occurs in the interval $(-3,-1)$.

54. $g(x)=x^3-4x^2-5x+13$

$$\begin{aligned} g(-1) &= (-1)^3-4(-1)^2-5(-1)+13 \\ &= -1-4+5+13 \\ &= 13 \end{aligned}$$

$$\begin{aligned} g(1) &= (1)^3-4(1)^2-5(1)+13 \\ &= 1-4-5+13 \\ &= 5 \end{aligned}$$

Since $g(-1)$ and $g(1)$ have the same sign, we cannot determine if a zero exists in the interval.

55. $g(x)=x^3-4x^2-5x+13$

$$\begin{aligned} g(1) &= (1)^3-4(1)^2-5(1)+13 \\ &= 1-4-5+13 \\ &= 5 \end{aligned}$$

$$\begin{aligned} g(5) &= (5)^3-4(5)^2-5(5)+13 \\ &= 125-100-25+13 \\ &= 13 \end{aligned}$$

Since $g(1)$ and $g(5)$ have the same sign, we cannot determine if a zero exists on the interval.

56. $g(x)=x^3-4x^2-5x+13$

$$\begin{aligned} g(4) &= (4)^3-4(4)^2-5(4)+13 \\ &= 64-64-20+13 \\ &= -7 \end{aligned}$$

$$\begin{aligned} g(5) &= (5)^3-4(5)^2-5(5)+13 \\ &= 125-100-25+13 \\ &= 13 \end{aligned}$$

Since $g(4)<0$ and $g(5)>0$, a zero occurs in the interval $(4,5)$.

57. $f(x)=x^3-x^2-8$

$$\begin{array}{r|rrrr} 2 & 1 & -1 & 0 & -8 \\ & & 2 & 2 & 4 \\ \hline & 1 & 1 & 2 & -4 \end{array}$$

Since the bottom row of synthetic division contains a negative value, we cannot conclude that 2 is an upper bound.

$$\begin{array}{r|rrrr} -1 & 1 & -1 & 0 & -8 \\ & & -1 & 2 & -2 \\ \hline & 1 & -2 & 2 & -10 \end{array}$$

Since the bottom row of synthetic division alternates between positive and negative values, we can conclude that -1 is a lower bound for the zeros. Since we cannot conclude that 2 is an upper bound for the zeros, we cannot use the boundness theorem to determine if the zeros are bounded by the interval $[-1,2]$.

58. $g(x)=x^3-x^2-11x-6$

3	1	−1	−11	−6
		3	6	−15
	1	2	−5	−21

Since the bottom row of synthetic division contains a negative value, we cannot conclude that 3 is an upper bound for the zeros.

−3	1	−1	−11	−6
		−3	12	−3
	1	−4	1	−9

Since the bottom of row of synthetic division alternates between positive and negative numbers, we can conclude that −3 is a lower bound for the zeros. Since we could not conclude that 3 was an upper bound for the zeros, we cannot use the boundness theorem to determine if the zeros are bounded by the interval $[-3,3]$.

59. $h(x)=x^4-5x^3+7x^2-12x-96$

6	1	−5	7	−12	−96
		6	6	78	396
	1	1	13	66	300

Since the bottom row of synthetic division is all positive numbers, we can conclude that 6 is an upper bound for the zeros.

−3	1	−5	7	−12	−96
		−3	24	−93	315
	1	−8	31	−105	219

Since the bottom row of synthetic division alternates between positive and negative numbers, we can conclude that −3 is a lower bound for the zeros. Since −3 is a lower bound and 6 is an upper bound, the zeros of $h(x)$ are bounded over the interval $[-3,6]$.

60. $f(x)=x^5-3x^4-5x^2+18x+3$

4	1	−3	0	−5	18	3
		4	4	16	44	248
	1	1	4	11	62	251

Since the bottom row of synthetic division is all positive numbers, we can conclude that 4 is an upper bound for the zeros.

−1	1	−3	0	−5	18	3
		−1	4	−4	9	−27
	1	−4	4	−9	27	−24

Since the bottom row of synthetic division alternates between positive and negative numbers, we can conclude that −1 is a lower bound for the zeros. Since −1 is a lower bound and 4 is an upper bound, the zeros of $f(x)$ are bounded over the interval $[-1,4]$.

61. $f(x)=2x^3+x^2+2x+1$

To find the y-intercept, evaluate $f(0)$.

$f(0)=2(0)^3+(0)^2+2(0)+1=1$

The y-intercept is $(0,1)$.

Since the coefficients of $f(x)$ have no sign change, $f(x)$ has no positive real zeros.

$$f(-x)=2(-x)^3+(-x)^2+2(-x)+1$$
$$=-2x^3+x^2-2x+1$$

Since the coefficients of $f(-x)$ have 3 sign changes, there are 3 or 1 negative real zeros. The leading coefficient is 2 and the constant term is 1, so the possible rational zeros are $\pm1,\pm\frac{1}{2}$.

Since there are no positive real zeros, we only need to consider the negatives.

−1	2	1	2	1
		−2	1	−3
	2	−1	3	−2

−1 is not a zero since the remainder is not 0.

$-\frac{1}{2}$	2	1	2	1
		−1	0	−1
	2	0	2	0

$-\frac{1}{2}$ is a zero so $x+\frac{1}{2}$ is a factor.

$$f(x)=\left(x+\frac{1}{2}\right)\left(2x^2+2\right)$$
$$=2\left(x+\frac{1}{2}\right)\left(x^2+1\right)$$

Since x^2+1 is never 0, there is only one real zero. This means that the only x-intercept is $\left(-\frac{1}{2},0\right)$. This breaks the graph into 2 regions.

Find a point in each region.

Interval	x	$f(x)$
$\left(-\infty,-\frac{1}{2}\right)$	-1	$f(-1)=-2$
$\left(-\frac{1}{2},\infty\right)$	1	$f(1)=6$

Graph the intercepts and the points from each region and connect them with a smooth curve.

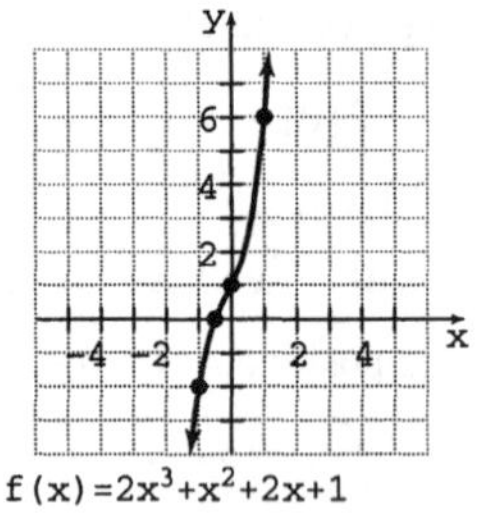

62. $g(x)=x^3+2x^2-19x-20$

To find the y-intercept, evaluate $g(0)$.

$g(0)=(0)^3+2(0)^2-19(0)-20=-20$

The y-intercept is $(0,-20)$.

Since the coefficients of $g(x)$ have one sign change, there is exactly one positive real zero.

$g(-x)=(-x)^3+2(-x)^2-19(-x)-20$

$=-x^3+2x^2+19x-20$

Singe the coefficients of $g(-x)$ have 2 sign changes, there are 2 or 0 negative real zeros. The leading coefficient is 1 and the constant term is -20, so the possible rational zeros are $\pm1,\pm2,\pm4,\pm5,\pm10,\pm20$

1	1	2	−19	−20
		1	3	−16
	1	3	−16	−36

1 is not a zero since the remainder is not 0.

2	1	2	−19	−20
		2	8	−22
	1	4	−11	−42

2 is not a zero since the remainder is not 0.

4	1	2	−19	−20
		4	24	20
	1	6	5	0

4 is a zero (the only positive zero), so $x-4$ is a factor.

$g(x)=(x-4)(x^2+6x+5)$

$=(x-4)(x+5)(x+1)$

The three zeros are $x=4$, $x=-5$, and $x=-1$. The three x-intercepts are $(4,0)$, $(-5,0)$, and $(-1,0)$. These break the graph into 4 regions. Find a point in each region.

Interval	x	$g(x)$
$(-\infty,-5)$	-6	$g(-6)=50$
$(-5,-1)$	-3	$g(-3)=28$
$(-1,4)$	2	$g(2)=-42$
$(4,\infty)$	5	$g(5)=60$

Graph the intercepts and the points from each region. Then connect them with a smooth curve.

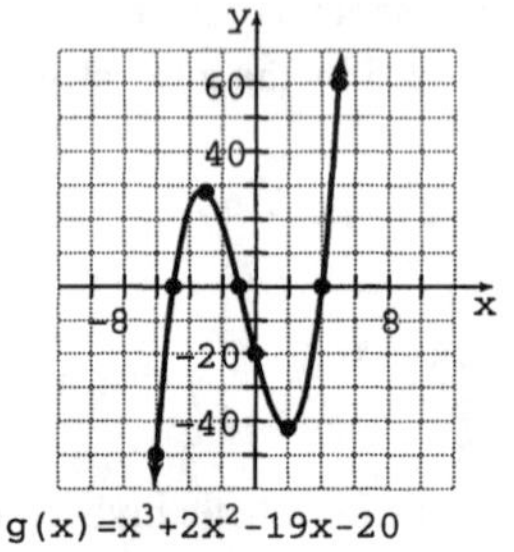

63. $h(x)=x^4-x^3-6x^2+4x+8$

Find the y-intercept by evaluating $h(0)$.

$h(0)=(0)^4-(0)^3-6(0)^2+4(0)+8=8$

The y-intercept is $(0,8)$.

Since the coefficients of $h(x)$ have 2 sign changes, there are 2 or 0 positive real zeros.

$h(-x)=(-x)^4-(-x)^3-6(-x)^2+4(-x)+8$

$=x^4+x^3-6x^2-4x+8$

Since the coefficients of $h(-x)$ have 2 sign changes, there are 2 or 0 negative real zeros. The leading coefficient is 1 and the constant terms is 8, so the potential rational zeros are $\pm1,\pm2,\pm4,\pm8$

1	1	−1	−6	4	8
		1	0	−6	−2
	1	0	−6	−2	6

1 is not a zero since the remainder is not 0.

$$\begin{array}{r|rrrrr} 2 & 1 & -1 & -6 & 4 & 8 \\ & & 2 & 2 & -8 & -8 \\ \hline & 1 & 1 & -4 & -4 & 0 \end{array}$$

2 is a zero, so $x-2$ is a factor.

$$\begin{aligned} h(x) &= (x-2)(x^3+x^2-4x-4) \\ &= (x-2)(x^2(x+1)-4(x+1)) \\ &= (x-2)(x+1)(x^2-4) \\ &= (x-2)(x+1)(x-2)(x+2) \\ &= (x-2)^2(x+1)(x+2) \end{aligned}$$

The zeros are $x=2$ (with multiplicity 2), $x=-1$, and $x=-2$. The x-intercepts are $(2,0)$, $(-1,0)$, and $(-2,0)$. They break the graph into 4 regions. Find a point in each region.

Interval	x	$h(x)$
$(-\infty,-2)$	-3	$h(-3)=50$
$(-2,-1)$	-1.5	$h(-1.5)=-3.0625$
$(-1,2)$	1	$h(1)=6$
$(2,\infty)$	3	$h(3)=20$

Graph the intercepts and the points from each region. Then connect them with a smooth curve.

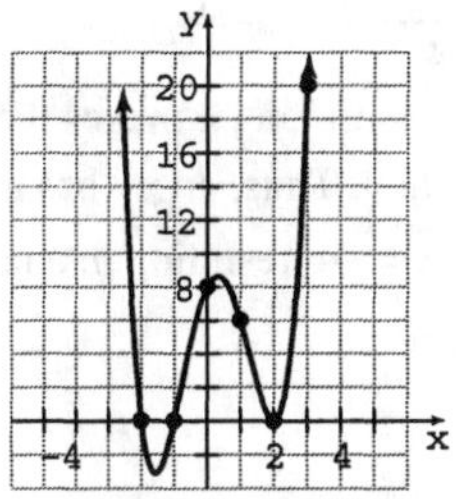

$h(x)=x^4-x^3-6x^2+4x+8$

64. $h(x)=4x^5+12x^4-x-3$

To find the y-intercept, evaluate $h(0)$.

$h(0)=4(0)^5+12(0)^4-(0)-3=-3$

The y-intercept is $(0,-3)$.

Since the coefficients of $h(x)$ have 1 sign change, there is exactly one positive real zero.

$$\begin{aligned} h(-x) &= 4(-x)^5+12(-x)^4-(-x)-3 \\ &= -4x^5+12x^4+x-3 \end{aligned}$$

Since the coefficients of $h(-x)$ have 2 sign changes, there are 2 or 0 negative real zeros. The leading coefficient is 4 and the constant term is -3, so the potential rational zeros are $\pm1, \pm3, \pm\frac{1}{2}, \pm\frac{3}{2}, \pm\frac{1}{4}, \pm\frac{3}{4}$

$$\begin{array}{r|rrrrrr} 1 & 4 & 12 & 0 & 0 & -1 & -3 \\ & & 4 & 16 & 16 & 16 & 15 \\ \hline & 4 & 16 & 16 & 16 & 15 & 12 \end{array}$$

1 is not a zero since the remainder is not 0.

$$\begin{array}{r|rrrrrr} \frac{3}{4} & 4 & 12 & 0 & 0 & -1 & -3 \\ & & 3 & \frac{45}{4} & \frac{135}{16} & \frac{405}{64} & \frac{1023}{256} \\ \hline & 4 & 15 & \frac{45}{4} & \frac{135}{16} & \frac{341}{64} & \frac{255}{256} \end{array}$$

$\frac{3}{4}$ is not a zero since the remainder is not 0.

$$\begin{array}{r|rrrrrr} \frac{1}{2} & 4 & 12 & 0 & 0 & -1 & -3 \\ & & 2 & 7 & \frac{7}{2} & \frac{7}{4} & \frac{3}{8} \\ \hline & 4 & 14 & 7 & \frac{7}{2} & \frac{3}{4} & -\frac{21}{8} \end{array}$$

$\frac{1}{2}$ is not a zero since the remainder is not 0.

Since $h\left(\frac{3}{4}\right)$ is positive and $h\left(\frac{1}{2}\right)$ is negative, there is a zero between $\frac{3}{4}$ and $\frac{1}{2}$. This is the only positive real zero.

$$\begin{array}{r|rrrrrr} -1 & 4 & 12 & 0 & 0 & -1 & -3 \\ & & -4 & -8 & 8 & -8 & 9 \\ \hline & 4 & 8 & -8 & 8 & -9 & 6 \end{array}$$

-1 is not a zero since the remainder is not 0.

$$\begin{array}{r|rrrrrr} -\frac{3}{4} & 4 & 12 & 0 & 0 & -1 & -3 \\ & & -3 & -\frac{27}{4} & \frac{81}{16} & -\frac{243}{64} & \frac{921}{256} \\ \hline & 4 & 9 & -\frac{27}{4} & \frac{81}{16} & -\frac{307}{64} & \frac{153}{256} \end{array}$$

$-\frac{3}{4}$ is not a zero since the remainder is not 0.

$$\begin{array}{r|rrrrrr} -\frac{1}{2} & 4 & 12 & 0 & 0 & -1 & -3 \\ & & -2 & -5 & \frac{5}{2} & -\frac{5}{4} & \frac{9}{8} \\ \hline & 4 & 10 & -5 & \frac{5}{2} & -\frac{9}{4} & -\frac{15}{8} \end{array}$$

$-\frac{1}{2}$ is not a zero since the remainder is not 0.

Since $h\left(-\frac{3}{4}\right)$ is positive and $h\left(-\frac{1}{2}\right)$ is negative, there is a zero between $-\frac{3}{4}$ and $-\frac{1}{2}$.

$$\begin{array}{r|rrrrrr} -3 & 4 & 12 & 0 & 0 & -1 & -3 \\ & & -12 & 0 & 0 & 0 & 3 \\ \hline & 4 & 0 & 0 & 0 & -1 & 0 \end{array}$$

-3 is a zero, so $x+3$ is a factor.

$h(x)=(x+3)(4x^4-1)$

$4x^4-1=0$

$4x^4=1$

$x^4=\dfrac{1}{4}$

$x=\pm\sqrt[4]{\dfrac{1}{4}}=\pm\sqrt{\dfrac{1}{2}}=\pm\dfrac{1}{\sqrt{2}}=\pm\dfrac{\sqrt{2}}{2}$

$\left(\dfrac{\sqrt{2}}{2}\approx 0.707\right)$

The zeros are $x=-3$, $x=-\dfrac{\sqrt{2}}{2}$, and $x=\dfrac{\sqrt{2}}{2}$.

These break the graph into 4 regions. Find a point in each region.

Interval	x	$h(x)$
$(-\infty,-3)$	-4	$h(-4)=-1023$
$\left(-3,-\dfrac{\sqrt{2}}{2}\right)$	-2	$h(-2)=63$
$\left(-\dfrac{\sqrt{2}}{2},\dfrac{\sqrt{2}}{2}\right)$	0	$h(0)=-3$
$\left(\dfrac{\sqrt{2}}{2},\infty\right)$	1	$h(1)=12$

Graph the intercepts and the points in each region. Then connect them with a smooth curve.

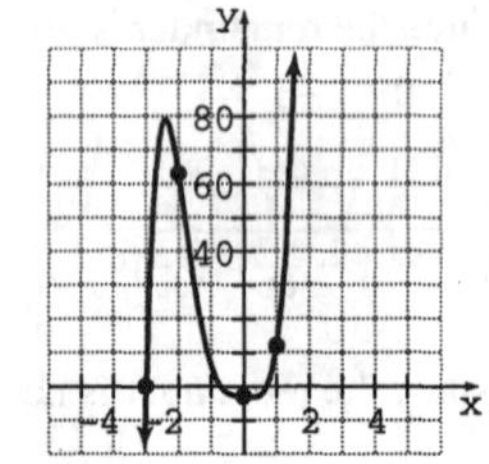

h(x)=4x^5+12x^4-x-3

65. $f(x)=\dfrac{4x}{x^2-25}=\dfrac{4x}{(x-5)(x+5)}$

Since f is in lowest terms, all real zeros of its denominator will result in points of discontinuity with vertical asymptotes.

$(x-5)(x+5)=0$

$x-5=0$ or $x+5=0$

$x=5$ or $x=-5$

$f(x)$ has a point of discontinuity at $x=5$ and $x=-5$, both resulting in vertical asymptotes.

66. $g(x)=\dfrac{x+5}{x^2-25}=\dfrac{x+5}{(x-5)(x+5)}$

Since g is not in lowest terms, we first not all values which make the denominator equal 0 and then simplify.

$(x-5)(x+5)=0$

$x-5=0$ or $x+5=0$

$x=5$ or $x=-5$

$g(x)=\dfrac{x+5}{(x-5)(x+5)}=\dfrac{1}{x-5}$

Notice that $x=-5$ no longer makes the denominator equal 0. Thus, $g(x)$ has a point of discontinuity at $x=-5$ resulting in a hole in the graph and a point of discontinuity at $x=5$ resulting in a vertical asymptote.

67. $h(x)=\dfrac{x^2-16}{x+4}=\dfrac{(x-4)(x+4)}{x+4}$

Since h is not in lowest terms, we first note the values that make the denominator equal 0 and then simplify.

$x+4=0$

$x=-4$

$h(x)=\dfrac{(x-4)(x+4)}{x+4}=x-4$

Notice that $x=-4$ no longer makes the denominator equal 0. Thus, $h(x)$ has a point of discontinuity at $x=-4$ resulting in a hole in the graph.

68. $f(x)=\dfrac{x+4}{x^2+16}$

Since f is in lowest terms, all real zeros of its denominator will result in points of discontinuity with vertical asymptotes.

$x^2+16=0$

$x^2=-16$

$x=\pm\sqrt{-16}$

$x=\pm 4i$

Since the denominator has no real zeros, there are no points of discontinuity.

69. $f(x)=\dfrac{16x^5+5}{4x^7-12}$

Since $f(x)$ is a proper rational function, it has a horizontal asymptote at $y=0$.

70. $g(x)=\dfrac{24x^2+x-10}{8x^2-12x+3}$

Since the degree of the numerator is the same as the degree of the denominator, $g(x)$ has a horizontal asymptote at the ratio of the leading coefficients, $y=\dfrac{24}{8}=3$.

71. $h(x)=\dfrac{6x^2-5x+7}{2x+3}$

Since the degree of the numerator is exactly one more than the degree of the denominator, $h(x)$ may have an oblique asymptote.

$$\begin{array}{r}
3x-7 \\
2x+3\overline{)6x^2-5x+7} \\
\underline{-(6x^2+9x)} \\
-14x+7 \\
\underline{-(-14x-21)} \\
28
\end{array}$$

Since the remainder is nonzero, the function has an oblique asymptote at $y=3x-7$.

72. $f(x)=\dfrac{x^2+9x+14}{x+2}$

Since the degree of the numerator is exactly one more than the degree of the denominator, $f(x)$ may have an oblique asymptote.

$$\begin{array}{r}
x+7 \\
x+2\overline{)x^2+9x+14} \\
\underline{-(x^2+2x)} \\
7x+14 \\
\underline{-(7x+14)} \\
0
\end{array}$$

Since the remainder is 0, the function does not have an oblique asymptote.

73. $g(x)=\dfrac{8x^2+6x-25}{4x+9}$

Since the degree of the numerator is exactly one more than the degree of the denominator, $g(x)$ may have an oblique asymptote.

$$\begin{array}{r}
2x-3 \\
4x+9\overline{)8x^2+6x-25} \\
\underline{-(8x^2+18x)} \\
-12x-25 \\
\underline{-(-12x-27)} \\
2
\end{array}$$

Since the remainder is not 0, the function has an oblique asymptote at $y=2x-3$.

74. $h(x)=\dfrac{18x^3-12x^2+7x}{12x^3-9x^2+4x}$

Since the degree of the numerator is the same as the degree of the denominator, the function has a horizontal asymptote at the ratio of the leading coefficients, $y=\dfrac{18}{12}=\dfrac{3}{2}$.

75. $f(x)=\dfrac{2}{x+2}$

Find the y-intercept by evaluating $f(0)$.

$f(0)=\dfrac{2}{0+2}=\dfrac{2}{2}=1$

The y-intercept is $(0,1)$.

Since the numerator is never equal to 0, there are no x-intercepts.

Since f is a proper rational function, there is a horizontal asymptote at $y=0$.

Find the points of discontinuity by finding the real zeros of the denominator.

$x+2=0$

$x=-2$

The function has a vertical asymptote at $x=-2$.

Evaluate $f(x)$ as $x\to-2$ from the right.

x	$f(x)$
-1	2
-1.5	4
-1.9	20

As $x\to-2$ from the right, $f(x)$ increases.

Evaluate $f(x)$ as $x\to-2$ from the left.

x	$f(x)$
-3	-2
-2.5	-4
-2.1	-20

As $x \to -2$ from the left, $f(x)$ decreases.

Graph the asymptotes with dashed lines and plot the points. Connect the points on either side of the vertical asymptote with a smooth curve that approaches the asymptotes.

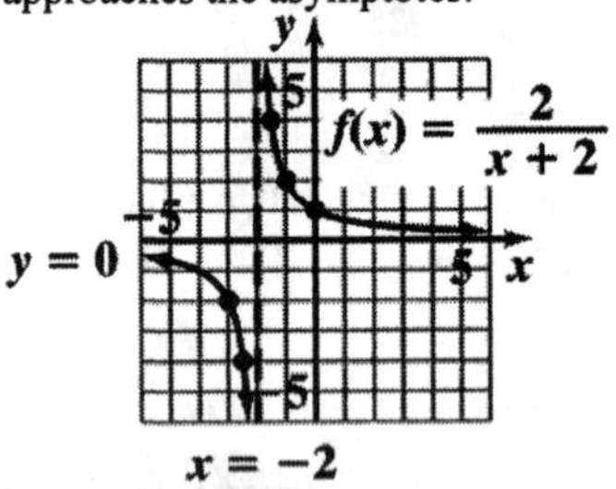

76. $g(x)=\dfrac{7x-5}{2x+1}$

Find the y-intercept by evaluating $g(0)$.

$$g(0)=\frac{7(0)-5}{2(0)+1}=\frac{-5}{1}=-5$$

The y-intercept is $(0,-5)$.

Find the x-intercepts by finding the real zeros of the numerator.

$$7x-5=0$$
$$7x=5$$
$$x=\frac{5}{7}$$

The only x-intercept is $\left(\frac{5}{7},0\right)$.

Since the degree of the numerator is the same as the degree of the denominator, the function has a horizontal asymptote at the ratio of the leading coefficients, $y=\frac{7}{2}$.

Find the points of discontinuity by finding the real zeros of the denominator.

$$2x+1=0$$
$$2x=-1$$
$$x=-\frac{1}{2}$$

The function has a vertical asymptote at $x=-\frac{1}{2}$.

Evaluate $g(x)$ as $x \to -\frac{1}{2}$ from the right.

x	$g(x)$
2	1.8
1	0.7
0	-5

As $x \to -\frac{1}{2}$ from the right, $g(x)$ decreases.

Evaluate $g(x)$ as $x \to -\frac{1}{2}$ from the left.

x	$g(x)$
-3	5.2
-2	6.3
-1	12

As $x \to -\frac{1}{2}$ from the left, $g(x)$ increases.

Graph the asymptotes as dashed lines and plot the points. Connect the points on either side of the vertical asymptote with a smooth curve that approaches the asymptotes.

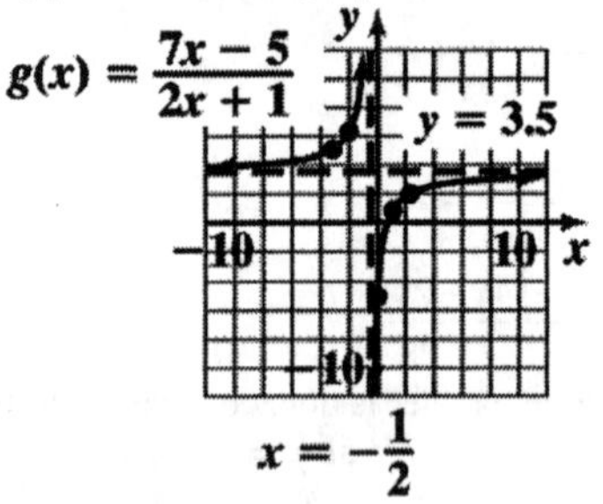

77. $h(x)=\dfrac{x^2+x-20}{x-4}$

$$=\frac{(x+5)(x-4)}{x-4}$$
$$=x+5, \quad (x \neq 4)$$

At $x=4$ both the numerator and denominator equal 0. Since $x=4$ does not make the denominator of the simplified function equal 0, there is a hole in the graph at $x=4$. Graph the line $y=x+5$, making a hole in the graph at $x=4$.

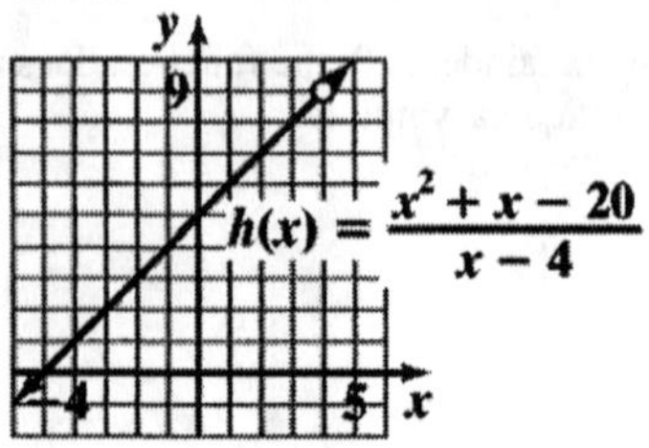

78. $f(x)=\dfrac{x-3}{x^2+x-2}=\dfrac{x-3}{(x+2)(x-1)}$

Find the y-intercept by evaluating $f(0)$.

$f(0)=\dfrac{0-3}{(0)^2+0-2}=\dfrac{-3}{-2}=\dfrac{3}{2}$

The y-intercept is $\left(0,\dfrac{3}{2}\right)$.

Find the x-intercepts by finding the real zeros of the numerator.

$x-3=0$

$x=3$

The only x-intercept is $(3,0)$.

Since *f* is a proper rational function, there is a horizontal asymptote at $y=0$. To find the points of discontinuity, find all the real zeros of the denominator.

$(x+2)(x-1)=0$

$x+2=0$ or $x-1=0$

$x=-2$ or $x=1$

The graph of $f(x)$ has vertical asymptotes at $x=-2$ and $x=1$.

Evaluate $f(x)$ as $x\to 1$ from the right.

x	$f(x)$
2	–0.25
1.5	–0.36
1.1	–6.13

As $x\to 1$ from the right, $f(x)$ decreases.

Evaluate $f(x)$ as $x\to 1$ from the left.

x	$f(x)$
0	1.5
0.5	2
0.9	7.24

As $x\to 1$ from the left, $f(x)$ increases.

Evaluate $f(x)$ as $x\to -2$ from the right.

x	$f(x)$
–1	2
–1.5	3.6
–1.9	16.89

As $x\to -2$ from the right, $f(x)$ increases.

Evaluate $f(x)$ as $x\to -2$ from the left.

x	$f(x)$
–3	–1.5
–2.5	–3.14
–2.1	–16.45

As $x\to -2$ from the left, $f(x)$ decreases.

Graph the asymptotes as dashed lines and plot the points. Connect the points in each region defined by the vertical asymptotes with a smooth curve that approaches the asymptotes.

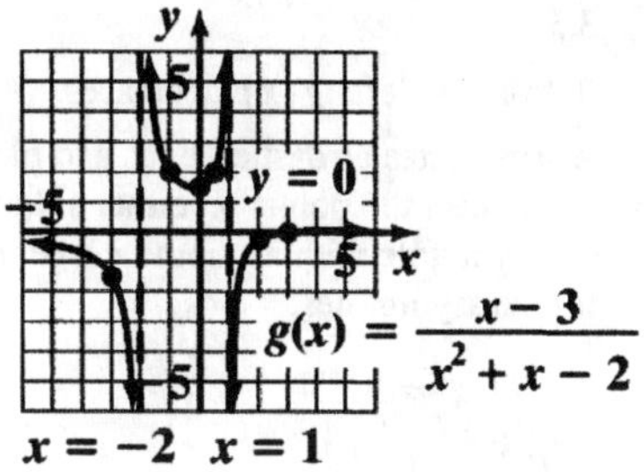

79. $g(x)=\dfrac{x^2+5x-12}{x-3}$

Find the y-intercept by evaluating $g(0)$.

$g(0)=\dfrac{(0)^2+5(0)-12}{0-3}=\dfrac{-12}{-3}=4$

The y-intercept is $(0,4)$.

Since the degree of the numerator is exactly one more than the degree of the denominator, the function may have an oblique asymptote.

$$\begin{array}{r} x+8 \\ x-3\overline{)x^2+5x-12} \\ \underline{-(x^2-3x)} \\ 8x-12 \\ \underline{-(8x-24)} \\ 12 \end{array}$$

Since the remainder is not 0, the function has an oblique asymptote at $y=x+8$.

To find the points of discontinuity, find the real zeros of the denominator.

$x-3=0$

$x=3$

The function has a vertical asymptote at $x=3$.

Evaluate $g(x)$ as $x \to 3$ from the right.

x	$g(x)$
5	19
4	24
3.5	35.5

As $x \to 3$ from the right, $g(x)$ increases.

Evaluate $g(x)$ as $x \to 3$ from the left.

x	$g(x)$
1	3
2	−2
2.5	−13.5

As $x \to 3$ from the left, $g(x)$ decreases.

Graph the asymptotes as dashed lines and plot the points. Connect the points on either side of the vertical asymptote with a smooth curve that approaches the asymptotes.

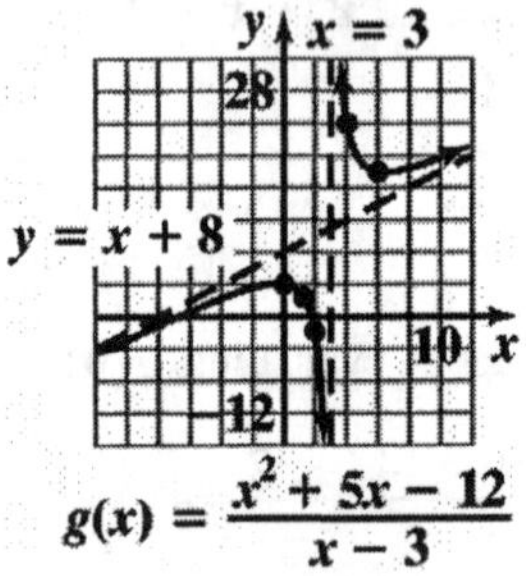

80. $h(x) = \dfrac{2x^2 - 9x + 15}{x-6}$

Find the y-intercept by evaluating $h(0)$.

$$h(0) = \frac{2(0)^2 - 9(0) + 15}{0-6} = \frac{15}{-6} = -\frac{5}{2}$$

The y-intercept is $\left(0, -\frac{5}{2}\right)$.

Since the degree of the numerator is exactly one more than the degree of the denominator, the function may have an oblique asymptote.

$$\begin{array}{r} 2x+3 \\ x-6\overline{)2x^2-9x+15} \\ \underline{-(2x^2-12x)} \\ 3x+15 \\ \underline{-(3x-18)} \\ 33 \end{array}$$

Since the remainder is not 0, the function has an oblique asymptote at $y = 2x+3$.

To find the points of discontinuity, find all the real zeros of the denominator.

$x - 6 = 0$

$x = 6$

The function has a vertical asymptote at $x = 6$.

Evaluate $h(x)$ as $x \to 6$ from the right.

x	$h(x)$
9	32
8	35.5
7	50
6.5	82

As $x \to 6$ from the right, $h(x)$ increase.

Evaluate $h(x)$ as $x \to 6$ from the left.

x	$h(x)$
3	−2
4	−5.5
5	−20
5.5	−52

As $x \to 6$ from the left, $h(x)$ decreases.

Graph the asymptotes as dashed lines and plot the points. Connect the points on either side of the vertical asymptote with a smooth curve that approaches the asymptotes.

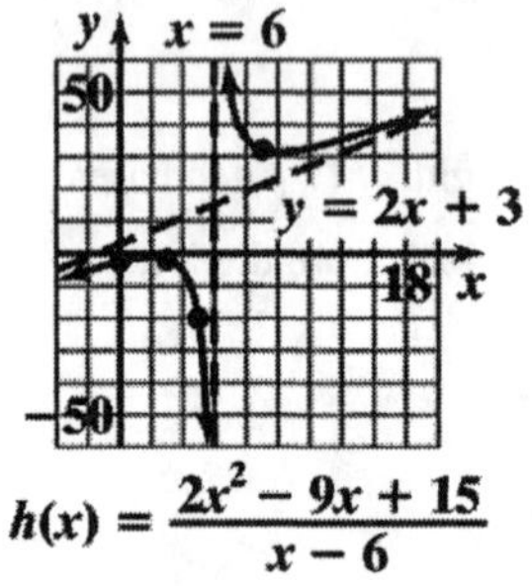

Chapter 10 Practice Test

1. a. Since f has an even degree and a negative leading coefficient, its graph is below the x-axis for large positive values of x.

   b. Since f has an even degree and a negative leading coefficient, its graph is below the x-axis for large negative values of x.

2. $f(x)=x(x-4)^3(2x+3)^2$

   $x(x-4)^3(2x+3)^2=0$

   $x=0$ or $x-4=0$ or $2x+3=0$

   $x=4 \qquad 2x=-3$

   $x=-\frac{3}{2}$

   The zeros are $x=0$, $x=4$ with multiplicity 3, and $x=-\frac{3}{2}$ with multiplicity 2.

3. Since $f(x)=3x^7-12x^5+x^3-x^2-11$ is a polynomial of degree 7, its graph has a maximum of $7-1=6$ turning points.

4. Since the graph has 3 turning points, the degree of the function must be at least $3+1=4$.

5. Since $f(x)=3x^5+2x^4-20x^3-12x^2+x+15$ is a polynomial of degree 5, its graph has a maximum of 5 x-intercepts. Since it degree is odd, the ends go in opposite directions. Thus, its graph will have at least one x-intercept.

6. a. $f(x)=x^5-3x^4-7x^3-x^2+6$

   Since the coefficients of $f(x)$ have 2 sign changes, there are 2 or 0 positive real zeros.

   b. $f(-x)=(-x)^5-3(-x)^4-7(-x)^3-(-x)^2+6$

   $=-x^5-3x^4+7x^2-x+6$

   Since the coefficients of $f(-x)$ have 3 sign changes, there are 3 or 1 negative real zeros.

7. $g(x)=4x^6+5x^4-12x^3-18x-6$

   The potential rational zeros of $g(x)$ are of the form $\frac{p}{q}$ where p is a factor of –6 and q is a factor of 4.

   Factors of –6: $\pm1,\pm2,\pm3,\pm6$

   Factors of 4: $\pm1,\pm2,\pm4$

   Potential rational zeros:

   $\pm1,\pm2,\pm3,\pm6,\pm\frac{1}{2},\pm\frac{3}{2},\pm\frac{1}{4},\pm\frac{3}{4}$

8. $f(x)=x^4+x^3-3x^2+9x-108$

   There are 3 sign changes in $f(x)$, so there are 3 or 1 positive real zeros.

   $f(-x)=x^4-x^3-3x^2-9x-108$

   There is one sign change in $f(-x)$, so there is exactly one negative real zero.

   Potential rational zeros:

   $\pm1,\pm2,\pm3,\pm4,\pm6,\pm9,\pm12,\pm18,\pm27,\pm36,\pm54,\pm108$

   $$\begin{array}{r|rrrrr} -1 & 1 & 1 & -3 & 9 & -108 \\ & & -1 & 0 & 3 & -12 \\ \hline & 1 & 0 & -3 & 12 & -120 \end{array}$$

   -1 is not a zero since the remainder is not 0.

   $$\begin{array}{r|rrrrr} -2 & 1 & 1 & -3 & 9 & -108 \\ & & -2 & 2 & 2 & -22 \\ \hline & 1 & -1 & -1 & 11 & -130 \end{array}$$

   -2 is not a zero since the remainder is not 0.

   $$\begin{array}{r|rrrrr} -3 & 1 & 1 & -3 & 9 & -108 \\ & & -3 & 6 & -9 & 0 \\ \hline & 1 & -2 & 3 & 0 & -108 \end{array}$$

   -3 is not a zero since the remainder is not 0.

   $$\begin{array}{r|rrrrr} -4 & 1 & 1 & -3 & 9 & -108 \\ & & -4 & 12 & -36 & 108 \\ \hline & 1 & -3 & 9 & -27 & 0 \end{array}$$

   -4 is a zero, so $x+4$ is a factor.

   $f(x)=(x+4)(x^3-3x^2+9x-27)$

   $=(x+4)(x^2(x-3)+9(x-3))$

   $=(x+4)(x-3)(x^2+9)$

   $(x+4)(x-3)(x^2+9)=0$

   $x+4=0$ or $x-3=0$ or $x^2+9=0$

   $x=-4 \qquad x=3 \qquad x^2=-9$

   $x=\pm\sqrt{-9}$

   $x=\pm3i$

   The zeros are $x=-4, x=3, x=-3i$, and $x=3i$.

9. a. $f(x)=x^3-5x^2-4x+8$

$$f(2)=(2)^3-5(2)^2-4(2)+8$$
$$=8-20-8+8$$
$$=-12$$
$$f(5)=(5)^3-5(5)^2-4(5)+8$$
$$=125-125-20+8$$
$$=-12$$

Since $f(2)$ and $f(5)$ have the same sign, we cannot determine if a zero exists on the interval $(2,5)$.

b. $f(x)=x^3-5x^2-4x+8$

$$f(-2)=(-2)^3-5(-2)^2-4(-2)+8$$
$$=-8-20+8+8$$
$$=-12$$
$$f(0)=(0)^3-5(0)^2-4(0)+8$$
$$=0+0+0+8$$
$$=8$$

Since $f(-2)$ and $f(0)$ have opposite signs (and the function is continuous), there is at least one real zero in the interval $(-2,0)$.

10. $f(x)=x^4-5x^3-4x^2+6x-10$

5	1	−5	−4	6	−10
		5	0	−20	−70
	1	0	−4	−14	−80

Since the bottom row of synthetic division contains a negative, we cannot conclude that 5 is an upper bound for the zeros.

−2	1	−5	−4	6	−10
		−2	14	−20	28
	1	−7	10	−14	18

Since the bottom row of synthetic division alternates between positive and negative values, we can conclude that −2 is a lower bound for the zeros. Since we cannot determine if 5 is an upper bound, we also cannot determine if the real zeros are bounded over the interval $[-2,5]$.

11. $f(x)=\dfrac{x^2+2x-8}{x+4}=\dfrac{(x+4)(x-2)}{x+4}$

Points of discontinuity occur where the denominator equals 0.

$$x+4=0$$
$$x=-4$$

The function has a discontinuity at $x=-4$. Simplifying the function yields

$$f(x)=\frac{(x+4)(x-2)}{x+4}=x-2$$

Since $x=-4$ no longer makes the denominator equal 0, the function has a hole in the graph at $x=-4$.

12. $g(x)=\dfrac{x+5}{x-5}$

Points of discontinuity occur where the denominator equals 0.

$$x-5=0$$
$$x=5$$

There is a point of discontinuity at $x=5$. Since the function is in lowest terms, there is a vertical asymptote at $x=5$.

13. $h(x)=\dfrac{x^2+3x-10}{2x^2+3x}=\dfrac{(x+5)(x-2)}{x(2x+3)}$

Points of discontinuity occur where the denominator equals 0.

$$x(2x+3)=0$$
$$x=0 \text{ or } 2x+3=0$$
$$2x=-3$$
$$x=-\frac{3}{2}$$

There are points of discontinuity at $x=0$ and $x=-\frac{3}{2}$. Since the function is in lowest terms, both result in vertical asymptotes.

14. $f(x)=\dfrac{15x^2-7x+3}{3x^2+4}$

Since the degree of the numerator is the same as the degree of the denominator, the function has a horizontal asymptote at the ratio of the leading coefficients, $y=\frac{15}{3}=5$.

15. $g(x)=\dfrac{x^2-6x+9}{x-3}$

Since the degree of the numerator is exactly one more than the degree of the denominator, the function may have an oblique asymptote.

$$\begin{array}{r} x-3 \\ x-3\overline{)x^2-6x+9} \\ \underline{-(x^2-3x)} \\ -3x+9 \\ \underline{-(-3x+9)} \\ 0 \end{array}$$

Since the remainder is 0, the function does not have an oblique asymptote. (There is a hole in the graph at $x=3$.)

16. $h(x)=\dfrac{4x^2+9x-7}{x+3}$

Since the degree of the numerator is exactly one more than the degree of the denominator, the function may have an oblique asymptote.

$$\begin{array}{r} 4x-3 \\ x+3\overline{)4x^2+9x-7} \\ \underline{-(4x^2+12x)} \\ -3x-7 \\ \underline{-(-3x-9)} \\ 2 \end{array}$$

Since the remainder is not 0, the function has an oblique asymptote at $y=4x-3$.

17. $f(x)=\dfrac{4x^3-5x^2+64}{x^5-4x^4+16x+32}$

Since this is a proper rational function, there is a horizontal asymptote at $y=0$.

18. $f(x)=\begin{cases} 2x-3 & \text{if } x\le -2 \\ -2x+3 & \text{if } x>-2 \end{cases}$

For $x\le -2$, graph $f(x)=2x-3$.

$f(-3)=2(-3)-3=-9$

$f(-2)=2(-2)-3=-7$

Since this piece is defined for $x=-2$, make a solid dot at $(-2,-7)$.

For $x>-2$, graph $f(x)=-2x+3$.

$f(-2)=-2(-2)+3=7$

$f(0)=-2(0)+3=3$

Since this piece is not defined for $x=-2$, make an open circle at $(-2,7)$.

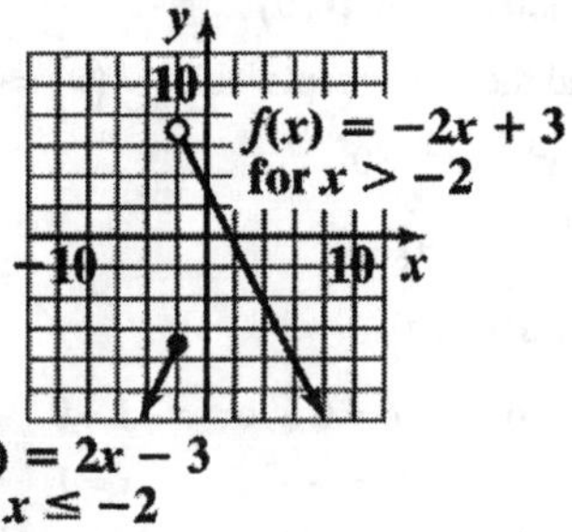

19. $g(x)=\left[\!\left[\dfrac{x-1}{4}\right]\!\right]$

Find the closed endpoint value for $g(x)=0$:

$0=\dfrac{x-1}{4}$

$0=x-1$

$1=x$

Find the closed endpoint value for $g(x)=1$:

$1=\dfrac{x-1}{4}$

$4=x-1$

$5=x$

Find the closed endpoint value for $g(x)=2$:

$2=\dfrac{x-1}{4}$

$8=x-1$

$9=x$

From these points, we know the interval for each line segment spans 4 units.

Interval	$g(x)$
$[1,5)$	0
$[5,9)$	1
$[9,13)$	2

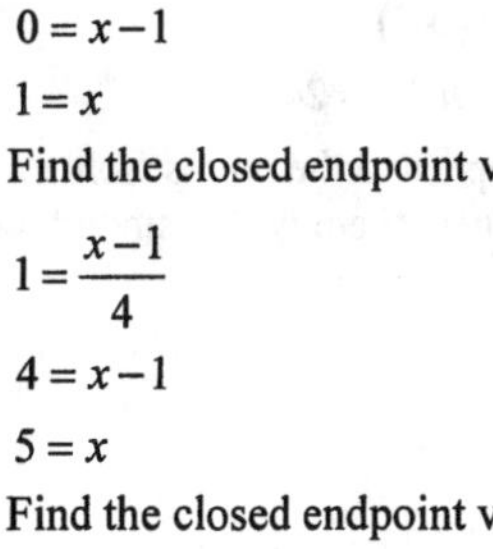

20. $h(x)=x^4+3x^3-4x^2$

Find the y-intercept by evaluating $h(0)$.

$h(0)=(0)^4+3(0)^3-4(0)^2=0$

The y-intercept is $(0,0)$.

To find the x-intercepts, solve $h(x)=0$.

$x^4+3x^3-4x^2=0$

$x^2(x^2+3x-4)=0$

$x^2(x+4)(x-1)=0$

$x^2=0$ or $x+4=0$ or $x-1=0$

$x=0$ $\quad x=-4$ $\quad x=1$

The x-intercepts are $(0,0),(-4,0)$ and $(1,0)$.

These break the graph into 4 regions. Find a point in each region.

Interval	x	$h(x)$
$(-\infty,-4)$	-5	$h(-5)=150$
$(-4,0)$	-2	$h(-2)=-24$
$(0,1)$	0.5	$h(0.5)=-0.6$
$(1,\infty)$	2	$h(2)=24$

Graph the intercepts and the points from each region. Then connect them with a smooth curve.

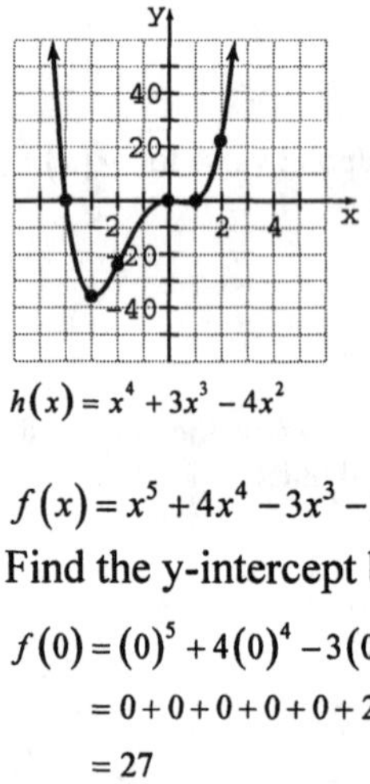

$h(x)=x^4+3x^3-4x^2$

21. $f(x)=x^5+4x^4-3x^3-15x^2+18x+27$

Find the y-intercept by evaluating $f(0)$.

$$f(0)=(0)^5+4(0)^4-3(0)^3-15(0)^2+18(0)+27$$
$$=0+0+0+0+0+27$$
$$=27$$

The y-intercept is $(0,27)$.

Since the coefficients of $f(x)$ have 2 sign changes, there are 2 or 0 positive real zeros.

$f(-x)=-x^5+4x^4+3x^3-15x^2-18x+27$

Since the coefficients of $f(-x)$ have 3 sign changes, there are 3 or 1 negative real zeros.

The leading coefficient is 1 and the constant term is 27, so the potential rational zeros are $\pm1,\pm3,\pm9,\pm27$

Since there is at least one negative real zero, start by checking the negative possibilities.

$$\begin{array}{r|rrrrrr} -1 & 1 & 4 & -3 & -15 & 18 & 27 \\ & & -1 & -3 & 6 & 9 & -27 \\ \hline & 1 & 3 & -6 & -9 & 27 & 0 \end{array}$$

-1 is a zero, so $x+1$ is a factor.

$f(x)=(x+1)(x^4+3x^3-6x^2-9x+27)$

To find the other zeros, we work with the reduced polynomial

$Q(x)=x^4+3x^3-6x^2-9x+27$ which has the same potential rational zeros as $f(x)$.

$$\begin{array}{r|rrrrr} -1 & 1 & 3 & -6 & -9 & 27 \\ & & -1 & -2 & 8 & 1 \\ \hline & 1 & 2 & -8 & -1 & 28 \end{array}$$

-1 is not a zero since the remainder is not 0.

$$\begin{array}{r|rrrrr} -3 & 1 & 3 & -6 & -9 & 27 \\ & & -3 & 0 & 18 & -27 \\ \hline & 1 & 0 & -6 & 9 & 0 \end{array}$$

-3 is a zero, so $x+3$ is a factor.

$f(x)=(x+1)(x+3)(x^3-6x+9)$

We continue working with the reduced polynomial $R(x)=x^3-6x+9$. The potential rational zeros are $\pm1,\pm3,\pm9$. Since we eliminated -1 as a zero, we can start with -3.

$$\begin{array}{r|rrrr} -3 & 1 & 0 & -6 & 9 \\ & & -3 & 9 & -9 \\ \hline & 1 & -3 & 3 & 0 \end{array}$$

-3 is a zero, so $x+3$ is a factor.

$f(x)=(x+1)(x+3)^2(x^2-3x+3)$

We can now use the quadratic formula to find the zeros of x^2-3x+3.

$a=1, b=-3, c=3$

$$x=\frac{-b\pm\sqrt{b^2-4ac}}{2a}$$
$$=\frac{-(-3)\pm\sqrt{(-3)^2-4(1)(3)}}{2(1)}$$
$$=\frac{3\pm\sqrt{9-12}}{2}=\frac{3\pm\sqrt{-3}}{2}$$
$$=\frac{3\pm i\sqrt{3}}{2}$$

The last two zeros are complex numbers. The graph only has two x-intercepts: $(-3,0)$ and $(-1,0)$. They break the graph into 3 regions. Find a point in each region.

Interval	x	$f(x)$
$(-\infty,-3)$	-4	$f(-4)=-93$
$(-3,-1)$	-2	$f(-2)=-13$
$(-1,\infty)$	1	$f(1)=32$

Graph the intercepts and the points. Then connect them with a smooth curve.

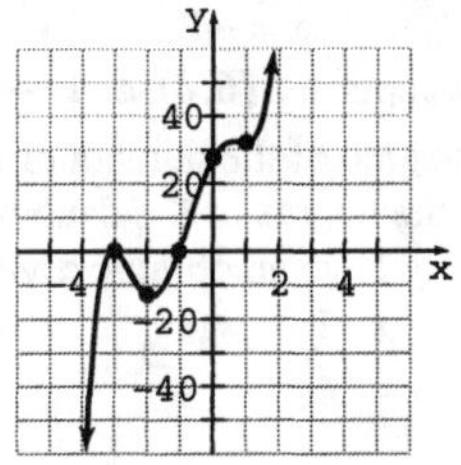

$f(x)=x^5+4x^4-3x^3-15x^2+18x+27$

22. $g(x)=x^3-9x^2+6x+16$

Find the y-intercept by evaluating $g(0)$.

$g(0)=(0)^3-9(0)^2+6(0)+16=16$

The y-intercept is $(0,16)$.

Since the coefficients of $g(x)$ have 2 sign changes, there are 2 or 0 positive real zeros.

$g(-x)=-x^3-9x^2-6x+16$

Since the coefficients of $g(-x)$ have 1 sign change, there is exactly one negative real zero. The leading coefficient is 1 and the constant term is 16, so the potential rational zeros are $\pm1,\pm2,\pm4,\pm8,\pm16$

$$\begin{array}{r|rrrr} -1 & 1 & -9 & 6 & 16 \\ & & -1 & 10 & -16 \\ \hline & 1 & -10 & 16 & 0 \end{array}$$

-1 is a zero, so $x+1$ is a factor.

$g(x)=(x+1)(x^2-10x+16)$
$=(x+1)(x-8)(x-2)$

The x-intercepts are $(-1,0)$, $(8,0)$, and $(2,0)$. These break the graph into 4 regions. Find a point in each region.

Interval	x	$g(x)$
$(-\infty,-1)$	-2	$g(-2)=-40$
$(-1,2)$	1	$g(1)=14$
$(2,8)$	5	$g(5)=-54$
$(8,\infty)$	9	$g(9)=70$

Graph the intercepts and the points from each region. Then connect them with a smooth curve.

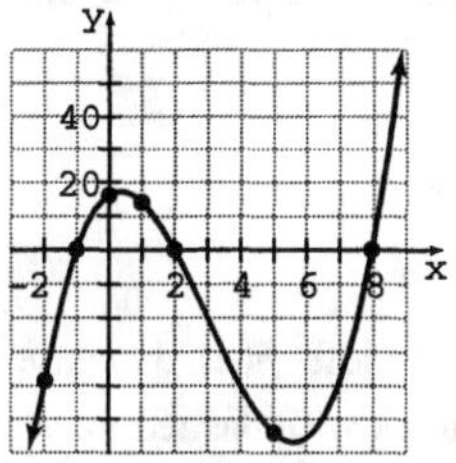

$g(x)=x^3-9x^2+6x+16$

23. $h(x)=\dfrac{3x}{x^2-2x-15}=\dfrac{3x}{(x-5)(x+3)}$

Find the y-intercept by evaluating $h(0)$.

$h(0)=\dfrac{3(0)}{(0)^2-2(0)-15}=0$

The y-intercept is $(0,0)$.

Since $h(x)$ is a proper rational function, there is a horizontal asymptote at $y=0$.

Points of discontinuity occur where the denominator equals 0.

$(x-5)(x+3)=0$

$x-5=0$ or $x+3=0$

$x=5$ $\quad x=-3$

Since the function is in lowest terms, there are vertical asymptotes at $x=5$ and $x=-3$.

Evaluate $h(x)$ as $x\to5$ from the right.

x	$h(x)$
7	1.05
6	2
5.5	3.8824

As $x\to5$ from the right, $h(x)$ increases.

Evaluate $h(x)$ as $x\to5$ from the left.

x	$h(x)$
3	-0.75
4	-1.71
4.5	-3.6

As $x\to5$ from the left, $h(x)$ decreases.

Evaluate $h(x)$ as $x\to-3$ from the right.

x	$h(x)$
-1	0.25
-2	0.86
-2.5	2

As $x \to -3$ from the right, $h(x)$ increases.

Evaluate $h(x)$ as $x \to -3$ from the left.

x	$h(x)$
–5	–0.75
–4	–1.33
–3.5	–2.47

As $x \to -3$ from the left, $h(x)$ decreases.

Graph the asymptotes with dashed lines and plot the points. Connect the points within each region defined by the vertical asymptotes with a smooth curve that approaches the asymptotes.

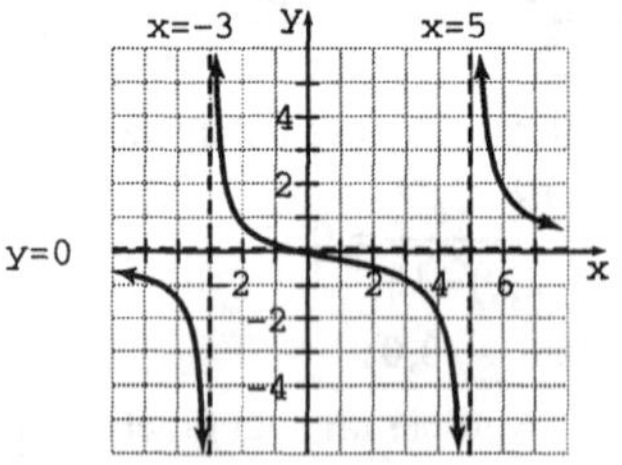

$$h(x) = \frac{3x}{x^2 - 2x - 15}$$

24. $f(x) = \dfrac{12x+9}{4x+3} = \dfrac{3(4x+3)}{4x+3} = 3, \quad \left(x \neq -\dfrac{3}{4}\right)$

At $x = -\frac{3}{4}$ both the numerator and denominator have a value of 0. Since $x = -\frac{3}{4}$ does not make the denominator of the simplified function equal 0, there is a hole in the graph at $x = -\frac{3}{4}$. Graph the line $y = 3$, making a hole in the graph at $x = -\frac{3}{4}$.

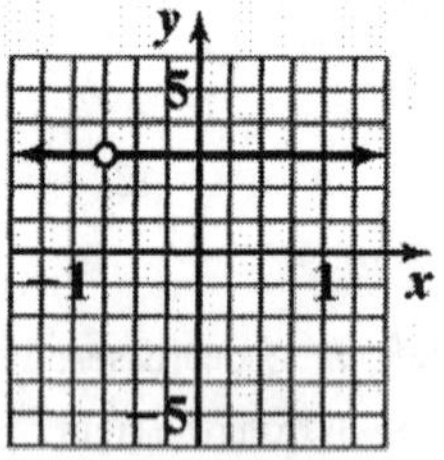

$$f(x) = \frac{12x+9}{4x+3}$$

25. $g(x) = \dfrac{x^2 - 2x - 24}{x - 1} = \dfrac{(x-6)(x+4)}{x-1}$

Find the y-intercept by evaluating $g(0)$.

$$g(0) = \frac{(0)^2 - 2(0) - 24}{0 - 1} = \frac{-24}{-1} = 24$$

The y-intercept is $(0, 24)$.

Find the x-intercepts by finding values where the numerator equals 0.

$(x-6)(x+4) = 0$

$x - 6 = 0$ or $x + 4 = 0$

$x = 6 \qquad x = -4$

The x-intercepts are $(6, 0)$ and $(-4, 0)$.

Since the degree of the numerator is exactly one more than the degree of the denominator, the function may have an oblique asymptote.

$$\begin{array}{r} x-1 \\ x-1\overline{)x^2-2x-24} \\ \underline{-(x^2-x)} \\ -x-24 \\ \underline{-(-x+1)} \\ -25 \end{array}$$

Since the remainder is not 0, the function has an oblique asymptote at $y = x - 1$.

Points of discontinuity occur where the denominator equals 0.

$x - 1 = 0$

$x = 1$

The function has a vertical asymptote at $x = 1$.

Evaluate $g(x)$ as $x \to 1$ from the right.

x	$g(x)$
3	–10.5
2	–24
1.5	–49.5

As $x \to 1$ from the right, $g(x)$ decreases.

Evaluate $g(x)$ as $x \to 1$ from the left.

x	$g(x)$
–1	10.5
0	24
0.5	49.5

As $x \to 1$ from the left, $g(x)$ increases.

Graph the asymptotes with dashed lines and plot the points. Connect the points on either side of the vertical asymptote with a smooth curve that approaches the asymptotes.

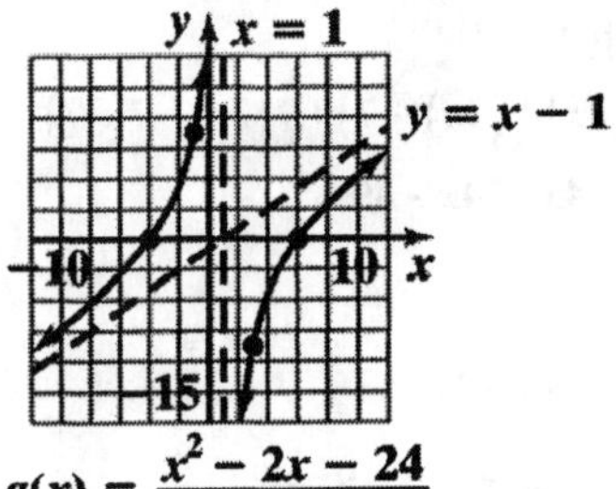

$g(x) = \dfrac{x^2 - 2x - 24}{x - 1}$

Chapter 10 Cumulative Review Test

1. $$\begin{aligned} -6^2 + 12 \div 3 - 4 \cdot 6 &= -36 + 12 \div 3 - 4 \cdot 6 \\ &= -36 + 4 - 24 \\ &= -32 - 24 \\ &= -56 \end{aligned}$$

2. $$\begin{aligned} \frac{5.2 \times 10^7}{1.3 \times 10^{-3}} &= \frac{5.2}{1.3} \times \frac{10^7}{10^{-3}} \\ &= 4 \times 10^{7-(-3)} \\ &= 4 \times 10^{10} \\ &= 40{,}000{,}000{,}000 \end{aligned}$$

3. $$\begin{aligned} -\frac{1}{3}(9x - 4) &= \frac{1}{2}(2 - 4x) \\ -3x + \frac{4}{3} &= 1 - 2x \\ -x &= -\frac{1}{3} \\ x &= \frac{1}{3} \end{aligned}$$

4. $|3x - 5| > 7$

$3x - 5 < -7$ or $3x - 5 > 7$

$3x < -2$ $\quad$ $3x > 12$

$x < -\frac{2}{3}$ $\quad$ $x > 4$

$\left\{ x \mid x < -\frac{2}{3} \text{ or } x > 4 \right\}$

5. $y = \frac{1}{4}x - 3$

y-intercept: $(0, -3)$

slope: $\frac{1}{4}$

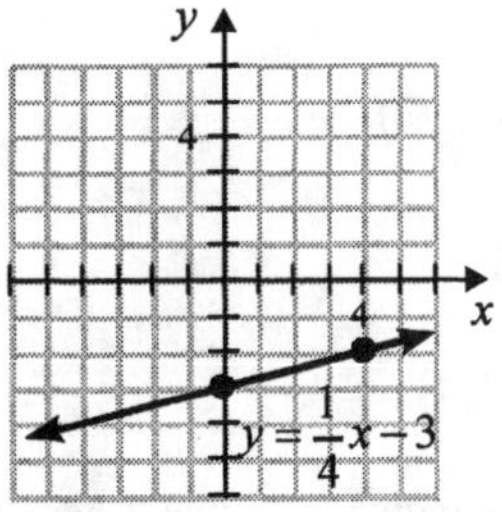

6. Start by finding the slope of the line $4x - 6y = 7$.

$$\begin{aligned} 4x - 6y &= 7 \\ -6y &= -4x + 7 \\ y &= \frac{-4x + 7}{-6} \\ y &= \frac{2}{3}x - \frac{7}{6} \end{aligned}$$

The slope of the given line is $\frac{2}{3}$. Since the desired line is perpendicular to the given line, its slope must be the negative-reciprocal of this slope.

$$\frac{2}{3}m = -1$$

$$m = -\frac{3}{2}$$

Now find the equation of the line with slope $m = -\frac{3}{2}$ that passes through the point $(-4, 3)$.

$$\begin{aligned} y - y_1 &= m(x - x_1) \\ y - 3 &= -\frac{3}{2}(x + 4) \\ y - 3 &= -\frac{3}{2}x - 6 \\ y &= -\frac{3}{2}x - 3 \end{aligned}$$

7. $3x + y = 10$ (1)

$y = 3x - 2$ (2)

Substitute $3x - 2$ for y in (1).

$$3x + (3x - 2) = 10$$
$$6x - 2 = 10$$
$$6x = 12$$
$$x = 2$$

Substitute 2 for x in (2).

$$y = 3(2) - 2$$
$$= 6 - 2$$
$$= 4$$

The solution is $(2, 4)$.

8. Graph $y = -x + 6$ with a solid line. Check if the point $(0,0)$ is a solution to the strict inequality.

$$y < -x + 6$$
$$0 < -(0) + 6$$
$$0 < 6 \quad \text{TRUE}$$

Since a true statement resulted, shade the region containing the point $(0,0)$.

Graph $y = \frac{2}{3}x - 1$ with a dashed line since the inequality is strict. Check if the point $(0,0)$ is a solution the inequality.

$$y > \frac{2}{3}x - 1$$
$$0 > \frac{2}{3}(0) - 1$$
$$0 > -1 \quad \text{TRUE}$$

Since a true statement resulted, shade the region containing the point $(0,0)$.

The solution is the set of points in the region that contains both shadings.

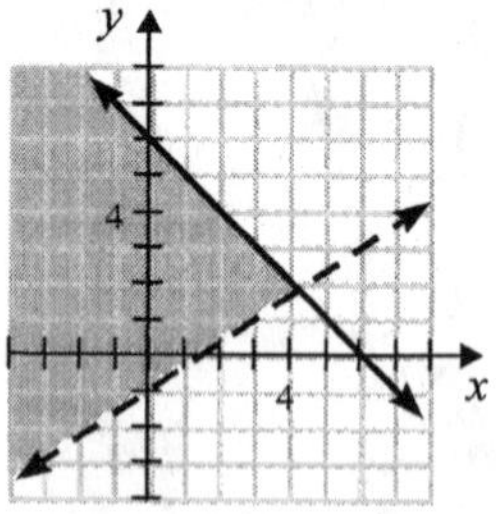

9. $(2x + 7)(2x - 7)$

$$= (2x)(2x) + (2x)(-7) + 7(2x) + 7(-7)$$
$$= 4x^2 - 14x + 14x - 49$$
$$= 4x^2 - 49$$

10. $(x^4 - 11) \div (x - 3)$

$$\begin{array}{r} x^3 + 3x^2 + 9x + 27 \\ x-3 \overline{)\, x^4 + 0x^3 + 0x^2 + 0x - 11} \\ \underline{-(x^4 - 3x^3)} \quad\quad\quad\quad\quad\quad \\ 3x^3 + 0x^2 \quad\quad\quad\quad \\ \underline{-(3x^3 - 9x^2)} \quad\quad\quad\quad \\ 9x^2 + 0x \quad\quad \\ \underline{-(9x^2 - 27x)} \quad\quad \\ 27x - 11 \\ \underline{-(27x - 81)} \\ 70 \end{array}$$

$$(x^4 - 11) \div (x - 3) = x^3 + 3x^2 + 9x + 27 + \frac{70}{x - 3}$$

11. $6x^2 + 9xy + 4xy + 6y^2$

$$= 3x(2x + 3y) + 2y(2x + 3y)$$
$$= (2x + 3y)(3x + 2y)$$

12. $\frac{7}{10x} + \frac{2}{3x^2} = \frac{7}{10x} \cdot \frac{3x}{3x} + \frac{2}{3x^2} \cdot \frac{10}{10}$

$$= \frac{21x}{30x^2} + \frac{20}{30x^2}$$
$$= \frac{21x + 20}{30x^2}$$

13. $$\frac{15}{x}+\frac{9x-7}{x+2}=9$$
$$\frac{15}{x}\cdot\frac{(x+2)}{(x+2)}+\frac{(9x-7)}{(x+2)}\cdot\frac{x}{x}=\frac{9}{1}\cdot\frac{x(x+2)}{x(x+2)}$$
$$\frac{15(x+2)}{x(x+2)}+\frac{x(9x-7)}{x(x+2)}=\frac{9x(x+2)}{x(x+2)}$$
Restricted values: $x=0, x=-2$
$$15(x+2)+x(9x-7)=9x(x+2)$$
$$15x+30+9x^2-7x=9x^2+18x$$
$$9x^2+8x+30=9x^2+18x$$
$$8x+30=18x$$
$$-10x=-30$$
$$x=3$$
Since this is not a restricted value, $x=3$ is the solution.

14. a. $f(x)=\sqrt{x+3}$
$$x+3\ge 0$$
$$x\ge -3$$
The domain is all real numbers $x\ge -3$.
$\{x\mid x\ge -3\}$

b.

x	$f(x)$
-3	$f(-3)=0$
-2	$f(-2)=1$
1	$f(1)=2$
6	$f(6)=3$

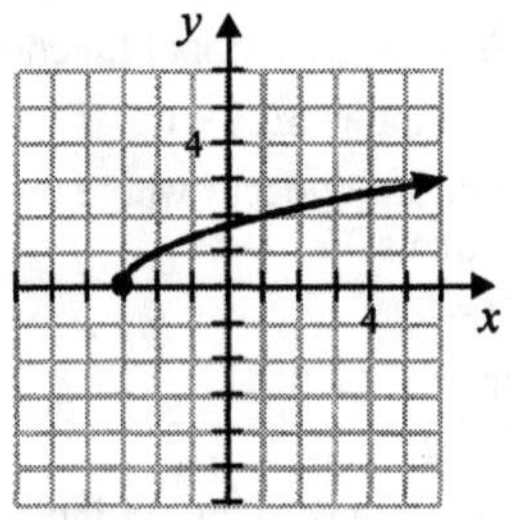

$f(x)=\sqrt{x+3}$

15. $$\frac{4}{3+\sqrt{5}}=\frac{4}{3+\sqrt{5}}\cdot\frac{3-\sqrt{5}}{3-\sqrt{5}}$$
$$=\frac{4(3-\sqrt{5})}{(3+\sqrt{5})(3-\sqrt{5})}$$
$$=\frac{12-4\sqrt{5}}{9-5}$$
$$=\frac{12-4\sqrt{5}}{4}$$
$$=3-\sqrt{5}$$

16. $x^2-14x=-9$
$$x^2-14x+9=0$$
$$a=1, b=-14, c=9$$
$$x=\frac{-b\pm\sqrt{b^2-4ac}}{2a}$$
$$=\frac{-(-14)\pm\sqrt{(-14)^2-4(1)(9)}}{2(1)}$$
$$=\frac{14\pm\sqrt{196-36}}{2}$$
$$=\frac{14\pm\sqrt{160}}{2}$$
$$=\frac{14\pm4\sqrt{10}}{2}$$
$$=7\pm2\sqrt{10}$$

17. $4\log_3 81=4\log_3 3^4$
$$=4\cdot4\log_3 3$$
$$=4\cdot4\cdot1$$
$$=16$$

18. $\log_4(x+16)=3$
$$4^3=x+16$$
$$64=x+16$$
$$48=x$$

19. $P(x)=2x^3-3x^2-59x+30$

Find the y-intercept by evaluating $P(0)$.

$P(0)=2(0)^3-3(0)^2-59(0)+30=30$

The y-intercept is $(0,30)$.

There are 2 sign changes in the coefficients of $P(x)$ so there are 2 or 0 positive real zeros.

$P(-x)=-2x^3-3x^2+59x+30$

There is one sign change in the coefficients of $P(-x)$ so there is exactly one negative real zero.

The leading coefficient is 2 and the constant term is 30, so the potential rational zeros are

$\pm1,\pm2,\ \pm3,\pm5,\pm6,\pm10,\pm15,\pm30,\pm\frac{1}{2},\pm\frac{3}{2},\pm\frac{5}{2},$

and $\pm\frac{15}{2}$.

$$\begin{array}{r|rrrr} -1 & 2 & -3 & -59 & 30 \\ & & -2 & 5 & 54 \\ \hline & 2 & -5 & -54 & 84 \end{array}$$

-1 is not a zero since the remainder is not 0.

$$\begin{array}{r|rrrr} -2 & 2 & -3 & -59 & 30 \\ & & -4 & 14 & 90 \\ \hline & 2 & -7 & -45 & 120 \end{array}$$

-2 is not a zero since the remainder is not 0.

$$\begin{array}{r|rrrr} -3 & 2 & -3 & -59 & 30 \\ & & -6 & 27 & 96 \\ \hline & 2 & -9 & -32 & 126 \end{array}$$

-3 is not a zero since the remainder is not 0.

$$\begin{array}{r|rrrr} -5 & 2 & -3 & -59 & 30 \\ & & -10 & 65 & -30 \\ \hline & 2 & -13 & 6 & 0 \end{array}$$

-5 is a zero, so $x+5$ is a factor.

$$\begin{aligned} P(x)&=(x+5)(2x^2-13x+6) \\ &=(x+5)(2x-1)(x-6) \end{aligned}$$

The x-intercepts are $(-5,0),\left(\frac{1}{2},0\right)$, and $(6,0)$.

They break the graph into 4 regions. Find a point in each region.

Interval	x	$P(x)$
$(-\infty,-5)$	-6	$P(-6)=-156$
$\left(-5,\frac{1}{2}\right)$	0	$P(0)=30$
$\left(\frac{1}{2},6\right)$	4	$P(4)=-126$
$(6,\infty)$	7	$P(7)=156$

Graph the intercepts and the points from each region. Then connect the points with a smooth curve.

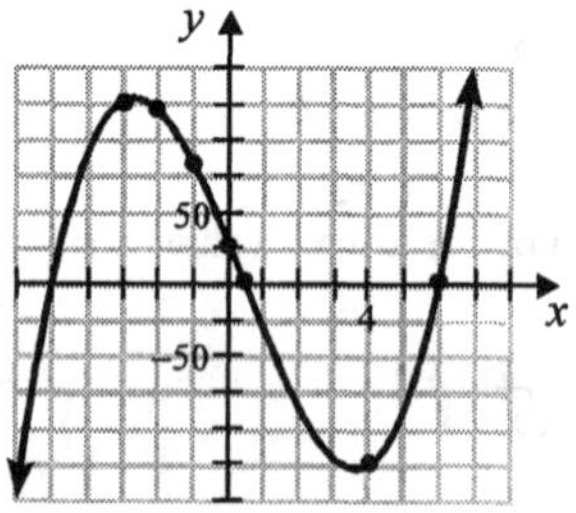

$P(x)=2x^3-3x^2-59x+30$

20. $f(x)=\dfrac{x+3}{x^2+5x+6}=\dfrac{x+3}{(x+3)(x+2)}$

Find the y-intercept by evaluating $f(0)$.

$f(0)=\dfrac{0+3}{(0)^2+5(0)+6}=\dfrac{3}{6}=\dfrac{1}{2}$

The y-intercept is $\left(0,\frac{1}{2}\right)$.

Since $f(x)$ is a proper rational function, there is a horizontal asymptote at $y=0$.

Points of discontinuity occur where the denominator equals 0.

$(x+3)(x+2)=0$

$x+3=0$ or $x+2=0$

$x=-3 \qquad x=-2$

The function has points of discontinuity at $x=-3$ and $x=-2$. Simplifying the function gives

$f(x)=\dfrac{x+3}{(x+3)(x+2)}=\dfrac{1}{x+2}$

Since $x=-3$ does not make the denominator of the simplified function equal 0, the function has a hole in the graph at $x=-3$. The function has a vertical asymptote at $x=-2$.

Evaluate $f(x)$ as $x\to-2$ from the right.

x	$f(x)$
−1	1
−1.5	2
−1.9	10

As $x \to -2$ from the right, $f(x)$ increases.

Evaluate $f(x)$ as $x \to -2$ from the left.

x	$f(x)$
−4	−0.5
−3	undefined
−2.5	−2
−2.1	−10

As $x \to -2$ from the left, $f(x)$ decreases.

Graph the asymptotes with dashed lines and plot the points. Connect the points on either side of the vertical asymptote with a smooth curve that approaches the asymptotes. Make a hole in the graph at $x = -3$.

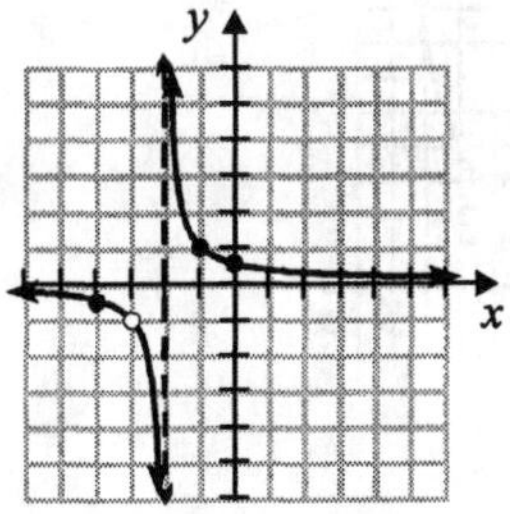

$$f(x) = \frac{x+3}{x^2+5x+6}$$

Chapter 11

Exercise Set 11.1

1.

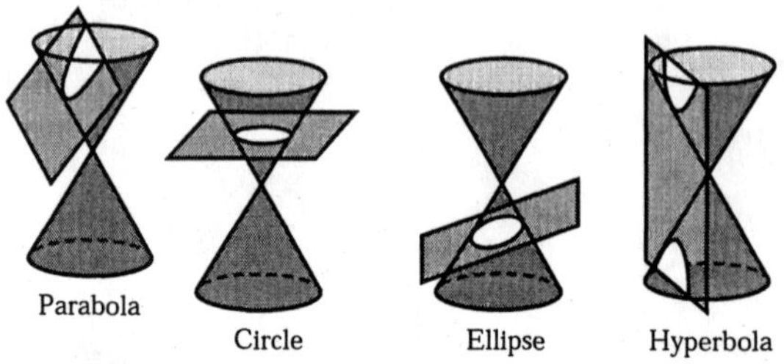

3. Yes, any parabola in the form $y = a(x-h)^2 + k$ is a function because each value of x corresponds to only one value of y. The domain is $\mathbb{R}$, the set of all real numbers. Since the vertex is at (h, k) and $a > 0$, the range is $\{y | y \geq k\}$.

5. The graphs have the same vertex, (3, 4). The first graph opens upward, and the second one opens downward.

7. The distance is always a positive number because both distances are squared and we use the principal square root.

9. A circle is the set of all points in a plane that are the same distance from a fixed point.

11. No, the coefficients of the y^2- term and the x^2- term must both be the same.

13. No, the coefficients of the y^2- term and the x^2- term must both be the same.

15. No, equations of parabolas do not include both x^2- and y^2- terms.

17. $y = (x-2)^2 + 3$

This is a parabola in the form $y = a(x-h)^2 + k$ with $a = 1$, $h = 2$ and $k = 3$. Since $a > 0$, the parabola opens upward. The vertex is (2, 3). The y-intercept is (0, 7). There are no x-intercepts.

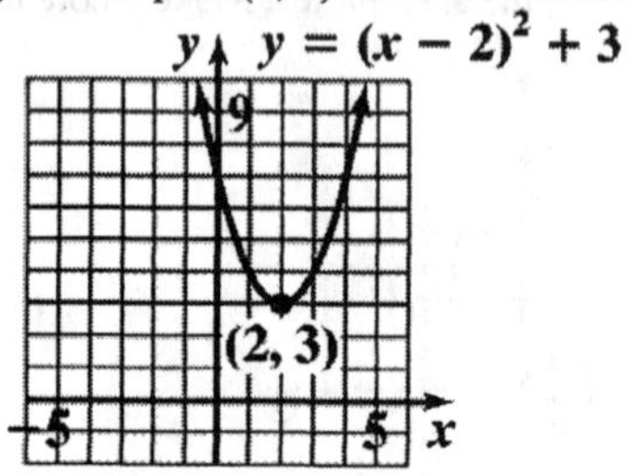

19. $y = (x+3)^2 + 2$

This is a parabola in the form $y = a(x-h)^2 + k$ with $a = 1$, $h = -3$ and $k = 2$. Since $a > 0$, the parabola opens upward. The vertex is (−3, 2). The y-intercept is (0, 11). There are no x-intercepts.

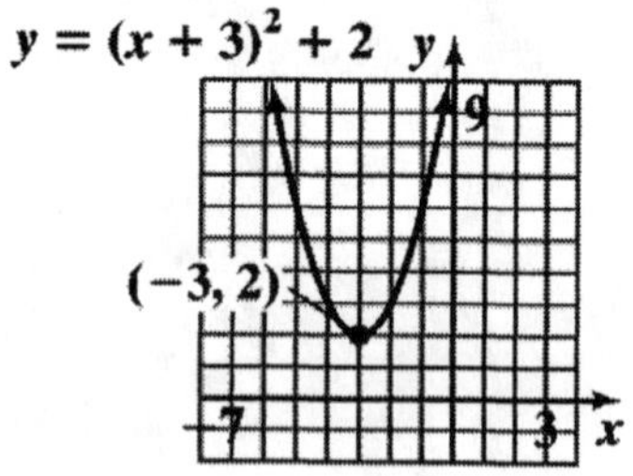

21. $y = (x-2)^2 - 1$

This is a parabola in the form $y = a(x-h)^2 + k$ with $a = 1$, $h = 2$ and $k = -1$. Since $a > 0$, the parabola opens upward. The vertex is (2, –1). The y-intercept is (0, 3). The x-intercepts are (1, 0) and (3, 0).

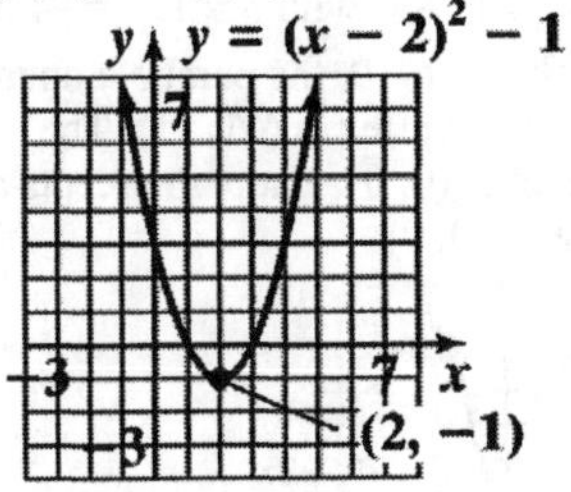

23. $y = -(x-1)^2 + 1$

This is a parabola in the form $y = a(x-h)^2 + k$ with $a = -1$, $h = 1$ and $k = 1$. Since $a < 0$, the parabola opens downward. The vertex is (1, 1). The y-intercept is (0, 0). The x-intercepts are (0, 0) and (2, 0).

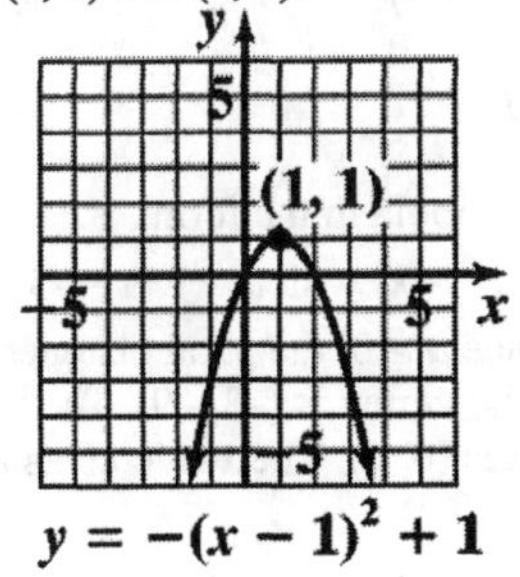

25. $y = -(x+3)^2 + 4$

This is a parabola in the form $y = a(x-h)^2 + k$ with $a = -1$, $h = -3$ and $k = 4$. Since $a < 0$, the parabola opens downward. The vertex is (–3, 4). The y-intercept is (0, –5). The x-intercepts are (–5, 0) and (–1, 0).

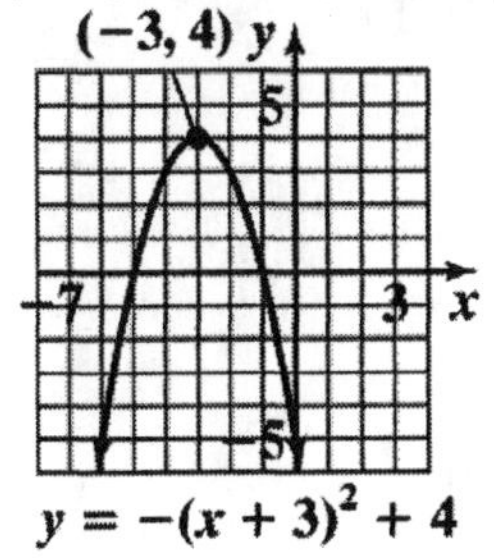

27. $y = -3(x-5)^2 + 3$

This is a parabola in the form $y = a(x-h)^2 + k$ with $a = -3$, $h = 5$ and $k = 3$. Since $a < 0$, the parabola opens downward. The vertex is (5, 3). The y-intercept is (0, –72). The x-intercepts are (4, 0) and (6, 0).

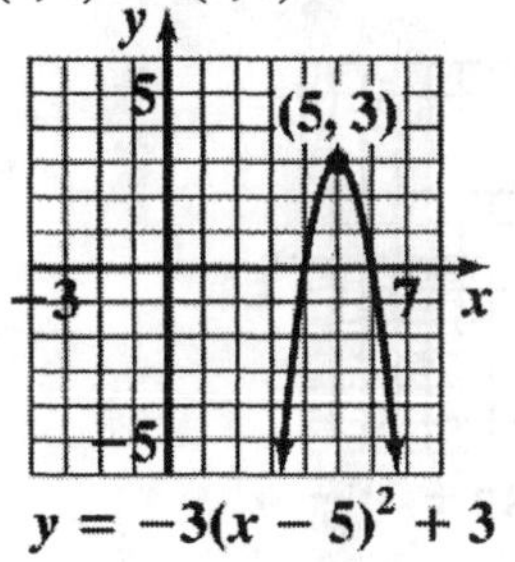

29. $x = (y-4)^2 - 3$

This is a parabola in the form $x = a(y-k)^2 + h$ with $a = 1$, $h = -3$ and $k = 4$. Since $a > 0$, the parabola opens to the right. The vertex is (–3, 4). The y-intercepts are about (0, 2.27) and (0, 5.73). The x-intercept is (13, 0).

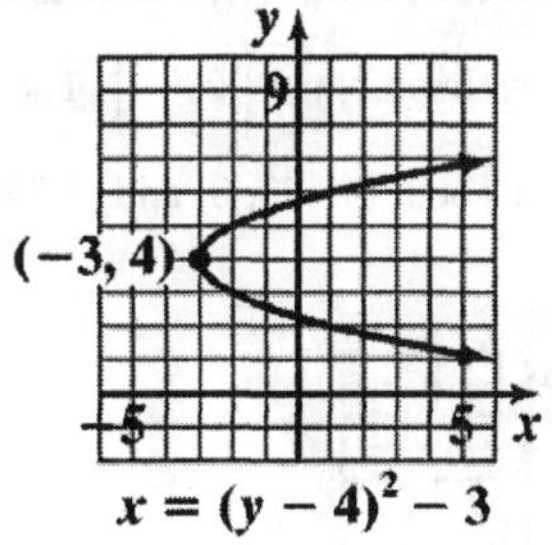

31. $x = -(y-5)^2 + 4$

This is a parabola in the form $x = a(y-k)^2 + h$ with $a = -1$, $h = 4$ and $k = 5$. Since $a < 0$, the parabola opens to the left. The vertex is (4, 5). The y-intercepts are (0, 3) and (0, 7). The x-intercept is (–21, 0).

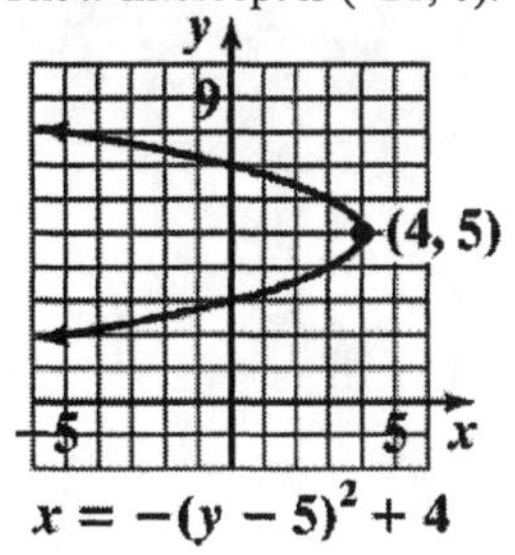

33. $x = -5(y+3)^2 - 6$

This is a parabola in the form $x = a(y-k)^2 + h$ with $a = -5$, $h = -6$ and $k = -3$. Since $a < 0$, the parabola opens to the left. The vertex is (–6, –3). There are no y-intercepts.
The x-intercept is (–51, 0)

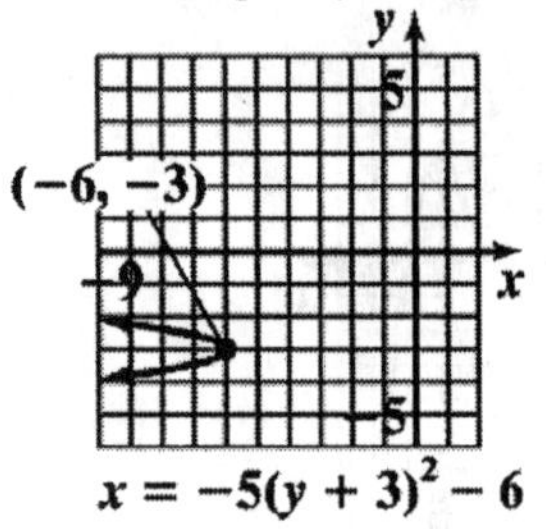

35. $y = -2\left(x + \frac{1}{2}\right)^2 + 6$

This is a parabola in the form $y = a(x-h)^2 + k$ with $a = -2$, $h = -\frac{1}{2}$ and $k = 6$. Since $a < 0$, the parabola opens downward. The vertex is $\left(-\frac{1}{2}, 6\right)$. The y-intercept is $\left(0, \frac{11}{2}\right)$. The x-intercepts are about (–2.23, 0) and (1.23, 0).

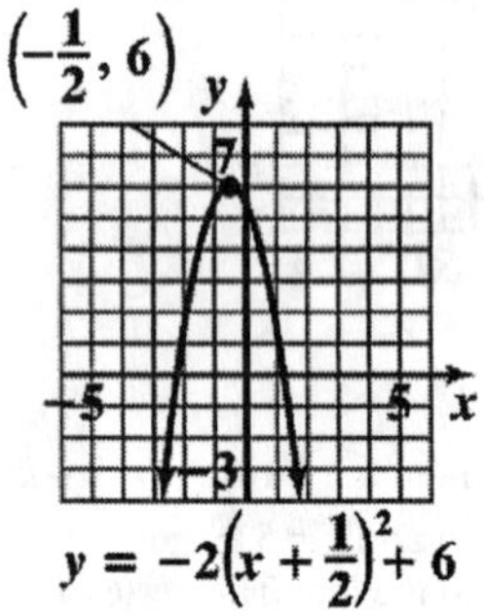

37. a. $y = x^2 + 2x$

$y = (x^2 + 2x + 1) - 1$

$y = (x+1)^2 - 1$

b. This is a parabola in the form $y = a(x-h)^2 + k$ with $a = 1$, $h = -1$ and $k = -1$. Since $a > 0$, the parabola opens upward. The vertex is (–1, –1). The y-intercept is (0, 0). The x-intercepts are (–2, 0) and (0, 0).

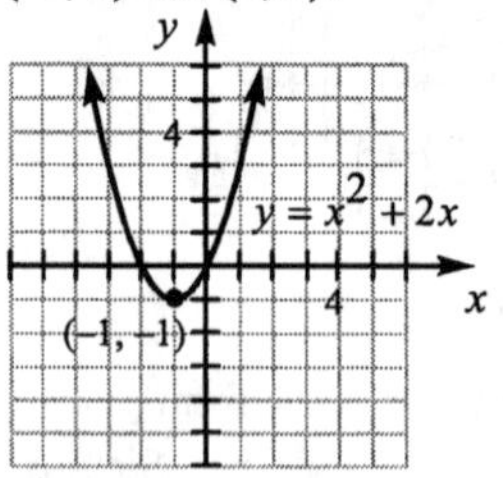

39. a. $y = x^2 + 6x$

$y = (x^2 + 6x + 9) - 9$

$y = (x+3)^2 - 9$

b. This is a parabola in the form $y = a(x-h)^2 + k$ with $a = 1$, $h = -3$ and $k = -9$. Since $a > 0$, the parabola opens upward. The vertex is (–3, –9). The y-intercept is (0, 0). The x-intercepts are (–6, 0) and (0, 0).

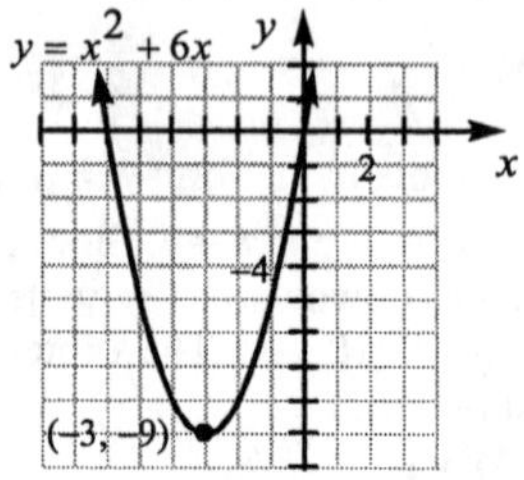

41. a. $x = y^2 + 4y$

$x = (y^2 + 4y + 4) - 4$

$x = (y + 2)^2 - 4$

b. This is a parabola in the form $x = a(y - k)^2 + h$ with $a = 1$, $h = -4$ and $k = -2$. Since $a > 0$, the parabola opens to the right. The vertex is (–4, –2). The y-intercepts are (0, –4) and (0, 0). The x-intercept is (0, 0).

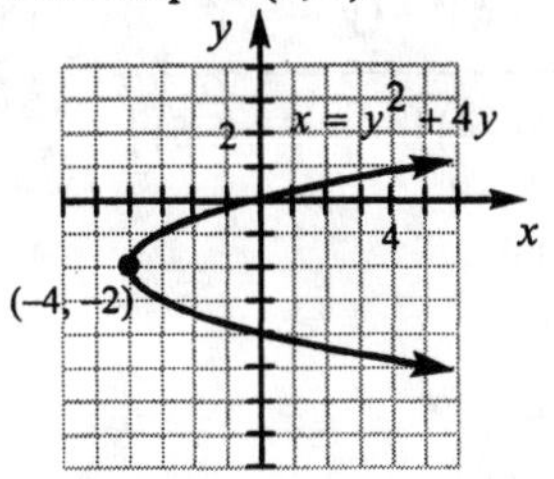

43. a. $y = x^2 + 7x + 10$

$y = \left(x^2 + 7x + \frac{49}{4}\right) - \frac{49}{4} + 10$

$y = \left(x + \frac{7}{2}\right)^2 - \frac{9}{4}$

b. This is a parabola in the form $y = a(x - h)^2 + k$ with $a = 1$, $h = -\frac{7}{2}$ and $k = -\frac{9}{4}$. Since $a > 0$, the parabola opens upward. The vertex is $\left(-\frac{7}{2}, -\frac{9}{4}\right)$. The y-intercept is (0, 10). The x-intercepts are (–5, 0) and (–2, 0).

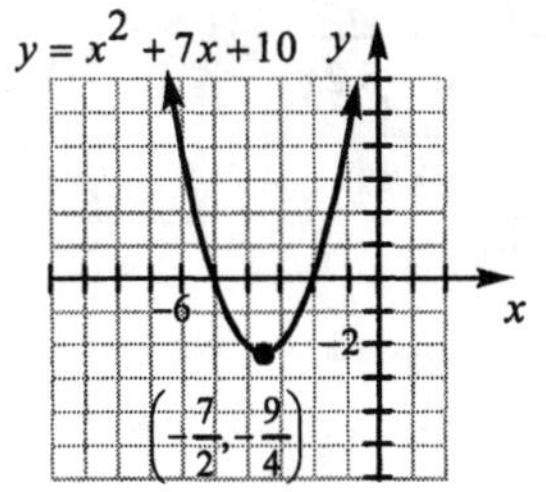

45. a. $x = -y^2 + 6y - 9$

$x = -(y^2 - 6y) - 9$

$x = -(y^2 - 6y + 9) + 9 - 9$

$x = -(y - 3)^2$

b. This is a parabola in the form $x = a(y - k)^2 + h$ with $a = -1$, $h = 0$ and $k = 3$. Since $a < 0$, the parabola opens to the left. The vertex is (0, 3). The y-intercept is (0, 3). The x-intercept is (–9, 0).

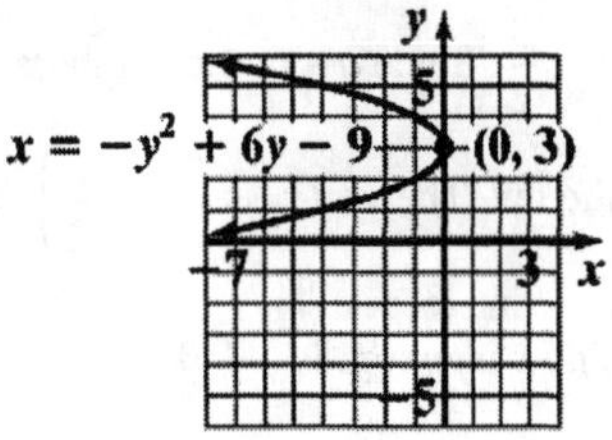

47. a. $y = -x^2 + 4x - 4$

$y = -(x^2 - 4x) - 4$

$y = -(x^2 - 4x + 4) + 4 - 4$

$y = -(x - 2)^2$

b. This is a parabola in the form $y = a(x - h)^2 + k$ with $a = -1$, $h = 2$ and $k = 0$. Since $a < 0$, the parabola opens downward. The vertex is (2, 0). The y-intercept is (0, –4). The x-intercept is (2, 0).

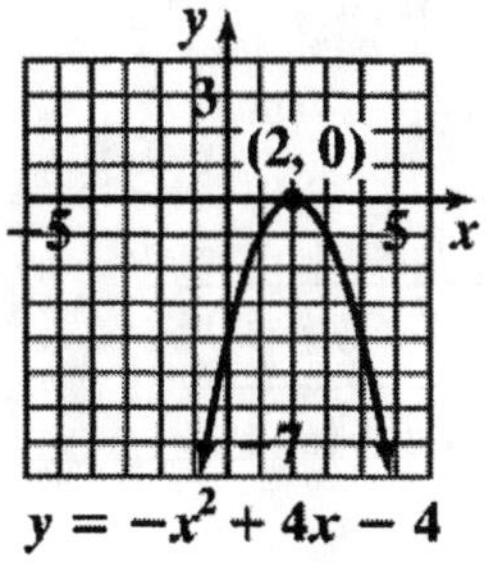

49. a. $x=-y^2+3y-4$

$x=-\left(y^2-3y\right)-4$

$x=-\left(y^2-3y+\frac{9}{4}\right)+\frac{9}{4}-4$

$x=-\left(y-\frac{3}{2}\right)^2-\frac{7}{4}$

b. This is a parabola in the form $x=a(y-k)^2+h$ with $a=-1$, $h=-\frac{7}{4}$ and $k=\frac{3}{2}$. Since $a<0$, the parabola opens to the left. The vertex is $\left(-\frac{7}{4},\frac{3}{2}\right)$. There are no y-intercepts.
The x-intercept is $(-4, 0)$.

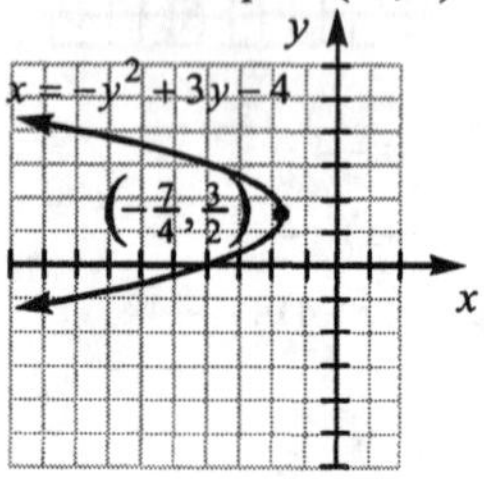

51. $d=\sqrt{(x_2-x_1)^2+(y_2-y_1)^2}$

$=\sqrt{(5-5)^2+[-6-(-1)]^2}$

$=\sqrt{0^2+(-5)^2}$

$=\sqrt{0+25}$

$=\sqrt{25}$

$=5$

53. $d=\sqrt{(x_2-x_1)^2+(y_2-y_1)^2}$

$=\sqrt{[8-(-1)]^2+(6-6)^2}$

$=\sqrt{9^2+0^2}$

$=\sqrt{81+0}$

$=\sqrt{81}$

$=9$

55. $d=\sqrt{(x_2-x_1)^2+(y_2-y_1)^2}$

$=\sqrt{[4-(-1)]^2+[9-(-3)]^2}$

$=\sqrt{5^2+12^2}$

$=\sqrt{25+144}$

$=\sqrt{169}$

$=13$

57. $d=\sqrt{(x_2-x_1)^2+(y_2-y_1)^2}$

$=\sqrt{[5-(-4)]^2+[-2-(-5)]^2}$

$=\sqrt{9^2+3^2}$

$=\sqrt{81+9}$

$=\sqrt{90}\approx 9.49$

59. $d=\sqrt{(x_2-x_1)^2+(y_2-y_1)^2}$

$=\sqrt{\left(\frac{1}{2}-3\right)^2+[4-(-1)]^2}$

$=\sqrt{\left(-\frac{5}{2}\right)^2+5^2}$

$=\sqrt{\frac{25}{4}+25}$

$=\sqrt{\frac{125}{4}}$

≈ 5.59

61. $d=\sqrt{(x_2-x_1)^2+(y_2-y_1)^2}$

$=\sqrt{[-4.3-(-1.6)]^2+(-1.7-3.5)^2}$

$=\sqrt{(-2.7)^2+(-5.2)^2}$

$=\sqrt{7.29+27.04}$

$=\sqrt{34.33}$

≈ 5.86

63. $d=\sqrt{(x_2-x_1)^2+(y_2-y_1)^2}$
$=\sqrt{\left(0-\sqrt{7}\right)^2+\left[0-\sqrt{3}\right]^2}$
$=\sqrt{\left(-\sqrt{7}\right)^2+\left(\sqrt{3}\right)^2}$
$=\sqrt{7+3}$
$=\sqrt{10}$
≈ 3.16

65. $\text{Midpoint}=\left(\frac{x_1+x_2}{2},\frac{y_1+y_2}{2}\right)$
$=\left(\frac{1+5}{2},\frac{9+3}{2}\right)$
$=(3,6)$

67. $\text{Midpoint}=\left(\frac{x_1+x_2}{2},\frac{y_1+y_2}{2}\right)$
$=\left(\frac{-7+7}{2},\frac{2+(-2)}{2}\right)$
$=(0,0)$

69. $\text{Midpoint}=\left(\frac{x_1+x_2}{2},\frac{y_1+y_2}{2}\right)$
$=\left(\frac{-1+4}{2},\frac{4+6}{2}\right)$
$=\left(\frac{3}{2},5\right)$

71. $\text{Midpoint}=\left(\frac{x_1+x_2}{2},\frac{y_1+y_2}{2}\right)$
$=\left(\frac{3+2}{2},\frac{\frac{1}{2}+(-4)}{2}\right)$
$=\left(\frac{5}{2},-\frac{7}{4}\right)$

73. $\text{Midpoint}=\left(\frac{x_1+x_2}{2},\frac{y_1+y_2}{2}\right)$
$=\left(\frac{\sqrt{3}+\sqrt{2}}{2},\frac{2+7}{2}\right)$
$=\left(\frac{\sqrt{3}+\sqrt{2}}{2},\frac{9}{2}\right)$

75. $(x-h)^2+(y-k)^2=r^2$
$(x-0)^2+(y-0)^2=4^2$
$x^2+y^2=16$

77. $(x-h)^2+(y-k)^2=r^2$
$(x-2)^2+(y-0)^2=5^2$
$(x-2)^2+y^2=25$

79. $(x-h)^2+(y-k)^2=r^2$
$(x-0)^2+(y-5)^2=1^2$
$x^2+(y-5)^2=1$

81. $(x-h)^2+(y-k)^2=r^2$
$(x-3)^2+(y-4)^2=(8)^2$
$(x-3)^2+(y-4)^2=64$

83. $(x-h)^2+(y-k)^2=r^2$
$(x-7)^2+\left[y-(-6)\right]^2=10^2$
$(x-7)^2+(y+6)^2=100$

85. $(x-h)^2+(y-k)^2=r^2$
$(x-1)^2+(y-2)^2=\left(\sqrt{5}\right)^2$
$(x-1)^2+(y-2)^2=5$

87. The center is (0, 0) and the radius is 4.
$(x-h)^2+(y-k)^2=r^2$
$(x-0)^2+(y-0)^2=4^2$
$x^2+y^2=16$

89. The center is (3, –2) and the radius is 3.
$(x-h)^2+(y-k)^2=r^2$
$(x-3)^2+[y-(-2)]^2=3^2$
$(x-3)^2+(y+2)^2=9$

91. $x^2 + y^2 = 16$

$x^2 + y^2 = 4^2$

The graph is a circle with its center at the origin and radius 4.

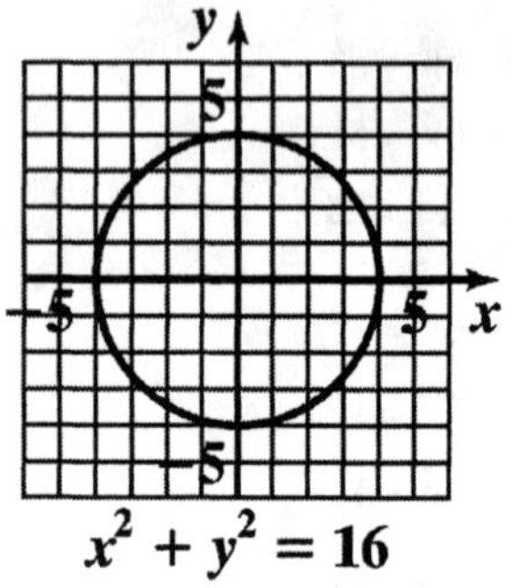

93. $x^2 + y^2 = 10$

$x^2 + y^2 = \left(\sqrt{10}\right)^2$

The graph is a circle with its center at the origin and radius $\sqrt{10}$.

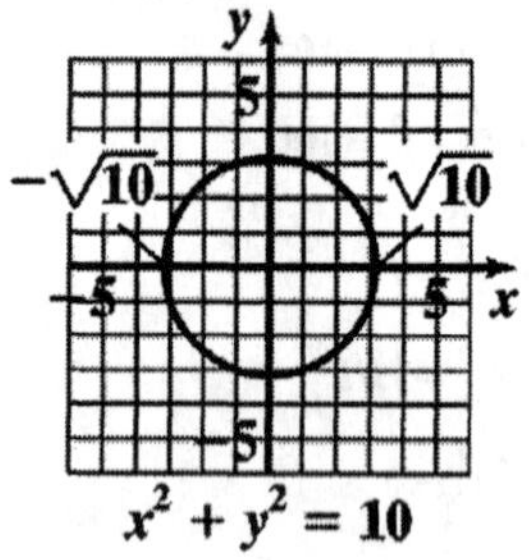

95. $(x+4)^2 + y^2 = 25$

$(x+4)^2 + (y-0)^2 = 5^2$

The graph is a circle with its center at (–4, 0) and radius 5.

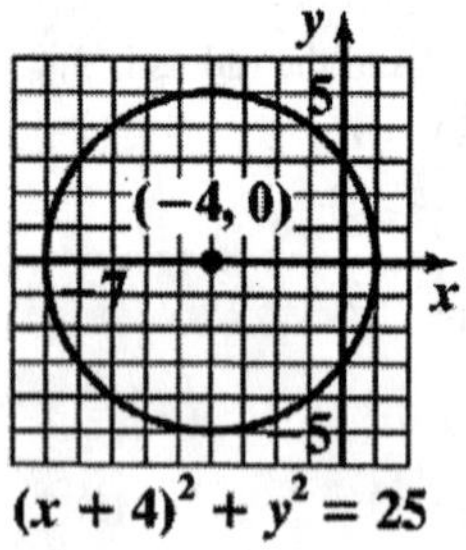

97. $x^2 + \left(y-3\right)^2 = 4$

$(x-0)^2 + (y-3)^2 = \left(2\right)^2$

The graph is a circle with its center at (0, 3) and radius 2.

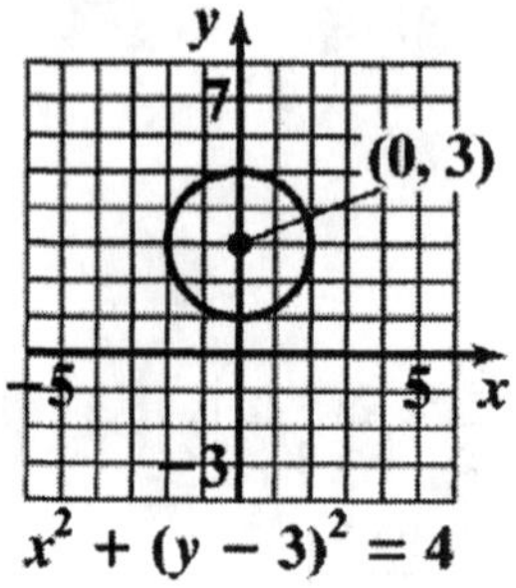

99. $(x+8)^2 + (y+2)^2 = 9$

$(x+8)^2 + (y+2)^2 = 3^2$

The graph is a circle with its center at (–8, –2) and radius 3.

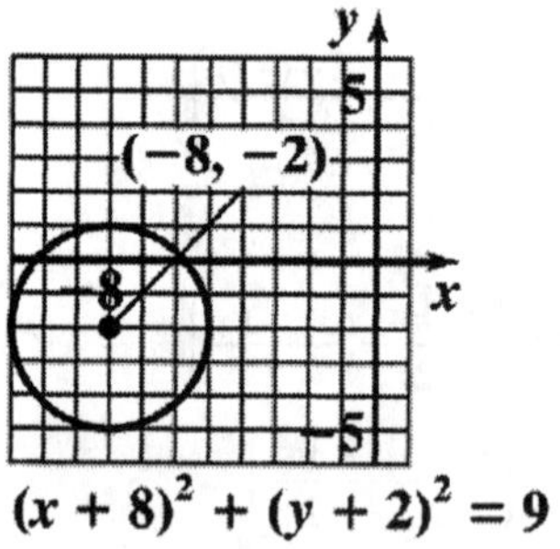

101. $y = \sqrt{25 - x^2}$

If we solve $x^2 + y^2 = 25$ for y, we obtain $y = \pm\sqrt{25 - x^2}$. Therefore, the graph of $y = \sqrt{25 - x^2}$ is the upper half ($y \geq 0$) of a circle with its center at the origin and radius 5.

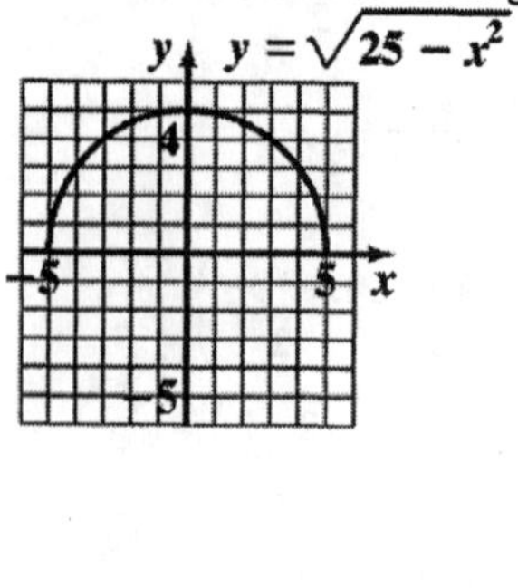

103. $y = -\sqrt{4 - x^2}$

If we solve $x^2 + y^2 = 4$ for *y*, we obtain $y = \pm\sqrt{4 - x^2}$. Therefore, the graph of $y = -\sqrt{4 - x^2}$ is the lower half ($y \leq 0$) of a circle with its center at the origin and radius 2.

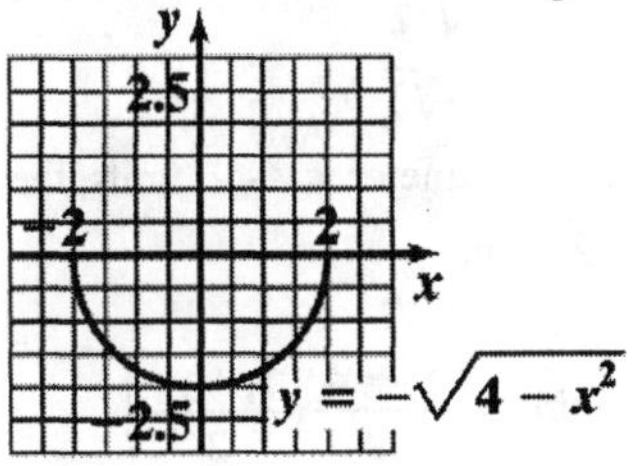

105. a.

$$x^2 + y^2 + 8x + 15 = 0$$
$$x^2 + 8x + y^2 = -15$$
$$(x^2 + 8x + 16) + y^2 = -15 + 16$$
$$(x+4)^2 + y^2 = 1$$
$$(x+4)^2 + y^2 = 1^2$$

b. The graph is a circle with center (–4, 0) and radius 1.

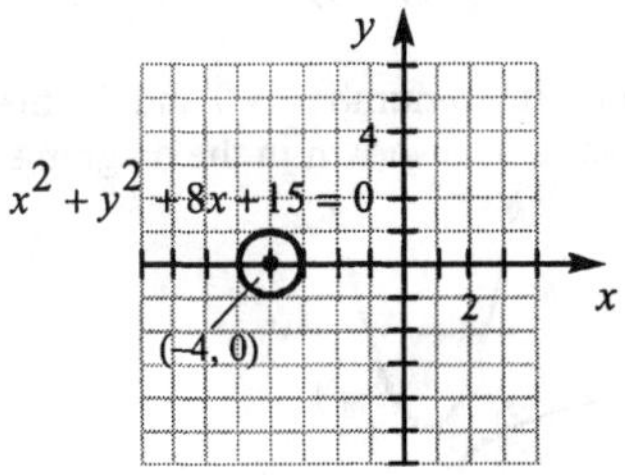

107. a.

$$x^2 + y^2 + 6x - 4y + 4 = 0$$
$$x^2 + 6x + y^2 - 4y = -4$$
$$(x^2 + 6x + 9) + (y^2 - 4y + 4) = -4 + 9 + 4$$
$$(x+3)^2 + (y-2)^2 = 9$$
$$(x+3)^2 + (y-2)^2 = 3^2$$

b. The graph is a circle with center (–3, 2) and radius 3.

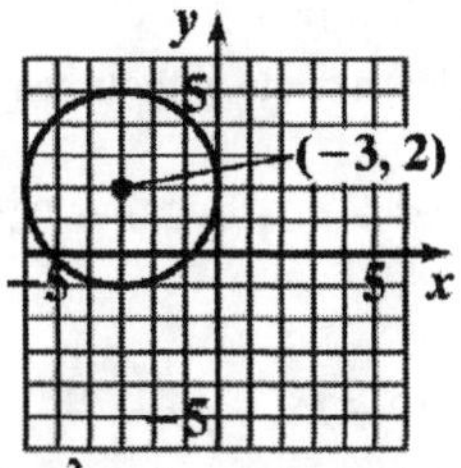

$x^2 + y^2 + 6x - 4y + 4 = 0$

109. a.

$$x^2 + y^2 + 6x - 2y + 6 = 0$$
$$x^2 + 6x + y^2 - 2y = -6$$
$$(x^2 + 6x + 9) + (y^2 - 2y + 1) = -6 + 9 + 1$$
$$(x+3)^2 + (y-1)^2 = 4$$
$$(x+3)^2 + (y-1)^2 = 2^2$$

b. The graph is a circle with center (–3, 1) and radius 2.

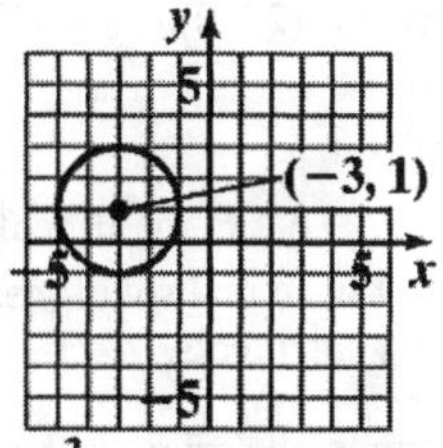

$x^2 + y^2 + 6x - 2y + 6 = 0$

111. a.

$$x^2 + y^2 - 8x + 2y + 13 = 0$$
$$x^2 - 8x + y^2 + 2y = -13$$
$$(x^2 - 8x + 16) + (y^2 + 2y + 1) = -13 + 16 + 1$$
$$(x-4)^2 + (y+1)^2 = 4$$
$$(x-4)^2 + (y+1)^2 = 2^2$$

b. The graph is a circle with center (4, –1) and radius 2.

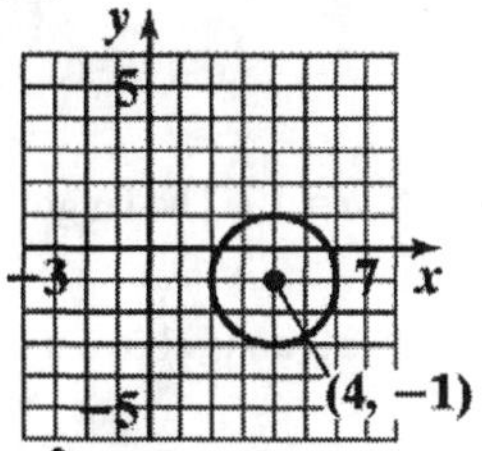

$x^2 + y^2 - 8x + 2y + 13 = 0$

113. $(x+4)^2+y^2=25$

$(x+4)^2+(y-0)^2=5^2$

The graph is a circle with its center at (–4, 0) and radius 5.

Area $=\pi r^2=\pi(5)^2=25\pi\approx 78.5$ sq. units

115. x-intercept:

$x=0^2-6(0)-7$

$x=-7$

The x-intercept is (–7, 0)

y-intercepts:

$0=y^2-6y-7$

$0=(y+1)(y-7)$

$y=-1$ or $y=7$

The y-intercepts are (0, –1) and (0, 7).

117. x-intercept:

$x=2(0-3)^2+6$

$x=24$

The x-intercept is (24, 0).

y-intercepts:

$0=2(y-3)^2+6$

Since $2(y-3)^2+6\ge 6$ for all real values of y, this equation has no real solutions.

There are no y-intercepts.

119. No. For example, the origin is the midpoint of both the segment from (1, 1) to (–1, –1) and the segment from (2, 2) to (–2, –2), but these segments have different lengths.

121. The distance from the midpoint (4, –6) to the endpoint (7, –2) is half the length of the line segment.

$$\frac{d}{2}=\sqrt{(7-4)^2+[-2-(-6)]^2}$$
$$=\sqrt{3^2+4^2}$$
$$=\sqrt{25}$$
$$=5$$

Since $\frac{d}{2}=5$, $d=10$. The length is 10 units.

123. Since (–6, 2) is 2 units above the x-axis, the radius is 2.

$(x-h)^2+(y-k)^2=r^2$

$(x+6)^2+(y-2)^2=2^2$

$(x+6)^2+(y-2)^2=4$

125. a. Diameter $=\sqrt{(x_2-x_1)^2+(y_2-y_1)^2}$

$=\sqrt{(9-5)^2+(8-4)^2}$

$=\sqrt{4^2+4^2}$

$=\sqrt{16+16}$

$=\sqrt{32}$

$=4\sqrt{2}$

Since the diameter is $4\sqrt{2}$ units, the radius is $2\sqrt{2}$ units.

b. Midpoint $=\left(\frac{x_1+x_2}{2},\frac{y_1+y_2}{2}\right)$

$=\left(\frac{5+9}{2},\frac{4+8}{2}\right)$

$=(7,6)$

The center is (7, 6).

c. $(x-h)^2+(y-k)^2=r^2$

$(x-7)^2+(y-6)^2=\left(2\sqrt{2}\right)^2$

$(x-7)^2+(y-6)^2=8$

127. The minimum number is 0 and the maximum number is 4 as shown in the diagrams.

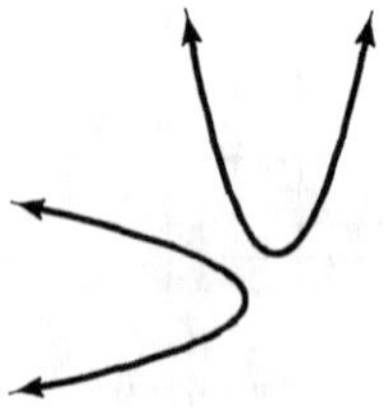

No points of intersection

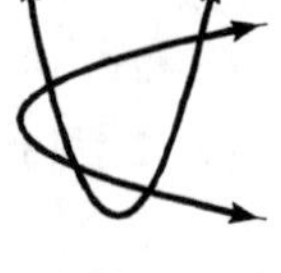

Four points of intersection

129. a. Since $150-2(68.2)=13.6$, the clearance is 13.6 feet.

b. Since $150-68.2=81.8$, the center of the wheel is 81.8 feet above the ground.

c. $(x-h)^2+(y-k)^2=r^2$

$(x-0)^2+(y-81.8)^2=68.2^2$

$x^2+(y-81.8)^2=68.2^2$

$x^2+(y-81.8)^2=4651.24$

131. a. The center of the blue circle is the origin, and the radius is 4.

$x^2+y^2=r^2$

$x^2+y^2=4^2$

$x^2+y^2=16$

b. The center of the red circle is (2, 0), and the radius is 2.

$(x-h)^2+(y-k)^2=r^2$

$(x-2)^2+(y-0)^2=2^2$

$(x-2)^2+y^2=4$

c. The center of the green circle is (–2, 0), and the radius is 2.

$(x-h)^2+(y-k)^2=r^2$

$[x-(-2)]^2+(y-0)^2=2^2$

$(x+2)^2+y^2=4$

d. Shaded area = (blue circle area) – (red circle area) – (green circle area)

$=\pi(4^2)-\pi(2^2)-\pi(2^2)$

$=16\pi-4\pi-4\pi$

$=8\pi$

133. The radii are 4 and 8, respectively. So, the area between the circles is

$\pi(8)^2-\pi(4)^2=64\pi-16\pi=48\pi$ square units.

136. $\dfrac{6x^{-3}y^4}{18x^{-2}y^3}=\dfrac{6}{18}x^{-3-(-2)}y^{4-3}$

$=\dfrac{1}{3}x^{-3+2}y^1$

$=\dfrac{1}{3}x^{-1}y$ or $\dfrac{y}{3x}$

137. $-4<3x-4<17$

$-4+4<3x-4+4<17+4$

$0<3x<21$

$\dfrac{0}{3}<\dfrac{3x}{3}<\dfrac{21}{3}$

$0<x<7$

In interval notation: $(0, 7)$

138. $\begin{vmatrix}4 & 0 & 3\\5 & 2 & -1\\3 & 6 & 4\end{vmatrix}=4\begin{vmatrix}2 & -1\\6 & 4\end{vmatrix}-5\begin{vmatrix}0 & 3\\6 & 4\end{vmatrix}+3\begin{vmatrix}0 & 3\\2 & -1\end{vmatrix}$

$=4(8+6)-5(0-18)+3(0-6)$

$=4(14)-5(-18)+3(-6)$

$=56+90-18$

$=128$

139. a. area 1: a^2 area 2: ab

area 3: ab area 4: b^2

b. $a^2+ab+ab+b^2=a^2+2ab+b^2$

$=(a+b)^2$

140. This is a parabola in the form $y=a(x-h)^2+k$ with $a=1$, $h=4$ and $k=1$. Since $a>0$, the parabola opens upward. The vertex is (4, 1). The y-intercept is (0, 17). There are no x-intercepts.

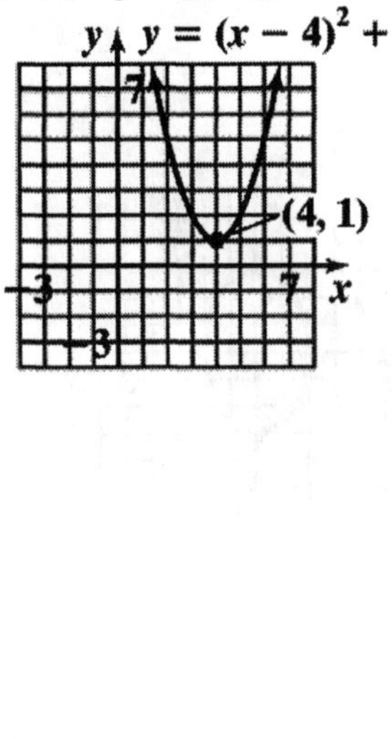

Exercise Set 11.2

1. An ellipse is a set of points in a plane, the sum of whose distances from two fixed points is constant.

3. $\frac{(x-h)^2}{a^2}+\frac{(y-k)^2}{b^2}=1$

5. If $a=b$, the formula for a circle is obtained.

7. First divide both sides by 180.

9. No, the sign in front of the y^2 is negative.

11. $\frac{x^2}{4}+\frac{y^2}{1}=1$

Since $a^2=4, a=2.$

Since $b^2=1, b=1.$

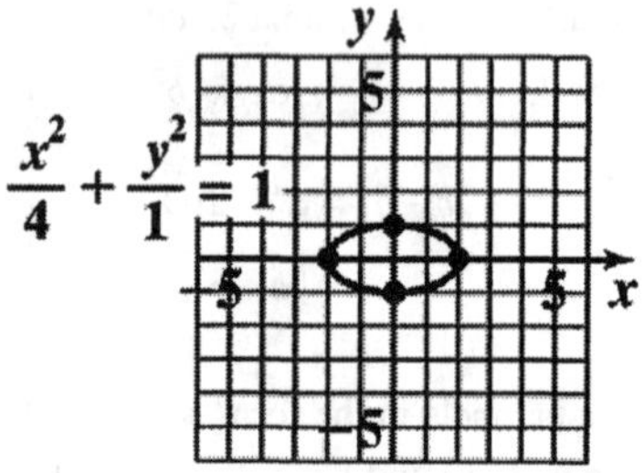

13. $\frac{x^2}{4}+\frac{y^2}{9}=1$

Since $a^2=4, a=2.$

Since $b^2=9, b=3.$

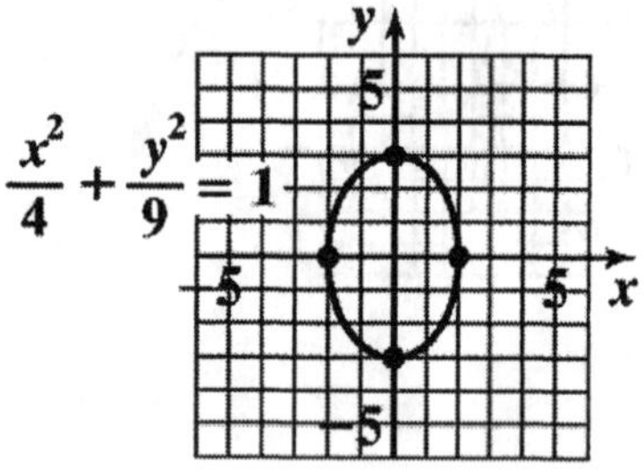

15. $\frac{x^2}{25}+\frac{y^2}{9}=1$

Since $a^2=25, a=5.$

Since $b^2=9, b=3.$

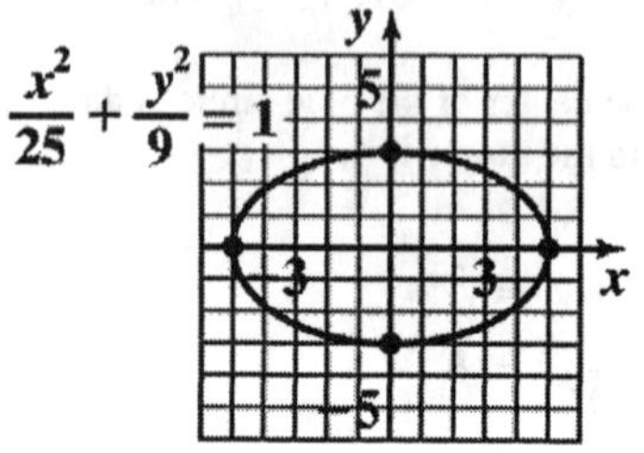

17. $\frac{x^2}{16}+\frac{y^2}{25}=1$

Since $a^2=16, a=4.$

Since $b^2=25, b=5.$

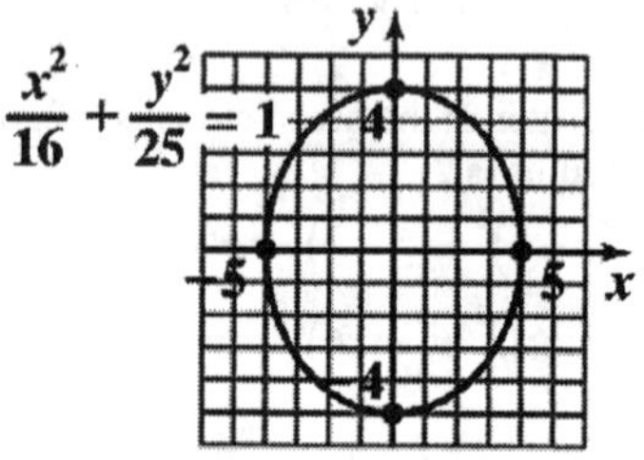

19. $x^2+16y^2=16$

$\frac{x^2}{16}+\frac{16y^2}{16}=1$

$\frac{x^2}{16}+\frac{y^2}{1}=1$

Since $a^2=16, a=4w.$

Since $b^2=1, b=1.$

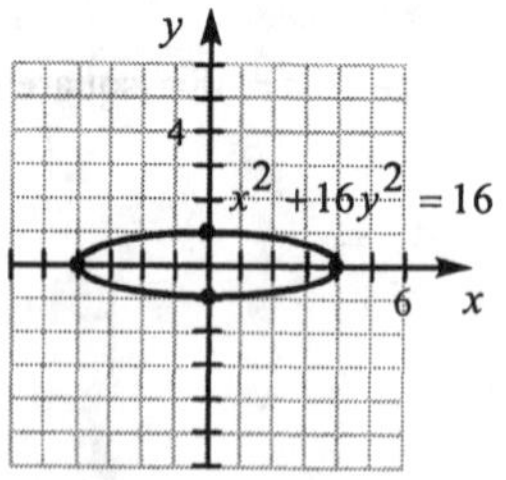

21. $49x^2 + y^2 = 49$

$$\frac{49x^2}{49} + \frac{y^2}{49} = 1$$

$$\frac{x^2}{1} + \frac{y^2}{49} = 1$$

Since $a^2 = 1, a = 1$.

Since $b^2 = 49, b = 7$.

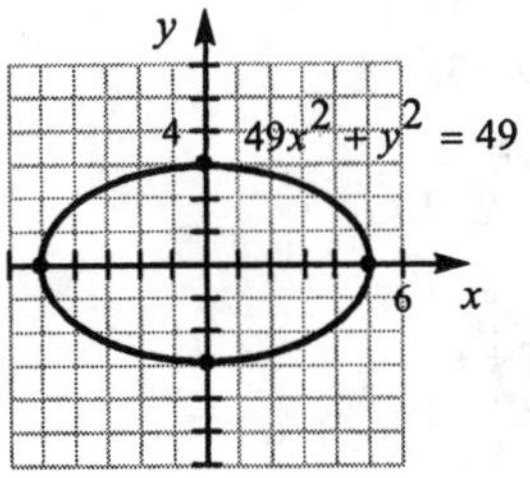

23. $9x^2 + 16y^2 = 144$

$$\frac{9x^2}{144} + \frac{16y^2}{144} = 1$$

$$\frac{x^2}{16} + \frac{y^2}{9} = 1$$

Since $a^2 = 16, a = 4$.

Since $b^2 = 9, b = 3$.

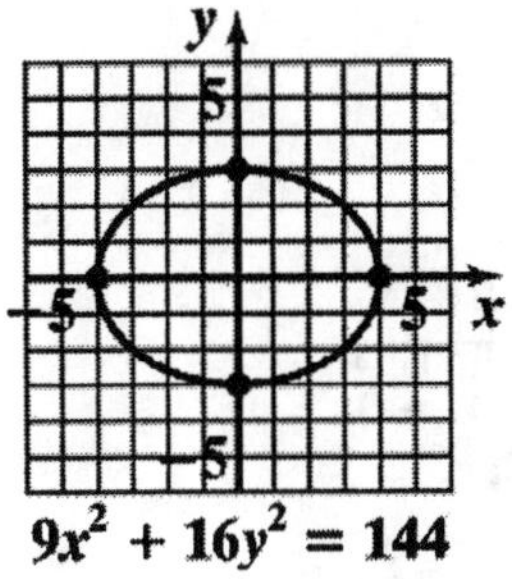

25. $25x^2 + 100y^2 = 400$

$$x^2 + 4y^2 = 16$$

$$\frac{x^2}{16} + \frac{4y^2}{16} = 1$$

$$\frac{x^2}{16} + \frac{y^2}{4} = 1$$

Since $a^2 = 16, a = 4$.

Since $b^2 = 4, b = 2$.

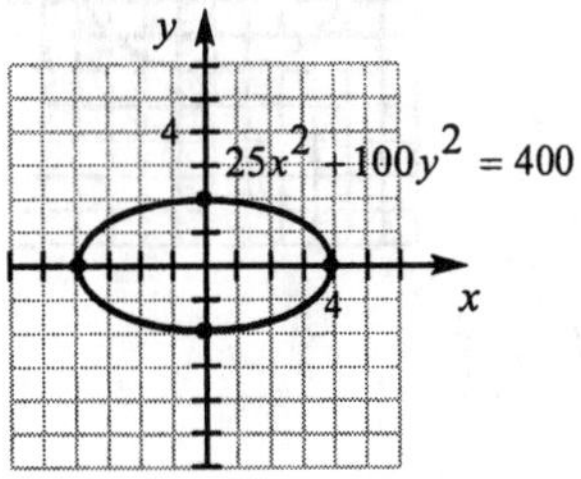

27. $x^2 + 2y^2 = 8$

$$\frac{x^2}{8} + \frac{2y^2}{8} = 1$$

$$\frac{x^2}{8} + \frac{y^2}{4} = 1$$

Since $a^2 = 8, a = \sqrt{8} = 2\sqrt{2} \approx 2.83$.

Since $b^2 = 4, b = 2$.

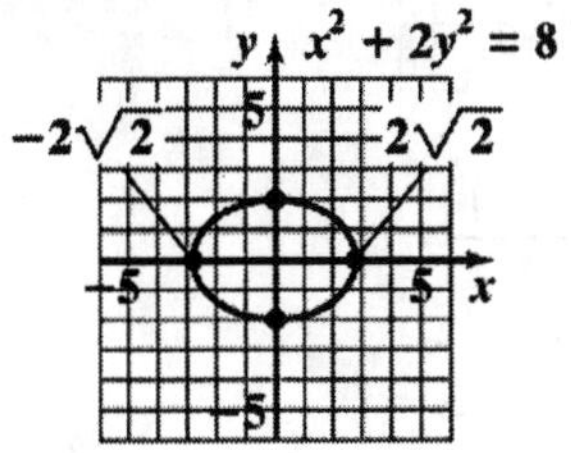

29. $\dfrac{x^2}{16}+\dfrac{(y-2)^2}{9}=1$

The center is (0, 2).

Since $a^2=16, a=4$.

Since $b^2=9, b=3$.

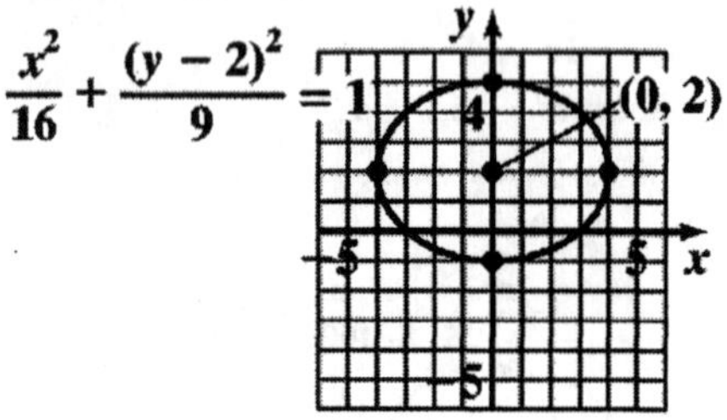

31. $\dfrac{(x-4)^2}{9}+\dfrac{(y+3)^2}{25}=1$

The center is (4, −3).

Since $a^2=9, a=3$.

Since $b^2=25, b=5$.

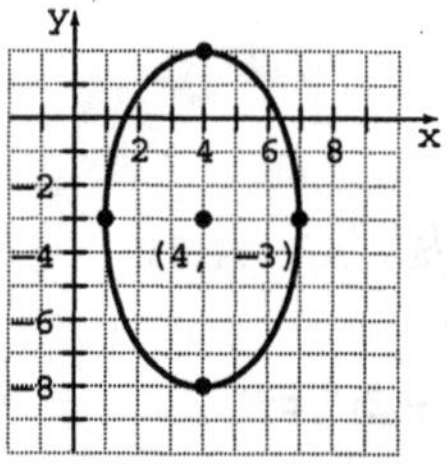

33. $\dfrac{(x+1)^2}{9}+\dfrac{(y-2)^2}{4}=1$

The center is (−1, 2).

Since $a^2=9, a=3$.

Since $b^2=4, b=2$.

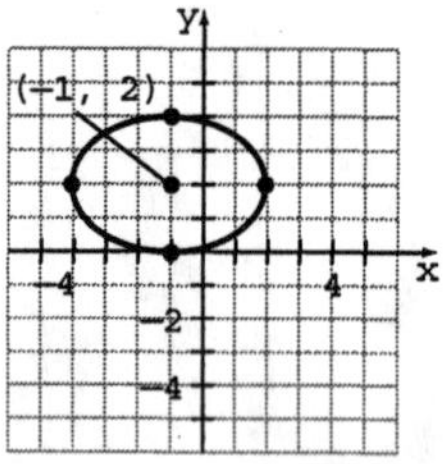

35. $(x+3)^2+9(y+1)^2=81$

$\dfrac{(x+3)^2}{81}+\dfrac{9(y+1)^2}{81}=1$

$\dfrac{(x+3)^2}{81}+\dfrac{(y+1)^2}{9}=1$

The center is (−3, −1).

Since $a^2=81, a=9$.

Since $b^2=9, b=3$.

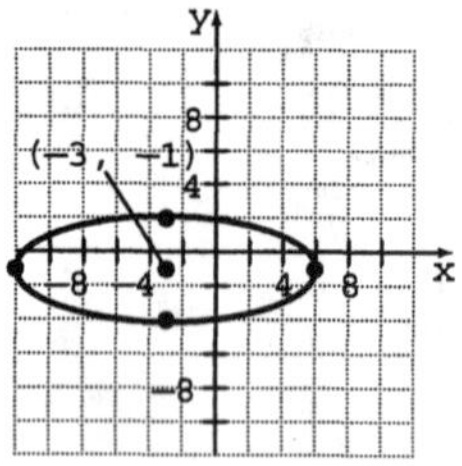

37. $(x-5)^2+4(y+4)^2=4$

$\dfrac{(x-5)^2}{4}+\dfrac{4(y+4)^2}{4}=1$

$\dfrac{(x-5)^2}{4}+\dfrac{(y+4)^2}{1}=1$

The center is (5, −4).

Since $a^2=4, a=2$.

Since $b^2=1, b=1$.

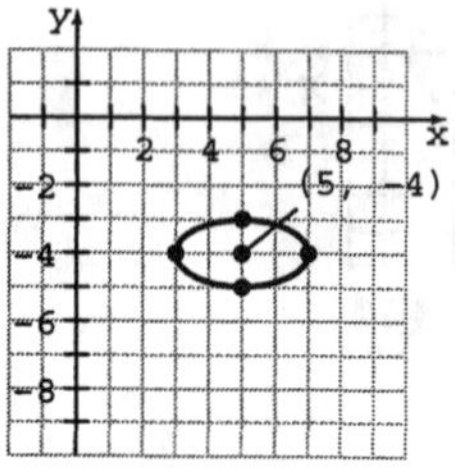

39. $12(x+4)^2+3(y-1)^2=48$

$$\frac{12(x+4)^2}{48}+\frac{3(y-1)^2}{48}=1$$

$$\frac{(x+4)^2}{4}+\frac{(y-1)^2}{16}=1$$

The center is (–4, 1).

Since $a^2=4, a=2$.

Since $b^2=16, b=4$.

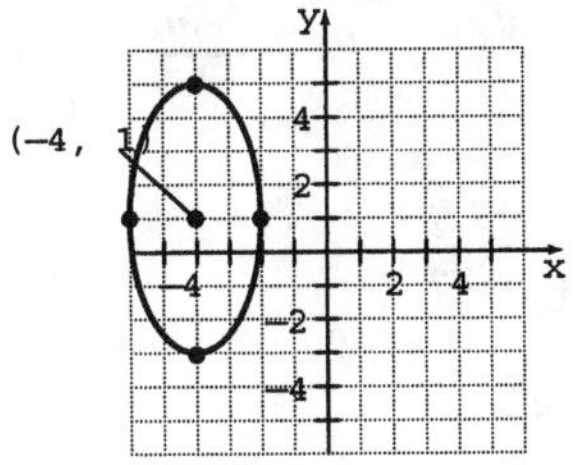

41. $\frac{x^2}{4}+\frac{y^2}{1}=1$

Since $a^2=4, a=2$; Since $b^2=1, b=1$.

Area $=\pi ab$
$=\pi(2)(1)$
$=2\pi\approx 6.3$ square units

43. There is only one point, at (0, 0). Two non-negative numbers can only sum to 0 if they are both 0.

45. The center is the origin, $a=3$, and $b=4$.

$$\frac{x^2}{a^2}+\frac{y^2}{b^2}=1$$

$$\frac{x^2}{3^2}+\frac{y^2}{4^2}=1$$

$$\frac{x^2}{9}+\frac{y^2}{16}=1$$

47. The center is the origin, $a=2$, and $b=3$.

$$\frac{x^2}{a^2}+\frac{y^2}{b^2}=1$$

$$\frac{x^2}{2^2}+\frac{y^2}{3^2}=1$$

$$\frac{x^2}{4}+\frac{y^2}{9}=1$$

49. There are no points of intersection, because the ellipse with $a=4$ and $b=5$ is completely inside the circle of radius 7.

51. $x^2+4y^2+6x+16y-11=0$

$$x^2+6x+4y^2+16y=11$$

$$(x^2+6x+9)+4(y^2+4y+4)=11+9+16$$

$$(x+3)^2+4(y+2)^2=36$$

$$\frac{(x+3)^2}{36}+\frac{(y+2)^2}{9}=1$$

The center is (–3, –2).

53. Since 90.2 – 20.7 = 69.5, the distance between the foci is 69.5 feet.

55. **a.** Consider the ellipse to be centered at the origin, (0, 0). Here, $a=10$ and $b=24$. The equation is

$$\frac{x^2}{10^2}+\frac{y^2}{24^2}=1 \Rightarrow \frac{x^2}{100}+\frac{y^2}{576}=1.$$

b. Area $=\pi ab$
$=\pi(10)(24)$
$=240\pi\approx 753.98$ square feet

c. Area of opening is half the area of ellipse

$$=\frac{\pi ab}{2}$$

$$=\frac{\pi(10)(24)}{2}$$

$=120\pi\approx 376.99$ square feet

57. Using $a = 3$ and $b = 2$, we may assume that the ellipse has the equation $\frac{x^2}{9} + \frac{y^2}{4} = 1$ and that the foci are located at $(\pm c, 0)$. Apply the definition of an ellipse using the points (3, 0) and (0, 2). That is, the distance from (3, 0) to $(-c, 0)$ plus the distance from (3, 0) to $(c, 0)$ is the same as the sum of the distance from (0, 2) to $(-c, 0)$ and the distance from (0, 2) to $(c, 0)$.

$$\sqrt{[3-(-c)]^2 + (0-0)^2} + \sqrt{(3-c)^2 + (0-0)^2} = \sqrt{[0-(-c)]^2 + (2-0)^2} + \sqrt{(0-c)^2 + (2-0)^2}$$

$$|3+c| + |3-c| = \sqrt{c^2+4} + \sqrt{(-c)^2+4}$$

Note that the foci are inside the ellipse, so $3+c>0$ and $3-c>0$. So, $|3+c| = 3+c$ and $|3-c| = 3-c$.

$$(3+c)+(3-c) = 2\sqrt{c^2+4}$$
$$6 = 2\sqrt{c^2+4}$$
$$3 = \sqrt{c^2+4}$$
$$9 = c^2+4$$
$$5 = c^2$$
$$c = \pm\sqrt{5}$$

The foci are located at $(\pm\sqrt{5}, 0)$. That is, the foci are $\sqrt{5} \approx 2.24$ feet, in both directions, from the center of the ellipse, along the major axis.

59. Answers will vary.

61. Answers will vary.

63. $h = \frac{-3+11}{2} = \frac{8}{2} = 4$; $k = \frac{5+(-1)}{2} = \frac{4}{2} = 2$

Thus, the center is (4, 2).

$a = \frac{11-(-3)}{2} = \frac{14}{2} = 7$; $b = \frac{5-(-1)}{2} = \frac{6}{2} = 3$

$$\frac{(x-h)^2}{a^2} + \frac{(y-k)^2}{b^2} = 1$$
$$\frac{(x-4)^2}{7^2} + \frac{(y-2)^2}{3^2} = 1$$
$$\frac{(x-4)^2}{49} + \frac{(y-2)^2}{9} = 1$$

65. $S = \frac{n}{2}(f+l)$, for l

$$S = \frac{nf}{2} + \frac{nl}{2}$$
$$S - \frac{nf}{2} = \frac{nl}{2}$$
$$2\left(S - \frac{nf}{2}\right) = 2\left(\frac{nl}{2}\right)$$
$$2S - nf = nl$$
$$\frac{2S-nf}{n} = \frac{nl}{n} \Rightarrow l = \frac{2S-nf}{n}$$

66.

$$\begin{array}{r} x+\frac{5}{2} \\ 2x-3\overline{)2x^2+2x-7} \\ \underline{2x^2-3x} \\ 5x-7 \\ \underline{5x-\frac{15}{2}} \\ \frac{1}{2} \end{array}$$

$$\frac{2x^2+2x-7}{2x-3} = x + \frac{5}{2} + \frac{1}{2(2x-3)}$$

67. $\sqrt{3b-2} = 10-b$

$$3b-2 = (10-b)^2$$
$$3b-2 = 100-20b+b^2$$
$$b^2-23b+102 = 0$$
$$(b-6)(b-17) = 0$$
$$b-6=0 \quad \text{or} \quad b-17=0$$
$$b=6 \qquad\qquad b=17$$

Upon checking $b = 17$ is extraneous. The solution is $b = 6$.

68. $\frac{3x+5}{x-4} \le 0$

$$3x+5=0 \Rightarrow x = -\frac{5}{3}$$
$$x-4=0 \Rightarrow x=4$$
$$\frac{3x+5}{x-4} \le 0 \Rightarrow -\frac{5}{3} \le x < 4 \text{ or } \left[-\frac{5}{3}, 4\right)$$

69. $\log_8 321 = \frac{\ln 321}{\ln 8} \approx 2.7755$

Mid-Chapter Test: 11.1 – 11.2

1. This is a parabola in the form $y = a(x-h)^2 + k$ with $a = 1$, $h = 2$, and $k = -1$. Since $a > 0$, the parabola opens upward. The vertex is $(2,8)$. The y-intercept is $(0,3)$. The x-intercepts are $(1,0)$ and $(3,0)$.

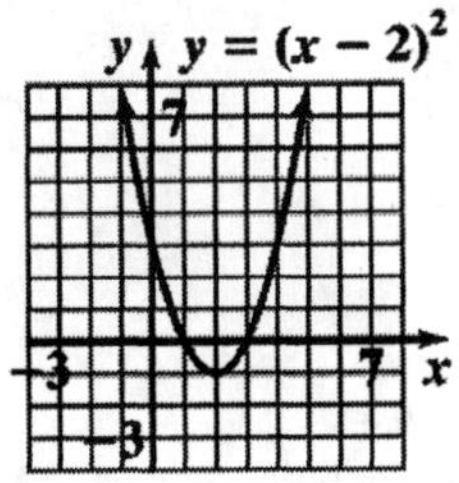

2. This is a parabola in the form $y = a(x-h)^2 + k$ with $a = -1$, $h = -1$, and $k = 3$. Since $a < 0$, the parabola opens downward. The vertex is $(-1,3)$. The y-intercept is $(0,2)$. The x-intercepts are about $(-2.732,0)$ and $(0.732,0)$.

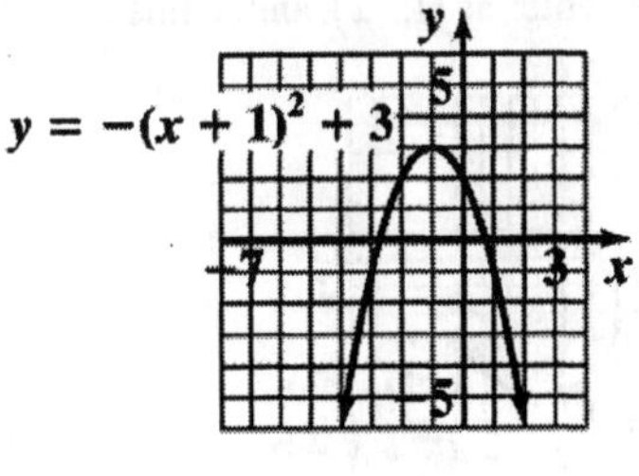

3. This is a parabola in the form $x = a(y-k)^2 + h$ with $a = -1$, $h = 1$, and $k = 4$. Since $a < 0$, the parabola opens to the left. The vertex is $(1,4)$. The x-intercept is $(-15,0)$. The y-intercepts are $(0,5)$ and $(0,3)$.

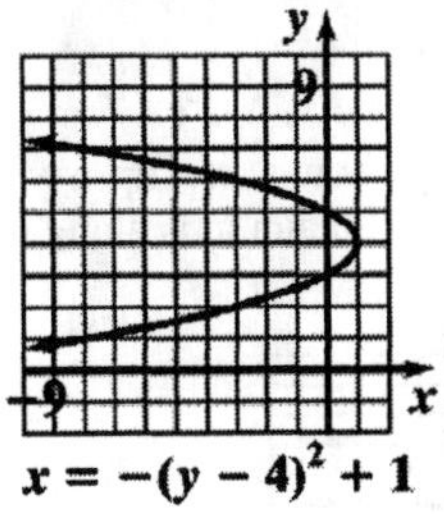

4. This is a parabola in the form $x = a(y-k)^2 + h$ with $a = 2$, $h = -2$, and $k = -3$. Since $a > 0$, the parabola opens to the right. The vertex is $(-2,-3)$. The x-intercept is $(16,0)$. The y-intercepts are $(0,-4)$ and $(0,-2)$.

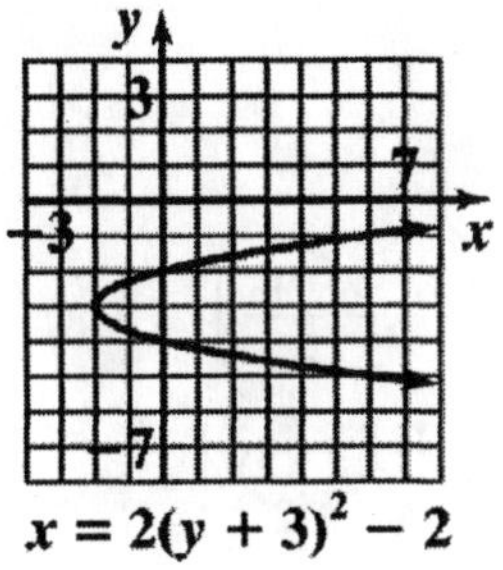

5.
$$\begin{aligned} y &= x^2 + 6x + 10 \\ &= \left(x^2 + 6x\right) + 10 \\ &= \left(x^2 + 6x + 9\right) + 10 - 9 \\ &= (x+3)^2 + 1 \end{aligned}$$

This is a parabola in the form $y = a(x-h)^2 + k$ with $a = 1$, $h = -3$, and $k = 1$. Since $a > 0$, the parabola opens upward. The vertex is $(-3,1)$. The y-intercept is $(0,10)$. There are no x-intercepts.

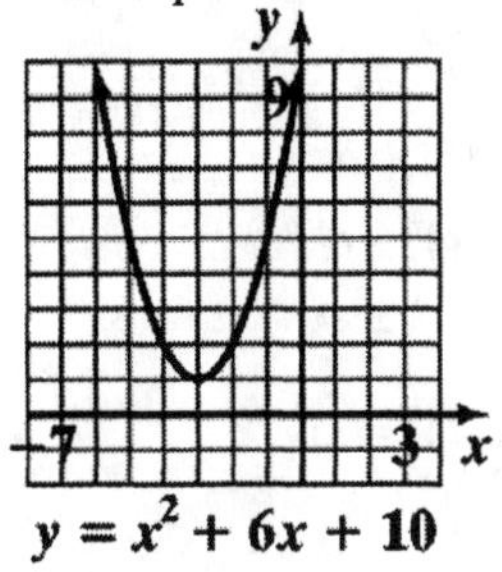

6.
$$\begin{aligned} d &= \sqrt{(x_2 - x_1)^2 + (y_2 - y_1)^2} \\ &= \sqrt{(-2-(-7))^2 + (-8-4)^2} \\ &= \sqrt{(5)^2 + (-12)^2} \\ &= \sqrt{25 + 144} \\ &= \sqrt{169} \\ &= 13 \end{aligned}$$

7. $d=\sqrt{(x_2-x_1)^2+(y_2-y_1)^2}$
$=\sqrt{(2-5)^2+(9-(-3))^2}$
$=\sqrt{(-3)^2+(12)^2}$
$=\sqrt{9+144}$
$=\sqrt{153}$
≈ 12.37

8. Midpoint $=\left(\dfrac{x_1+x_2}{2},\dfrac{y_1+y_2}{2}\right)$
$=\left(\dfrac{9+(-11)}{2},\dfrac{-1+6}{2}\right)$
$=\left(\dfrac{-2}{2},\dfrac{5}{2}\right)$
$=\left(-1,\dfrac{5}{2}\right)$

9. Midpoint $=\left(\dfrac{x_1+x_2}{2},\dfrac{y_1+y_2}{2}\right)$
$=\left(\dfrac{-\frac{5}{2}+8}{2},\dfrac{7+\frac{1}{2}}{2}\right)$
$=\left(\dfrac{11}{4},\dfrac{15}{4}\right)$

10. $(h,k)=(-3,2);\ r=5$
$(x-h)^2+(y-k)^2=r^2$
$(x-(-3))^2+(y-2)^2=5^2$
$(x+3)^2+(y-2)^2=25$

11. $x^2+(y-1)^2=16$
$x^2+(y-1)^2=4^2$
The graph is a circle with its center at $(0,1)$ and radius 4.

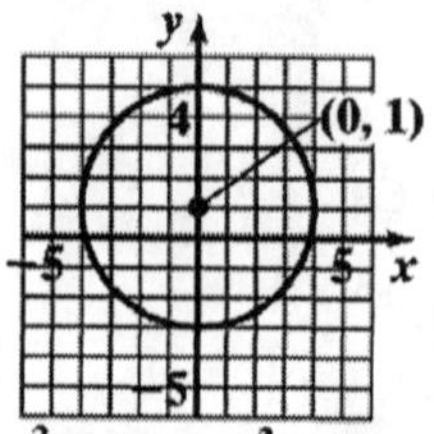

$x^2+(y-1)^2=16$

12. $y=\sqrt{36-x^2}$
If we solve $x^2+y^2=36$ for *y*, we obtain $y=\pm\sqrt{36-x^2}$. Therefore, the graph of $y=\sqrt{36-x^2}$ is the upper half ($y \geq 0$) of a circle with its center at the origin and radius 6.

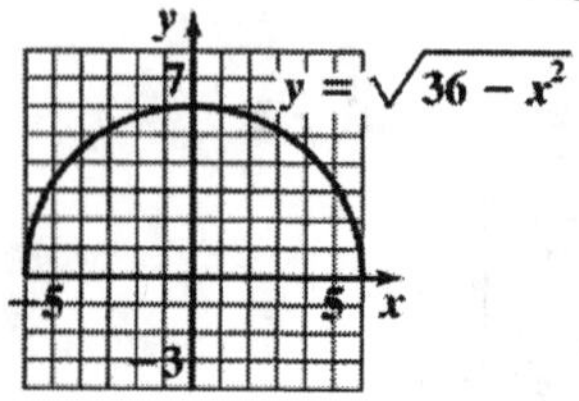

13. $x^2+y^2-2x+4y-4=0$
Complete the square in both *x* and *y*.
$(x^2-2x)+(y^2+4y)=4$
$(x^2-2x+1)+(y^2+4y+4)=4+1+4$
$(x-1)^2+(y+2)^2=9$
The graph of $x^2+y^2-2x+4y-4=0$ is a circle with its center at $(1,-2)$ and radius 3.

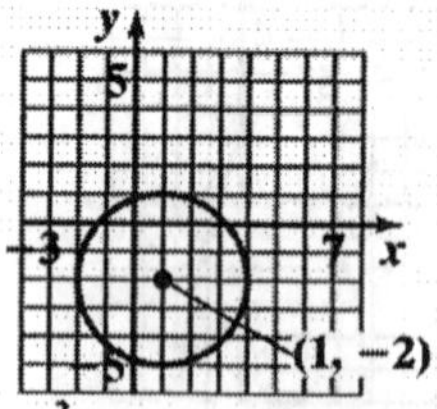

$x^2+y^2-2x+4y-4=0$

14. A circle is defined to be the set of points that are equidistant from a fixed point. This distance is called the radius and the fixed point is called the center.

15. $\dfrac{x^2}{4}+\dfrac{y^2}{9}=1$
$a^2=4\rightarrow a=2;\ b^2=9\rightarrow b=3$
The graph of this equation is an ellipse with center at the origin and major axis along the *y*-axis.

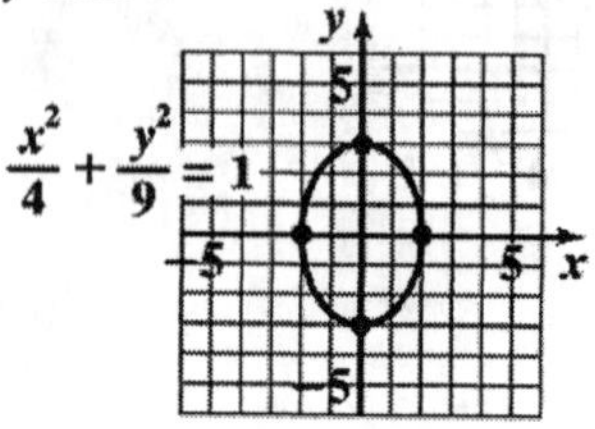

16. $\frac{x^2}{81}+\frac{y^2}{25}=1$

$a^2=81 \to a=9$; $b^2=25 \to b=5$

The graph of this equation is an ellipse with center at the origin and major axis along the x-axis.

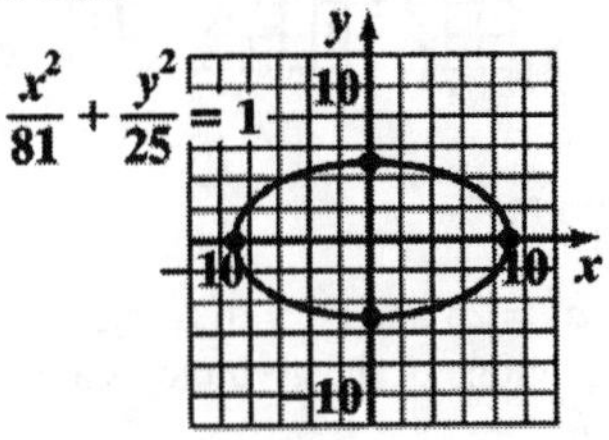

17. $\frac{(x-1)^2}{49}+\frac{(y+2)^2}{4}=1$

$a^2=49 \to a=7$; $b^2=4 \to b=2$

The graph of this equation is an ellipse with center at $(1,-2)$ and major axis parallel to the x-axis.

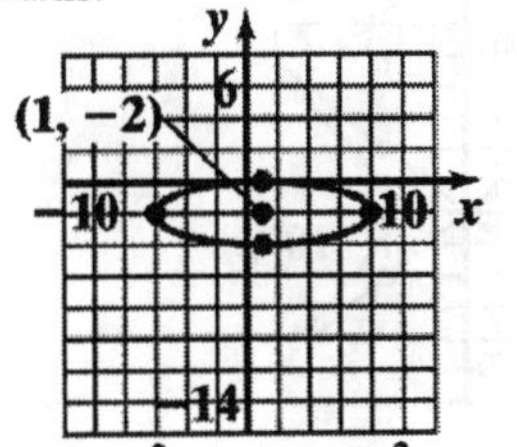

$\frac{(x-1)^2}{49}+\frac{(y+2)^2}{4}=1$

18. $36(x+3)^2+(y-4)^2=36$

$\frac{36(x+3)^2}{36}+\frac{(y-4)^2}{36}=\frac{36}{36}$

$\frac{(x+3)^2}{1}+\frac{(y-4)^2}{36}=1$

$a^2=1 \to a=1$; $b^2=36 \to b=6$

The graph of this equation is an ellipse with center at $(-3,4)$ and major axis parallel to the y-axis.

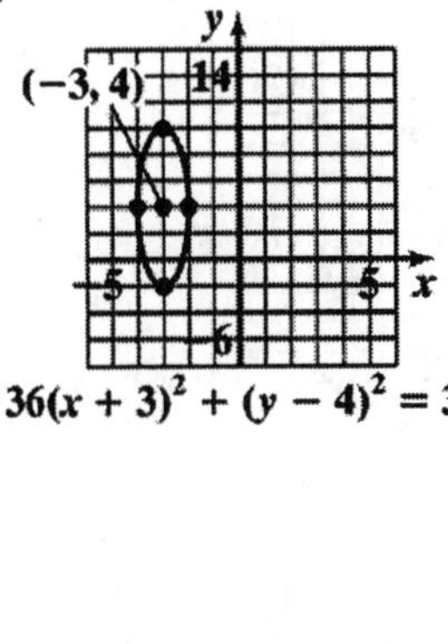

$36(x+3)^2+(y-4)^2=36$

19. $\frac{x^2}{4}+\frac{y^2}{9}=1$

Since $a^2=4, a=2$; Since $b^2=9, b=3$.

Area $= \pi ab = \pi(2)(3)$

$= 6\pi \approx 18.85$ square units

20. The center of the ellipse is the origin, $(0,0)$. We have $a=8$ and $b=5$ so $a^2=64$ and $b^2=25$. Thus, the equation of the ellipse is

$\frac{(x-0)^2}{8^2}+\frac{(y-0)^2}{5^2}=1$

$\frac{x^2}{64}+\frac{y^2}{25}=1$

Exercise Set 11.3

1. A hyperbola is the set of points in a plane, the difference of whose distances from two fixed points (called foci) is a constant.

3. The graph of $\frac{x^2}{a^2}-\frac{y^2}{b^2}=1$ is a hyperbola with vertices at $(a, 0)$ and $(-a, 0)$. Its transverse axis lies along the x-axis. The asymptotes are $y=\pm\frac{b}{a}x$.

5. No, equations of hyperbolas have one positive square term and one negative square term. This equation has two positive square terms.

7. Yes, equations of hyperbolas have one positive square term and one negative square term. This equation satisfies that condition.

9. The first step is to divide both sides by 81 in order to make the right side equal to 1.

11. a. $\frac{x^2}{9}-\frac{y^2}{4}=1$

Since $a^2=9$ and $b^2=4$, $a=3$ and $b=2$. The equations of the asymptotes are $y=\pm\frac{b}{a}x$, or $y=\pm\frac{2}{3}x$.

b. To graph the asymptotes, plot the points (3, 2), (–3, 2), (3, –2), and (–3, –2). The graph intersects the x-axis at (–3, 0) and (3, 0).

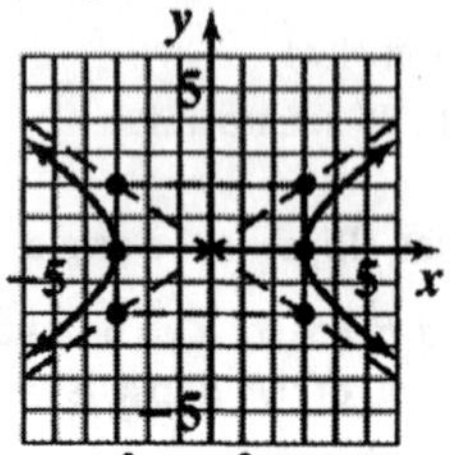

$$\frac{x^2}{9}-\frac{y^2}{4}=1$$

13. a. $\dfrac{x^2}{4}-\dfrac{y^2}{1}=1$

Since $a^2=4$ and $b^2=1$, $a=2$ and $b=1$. The equations of the asymptotes are $y=\pm\dfrac{b}{a}x$, or $y=\pm\dfrac{1}{2}x$.

b. To graph the asymptotes, plot the points (2, 1), (–2, 1), (2, –1), and (–2, –1). The graph intersects the x-axis at (–2, 0) and (2, 0).

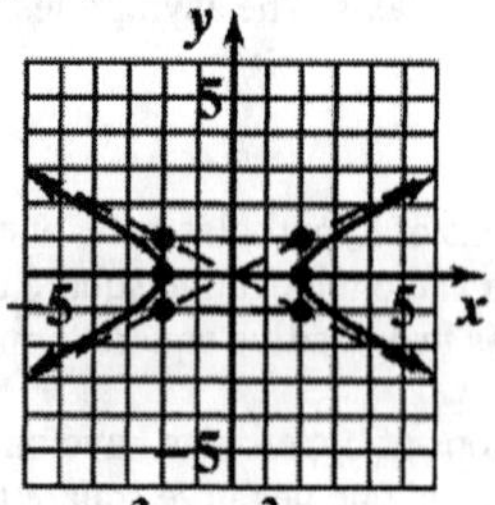

$$\frac{x^2}{4}-\frac{y^2}{1}=1$$

15. a. $\dfrac{x^2}{9}-\dfrac{y^2}{25}=1$

Since $a^2=9$ and $b^2=25$, $a=3$ and $b=5$. The equations of the asymptotes are $y=\pm\dfrac{b}{a}x$, or $y=\pm\dfrac{5}{3}x$.

b. To graph the asymptotes, plot the points (3, 5), (–3, 5), (3, –5), and (–3, –5). The graph intersects the x-axis at (–3, 0) and (3, 0).

$$\frac{x^2}{9}-\frac{y^2}{25}=1$$

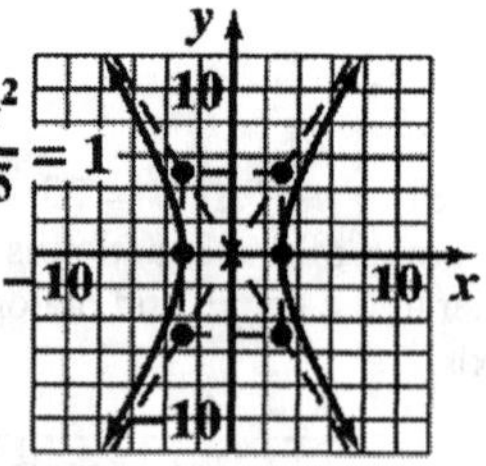

17. a. $\dfrac{x^2}{25}-\dfrac{y^2}{16}=1$

Since $a^2=25$ and $b^2=16$, $a=5$ and $b=4$. The equations of the asymptotes are $y=\pm\dfrac{b}{a}x$, or $y=\pm\dfrac{4}{5}x$.

b. To graph the asymptotes, plot the points (5, 4), (–5, 4), (5, –4), and (–5, –4). The graph intersects the x-axis at (–5, 0) and (5, 0).

$$\frac{x^2}{25}-\frac{y^2}{16}=1$$

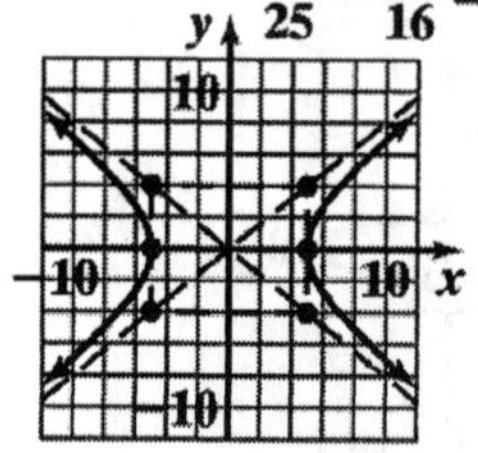

19. a. $\dfrac{y^2}{25}-\dfrac{x^2}{36}=1$

Since $a^2=36$ and $b^2=25$, $a=6$ and $b=5$. The equations of the asymptotes are $y=\pm\dfrac{b}{a}x$, or $y=\pm\dfrac{5}{6}x$.

b. To graph the asymptotes, plot the points (6, 5), (–6, 5), (6, –5), and (–6, –5). The graph intersects the y-axis at (0, –5) and (0, 5).

$$\frac{y^2}{25}-\frac{x^2}{36}=1$$

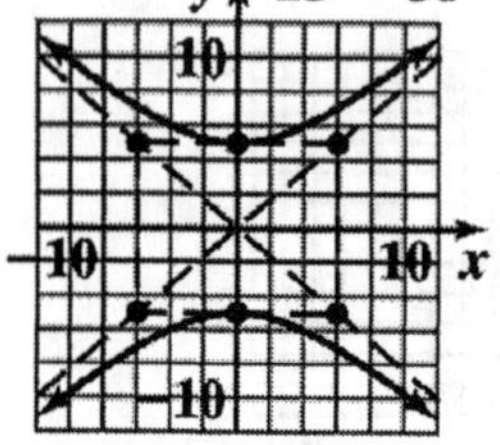

21. a. $\frac{y^2}{9}-\frac{x^2}{16}=1$

Since $a^2=16$ and $b^2=9$, $a=4$ and $b=3$. The equations of the asymptotes are $y=\pm\frac{b}{a}x$, or $= y\pm\frac{3}{4}x$.

b. To graph the asymptotes, plot the points (4, 3), (–4, 3), (4, –3) and (–4, –3). The graph intersects the y-axis at (0, –3) and (0, 3).

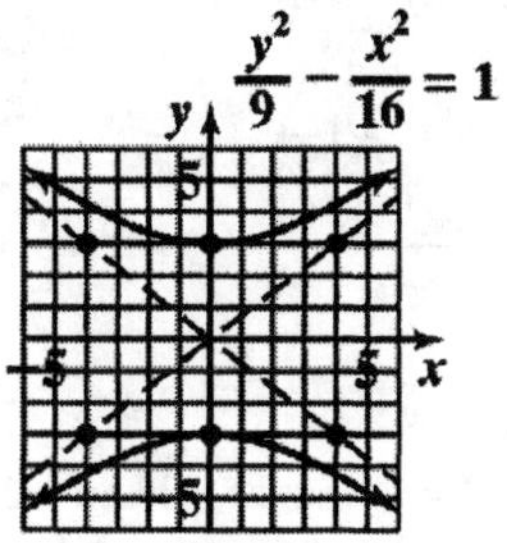

23. a. $\frac{y^2}{25}-\frac{x^2}{4}=1$

Since $a^2=4$ and $b^2=25$, $a=2$ and $b=5$. The equations of the asymptotes are $y=\pm\frac{b}{a}x$ or $y=\pm\frac{5}{2}x$.

b. To graph the asymptotes, plot the points (2, 5), (–2, 5), (2, –5) and (–2, –5). The graph intersects the y-axis at (0, –5) and (0, 5).

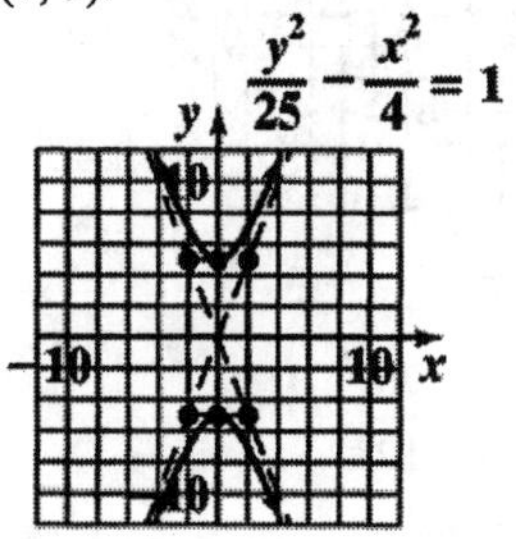

25. a. $\frac{x^2}{81}-\frac{y^2}{16}=1$

Since $a^2=81$ and $b^2=16$, $a=9$ and $b=4$. The equations of the asymptotes are $y=\pm\frac{b}{a}x$, or $y=\pm\frac{4}{9}x$.

b. To graph the asymptotes, plot the points (9, 4), (–9, 4), (9, –4), and (–9, –4). The graph intersects the x-axis at (–9, 0) and (9, 0).

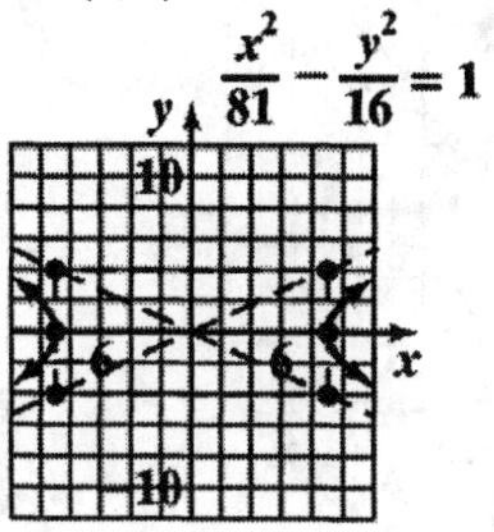

27. a. $x^2-25y^2=25$

$\frac{x^2}{25}-\frac{25y^2}{25}=1$

$\frac{x^2}{25}-\frac{y^2}{1}=1$

Since $a^2=25$ and $b^2=1$, $a=5$ and $b=1$. The equations of the asymptotes are $y=\pm\frac{b}{a}x$, or $y=\pm\frac{1}{5}x$.

b. To graph the asymptotes, plot the points (5, 1), (–5, 1), (5, –1), and (–5, –1). The graph intersects the x-axis at (–5, 0), and (5, 0).

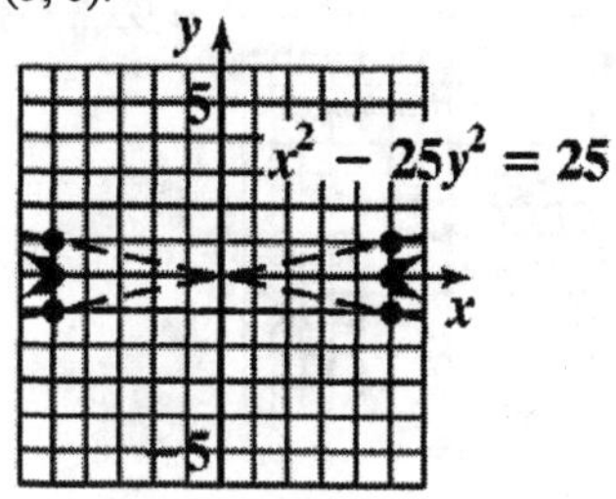

29. a. $4y^2-16x^2=64$

$\frac{4y^2}{64}-\frac{16x^2}{64}=1$

$\frac{y^2}{16}-\frac{x^2}{4}=1$

Since $a^2=4$ and $b^2=16$, $a=2$ and $b=4$. The equations of the asymptotes are $y=\pm\frac{b}{a}x$, or $y=\pm 2x$.

b. To graph the asymptotes, plot (2, 4), (–2, 4), (2 –4), and (–2, –4). The graph intersects the *y*-axis at (0, –4) and (0, 4).

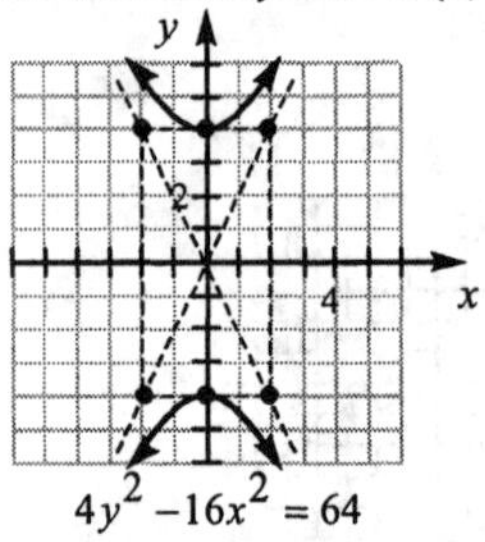

31. a. $9y^2 - x^2 = 9$

$$\frac{9y^2}{9} - \frac{x^2}{9} = 1$$

$$\frac{y^2}{1} - \frac{x^2}{9} = 1$$

Since $a^2 = 9$ and $b^2 = 1$, $a = 3$ and $b = 1$. The equations of the asymptotes are $y = \pm\frac{b}{a}x$, or $\pm\frac{1}{3}x$.

b. To graph the asymptotes, plot the points (3, 1), (–3, 1), (3, –1), and (–3, –1). The graph intersects the *y*-axis at (0, –1) and (0, 1).

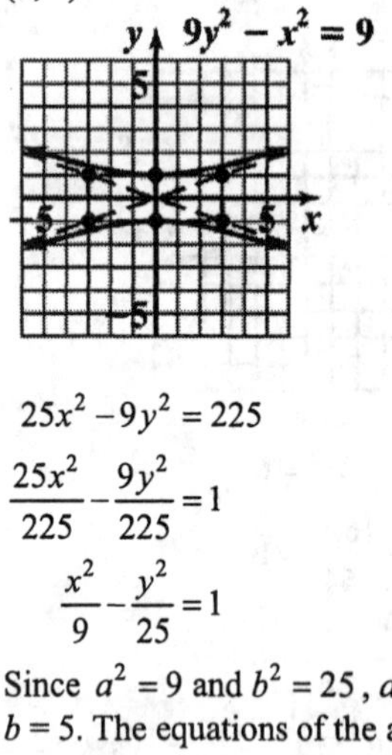

33. a. $25x^2 - 9y^2 = 225$

$$\frac{25x^2}{225} - \frac{9y^2}{225} = 1$$

$$\frac{x^2}{9} - \frac{y^2}{25} = 1$$

Since $a^2 = 9$ and $b^2 = 25$, $a = 3$ and $b = 5$. The equations of the asymptotes are $y = \pm\frac{b}{a}x$, or $y = \pm\frac{5}{3}x$.

b. To graph the asymptotes, plot the points (3, 5), (–3, 5), (3, –5), and (–3, –5). The graph intersects the *x*-axis at (–3, 0) and (3, 0).

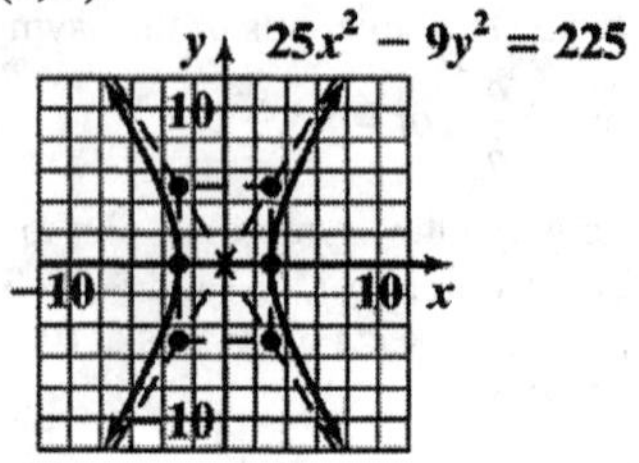

35. a. $4y^2 - 36x^2 = 144$

$$\frac{4y^2}{144} - \frac{36x^2}{144} = \frac{144}{144}$$

$$\frac{y^2}{36} - \frac{x^2}{4} = 1$$

Since $a^2 = 4$ and $b^2 = 36$, $a = 2$ and $b = 6$. The equations of the asymptotes are $y = \pm\frac{b}{a}x$, or $y = \pm 3x$.

b. To graph the asymptotes, plot the points (2, 6), (–2, 6), (2, –6), and (–2, –6). The graph intersects the *y*-axis at (0, –6) and (0, 6).

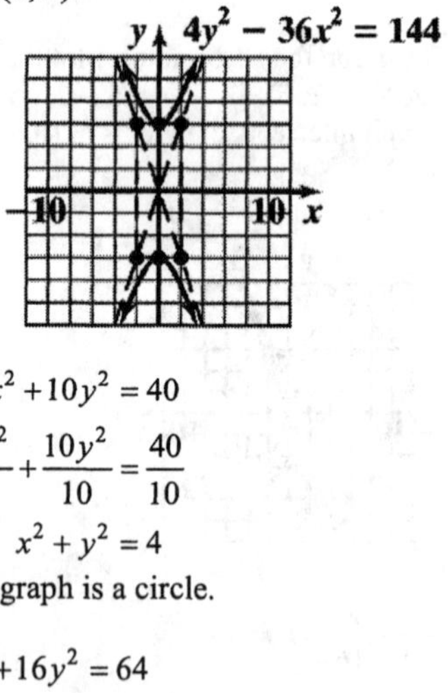

37. $10x^2 + 10y^2 = 40$

$$\frac{10x^2}{10} + \frac{10y^2}{10} = \frac{40}{10}$$

$$x^2 + y^2 = 4$$

The graph is a circle.

39. $x^2 + 16y^2 = 64$

$$\frac{x^2}{64} + \frac{16y^2}{64} = \frac{64}{64}$$

$$\frac{x^2}{64} + \frac{y^2}{4} = 1$$

The graph is an ellipse.

41. $4x^2-4y^2=29$

$$\frac{4x^2}{29}-\frac{4y^2}{29}=\frac{29}{29}$$

$$\frac{x^2}{\frac{29}{4}}-\frac{y^2}{\frac{29}{4}}=1$$

The graph is a hyperbola.

43. $2y=12x^2-8x+16$

$$y=6x^2-4x+8$$

The graph is a parabola.

45. $6x^2+9y^2=54$

$$\frac{6x^2}{54}+\frac{9y^2}{54}=\frac{54}{54}$$

$$\frac{x^2}{9}+\frac{y^2}{6}=1$$

The graph is an ellipse.

47. $3x=-2y^2+9y-15$

$$x=-\frac{2}{3}y^2+3y-5$$

The graph is a parabola.

49. $6x^2+6y^2=36$

$$\frac{6x^2}{6}+\frac{6y^2}{6}=\frac{36}{6}$$

$$x^2+y^2=6$$

The graph is a circle.

51. $14y^2=7x^2+35$

$$14y^2-7x^2=35$$

$$\frac{14y^2}{35}-\frac{7x^2}{35}=\frac{35}{35}$$

$$\frac{y^2}{\frac{5}{2}}-\frac{x^2}{5}=1$$

The graph is a hyperbola.

53. $x+y=2y^2+6$

$$x=2y^2-y+6$$

The graph is a parabola.

55. $12x^2=4y^2+48$

$$12x^2-4y^2=48$$

$$\frac{12x^2}{48}-\frac{4y^2}{48}=\frac{48}{48}$$

$$\frac{x^2}{4}-\frac{y^2}{12}=1$$

The graph is a hyperbola.

57. $y-x+4=x^2$

$$y=x^2+x-4$$

The graph is a parabola.

59. $-3x^2-3y^2=-27$

$$\frac{-3x^2}{-3}+\frac{-3y^2}{-3}=\frac{-27}{-3}$$

$$x^2+y^2=9$$

The graph is a circle.

61. Since the vertices are (0, ±2), the hyperbola is of the form $\frac{y^2}{b^2}-\frac{x^2}{a^2}=1$ with $b=2$. Since the asymptotes are $y=\pm\frac{1}{2}x$, we have $\frac{b}{a}=\frac{1}{2}$. Therefore, $\frac{2}{a}=\frac{1}{2}$, so $a=4$. The equation of the hyperbola is $\frac{y^2}{2^2}-\frac{x^2}{4^2}=1$, or $\frac{y^2}{4}-\frac{x^2}{16}=1$.

63. Since the vertices are $(0,\pm 3)$, the hyperbola is of the form $\frac{x^2}{a^2}-\frac{y^2}{b^2}=1$ with $a=3$. Since the asymptotes are $y=\pm 2x$, we have $\frac{b}{a}=2$. Therefore, $\frac{b}{3}=2$, so $b=6$. The equation of the hyperbola is $\frac{x^2}{3^2}-\frac{y^2}{6^2}=1$, or $\frac{x^2}{9}-\frac{y^2}{36}=1$.

65. Since the transverse axis is along the x-axis, the equation is of the form $\frac{x^2}{a^2}-\frac{y^2}{b^2}=1$ Since the asymptotes are $y=\pm\frac{5}{3}x$, we require $\frac{b}{a}=\frac{5}{3}$.
Using $a=3$ and $b=5$, the equation of the hyperbola is $\frac{x^2}{3^2}-\frac{y^2}{5^2}=1$, or $\frac{x^2}{9}-\frac{y^2}{25}=1$.
No, this is not the only possible answer, because a and b are not uniquely determined.
$\frac{x^2}{18}-\frac{y^2}{50}=1$ and others will also work.

67. No, for each value of x with $|x|>a$, there are 2 possible values of y.

69. $\frac{x^2}{25}-\frac{y^2}{4}=1$. This hyperbola has its transverse axis along the x-axis with vertices at $(\pm 5,0)$.
Domain: $(-\infty,-5]\cup[5,\infty)$
Range: $\mathbb{R}$

71. The equation is changed from $\frac{x^2}{a^2}-\frac{y^2}{b^2}=1$ to $\frac{x^2}{b^2}-\frac{y^2}{a^2}=1$. Both graphs have a transverse axis along the x-axis. The vertices of the second graph will be closer to the origin, at $(\pm b,0)$ instead of $(\pm a,0)$. The second graph will open wider.

73. Answers will vary.

75. The points are $(-6,4)$ and $(-2,2)$.

$$m=\frac{y_2-y_1}{x_2-x_1}=\frac{2-4}{-2-(-6)}=\frac{-2}{4}=-\frac{1}{2}$$

Use $y-y_1=m(x-x_1)$, with $m=-\frac{1}{2}$, $(-2,2)$

$$y-2=-\frac{1}{2}(x-(-2))$$
$$y-2=-\frac{1}{2}x-1$$
$$y=-\frac{1}{2}x+1$$

76. $f(x)=3x^2-x+5,\ g(x)=6-4x^2$

$$(f+g)(x)=(3x^2-x+5)+(6-4x^2)$$
$$=-x^2-x+11$$

77.

$$5(-4x+9y=7) \Rightarrow -20x+45y=35$$
$$4(5x+6y=-3) \Rightarrow \underline{20x+24y=-12}$$
$$69y=23$$
$$y=\frac{1}{3}$$

$$5x+6\left(\frac{1}{3}\right)=-3 \Rightarrow 5x+2=-3 \Rightarrow x=-1$$

The solution is $\left(-1,\frac{1}{3}\right)$.

78.

$$\frac{3x}{2x-3}+\frac{2x+4}{2x^2+x-6}$$
$$=\frac{3x}{2x-3}+\frac{2x+4}{(2x-3)(x+2)}$$
$$=\frac{3x}{2x-3}\cdot\frac{x+2}{x+2}+\frac{2x+4}{(2x-3)(x+2)}$$
$$=\frac{3x^2+6x+2x+4}{(2x-3)(x+2)}$$
$$=\frac{3x^2+8x+4}{(2x-3)(x+2)}$$
$$=\frac{(3x+2)(x+2)}{(2x-3)(x+2)}$$
$$=\frac{3x+2}{2x-3}$$

79. $E=\frac{1}{2}mv^2$, for v

$$2E=mv^2$$
$$\frac{2E}{m}=v^2$$
$$\sqrt{\frac{2E}{m}}=v \quad\text{or}\quad v=\sqrt{\frac{2E}{m}}$$

80. $\log(x+4) = \log 5 - \log x$

$\log(x+4) = \log\frac{5}{x}$

$x+4 = \frac{5}{x}$

$x(x+4) = 5$

$x^2 + 4x - 5 = 0$

$(x+5)(x-1) = 0$

$x+5=0$ or $x-1=0$

$x=-5$ $\quad x=1$

Upon checking, $x = -5$ does not satisfy the equation. The solution is $x = 1$.

Exercise Set 11.4

1. A nonlinear system of equations is a system in which at least one equation is nonlinear.

3. Yes

5. Yes

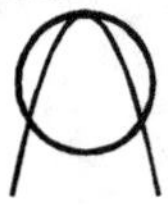

7. $x^2 + y^2 = 18$

$x + y = 0$

Solve $x+y=0$ for x: $x=-y$.

Substitute $x=-y$ for x in $x^2+y^2=18$.

$x^2 + y^2 = 18$

$(-y)^2 + y^2 = 18$

$y^2 + y^2 = 18$

$2y^2 = 18$

$y^2 = 9$

$y = \pm 3$

$y=3$ or $y=-3$

$x=-3$ $\quad x=3$

The solutions are (3, –3) and (–3, 3).

9. $x^2 + y^2 = 9$

$x + 2y = 3$

Solve $x+2y=3$ for x: $x = 3-2y$.

Substitute $3-2y$ for x in $x^2+y^2=9$.

$x^2 + y^2 = 9$

$(3-2y)^2 + y^2 = 9$

$9 - 12y + 4y^2 + y^2 = 9$

$5y^2 - 12y = 0$

$y(5y-12) = 0$

$y=0$ or $y = \frac{12}{5}$

$x = 3-2y$

$x = 3-2(0)$

$x = 3$

$x = 3-2y$

$x = 3-2\left(\frac{12}{5}\right)$

$x = -\frac{9}{5}$

The solutions are (3, 0) and $\left(-\frac{9}{5}, \frac{12}{5}\right)$.

11. $y = x^2 - 5$

$3x + 2y = 10$

Substitute $x^2 - 5$ for y in $3x+2y=10$

$3x + 2y = 10$

$3x + 2(x^2-5) = 10$

$3x + 2x^2 - 10 = 10$

$2x^2 + 3x - 20 = 0$

$(x+4)(2x-5) = 0$

$x=-4$ or $x = \frac{5}{2}$

$y = x^2 - 5$

$y = (-4)^2 - 5$

$y = 11$

$y = x^2 - 5$

$y = \left(\frac{5}{2}\right)^2 - 5$

$y = \frac{5}{4}$

The solutions are $(-4, 11)$ and $\left(\frac{5}{2}, \frac{5}{4}\right)$.

13. $x^2 + y = 6$

$y = x^2 + 4$

Substitute $x^2 + 4$ for y in $x^2 + y = 6$

$x^2 + x^2 + 4 = 6$

$2x^2 = 2$

$x^2 = 1$

$x = \pm 1$

$y = (1)^2 + 4$ or $y = (-1)^2 + 4$

$y = 1 + 4$ $\quad$ $y = 1 + 4$

$y = 5$ $\quad$ $y = 5$

The solutions are (1, 5) and (–1, 5).

15. $2x^2 + y^2 = 16$

$x^2 - y^2 = -4$

Solve $x^2 - y^2 = -4$ for y^2: $y^2 = x^2 + 4$

Substitute $x^2 + 4$ for y^2 in $2x^2 + y^2 = 16$.

$2x^2 + y^2 = 16$

$2x^2 + (x^2 + 4) = 16$

$3x^2 + 4 = 16$

$3x^2 = 12$

$x^2 = 4$

$x = 2$ or $x = -2$

$y^2 = x^2 + 4$ $\quad$ $y^2 = x^2 + 4$

$y^2 = 2^2 + 4$ $\quad$ $y = (-2)^2 + 4$

$y^2 = 8$ $\quad$ $y = 8$

$y = \pm\sqrt{8} = \pm 2\sqrt{2}$ $\quad$ $y = \pm\sqrt{8} = \pm 2\sqrt{2}$

The solutions are $\left(2, 2\sqrt{2}\right), \left(2, -2\sqrt{2}\right)$, $\left(-2, 2\sqrt{2}\right)$, and $\left(-2, -2\sqrt{2}\right)$.

17. $x^2 + y^2 = 4$

$y = x^2 - 6 \Rightarrow x^2 = y + 6$

Substitute $y + 6$ for x^2 in $x^2 + y^2 = 4$.

$x^2 + y^2 = 4$

$(y + 6) + y^2 = 4$

$y^2 + y + 2 = 0$

$$y = \frac{-1 \pm \sqrt{1^2 - 4(1)(2)}}{2(1)}$$

$$= \frac{-1 \pm \sqrt{-7}}{2}$$

$$= \frac{-1 \pm i\sqrt{31}}{2}$$

There is no real solution.

19. $x^2 + y^2 = 9$

$y = x^2 - 3$

Solve $y = x^2 - 3$ for x^2: $x^2 = y + 3$.

Substitute $y + 3$ for x^2 in $x^2 + y^2 = 9$.

$x^2 + y^2 = 9$

$(y + 3) + y^2 = 9$

$y^2 + y - 6 = 0$

$(y - 2)(y + 3) = 0$

$y = 2$ or $y = -3$

$x^2 = y + 3$ $\quad$ $x^2 = y + 3$

$x^2 = 2 + 3$ $\quad$ $x^2 = -3 + 3$

$x^2 = 5$ $\quad$ $x^2 = 0$

$x = \pm\sqrt{5}$ $\quad$ $x = 0$

The solutions are $(0, -3)$, $\left(\sqrt{5}, 2\right)$, and $\left(-\sqrt{5}, 2\right)$.

21. $2x^2 - y^2 = -8$

$x - y = 6$

Solve the second equation for y: $y = x - 6$.

Substitute $x - 6$ for y in $2x^2 - y^2 = -8$.

$$2x^2 - y^2 = -8$$
$$2x^2 - (x-6)^2 = -8$$
$$2x^2 - (x^2 - 12x + 36) = -8$$
$$2x^2 - x^2 + 12x - 36 = -8$$
$$x^2 + 12x - 28 = 0$$
$$(x-2)(x+14) = 0$$

or

$x = 2$ $\qquad x = -14$

$y = x - 6$ $\qquad y = x - 6$

$y = 2 - 6$ $\qquad y = -14 - 6$

$y = -4$ $\qquad y = -20$

The solutions are (2, –4) and (–14, –20).

23. $x^2 - y^2 = 4$

$2x^2 + y^2 = 8$

$3x^2 \quad = 12$

$x^2 = 4$

$x = 2$ or $x = -2$

$x^2 - y^2 = 4$ $\qquad x^2 - y^2 = 4$

$2^2 - y^2 = 4$ $\qquad (-2)^2 - y^2 = 4$

$y^2 = 0$ $\qquad y^2 = 0$

$y = 0$ $\qquad y = 0$

The solutions are (2, 0) and (–2, 0).

25. $x^2 + y^2 = 16$ (1)

$2x^2 - 5y^2 = 25$ (2)

$-2x^2 - 2y^2 = -32$ (1) multiplied by –2

$2x^2 - 5y^2 = 25$ (2)

$-7y^2 = -7$

$y^2 = 1$

$y = 1$ or $y = -1$

$x^2 + y^2 = 16$ $\qquad x^2 + y^2 = 16$

$x^2 + 1^2 = 16$ $\qquad x^2 + (-1)^2 = 16$

$x^2 = 15$ $\qquad x^2 = 15$

$x = \pm\sqrt{15}$ $\qquad x = \pm\sqrt{15}$

The solutions are $\left(-\sqrt{15}, 1\right)$, $\left(-\sqrt{15}, -1\right)$, $\left(\sqrt{15}, -1\right)$, and $\left(\sqrt{15}, 1\right)$.

27. $3x^2 - y^2 = 4$ (1)

$x^2 + 4y^2 = 10$ (2)

$12x^2 - 4y^2 = 16$ (1) multiplied by 4

$x^2 + 4y^2 = 10$ (2)

$13x^2 = 26$

$x^2 = 2$

$x = \sqrt{2}$ or $x = -\sqrt{2}$

$3x^2 - y^2 = 4$ $\qquad 3x^2 - y^2 = 4$

$3(\sqrt{2})^2 - y^2 = 4$ $\qquad 3(-\sqrt{2})^2 - y^2 = 4$

$6 - y^2 = 4$ $\qquad 6 - y^2 = 4$

$y^2 = 2$ $\qquad y^2 = 2$

$y = \pm\sqrt{2}$ $\qquad y = \pm\sqrt{2}$

The solutions are $\left(\sqrt{2}, \sqrt{2}\right)$, $\left(\sqrt{2}, -\sqrt{2}\right)$, $\left(-\sqrt{2}, \sqrt{2}\right)$, and $\left(-\sqrt{2}, -\sqrt{2}\right)$.

29. $4x^2+9y^2=36$

$2x^2-9y^2=18$

$6x^2=54$

$x^2=9$

$x=3$ or $x=-3$

$4x^2+9y^2=36$ $\quad$ $4x^2+9y^2=36$

$4(3)^2+9y^2=36$ $\quad$ $4(-3)^2+9y^2=36$

$9y^2=0$ $\quad$ $9y^2=0$

$y^2=0$ $\quad$ $y^2=0$

$y=0$ $\quad$ $y=0$

The solutions are (3, 0) and (–3, 0).

31. $2x^2-y^2=7$ (1)

$x^2+2y^2=6$ (2)

$4x^2-2y^2=14$ (1) multiplied by 2

$x^2+2y^2=6$ (2)

$5x^2=20$

$x^2=4$

$x=2$ or $x=-2$

$2(2)^2-y^2=7$ $\quad$ $2(-2)^2-y^2=7$

$8-y^2=7$ $\quad$ $8-y^2=7$

$y^2=1$ $\quad$ $y^2=1$

$x=\pm1$ $\quad$ $x=\pm1$

The solutions are (2, 1), (2, –1), (–2, 1), and (–2, –1).

33. $x^2+y^2=25$ (1)

$2x^2-3y^2=-30$ (2)

$3x^2+3y^2=75$ (1) multiplied by 3

$2x^2-3y^2=-30$ (2)

$5x^2=45$

$x^2=9$

$x=3$ or $x=-3$

$x^2+y^2=25$ $\quad$ $x^2+y^2=25$

$(3)^2+y^2=25$ $\quad$ $(-3)^2+y^2=25$

$y^2=16$ $\quad$ $y^2=16$

$y=\pm4$ $\quad$ $y=\pm4$

The solutions are(3, 4), (3, –4), (–3, 4), and (–3, –4).

35. $x^2+y^2=9$ (1)

$16x^2-4y^2=64$ (2)

$4x^2+4y^2=36$ (1) multiplied by 4

$16x^2-4y^2=64$ (2)

$20x^2=100$

$x^2=5$

$x=\sqrt{5}$ or $x=-\sqrt{5}$

$x^2+y^2=9$ $\quad$ $x^2+y^2=9$

$(\sqrt{5})^2+y^2=9$ $\quad$ $(-\sqrt{5})^2+y^2=9$

$y^2=4$ $\quad$ $y^2=4$

$y=\pm2$ $\quad$ $y=\pm2$

The solutions are $(\sqrt{5},2)$, $(\sqrt{5},-2)$, $(-\sqrt{5},2)$, and $(-\sqrt{5},-2)$.

37. $x^2+y^2=4 \quad (1)$

$16x^2+9y^2=144 \quad (2)$

Multiply the first equation by -16 and add.

$$\begin{aligned}-16x^2-16y^2&=-164\\ 16x^2+9y^2&=144\\ \hline 7y^2&=-20\\ y^2&=-\frac{20}{7}\end{aligned}$$

This equation has no real solutions so the system has no real solutions.

39. $x^2+4y^2=4$

$10y^2-9x^2=90$

Write the equations in standard form.

$$\begin{aligned}x^2+4y^2&=4\\ -9x^2+10y^2&=90\end{aligned}$$

Multiply the first equation by -2 and the second equation by 2, then add.

$$\begin{aligned}-5x^2-20y^2&=-20\\ -18x^2+20y^2&=180\\ \hline -23x^2&=160\\ x^2&=-\frac{160}{23}\end{aligned}$$

This equation has no real solutions so the system has no real solutions.

41. Answers will vary.

43. Let x = length; y = width

$xy=440$

$2x+2y=84$

Solve $2x+2y=84$ for y: $y=42-x$.

Substitute $42-x$ for y in $xy=440$.

$$\begin{aligned}xy&=440\\ x(42-x)&=440\\ 42x-x^2&=440\end{aligned}$$

$x^2-42x+440=0$

$(x-20)(x-22)=0$

$x-20=0$ or $x-22=0$

$x=20 \qquad x=22$

$y=42-x$ or $y=42-22$

$y=42-20 \qquad y=42-22$

$y=22 \qquad y=20$

The solutions are (20, 22) and (22, 20).

The dimensions of the floor are 20 m by 22 m.

45. Let x = length

y = width

$xy=270$

$2x+2y=78$

Solve $2x+2y=78$ for y: $y=39-x$.

Substitute $39-x$ for y in $xy=270$.

$$\begin{aligned}xy&=270\\ x(39-x)&=270\\ 39x-x^2&=270\end{aligned}$$

$x^2-39x+270=0$

$(x-9)(x-30)=0$

$x-9=0$ or $x-30=0$

$x=9 \qquad x=30$

$y=39-x \qquad y=39-x$

$y=39-9 \qquad y=39-30$

$y=30 \qquad y=9$

The solutions are (9, 30) and (30, 9). The dimensions of the garden are 9 ft by 30 ft.

47. Let x = length; y = width

$xy=112$

$x^2+y^2=\left(\sqrt{260}\right)^2$

Solve $xy=112$ for y: $y=\frac{112}{x}$.

$$\begin{aligned}x^2+y^2&=260\\ x^2+\left(\frac{112}{x}\right)^2&=260\\ x^2+\frac{12{,}544}{x^2}&=260\\ x^4+12{,}544&=260x^2\end{aligned}$$

$x^4-260x^2+12{,}544=0$

$(x^2-64)(x^2-196)=0$

$x^2-64=0$ or $x^2-196=0$

$x^2=64 \qquad x^2=196$

$x=\pm 8 \qquad x=\pm 14$

Since x must be positive, $x=8$ or $x=14$.

If $x=8$, then $y=\frac{112}{8}=14$.

If $x=14$, then $y=\frac{112}{14}=8$.

The dimensions of the new bill are 8 cm by 14 cm.

49. Let x = length

y = width

$$x^2 + y^2 = 34^2$$

$$x + y + 34 = 80$$

Solve $x + y + 34 = 80$ for y: $y = 46 - x$.

Substitute $46 - x$ for y in $x^2 + y^2 = 34^2$.

$$x^2 + y^2 = 34^2$$

$$x^2 + (46 - x)^2 = 34^2$$

$$x^2 + (2116 - 92x + x^2) = 1156$$

$$2x^2 - 92 + 2116 = 1156$$

$$2x^2 - 92x + 960 = 0$$

$$x^2 - 46x + 480 = 0$$

$$(x - 16)(x - 30) = 0$$

$x - 16 = 0$ or $x - 30 = 0$

$x = 16$ $\quad$ $x = 30$

$y = 46 - x$ $\quad$ $y = 46 - x$

$y = 46 - 16$ $\quad$ $y = 46 - 30$

$y = 30$ $\quad$ $y = 16$

The solutions are (16, 30) and (30, 16).
The dimensions of the piece of wood are 16 in. by 30 in.

51. $d = -16t^2 + 64t$

$d = -16t^2 + 16t + 80$

Substitute $-16t^2 + 64t$ for d in $d = -16t^2 + 16t + 80$.

$$d = -16t^2 + 16t + 80$$

$$-16t^2 + 64t = -16t^2 + 16t + 80$$

$$64t = 16t + 80$$

$$48t = 80$$

$$t = \frac{80}{48} = \frac{5}{3} \approx 1.67$$

The balls are the same height above the ground at $t \approx 1.67$ sec.

53. Since $t = 1$ year, we may write the formula as $i = pr$.

$7.50 = pr$

$7.50 = (p + 25)(r - 0.01)$

Rewrite the second equation by multiplying the binomials. Then substitute 7.50 for pr and solve for r.

$$7.50 = (p + 25)(r - 0.01)$$

$$7.50 = pr - 0.01p + 25r - 0.25$$

$$7.50 = 7.50 - 0.01p + 25r - 0.25$$

$$0 = -0.01p + 25r - 0.25$$

$$0.01p + 0.25 = 25r$$

$$\frac{0.01p}{25} + \frac{0.25}{25} = \frac{25r}{25}$$

$$r = 0.0004p + 0.01$$

Substitute $0.0004p + 0.01$ for r in $7.50 = pr$.

$$7.50 = pr$$

$$7.50 = p(0.0004p + 0.01)$$

$$7.50 = 0.0004p^2 + 0.01p$$

$$0 = 0.0004p^2 + 0.01p - 7.50$$

$$0 = p^2 + 25p - 18{,}750$$

$$0 = (p - 125)(p + 150)$$

$p - 125 = 0$ or $p + 150 = 0$

$p = 125$ $\quad$ $p = -150$

Since the principal must be positive, use $p = 125$.

$r = 0.0004p + 0.01$

$r = 0.0004(125) + 0.01$

$r = 0.06$

The principal is \$125 and the interest rate is 6%.

55. $C = 10x + 300$

$R = 30x - 0.1x^2$

$$C = R$$

$$10x + 300 = 30x - 0.1x^2$$

$$0.1x^2 - 20x + 300 = 0$$

$$x = \frac{-b \pm \sqrt{b^2 - 4ac}}{2a}$$

$$= \frac{-(-20) \pm \sqrt{(-20)^2 - 4(0.1)(300)}}{2(0.1)}$$

$$= \frac{20 \pm \sqrt{280}}{0.2}$$

$$x = \frac{20 + \sqrt{280}}{0.2} \approx 183.7$$

or

$$x = \frac{20 - \sqrt{280}}{0.2} \approx 16.3$$

The break-even points are ≈ 16 and ≈ 184.

57. $C = 12.6x + 150$

$R = 42.8x - 0.3x^2$

$$C = R$$

$$12.6x + 150 = 42.8x - 0.3x^2$$

$$0.3x^2 - 30.2x + 150 = 0$$

$$x = \frac{-b \pm \sqrt{b^2 - 4ac}}{2a}$$

$$= \frac{-(-30.2) \pm \sqrt{(-30.2)^2 - 4(0.3)(150)}}{2(0.3)}$$

$$= \frac{30.2 \pm \sqrt{732.04}}{0.6}$$

$$x = \frac{30.2 + \sqrt{732.04}}{0.6} \approx 95.4$$

or

$$x = \frac{30.2 - \sqrt{732.04}}{0.6} \approx 5.2$$

The break-even points are ≈ 5 and ≈ 95.

59. Solve each equation for y.

$$3x - 5y = 12 \qquad x^2 + y^2 = 10$$

$$-5y = -3x + 12 \qquad y^2 = 10 - x^2$$

$$y = \frac{3}{5}x - \frac{12}{5} \qquad y = \pm\sqrt{10 - x^2}$$

Use $y_1 = \frac{3}{5}x - \frac{12}{5}$, $y_2 = \sqrt{10 - x^2}$, and $y_2 = -\sqrt{10 - x^2}$.

-9.4, 9.4, 1, -6.2, 6.2, 1

Approximate solutions: (–1, –3), (3.12, –0.53)

61. Let x = length of one leg

y = length of other leg

$$x^2 + y^2 = 26^2$$

$$\frac{1}{2}xy = 120$$

Solve $\frac{1}{2}xy = 120$ for y: $y = \frac{240}{x}$.

Substitute $\frac{240}{x}$ for y in $x^2 + y^2 = 26^2$

$$x^2 + y^2 = 26^2$$

$$x^2 + \left(\frac{240}{x}\right)^2 = 676$$

$$x^2 + \frac{57,600}{x^2} = 676$$

$$x^4 + 57,600 = 676x^2$$

$$x^4 - 676x^2 + 57,600 = 0$$

$$(x^2 - 100)(x^2 - 576) = 0$$

$$x^2 - 100 = 0 \quad \text{or} \quad x^2 - 576 = 0$$

$$x^2 = 100 \qquad x^2 = 576$$

$$x = \pm 10 \qquad x = \pm 24$$

Since x is a length, x must be positive.

If $x = 10$, then $y = \frac{240}{10} = 24$. If $x = 24$, then $y = \frac{240}{24} = 10$. The legs have lengths 10 yards and 24 yards.

63. The operations are evaluated in the following order: parentheses, exponents, multiplication or division, addition or subtraction.

64. $(x+1)^3 + 1$

$$= (x+1)^3 + (1)^3$$

$$= (x+1+1)\left((x+1)^2 - (x+1)(1) + (1)^2\right)$$

$$= (x+2)(x^2 + 2x + 1 - x - 1 + 1)$$

$$= (x+2)(x^2 + x + 1)$$

65. $x = \frac{k}{P^2}$

$10 = \frac{k}{6^2} \Rightarrow k = 360 \Rightarrow x = \frac{360}{P^2}$

$x = \frac{360}{20^2} = \frac{360}{400} = \frac{9}{10}$ or 0.9

66. $\frac{5}{\sqrt{x+2}-3} = \frac{5}{\sqrt{x+2}-3} \cdot \frac{\sqrt{x+2}+3}{\sqrt{x+2}+3}$

$= \frac{5\sqrt{x+2}+15}{x+2+3\sqrt{x+2}-3\sqrt{x+2}-9}$

$= \frac{5\sqrt{x+2}+15}{x-7}$

67. $A = A_0 e^{kt}$, for k

$\frac{A}{A_0} = e^{kt}$

$\ln \frac{A}{A_0} = \ln e^{kt}$

$\ln A - \ln A_0 = kt$

$\frac{\ln A - \ln A_0}{t} = k$

Exercise Set 11.5

1. When graphing an inequality containing > or <, the boundary line is graphed with a dashed line because the boundary line is not part of the solution.

3. The graph of a circle divides the xy-plane into 2 regions: the region inside the circle and the region outside the circle.

5. $y < (x-3)^2 + 2$

Graph $y = (x-3)^2 + 2$ with a dashed line. Select a point not on the graph and determine if the point satisfies the inequality.

$(0,0)$

$y < (x-3)^2 + 2$

$0 < (0-3)^2 + 2$

$0 < 9 + 2$

$0 < 11$ TRUE

The point (0, 0) is a solution. Shade the region containing this point.

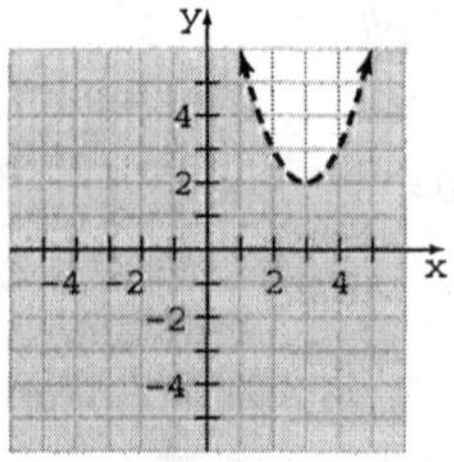

7. $x \ge (y-5)^2 - 3$

Graph $x = (y-5)^2 - 3$ with a solid line. Select a point not on the graph and determine if this point satisfies the inequality.

$(0,0)$

$x \ge (y-5)^2 - 3$

$0 \ge (0-5)^2 - 3$

$0 \ge 25 - 3$

$0 \ge 22$ FALSE

The point $(0,0)$ is not a solution. Shade the region that does not contain $(0,0)$.

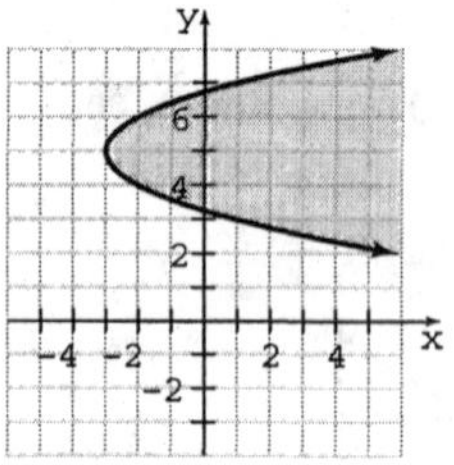

9. $y < -x^2 + 4x - 4$

Graph $y = -x^2 + 4x - 4$ with a dashed line.
Select a point not on the graph and determine if this point satisfies the inequality.

$(0,0)$

$y < -x^2 + 4x - 4$

$0 < -(0)^2 + 4(0) - 4$

$0 < -4$ TRUE

The point $(0,0)$ is not a solution. Shade the region that does not contain $(0,0)$.

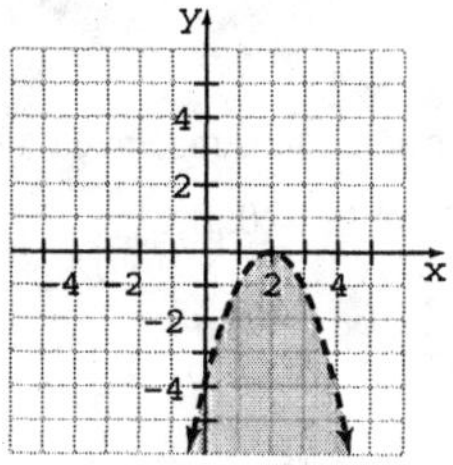

11. $x^2 + y^2 > 9$

Graph $x^2 + y^2 = 9$ with a dashed line. Select a point not on the graph and determine if this point satisfies the inequality.

$(0,0)$

$0^2 + 0^2 > 9$

$0 > 9$ FALSE

The point $(0,0)$ is not a solution. Shade the region that does not contain $(0,0)$.

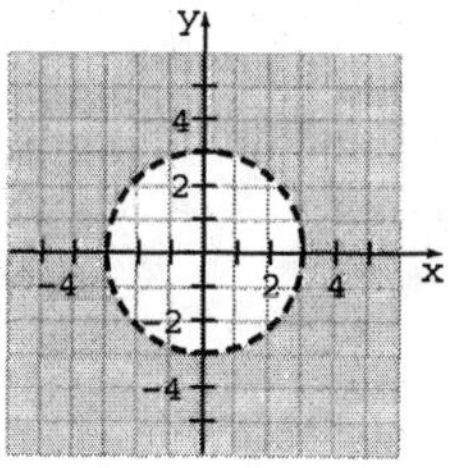

13. $(x+2)^2 + (y-1)^2 \geq 25$

Graph $(x+2)^2 + (y-1)^2 = 25$ with a solid line.
Select a point not on the graph and determine if this point satisfies the inequality.

$(0,0)$

$(x+2)^2 + (y-1)^2 \geq 25$

$(0+2)^2 + (0-1)^2 \geq 25$

$4 + 1 \geq 25$

$5 \geq 25$ FALSE

The point $(0,0)$ is not a solution. Shade the region that does not contain $(0,0)$.

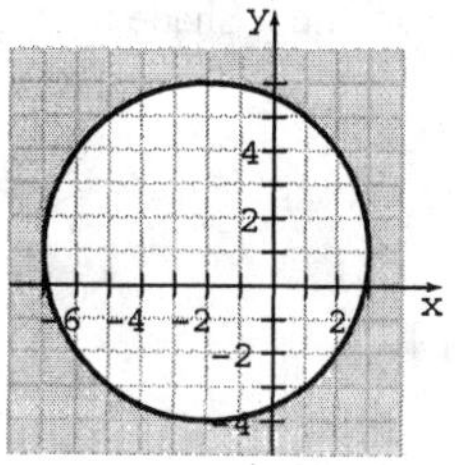

15. $\dfrac{x^2}{1} + \dfrac{y^2}{4} \leq 1$

Graph $\dfrac{x^2}{1} + \dfrac{y^2}{4} = 1$ with a solid line. Select a point not on the graph and determine if this point satisfies the inequality.

$(0,0)$

$\dfrac{x^2}{1} + \dfrac{y^2}{4} \leq 1$

$\dfrac{0^2}{1} + \dfrac{0^2}{4} \leq 1$

$0 + 0 \leq 1$

$0 \leq 1$ TRUE

The point $(0,0)$ is a solution. Shade the region that contains $(0,0)$.

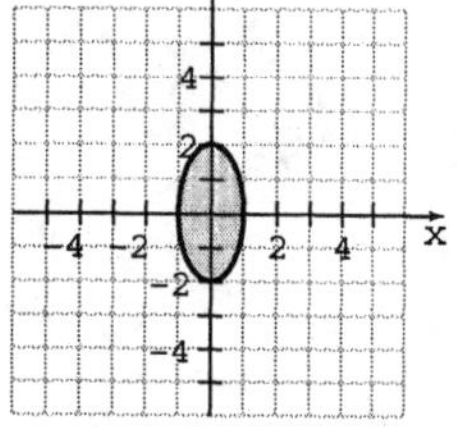

17. $\frac{x^2}{64}+\frac{y^2}{49}<1$

Graph $\frac{x^2}{64}+\frac{y^2}{49}=1$ with a dashed line. Select a point not on the graph and determine if this point satisfies the inequality.

$(0,0)$

$\frac{x^2}{64}+\frac{y^2}{49}<1$

$\frac{0^2}{64}+\frac{0^2}{49}<1$

$0+0<1$

$0<1$ TRUE

The point $(0,0)$ is a solution. Shade the region that contains $(0,0)$.

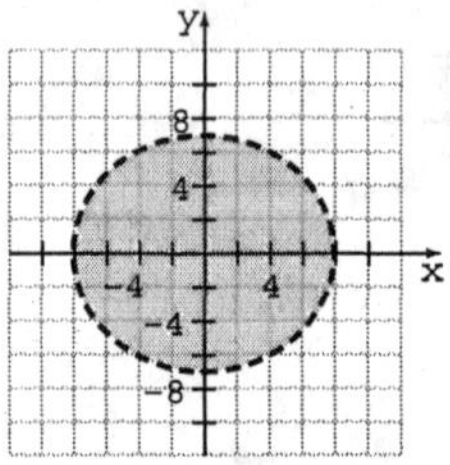

19. $16x^2+4y^2\ge 64$

Graph $16x^2+4y^2=64$ with a solid line. Select a point not on the graph and determine if this point satisfies the inequality.

$(0,0)$

$16x^2+4y^2\ge 64$

$16(0)^2+4(0)^2\ge 64$

$0\ge 64$ FALSE

The point $(0,0)$ is not a solution. Shade the region that does not contain $(0,0)$.

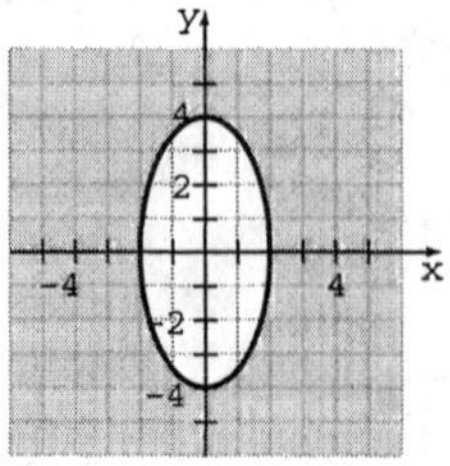

21. $\frac{y^2}{9}-\frac{x^2}{25}>1$

Graph $\frac{y^2}{9}-\frac{x^2}{25}=1$ with a dashed line. Select a point not on the graph and determine if this point satisfies the inequality.

$(0,0)$

$\frac{y^2}{9}-\frac{x^2}{25}>1$

$\frac{0^2}{9}-\frac{0^2}{25}>1$

$0-0>1$

$0>1$ FALSE

The point $(0,0)$ is not a solution. Shade the region that does not contain (0, 0).

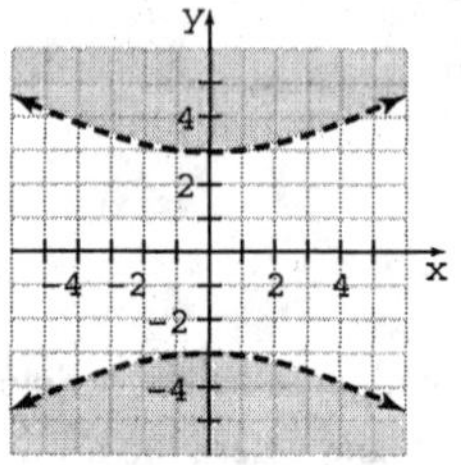

23. $\frac{x^2}{4}-\frac{y^2}{36}\le 1$

Graph $\frac{x^2}{4}-\frac{y^2}{36}=1$ with a solid line. Select a point not on the graph and determine if this point satisfies the inequality.

(0, 0)

$\frac{x^2}{4}-\frac{y^2}{36}\le 1$

$\frac{0^2}{4}-\frac{0^2}{36}\le 1$

$0-0\le 1$

$0\le 1$ TRUE

The point (0, 0) is a solution. Shade the region that contains (0, 0).

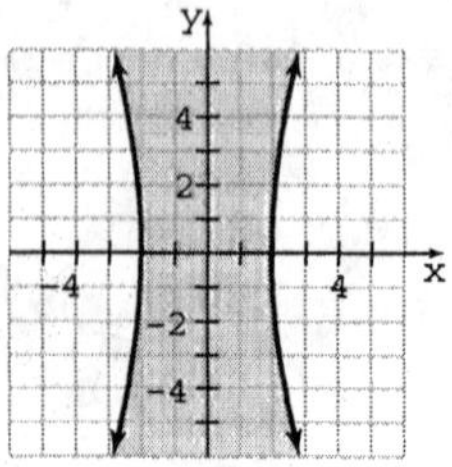

25. $x^2 + y^2 \geq 16$

$x + y < 3$

Graph $x^2 + y^2 \geq 16$ with a solid line. Select a point not on the graph and determine if it satisfies the inequality.

$(0,0)$

$x^2 + y^2 \geq 16$

$0^2 + 0^2 \geq 16$

$0 + 0 \geq 16$

$0 \geq 16$ FALSE

The point $(0,0)$ is not a solution. Shade the region that does not contain $(0,0)$.

On the same axes, graph $x + y < 3$ with a dashed line. Select a point not on the graph and determine if it satisfies the inequality.

$(0,0)$

$x + y \leq 3$

$0 + 0 < 3$

$0 < 3$ TRUE

The point $(0,0)$ is a solution. Using a different shading, shade the region containing $(0,0)$. The solution is the area containing both shadings.

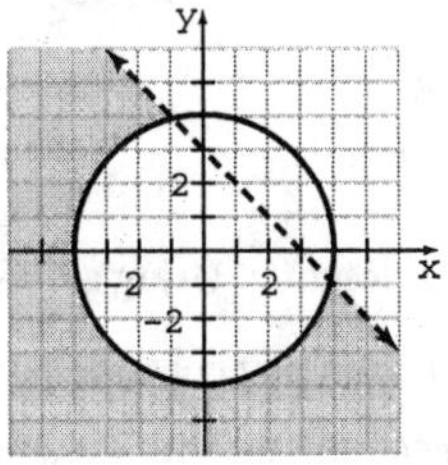

27. $4x^2 + y^2 > 16$

$y \geq 2x + 2$

Graph $4x^2 + y^2 = 16$ with a dashed line. Select a point not on the graph and determine if it satisfies the inequality.

$(0,0)$

$4x^2 + y^2 > 16$

$4(0)^2 + 0^2 > 16$

$0 > 16$ FALSE

The point $(0,0)$ is not a solution. Shade the region that does not contain $(0,0)$. On the same axes, graph $y = 2x + 2$ with a solid line. Select a point not on the graph and determine if it satisfies the inequality.

$(0,0)$

$y \geq 2x + 2$

$0 \geq 2(0) + 2$

$0 \geq 2$ FALSE

The point $(0,0)$ is not a solution. Using a different shading, shade the region that does not contain $(0,0)$. The solution is the area containing both shadings.

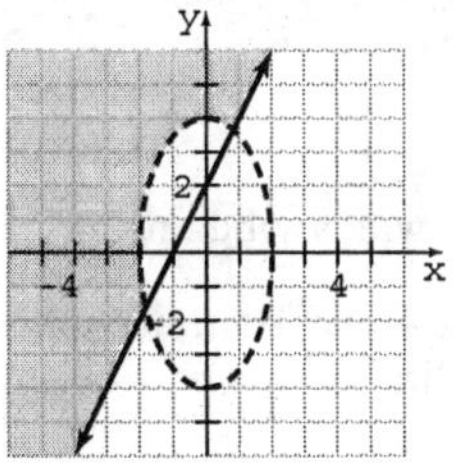

29. $x^2 + y^2 \leq 36$

$y < (x+1)^2 - 5$

Graph $x^2 + y^2 = 36$ with a solid line. Select a point not on the graph and determine if it satisfies the inequality.

$(0,0)$

$x^2 + y^2 \leq 36$

$0^2 + 0^2 \leq 36$

$0 \leq 36$ TRUE

The point $(0,0)$ is a solution. Shade the region that contains $(0,0)$. On the same axes, graph $y = (x+1)^2 - 5$ with a dashed line. Select a point not on the graph and determine if it satisfies the inequality.

$(0,0)$

$y < (x+1)^2 - 5$

$0 < (0+1)^2 - 5$

$0 < 1 - 5$

$0 < -4$ FALSE

The point $(0,0)$ is not a solution. Using a different shading, shade the region that does not contain $(0,0)$. The solution is the area containing both shadings.

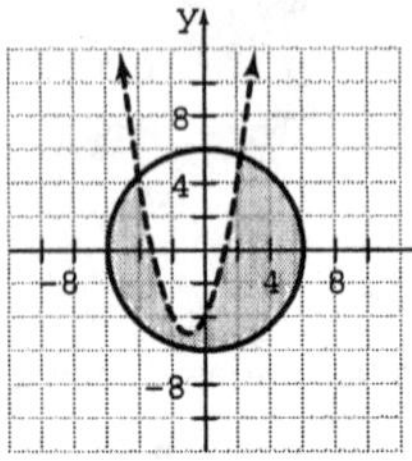

31. $\frac{y^2}{16}-\frac{x^2}{4}\ge 1$

$\frac{x^2}{4}-\frac{y^2}{1}<1$

Graph $\frac{y^2}{16}-\frac{x^2}{4}=1$ with a solid line. Select a point not on the graph and determine if it satisfies the inequality.

$(0,0)$

$\frac{y^2}{16}-\frac{x^2}{4}\ge 1$

$\frac{0^2}{16}-\frac{0^2}{4}\ge 1$

$0\ge 1$ FALSE

The point $(0,0)$ is not a solution. Shade the region that does not contain $(0,0)$. On the same axes, graph $\frac{x^2}{4}-\frac{y^2}{1}=1$ with a dashed line. Select a point not on the graph and determine if it satisfies the inequality.

$(0,0)$

$\frac{x^2}{4}-\frac{y^2}{1}<1$

$\frac{0^2}{4}-\frac{0^2}{1}<1$

$0<1$ TRUE

The point $(0,0)$ is a solution. Using a different shading, shade the region that contains $(0,0)$. The solution is the area containing both shadings.

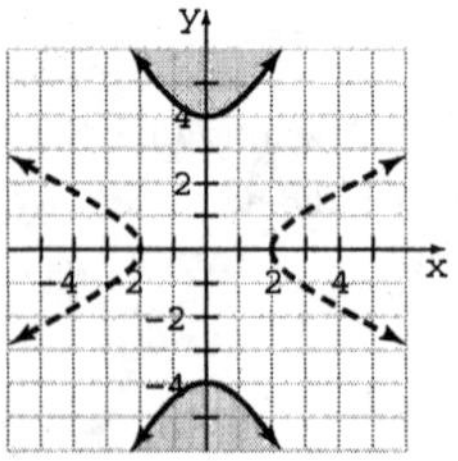

33. $xy\le 6$

$2x-y\le 8$

Graph $xy=6$ with a solid line. Select a point not on the graph and determine if it satisfies the inequality.

$(0,0)$

$xy\le 6$

$(0)(0)\le 6$

$0\le 6$ TRUE

The point $(0,0)$ is a solution. Shade the region that contains $(0,0)$. On the same axes, graph $2x-y=8$ with a solid line. Select a point not on the graph and determine if it satisfies the inequality.

$(0,0)$

$2x-y\le 8$

$2(0)-(0)\le 8$

$0\le 8$ TRUE

The point $(0,0)$ is a solution. Using a different shading, shade the region that contains $(0,0)$. The solution is the area containing both shadings.

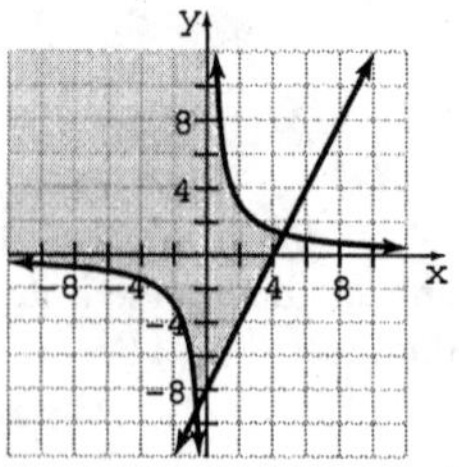

35. $(x-3)^2+(y+2)^2 \ge 16$
$y \le 4x-2$
Graph $(x-3)^2+(y+2)^2=16$ with a solid line. Select a point not on the graph and determine if it satisfies the inequality.
$(0,0)$
$(x-3)^2+(y+2)^2 \ge 16$
$(0-3)^2+(0+2)^2 \ge 16$
$9+4 \ge 16$
$13 \ge 16$ FALSE
The point $(0,0)$ is not a solution. Shade the region that does not contain $(0,0)$. On the same axes, graph $y=4x-2$ with a solid line. Select a point not on the graph and determine if it satisfies the inequality.
$(0,0)$
$y \le 4x-2$
$0 \le 4(0)-2$
$0 \le 0-2$
$0 \le -2$ FALSE
The point $(0,0)$ is not a solution. Using a different shading, shade the region that does not contain $(0,0)$. The solution is the area containing both shadings.

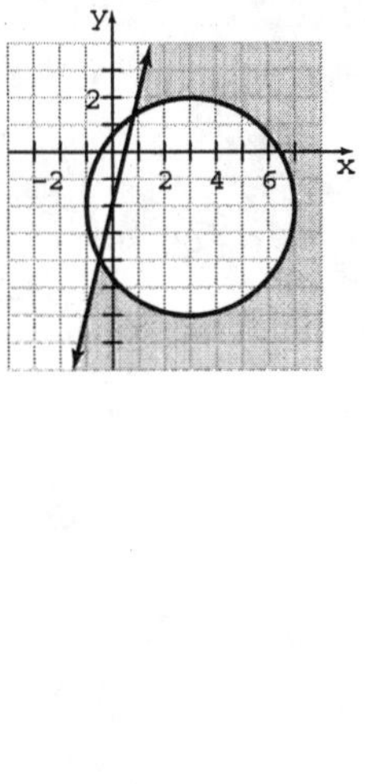

37. a. Yes, the solution to a system of linear inequalities containing two linear inequalities can contain no points. For instance, if the lines are parallel and the top line was shaded above and the bottom line was shaded below, the shaded regions would not overlap. Thus, the solution would contain no points.

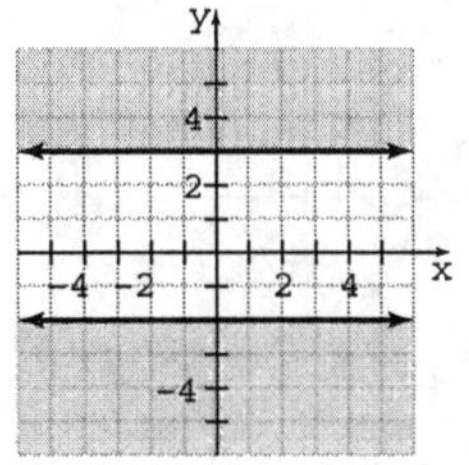

b. No, the solution to a system of linear inequalities containing two linear inequalities cannot contain exactly one point. If there is a solution, either the lines overlap completely or the shaded regions overlap. In either case, more than 1 solution exists.

c. No, the solution to a system of linear inequalities containing two linear inequalities cannot contain exactly 2 points. If there is a solution, either the lines overlap completely or the shaded regions overlap. In either case, more than 2 solutions exist.

d. No, the solution to a system of linear inequalities containing two linear inequalities cannot contain all points on the *xy*-plane. The solution to a system of inequalities is the intersection of the solutions to each inequality. Since the solution to an inequality cannot contain all points in the *xy*-plane, the intersection of two inequalities cannot contain all points in the *xy*-plane.

39. a. Yes, the solution to a system of inequalities containing two second-degree inequalities can contain no points. For instance, two circles that do not overlap and are shaded within have no overlapping regions. Thus, the solution would contain no points.

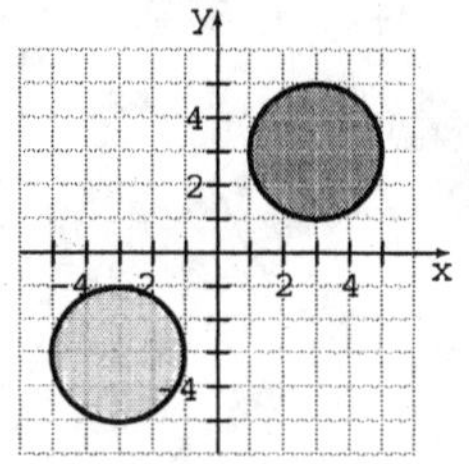

b. Yes, the solution to a system of inequalities containing two second-degree inequalities can contain exactly one point. For instance, two circles that are shaded within and intersect at exactly one point would have exactly one point in common. Thus, the solution would contain exactly one point.

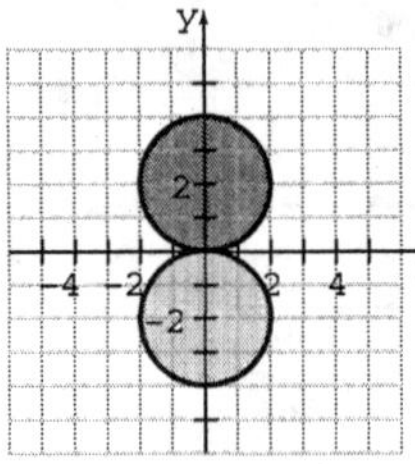

c. Yes, the solution to a system of inequalities containing two second-degree inequalities can contain exactly two points. For instance, an ellipse and a circle can intersect in exactly 2 points. If the ellipse is within the circle and its interior is shaded and the circle's exterior is shaded, they intersect in exactly two points. Thus, the solution would contain exactly two points.

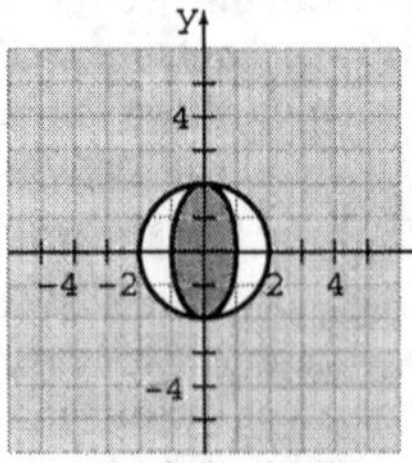

d. No, the solution to a system of inequalities containing two second degree inequalities cannot contain all points in the *xy*-plane. The solution to a system of inequalities is the intersection of the solutions to each inequality. Since the solution to an inequality cannot contain all points in the *xy*-plane, the intersection of two inequalities cannot contain all points in the *xy*-plane.

41. **a.**

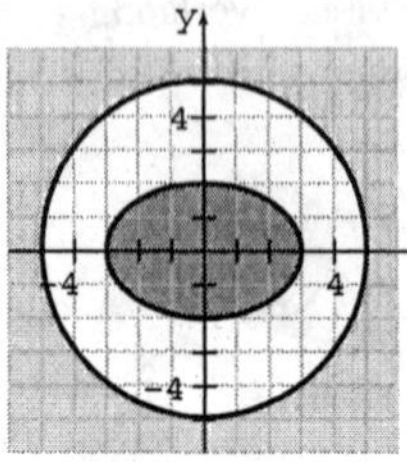

b.

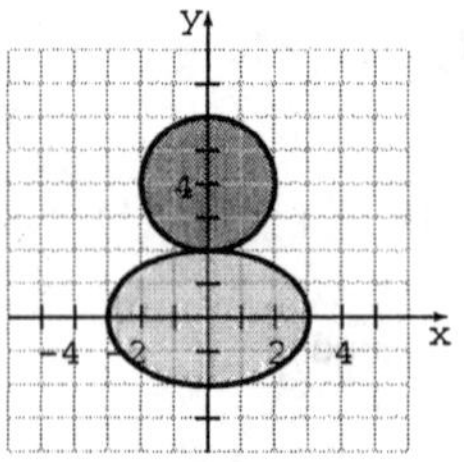

c.

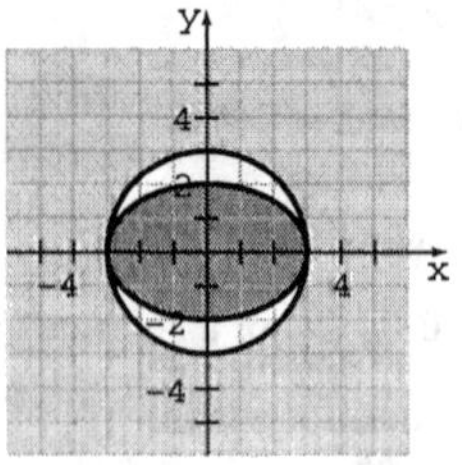

45. $\left(\frac{4x^{-2}y^3}{2xy^{-4}}\right)^2\left(\frac{3xy^{-1}}{6x^4y^{-3}}\right)^{-2}$

$\left(2x^{-2-1}y^{3-(-4)}\right)^2\left(\frac{1}{2}x^{1-4}y^{-1-(-3)}\right)^{-2}$

$\left(2x^{-3}y^7\right)^2\left(\frac{1}{2}x^{-3}y^2\right)^{-2}$

$\left(2^2x^{-3(2)}y^{7(2)}\right)\left(\left(\frac{1}{2}\right)^{-2}x^{-3(-2)}y^{2(-2)}\right)$

$\left(4x^{-6}y^{14}\right)\left(4x^6y^{-4}\right)$

$4\cdot 4x^{-6+6}y^{14-4}$

$16x^0y^{10}$

$16y^{10}$

46. $x^2-2x-4=0$

$x^2-2x=4$

$x^2-2x+1=4+1$

$(x-1)^2=5$

$x-1=\pm\sqrt{5}$

$x=1\pm\sqrt{5}$

47. $x^2 - 2x - 4 = 0$

$a = 1, b = -2, c = -4$

$$x = \frac{-b \pm \sqrt{b^2 - 4ac}}{2a}$$

$$= \frac{-(-2) \pm \sqrt{(-2)^2 - 4(1)(-4)}}{2(1)}$$

$$= \frac{2 \pm \sqrt{4+16}}{2} = \frac{2 \pm \sqrt{20}}{2}$$

$$= \frac{2 \pm 2\sqrt{5}}{2} = 1 \pm \sqrt{5}$$

48. $(x+4)(x-2)(x-4) \le 0$

$(x+4)(x-2)(x-4) = 0$

$x+4=0$ or $x-2=0$ or $x-4=0$

$x=-4$ $\quad x=2$ $\quad x=4$

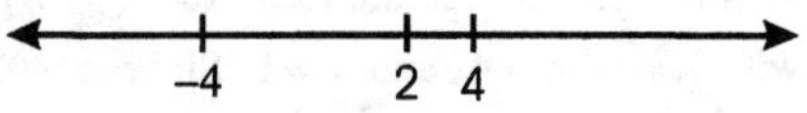

Interval	Test Value	$(x+4)(x-2)(x-4)$	≤ 0
A	–5	–63	TRUE
B	0	32	FALSE
C	3	–7	TRUE
D	5	27	FALSE

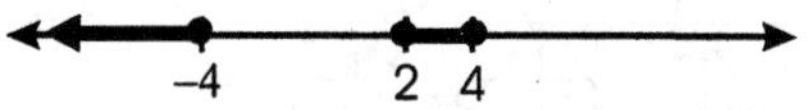

Chapter 11 Review Exercises

1. $d = \sqrt{(x_2 - x_1)^2 + (y_2 - y_1)^2}$

$= \sqrt{(5-0)^2 + (-12-0)^2}$

$= \sqrt{5^2 + (-12)^2}$

$= \sqrt{25 + 144}$

$= \sqrt{169}$

$= 13$

$\text{Midpoint} = \left(\frac{x_1 + x_2}{2}, \frac{y_1 + y_2}{2}\right)$

$= \left(\frac{0+5}{2}, \frac{0+(-12)}{2}\right)$

$= \left(\frac{5}{2}, -6\right)$

2. $d = \sqrt{(x_2 - x_1)^2 + (y_2 - y_1)^2}$

$= \sqrt{(-1-(-4))^2 + (5-1)^2}$

$= \sqrt{3^2 + 4^2}$

$= \sqrt{9 + 16}$

$= \sqrt{25}$

$= 5$

$\text{Midpoint} = \left(\frac{x_1 + x_2}{2}, \frac{y_1 + y_2}{2}\right)$

$= \left(\frac{-4+(-1)}{2}, \frac{1+5}{2}\right)$

$= \left(-\frac{5}{2}, 3\right)$

3. $d=\sqrt{\left(x_2-x_1\right)^2+\left(y_2-y_1\right)^2}$

$=\sqrt{\left[-1-(-9)\right]^2+\left[10-(-5)\right]^2}$

$=\sqrt{(8)^2+15^2}$

$=\sqrt{64+225}$

$=\sqrt{289}$

$=17$

$\text{Midpoint}=\left(\frac{x_1+x_2}{2},\frac{y_1+y_2}{2}\right)$

$=\left(\frac{-9+(-1)}{2},\frac{-5+10}{2}\right)$

$=\left(-5,\frac{5}{2}\right)$

4. $d=\sqrt{\left(x_2-x_1\right)^2+\left(y_2-y_1\right)^2}$

$=\sqrt{\left[-2-(-4)\right]^2+(5-3)^2}$

$=\sqrt{2^2+2^2}$

$=\sqrt{4+4}$

$=\sqrt{8}$

≈ 2.83

$\text{Midpoint}=\left(\frac{x_1+x_2}{2},\frac{y_1+y_2}{2}\right)$

$=\left(\frac{-4+(-2)}{2},\frac{3+5}{2}\right)$

$=(-3,4)$

5. $y=(x-2)^2+1$

This is a parabola in the form $y=a(x-h)^2+k$ with $a=1$, $h=2$, and $k=1$. Since $a>0$, the parabola opens upward. The vertex is (2, 1). The y-intercept is (0, 5).
There are no x-intercepts.

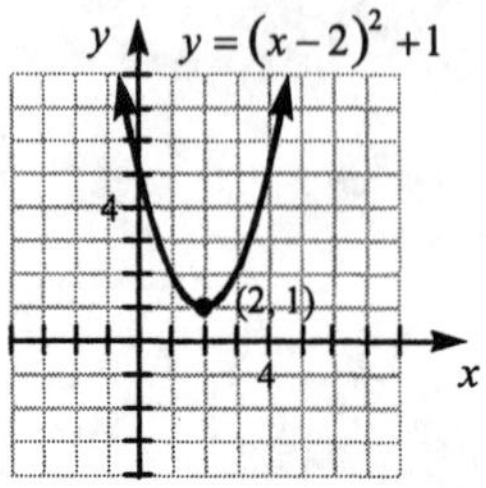

6. $y=(x+3)^2-4$

This is a parabola in the form $y=a(x-h)^2+k$ with $a=1$, $h=-3$, and $k=-4$. Since $a>0$, the parabola opens upward. The vertex is (–3, –
4). The
y-intercept is (0, 5).
The x-intercepts are about (–5, 0) and
(–1, 0).

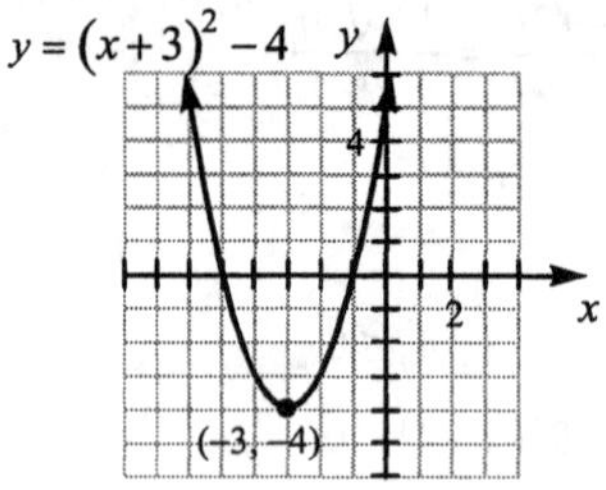

7. $x=(y-1)^2+4$

This is a parabola in the form $x=a(y-k)^2+h$ with $a=1$, $h=4$, and $k=1$. Since $a>0$, the parabola opens to the right. The vertex is (4, 1). There are no
y-intercepts.
The x-intercept is (5, 0).

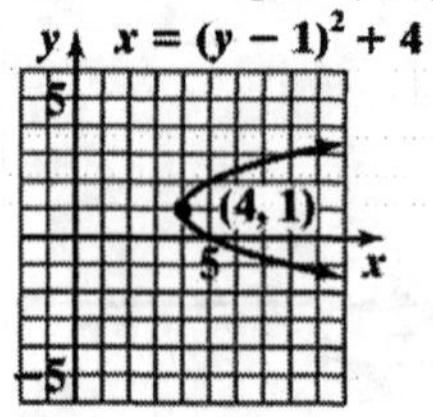

8. $x=-2(y+4)^2-3$

This is a parabola in the form $x=a(y-k)^2+h$ with $a=-2$, $h=-3$, and $k=-4$. Since $a<0$, the parabola opens to the left. The vertex is (–3, –4). There are no y-intercepts.
The x-intercept is (–35, 0).

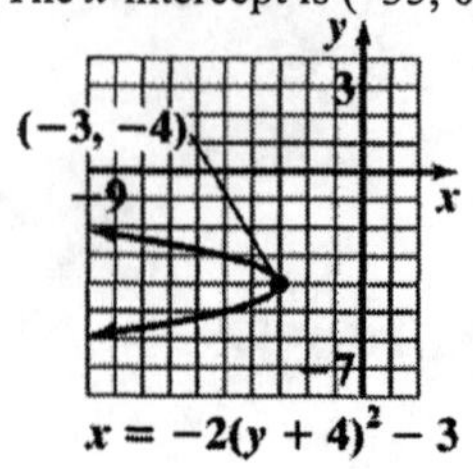

9. a. $y = x^2 - 8x + 22$

$y = \left(x^2 - 8x + 16\right) + 22 - 16$

$y = (x-4)^2 + 6$

b. This is a parabola in the form $y = a(x-h)^2 + k$ with $a = 1$, $h = 4$, and $k = 6$. Since $a > 0$, the parabola opens upward. The vertex is (4, 6). The y-intercept is (0, 22). There are no y-intercepts.

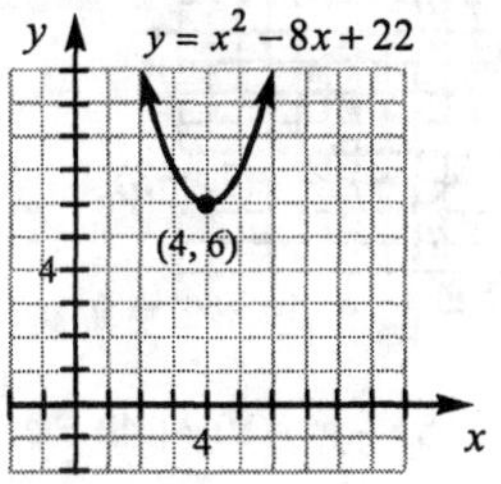

10. a. $x = -y^2 - 2y + 5$

$x = -(y^2 + 2y) + 5$

$x = -(y^2 + 2y + 1) + 1 + 5$

$x = -(y+1)^2 + 6$

b. This is a parabola in the form $x = a(y-k)^2 + h$ with $a = -1$, $h = 6$, and $k = -1$. Since $a < 0$, the parabola opens to the left. The vertex is (6, –1). The y-intercepts are about (0, –3.45) and (0, 1.45). The x-intercept is (5, 0).

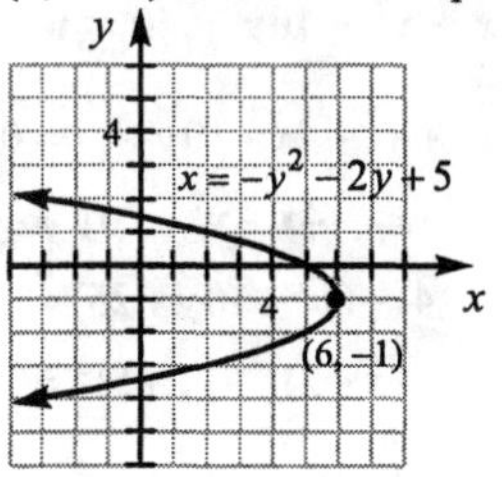

11. a. $x = y^2 + 5y + 4$

$x = \left(y^2 + 5y + \frac{25}{4}\right) - \frac{25}{4} + 4$

$x = \left(y + \frac{5}{2}\right)^2 - \frac{9}{4}$

b. This is a parabola in the form $x = a(y-k)^2 + h$ with $a = 1$, $h = -\frac{9}{4}$, and $k = -\frac{5}{2}$. Since $a > 0$, the parabola opens to the right. The vertex is $\left(-\frac{9}{4}, -\frac{5}{2}\right)$.

The y-intercepts are (0, –4) and (0, –1). The x-intercept is (4, 0).

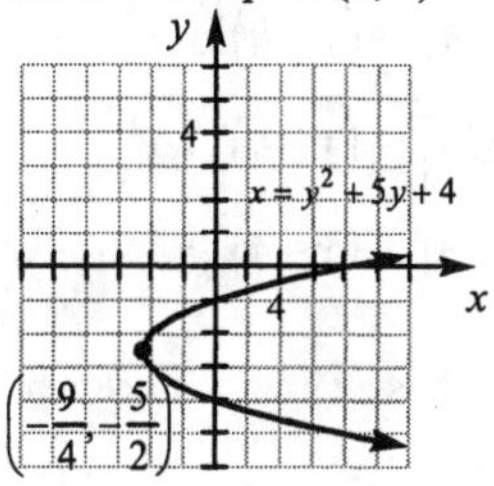

12. a. $y = 2x^2 - 8x - 24$

$y = 2(x^2 - 4x) - 24$

$y = 2(x^2 - 4x + 4) - 8 - 24$

$y = 2(x-2)^2 - 32$

b. This is a parabola in the form $y = a(x-h)^2 + k$ with $a = 2$, $h = 2$, and $k = -32$. Since $a > 0$, the parabola opens upward. The vertex is (2, –32). The y-intercept is (0, –24). The x-intercepts are (–2, 0) and (6, 0).

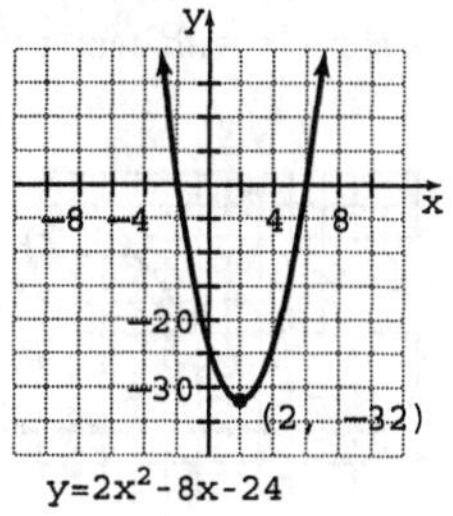

13. a.
$$(x-h)^2+(y-k)^2=r^2$$
$$(x-0)^2+(y-0)^2=4^2$$
$$x^2+y^2=4^2$$

b. The graph is a circle with center $(0,0)$ and radius 4.

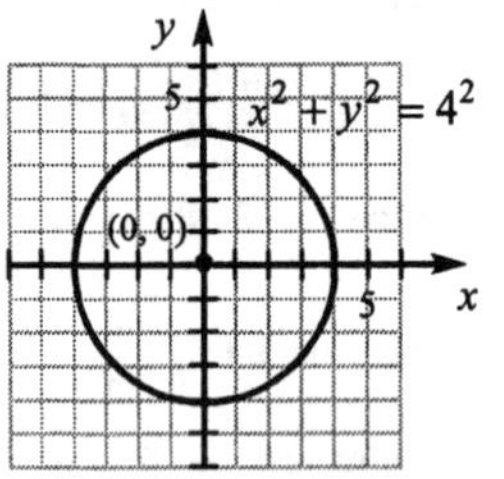

14. a.
$$(x-h)^2+(y-k)^2=r^2$$
$$[x-(-3)]^2+(y-4)^2=1^2$$
$$(x+3)^2+(y-4)^2=1^2$$

b. The graph is a circle with center $(-3,4)$ and radius 1.

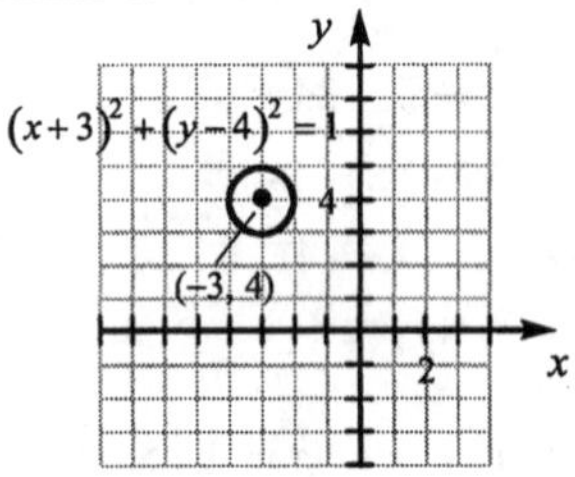

15. a.
$$x^2+y^2-4y=0$$
$$x^2+(y^2-4y+4)=4$$
$$x^2+(y-2)^2=2^2$$

b. The graph is a circle with center (0, 2) and radius 2.

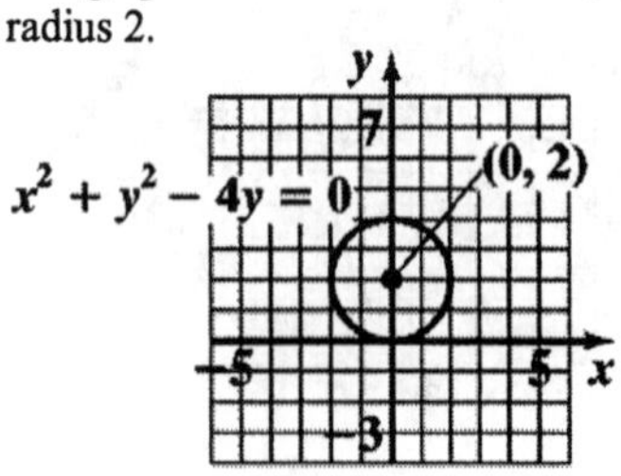

16. a.
$$x^2+y^2-2x+6y+1=0$$
$$x^2-2x+y^2+6y=-1$$
$$(x^2-2x+1)+(y^2+6y+9)=-1+1+9$$
$$(x-1)^2+(y+3)^2=9$$
$$(x-1)^2+(y+3)^2=3^2$$

b. The graph is a circle with center (1, –3) and radius 3.

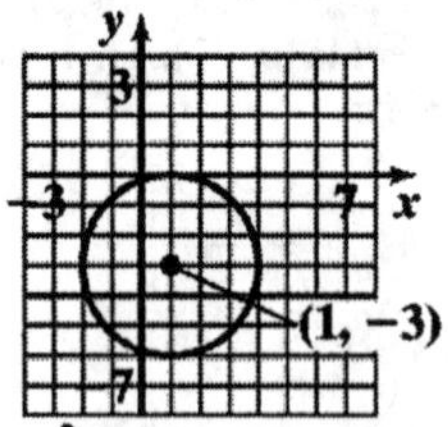

$x^2+y^2-2x+6y+1=0$

17. a.
$$x^2-8x+y^2-10y+40=0$$
$$(x^2-8x+16)+(y^2-10y+25)=-40+16+25$$
$$(x-4)^2+(y-5)^2=1$$
$$(x-4)^2+(y-5)^2=1^2$$

b. The graph is a circle with center (4, 5) and radius 1.

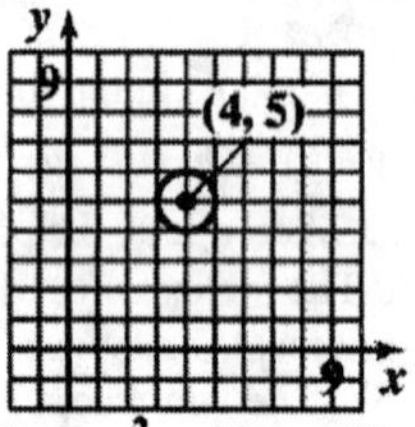

$x^2-8x+y^2-10y+40=0$

18. a.
$$x^2+y^2-4x+10y+17=0$$
$$x^2-4x+y^2+10y=-17$$
$$(x^2-4x+4)+(y^2+10y+25)=-17+4+25$$
$$(x-2)^2+(y+5)^2=12$$
$$(x-2)^2+(y+5)^2=\left(\sqrt{12}\right)^2$$

b. The graph is a circle with center (2, –5) and radius $\sqrt{12} \approx 3.46$.

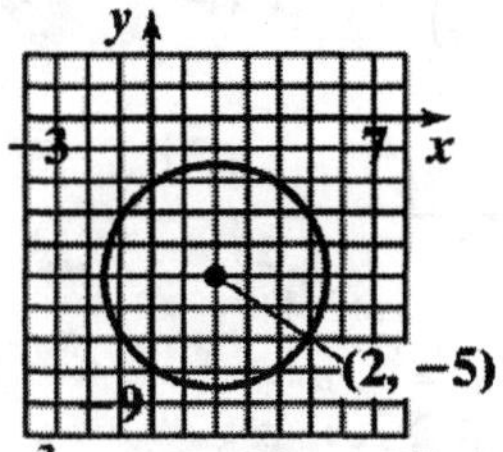

$x^2 + y^2 - 4x + 10y + 17 = 0$

19. $y = \sqrt{9 - x^2}$

If we solve $x^2 + y^2 = 9$ for *y*, we obtain $y = \pm\sqrt{9 - x^2}$. Therefore, the graph of $y = \sqrt{9 - x^2}$ is the upper half $(y \ge 0)$ of a circle with its center at the origin and radius 4.

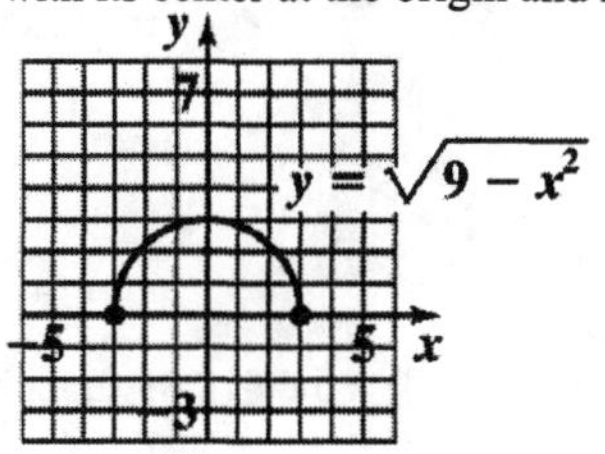

20. $y = -\sqrt{36 - x^2}$

If we solve $x^2 + y^2 = 36$ for *y*, we obtain $y = \pm\sqrt{36 - x^2}$. Therefore, the graph of $y = -\sqrt{36 - x^2}$ is the lower half $(y \le 0)$ of a circle with its center at the origin and radius 6.

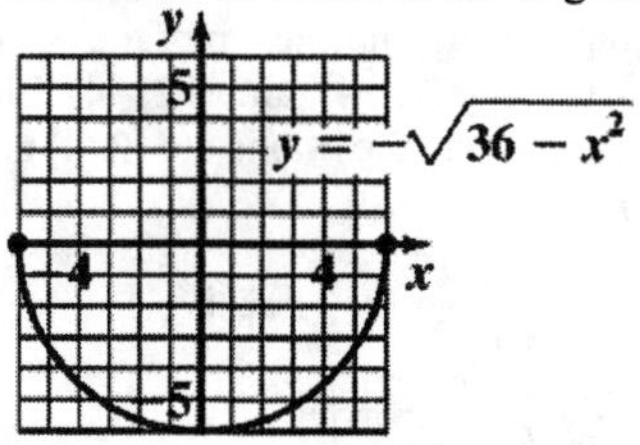

21. The center is (–1, 1) and the radius is 2.

$$(x-h)^2 + (y-k)^2 = r^2$$

$$[x-(-1)]^2 + (y-1)^2 = 2^2$$

$$(x+1)^2 + (y-1)^2 = 4$$

22. The center is (5, –3) and the radius is 3.

$$(x-h)^2 + (y-k)^2 = r^2$$

$$(x-5)^2 + [y-(-3)]^2 = 3^2$$

$$(x-5)^2 + (y+3)^2 = 9$$

23. $\dfrac{x^2}{4} + \dfrac{y^2}{9} = 1$

Since $a^2 = 4, a = 2$.

Since $b^2 = 9, b = 3$.

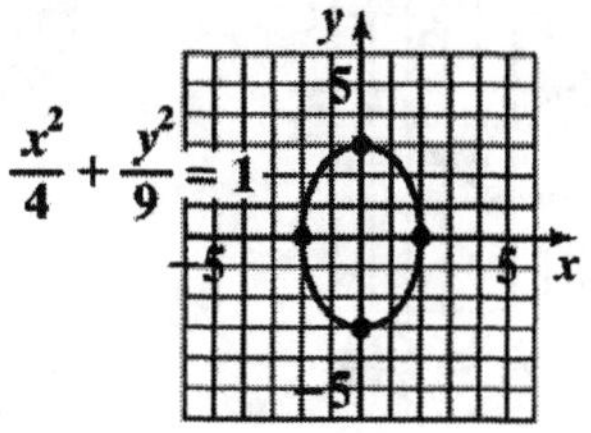

24. $\dfrac{x^2}{36} + \dfrac{y^2}{64} = 1$

Since $a^2 = 36, a = 6$.

Since $b^2 = 64$, $b = 8$.

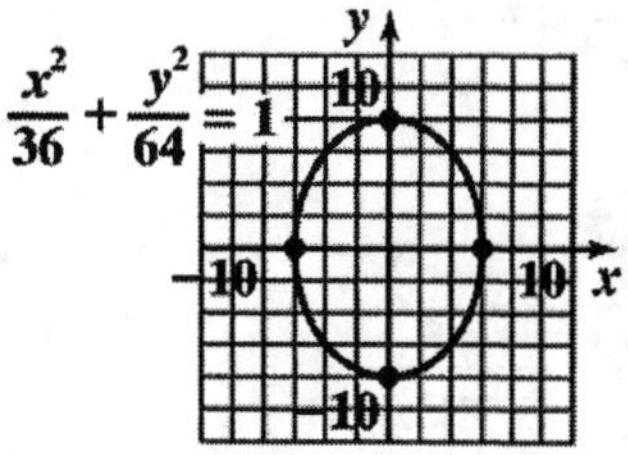

25. $4x^2 + 9y^2 = 36$

$$\frac{4x^2}{36} + \frac{9y^2}{36} = 1$$

$$\frac{x^2}{9} + \frac{y^2}{4} = 1$$

Since $a^2 = 9$, $a = 3$.

Since $b^2 = 4$, $b = 2$.

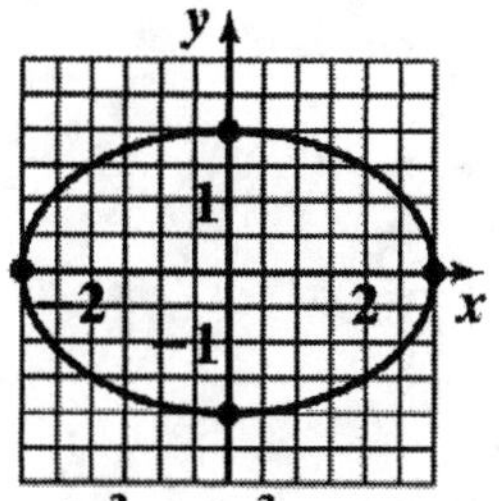

$4x^2 + 9y^2 = 36$

26. $9x^2 + 16y^2 = 144$

$$\frac{9x^2}{144} + \frac{16y^2}{144} = 1$$

$$\frac{x^2}{16} + \frac{y^2}{9} = 1$$

Since $a^2 = 16$, $a = 4$.

Since $b^2 = 9$, $b = 3$.

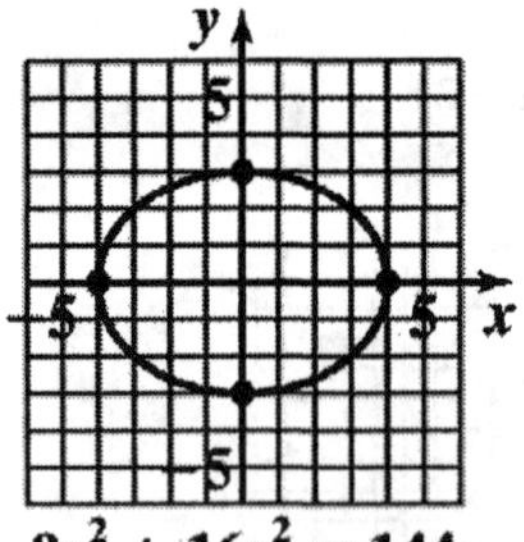

$\mathbf{9x^2 + 16y^2 = 144}$

27. $\frac{(x-3)^2}{16} + \frac{(y+2)^2}{4} = 1$

The center is (3, –2).

Since $a^2 = 16$, $a = 4$.

Since $b^2 = 4$, $b = 2$.

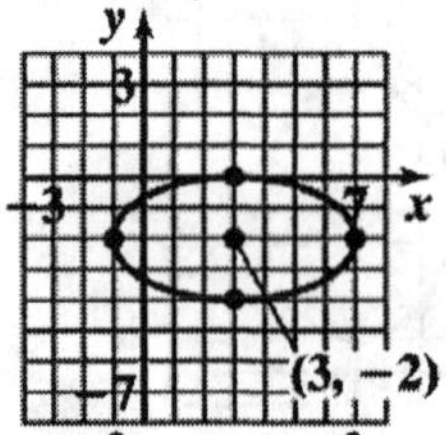

$$\frac{(x-3)^2}{16} + \frac{(y+2)^2}{4} = 1$$

28. $\frac{(x+3)^2}{9} + \frac{y^2}{25} = 1$

The center is (–3, 0).

Since $a^2 = 9$, $a = 3$.

Since $b^2 = 25$, $b = 5$.

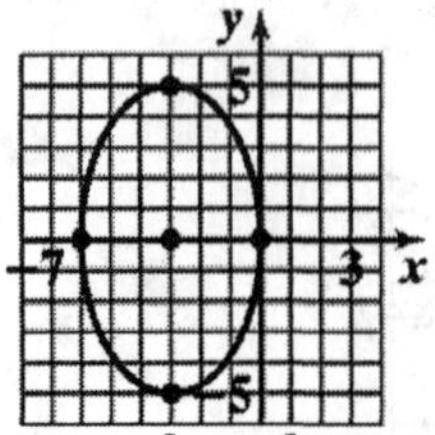

$$\frac{(x+3)^2}{9} + \frac{y^2}{25} = 1$$

29. $25(x-2)^2 + 9(y-1)^2 = 225$

$$\frac{25(x-2)^2}{225} + \frac{9(y-1)^2}{225} = 1$$

$$\frac{(x-2)^2}{9} + \frac{(y-1)^2}{25} = 1$$

The center is (2, 1).

Since $a^2 = 9$, $a = 3$.

Since $b^2 = 25$, $b = 5$.

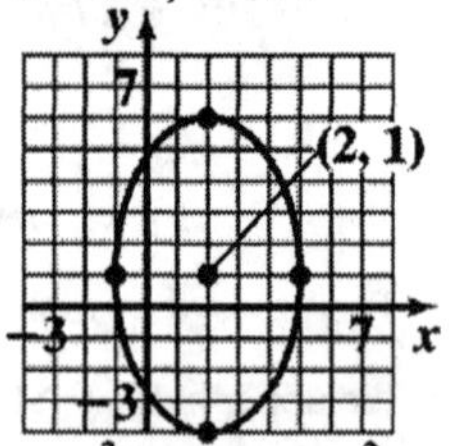

$\mathbf{25(x-2)^2 + 9(y-1)^2 = 225}$

30. $\frac{x^2}{4} + \frac{y^2}{9} = 1$

Since $a^2 = 4$, $a = 2$.

Since $b^2 = 9$, $b = 3$.

Area $= \pi ab = \pi(2)(3) = 6\pi \approx 18.85$ sq. units

31. a. $\frac{x^2}{4} - \frac{y^2}{16} = 1$

Since $a^2 = 4$ and $b^2 = 16$, $a = 2$ and $b = 4$. The equations of the asymptotes are $y = \pm\frac{b}{a}x$, or $y = \pm 2x$.

b. To graph the asymptotes, plot the points (2, 4), (–2, 4), (2, –4), and (–2, –4). The graph intersects the x-axis at (–2, 0) and (2, 0).

$$\frac{x^2}{4} - \frac{y^2}{16} = 1$$

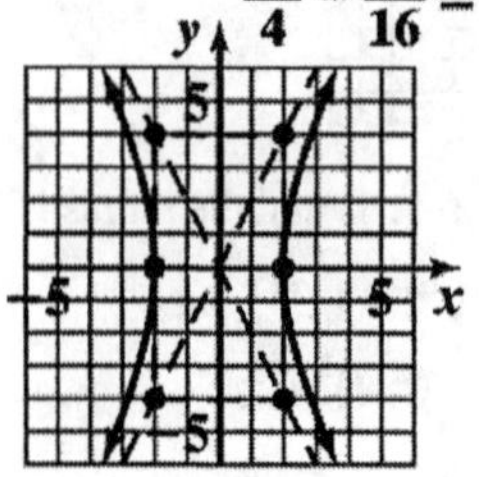

32. a. $\frac{x^2}{4} - \frac{y^2}{4} = 1$

Since $a^2 = 4$ and $b^2 = 4$, $a = 2$ and $b = 2$. The equations of the asymptotes are

$y = \pm\frac{b}{a}x$, or $y = \pm\frac{2}{2}x = \pm x$.

b. To graph the asymptotes, plot the points (2, 2), (–2, 2), (2, –2), and (–2, –2). The graph intersects the x-axis at (–2, 0) and (2, 0).

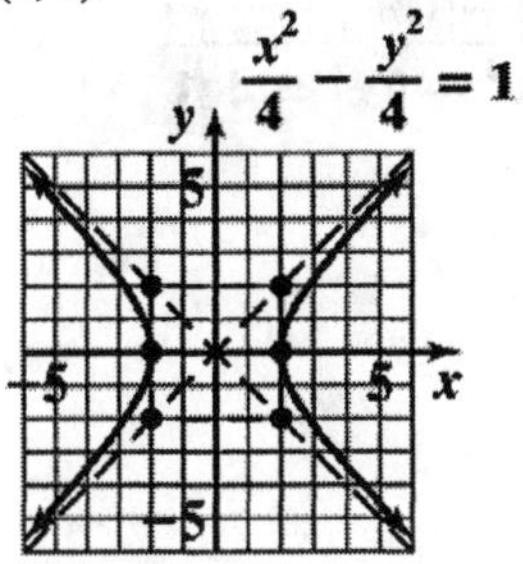

33. a. $\frac{y^2}{4} - \frac{x^2}{36} = 1$

Since $a^2 = 36$ and $b^2 = 4$, $a = 6$ and $b = 2$. The equations of the asymptotes are

$y = \pm\frac{b}{a}x$, or $y = \pm\frac{1}{3}$.

b. To graph the asymptotes, plot the points (6, 2), (6, –2), (–6, 2), and (–6, –2). The graph intersects the y-axis at (0,–2) and (0, 2).

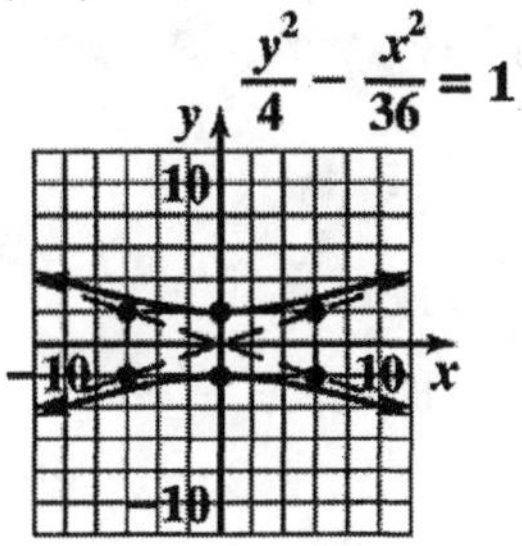

34. a. $\frac{y^2}{25} - \frac{x^2}{16} = 1$

Since $a^2 = 16$ and $b^2 = 25$, $a = 4$ and $b = 5$. The equations of the asymptotes are

$y = \pm\frac{b}{a}x$, or $y = \pm\frac{5}{4}x$.

b. To graph the asymptotes, plot the points (4, 5), (4, –5), (–4, 5), and (–4, –5). The graph intersects the y-axis at (0, –5) and (0, 5).

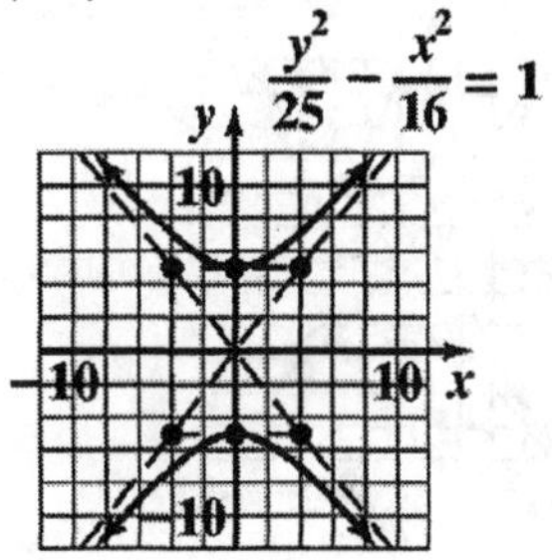

35. a. $x^2 - 9y^2 = 9$

$\frac{x^2}{9} - \frac{9y^2}{9} = 1$

$\frac{x^2}{9} - \frac{y^2}{1} = 1$

b. Since $a^2 = 9$ and $b^2 = 1$, $a = 3$ and $b = 1$. The equations of the asymptotes are

$y = \pm\frac{b}{a}x$, or $y = \pm\frac{1}{3}x$.

c. To graph the asymptotes, plot the points (3, 1), (–3, 1), (3, –1), and (–3, –1). The graph intersects the x-axis at (–3, 0) and (3, 0).

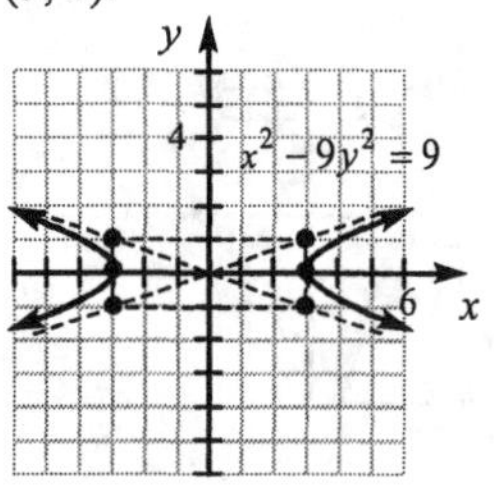

36. a. $25x^2 - 16y^2 = 400$

$\frac{25x^2}{400} - \frac{16y^2}{400} = 1$

$\frac{x^2}{16} - \frac{y^2}{25} = 1$

b. Since $a^2 = 16$ and $b^2 = 25$, $a = 4$ and $b = 5$. The equations of the asymptotes are

$y = \pm\frac{b}{a}x$, or $y = \pm\frac{5}{4}x$.

c. To graph the asymptotes, plot the points (4, 5), (–4, 5), (4, –5), and (–4, –5). The graph intersects the *x*-axis at (–4, 0) and (4, 0).

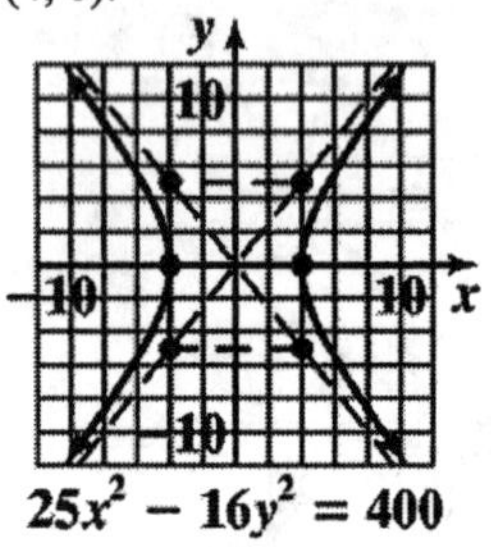

37. a. $4y^2 - 25x^2 = 100$

$$\frac{4y^2}{100} - \frac{25x^2}{100} = 1$$

$$\frac{y^2}{25} - \frac{x^2}{4} = 1$$

b. Since $a^2 = 4$ and $b^2 = 25$, $a = 2$ and $b = 5$. The equations of the asymptotes are $y = \pm\frac{b}{a}x$, or $y = \pm\frac{5}{2}x$.

c. To graph the asymptotes, plot the points (2, 5), (2, –5), (–2, 5), and (–2, –5). The graph intersects the *y*-axis at (0, –5) and (0, 5).

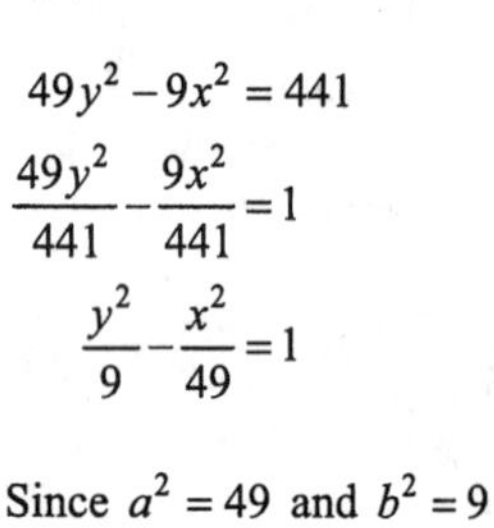

38. a. $49y^2 - 9x^2 = 441$

$$\frac{49y^2}{441} - \frac{9x^2}{441} = 1$$

$$\frac{y^2}{9} - \frac{x^2}{49} = 1$$

b. Since $a^2 = 49$ and $b^2 = 9$, $a = 7$ and $b = 3$. The equations of the asymptotes are $y = \pm\frac{b}{a}x$, or $y = \pm\frac{3}{7}x$.

c. To graph the asymptotes, plot the points (7, 3), (–7, 3), (7, –3), and (–7, –3). The graph intersects the *y*-axis at (0, –3) and (0, 3).

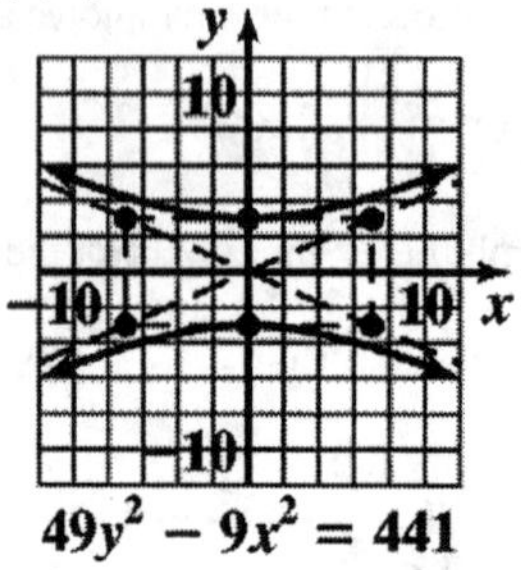

39. $\frac{x^2}{49} - \frac{y^2}{16} = 1$

The graph is a hyperbola.

40. $4x^2 + 8y^2 = 32$

$$\frac{4x^2}{32} + \frac{8y^2}{32} = \frac{32}{32}$$

$$\frac{x^2}{8} + \frac{y^2}{4} = 1$$

The graph is an ellipse.

41. $5x^2 + 5y^2 = 125$

$$\frac{5x^2}{5} + \frac{5y^2}{5} = \frac{125}{5}$$

$$x^2 + y^2 = 25$$

The graph is a circle.

42. $4x^2 - 25y^2 = 25$

$$\frac{4x^2}{25} - \frac{25y^2}{25} = \frac{25}{25}$$

$$\frac{x^2}{6.25} - \frac{y^2}{1} = 1$$

The graph is a hyperbola.

43. $\frac{x^2}{18} + \frac{y^2}{9} = 1$

The graph is an ellipse.

44. $y = (x-2)^2 + 1$

The graph is a parabola.

45. $12x^2 + 9y^2 = 108$

$$\frac{12x^2}{108} + \frac{9y^2}{108} = \frac{108}{108}$$

$$\frac{x^2}{9} + \frac{y^2}{12} = 1$$

The graph is an ellipse.

46. $x = -y^2 + 8y - 9$

The graph is a parabola.

47. $x^2 + 2y^2 = 25$

$x^2 - 3y^2 = 25 \Rightarrow x^2 = 3y^2 + 25$

Substitute $3y^2 + 25$ for x^2 in $x^2 + 2y^2 = 25$.

$$x^2 + 2y^2 = 25$$
$$3y^2 + 25 + 2y^2 = 25$$
$$5y^2 = 0$$
$$y^2 = 0$$

Substitute 0 for y^2 in $x^2 - 3y^2 = 25$

$x^2 - 0 = 25 \Rightarrow x = \pm 5$

The solutions are (5, 0) and (–5, 0).

48. $x^2 = y^2 + 4$

$x + y = 4$

Solve $x + y = 4$ for y: $y = 4 - x$.

Substitute $4 - x$ for y in $x^2 = y^2 + 4$.

$$x^2 = y^2 + 4$$
$$x^2 = (4 - x)^2 + 4$$
$$x^2 = (16 - 8x + x^2) + 4$$
$$8x - 16 = 4$$
$$8x = 20$$
$$x = \frac{5}{2}$$

$$y = 4 - x$$
$$y = 4 - \frac{5}{2}$$
$$y = \frac{3}{2}$$

The solution is $\left(\frac{5}{2}, \frac{3}{2}\right)$.

49. $x^2 + y^2 = 9$

$y = 3x + 9$

Substitute $3x + 9$ for y in $x^2 + y^2 = 9$.

$$x^2 + y^2 = 9$$
$$x^2 + (3x + 9)^2 = 9$$
$$x^2 + 9x^2 + 54x + 81 = 9$$
$$10x^2 + 54x + 72 = 0$$
$$5x^2 + 27x + 36 = 0$$
$$(x + 3)(5x + 12) = 0$$

$x + 3 = 0$ or $5x + 12 = 0$

$x = -3$ $\quad x = -\frac{12}{5}$

$y = 3x + 9$

$y = 3(-3) + 9$

$y = 0$

$y = 3x + 9$

$y = 3\left(-\frac{12}{5}\right) + 9$

$y = \frac{9}{5}$

The solutions are (–3, 0) and $\left(-\frac{12}{5}, \frac{9}{5}\right)$.

50. $x^2 + 2y^2 = 9$

$x^2 - 6y^2 = 36$

Solve $x^2 + 2y^2 = 9$ for x^2: $x^2 = 9 - 2y^2$.

Substitute $9 - 2y^2$ for x^2 in $x^2 - 6y^2 = 36$.

$$x^2 - 6y^2 = 36$$
$$(9 - 2y^2) - 6y^2 = 36$$
$$9 - 8y^2 = 36$$
$$-8y^2 = 27$$
$$y^2 = -\frac{27}{8}$$

There is no real solution to this equation so the system has no real solution.

51. $x^2+y^2=36$
$x^2-y^2=36$
$2x^2=72$
$x^2=36$
$x=6$ or $x=-6$

$x^2+y^2=36$ | $x^2+y^2=36$
$6^2+y^2=36$ | $(-6)^2+y^2=36$
$y^2=0$ | $y^2=0$
$y=0$ | $y=0$

The solutions are (6, 0) and (–6, 0).

52. $x^2+y^2=25$ (1)
$x^2-2y^2=-2$ (2)
$2x^2+2y^2=50$ (1) multiplied by 2
$x^2-2y^2=-2$ (2)
$3x^2=48$
$x^2=16$
$x=4$ or $x=-4$

$x^2+y^2=25$ | $x^2+y^2=25$
$4^2+y^2=25$ | $(-4)^2+y^2=25$
$y^2=9$ | $y^2=9$
$y=\pm3$ | $y=\pm3$

The solutions are (4, 3), (4, –3), (–4, 3) and (–4, –3).

53. $-4x^2+y^2=-15$ (1)
$8x^2+3y^2=-5$ (2)
$-8x^2+2y^2=-30$ (1) multiplied by 2
$8x^2+3y^2=-5$ (2)
$5y^2=-35$
$y^2=-7$

This equation has no real solution so there is no real solution to the system.

54. $3x^2+2y^2=6$ (1)
$4x^2+5y^2=15$ (2)
$-12x^2-8y^2=-24$ (1) multiplied by -4
$12x^2+15y^2=45$ (2) multiplied by 3
$7y^2=21$
$y^2=3$
$y=\pm\sqrt{3}$
$3x^2+2y^2=6$
$3x^2+2(3)=6$
$3x^2=0$
$x^2=0$

The solutions are $(0,\sqrt{3})$ and $(0,-\sqrt{3})$.

55. Let $x=$ length
$y=$ width
$xy=45$
$2x+2y=28$
Solve $2x+2y=28$ for y: $y=14-x$.
Substitute $14-x$ for y in $xy=45$.
$xy=45$
$x(14-x)=45$
$14x-x^2=45$
$x^2-14x+45=0$
$(x-5)(x-9)=0$
$x-5=0$ or $x-9=0$
$x=5$ | $x=9$

$y=14-x$ | $y=14-x$
$y=14-5$ | $y=14-9$
$y=9$ | $y=5$

The solutions are (5, 9) and (9, 5).
The dimensions of the pool table are 5 feet by 9 feet.

56. $C = 20.3x + 120$

$R = 50.2x - 0.2x^2$

$$C = R$$

$$20.3x + 120 = 50.2x - 0.2x^2$$

$0.2x^2 - 29.9 + 120 = 0$

$$x = \frac{29.9 \pm \sqrt{(-29.9)^2 - 4(0.2)(120)}}{2(0.2)}$$

$$x = \frac{29.9 \pm \sqrt{798.01}}{0.4}$$

$x \approx 145$ or 4

The company must sell either 4 bottles or 145 bottles to break even.

57. Since $t = 1$ year, we may rewrite the formula $i = prt$ as $i = pr$.

$120 = pr$

$120 = (p + 2000)(r - 0.01)$

Rewrite the second equation by multiplying the binomials. Then substitute 120 for pr and solve for r.

$$120 = pr - 0.01p + 2000r - 20$$

$$120 = 120 - 0.01p + 2000r - 20$$

$$0 = -0.01p + 2000r - 20$$

$$0.01p + 20 = 2000r$$

$$\frac{0.01p}{2000} + \frac{20}{2000} = r$$

$$r = 0.000005p + 0.01$$

Substitute $0.000005p + 0.01$ for r in $120 = pr$.

$120 = pr$

$120 = p(0.000005p + 0.01)$

$120 = 0.000005p^2 + 0.01p$

$0 = 0.000005p^2 + 0.01p - 120$

$0 = p^2 + 2000p - 24{,}000{,}000$

$0 = (p - 4000)(p + 6000)$

$p - 4000 = 0$ or $p + 6000 = 0$

$p = 4000$ $\quad p = -6000$

The principal must be positive, so use $p = 4000$.

$r = 0.000005p + 0.01$

$r = 0.000005(4000) + 0.01$

$r = 0.03$

The principal is \$4000 and the rate is 3%.

58. Let x = length; y = width

$xy = 300$

$x^2 + y^2 = 25^2$

Solve $xy = 300$ for y: $y = \dfrac{300}{x}$.

Substitute $\dfrac{300}{x}$ for y in $x^2 + y^2 = 25^2$.

$x^2 + y^2 = 25^2$

$x^2 + \left(\dfrac{300}{x}\right)^2 = 625$

$x^2 + \dfrac{90{,}000}{x^2} = 625$

$x^4 + 90{,}000 = 625x^2$

$x^4 - 625x^2 + 90{,}000 = 0$

$(x^2 - 225)(x^2 - 400) = 0$

$x^2 - 225 = 0$ or $x^2 - 400 = 0$

$x^2 = 225$ $\quad x^2 = 400$

$x = \pm 15$ $\quad x = \pm 20$

Since x must be positive, $x = 15$ or $x = 20$.

If $x = 15$, then $y = \dfrac{300}{15} = 20$.

If $x = 20$, then $y = \dfrac{300}{20} = 15$.

The dimensions of the carpet are 15 ft by 20 ft.

59. $y > -x^2 + 6x - 8$

Graph $y = -x^2 + 6x - 8$ with a dashed line.

Select a point not on the graph and determine if it satisfies the inequality.

$(0, 0)$

$y > -x^2 + 6x - 8$

$0 > -(0)^2 + 6(0) - 8$

$0 > -8$ TRUE

The point $(0, 0)$ is a solution. Shade the region containing (0, 0).

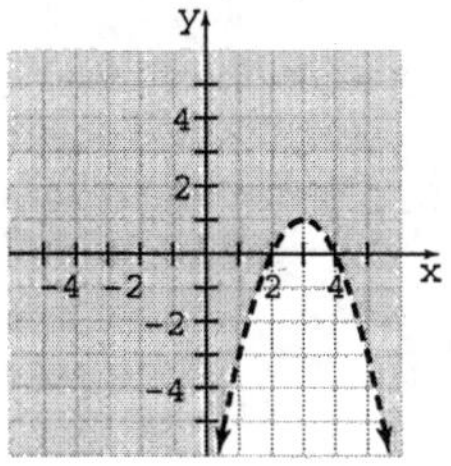

60. $(x+3)^2+(y-2)^2 \ge 16$

Graph $(x+3)^2+(y-2)^2=16$ with a solid line.

Select a point not on the graph and determine if it satisfies the inequality.

$(0,0)$

$(x+3)^2+(y-2)^2 \ge 16$

$(0+3)^2+(0-2)^2 \ge 16$

$9+4 \ge 16$

$13 \ge 16$ FALSE

The point (0, 0) is not a solution. Shade the region that does not contain (0, 0).

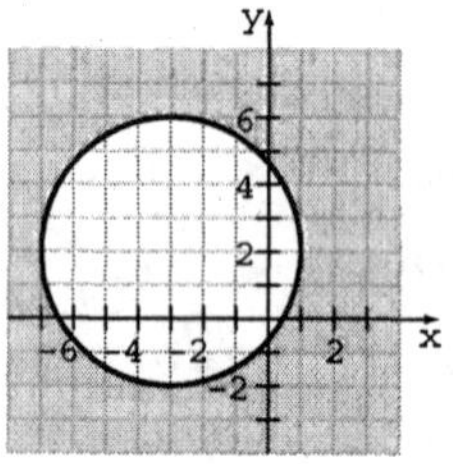

61. $\dfrac{(x-1)^2}{36}+\dfrac{(y+3)^2}{16} \le 1$

Graph $\dfrac{(x-1)^2}{36}+\dfrac{(y+3)^2}{16}=1$ with a solid line.

Select a point not on the graph and determine if it satisfies the inequality.

$(0,0)$

$\dfrac{(x-1)^2}{36}+\dfrac{(y+3)^2}{16} \le 1$

$\dfrac{(0-1)^2}{36}+\dfrac{(0+3)^2}{16} \le 1$

$\dfrac{1}{36}+\dfrac{9}{16} \le 1$

$\dfrac{85}{144} \le 1$ TRUE

The point (0, 0) is a solution. Shade the region that contains $(0,0)$.

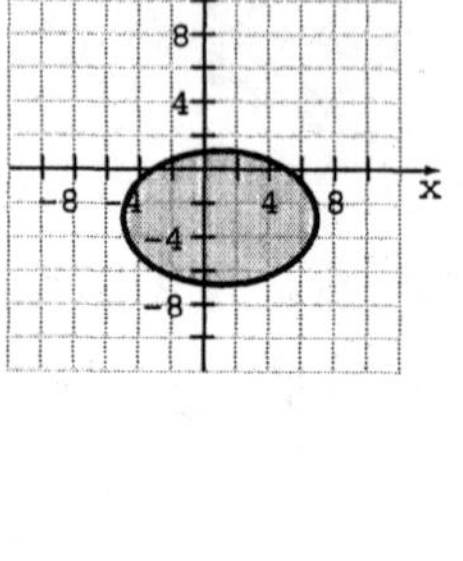

62. $\dfrac{y^2}{25}-\dfrac{x^2}{49} \ge 1$

Graph $\dfrac{y^2}{25}-\dfrac{x^2}{49}=1$ with a solid line. Select a point not on the graph and determine if it satisfies the inequality.

$(0,0)$

$\dfrac{y^2}{25}-\dfrac{x^2}{49} \ge 1$

$\dfrac{0^2}{25}-\dfrac{0^2}{49} \ge 1$

$0 \ge 1$ FALSE

The point $(0,0)$ is not a solution. Shade the region that does not contain $(0,0)$.

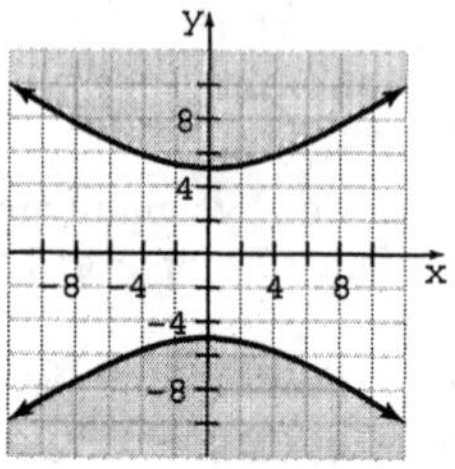

63. $2x+y \ge 6$

$x^2+y^2 < 9$

Graph $2x+y=6$ with a solid line. Select a point not on the graph and determine if it satisfies the inequality.

$(0,0)$

$2x+y \ge 6$

$2(0)+0 \ge 6$

$0 \ge 6$ FALSE

The point $(0,0)$ is not a solution. Shade the region that does not contain $(0,0)$. On the same axes, graph $x^2+y^2=9$ with a dashed line. Select a point not on the graph and determine if it satisfies the inequality.

$(0,0)$

$x^2+y^2 < 9$

$0^2+0^2 < 9$

$0 < 9$ TRUE

The point $(0,0)$ is a solution. Using a different

shading, shade the region that contains $(0,0)$.
The solution is the area that contains both shadings.

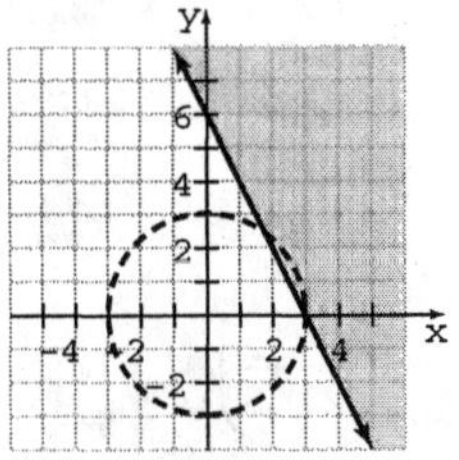

64. $xy > 5$
$y < 3x + 4$
Graph $xy = 5$ with a dashed line. Select a point not on the graph and determine if it satisfies the inequality.
$(0,0)$
$xy > 5$
$(0)(0) > 5$
$0 > 5$ FALSE
The point $(0,0)$ is not a solution. Shade the region that does not contain $(0,0)$. On the same axes, graph $y = 3x + 4$ with a dashed line. Select a point not on the graph and determine if it satisfies the inequality.
$(0,0)$
$y < 3x + 4$
$0 < 3(0) + 4$
$0 < 4$ TRUE
The point $(0,0)$ is a solution. Using a different shading, shade the region that contains $(0,0)$.
The solution is the area that contains both shadings.

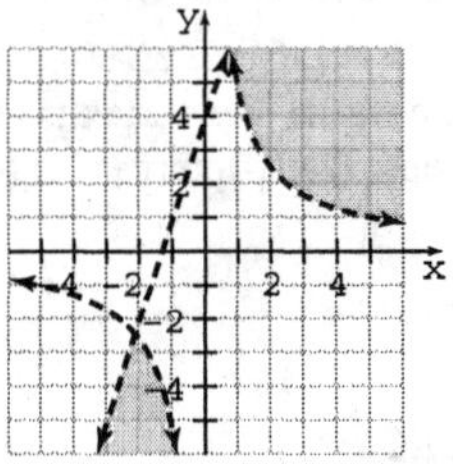

65. $4x^2 + 9y^2 \le 36$
$x^2 + y^2 > 25$
Graph $4x^2 + 9y^2 = 36$ with a solid line. Select a point not on the graph and determine if it satisfies the inequality.
$(0,0)$
$4x^2 + 9y^2 \le 36$
$4(0)^2 + 9(0)^2 \le 36$
$0 \le 36$ TRUE
The point $(0,0)$ is a solution. Shade the region that contains $(0,0)$. On the same axes, graph $x^2 + y^2 = 25$ with a dashed line. Select a point not on the graph and determine if it satisfies the inequality.
$(0,0)$
$x^2 + y^2 > 25$
$0^2 + 0^2 > 25$
$0 > 25$ FALSE
The point $(0,0)$ is not a solution. Using a different shading, shade the region that does not contain $(0,0)$. The solution is the area that contains both shadings.

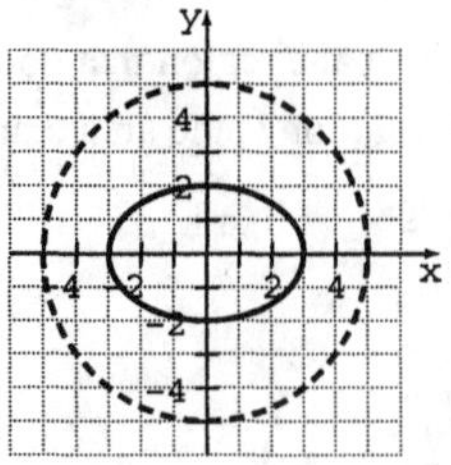

66. $\frac{x^2}{4}-\frac{y^2}{9}>1$

$y \ge (x-3)^2-4$

Graph $\frac{x^2}{4}-\frac{y^2}{9}=1$ with a dashed line. Select a point not on the graph and determine if it satisfies the inequality.

$(0,0)$

$\frac{x^2}{4}-\frac{y^2}{9}>1$

$\frac{0^2}{4}-\frac{0^2}{9}>1$

$0>1$ FALSE

The point $(0,0)$ is not a solution. Shade the region that does not contain $(0,0)$. On the same axes, graph $y=(x-3)^2-4$ with a solid line. Select a point not on the graph and determine if it satisfies the inequality.

$(0,0)$

$y \ge (x-3)^2-4$

$0 \ge (0-3)^2-4$

$0 \ge 5$ FALSE

The point $(0,0)$ is not a solution. Using a different shading, shade the region that does not contain $(0,0)$. The solution is the area that contains both shadings.

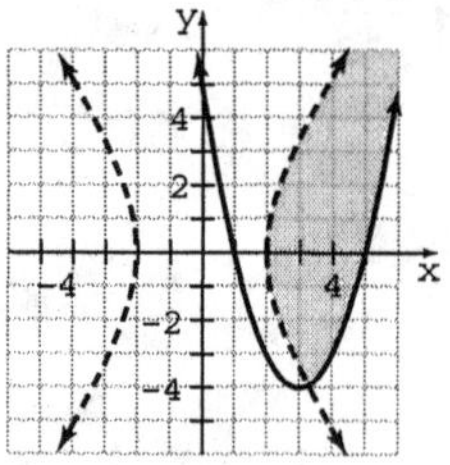

Chapter 11 Practice Test

1. They are formed by cutting a cone or pair of cones.

2. $d=\sqrt{(x_2-x_1)^2+(y_2-y_1)^2}$
$=\sqrt{[6-(-1)]^2+(7-8)^2}$
$=\sqrt{7^2+(-1)^2}$
$=\sqrt{50}$
The length is $\sqrt{50}\approx 7.07$ units.

3. Midpoint $=\left(\frac{x_1+x_2}{2},\frac{y_1+y_2}{2}\right)$
$=\left(\frac{-9+7}{2},\frac{4+(-1)}{2}\right)$
$=\left(-1,\frac{3}{2}\right)$

4. $y=-2(x+3)^2+1$
This is a parabola in the form $y=a(x-h)^2+k$ with $a=-2$, $h=-3$, and $k=1$. Since $a<0$, the parabola opens downward. The vertex is $(-3, 1)$. The y-intercept is $(0, -17)$. The x-intercepts are about $(-3.71, 0)$ and $(-2.29, 0)$.

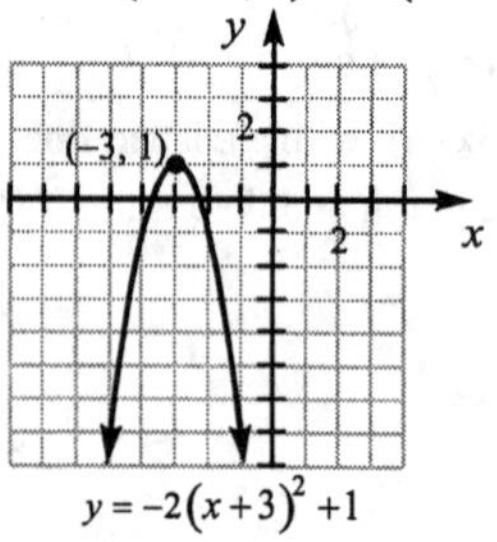

5. $x=y^2-2y+4$
$x=(y^2-2y+1)-1+4$
$x=(y-1)^2+3$
This is a parabola in the form $x=a(y-k)^2+h$ with $a=1$, $h=3$ and $k=1$. Since $a>0$, the parabola opens to the right. The vertex is $(3, 1)$. There is no y-intercept. The x-intercept is $(4, 0)$.

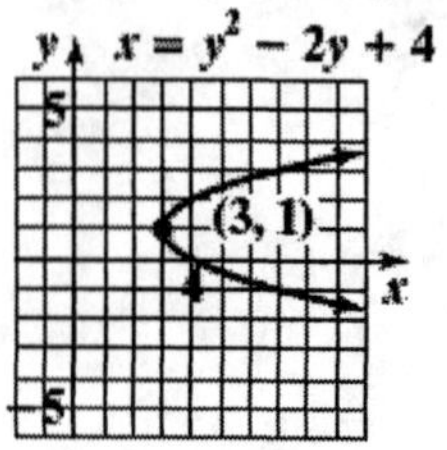

6. $x = -y^2 - 4y - 5$

$x = -(y^2 + 4y) - 5$

$x = -(y^2 + 4y + 4) + 4 - 5$

$x = -(y + 2)^2 - 1$

This is a parabola in the form $x = a(y-k)^2 + h$ with $a = -1$, $h = -1$, and $k = -2$. Since $a < 0$, the parabola opens to the left. The vertex is (–1, –2). There are no y-intercepts. The x-intercept is (–5,0).

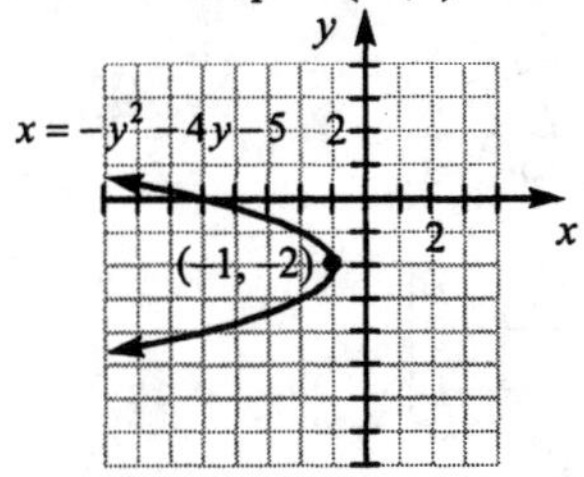

7. $(x-h)^2 + (y-k)^2 = r^2$

$[x-2]^2 + [y-4]^2 = 3^2$

$(x-2)^2 + (y-4)^2 = 9$

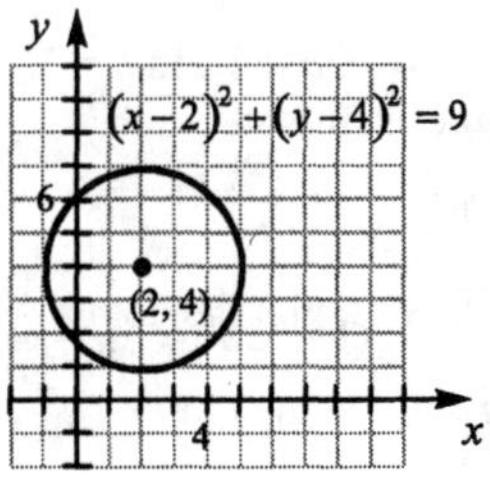

8. $(x+2)^2 + (y-8)^2 = 9$. The graph of this equation is a circle with center (–2, 8) and radius 3.

Area $= \pi r^2$

$= \pi 3^2 = 9\pi \approx 28.27$ sq. units

9. The center is (3, –1) and the radius is 4.

$(x-h)^2 + (y-k)^2 = r^2$

$(x-3)^2 + [y-(-1)]^2 = 4^2$

$(x-3)^2 + (y+1)^2 = 16$

10. $y = -\sqrt{16 - x^2}$

If we solve $x^2 + y^2 = 16$ for y, we obtain $y = \pm\sqrt{16 - x^2}$. Therefore, the graph of $y = -\sqrt{16 - x^2}$ is the lower half ($y \le 0$) of a circle with its center at the origin and radius 4.

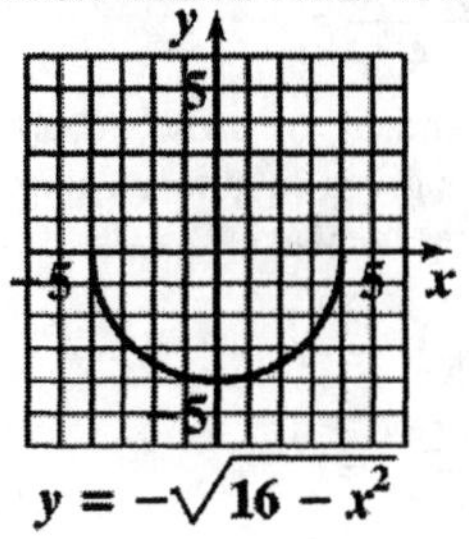

11.

$x^2 + y^2 + 2x - 6y + 1 = 0$

$x^2 + 2x + y^2 - 6y = -1$

$(x^2 + 2x + 1) + (y^2 - 6y + 9) = -1 + 1 + 9$

$(x+1)^2 + (y-3)^2 = 9$

The graph is a circle with center (1, 3) and radius 3.

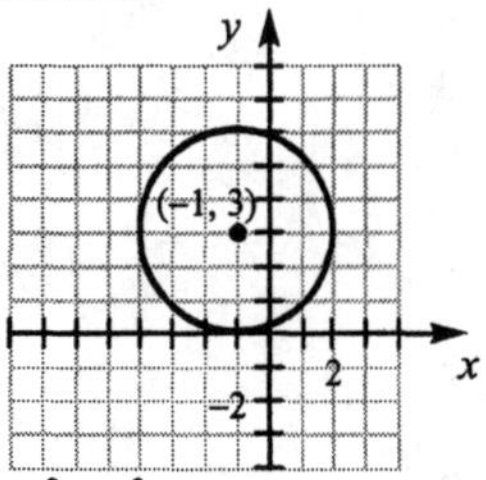

$x^2 + y^2 + 2x - 6y + 1 = 0$

12. $4x^2 + 25y^2 = 100$

$\frac{4x^2}{100} + \frac{25y^2}{100} = 1$

$\frac{x^2}{25} + \frac{y^2}{4} = 1$

Since $a^2 = 25$, $a = 5$

Since $b^2 = 4$, $b = 2$.

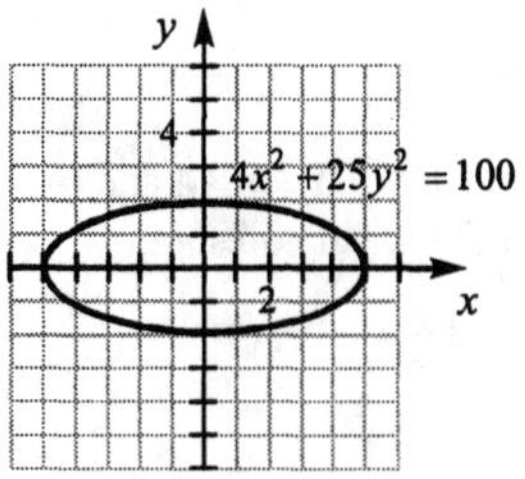

13. The center is (–2, –1), $a = 4$, and $b = 2$.

$$\frac{(x-h)^2}{a^2}+\frac{(y-k)^2}{b^2}=1$$

$$\frac{[x-(-2)]^2}{4^2}+\frac{[y-(-1)]^2}{2^2}=1$$

$$\frac{(x+2)^2}{16}+\frac{(y+1)^2}{4}=1$$

The values of a^2 and b^2 are switched, so this is not the graph of the given equation. The major axis should be along the y-axis.

14. $4(x-4)^2+36(y+2)^2=36$

$$\frac{4(x-4)^2}{36}+\frac{36(y+2)^2}{36}=1$$

$$\frac{(x-4)^2}{9}+\frac{(y+2)^2}{1}=1$$

The center is (4, –2). Since $a^2 = 9$, $a = 3$

Since $b^2 = 1$, $b = 1$

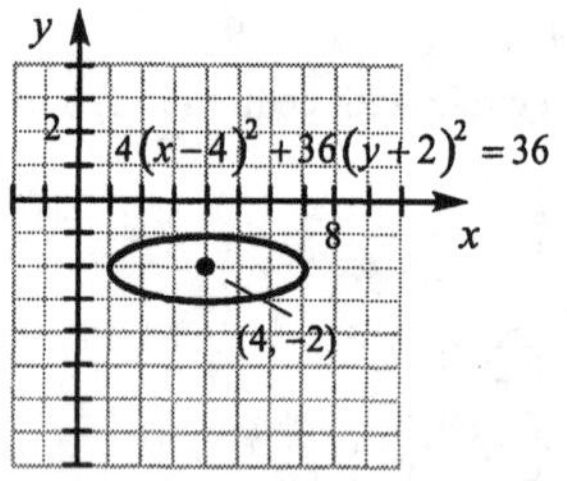

15. $3(x-8)^2+6(y+7)^2=18$

$$\frac{3(x-8)^2}{18}+\frac{6(y+7)^2}{18}=\frac{18}{18}$$

$$\frac{(x-8)^2}{6}+\frac{(y+7)^2}{3}=1$$

The center is $(8, -7)$.

16. The transverse axis lies along the axis corresponding to the positive term of the equation in standard form.

17. $\frac{x^2}{16}-\frac{y^2}{49}=1$

Since $a^2 = 16$ and $b^2 = 49$, $a = 4$ and $b = 7$. The equations of the asymptotes are

$y=\pm\frac{b}{a}x$, or $y=\pm\frac{7}{4}x$.

18. $\frac{y^2}{25}-\frac{x^2}{1}=1$

Since $a^2 = 1$ and $b^2 = 25$, $a = 1$ and $b = 5$. The equations of the asymptotes are

$y=\pm\frac{b}{a}x$, or $y=\pm 5x$.

To graph the asymptotes, plot the points (1, 5),
(–1, 5), (1, –5), and (–1, –5).
The graph intersects the y-axis at (0, –5) and (0, 5).

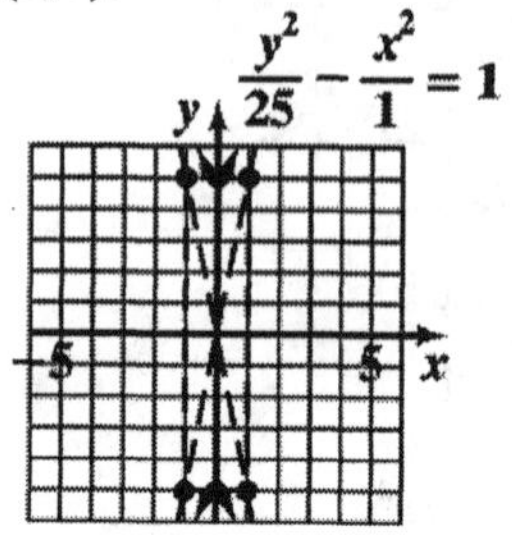

19. $\frac{x^2}{4}-\frac{y^2}{9}=1$

Since $a^2 = 4$ and $b^2 = 9$, $a = 2$ and $b = 3$. The equations of the asymptotes are

$y=\pm\frac{b}{a}x$, or $y=\pm\frac{3}{2}x$.

To graph the asymptotes, plot the points (2, 3), (–2, 3), (2, –3), and (–2, –3). The graph intersects the x-axis at (–2, 0) and (2, 0).

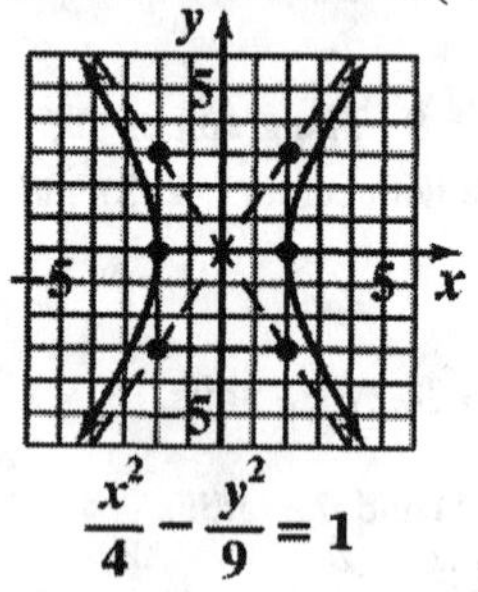

20. $4x^2 - 15y^2 = 30$

$$\frac{4x^2}{30} - \frac{15y^2}{30} = \frac{30}{30}$$

$$\frac{x^2}{\frac{15}{2}} - \frac{y^2}{2} = 1$$

Since the equation is of the form $\frac{x^2}{a^2} - \frac{y^2}{b^2} = 1$, the graph is a hyperbola.

21. $25x^2 + 4y^2 = 100$

$$\frac{25x^2}{100} + \frac{4y^2}{100} = \frac{100}{100}$$

$$\frac{x^2}{4} + \frac{y^2}{25} = 1$$

Since the equation is of the form $\frac{x^2}{a^2} + \frac{y^2}{b^2} = 1$, the graph is an ellipse.

22. $x^2 + y^2 = 7 \overset{\times 3}{\Rightarrow} 3x^2 + 3y^2 = 21$

$2x^2 - 3y^2 = -1 \qquad 2x^2 - 3y^2 = -1$

$$5x^2 = 20$$

$$x^2 = 4$$

$x = 2$ or $x = -2$

$x = 2$	$x = -2$
$x^2 + y^2 = 7$	$x^2 + y^2 = 7$
$(2)^2 + y^2 = 7$	$(-2)^2 + y^2 = 7$
$4 + y^2 = 7$	$4 + y^2 = 7$
$y^2 = 3$	$y^2 = 3$
$y = \pm\sqrt{3}$	$y = \pm\sqrt{3}$

The solutions are $\left(2, \sqrt{3}\right), \left(2, -\sqrt{3}\right)$, $\left(-2, \sqrt{3}\right)$, and $\left(-2, -\sqrt{3}\right)$.

23. $x + y = 8$

$x^2 + y^2 = 4$

Solve $x + y = 8$ for y: $y = 8 - x$.

Substitute $8 - x$ for y in $x^2 + y^2 = 4$.

$$x^2 + y^2 = 4$$

$$x^2 + (8 - x)^2 = 4$$

$$x^2 + 64 - 16x + x^2 = 4$$

$$2x^2 - 16x + 60 = 0$$

$$x^2 - 8x + 30 = 0$$

$$x = \frac{-(-8) \pm \sqrt{(-8)^2 - 4(1)(30)}}{2(1)}$$

$$= \frac{8 \pm \sqrt{-56}}{2} = 4 \pm i\sqrt{14}$$

There is no real solution.

24. Let x = length, y = width.

$xy = 1500$

$2x + 2y = 160$

Solve $2x + 2y = 160$ for y: $y = 80 - x$.

Substitute $80 - x$ for y in $xy = 1500$.

$$xy = 1500$$

$$x(80 - x) = 1500$$

$$80x - x^2 = 1500$$

$x^2 - 80x + 1500 = 0$

$(x - 30)(x - 50) = 0$

$x - 30 = 0$ or $x - 50 = 0$

$x = 30$	$x = 50$
$y = 80 - 30$	$y = 80 - 50$
$y = 50$	$y = 30$

The solutions are (30, 50) and (50, 30).

The dimensions are 30 m by 50 m.

25. $\frac{x^2}{9}-\frac{y^2}{25}<1$

$x^2+y^2\le 25$

Graph $\frac{x^2}{9}-\frac{y^2}{25}=1$ with a dashed line. Select a point not on the graph and determine if it satisfies the inequality.

$(0,0)$

$\frac{x^2}{9}-\frac{y^2}{25}<1$

$\frac{0^2}{9}-\frac{0^2}{25}<1$

$0<1$ TRUE

The point $(0,0)$ is a solution. Shade the region that contains $(0,0)$. On the same axes, graph $x^2+y^2=25$ with a solid line. Select a point not on the graph and determine if it satisfies the inequality.

$(0,0)$

$x^2+y^2\le 25$

$0^2+0^2\le 25$

$0\le 25$ TRUE

The point $(0,0)$ is a solution. Using a different shading, shade the region that contains $(0,0)$. The solution is the area that contains both shadings.

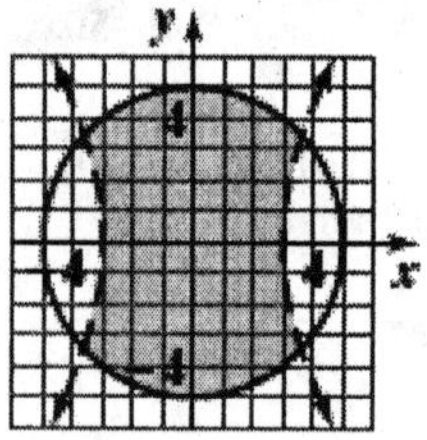

Chapter 11 Cumulative Review Test

1. $(9x^2y^5)(-3xy^4)=(9)(-3)x^{2+1}y^{5+4}=-27x^3y^9$

2. $4x-2(3x-7)=2x-5$

$4x-6x+14=2x-5$

$-2x+14=2x-5$

$-4x=-19$

$x=\frac{-19}{-4}=\frac{19}{4}$

3. $2(x-5)+2x=4x-7$

$2x-10+2x=4x-7$

$4x-10=4x-7$

$-10=-7$

This is a contradiction so the solution set is $\varnothing$.

4. $|3x+1|>4$

$3x+1>4 \qquad 3x+1<-4$

$3x>3 \qquad 3x<-5$

$x>1 \qquad x<-\frac{5}{3}$

The solution set is $\left\{x \,\middle|\, x<-\frac{5}{3} \text{ or } x>1\right\}$

5. $y=-2x+2$

The slope is –2 and the y-intercept is $(0, 2)$.

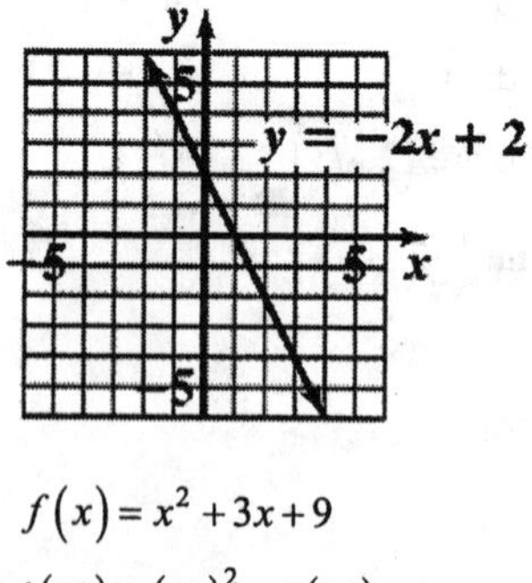

6. $f(x)=x^2+3x+9$

$f(10)=(10)^2+3(10)+9$

$=100+30+9$

$=139$

7. $\frac{1}{2}x - \frac{1}{3}y = 2 \overset{\times(-6)}{\Rightarrow} -3x + 2y = -12$

$\frac{1}{4}x + \frac{2}{3}y = 6 \overset{\times 12}{\Rightarrow} \quad \underline{3x + 8y = 72}$

$10y = 60$

$y = 6$

Substitute $y = 6$ into $-3x + 2y = -12$

$-3x + 2(6) = -12$

$-3x + 12 = -12$

$-3x = -24 \Rightarrow x = 8$

The solution is (8, 6).

8. $x^4 - x^2 - 42$ let $u = x^2$

$u^2 - u - 42$

$(u-7)(u+6)$

$(x^2-7)(x^2+6)$

9. Let x be the base of the sign. Then the height can be expressed as $x - 6$.

Area $= \frac{1}{2}$ (base × height)

$56 = \frac{1}{2}(x-6)(x)$

$112 = x^2 - 6x$

$0 = x^2 - 6x - 112$

$0 = (x-14)(x+8)$

$x - 14 = 0$ or $x + 8 = 0$

$x = 14 \qquad x = -8$

Disregard the negative value. The base of the sign is 14 feet and the height is $14 - 6 = 8$ ft.

10. $\frac{3x^2 - x - 4}{4x^2 + 7x + 3} \cdot \frac{2x^2 - 5x - 12}{6x^2 + x - 12}$

$\frac{(3x-4)(x+1)}{(4x+3)(x+1)} \cdot \frac{(2x+3)(x-4)}{(3x-4)(2x+3)}$

$\frac{(3x-4)(x+1)(2x+3)(x-4)}{(4x+3)(x+1)(3x-4)(2x+3)}$

$\frac{x-4}{4x+3}$

11. $\frac{x}{x+3} - \frac{x+5}{2x^2 - 2x - 24}$

$= \frac{x}{x+3} - \frac{x+5}{2(x+3)(x-4)}$

$= \frac{2x(x-4)}{2(x+3)(x-4)} - \frac{x+5}{2(x+3)(x-4)}$

$= \frac{2x^2 - 8x - (x+5)}{2(x+3)(x-4)}$

$= \frac{2x^2 - 9x - 5}{2(x+3)(x-4)}$

12. $\frac{3}{x+3} + \frac{5}{x+4} = \frac{12x+19}{x^2 + 7x + 12}$

$\left(\frac{3}{x+3} + \frac{5}{x+4} = \frac{12x+19}{(x+3)(x+4)}\right)(x+3)(x+4)$

$3(x+4) + 5(x+3) = 12x + 19$

$3x + 12 + 5x + 15 = 12x + 19$

$8x + 27 = 12x + 19$

$-4x = -8$

$x = 2$

13. $\left(\frac{18x^{1/2}y^3}{2x^{3/2}}\right)^{1/2} = \left(\frac{18}{2}x^{1/2-3/2}y^3\right)^{1/2}$

$= (9x^{-1}y^3)^{1/2}$

$= 9^{1/2}x^{-1/2}y^{3/2}$

$= \sqrt{9} \cdot x^{-1/2}y^{3/2}$

$= \frac{3y^{3/2}}{x^{1/2}}$

14. $\frac{6\sqrt{x}}{\sqrt{x} - y} = \frac{6\sqrt{x}}{\sqrt{x} - y} \cdot \frac{\sqrt{x} + y}{\sqrt{x} + y}$

$= \frac{6x + 6y\sqrt{x}}{x + y\sqrt{x} - y\sqrt{x} - y^2}$

$= \frac{6x + 6y\sqrt{x}}{x - y^2}$

15. $3\sqrt[3]{2x+2} = \sqrt[3]{80x-24}$

$\left(3\sqrt[3]{2x+2}\right)^3 = \left(\sqrt[3]{80x-24}\right)^3$

$27(2x+2) = (80x-24)$

$54x + 54 = 80x - 24$

$54x = 80x - 78$

$-26x = -78$

$x = 3$

Check: $3\sqrt[3]{2(3)+2} \stackrel{?}{=} \sqrt[3]{80(3)-24}$

$3\sqrt[3]{8} \stackrel{?}{=} \sqrt[3]{216}$

$3 \cdot 2 = 6$ True

The solution is 3.

16. $3x^2 - 4x + 5 = 0$

$x = \dfrac{-b \pm \sqrt{b^2 - 4ac}}{2a}$

$= \dfrac{-(-4) \pm \sqrt{(-4)^2 - 4(3)(5)}}{2(3)}$

$= \dfrac{4 \pm \sqrt{-44}}{6}$

$= \dfrac{4 \pm 2i\sqrt{11}}{6}$

$= \dfrac{2 \pm i\sqrt{11}}{3}$

17. $\log(3x-4) + \log(4) = \log(x+6)$

$\log 4(3x-4) = \log(x+6)$

$\log(12x-16) = \log(x+6)$

$12x - 16 = x + 6$

$12x = x + 22$

$11x = 22$

$x = 2$

Check: $\log[3(2)-4] + \log 4 \stackrel{?}{=} \log(2+6)$

$\log 2 + \log 4 \stackrel{?}{=} \log 8$

$\log(2 \cdot 4) = \log(8)$ True

The solution is 2.

18. $f(x) = 4x^4 - 8x^3 - 7x^2 + 17x - 6$

The possible rational zeros are of the form $\frac{p}{q}$ where p is a factor of –6 and q is a factor of 4.

Factors of –6: $\pm1, \pm2, \pm3, \pm6$

Factors of 4: $\pm1, \pm2, \pm4$

Possible rational zeros:

$\pm1, \pm2, \pm3, \pm6, \pm\frac{1}{2}, \pm\frac{3}{2}, \pm\frac{1}{4}, \pm\frac{3}{4}$

Since the coefficients of $f(x)$ have 3 sign changes, there are 3 or 1 positive real zeros.

$f(-x) = 4x^4 + 8x^3 - 7x^2 - 17x - 6$

Since the coefficients of $f(-x)$ have 1 sign change, there is exactly 1 negative real zero.

–1	4	–8	–7	17	–6
		–4	12	–5	–12
	4	–12	5	12	–18

–1 is not a zero since the remainder is not 0.

–2	4	–8	–7	17	–6
		–8	32	–50	66
	4	–16	25	–33	60

–2 is not a zero since the remainder is not zero. Since the last row of synthetic division alternate between positive and negative values, –2 is a lower bound on the zeros.

$-\frac{3}{2}$	4	–8	–7	17	–6
		–6	21	–21	6
	4	–14	14	–4	0

$-\frac{3}{2}$ is a zero and $f(x)$ can be rewritten in factored form.

$f(x) = \left(x + \frac{3}{2}\right)\left(4x^3 - 14x^2 + 14x - 4\right)$

The other rational zeros are the zeros of $Q(x) = 4x^3 - 14x^2 + 14x - 4$

The possible rational zeros of $Q(x)$ are of the form $\frac{p}{q}$ where p is a factor of –4 and q is a factor of 4.

Factors of –4: $\pm1, \pm2, \pm4$

Factors of 4: $\pm1, \pm2, \pm4$

Possible rational zeros: $\pm1, \pm2, \pm4, \pm\frac{1}{2}, \pm\frac{1}{4}$

Since we found the only negative real zero, we only need to consider positive values now. The only possible rational zeros are $1, 2, 4, \frac{1}{2}$, and $\frac{1}{4}$.

1|	4	−14	14	−4
		4	−10	4
	4	−10	4	0

1 is a zero and $f(x)$ can be factored further.

$$f(x) = \left(x + \frac{3}{2}\right)(x-1)\left(4x^2 - 10x + 4\right)$$

$$f(x) = \left(x + \frac{3}{2}\right)(x-1)2\left(2x^2 - 5x + 2\right)$$

$$f(x) = 2\left(x + \frac{3}{2}\right)(x-1)(x-2)(2x-1)$$

The rational zeros are:

$x = -\frac{3}{2}, x = 1, x = 2,$ and $x = \frac{1}{2}$

19. $9x^2 + 4y^2 = 36$

$$\frac{9x^2}{36} + \frac{4y^2}{36} = 1$$

$$\frac{x^2}{4} + \frac{y^2}{9} = 1$$

Since $a^2 = 4$, $a = 2$.

Since $b^2 = 9$, $b = 3$.

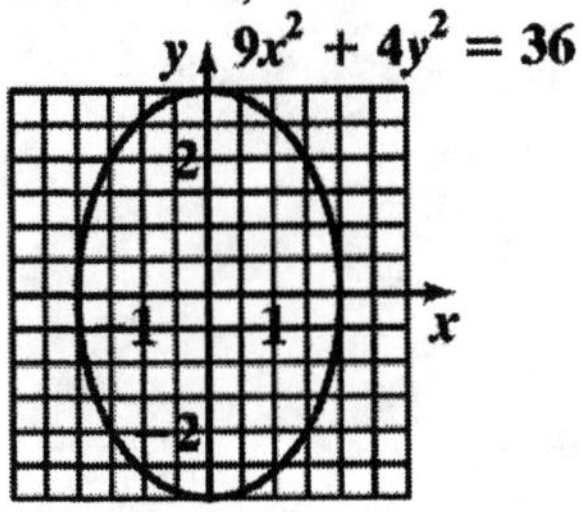

20. $\frac{y^2}{25} - \frac{x^2}{16} = 1$

Since $a^2 = 25$ and $b^2 = 16$, $a = 5$ and $b = 4$

The equations of the asymptotes are $y = \pm\frac{b}{a}x$, or $y = \pm\frac{5}{4}x$. To graph the asymptotes, plot the points (4, 5), (4, −5), (−4, 5), and (−4, −5). The graph intersects the y-axis at (0, −5) and (0, 5).

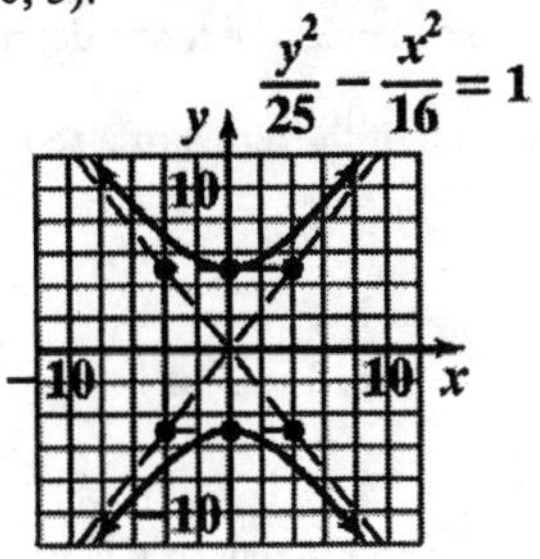

Chapter 12

Exercise Set 12.1

1. A sequence is a list of numbers arranged in a specific order.

3. A finite sequence is a function whose domain includes only the first n natural numbers.

5. In a decreasing sequence, the terms decrease.

7. A series is the sum of the terms of a sequence.

9. $\sum_{i=1}^{5}(i+4)$

The sum as i goes from 1 to 5 of $i+4$.

11. The sequence $a_n = 2n-1$ is an increasing sequence since the coefficient of n is positive.

13. Yes, $a_n = 1+(-2)^n$ is an alternating sequence since $(-2)^n$ alternates between positive and negative values as n alternates from odd to even.

15. $a_n = 6n$

$a_1 = 6(1) = 6$
$a_2 = 6(2) = 12$
$a_3 = 6(3) = 18$
$a_4 = 6(4) = 24$
$a_5 = 6(5) = 30$

The terms are 6, 12, 18, 24, 30.

17. $a_n = 4n-1$

$a_1 = 4(1)-1 = 3$
$a_2 = 4(2)-1 = 7$
$a_3 = 4(3)-1 = 11$
$a_4 = 4(4)-1 = 15$
$a_5 = 4(5)-1 = 19$

The terms are 3, 7, 11, 15, 19.

19. $a_n = \frac{7}{n}$

$a_1 = \frac{7}{1} = 7$
$a_2 = \frac{7}{2}$
$a_3 = \frac{7}{3}$
$a_4 = \frac{7}{4}$
$a_5 = \frac{7}{5}$

The terms are 7, $\frac{7}{2}$, $\frac{7}{3}$, $\frac{7}{4}$, and $\frac{7}{5}$.

21. $a_n = \frac{n+2}{n+1}$

$a_1 = \frac{1+2}{1+1} = \frac{3}{2}$
$a_2 = \frac{2+2}{2+1} = \frac{4}{3}$
$a_3 = \frac{3+2}{3+1} = \frac{5}{4}$
$a_4 = \frac{4+2}{4+1} = \frac{6}{5}$
$a_5 = \frac{5+2}{5+1} = \frac{7}{6}$

The terms are $\frac{3}{2}$, $\frac{4}{3}$, $\frac{5}{4}$, $\frac{6}{5}$, and $\frac{7}{6}$.

23. $a_n = (-1)^n$

$a_1 = (-1)^1 = -1$
$a_2 = (-1)^2 = 1$
$a_3 = (-1)^3 = -1$
$a_4 = (-1)^4 = 1$
$a_5 = (-1)^5 = -1$

The terms are $-1, 1, -1, 1, -1$.

25. $a_n = (-2)^{n+1}$

$a_1 = (-2)^{1+1} = (-2)^2 = 4$
$a_2 = (-2)^{2+1} = (-2)^3 = -8$
$a_3 = (-2)^{3+1} = (-2)^4 = 16$
$a_4 = (-2)^{4+1} = (-2)^5 = -32$
$a_5 = (-2)^{5+1} = (-2)^6 = 64$

The terms are 4, -8, 16, -32, 64.

27. $a_n = 2n+7$

$a_{12} = 2(12)+7 = 24+7 = 31$

29. $a_n = \frac{n}{4}+8$

$a_{16} = \frac{16}{4}+8 = 4+8 = 12$

31. $a_n = (-1)^n$

$a_8 = (-1)^8 = 1$

33. $a_n = n(n+2)$

$a_9 = 9(9+2) = 9(11) = 99$

35. $a_n = \frac{n^2}{2n+7}$

$a_9 = \frac{9^2}{2(9)+7} = \frac{81}{18+7} = \frac{81}{25}$

37. $a_n = 3n-1$

$a_1 = 3(1)-1 = 3-1 = 2$

$a_2 = 3(2)-1 = 6-1 = 5$

$a_3 = 3(3)-1 = 9-1 = 8$

$s_1 = a_1 = 2$

$s_3 = a_1 + a_2 + a_3 = 2+5+8 = 15$

39. $a_n = 2^n + 1$

$a_1 = 2^1 + 1 = 2+1 = 3$

$a_2 = 2^2 + 1 = 4+1 = 5$

$a_3 = 2^3 + 1 = 8+1 = 9$

$s_1 = a_1 = 3$

$s_3 = a_1 + a_2 + a_3 = 3+5+9 = 17$

41. $a_n = \frac{n-1}{n+2}$

$a_1 = \frac{1-1}{1+2} = \frac{0}{3} = 0$

$a_2 = \frac{2-1}{2+2} = \frac{1}{4}$

$a_3 = \frac{3-1}{3+2} = \frac{2}{5}$

$s_1 = 0$

$s_3 = 0 + \frac{1}{4} + \frac{2}{5} = \frac{5}{20} + \frac{8}{20} = \frac{13}{20}$

43. $a_n = (-1)^n$

$a_1 = (-1)^1 = -1$

$a_2 = (-1)^2 = 1$

$a_3 = (-1)^3 = -1$

$s_1 = a_1 = -1$

$s_3 = a_1 + a_2 + a_3 = -1+1+-1 = -1$

45. $a_n = \frac{n^2}{2}$

$a_1 = \frac{1^2}{2} = \frac{1}{2}$

$a_2 = \frac{2^2}{2} = \frac{4}{2} = 2$

$a_3 = \frac{3^2}{2} = \frac{9}{2}$

$s_1 = a_1 = \frac{1}{2}$

$s_3 = a_1 + a_2 + a_3 = \frac{1}{2} + \frac{4}{2} + \frac{9}{2} = \frac{14}{2} = 7$

47. Each term is twice the preceding term. The next three terms are 64, 128, 256.

49. Each term is two more than the preceding term. The next three terms are 17, 19, 21.

51. Each denominator is one more than the preceding one while each numerator is one. The next three terms are $\frac{1}{6}, \frac{1}{7}, \frac{1}{8}$.

53. Each term is –1 times the previous term. The next three terms are 1, –1, 1.

55. Each denominator is three times the previous one while each numerator is one. The next three terms are $\frac{1}{81}, \frac{1}{243}, \frac{1}{729}$.

57. Each term is $-\frac{1}{2}$ times the preceding term. The next three terms are $\frac{1}{16}, -\frac{1}{32}, \frac{1}{64}$.

59. Each term is 5 less than the preceding term. The next three terms are 17, 12, 7.

61. $\sum_{i=1}^{5}(3i-1) = [3(1)-1]+[3(2)-1]+[3(3)-1]+[3(4)-1]+[3(5)-1]$
$= 2+5+8+11+14$
$= 40$

63. $\sum_{i=1}^{6}(i^2+1) = (1^2+1)+(2^2+1)+(3^2+1)+(4^2+1)+(5^2+1)+(6^2+1)$
$= (1+1)+(4+1)+(9+1)+(16+1)+(25+1)+(36+1)$
$= 2+5+10+17+26+37$
$= 97$

65. $\sum_{i=1}^{4}\frac{i^2}{2} = \frac{1^2}{2}+\frac{2^2}{2}+\frac{3^2}{2}+\frac{4^2}{2} = \frac{1}{2}+\frac{4}{2}+\frac{9}{2}+\frac{16}{2} = \frac{1}{2}+2+\frac{9}{2}+8 = \frac{30}{2} = 15$

67. $\sum_{i=4}^{9}\frac{i^2+i}{i+1} = \frac{4^2+4}{4+1}+\frac{5^2+5}{5+1}+\frac{6^2+6}{6+1}+\frac{7^2+7}{7+1}+\frac{8^2+8}{8+1}+\frac{9^2+9}{9+1}$
$= \frac{20}{5}+\frac{30}{6}+\frac{42}{7}+\frac{56}{8}+\frac{72}{9}+\frac{90}{10}$
$= 4+5+6+7+8+9$
$= 39$

---

69. $a_n = n+8$

The fifth partial sum is $\sum_{i=1}^{5}(i+8)$.

71. $a_n = \frac{n^2}{4}$

The third partial sum is $\sum_{i=1}^{3}\frac{i^2}{4}$.

73. $\sum_{i=1}^{5}x_i = x_1+x_2+x_3+x_4+x_5$
$= 2+3+5+(-1)+4$
$= 13$

75. $\left(\sum_{i=1}^{5}x_i\right)^2 = (x_1+x_2+x_3+x_4+x_5)^2$
$= (2+3+5+(-1)+4)^2$
$= 13^2$
$= 169$

77. $\sum_{i=1}^{5}x_i^2 = x_1^2+x_2^2+x_3^2+x_4^2+x_5^2$
$= 2^2+3^2+5^2+(-1)^2+4^2$
$= 55$

79. $\bar{x} = \frac{15+20+25+30+35}{5} = \frac{125}{5} = 25$

81. $\bar{x} = \frac{72+83+4+60+18+20}{6} = \frac{257}{6} \approx 42.83$

83. a. Perimeter of rectangle: $p = 2l+2w$
$p_1 = 2(1)+2(2\cdot1) = 2+4 = 6$
$p_2 = 2(2)+2(2\cdot2) = 4+8 = 12$
$p_3 = 2(3)+2(2\cdot3) = 6+12 = 18$
$p_4 = 2(4)+2(2\cdot4) = 8+16 = 24$

b. $p_n = 2n+2(2n) = 2n+4n = 6n$

85 – 87. Answers will vary.

89. $\bar{x} = \frac{\sum x}{n}$
$n\bar{x} = n\cdot\frac{\sum x}{n}$
$n\bar{x} = \sum x$ or $\sum x = n\bar{x}$

91. Yes, $\sum_{i=1}^{n} 4x_i = 4\sum_{i=1}^{n} x_i$. Examples will vary.

93. **a.** $\sum x = x_1 + x_2 + x_3 = 3+5+2 = 10$

b. $\sum y = y_1 + y_2 + y_3 = 4+1+6 = 11$

c. $\sum x \cdot \sum y = 10 \cdot 11 = 110$

d. $\sum xy = x_1y_1 + x_2y_2 + x_3y_3$
$= 3(4)+5(1)+2(6)$
$= 12+5+12$
$= 29$

e. No, $\sum xy \neq \sum x \cdot \sum y$.

94. $$\left|\frac{1}{2}x+\frac{3}{5}\right| = \left|\frac{1}{2}x-1\right|$$

$\frac{1}{2}x+\frac{3}{5}=\frac{1}{2}x-1$ or $\frac{1}{2}x+\frac{3}{5}=-\left(\frac{1}{2}x-1\right)$

$\frac{3}{5}=-1$ $\qquad \frac{1}{2}x+\frac{3}{5}=-\frac{1}{2}x+1$

$\varnothing$ $\qquad x=1-\frac{3}{5}$

$x=\frac{2}{5}$

The solution is $x=\frac{2}{5}$.

95. $8y^3-64x^6 = 8(y^3-8x^6)$
$= 8\left[(y)^3-(2x^2)^3\right]$
$= 8\left[(y-2x^2)(y^2+2x^2y+(2x^2)^2)\right]$
$= 8\left[(y-2x^2)(y^2+2x^2y+4x^4)\right]$

96.
$$\sqrt{x+5}-1=\sqrt{x-2}$$
$$\left(\sqrt{x+5}-1\right)^2 = x-2$$
$$\left(\sqrt{x+5}\right)^2-2\left(\sqrt{x+5}\right)(1)+(1)^2 = x-2$$
$$x+5-2\sqrt{x+5}+1 = x-2$$
$$-2\sqrt{x+5}=-8$$
$$\sqrt{x+5}=4$$
$$x+5=4^2$$
$$x+5=16$$
$$x=11$$

Since we have raised both sides of the equation to an even-numbered power, we must check for extraneous solutions:

$\sqrt{11+5}-1=\sqrt{11-2}$?
$\sqrt{16}-1=\sqrt{9}$?
$4-1=3$?
$3=3$ True

Thus, $x = 11$ is the solution to the equation.

97. $V=\pi r^2 h$, for r
$$\frac{V}{\pi h}=\frac{\pi r^2 h}{\pi h}$$
$$\frac{V}{\pi h}=r^2$$
$$\sqrt{\frac{V}{\pi h}}=r$$

Exercise Set 12.2

1. In an arithmetic sequence, each term differs by a constant amount.

3. It is called the common difference.

5. The common difference, d, must be a positive number.

7. Yes. For example, –1, –2, –3, –4, ... is an arithmetic sequence with $a_1=-1$ and $d=-1$.

9. Yes. For example, 2, 4, 6, 8, ... is an arithmetic sequence with $a_1=2$ and $d=2$.

11. $a_1=4$
$a_2=4+(2-1)(3)=4+3=7$
$a_3=4+(3-1)(3)=4+2(3)=4+6=10$
$a_4=4+(4-1)(3)=4+3(3)=4+9=13$
$a_5=4+(5-1)(3)=4+4(3)=4+12=16$
The terms are 4, 7, 10, 13, 16. The general term is $a_n=4+(n-1)3$ or $a_n=3n+1$.

13. $a_1=7$
$a_2=7+(2-1)(-2)=7-2=5$
$a_3=7+(3-1)(-2)=7+2(-2)=7-4=3$
$a_4=7+(4-1)(-2)=7+3(-2)=7-6=1$
$a_5=7+(5-1)(-2)=7+4(-2)=7-8=-1$
The terms are 7, 5, 3, 1, –1. The general term is $a_n=7+(n-1)(-2)$ or $a_n=-2n+9$.

15. $a_1 = \frac{1}{2}$

$$a_2 = \frac{1}{2} + (2-1)\left(\frac{3}{2}\right) = \frac{1}{2} + \frac{3}{2} = \frac{4}{2} = 2$$

$$a_3 = \frac{1}{2} + (3-1)\left(\frac{3}{2}\right) = \frac{1}{2} + 2\left(\frac{3}{2}\right) = \frac{1}{2} + \frac{6}{2} = \frac{7}{2}$$

$$a_4 = \frac{1}{2} + (4-1)\left(\frac{3}{2}\right) = \frac{1}{2} + 3\left(\frac{3}{2}\right) = \frac{1}{2} + \frac{9}{2} = \frac{10}{2} = 5$$

$$a_5 = \frac{1}{2} + (5-1)\left(\frac{3}{2}\right) = \frac{1}{2} + 4\left(\frac{3}{2}\right) = \frac{1}{2} + \frac{12}{2} = \frac{13}{2}$$

The terms are $\frac{1}{2}, 2, \frac{7}{2}, 5, \frac{13}{2}$. The general term is $a_n = \frac{1}{2} + (n-1)\frac{3}{2}$ or $a_n = \frac{3}{2}n - 1$.

17. $a_1 = 100$

$$a_2 = 100 + (2-1)(-5) = 100 + (-5) = 100 - 5 = 95$$

$$a_3 = 100 + (3-1)(-5) = 100 + 2(-5) = 100 - 10 = 90$$

$$a_4 = 100 + (4-1)(-5) = 100 + 3(-5) = 100 - 15 = 85$$

$$a_5 = 100 + (5-1)(-5) = 100 + 4(-5) = 100 - 20 = 80$$

The terms are 100, 95, 90, 85, 80. The general term is $a_n = 100 + (n-1)(-5)$ or $a_n = -5n + 105$.

19. $a_n = a_1 + (n-1)d$

$$a_4 = 5 + (4-1)3 = 5 + 3 \cdot 3 = 5 + 9 = 14$$

21. $a_n = a_1 + (n-1)d$

$$a_{10} = -9 + (10-1)(4) = -9 + 9(4) = -9 + 36 = 27$$

23. $a_n = a_1 + (n-1)d$

$$a_{13} = -8 + (13-1)\left(\frac{5}{3}\right) = -8 + 12\left(\frac{5}{3}\right) = -8 + 20 = 12$$

25. $a_n = a_1 + (n-1)d$

$$27 = 11 + (9-1)d$$
$$27 = 11 + 8d$$
$$16 = 8d$$
$$2 = d$$

27. $a_n = a_1 + (n-1)d$

$$28 = 4 + (n-1)(3)$$
$$28 = 4 + 3n - 3$$
$$28 = 1 + 3n$$
$$27 = 3n$$
$$9 = n$$

29. $a_n = a_1 + (n-1)d$

$$42 = 82 + (n-1)(-8)$$
$$42 = 82 - 8n + 8$$
$$42 = 90 - 8n$$
$$-48 = -8n$$
$$6 = n$$

31. $s_{10} = \frac{10(a_1 + a_{10})}{2} = \frac{10(1+19)}{2} = 5(20) = 100$

$$a_{10} = a_1 + (10-1)d$$
$$a_{10} = a_1 + 9d$$
$$19 = 1 + 9d$$
$$18 = 9d$$
$$2 = d$$

33. $s_8 = \frac{8(a_1 + a_8)}{2} = \frac{8(\frac{3}{5}+2)}{2} = 4\left(\frac{3}{5}+2\right) = 4\left(\frac{3}{5}+\frac{10}{5}\right) = 4\left(\frac{13}{5}\right) = \frac{52}{5}$

$$a_8 = a_1 + (8-1)d$$
$$a_8 = a_1 + 7d$$
$$2 = \frac{3}{5} + 7d$$
$$\frac{7}{5} = 7d$$
$$d = \frac{1}{7} \cdot \frac{7}{5} = \frac{1}{5}$$

35. $s_6 = \frac{6(a_1 + a_6)}{2} = \frac{6(-5+13.5)}{2} = \frac{6(8.5)}{2} = 25.5$

$$a_6 = a_1 + (6-1)d$$
$$a_6 = a_1 + 5d$$
$$13.5 = -5 + 5d$$
$$18.5 = 5d$$
$$3.7 = d$$

37. $s_{11} = \frac{11(a_1 + a_{11})}{2} = \frac{11(7+67)}{2} = \frac{11(74)}{2} = 407$

$a_{11} = a_1 + (11-1)d$

$a_{11} = a_1 + 10d$

$67 = 7 + 10d$

$60 = 10d$

$6 = d$

39. $a_1 = 4$

$a_2 = 4 + (2-1)(3) = 4 + 3 = 7$

$a_3 = 4 + (3-1)(3) = 4 + 2(3) = 4 + 6 = 10$

$a_4 = 4 + (4-1)(3) = 4 + 3(3) = 4 + 9 = 13$

The terms are 4, 7, 10, 13.

$a_{10} = 4 + (10-1)(3) = 4 + 9(3) = 4 + 27 = 31$

$s_{10} = \frac{10(4+31)}{2} = \frac{10(35)}{2} = 175$

41. $a_1 = -6$

$a_2 = -6 + (2-1)(2) = -6 + 1(2) = -6 + 2 = -4$

$a_3 = -6 + (3-1)(2) = -6 + 2(2) = -6 + 4 = -2$

$a_4 = -6 + (4-1)(2) = -6 + 3(2) = -6 + 6 = 0$

The terms are –6, –4, –2, 0.

$a_{10} = -6 + (10-1)(2) = -6 + 9(2) = -6 + 18 = 12$

$s_{10} = \frac{10(-6+12)}{2} = \frac{10(6)}{2} = \frac{60}{2} = 30$

43. $a_1 = -8$

$a_2 = -8 + (2-1)(-5) = -8 - 5 = -13$

$a_3 = -8 + (3-1)(-5) = -8 + 2(-5) = -8 - 10 = -18$

$a_4 = -8 + (4-1)(-5) = -8 + 3(-5) = -8 - 15 = -23$

The terms are –8, –13, –18, –23.

$a_{10} = -8 + (10-1)(-5)$

$= -8 + 9(-5) = -8 - 45 = -53$

$s_{10} = \frac{10[-8+(-53)]}{2} = \frac{10(-61)}{2} = \frac{-610}{2} = -305$

45. $a_1 = \frac{7}{2}$

$a_2 = \frac{7}{2} + (2-1)\left(\frac{5}{2}\right) = \frac{7}{2} + 1\left(\frac{5}{2}\right) = \frac{7}{2} + \frac{5}{2} = \frac{12}{2} = 6$

$a_3 = \frac{7}{2} + (3-1)\left(\frac{5}{2}\right) = \frac{7}{2} + 2\left(\frac{5}{2}\right) = \frac{7}{2} + \frac{10}{2} = \frac{17}{2}$

$a_4 = \frac{7}{2} + (4-1)\left(\frac{5}{2}\right) = \frac{7}{2} + 3\left(\frac{5}{2}\right) = \frac{7}{2} + \frac{15}{2} = \frac{22}{2} = 11$

The terms are $\frac{7}{2}$, 6, $\frac{17}{2}$, 11.

$a_{10} = \frac{7}{2} + (10-1)\left(\frac{5}{2}\right)$

$= \frac{7}{2} + 9\left(\frac{5}{2}\right) = \frac{7}{2} + \frac{45}{2} = \frac{52}{2} = 26$

$s_{10} = \frac{10(3.5+26)}{2} = \frac{10(29.5)}{2} = 147.5$

47. $a_1 = 100$

$a_2 = 100 + (2-1)(-7)$

$= 100 + 1(-7) = 100 - 7 = 93$

$a_3 = 100 + (3-1)(-7)$

$= 100 + 2(-7) = 100 - 14 = 86$

$a_4 = 100 + (4-1)(-7)$

$= 100 + 3(-7) = 100 - 21 = 79$

The terms are 100, 93, 86, 79.

$a_{10} = 100 + (10-1)(-7)$

$= 100 + 9(-7) = 100 - 63 = 37$

$s_{10} = \frac{10(100+37)}{2} = \frac{10(137)}{2} = 685$

49. $d = 4 - 1 = 3$

$a_n = a_1 + (n-1)d$

$43 = 1 + (n-1)(3)$

$43 = 1 + 3n - 3$

$43 = -2 + 3n$

$45 = 3n$

$15 = n$

$s_{15} = \frac{15(a_1 + a_{15})}{2} = \frac{15(1+43)}{2} = \frac{15(44)}{2} = 330$

51. $d = -5 - (-9) = -5 + 9 = 4$

$a_n = a_1 + (n-1)d$

$31 = -9 + (n-1)(4)$

$31 = -9 + 4n - 4$

$31 = -13 + 4n$

$44 = 4n$

$11 = n$

$s_{10} = \frac{11(a_1 + a_{10})}{2} = \frac{11(-9+31)}{2} = \frac{11(22)}{2} = 121$

53. $d = \frac{2}{2} - \frac{1}{2} = \frac{1}{2}$

$a_n = \frac{1}{2} + (n-1)\left(\frac{1}{2}\right)$

$\frac{17}{2} = \frac{1}{2} + \frac{1}{2}n - \frac{1}{2}$

$\frac{17}{2} = \frac{1}{2}n$

$17 = n$

$s_{17} = \frac{17(a_1 + a_{17})}{2}$

$= \frac{17\left(\frac{1}{2} + \frac{17}{2}\right)}{2} = \frac{17\left(\frac{18}{2}\right)}{2} = \frac{17(9)}{2} = \frac{153}{2}$

55. $d = 10 - 7 = 3$

$a_n = a_1 + (n-1)d$

$91 = 7 + (n-1)(3)$

$91 = 7 + 3n - 3$

$91 = 4 + 3n$

$87 = 3n$

$29 = n$

$s_{29} = \frac{29(a_1 + a_{29})}{2} = \frac{29(7+91)}{2} = \frac{29(98)}{2} = 1421$

57. $s_n = \frac{n(a_1 + a_n)}{2}$

$s_{50} = \frac{50(1+50)}{2} = \frac{50(51)}{2} = 1275$

59. $s_n = \frac{n(a_1 + a_n)}{2}$

$s_{50} = \frac{50(1+99)}{2} = \frac{50(100)}{2} = 2500$

61. $s_n = \frac{n(a_1 + a_n)}{2}$

$s_{30} = \frac{30(3+90)}{2} = \frac{30(93)}{2} = 1395$

63. The smallest number greater than 7 that is divisible by 6 is 12. The largest number less than 1610 that is divisible by 6 is 1608. Now find n in the equation $a_n = a_1 + (n-1)d$.

$1608 = 12 + (n-1)6$

$1596 = 6(n-1)$

$266 = n - 1$

$267 = n$

There are 267 numbers between 7 and 1610 that are divisible by 6.

65. $a_1 = 20,\ d = 2,\ n = 12$

$a_n = a_1 + (n-1)d$

$a_{12} = 20 + (12-1)(2) = 20 + 11(2) = 20 + 22 = 42$

$s_n = \frac{n(a_1 + a_n)}{2}$

$s_{12} = \frac{12(20+42)}{2} = \frac{12(62)}{2} = \frac{744}{2} = 372$

There are 42 seats in the twelfth row and 372 seats in the first twelve rows.

67. $26 + 25 + 24 + \cdots + 1$ or $1 + 2 + 3 + \cdots + 26$

$s_n = \frac{n(a_1 + a_n)}{2}$

$s_{26} = \frac{26(1+26)}{2} = \frac{26(27)}{2} = \frac{702}{2} = 351$

There are 351 logs in the pile.

69. $a_1 = 1,\ d = 2,\ n = 14$

$a_n = a_1 + (n-1)d$

$a_{14} = 1 + (14-1)(2) = 1 + 13(2) = 1 + 26 = 27$

$s_n = \frac{n(a_1 + a_n)}{2}$

$s_{14} = \frac{14(1+27)}{2} = \frac{14(28)}{2} = \frac{392}{2} = 196$

There are 27 glasses in the 14th row and 196 glasses in all.

71. $1 + 2 + 3 + \cdots + 100$

$= (1+100) + (2+99) + \cdots + (50+51)$

$= 101 + 101 + \cdots + 101$

$= 50(101)$

$= 5050$

73. $s_n = \frac{n(a_1 + a_n)}{2}$

$= \frac{n[1 + (2n-1)]}{2} = \frac{n(2n)}{2} = \frac{2n^2}{2} = n^2$

75. a. $a_1 = 22,\ d = -\frac{1}{2},\ n = 7$

$a_n = a_1 + (n-1)d$

$a_7 = 22 + (7-1)\left(-\frac{1}{2}\right) = 22 - 3 = 19$

Her seventh swing is 19 feet.

b. $s_n = \frac{n(a_1 + a_n)}{2}$

$s_7 = \frac{7(22+19)}{2} = 143.5$

She travels 143.5 feet during the seven swings.

77. $d = -6 \text{ in.} = -\frac{1}{2} \text{ ft}, a_1 = 6$

$a_n = a_1 + (n-1)d$

$a_9 = 6 + (9-1)\left(-\frac{1}{2}\right) = 6 + 8\left(-\frac{1}{2}\right) = 6 - 4 = 2$

The ball bounces 2 feet on the ninth bounce.

79. a. Note that if March17th is day 1, then March 22nd is day 6.

$a_1 = 105, d = 10, n = 6$

$a_n = a_1 + (n-1)d$

$a_6 = 105 + (6-1)(10)$

$= 105 + 5(10) = 105 + 50 = 155$

He can prepare 155 packages for shipment on March 22nd.

b. $s_n = \frac{n(a_1 + a_n)}{2}$

$s_5 = \frac{5(105+155)}{2} = \frac{5(260)}{2} = \frac{1300}{2} = 650$

He can prepare 650 packages for shipment from March 17th through March 22nd.

81. $s_n = \frac{n(a_1 + a_n)}{2}$

$s_{31} = \frac{31(1+31)}{2} = \frac{31(32)}{2} = 496$

On day 31, Craig will have saved \$496.

83. a. $a_{10} = 42{,}000 + (10-1)(400) = 45{,}600$

She will receive \$45,600 in her tenth year of retirement.

b. $s_{10} = \frac{10(42{,}000 + 45{,}600)}{2}$

$= \frac{10(87{,}600)}{2}$

$= 438{,}000$

In her first 10 years, she will receive a total of \$438,000.

85. $360 - 180 = 180$

$540 - 360 = 180$

$720 - 540 = 180$

The terms form an arithmetic sequence with $d = 180$ and $a_3 = 180$.

$a_n = a_1 + (n-1)d$

$180 = a_1 + (3-1)180$

$180 = a_1 + 360$

$-180 = a_1$

$a_n = a_1 + (n-1)d$

$a_n = -180 + (n-1)(180)$

$= -180 + 180n - 180$

$= 180n - 360$

$= 180(n-2)$

93. $A = P + Prt$

$A - P = Prt$

$\frac{A-P}{Pt} = r$ or $r = \frac{A-P}{Pt}$

94. $y = 2x + 1$

$3x - 2y = 1$

Substitute $2x+1$ for y in the second equation.

$3x - 2y = 1$

$3x - 2(2x+1) = 1$

$3x - 4x - 2 = 1$

$-x - 2 = 1$

$-x = 3$

$x = -3$

Substitute -3 for x in the first equation.

$y = 2(-3) + 1 = -6 + 1 = -5$

The solution is $(-3, -5)$.

95. $12n^2 - 6n - 30n + 15$

$= 3(4n^2 - 2n - 10n + 5)$

$= 3[(4n^2 - 2n) - (10n - 5)]$

$= 3[2n(2n-1) - 5(2n-1)]$

$= 3[(2n-1)(2n-5)]$

$= 3(2n-1)(2n-5)$

96. $(x+4)^2 + y^2 = 25$

$(x+4)^2 + y^2 = 5^2$

The center is $(-4, 0)$ and the radius is 5.

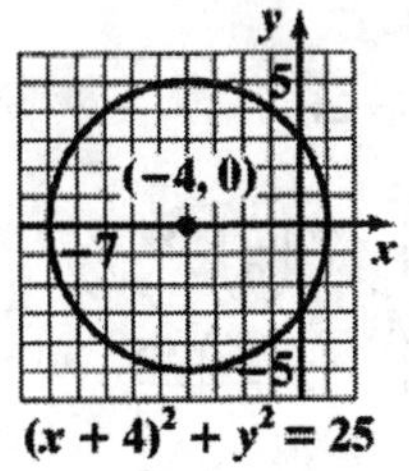

Exercise Set 12.3

1. A geometric sequence is a sequence in which each term after the first is the same multiple of the preceding term.

3. To find the common ratio, take any term except the first and divide by the term that precedes it.

5. r^n approaches 0 as n gets larger and larger when $|r|<1$.

7. Yes

9. Yes, s_∞ exists. $s_\infty = \frac{a_1}{1-r} = \frac{6}{1-\frac{1}{4}} = \frac{6}{\frac{3}{4}} = 6\cdot\frac{4}{3} = 8$

This is true since $|r|<1$.

11. $a_1 = 2$

$a_2 = 2(3)^{2-1} = 2(3) = 6$

$a_3 = 2(3)^{3-1} = 2(3)^2 = 2(9) = 18$

$a_4 = 2(3)^{4-1} = 2(3)^3 = 2(27) = 54$

$a_5 = 2(3)^{5-1} = 2(3)^4 = 2(81) = 162$

The terms are 2, 6, 18, 54, 162.

13. $a_1 = 6$

$a_2 = 6\left(-\frac{1}{2}\right)^{2-1} = 6\left(-\frac{1}{2}\right) = -3$

$a_3 = 6\left(-\frac{1}{2}\right)^{3-1} = 6\left(-\frac{1}{2}\right)^2 = 6\left(\frac{1}{4}\right) = \frac{3}{2}$

$a_4 = 6\left(-\frac{1}{2}\right)^{4-1} = 6\left(-\frac{1}{2}\right)^3 = 6\left(-\frac{1}{8}\right) = -\frac{3}{4}$

$a_5 = 6\left(-\frac{1}{2}\right)^{5-1} = 6\left(-\frac{1}{2}\right)^4 = 6\left(\frac{1}{16}\right) = \frac{3}{8}$

The terms are $6, -3, \frac{3}{2}, -\frac{3}{4}, \frac{3}{8}$.

15. $a_1 = 72$

$a_2 = 72\left(\frac{1}{3}\right)^{2-1} = 72\left(\frac{1}{3}\right) = 24$

$a_3 = 72\left(\frac{1}{3}\right)^{3-1} = 72\left(\frac{1}{3}\right)^2 = 72\left(\frac{1}{9}\right) = 8$

$a_4 = 72\left(\frac{1}{3}\right)^{4-1} = 72\left(\frac{1}{3}\right)^3 = 72\left(\frac{1}{27}\right) = \frac{8}{3}$

$a_5 = 72\left(\frac{1}{3}\right)^{5-1} = 72\left(\frac{1}{3}\right)^4 = 72\left(\frac{1}{81}\right) = \frac{8}{9}$

The terms are 72, 24, 8, $\frac{8}{3}, \frac{8}{9}$.

17. $a_1 = 90$

$a_2 = 90\left(-\frac{1}{3}\right)^{2-1} = 90\left(-\frac{1}{3}\right) = -30$

$a_3 = 90\left(-\frac{1}{3}\right)^{3-1} = 90\left(-\frac{1}{3}\right)^2 = 90\left(\frac{1}{9}\right) = 10$

$a_4 = 90\left(-\frac{1}{3}\right)^{4-1} = 90\left(-\frac{1}{3}\right)^3 = 90\left(-\frac{1}{27}\right) = -\frac{10}{3}$

$a_5 = 90\left(-\frac{1}{3}\right)^{5-1} = 90\left(-\frac{1}{3}\right)^4 = 90\left(\frac{1}{81}\right) = \frac{10}{9}$

The terms are 90, –30, 10, $-\frac{10}{3}, \frac{10}{9}$.

19. $a_1 = -1$

$a_2 = -1(3)^{2-1} = -1(3) = -3$

$a_3 = -1(3)^{3-1} = -1(3)^2 = -1(9) = -9$

$a_4 = -1(3)^{4-1} = -1(3)^3 = -1(27) = -27$

$a_5 = -1(3)^{5-1} = -1(3)^4 = -1(81) = -81$

The terms are –1, –3, –9, –27, –81.

21. $a_1 = 5$

$a_2 = 5(-2)^{2-1} = 5(-2)^1 = 5(-2) = -10$

$a_3 = 5(-2)^{3-1} = 5(-2)^2 = 5(4) = 20$

$a_4 = 5(-2)^{4-1} = 5(-2)^3 = 5(-8) = -40$

$a_5 = 5(-2)^{5-1} = 5(-2)^4 = 5(16) = 80$

The terms are 5, –10, 20, –40, 80.

23. $a_1 = \frac{1}{3}$

$a_2 = \frac{1}{3}\left(\frac{1}{2}\right)^{2-1} = \frac{1}{3}\left(\frac{1}{2}\right) = \frac{1}{6}$

$a_3 = \frac{1}{3}\left(\frac{1}{2}\right)^{3-1} = \frac{1}{3}\left(\frac{1}{2}\right)^2 = \frac{1}{3}\left(\frac{1}{4}\right) = \frac{1}{12}$

$a_4 = \frac{1}{3}\left(\frac{1}{2}\right)^{4-1} = \frac{1}{3}\left(\frac{1}{2}\right)^3 = \frac{1}{3}\left(\frac{1}{8}\right) = \frac{1}{24}$

$a_5 = \frac{1}{3}\left(\frac{1}{2}\right)^{5-1} = \frac{1}{3}\left(\frac{1}{2}\right)^4 = \frac{1}{3}\left(\frac{1}{16}\right) = \frac{1}{48}$

The terms are $\frac{1}{3}, \frac{1}{6}, \frac{1}{12}, \frac{1}{24}, \frac{1}{48}$.

25. $a_1 = 3$

$a_2 = 3\left(\frac{3}{2}\right)^{2-1} = 3\left(\frac{3}{2}\right) = \frac{9}{2}$

$a_3 = 3\left(\frac{3}{2}\right)^{3-1} = 3\left(\frac{3}{2}\right)^2 = 3\left(\frac{9}{4}\right) = \frac{27}{4}$

$a_4 = 3\left(\frac{3}{2}\right)^{4-1} = 3\left(\frac{3}{2}\right)^3 = 3\left(\frac{27}{8}\right) = \frac{81}{8}$

$a_5 = 3\left(\frac{3}{2}\right)^{5-1} = 3\left(\frac{3}{2}\right)^4 = 3\left(\frac{81}{16}\right) = \frac{243}{16}$

The terms are 3, $\frac{9}{2}, \frac{27}{4}, \frac{81}{8}, \frac{243}{16}$.

27. $a_6 = a_1 r^{6-1}$

$a_6 = 4(2)^{6-1} = 4(2)^5 = 4(32) = 128$

29. $a_9 = a_1 r^{9-1}$

$a_9 = -12\left(\frac{1}{2}\right)^{9-1} = -12\left(\frac{1}{2}\right)^8 = -12\left(\frac{1}{256}\right) = -\frac{3}{64}$

31. $a_{10} = a_1 r^{10-1}$

$a_{10} = \frac{1}{4}(2)^{10-1} = \frac{1}{4}(2)^9 = \frac{1}{4}(512) = 128$

33. $a_{12} = a_1 r^{12-1}$

$a_{12} = -3(-2)^{12-1} = -3(-2)^{11} = -3(-2048) = 6144$

35. $a_8 = a_1 r^{8-1}$

$a_8 = 2\left(\frac{1}{2}\right)^{8-1} = 2\left(\frac{1}{2}\right)^7 = 2\left(\frac{1}{128}\right) = \frac{1}{64}$

37. $a_7 = a_1 r^{7-1}$

$a_7 = 50\left(\frac{1}{3}\right)^{7-1} = 50\left(\frac{1}{3}\right)^6 = 50\left(\frac{1}{729}\right) = \frac{50}{729}$

39. $s_5 = \frac{a_1(1-r^5)}{1-r}$

$s_5 = \frac{5(1-2^5)}{1-2} = \frac{5(1-32)}{-1} = \frac{5(-31)}{-1} = \frac{-155}{-1} = 155$

41. $s_6 = \frac{a_1(1-r^6)}{1-r}$

$s_6 = \frac{2(1-5^6)}{1-5} = \frac{2(1-15{,}625)}{-4} = \frac{2(-15{,}624)}{-4} = 7812$

43. $s_7 = \frac{a_1(1-r^7)}{1-r}$

$s_7 = \frac{80(1-2^7)}{1-2}$

$= \frac{80(1-128)}{-1} = \frac{80(-127)}{-1} = \frac{-10{,}160}{-1} = 10{,}160$

45. $s_9 = \frac{a_1(1-r^9)}{1-r}$

$s_9 = \frac{-15\left[1-\left(-\frac{1}{2}\right)^9\right]}{1-\left(-\frac{1}{2}\right)}$

$= \frac{-15\left[1-\left(-\frac{1}{512}\right)\right]}{\frac{3}{2}}$

$= \frac{-15\left(1+\frac{1}{512}\right)}{\frac{3}{2}}$

$= \frac{-15\left(\frac{513}{512}\right)}{\frac{3}{2}}$

$= -15\left(\frac{513}{512}\right)\left(\frac{2}{3}\right)$

$= -\frac{2565}{256}$

47. $s_5 = \frac{a_1(1-r^5)}{1-r}$

$s_5 = \frac{-9\left[1-\left(\frac{2}{5}\right)^5\right]}{1-\frac{2}{5}} = \frac{-9\left(1-\frac{32}{3125}\right)}{\frac{3}{5}}$

$= \frac{-9\left(\frac{3093}{3125}\right)}{\frac{3}{5}} = -9\left(\frac{3093}{3125}\right)\left(\frac{5}{3}\right) = -\frac{9279}{625}$

49. $r = \frac{3}{2} \div 3 = \frac{3}{2} \cdot \frac{1}{3} = \frac{1}{2}$

$a_n = 3\left(\frac{1}{2}\right)^{n-1}$

51. $r = 18 \div 9 = 2$

$a_n = 9(2)^{n-1}$

53. $r = -6 \div 2 = -3$

$a_n = 2(-3)^{n-1}$

55. $r = \frac{1}{2} \div \frac{3}{4} = \frac{1}{2} \cdot \frac{4}{3} = \frac{2}{3}$

$a_n = \frac{3}{4}\left(\frac{2}{3}\right)^{n-1}$

57. $r = \frac{1}{2} \div 1 = \frac{1}{2}$; $S_\infty = \frac{1}{1-\frac{1}{2}} = \frac{1}{\frac{1}{2}} = 1\left(\frac{2}{1}\right) = 2$

59. $r = \frac{1}{5} \div 1 = \frac{1}{5}$; $S_\infty = \frac{1}{1-\frac{1}{5}} = \frac{1}{\frac{4}{5}} = 1\left(\frac{5}{4}\right) = \frac{5}{4}$

61. $r = 3 \div 6 = \frac{1}{2}$; $S_\infty = \frac{6}{1-\frac{1}{2}} = \frac{6}{\frac{1}{2}} = 6\left(\frac{2}{1}\right) = 12$

63 $r = 2 \div 5 = \frac{2}{5}$; $S_\infty = \frac{5}{1-\frac{2}{5}} = \frac{5}{\frac{3}{5}} = 5\left(\frac{5}{3}\right) = \frac{25}{3}$

65.
$$S_n = \frac{a_1(1-r^n)}{1-r}$$
$$93 = \frac{3(1-2^n)}{1-2}$$
$$93 = \frac{3(1-2^n)}{-1}$$
$$93 = -3(1-2^n)$$
$$-31 = 1-2^n$$
$$-32 = -2^n$$
$$32 = 2^n$$
$$n = 5 \text{ since } 2^5 = 32$$

67.
$$S_n = \frac{a_1(1-r^n)}{1-r}$$
$$\frac{189}{32} = \frac{3\left[1-\left(\frac{1}{2}\right)^n\right]}{1-\frac{1}{2}}$$
$$\frac{189}{32} = \frac{3\left[1-\left(\frac{1}{2}\right)^n\right]}{\frac{1}{2}}$$
$$\frac{1}{2}\cdot\frac{1}{3}\cdot\frac{189}{32} = 1-\left(\frac{1}{2}\right)^n$$
$$\frac{63}{64} = 1-\left(\frac{1}{2}\right)^n$$
$$-\frac{1}{64} = -\left(\frac{1}{2}\right)^n$$
$$\frac{1}{64} = \left(\frac{1}{2}\right)^n$$
$$n = 6 \text{ since } \left(\frac{1}{2}\right)^6 = \frac{1}{64}$$

69. $r = 1 \div 2 = \frac{1}{2}$
$$S_\infty = \frac{2}{1-\frac{1}{2}} = \frac{2}{\frac{1}{2}} = 2\left(\frac{2}{1}\right) = 4$$

71. $r = \frac{16}{3} \div 8 = \frac{16}{3}\left(\frac{1}{8}\right) = \frac{2}{3}$
$$S_\infty = \frac{8}{1-\frac{2}{3}} = \frac{8}{\frac{1}{3}} = 8\left(\frac{3}{1}\right) = 24$$

73. $r = 20 \div -60 = \frac{20}{-60} = -\frac{1}{3}$
$$S_\infty = \frac{-60}{1-\left(-\frac{1}{3}\right)} = \frac{-60}{1+\frac{1}{3}} = \frac{-60}{\frac{4}{3}} = -60\left(\frac{3}{4}\right) = -45$$

75. $r = -\frac{12}{5} \div -12 = -\frac{12}{5}\left(-\frac{1}{12}\right) = \frac{1}{5}$
$$S_\infty = \frac{-12}{1-\frac{1}{5}} = \frac{-12}{\frac{4}{5}} = -12\left(\frac{5}{4}\right) = -15$$

77. $0.242424\ldots = 0.24 + 0.0024 + 0.000024 + \cdots$
$$= 0.24 + 0.24(0.01) + 0.24(0.01)^2 + \cdots$$
$r = 0.01$ and $a_1 = 0.24$
$$S_\infty = \frac{0.24}{1-0.01} = \frac{0.24}{0.99} = \frac{24}{99} = \frac{8}{33}$$

79. $0.8888\ldots = 0.8 + 0.08 + 0.008 + \cdots$
$$= 0.8 + 0.8(0.1) + 0.8(0.1)^2 + \cdots$$
$r = 0.1$ and $a_1 = 0.8$
$$S_\infty = \frac{0.8}{1-0.1} = \frac{0.8}{0.9} = \frac{8}{9}$$

81. $0.515151\cdots = 0.51 + 0.0051 + 0.000051 + \cdots$
$$= 0.51 + 0.51(0.01) + 0.51(0.01)^2 + \cdots$$
$r = 0.01$ and $a_1 = 0.51$
$$S_\infty = \frac{0.51}{1-0.01} = \frac{0.51}{0.99} = \frac{51}{99} = \frac{17}{33}$$

83. Consider a new series $b_1, b_2, b_3, \ldots$ where $b_1 = 15$ and $b_4 = 405$. Now $b_4 = b_1 r^{4-1}$ becomes
$$405 = 15r^3$$
$$\frac{405}{15} = r^3$$
$$27 = r^3$$
$$\sqrt[3]{27} = r$$
$$3 = r$$
From the original series, $a_1 = \frac{a_2}{r} = \frac{15}{3} = 5.$

85. Consider a new series $b_1, b_2, b_3, \ldots$ where $b_1 = 28$ and $b_3 = 112$. Now $b_3 = b_1 r^{3-1}$ becomes

$$112 = 28r^2$$
$$\frac{112}{28} = r^2$$
$$4 = r^2$$

so that $r = 2$ or $r = -2$. From the original series $a_1 = \frac{a_3}{r^2} = \frac{28}{4} = 7$.

87. $a_1 = 1.40$, $n = 9$, $r = 1.03$

$$a_n = a_1 r^{n-1}$$
$$a_9 = 1.4(1.03)^{9-1} = 1.4(1.03)^8 \approx 1.77$$

In 8 years, a loaf of bread would cost \$1.77.

89. $r = \frac{1}{2}$. Let a_n be the amount left after the nth day. After 1 day there are $600\left(\frac{1}{2}\right) = 300$ grams left, so $a_1 = 300$.

a. $37.5 = 300\left(\frac{1}{2}\right)^{n-1}$

$$\frac{37.5}{300} = \left(\frac{1}{2}\right)^{n-1}$$
$$\frac{1}{8} = \left(\frac{1}{2}\right)^{n-1}$$
$$\left(\frac{1}{2}\right)^3 = \left(\frac{1}{2}\right)^{n-1}$$
$$n - 1 = 3$$
$$n = 4$$

37.5 grams are left after 4 days.

b. $a_9 = 300\left(\frac{1}{2}\right)^{9-1} = 300\left(\frac{1}{256}\right) \approx 1.172$

After 9 days, about 1.172 grams of the substance remain.

91. After ten years will be at the beginning of the eleventh year. If the population increases by 1.1% each year, then the population at the beginning of a year will be 1.011 times the population at the beginning of the previous year.

a. Use $a_1 = 296.5$, $r = 1.011$, and $n = 11$.

$$a_{11} = a_1 r^{11-1} = 296.5(1.011)^{10} \approx 330.78$$

After ten years (the beginning of the eleventh year), the population is about 330.78 million people.

b. $a_n = 2(296.5) = 593$

Now, use $a_n = a_1 r^{n-1}$

$$593 = 296.5(1.011)^{n-1}$$
$$\frac{593}{296.5} = (1.011)^{n-1}$$
$$2 = (1.011)^{n-1}$$

Now, use logarithms.

$$\log 2 = \log(1.011)^{n-1}$$
$$\log 2 = (n-1)\log 1.011$$
$$\frac{\log 2}{\log 1.011} = n - 1$$
$$63.4 \approx n - 1$$
$$64.4 \approx n$$

The population will be double at the beginning of the 64.4 year, which is the end of the 63.4 year.

93. a. After 1 meter there is $\frac{1}{2}$ of the original light remaining, so $a_1 = \frac{1}{2}$.

$$a_1 = \frac{1}{2}$$
$$a_2 = \frac{1}{2}\left(\frac{1}{2}\right)^{2-1} = \left(\frac{1}{2}\right)\left(\frac{1}{2}\right) = \frac{1}{4}$$
$$a_3 = \frac{1}{2}\left(\frac{1}{2}\right)^{3-1} = \left(\frac{1}{2}\right)\left(\frac{1}{4}\right) = \frac{1}{8}$$
$$a_4 = \frac{1}{2}\left(\frac{1}{2}\right)^{4-1} = \left(\frac{1}{2}\right)\left(\frac{1}{8}\right) = \frac{1}{16}$$
$$a_5 = \frac{1}{2}\left(\frac{1}{2}\right)^{5-1} = \left(\frac{1}{2}\right)\left(\frac{1}{16}\right) = \frac{1}{32}$$

b. $a_n = \frac{1}{2}\left(\frac{1}{2}\right)^{n-1} = \left(\frac{1}{2}\right)^n$

c. $a_7 = \left(\frac{1}{2}\right)^7 = \frac{1}{128} \approx 0.0078$ or 0.78%

95. After 8 years will be the beginning of the ninth year. In any given year (except the first) there will be 106% of the previous amount in the account. Use $a_1 = 10{,}000$, $r = 1.06$, and $n = 9$

$$a_9 = 10{,}000(1.06)^{9-1}$$
$$\approx 10{,}000(1.5938481)$$
$$\approx 15{,}938.48$$

At the end of 8 years, there is \$15,938.48 in the account.

97. a. $a_1 = 0.6(220) = 132$, $r = 0.6$

$a_n = a_1 r^{n-1}$

$a_4 = 132(0.6)^{4-1} = 132(0.6)^3 = 28.512$

The height of the fourth bounce is 28.512 feet.

b. $s_\infty = \frac{a_1}{1-r} = \frac{220}{1-0.6} = \frac{220}{0.4} = 550$

She travels a total of 550 feet in the downward direction.

99. a. $a_1 = 30(0.7) = 21$, $r = 0.7$

$a_n = a_1 r^{n-1}$

$a_3 = 21(0.7)^{3-1} = 21(0.7)^2 = 10.29$

The ball will bounce 10.29 inches on the third bounce.

b. $s_\infty = \frac{a_1}{1-r} = \frac{30}{1-0.7} = \frac{30}{0.3} = 100$

The ball travels a total of 100 inches in the downward direction.

101. Blue: $a_1 = 1$, $r = 2$

Red: $a_1 = 1$, $r = 3$

$a_6 = a_1 r^{6-1} = a_1 r^5$

Blue: $a_6 = 1(2)^5 = 32$

Red: $a_6 = 1(3)^5 = 243$

$243 - 32 = 211$

There are 211 more chips in the sixth stack of red chips.

103. Let a_n = value left after the *n*th year. After the first year there is $15000\left(\frac{4}{5}\right) = 12000$ of the value left, so $a_1 = 12000$, $r = \frac{4}{5}$.

$a_2 = 12000\left(\frac{4}{5}\right)^{2-1} = 12000\left(\frac{4}{5}\right) = 9600$

$a_3 = 12000\left(\frac{4}{5}\right)^{3-1} = 12000\left(\frac{16}{25}\right) = 7680$

$a_4 = 12000\left(\frac{4}{5}\right)^{4-1} = 12000\left(\frac{64}{125}\right) = 6144$

a. \$12,000, \$9,600, \$7,680, \$6,144

b. $a_n = 12000\left(\frac{4}{5}\right)^{n-1}$

c. $a_5 = 12000\left(\frac{4}{5}\right)^{5-1} = 12000\left(\frac{256}{625}\right) = 4915.20$

After 5 years, the value of the car is \$4,915.20.

105. Each time the ball bounces it goes up and then comes down the same distance. Therefore, the total vertical distance will be twice the height it rises after each bounce plus the initial 10 feet. The heights after each bounce form an infinite geometric sequence with $r = 0.9$ and $a_1 = 9$.

$s_\infty = \frac{9}{1-0.9} = \frac{9}{0.1} = 90$

Total distance: 6

$2(s_\infty) + 10 = 2(90) + 10 = 190$

The total vertical distance is 190 feet.

107. a. y_2 goes up more steeply.

b.

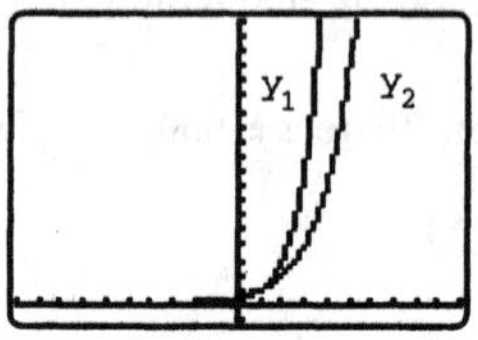

-10, 10, 1, -1, 19, 1

From the graph y_2 goes up more steeply, which agrees with our answer to part (a).

109. This is a geometric sequence with $r = \frac{2}{1} = 2$.

Also, $a_n = 1{,}048{,}576 = 2^{20}$. Using $a_n = a_1 r^{n-1}$ gives

$2^{20} = 1(2)^{n-1}$

$2^{20} = 2^{n-1}$

$20 = n - 1$

$21 = n$

Thus, there are 21 terms in the sequence.

$$s_{21} = \frac{a_1(1-r^{21})}{1-r} = \frac{1(1-2^{21})}{1-2}$$

$$= \frac{1-2{,}097{,}152}{-1} = \frac{-2{,}097{,}151}{-1} = 2{,}097{,}151$$

110. $f(x) = x^2 - 4$, $g(x) = x - 3$

$(f \cdot g)(4) = f(4) \cdot g(4)$

$= (4^2 - 4) \cdot (4 - 3)$

$= (16 - 4)(1)$

$= 12 \cdot 1$

$= 12$

111.

$$\begin{array}{r} 3x^2+4xy-2y^2 \\ \times \qquad 2x-3y \\ \hline -9x^2y-12xy^2+6y^3 \\ 6x^3+8x^2y-4xy^2 \qquad \\ \hline 6x^3-x^2y-16xy^2+6y^3 \end{array}$$

112. $S=\dfrac{2a}{1-r}$, for r

$S(1-r)=2a$

$1-r=\dfrac{2a}{S}$

$-r=\dfrac{2a}{S}-1$

$r=1-\dfrac{2a}{S}$ or $r=\dfrac{S-2a}{S}$

113. $g(x)=x^3+9$

$y=x^3+9$

$x=y^3+9$

$x-9=y^3$

$\sqrt[3]{x-9}=y$

$g^{-1}(x)=\sqrt[3]{x-9}$

114. $\log x+\log(x-1)=\log 20$

$\log[x(x-1)]=\log 20$

$x(x-1)=20$

$x^2-x-20=0$

$(x+4)(x-5)=0$

$x+4=0$ or $x-5=0$

$x=-4$ $\quad x=5$

$x=-4$ is an extraneous solution. The solution is $x=5$.

115. Let x and y be the lengths of the legs. Then

$x^2+y^2=15^2$

$x+y+15=36 \Rightarrow y=21-x$

$x^2+(21-x)^2=225$

$x^2+441-42x+x^2=225$

$2x^2-42x+216=0$

$2(x^2-21x+108)=0$

$2(x-9)(x-12)=0$

$x-9=0$ or $x-12=0$

$x=9$ $\quad x=12$

$y=12$ $\quad y=9$

The solutions are (9, 12) and (12, 9). The legs are 9 meters and 12 meters.

Mid-Chapter Test: 12.1-12.3

1. $a_n=-3n+5$

$a_1=-3(1)+5=-3+5=2$

$a_2=-3(2)+5=-6+5=-1$

$a_3=-3(3)+5=-9+5=-4$

$a_4=-3(4)+5=-12+5=-7$

$a_5=-3(5)+5=-15+5=-10$

The terms are 2, –1, –4, –7, –10.

2. $a_n=n(n+6)$

$a_7=7(7+6)=7(13)=91$

3 $a_n=2^n-1$

$a_1=2^1-1=2-1=1$

$a_2=2^2-1=4-1=3$

$a_3=2^3-1=8-1=7$

$s_1=a_1=1$

$s_3=a_1+a_2+a_3=1+3+7=11$

4. Each term is 4 less than the previous term. The next three terms are $-15,-19,-23$.

5. $\sum_{i=1}^{5}(4i-3)$

$=[4(1)-3]+[4(2)-3]+[4(3)-3]+[4(4)-3]+[4(5)-3]$

$=[4-3]+[8-3]+[12-3]+[16-3]+[20-3]$

$=1+5+9+13+17$

$=45$

6. $a_n=\dfrac{1}{3}n+7$

The fifth partial sum is $\sum_{i=1}^{5}\left(\dfrac{1}{3}i+7\right)$.

7. $a_1=-6$

$a_2=-6+(2-1)(5)=-6+1(5)=-6+5=-1$

$a_3=-6+(3-1)(5)=-6+2(5)=-6+10=4$

$a_4=-6+(4-1)(5)=-6+3(5)=-6+15=9$

$a_5=-6+(5-1)(5)=-6+4(5)=-6+20=14$

The terms are –6, –1, 4, 9, 14. The general term is $a_n=-6+(n-1)5$ or $a_n=5n-11$.

8. $a_n = a_1 + (n-1)d$

$-\frac{1}{2} = \frac{11}{2} + (7-1)d$

$-\frac{1}{2} = \frac{11}{2} + 6d$

$-\frac{12}{2} = 6d$

$-6 = 6d$

$-1 = d$

9. $a_n = a_1 + (n-1)d$

$-3 = 22 + (n-1)(-5)$

$-3 = 22 - 5n + 5$

$-3 = 27 - 5n$

$-30 = -5n$

$6 = n$

10. $a_n = a_1 + (n-1)d$

$7 = -8 + (6-1)d$

$7 = -8 + 5d$

$15 = 5d$

$3 = d$

$$s_6 = \frac{6(a_1 + a_6)}{2} = \frac{6(-8+7)}{2} = \frac{6(-1)}{2} = \frac{-6}{2} = -3$$

11. $a_n = a_1 + (n-1)d$

$$a_{10} = \frac{5}{2} + (10-1)\left(\frac{1}{2}\right) = \frac{5}{2} + 9\left(\frac{1}{2}\right) = \frac{5}{2} + \frac{9}{2} = \frac{14}{2} = 7$$

$$s_{10} = \frac{10(a_1 + a_{10})}{2}$$

$$= \frac{10(\frac{5}{2}+7)}{2} = \frac{25+70}{2} = \frac{95}{2} = 47\frac{1}{2}$$

12. $d = 0 - (-7) = 0 + 7 = 7$

$a_n = a_1 + (n-1)d$

$63 = -7 + (n-1)(7)$

$63 = -7 + 7n - 7$

$63 = 7n - 14$

$77 = 7n$

$11 = n$

There are 11 terms in the sequence.

13. $16+15+14+\cdots+1$ or $1+2+3+\cdots+16$

$$s_n = \frac{n(a_1 + a_n)}{2}$$

$$s_{16} = \frac{16(1+16)}{2} = \frac{16(17)}{2} = \frac{272}{2} = 136$$

There are 136 logs in the pile.

14. $a_1 = 80$

$$a_2 = 80\left(-\frac{1}{2}\right)^{2-1} = 80\left(-\frac{1}{2}\right) = -40$$

$$a_3 = 80\left(-\frac{1}{2}\right)^{3-1} = 80\left(-\frac{1}{2}\right)^2 = 80\left(\frac{1}{4}\right) = 20$$

$$a_4 = 80\left(-\frac{1}{2}\right)^{4-1} = 80\left(-\frac{1}{2}\right)^3 = 80\left(-\frac{1}{8}\right) = -10$$

$$a_5 = 80\left(-\frac{1}{2}\right)^{5-1} = 80\left(-\frac{1}{2}\right)^4 = 80\left(\frac{1}{16}\right) = 5$$

The terms are 80, –40, 20, –10, 5.

15. $a_7 = a_1 r^{7-1}$

$$a_7 = 81\left(\frac{1}{3}\right)^{7-1} = 81\left(\frac{1}{3}\right)^6 = 81\left(\frac{1}{729}\right) = \frac{1}{9}$$

16. $s_6 = \frac{a_1(1-r^6)}{1-r}$

$$s_6 = \frac{5(1-2^6)}{1-2} = \frac{5(1-64)}{-1} = \frac{5(-63)}{-1} = 315$$

17. $r = -\frac{16}{3} \div 8 = -\frac{16}{3} \cdot \frac{1}{8} = -\frac{2}{3}$

18. $r = 4 \div 12 = \frac{4}{12} = \frac{1}{3}$

$$s_\infty = \frac{a_1}{1-r} = \frac{12}{1-\frac{1}{3}} = \frac{12}{\frac{2}{3}} = 12\left(\frac{3}{2}\right) = 18$$

19. $0.878787\ldots = 0.87 + 0.0087 + 0.000087 + \cdots$

$= 0.87 + 0.87(0.01) + 0.87(0.01)^2 + \cdots$

$r = 0.01$ and $a_1 = 0.87$

$$s_\infty = \frac{0.87}{1-0.01} = \frac{0.87}{0.99} = \frac{87}{99} = \frac{29}{33}$$

20. a. A sequence is a list of numbers arranged in a specific order.

b. An arithmetic sequence is a sequence where each term differs by a constant amount.

c. A geometric sequence is a sequence in which each term after the first is the same multiple of the preceding term.

d. A series is the sum of the terms of a sequence.

Exercise Set 12.4

1. No. When we test a case, we are only able to prove that one particular case. If there are an infinite number of cases to be tested, we will never be able to prove that all are true by testing a finite number of cases.

3. S_n is proven to be true for all *n*, if (1) it is proven that S_1 is true and (2) if S_k is true, then S_{k+1} is true.

---

5. Prove $1+5+9+\cdots+(4n-3)=n(2n-1)$ for all positive integers *n*.

Condition 1:
For $n=1$,

$$1\stackrel{?}{=}1(2\cdot 1-1)$$
$$1\stackrel{?}{=}1(2-1)$$
$$1\stackrel{?}{=}1(1)$$
$$1=1 \leftarrow \text{True}$$

Condition 2:
Assume $1+5+9+\cdots+(4k-3)=k(2k-1)$ for some positive integer *k*. Then

$$1+5+9+\cdots+(4k-3)+[4(k+1)-3]=k(2k+1)+[4(k+1)-3]$$
$$1+5+9+\cdots+(4k-3)+[4(k+1)-3]=2k^2-k+4k+4-3$$
$$1+5+9+\cdots+(4k-3)+[4(k+1)-3]=2k^2+3k+1$$
$$1+5+9+\cdots+(4k-3)+[4(k+1)-3]=(k+1)(2k+1)$$
$$1+5+9+\cdots+(4k-3)+[4(k+1)-3]=(k+1)(2k+2-1)$$
$$1+5+9+\cdots+(4k-3)+[4(k+1)-3]=(k+1)[2(k+1)-1] \quad \leftarrow S_{k+1}$$

Since Conditions 1 and 2 hold, the above formula is true for all positive integers (by mathematical induction).

7. Prove $2+4+6+\cdots+2n=n(n+1)$ for all positive integers *n*.

Condition 1:
For $n=1$,

$$2\stackrel{?}{=}1(1+1)$$
$$2\stackrel{?}{=}1(2)$$
$$2=2 \leftarrow \text{True}$$

Condition 2:
Assume $2+4+6+\cdots+2k=k(k+1)$ for some positive integer *k*. Then

$$2+4+6+\cdots+2k+2(k+1)=k(k+1)+2(k+1)$$
$$2+4+6+\cdots+2k+2(k+1)=k^2+k+2k+2$$
$$2+4+6+\cdots+2k+2(k+1)=k^2+3k+2$$
$$2+4+6+\cdots+2k+2(k+1)=(k+1)(k+2)$$
$$2+4+6+\cdots+2k+2(k+1)=(k+1)[(k+1)+1] \quad \leftarrow S_{k+1}$$

Since Conditions 1 and 2 hold, the above formula is true for all positive integers (by mathematical induction).

9. Prove $3+4+5+\cdots+(n+2)=\frac{n}{2}(n+5)$ for all positive integers *n*.

Condition 1:
For $n=1$,

$$3\stackrel{?}{=}\frac{1}{2}(1+5)$$
$$3\stackrel{?}{=}\frac{1}{2}(6)$$
$$3=3 \leftarrow \text{True}$$

Condition 2:
Assume $3+4+5+\cdots+(k+2)=\frac{k}{2}(k+5)$ for some positive integer *k*. Then

$$3+4+5+\cdots+(k+2)+[(k+1)+2]=\frac{k}{2}(k+5)+[(k+1)+2]$$
$$3+4+5+\cdots+(k+2)+[(k+1)+2]=\frac{k^2}{2}+\frac{5k}{2}+k+3$$
$$3+4+5+\cdots+(k+2)+[(k+1)+2]=\frac{k^2+7k+6}{2}$$
$$3+4+5+\cdots+(k+2)+[(k+1)+2]=\frac{(k+1)(k+6)}{2}$$
$$3+4+5+\cdots+(k+2)+[(k+1)+2]=\frac{k+1}{2}[(k+1)+5] \quad \leftarrow S_{k+1}$$

Since Conditions 1 and 2 hold, the above formula is true for all positive integers (by mathematical induction).

11. Prove $3+6+9+\cdots+3n=\dfrac{3n(n+1)}{2}$ for all positive integers *n*.

Condition 1:

For $n=1$,

$$3\overset{?}{=}\frac{3\cdot1(1+1)}{2}$$
$$3\overset{?}{=}\frac{3(2)}{2}$$
$$3=3 \leftarrow \text{True}$$

Condition 2:

Assume $3+6+9+\cdots+3k=\dfrac{3k(k+1)}{2}$ for some positive integer *k*. Then

$$3+6+9+\cdots+3k+3(k+1)=\frac{3k(k+1)}{2}+3(k+1)$$
$$3+6+9+\cdots+3k+3(k+1)=\frac{3k^2}{2}+\frac{3k}{2}+3k+3$$
$$3+6+9+\cdots+3k+3(k+1)=\frac{3k^2+9k+6}{2}$$
$$3+6+9+\cdots+3k+3(k+1)=\frac{3(k^2+3k+2)}{2}$$
$$3+6+9+\cdots+3k+3(k+1)=\frac{3(k+1)(k+2)}{2}$$
$$3+6+9+\cdots+3k+3(k+1)=\frac{3(k+1)[(k+1)+1]}{2} \quad \leftarrow S_{k+1}$$

Since Conditions 1 and 2 hold, the above formula is true for all positive integers (by mathematical induction).

13. Prove $5+10+15+\cdots+5n=\dfrac{5n(n+1)}{2}$ for all positive integers *n*.

Condition 1:

For $n=1$,

$$5\overset{?}{=}\frac{5\cdot1(1+1)}{2}$$
$$5\overset{?}{=}\frac{5(2)}{2}$$
$$5=5 \leftarrow \text{True}$$

Condition 2:

Assume $5+10+15+\cdots+5k=\dfrac{5k(k+1)}{2}$ for some positive integer *k*. Then

$$5+10+15+\cdots+5k+5(k+1)=\frac{5k(k+1)}{2}+5(k+1)$$
$$5+10+15+\cdots+5k+5(k+1)=\frac{5k^2}{2}+\frac{5k}{2}+5k+5$$
$$5+10+15+\cdots+5k+5(k+1)=\frac{5k^2+15k+10}{2}$$
$$5+10+15+\cdots+5k+5(k+1)=\frac{5(k^2+3k+2)}{2}$$
$$5+10+15+\cdots+5k+5(k+1)=\frac{5(k+1)(k+2)}{2}$$
$$5+10+15+\cdots+5k+5(k+1)=\frac{5(k+1)[(k+1)+1]}{2} \quad \leftarrow S_{k+1}$$

Since Conditions 1 and 2 hold, the above formula is true for all positive integers (by mathematical induction).

15. Prove $1+2+2^2+\cdots+2^{n-1}=2^n-1$ for all positive integers *n*.

Condition 1:

For $n=1$,

$$1\overset{?}{=}2^1-1$$
$$1\overset{?}{=}2-1$$
$$1=1 \leftarrow \text{True}$$

Condition 2:

Assume $1+2+2^2+\cdots+2^{k-1}=2^k-1$ for some positive integer *k*. Then

$$1+2+2^2+\cdots+2^{k-1}+2^{(k+1)-1}=2^k-1+2^{(k+1)-1}$$
$$1+2+2^2+\cdots+2^{k-1}+2^{(k+1)-1}=2^k-1+2^k$$
$$1+2+2^2+\cdots+2^{k-1}+2^{(k+1)-1}=2^k+2^k-1$$
$$1+2+2^2+\cdots+2^{k-1}+2^{(k+1)-1}=2(2^k)-1$$
$$1+2+2^2+\cdots+2^{k-1}+2^{(k+1)-1}=2^{k+1}-1 \quad \leftarrow S_{k+1}$$

Since Conditions 1 and 2 hold, the above formula is true for all positive integers (by mathematical induction).

17. Prove $1+4+4^2+\cdots+4^{n-1}=\dfrac{4^n-1}{3}$ for all positive integers *n*.

Condition 1:
For $n=1$,

$$1\stackrel{?}{=}\frac{4^1-1}{3}$$
$$1\stackrel{?}{=}\frac{4-1}{3}$$
$$1\stackrel{?}{=}\frac{3}{3}$$
$$1=1 \leftarrow \text{True}$$

Condition 2:
Assume $1+4+4^2+\cdots+4^{k-1}=\dfrac{4^k-1}{3}$ for some positive integer *k*. Then

$$1+4+4^2+\cdots+4^{k-1}+4^{(k+1)-1}=\frac{4^k-1}{3}+4^{(k+1)-1}$$
$$1+4+4^2+\cdots+4^{k-1}+4^{(k+1)-1}=\frac{4^k-1}{3}+\frac{3\cdot 4^k}{3}$$
$$1+4+4^2+\cdots+4^{k-1}+4^{(k+1)-1}=\frac{4^k+3\cdot 4^k-1}{3}$$
$$1+4+4^2+\cdots+4^{k-1}+4^{(k+1)-1}=\frac{4^k(1+3)-1}{3}$$
$$1+4+4^2+\cdots+4^{k-1}+4^{(k+1)-1}=\frac{4^k(4)-1}{3}$$
$$1+4+4^2+\cdots+4^{k-1}+4^{(k+1)-1}=\frac{4^{k+1}-1}{3} \leftarrow S_{k+1}$$

Since Conditions 1 and 2 hold, the above formula is true for all positive integers (by mathematical induction).

19. Prove $2+2^2+2^3+\cdots+2^n=2^{n+1}-2$ for all positive integers *n*.

Condition 1:
For $n=1$,

$$2\stackrel{?}{=}2^{1+1}-2$$
$$2\stackrel{?}{=}2^2-2$$
$$2\stackrel{?}{=}4-2$$
$$2=2 \leftarrow \text{True}$$

Condition 2:
Assume $2+2^2+2^3+\cdots+2^k=2^{k+1}-2$ for some positive integer *k*. Then

$$2+2^2+2^3+\cdots+2^k+2^{k+1}=2^{k+1}-2+2^{k+1} \updownarrow$$
$$2+2^2+2^3+\cdots+2^k+2^{k+1}=2^{k+1}+2^{k+1}-2$$
$$2+2^2+2^3+\cdots+2^k+2^{k+1}=2\cdot 2^{k+1}-2$$
$$2+2^2+2^3+\cdots+2^k+2^{k+1}=2^{(k+1)+1}-2 \leftarrow S_{k+1}$$

Since Conditions 1 and 2 hold, the above formula is true for all positive integers (by mathematical induction).

21. Prove $4+4^2+4^3+\cdots+4^n=4\left(\dfrac{4^n-1}{3}\right)$ for all positive integers *n*.

Condition 1:
For $n=1$,

$$4\stackrel{?}{=}4\left(\frac{4^1-1}{3}\right)$$
$$4\stackrel{?}{=}4\left(\frac{4-1}{3}\right)$$
$$4\stackrel{?}{=}4\left(\frac{3}{3}\right)$$
$$4=4 \leftarrow \text{True}$$

Condition 2:
Assume $4+4^2+4^3+\cdots+4^k=4\left(\dfrac{4^k-1}{3}\right)$ for some positive integer *k*. Then

$$4+4^2+4^3+\cdots+4^k+4^{k+1}=4\left(\frac{4^k-1}{3}\right)+4^{k+1}$$
$$4+4^2+4^3+\cdots+4^k+4^{k+1}=\frac{4(4^k-1)}{3}+\frac{3\cdot 4\cdot 4^k}{3}$$
$$4+4^2+4^3+\cdots+4^k+4^{k+1}=\frac{4\left[(4^k-1)+3\cdot 4^k\right]}{3}$$
$$4+4^2+4^3+\cdots+4^k+4^{k+1}=\frac{4(4^k+3\cdot 4^k-1)}{3}$$
$$4+4^2+4^3+\cdots+4^k+4^{k+1}=\frac{4\left[4^k(1+3)-1\right]}{3}$$
$$4+4^2+4^3+\cdots+4^k+4^{k+1}=\frac{4(4^k\cdot 4-1)}{3}$$
$$4+4^2+4^3+\cdots+4^k+4^{k+1}=4\left(\frac{4^{k+1}-1}{3}\right) \leftarrow S_{k+1}$$

Since Conditions 1 and 2 hold, the above formula is true for all positive integers (by mathematical induction).

23. Prove $1^3+2^3+3^3+\cdots+n^3=\left[\dfrac{n(n+1)}{2}\right]^2$ for all positive integers n.

Condition 1:
For $n=1$,

$$1^3 \stackrel{?}{=} \left[\frac{1(1+1)}{2}\right]^2$$

$$1 \stackrel{?}{=} \left[\frac{1(2)}{2}\right]^2$$

$$1 \stackrel{?}{=} (1)^2$$

$$1=1 \leftarrow \text{True}$$

Condition 2:
Assume $1^3+2^3+3^3+\cdots+k^3=\left[\dfrac{k(k+1)}{2}\right]^2$ for some positive integer k. Then

$$1^3+2^3+3^3+\cdots+k^3+(k+1)^3=\left[\frac{k(k+1)}{2}\right]^2+(k+1)^3$$

$$1^3+2^3+3^3+\cdots+k^3+(k+1)^3=\frac{k^2(k+1)^2}{4}+\frac{4(k+1)^2(k+1)}{4}$$

$$1^3+2^3+3^3+\cdots+k^3+(k+1)^3=\frac{(k+1)^2\left[k^2+4(k+1)\right]}{4}$$

$$1^3+2^3+3^3+\cdots+k^3+(k+1)^3=\frac{(k+1)^2(k^2+4k+4)}{4}$$

$$1^3+2^3+3^3+\cdots+k^3+(k+1)^3=\frac{(k+1)^2(k+2)^2}{2^2}$$

$$1^3+2^3+3^3+\cdots+k^3+(k+1)^3=\left[\frac{(k+1)[(k+1)+1]}{2}\right]^2 \quad \leftarrow S_{k+1}$$

Since Conditions 1 and 2 hold, the above formula is true for all positive integers (by mathematical induction).

25. Prove $\dfrac{1}{1\cdot 2}+\dfrac{1}{2\cdot 3}+\dfrac{1}{3\cdot 4}+\cdots+\dfrac{1}{n(n+1)}=\dfrac{n}{n+1}$ for all positive integers n.

Condition 1:
For $n=1$,

$$\frac{1}{1\cdot 2} \stackrel{?}{=} \frac{1}{1+1}$$

$$\frac{1}{2}=\frac{1}{2} \leftarrow \text{True}$$

Condition 2:
Assume $\dfrac{1}{1\cdot 2}+\dfrac{1}{2\cdot 3}+\dfrac{1}{3\cdot 4}+\cdots+\dfrac{1}{k(k+1)}=\dfrac{k}{k+1}$ for some positive integer k. Then

$$\frac{1}{1\cdot 2}+\frac{1}{2\cdot 3}+\frac{1}{3\cdot 4}+\cdots+\frac{1}{k(k+1)}+\frac{1}{(k+1)[(k+1)+1]}=\frac{k}{k+1}+\frac{1}{(k+1)[(k+1)+1]}$$

$$\frac{1}{1\cdot 2}+\frac{1}{2\cdot 3}+\frac{1}{3\cdot 4}+\cdots+\frac{1}{k(k+1)}+\frac{1}{(k+1)[(k+1)+1]}=\frac{k[(k+1)+1]+1}{(k+1)[(k+1)+1]}$$

$$\frac{1}{1\cdot 2}+\frac{1}{2\cdot 3}+\frac{1}{3\cdot 4}+\cdots+\frac{1}{k(k+1)}+\frac{1}{(k+1)[(k+1)+1]}=\frac{k^2+2k+1}{(k+1)(k+2)}$$

$$\frac{1}{1\cdot 2}+\frac{1}{2\cdot 3}+\frac{1}{3\cdot 4}+\cdots+\frac{1}{k(k+1)}+\frac{1}{(k+1)[(k+1)+1]}=\frac{(k+1)^2}{(k+1)(k+2)}$$

$$\frac{1}{1\cdot 2}+\frac{1}{2\cdot 3}+\frac{1}{3\cdot 4}+\cdots+\frac{1}{k(k+1)}+\frac{1}{(k+1)[(k+1)+1]}=\frac{k+1}{k+2}$$

$$\frac{1}{1\cdot 2}+\frac{1}{2\cdot 3}+\frac{1}{3\cdot 4}+\cdots+\frac{1}{k(k+1)}+\frac{1}{(k+1)[(k+1)+1]}=\frac{k+1}{(k+1)+1} \quad \leftarrow S_{k+1}$$

Since Conditions 1 and 2 hold, the above formula is true for all positive integers (by mathematical induction).

27. Prove $1+\frac{1}{2}+\frac{1}{2^2}+\cdots+\frac{1}{2^{n-1}}=2\left(1-\frac{1}{2^n}\right)$ for all positive integers n.

Condition 1:

For $n=1$,

$$1\overset{?}{=}2\left(1-\frac{1}{2^1}\right)$$

$$1\overset{?}{=}2\left(1-\frac{1}{2}\right)$$

$$1\overset{?}{=}2\left(\frac{1}{2}\right)$$

$1=1 \quad \leftarrow$ True

Condition 2:

Assume $1+\frac{1}{2}+\frac{1}{2^2}+\cdots+\frac{1}{2^{k-1}}=2\left(1-\frac{1}{2^k}\right)$ for some positive integer k. Then

$$1+\frac{1}{2}+\frac{1}{2^2}+\cdots+\frac{1}{2^{k-1}}+\frac{1}{2^{(k+1)-1}}=2\left(1-\frac{1}{2^k}\right)+\frac{1}{2^{(k+1)-1}}$$

$$1+\frac{1}{2}+\frac{1}{2^2}+\cdots+\frac{1}{2^{k-1}}+\frac{1}{2^{(k+1)-1}}=2-\frac{2}{2^k}+\frac{1}{2^k}$$

$$1+\frac{1}{2}+\frac{1}{2^2}+\cdots+\frac{1}{2^{k-1}}+\frac{1}{2^{(k+1)-1}}=2-\frac{1}{2^k}$$

$$1+\frac{1}{2}+\frac{1}{2^2}+\cdots+\frac{1}{2^{k-1}}+\frac{1}{2^{(k+1)-1}}=2-\frac{2}{2\cdot 2^k}$$

$$1+\frac{1}{2}+\frac{1}{2^2}+\cdots+\frac{1}{2^{k-1}}+\frac{1}{2^{(k+1)-1}}=2-\frac{2}{2^{k+1}}$$

$$1+\frac{1}{2}+\frac{1}{2^2}+\cdots+\frac{1}{2^{k-1}}+\frac{1}{2^{(k+1)-1}}=2\left(1-\frac{1}{2^{k+1}}\right) \quad \leftarrow S_{k+1}$$

Since Conditions 1 and 2 hold, the above formula is true for all positive integers (by mathematical induction).

29. Prove $1+\frac{1}{4}+\frac{1}{4^2}+\cdots+\frac{1}{4^{n-1}}=\frac{4}{3}\left(1-\frac{1}{4^n}\right)$ for all positive integers n.

Condition 1:

For $n=1$,

$$1\overset{?}{=}\frac{4}{3}\left(1-\frac{1}{4^1}\right)$$

$$1\overset{?}{=}\frac{4}{3}\left(1-\frac{1}{4}\right)$$

$$1\overset{?}{=}\frac{4}{3}\left(\frac{3}{4}\right)$$

$1=1 \quad \leftarrow$ True

Condition 2:

Assume $1+\frac{1}{4}+\frac{1}{4^2}+\cdots+\frac{1}{4^{k-1}}=\frac{4}{3}\left(1-\frac{1}{4^k}\right)$ for some positive integer k. Then

$$1+\frac{1}{4}+\frac{1}{4^2}+\cdots+\frac{1}{4^{k-1}}+\frac{1}{4^{(k+1)-1}}=\frac{4}{3}\left(1-\frac{1}{4^k}\right)+\frac{1}{4^{(k+1)-1}}$$

$$1+\frac{1}{4}+\frac{1}{4^2}+\cdots+\frac{1}{4^{k-1}}+\frac{1}{4^{(k+1)-1}}=\frac{4}{3}-\frac{4}{3\cdot 4^k}+\frac{1}{4^k}$$

$$1+\frac{1}{4}+\frac{1}{4^2}+\cdots+\frac{1}{4^{k-1}}+\frac{1}{4^{(k+1)-1}}=\frac{4}{3}-\frac{4}{3\cdot 4^k}+\frac{3}{3\cdot 4^k}$$

$$1+\frac{1}{4}+\frac{1}{4^2}+\cdots+\frac{1}{4^{k-1}}+\frac{1}{4^{(k+1)-1}}=\frac{4}{3}-\frac{1}{3\cdot 4^k}$$

$$1+\frac{1}{4}+\frac{1}{4^2}+\cdots+\frac{1}{4^{k-1}}+\frac{1}{4^{(k+1)-1}}=\frac{4}{3}-\frac{4}{3\cdot 4\cdot 4^k}$$

$$1+\frac{1}{4}+\frac{1}{4^2}+\cdots+\frac{1}{4^{k-1}}+\frac{1}{4^{(k+1)-1}}=\frac{4}{3}-\frac{4}{3\cdot 4^{k+1}}$$

$$1+\frac{1}{4}+\frac{1}{4^2}+\cdots+\frac{1}{4^{k-1}}+\frac{1}{4^{(k+1)-1}}=\frac{4}{3}\left(1-\frac{1}{4^{k+1}}\right) \quad \leftarrow S_{k+1}$$

Since Conditions 1 and 2 hold, the above formula is true for all positive integers (by mathematical induction).

31. Prove $1+\frac{3}{4}+\left(\frac{3}{4}\right)^2+\cdots+\left(\frac{3}{4}\right)^{n-1}=4\left[1-\left(\frac{3}{4}\right)^n\right]$ for all positive integers n.

Condition 1:
For $n=1$,

$1\stackrel{?}{=}4\left[1-\left(\frac{3}{4}\right)^1\right]$

$1\stackrel{?}{=}4\left(1-\frac{3}{4}\right)$

$1\stackrel{?}{=}4\left(\frac{1}{4}\right)$

$1=1 \quad \leftarrow \text{True}$

Condition 2:
Assume $1+\frac{3}{4}+\left(\frac{3}{4}\right)^2+\cdots+\left(\frac{3}{4}\right)^{k-1}=4\left[1-\left(\frac{3}{4}\right)^k\right]$ for some positive integer k. Then

$1+\frac{3}{4}+\left(\frac{3}{4}\right)^2+\cdots+\left(\frac{3}{4}\right)^{k-1}+\left(\frac{3}{4}\right)^{(k+1)-1}=4\left[1-\left(\frac{3}{4}\right)^k\right]+\left(\frac{3}{4}\right)^{(k+1)-1}$

$1+\frac{3}{4}+\left(\frac{3}{4}\right)^2+\cdots+\left(\frac{3}{4}\right)^{k-1}+\left(\frac{3}{4}\right)^{(k+1)-1}=4-4\left(\frac{3}{4}\right)^k+\left(\frac{3}{4}\right)^k$

$1+\frac{3}{4}+\left(\frac{3}{4}\right)^2+\cdots+\left(\frac{3}{4}\right)^{k-1}+\left(\frac{3}{4}\right)^{(k+1)-1}=4-3\left(\frac{3}{4}\right)^k$

$1+\frac{3}{4}+\left(\frac{3}{4}\right)^2+\cdots+\left(\frac{3}{4}\right)^{k-1}+\left(\frac{3}{4}\right)^{(k+1)-1}=4-4\cdot\frac{3}{4}\left(\frac{3}{4}\right)^k$

$1+\frac{3}{4}+\left(\frac{3}{4}\right)^2+\cdots+\left(\frac{3}{4}\right)^{k-1}+\left(\frac{3}{4}\right)^{(k+1)-1}=4-4\left(\frac{3}{4}\right)^{k+1}$

$1+\frac{3}{4}+\left(\frac{3}{4}\right)^2+\cdots+\left(\frac{3}{4}\right)^{k-1}+\left(\frac{3}{4}\right)^{(k+1)-1}=4\left[1-\left(\frac{3}{4}\right)^{k+1}\right] \quad \leftarrow S_{k+1}$

Since Conditions 1 and 2 hold, the above formula is true for all positive integers (by mathematical induction).

33. Prove $\left(\frac{3}{4}\right)^{n+1}<\frac{3}{4}$ for all positive integers n.

Condition 1:
For $n=1$,

$\left(\frac{3}{4}\right)^{1+1}\stackrel{?}{<}\frac{3}{4}$

$\left(\frac{3}{4}\right)^{2}\stackrel{?}{<}\frac{3}{4}$

$\frac{9}{16}<\frac{3}{4} \quad \leftarrow \text{True}$

Condition 2:
Assume $\left(\frac{3}{4}\right)^{k+1}<\frac{3}{4}$ for some positive integer k. Then

$\frac{3}{4}\left(\frac{3}{4}\right)^{k+1}<\frac{3}{4}\cdot\frac{3}{4}$

$\left(\frac{3}{4}\right)^{(k+1)+1}<\frac{9}{16}<\frac{3}{4}$

Since Conditions 1 and 2 hold, the above inequality is true for all positive integers (by mathematical induction).

35. Prove $\left(\frac{1}{4}\right)^{n+1}<\frac{1}{4}$ for all positive integers n.

Condition 1:
For $n=1$,

$\left(\frac{1}{4}\right)^{1+1}\stackrel{?}{<}\frac{1}{4}$

$\left(\frac{1}{4}\right)^{2}\stackrel{?}{<}\frac{1}{4}$

$\frac{1}{16}<\frac{1}{4} \quad \leftarrow \text{True}$

Condition 2:
Assume $\left(\frac{1}{4}\right)^{k+1}<\frac{1}{4}$ for some positive integer k. Then

$\frac{1}{4}\left(\frac{1}{4}\right)^{k+1}<\frac{1}{4}\cdot\frac{1}{4}$

$\left(\frac{1}{4}\right)^{(k+1)+1}<\frac{1}{16}<\frac{1}{4}$

Since Conditions 1 and 2 hold, the above inequality is true for all positive integers (by mathematical induction).

37. Prove $3^n \geq 2n+1$ for all positive integers *n*.

Condition 1:

For $n=1$,

$3^1 \overset{?}{\geq} 2(1)+1$

$3 \overset{?}{\geq} 2+1$

$3 \geq 3 \quad \leftarrow$ True

Condition 2:

Assume $3^k \geq 2k+1$ for some positive integer *k*. Then

$3 \cdot 3^k \geq 3(2k+1)$

$3^{k+1} > 6k+3$

$3^{k+1} > 4k+2k+2+1$

$3^{k+1} > 4k+2(k+1)+1 \geq 2(k+1)+1$

Since Conditions 1 and 2 hold, the above inequality is true for all positive integers (by mathematical induction).

39. Prove n^2+n is even for all positive integers *n*.

If n^2+n is even, then it will be divisible by 2 and $(n^2+n) \div 2$ must be an integer.

Condition 1:

For $n=1$,

$(1^2+1) \div 2$

$= (1+1) \div 2$

$= 2 \div 2$

$= 1 \quad \leftarrow$ an integer

Condition 2:

Assume $(k^2+k) \div 2 = m$ for some positive integer *m*. Then

$$\begin{aligned}\left[(k+1)^2+(k+1)\right] \div 2 &= (k^2+2k+1+k+1) \div 2 \\ &= (k^2+3k+2) \div 2 \\ &= (k^2+k+2k+2) \div 2 \\ &= (k^2+k) \div 2 + (2k+2) \div 2 \\ &= m+k+1 \quad \leftarrow \text{an integer}\end{aligned}$$

Thus, $\left[(k+1)^2+(k+1)\right] \div 2$ must be an integer.

Since Conditions 1 and 2 hold, n^2+n must be an even for all positive integers *n* (by mathematical induction).

41. Prove n^3+2n is divisible by 3 for all positive integers *n*.

If n^3+2n is divisible by 3, then $(n^3+2n) \div 3$ must be an integer.

Condition 1:

For $n=1$,

$(1^3+2 \cdot 1) \div 3$

$= (1+2) \div 3$

$= 3 \div 3$

$= 1 \quad \leftarrow$ an integer

Condition 2:

Assume $(k^3+2k) \div 3 = m$ for some positive integer *m*. Then

$$\begin{aligned}\left[(k+1)^3+2(k+1)\right] \div 3 &= \left[k^3+k^2+2k^2+2k+k+1+2k+2\right] \div 3 \\ &= (k^3+3k^2+5k+3) \div 3 \\ &= (k^3+2k+3k^2+3k+3) \div 3 \\ &= (k^3+2k) \div 3 + (3k^2+3k+3) \div 3 \\ &= m+k^2+k+1 \quad \leftarrow \text{an integer}\end{aligned}$$

Thus, $\left[(k+1)^3+2(k+1)\right] \div 3$ must be an integer

Since Conditions 1 and 2 hold, n^3+2n must be divisible by 3 for all positive integers *n*.

43. Prove $a+ar+ar^2+\cdots+ar^{n-1}=\dfrac{a(1-r^n)}{1-r}$ for all positive integers n and $r\neq 1$.

Condition 1:

For $n=1$,

$$a\overset{?}{=}\frac{a(1-r^1)}{1-r}$$

$$a\overset{?}{=}\frac{a(1-r)}{1-r}$$

$$a=a \quad \leftarrow \text{True}$$

Condition 2:

Assume $a+ar+ar^2+\cdots+ar^{k-1}=\dfrac{a(1-r^k)}{1-r}$ for some positive integer k. Then

$$a+ar+ar^2+\cdots+ar^{k-1}+ar^{(k+1)-1}=\frac{a(1-r^k)}{1-r}+ar^{(k+1)-1}$$

$$a+ar+ar^2+\cdots+ar^{k-1}+ar^{(k+1)-1}=\frac{a-ar^k}{1-r}+\frac{ar^k(1-r)}{1-r}$$

$$a+ar+ar^2+\cdots+ar^{k-1}+ar^{(k+1)-1}=\frac{a-ar^k+ar^k-ar^{k+1}}{1-r}$$

$$a+ar+ar^2+\cdots+ar^{k-1}+ar^{(k+1)-1}=\frac{a-ar^{k+1}}{1-r}$$

$$a+ar+ar^2+\cdots+ar^{k-1}+ar^{(k+1)-1}=\frac{a(1-r^{k+1})}{1-r} \quad \leftarrow S_{k+1}$$

Since Conditions 1 and 2 hold, the above formula is true for all positive integers (by mathematical induction).

---

45. Use $1+2+3+\cdots+n=\dfrac{n(n+1)}{2}$ with $n=8$.

$$1+2+3+\cdots+8=\frac{8(8+1)}{2}=\frac{8(9)}{2}=36$$

47. Use $1+2+3+\cdots+n=\dfrac{n(n+1)}{2}$ with $n=22$.

$$1+2+3+\cdots+22=\frac{22(22+1)}{2}$$
$$=\frac{22(23)}{2}$$
$$=253$$

49. Use $1+2+3+\cdots+n=\dfrac{n(n+1)}{2}$ with $n=32$.

$$1+2+3+\cdots+32=\frac{32(32+1)}{2}$$
$$=\frac{32(33)}{2}$$
$$=528$$

51. Use $1+2+3+\cdots+n=\dfrac{n(n+1)}{2}$ with $n=45$.

$$1+2+3+\cdots+45=\frac{45(45+1)}{2}$$
$$=\frac{45(46)}{2}$$
$$=1035$$

53. Use $1+2+3+\cdots+n=\dfrac{n(n+1)}{2}$ with $n=70$.

$$1+2+3+\cdots+70=\frac{70(70+1)}{2}$$
$$=\frac{70(71)}{2}$$
$$=2485$$

57. $(2x+3y)^2=(2x)^2+2(2x)(3y)+(3y)^2$
$$=4x^2+12xy+9y^2$$

58. $(5x^2-2y)^2=(5x^2)^2-2(5x^2)(2y)+(2y)^2$
$$=25x^4-20x^2y+4y^2$$

59. $[4x+(y-3)]^2=(4x)^2+2(4x)(y-3)+(y-3)^2$
$$=16x^2+8x(y-3)+y^2-2(y)(3)+3^2$$
$$=16x^2+8xy-24x+y^2-6y+9$$

60. $(2x+3y)^3$
$$=(2x+3y)^2(2x+3y)$$
$$=\left[(2x)^2+2(2x)(3y)+(3y)^2\right](2x+3y)$$
$$=(4x^2+12xy+9y^2)(2x+3y)$$
$$=4x^2(2x)+4x^2(3y)+12xy(2x)+12xy(3y)+9y^2(2x)+9y^2(3y)$$
$$=8x^3+36x^2y+54xy^2+27x^3$$

Exercise Set 12.5

1. Answers will vary. One possibility follows. The first and last numbers in each row are 1 and the inner numbers are obtained by adding the two numbers in the row above (to the right and left).

$$\begin{array}{ccccccccc} & & & & 1 & & & & \\ & & & 1 & & 1 & & & \\ & & 1 & & 2 & & 1 & & \\ & 1 & & 3 & & 3 & & 1 & \\ 1 & & 4 & & 6 & & 4 & & 1 \end{array}$$

3. $1! = 1$

5. No. Factorials are only defined for nonnegative integers.

7. The expansion of $(a+b)^{13}$ has 14 terms, one more than the power to which the binomial is raised.

9.
$$\begin{aligned}\binom{5}{2} &= \frac{5!}{2!\cdot(5-2)!}\\ &= \frac{5!}{2!\,3!}\\ &= \frac{5\cdot 4\cdot \cancel{3\cdot 2\cdot 1}}{(2\cdot 1)(\cancel{3\cdot 2\cdot 1})}\\ &= \frac{20}{2}\\ &= 10\end{aligned}$$

11. $\binom{5}{5} = \frac{5!}{5!(5-5)!} = \frac{\cancel{5!}}{\cancel{5!}\cdot 0!} = \frac{1}{0!} = \frac{1}{1} = 1$

13. $\binom{7}{0} = \frac{7!}{0!(7-0)!} = \frac{7!}{0!\cdot 7!} = \frac{\cancel{7!}}{0!\cdot\cancel{7!}} = \frac{1}{0!} = \frac{1}{1} = 1$

15.
$$\begin{aligned}\binom{8}{4} &= \frac{8!}{4!\cdot(8-4)!}\\ &= \frac{8!}{4!\,4!}\\ &= \frac{8\cdot 7\cdot 6\cdot 5\cdot \cancel{4\cdot 3\cdot 2\cdot 1}}{(4\cdot 3\cdot 2\cdot 1)(\cancel{4\cdot 3\cdot 2\cdot 1})}\\ &= \frac{1680}{24}\\ &= 70\end{aligned}$$

17.
$$\begin{aligned}\binom{8}{2} &= \frac{8!}{2!\cdot(8-2)!}\\ &= \frac{8!}{2!\,6!}\\ &= \frac{8\cdot 7\cdot \cancel{6\cdot 5\cdot 4\cdot 3\cdot 2\cdot 1}}{(2\cdot 1)\cdot(\cancel{6\cdot 5\cdot 4\cdot 3\cdot 2\cdot 1})}\\ &= \frac{56}{2}\\ &= 28\end{aligned}$$

19.
$$\begin{aligned}(x+4)^3 &= \binom{3}{0}x^3 4^0 + \binom{3}{1}x^2 4^1 + \binom{3}{2}x^1 4^2 + \binom{3}{3}x^0 4^3\\ &= 1x^3(1) + 3x^2(4) + 3x(16) + 1(1)64\\ &= x^3 + 12x^2 + 48x + 64\end{aligned}$$

21.
$$\begin{aligned}(2x-3)^3 &= \binom{3}{0}(2x)^3(-3)^0 + \binom{3}{1}(2x)^2(-3)^1 + \binom{3}{2}(2x)^1(-3)^2 + \binom{3}{3}(2x)^0(-3)^3\\ &= 1(8x^3)(1) + 3(4x^2)(-3) + 3(2x)(9) + 1(1)(-27)\\ &= 8x^3 - 36x^2 + 54x - 27\end{aligned}$$

23.
$$\begin{aligned}(a-b)^4 &= \binom{4}{0}a^4(-b)^0 + \binom{4}{1}a^3(-b)^1 + \binom{4}{2}a^2(-b)^2 + \binom{4}{3}a^1(-b)^3 + \binom{4}{4}a^0(-b)^4\\ &= 1a^4(1) + 4a^3(-b) + 6a^2b^2 + 4a(-b^3) + 1(1)b^4\\ &= a^4 - 4a^3b + 6a^2b^2 - 4ab^3 + b^4\end{aligned}$$

25. $(3a-b)^5 = \binom{5}{0}(3a)^5(-b)^0 + \binom{5}{1}(3a)^4(-b)^1 + \binom{5}{2}(3a)^3(-b)^2 + \binom{5}{3}(3a)^2(-b)^3 + \binom{5}{4}(3a)^1(-b)^4 + \binom{5}{5}(3a)^0(-b)^5$

$= 1(243a^5)(1) + 5(81a^4)(-b) + 10(27a^3)b^2 + 10(9a^2)(-b^3) + 5(3a)b^4 + 1(1)(-b^5)$

$= 243a^5 - 405a^4b + 270a^3b^2 - 90a^2b^3 + 15ab^4 - b^5$

27. $\left(2x+\frac{1}{2}\right)^4 = \binom{4}{0}(2x)^4\left(\frac{1}{2}\right)^0 + \binom{4}{1}(2x)^3\left(\frac{1}{2}\right)^1 + \binom{4}{2}(2x)^2\left(\frac{1}{2}\right)^2 + \binom{4}{3}(2x)^1\left(\frac{1}{2}\right)^3 + \binom{4}{4}(2x)^0\left(\frac{1}{2}\right)^4$

$= 1(16x^4)(1) + 4(8x^3)\left(\frac{1}{2}\right) + 6(4x^2)\left(\frac{1}{4}\right) + 4(2x)\left(\frac{1}{8}\right) + 1(1)\left(\frac{1}{16}\right)$

$= 16x^4 + 16x^3 + 6x^2 + x + \frac{1}{16}$

29. $\left(\frac{x}{2}-3\right)^4 = \binom{4}{0}\left(\frac{x}{2}\right)^4(-3)^0 + \binom{4}{1}\left(\frac{x}{2}\right)^3(-3)^1 + \binom{4}{2}\left(\frac{x}{2}\right)^2(-3)^2 + \binom{4}{3}\left(\frac{x}{2}\right)^1(-3)^3 + \binom{4}{4}\left(\frac{x}{2}\right)^0(-3)^4$

$= 1\left(\frac{x^4}{16}\right)(1) + 4\left(\frac{x^3}{8}\right)(-3) + 6\left(\frac{x^2}{4}\right)(9) + 4\left(\frac{x}{2}\right)(-27) + 1(1)(81)$

$= \frac{x^4}{16} - \frac{3x^3}{2} + \frac{27x^2}{2} - 54x + 81$

31. $(x+10)^{10} = \binom{10}{0}x^{10}(10)^0 + \binom{10}{1}x^9(10)^1 + \binom{10}{2}x^8(10)^2 + \binom{10}{3}x^7(10)^3 + \cdots$

$= 1x^{10}(1) + \frac{10}{1}x^9(10) + \frac{10\cdot 9}{2\cdot 1}x^8(100) + \frac{10\cdot 9\cdot 8}{3\cdot 2\cdot 1}x^7(1000) + \cdots$

$= x^{10} + 100x^9 + 4{,}500x^8 + 120{,}000x^7 + \cdots$

33. $(3x-y)^7 = \binom{7}{0}(3x)^7(-y)^0 + \binom{7}{1}(3x)^6(-y)^1 + \binom{7}{2}(3x)^5(-y)^2 + \binom{7}{3}(3x)^4(-y)^3 + \cdots$

$= 1(2187x^7)(1) + \frac{7}{1}(729x^6)(-y) + \frac{7\cdot 6}{2\cdot 1}(243x^5)(y^2) + \frac{7\cdot 6\cdot 5}{3\cdot 2\cdot 1}(81x^4)(-y^3) + \cdots$

$= 2187x^7 + 7(729x^6)(-y) + 21(243x^5)(y^2) + 35(81x^4)(-y^3) + \cdots$

$= 2187x^7 - 5103x^6y + 5103x^5y^2 - 2835x^4y^3 + \cdots$

35. $(x^2-3y)^8 = \binom{8}{0}(x^2)^8(-3y)^0 + \binom{8}{1}(x^2)^7(-3y)^1 + \binom{8}{2}(x^2)^6(-3y)^2 + \binom{8}{3}(x^2)^5(-3y)^3 + \cdots$

$= 1(x^{16})(1) + \frac{8}{1}(x^{14})(-3y) + \frac{8\cdot 7}{2\cdot 1}(x^{12})(9y^2) + \frac{8\cdot 7\cdot 6}{3\cdot 2\cdot 1}(x^{10})(-27y^3) + \cdots$

$= x^{16} + 8(x^{14})(-3y) + 28(x^{12})(9y^2) + 56(x^{10})(-27y^3) + \cdots$

$= x^{16} - 24x^{14}y + 252x^{12}y^2 - 1512x^{10}y^3 + \cdots$

37. Yes, $n! = n\cdot(n-1)!$

$$\begin{aligned}4! &= 4\cdot3\cdot2\cdot1\\ &= 4\cdot(3\cdot2\cdot1)\\ &= 4\cdot(3)!\\ &= 4\cdot(4-1)!\end{aligned}$$

39. Yes, $(n-3)! = (n-3)(n-4)(n-5)!$ for $n \ge 5$.
Let $n = 7$:
$(7-3)! = (7-3)(7-4)(7-5)!$ or
$4! = 4\cdot3\cdot2! = 4\cdot3\cdot2\cdot1 = 4!$

41. $\binom{n}{m} = 1$ when either $n = m$ or $m = 0$.

43. $(x+3)^8$

First term is $\binom{8}{0}(x)^8(3)^0 = 1(x^8)(1) = x^8$.

Second term is $\binom{8}{1}(x)^7(3)^1 = 8(x^7)(3) = 24x^7$.

Next to last term is

$\binom{8}{7}(x)^1(3)^7 = 8(x)(2187) = 17{,}496x$.

Last term is $\binom{8}{8}(x)^0(3)^8 = 1(1)(6561) = 6561$.

45. $(a+b)^n = \sum_{i=0}^{n}\binom{n}{i}a^{n-i}b^i$

47. Let $x = 0$.

$$\begin{aligned}2x+y &= 10\\ 2(0)+y &= 6\\ y &= 6\end{aligned}$$

The y-intercept is (0, 10).

48. $\frac{1}{5}x+\frac{1}{2}y = 4 \overset{\times(-10)}{\Rightarrow} -2x-5y = -40$ (1)

$\frac{2}{3}x - y = \frac{8}{3} \overset{\times 3}{\Rightarrow} 2x-3y = 8$ (2)

$$\begin{aligned}-8y &= -32\\ y &= 4\end{aligned}$$

Substitute $y = 4$ into equation (2):

$$\begin{aligned}2x-3(4) &= 8\\ 2x &= 20\\ x &= 10\end{aligned}$$

The solution is $(10, 4)$.

49.

$$\begin{aligned}x(x-11) &= -18\\ x^2-11x+18 &= 0\\ (x-9)(x-2) &= 0\end{aligned}$$

$x-9=0$ or $x-2=0$

$x=9$ $\quad$ $x=2$

50. $\sqrt{20xy^4}\sqrt{6x^5y^7} = \sqrt{120x^6y^{11}}$

$$= \sqrt{4x^6y^{10}\cdot30y}$$
$$= 2x^3y^5\sqrt{30y}$$

51.

$$\begin{aligned}f(x) &= 3x+8\\ y &= 3x+8\\ x &= 3y+8\\ 3y &= x-8\\ y &= \frac{x-8}{3}\\ f^{-1}(x) &= \frac{x-8}{3}\end{aligned}$$

Exercise Set 12.6

1. A permutation is an ordered arrangement of a given set of objects.

3. ${}_nP_r$ is a symbol used to represent the number of arrangements of r objects that can be made from a set of n objects.

5. Answers will vary. Possible answer: Find the product of all positive integers less than or equal to n.

7. Answers will vary.

9. $7! = 7\cdot6\cdot5\cdot4\cdot3\cdot2\cdot1 = 5040$

11. $\frac{12!}{8!} = \frac{12\cdot11\cdot10\cdot9\cdot\cancel{8!}}{\cancel{8!}} = 11{,}880$

13. $\frac{13!}{(13-4)!} = \frac{13!}{9!} = \frac{13\cdot12\cdot11\cdot10\cdot\cancel{9!}}{\cancel{9!}} = 17{,}160$

15. $\frac{9!}{2!4!} = \frac{9\cdot8\cdot7\cdot6\cdot5\cdot\cancel{4!}}{2\cdot1\cdot\cancel{4!}} = \frac{15{,}120}{2} = 7560$

17. ${}_5P_0 = \frac{5!}{(5-0)!} = \frac{5!}{5!} = 1$

19. ${}_5P_2 = \frac{5!}{(5-2)!} = \frac{5!}{3!} = \frac{5\cdot 4\cdot \cancel{3!}}{\cancel{3!}} = 20$

21. ${}_5P_4 = \frac{5!}{(5-4)!} = \frac{5!}{1!} = \frac{5\cdot 4\cdot 3\cdot 2\cdot 1}{1} = 120$

23. ${}_6P_4 = \frac{6!}{(6-4)!} = \frac{6!}{2!} = \frac{6\cdot 5\cdot 4\cdot 3\cdot \cancel{2!}}{\cancel{2!}} = 360$

25. ${}_9P_4 = \frac{9!}{(9-4)!} = \frac{9!}{5!} = \frac{9\cdot 8\cdot 7\cdot 6\cdot \cancel{5!}}{\cancel{5!}} = 3024$

27. ${}_{10}P_0 = \frac{10!}{(10-0)!} = \frac{10!}{10!} = 1$

29. ${}_5C_0 = \frac{5!}{(5-0)!0!} = \frac{5!}{5!0!} = \frac{\cancel{5!}}{\cancel{5!}\cdot 1} = 1$

31. ${}_5C_2 = \frac{5!}{(5-2)!2!} = \frac{5!}{3!2!} = \frac{5\cdot 4\cdot \cancel{3!}}{\cancel{3!}\cdot 2\cdot 1} = \frac{20}{2} = 10$

33. ${}_5C_4 = \frac{5!}{(5-4)!4!} = \frac{5!}{1!4!} = \frac{5\cdot \cancel{4!}}{1\cdot \cancel{4!}} = 5$

35. ${}_7C_3 = \frac{7!}{(7-3)!3!} = \frac{7!}{4!3!} = \frac{7\cdot 6\cdot 5\cdot \cancel{4!}}{\cancel{4!}\cdot 3\cdot 2\cdot 1} = \frac{210}{6} = 35$

37. ${}_9C_6 = \frac{9!}{(9-6)!6!} = \frac{9!}{3!6!} = \frac{9\cdot 8\cdot 7\cdot \cancel{6!}}{3\cdot 2\cdot 1\cdot \cancel{6!}} = \frac{504}{6} = 84$

39. ${}_7C_7 = \frac{7!}{(7-7)!7!} = \frac{7!}{0!7!} = \frac{\cancel{7!}}{1\cdot \cancel{7!}} = 1$

41. There are 7 possible winners of the first race and 8 possible winners of the second race. By the counting principle, there are $7\cdot 8 = 56$ possible outcomes for the daily double. Thus, 56 daily double tickets must be purchased to guarantee a win.

43. Since repetition is allowed, there are 10 choices for each of the six buttons in the code. Thus, by the counting principle, there are $10\cdot 10\cdot 10\cdot 10\cdot 10\cdot 10 = 10^6 = 1{,}000{,}000$ different codes possible.

45. Since repetition is allowed, there are 10 choices for each of the nine digits in the number. Thus, by the counting principle, there are
$10\cdot 10\cdot 10\cdot 10\cdot 10\cdot 10\cdot 10\cdot 10\cdot 10 = 10^9$
$= 1{,}000{,}000{,}000$
different social security numbers possible.

47. By the counting principle, there are $10\cdot 12\cdot 9 = 1080$ different sound systems that can be advertised.

49. Since the ordering of the 3 gifts among the 7 people is important, the number of ways in which the gifts can be awarded is
$${}_7P_3 = \frac{7!}{(7-3)!} = \frac{7!}{4!} = \frac{7\cdot 6\cdot 5\cdot \cancel{4!}}{\cancel{4!}} = 210.$$

51. Since the ordering of the 6 prizes among the 30 students is important, the number of ways in which the offices can be awarded is
$$\begin{aligned}{}_{30}P_6 &= \frac{30!}{(30-6)!}\\ &= \frac{30!}{24!}\\ &= \frac{30\cdot 29\cdot 28\cdot 27\cdot 26\cdot 25\cdot \cancel{24!}}{\cancel{24!}}\\ &= 427{,}518{,}000.\end{aligned}$$

53. Since the ordering of the visits is important, the number of ways in which the guard can visit the 8 offices is $8! = 40{,}320$.

55. The number of permutations of the 5 distinct letters is $5! = 120$.

57. The number of arrangements of the 10 distinct answers is $10! = 3{,}628{,}800$.

59. Since the ordering of the 4 selected books in not important, the number of possible selections is
$$\begin{aligned}{}_{12}C_4 &= \frac{12!}{(12-4)!4!}\\ &= \frac{12!}{8!4!}\\ &= \frac{12\cdot 11\cdot 10\cdot 9\cdot \cancel{8!}}{\cancel{8!}\cdot 4\cdot 3\cdot 2\cdot 1}\\ &= \frac{11{,}880}{24}\\ &= 495.\end{aligned}$$

61. Since the prizes are identical, the ordering of the 6 prize winners from the 18 students is not important. Thus, the number of ways in which the prizes can be given is

$$_{18}C_6 = \frac{18!}{(18-6)!6!} = \frac{18!}{12!6!}$$
$$= \frac{18\cdot 17\cdot 16\cdot 15\cdot 14\cdot 13\cdot \cancel{12!}}{\cancel{12!}\cdot 6\cdot 5\cdot 4\cdot 3\cdot 2\cdot 1}$$
$$= \frac{13,366,080}{720}$$
$$= 18,564.$$

63. Since the order of the 3 selections from the 6 books is not important, the number of ways in which the selections can be made is

$$_6C_3 = \frac{6!}{(6-3)!3!} = \frac{6!}{3!3!} = \frac{6\cdot 5\cdot 4\cdot \cancel{3!}}{\cancel{3!}\cdot 3\cdot 2\cdot 1} = \frac{120}{6} = 20.$$

65. Since the order of the 3 awardees from the 8 finalists is not important, the number of ways in which the selections can be made is

$$_8C_3 = \frac{8!}{(8-3)!3!} = \frac{8!}{5!3!} = \frac{8\cdot 7\cdot 6\cdot \cancel{5!}}{\cancel{5!}\cdot 3\cdot 2\cdot 1} = \frac{336}{6} = 56.$$

67. Since the order of the 6 students receiving A's from the 30 students in the class is not important, the number of different combinations possible is

$$_{30}C_6 = \frac{30!}{(30-6)!6!} = \frac{30!}{24!6!}$$
$$= \frac{30\cdot 29\cdot 28\cdot 27\cdot 26\cdot 25\cdot \cancel{24!}}{\cancel{24!}\cdot 6\cdot 5\cdot 4\cdot 3\cdot 2\cdot 1}$$
$$= \frac{427,518,000}{720}$$
$$= 593,775.$$

69. Since the order of 3 selections from the 6 limousines is not important, the number of different selections is

$$_6C_3 = \frac{6!}{(6-3)!3!} = \frac{6!}{3!3!} = \frac{6\cdot 5\cdot 4\cdot \cancel{3!}}{3\cdot 2\cdot 1\cdot \cancel{3!}} = \frac{120}{6} = 20.$$

71. Since repetition is allowed, there are 10 choices for each digit of the 10-digit number. By the counting principle, the number of different ISBN numbers possible is

$$10\cdot 10\cdot 10\cdot 10\cdot 10\cdot 10\cdot 10\cdot 10\cdot 10\cdot 10$$
$$= 10^{10} = 10,000,000,000.$$

73. The number of permutations of the 6 distinct letters is $6! = 720$.

75. a. Because the order of the 3-number sequence is important, this is more like a permutation problem than a combination problem. However, it is not a true permutation problem because repetition is permitted.

b. Since repetition is permitted, there are 40 possible choices for each of the 3 numbers. By the counting principle, the number of possible 3-number arrangements is

$$40\cdot 40\cdot 40 = 40^3 = 64,000.$$

c. Since repetition is no longer permitted, there are 40 choices for the first number, 39 choices for the second number, and 38 choices for the third number. By the counting principle, the number of possible 3-number arrangements is $40\cdot 39\cdot 38 = 59,280$.

77. a. Since the first digit cannot be 0 or 1, there are 8 choices for the first digit. Since repetition is permitted, there are 10 choices for each of the remaining 6 digits. Thus, there will be $8\cdot 10\cdot 10\cdot 10\cdot 10\cdot 10\cdot 10 = 8,000,000$ different telephone numbers possible.

b. For the area code, there are 8 choices for the first digit and 10 choices for the remaining two digits. Thus, the number of distinct telephone numbers possible will be $8\cdot 10\cdot 10\cdot 8,000,000 = 6,400,000,000$

c. Adding 4 more digits with 10 choices for each one results in

$$6,400,000,000\cdot 10\cdot 10\cdot 10\cdot 10$$
$$= 64,000,000,000,000$$

distinct telephone numbers.

79. a. The number of arrangements of the 10 people is $10! = 3,628,800$.

b. There is 1 way to arrange Mr. and Mrs. Doan and 8! ways to arrange the remaining 8 people. Thus, there are $1\cdot 8! = 40,320$ arrangements possible.

c. The person next to Mrs. Doan must be male. There are 4 possibilities. The person next to him must be a female. There are 4 possibilities. The person next to her must be male. There are 3 possibilities, and so on. The number of possible arrangements is $1\cdot 1\cdot 4\cdot 4\cdot 3\cdot 3\cdot 2\cdot 2\cdot 1\cdot 1 = 576$.

81. For Part 1 of the exam, the number of combinations of questions Frank can answer is

$${}_4C_3 = \frac{4!}{(4-3)!3!} = \frac{4!}{1!3!} = \frac{4 \cdot \cancel{3!}}{1 \cdot \cancel{3!}} = 4.$$

For Part 2, the number of combinations of questions he can answer is

$${}_5C_3 = \frac{5!}{(5-3)!3!} = \frac{5!}{2!3!} = \frac{5 \cdot 4 \cdot \cancel{3!}}{2 \cdot 1 \cdot \cancel{3!}} = \frac{20}{2} = 10.$$

By the counting principle, Frank has a total of $4 \cdot 10 = 40$ combinations of questions he can answer.

83. The number of combinations of types of faculty is ${}_5C_3 = \frac{5!}{(5-3)!3!} = \frac{5!}{2!3!} = \frac{5 \cdot 4 \cdot \cancel{3!}}{2 \cdot 1 \cdot \cancel{3!}} = 10.$

The number of combinations of students is

$$\begin{aligned} {}_{30}C_3 &= \frac{30!}{(30-3)!3!} \\ &= \frac{30!}{27!3!} \\ &= \frac{30 \cdot 29 \cdot 28 \cdot \cancel{27!}}{\cancel{27!} \cdot 3 \cdot 2 \cdot 1} \\ &= 4060. \end{aligned}$$

By the counting principle, there are $10 \cdot 4060 = 40{,}600$ committees possible.

85. The number of combinations of oat cereals is

$${}_6C_3 = \frac{6!}{(6-3)!3!} = \frac{6!}{3!3!} = \frac{6 \cdot 5 \cdot 4 \cdot \cancel{3!}}{3 \cdot 2 \cdot 1 \cdot \cancel{3!}} = 20.$$

The number of combinations of wheat cereals is

$${}_5C_2 = \frac{5!}{(5-2)!2!} = \frac{5!}{3!2!} = \frac{5 \cdot 4 \cdot \cancel{3!}}{\cancel{3!} \cdot 2 \cdot 1} = 10.$$

The number of combinations of rice cereals is

$${}_4C_2 = \frac{4!}{(4-2)!2!} = \frac{4!}{2!2!} = \frac{4 \cdot 3 \cdot \cancel{2!}}{\cancel{2!} \cdot 2 \cdot 1} = 6.$$

By the counting principle, there are $20 \cdot 10 \cdot 6 = 1200$ different combinations.

87. A pizza could contain 0, 1, 2, 3, 3, 5, or 6 of the 6 toppings available. The number of different kinds of pizza that can be ordered is

$$\begin{aligned} &{}_6C_0 + {}_6C_1 + {}_6C_2 + {}_6C_3 + {}_6C_4 + {}_6C_5 + {}_6C_6 \\ &= \frac{6!}{6!0!} + \frac{6!}{5!1!} + \frac{6!}{4!2!} + \frac{6!}{3!3!} + \frac{6!}{2!4!} + \frac{6!}{1!5!} + \frac{6!}{0!6!} \\ &= 1 + 6 + 15 + 20 + 15 + 6 + 1 \\ &= 64 \end{aligned}$$

92. $\frac{3}{4}x - 2 = 4$

$$\frac{3}{4}x = 6$$

$$x = \frac{4}{3} \cdot 6 = 8$$

93.

$$\begin{aligned} 3x^2 &= 18x \\ 3x^2 - 18x &= 0 \\ 3x(x-6) &= 0 \end{aligned}$$

$3x = 0$ or $x - 6 = 0$

$x = 0$ $\quad$ $x = 6$

94.

$$\frac{x^2}{x+5} = \frac{25}{x+5}$$

$$(x+5)\left(\frac{x^2}{x+5}\right) = (x+5)\left(\frac{25}{x+5}\right)$$

$$x^2 = 25$$

$$x = \pm\sqrt{25}$$

$$x = \pm 5$$

Check the possible answers:

$\frac{(-5)^2}{(-5)+5} \stackrel{?}{=} \frac{25}{(-5)+5}$ $\qquad$ $\frac{5^2}{5+5} \stackrel{?}{=} \frac{25}{5+5}$

$\frac{25}{0} \stackrel{?}{=} \frac{25}{0}$ ← undefined $\qquad$ $\frac{25}{10} = \frac{25}{10}$ ← True

The equation is undefined for $x = -5$. Thus, $x = -5$ is not a solution. The equation is true for $x = 5$. Thus, the only solution to this equation is $x = 5$.

95. $\sqrt{x-8} + 12 = 19$

$$\sqrt{x-8} = 7$$

$$x - 8 = 7^2$$

$$x - 8 = 49$$

$$x = 57$$

Check the possible answer:

$\sqrt{57-8} + 12 \stackrel{?}{=} 19$

$\sqrt{49} + 12 \stackrel{?}{=} 19$

$7 + 12 \stackrel{?}{=} 19$

$19 = 19$ ← True

Thus, the solution to the equation is $x = 57$.

Exercise Set 12.7

1. Equally likely outcomes are outcomes that have the same chance of occurring.

3. Since event A will either occur or not occur, $P(A)+P(\text{not } A)=1$.

5. Two events that are mutually exclusive cannot occur simultaneously.

7. All probabilities must be between 0 and 1, inclusive.

9. $P(\text{a red marble is chosen})$

$$=\frac{\text{Number of red marbles}}{\text{Total number of marbles}}=\frac{15}{75}=\frac{1}{5}$$

11. $P(\text{a green marble is chosen})$

$$=\frac{\text{Number of green marbles}}{\text{Total number of marbles}}=\frac{10}{75}=\frac{2}{15}$$

13. $P(\text{a red marble is } \textit{not} \text{ chosen})$

$$=1-P(\text{a red marble } \textit{is} \text{ chosen})$$

$$=1-\frac{\text{Number of red marbles}}{\text{Total number of marbles}}=1-\frac{15}{75}=1-\frac{1}{5}=\frac{4}{5}$$

---

15. $P(\text{a blue } \textit{or} \text{ a green marble is chosen})$

$$=P(\text{a blue marble is chosen})+P(\text{a green marble is chosen})-P(\text{a blue and green marble is chosen})$$

$$=\frac{\text{Number of blue marbles}}{\text{Total number of marbles}}+\frac{\text{Number of green marbles}}{\text{Total number of marbles}}-\frac{\text{Number of blue and green marbles}}{\text{Total number of marbles}}$$

$$=\frac{20}{75}+\frac{10}{75}-\frac{0}{75}=\frac{30}{75}=\frac{2}{5}$$

---

17. $P(\text{selecting a 9})$

$$=\frac{\text{Number of 9's}}{\text{Total number of cards}}=\frac{4}{52}=\frac{1}{13}$$

19. $P(\text{selecting the ace of spades})$

$$=\frac{\text{Number of aces of spades}}{\text{Total number of cards}}=\frac{1}{52}$$

21. $P(\text{selecting a red card})$

$$=\frac{\text{Number of red cards}}{\text{Total number of cards}}=\frac{26}{52}=\frac{1}{2}$$

23. $P(\text{selecting a card that is not a king})$

$$=1-P(\text{selecting a king})$$

$$=1-\frac{\text{Number of kings}}{\text{Total number of cards}}=1-\frac{4}{52}=1-\frac{1}{13}=\frac{12}{13}$$

---

25. $P(\text{selecting a card that is a 9 or a 10})=P(\text{selecting a 9})+P(\text{selecting a 10})-P(\text{selecting a card that is a 9 and a 10})$

$$=\frac{\text{Number of 9's}}{\text{Total number of cards}}+\frac{\text{Number of 10's}}{\text{Total number of cards}}-\frac{\text{Number of cards that are a 9 and a 10}}{\text{Total number of cards}}$$

$$=\frac{4}{52}+\frac{4}{52}-\frac{0}{52}=\frac{8}{52}=\frac{2}{13}$$

27. $P(\text{selecting a card that is a red card or a king})$

$$=P(\text{selecting a red card})+P(\text{selecting a king})-P(\text{selecting a card that is red and a king})$$

$$=\frac{\text{Number of red cards}}{\text{Total number of cards}}+\frac{\text{Number of kings}}{\text{Total number of cards}}-\frac{\text{Number of cards that are red and a king}}{\text{Total number of cards}}$$

$$=\frac{26}{52}+\frac{4}{52}-\frac{2}{52}=\frac{28}{52}=\frac{7}{13}$$

29. $P(\text{selecting a rock/pop song})$
$$=\frac{\text{Number of rock/pop songs}}{\text{Total number of songs}}=\frac{40}{70}=\frac{4}{7}$$

31. $P(\text{selecting a classical piece})$
$$=\frac{\text{Number of classical pieces}}{\text{Total number of songs}}=\frac{3}{70}$$

---

33. $P(\text{selecting a country song } or \text{ a classical piece})$
$$=P(\text{selecting a country song})+P(\text{selecting a classical piece})-P(\text{selecting a song that is country and classical})$$
$$=\frac{\text{Number of country song}}{\text{Total number of songs}}+\frac{\text{Number of classical pieces}}{\text{Total number of songs}}-\frac{\text{Number of songs that are country and classical}}{\text{Total number of songs}}$$
$$=\frac{27}{70}+\frac{3}{70}-\frac{0}{70}=\frac{30}{70}=\frac{3}{7}$$

---

35. The spinner is divided into 4 equal pieces.

a. $P(\text{yellow})=\frac{2}{4}=\frac{1}{2}$

b. $P(\text{purple})=\frac{1}{4}$

c. $P(\text{green})=\frac{1}{4}$

d. $P(\text{yellow } or \text{ purple})$
$$=P(\text{yellow})+P(\text{purple})-P(\text{yellow } and \text{ purple})$$
$$=\frac{2}{4}+\frac{1}{4}-\frac{0}{4}$$
$$=\frac{3}{4}$$

e. $P(not\text{ purple})=1-P(\text{purple})$
$$=1-\frac{1}{4}$$
$$=\frac{3}{4}$$

37. The spinner is divided into 3 pieces. The purple piece is $\frac{1}{2}$ of the whole, and the yellow and green pieces are each $\frac{1}{4}$ of the whole.

a. $P(\text{yellow})=\frac{1}{4}$

b. $P(\text{purple})=\frac{1}{2}$

c. $P(\text{green})=\frac{1}{4}$

d. $P(\text{yellow } or \text{ purple})$
$$=P(\text{yellow})+P(\text{purple})-P(\text{yellow } and \text{ purple})$$
$$=\frac{1}{4}+\frac{1}{2}-\frac{0}{4}$$
$$=\frac{3}{4}$$

e. $P(not\text{ purple})=1-P(\text{purple})=1-\frac{1}{2}=\frac{1}{2}$

39. The spinner is divided into 5 pieces. Four of the pieces are each $\frac{1}{6}$ of the whole, and one piece is $\frac{1}{3}$ of the whole.

a. $P(\text{yellow})=\frac{2}{6}=\frac{1}{3}$

b. $P(\text{purple})=\frac{2}{6}=\frac{1}{3}$

c. $P(\text{green})=\frac{1}{3}$

d. $P(\text{yellow } or \text{ purple})$
$$=P(\text{yellow})+P(\text{purple})-P(\text{yellow } and \text{ purple})$$
$$=\frac{2}{6}+\frac{2}{6}-\frac{0}{6}$$
$$=\frac{4}{6}$$
$$=\frac{2}{3}$$

e. $P(not\text{ purple})=1-P(\text{purple})=1-\frac{1}{3}=\frac{2}{3}$

41. $P(\text{the light is red})=\dfrac{\text{Time the light is red}}{\text{Total time}}=\dfrac{40\text{ seconds}}{110\text{ seconds}}=\dfrac{4}{11}$

43. $P(\text{the light is green})=\dfrac{\text{Time the light is green}}{\text{Total time}}=\dfrac{60\text{ seconds}}{110\text{ seconds}}=\dfrac{6}{11}$

---

45. $P(\text{the light is yellow } \textit{or} \text{ green}) = P(\text{the light is yellow})+P(\text{the light is green})-P(\text{the light is yellow } \textit{and} \text{ green})$

$$=\frac{\text{Time the light is yellow}}{\text{Total time}}+\frac{\text{Time the light is green}}{\text{Total time}}-\frac{\text{Time the light is yellow and green}}{\text{Total time}}$$

$$=\frac{10\text{ seconds}}{110\text{ seconds}}+\frac{60\text{ seconds}}{110\text{ seconds}}-\frac{0\text{ seconds}}{110\text{ seconds}}$$

$$=\frac{70\text{ seconds}}{110\text{ seconds}}$$

$$=\frac{7}{11}$$

---

47. $P(A \text{ or } B)=P(A)+P(B)-P(A \text{ and } B)$
$=0.4+0.5-0.2$
$=0.7$

49. $P(A \text{ or } B)=P(A)+P(B)-P(A \text{ and } B)$
$=0.8+0.2-0.1$
$=0.9$

51. a. $P(\text{white})=\dfrac{\text{Number of white blouses}}{\text{Total number of blouses}}=\dfrac{4}{12}=\dfrac{1}{3}$

b. $P(\textit{not}\text{ white})=1-P(\text{white})=1-\dfrac{1}{3}=\dfrac{2}{3}$

c. Odds against it being white $=\dfrac{P(\textit{not}\text{ white})}{P(\text{white})}=\dfrac{\frac{2}{3}}{\frac{1}{3}}=\dfrac{2}{3}\cdot\dfrac{3}{1}=\dfrac{2}{1}$ or 2:1

d. Odds in favor of it being white $=\dfrac{P(\text{white})}{P(\textit{not}\text{ white})}=\dfrac{\frac{1}{3}}{\frac{2}{3}}=\dfrac{1}{3}\cdot\dfrac{3}{2}=\dfrac{1}{2}$ or 1:2

53. Odds against rolling a 4 $=\dfrac{P(\textit{not}\text{ rolling a 4})}{P(\text{rolling a 4})}=\dfrac{\frac{5}{6}}{\frac{1}{6}}=\dfrac{5}{6}\cdot\dfrac{6}{1}=\dfrac{5}{1}$ or 5:1

55. Odds against rolling a number greater than 5

$$=\frac{P(\textit{not}\text{ rolling a number greater than 5})}{P(\text{rolling a number greater than 5})}=\frac{\frac{5}{6}}{\frac{1}{6}}=\frac{5}{6}\cdot\frac{6}{1}=\frac{5}{1}\text{ or }5:1$$

57. Odds against selecting a 9 $= \dfrac{P(\textit{not}\text{ selecting } 9)}{P(\text{selecting a } 9)}$

$$= \frac{\frac{48}{52}}{\frac{4}{52}}$$
$$= \frac{48}{52} \cdot \frac{52}{4}$$
$$= \frac{48}{4}$$
$$= \frac{12}{1} \text{ or } 12{:}1$$

59. Odds in favor of selecting a red card

$$= \frac{P(\text{selecting a red card})}{P(\textit{not}\text{ selecting a red card})}$$
$$= \frac{\frac{26}{52}}{\frac{26}{52}}$$
$$= \frac{26}{52} \cdot \frac{52}{26}$$
$$= \frac{1}{1} \text{ or } 1{:}1$$

61. a. Odds against winning $= \dfrac{P(\textit{not}\text{ winning})}{P(\text{winning})}$

$$= \frac{\frac{999{,}999}{1{,}000{,}000}}{\frac{1}{1{,}000{,}000}}$$
$$= \frac{999{,}999}{1{,}000{,}000} \cdot \frac{1{,}000{,}000}{1}$$
$$= \frac{999{,}999}{1} \text{ or } 999{,}999{:}1$$

b. Odds against winning $= \dfrac{P(\textit{not}\text{ winning})}{P(\text{winning})}$

$$= \frac{\frac{999{,}990}{1{,}000{,}000}}{\frac{10}{1{,}000{,}000}}$$
$$= \frac{999{,}990}{1{,}000{,}000} \cdot \frac{1{,}000{,}000}{10}$$
$$= \frac{999{,}990}{10}$$
$$= \frac{99{,}990}{1} \text{ or } 99{,}990{:}1$$

63. a. Odds in favor of an event

$$= \frac{P(\text{event occurs})}{P(\text{event does not occur})}$$
$$= \frac{P(\text{event occurs})}{1 - P(\text{event occurs})}$$

Let $x = P(\text{event occurs})$.

$$\frac{7}{3} = \frac{x}{1-x}$$
$$7(1-x) = 3x$$
$$7 - 7x = 3x$$
$$7 = 10x$$
$$\frac{7}{10} = x$$
$$P(\text{event occurs}) = \frac{7}{10}$$

b. $P(\text{event does not occur}) = 1 - P(\text{event occurs})$

$$= 1 - \frac{7}{10}$$
$$= \frac{3}{10}$$

65. Odds against Jamal getting promoted

$$= \frac{P(\text{Jamal does not get promoted})}{P(\text{Jamal gets promoted})}$$
$$= \frac{1 - P(\text{Jamal gets promoted})}{P(\text{Jamal gets promoted})}$$

Let $x = P(\text{Jamal gets promoted})$.

$$\frac{3}{7} = \frac{1-x}{x}$$
$$3x = 7(1-x)$$
$$3x = 7 - 7x$$
$$10x = 7$$
$$x = \frac{7}{10}$$
$$P(\text{Jamal gets promoted}) = \frac{7}{10}$$

67. Odds against buying a house this week

$$= \frac{P(\text{not buying a house this week})}{P(\text{buying a house this week})}$$
$$= \frac{1 - P(\text{buying a house this week})}{P(\text{buying a house this week})}$$
$$= \frac{1-0.6}{0.6}$$
$$= \frac{0.4}{0.6}$$
$$= \frac{2}{3} \text{ or } 2{:}3$$

69. Odds against water heater being fixed the first time

$$= \frac{P(\text{water heater is fixed the first time})}{P(\text{water heater is not fixed the first time})}$$
$$= \frac{1 - P(\text{water heater is fixed the first time})}{P(\text{water heater is fixed the first time})}$$
$$= \frac{1-0.7}{0.7}$$
$$= \frac{0.3}{0.7}$$
$$= \frac{3}{7} \text{ or } 3{:}7$$

72. $12x - 16y = 48$

Find the x-intercept by letting $y = 0$.

$$12x - 16(0) = 48$$
$$12x = 48$$
$$x = 4$$

The x-intercept is (4, 0).

Find the y-intercept by letting $x = 0$.

$$12(0) - 16y = 48$$
$$-16x = 48$$
$$x = -3$$

The y-intercept is (0, –3).

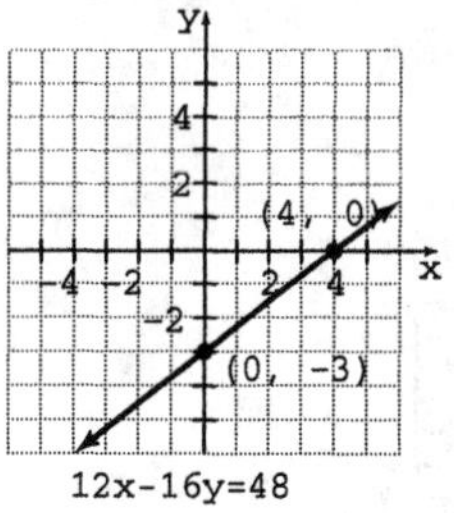

73. $P(x) = x^4 - 2x^3 + 4x^2 - 14x - 21$

To find the y-intercept, substitute 0 for x: $P(0) = -21$. The y-intercept is $(0, -21)$.

Since the coefficients of $P(x)$ have three variations in signs, it has either 3 or 1 positive real zeros. Since the coefficients of $P(-x) = x^4 + 2x^3 + 4x^2 + 14x - 21$ have one variation in signs, P has 1 negative real zero. The possible rational zeros of P are of the form $\frac{p}{q}$ where p is a factor of -21 and q is a factor of 1.

Factors of -21: $\pm1, \pm3, \pm7, \pm21$

Factors of 1: ±1

Possible rational zeros: $\pm1, \pm3, \pm7, \pm21$

Using synthetic division to find the one negative real zero and checking the possibility of -1 being that zero, we get

-1	1	-2	4	-14	-21
		-1	3	-7	21
	1	-3	7	-21	0

Since the remainder is 0, -1 is a zero of P and $x+1$ is a factor of P. Thus, P can be written as follows:

$$P(x) = x^4 - 2x^3 + 4x^2 - 14x - 21$$
$$= (x+1)(x^3 - 3x^2 + 7x - 21)$$

We now need to determine any positive real zeros. To do this, focus on the polynomial. $Q(x) = x^3 - 3x^2 + 7x - 21$. Using synthetic division and checking the possibility of 3 being a zero, we get

3	1	-3	7	-21
		3	0	21
	1	0	7	0

Since the remainder is 0, 3 is a real zero of P and $x-3$ is a factor of P. Writing P in factored form, we get

$$P(x) = x^4 - 2x^3 + 4x^2 - 14x - 21$$
$$= (x+1)(x^3 - 3x^2 + 7x - 21)$$
$$= (x+1)(x-3)(x^2+7)$$

$x + 1 = 0$ or $x - 3 = 0$ or $x^2 + 7 = 0$

$x = -1$ $\quad$ $x = 3$ $\quad$ $x^2 = -7$

$$x = \pm\sqrt{-7}$$
$$x = \pm i\sqrt{7}$$

Thus, the four zeros are -1, 3, $i\sqrt{7}$, and $-i\sqrt{7}$. Since P only has two real zeros, it has only two x-intercepts which divide the graph into three regions. We will find a point in each region:

Interval	x	Value of $h(x)$
$(-\infty,-1)$	-2	$P(-2)=55$
$(-1,3)$	2	$P(2)=-33$
$(3,\infty)$	4	$P(4)=115$

Graph the intercepts and points from each region. Then connect them with a smooth curve.

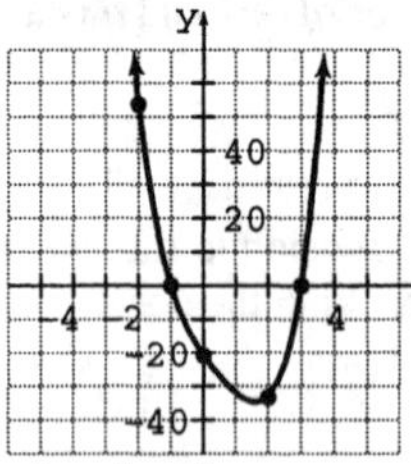

P(x) $=x^4-2x^3+4x^2-14x-21$

74. $f(x)=\dfrac{3x^2-9x-12}{2x^2-10x+8}$

To find the y-intercept, substitute 0 for x:

$f(0)=\dfrac{-12}{8}=-\dfrac{3}{2}$. The y-intercept is $\left(0,-\dfrac{3}{2}\right)$.

Since the degree of the numerator is the same as the degree of the denominator, there is a horizontal asymptote at the ratio of their leading coefficients: $y=\dfrac{3}{2}$.

To find the points of discontinuity, find the zeros of the denominator:

$$2x^2-10x+8=0$$
$$2(x^2-5x+4)=0$$
$$2(x-4)(x-1)=0$$
$$x-4=0 \quad \text{or} \quad x-1=0$$
$$x=4 \qquad\qquad x=1$$

Points of discontinuity occur at $x=4$ and $x=1$. Now simplify the function:

$$f(x)=\frac{3x^2-9x-12}{2x^2-10x+8}$$
$$=\frac{3(x^2-3x-4)}{2(x^2-5x+4)}$$
$$=\frac{3(x-4)(x+1)}{2(x-4)(x-1)}$$
$$=\frac{3(x+1)}{2(x-1)}$$

Since $x=4$ does not make the denominator of the simplified function 0, there is a hole at $x=4$. Since $x=1$ makes the denominator of the simplified function 0, there is a vertical asymptote ate $x=1$.

Evaluate $f(x)$ as $x\to1$ from both the right and left:

x	$f(x)$
5	2.25
4	undefined
3	3
2	4.5

x	$f(x)$
-2	0.5
-1	0
0	-1.5

As $x\to1$ from the right, $f(x)$ gets smaller and smaller. As $x\to1$ from the left, $f(x)$ gets larger and larger.

Connect the points on either side of the vertical asymptote with a smooth curve that approaches the asymptotes. Make a hole in the graph at $x=4$.

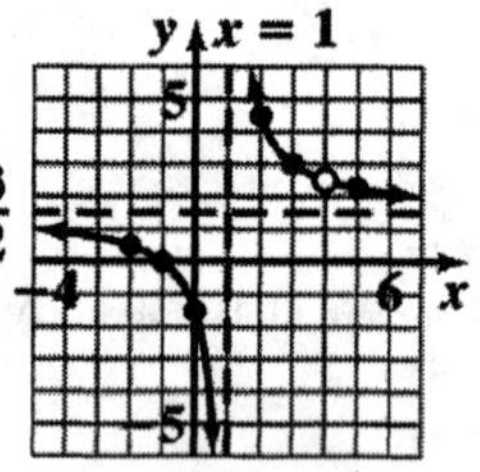

$f(x)=\dfrac{3x^2-9x-12}{2x^2-10x+8}$

75. $\dfrac{(x+3)^2}{25}+\dfrac{(y-5)^2}{9}=1$

$\dfrac{(x+3)^2}{5^2}+\dfrac{(y-5)^2}{3^2}=1$

Center at $(-3, 5)$.

$a=5$ and $b=3$.

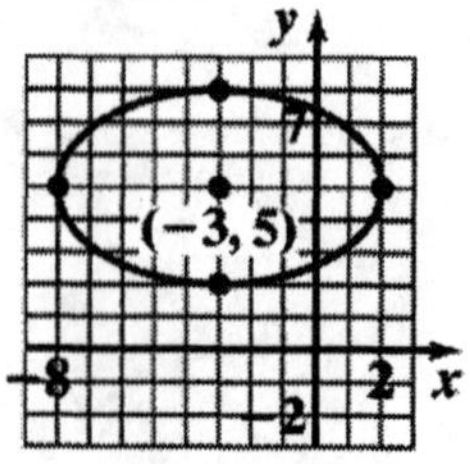

$\dfrac{(x+3)^2}{25}+\dfrac{(y-5)^2}{9}=1$

Chapter 12 Review Exercises

1. $a_n = n+5$

$a_1 = 1+5 = 6$

$a_2 = 2+5 = 7$

$a_3 = 3+5 = 8$

$a_4 = 4+5 = 9$

$a_5 = 5+5 = 10$

The terms are 6, 7, 8, 9, 10.

2. $a_n = n^2 + n - 3$

$a_1 = 1^2 + 1 - 3 = -1$

$a_2 = 2^2 + 2 - 3 = 3$

$a_3 = 3^2 + 3 - 3 = 9$

$a_4 = 4^2 + 4 - 3 = 17$

$a_5 = 5^2 + 5 - 3 = 27$

The terms are -1, 3, 9, 17, 27.

3. $a_n = \frac{6}{n}$

$a_1 = \frac{6}{1} = 6$

$a_2 = \frac{6}{2} = 3$

$a_3 = \frac{6}{3} = 2$

$a_4 = \frac{6}{4} = \frac{3}{2}$

$a_5 = \frac{6}{5}$

The terms are 6, 3, 2, $\frac{3}{2}$, $\frac{6}{5}$.

4. $a_n = \frac{n^2}{n+4}$

$a_1 = \frac{1^2}{1+4} = \frac{1}{5}$

$a_2 = \frac{2^2}{2+4} = \frac{4}{6} = \frac{2}{3}$

$a_3 = \frac{3^2}{3+4} = \frac{9}{7}$

$a_4 = \frac{4^2}{4+4} = \frac{16}{8} = 2$

$a_5 = \frac{5^2}{5+4} = \frac{25}{9}$

The terms are $\frac{1}{5}, \frac{2}{3}, \frac{9}{7}, 2, \frac{25}{9}$.

5. $a_n = 3n - 10$

$a_7 = 3(7) - 10 = 21 - 10 = 11$

6. $a_n = (-1)^n + 5$

$a_7 = (-1)^7 + 5 = -1 + 5 = 4$

7. $a_n = \frac{n+17}{n^2}$

$a_9 = \frac{9+17}{9^2} = \frac{26}{81}$

8. $a_n = (n)(n-3)$

$a_{11} = (11)(11-3) = (11)(8) = 88$

9. $a_n = 2n+5$

$a_1 = 2(1) + 5 = 2 + 5 = 7$

$a_2 = 2(2) + 5 = 4 + 5 = 9$

$a_3 = 2(3) + 5 = 6 + 5 = 11$

$s_1 = a_1 = 7$

$s_3 = a_1 + a_2 + a_3 = 7 + 9 + 11 = 27$

10. $a_n = n^2 + 8$

$a_1 = (1)^2 + 8 = 1 + 8 = 9$

$a_2 = (2)^2 + 8 = 4 + 8 = 12$

$a_3 = (3)^2 + 8 = 9 + 8 = 17$

$s_1 = a_1 = 9$

$s_3 = a_1 + a_2 + a_3 = 9 + 12 + 17 = 38$

11. $a_n = \frac{n+3}{n+2}$

$a_1 = \frac{1+3}{1+2} = \frac{4}{3}$

$a_2 = \frac{2+3}{2+2} = \frac{5}{4}$

$a_3 = \frac{3+3}{3+2} = \frac{6}{5}$

$s_1 = a_1 = \frac{4}{3}$

$$\begin{aligned} s_3 &= a_1 + a_2 + a_3 \\ &= \frac{4}{3} + \frac{5}{4} + \frac{6}{5} \\ &= \frac{80}{60} + \frac{75}{60} + \frac{72}{60} \\ &= \frac{227}{60} \end{aligned}$$

12. $a_n = (-1)^n(n+8)$

$a_1 = (-1)^1(1+8) = (-1)(9) = -9$

$a_2 = (-1)^2(2+8) = 1(10) = 10$

$a_3 = (-1)^3(3+8) = -1(11) = -11$

$s_1 = a_1 = -9$

$s_3 = a_1 + a_2 + a_3 = -9 + 10 - 11 = -10$

13. This is a geometric sequence with $r = 4 \div 2 = 2$ and $a_1 = 2$.

$a_5 = 2(2)^{5-1} = 2 \cdot 2^4 = 32$

$a_6 = 2(2)^{6-1} = 2 \cdot 2^5 = 64$

$a_7 = 2(2)^{7-1} = 2 \cdot 2^6 = 128$

The terms are 32, 64, 128.

$a_n = 2(2)^{n-1} = 2^1 2^{n-1} = 2^n$

14. This is a geometric sequence with

$r = 9 \div (-27) = \frac{9}{-27} = -\frac{1}{3}$ and $a_1 = -27$.

$$a_5 = -27\left(-\frac{1}{3}\right)^4 = -27\left(\frac{1}{81}\right) = -\frac{1}{3}$$

$$a_6 = -27\left(-\frac{1}{3}\right)^5 = -27\left(-\frac{1}{243}\right) = \frac{1}{9}$$

$$a_5 = -27\left(-\frac{1}{3}\right)^6 = -27\left(\frac{1}{729}\right) = -\frac{1}{27}$$

The terms are $-\frac{1}{3}, \frac{1}{9}, -\frac{1}{27}$.

$$a_n = -27\left(-\frac{1}{3}\right)^{n-1} \text{ or } a_n = (-1)^n\left(3^{4-n}\right)$$

15. This is a geometric sequence with

$r = \frac{2}{7} \div \frac{1}{7} = \frac{2}{7} \cdot \frac{7}{1} = 2$ and $a_1 = \frac{1}{7}$.

$$a_5 = \frac{1}{7}(2)^{5-1} = \frac{2^4}{7} = \frac{16}{7}$$

$$a_6 = \frac{1}{7}(2)^{6-1} = \frac{2^5}{7} = \frac{32}{7}$$

$$a_7 = \frac{1}{7}(2)^{7-1} = \frac{2^6}{7} = \frac{64}{7}$$

The terms are $\frac{16}{7}, \frac{32}{7}, \frac{64}{7}$.

$$a_n = \frac{1}{7}(2)^{n-1} = \frac{2^{n-1}}{7}$$

16. This is an arithmetic sequence with $d = 9 - 13 = -4$ and $a_1 = 13$.

$a_5 = 13 + (5-1)(-4) = 13 - 16 = -3$

$a_6 = 13 + (6-1)(-4) = 13 - 20 = -7$

$a_7 = 13 + (7-1)(-4) = 13 - 24 = -11$

The terms are $-3, -7, -11$.

$$\begin{aligned} a_n &= a_1 + (n-1)d \\ &= 13 + (n-1)(-4) \\ &= 13 - 4n + 4 \\ &= 17 - 4n \end{aligned}$$

17.
$$\begin{aligned} \sum_{i=1}^{3}(i^2+9) &= (1^2+9)+(2^2+9)+(3^2+9) \\ &= (1+9)+(4+9)+(9+9) \\ &= 10+13+18 \\ &= 41 \end{aligned}$$

18.
$$\begin{aligned} &\sum_{i=1}^{4} i(i+5) \\ &= 1(1+5)+2(2+5)+3(3+5)+4(4+5) \\ &= 1(6)+2(7)+3(8)+4(9) \\ &= 6+14+24+36 \\ &= 80 \end{aligned}$$

19.
$$\begin{aligned} \sum_{i=1}^{5}\frac{i^2}{6} &= \frac{1^2}{6}+\frac{2^2}{6}+\frac{3^2}{6}+\frac{4^2}{6}+\frac{5^2}{6} \\ &= \frac{1}{6}+\frac{4}{6}+\frac{9}{6}+\frac{16}{6}+\frac{25}{6} \\ &= \frac{55}{6} \end{aligned}$$

20.
$$\begin{aligned} \sum_{i=1}^{4}\frac{i}{i+1} &= \frac{1}{1+1}+\frac{2}{2+1}+\frac{3}{3+1}+\frac{4}{4+1} \\ &= \frac{1}{2}+\frac{2}{3}+\frac{3}{4}+\frac{4}{5} \\ &= \frac{163}{60} \end{aligned}$$

21. $\sum_{i=1}^{4} x_i = x_1 + x_2 + x_3 + x_4 = 3 + 9 + 7 + 10 = 29$

22.
$$\begin{aligned} \sum_{i=1}^{4}(x_i)^2 &= x_1^2 + x_2^2 + x_3^2 + x_4^2 \\ &= 3^2 + 9^2 + 7^2 + 10^2 \\ &= 9 + 81 + 49 + 100 \\ &= 239 \end{aligned}$$

23. $\sum_{i=2}^{3}(x_i^2+1)=(x_2^2+1)+(x_3^2+1)$
$=(9^2+1)+(7^2+1)$
$=(81+1)+(49+1)$
$=82+50$
$=132$

24. $\left(\sum_{i=1}^{4}x_i\right)^2=(x_1+x_2+x_3+x_4)^2$
$=(3+9+7+10)^2$
$=(29)^2$
$=841$

25. a. perimeter of rectangle: $p=2l+2w$
$p_1=2(1)+2(1+3)=2+8=10$
$p_2=2(2)+2(2+3)=4+10=14$
$p_3=2(3)+2(3+3)=6+12=18$
$p_4=2(4)+2(4+3)=8+14=22$

b. $p_n=2n+2(n+3)=2n+2n+6=4n+6$

26. a. area of rectangle: $a=l\cdot w$
$a_1=1(1+3)=4$
$a_2=2(2+3)=10$
$a_3=3(3+3)=18$
$a_4=4(4+3)=28$

b. $a_n=n(n+3)=n^2+3n$

27. $a_1=5$
$a_2=5+(2-1)(3)=5+1(3)=5+3=8$
$a_3=5+(3-1)(3)=5+2(3)=5+6=11$
$a_4=5+(4-1)(3)=5+3(3)=5+9=14$
$a_5=5+(5-1)(3)=5+4(3)=5+12=17$
The terms are 5, 7, 9, 11, 13.

28. $a_1=5$
$a_2=5+(2-1)\left(-\frac{1}{3}\right)=5+1\left(-\frac{1}{3}\right)=\frac{15}{3}-\frac{1}{3}=\frac{14}{3}$
$a_3=5+(3-1)\left(-\frac{1}{3}\right)=5+2\left(-\frac{1}{3}\right)=\frac{15}{3}-\frac{2}{3}=\frac{13}{3}$
$a_4=5+(4-1)\left(-\frac{1}{3}\right)=5+3\left(-\frac{1}{3}\right)=5-1=4$
$a_5=5+(5-1)\left(-\frac{1}{3}\right)=5+4\left(-\frac{1}{3}\right)=\frac{15}{3}-\frac{4}{3}=\frac{11}{3}$
The terms are 5, $\frac{14}{3}$, $\frac{13}{3}$, 4, $\frac{11}{3}$.

29. $a_1=\frac{1}{2}$
$a_2=\frac{1}{2}+(2-1)(-2)=\frac{1}{2}-2=-\frac{3}{2}$
$a_3=\frac{1}{2}+(3-1)(-2)=\frac{1}{2}-4=-\frac{7}{2}$
$a_4=\frac{1}{2}+(4-1)(-2)=\frac{1}{2}-6=-\frac{11}{2}$
$a_5=\frac{1}{2}+(5-1)(-2)=\frac{1}{2}-8=-\frac{15}{2}$
The terms are $\frac{1}{2},-\frac{3}{2},-\frac{7}{2},-\frac{11}{2},-\frac{15}{2}$.

30. $a_1=-100$
$a_2=-100+(2-1)\left(\frac{1}{5}\right)=-100+\frac{1}{5}=-\frac{499}{5}$
$a_3=-100+(3-1)\left(\frac{1}{5}\right)=-100+\frac{2}{5}=-\frac{498}{5}$
$a_4=-100+(4-1)\left(\frac{1}{5}\right)=-100+\frac{3}{5}=-\frac{497}{5}$
$a_5=-100+(5-1)\left(\frac{1}{5}\right)=-100+\frac{4}{5}=-\frac{496}{5}$
The terms are $-100,-\frac{499}{5},-\frac{498}{5},-\frac{497}{5},-\frac{496}{5}$.

31. $a_9=a_1+(9-1)d$
$a_9=6+(9-1)(3)=6+8(3)=6+24=30$

32. $a_7=a_1+(7-1)d$
$-14=10+6d$
$-24=6d$
$-4=d$

33. $a_{11}=a_1+(11-1)d$
$2=-3+10d$
$5=10d$
$d=\frac{5}{10}=\frac{1}{2}$

34. $a_n=a_1+(n-1)d$
$-3=22+(n-1)(-5)$
$-3=22-5n+5$
$-3=27-5n$
$-30=-5n$
$6=n$

35. $a_8 = a_1 + (8-1)d$

$21 = 7 + 7d$

$14 = 7d$

$2 = d$

$$s_8 = \frac{8(a_1 + a_8)}{2} = \frac{8(7+21)}{2} = \frac{8(28)}{2} = \frac{224}{2} = 112$$

36. $a_7 = a_1 + (7-1)d$

$-48 = -12 + 6d$

$-36 = 6d$

$-6 = d$

$$s_7 = \frac{7(a_1 + a_7)}{2} = \frac{7(-12-48)}{2} = \frac{7(-60)}{2} = -210$$

37. $a_6 = a_1 + (6-1)d$

$\frac{13}{5} = \frac{3}{5} + 5d$

$\frac{13}{5} - \frac{3}{5} = 5d$

$\frac{10}{5} = 5d$

$2 = 5d$

$\frac{2}{5} = d$

$$s_6 = \frac{6(a_1 + a_6)}{2}$$

$$= \frac{6\left(\frac{3}{5} + \frac{13}{5}\right)}{2} = \frac{6\left(\frac{16}{5}\right)}{2} = 6\left(\frac{16}{5}\right)\left(\frac{1}{2}\right) = \frac{48}{5}$$

38. $a_9 = a_1 + (9-1)d$

$-6 = -\frac{10}{3} + 8d$

$-6 + \frac{10}{3} = 8d$

$-\frac{8}{3} = 8d$

$d = \frac{1}{8}\left(-\frac{8}{3}\right) = -\frac{1}{3}$

$$s_9 = \frac{9(a_1 + a_n)}{2}$$

$$= \frac{9\left(-\frac{10}{3} - 6\right)}{2} = \frac{9\left(-\frac{28}{3}\right)}{2} = 9\left(-\frac{28}{3}\right)\left(\frac{1}{2}\right) = -42$$

39. $a_1 = -7$

$a_2 = -7 + (2-1)(4) = -7 + 1(4) = -7 + 4 = -3$

$a_3 = -7 + (3-1)(4) = -7 + 2(4) = -7 + 8 = 1$

$a_4 = -7 + (4-1)(4) = -7 + 3(4) = -7 + 12 = 5$

The terms are −7, −3, 1, 5.

$a_{10} = -7 + (10-1)(4) = -7 + 9(4) = -7 + 36 = 29$

$$s_{10} = \frac{10(-7+29)}{2} = \frac{10(22)}{2} = \frac{220}{2} = 110$$

40. $a_1 = 4$

$a_2 = 4 + (2-1)(-3) = 4 + 1(-3) = 4 - 3 = 1$

$a_3 = 4 + (3-1)(-3) = 4 + 2(-3) = 4 - 6 = -2$

$a_4 = 4 + (4-1)(-3) = 4 + 3(-3) = 4 - 9 = -5$

The terms are 4, 1, −2, −5.

$a_{10} = 4 + (10-1)(-3) = 4 + 9(-3) = 4 - 27 = -23$

$$s_{10} = \frac{10[4 + (-23)]}{2} = \frac{10(-19)}{2} = \frac{-190}{2} = -95$$

41. $a_1 = \frac{5}{6}$

$a_2 = \frac{5}{6} + (2-1)\left(\frac{2}{3}\right) = \frac{5}{6} + \frac{2}{3} = \frac{9}{6} = \frac{3}{2}$

$a_3 = \frac{5}{6} + (3-1)\left(\frac{2}{3}\right) = \frac{5}{6} + 2\left(\frac{2}{3}\right) = \frac{5}{6} + \frac{4}{3} = \frac{13}{6}$

$a_4 = \frac{5}{6} + (4-1)\left(\frac{2}{3}\right) = \frac{5}{6} + 3\left(\frac{2}{3}\right) = \frac{5}{6} + \frac{6}{3} = \frac{17}{6}$

The terms are $\frac{5}{6}, \frac{3}{2}, \frac{13}{6}, \frac{17}{6}$.

$a_{10} = \frac{5}{6} + (10-1)\left(\frac{2}{3}\right) = \frac{5}{6} + 9\left(\frac{2}{3}\right) = \frac{5}{6} + 6 = \frac{41}{6}$

$$s_{10} = \frac{10\left(\frac{5}{6} + \frac{41}{6}\right)}{2} = \frac{10\left(\frac{46}{6}\right)}{2} = 5\left(\frac{46}{6}\right) = 5\left(\frac{23}{3}\right) = \frac{115}{3}$$

42. $a_1 = -60$

$a_2 = -60 + (2-1)(5) = -60 + 1(5) = -60 + 5 = -55$

$a_3 = -60 + (3-1)(5) = -60 + 2(5) = -60 + 10 = -50$

$a_4 = -60 + (4-1)(5) = -60 + 3(5) = -60 + 15 = -45$

The terms are −60, −55, −50, −45.

$a_{10} = -60 + (10-1)(5) = -60 + 45 = -15$

$$s_{10} = \frac{10(-60-15)}{2} = 5(-75) = -375$$

43. $d = 9 - 4 = 5$

$a_n = a_1 + (n-1)d$

$64 = 4 + (n-1)5$

$64 = 4 + 5n - 5$

$64 = 5n - 1$

$65 = 5n$

$13 = n$

$s_{13} = \dfrac{13(a_1 + a_{13})}{2} = \dfrac{13(4+64)}{2} = \dfrac{13(68)}{2} = 442$

44. $d = -4 - (-7) = -4 + 7 = 3$

$a_n = a_1 + (n-1)d$

$11 = -7 + (n-1)3$

$11 = -7 + 3n - 3$

$11 = 3n - 10$

$21 = 3n$

$7 = n$

$s_7 = \dfrac{7(a_1 + a_7)}{2} = \dfrac{7(-7+11)}{2} = \dfrac{7(4)}{2} = \dfrac{28}{2} = 14$

45. $d = \dfrac{9}{10} - \dfrac{6}{10} = \dfrac{3}{10}$

$a_n = a_1 + (n-1)d$

$\dfrac{36}{10} = \dfrac{6}{10} + (n-1)\dfrac{3}{10}$

$\dfrac{36}{10} = \dfrac{6}{10} + \dfrac{3}{10}n - \dfrac{3}{10}$

$\dfrac{36}{10} = \dfrac{3}{10} + \dfrac{3}{10}n$

$\dfrac{33}{10} = \dfrac{3}{10}n$

$n = \dfrac{10}{3}\left(\dfrac{33}{10}\right) = 11$

$s_{11} = \dfrac{11(a_1 + a_{11})}{2}$

$= \dfrac{11\left(\frac{6}{10} + \frac{36}{10}\right)}{2} = \dfrac{11\left(\frac{42}{10}\right)}{2} = 11\left(\dfrac{42}{10}\right)\left(\dfrac{1}{2}\right) = \dfrac{231}{10}$

46. $d = -3 - (-9) = -3 + 9 = 6$

$a_n = a_1 + (n-1)d$

$45 = -9 + (n-1)6$

$45 = -9 + 6n - 6$

$45 = -15 + 6n$

$60 = 6n$

$10 = n$

$s_{10} = \dfrac{10(a_1 + a_{10})}{2} = \dfrac{10(-9+45)}{2} = \dfrac{10(36)}{2} = 180$

47. $a_1 = 6$

$a_2 = 6(2)^{2-1} = 6(2) = 12$

$a_3 = 6(2)^{3-1} = 6(2)^2 = 6(4) = 24$

$a_4 = 6(2)^{4-1} = 6(2)^3 = 6(8) = 48$

$a_5 = 6(2)^{5-1} = 6(2)^4 = 6(16) = 96$

The terms are 6, 12, 24, 48, 96.

48. $a_1 = -12$

$a_2 = -12\left(\dfrac{1}{2}\right)^{2-1} = -12\left(\dfrac{1}{2}\right) = -6$

$a_3 = -12\left(\dfrac{1}{2}\right)^{3-1} = -12\left(\dfrac{1}{4}\right) = -3$

$a_4 = -12\left(\dfrac{1}{2}\right)^{4-1} = -12\left(\dfrac{1}{8}\right) = -\dfrac{3}{2}$

$a_5 = -12\left(\dfrac{1}{2}\right)^{5-1} = -12\left(\dfrac{1}{16}\right) = -\dfrac{3}{4}$

The terms are $-12, -6, -3, -\dfrac{3}{2}, -\dfrac{3}{4}$.

49. $a_1 = 20$

$a_2 = 20\left(-\dfrac{2}{3}\right)^{2-1} = 20\left(-\dfrac{2}{3}\right) = -\dfrac{40}{3}$

$a_3 = 20\left(-\dfrac{2}{3}\right)^{3-1} = 20\left(-\dfrac{2}{3}\right)^2 = 20\left(\dfrac{4}{9}\right) = \dfrac{80}{9}$

$a_4 = 20\left(-\dfrac{2}{3}\right)^{4-1} = 20\left(-\dfrac{2}{3}\right)^3 = 20\left(-\dfrac{8}{27}\right) = -\dfrac{160}{27}$

$a_5 = 20\left(-\dfrac{2}{3}\right)^{5-1} = 20\left(-\dfrac{2}{3}\right)^4 = 20\left(\dfrac{16}{81}\right) = \dfrac{320}{81}$

The terms are $20, -\dfrac{40}{3}, \dfrac{80}{9}, -\dfrac{160}{27}, \dfrac{320}{81}$.

50. $a_1 = -20$

$a_2 = -20\left(\dfrac{1}{5}\right)^{2-1} = -20\left(\dfrac{1}{5}\right) = -4$

$a_3 = -20\left(\dfrac{1}{5}\right)^{3-1} = -20\left(\dfrac{1}{25}\right) = -\dfrac{4}{5}$

$a_4 = -20\left(\dfrac{1}{5}\right)^{4-1} = -20\left(\dfrac{1}{125}\right) = -\dfrac{4}{25}$

$a_5 = -20\left(\dfrac{1}{5}\right)^{5-1} = -20\left(\dfrac{1}{625}\right) = -\dfrac{4}{125}$

The terms are $-20, -4, -\dfrac{4}{5}, -\dfrac{4}{25}, -\dfrac{4}{125}$.

51. $a_5 = a_1 \cdot r^{5-1} = 6\left(\frac{1}{3}\right)^{5-1} = 6\left(\frac{1}{3}\right)^4 = 6\left(\frac{1}{81}\right) = \frac{2}{27}$

52. $a_6 = a_1 \cdot r^{6-1} = 15(2)^{6-1} = 15(2)^5 = 15(32) = 480$

53. $a_4 = -8(-3)^{4-1} = -8(-3)^3 = -8(-27) = 216$

54. $a_5 = \frac{1}{12}\left(\frac{2}{3}\right)^{5-1} = \frac{1}{12}\left(\frac{2}{3}\right)^4 = \frac{1}{12}\left(\frac{16}{81}\right) = \frac{4}{243}$

55. $s_6 = \frac{6(1-2^6)}{1-2} = \frac{6(1-64)}{-1} = \frac{6(-63)}{-1} = 378$

56.
$$s_5 = \frac{-84\left[1-\left(-\frac{1}{4}\right)^5\right]}{1-\left(-\frac{1}{4}\right)}$$
$$= \frac{-84\left[1-\left(-\frac{1}{1024}\right)\right]}{1+\frac{1}{4}}$$
$$= \frac{-84\left(1+\frac{1}{1024}\right)}{\frac{5}{4}}$$
$$= -84 \cdot \left(\frac{1025}{1024}\right) \cdot \frac{4}{5}$$
$$= -\frac{4305}{64}$$

57. $s_4 = \frac{9\left[1-\left(\frac{3}{2}\right)^4\right]}{1-\frac{3}{2}} = \frac{9\left(1-\frac{81}{16}\right)}{-\frac{1}{2}} = 9\left(-\frac{65}{16}\right)\left(-\frac{2}{1}\right) = \frac{585}{8}$

58. $s_7 = \frac{8\left[1-\left(\frac{1}{2}\right)^7\right]}{1-\frac{1}{2}} = \frac{8\left(1-\frac{1}{128}\right)}{\frac{1}{2}} = 8\left(\frac{127}{128}\right)\left(\frac{2}{1}\right) = \frac{127}{8}$

59. $r = 12 \div 6 = 2$
$a_n = 6(2)^{n-1}$

60. $r = -20 \div (-4) = 5$
$a_n = -4(5)^{n-1}$

61. $r = \frac{10}{3} \div 10 = \frac{10}{3} \cdot \frac{1}{10} = \frac{1}{3}$
$a_n = 10\left(\frac{1}{3}\right)^{n-1}$

62. $r = \frac{18}{15} \div \frac{9}{5} = \frac{18}{15} \cdot \frac{5}{9} = \frac{2}{3}$
$a_n = \frac{9}{5}\left(\frac{2}{3}\right)^{n-1}$

63. $r = \frac{5}{2} \div 5 = \frac{5}{2} \cdot \frac{1}{5} = \frac{1}{2}$
$s_\infty = \frac{5}{1-\frac{1}{2}} = \frac{5}{\frac{1}{2}} = 5\left(\frac{2}{1}\right) = 10$

64. $r = 1 \div \frac{5}{2} = 1\left(\frac{2}{5}\right) = \frac{2}{5}$
$s_\infty = \frac{\frac{5}{2}}{1-\frac{2}{5}} = \frac{\frac{5}{2}}{\frac{3}{5}} = \frac{5}{2}\left(\frac{5}{3}\right) = \frac{25}{6}$

65. $r = \frac{8}{3} \div (-8) = \frac{8}{3}\left(-\frac{1}{8}\right) = -\frac{1}{3}$
$s_\infty = \frac{-8}{1-\left(-\frac{1}{3}\right)} = \frac{-8}{\frac{4}{3}} = -8\left(\frac{3}{4}\right) = -6$

66. $r = -4 \div -6 = \frac{-4}{-6} = \frac{2}{3}$
$s_\infty = \frac{-6}{1-\frac{2}{3}} = \frac{-6}{\frac{1}{3}} = -6\left(\frac{3}{1}\right) = -18$

67. $r = 8 \div 16 = \frac{8}{16} = \frac{1}{2}$
$s_\infty = \frac{16}{1-\frac{1}{2}} = \frac{16}{\frac{1}{2}} = 16\left(\frac{2}{1}\right) = 32$

68. $r = \frac{9}{3} \div 9 = \frac{9}{3} \cdot \frac{1}{9} = \frac{1}{3}$
$s_\infty = \frac{9}{1-\frac{1}{3}} = \frac{9}{\frac{2}{3}} = 9\left(\frac{3}{2}\right) = \frac{27}{2}$

69. $r = -1 \div 5 = -\frac{1}{5}$
$s_\infty = \frac{5}{1-\left(-\frac{1}{5}\right)} = \frac{5}{\frac{6}{5}} = 5\left(\frac{5}{6}\right) = \frac{25}{6}$

70. $r = -\frac{8}{3} \div -4 = -\frac{8}{3}\left(-\frac{1}{4}\right) = \frac{2}{3}$
$s_\infty = \frac{-4}{1-\frac{2}{3}} = \frac{-4}{\frac{1}{3}} = -4\left(\frac{3}{1}\right) = -12$

71. $0.363636\ldots = 0.36 + 0.0036 + 0.000036 + \ldots$
$= 0.36 + 0.36(0.01) + 0.36(0.01)^2 + \cdots$
$a_1 = 0.36$ and $r = 0.01$
$s_\infty = \frac{0.36}{1-0.01} = \frac{0.36}{0.99} = \frac{36}{99} = \frac{4}{11}$

72. $0.621621... = 0.621 + 0.000621 + 0.000000621 + \cdots = 0.621 + 0.621(0.001) + 0.621(0.001)^2 + \cdots$

$a_1 = 0.621$ and $r = 0.001$

$$s_\infty = \frac{0.621}{1-0.001} = \frac{0.621}{0.999} = \frac{621}{999} = \frac{23}{37}$$

73. Prove $2+5+8+\cdots+(3n-1) = \dfrac{n(3n+1)}{2}$ for all positive integers n.

Condition 1:
For $n = 1$,

$$2 \stackrel{?}{=} \frac{1(3\cdot1+1)}{2}$$
$$2 \stackrel{?}{=} \frac{1(3+1)}{2}$$
$$2 \stackrel{?}{=} \frac{1(4)}{2}$$
$$2 \stackrel{?}{=} \frac{4}{2}$$
$$2 = 2 \leftarrow \text{True}$$

Condition 2:
Assume $2+5+8+\cdots+(3k-1) = \dfrac{k(3k+1)}{2}$ for some positive integer k. Then

$$2+5+8+\cdots+(3k-1)+[3(k+1)-1] = \frac{k(3k+1)}{2} + [3(k+1)-1]$$
$$2+5+8+\cdots+(3k-1)+[3(k+1)-1] = \frac{3k^2+k}{2} + 3k+3-1$$
$$2+5+8+\cdots+(3k-1)+[3(k+1)-1] = \frac{3k^2+k+6k+6-2}{2}$$
$$2+5+8+\cdots+(3k-1)+[3(k+1)-1] = \frac{3k^2+7k+4}{2}$$
$$2+5+8+\cdots+(3k-1)+[3(k+1)-1] = \frac{(k+1)(3k+4)}{2}$$
$$2+5+8+\cdots+(3k-1)+[3(k+1)-1] = \frac{(k+1)[3(k+1)+1]}{2} \quad \leftarrow S_{k+1}$$

Since Conditions 1 and 2 hold, the above formula is true for all positive integers (by mathematical induction).

74. Prove $2^2+4^2+6^2+\cdots+(2n)^2 = \dfrac{2n(2n+1)(n+1)}{3}$ for all positive integers n.

Condition 1:
For $n = 1$,

$$2^2 \stackrel{?}{=} \frac{2\cdot1(2\cdot1+1)(1+1)}{3}$$
$$4 \stackrel{?}{=} \frac{2(2+1)(2)}{3}$$
$$4 \stackrel{?}{=} \frac{2(3)(2)}{3}$$
$$4 \stackrel{?}{=} \frac{12}{3}$$
$$4 = 4 \leftarrow \text{True}$$

Condition 2:
Assume $2^2+4^2+6^2+\cdots+(2k)^2 = \dfrac{2k(2k+1)(k+1)}{3}$ for some positive integer k.

Then

$$2^2+4^2+6^2+\cdots+(2k)^2+[2(k+1)]^2 = \frac{2k(2k+1)(k+1)}{3} + [2(k+1)]^2$$
$$2^2+4^2+6^2+\cdots+(2k)^2+[2(k+1)]^2 = \frac{2k(2k^2+3k+1)}{3} + (2k+2)^2$$
$$2^2+4^2+6^2+\cdots+(2k)^2+[2(k+1)]^2 = \frac{4k^3+6k^2+2k}{3} + 4k^2+8k+4$$
$$2^2+4^2+6^2+\cdots+(2k)^2+[2(k+1)]^2 = \frac{4k^3+6k^2+2k+12k^2+24k+12}{3}$$
$$2^2+4^2+6^2+\cdots+(2k)^2+[2(k+1)]^2 = \frac{4k^3+18k^2+26k+12}{3}$$
$$2^2+4^2+6^2+\cdots+(2k)^2+[2(k+1)]^2 = \frac{2(2k^3+9k^2+13k+6)}{3}$$
$$2^2+4^2+6^2+\cdots+(2k)^2+[2(k+1)]^2 = \frac{2(2k^2+5k+3)(k+2)}{3}$$
$$2^2+4^2+6^2+\cdots+(2k)^2+[2(k+1)]^2 = \frac{2(k+1)(2k+3)(k+2)}{3}$$
$$2^2+4^2+6^2+\cdots+(2k)^2+[2(k+1)]^2 = \frac{2(k+1)[2(2k+1)+1][(k+1)+1]}{3} \quad \leftarrow S_{k+1}$$

Since Conditions 1 and 2 hold, the above formula is true for all positive integers (by mathematical induction).

75. Prove $1\cdot 2+2\cdot 3+3\cdot 4+\cdots+n(n+1)=\dfrac{n(n+1)(n+2)}{3}$ for all positive integers n.

Condition 1:
For $n=1$,

$$1\cdot 2\overset{?}{=}\frac{1(1+1)(1+2)}{3}$$

$$2\overset{?}{=}\frac{1(2)(3)}{3}$$

$$2\overset{?}{=}\frac{6}{3}$$

$$2=2 \leftarrow \text{True}$$

Condition 2:
Assume $1\cdot 2+2\cdot 3+3\cdot 4+\cdots+k(k+1)=\dfrac{k(k+1)(k+2)}{3}$ for some positive integer k. Then

$$1\cdot 2+2\cdot 3+3\cdot 4+\cdots+k(k+1)+(k+1)\left[(k+1)+1\right]=\frac{k(k+1)(k+2)}{3}+(k+1)\left[(k+1)+1\right]$$

$$1\cdot 2+2\cdot 3+3\cdot 4+\cdots+k(k+1)+(k+1)\left[(k+1)+1\right]=\frac{k(k^2+3k+2)}{3}+(k+1)(k+2)$$

$$1\cdot 2+2\cdot 3+3\cdot 4+\cdots+k(k+1)+(k+1)\left[(k+1)+1\right]=\frac{k^3+3k^2+2k}{3}+k^2+3k+2$$

$$1\cdot 2+2\cdot 3+3\cdot 4+\cdots+k(k+1)+(k+1)\left[(k+1)+1\right]=\frac{k^3+3k^2+2k+3k^2+9k+6}{3}$$

$$1\cdot 2+2\cdot 3+3\cdot 4+\cdots+k(k+1)+(k+1)\left[(k+1)+1\right]=\frac{k^3+6k^2+11k+6}{3}$$

$$1\cdot 2+2\cdot 3+3\cdot 4+\cdots+k(k+1)+(k+1)\left[(k+1)+1\right]=\frac{(k^2+3k+2)(k+3)}{3}$$

$$1\cdot 2+2\cdot 3+3\cdot 4+\cdots+k(k+1)+(k+1)\left[(k+1)+1\right]=\frac{(k+1)(k+2)(k+3)}{3}$$

$$1\cdot 2+2\cdot 3+3\cdot 4+\cdots+k(k+1)+(k+1)\left[(k+1)+1\right]=\frac{(k+1)\left[(k+1)+1\right]\left[(k+1)+2\right]}{3} \quad \leftarrow S_{k+1}$$

Since Conditions 1 and 2 hold, the above formula is true for all positive integers (by mathematical induction).

76. Prove $2+6+10+\cdots+(4n-2)=2n^2$ for all positive integers n.

Condition 1:
For $n=1$,

$$2\overset{?}{=}2(1)^2$$

$$2\overset{?}{=}2(1)$$

$$2=2 \leftarrow \text{True}$$

Condition 2:
Assume $2+6+10+\cdots+(4k-2)=2k^2$ for some positive integer k. Then

$$2+6+10+\cdots+(4k-2)+\left[4(k+1)-2\right]=2k^2+\left[4(k+1)-2\right]$$

$$2+6+10+\cdots+(4k-2)+\left[4(k+1)-2\right]=2k^2+4k+4-2$$

$$2+6+10+\cdots+(4k-2)+\left[4(k+1)-2\right]=2k^2+4k+2$$

$$2+6+10+\cdots+(4k-2)+\left[4(k+1)-2\right]=2(k^2+2k+1)$$

$$2+6+10+\cdots+(4k-2)+\left[4(k+1)-2\right]=2(k+1)^2 \quad \leftarrow S_{k+1}$$

Since Conditions 1 and 2 hold, the above formula is true for all positive integers (by mathematical induction).

77.
$$(3x+y)^4=\binom{4}{0}(3x)^4(y)^0+\binom{4}{1}(3x)^3(y)^1+\binom{4}{2}(3x)^2(y)^2+\binom{4}{3}(3x)^1(y)^3+\binom{4}{4}(3x)^0(y)^4$$
$$=1(81x^4)(1)+4(27x^3)(y)+6(9x^2)(y^2)+4(3x)(y^3)+1(1)(y^4)$$
$$=81x^4+108x^3y+54x^2y^2+12xy^3+y^4$$

78.
$$(2x-3y^2)^3=\binom{3}{0}(2x)^3(-3y^2)^0+\binom{3}{1}(2x)^2(-3y^2)^1+\binom{3}{2}(2x)^1(-3y^2)^2+\binom{3}{3}(2x)^0(-3y^2)^3$$
$$=1(8x^3)(1)+3(4x^2)(-3y^2)+3(2x)(9y^4)+1(1)(-27y^6)$$
$$=8x^3-36x^2y^2+54xy^4-27y^6$$

79. $(x-2y)^9 = \binom{9}{0}(x)^9(-2y)^0 + \binom{9}{1}(x)^8(-2y)^1 + \binom{9}{2}(x)^7(-2y)^2 + \binom{9}{3}(x)^6(-2y)^3 + \cdots$

$= 1(x^9)(1) + 9(x^8)(-2y) + 36(x^7)(4y^2) + 84(x^6)(-8y^3) + \cdots$

$= x^9 - 18x^8y + 144x^7y^2 - 672x^6y^3 + \cdots$

80. $(2a^2+3b)^8 = \binom{8}{0}(2a^2)^8(3b)^0 + \binom{8}{1}(2a^2)^7(3b)^1 + \binom{8}{2}(2a^2)^6(3b)^2 + \binom{8}{3}(2a^2)^5(3b)^3 + \cdots$

$= 1(256a^{16})(1) + 8(128a^{14})(3b) + 28(64a^{12})(9b^2) + 56(32a^{10})(27b^3) + \cdots$

$= 256a^{16} + 3072a^{14}b + 16{,}128a^{12}b^2 + 48{,}384a^{10}b^3 + \cdots$

---

81. This is an arithmetic series with $d = 1$, $a_1 = 101$, and $a_n = 200$.

$a_n = a_1 + (n-1)d$

$200 = 101 + (n-1)(1)$

$200 = 101 + n - 1$

$200 = n + 100$

$100 = n$

The sum is

$s_{100} = \dfrac{n(a_1 + a_{100})}{2}$

$= \dfrac{100(101+200)}{2} = \dfrac{100(301)}{2} = 15{,}050$

82. $21 + 20 + 19 + \cdots + 1$ or $1 + 2 + 3 + \cdots + 21$

$s_n = \dfrac{n(a_1 + a_n)}{2}$

$s_{21} = \dfrac{21(1+21)}{2} = \dfrac{21(22)}{2} = \dfrac{462}{2} = 231$

There are 231 barrels in the stack.

83. This is an arithmetic sequence with $d = 1000$

a. $a_1 = 36{,}000$

$a_2 = 36{,}000 + (2-1)(1000)$

$= 36{,}000 + 1000$

$= 37{,}000$

$a_3 = 36{,}000 + (3-1)(1000)$

$= 36{,}000 + 2000$

$= 38{,}000$

$a_4 = 36{,}000 + (4-1)(1000)$

$= 36{,}000 + 3000$

$= 39{,}000$

Ahmed's salaries for the first four years are $36,000, $37,000, $38,000, and $39,000.

b. $a_n = 36{,}000 + (n-1)(1000)$

$= 36{,}000 + 1000n - 1000 = 35{,}000 + 1000n$

c. $a_6 = 35{,}000 + 1000(6)$

$= 35{,}000 + 6{,}000 = 41{,}000$

In the 6th year, his salary would be $41,000.

d. $a_{11} = 36{,}000 + (11-1)1000 = 46{,}000$

$s_n = \dfrac{n(a_1 + a_n)}{2}$

$s_{11} = \dfrac{11(36{,}000 + 46{,}000)}{2}$

$= \dfrac{11(82{,}000)}{2} = 451{,}000$

Ahmed will make a total of $451,000 in his first 11 years.

84. This is a geometric series with $r = 2$, $a_1 = 100$, and $n = 11$. (Note: There are 11 terms here since 200 represents the first doubling.)

$a_{11} = a_1 \cdot r^{11-1}$

$= 100(2)^{11-1} = 100(2)^{10} = 100(1024) = 102{,}400$

You would have $102,400.

85. $a_1 = 1600$, $r = 1.04$, $a_n = a_1 r^{n-1}$

a. July is the 7th month, so $n = 7$.

$a_7 = 1600(1.04)^{7-1} = 1600(1.04)^6 \approx 2024.51$

Gertrude's salary will be $2,024.51 in July.

b. December is the 12th month, so $n = 12$.

$a_{12} = 1600(1.04)^{12-1} = 1600(1.04)^{11} \approx 2463.13$

Gertrude's salary will be $2,463.13 in December.

c. $s_n = \dfrac{a_1(1-r^n)}{1-r}$, $n = 12$

$s_{12} = \dfrac{1600\left[1-(1.04)^{12}\right]}{1-1.04} \approx 24{,}041.29$

Gertrude will make $24,041.29 in 2006.

86. This is an infinite geometric series with $r = 0.92$ and $a_1 = 12$.

$$s_\infty = \frac{12}{1-0.92} = \frac{12}{0.08} = 150$$

The pendulum travels a total distance of 150 feet.

87. By the counting principle, Jeff can $7 \cdot 5 = 35$ different paper and envelope color sets.

88. Since order of the seating is important, the eight people can be arranged in $8! = 40{,}320$ ways.

89. Since order of the gifts is important, they can distributed in $12! = 479{,}001{,}600$ ways.

90. Since the ordering of the 7 selected huskies from the 15 choices is important, the number of arrangements is

$$_{15}P_7 = \frac{15!}{(15-7)!} = \frac{15!}{8!}$$
$$= \frac{15 \cdot 14 \cdot 13 \cdot 12 \cdot 11 \cdot 10 \cdot 9 \cdot \cancel{8!}}{\cancel{8!}} = 32{,}432{,}400.$$

91. Since the ordering of the 2 prize winners from the 5 finalists is important, the number of arrangements is

$$_5P_2 = \frac{5!}{(5-2)!} = \frac{5!}{3!} = \frac{5 \cdot 4 \cdot \cancel{3!}}{\cancel{3!}} = 20.$$

92. Since the ordering of the selected group of 4 from the 37 employees is not important, the number of possible groups is

$$_{37}C_4 = \frac{37!}{(37-4)!4!} = \frac{37!}{33!4!}$$
$$= \frac{37 \cdot 36 \cdot 35 \cdot 34 \cdot \cancel{33!}}{\cancel{33!} \cdot 4 \cdot 3 \cdot 2 \cdot 1} = \frac{1{,}585{,}080}{24} = 66{,}045.$$

93. The number of possible combination of males is

$$_6C_3 = \frac{6!}{(6-3)!3!} = \frac{6!}{3!3!} = \frac{6 \cdot 5 \cdot 4 \cdot \cancel{3!}}{\cancel{3!} \cdot 3 \cdot 2 \cdot 1} = \frac{120}{6} = 20.$$

The number of possible combination of females is

$$_8C_3 = \frac{8!}{(8-3)!3!} = \frac{8!}{5!3!} = \frac{8 \cdot 7 \cdot 6 \cdot \cancel{5!}}{\cancel{5!} \cdot 3 \cdot 2 \cdot 1} = \frac{336}{6} = 56.$$

By the counting principle, the number of combinations of males and females is $20 \cdot 56 = 1120$.

94. Since 1, 2, or 3 books may be checked out, we sum the number of combination for each number of books checked out:

$$_{10}C_1 + {}_{10}C_2 + {}_{10}C_3$$
$$= \frac{10!}{(10-1)!1!} + \frac{10!}{(10-2)!2!} + \frac{10!}{(10-3)!3!}$$
$$= \frac{10!}{9!1!} + \frac{10!}{8!2!} + \frac{10!}{7!3!}$$
$$= \frac{10 \cdot \cancel{9!}}{\cancel{9!} \cdot 1} + \frac{10 \cdot 9 \cdot \cancel{8!}}{\cancel{8!} \cdot 2 \cdot 1} + \frac{10 \cdot 9 \cdot 8 \cdot \cancel{7!}}{\cancel{7!} \cdot 3 \cdot 2 \cdot 1}$$
$$= 10 + 45 + 120$$
$$= 175$$

95. $P(\text{number is odd})$

$$= \frac{\text{Number of odd numbers}}{\text{Total number of numbers}} = \frac{5}{10} = \frac{1}{2}$$

96. $P(\text{number is even and greater than 4})$

$$= \frac{\text{Number of even numbers greater than 4}}{\text{Total number of numbers}} = \frac{2}{10} = \frac{1}{5}$$

---

97. $P(\text{even or greater than 4})$

$$= P(\text{even}) + P(\text{greater than 4}) - P(\text{even and greater than 4})$$
$$= \frac{\text{Number of even numbers}}{\text{Total number of numbers}} + \frac{\text{Number of numbers greater than 4}}{\text{Total number of numbers}} - \frac{\text{Number of even numbers greater than 4}}{\text{Total number of numbers}}$$
$$= \frac{5}{10} + \frac{5}{10} - \frac{2}{10} = \frac{8}{10} = \frac{4}{5}$$

98. $$P(\text{number is greater than 6 and less than 4}) = \frac{\text{Number of numbers that are greater than 6 and less than 4}}{\text{Total number of numbers}} = \frac{0}{10} = 0$$

99. $P(\text{number is a multiple of 3})$

$$= \frac{\text{Number of numbers that are multiples of 3}}{\text{Total number of numbers}}$$

$$= \frac{3}{10}$$

100. $P(\text{number is less than 10})$

$$= \frac{\text{Number of numbers that are less than 10}}{\text{Total number of numbers}}$$

$$= \frac{10}{10}$$

$$= 1$$

101. $\text{Odds against rain} = \frac{P(\text{no rain})}{P(\text{rain})}$

$$= \frac{1 - P(\text{rain})}{P(\text{rain})}$$

$$= \frac{1 - 0.75}{0.75}$$

$$= \frac{0.25}{0.75}$$

$$= \frac{25}{75}$$

$$= \frac{1}{3} \text{ or } 1:3$$

102. $\text{Odds against winning} = \frac{P(\text{not winning})}{P(\text{winning})}$

$$= \frac{1 - P(\text{winning})}{P(\text{winning})}$$

Let $x = P(\text{winning})$.

$$\frac{5}{4} = \frac{1-x}{x}$$

$$5x = 4(1-x)$$

$$5x = 4 - 4x$$

$$9x = 4$$

$$x = \frac{4}{9}$$

$$P(\text{winning}) = \frac{4}{9}$$

Chapter 12 Practice Test

1. **a.** This sequence in neither arithmetic nor geometric because the terms do not differ by a constant amount nor by a common multiple.

b. This sequence is arithmetic because the terms differ by –3.

c. This sequence is geometric because the terms differ by a multiple of $-\frac{1}{2}$.

2. $a_n = \frac{n-2}{3n}$

$$a_1 = \frac{1-2}{3(1)} = \frac{-1}{3} = -\frac{1}{3}$$

$$a_2 = \frac{2-2}{3(2)} = \frac{0}{6} = 0$$

$$a_3 = \frac{3-2}{3(3)} = \frac{1}{9}$$

$$a_4 = \frac{4-2}{3(4)} = \frac{2}{12} = \frac{1}{6}$$

$$a_5 = \frac{5-2}{3(5)} = \frac{3}{15} = \frac{1}{5}$$

The terms are $-\frac{1}{3}, 0, \frac{1}{9}, \frac{1}{6}, \frac{1}{5}$.

3. $a_n = \frac{2n+1}{n^2}$

$$a_1 = \frac{2(1)+1}{1^2} = \frac{2+1}{1} = 3$$

$$a_2 = \frac{2(2)+1}{2^2} = \frac{4+1}{4} = \frac{5}{4}$$

$$a_3 = \frac{2(3)+1}{3^2} = \frac{6+1}{9} = \frac{7}{9}$$

$$s_1 = a_1 = 3$$

$$s_3 = a_1 + a_2 + a_3 = 3 + \frac{5}{4} + \frac{7}{9} = \frac{181}{36}$$

4. $\sum_{i=1}^{5} (2i^2 + 3)$

$$= [2(1)^2 + 3] + [2(2)^2 + 3] + [2(3)^2 + 3] + [2(4)^2 + 3] + [2(5)^2 + 3]$$

$$= (2+3) + (8+3) + (18+3) + (32+3) + (50+3)$$

$$= 5 + 11 + 21 + 35 + 53$$

$$= 125$$

5. $\sum_{i=1}^{4}(x_i)^2 = x_1^2 + x_2^2 + x_3^2 + x_4^2$
$= 4^2 + 2^2 + 8^2 + 10^2$
$= 16 + 4 + 64 + 100$
$= 184$

6. $d = \frac{2}{3} - \frac{1}{3} = \frac{1}{3}$
$a_n = a_1 + (n-1)d$
$= \frac{1}{3} + (n-1)\left(\frac{1}{3}\right)$
$= \frac{1}{3} + \frac{1}{3}n - \frac{1}{3}$
$= \frac{1}{3}n$

7. $r = 10 \div 5 = \frac{10}{5} = 2$
$a_n = a_1 r^{n-1} = 5(2)^{n-1}$

8. $a_1 = 15$
$a_2 = 15 + (2-1)(-6) = 15 + 1(-6) = 15 - 6 = 9$
$a_3 = 15 + (3-1)(-6) = 15 + 2(-6) = 15 - 12 = 3$
$a_4 = 15 + (4-1)(-6) = 15 + 3(-6) = 15 - 18 = -3$
The terms are 15, 9, 3, -3.

9. $a_1 = \frac{5}{12}$
$a_2 = a_1 r^1 = \frac{5}{12}\left(\frac{2}{3}\right)^1 = \frac{5}{12}\left(\frac{2}{3}\right) = \frac{5}{18}$
$a_3 = a_1 r^2 = \frac{5}{12}\left(\frac{2}{3}\right)^2 = \frac{5}{12}\left(\frac{4}{9}\right) = \frac{5}{27}$
$a_4 = a_1 r^3 = \frac{5}{12}\left(\frac{2}{3}\right)^3 = \frac{5}{12}\left(\frac{8}{27}\right) = \frac{10}{81}$
The terms are $\frac{5}{12}, \frac{5}{18}, \frac{5}{27}, \frac{10}{81}$.

10. $a_{11} = a_1 + 10d$
$= 40 + (10)(-8)$
$= 40 - 80$
$= -40$

11. $s_8 = \frac{8(a_1 + a_8)}{2}$
$= \frac{8[7 + (-12)]}{2}$
$= \frac{8(-5)}{2}$
$= -20$

12. $d = -16 - (-4) = -16 + 4 = -12$
$a_n = a_1 + (n-1)d$
$-136 = -4 + (n-1)(-12)$
$-136 = -4 - 12n + 12$
$-136 = 8 - 12n$
$-144 = -12n$
$12 = n$

13. $a_6 = a_1 r^5 = 8\left(\frac{2}{3}\right)^5 = 8\left(\frac{32}{243}\right) = \frac{256}{243}$

14. $s_7 = \frac{a_1(1 - r^7)}{1 - r}$
$= \frac{\frac{3}{5}[1 - (-5)^7]}{1 - (-5)}$
$= \frac{\frac{3}{5}[1 - (-78{,}125)]}{1 - (-5)}$
$= \frac{\frac{3}{5}(1 + 78{,}125)}{1 + 5}$
$= \frac{\frac{3}{5}(78{,}126)}{6}$
$= \frac{3(78{,}126)}{5 \cdot 6}$
$= \frac{78{,}126}{5 \cdot 2}$
$= \frac{39{,}063}{5}$

15. $r = \frac{8}{3} \div 4 = \frac{8}{3} \cdot \frac{1}{4} = \frac{2}{3}$
$s_\infty = \frac{a_1}{1 - r} = \frac{4}{1 - \frac{2}{3}} = \frac{4}{\frac{1}{3}} = 4 \cdot \frac{3}{1} = 12$

16. Prove $5+10+15+\cdots+5n=\dfrac{5n(n+1)}{2}$ for all positive integers n.

Condition 1:

For $n=1$,

$5\overset{?}{=}\dfrac{5\cdot1(1+1)}{2}$

$5\overset{?}{=}\dfrac{5(2)}{2}$

$5=5 \leftarrow$ True

Condition 2:

Assume $5+10+15+\cdots+5k=\dfrac{5k(k+1)}{2}$ for some positive integer k. Then

$$5+10+15+\cdots+5k+5(k+1)=\frac{5k(k+1)}{2}+5(k+1)$$
$$5+10+15+\cdots+5k+5(k+1)=\frac{5k^2}{2}+\frac{5k}{2}+5k+5$$
$$5+10+15+\cdots+5k+5(k+1)=\frac{5k^2+15k+10}{2}$$
$$5+10+15+\cdots+5k+5(k+1)=\frac{5(k^2+3k+2)}{2}$$
$$5+10+15+\cdots+5k+5(k+1)=\frac{5(k+1)(k+2)}{2}$$
$$5+10+15+\cdots+5k+5(k+1)=\frac{5(k+1)[(k+1)+1]}{2} \quad \leftarrow S_{k+1}$$

Since Conditions 1 and 2 hold, the above formula is true for all positive integers (by mathematical induction).

17. $(x+2y)^4=\binom{4}{0}(x)^4(2y)^0+\binom{4}{1}(x)^3(2y)^1+\binom{4}{2}(x)^2(2y)^2+\binom{4}{3}(x)^1(2y)^3+\binom{4}{4}(x)^0(2y)^4$

$=1(x^4)(1)+4(x^3)(2y)+6(x^2)(4y^2)+4(x)(8y^3)+1(1)(16y^4)$

$=x^4+8x^3y+24x^2y^2+32xy^3+16y^4$

18. By the counting principle, $3\cdot2\cdot4=24$ different sandwiches can be prepared.

19. Since order in which the letters are opened is important, the seven letters can be opened in $7!=5040$ ways.

20. Since the order of the 4 selected chores from the 8 choices is not important, the number of options is

$${}_8C_2=\frac{8!}{(8-2)!2!}=\frac{8!}{6!2!}=\frac{8\cdot7\cdot\cancel{6!}}{\cancel{6!}\cdot2\cdot1}=\frac{56}{2}=28.$$

21. Since the order of the 4 selected photos from the 11 choices is not important, the number of ways the photos can be chosen is ${}_{11}C_4=\dfrac{11!}{(11-4)!4!}=\dfrac{11!}{7!4!}=\dfrac{11\cdot10\cdot9\cdot8\cdot\cancel{7!}}{\cancel{7!}\cdot4\cdot3\cdot2\cdot1}=\dfrac{7920}{24}=330$.

22. $P(\text{multiple of 2 } or \text{ multiple of 3})=P(\text{multiple of 2})+P(\text{multiple of 3})-P(\text{multiples of 2 and 3})$

$=\dfrac{\text{Number of multiples of 2}}{\text{Total number of numbers}}+\dfrac{\text{Number of multiples of 3}}{\text{Total number of numbers}}-\dfrac{\text{Number of multiple of 2 and 3}}{\text{Total number of numbers}}$

$=\dfrac{3}{6}+\dfrac{2}{6}-\dfrac{1}{6}$

$=\dfrac{4}{6}$

$=\dfrac{2}{3}$

23. $\text{Odds against winning} = \dfrac{P(\text{not winning})}{P(\text{winning})}$

$= \dfrac{1-P(\text{winning})}{P(\text{winning})}$

$= \dfrac{1-\frac{1}{500}}{\frac{1}{500}}$

$= \dfrac{\frac{499}{500}}{\frac{1}{500}}$

$= \dfrac{499}{500}\cdot\dfrac{500}{1}$

$= \dfrac{499}{1}$ or 499:1

24. $a_1 = 1000$, $n = 20$

$a_n = a_1 + (n-1)d$

$a_{20} = 1000 + (20-1)(1000)$

$= 20{,}000$

$s_{20} = \dfrac{20(1000 + 20{,}000)}{2}$

$= \dfrac{20(21{,}000)}{2}$

$= 210{,}000$

After 20 years, she will have saved \$210,000.

25. $r = 3,\ a_1 = 500(3) = 1500$

$a_6 = 1500(3)^{6-1} = 1500(3)^5 = 364{,}500$

At the end of the sixth hour, there will be 364,500 bacteria in the culture.

Chapter 12 Cumulative Review Test

1. $A = \frac{1}{2}bh$, for b

$2A = bh$

$\dfrac{2A}{h} = \dfrac{bh}{h}$

$\dfrac{2A}{h} = b$

2. $m = \dfrac{y_2 - y_1}{x_2 - x_1} = \dfrac{9-(-2)}{1-4} = \dfrac{11}{-3} = -\dfrac{11}{3}$

Use the point-slope equation with $m = -\frac{11}{3}$ and $(4,-2)$.

$y - y_1 = m(x - x_1)$

$y - (-2) = -\frac{11}{3}(x-4)$

$y + 2 = -\frac{11}{3}x + \frac{44}{3}$

$y = -\frac{11}{3}x + \frac{44}{3} - 2$

$y = -\frac{11}{3}x + \frac{38}{3}$

3. $x + y + z = 1$ (1)

$2x + 2y + 2z = 2$ (2)

$3x + 3y + 3z = 3$ (3)

Notice that if you multiply equation (1) by 2, the result is exactly the same as equation (2). This implies that this is a dependent system and therefore has infinitely many solutions.

4.

$$\begin{array}{r} 5x^3 + 4x^2 - 6x + 2 \\ x + 5 \\ \hline 5x^4 + 4x^3 - \ \ 6x^2 + \ \ 2x \\ 25x^3 + 20x^2 - 30x + 10 \\ \hline 5x^4 + 29x^3 + 14x^2 - 28x + 10 \end{array}$$

5. $x^3 + 2x - 5x^2 - 10 = (x^3 + 2x) - (6x^2 + 12)$

$= x(x^2+2) - 6(x^2+2)$

$= (x^2+2)(x-6)$

6. $(a+b)^2 + 8(a+b) + 16$ [Let $u = a + b$.]

$= u^2 + 8u + 16$

$= (u+4)(u+4)$

$= (u+4)^2$ [Back substitute $a+b$ for u.]

$= (a+b+4)^2$

7. $5 - \dfrac{x-1}{x^2+3x-10} = 5 - \dfrac{x-1}{(x+5)(x-2)}$

$= \dfrac{5(x+5)(x-2)}{(x+5)(x-2)} - \dfrac{x-1}{(x+5)(x-2)}$

$= \dfrac{5x^2+15x-50}{(x+5)(x-2)} - \dfrac{x-1}{(x+5)(x-2)}$

$= \dfrac{5x^2+15x-50-(x-1)}{(x+5)(x-2)}$

$= \dfrac{5x^2+15x-50-x+1}{(x+5)(x-2)}$

$= \dfrac{5x^2+14x-49}{(x+5)(x-2)}$

8. $y = kz^2$

$80 = k(20)^2$

$80 = 400k$

$k = \frac{80}{400} = 0.2$

$y = 0.2z^2$

Let $z = 50$: $y = 0.2(50)^2 = 0.2(2500) = 500$

9. $f(x) = 3\sqrt[3]{x-2},\ g(x) = \sqrt[3]{7x-14}$

$f(x) = g(x)$

$2\sqrt[3]{x-3} = \sqrt[3]{5x-15}$

$\left(2\sqrt[3]{x-3}\right)^3 = \left(\sqrt[3]{5x-15}\right)^3$

$8(x-3) = 5x-15$

$8x-24 = 5x-15$

$3x = 9$

$x = 3$

10. $\sqrt{6x-5} - \sqrt{2x+6} - 1 = 0$

$\sqrt{6x-5} = 1 + \sqrt{2x+6}$

$\left(\sqrt{6x-5}\right)^2 = \left(1+\sqrt{2x+6}\right)^2$

$6x-5 = 1 + 2\sqrt{2x+6} + 2x + 6$

$4x - 12 = 2\sqrt{2x+6}$

$2x - 6 = \sqrt{2x+6}$

$(2x-6)^2 = \left(\sqrt{2x+6}\right)^2$

$4x^2 - 24x + 36 = 2x + 6$

$4x^2 - 26x + 30 = 0$

$2\left(2x^2 - 13x + 15\right) = 0$

$2(2x-3)(x-5) = 0$

$2x - 3 = 0$ or $x - 5 = 0$

$x = \frac{3}{2}$ $\qquad x = 5$

Upon checking, $x = \frac{3}{2}$ is an extraneous solution.

The solution is $x = 5$.

11. $x^2 + 2x + 15 = 0$

$x^2 + 2x = -15$

$x^2 + 2x + 1 = -15 + 1$

$(x+1)^2 = -14$

$x + 1 = \pm\sqrt{-14}$

$x + 1 = \pm i\sqrt{14}$

$x = -1 \pm i\sqrt{14}$

12. $x^2 - \frac{x}{5} - \frac{1}{3} = 0$

$15\left(x^2 - \frac{x}{5} - \frac{1}{3}\right) = 15(0)$

$15x^2 - 3x - 5 = 0$

$x = \frac{-b \pm \sqrt{b^2 - 4ac}}{2a}$

$= \frac{-(-3) \pm \sqrt{(-3)^2 - 4(15)(-5)}}{2(15)}$

$= \frac{3 \pm \sqrt{9 + 300}}{30}$

$= \frac{3 \pm \sqrt{309}}{30}$

13. Let x be the number. Then

$2x^2 - 9x = 5$

$2x^2 - 9x - 5 = 0$

$(2x+1)(x-5) = 0$

$2x + 1 = 0$ or $x - 5 = 0$

$x = -\frac{1}{2}$ $\qquad x = 5$

Since the number must be positive, disregard $x = -\frac{1}{2}$. The number is 5.

14. $y = x^2 - 4x$

$y = \left(x^2 - 4x + 4\right) - 4$

$y = (x-2)^2 - 4$

The vertex is $(2, -4)$.

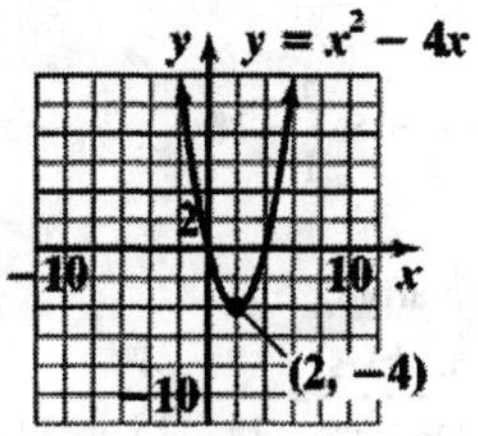

15. $\log_a \frac{1}{64} = 6$

$$a^6 = \frac{1}{64}$$
$$a^6 = \left(\frac{1}{2}\right)^6$$
$$a = \frac{1}{2}$$

16. $y = 2^x - 1$

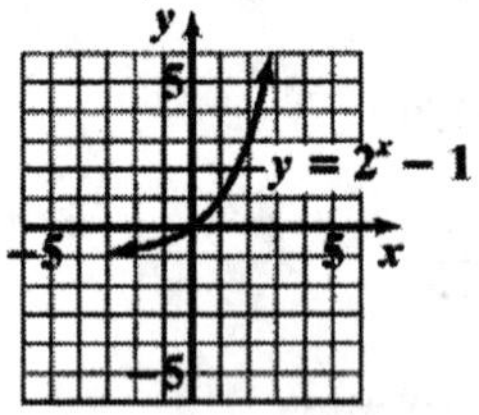

17. $(h, k) = (-6, 2),\ r = 7$

$$(x-h)^2 + (y-k)^2 = r^2$$
$$(x-(-6))^2 + (y-2)^2 = 7^2$$
$$(x+6)^2 + (y-2)^2 = 49$$

18. $9x^2 + 16y^2 = 144$

$$\frac{9x^2}{144} + \frac{16y^2}{144} = 1$$
$$\frac{x^2}{16} + \frac{y^2}{9} = 1$$
$$\frac{x^2}{4^2} + \frac{y^2}{3^2} = 1 \Rightarrow a = 4 \text{ and } b = 3$$

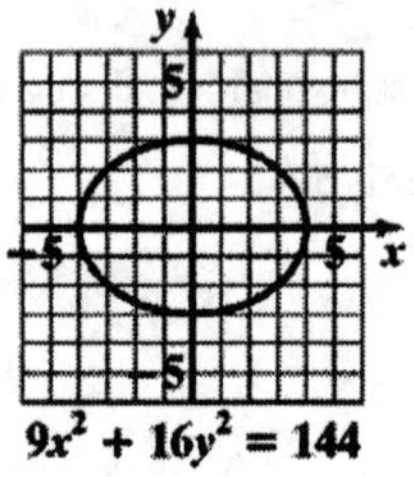

19. $r = 4 \div 6 = \frac{4}{6} = \frac{2}{3}$

$$s_\infty = \frac{a_1}{1-r} = \frac{6}{1-\frac{2}{3}} = \frac{6}{\frac{1}{3}} = 6 \cdot \frac{3}{1} = 18$$

20. $P(\text{selecting a blue marble})$

$$= \frac{\text{Number of blue marbles}}{\text{Total number of marbles}} = \frac{20}{75} = \frac{4}{15}$$